The Agricultural Notebook

Primrose McConnell compiled the first edition of The Agricultural Notebook in 1883. This edition has been completely revised and updated by the contributors under the editorship of Richard J. Soffe.

Other titles of interest

Fream's Principles of Food and Agriculture
17th edition
Edited by C.R.W. Spedding
0 632 02978 1

Farm Machinery
12th edition
Claude Culpin
0 632 03159 X

Profitable Organic Farming
J. Newton
0 632 03929 9

Understanding The Dairy Cow
2nd edition
John Webster
0 632 03438 6

Poultry Heath and Management
3rd edition
David Sainsbury
0 632 03325 8

Planned Sheep Production
2nd edition
David Croston and Geoff Pollott
0 632 03576 5

Animal Health
2nd edition
David Sainsbury
0 632 03888 8

Aquaculture: Principles and Practices
T.V.R. Pillay
0 85238 202 2

Horse and Stable Management
(incorporating *Horse Care*)
3rd edition
Jeremy Houghton Brown,
Vincent Powell-Smith and
Sarah Pilliner
0 632 04152 8

Primrose McConnell's

The Agricultural Notebook

19th Edition

Edited by

Richard J. Soffe

Seale-Hayne Faculty of Agriculture,
Food and Land Use
University of Plymouth

Blackwell
Science

Editorial Offices:
Osney Mead, Oxford OX2 0EL
25 John Street, London WC1N 2BL
23 Ainslie Place, Edinburgh EH3 6AJ
350 Main Street, Malden
MA 02148 5018, USA
54 University Street, Carlton
Victoria 3053, Australia

Other Editorial Offices:
Blackwell Wissenschafts-Verlag GmbH
Kurfürstendamm 57
10707 Berlin, Germany

Zehetnergasse 6
A-1140 Wien
Austria

First edition published 1883
18th edition published 1988
Reissued in paperback 1992
Reprinted 1994
19th edition published 1995
Reprinted 1996, 1977

Set in Ehrhardt 8.5 on 9 pt
by Setrite Typesetters, Hong Kong
Printed and bound in Great Britain
at the University Press, Cambridge

DISTRIBUTORS

Marston Book Services Ltd
PO Box 269
Abingdon
Oxon OX14 4YN
(*Orders*: Tel: 01235 465500
Fax: 01235 465555)

USA
Blackwell Science, Inc.
Commerce Place
350 Main Street
Malden, MA 02148 5018
(*Orders*: Tel: 800 759 6102
617 388 8250
Fax: 617 388 8255)

Canada
Copp Clark Professional
200 Adelaide Street, West, 3rd Floor
Toronto, Ontario M5H 1W7
(*Orders*: Tel: 416 597-1616
800 815-9417
Fax: 416 597-1617)

Australia
Blackwell Science Pty Ltd
54 University Street
Carlton, Victoria 3053
(*Orders*: Tel: 03 9347-0300
Fax: 03 9347-5001)

A catalogue record for this title
is available from the British Library

ISBN 0 632 03643 5

Library of Congress
Cataloging-in-Publication Data

McConnell, Primrose.
[Agricultural notebook]
Primrose McConnell's The agricultural notebook. – 19th ed./
edited by R.J. Soffe.
p. cm.
Includes bibliographical references (p.) and index.
ISBN 0-632-03643-5
1. Agriculture – Handbooks, manuals, etc. 2. Agriculture – Great Britain – Handbooks, manuals, etc. I. Soffe, R.J. II. Title
III. Title: Agricultural handbook.
S513.2.M33 1995
630′.941 – dc20 94-41071
CIP

Contents

Preface to 19th Edition xi

Primrose McConnell: a brief biographical sketch xiii

Contributors xv

Part 1 Farm Management 1

1 The Common Agricultural Policy of the European Union 3
Background, institutions and the legislative process 3
The formation and development of the Common Agricultural Policy 5
CAP price mechanisms 9
Monetary arrangements in the CAP 13
Structural and environmental policy 14
Problems and developments in the CAP 16

2 Farm business management 17
What is business management? 17
Setting primary objectives 17
Planning to achieve primary objectives 18
Information for financial management 19

3 Farm staff management 39
Introduction 39
Job analysis and job design 39
Recruitment and pay 41
Interviewing and discrimination 42
Control, guidance and negotiation 44
Training and development 47
Maintaining and increasing performance 49
The employment environment 51

4 Agricultural law 53
Introduction 53
The English legal system 53
Legal aspects of the ownership, possession and occupation of agricultural land 55
Legal constraints on the development and use of agricultural land 58
Employer's liabilities 61
Conclusions 62

5 **Health and safety in agriculture** **64**
Introduction 64
The Health and Safety at Work, etc. Act 1974 64
Some of the statutory requirements of the HSW Act 66
Regulations applying to agriculture activities 67
Food and Environment Protection Act 1985 69
Conclusions 69

6 **Alternative enterprises** **71**
Defining alternatives 71
The policy climate 72
Types of alternative enterprise 74
Management of alternative enterprises 78
Two examples of diversified enterprises 80
Summary 83

Part 2 Crop Production **85**

7 **Soil management** **87**
Introduction 87
Soil physical properties 87
Drainage 94
Soil chemical properties 100
Soil biological properties 108
Soil classification and analysis 111
Soils and sustainable production 114

8 **Crop physiology** **118**
Fundamental physiological processes 118
Growth 120
Crop development cycle 121
Environmental influences on growth and development 123
Competition 127
Modification of crops by growth regulators and breeding 129

9 **Crop nutrition** **130**
Summary of soil-supplied elements 130
Organic fertilizers and manures 134
Inorganic fertilizers 138
Forms of fertilizer available 141
Compound fertilizers 143
Fertilizer placement 143
Fertilizers and the environment 144
Optimum fertilizer rates 144
Fertilizer recommendations 146
Liming 147

10 **Arable cropping** **150**
Arable crop planning and production practice 151
Choice of arable crop 151
Choice of cropping system 152
Enterprise planning 154

Arable crop enterprises 157
Combinable crops 157
Root and vegetable crops 174
Fodder and forage crops 187
Fibre, fuel and other non-food crops 192

11 **Grassland** 194
Distribution and purpose of grassland in the UK 195
Grassland improvement 196
Characteristics of agricultural grasses 200
Herbage legumes 202
Herbage seed production 203
Basis for seeds mixtures 203
Fertilizers for grassland 206
Patterns of grassland production 210
Expression of grassland output on the farm 211
Output from grazing animals 213
Grazing systems 214
Conservation of grass 216
Grassland farming and the environment 223
Glossary of grassland terms 224

12 **Farm woodland management** 227
Setting management objectives 227
Forest products and markets 227
Silvicultural options for existing woodlands 228
Improvement of neglected woodland 230
Forest measurement 230
Thinning 232
Harvesting 234
New farm woodlands 235
Forest protection 246
Landscape design for farm woodlands 247
Forest management for non-timber uses 248
Forest investment 249
Integration of agriculture and forestry 249
Grants 250

13 **Crop health** 252
Weed control 252
Occurrence of weed problems 252
General control measures – applicable to all systems 253
Principles of chemical (herbicidal) weed control 253
Choice of herbicide 255
Arable crops and seedling leys 257
Weed control in cereals 258
Root crops, row crops and other broadleaved crops 262
Weeds from shed crop seed or groundkeepers 262
Weed control in established grassland 268
Weed control in forest and woodlands 270
Treatment of individual large trees 275
Weed control (unselective) in non-crop situations 276

Pests of crops 277
Cereal pests 277
Potato pests 282
Sugar beet pests 285
Oilseed rape pests 290
Kale pests 293
Pea and bean pests 294
Field vegetable pests 296
Schemes, regulations and acts relating to the use of insecticides 297

Diseases of crops 298
Air-borne diseases 298
Soil-borne diseases 300
Seed-borne and inflorescence diseases 301
Vector-borne diseases 301

Diseases of wheat 302
Diseases of barley 305
Diseases of oat 307
Diseases of rye/triticale 308
Diseases of forage maize and sweet corn 308
Diseases of potato 309
Diseases of brassicas 312
Diseases of sugar beet, fodder beet, mangold 315
Diseases of peas and beans (field, broad/dwarf, French) 316
Diseases of linseed 317
Diseases of carrots 317
Diseases of onions and leeks 318
Diseases of grasses and herbage legumes 319

14 Grain preservation and storage 321
Conditions for the safe storage of living grain 321
The time factor in grain storage 322
Grain moisture content and its measurement 322
Drying grain 323
Fans for grain stores 327
Grain cleaners 328
Management of grain in store 328
Handling grain 329
Storage of grain in sealed containers 330
Use of chemical preservatives to store grain 331

Part 3 Animal Production 333

15 Animal physiology and nutrition 335
Regulation of body function 335
Chemical composition of the animal and its food 337
Digestion 343
Metabolism 348
Voluntary food intake 360
Reproduction 361
Lactation 373
Growth 379
Environmental physiology 382

16 Animal welfare 385
Background 385
What is welfare? 385
How do we assess welfare? 389
What is an acceptable level of welfare? 392

17 Livestock feeds and feeding 394
Nutrient evaluation of feeds 394
Raw materials for diet and ration formulation 401
Nutrient requirements 414

18 Cattle 421
Definitions of common cattle terminology 422
Calf rearing 424
Management of breeding stock replacements 427
Feeding dairy and beef cattle 429
Beef production 432
Beef production systems 434
Dairying 437
Cattle breeding 443

19 Sheep and goats 445
Sheep 445
Goats 460
Appendix 1: Body condition scores 463
Appendix 2: Calendar for frequent-lambing flock 463

20 Pig production **464**
Introduction 464
The structure of the UK pig industry 464
Pig housing and animal welfare 466
Genetics and pig improvement 468
Sow and gilt reproduction 472
Weaner piglets 478
Finishing pigs 480
Conclusions 485

21 Poultry **486**
Meat production 486
Commercial egg production 489
Broiler breeding industry 491
Environment of layers and broilers 491
Turkeys and waterfowl 492
Poultry welfare 496

22 Animal health **497**
Introduction 497
Disease and immunity 501
The health of cattle 505
The health of pigs 517
The health of sheep 521

Part 4 Farm Equipment 525

23 Services **527**
Energy 527
Chases and conduits 530
Shafts and ducts 531
Drainage 531
Rainwater goods 532
Water supplies 532
Fire fighting 534
Security 534

24 Farm machinery **536**
The agricultural tractor 536
Cultivation machines 541
Fertilizer application machines 544
Planting machine 545
Crop sprayers 549
Harvesting forage crops 552
Grain harvesting 558
Root harvesting 561
Irrigation 562
Feed preparation machinery 568
Milking equipment 569
Manure handling and spreading 571

25 Farm buildings **575**
Design criteria (BS 5502, Part 21, refers) 575
Stages in constructing a building 580
Building materials and techniques 584
Environmental control 592
Water pollution control 598
Specific purpose buildings 601

Glossary of units 626

Index 629

Preface to 19th Edition

The Agricultural Notebook has come a long way since 1883 when Primrose McConnell wrote, 'To dig with a spade an acre of land 9 inches deep, a man will take from 14 to 21 days'. The current edition indicates less than one hour for a tractor to do the same. Even in the seven years between the 18th edition and this edition, agriculture and agricultural developments, both at governmental level and at a technical field level, have moved apace.

Key areas of change included in this edition are: developments in the European Union; reform of the Common Agricultural Policy; increasing awareness and action regarding animal welfare; continued emphasis on pollution control and the environment; and the steady increase in sustainable farming.

The chapters on traditional enterprises, which remain the backbone of agriculture, have all been updated and rewritten where appropriate, and are supplemented by new contributors in a number of areas, and an additional chapter on animal welfare has been included. All the chapters have a section on references and further reading which has proved popular with readers.

The result I hope is an edition that gives those in the agricultural and related industries the best single agricultural reference source contained within one cover. My grateful thanks to all the contributors for their time, effort and enthusiasm.

Primrose McConnell: a brief biographical sketch

P.W. Brassley

'I wish I had not been born for a hundred years to come, for there will be so many things found out after I am done with . . .', wrote Primrose McConnell in the spring of 1906. It was typical of him: always fascinated by the latest discoveries and inventions, but concerned to test them against his own extensive knowledge and experience.

McConnell was born at Lesnessock Farm, near Ochiltree in Ayrshire, on 11 April 1856, the son and grandson of tenant farmers. After leaving Ayr Academy he was apprenticed to a Glasgow engineering firm, but subsequently, in the 1870s, went to the University of Edinburgh, which did not then offer degrees in agriculture but prepared students for the diploma examinations of the Highland and Agricultural Society. McConnell passed in 1878, and a little later also passed the certificate examinations of the Royal Agricultural Society of England. When Edinburgh eventually introduced a degree, he returned, and was the second student to be awarded the BSc in Agriculture in 1886.

The *Notebook* is his lasting memorial, but his writing was based on the foundation of his other activities, as farmer, scientist, engineer and inventor, traveller, lecturer and all-round man of agricultural affairs. Farming formed the foundation of his life. He first rented the 636-acre Ongar Park Hall Farm, about 20 miles from London, near Epping in Essex, initially in partnership with his father. When he began farming, half of the land was in arable, but this was in 1883, when cheap grain from the new world was beginning to make life difficult for corn producers on stiff London clays. He therefore grassed down about 200 acres and based his farming upon dairying (he had about 60 cows in milk at any one time) and feeding bullocks, heifers and sheep. In 1905 he moved. Why? 'For the very good reason that I was losing more money than I could afford', he wrote soon afterwards, 'but also for various other reasons', which in fact centred on a dispute with his landlord over how he was to be compensated for capital invested in buildings. The case went to court, and McConnell had the better of the legal arguments, but he was left with a jaundiced view of landlords. He was the owner-occupier of his new farm, North Wycke, 500 acres of land 'as flat as a table' near Southminster in Essex, between the Crouch and Blackwater estuaries, with 'nothing higher than a tree or a house between me and the Ural Mountains'. When he began to farm there, it was half arable and half grass, and he kept 80 cows, nine work horses, a pony, two dogs, three tomcats and '135 head of poultry of all breeds under the sun, including those that do not lay in winter'. He was understocked, he knew, but he was too short of capital to buy any more. He continued to farm there until he died in 1931.

McConnell farmed to make money, but not only to make money. Fascinated by the scientific and technical problems involved, he attempted to deal with them professionally, as befitted an agricultural graduate. He tried machine milking, found that it resulted in decreased yields, and so went back to hand milking. Then, after considering his experiences for a year or so, he wrote a detailed article for the *Agricultural Gazette* setting out the costs, technical details, yield changes and probable explanations, before concluding that 'It is rather a dangerous thing to prophesy as to future inventions, and we do not know what mankind may accomplish in another generation. We may, therefore, still see a successful milking machine, but it has not arrived yet.' He was an early advocate of milk recording and kept a Gerber fat testing machine in his dairy. He experimented with silage and he designed his own elevator. The string-binder, he thought, was the greatest invention of the nineteenth century. He had 'an outfit of every possible kind of tool in my workshop on the farm that is likely to be of use', and was 'never ... happier than when at the bench or vice'. When he wanted to try out a new plough, he would use it himself for a day, with a dynamometer between the horses and the plough. Not surprisingly, the shortcomings of farm machinery provoked some of his more vitriolic comments.

McConnell also led a busy life away from the farm. He lectured, at various times, at the Glasgow Veterinary College (where he was appointed Professor of Agriculture at the age of 24) and Oxford and Edinburgh universities and the Essex Winter School of Agriculture (the forerunner of Writtle College), and examined at Reading, Wye and the Royal Agricultural College. He was a Fellow of the Geological Society. He was on the Council of the Dairy Show and a milking judge there, and involved with the British Dairy Farmers' Association and the Eastern Counties Co-operative Dairy Farmers' Association. He visited farms in Holland in 1899 and made at least two trips to North America, in 1890 and 1893, on one occasion meeting some of the Sioux who had taken part in the Custer massacre in 1876. He was also one of the pioneers in the migration of farmers from Scotland to Essex in the late nineteenth century. No sooner had he found his farm at Ongar Park Hall than he was writing articles about the potential of Essex farms for the *North British Agriculturalist*.

And it is as a writer that he is now best remembered. 'I began to write to the farm papers at the age of eighteen, when first learning to hold the plough', he recalled, and he produced eleven editions of the *Notebook* between 1883, when he was 27, and 1930, the year before his death at the age of 75. He also wrote *The Elements of Farming* (1896), an elementary textbook, *The Elements of Agricultural Geology* (1902), *The Diary of a Working Farmer* (1906) and *The Complete Farmer* (1908), in addition to articles in the journals of the Royal Agricultural Society of England and the Bath and West Society. Later he spent many years as dairy editor of the *Agricultural Gazette* and editor of *Farm Life*.

Thus he was academically successful, and he clearly enjoyed writing. But it is also evident from his diary that he enjoyed physical work too: he writes enthusiastically of making his own cheese, digging, ploughing and broadcasting, and stooking even though the sheaves are drawing blood from his forearms. He was a teetotaller and a dissenting churchman (his wife Katherine was the daughter of a Free Church minister) who took a five-day study tour with the British Dairy Farmers' Association as his annual holidays. He took an unsentimental attitude to landscape: 'in a level district you get a great wide sky, and the sun shines longer'. But if this suggests the stereotypical dour Scot, then his irascibility, his sense of humour and his benevolent interest in the world around him keep breaking through, as does his pride in his family, when he mentions that his daughter Ann is an accredited dairymaid, and prints a photograph of his sons Archibald and Primrose (who was to be killed in the last days of the First World War). Farming, and thinking and writing about farming, provided him with stimulation and satisfaction. Fortunately, there was always something new to learn: 'Agriculture is a very wide subject, and no one can master it all within the limits of an ordinary lifetime'.

Sources

Most of the material used here is taken from McConnell's *The Diary of a Working Farmer* (1906) and from his obituary, published in the [Essex] *Weekly News* of Friday, 10 July 1931. For the latter, an enormous amount of other biographical material on McConnell, and many perceptive editorial comments, the author is indebted to Elizabeth Sellers of Chelmsford.

Contributors

P.H. Bomford, BSc (Hons), MSc, CEng, MIAgrE.

Retired; formerly Senior Lecturer, Engineering Department, Seale-Hayne Faculty, University of Plymouth.

Educated at Reading University and the University of Newcastle-upon-Tyne. Previous appointments have included assistant lecturer in Farm Mechanization, Wye College, University of London, lecturer in farm mechanization, Essex Institute of Agriculture, and head of Engineering Department, Ridgetown College of Agricultural Technology (Canada). Directed and carried out research into field drainage, environmental control, vegetable crops mechanization, farm machine performance, reversible tractors, maize harvesting, mulch laying. Recent professional interests were cereal and forage harvesting and storage, tractors and power units and alternative sources of energy.

Carl Boswell, CBE, TD, CEng, MIStruct E, FIAgrE, Sen MWeldI.

Formerly HM Chief Agricultural Inspector and Regional Director, Health and Safety Executive.

Worked as a Director of an Engineering Company and then as Chief Structural Engineer in the CWS Ltd before starting a career in Health and Safety in 1968. Served as HM Chief Agricultural Inspector 1984–93 and as a Regional Director of HSE 1989–93. A member of the Governing Body of Silsoe Research Institute. Past Chairman of the Agricultural Industry Advisory Committee and EC, ISO and BSI Committees and member of the International Social Security Association Advisory Board for Agriculture.

Paul Brassley, BSc (Hons), BLitt.

Present appointment: Senior Lecturer Agricultural Economics, Seale-Hayne Faculty, University of Plymouth.

After reading agriculture and agricultural economics at the University of Newcastle-upon-Tyne went on to research in agricultural history at Oxford. Since his appointment to his present post at Seale-Hayne has taught agricultural economics and policy and researched agricultural history from the 17th to the 20th centuries.

J.S. Brockman, BSc, PhD.

Present appointment: Principal Lecturer, Agriculture, Seale-Hayne Faculty, University of Plymouth.

After obtaining a PhD for a residual study of the classic Treefield experiment at Cockle Park, helped to found Fisons North Wyke Grassland Research Centre in Devon. Has published many papers on the nutrition and utilization of grassland. Since 1978 has been at Seale-Hayne. President of the British Grassland Society, 1982–3. Associate Editor of *Grass and Forage Science*; Editorial Board of *Plant Varieties and Seeds*; Chairman of British Grassland Society Publications Committee, 1984–93.

J.L. Carpenter, BSc, MSc, CEng, NDAgrE, FIAgrE.

Retired; formerly Senior Lecturer, Agricultural Structures, Seale-Hayne Faculty, University of Plymouth.

Educated at Leeds and Newcastle-upon-Tyne Universities. Taught horticultural mechanization and farm buildings at Writtle Agricultural College, followed by farm structures design and construction at Seale-Hayne. Research interests and papers published include dairy energy use, general water utilization on farms, the provision of water for pigs and the economics of farm utility vehicles. Currently a director of a company advising on water supply equipment for farms.

Adam Carter, BSc (Hons), MSc.

Present appointment: Lecturer in Rural Resource Management, Seale-Hayne Faculty, University of Plymouth.

Graduated in Rural Environment Studies at Wye College, University of London, before completing an MSc in Forestry and its relation to land use at the Oxford Forestry Institute, University of Oxford. Previous experience in British silviculture and forestry extension work in Northern Sudan with Voluntary Services Overseas. Current research interests in forest ecology and silviculture.

Richard Coates, FRICS, MRAC.

Present appointments: Associate Senior Lecturer, Agricultural Buildings, Seale-Hayne Faculty, University of Plymouth. Sole Principal of Rural Design Consultancy, Lemprice Farm, Budleigh Salterton, Devon.

Chartered Building Surveyor and Land Agent with a special interest in design in the countryside – both in agricultural construction and other rural building projects. Thirty years experience in field including 18 years as Resident Building Surveyor to Clinton Devon Estates. Only designer to have won the coveted CLA Farm Buildings Award three times.

R.A. Cooper, CDA (Hons), NDA, MSc, PhD.

Present appointment: Principal Lecturer, Animal Production, Seale-Hayne Faculty, University of Plymouth.

Taught animal production at Shropshire Farm Institute before moving to University of Malawi as lecturer in animal husbandry and assistant farms director. Joined Seale-Hayne in 1974 following MSc at Reading University and completed PhD on interactions between growth promoters and reproductive physiology in ewe lambs in 1982. Main research interests in aspects of goat-production, particularly in milking does, and in water intake studies.

J.C. Eddison, BSc, PhD, CBiol, MIBiol.

Present appointment: Senior Lecturer, Seale-Hayne Faculty, University of Plymouth.

Graduated in Zoology from Leeds University and then conducted PhD research at Aberdeen University into the effect of environmental variation on ecological communities. Post-doctoral research followed at the Edinburgh School of Agriculture investigating the grazing behaviour and ecology of hill sheep. Took up his position in 1983 and now has a number of research projects on the behaviour and welfare of farm animals.

Tim Felton, LLb (Hons) of the Middle Temple, Barrister at Law.

Present appointment: Senior Lecturer, Law and Business Management, Seale-Hayne Faculty, University of Plymouth.

After reading law at Leeds University was called to the Bar of the Middle Temple. Following a period of legal practice he took up a career in practical farming and obtained a Diploma in Farm Management from Seale-Hayne. Prior to taking his present position he was share farming a mixed dairy, arable and beef farm at Tiverton, Devon of 185 hectares. Particular interests include access to the countryside and employment law issues in the rural environment.

Michael P. Fuller, BSc, PhD.

Present appointment: Principal Lecturer, Crop Physiology, Seale-Hayne Faculty, University of Plymouth.

Graduated from Leicester University and then completed a PhD at the Welsh Plant Breeding Station on frost resistance in grasses. Employed as a research demonstrator at Leeds University and as a Research Trials Officer at the Sports Turf Research Institute before taking up his present position in 1980. He currently has research programmes on biotechnological manipulation of cauliflowers, frost resistance and field evaluation of composted domestic refuse and sewage sludge.

Richard P. Heath, BA, NDA, NDAgrE, IEng, MIAgrE, Cert Ed.

Present appointment: Senior Lecturer, Mechanization, Seale-Hayne Faculty, University of Plymouth.

Trained at Shuttleworth and Writtle Colleges. Joined the Engineering Department staff in 1970, teaching TEC diploma, higher diploma and degree levels. Interests are mechanization aspects of crop production and dairying. Research and investigation extend his teaching interests. A member of the Institution of Agricultural Engineers, he has held branch and national officer positions.

Anita J. Jellings, BSc (Hons), PhD.

Present appointment: Senior Lecturer, Seale-Hayne Faculty, University of Plymouth.

After reading Applied Biology at the University of Bath, completed at PhD at the Plant Breeding Institute with a cytogenetic study of cereal ovules. This was followed by a research fellowship at the University of York investigating constraints on photosynthetic capacity. Current interests include cellular development of *in vitro* cultures, integrated crop management and sustainable agricultural systems, and brassica crop improvement.

J.A. Kirk, BSc, PhD.

Present appointment: Principal Lecturer, Animal Production, Seale-Hayne Faculty, University of Plymouth.

Spent nine years farming before reading agriculture with agricultural economics at the University College of North Wales, Bangor. Postgraduate research on the growth and development of Welsh Mountain lambs led to a PhD. Current research interests are the growth and development of animals and the improvement of meat quality.

Graham Moule, BSc (Hons), MSc.

Present appointment: Senior Lecturer, Crop Production and Forestry, Seale-Hayne Faculty, University of Plymouth.

Trained at London and Exeter Universities before taking up present appointment in 1971. Professional interests include crop health, the epidemiology of crop pathogens and tree establishment techniques.

R.M. Orr, BSc, PhD, MIBiol.

Present appointment: Senior Lecturer, Seale-Hayne Faculty, University of Plymouth.

Read agriculture with animal science at the University of Aberdeen. Postgraduate research at Edinburgh University on appetite regulation led to a PhD. Teaching and research interests include food science and nutrition.

Robert Parkinson, BSc (Hons), MSc, PhD.

Present appointment: Principal Lecturer, Soil Science, Seale-Hayne Faculty, University of Plymouth.

Studied at Leeds and Reading Universities before taking up a post as Research Officer at Birkbeck College, University of London, working on soil water dynamics and agricultural drainage. This research subsequently led to the award of PhD. On moving to Seale-Hayne, research interests have broadened to include efficiency of drainage systems, fate of animal and human wastes recycled to agricultural soils, interactions between cultivation systems and the soil, and wider aspects of the environmental impact of agricultural practices.

John I. Portsmouth, NCP, NDP, DipPoult, NDR, FPH.

Present appointment: Managing Director, JP Enterprises, Consultant Nutritionist, Tremaen, Maenporth Hill, Falmouth, Cornwall.

Began his own private consulting service in 1991 following some thirty years in animal nutrition and animal health industry. Author of several books and many technical articles on poultry nutrition and management. Specialist subjects are calcium metabolism in laying hens and micro-nutrient requirements of poultry. In 1988 given a distinguished service award for services in nutrition to the UK poultry industry.

David Sainsbury, MA, PhD, BSc, MRCVS, FRSH, CBiol, FIBiol.

Present appointment: Director of The Cambridge Centre for Animal Health and Welfare, 101 Madingley Road, Cambridge.

After a period of research and lecturing at the Department of Veterinary Hygiene, Royal Veterinary College, London, until 1993 worked at the Department of Clinical Veterinary Medicine, University of Cambridge in the Division of Animal Health. Main interests have been concerned with the health and well-being of farm livestock especially under intensively managed conditions. Author of five textbooks and some 130 scientific papers.

R.W. Slee, MA, PhD, Dip Land Economy.

Present appointment: Senior Lecturer Rural Economics, Department of Agriculture, University of Aberdeen.

Before moving to Aberdeen in 1989, spent over ten years as a lecturer at Seale Hayne Faculty, University of Plymouth. Has written widely on rural economic change, including contributions to *The Changing Countryside* (eds J. Blunden and N. Curry, Croom Helm, 1985) and *A Future for Our Countryside* (eds J. Blunden and N. Curry, Blackwells, 1988) Author of *Alternative Farm Enterprises* (2nd edn, Farming Press, 1989). Current research interests include agri-environmental policy questions and the economic impact of rural tourism.

Richard J. Soffe, HND, MPhil, MCIM, MBIM, M Inst M.

Present appointment: Senior Lecturer, Farm Business Management, Seale-Hayne Faculty, University of Plymouth.

Previous appointments have included a lectureship at Sparsholt College, Winchester and assistant farm manager of a large estate in Hampshire. Research interests include financial management and the marketing of financial products to farmers. He was co-editor of the eighteenth edition of *The Agricultural Notebook* and is editor of the current edition.

Mark A.H. Stone, BSc (Hons), MSc, (Econ), MIPD.

Present appointment: Senior Lecturer, Human Resource Management, Seale-Hayne Faculty, University of Plymouth.

After reading economics and politics at the University of Central Lancashire, went on to a masters degree in employment studies at University College, Cardiff. After working short-term for the Barry Development Partnership and Portsmouth City Council, spent four years working in industry. During this time he achieved membership status within the Institute of Personnel and Development (previously I.P.M.). Current research interests include the use of information technology to allow individuals within rural organizations/ businesses to update and develop their management skills.

M.A. Varley, BSc, PhD, MIBiol, CBiol.

Present appointment: Lecturer, Department of Animal Physiology and Nutrition, University of Leeds.

Before his present position, spent four years as a lecturer in animal science at Seale-Hayne Faculty, University of Plymouth, and then moved to the Rowett Research Institute where he was a Senior Scientific officer for six years. Research interests include reproductive physiology of the sow, neonatal immunology and animal behaviour and welfare.

Martyn Warren, BSc (Hons), MSc.

Present appointment: Head of Land Use and Rural Management Department, Seale-Hayne Faculty, University of Plymouth.

Started teaching farm management at Harper Adams Agricultural College in 1974 before moving to Seale-Hayne where for several years he was Course Manager of the Diploma in Farm Management (now the Diploma/MSc in Agricultural Business Management). An experienced trainer in management for rural businesses, his publications include various studies of agricultural change and the resulting training needs. His book *Financial Management for Farmers*, now in its third edition, has become a standard student textbook in the UK and abroad.

Part 1

Farm Management

1

The Common Agricultural Policy of the European Union

P.W. Brassley

Background, institutions and the legislative process*

The need for a Common Agricultural Policy

When the European Community was created, it needed a Common Agricultural Policy for two reasons: first, the agricultural industries in each of the member states were subject to government intervention, and, second, if this intervention were to continue, it had to be compatible with the other provisions of the Community.

Left to themselves, in the absence of government intervention, the markets for agricultural products in developed economies will produce:

- fluctuating prices, and, as a result, fluctuating incomes for producers; and
- gradually declining prices, in terms of purchasing power, and, as a result, declining real incomes for producers in the long run.

The price of agricultural products, like that of any other product in a free market, depends upon the balance of demand and supply. If demand increases less than supply, the price will fall.

The demand for agricultural products in a developed country does not usually increase rapidly, unless those products can be sold cheaply on the world market. Technical change produces increased output per hectare and per unit of labour, so supplies tend to increase more rapidly than demand. Thus, in the long run, prices tend to fall. Demand is also relatively unresponsive to price changes, so short-run supply variations, caused by changes in weather and disease, produce short-run price fluctuations.

Short-run price fluctuations in a free market may result in the demise of businesses which would be viable at average levels of input and output prices, and the possibility of this event creates a disincentive to investment on such farms. In the long run, falling prices produce lower incomes for those farms which are unable either to reduce costs or to increase output. If such low incomes are unacceptable, they may cease trading altogether. It is usually found that farm incomes have to fall to very low levels before farmers leave the agricultural industry, and in any case the land that they no longer farm is often taken over by another farmer. The total output of the industry is therefore maintained, so there is no tendency for prices to rise. While farmers with low incomes may be found in all areas, some regions may be particularly disadvantaged by physiographical or structural factors (i.e. infertile land and small farms), and if agriculture is the major industry the whole region may be affected by income and outmigration problems.

In a free-trading economy, farm income problems may be further intensified by the availability of imported agricultural products at prices below the domestic supply price. Imports will not only restrict domestic price rises but will also increase the total import bill, which may be considered a problem in countries with balance of payments deficit problems. However, the government of a country which has no difficulty in exporting enough to pay for food imports may still consider it unwise to be totally reliant on imported food supplies: unforeseen circumstances may give rise to difficulties in obtaining supplies, or something may reduce the price advantage of imports in the long run. A capacity for rapid expansion of domestic agricultural output may therefore be desirable even though considerable reliance is placed on imports. In short, free markets for agricultural products in developed free market economies may produce low farm income, balance of payments and potential supply security problems.

By the middle of the 1950s the agricultural industries in all western European countries faced these problems to some extent, and governments had evolved a number of policies for dealing with them. In the continental European countries, agricultural policies usually involved a system of import controls and price support, which kept price levels higher than they would have been in a free market and so maintained agricultural incomes and self-sufficiency levels. When the discussions which led to the formation of the Community were held in 1955–6, it was agreed that the exclusion of agriculture from the general common market was impossible: without free trade in farm products, national price levels could differ and those countries with the lowest food prices would

* Since the Maastricht Treaty came into force on 1 November 1993, the CAP has been under the control of the European Union. Therefore, except when discussing the actions of the European Community, the following pages will refer to the European Union or the EU.

have the lowest industrial costs, thus undermining the common policies which would be introduced for other industries. Although there was a general similarity between the existing agricultural policies, the differences in detail were so numerous that the simple continuation of existing policies would have been impossible. Not only did the Community countries require an agricultural policy, they required a *common* agricultural policy.

The process of decision making in the European Union

The European Union must make decisions about many areas of policy without, on the one hand, being bogged down in the process of consultation, or, on the other, neglecting the needs and views of fifteen member states, over 300 million individual citizens, and numerous lobbies and interest groups. Decisions must be made about what policy shall be (primary legislation) and how it should work in detail (secondary legislation). Several institutions are involved in this process, of which the two most important are the Commission and the Council.

The *European Commission* is presided over by 20 Commissioners. They are appointed by agreement between the member governments, but are required to act in the interest of the EU and not of individual governments. From 1995 they will serve for terms of five years. The Commission employs roughly 16 000 officials, of whom about a quarter are translators, and they work in 14 special services such as the legal department and the statistical office, and 23 Directorates General (DGs) which are concerned with policy areas. Thus DG II deals with Economic and Financial Affairs, DG VI with agriculture, DG XI with the environment, DG XIV with fisheries, and so on.

Individual Commissioners have responsibility for one or more DGs. Therefore in some ways the Commission is like a national civil service, but it also has extra functions: it is responsible for proposing primary legislation; it handles the day-to-day administration of Community laws and policies resulting from this primary legislation, and may enact secondary legislation in order to do so; and it represents the Community in its relations with non-member states. Thus Commissioners are responsible for representing the Community in the negotiations under the terms of the General Agreement on Tariffs and Trade (the GATT negotiations).

The *Council of the European Union* is the major legislative body of the Community, and consists of a minister from each member state, depending on the subject under discussion: agriculture ministers for agricultural matters, finance ministers for financial matters, and so on. The chair is held by an individual country for a six-month period, and rotates from country to country in alphabetical order, according to the name of each country in its national language: Greece (E for Ellas) therefore follows Denmark. It is normally the final decision-making body for all primary legislation, although major constitutional issues or especially insoluble problems may be passed on to the *European Council*, a meeting of the heads of government of the member states which has no legal basis in any treaty and which developed from earlier summit meetings. Its decisions have to be passed back to the Council of the European Union to be given legal validity.

Since the Community is a partnership of the member states, it makes its decisions by negotiations in a series of committees. The annual review of agricultural prices is a good illustration of this process. The initiation of *primary legislation* is normally the responsibility of the Commission. The first moves are made by the appropriate department in DG VI, usually in consultation with national civil servants, members of trade associations, and independent experts, all of whom may serve on expert working groups chaired by a member of the Commission staff. Other Commission departments with a legitimate interest in the proposals are consulted at an early stage, since agricultural policy measures might affect negotiations on, for example, external trade or monetary affairs. By the time the draft proposals are nearing completion the pressure groups make their views known to the Commission. There is a wide variety of these groups, from BEUC, the European Consumer organization, to the European Environmental Bureau (EEB), and various European trade associations such as COCERAL (the grain trade organization), ASSILEC (the dairy industry), CIAA (the food industry), and, perhaps most influential, COPA, the Committee of Professional Agricultural Organizations, which acts for farmers' unions in the Community. When DG VI consider that the draft is complete, they submit it to a full meeting of the 20 Commissioners for their approval.

Once this approval is given, the draft becomes a Commission proposal and is submitted to the Council of the European Union and consultative bodies such as the European Parliament, the Economic and Social Committee and the relevant Management Committees (which are mostly concerned with secondary legislation – see below).

For most proposals the Council, before it makes a decision, is required to receive the opinion of the *European Parliament*. The Parliament carries out this part of its work by nominating a committee which produces a report on the proposal which may then be debated by a plenary session of Parliament before becoming the Opinion which is passed on to the Council. In practice this opinion carries little weight and may often be ignored. The Parliament also has considerable indirect impact on the formulation of policy through its questioning of members of the Commission, both formally and informally.

The Council must also receive the Opinion of the *Economic and Social Committee* (ESC), which is a body set up under the Treaty of Rome to advise the Council and the Commission. It consists of members who are nominated by national governments in their personal capacities, often because of their experience or knowledge as employers, trade unionists or consumers.

Meetings of the Council are often complicated and lengthy, and so require detailed preparation. This preparation is the task of the *Committee of Permanent Representatives* (COREPER) for non-agricultural matters, and of the *Special Committee for Agriculture* (SCA) for agricultural legislation. Both of these committees consist of senior civil servants from the member states, meeting with members of the Commission. Detailed consideration of the proposals is carried out in working parties of national civil servants. The purpose of this procedure is to identify and, if possible, to resolve points of conflict. If it is possible to produce a draft proposal which is acceptable to all member states, it is returned to the Council of Ministers on the 'A' list, which can be passed by the

Council with no further discussion. Otherwise it returns to the Council as a 'B' point, for further negotiation.

While this process of sorting out the agenda for the Council meeting is going on, the business of lobbying continues. Pressure groups, national government ministers and other politicians express their views on the Commission proposals, both at a European and at a national level. The President of the Council of Ministers is required to hold meetings with both COPA and, since 1980, BEUC, and to report on them to the rest of the Council. Nevertheless, most of the political pressure on individual ministers in the Council comes through national lobbying channels, so that national political problems, such as a forthcoming national or even local election, can have significant effects on the decisions made.

Thus any proposal, before a decision is taken on it in the Council, will have been the subject of comment by a wide variety of formal and informal, corporate and individual, Community and national, expert and lay sources. The issues which remain to be resolved by the Council should be the basic political ones. After debating a proposal, a decision must be made. Some legislation requires the unanimous approval of the Council, and a quasi-formal agreement, known as the 'Luxemburg Compromise', provides for a member state to exercise a veto if it believes that its 'vital national interests' are at stake. But increasingly, and especially since the passage of the Single European Act, most decisions are taken on the basis of a *qualified majority*, which requires 62 or more votes in favour. Each country has the number of votes shown in Table 1.1.

By this process the Council is said to 'adopt a common position'. Since the adoption of the Single European Act this is not the end of the decision-making process. The common position is then sent to the Parliament, which has three months to carry out a second reading and make its opinion known. It may:

(1) approve or take no decision, in which case the Council adopts the measure in question and it is effectively passed;
(2) reject the common position by an absolute majority, in which case the Council may maintain it and adopt it, but only if it can do so unanimously; or
(3) amend the common position by an absolute majority, in which case the Commission revises its proposals within one month and re-submits them to the Council, which may then:

- adopt the Commission proposal without change by a qualified majority;
- adopt the Commission proposal after amendment, but only if it can do so unanimously; or
- reject the proposal, in which case it lapses.

Table 1.1 Votes in the Council of Ministers

Country	*Number of votes*
France, Germany, Italy, UK	10 each
Spain	8
Belgium, Greece, The Netherlands, Portugal	5 each
Austria, Sweden	4 each
Denmark, Ireland, Finland	3 each
Luxemburg	2

The proposals thus adopted fall into three categories. A *Regulation* is directly applicable to all people and governments in all member states. A *Directive* is binding as to its intention on governments, which must then pass legislation to give it effect, so that it is more flexible than a Regulation. A *Decision* is as binding and immediately applicable as a Regulation, but only on the people or governments to whom it is addressed.

Secondary legislation, which is concerned with the day-to-day running of the CAP (e.g. the administration of tenders to intervention stores, or setting the level of export refunds) is largely the responsibility of the Commission. In formulating its proposals, it may or may not take the advice of an advisory committee of representatives from all sides of the appropriate industry. The proposal is then considered by a *Management Committee* made up of national civil servants from the relevant division of the national ministries and Commission officials from the relevant Directorate General. There is a Management Committee for each of the main commodities, and others for agri-monetary affairs, plant health, structural policy, and so on. The frequency of the meetings reflects the amount of work to be done: the beef committee may meet every fortnight, whereas the research committee may only meet a few times a year. The Committee gives its opinion on the Commission proposal using the same qualified majority system as the Council. After a proposal has passed through the Management Committee and been adopted by the Commission it becomes, in effect, law.

Community law is thus the primary and secondary legislation produced by the Council and the Commission, together with the law embodied in the Treaty of Rome and the various treaties of accession to the Community. It takes precedence over national law. It is the responsibility of the *Court of Justice* to decide whether or not Community legislation has been correctly applied, is being applied, or is flawed. Some of the decisions of the Court have had a substantial effect on the way in which the Community is run, from deciding upon the powers of the Parliament to defining what means may legitimately be used to prevent trade in food products.

The *Court of Auditors* exists to audit the expenditure of the Community and to ensure that its finances are properly managed. In this context it has published a number of reports critical of the operation of the CAP and has highlighted the problems of CAP fraud.

The formation and development of the Common Agricultural Policy

The origins and development of the European Union

The treaty which established the European Community was signed in Rome on 25 March 1957, and the Community came into being at the beginning of 1958. The main objective of the original six member states was to promote economic development by removing the barriers to trade between them. As they gradually succeeded in bringing this about they were joined by other countries:

Date of accession	*Member states*
1958	Belgium, France, West Germany, Italy, Luxemburg, The Netherlands
1973	Denmark, Ireland, UK
1981	Greece
1986	Spain, Portugal
1995	Austria, Sweden, Finland

By 1993 most of the other countries of northern and western Europe had made formal applications for membership, and many eastern and southern European countries were seriously discussing the possibility of applying. Since agriculture in the northern European countries is highly protected, and in the eastern European countries is potentially very productive, this expansion has major implications for the future of the CAP.

The integration of the EU was further developed by the signing of the *Single European Act*, which came into force in July 1987 and provided for the removal of all tariff barriers by the end of 1992, moves towards monetary union, extended use of majority voting in meetings of the Council, and further powers for the European Parliament. Further progress towards economic and political union was brought about by the Maastricht Agreement of late 1991.

The origins of the CAP

Articles 38–47 of the Treaty of Rome apply directly to agriculture. The products covered by the CAP are listed in Annex II of the Treaty. In practice, some products not listed there (and so called 'non-Annex II products') are sometimes affected by CAP rules. Article 39.1 lays down the objectives of the CAP:

(1) to increase agricultural productivity by promoting technical progress and by ensuring the rational development of agricultural production and the optimum utilization of the factors of production, in particular labour;
(2) thus to ensure a fair standard of living for the agricultural community, in particular by increasing the individual earnings of persons engaged in agriculture;
(3) to stabilize markets;
(4) to ensure the availability of supplies;
(5) to ensure that supplies reach consumers at reasonable prices.

Article 40 lays down the broad guidelines for the various policy instruments by which these objectives are to be achieved, and Article 43 indicates the procedure to be followed in reaching agreement on the detailed provisions of the CAP and requires the Commission to submit detailed proposals to the Council of Ministers. The Commission produced these proposals in 1960, and although altered in detail, and changed more substantially in 1992, in many ways they continue to form the basis of the CAP described below. It is interesting to note that three principles which are often assumed to lie behind the CAP system (a single market, joint financing, and Community preference) appear neither in the Treaty of Rome nor in the Commission proposals, but seem to have appeared early in the working life of the CAP. This was in the 1960s: the first products were subjected to a common set of market rules in 1962, and common prices were applied in 1968.

An outline of the original price mechanisms

The underlying principle of the CAP price mechanism was that the producer should receive a price determined by market forces. But these market forces were controlled so that the market price fluctuated only between predetermined upper and lower limits. Thus the farmer was protected against excessively low prices, and the consumer against excessively high prices.

This price support mechanism recognized that farm products might be produced both within and outside the EU. The demand for farm products was (and is) relatively constant, so that if the market were supplied only from within the EU the major reason for low prices would be an excess supply. The effects of this could be mitigated by artificially increasing demand, so the EC entered the market to buy up farm products for storage when prices were low, a process known as *intervention*. When prices subsequently rose, this stored produce could be released on to the market again. If prices remained low, intervention stocks could be sold on the world market with the aid of subsidies known as *export restitutions* or refunds.

Many farm products might also be supplied from countries outside the EU (known as *third countries*). If these imports were available at less than the EU market price, that market price would be reduced, so they were subject to *variable levies* to raise the price at which they might profitably be sold on the internal EC market. Conversely, when EC prices were high, perhaps as a result of supply shortages, the entry of produce from third countries would serve to reduce prices by increasing supplies.

In the UK, a national agency, the *Intervention Board for Agricultural Products* (IBAP), was established to deal with intervention, import levies, export restitutions and all the payments made under the Guarantee section of the EAGGF (see below). It did not itself physically handle any goods, but supervised the work of commercial intervention store operators and bodies such as the Home Grown Cereals Authority. The financial resources required to pay for intervention, export refunds, and so on, were obtained from the *European Agricultural Guidance and Guarantee Fund* (EAGGF – often referred to by its French acronym FEOGA). Likewise, import levies and customs duties on imports of farm products were paid into it.

This basic system was used for regulating markets in most of the major commodities produced by EC farmers, including cereals, milk products, sugar, beef and veal, pigmeat, poultrymeat and eggs. Different systems were used for fruit and vegetables and other crops. A sheepmeat regime was introduced later, although there was no regime for wool, which was treated as an industrial product. Neither was there a potato regime, although at the time of writing there are proposals to introduce one. Although there have been amendments and further developments, this system is still recognizable as the basis of the commodity regimes in force today.

The problems of the CAP

Initially, although not without difficulties, the CAP was reasonably successful. It soon became clear that open-ended price guarantees would result in increased output (self-sufficiency in wheat, for example, rose from 89% in the late 1950s to 101% in the mid-1960s (Fearne, 1991), but initially this might have appeared to be consistent with the objectives of the CAP. When, a little later, it began to present problems, the Commission soon recognized them, and put forward proposals for structural change. However, they were watered down by the Council, and eventually resulted only in the three structural directives of 1972 (see below). The potential crisis caused by the breakdown of the post-war system of fixed exchange rate was met by the development of monetary compensatory amounts, although they soon began to require a significant proportion of the agricultural budget.

But the original six member states all had roughly similar agricultural structures and agricultural histories, and had been used to a policy of agricultural self-sufficiency for a century or more. Hence all had roughly similar agricultural policies which did not have to be changed very much to conform with the new common policy. If they had difficulties in making the new common policy work, how much greater might the difficulties be if the Community were expanded to 12 states, some traditional exporters of farm products, some traditional importers, some technically advanced, and some backward? And all spread across Europe from the north west to the south east, from the northern shores of the Atlantic to the eastern end of the Mediterranean.

Consequently, by the middle of the 1980s the CAP was beset by major problems. First, it was producing surpluses of agricultural products which were expensive to store. In 1985/6 the level of self-sufficiency in the 10 member states was:

	per cent
Cereals	121
Vegetables	101
Butter	133
Cheese	107
Beef and veal	108
Poultrymeat	107
Sheepmeat	76

Source: Commission, 1993.

The accession of Spain and Portugal in 1986 produced a small reduction in these figures, but they remained higher than could be justified by the demands of the commercial export market or the requirement to maintain strategic stocks. In part surpluses were the result of technical change and its effect on yields. The table below shows how yields changed in France between 1970 and 1990:

	1970	*1990*
Cereals (t per ha)	3.4	6.1
Sugar (t per ha)	6.7	9.5
Rape (t per ha)	1.8	2.8
Potatoes (t per ha)	14.0	29.0
Milk ('000 kg per cow)	3.1	4.6

Source: Commission, 1993.

For the community as a whole, output has generally risen by about 2% per year while demand has been generally static, and for red meat and some dairy products, as a result of changing consumer tastes, has actually fallen. In an unsupported market this would have brought about price decreases and so reduced the incentive to produce, but Community prices have not been unsupported. Moreover, in theory at least, the existence of a single market price prevailing throughout the Community should have produced a pattern of production which reflected the physical, structural and economic strengths and weaknesses of each of the farming regions. Regions such as East Anglia and the Paris Basin would be expected to specialize in cereals, while Mediterranean regions would produce fruit and vegetables and the cooler, wetter north west of the Community would produce grass-based products.

In practice, Intra-Community trade has indeed increased. Allowing for changes in the value of money and the agricultural area of the EC, intra-EC trade in 1990 was (very roughly) about four times greater than the same figure for 1968. As an alternative measure, it may be compared with the level of spending on farm income support: the intra-EC trade figure was about twice the support figure in 1968 but more than three times the support figure in 1990 (Fearne, 1991; Commission, 1993). But these changes might have been greater still in the absence of price support. The maintenance of high levels of support prices has meant that those producers with relatively high production costs have not been forced to cease production, while low-cost producers have been able to make high profits. In addition, the green money system, the continuation of national support measures, and the existence of non-tariff barriers to trade, such as health and veterinary regulations (some of which were not easy to justify on technical grounds), prevented the realization of the ideal of a community-wide market, and again kept high-cost producers in business. Thus surplus production was a constant feature of Community agriculture.

The surpluses had to go somewhere. Some were simply destroyed. Newspapers printed pictures of tractors rolling over heaps of luscious peaches and juicy tomatoes before ploughing them into the ground. Some surpluses were converted into lower-return products with the aid of subsidies: fruit was juiced and wine converted into industrial alcohol. But much of the excess production was exported to the world market with the aid of export refunds. After allowing for changes in the value of money, exports to non-member states roughly quadrupled while imports remained about constant over the period 1968–90. And the composition of the imports changed. CAP price mechanisms excluded cereals more than oilseeds, protein meals and cereal substitutes such as manioc. Although the EC market for compound feeds expanded, the use of cereals therein decreased, so contributing to the cereals surplus. Thus, despite EC encouragement of domestic production of rapeseed, sunflower, dried fodder and peas and beans, it was imports of soya and cereal substitutes which increased, at the expense of traditional cereal imports such as maize. These trends eventually antagonized the traditional food exporting countries, so that the Uruguay Round of negotiations under the General Agreement on Tariffs and Trade (GATT) were characterized by interminable arguments over agricultural policy.

In addition to the political costs of the surpluses, the cost of the Community's budget was not insignificant. For most of the 1980s, spending on agriculture accounted for at least 60% of total budget spending. By 1990 it was over 26 billion ECU, which in real terms, and after allowing for the expansion in the size of the Community, represented more than twice the level of expenditure in the late 1960s (Commission, 1993).

Environmental problems were another result of farm income support through the price mechanism. The more farmers produced the more support they received. Consequently agricultural production was encouraged on land which would otherwise have been left uncultivated, or only extensively cultivated. Hedges were removed, wetlands drained, moors and heaths reclaimed, and so on. The returns to applications of fertilizers and pesticides were increased, so provoking their greater use. The land affected by these changes was no longer so valuable as a wildlife habitat. This might have been acceptable had food been in short supply; at a time of surpluses, when increasing numbers of people were interested in wildlife and landscape for recreational purposes, it was seen as a misuse of resources.

Nevertheless, it might have been justifiable to some extent if it had brought about a more desirable distribution of income between individuals, regions, or even member states. But the CAP raised food prices, because the consumer was no longer free to buy from the world market, where prices were generally lower. Rich regions benefited more than poor, because most of them happened to be in the north of the Community, and northern products were supported more than Mediterranean products. And the countries which gained most from the CAP were not necessarily the poorest, but simply those which produced most farm products and imported least. Perhaps most importantly, all this expenditure did not go to help the poorer farmers and farm workers. In 1992 the Commission estimated that 80% of EC farm spending went to only 20% of the farmers, and that these were 'generally the bigger and more efficient ones' (Commission, 1993). Between 1960 and 1990 the number of farmers and farm workers in the original six member states of the Community fell from more than 10 million to less than 5 million. The number of farms in these countries decreased from 5.9 million in 1970 to 4.7 million in 1987 (Commission, 1993). By the end of the 1970s it was increasingly accepted that the CAP was not meeting its objectives.

Reforming the CAP

The first attempts to change the CAP came in the late 1970s. The most immediate problem at that time was the surplus of milk products, and so in 1979 a co-responsibility levy was imposed on milk producers to help meet the cost of disposing of the surplus and to reduce the incentive to produce more milk. It failed to bring about balance in the market, and in 1984 milk quotas were introduced, although the total quota allowed the production of more milk than was consumed in the Community. In 1984 a ceiling on agricultural spending was introduced, whereby the increase in the farm budget was limited to 74% of the rate of economic growth in the EC. At the same time the stabilizer system was introduced, which provided for reductions in support prices if a maximum guaranteed quantity of production was exceeded. But these changes were not enough. Surpluses remained, the cost of export subsidies was increased by the falling value of the dollar in the late 1980s, rising expenditure put pressure on the agricultural budget, sales to the former Soviet Union decreased as its economic difficulties increased, sales to the Middle East were affected by the Gulf War, and the USA and most of the rest of the world in the GATT negotiations were arguing for major changes in the CAP. The economic and political pressures for reform of the CAP were continuing to build up.

The Commission recognized the need for more radical reform, and in 1991 presented proposals to this effect. The objectives of the reform package were subsequently stated (Commission, 1993) by the Commission to be:

(1) to maintain the Community's position as a major agricultural producer and exporter by making its farmers more competitive in home and export markets;
(2) to bring production down to levels more in line with market demand;
(3) to focus support for farmers' incomes where it is most needed;
(4) to encourage farmers to remain on the land;
(5) to protect the environment and develop the natural potential of the countryside.

However, it is interesting to note that in the objectives in its original proposals the Commission listed encouragement of the non-food use of agricultural products and better incentives for farmers to take early retirement (Commission, 1991). Whether these proposals are supposed to be additional to the objectives stated in the Treaty of Rome, or a replacement for them, has never been clearly stated, but there are some clear conflicts between the two lists.

The reform package finally adopted in 1992 – sometimes known as the MacSharry package after the Commissioner in charge of agriculture at the time – was based on:

- cuts over a 3-year period in the prices of the main products in surplus, in order to bring EC market prices closer to world prices. It was originally proposed by the Commission that the threshold price should be set at or around the world price, but by the time the package was agreed by the Council the threshold price for cereals was to be significantly above the world price;
- set-aside, or the withdrawal of land from production, and the encouragement of extensive beef production, in order to bring about as rapid a reduction as possible in the level of output;
- compensation for the producer for the resultant fall in income by area or headage payments available only to farmers participating in the set-aside schemes;
- measures to improve the environment and encourage forestry.

The milk regime, having undergone major change with the introduction of quotas, was only changed in detail, and the sugar regime was left unchanged, to be dealt with on a future date. Contrary to the original Commission proposals, the Council eventually agreed that compensation would be paid on all set-aside, and that big sheep flocks would be paid at 50% of the full rate for ewes in excess of quota, both of which decisions were felt to be favourable to countries with many big farms. Subsequently, at the European Council of December 1992, it was agreed to maintain the limit on CAP spending to a maximum of 34.8 billion ECU in 1999.

The details of the new commodity regimes are described in the following section on CAP price mechanisms.

CAP price mechanisms

Cereals

Before the reform of the CAP in 1992, the incomes of cereal producers in the EU were maintained by supporting the prices they received for their products at a level higher than that which would have been produced by unregulated market forces. This system remains in force after the reforms, but at a lower level of prices. Producers who put a proportion of their land into a set-aside scheme also receive compensation payments, to compensate them for the lower price levels and the loss in output from the land set aside. This regime applies to cereal crops such as wheat, durum wheat, rye, barley, oats and maize; oilseeds such as soya beans, rape and sunflower; and protein crops such as peas, field beans and lupins.

The price support system distinguishes between cereals produced within the Community and those imported. The terms relating to each are set out in Fig. 1.1.

The intervention price is set at a level that is roughly similar to what the EU believes the world price is likely to be when the system is fully operational, with the target price being roughly 10% higher. The underlying principle is that the possibility of sales into intervention stores should prevent the domestic market price for cereals falling much below the intervention price.

In practice, intervention buying is controlled by several detailed regulations which usually modify the intervention price. For example, there are monthly increments which are paid in equal amounts between November and May, to compensate for the cost of storing grain.

Intervention stores are only open between November and May in northern member states (and August to April in southern member states). Not all grain is acceptable for intervention. Although intervention is mandatory for common wheat, durum wheat, barley, grain maize and rye, not all wheat qualifies as common wheat. To qualify for the full common wheat intervention price, wheat must have a Hagberg falling number of at least 220 seconds, a Zeleny index of at least 20, a protein content of at least 11.5% and, if the Zeleny index is less than 30, the ability to pass the dough machinability test. These are milling quality standards. Other intervention standards

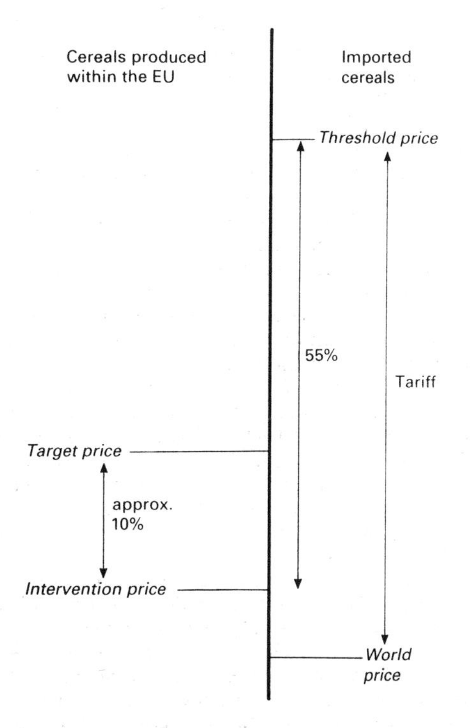

Fig. 1.1 The cereal price mechanism.

change from time to time, but they are normally specified in terms of moisture content, colour, smell, impurities, freedom from pests, and specific weight.

There are also minimum quantities which may be offered into intervention. The EU standard is a minimum offer of 100 t, but individual member states may adopt higher standards. For example, the UK has a minimum quantity of 500 t of wheat and 100 t of other grains.

Thus the farm-gate price of cereals may be below intervention price, because of adjustments for quality and transport costs from the farm to the intervention store and for the 30- to 35-day delay in payments for grain sold to intervention stores. Having established the appropriate price in ECU, it must then be converted into national currencies. This conversion is carried out at the current green rate, so that the intervention price in national currency terms may change as often as the green rate changes.

The current state of the internal community market is taken into account by the Commission when deciding whether or not to make sales from intervention. Grain may be sold either for export or to the internal market. Traders offer tenders for these sales to the domestic market, and these tender prices must not undercut the current market price and must be higher than the intervention price. Export sales normally qualify for *export refunds*, the size of which is determined by the difference between the EU market price and the world price.

Since the price on the world market could be, and often has been, lower than the intervention price, it would be worth while for traders to import grain simply for sale into intervention in the absence of any controls. In fact imports are controlled by the issue of import licences, and when world prices are lower than the threshold price a tariff is payable. From 1995 to 1996 the level of the threshold price will be equal to the target price plus the compensation payment expressed on an ECU per tonne basis. In other words, it will be significantly (55%) higher than the intervention price. The size of the tariff, according to the terms of the settlement of the GATT negotiations of 1993, is in theory determined by the difference between the world market price and 110% of the intervention price. This tariff was then to be reduced by at least 15% for any one commodity, and an average of 36% for all commodities, over a period of six years from the base year. In practice, the intervention price in the chosen base year (1986–8) was higher than the threshold price under the reformed CAP. It was also agreed in the 1993 GATT settlement that the minimum import price should not be higher than 155% of the intervention price, so in practice it is the difference between the threshold price and the world price which determines the size of the tariff.

In addition to this manipulation of the cereals market, the cereals regime provides for compensation payments to cereal producers who set aside a proportion of their land (see below). The logic of the system is that the compensation (in ECU per tonne) plus the target price will be equal to the threshold price when the reformed regime is fully operational from 1995 to 1996. But although compensation payments are defined in ECU per tonne they are paid on an area basis. The compensation per tonne is multiplied by the average historic yield for the region in which the farm is located to give a compensation payment per hectare in ECU.

For example, for England in 1993–4 the compensation payment was 25 ECU per t, the regional average yield was 5.93 t per ha, so the compensation payment was 148.25 ECU per ha of cereals.

This ECU figure is then converted into the appropriate national currency (according to Regulation 1068/93) at the green rate applying at the start of the marketing year (which in the case of cereals is 1 July). If the currency has been revalued, so that the value of the compensation payment in national currency terms is lower than it was in the previous year, the member state concerned may ask for an increase in the ECU amount of the payment, and the new figure will apply to all member states.

Each member state decides upon the number of regions for compensation payments. In the UK there are five regions (England, Scotland non-LFA, Scotland LFA, Wales and Northern Ireland), whereas in France there are 90 regions. Regional average yields for cereals are also decided by the member state, subject to the approval of the Commission. In the UK they vary from a maximum of 5.93 t per ha for England to a minimum of 4.65 t per ha for Wales. These figures are generally lower than the yield actually produced. There are different compensation rates and average yields for oilseeds and protein crops. A maximum total of area payments for each region (known as the *regional base area*) is determined on the basis of the hectares sown to each crop in the period 1989–91: exceeding the maximum results in a pro-rata reduction in all payments for the year, and the obligation to increase the set-aside area without any extra set-aside payments.

Farmers who produce more than 92 t (equivalent to about 15.5 ha at English regional average yields) of arable crops are only eligible for compensation payments if they *set aside* a proportion of their arable land. If land is set aside

- on a *rotational* basis, 15% of the total area on which Arable Area Payments (i.e. compensation payments) are claimed must be set aside between January and August of the year in question, and the same land can only be put into set-aside once every six years;
- on a *non-rotational* basis, it must remain so for five years, and 18% of the area on which Arable Area Payments are claimed must be set aside.

There is a simplified scheme for farmers who produce less than 92 t of cereals which does not require them to set any land aside. They receive area payments at the cereal rate. The purpose of this scheme is clearly to absolve small farmers from the necessity of set-aside. In fact any farmer may join this scheme, but of course compensation payments will be limited to those payable on the first 92 t expressed at an area rate (i.e. about 15.5 ha in England). For producers with large acreages, only those who could virtually guarantee high yields sold at good prices would therefore find this option attractive. Under certain circumstances it is possible to grow non-food crops such as short-rotation trees or high erucic acid rape on set-aside land.

In some circumstances it may be possible to have a mixture of rotational and non-rotational set-aside. The precise regulations about what land can be set aside, what is eligible, and what can be done with the set-aside land are contained in the explanatory documents produced by MAFF on the Integrated Administration and Control System (IACS) and Arable Area Payments. Since the details of these schemes can change from year to year it is vital to read the current editions of these carefully before making any significant farm management decisions.

Oilseeds

The system of compensation or area payments payable to cereal producers, and the accompanying set-aside regulations, also applies to oilseeds, but there are differences between the cereals and oilseeds regimes because the structures of the oilseeds and cereal markets are different. Whereas the Community is self-sufficient in most cereals, it is not so in oilseeds, with the exception of olive oil. In the early 1990s the EU produced about 60–65% of its consumption of vegetable oils (including olive oil) and 20–25% of its consumption of meals derived from oilseeds. Therefore considerable quantities of oilseeds are imported to supply the domestic crushing industry, as well as large amounts of oils and oil meals, mostly for use as protein feeds. Consequently a support regime which allows oilseeds to be traded at (or close to) world price levels is required.

For northern European producers the important regime is the one which applies to rapeseed, sunflowerseed and soya beans. Growers of these crops receive a return from selling on the open market (which, since there are no barriers to imports, operates at world market prices) and from area payments. The area payments are calculated according to the following procedure:

(1) The Commission believes that oilseed and cereal production will be equally profitable when the oilseeds price is 2.1 times higher than the cereal price. Therefore if the cereal price were 150 ECU per tonne, an oilseed support price would be set at $150 \times 2.1 = 315$ ECU per tonne.
(2) The Commission sets a projected reference price (PRP), which is effectively its estimate of the world price for the coming marketing year, and thus the price likely to be received from the market by the producer. This is subtracted from the support price to determine the compensation or support to be paid per tonne. So if the PRP were set at 163 ECU per tonne, the support would be $315 - 163 = 152$ ECU per tonne.
(3) The support per tonne is then multiplied by the EU average yield to give the EU Reference Amount per hectare. If the average yield were 2.36 tonnes per hectare the Reference Amount would therefore be $152 \times 2.36 = 358.72$ ECU per ha.
(4) The Reference Amount is then modified by the regional average yield to give the regional payment per hectare. In the UK the regions for oilseeds purposes are the same as those used for cereals compensation payments. Thus if the regional average yield were 3.14 t per ha the Reference Amount would be increased by 3.14/2.36, so the regional payment per ha would be $358.72 \times (3.14/2.36) = 477.28$ ECU per ha.

These regional reference amounts are paid in two instalments, the first after planting, and the second (which may be reduced if the actual reference price has risen by more than 8% of the PRP) after harvest. The green rate used for conversion of these ECU amounts into national currencies is the one applying on 1 July of each year.

A maximum guaranteed area (MGA) applies to oilseeds. If the area planted exceeds the MGA, the area payments are reduced by 1% for each 1% excess planting in member states exceeding their own national reference areas, and the adjustment is made to the second instalment of the area payment.

Protein crops

The regime for protein crops – peas (including vining peas), field beans, linseed and sweet lupins, whether grown for grain, seed or fodder – is similar to that for oilseeds in that there is no support for the market price but there is an area payment and a set-aside requirement. The area payment is calculated by multiplying the notional compensation figure (in ECU per t) by the regional average yield.

For example, in 1993–4 the notional compensation was 65 ECU per t, and the regional average yield was 5.93 t per ha, so the compensation payment was $65 \times 5.93 = 385.45$ ECU per ha.

Sugar

Target, intervention and threshold prices exist for sugar as for cereals, but there are some significant differences from the cereals regime: intervention applies to the processed product and not to sugar beet or cane; intervention is limited by quotas; and there is guaranteed access for an agreed quantity of imports from ACP (African, Caribbean and Pacific) countries, mostly former colonies of EU member states).

The important price for farmers is the minimum beet price, which is the price which processors are required to pay to producers for production within the A quota. It is 98% of the basic beet price, which is derived from the intervention price, adjusted for the processing margin, the sugar yield of the beet, receipts from the sale of molasses, and costs of delivery to processors. The A quota for the whole European Union is approximately equal to domestic consumption. There is also a B quota, and production within the B quota is subject to a levy which has in the past varied from almost 40% to about 23% of the A quota price. No minimum beet price is set for beet used to produce C quota sugar, which must be sold on the world market at world market prices. The minimum beet price, as defined in ECU terms, is then converted into national currencies at the current green rate, and so can vary through the marketing year.

The calculation of the threshold price for white sugar imports is different from that for cereals. The cost of transporting sugar from the area of greatest surplus – northern France – to the area of greatest deficit – Sicily – is added to a storage levy, and the total then added to the target price to give the threshold price. An import levy is then applied to raise the import offer price to this threshold price.

Prices for industrially produced sugar substitutes such as isoglucose and aspartame are also regulated by the sugar regime.

It is possible that a new sugar regime may come into force but until that is agreed the existing system remains in force. No details of this new regime are available at the time of writing, but there is some political pressure to reduce EU sugar output. The Court of Auditors has argued that, since it results in the production of about one-third more sugar than the EU consumes, the effect of the existing regime is to depress world prices. Suggested changes have included the removal of support from B quota production.

Milk and milk products

The basic support system for milk and dairy products is essentially similar to that for the arable products, although it is more complex because of the greater range of products involved and the measures which have been introduced to attempt to ameliorate the problems of surplus which continue to exist in the EU milk market.

There is a target price for milk, but support for the milk market price is through intervention buying of butter and skimmed milk powder (SMP). Milk products offered to intervention must meet certain quality criteria, and traders must offer to intervention stores at a tender price between 90 and 92% of the intervention price. SMP is not accepted into intervention between the beginning of October and the end of February. From the prices paid for these intervention products an *Intervention Milk Price Equivalent* (IMPE) is calculated, by subtracting a processing margin from the tender price and dividing by the amount of milk required to produce a kilogram of the product. The receipts from butter and SMP are then added together to make the IMPE. This therefore acts as a floor price in the market, although the price received by the farmer will then depend on the commercial judgement of the first buyer of the milk and be influenced by the proportion sold to various segments of the market, such as the liquid trade and the various manufacturing outlets. Internal subsidies have also been introduced to reduce intervention stocks by promoting sales of butter to the bakery and ice-cream trades and of SMP to animal feed manufacturers.

Milk products may only be imported from third countries at or above the threshold price, so import levies are applied to such imports when the world price is below the threshold price. Threshold prices are set for 12 different product types, such as butter, SMP and various cheeses. In 1992–3, for example, the butter threshold price was roughly 12% higher than the intervention price. Export levies are also available to enable EC-produced milk products to be sold on the lower-priced world market.

The main mechanism used to reduce supplies of milk is the quota. Quotas are set for each member state, and levies are payable to the Commission if the country's total quota is exceeded. Each member state may then decide to set individual quotas at the level of the milk producer (as in Belgium, Germany and Italy) or the purchasing dairy, which then applies the system to the individual producer. The size of the quota is specified not only in litres, but also in terms of fat content. If the national quota is exceeded, all producers who exceeded their individual quotas are liable for a levy of 115% of the target price for their excess production. Quotas may be transferred between producers by sale or, in some cases, by temporary leasing arrangements.

Beef and veal

The changes to the CAP agreed in 1992 shifted the balance of support for beef away from price support for the product towards income support for the beef producer. As with other products, intervention buying remains in force but at a reduced price and for limited quantities: by 1997 the maximum quantity to be bought into intervention will be set at 350 000 tonnes. In addition to this tonnage, safety-net intervention at times when the market price is less than 60% of the intervention price will be allowed. Imports from non-member states are subject to customs duties and tariffs.

To compensate producers for the decreased prices which will presumably result from these changes, the Beef Special Premium Scheme and the Suckler Cow Premium Scheme have been introduced.

The Beef Special Premium is paid only on male cattle, at a rate increasing from 60 ECU per head in 1993 to 90 ECU per head in 1995. These sums are converted into national currencies at a rate which is fixed for any one year and may change on 1 January depending on the green rate then in force. Animals qualify for premium on attaining the age of 8 months and again at 21 months, provided in each case that they are retained on the claimant's holding for two months after the claim has been made. Each animal over three months old must have a Cattle Identification Document (CID), the colour of which indicates the number of payments which have been made on the animal. It is an offence to buy or sell male animals without these documents, and premiums cannot be paid without them so they are valuable, and the producer is responsible for their safe keeping. Each producer is allowed to claim for up to 90 animals in each age category. However, if the number of premiums claims exceeds a regional ceiling figure, all claims for both first and second premiums are reduced in proportion to the excess. There are also stocking density limits which must be observed (see below).

The Suckler Cow Premium is paid on cows kept for producing beef calves at a rate increasing from 70 ECU per head in 1993 to 120 ECU per head in 1995. Member states may also choose to pay an additional sum, up to 25 ECU, from national funds, although the UK government has declined to do so. These sums are converted into the appropriate national currency using the green rate in force on 1 January, and so may change from year to year. There are several restrictions on the claims that may be made: cows must be retained on the farm for six months after the date of application for the premium; there is an individual quota on premiums, which are only payable on the number of cows for which premium was claimed in the reference year (1992 for the UK), although quota may be transferred between producers subject to siphoning off into a national reserve; and dairy farmers may only apply for premium on any suckler cows they may have if their milk quota is less than 120 000 kg (i.e. 116 540 litres). The stocking density limits must also be observed.

The stocking density limits apply to both the Beef Special Premium and the Suckler Cow Premium schemes. The maximum stocking density allowed will be reduced from 3.5 Livestock Units (LU) per hectare of forage area in 1993 to 2 LU per hectare in 1996. Farms with low stocking densities (less than 1.4 LU per forage hectare) also qualify for an Extensification Premium (of 30 ECU per head in 1993, converted to national currencies using the 1 January green rate). The total livestock units are calculated from Table 1.2.

In each case these figures apply to animals on which premium has been claimed, and animals on which no premium is claimed do not count as livestock units for the purposes of this calculation. The forage area of the farm is, essentially, the total area minus the arable area, the set-aside area, and the area of buildings, woods, paths and ponds, although more precise definitions are used for the purposes of making a claim. Thus it might be possible to keep a suckler cow, her calf, now a 25-month-old bullock, and her succeeding calf, now 12 months old, on a hectare of grass, but from 1996 it would only be possible to claim premiums for the cow and the bullock. However, producers with less than 15 LU are exempt from these stocking density rules.

Table 1.2 Livestock units for stocking density calculations

	LU
Dairy cows	1.0
Breeding ewes	0.15
Male cattle under 2 years	0.6
Male cattle over 2 years	1.0
Suckler cows	1.0

Sheepmeat

Sheepmeat production in the EU is supported by aids to private storage and the Sheep Annual Premium Scheme. A basic price is set as part of the annual price negotiations and adjusted weekly. If the market price in both the EU and the region for standard quality lambs falls below 90% of the seasonally adjusted price, or 85% of the basic price, traders may tender for aids to private storage. If the price in the region falls to less than 70% of the basic price for more than two weeks, they may do so regardless of the EU price.

The size of the annual ewe premium depends on the difference between the EU average market price and the basic price, and so varies from year to year. There is an additional flat rate premium for ewes in Less Favoured Areas (LFAs). Both of these payments are subject to headage quotas. Normally, a producer may only claim for the number of ewes on which premium was paid in 1991, up to a maximum of 1000 ewes in LFAs and 500 ewes in other areas. Ewes in excess of these numbers receive headage payments at 50% of the full rate. In some circumstances, additional quota may be available for new entrants to the industry or producers whose flock sizes were reduced by abnormal circumstances in 1991. Quota rights may be transferred or leased, subject to certain conditions.

There is no CAP regime for wool, which is treated as an industrial product.

Pigmeat

The pigmeat regime is based on aids to private storage to maintain domestic prices, sluicegate prices and import levies to prevent the EU market price being undercut by imports from countries where grain is cheaper, and export refunds to prevent the build-up of stocks.

In theory *aids to private storage* may be introduced when the *reference price* (i.e. the weighted average EU price) falls below 103% of the *basic price*, but following increases in the basic price up to the mid-1980s, this is in practice always the case. Therefore the pigmeat Management Committee introduce private storage schemes when, in their view, they are warranted by the situation in the market. Traders are paid to store their meat for a specified period, normally between four and seven months. The aid payable, the storage time and the eligible cuts are decided by the Management Committee, which also decides on the level of *export refunds* for sales of pigmeat to non-member states. These refunds are denominated in ECU and converted at the green rates in force at the time of export, although there are provisions for prefixing green rates for up to three months.

Pigmeat imports from third countries are controlled by sluicegate prices and basic import levies. The *sluicegate price* is an estimate of world production costs made up of cereal and other feed costs together with overhead costs. The *basic import levy* comprises a variable element, reflecting the difference between world and EU cereal (threshold) prices, and a fixed element, of 7% of the sluicegate price, to protect the domestic industry. Thus the minimum price at which imported pigmeat may be sold within the EU is the sluicegate price plus the basic import levy, since a *supplementary levy* is charged on pigmeat offered at less than the sluicegate price.

Eggs and poultrymeat

There are no internal market support arrangements for eggs or poultrymeat, but producers are protected from competition from third countries by the same sort of sluicegate price and levy arrangements as apply to pigmeat. Export refunds are also available to promote exports to third countries. Association agreements have been concluded with some eastern European countries which allow the import of specified quantities of eggs and poultrymeat free of levy.

Fruit and vegetables

Prices for fruit and vegetables are supported by compensating growers for withdrawing produce from the market. In some ways this system is not unlike intervention, but the compensation received for the withdrawn produce is far below the market price, and most of it is destroyed, although some may be used for animal feed or processed.

The withdrawal prices are paid to producer organizations. They are calculated from basic and buying-in prices which are determined by the Council. The buying-in prices range from 30% to 65% of the basic price, depending on the product in question, and the compensation for withdrawal is then determined from this price, modified by quality, variety and size coefficients. There are different withdrawal seasons for different crops. Imports from third countries are subject to *ad valorem* common customs tariffs.

Other products

The EU also has regulations pertaining to markets in wine, raw tobacco, hops, seeds, flowers and ornamental plants, silkworms, and fibre plants. Of these, the last category is probably the most significant for UK farmers, for it provides for the payment of a flat-rate production aid. In the 1992/3 marketing year this was 374.36 ECU per hectare for fibre flax and 339.42 ECU per hectare for approved varieties of hemp.

Monetary arrangements in the CAP

Institutional prices in the CAP, such as intervention prices, and compensation payments, are denominated in European Currency Units (ECU). The ECU is a basket currency comprised of fixed amounts of the currencies of member states of the Community. For the purposes of every-day transactions, such as sales into intervention, ECU amounts need to be converted into their equivalents in national currencies. If exchange rates were fixed for significant periods of time, as they were when the CAP was set up, this would present no problems. It would be sufficient to multiply the ECU price by the fixed ECU/national currency exchange rate to determine the price in terms of the national currency.

In practice, exchange rates are free to move, to some extent at least. Most member states of the EU declare an exchange rate, known as the 'central rate', between each of their national currencies and the ECU, and then undertake to maintain the cross rates based on these central rates within a margin of plus or minus 15%. These are members of the Exchange Rate Mechanism (ERM) of the European Monetary System (EMS). (Until August 1993 the permitted deviation from the central rate was much smaller for some countries. Whether it is so again will depend upon future progress towards European monetary union.)

The values of other currencies, which are not members of the ERM, are allowed to change in response to the pressures of market forces, and so are known as 'floating' currencies. However, given the wide permitted deviations from central rates, all EU currencies are effectively floating. Thus the value, in national currency terms, of any price or payment denominated in ECU would vary from hour to hour in response to changing currency exchange rates on the international money markets. The result would be chaos. Therefore, for agricultural purposes, a fixed exchange rate is used, which is known as the representative or green rate, e.g. the green pound.

A change in the green rate will affect the price received by farmers. If the green rate is *devalued* (so that more pounds are required to buy one ECU) prices will *rise*, and vice versa. For example, for a commodity with an intervention price of 120 ECU per tonne:

Green rate (1 ECU =)	× *Intervention price in ECU*	= *Intervention price in £ sterling*
0.77	120	92.4
0.77 (devaluation)	120	94.8
0.75 (revaluation)	120	90.0

Since green rates are fixed but market rates of exchange may vary, it is possible for a gap – the monetary gap – to develop between the green and market rates. The size of the monetary gap is calculated using the expression:

$$\text{Monetary gap} = \frac{\text{Green rate} - \text{Market rate}}{\text{Green rate}} \times 100$$

Thus if the green rate is 1 ECU = £0.77 012, and the market rate is 1 ECU = £0.74, the monetary gap will be

$$\frac{0.77\,012 - 0.74}{0.77\,012} = +3.91\,107$$

But if sterling devalues, so that the market rate is 1 ECU = £0.79, the monetary gap will be negative, i.e.

$$\frac{0.77\,012 - 0.79}{0.77\,012} = -2.58\,141$$

Therefore the monetary gaps of individual currencies may be positive or negative, and so the *spread* of monetary gaps, which is the difference between any individual gap and the highest positive gap, may be calculated.

All this means that farmers could receive different prices in different member states for the same product. Consequently normal trade patterns could be disrupted. The EU believes that this would be undesirable, and that large monetary gaps should be corrected. The correction involves a change in the green rate, triggered by the monetary gap for an individual currency becoming too big, or by the spread of gaps becoming too big.

For example, the monetary gap of +3.363 calculated above would breach regulations which allow gaps of between −2 and +3% (these limits are called *franchises*), so the green rate would be adjusted using the formula

$$\text{Market rate} \times \frac{100}{(100 - \text{monetary gap}/2)} = \text{New green rate}$$

$$\text{i.e. } 0.74 \times \frac{100}{100 - \left(\dfrac{3.91\,107}{2}\right)} = 0.75\,476$$

It is important to remember that this is only *one* of the mechanisms for triggering green rate changes: if a currency continues to appreciate, a 'floating' franchise is allowed to rise to +5% as long as the negative gaps have been reduced to zero. Once the 5% limit has been breached there must be a revaluation of the green rate concerned. All of these rules have been changed from time to time by decisions of the Council or the Commission.

The Commission, operating in the Agri-monetary Management Committee, usually tries to reduce gaps or spreads by devaluing weak currency green rates, since this gives rise to institutional price increases (e.g. in intervention prices) and so is thought to be less politically sensitive than revaluation of strong currency green rates, which would decrease institutional prices.

In early 1995 the switchover mechanism, which had involved the use of an artificially inflated 'green' ECU, was discontinued, and the ordinary ECU was employed for agricultural purposes. Consequently the green rate was reduced by some 20.8%. Simultaneously all ECU prices and compensatory payments were increased in the same proportion, so that the net effect in national currency terms was zero.

It is important to remember that the detailed rules for the operation of the green money system may be adjusted from time to time. This may not make very much difference to the day-to-day decisions of farmers, but it may have a significant effect on the operations of traders who are engaged in international trade. Detailed information on current green rates may be obtained from organizations such as the Home Grown Cereals Authority and the Grain and Feed Trade Association, and market rates, current and predicted green rates, monetary gaps and spreads are reported each week in *Agra Europe*.

Structural and environmental policy

Structural measures

In addition to the common price and market measures financed through the guarantee section of FEOGA, the CAP also includes a number of structural measures financed through the *guidance* section of FEOGA. In this context the term 'structural' refers to aspects of agricultural policy which are concerned with farm size, problem areas, investment and marketing.

The major structural problem in the European Union arises from the existence of many small farm businesses. This is not to imply that the large business is necessarily more efficient than the small, but simply that the larger business can make lower profits per unit of output and survive because it produces more units of output to cover fixed costs. Consequently small farm businesses are often forced to concentrate on the more intensive crop and livestock enterprises, and may have difficulty in acquiring capital. The price and market policies of the CAP are not designed to solve these problems. Indeed, insofar as they support farm incomes and so prevent the disappearance of small businesses from the farming industry, they perpetuate them, without necessarily solving small farm income problems. By the late 1960s it was therefore felt that measures designed to decrease the number of small farms and increase the viability of those remaining were required.

The original Commission proposals made in response to this view led in 1972 to three directives on the modernization of farms, payments to outgoers, provision of socio–economic guidance, and, later, on special measures to ameliorate the problems of farming in mountain, hill and other less favoured areas. Following reform of the structural policy in 1988 and after, all these directives have been replaced by the provisions of Regulation 2328/91 and other regulations (2078/92 on environmental measures, 2079/92 on early retirement, and 2080/92 on afforestation) introduced as part of the MacSharry reforms of 1992.

The main function of these regulations is to provide funds for co-operatives and producer groups, improvements in marketing and processing, technical and business training, incentives for early retirement, compensatory payments for farmers in Less Favoured Areas (LFAs), and investment grants (not only for agricultural purposes, but also for farm diversification into tourism or crafts). In most cases the member state introduces legislation or schemes to put the policy into practical effect, and pays part of the cost. The level of EU funding may thus vary between 25 and 75% of the total grant disbursed. In order to maintain extensive livestock production in LFAs in the UK, for example, *Hill Livestock Compensatory Allowances* are paid on breeding cattle and sheep in regular breeding herds and flocks where stocking density is no more than 1.4 livestock units per forage hectare.

Early structural policies were criticized on the grounds that they concentrated too much on agricultural investment in the northern member states. Subsequently, therefore, attempts were made to divert more of the funds available to the Mediterranean countries, and to recognize that development in rural areas might involve other industries as well as agriculture. In 1981, for example, there was an 'integrated programme' for the Western Isles of Scotland, and in 1985 'integrated Mediterranean programmes'. The 1988 reform of structural funding carries this principle further, to the point of combining spending from the European Regional Development Fund and the European Social Fund with FEOGA guidance section money to attempt to achieve five objectives, only some of which are relevant to rural areas.

- *Objective 1* regions are those in which income per head is less than 75% of the EU average. They include the whole of Greece, Portugal and Ireland (including Northern Ireland), and parts of other countries. Here the EU funds up to 75% of the cost of 'Operational Programmes', involving, amongst other things, rural infrastructure and irrigation schemes.
- *Objective 5b* areas are basically agricultural, with low income levels and low levels of socio-economic development. Parts of Devon and Cornwall, mid-Wales, and south-west and north-west Scotland are included in this category. Again the EU helps to finance Operational Programmes, albeit contributing a smaller proportion of the total cost.
- *Objective 5a* is concerned with speeding up the adjustment of agricultural structures. It uses only FEOGA guidance section funds, and does not operate on a regional basis but across the EU as a whole.

One of the aims of the Maastricht Treaty was to reduce economic disparities between regions, including rural regions. It also provided for the establishment of a Cohesion Fund to make extra resources available to Greece, Spain, Portugal and Ireland, the four poorest member states. Consequently, a substantial part of FEOGA guidance money is now spent on a regional basis, either on the operational programmes or in Less Favoured Areas. The European Council, at its meeting in Edinburgh in December 1992, decided to increase structural spending as a whole up to a limit of 30 billion ECU in 1999, and, in addition, allocated a further 15.15 billion ECU to the Cohesion Fund over a seven-year period.

At a national level, EU structural funds are channelled through national schemes such as the Farm and Conservation Grant Scheme in the UK, which provides assistance for various types of investment from reseeding of grassland and thermal insulation of buildings to hedge laying and establishment of shelter belts. The rates of grant vary from as little as 15% to as much as 50%, and are usually higher in Less Favoured Areas. Further details of these schemes are available from MAFF Regional Service Centres.

The CAP and the environment

Initially, the CAP contained virtually no provision for anything concerned with the environment, but as the policy has developed and public concern has increased, environmental issues have become increasingly important. The Single European Act of 1986 provided for the addition of environmental objectives to the Treaty of Rome, but even before that some aspects of structural policy had been used for environmental purposes. In the structural regulations of 1985 the establishment of *environmentally sensitive areas* (ESAs) was permitted, and later some EC funding was made available for such schemes. ESAs are areas in which nationally important landscapes or habitats have been produced by traditional farming practices, and farmers are given incentives to maintain these practices. The UK was instrumental in the establishment of the scheme, but it is now applied in at least six other member states.

In 1988 a *voluntary set-aside* scheme was introduced as part of the stabilizer scheme. Producers were compensated for withdrawing at least 20% of their arable area from production for at least five years. The Nitrate Sensitive Areas Scheme introduced into the UK in 1990 also preceded its EC equivalent, the 1991 *Nitrates Directive* (91/676), which required member states to designate vulnerable zones in which fertilizer application would be restricted. At the same time an *organic farming* regulation (2092/91) was also introduced which defined the cultivation practices required for plant products to be labelled as organic.

The MacSharry reforms of 1992 also included a package of environmental measures (Regulation 2078/92, sometimes known as the Agri-Environment Regulation) which allowed for payment of aids for water protection, reconversion of arable to extensive pasture land, organic farming, extensification of livestock enterprises, preservation of environmentally friendly production practices, and public access and recreation. As with many of the other environmental measures, this was clearly designed to limit the output of products in surplus as well as producing environmental benefits. The same considerations influenced provisions for early retirement (Regulation 2079/92) and *afforestation on farms* (Regulation 2080/92). These regulations are essentially enabling legislation, in that the individual member state does not have to introduce all of their provisions (e.g. the early retirement regulation does not apply at all in the UK) or apply them over the whole country. In the UK, the afforestation regulation is applied through the Woodland Grant Scheme of the Forestry Authority: 50% of the money disbursed through this scheme is recovered from the EU. The Agri-Environment regulation is applied through a variety of schemes, including Countryside Stewardship, Environmentally Sensitive Areas, the Moorland scheme, the Habitat scheme, Nitrate Sensitive Areas, the Organic Aid scheme, and the Countryside Access scheme. Details of all of these are available from MAFF offices.

Problems and developments in the CAP

It is unlikely that the MacSharry reforms of 1992 will solve all the problems of the CAP, but they clearly represent a significant change. They made settlement of the GATT trade negotiations easier, and they are likely to reduce output and support costs in the long run. In the short run any saving in the cost of disposing of surplus is likely to be outweighed by the cost of area and headage payments, but eventually, in theory at least, transferring part of the support from price to the producer and compulsory set-aside will reduce the incentive to produce, so cutting the cost of surplus disposal. Thus the cost of the CAP is more likely to remain within the budget guidelines, and support spending is concentrated on those who produce least. Moreover, the measures now introduced provide some flexibility for the future, in that quotas may be cut and the proportion set aside raised if surpluses remain. Conversely, the European Union retains the capacity to expand production fairly rapidly if market conditions make it desirable to do so. And, finally, the new policy seems more likely to assist in the production of environmental goods than the old one.

Nevertheless, it is clear that many problems remain. Technical change, the effects of partial fallowing, and the exemption of sub-92-tonne producers will reduce the impact of set-aside, and its environmental benefits are reduced by its rotational nature. Some of the contradictions of the CAP remain: output-increasing technical change is maintained at the same time as supply restriction, and there is a conflict between efficiency and conservation. Although there is still a surplus of milk in the EU as a whole, processors in the UK may be starved of raw material because quota may not be sold across national boundaries. The impact of lower prices in ECU terms is reduced by fluctuations in exchange rates. Some commentators have suggested that high levels of debt encourage the maintenance of output despite lower prices. Area payments may not be as regressive as price support, but those with the greater areas or numbers of livestock still get more support than those with only a few hectares of corn or a few animals. The farm size structure of the EU will remain diverse even though the number of farms and farmers will decrease, so a common price level will still produce different profit levels.

Few people expected the MacSharry reforms to be the last word in the evolution of the CAP; what matters is how long they will last. Some have argued that they are designed to fail fairly rapidly in order to strengthen the political will to introduce really radical change. Much probably depends on the evolution of world price levels: the closer they are to CAP prices the longer the existing CAP can be sustained.

References and further reading

Agra Europe *The CAP Monitor* (continuously updated).
Agra Europe (weekly).
Commission of the European Communities. (1990) *Agriculture and the Reform of the Structural Funds.*
Commission of the European Communities (1991) *The Development and Future of the Common Agricultural Policy.* COM (91) 258.
Commission of the European Communities (1993) *Our Farming Future.*
Fearne, A. (1991) The history and development of the CAP 1945–85. In: *The CAP and the World Economy: Essays in Honour of John Ashton* (eds C. Ritson & D. Harvey). CAB, Wallingford.
Fennell, R. (1991) *The Common Agricultural Policy of the European Community.* Blackwell Science, Oxford.
Neville, W. & Mordaunt, F. (1993) *A Guide to the Reformed Common Agricultural Policy.* Estates Gazette, London.
Tracy, M. (1993) *Food and Agriculture in a Market Economy.* APS, La Hutte, Belgium.

2

Farm Business Management

M.F. Warren

What is business management?

Business management is difficult to define precisely, and is best described in terms of the activities performed by the manager of a business – whether employee or owner.

A manager's job is to make decisions – not just the day-to-day ones that everyone has to make, but decisions concerning the next month, the next year, or even the next ten years. The manager is deciding how best to use the resources at his or her disposal in trying to achieve the objectives of the business.

A manager's job is also concerned with ensuring that his decisions are having the desired effect. The outcome of each decision must be measured and checked to see that it matches up to expectations. If actual progress does not compare well with the expected performance, the manager should take the appropriate action to bring the two into line.

Viewed in this way, the manager's job can be seen as *planning* (making decisions which affect the future operation of the business) and *control* (monitoring progress of the decision and taking necessary corrective action), in order to achieve the *objectives* of the business.

In a one-person business, where the owner is also the manager and the workforce, the business has only to satisfy the owner's objectives. In most farm businesses, however, there are other people involved – family, partners, employees, maybe even a paid manager. These people have their own objectives, and this adds another dimension to the manager's job – making sure that everyone is working towards the same end. A major objective of many farm workers, for instance, is to do a job to be proud of. The task of the manager is to use that fact to help the business achieve its goals rather than allow it to cause a nuisance.

This dimension of the manager's job – marshalling resources, coordinating and harmonizing objectives and efforts – can be summed up as *organization*. Management thus becomes a process of planning, organizing and controlling (Fig. 2.1).

This chapter is concerned largely with the planning and control functions: the following chapter includes some of the skills needed in organization.

Fig. 2.1 The management process.

Setting primary objectives

Setting objectives is the key to the management process. Unless the manager is sure what he or she is trying to achieve, the business can have no clear direction.

Objectives can be roughly grouped into primary and secondary objectives. Primary objectives are the overall aims of the business – the long-term goals for which the business is striving. Secondary objectives are the shorter-term goals that the business needs to meet in order to achieve the primary objectives.

A primary objective might be, for instance, to increase the wealth of the business by 10% in the next five years; one of the associated secondary objectives might be to achieve an average wheat yield of 10 t/ha over this period.

Primary objectives are reflections of the long-term expectations of the manager. These expectations will range far and wide, but for convenience can be classified into personal, physical, responsibility and financial.

(1) *Personal expectations* are those which are concerned with the lifestyle and personal preferences of the manager. They often include time for family, social and leisure activity, and acquisition of status or

influence through, for instance, being a magistrate or county chairman of the NFU.

(2) *Physical expectations* relate to the productivity of the farm, to innovation, and to general technical prowess.

(3) *Responsibility expectations* involve satisfaction of the responsibility the manager feels towards other groups in society. These can include employees, customers, neighbours and the local community.

(4) *Financial expectations* relate to the generation of income and wealth. These are the crucial expectations, not because they are particularly important in themselves (although some people are highly motivated by the acquisition of money for its own sake), but rather because they govern the ability of the business to satisfy the personal, physical and responsibility expectations.

It is because of the importance of financial expectations that the following sections relate particularly to financial management.

A further reason is that, to plan and control effectively, a manager needs a widely accepted and flexible measure by which to judge performance. Money provides that measure.

Financial measures

Where the finances of a farm are concerned, three factors are crucial: survival, efficiency and stability. These will dictate the financial objectives of the business, and the way in which the manager plans and monitors progress.

(1) Survival, in financial terms, is a matter of having sufficient cash available at any time to pay for essential inputs, such as labour, fertilizers, diesel, finance charges and so on. Cash availability is measured through net *cash flow* – the difference between cash coming into the business and cash going out.

(2) Operational efficiency is also reflected in the money coming in and going out of the business. There are other factors to take into account, however, such as changes in the levels of stocks and of bills outstanding. The financial measure that attempts to measure efficiency by taking these and other factors into account is known as net *profit*.

(3) Financial stability is imparted by wealth. Wealth implies ownership of assets – money and things of value – but it must also take account of outstanding debts, or liabilities. A business which has £1 000 000 assets and £999 000 debts is likely to be less stable than one with £10 000 assets and £50 debts. The appropriate measure of financial stability is thus net *capital* – the difference between assets and liabilities.

Planning to achieve primary objectives

Once the primary objectives have been established, the planning process can get under way. The first, and usually most protracted, stage is to formulate an overall or long-term plan. The time period covered by such a plan will be one or more years, and it will indicate the decisions and actions needed for the business to have the best possible chance of achieving the primary objectives.

This process relies much on subjective judgement and can never be made foolproof. A structured approach can, however, minimize the risks of making grossly erratic judgements. One such approach is as follows:

(1) Assess the farm and its resources.

Take a hard look at the *strengths* and *weaknesses* of the business. Assess the quantity and quality of the resources available, such as the land, buildings, expertise of the employees, ease of access to markets, and so on. Take account of factors that could limit freedom of action; not only physical and financial factors such as the slope of the ground and the amount of capital available, but also legal and administrative constraints such as town and country planning laws, restrictive clauses in a lease, or the likelihood of obtaining a quota or contract.

(2) Appraise the present and future environment of the business.

Identify the *opportunities* that are open to the business, and the *threats* to its welfare. There is more to this than keeping abreast of technological change. It involves being aware of the economic and political climate affecting the business, and making judgements about how that climate is likely to change during the period planned for (see Chapter 1). It also requires an awareness of changes in the habits and attitudes of society in general (e.g. food consumption habits and attitudes towards farming). The likely actions of competitors must be anticipated. The search for opportunities should not be confined to farming, but should include all possibilities for using the business resources effectively.

(3) Formulate alternative plans.

List all the possible ways of using the resources of the business to achieve its primary objectives. 'Possible', in this context, must be judged by reference to the strengths and weaknesses, opportunities and threats listed earlier. It is essential to look further than the 'obvious' solution, and consider as wide a range of alternatives as possible. This is, in many ways, the most difficult part of the planning process: it is hard to be objective and imaginative about a business which one knows intimately. One way of overcoming this problem is to enlist the help of family and/or friends in a 'talking shop' aimed at stimulating new ideas.

(4) Appraise the alternative plans and select the one most likely to succeed.

The likely outcomes of the alternative plans must be measured and compared. Given the importance of financial objectives, and the fact that money provides a common unit of measurement, it is critical at this stage to prepare budgets. A budget is a calculation of the financial effects of a course of action, and given that the financial objectives of a business can be measured in terms of cash flow, profit and capital, budgets should be expressed in these terms.

The plan whose budgets indicate the nearest

result to the financial objectives will then be chosen, unless another plan more nearly achieves the non-financial objectives specified, and the manager is content to forgo the extra financial benefit for the sake of achieving non-financial expectations. He or she may happily put up with a low-key, low-profit system for the sake of minimizing mental and physical strain.

The resulting overall plan can then be used as a focus for short-term planning. The manager will need to:

(1) set secondary objectives. These are the targets which must be achieved for the overall plan to work. They may be expressed purely in financial terms, as the margin which a particular enterprise should achieve in a given year. They may also include key physical targets, such as the yield produced by a major enterprise.
(2) formulate short-term plans. This is a continual process aimed at enabling these targets to be achieved. It ranges from formal budgeting of enterprise margins for a season, or cash flow for a month, down to day-to-day decisions about organization of the workforce.

Once plans have been formulated and acted upon, the control process can start. On a week-by-week, month-by-month basis, progress can be monitored by comparing key results with the short-term targets (most valuable in dairy and intensive livestock enterprises).

On an annual basis, control will involve monitoring the progress of overall plans. Given the importance of financial objectives at this level, this implies comparing financial expectations with what has actually happened. This *budgetary control* can only take place if the manager has available two key sets of information – the *budgets* which show the detail of financial expectations, and the *accounts* which record the actual performance. Both budgets and accounts use the same measures – cash flow, profit and capital.

Information for financial management

Budgets

Budgeting for profit

A profit budget in conventional form

A budget shows the expected financial performance of a business over a future time period. This period is normally a year and, for convenience, is usually the financial year used by the farm's accountant. Many farm financial years start from Michaelmas (end of September) or Lady Day (end of March): other popular dates are the beginning of January and the beginning of April.

The easiest way of beginning the budgeting process is to start from the practical basics of the farming system – how many hectares of crops are to be grown during the year in question, how many head of livestock are to be kept?

On this framework can be built a detailed picture, by estimating the quantities of inputs which are likely to be consumed and their prices, and the quantities of products which are likely to be produced and their prices.

To reflect fully the business performance of the farm, though, some adjustments need to be made to these figures. The first is that allowance should be made for any goods or livestock likely to be on the farm at the end of year. Although these 'stocks' will not yet have been sold, they will have been produced or purchased by the business during the year, and thus ought to be included. Similarly, any bills yet to be paid at the end of the year (creditors) or income due from goods which have been sold and not yet paid for (debtors) must be included as 'belonging' to the budget year. Conversely, any stocks, creditors or debtors likely to be outstanding at the *beginning* of the budget year are taken out of the calculation – they 'belong' to last year's production.

The second type of adjustment is to remove any payment or receipt concerned with personal, tax or capital matters. The amount the owner of a business takes for his or her own use is not determined by the efficiency of the business, but by lifestyle: the same applies to tax payments, which also vary according to how good the accountant is and the policy of the prevailing government. Capital items relate to lump sums, such as the receipt and repayment of loans, and purchase of land, buildings and machinery. Since these do not arise out of the normal trading of the business, they are left out of the calculation with the exception of 'wasting assets' such as buildings and machinery. The cost of these is spread over their useful life in the business: the resulting annual cost is known as 'depreciation' (see 'Estimating outputs and costs' below).

Finally, any farm products consumed without payment by the farmer, family and employees should be added in. These 'benefits in kind' or 'household consumption' are just as much part of the farm production as those sold. Similarly, goods and services paid for on the farm account but consumed without payment should be credited back to the farm.

The result (shown in Fig. 2.2) is a budget showing the expected net profit for the year concerned.

Enterprise budgeting

A profit budget of the type just illustrated is useful in a limited way. One can tell whether the operation of the farm is likely to make a positive or negative contribution to the owner's financial welfare (i.e. profit or loss). It is also possible to see whether the result represents an improvement on previous years. It is difficult, however, to see how particular costs and returns relate to each other, and to see how the financial quantities reflect physical performance. The *value* of milk produced is easily apparent, for instance, but not the *amount* of milk produced in total and, more important, per cow.

To overcome these problems, it is possible to rearrange the information in the budget on an enterprise basis. Conceptually, the easiest way of doing this is by calculating a net profit for each enterprise (an enterprise being an output-producing part of the business, such as a particular crop or livestock type, or a service such as holiday accommodation or contracting).

The form of enterprise accounting most commonly found in UK farming is the gross margin system. This involves identifying for each enterprise the value produced (the *output*) and the *variable costs*. The latter are costs

Trading and profit and loss account, year ended 31 March 199x

	This year (£)		*Last year (£)*	
Farm income				
Milk	98 159		88 521	
Calves	7 700		11 569	
Cull cows	8 750		17 580	
Grain	22 400		512	
Straw	2 500		335	
Area payments	7 600		5 642	
Miscellaneous	1 500		500	
		148 609		124 659
Less: Purchases				
Feed	7 200		26 513	
Livestock	14 400		725	
Vet. and med., AI, etc.	4 800		4 958	
Fertilizer	14 740		2 160	
Seed	3 590		12 589	
Sprays	4 120		154	
Regular labour	20 000		2 561	
Machinery running costs	15 000		1 583	
Rent and rates	14 000		13 521	
Miscellaneous costs	10 650		9 852	
Bank interest	9 243		10 256	
Machinery depreciation	9 000		8 250	
		126 743		93 122
Benefits in kind		1 151		958
Increase in valuation of stocks		50		1 361
Net profit		23 067		33 856

Fig. 2.2 Example of a profit/loss statement.

which both are easily attributable to the enterprise concerned, and also vary in direct proportion to small changes in the scale of the enterprise. Thus, for a livestock enterprise, livestock purchase, feed, veterinary and medical costs, transport and so on are regarded as variable costs. Each one is readily identifiable with the enterprise, and a 5% increase in the size of the flock or herd, for example, would tend to increase each of these by a similar proportion. Typical variable costs for a crop enterprise include seed, fertilizer and spray costs.

Subtracting the variable costs from the output gives, for each enterprise, a *gross margin*. When the gross margins from the various enterprises are added together, the farm overhead or *fixed costs* can be deducted to give the net profit for the business (Fig. 2.3). This net profit should be the same as that shown by the budget in conventional form described above – it is only the layout of the information that has changed, not the information itself.

The main advantage of this method is that it avoids the allocation of the fixed costs between enterprises. Such allocation can involve difficult and arbitrary decisions, particularly when general farm costs are concerned – such as the cost of the farm Land Rover, office expenses, finance charges, etc. Gross margins are also useful when considering small changes to the farm system. The

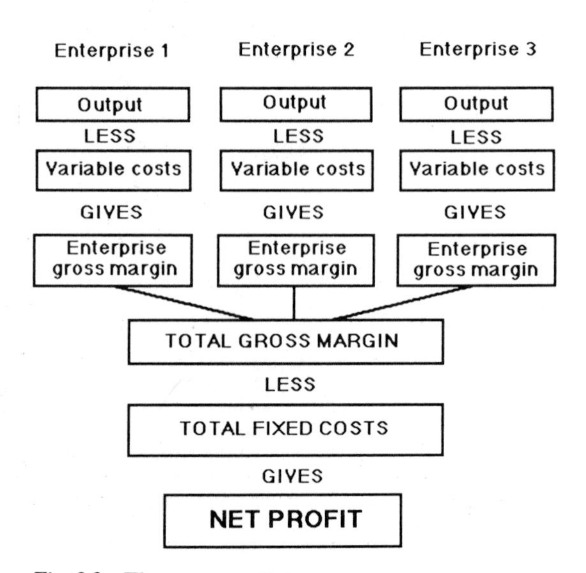

Fig. 2.3 The gross margin system.

	£
Gross margin gained: 10 ha wheat at £645.50/ha	6 455
Gross margin lost: 10 ha barley at £527.50/ha	5 275
Net gain	1 180

Fig. 2.4 Net gain from substituting 10 ha winter wheat for spring barley.

financial effects of switching one field from spring barley to winter wheat can be seen by substituting the gross margin of one crop for that of the other in the budget (Fig. 2.4).

There are situations when the gross margin method is insufficient. If it is important to know exactly how much a particular enterprise contributes to the business (when pricing a contract, for instance), it will be necessary to calculate a *net margin*. In its extreme form, this involves allocating all the costs of the business (including the general overheads mentioned earlier) between the enterprises. A useful compromise is to allocate to enterprises all those fixed costs which can easily be attributed, leaving only such costs as general overheads and finance charges to be deducted on a whole-business basis (Giles, 1986).

Where the farm is large enough to support its own secretarial services and/or its enterprises are fairly self-contained, the benefits of using this system can outweigh the effort involved in allocating labour and machinery costs (including analysis of timesheets when recording). For most general-purpose management accounting, though, the gross margin system remains a valuable compromise, provided that fixed cost implications are not overlooked.

Preparing an enterprise gross margin budget
Estimating an enterprise gross margin involves two tasks – identifying the information needed, and organizing the information so that it can easily be used.

Calculating the likely output of an enterprise needs information about any value produced by that enterprise. As well as sales (adjusted for changes in debtors), therefore, the calculation must include transfers of products between enterprises within the business. Thus home-grown barley fed to livestock enterprises must be valued at market value at time of transfer, and credited to the barley enterprise as part of its output. In addition, any produce consumed as benefits in kind (milk consumed in the farmhouse, for instance) should be counted as output. Finally, allowance should be made for any increase or decrease in the valuation of stocks of output items (crops in store or in ground, livestock and livestock products).

Similar conditions apply to the variable costs. Allowance must be made for transfers of home-grown inputs from other enterprises, for household consumption, and for changes in stock valuation of each input.

The manner of organizing this information is partly a matter of taste, but the important criteria are that the result should be easy to read, and show readily the type of information that is important to the manager. One way of laying out a crop gross margin is shown in Fig. 2.5. Note that the five columns give information not just on financial amounts in total and per unit of production, but also physical information such as yields and usage per head or hectare. This degree of detail is necessary for the budget to be used effectively in budgetary control, as it enables the manager to identify discrepancies between actual and planned results, and also to explain how they have arisen.

Figure 2.6 shows a similar calculation for a breeding livestock enterprise. Note that the cost of buying and

Ha 40	*Total Quantity*	*Tonnes/ Hectare*	*Price/ Tonne*	*Total £*	*£/Ha*
Grain					
Sales	220	5.5	102	22400	560.00
Transfers out	25	0.6	90	2250	56.25
Valuation change	-10	-0.3	90	-900	-22.50
Private consumption	0				
Total grain	235	5.9		23750	593.75
Straw					
Sales	80	2.0	31	2500	62.50
Transfers out	25	0.6	25	625	15.63
Valuation change	+5	0.1	25	125	3.13
Private consumption	0				
Total straw	110	2.8		3250	81.25
Area payments				7600	190.00
ENTERPRISE OUTPUT				34600	865.00
	(tonnes)	*(kg)*	*(£)*	*(£)*	*(£)*
Fertiliser (nitrogen)	11	275	100	1100	27.50
Fertiliser (compound)	20	500	120	2400	60.00
Seed	9	225	250	2250	56.25
Sprays				2800	70.00
Other costs				230	5.75
VARIABLE COSTS				8780	219.50
GROSS MARGIN				25820	645.50

Fig. 2.5 Example of gross margin calculation for winter wheat.

Average numbers 70	Total Quantity	Quantity/ Head	Price/ Unit	Total £	£/Head
Milk					
Sales	455000	6500	0.215	97803	1397.19
Private consumption	1655	24	0.215	356	5.08
Total milk	456655	6524	0.215	98159	1402.27
Calves					
Sales	65	0.9	118	7700	110.00
Transfers out	0				
Valuation change	0				
Total calves	65	0.9	118	7700	110.00
Cull sales					
Sales	15	0.2	583	8750	125.00
Valuation change	1			800	11.43
Less l/stock purchase	-16	-0.2	900	-14400	-205.71
Net replacement cost	0			-4850	-69.29
ENTERPRISE OUTPUT				101009	1442.98
Concentrate feed: homegrown	25	0.36	90	2250	32.15
Concentrate feed: bought	48	0.69	150	7200	102.86
Total concentrates	73	1.04	129	9450	135.01
Vet and med, AI				4800	68.57
Bedding: home-grown	25	0.36	25	625	8.93
Other				2160	30.86
Forage costs				5890	84.14
VARIABLE COSTS				22925	327.51
GROSS MARGIN				78083	1115.48
Stocking rate (head/ha)		2.0	**Gross margin/ha**		2230.95

Fig. 2.6 Example of gross margin calculation for a dairy enterprise.

transferring in livestock, though strictly a variable cost, is deducted in calculating the enterprise output. There is no particular reason why you should follow this practice, but it has become an established convention in agricultural management and most published gross margin information is expressed in this way.

When budgeted gross margins have been prepared for each of the planned enterprises, the net profit can be calculated by totalling the gross margins and deducting the fixed costs. The major fixed costs are likely to be those of regular labour, machinery running costs and depreciation, rent and/or property depreciation and maintenance costs, and finance costs such as interest and leasing charges. Overdraft interest charges, being dependent on fluctuating overdraft levels, can be left on one side for now. This results in a budgeted net profit before overdraft interest; the final net profit can be calculated after using the cash flow budget to estimate interest costs.

Estimating outputs and costs

The value of a budget depends on the quality of the information used in its compilation. Estimates of outputs and variable costs can be built up from forecasts of prices and quantities. The best guide for yields and quantities of inputs is usually the performance of previous years, taking bad years with the good. A good farm management handbook will also help, particularly where the manager has little previous experience of a particular enterprise (see, for instance, textbooks mentioned under 'Farm management data' at the end of this chapter. The biggest danger in estimating is that of being too optimistic – the most realistic estimate is usually produced by incorporating a healthy pessimism into budget figures.

Prices are affected by a great many factors, including the state of world trade, the political climate, changes in consumption habits, and quality of the product. The individual manager is unlikely to have the time or the expertise to weigh up all these factors for every commodity. The best procedure is to make use of the opinions of those who *do* have the time and expertise. Economic researchers are employed by a number of organizations: their conclusions are frequently reported in the farming and other media. The job of the manager then becomes one of locating the various price forecasts, reconciling any conflicts between them, and relating them to the business in question.

In estimating fixed costs, previous years can again be a useful guide, allowing for intervening wage awards and increases in prices. The handbooks mentioned above contain guidance where previous years' figures are not available or not relevant (where the system has been changed, for instance). Tables show the likely number of man- and tractor-hours needed for particular enterprises, from which can be calculated likely running costs.

Two methods are in common use for estimating depreciation. The straight-line method takes the original cost of the asset (machine, building or property improvement) divided by the likely life of the asset. Thus, an asset costing £10 000 and expected to last 20 years would have an annual depreciation of £500 per year. This method is useful for budgeting several years in advance, the constant annual payments making estimation quick and easy. It does not, however, result in a realistic estimate of the depreciation of machines, which tend to lose value more quickly in the first years of their lives than in the later. To allow for the latter effect, the diminishing balance method can be used. Here a constant annual percentage depreciation rate is applied to the depreciated value from the end of the previous year. Figure 2.7 shows the depreciation pattern of an asset costing £10 000, depreciated at a rate of 20% per year.

Ideally, the depreciation rate should be assessed individually for each machine, taking account of its likely life and resale value. Some farm management handbooks contain tables of appropriate rates. In practice, this is too complicated a procedure for most management purposes, and in both budgets and accounts, machines and buildings tend to be grouped or 'pooled' with other similar assets. A depreciation rate is chosen for the type of assets in the 'pool' (e.g. tractors), and is applied to the written-down value from the last year, adjusted for sales and additions during the current year Figure 2.8 illustrates the process.

Assessing the profit budget

Before moving on to the cash flow budget, it is important to make an initial assessment of the implications of the profit budget. The first obvious check is whether the budget shows a net profit or a loss, and whether that figure matches up to the profit objective. Another useful check is whether the profit is sufficient to cover basic

Year	*Opening value*	*Depreciation*	*Closing value*
1	10 000	2 000	8 000
2	8 000	1 600	6 400
3	6 400	1 280	5 120
4	5 120	1 024	4 096
5	4 096	819	3 277
6	3 277	655	2 621
7	2 621	524	2 097
8	2 097	419	1 678
9	1 678	336	1 342
10	1 342	268	1 074
11	1 074		

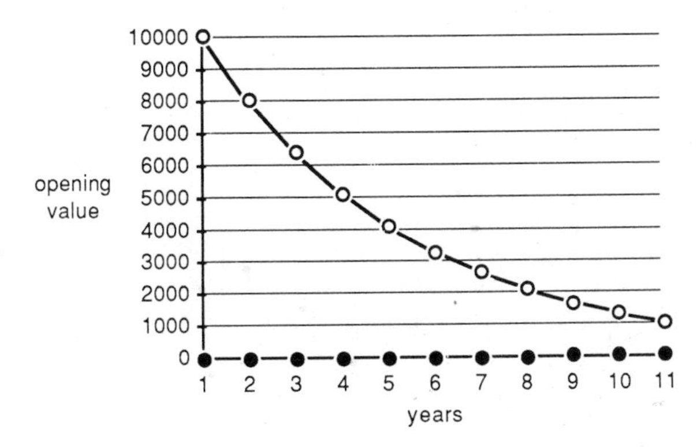

Fig. 2.7 Example of diminishing balance depreciation.

	£
Value of machinery at beginning of year	30 000
plus purchases of machinery	20 000
less sales of machinery	−5 000
Value before depreciation	45 000
Depreciation at 20%	9 000
Value of machinery at end of year	36 000

Fig. 2.8 Example of 'pool' calculation.

family drawings and known tax commitments. A comparison between the result of the budget and known results for previous years will help to give a feel as to whether the budget is realistic (it is very easy to be overoptimistic in budgeting).

It is likely that such an assessment of the budget will indicate deficiencies, and encourage adjustments in the projected gross margins as a result. Indeed, several such adjustment stages may be necessary before the manager is sufficiently satisfied (or convinced that no more profitable plan is possible) and moves on to test the effects on cash flow.

Budgeting for cash flow

Once a profit budget in gross margin form has been compiled, preparation of a cash flow budget is relatively easy. The aim is to present a picture of the flow of cash in and out of the business over the budget year.

An outline cash flow budget is illustrated in Fig. 2.9. For each month, the likely payments and receipts of cash are estimated. By taking the total cash payments from the total cash receipts for the month, an estimate is formed of the likely net cash flow for that month. When the net cash flow is added to the bank balance at the beginning of the month, the estimated bank balance at the end of the month is achieved. If this is done for each month of the budget year, the result is a picture of the way in which the bank balance of the business is likely to vary during that year.

The first stage in compiling a cash flow budget is to make a list of cash receipts and cash payments likely to affect the business bank balance during the budget year. A profit budget, especially one in gross margin form, makes an excellent basis for this list. It is important to remember, however, that one is now concerned only with cash items. Changes in debtors, creditors and stock valuations must be excluded, as must depreciation and benefits in kind. All personal, tax or capital receipts or payments must be included.

Once the list is complete, the pattern of timing of receipts and payments can be worked out. Each type of receipt is examined in turn, and a judgement made concerning the way in which that receipt will be split between the months of the year. With many items, this will require careful estimation of the pattern of physical production – the way in which milk receipts will occur during the year, for instance, will be governed by the combined lactation curves of the cows in the herd.

It is a normal trade practice to alow between two and four weeks' credit on sales, so many cash receipts will be paid into the bank the month after the actual sale has taken place. This must be allowed for in deciding in

	TOTAL	Oct	Nov	Dec	Jan	Feb	Mar	Apr	May	June	July	Aug	Sep
					12500								
Income													
Milk	97803	8910	8894	8823	8751	8680	8609	8446	7705	6530	6417	7430	8608
Calves	7700	2100	1400	1050	700	350						700	1400
Cull cows	8750			350		350		700	350	700	3850	2450	
Wheat	22400				22400								
Wheat straw	2500	2500											
Area payments	12500				12500								
Sundries	1500		300			200		500		200		300	
VAT charged on outputs	0												
VAT refunds	0												
Capital - grants	0												
machinery sales	5000				5000								
Personal receipts	0												
SUB-TOTAL	158153	13510	10594	10223	49351	9580	8609	9646	8055	7430	10267	10880	10008
Payments													
Feed: cows	7200	3000		2500		1000		700					
Vet and Med	4800	500	700	600	600	400	300	200	200	200	200	400	500
Livestock purchases: cows	14400										6750	3600	4050
Misc. dairy costs	2160	180	180	180	180	180	180	180	180	180	180	180	180
Seed	3590	2420					1170						
Fertiliser	14590		14590										
Spray	4120		4120										
Misc. crop variable costs	490								490				
Wages - permanent	20000	2000	1500	1500	1500	1500	2000	1500	1500	1500	1500	2000	2000
Power and machinery	15000	2000	1000	1000	1000	1000	1000	2000	1000	1000	1000	1000	2000
Rent and rates	14000						7000						7000
Misc. fixed costs	8000	650	750	650	650	650	650	750	650	650	650	650	650
Overdraft interest	4967	417	408	541	554	378	351	395	373	353	380	415	403
Loan interest	4276		2138							2138			
Capital - buildings	0												
machinery	20000				20000								
capital repayments	7000			3500						3500			
	0												
Personal - drawings	15000	1250	1250	1250	1250	1250	1250	1250	1250	1250	1250	1250	1250
tax	5000				2500						2500		
SUB-TOTAL	164593	12417	26636	11721	28234	6358	13901	6975	5643	10771	14410	9495	18033
NET CASH FLOW	-6440	1093	-16042	-1498	21117	3222	-5292	2671	2412	-3341	-4143	1385	-8025
OPENING BANK BALANCE	-50000	-50000	-48907	-64948	-66446	-45329	-42107	-47399	-44728	-42316	-45656	-49800	-48415
CLOSING BANK BALANCE	-56440	-48907	-64948	-66446	-45329	-42107	-47399	-44728	-42316	-45656	-49800	-48415	-56440

Fig. 2.9 Example of a cash flow budget.

which month the cash is likely to enter the bank account. Payments by opening debtors should be assumed to be received in the first month of the year.

The process is repeated with each payment item, again ensuring that allowance is made for credit taken. It is sometimes possible to negotiate extended credit (three months, for instance) with suppliers of items such as fertilizer and machinery. This must be allowed for in deciding timings.

The budget can now be completed by entering receipts and payments in the body of the table according to the timings decided, and calculating monthly bank balances. Beginning with the column for the first month of the budget year, total payments for the month are subtracted from total receipts to give the month's net cash flow. When added to the bank balance expected for the beginning of the year, this gives the expected bank balance for the end of the first month. A positive figure denotes money in the bank: a negative figure indicates an overdraft.

This closing bank balance for the first month becomes the opening balance for the second month, and the above calculation is repeated for this and each of the remaining ten months of the year. If the closing balance of any month is negative (i.e. shows an overdraft), the interest due for that month can be estimated by multiplying the balance by one-twelfth of the likely overdraft rate. The result is entered under 'overdraft interest' in the following month's column.

A vital part of the budget is the 'total' column, used to summarize the cash flow for the whole year. This is compiled by adding the contents of each line across the sheet, and then finding the annual net cash flow and closing balance as for a monthly column (the opening balance being that used in the first month's column). As well as providing a useful summary of the year's expected cash flow, and a means of cross-checking with the contents of the profit budget, it provides an important internal check. If the closing balance calculated in this manner matches that shown in the final month's column, one can be confident that the arithmetic in the budget is correct. If it does not, the calculations must be checked and rechecked until the error is located.

A valuable aid to the interpretation of the cash flow budget is a graph of the monthly closing bank balances, as shown in Fig. 2.10.

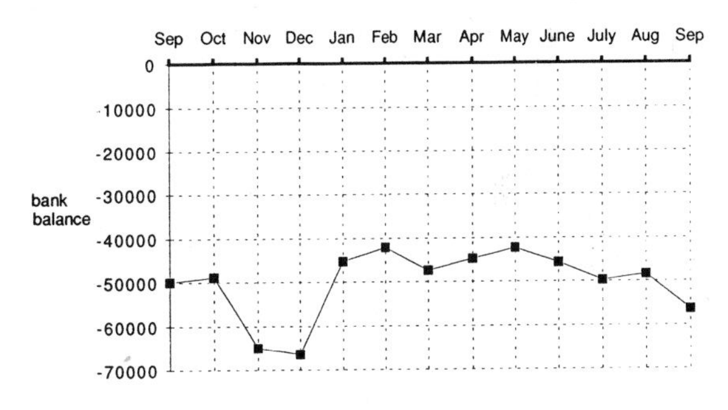

Fig. 2.10 Cash flow chart.

Assessing the cash flow budget

Before moving on to the budget for capital, it is important to weigh up the implications of the cash flow budget. As before, the main test is whether the budget indicates that the cash flow objective (probably expressed as the maximum tolerable overdraft) will be achieved. A parallel test is whether the overdraft at any time is likely to exceed the existing limit agreed with the bank manager. The next concern should be to check for significant variations in bank balance between months, and for any unusual movements in key items.

If these checks indicate potential problems, there are four possible strategies open to the manager. Ways can be sought of *increasing cash flow*, and thus lifting the overdraft away from problems. Possibilities include cutting or postponing private drawings and capital expenditure, and a general reduction of overhead costs. Another possibility is that enterprise gross margins could be examined for possibilities of increasing output and/or reducing variable costs, although most avenues should have been explored when assessing and adjusting the profit budget.

The second opportunity is that of *delaying payments* to reduce overdraft peaks. Buying fertilizer as it is needed, for instance, or negotiating longer trade credit from a machinery supplier, could avoid the overdraft limit being exceeded and help to cut interest charges.

Another way of reducing overdraft peaks is by *advancing receipts*, such as selling the wheat off the combine rather than storing, or pressing debtors for earlier payment.

These possibilities should be tested by reworking the cash flow budget and seeing the effect on the cash flow. Each change will have a cost – lost discounts on fertilizer, lower prices on corn sold early – and the costs and benefits of the change should be examined (a partial budget is an ideal tool for this – see 'Planning for change' later in this chapter). Some changes antagonistic to the manager's objectives (for instance pressing debtors for prompt payment, or cutting personal expenditure) may be totally unacceptable.

The final possibility, when all other options have been considered and the overdraft is still over the bank's existing limit, is to seek to negotiate a new overdraft limit. The existence of a carefully worked and reworked cash flow budget will be of great help in this process.

If this fails, and/or the new limit goes against personal objectives with respect to the maximum borrowing, a complete rethink of the business objectives and plan is necessary.

Otherwise, the next stage is to prepare a budgeted balance sheet.

Budgeting for capital

The budgets for profit and cash flow provide the raw material for a projected balance sheet. This is a statement the wealth of the business at a particular time. Wealth denotes ownership of assets – money and other things of value, such as stocks, machinery and land. Against these assets must be set liabilities, however – debts such as outstanding trade creditors, overdraft and loans.

A typical farm balance sheet is shown in Fig. 2.11. The difference between assets and liabilities is shown as *net capital* (also known as 'net worth' or 'owner equity'). This figure represents the net value of the business to its owner(s). In a company business net capital will be expressed in a slightly different form, as a combination of share capital and reserves.

The figure used for bank balance or overdraft is the

Projected balance sheet at 31 March 199x	*This year* £	£	*Last year* £	£
Fixed assets				
Bungalow (at cost 199x)	70 000		70 000	
Plant, machinery and vehicles	36 000		30 000	
Breeding livestock (dairy herd)	50 800	156 800	50 000	150 000
Current assets				
Trading livestock (calves)	280		280	
Crops in store	41 950		42 700	
Crops in ground	3 000		3 000	
Miscellaneous stores	9 000		9 000	
Trade debtors	9 265		8 910	
Cash	0		0	
	63 495		63 890	
Current liabilities (due within one year)				
Trade creditors	1 349		1 200	
Bank overdraft	56 440		50 000	
	57 789		51 200	
Net current assets (liabilities)		5 706		12 690
Total assets less current liabilities		162 506		162 690
Deferred liabilities (due after one year)				
Bank loan		25 000		27 100
Capital account				
Opening net capital	135 590			
Add: net profit	23 067			
	158 657			
Less: Private drawings	15 000			
Income tax	5 000			
Goods for own use	1 151			
Net capital		137 506		135 590
Capital employed		162 506		162 690

Fig. 2.11 Example of a balance sheet (tenant).

closing balance from the last column of the cash flow budget. The debtors and stock valuations are those used previously in compiling the profit budget. The values of machinery and buildings are those after depreciation (used in the profit budget). The value of land can be taken as that at the beginning of the year, plus or minus any sales or additions during the year (which will be shown in the cash flow budget).

Apart from the overdraft, the liabilities include creditors, which have already been determined in compiling the profit budget. The value of loans can be taken as the amounts outstanding at the beginning of the year, plus or minus any additions or repayments during the year (shown in the cash flow budget).

Figure 2.11 also shows the previous year's balance sheet figures in a separate column, thus enabling an easy assessment of likely changes over the year.

A further refinement is the inclusion of a capital account. This traces the growth or otherwise of net capital over the year, by reference to profit and injections of money from outside the business (both adding to net capital), and private drawings, tax payments and benefits in kind (all reducing net capital). As well as showing the relationship between profit and drawings, the capital account acts as a useful arithmetical check. The result of the calculation should be identical to the closing net capital shown by the balance sheet: if it is not, the budget calculations should be checked and rechecked until the error is found.

Assessing the projected balance sheet

Much of the message of the projected balance sheet can be seen at a glance. A positive net capital (i.e. assets more than liabilities) confirms that the business is still likely to be solvent (in business) at the end of the year. Comparison of the opening and closing balance sheets shows how the financial position of the business is likely to change over the budget year – whether net capital is likely to increase, how much total assets and total liabilities are likely to change, and to what degree major changes in individual items are responsible.

Additional clues can be derived from the application of various ratios. Space precludes mention of all such, and indeed it is doubtful whether it is useful to know more than a few easily remembered calculations (but see Bull (1990), Chapter 2, and Warren (1992) for more detailed discussion). To check on overall stability of the finances, the *percentage owned* can be calculated (net capital as a percentage of total assets). This both shows the owner's stake in the business, and also gives an impression of the 'safety margin' enjoyed by the business. A business with 40% owned has rather less buffer against changes in financial fortunes than one with 80% (typical of many long-established owner-occupied farms). The degree to which this percentage is likely to change during the course of the budget year is a vital indication of the financial health of the business. The example farm (Fig. 2.11) has a low and declining percentage owned of 62%.

An additional overall check is provided by the *long-term debt to equity ratio*. This expresses long-term borrowing (loans and any other finance with a fixed repayment pattern) as a percentage of the net capital. Such borrowing imposes demands for both interest and repayment, irrespective of the general cash flow of the business. Thus the higher the ratio, the higher the risk faced by the business, and the greater the need for a high return on capital (see below) to cope with these demands. In an occupation such as farming, which commonly gives low returns on capital, a ratio of above about 25% would give rise to concern – but this is only a very general guide. The example farm has a ratio of 18%.

Short-term stability can be measured by the ratio between the liquid assets of the business (very short-term assets such as cash and trade debtors) and current liabilities (those at short call such as overdraft and trade creditors). This is the *acid-test ratio*. A ratio of 100% indicates that even if the current liabilities were suddenly called in, the business would have enough short-term reserves to cope. The lower this ratio, the more susceptible is the business to running out of cash. As before, changes over the year are more important than static measures – a declining ratio is a warning signal. The example farm has a ratio of 110% – down on the opening figure of 125%.

Finally, budgeted net profit and net capital may be linked by calculating a return on capital, such as *return on owner equity*. This is net profit expressed as a percentage of net capital, and indicates how efficiently the owner's capital is used within the business. The farmer who finds that his return on net capital will be less than 3% while he could get an assured 10% by investing in government securities (say), needs to ask himself some searching questions – such as whether satisfying personal objectives (for instance pleasure in running a farm) is worth the financial cost involved. Profits can be highly variable between years, and before making a decision based on return on capital, outline budgets should be used to check likely trends over the following two or three years. The example farm has a return on net capital of 16.8%, which will go some way to compensating for the low stability measures.

As with the other budgets, if the picture indicated by the assessment of the projected balance sheet does not meet the objectives of the owner, alternative plans must be investigated and budgeted. First, though, the valuations of the assets in the balance sheet should be checked. If these are based on past balance sheets, asset values may be out of date. This should be checked, assets conservatively revalued, and ratios recalculated, before any decision to rebudget is made.

Accounts

While budgets estimate the likely future effect of decisions, accounts provide the historic information to show the actual effects of those decisions.

Recording cash flow

The primary source of historic information about a business over a year is the record of the business transactions made during that year. Various methods exist of making this record, ranging from merely collecting invoices in a cardboard box, to using a computer accounting system. That which is most commonly found on farms is the *cash analysis* system of book-keeping.

As its name suggests, this is a method of recording net cash flow. It is generally based on monthly intervals, recording expenditure and receipts as the payment is made or received. The main features of a cash analysis book are shown in the example of a monthly payments and receipts analysis shown in Fig. 2.12. The broad pages accommodate a large number of columns. Those at the left are used to record the essential information about each transaction: date, details, cheque number (paying-in slip number for receipts) and the amount of money involved.

The remaining columns allow the amount paid or received to be categorized, ideally by reference to eventual use in gross margin accounts. Thus payment analysis columns could be divided into variable and fixed cost groups, with the former including columns for feed, livestock purchase, veterinary and medical, seeds, sprays, fertilizers, contract hire and so on, and the fixed cost columns including those for regular labour, machinery running costs, rent and rates, finance charges, and miscellaneous overheads. In addition to these 'trading' items, some of the transactions will relate to personal, capital or tax items (VAT, income and capital taxes): hence the columns at the right of the sheet.

The receipts sheet is laid out in a similar manner, with the analysis columns categorizing receipts by type of output. The extent of the division is limited partly by the number of columns on the sheet, and partly by the need for clarity and simplicity.

At the end of each month, a line should be drawn beneath the last entry on each page, and the total found for each column. Taking the total of the payments page 'bank' column from the total of that on the 'receipts' page gives the bank account at the end of the month – a negative balance denoting an overdrawn account. The accuracy of this balance (and thus the recording for the whole month) can be checked by comparing the balance with that shown on the bank statement (which should be requested to show the cleared transactions up to the end of each month).

The two figures will rarely be identical, due to delays between cheques being written and their reaching the bank, and to non-cheque items such as standing orders and direct debits and credits, which may not have been recorded in the cash analysis book. If, though, after making adjustment for these items, the balances in cash analysis book and bank statement still disagree, a mistake is indicated (either at the bank or in the book-keeping) and must be located before proceeding.

Monitoring cash flow

The totals of the analysis columns can now be transferred to the 'actual' column for that month in the cash flow budget (see 'Budgeting for cash flow' above). Ideally the headings used in both cash analysis book and cash flow budget will be the same, making the transfer of information between the two as easy as possible. Once this is

PAYMENT ANALYSIS

Date	Details	stub no.	Bank	Dairy feed	Vet and med	L/stock purchase	Misc. dairy	Crop costs	Wages	Power	Rent,rates, misc o/hds	Interest & charges		Capital	Personal and tax	VAT paid
1/10	Opening bank balance b/fwd (o/d)		50000.00													
5/10	Wages	120	438.40						438.40							
5/10	A. Dealer: tractor repair	121	575.31							489.63						85.68
5/10	Farmco Ltd: 25t dairy feed @ £142.50/t	123	2850.00	2850.00												
9/10	Adams: vet bill	124	405.38		345.00											60.38
13/10	Newgro Ltd.: corn & grass seeds	125	2560.32					2560.32								
13/10	Fastfuel Ltd.:tractor diesel	126	532.58							532.58						
13/10	telephone bill	127	162.85								138.59					24.26
13/10	wages	128	484.56						484.56							
20/10	wages	129	523.87						523.87							
20/10	High St. Garage: petrol & car service	130	300.66							255.88						44.78
20/10	Browns: dairy supplies	131	550.83				468.79									82.04
28/10	wages	131	427.35						427.35							
28/10	fencing wire	132	447.06								380.48					66.58
29/10	A Dealer: tractor	133	21737.50											18500.00		3237.50
29/10	Transfer to personal a/c	134	1250.00												1250.00	
	TOTAL PAYMENTS		83246.66	2850.00	345.00		468.79	2560.32	1874.18	1278.09	519.07			18500.00	1250.00	3601.21

INCOME ANALYSIS

Date	Details	stub no.	Bank	Milk	Calves & culls	Crop sales	Sundry income		Capital	Personal	VAT charged	VAT refunds
5/10	J. Smith: 15 calves	927	1568.00		1568.00							
20/10	M.M.B. : milk sales (direct credit)		9234.32	9234.32								
20/10	F. Bloggs: contract ploughing	928	327.83				279.00				48.83	
29/10	A. Dealer: sale of tractor	929	7637.50						6500.00		1137.50	
	TOTAL INCOME		18767.65	9234.32	1568.00		279.00		6500.00		1186.33	
	BALANCE CARRIED FORWARD (0/D)		64479.01									
			83246.66									

Fig. 2.12 Example of cash analysis for one month.

done, and the actual closing balance entered at the bottom of the column, the two sets of information can be compared to detect discrepancies. The aim in this comparison should be to establish the main causes of discrepancy, to anticipate the likely effect over the remainder of the year, and to identify appropriate remedial action (see Fig. 2.13).

The cash flow budget is an essential tool for the latter two tasks. If there is little that can be done to bring actual performance into line with the budget, it may be time to revise the budget for the rest of the year. Should this revision imply a higher overdraft facility than has been agreed with the bank, it can be used to good effect in renegotiating that facility well before it is actually needed.

For a more complete description of the mechanics of cash recording and control, a specialist text is recommended such as Atkinson *et al.* (1990), Hosken & Brown (1991), or Warren (1992).

	OCTOBER	
	Budget	Actual
Income		
Milk	8910	9234
Calves	2100	1568
Cull cows		
Wheat		
Wheat straw	2500	
Sundries		279
VAT charged on outputs		1186
VAT refunds		
Capital - grants		
machinery sales		6500
Personal receipts		
SUB-TOTAL	13510	18767
Payments		
Feed: cows	3000	2850
Vet.and Med.	500	345
Livestock purchases: cows		
misc. dairy costs	180	469
Seed	2420	2560
Fertiliser		0
Spray		0
misc. crop variable costs		
Wages	2000	1874
Power and machinery	2000	1278
Rent and rates		
Misc. fixed costs	650	520
Overdraft interest		
Loan interest		0
VAT payments		3601
Capital - buildings		
machinery		18500
capital repayments		
Personal - drawings	1250	1250
tax		
SUB-TOTAL	12000	33247
NET CASH FLOW	1510	-14480
OPENING BANK BALANCE	-50000	-50000
CLOSING BANK BALANCE	-48490	-64480

Fig. 2.13 Example of cash flow monitoring.

Recording profit and capital

Actual profits or losses, and actual net capital, are measured by exactly the same methods as used in the budgets. Thus a profit and loss account, in conventional or gross margin form, is used to measure net profit or loss, and a balance sheet to measure the net capital.

It was suggested above that the selection of headings in the cash analysis book should be influenced by the needs of gross margin accounts, with payments divided into fixed and variable categories. The result will be that, at the end of a year, the monthly totals of each of the 'trading' analysis columns (i.e. excluding capital, personal and tax columns) can be accumulated to give the raw material for the gross margin enterprise accounts. Given the limitations on the number of analysis columns in the cash analysis book, it may be necessary to use a simple coding system during the year to distinguish between items within the same column. A letter 'P' could be written alongside appropriate items in the 'feed' payments column, for instance, to indicate pig feed as opposed to dairy or sheep feed.

From this point, it is a relatively simple matter to compile gross margin accounts for the year, using the procedure described for the preparation of budgeted gross margins. Fixed costs are likewise derived from the relevant analysis column totals, and deducted from total gross margin to arrive at an actual net profit.

Analysis and interpretation of profit and loss accounts

Interpretation of a profit statement depends on comparison: with previous years' figures, with other businesses, and with the relevant budget. All of these methods can be used with a profit and loss account in conventioonal form, but provide much more useful management information when used with a profit statement in enterprise account form, such as the gross margin accounts discussed above.

Comparison with previous years gives a valuable picture of trends within the business, and management 'alarm bells' will be rung by major differences between the performance of the immediate past year and those preceding it. As always, the comparison should start with the 'bottom line' – the net profit – and work back through total revenues, expenses and stock valuations to build up a picture of cause and effect concerning the variations. Given the physical and financial detail shown in each enterprise gross margin account, such comparisons can be used in building up a list of clues to help management explain how discrepancies arose, and how to correct them in the future (or capitalize on them, where performance is better than before).

Previous-year comparisons are limited by variations in the general environment (e.g. weather, prices) between years, and are also insular, relating as they do to just one business. To avoid the latter problem in particular, comparisons can be made with other businesses. Various comparison media are available in UK agriculture. Some are based on surveys, such as the annual Farm Business Survey (FBS) conducted by Regional Agricultural Economics Centres. The results of this survey are published in regional reports, most of which present their results in enterprise account as well as whole-farm form. Comparing farm gross margin results to these survey results thus gives rise to a number of new clues as to where farm profitability could be improved.

The value of survey results is often limited by the time taken to collect and process the information, so that the latest FBS results available in a region are likely to be at least one year out of date at publication. Survey regions are large, and the number of farms of a particular type in each sample usually relatively small. A further drawback is that such surveys are often not designed to collect detailed physical enterprise information: this inevitably limits the diagnostic possibilities of comparisons.

An alternative source of comparison information is a costing service. Such services are offered to farmers by a number of organizations, including quasi-government agencies (such as the MLC Pigplan service), commercial companies and other agricultural institutions. These services, particularly those relating to dairy and other intensive livestock, usually depend on monthly updating of information from the farmer, and use computers for the collation and analysis of that information.

As a result they can process information rapidly and provide reasonably up-to-date comparisons. Moreover, being designed specifically as management aids for the farmer, they usually contain a thorough breakdown of the most important physical information concerning the enterprises concerned. Thus a farmer using MLC's Pigplan can trace a low gross margin back to such details as low gradings, high feed used, low liveweight gain, poor market prices, and so on. In these respects, the use of comparisons via a costing scheme has advantages over the use of survey data. On the other hand, it must be remembered that the results of a costing scheme relate to a self-selected sample of farmers, and that the accuracy of the information fed into the costing scheme by farmers is not subject to the same rigorous control over accuracy as would be the data collected for survey purposes. As long as the user remembers that *any* form of comparison is merely a process of looking for clues rather than answers, this should not pose too great a problem.

The third, and potentially most useful, method of comparison is that with budgets. This is part of the budgetary control described earlier as an integral part of the management process, and is analogous to the budget/actual comparisons explained in the section on monitoring cash flow. The process of comparison is as before, but this time using the budgets prepared at the beginning of the year in question, rather than previous years' figures, or those derived from other farms. This form of comparison has the advantage of being specific to the farm in question, and relating to the most recent year for which results are available. Most importantly, it measures the farm's performance against the objectives set at the beginning of the year, as part of the planning process.

Similar strictures apply to this as to the other forms of comparison – that it is concerned with collecting clues rather than answers, and that to follow up the clues needs the application of down-to-earth farming knowledge, such as that provided by the other chapters in this volume.

Recording and analysis of net capital

Once cash analysis book and profit statement have been completed for a particular year, the compilation of a closing balance sheet is a simple matter, using information contained in those accounts and the methods described in the earlier section on budgets.

Similarly, analysis of the balance sheet is exactly as described for the projected balance sheet, with the sole additional facility of comparison between budgeted and actual balance sheets.

In summary, effective control of a business depends on *budgets* anticipating the outcome of decisions in terms of cash, profit and capital. It also requires *accounts* to show actual performance in the same terms, allowing the manager to monitor progress and take corrective action.

Physical information

Calculation of reliable gross margin accounts, particularly in the detail suggested above, depends on good physical records. Examples include records of output, such as milk yields, pig liveweights, and grain yields; and records of inputs such as amounts of feed consumed and fertilizer applied.

Physical records are important for other reasons. Breeding records and charts, for instance, are vital in maintaining and improving the genetic potential of a breeding livestock enterprise. Short-term control often depends on the use of physical records. While the cost of water may not be sufficiently large to justify recording solely for enterprise accounts, for example, water records can give valuable advance warning of changes in the health of an intensive poultry flock or pig herd. Other physical records, such as those of labour use and management time, may be useful in making the use of time more efficient.

There is no 'right' way of keeping physical records, but certain guidelines should be borne in mind:

(1) Never record for the sake of recording – be sure that the benefit gained from using each record will justify the trouble taken. It is easy to find oneself keeping records that are rarely, if ever, used, and which could be safely discontinued.
(2) Make it as easy as possible for the records to be entered. This implies careful design of recording sheets, but also careful positioning of the sheets, near to the place where measurement is made, well-lit and with a pencil handy.
(3) Make it as easy as possible for the records to be interpreted. This is particularly important where the records are used in day-to-day control: key information should be highlighted and use made of graphs so that problems are clearly and rapidly shown up. It also applies where records are used as the basis for financial accounts. a little thought at the design stage about the way in which the information is to be used could save a great deal of trouble later.

For more information on the keeping of physical records, Hosken & Brown (1991) is recommended.

Planning for change

There are times when to use a full-blown set of budgets would be inappropriate. An example is where a number of partial changes to an established system are being considered. To avoid unnecessary work and potential confusion, a device is needed which will allow consideration of only the *net* financial effects of the change to be calculated. Such a device is the partial budget.

Partial budget principles can be applied to any of the three financial measures – profit, cash flow and capital – but are most commonly used to test the effects on profitability of a given change. These effects are grouped under four headings, revenue gained; costs saved; revenue lost; and extra costs.

The result is a calculation such as that shown in Fig. 2.14. This is a budget for a 'normal' year – in other words, when the change has been fully established. It is possible to compile a partial budget for a specific year, such as the first year the change is implemented, if profitability is likely to vary significantly between years. The normal year budget is the most important, though: if this does not show a benefit, there is no point in looking more closely at particular years.

Note that all costs and revenues are expressed on an annual basis. Thus expenditure made (or saved) on buildings and machinery is represented by depreciation (initial cost divided by the useful life in years), and the cost of building up (or reducing) a breeding herd or flock of livestock is shown by the extra annual costs of replacement incurred by those animals.

Note also that the budget is kept clear and uncluttered by including only those items that are likely to change. If a change is likely to reduce workloads, but it is unlikely that a worker can be laid off or overtime reduced, wages costs can safely be omitted.

The best use of a partial budget of this sort is as an 'initial screening' device, allowing a quick and simple check of the relative merits of a number of alternative changes to the farm system. Those changes which partial budgets show to be likely to lose money, or make insufficient profit to justify the effort and inconvenience caused, can be scrapped without further ado, and the remainder built in to revised whole-farm budgets to test the effects on cash flow and balance sheet, as well as profit.

Where the change requires investment, it is important to relate the extra profit to the capital used. There are, broadly, two ways of doing this – by including an interest charge in the partial budget, and by calculating a marginal return on capital.

The ideal way of achieving the former is to compile a cash flow budget. For this, however, the cash flow budget would have to relate to a particular year, rather than a 'normal' year, and its complexity of calculation militates against its convenient use in initial screening. The alternative is to make an estimate of the amount borrowed, and apply to this an interest rate.

Rather than estimate the amount of borrowing outstanding year by year (which again would lose the simplicity that makes the partial budget so effective), a common practice is to take the initial amount borrowed and assume it will be paid off in equal instalments over the life of the project. The average borrowing outstanding in a 'normal' year is then taken as the initial borrowing divided by two. At worst it is possible to assume that all the net initial capital is to be borrowed, but only if no firm information is available concerning the amount that is to be borrowed.

Net initial capital is the amount of money to be invested in the first instance. It includes both fixed capital (items such as property, machines, breeding stock, etc.) and working capital (the cash needed to finance the running costs of the change until it begins to bring cash in to defray those costs). It also takes account of any fixed and working capital set free by the introduction of the change. An example, related to the earlier partial budget, is shown in Fig. 2.15.

Using this example, and assuming all the capital needs to be borrowed, the average borrowing would be £5078 ÷ 2 = £2539. At an interest rate of, say, 10%, the interest charge in a 'normal' year would be £254, and the extra profit after interest would be 750 – 254 = £496.

There are various problems with this type of calcu-

COSTS OF CHANGE	£	BENEFITS OF CHANGE	£
Revenue Lost		***Extra Revenue***	
cows: milk: 10 @ 6524 litres		milk: 60 ewes @350l. @65p/l.	13650
@ £0.215/litre	14023	lambs: £50/ewe	3000
culls: 2/year @ £583	1166	cull ewes:18/year @ £10	180
calves: 0.95/cow @ £118	112	cull rams: 1/year @ £20	20
		wool: 2kg/ewe @ £0.30/kg	36
		ewe premium: £15/ewe	900
Sub-total	15301	Sub-total	17786
Extra Costs		***Costs Saved***	
ewe purch: 18/year @ £60	1080	cows: concs. (1.04t/cow)	1350
ram purch: 1/year @ 500	500	vet & med & AI etc.	686
concs.:200kg/ewe @ £160/t	1680	other	398
vet & med	500	forage: £198/ha x 5 ha	841
other	450	replacements: 2 @ £900	1800
forage	600		
extra labour:	1000		
capital: 5000 over 5 years	1000		
Sub-total	6810	Sub-total	5075
TOTAL COSTS	22111	TOTAL BENEFITS	22861
BALANCE -EXTRA PROFIT	**750**	BALANCE - LOSS OF PROFIT	0

Fig. 2.14 Partial budget for reducing the dairy herd by ten cows, and releasing 5 ha for 50 milking ewes and four rams.

Fixed capital		
Building and plant	5000	
Breeding stock – ewes	3600	
rams	2000	
Less sale of cows: 10 @ £600	–6000	
		4600
Working capital		
Half running costs of sheep	2115	
Less half running costs of cows	–1638	
		478
Total net initial capital		5078

Fig. 2.15 Calculation of net initial capital.

lation, not least because of the many assumptions and generalizations which have to be made in the process. Moreover, it only measures the *cash* cost of borrowing, and puts no value on any non-borrowed capital used.

An alternative is to relate profit to capital through calculating return on capital. The appropriate measure here is *marginal return on capital*, calculated by dividing the extra profit before interest (from the partial budget) by the net initial capital required, and multiplying by 100. In the example above, for instance, the rate of return would be $750 \div 5078 \times 100 = 15\%$.

The resulting percentage rate can firstly be compared with interest rates payable on borrowed money. If the rate of return is significantly more than the interest rate, it looks financially worth while; if less, the change is a non-starter from a financial point of view – the finance costs will outweigh the return. If the rate of return is only marginally more than interest rates, the farmer must decide whether it is worth the risk that profit might be less than forecast (or capital needs greater).

If the change passes this test, its marginal return on capital can be compared with those of other possible changes as a basis for selection between projects.

Among the drawbacks of this use of marginal return on capital are that it rests on large assumptions, takes no account of the pattern of costs and returns either within or between years and can give pessimistic results. The latter can be countered by using average rather than initial capital in calculating the return, but this tends to give overoptimistic results – it is perhaps better to tolerate a pessimistic result than to risk making decisions based on a result which may never be achieved.

More sophistication is available in the shape of 'discounted cash flow' techniques. These are essentially partial budgets which take account of differences in costs and returns between years of a project, and allow flows of money in the near future to be weighted more heavily than those in the far future. Tax payments can also be incorporated. The result is a 'net present value' showing the consolidated margin of future cash flows over the amount invested, after allowing for the time factor. This can in turn be used in calculating an 'internal rate of return', which can be used in the same way as the marginal return on capital described above.

The very sophistication of the techniques gives rise to difficulties in application, calculation and interpretation. Without training, and preferably a computer, it is likely to be better to use the crude but simple devices described earlier in this section. If these are treated purely as initial screening measures, to be followed by full budgets where these appear to be warranted, the crudity is acceptable. Further reading is suggested at the end of the chapter.

One additional device can be useful in planning for change. This is *payback* – the number of years needed for a change to repay the net initial capital.

For this it is necessary to estimate the net cash inflow for each year of the project's life. One way of doing this is merely to take the extra profit from the partial budget, exclude depreciation (the main non-cash expense) and include interest. It is possible to make more precise estimates using simple cash flow budgets. The net cash flows are then progressively totalled, starting from year one (allowing for negative flows), until they accumulate to the net initial capital invested. The payback of the project is the number of years it takes to get to this point. A quicker and cruder method is shown in Fig. 2.16, where a 'normal year' cash flow is calculated and divided by the net initial capital.

Payback is of use in its own right, in giving a 'feel' of the worthwhileness of the project. Compared with that of other possible projects, it gives a measure, though imprecise, of the relative riskiness of the projects. If size, marginal returns and personal preferences were equal between three projects with paybacks of three, six and ten years, the rational decision-maker would go for that with the lowest payback.

Whatever partial budgeting devices are used for the testing of the effects of a change from an established system or a whole-farm budget, it will be necessary, for all but the smallest of changes, to eventually test the effects of the change using whole-farm budgets for profit, cash flow and capital.

Extra profit from change after interest	496
Add back depreciation	1000
Annual cash surplus	**1496**
Net initial capital	5078
Payback (5078/1496): 3.39 years	

Fig. 2.16 Example of simple payback calculation.

Sources of capital

To put a business plan into operation (and to maintain it) usually needs investment of lump sums of money – capital. That capital may be wholly or partly on hand, in the form of savings, retained profits or the proceeds of the sale of assets. The farmer may be lucky enough to be able to claim grants to defray some of the costs of investments, or even to have relatives or other well-wishers willing and able to make outright gifts of money. Though such capital may appear to be free of charge, there are costs attached to the use of any capital sum. If nothing else, there is the opportunity cost arising from the lost benefits from using the money in some other way. Other, non-financial costs arise from legal obligations, in the case of a grant, or 'moral' obligations in the case of a gift, for instance. Nevertheless, the costs are usually relatively low, and these sources should be exploited to the full before turning to raising money from outside the business through borrowing or shared ownership.

Borrowing

The most common alternative is to borrow money. Borrowed money incurs opportunity costs and obligation costs as described above. In addition it incurs a direct cost – that of interest (the exception being some private loans from relatives). This is a fee charged by the lender of the money, expressed as an annual percentage of the amount borrowed. The rate of interest depends on both external factors such as the state of the economy, and internal factors such as the security offered and the apparent creditworthiness of the business.

Interest rates are often expressed in relation to the clearing banks' base rate, which reflects the level of interest in the economy as a whole and is published in the financial sections of quality newspapers. An interest rate may be variable (i.e. it fluctuates over time with changes in the base rate) or fixed (i.e. it stays the same over the length of the borrowing term).

Rates can be applied in various ways. For instance they can be related to the daily balance outstanding (as with an overdraft), to the amount outstanding at the beginning of a particular year, or applied to the initial amount borrowed ('flat rate'). To allow the cost of different forms of borrowing to be compared, each rate should be expressed as an Annual Percentage Rate (APR). The APR on an overdraft is normally close to its stated rate, while that of a flat rate can be nearly double the expressed rate.

A lender will usually require some form of security against default on payments by the borrower. This security can be in the form of title to specific fixed assets, enabling the bank to sell those assets to recover the debt if necessary. A 'floating charge', commonly used for overdrafts, is a generalized version including more movable assets of the business. Alternatively, security can be provided in the form of a guarantee from an individual or an institution that, if the borrower defaults, the guarantor will repay the money outstanding.

Sources of borrowed capital

It is difficult to avoid generalizing when referring to the characteristics of various sources of borrowing. The following notes should be judged in this context.

The high street banks are the main source of borrowed capital for farming, and the largest part of this is in the form of *overdrafts*. An overdraft is created when more money is drawn out of an ordinary bank current account than was deposited there in the first instance. Interest rates are low (typically 2–3% above base rate), and are applied to the daily balance outstanding. Unauthorized overdrafts are charged a penal rate which can be more than 15% above the base rate, so it is important to agree an overdraft facility with the bank manager in advance of needing it. When an overdraft facility is negotiated or reviewed, an arrangement fee may be charged. This fee can be as much as 1% of the overdraft facility.

Overdrafts are intended for short-term finance, and the bank will normally expect the account to return to credit at some stage during the year. It is normally secured by a 'floating charge' on all the assets of the business, and is technically repayable on demand.

Bank loans, on the other hand, are subject to a contract binding each party. When a loan is arranged, the borrower is given the amount of the loan as a lump sum. This amount is repaid over time in regular instalments, with interest payments (usually at a fixed rate) reflecting the decline in the loan outstanding. As long as these instalments are paid, and the other terms of the contract are met by the farmer, a loan cannot be called in before it is due (or the business becomes insolvent). To protect itself in the event of insolvency, the bank will normally expect the loan to be 'secured' on a specific asset – in other words, if the business defaults on the loan, the bank has a legally valid claim to the title of the asset, and can sell it in order to recover the money owed.

A loan is normally intended for longer-term finance than an overdraft. Interest rates are higher (e.g. 3–6% above base rate) and arrangement fees are again likely to be payable. Against the higher costs must be set the greater security of the loan and the regular and predictable interest and capital payments, making budgeting easier.

The high street banks are not the only institutions offering loans to farmers. Loans secured on land and property ('mortgages') are available from a variety of sources, including building societies (for house purchase and building) and the Agricultural Mortgage Corporation (for land purchase, buildings, and other longer-term needs).

Trade credit is a popular form of short-term finance, being quick and easy. It is obtained by not paying bills immediately. Most suppliers will allow a period of grace for payment of invoices, usually two to four weeks, though sometimes as short as seven days. Using this period to the full can help keep down the business overdraft. Extending credit beyond this limit will, however, incur penalties – at the least, loss of goodwill from the supplier, and at the worst a credit charge or 'loss of cash discount'. These charges are invariably much more expensive than overdraft finance.

Two forms of borrowing are particularly relevant to machinery purchase. *Hire purchase* or lease purchase uses the asset purchased as the security for the borrowing: repayments are arranged so that the recoverable value of the asset is always greater than the debt outstanding. It is convenient, as purchase and finance can be arranged in one operation, but it can be considerably more expensive than bank finance.

Finance leasing involves a machine (or other asset) being bought by a finance company (often a subsidiary of a major bank) and rented to the farmer. After the initial rental period of two to five years (depending on the life of the asset) the farmer may be able to continue the lease into a 'secondary' rental at a nominal rent, or even to buy it. If the machine is sold on behalf of the finance company, he is usually able to retain a large proportion of the proceeds. Maintenance, servicing and insurance are all the farmer's responsibility. Thus the effect is of purchase, even though the arrangement is technically a lease. Rentals are quoted in terms of interest rates, and the finance companies' ability to exploit tax concessions helps to keep rates low even compared with overdraft finance. An attractive characteristic is that spreading the cost over several years enables machines to be acquired without immediate effects on the overdraft. This can be dangerous if done to deceive the bank manager as to the true state of affairs (at the time of writing it is not obligatory for sole traders and partnership to show leased assets on their balance sheets).

Calculating the cost of borrowing
It is crucial, if considering borrowing capital, to be able to estimate the likely costs to the business. Apart from administration costs (such as arrangement fees), the annual costs are composed of two elements – the repayment of the amount borrowed (needed for the cash flow calculations), and interest charges (needed for both the profit and the cash flow budgets).

In calculating the financial commitments arising from an overdraft, there is no substitute for a cash flow budget, as described earlier in this chapter.

Where interest on loans or other forms of finance is quoted at a 'flat' rate, the annual interest charge is given by the initial amount borrowed multiplied by the flat interest rate. The annual capital repayment is given by the initial amount borrowed divided by the number of years of the term. Thus £10 000 borrowed over a term of five years at a flat rate of 8% will give rise to interest payments of £800 per year (10 000 × 8%) and capital repayments of £2000 per year (10 000 ÷ 5 years).

A very common method of repayment is that known as the 'annuity' or 'normal' method. This adjusts the proportions of interest and repayment throughout the term of borrowing, to keep the combined annual payment constant while allowing interest payments to reflect the gradual repayment of the capital. Tables are provided in most farm management handbooks and financial textbooks to enable estimation of interest and repayment charges for a particular year in the life of a loan. Part of such a table is shown in Fig. 2.17. From this it can be calculated that in year 3 of a 5-year loan of £10 000 at 12% a business will have to pay £800 interest (80 × 10) and £1970 capital repayment (197 × 10) (see Fig. 2.17).

Shared ownership

An alternative to borrowing is to share all or part of the business with others, in return for an injection of capital. The simplest form of sharing is a *partnership*, where two or more people agree to run the business in common, sharing profits and control in return for investment of capital.

Each of the partners is jointly and severally liable for debts incurred by the other partners as well as by himself. In other words, if the business fails, the partners may have to sell all their personal assets if that is the only way that the creditors can be repaid – even if the cause of the failure was the action of only one of the partners. This is the motivation behind the formation of many private *limited companies*. A company can be formed with as few as two people, and has the advantage of liability being limited to the capital invested in the company – at the expense of higher formation and administration costs, and a certain lack of privacy.

Formation of a *cooperative* can allow shared investment in just part of a business, ranging from a single machine to a complete marketing and distribution service for produce. Liability is limited, and formation costs are similar to those of a company.

Any form of shared ownership raises complex and difficult questions, particularly where taxation is concerned. Generalization is impossible, and a reputable solicitor and accountant should be consulted before entering into any such arrangement (Panes (1980); Centre for Management in Agriculture (1986); Furlong (1987); and Wright (1989)).

Coping with risk and uncertainty

Implicit in all the above discussion is the difficulty of planning and budgeting for the future. This arises from the fact that nothing about the future is certain.

The sources of this uncertainty, and the consequent risk to the business owner on making decisions about future events, are many and varied. In farming, physical factors are prominent, such as the possibility of disease, pest damage, drought, flood, fire, health of farmer and employees, and so on. Economic fluctuations also have effects – the state of world trade, the strength of the pound, and the growth of incomes of the population as a whole are some of the influences on product and input prices at the farm gate.

Just as important can be political factors, affecting such things as tax allowances, agricultural support prices and quotas on production. Social and personal aspects also have an effect – such as changes in general food consumption habits, hardening attitudes of non-farming neighbours to noisy or smelly activities, marriage of the farmer's son or daughter, or the divorce of the farmer.

There are ways of organizing the business to minimize some of the effects of uncertainty. Some of the factors mentioned can be insured against (fire, for instance). Prices of some products and inputs can be protected by selling and buying forward, or by using the futures

	Interest rate					
	8%		10%		12%	
Year	Interest	Repayment	Interest	Repayment	Interest	Repayment
1	80	170	100	164	120	157
2	166	184	84	180	101	176
3	52	199	66	198	80	197
4	36	215	46	218	56	222
5	19	232	24	240	30	248
	£ Payment per £1000 borrowed					

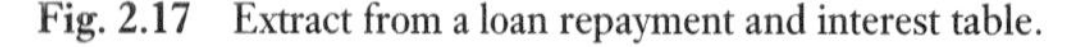

Fig. 2.17 Extract from a loan repayment and interest table.

markets. The farm system can be designed with a wide spread of different enterprises, so that if one enterprise performs badly, the others can compensate. The farmer can avoid trying any new products or production methods to avoid the risk of their failure.

Not all risks can be avoided by such methods, and reducing risk usually has a cost. A farmer may prefer to take the risk rather than incur that cost. Insurance incurs the cost of the premium. Forward buying and futures trading incur the cost of losing potential revenue if spot market prices improve and the farm is tied to a contract price. Diversification and conservatism involve lost opportunities to increase profits – either by losing the chance of specializing in products that the farm and farmer are good at producing, or by forgoing the uptake of new, potentially profitable enterprises and production methods.

If risk cannot be limited beyond a certain point, it can at least be allowed for in the planning and control process. Reference has earlier been made to the importance of erring on the pessimistic side when budgeting, and to the need for adequate research to ensure that the forecasts on which the budgets are based are as accurate as possible. Most important of all is the process of budgetary control, whereby the effects of a deviation from the expected performance can be readily identified and appropriate action taken.

None of these methods gives the farmer much advance warning of *how much* risk he or she faces – and this can be crucial in making a decision. The key question concerns the 'downside risk': 'How much do I stand to lose if things go wrong?'

The simplest solution in concept is to prepare two sets of budgets, one based on the inputs and outputs that the farmer thinks are most likely, and the other based on the worst possible performance. The result is effective, with the farmer being able to see what he or she stands to lose if things go wrong and to decide whether or not to take this risk. Preparing two sets of budgets is time-consuming, though, unless one has available a suitably programmed computer. Even with a computer, interpreting the results can be difficult: so many variables may have been changed that it is impossible to pick out the factors most responsible for the variations.

The latter problem may be overcome by the calculation of *breakeven points*. Those factors which seem most likely to affect the outcome of the profit budget are identified. Two such factors on a dairy/arable farm might be the amount of concentrate fed to the cows, and the likely price for winter wheat. The budgetary margin (net profit or loss in a whole-farm budget, extra profit or loss in a partial budget) is then expressed in terms of that factor. For instance, in the partial budget example used earlier, the extra profit estimated was £496 after interest. At a price of 65 p/litre of ewe's milk, this represents $496 \div 0.65 = 2712$ litres (45 litres/ewe). This is the amount by which the milk yield would have to fall before the extra profit started to become a loss of profit. The breakeven milk yield is thus $350 - 45 = 305$ litres/ewe.

Working out breakeven points for various components of a budget does two things. By looking at each breakeven point, the farmer can make subjective judgements as to which of the breakeven points is most likey to be reached: this gives an indication of those items which most contribute to risk, and which thus should be monitored most closely. It also helps the farmer to judge whether or not the risk is worth taking. In the example above, for instance, the farmer can ask himself, 'How likely is it that I will achieve only 341 litres/ewe, and am I prepared to take that risk?'

The strength of the breakeven point technique – that it allows the effect of particular factors to be singled out – is also its weakness. It is not possible to identify the combined effect of more than one factor varying at the same time. While this may not be a huge problem in a simple partial budget, it is a severe limitation in budgets with many component variables.

An alternative is a form of *sensitivity analysis*. This uses tables to show the way in which the budgetary margin is likely to vary with changes in the value of one component of the budget. The example in Fig. 2.18 is based on the partial budget from earlier in the chapter. It shows how the extra profit is likely to change from the value in the budget (£496) if ewe milk yield were to vary from the 'most likely' assumption of 350 litres used in that budget. The lower figure is, in the farmer's view, the worst likely outcome, and the higher figure is the best that he or she feels is possible.

This gives the farmer a feel for the 'sensitivity' of the budgetary margin to variations of a particular component of the budget, and thus the degree to which the successful outcome of the plan depends on the farmer's ability to achieve this level of performance. Better still, this table can be combined with others to show the effect of interaction between components. In Fig. 2.19, for instance, the milk yield table has been extended to incorporate variations in milk price. Here the decision-maker has the opportunity to see what the consequences would be of poor performance on both fronts – and to decide whether or not to take the risk. At the cost of simplicity and clarity, such a table can be extended to show the combined effects of variation in three or even four budget components.

The methods described merely present the farmer with information about the likely consequences of deviation from the forecast performance assumed in the budgets. They leave the farmer to use subjective judgement as to how likely these deviations are to occur. Other techniques exist which allow probabilities of different

Milk yield per ewe		
300 litres	350 litres	400 litres
−1454	496	2446
Extra profit from change, after interest		

Fig. 2.18 Sensitivity analysis – one-way table.

Milk yield per ewe			
Milk price	300 litres	350 litres	400 litres
55p/litre	−3554	−1604	346
65p/litre	−1454	496	2446
70p/litre	−404	1546	3496
	Extra profit from change, after interest		

Fig. 2.19 Sensitivity analysis – two-way table.

outcomes to be incorporated in the calcuation, and these are described in texts such as Barnard & Nix (1979) and McCrimmon & Wehrung (1986). For most practical purposes in farm management the simpler devices described above will be sufficient, particularly if combined with a rigorous system of budgetary control.

Marketing

The welfare of a business depends on its being able to sell the goods and services produced at a profit. Crucial to this is the process of marketing, which can be conveniently, if predictably, divided into four functions:

(1) determining the *products* to be produced;
(2) deciding the *price* at which they should be sold;
(3) deciding the *place* at which they would best be sold; and
(4) determining the most appropriate forms of *promotion* to use in persuading potential customers to buy.

For these functions to be performed effectively, they must be considered from the viewpoint of the consumer of the product. The business needs a product wanted by consumers at the price and location wanted by consumers, and promotion which will appeal to the consumer. A 'market-orientated' business such as this has a far better chance of maximizing profitability than one whose management is totally concerned with getting production techniques right (the 'production-orientated' business).

For many farm products, the farmer has traditionally been a 'price-taker'. One farm's products are very similar to another's, and any individual business is such a small part of a very large market that it can have little or no influence over the price. In such circumstances marketing decisions at farm level are limited to deciding the products to be produced: these are then sold automatically to an institution such as a dairy company or a merchant, which then makes decisions about place, price and promotion. Even here, though, it is important that the business is market-orientated, responding to the information relayed back from the consumer by the wholesaling institution. A classic example is the way in which some sheep farmers have been able to boost profits of their flocks by producing the leaner meat requested by supermarkets, who are themselves responding to the wishes of their customers.

Increasingly, farmers are having to look beyond these traditional markets in a quest for profitability (see Chapter 6). This brings them to consider products which are particular to that business – for instance contracting, special food products, tourist facilities, farmed deer and trout, and so on. Here the market is small enough to be influenced by the farmer, who then has to consider all aspects of the 'marketing mix'.

If a business in this position is to supply what consumers want, it must first find out some of the characteristics of those consumers. This implies some form of market research. The aim is not only to identify possible new products, but also to establish facts such as who is likely to buy a particular type of product, the characteristics they would like that product to have, and the price they would be prepared to pay. Once these facts have been ascertained, decisions can be made about the exact form of the product and how best to promote it.

The term 'market research' implies large and complicated surveys, but can be applied to any ordered and critical examination of the likely market for a product. It should start at the simplest level, with the questions, 'why should anyone want to buy my product rather than those of my competitors?' and, if there are not apparent competitors, 'why is no-one supplying this product already?' Honest, objective answers to such questions may avoid the danger of the emotional decision: the obsessive attachment to a particular idea coupled with a refusal to give full consideration to its limitations.

The next stage is to consider the factors which are likely to affect the demand and supply of the product concerned, since these factors govern the prices which can be charged.

From the *supply* point of view, the most important factor is likely to be the number and type of competitors. It is essential to consider all possible competitors, rather than just those who are providing direct substitutes. For instance, a contractor needs to take account of the presence in the area of an active machinery leasing firm; a pick-your-own enterprise needs to consider greengrocers and supermarkets in the towns he or she hopes to service, and a day-visitor enterprise needs to bear in mind the proximity of National Parks and other areas which can be visited without charge.

It is vital to anticipate the way in which the identified competitors are likely to react to the new challenge. A competitor with substantial financial reserves may adopt aggressive pricing policies in order to safeguard his or her markets; thus many producer-retailers of milk have been forced out of operation by large dairies which can afford to give loss-making discounts for long periods. A long-established firm with a large share of the market may be more complacent about competition than a relatively new business which has built up its trade by hard graft and has no intention of giving any of it up.

Among the factors which may affect the *demand* for a product are changes in the number of potential customers within the area of sale, changes in the level and distribution of their incomes, and changes in their tastes and lifestyles. Other influences will be the presence of businesses producing complementary products, the promotion of the product, and the level at which prices are set in comparison to those of competitors.

Clues to many of these factors can be found in local libraries: residential population and incomes information is published by both central and local government; seasonal population by many of the regional tourist boards; changes in consumption (e.g. the types of food, or cars, or freezers) in the National Incomes and Expenditure Survey (Her Majesty's Stationery Office). Other sources of information may come to light when pressure is put on organizations and individuals with particular interests. It is possible, for instance, that the county council has recently measured the flow of traffic along nearby roads, and is prepared to release the results. For a visitor enterprise, this could be crucial information.

Published information can be supplemented by personal investigation, given common sense and a little imagination. Traffic levels can be assessed roughly by first-hand observation. The Yellow Pages in the telephone directory, together with the advertisement pages of the local newspaper, can be used to identify potential competitors, whose prices can be determined by making telephone enquiries as if from a potential customer. The same sources may be used to identify complementary rather than competitive businesses. The golf club down the road could encourage the development of a sport fishery. One contractor in a locality might be geared up for operations for which his neighbour is not, and vice versa. The presence of a neighbouring caravan site could provide a ready-made market for a pick-your-own enterprise. At the least such information should influence the way in which the enterprise is promoted; at the best it could be an avenue to profitable cooperation.

This basic research may give a rough idea of the numbers of people who come into the target population of the enterprise, but will not give a reliable estimate of the proportion who might be tempted to buy the produce on offer. The market must be tested in some way in order to see whether a reasonable number of people are interested. One way of doing this is to start the enterprise in a small way, with little commitment of capital. This is easy in some enterprises, such as bed and breakfast in a half-empty farmhouse, or using underemployed men and machines for contracting. In other cases, however, the effort and expense of establishing even a small enterprise are such that full commitment is needed from the start; where planning permission and/or licences are necessary, for instance, or where major capital works are called for. In these circumstances, a market survey of some form is needed. At its simplest, this can take the form of a series of informal discussions with potential customers, or with people who are likely to be 'in the know', such as tourist information officers or officials of the local chamber of commerce (as long as they are not likely to be competitors themselves). If the risks are great and the financial commitment large, it may be worth employing a market research bureau to conduct a formal survey among potential customers.

Once the enterprise has begun operation, a check can be kept on the market by various types of feedback from customers, ranging from a verbal enquiry 'was everything all right?', through visitors' books and suggestion boxes, to analysis of sales records in detecting significant variations from target.

Promotion

Promotion can be defined as any tactic which is used in attempting to stimulate demand for a given product, and can include devices such as free gifts and special offers as well as advertising. Most forms of promotion cost money and effort, and so discrimination is needed with respect to the amount and type of promotion that is employed. Generally speaking, the effectiveness of promotion depends on the degree of product differentiation that exists. If a business is offering an identical product to those of several competitors in the same area, its promotion may benefit its competitors as much as, or even more than, itself. If, on the other hand, the business can convince the customers that it is offering something unique, the chances of promotion being successful are much higher. This uniqueness can be brought about in a variety of ways, including developing an 'own brand', growing exotic or out-of-season produce, offering a high quality of product and/or service, or providing a particularly attractive environment.

The cheapest and possibly most effective form of advertising is that of word of mouth. Since this takes some time to build up, however, the new enterprise has to use other media. These can include press, TV and radio advertisements and editorial, direct mailing and hand-posting of leaflets, personal calls, membership of professional associations, directories and guide-books, printed tee-shirts and sponsorship of sports and cultural events.

An important rule in advertising is to match the medium to the market. Market research of even the simplest form will indicate the target population of the enterprise, and it remains to decide which form of advertising would be most effective in reaching that population. A pick-your-own enterprise relies on local trade, so media such as local newspapers and leaflets are appropriate, while national newspapers are not. The reverse is true of a holiday accommodation enterprise. A farm contractor would probably get better results from use of a circular letter followed by a telephone call, than from an advertisement tucked away in the classified advertisement page of the local paper.

Next it is vital to plan the advertising carefully, leaving nothing to chance. The strengths of the product should be assessed so that they may be brought out in the advertisement. The wording should be pared down to leave a clear message. This message should be hard-hitting, even brash. Good taste does not always win customers. It is worth taking pains over design of advertisements, making several mock-ups before deciding on the final version.

Some of the most effective advertising is free, or at least comparatively cheap. This applies particularly to editorial in newspapers and the broadcast media: a news-hungry editor can often be persuaded to publish articles about local businesses. The key lies in developing an interesting story – a grand opening, for instance, or a free afternoon for the 'Over-Sixties Club'.

Planning alone is not enough to get the best out of advertising – its effectiveness must be continually monitored. Where orders or bookings are made by post, this is easily effected by means of a code inserted in the address quoted in the advertisement. Thus respondents to two separate advertisements might be asked to write to M. *A.* Warren and M. *B.* Warren respectively, where the *A* represents an advertisement in the *Barset Chronicle* and the *B* that in the *Barset Advertiser*.

Pricing

Two landmarks govern the process of pricing a product: the cost of production of the goods or service, and the prices charged by competitors for similar products. These landmarks must be determined, which implies careful research and budgeting. In particular it is important to be able to distinguish between fixed and variable costs, since in the short term it may be appropriate to set prices to cover variable costs only.

How the prices are set in relation to these landmarks will depend on one's own estimation of how the market will react, and one's own promotional objectives. The market reaction is dependent on the responses of both potential customers and competitors. If the competition is weak, and the customers are likely to be relatively insensitive to price, 'skimming pricing' can be used, setting an initially high price which can be progressively dropped as new competitors come on the scene (by which time one's own enterprise should be well established). If, on the other hand, competition is strong and customers are likely to be responsive to price levels, the new enterprise can carve out a market by 'penetration pricing'. This is charging low prices at first in order to attract custom away from the established competition. Use of this tactic requires reserves, in the form of cash or the support of other business enterprises, to enable survival in the initial stages of low or negative profits.

Consistently low prices may be used as a long-term strategy, if the objective is to generate high-volume sales which can withstand low margins per unit sold. On the other hand, it may be appropriate to set prices high to establish a high-quality image.

The owner of a well established enterprise might be tempted to rely on a mark-up method of pricing, applying a constant percentage margin on top of production costs in order to keep the pricing decision simple. The rigidity imposed by this sort of formula reduces the ability of the enterprise to adapt to changing market conditions, however, and could lead to a dangerous decline in competitiveness.

Conclusion

'Management' was described at the beginning of this chapter as a process of making decisions, and monitoring their effect in order to make better decisions in the future. Viewed from some standpoints, this could be considered to need merely the application of common sense. It is to be hoped, however, that this chapter has shown on the one hand the potential complexity of the management task, and on the other the way in which various techniques and ways of thinking can clarify issues and aid decision-making.

Sometimes common sense is not enough.

References and further reading

Atkinson, J.M., Hastings, M.R. & Warren, M.F. (1986, revised 1990) *Using Your Accounts*. Agricultural Training Board/Seale Hayne College/Manpower Services Commission (The Family Farm Series).

Barnard, C.S. & Nix, J.S. (1979) *Farm Planning and Control*, 2nd edn. Cambridge University Press, London.

Berry, A. & Jarvis, R. (1991) *Accounting in a Business Context*. Chapman & Hall, London.

Bull, R.J. (1990) *Accounting in Business*, 6th edn. Butterworths, London.

Centre for Management in Agriculture (1986) *Practical Share Farming*. British Institute of Management, Corby.

Furlong, L.A.C. (1987) Farming as a limited company. *Farm Management* **6**(5), 213–220.

Giles, A.K. (1986) *Net Margins and All That*. Study No. 9, The Farm Management Unit, University of Reading.

Giles, A.K. & Stansfield, J.M. (1990) *The Farmer as Manager*, 2nd edn. CAB International, Oxford.

Hosken, M. & Brown, D. (1991) *The Farm Office*, 4th edn. Farming Press, Ipswich.

Lumby, S. (1991) *Investment Appraisal and Financing Decisions*, 4th edn. Chapman & Hall, London.

McCrimmon, K.R. & Wehrung, D.A. (1986) *Taking Risks*. Collier Macmillan, London.

Mitchell, W. (1990) Passing on assets to the next generation. *Farm Management*, **7**(6), 277–284.

Mott, G. (1989) *Investment Appraisal*. Pitman, London.

Nix, J.S., Hill, G.P. & Williams, N. (1989) *Land and Estate Management*, 2nd edn. Packard Publishing, Chichester.

Panes, C. (1980) Farm partnerships. *Farm Management*, **4**(3), 108–116.

Warren, M.F. (1993) *Financial Management for Farmers*, 3rd edn. Stanley Thornes, Cheltenham.

Wright, D. (1989) Contract farming agreements. *Farm Management*, **7**(4), 177–184.

Farm management data

Agricultural Business Consultants, Ltd. *The Agricultural Budgeting and Costing Book*. ABC, Atherstone (published twice-yearly).

Nix, J.S. *The Farm Management Pocketbook*. Farm Business Unit, Wye College, Kent (published annually).

The Scottish Agricultural Colleges, *Farm Management Handbook* (published annually).

Each Regional Agricultural Economics Centre (usually based at a prominent university or college) also publishes annual reports and handbooks.

3

Farm Staff Management

M.A.H. Stone

Introduction

Between 1988 and 1992 the total labour force engaged in farming in Britain declined by 34 000, continuing a long established trend. However, farming still makes significant use of labour, providing work for 622 000 men and women (including spouses) in 1992 (MAFF, 1993). This represented 2.1% of the working population. This large total labour force is spread between 242 000 farm holdings, the majority of which are under 100 ha in size (MAFF, 1993).

While it is important for all enterprises to make the best use of their staff, effective use of labour is essential for the large proportion of farm enterprises using very small amounts of labour. It is very much with these farms in mind that the following material is presented. On those farms which employ only a small amount of labour it is likely that many farmers and managers will not have the benefit of frequent practice in the wide range of people management skills described below. Therefore the approach has been to break down each process into stages, to define the terms used and to detail the options.

There are three Ps which support any business; People, Product and Profit: all the three Ps need to be adequately addressed for the business to be successful. People are highly flexible but highly complex. Dealing with this resource will be dealt with under the following headings:

- Job analysis and job design
- Recruitment and pay
- Interviewing and discrimination
- Control, guidance and negotiation
- Training and development
- Maintaining and increasing performance
- The employment environment
- References and further reading
- Glossary of terms used

Job analysis and job design

Systems and methods of work need to match both the goals of the business and the capacity of the workforce available. However, the use of labour is not based on rational economic decisions alone; there are many other influences. Social, cultural and political influences provide a frame of reference within which organizational decisions about labour use are made.

The demand for labour is rarely constant. Some of the many causes of change are:

- growth or decline in the business
- seasons
- cycles of production
- changes in demand
- changes over the product life
- change in the product
- changes in the manpower itself
- changes in methods.

Jobs are most commonly defined in functional terms, e.g. to do, provide, support, be responsible for. This is the first step, for until you have decided exactly what the job will be, you will be unable to determine how much labour you need, who is best to do the job, how the job can be done best or even the best format for the labour provision, e.g. full/part time. This process is essential to ensure you are getting value for money from your labour. Each time the business requires additional labour the work to be done must be analysed and then a job designed or redesigned. This process helps to clearly determine what the organization wants the job holder to do.

The majority of job vacancies are those replacing leavers or for additional staff doing the same or similar jobs. For this reason, the best source of information about what a job consists of is the person doing the job. The manager must carry out a *job study*, or in its more scientific form *work study*, to determine working methods and required outputs.

Actively seeking the involvement of employees in the design, evolution and redesign of jobs is to be encouraged. The outcomes of worker involvement in job analysis and design are outlined below.

- *Positive:* The job holder is close to the work; has evolved the process by learning from experience; can

gain recognition for their work and skill; involvement is fostered and health and safety risks can be assessed.
- *Negative:* The job holder may want to hide the ease of some jobs, or may fear management motives regarding pay, grading, or bonus payments.

The best results will be achieved in a climate of real dialogue between employee and employer.

Job analysis and work study allow jobs to be broken down into the technical/non-technical functions required, and worker skills to be compared to the job's requirements. This allows managers to set standards which can be built into a job description and person specification.

The *job description* is a working document which details a job's tasks, duties and responsibilities in a measurable way. The job description can be used to supply information to candidates, as a reminder at the interview and as a guide for induction, training and appraisal.

The *person specification* can be thought of as a 'photofit' picture of the ideal candidate for the job described; detailing the knowledge, skills and attitudes required. Prioritize the criteria to assist selection for interview and interviewing.

Job information can also be obtained from leavers at an *exit interview*; an interview with an employee before they leave, conducted for the benefit of the firm. The employer may discover frustrations, suggestions and details concerning employee relations that will aid future job planning. This is best done after you have written the leaver's reference.

Labour turnover datas allow the leavers in a set period to be compared to the total workforce, normally expressed as a percentage.

$$\frac{\text{Leavers in the period}}{\text{Staff employed}} \times 100$$

This allows the manager to check for trends which might highlight problems. Some turnover cannot be stopped, and it is not desirable to try, for new staff bring new ideas and approaches to a business. Turnover in jobs with no promotion or development prospects should be planned for, and such openings should perhaps be advertised as stepping-stone jobs.

If labour turnover proves a problem, you will need to know if it is your problem alone or a local/typical problem. You can find answers or help with this from such places as: local employers' groups, Department of Employment/Job Centres, Chambers of Commerce, and professional or industry groups.

Labour profiles can be calculated so that seasonal manpower requirements are known and may be plotted as a bar chart. While not a precise tool these charts are also extremely helpful when considering a change in the size of an enterprise or the introduction of a new one. A labour profile can also show casual labour requirements. When compiling these profiles or gang work day charts, a useful source of data on average and premium worker performance is Nix (1987).

Gang-work day charts

Gang-work day charts (Fig. 3.1) aim to ensure that enough labour and machines are available to carry out operations at the correct time and within the time available for the operation. It is seldom necessary to chart the whole year; only peak periods usually need be charted. These charts may be useful when considering the viability of annual hours contracts for farm staff.

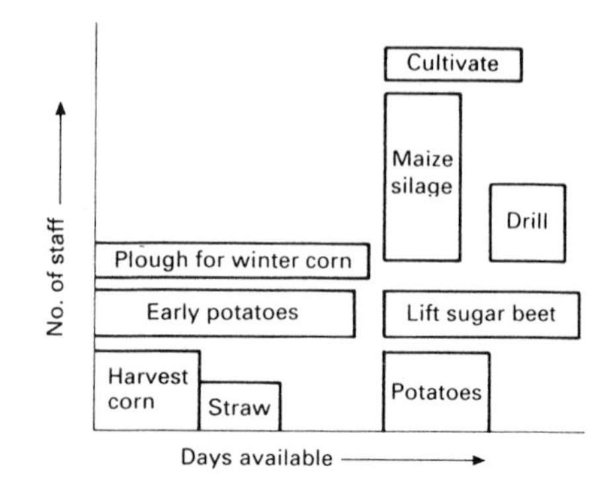

Fig. 3.1 An example gang-work day chart.

Construction of a gang-work day chart

(1) Collect data on:

- cropping and stocking
- regular labour employed
- machinery and implements available
- gang size and machines required for each operation
- rate of work of each operation
- earliest start and latest finish date for each operation
- number of days on which field work is possible in each month

(2) Tabulate the information for each operation, under the headings:

- Area
- Time period
- Working days available
- Work rate and gang size.

Then calculate the gang-work days required for each operation.

(3) Draw the chart.

Operations using large gangs are the least flexible and should be put in first. Initially assume operations can start at the earliest start date. Charting is a matter of trial and error, like a jigsaw. A chart with no spaces has no flexibility for poorer than average seasons or other delays. Casual labour may be written in or 'ballooned' on top.

One element in workforce planning often not recognized by farmers in time is that they and their workforce grow old together. The position then facing the heir or, equally, an appointed manager is that all the staff may have to be replaced fairly quickly with the consequent loss of knowledge and skill.

Recruitment and pay

Recruitment

Employers need to make quick, accurate, relevant and legally defensible decisions about people. Managers must plan and organize to get and keep labour. Re-recruitment most commonly occurs because of poor preparation for, or poor follow-up to, recruitment.

Constant recruitment when the business is not growing can quickly have a negative impact on the credibility and efficiency of the organization. For example:

- Too much concentrating on recruitment can become a vicious circle by lowering the morale of experienced staff who may then leave.
- Output can fall as initial training increases at the expense of output and developmental training.
- Short-term coverage can become the norm, leading to tiredness, increased accidents, errors of skill and judgement, withdrawal of good will and strained working relationships.
- Management credibility falls and with it falls the ability to motivate and control.

We all frequently make initial judgements (first impressions) on the basis of little evidence and at great speed. To be objective in recruitment we must be aware of our bias and suspend judgement until we have sufficient relevant evidence to make a choice.

The employer should concentrate on skills and behaviours because they are observable, describable and measurable. Past behaviour is the best guide to future behaviour that we have.

When measuring the individual against the criteria of the job, attention should also be paid to the future: the skills and behaviours that will aid the development of the business. The relationship of behaviour to the job is illustrated in Fig. 3.2.

Recruitment is a two-way selection process: potential employees too need information and time to make a rational decision.

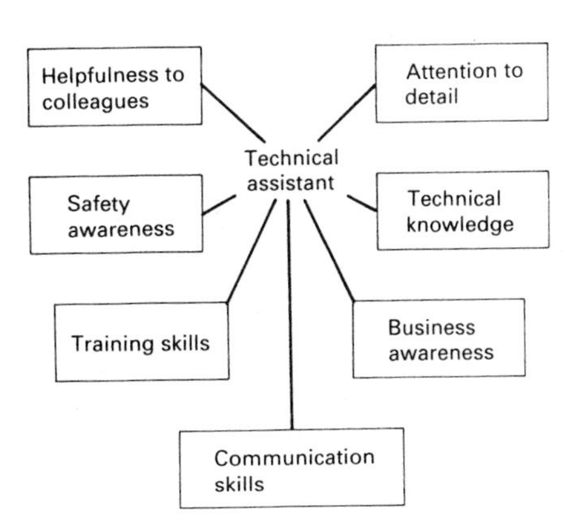

Fig. 3.2 Behavioural aspects of a technical assistant's job.

The process can be viewed as attempting to deter the unsuitable as well as attracting the suitable. The advantage of supplying negative as well as positive information to potential applicants is that selection can be made on the basis of employee/employer choice without any of the *sales pitch* element of recruitment. As an employer, try to build into jobs those elements that will appeal to your target audience as well as serving the needs of the business; clear responsibilities in jobs for recent graduates, for example.

Your own time is a commonly overlooked cost when running a business. If you do most jobs for yourself, you can get very caught up duplicating the work best done by others. Use the skill and experience of Job Centres, newspapers and agencies, much of which is free; e.g. use Job Centre application forms rather than design your own.

Advertising

The following are the advertising options open to you:

- Job Centres (national coverage)
- local papers
- national papers
- specialist journals (group or interest)
- agencies (recruitment and advertising)
- schools and colleges.

Pay and reward systems

These include:

- payment for an employee's time
- payment for an employee's output
- payment for an employee's service or loyalty.

The total reward package on offer, the rules covering eligibility and the period or date of review should be included in both the offer of employment and the written terms and conditions of service.

Time rate

This means payment for time worked, at a monetary rate per hour including provision for extra or differing working hours. Some of the options for the length of the working period are: day, week, month, season, year, salaried positions, fixed-hour contracts, flexi-time.

Production/performance rates

This refers to pay related wholly or in part to the achievement of production, sales, profit or other identified targets. For example:

- *Piece work*: pay is strictly limited to the quantity of output.
- *Commission*: this method of payment aims to give the employee motivation to sell, for each sale directly increases pay.

Payment based on performance or results relies on clear and effective measurement criteria. The manager must ensure work is available in sufficient and measurable quantity to allow workers to earn a reasonable and consistent wage. The rules must be clear from the start and not require quick or frequent adjustment. The system must have sufficient flexibility to deal with the changes in production and sales that may occur.

Most organizations generate a mixture of productive and non-productive work. It can be hard to measure accurately and consistently the contribution to profit or success of all employees.

Service and loyalty rewards

Many firms offer benefits in kind in addition to money to reward or try to encourage loyalty and long service, for example:

- subsidized housing
- child care
- food
- uniforms
- loans/mortgages
- non-contributory pension
- paid training expenses
- car
- membership fees.

The manager will need to measure the returns to the business of such rewards. These rewards are not of equal value to all employees or over time. One option may be to offer staff a choice from a range of benefits.

Establishing rates of pay and pay scales involves more than just establishing the going rate for a job: issues of power, personality, tradition and organizational culture are also involved. Pay levels are a way of establishing or reinforcing a hierarchy within an organization. As well as rewarding skill and experience, pay levels can signify levels of power and authority.

The opinion as to how fair a pay system is will depend upon the position of the individual within the system. For example, staff of differing ages offer different things to an organization but they may not fully understand the contribution of the other.

The effort–reward bargain

The effort–reward bargain is a personal perception of the exchange of effort, knowledge and time for the benefits of monetary and non-monetary reward. This perception will change over time and in comparison with the rewards of others.

Teams

Although many business goals are set and achieved through teams, managers must ensure that rewards to the individual have some relevance to individual performance.

Administration

Administration of pay and remuneration must be workable. Simplicity of measurement has meant that hourly pay is the most popular basis for reward. With this method, the control and measurement of output volume and quality will be a reflection of motivation and discipline. Before introducing a more complex form of payment, you must be sure that the new system's potential benefits outweigh the potential cost.

Interviewing and discrimination

Positive indicators to look for on application forms/CVs include:

- similar industry
- similar institution
- similar job
- stable employment history
- previous promotions
- positive leisure time
- appropriate educational background
- compatible wage/salary
- currently employed
- neat application.

The telephone screen

When screening a candidate by telephone:

- describe job requirements
- solicit candidate reaction
- check if experience is comparable
- check wage/salary requirements
- describe the selection process
- listen carefully.

The selection interview

The selection interview is a conversation with a purpose, to gain evidence of each applicant's suitability for the job, and to enable objective assessment to be made against the person specification.

Good interviews:

- Allow the participants to get acquainted.
- Help the organization make the best decision by collecting information in order to predict how successfully the individual would perform in the job.
- Help the applicant make the right decision by providing full details of the job and organization.
- Allow candidates to feel they have been given a fair hearing.

Preparing for the interview

Stress

A selection interview is not a natural situation and can put a lot of strain on the applicant; a potentially first-rate employee could be so nervous that you never discover how good they are. A good rule is that you should aim to stretch the candidate, but not to stress them.

Time

Give yourself enough time and tell interviewees how long you plan to keep them before they come.

Interruptions

Try to minimize them; they destroy concentration and do not allow for relaxation of interviewer or candidate.

The environment
If there is more than one applicant, make arrangements for their arrival and provide a place to wait, toilets and refreshments.

Documentation
You will need:

- person specification
- job description
- application form or CV
- interview plan/prepared questions/notes
- payment if travel expenses are to be paid
- notes from any references taken up.

Interview method

The acronym 'WASP' (Welcome, Acquire, Supply, Plan and part) is a useful reminder on interview procedure.

Welcome

- Establish rapport by adopting a friendly approach and showing interest in the candidate.
- Greet the applicant, giving your name and position.
- Explain the purposes of the interview.
- Give an outline of the interview's structure.
- Tell the candidate you will be taking notes.

Acquire

- Establish the applicant's experience, knowledge, skills and attitude.
- Ask questions using 'funnel technique', i.e. use
  - opening questions
  - probing questions
  - summarizing questions.
- Use a separate funnel of questions to deal with each new topic. (See 'Questions' below.)

Supply
Tell applicant about:

- the job, using the job description.
- the conditions of employment.
- the business/firm.
- the company benefits.
- career prospects,
- the down side of the job as well as the highlights.

Plan and part

- Tell the applicant when a decision will be made.
- Tell them how they will be informed.
- Pay travelling expenses.
- Thank applicant for attending interview.
- Pass to second interview/tour of the firm/arrange medical/testing.

Even if an applicant seems unsuitable, don't cut the interview abruptly short but consider all applicants fully and fairly for the job.

Even if you are impressed with an applicant, do not say anything that could be construed as a job offer. You may see a better applicant later, or a colleague may pick up an adverse point that you have missed. If the applicant thinks they have been offered the job and subsequently find this is not the case, they may have grounds for appeal to an Industrial Tribunal.

Questions

Avoid long or multiple questions; encourage the applicant to speak by keeping your questions short and simple. You should always be encouraging in your manner, give the applicant's answers your full attention and make use of verbal and non-verbal prompts. Try not to involve your own opinions on specific subjects mentioned.

A behavioural approach to interviewing (Figs 3.3 and 3.4) is based on gaining evidence of past performance behaviour as an indicator of future job performance regardless of background. Evidence of team-working behaviours and skills could be assessed by investigating any of the areas shown in Fig. 3.3.

General opening questions

- What opportunities have you had ...?
- Give me a recent example ...
- Describe a recent activity ...
- How did you go about ...?
- Tell me about the last occasion ...
- When have you been involved ...?
- What experience have you had ...?
- Describe a situation in which you ...?
- To what extent ...?

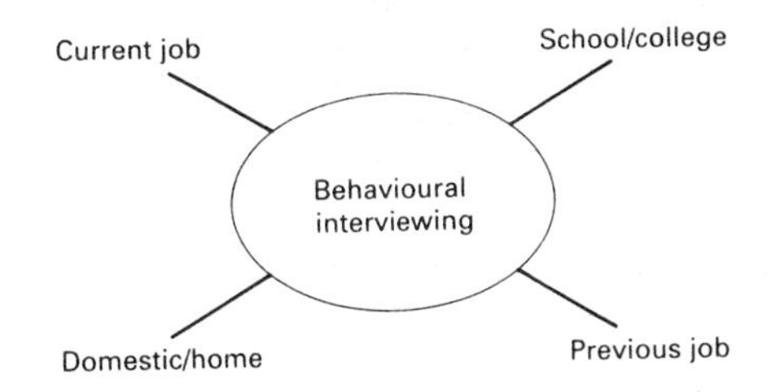

Fig. 3.3 The areas that behavioural questions can probe.

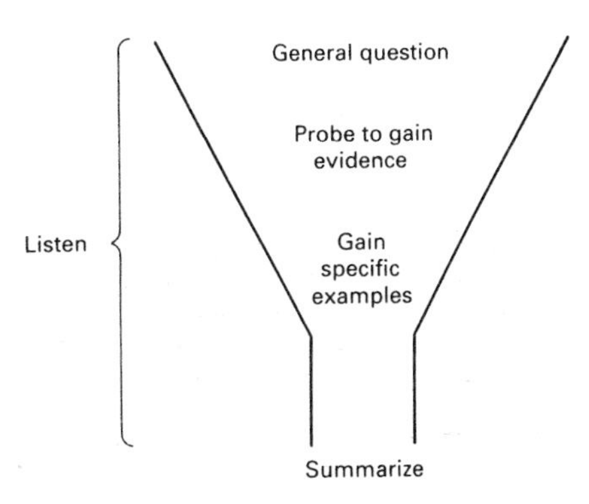

Fig. 3.4 Behavioural questions and the funnel.

Follow-up probes

- Describe specifically how . . .
- What exactly did you do . . .?
- Tell me more specifically . . .
- How did you go about . . .?
- What was your role . . .?
- What were your reactions . . .?
- Walk me through the process step by step . . .?
- How many times . . .?
- Describe exactly what you did in detail . . .
- How were you personally involved . . .?
- Would you choose to do it this way again?
- What went wrong?/What went well?
- How did you assess your success?
- How would you approach a similar situation again?

Summarize

- Am I right in saying . . .?
- So you are saying . . .?
- Is it fair to say . . .?
- What you are saying is . . .

Body language
Listen to the candidate and probe for more information where the body language signals do not seem consistent with the spoken message.

This questioning process is equally applicable to other types of interview situation e.g. counselling; disciplinary; grievance; appraisal; termination.

How to improve listening skills:
The acronym 'LISTEN' is a useful reminder about how to listen effectively:

- L*ook interested*: do not ignore candidate, yawn or look out of the window.
- Inquire with questions.
- Stay on target.
- Test your understanding.
- Evaluate the message.
- N*eutralize your feelings* – Do not discriminate or allow personal opinions to influence your decision.

Listening is not talking: resist the temptation to launch into a long discussion about your ideas or the company.

Questions you should not ask during an interview
You should not ask questions concerning:

- age
- race
- marital status (or living with someone)
- dependants
- child-care problems
- housing situation
- willingness to work different hours, unless this is a bona fide job requirement.

Minority/ethnic groups or females should not be asked questions not routinely put to white or male applicants.

Discrimination

In an employment context, discrimination is the application of unsound or irrelevant criteria to the process of selection, access and decision making.

Equal opportunities form part of good management practice. It is a reality that half the workforce is female, that significant numbers of people are disabled and that Britain is a multicultural country. These people can and want to make a full contribution. The number of school leavers is declining; employers who discriminate reduce this pool still further.

Skills and behaviour qualities are not found only in specific geographical areas, or among one sex or specific age or racial group; exclusion on these grounds is therefore not rational or legal.

Everyone experiences discrimination at some time, but most experiences do not last for life; those of race, colour and sex often do. These experiences will affect those discriminated against in their approach to getting a job. Discrimination in employment can persist even when members of these groups reach the interview stage owing to the absence of cultural or social reference points, i.e. the lack of common experience between interviewer and interviewee.

Fair, legal and efficient systems are those which look for the behaviours that will make the individual in your organization successful. This approach will enable you to deal positively with discrimination issues and assist the manager in making sound employment decisions about other unfamiliar groups, such as school leavers, the long-term unemployed or women returners to work.

Keep things simple: the more complex a recruitment system the more chance that it will not be carried out in full or to the same standard every time.

Psychometric tests are an increasingly common form of assessment. These tests are devised by professionals and therefore involve costs. The values of a chosen test must be compatible with those of the organization and its staff. Do not be tempted to use tests designed for other organizations.

Remember that all applicants including their relatives and friends, their contacts, official and commercial along with current employees, form attitudes and expectations based upon the publicity and promises delivered in the recruitment process. Those you deal with will have an opinion as to how they have been treated and so act as either good or bad ambassadors for you and your business.

Internal applicants should receive the same consideration and follow the same procedures as external candidates.

Disabilities

It is a legal requirement that 3% of staff should come from these groups unless an exemption certificate is in force.

Control, guidance and negotiation

Identifiable responsibilities in conjunction with house rules and disciplinary procedures provide a clear framework for the management and guidance of staff behaviour throughout the employment relationship.

House rules

House rules are guidelines, commonly written and displayed, that state the actions or behaviour demanded by the organization in set circumstances. They cover such things as the use of telephones or vehicles for private use, arrangements for visitors, uniform and safety equipment policy, and the right of search.

House rules often signal the organization's attitude towards such things as smoking and the protection of property and stock; drinking, on or prior to duty, especially for those who work with machinery; and security issues.

When small farms expand or develop, they can become vulnerable to dishonesty, from within and externally. *Audits* can help. In an employment context an audit can be seen as the scrutiny or review of records and systems, in the same way that a financial audit checks the books. Periodic audits of all systems and records can reduce the temptation of staff to break the rules through the knowledge that systems and records are checked. It is likely that, as an organization grows, more financial and access controls will be required.

The application of discipline and law; discipline can seem very legalistic, but there is a clear distinction between what constitutes a breach of workplace discipline and what constitutes a breach of law. For example, to discipline a member of staff for removing equipment, the manager does not need to prove a worker intended to steal from the firm but that equipment was taken off site without permission.

One approach is to set out the responsibility you expect each of your staff to take. Some firms will ask staff to sign statements to signify their understanding and agreement. This approach is commonly found within large established firms. These firms are not by nature less trusting, but have implemented these policies in the light of experience. The purpose of a system of workplace discipline is not punishment but behaviour change.

The legalistic way in which many of these situations are described and the emotions that can be generated often make the employee feel that punishment is at the root of the process in the same way that breach of school rules can lead to detention.

Staff are not always willing to see the person disciplining them as someone reasonable, for whom they may have caused problems. Individuals who are the subject of workplace discipline commonly feel threatened and many react accordingly.

Procedures

Establishing a system of discipline allows the manager or employer to introduce order into what can be situations of conflict and high emotion. Procedures help to set the agenda and provide the manager with a tool to:

- calm
- control
- gain consensus
- gain acceptance
- reduce uncertainty.

Steps to ensure that disciplinary procedures are seen as fair include:

- Publicity.
- Detail the procedures in the contract or terms and conditions of employment.
- Agree procedures with workers, their representatives, or organizations.
- Follow the 'best practice' advice of professional and advisory bodies.
- Always follow the procedures.
- Be seen to be consistent and impartial.

Remember that staff not involved in a disciplinary situation will be interested in the handling and outcome of the procedure out of curiosity and self-interest.

Investigation

This must be your first action. You do not have a disciplinary situation on the basis of one person's statement. If you start the disciplinary process by the investigation of incidents or accusations, you are immediately giving those concerned the chance to have their say (and let off steam). By not jumping to conclusions and by taking the time to check the situation, you are also allowing time for calming.

During an investigation the manager will need to:

- Explain the process.
- Talk to the relevant people in private.
- Try to ignore mere opinion.
- Make notes.
- Be aware of bias.
- Probe for the whole picture.
- *Remember: people do things for a reason.*

If the matter under consideration is serious and/or may take some time to investigate, you may wish to suspend the employee *on full pay*.

Threats are often associated with disciplinary situations. They come in the form of judgemental phrases, made on the spur of the moment. These do not help the orderly conduct of procedure, or the outcome. Do not endorse the threats or judgemental remarks of others, and if possible follow these up and prevent repetition.

Do not commit yourself to anything at the outset except to investigate as a preliminary to starting the disciplinary process.

Try to ensure that the person being disciplined understands why they are being disciplined; however, you may stress that time will be spent investigating the incident because you are interested in a positive outcome.

Time scales

Disciplinary issues should be resolved quickly. The uncertainty and stress of a long wait are unlikely to contribute positively to the business. The only legitimate forms of delay are:

- *For the manager*:
  - to carry out an investigation

  - to allow cooling off
  - time to consider your actions.
- *For the subject of the disciplinary process*:
  - to prepare evidence
  - to find a supportive colleague willing to attend
  - to consult with unions or other bodies.

Once a decision has been reached, a line should be drawn under events. The only circumstances in which the event can be relevantly revisited are those of repetition and only then in private. As with criminal matters, once a sentence has been served or a fine has been paid, the matter should be seen as closed.

Retrospective action

If a matter deserves investigation and/or implementation of disciplinary procedures, it must be fully dealt with at the time. If a blind eye is turned to a problem, it is not legitimate to use the incident as evidence in a later disciplinary situation. If you ignore something, you can be perceived as condoning it. This will express to staff how seriously the issue is taken; ignored more than once and it becomes custom and practice.

Fundamental steps in handling disciplinary situations

- Investigation
- Reference to personnel/responsible level
- Representation
- Notice of meeting
- Presentation of evidence
- Fair hearing
- Meeting/notes
- Pause for consideration
- Mitigation
- Make decision
- Right of appeal

Steps in a performance warning

- Explain what is being discussed.
- Restate the standard required.
- Outline the shortfall.
- Discuss the target standard.
- Give a time span for improvement.
- Provide assistance.
- Impose warning or offer guidance.

Steps in a conduct warning

- Explain what is being discussed.
- Detail the incident of poor conduct.
- State the standard expected.
- Outline the consequences of repetition.
- Give a time span for improvement.
- Provide assistance.
- Impose warning or offer guidance.

Dismissal

This means the involuntary termination of the employment relationship by the employer. Dismissal can be the culmination of the disciplinary process, or due to *gross misconduct*.

You can start the disciplinary process at any stage if the level of seriousness warrants this, but the common disciplinary actions short of dismissal are:

- an initial verbal warning;
- a follow-up written warning;
- the final written warning;
- dismissal.

Staff need to be aware of the actions that will result in disciplinary action and at what level this will be taken. Failure to make this clear could result in complaints of constructive dismissal.

Industrial tribunals

When the disciplinary system fails, the result may be an industrial tribunal. There the main concerns are to establish:

- any breach of employment law;
- whether the employer has acted in a way that a reasonable employer, of similar size/resources in a similar field would have acted.

Some possible outcomes of tribunals include:

- out-of-tribunal settlements
- compensation
- re-engagement
- reinstatement
- legal precedent.

Ill health

Termination on health grounds can be a delicate issue but failure to address the issue may prove costly, especially for small organizations. The common steps required prior to dismissal are to:

- ask for medical checks;
- seek independent advice;
- keep the employee informed of the likely outcomes.

Retirement

While it is sensible to plan for the changes in the level and type of input to the organization during working life, this is crucial for both employer and employee at retirement. Some options are:

- a stepped decline in effort/responsibility;
- Gradual handover;
- Consultancy.

The grievance interview

It is important to deal with grievances promptly to prevent them from escalating. The objectives are:

- to allow the employee to air the problem;
- to discover the causes of dissatisfaction/worry;
- to remove the problem/improve the situation.

When conducting the interview:

- Let employees state the case in their own way.
- Distinguish the facts and the feelings.
- Listen, don't argue.
- Get the employees to explore their own motives.
- Ask the employees what they think the solution is.
- Don't commit yourself to anything at this stage.
- Fix a date to come back to them.

Follow up

Investigate the facts, obtain information and communicate by the agreed date. Check that the situation has improved.

Negotiation and bargaining

Industrial relations issues are closely tied up with power, control, conflict and personal interest. What helps to cloud industrial relations issues further is their linkage to politics, ideology and propaganda. You as a manager must stick closely to the issues for negotiation. Any help you can give your employees and their representative groups to do the same will be rewarded.

Once one party deviates from the substantive issues, both sides can quickly take refuge behind the posturing that has characterized many famous disputes. Once this posturing takes place it allows those who have no part in or knowledge of the issues to stick their oar in, looking for proof that the 'Other Side' is abusing its position or being irresponsible: the very stuff of gossip and newspaper circulation. Once you have lost sight of the substantive issues it is very hard to re-focus the participants!

Types of negotiation and bargaining

Distributive bargaining
This is a fixed sum game which can be fine for simple issues. There are winners and losers, and the situation can be characterized by suspicion and defensiveness.

Integrative bargaining
This suits common problems that require a mutual solution. It is a positive sum game: the outcome can be more for all, because cooperation can increase the size of the cake. It also takes longer and requires greater honesty.

Procedure

Negotiation is not a one-off incident; there will always be some history to consider. Trust levels grow slowly and are best built up in low stress situations.

Opening moves
These are commonly a statement of your position and a rejection of the other side's position. This is one extreme of your bargaining position, the other being the resistance point beyond which you cannot go. For successful negotiation there needs to be some overlap in the bargaining range of each side.

Exploratory stage
Here you try to determine where your opponent stands. This process may consist of questions, threats and abuse. Frustration and time pressures are brought into play. Your aim is to move your opponent's resistance point.

Consolidation
Here a change of concentration occurs, moving from divergent issues to convergent issues. Contentious issues are avoided and conflict minimized.

Decision making
This starts when a compromise seems possible. Only then are concessions made to move from discussion to agreement. The goal changes from an ideal settlement to an acceptable one. It is sensible to help the other side accept the decision.

Negotiation has similarities to a game: it has rules, conventions, form and accepted moves, and there is usually an audience who are passionate about the result. Offers once made are normally not withdrawn and moves are normally towards the other side's position.

High initial demands may help to:

- widen the bargaining range in your favour;
- change the other side's expectation of what is reasonable;
- lead your opponent to raise their estimate of your resistance point;
- give you scope to make big concessions.

However, you may lose credibility if your opponents refuse to negotiate or the final settlement is small compared with the demand.

If you play to win in the short run, you may lose in the long run.

Training and development

Spending time and money on the training and development of staff is easy. Managing the investment to get value for money is not. Before resources are allocated to the development of people you need to know the level from which you are starting. This entails carrying out what is sometimes called a skills audit.

Skills audit

When looking for and measuring skill, the starting point should be goals and objectives to which the skills will be applied. In other words, what are your staff trying to achieve?

The next step is breaking down the defined goals and activities of the organization into measurable terms for each member of the team, including standards and objectives.

Once you have defined what the staff are trying to do

you can start to study the skills they employ to do this. A skills audit could also look for unused or under-used skills among the workforce – foreign language skills, for example.

Questions for a skills audit

- Are there defined standards of performance?
- Are the standards set at a correct level?
- Are these standards being met?
- Are our competitors achieving the same results by more simple/cheaper methods?
- Are our competitors achieving better results by the same/different means?

The skills audit will produce a comparison of the skills applied and the outcomes achieved. However, a shortfall in a standard of performance is not automatically a training need.

Training

A training need is an identifiable skill or body of knowledge, the learning or enhancement of which would generate positive, measurable benefits to the individual and/or the organization.

Expenditure to eliminate the training need should be actioned only when it is achievable and relevant and when the need for the training is clear to the trainer and trainee, for example to improve standards; to increase performance; to generate higher profits; in preparation for growth; to facilitate change; or to motivate.

A clear objective for any expenditure should have a quantifiable measure of success, i.e. a standard. Training to a standard helps you to:

- achieve continuity;
- let staff know what is expected;
- allow trainers to plan and be effective;
- set a measurable level of satisfactory performance;
- let the customer, both internal and external, know what to expect
- measure the standard so that you can then know when to stop training;
- focus competitive activity vs the standard (not staff vs staff).

Standards of performance

Ideally these should be written. If this is done, it is a good idea to make them available to the people who need them – the staff as well as the manager.

Setting standards and the recording of training

For many organizations the standards of performance training programmes and training records are all separate documents and systems. This need not be the case. The standards for the job can be both your training programme and record. A more detailed hard copy of standards may help trainers, but for the day-to-day monitoring and improvement of standards these are not required.

By establishing such systems you are providing the climate in which those who work for you can express their individual motivation to do a good job and improve their own and colleagues' skill. The manager will be facilitating by providing the tools for the job.

Sources of information on training needs

These include:

- job analysis
- appraisal
- workplace assessment
- changes in the nature of a job, equipment or markets
- general communication.

As well as providing guidance on training needs the Agricultural Training Board (ATB) through its network of trainers can provide training and development tailored to the specific needs of the individual farm enterprise.

Workplace assessment

A true measure of job performance can often only be gauged on the job. The person best suited to assess job performance is someone who:

- knows what the standards are;
- knows what the obstacles to achievement are;
- knows, or can see, ways of improvement.

Workplace assessment is undertaken to:

- give feedback on job performance;
- help measure and build consistency;
- assess the impact of pressure and environment on standard delivery;
- identify true long-term strengths and weaknesses of individuals and their environment.

An environment of trust and a collaborative approach to organizational development is required for best results. If standards of performance are outlined from the start, the regular assessment of the performance of starters will form a natural part of the training and induction process. After establishing such an environment, workplace assessment can focus on the introduction or change of standards and the upkeep of existing standards.

Workplace assessment is one way in which a manager can show they are committed to getting it right. However, this approach will only gain acceptance if shortfalls in standards that are the responsibility of the firm or management are pursued with the same vigour as shortfalls by staff, for example:

- lack of equipment
- lack of maintenance
- lack of staff
- poor planning
- poor communication from above.

For workplace assessment to work best, it must be seen as a collaborative process. Remember you are not there to conduct tests, but to facilitate staff achievement. During workplace assessment new or amended duties can be agreed, progress can be monitored and praise can be given. Problems or barriers to progress can be approached jointly.

Training should be carried out prior to the need for the skill. The work method can then be assessed in action to determine the need for additional training, further clarification of the standard, or amendment of the standard. This is a cyclical process.

The trainer also benefits from training because it:

- improves your handling of and confidence with people;
- makes the job easier;
- supports your staff while not doing their job for them;
- develops trust and respect;
- is part of managing instead of doing.

When offering training or development, be systematic; give the trainee a structure to their learning, such as the Kolb model in Fig. 3.5.

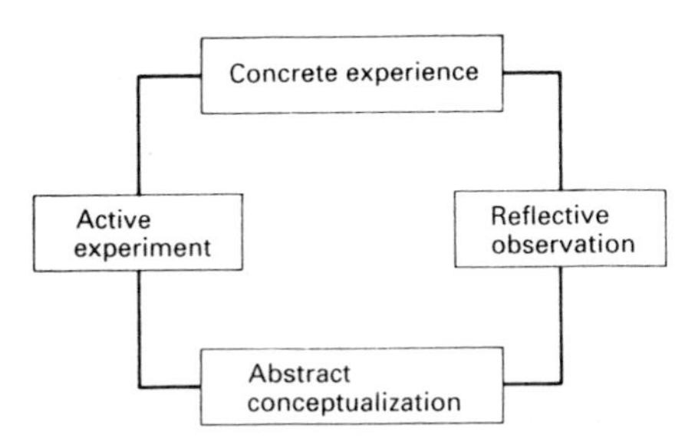

Fig. 3.5 The Kolb model of learning and development.

There are limits to learning and the acquisition of skill, imposed by:

- the situation and personal make-up of the person to be trained (learning style);
- the training techniques and the environment they are to be practised in;
- current skill limitations;
- intelligence limitations;
- lack of confidence;
- fear of the unknown;
- personal inhibitions;
- lack of practice;
- attitudes that are brought to the training;
- the effect of peer pressure or ridicule;
- the dynamics of the training group;
- the fact that people only learn what they want to learn;
- loss of concentration (after 20 min).

Aids to learning

There are a number of aids to learning including:

- Visual aids
- Mnemonics
- Use of more than one sense
- Build from known to unknown
- Show relevance to work situation (usefulness)
- Provide constructive feedback
- Use clear objectives
- Build on success

As well as technical competence a trainer needs to develop a working relationship with those to be trained, and to establish a climate in which learning can take place.

Learning is the gradual extension of knowledge, skill and attitudes.

The qualities that constitute a good trainer are those contained in the acronym 'TRAINER':

Tactfulness
Respectfulness
Accuracy
Imaginativeness
Neatness and tidyness
Enthusiasm
Reliability

The benefits of increased individual performance feed into company and personal career progression. These effects can be cumulative; development can progress faster if the individual is familiar with the methods and objectives of the organization.

Investing in people can be thought of as ensuring that the organization is covered by a maintenance and development contract for the human resource. Sir John Harvey Jones attacked British companies for investing more in plant and machinery than in their staff. Speaking at an 'Investors in People' conference, Harvey-Jones (1993) told business leaders to invest in their staff or risk failure in competitive markets.

Induction

Induction is a programme to bring about the smooth and effective introduction of people into an established working environment. Break the programme down according to what the employee needs to know by when. Use a variety of methods to allow a new employee to take on board potentially vast amounts of new information.

Special consideration will need to be given to different groups, such as *school leavers, the long-term unemployed, graduates, and women returning to work after a career break.*

Maintaining and increasing performance

Maintaining and guiding performance over time

Appraisal

An appraisal is a meeting between an individual and his/her immediate manager to discuss the job. The purpose is to maintain, guide and develop job performance. Best practice for appraisal suggests that past performance should not be looked at in a judgemental way, but reviewed to aid the process of work planning and goal setting for the period ahead.

More sophisticated systems make provision for the appraisee to carry out some self-assessment and provide comment and guidance to those who manage them (self and upward appraisal).

Appraisal benefits the manager by:

- allowing time to plan the work load and its allocation with his/her team for the period ahead;

- giving the manager feedback about his/her own performance;
- helping foster a joint approach to work;
- providing a recorded plan.

Appraisal benefits staff because:

- career planning and development can be discussed;
- feedback about performance can be gained;
- future work and priorities can be clarified;
- it enables the recording of a plan.

Job chats are an increasingly common alternative or supplement to the appraisal. They are shorter than the appraisal and less formal, but still provide a forum for the discussion of issues arising from the job, past, present and future.

The major threats to positive appraisal are:

- lack of trust;
- lack of understanding/knowledge of the job;
- lack of time;
- too close a link to pay: when appraisal is too closely linked to pay, individuals will concentrate on the assessment of past performance, negating the appraisal as a forward-looking collaborative exercise.

Activity plans

Typically activity plans are written plans for activities covering a fixed period, and include details of success criteria, support required and an interim review date. The plan should be reviewed at the conclusion of the period and a new one agreed. Plans can cover months or weeks and cover normal areas of work and development or specific projects/objectives.

Management by objective

This management tool aims to break down the organization's goals into subgoals and, in turn, individual goals. The starting point and focus is the goal and not the method. This opens up the possibility of more individuality of method, so long as the goals are common/compatible.

Feedback

Feedback is a powerful way of learning about ourselves and the impact we have on others. To make this process constructive requires skill and practice on the part of the giver and recipient of feedback.

The recipient needs something to build upon, positive or negative, and this information must be both understandable and usable.

We often fail to give positive feedback because we forget, we are embarrassed or we are worried that someone may become too conceited.

Negative feedback is rare because we fear the recipient may get upset and we are not ready or willing to deal with this. We may be worried that it will affect the relationship we have with that person in a lasting way, that they will not really understand what we are saying, or they will distort what we say. Finally we may say nothing because we believe saying something will have no effect.

Although not always a good idea, giving feedback does provide an opportunity for change.

To give effective feedback you must:

- be clear about what you want to say in advance;
- start with the positive;
- be specific;
- select priority areas;
- focus on the behaviour rather than the person;
- refer to behaviour which can be changed;
- offer alternatives;
- be descriptive rather than evaluative;
- own the feedback ('This is what I think');
- leave the recipient with a choice;
- think what it says about you;
- give the feedback as soon as you can after the event.

When receiving feedback:

- listen to the feedback rather than immediately rejecting or arguing with it;
- be clear about what is being said;
- check it out with others rather than relying on only one source;
- ask for the feedback you want but don't get;
- decide what you will do as a result of the feedback.

Coaching

Coaching is a method of developing and or correcting skills, attitudes, behaviours and techniques. With this approach the trainer/boss/mentor must have a basic agenda/outline of the subjects to be covered by the trainee.

The point is that situations that occur in the normal working day can be used as triggers for training to be carried out. One way to think of coaching is the use of mini case studies. In this way principles and applications can be looked at together and in a context that the trainee is more likely to understand.

It is important that the trainee understands the technique of coaching and is aware of when it is being practised, especially early in the coaching relationship so as to take advantage of the learning situations and not feel patronized.

Mentoring

Mentoring can work very much in the same way as coaching, but with mentoring the initiator of training or the coverage of issues is more likely to be the trainee. For best results the choice of mentor must at least partially lie with the mentee.

The mentor needs to be able to put him-/herself in the place of the trainee. The biggest expert is not always the best trainer. The mentor must have an understanding of mentee's personal goals and the need to make developmental activity personally relevant.

When looking at a scheme of mentoring you must decide what the basis of the relationship will be. Below are some popular models.

Description	*Application*
Patron – Protegé	Commercial
Expert – Trainee	Training
Role model – War	Compacts
Confidant – Colleague	Partnership
Educator – Disciple	Developmental

Poor standards

Poor standards or the failure to fulfil potential may be the result of:

- poor initial training
- forgetfulness
- cutting corners
- faulty/inadequate equipment
- lack of interest
- personal problems
- personality clashes.

The acronym 'OSCAR' provides an *aide mémoire* to the steps to corrective action:
Observe
Step in
Calm use of interpersonal skills
Ask questions
Repeat training

Incentives

Issues that the manager must bear in mind when considering the use of incentives include the following.

- The need to identify the aims of these incentives.
- Why are incentives needed over and above existing payment/remuneration policy?
- What happens when the scheme ends?
- What happens if the incentive does not work? There are numerous circumstances that may cause incentive schemes to fail, many of them outside an organization's control.

Incentives are introduced through:

- planning;
- consultation;
- explaining to staff why an incentive is needed and how the organization and the individual are going to benefit;
- trial;
- time to test the theory and your assumptions;
- review.

The goals of an incentive need to be precise. Having vague goals, such as to increase production or productivity, is the first step to failure. You need to know:

- how much
- to what quality
- for how long
- for what reward.

Fairness
An incentive scheme is a deal between you and your staff. The required effort must deliver the promised benefits, for both individuals and organization. It is rarely possible to apply the same scheme accurately to staff on different jobs. This is where life gets complicated. There are two main choices:

- development of additional criteria to allow more groups of staff to participate, while maintaining a benefit for the organization;
- exclusion from a scheme.

Teams
Incentives work best, administratively and practically, when applied to the individual because the motivation of a team is not the same as individual motivation. The evolution of team motivation is subject to group dynamics. The existing working relationships will need to be considered when designing incentives.

Quantity
Incentives which concentrate on increasing output (quantity) require safeguards for quality. These can be contentious issues, especially if quality problems result in scheme changes which lower income.

The desire to be flexible and to take time out for training and peripheral work can be reduced by incentives. What gets measured and paid for gets done. Short-term income generation can become the opportunity for the worker; while longer term concerns become problems for the manager. All rewards should be received as close to the effort that earned them as possible.

The employment environment

Data Protection Act

The Data Protection Act deals with automatically processed data relating to individuals and the provision of services in respect of such information (Crushway, 1992). The provisions of the act are administered by the Data Protection Registrar and Tribunal. Guidance on the provisions, exemptions and applicability of the Data Protection Act can be obtained from The Office of the Data Protection Registrar, Springfield House, Water Lane, Wilmslow, Cheshire SK8 5AX. This should be the first step before setting up a system.

Access to non-computerized employment and pay records is subject to agreement between employer and employee. However, with computerized records the individual has legal rights regarding the handling of information. To meet the terms of the act an employer must:

- register the data base/records;
- restrict access to authorized users;
- allow staff access to their records;
- change records when they are wrong;
- delete records no longer required;
- collect and use the data only for the purpose for which it was set up/registered.

Computerized records must be as secure and accurate as manual systems. When designing any pay or records system, the demands for information by government agencies, auditors and accountants should be considered.

Fire regulations

Fire regulations require:

- an established means of raising the alarm;

- an evacuation procedure;
- a safe and protected means of escape;
- an assembly area;
- records of training, equipment maintenance, fires and false alarms.

First aid

Legal and effective coverage requires more than the correct ratio of first aiders to staff, supported by the correct level and maintenance of equipment. Employers need to consider the geographical distribution of staff and to ensure that a first aider is available throughout the working day and week.

Health and safety at work

Health and safety are a vital area of farm staff management and should be a central consideration of job design, work planning, recruitment, training and the development of your staff. This complex area and its legal implications and effects upon efficiency are covered elsewhere in this book.

References and further reading

ACAS (1987) *Discipline At Work*. ACAS, London.
ACAS (1988) *Employing People*. ACAS, London.
ACAS (1990) *Employment Handbook*. ACAS, London.
Bartholomew, D.J. (Ed.) (1976) *Manpower Planning*. Penguin, Harmondsworth.
Courtis, J. (1988) *Interviews, Skills and Strategy*. IPM, London.
Crushway, B. (Ed.) (1992) *Essential Facts, Employment*. Gee Professional Publishing, London.
Farm Management Journal.
Fowler, A. (1990) *Negotiation Skills and Strategies*. IPM, London.
Handy, C. (1991) *Age of Unreason*. Arrow, London.
Harvey-Jones, J. (1993) *Personnel Today*, 9 Feb.
Honey, P. (1988) *Improve Your People Skills*. IPM, London.
HSE (1992a) *COSHH in Agriculture*. HMSO, London.
HSE (1992b) *Management of Health and Safety at Work (Regulations 1992): Approved Code of Practice*. HMSO, London.
Institute of Personnel Management (I.P.M.), Codes of Practice.
Keenoy, T. (1985) *An Invitation to Industrial Relations*, Blackwell, Oxford.
MAFF (1993) *Agriculture in the UK 1992*. HMSO, London.
Nix, J. (1987) *Farm Management Pocket Book*, 18th ed. Wye College – London University, Ashford.
Pearn, M. & Kandola, R. (1988) *Job Analysis (A Practical Guide for Managers)*. IPM, London.
Plumbly, P. (1985) *Recruitment and Selection*. IPM, London.
Pugh, D.S., Hickson, D.J. & Hinnings, C.R. (1989) *Writers on Organizations* 4th edn. Penguin, Harmondsworth.
Rae, L. (1983) *The Skills of Training*. Gower, Aldershot.
Robertson, I.T. & Smith, M. (1985) *Motivation and Job Design*. IPM, London.
Stewart, V. & Stewart, A. (1977) *Practical Performance Appraisal*. Gower, Aldershot.
Vroom, V.H. & Deci, E.L. (1992) *Management and Motivation*. Penguin, Harmondsworth.
Wood, S. (Ed.) (1988) *Continuous Development*. IPM, London.

4

Agricultural Law

T.J.F. Felton

Introduction

This chapter provides an outline of the law as it relates to the ownership, possession and use of agricultural land. The role of agriculture and agriculturalists in relation to methods, quality of production and ownership rights over land are matters of genuine public concern. Issues such as conservation, pollution of land, air, water, and access to the countryside remain high on the public agenda. The result has been the confirmation of agriculture's role in the area of environmental law. Europe in particular is a rich source of new law as it attempts both to protect the environment and restructure the CAP, with its knock-on effects as regards the ownership and transfer of quota.

It may be argued that in the past farmers have been slow to react to the changing perception of their industry and have paid the price with new laws replacing self-regulation; notable examples being the cases of straw and stubble burning and the ploughing of footpaths now controlled by the Environmental Protection Act 1990 and the Rights of Way Act 1990 respectively.

It is an imperative of modern farming that those involved in the industry take a positive attitude towards the legal framework that increasingly surrounds their everyday working lives so that they may operate most effectively within it. This chapter seeks to provide a starting point to the creation of such an attitude. It does not give, nor does it seek to provide, a comprehensive guide to all the rules and regulations that apply to the industry.

An understanding of the basic principles of the law as it affects the agricultural industry should be used in the following ways:

- to plan and manage effectively within the law;
- to understand how the law is developing, to monitor change and to react appropriately;
- to voice an opinion through the industry's pressure groups, for example the NFU, the CLA and the Rural, Agricultural and Allied Workers National Trade Group of the TGWU;
- to be alert to public attitudes and aspirations;
- to seek professional advice at the most appropriate and cost-effective time.

Advice should be sought to prevent problems arising, not as a damage limitation exercise. In selecting a solicitor to provide advice, care should be taken to make certain that your choice of individual or practice has the skills required to advise on rural business matters. The role of the solicitor may be explained as the general practitioner of the legal profession, but through them it is possible to obtain the services and advice of specialists in advocacy, drafting and specific areas of law. Members of this branch of the profession are known as barristers. In company with sound legal advice, farmers should also secure the services of competent insurance brokers to obtain cover against the risks involved.

The English legal system

Any greater understanding of the substantive law which controls an industry must commence with a review of the system that creates and administers that law. English law shares with our agriculture a long history. Many of its features retain traces of medieval origin, but in recent times large accretions and alterations have shown it to be as lively and innovative as modern farming. Our law is very different in substance and procedure from continental European practice – although this is not to say that ideas of justice differ – and it should be noted that in the ruling of the Court of Justice of the European Union there now exists a unifying factor of particular importance in relation to agricultural law. Agricultural law is not a separate part of English law; the same general principles apply in agricultural as in other areas.

Law consists of those rules of conduct that the courts will enforce. In this connection we include in the term 'court' all those bodies recognized by the judges as having an obligation to act judicially. So in addition to the civil and criminal courts we recognize various tribunals and other bodies set up by Act of Parliament (see below).

Civil courts decide disputes between fellow citizens where one, the plaintiff, alleges that he has suffered injury or loss by the unlawful act or omission of another, the defendant. If the defendant is adjudged legally responsible for the injury, he will be required to compensate the plaintiff, normally by a money payment called damages. When a state agency causes such an injury, an action brought against it will also be a matter of civil law.

The criminal courts decide cases where the state is involved as prosecutor against a citizen accused of committing a criminal offence. Accused persons, if found guilty, are punished by fine or imprisonment or both.

The two sets of courts are kept separate; Magistrates' Courts and the Crown Court hold criminal trials while County Courts and the High Court hear ordinary civil cases. Certain matters are, however, reserved for specialized tribunals of which the Agricultural Land Tribunal, the Lands Tribunal and Industrial Tribunals are particularly relevant to the agricultural industry. At the top of the hierarchy of courts the Court of Appeal and the House of Lords, sitting as a court, provide an appeal structure with the possibility of reference to the European Court on questions of European Union law.

Most hearings in court are concerned with matters of fact: in criminal trials the prosecution will have to prove 'beyond reasonable doubt' that the defendant has committed the crime of which he is accused; in civil trials the court will decide on the balance of probabilities, whether the events concerned took place as alleged by the plaintiff or not. But in some cases the decision depends on a question of law – for example the Dairy Produce Quotas Regulations 1984 refer to 'land used for milk production'. In *Puncknowle Farms Ltd* v *Kane* (1985)3 All ER 790, the High Court had to decide whether this phrase included only the farm area used for current milk production or whether it also included land used for dry cows and for young heifers, i.e. land used to support the dairy herd as a whole. The court decided that, in the context of apportionment of quota on partial transfer of a holding, the phrase included land used for the support of the dairy herd as a whole, and thus included the land used by followers, as well as by cows in-milk.

It is to be hoped that when an agriculturalist wishes to know the law on a certain topic, a source will already exist in clear terms and be readily available without the need for litigation.

There are two principal sources of law, decided cases and statute. Case law has been built up into a system termed the common law because, when a superior court makes a decision turning on a point of law, that decision becomes a precedent binding on judges dealing with subsequent cases involving the principle. Statute consists of Acts of Parliament and Statutory Instruments, i.e. orders and regulations made under the authority of Acts of Parliament by duly authorized Ministers. EC Treaty provisions are incorporated into English law by virtue of the statutory force to Regulations passed by the Council of the European Union. European Union Directives normally require legislative action by the UK Parliament before implementation but they are capable in certain circumstances of having direct effect. Decisions of the European Court also create precedents which UK courts must follow. For example, in *Von Menges (Klaus)* v *Land Nordrhein–Westfalen* No. 109/84 1986 CMLR 309, the plaintiff had kept a herd of dairy cows until 1980 when he obtained a premium for going out of milk production. He undertook, as required by Regulation 1078/77 EEC, not to market milk or milk products for five years. In 1981 he let his farm to another farmer who planned to use it for milk-sheep. He asked in the German courts for a declaration that the marketing of sheep's milk would not be contrary to the undertaking given. The local administrative court referred to the European Court under Art, 177 EEC, the question of the meaning of 'milk and milk production' in the Regulation. The court held that the words concerned ewe's milk and ewe's milk products as well as cow's milk and cow's milk products, since otherwise the premium payments would only encourage the replacement of dairy cows by milk-sheep leading to new surpluses. This decision is binding upon courts throughout the European Union which of course includes UK courts, who will in future follow this interpretation of 'milk and milk products' in this context.

A statutory provision overrules any common law precedents that directly conflict with it. Nowadays the definition of criminal offences and the powers of courts to penalize offenders depends almost exclusively upon statute, and a vast range of legislation covers fiscal and commercial matters together with the social activities of government, e.g. housing, employment, public health, and social security. Even in the field of property law, contract and tort (see below), formerly the domain of common law, parliament has codified or amended some of the rules developed in the courts.

Thus, to find the law relevant to a particular topic it is necessary to know if statutory rules apply. For instance, in respect of security of full-time farm workers occupying service cottages, the position is governed by one of two statutory codes: the Rent (Agriculture) Act 1976 for occupancies entered into prior to 15 January 1989, and the Housing Act Part 1 Chapter 3, which relates to occupancies that commenced on or after that date. Questions concerning the rights of the farmer and worker when employment comes to an end can only be resolved by reference to these acts.

When the topic is one that is covered by common law, the decisions of relevant cases can be found in the Law Reports, where decisions of significance are recorded. In practice, a legal practitioner will depend upon books of reference and texts dealing with specific areas of the law to enable him or her to find the statutes and cases that are relevant. The agriculturalist must know about changes in the law relating to his or her business. Details of many new statutory rules are publicized by government agencies, e.g. on employment law by the Department of Employment and on safety regulations by the Health and Safety Commission. But to keep up to date a farmer should read the professional journals.

Criminal offenders are liable to punishment; those committing civil wrongs are liable to pay damages or to have their activities stopped by an order termed an injunction made by a court. These are the means by which the law is enforced. By the standards of other legal systems enforcement of civil judgements in England is reasonably effective. Debtors may have their property sold up; bankrupt persons are subject to severe business disabilities. Flouters of injunctions may be imprisoned for contempt. Although legal procedures can be protracted, and nothing is to be gained by suing an impecunious defendant, the farmer should be prepared, when it is business-like to do so, to assert legal rights just as much as he or she should be careful to fulfil legal duties.

Legal aspects of the ownership, possession and occupation of agricultural land

Most agricultural enterprises are run by owner-occupiers or tenants of agricultural holdings. In many cases these will be single individuals or partnerships, and the persons involved will be legally responsible for any obligations that arise. Where the enterprise is run by a limited company, the company is recognized as a legal person with rights and duties separate from those of its members. The directors of an enterprise organized in this way must, however, recognize their obligations under the Companies Acts which contain provisions designed to prevent them from using the advantages of corporate personality to defraud creditors, members and employees of the company.

English law recognizes only two ways by which land may be held: freehold and leasehold. If it is desired to tie up freehold land in the ownership of succeeding generations within a family, this can only be done by the creation of a trust set up in accordance with certain rules – known as Equity – developed in the Court of Chancery, now a part of the High Court. This definitely requires professional expertise.

A freeholder has the largest possible freedom to decide how to use his land that the law recognizes. It has never been possible for freeholders to do whatever they pleased to the detriment of neighbours, but in addition to the limits already imposed by the law of nuisance, have been added the constraints of town and country planning and the compulsory acquisition of land. Public and private rights of way may also diminish a freeholder's privacy, and land can be lost by adverse possession. A freeholder's rights may be reduced by restrictive covenants and his obligations increased by a mortgage.

Although ownership of land normally includes rights over what lies in and beneath the soil, mineral rights may be held separately from the freehold, and coal and oil deposits belong to the State. Water may be taken from surface streams by riparian owners for normal agricultural purposes such as watering stock (not for aerial spraying), but it is generally the case that for other purposes it may only be taken under a licence from the National Rivers Authority and the same applies to most water taken from underground sources.

The freedom of landlords and tenants to negotiate the terms of which property is let has been greatly circumscribed by statute – in no area more so than agriculture. As a result there has been a fall-off in let land from some 90% in 1910 to 36% of the total agricultural land available in 1991. In 1991 the government produced a consultation paper entitled 'Agricultural Tenancy Law Proposals For Reform', the objectives of the reform being:

(1) to deregulate and simplify;
(2) to encourage the letting of land;
(3) to provide an enduring framework which can accommodate change.

Having undertaken a full consultation process, the government introduced the Agricultural Tenancies Bill to parliament in November 1994; it is intended that the new legislation should be in place by 1 September 1995. The government expressed themselves as still committed to reform. The proposals are that following legislation all new tenancies created would be 'farm business tenancies'. The terms of such tenancies would be freely negotiable between the landlord and tenant within a much simplified legal framework designed primarily to prevent disputes arising. The proposed legislation would:

(1) define farm business tenancies;
(2) require the service of notices to quit;
(3) prescribe arrangements for compensation;
(4) prescribe fall-back procedures for disputes;
(5) provide fall-back procedures on rent reviews.

All tenancies created under existing legislation would remain intact. As yet no clear indication has been given as to when the proposed legislation will reach the statute book, but as all interested parties appear committed to reform, the changes as outlined above appear inevitable barring a change of government.

An account of the present position is given below, but it should be emphasized that important regulations are made from time to time which adjust the details.

Both the owner–occupier and the tenant farmer have duties as occupiers towards neighbours, visitors and trespassers. These duties are in general the result of principles developed from case law, but they also arise as a result of the Environmental Pollution Act 1990 and from Health and Safety legislation. As mentioned above, recent legislation has continued to impose new constraints on agricultural activity and the use of land in the interests of production limitation, conservation and public access.

Farm tenancies

The law relating to the letting of agricultural property differs from the general law on leaseholds. It is therefore appropriate to consider it in more detail. In the first half of the twentieth century the normal agricultural tenancy developed under the common law was a tenancy from year to year terminable at six months' notice by landlord or tenant. Statutory provisions have now changed this law radically.

Security of tenure

Security for tenant farmers was first introduced as a permanent measure in the Agricultural Holdings Act 1948 by placing restrictions on the landlord's right to give notive to quit to his tenant. This Act together with all others dealing with agricultural tenancies has now been consolidated in the Agricultural Holdings Act 1986. If the landlord serves a 12 months' notice to quit and the tenant does nothing, the notice will be legally effective. However, the tenant may, within one month, serve a counter-notice if he does not wish to go. In this case the landlord's notice will not have effect unless the Agricultural Land Tribunal consents to its operation.

The tenant will be deprived of his security of tenure only on a limited number of grounds, i.e. that the purpose for which the landlord requires possession is:

(1) in the interests of good husbandry;
(2) in the interests of sound estate management;
(3) in the interests of agricultural research or education, or for the provision of smallholdings and allotments; or

(4) that greater hardship will be caused by the withholding than by the granting of consent.

If the Tribunal finds one or more of these grounds proven, it has nevertheless to refuse consent 'if in all the circumstances it appears to them that a fair and reasonable landlord would not insist on possession'.

The tenant's right to serve a counter-notice is excluded by a strictly limited list of cases covering consent by the Tribunal for the reasons given above, failure to remedy a breach of the contract of tenancy, or bankruptcy of the tenant, or bad husbandry, or failure to pay rent due by the tenant, also the intended use of the land for non-agricultural purposes for which planning permission has been granted or is not needed. Under the Act there are special rules regarding notices to remedy breaches of the tenant's obligations to keep fixed equipment including hedges, ditches, roads and ponds in good order, and the tenant may go to arbitration if he disputes the landlord's claim. An arbitrator can be chosen by agreement of the parties or one of them can ask the President of the Royal Institution of Chartered Surveyors to appoint one from an approved panel.

Until the Agriculture (Miscellaneous Provisions) Act 1976 the death of a tenant enabled the landlord to serve an incontestable notice to quit. This was then changed to enable security of tenure to be claimed for up to three generations including the first occupier, i.e. there can be two succession tenancies. In 1984 the law was changed to enable new tenancies to be created without such security, and the rules relating to succession are now included in the Agricultural Holdings Act 1986. This means that security of tenure for tenants' successors now applies in effect only to tenancies that were in existence during the period 1976 to 1984. Under these rules eligible persons can apply to the Tribunal for a tenancy of a holding whose tenant has died. To be eligible a person must be the widow, widower, brother, sister or child, natural or adopted, of the deceased; have derived his or her livelihood from agricultural work on the holding for five years out of the previous seven (up to three years spent at college or university will count) and not be the occupier of another viable commercial unit. An eligible applicant must also be found by the Tribunal to be suitable to take over the holding; suitability is judged by the agricultural experience, age, health and financial standing of the applicant.

If a landlord serves a notice to quit within three months of the death of a tenant and if no application is made by a suitable successor, this will end the tenancy. If an application is made, the landlord can dispute the application on grounds of unsuitability, ineligibility of the applicant, or for reasons of good estate management or hardship, but the 'fair and reasonable landlord test' still applies. Since 1984 succession is also possible subject to the above conditions on the retirement of a tenant aged 65 or over, either by agreement between landlord and tenant or on the tenant's application to the Agricultural Land Tribunal.

The law on security of tenure and succession is complicated, and there have been many rulings given by the courts on matters of detail. Moroever the issuing of the relevant notices must be made according to strict time limits and in prescribed form. Both landlord and tenant should insist upon skilled advice from qualified lawyers or surveyors.

Obligations of landlord and tenant

The landlord and tenant of an agricultural holding will be bound by the terms of their agreement, but there has been considerable intervention by statute so as to modify and extend their contractual obligations. Either party can insist on a written agreement, and terms may be fixed by arbitration. The parties can fix the rent at the start of the agreement and change it at any time by agreement. The 1986 Act provides for the rent reviews at three-year intervals and either party can demand an arbitration to fix it. The rules are to be found in Schedule 2: the rent has to be fixed by reference to a number of factors including the productive capacity and related earning capacity of the holding. Scarcity of lettings has to be disregarded when taking the rents of similar holdings into consideration as comparables.

Although the parties may agree their respective maintenance and insurance obligations, 'model repair clauses' are set out in the Agriculture (Maintenance, Repair and Insurance of Fixed Equipment) Regulations.

Improvements undertaken by the landlord with the tenant's agreement may lead to an increase in rent. If the tenant carries out improvements he will be entitled under certain circumstances to compensation at the end of the tenancy in accordance with the Act and relevant regulations. The Act also deals with the tenant's right to remove fixtures and the landlord's right to purchase them. When a tenancy comes to an end there will normally be a settlement of claims as between landlord and tenant, by arbitration if necessary, for disturbance and delapidations.

When milk quotas were introduced in 1984 it became evident that the value of a landlord's property could be greatly diminished if a tenant gave up a tenanted farm's quota. In general, however, for a tenant to do so now would be a breach of the contract of tenancy. Tenants meanwhile felt that, where the milk output of a farm had been increased during their term of occupation of a holding, the consequent size of the quota allocated should be reflected in compensation payable to them by the landlord when a tenancy was terminated. In the Agriculture Act 1986, S13 and Schedule 1, provision is made for such compensation for existing tenancies.

Further CAP reforms, including the introduction of sheep and beef quotas, set-aside and arable area payments, have made it a necessity for anyone considering entering into a tenancy agreement to undertake a full 'CAP assessment' of the land in question, so that a complete understanding is obtained of the commitments and opportunities available under the regime.

In general, subject to the comments made above, it may be said that a tenant farmer has freedom to crop a holding as he sees fit despite contrary indications in the tenancy agreement. There are, however, limits on a tenant's freedom in the last year of a tenancy when items of manurial value may not be sold or removed from the holding and the tenant may not establish a cropping scheme at any time that is not in accordance with the practice of good husbandry.

Occupation of agricultural land without security of tenure

If prior approval is given by the Minister of Agriculture, which will only be done for special reasons, an agreement for a letting for a specified period will be terminable when the period expires. By a quirk of the Agricultural

Holdings Act (but now endorsed by the courts in *EWP Ltd* v *Moore* (1992) QB 460), having a tenancy for more than one year but less than two does not have statutory protection (a *Gladstone* v *Bower* agreement), nor does a short grazing agreement. Thus an agreement for grazing or mowing, provided it is for a period of less than a year, gives no protection to the taker, but it must not permit ploughing or re-seeding or use of buildings unless they are simply shelters for grazing stock. Permission given to an employee to use agricultural land by a landlord employer during the period of employment does not create an agricultural lease. Partnerships between farmers and owners of agricultural land for agricultural business purposes do not create tenancies in favour of farmers. It must, however, be remembered that a partner is liable for his fellow partner's debts while a lease creates no such liability as between landlord and tenant. Share farming agreements are also possible but such contractual agreements require expert drafting.

Occupier's liability

Any person recognized by the law as the occupier of land owes duties towards persons who enter upon it and towards those who are in its vicinity. These duties are imposed by the law of tort which covers wrongs caused by a person's failure to carry out duties imposed by law as contrasted with duties imposed by a contract. The common law has evolved a number of such duties which have been recognized in judicial decisions as falling within the categories of trespass, nuisance, negligence and strict liability, to which further rules have been added by statute.

A farmer is more likely to suffer from trespassers against whom an action for damages or an injunction may be brought, than to be a trespasser himself. However, if his animals trespass on to neighbouring land, he will be liable for the damage they cause unless it was the fault of his neighbour or another person and could not have been reasonably anticipated. In this connection it should be remembered that a farmer is responsible for keeping his own stock in; he cannot complain if his animals escape through his neighbour's fence. Much of the law on liability for animals is covered by the Animals Act 1971 which also changed the rules about animals straying on to or off the highway. A person who negligently allows this to happen is now liable for damage done. But there is no duty to fence in animals grazed on common land by those who have the right to do so if it is customarily unfenced. The Act also permits a farmer to shoot dogs worrying livestock. However, he must notify the police within 48 hours in order to have a defence if he is sued by the dog's owner.

Persons entering land lawfully, known technically as 'visitors', as guests or for payment or because they have a statutory right of entry (e.g. Health and Safety Inspectors) are owed a duty by the occupier defined in the Occupiers Liability Act 1957 as 'a duty to take such care as in all the circumstances of the case is reasonable to see that the visitor shall be reasonably safe in using the premises for the purposes for which he is invited or permitted by the occupier to be there'. The Act permits the occupier to 'restrict, modify or exclude' the common duty of care by contract or adequate notice, but since the passing of the Unfair Contract Terms Act 1977 it is normally impossible for an occupier to exclude liability for personal injury caused by his own negligence.

Trespassers were regarded by the common law as entering upon other people's land at their peril until, in the 1970s (see *British Rail Board* v *Herrington* (1972) AC 877), the courts began to make rulings that implied that a farmer would be expected to take care to prevent child trespassers from encountering hazards. This idea has been confirmed and widened by the Occupiers Liability Act 1984 which imposes a duty of care on occupiers towards 'persons other than visitors' such as trespassers. The duty is to take reasonable care to give such persons protection against known dangers, for example by putting up warning notices, although this would probably not be sufficient where child trespassers are known to be at risk. Farmers allowing persons to enter their land for recreational or educational purposes are allowed to exclude their duty of care towards such persons so long as they are not allowed entry as part of the farmer's business enterprise.

'Mass trespass' has caused considerable loss to farmers in certain areas since the formation of 'hippy convoys'. Trespass is in itself not a crime but under the Public Order Act 1986, S39, police have powers to order trespassers to leave land where two or more have entered as such and done damage or used threatening behaviour or brought 12 vehicles on to the land. Failure to leave in response to such an order is an offence. The above legislation has proved inadequate to meet the needs of farmers who have suffered from the incursions of 'travellers', a phenomenon which has continued to persist throughout the 1980s and early 1990s.

New legislation to deal with this problem and the difficulties caused by those who unlawfully enter on to private land to take direct action against legitimate field sports is contained in the Criminal Justice and Public Order Act 1994. A new offence of 'aggravated trespass' by 'disruptive trespassers' has been created to deal with the latter situation. To bring a successful prosecution the following three elements have to be established: an act of trespass in the open air; the presence of a 'lawful activity' on that or adjoining land; and an intention to interfere with that activity. The Act also contains provision for local authorities and the Home Secretary to ban open-air assemblies of 20 or more persons which may lead to a 'serious disruption to the life of the community'. The purpose of this section is control the 'rave' culture.

As a result of the leading case of *Rylands* v *Fletcher* (1868) LR 3 HL 330, an occupier is strictly liable for the escape of any potentially dangerous thing kept on his land if it escapes and injures neighbouring property or people. Defences to such a claim are very limited and the occupier is liable even if the escape was caused by a contractor. The liability is, however, limited to non-natural uses of land. Both the doctrine of *Rylands* v *Fletcher* and the tort of nuisance were extensively reviewed by the House of Lords in the 1993 case of the *Cambridge Water Company* v *Eastern Counties Leather plc* 'the Cambridge Water Case'. Two important points to note from the judgement are:

- Foreseeability of harm of the type suffered is a requirement of liability under the rule in *Rylands* v *Fletcher*.
- The fact alone that usage of a particular item is common to an industry does not bring that use within

the definition of 'natural' for the purposes of the rule in *Rylands* v *Fletcher*.

A farmer will cover his liabilities in tort (including common law nuisance) as occupier by insurance but it should be remembered that insurance covers only for legally enforceable claims. Agriculture as an industry has a bad record so far as accidents are concerned. The utmost vigilance is required, for the law does not attempt to compensate for the incompensatable such as the death of a child drowned in a slurry pit, and damages cannot make good injury caused to an active adult crushed under a runaway machine.

Legal constraints on the development and use of agricultural land

Although some farmers have reacted positively to public opinion on the environment, concern remains as to the nature of modern farming methods and their effects on wildlife, human health and public amenity. The protection of water supplies remains high on the agenda and the momentum for greater public access to the countryside also continues. On this last issue a small amount of detailed knowledge concerning public rights of way legislation can be rewarded by a far better relationship with the general public.

Access to the countryside and public rights of way

There are no indications that the continued pressure for increased public access to the countryside will diminish in the foreseeable future. Indeed, the opposite appears to be the case, and further emphasis has been placed on the public rights of way system by the Countryside Commission's stated 'national target' of having all rights of way legally defined, properly maintained and well publicized by the year 2000. The past failures of both farmers and managers of land to respect public rights of way have resulted in further legislation to maintain and protect the network and to regularize agricultural activities that interfere with these rights, notably the preparation of ground for and the sowing of arable crops. In August 1990 the Rights of Way Act 1990 came into force. This legislation details the circumstances in which a farmer can exercise the statutory right to plough or otherwise disturb ground (the word 'ploughing' being replaced by the term 'disturbing the surface'); that which must be done to make good the ground; the time limits within which this must be achieved; the role of highway authorities in supervising the activities of farmers; and the results of non-compliance with the Act which may, in some circumstances, result in criminal prosecution.

The legal provisions that relate to public rights of way occur as the result of both common law and statute, are numerous and can be complicated. But if farmers are to maintain any credibility as 'stewards of the land' with those who seek to gain access to it, they must demonstrably be aware of their legal duties. Below is an outline of the main responsibilities of farmers and landowners with regard to public rights of way over their land.

(1) Obstructions

No person may wilfully obstruct the free passage along a highway (highway includes both footpaths and bridleways): S137 Highways Act 1980. This means that if a person 'without lawful authority or excuse, intentionally as opposed to accidentally, that is, by an exercise of their own free will, does something or omits to do something which will cause an obstruction, he or she is guilty of an offence'. Parker LCJ *Arrowsmith* v *Jenkins* (1963).

(2) Gates and stiles

The maintenance of a gate or stile that crosses a footpath or bridleway is the responsibility of the landowner: S146 Highways Act 1980. It must be maintained in a safe condition so that there is no unreasonable interference with the public's rights of access.

(3) Overhanging vegetation

Where any overgrowth interferes with a public right of access, the Highway Authority or District Council may serve a notice under S154 of the Highways Act 1980 requiring the owner or occupier to cut back the growth. In order to provide a way for pedestrians, clearance would normally be required to a height of 6 ft.

(4) Bulls: Wild Life and Countryside Act 1981 S59

Bulls may not be kept in fields crossed by rights of way unless the following conditions are met:

(a) The bull is less than ten months old, and
(b) It is not a recognized dairy breed and it is running with cows or heifers. Dairy breeds include Ayrshire, British Friesian, British Holstein, Dairy Shorthorn, Guernsey, Jersey and Kerry.

(5) Misleading notices

S57 National Parks and Access to the Countryside Act 1949 makes it an offence for any person to place or maintain, on or near any way shown on the definitive map, a notice which contains false or misleading statements which are calculated to deter the public form using the way.

(6) Barbed wire: Highways Act 1980 S164

Where barbed wire is placed on land adjoining a highway in a position in which it is likely to be injurious to persons or animals using the highway, the Highway Authority may serve a notice on the occupier of the land requiring him to remove the nuisance.

Farmers and landowners should also be aware that other activities that interfere with the public rights of access may be a nuisance at common law or under statute. In recent years the Environmental Protection Act 1990 Part 3 S79 has added to the list of statutory nuisances, which a farmer might breach when hindering the public's rights of access.

(7) Chief obligations imposed by the 1990 Rights of Way Act

Under the Act farmers may plough or disturb the surface of a crossfield, footpath or way as long as it is not convenient to avoid it. Farmers may not disturb field edge paths.

When the surface of a public right of way has been

'disturbed' (disturbance including all necessary operations for cultivation):

(a) The period allowed for restoration of the surface is 24 hours; however, where there is a sequence of operations leading up to sowing, a period of up to 14 days is allowed.
(b) When restoring the path or way the farmer is required to indicate the line of the route on the ground.
(c) A specific duty is imposed on farmers to prevent crops from encroaching on to paths by:
 – growing through the surface;
 – overgrowing on to them from the sides.
(d) Paths and ways which have been disturbed by cultivations must be restored to the following minimum widths:
 crossfield path 1 m
 crossfield bridleways 2.5 m

The minimum width for a field edge path is 1.5 m.
(e) Offences committed under the Act are punishable by fine which may be increased when an offence is repeated.
(f) If a farmer fails in his obligations under the Act, the local authority may act in default and seek reimbursement from the farmer.
(g) Excavations and engineering activities that disturb the surface of a public right of way may only be carried out if you first get written permission from the highway authority.

As indicated the above is merely an outline of the obligations placed upon those who have public rights of way over their land. However, likely future pressures suggest that it is the wise farmer who has a complete knowledge of the law as it affects his particular circumstances.

Legal constraints on the development of land.

The early 1990s saw a radical reshaping and the introduction of major changes to the statutory planning legislation in this country. This included the closer integration of agriculture into the planning system. In 1990 the Town and Country Planning Act (TCPA 1990) came into force consolidating previous legislation. On 25 July 1991 the Planning and Compensation Act (PCA 1991) received the royal assent. This act is extensive in the number of areas it effects and fundamental in a number of the changes it has brought about. The 1991 Act is an enabling act, i.e. it gives the power to the Secretary of State to introduce secondary or delegated legislation which is used to complete, amend and update the primary legislation. It is therefore extremely important when considering planning law to make certain you are working with up to date information. The planning legislation is complemented by government circulars and Planning Policy Guidance Notes (PPGs) which detail national policies. These do not have the force of law but are material factors to be taken into account when, for instance, planning applications are considered. For example PPG 7, the Countryside and the Rural Community, and PPG 2, Green Belts, are those of most relevance to farmers.

Use of land for agriculture is permitted because this activity does not constitute a development under the TPCA 1990. Other activities, i.e. building, mining, quarrying or the carrying out of engineering works, are development, as is a material change of use, e.g. the conversion of farm buildings to dwellings or for light industry. In such cases planning permission must be obtained from the local planning authority – normally the district council.

Under the General Development Order 1988 (GDO) certain agricultural developments are deemed not to require planning permission. However, farmers should ensure that the development complies with all the requirements of the GDO, e.g. the building does not exceed 465 m^2 subject to conditions as to its siting in relation to other buildings, other recent constructions, roads and its height. An amendment to the GDO in January 1992 made a number of significant changes to the permitted rights for agriculture and should be noted:

(a) Agricultural holdings of less than 5 ha no longer benefit from previous permitted development rights to construct farm buildings.
(b) More limited development rights have been introduced for agricultural holdings of less than 5 ha.
(c) Farmers should now check, before making use of agricultural permitted development rights, with their local planning authority as to whether the authority require to give prior approval for certain details of the development. In the case of an agricultural building this would include siting, design and external appearance. Under this 'determination' procedure the planning authority has 28 days to decide whether prior approval will be required for:
 (i) the siting, design and external appearance of agricultural and forestry buildings;
 (ii) the siting and means of construction of a private way;
 (iii) the siting of excavations or waste deposits with an area exceeding 0.5 ha;
 (iv) the siting and appearance of fish tanks.

If no response is forthcoming from the planning authority within 28 days, the farmer is entitled to proceed.

It is suggested that before starting any new 'development' project farmers and managers should obtain the answers to the following questions:

(a) Does the project involve 'development' within the definition of the 1990 Act, i.e. building, engineering, mining or other operations in or over land or a material change of use of the land, e.g. conversion of farm buildings to dwellings?
(b) If yes, are there permitted development rights available under the GDO? If yes, contact local planning authority to determine whether approval is required for some details. If no, planning permission should be sought in the normal way.

The value of a careful preparation to the application cannot be overstressed. This should include the canvassing of neighbours and local opinion and an understanding of the factors, e.g. development plans, PPGs

which the planning authority will take into account when reaching their decision. If the application is successful, it may well have conditions attached. These conditions:
(a) must relate to the permitted development;
(b) must serve some useful planning purpose;
(c) may not be manifestly unreasonable;
(d) may be declared invalid for uncertainty.

Agriculture no longer holds the privileged position it once did in the planning world, and this fact must be taken on board by farmers and managers of agricultural land if further restrictions are not to be imposed.

Nature conservation and production methods

The Wild Life and Countryside Acts 1981 and 1985 provide the cornerstone for the protection of birds and wild animals. Apart from the destruction of common pest birds it is an offence to kill wild birds, destroy their nests or disturb them near their nests. The Secretary of State for the Environment, working with English Nature and the Countryside Commission for Wales, has powers to make orders to protect Sites of Special Scientific Interest, and either the Minister or a local authority may enter into a management agreement with owners and occupiers of land to preserve and enhance the natural beauty of the countryside. The land involved will be subject to restrictions as to the use for which compensation is payable, that will be binding not only upon the owner or occupier who makes the management agreement, but also upon successors in title.

The EC Council Directive on the Conservation of Natural Habitats and of Wild Fauna and Flora (Directive 92/43/EEC) – the Habitats Directive – came into force in June 1994. The following designations are provided for:

- Special Areas of Conservation – SACs.
- Special Protection Areas – SPAs.

Sites designated as such will be subject to restrictions when a proposed development is likely to have a significant effect on the area. The local planning authority will be required to assess the impact of a development and respond accordingly. PPG Note 9 has been issued to advise on the planning implications of regulations which implement the EC Habitats Directive.

Limitation of production is intimately related to the Common Agricultural Policy. The introduction of Milk Quotas by EEC Council Regulations in 1984 introduced a new and complicated area of law to the agricultural scene. This scenario has had further complexities added in the form of Suckler Cow Premium Rights and Ewe Premium Rights. It should be noted that they differ from milk quota in two basic respects. First, they are owned by the producer and, secondly, they are a right to receive premium each year and not to produce. With the continuously shifting sands and perilous nature of this area of law, those involved in land and/or quota transfers are respectfully referred to take expert advice.

Another approach to production limitation coupled with conservation is to be found in the Environmentally Sensitive Areas Scheme (introduced by the Agriculture Act 1986) of which there are 16 in England. Payments are made to farmers in such areas for farming in accordance with traditional methods, thus maintaining established landscape patterns. This 'carrot' approach to the protection and enhancement of the countryside has also manifested itself in schemes such as that for 'Countryside Stewardship' and the introduction of grants for the management of hedgerows. Such initiatives are to be complemented by further schemes that will reward farmers who take positive action to conserve and enhance the rural countryside. Six such schemes, which are a response to the EU Agri–Environment Regulation (EC 2078/92) were implemented in 1994:

- a new ESA Scheme with public access options;
- a new Moorland Scheme to protect and improve heather and other scrubby moorland;
- a new Habitat Scheme aimed at improving wildlife habitats over a 20-year period;
- a new Organic Scheme to aid those farmers converting their means of production;
- a new Countryside Access Scheme for land in non-rotational set-aside;
- thirty new Nitrate Sensitive Areas to protect selected ground water sources.

Farmers would do well to contrast the approach that has been adopted with the more prescriptive alternatives.

Pesticides and pollution

The law relating to the use, supply and storage of pesticides is controlled by the Control of Pesticides Regulations 1986 (SI 1986 No. 1510) made under the Food and Environment Protection Act 1985, and the Control of Substances Hazardous to Health Regulations 1988 (SI 1988 No. 1657) more familiarly known as COSHH, made under the Health and Safety at Work Act 1974 (see Chapter 5 of this book). The term pesticide includes herbicides, insecticides and fungicides but does not include those substances applied directly to livestock (e.g. sheep dips which are regulated by separate legislation, namely the Medicines Act 1968). The legislation seeks to protect both human health and the environment by requiring Ministry approval for all pesticides and prescribing the training necessary for the use of such products. Those farmers or workers seeking to use pesticides who were born after 31 December 1964 must obtain a certificate of competence before using an approved pesticide. Failing this they must be directly supervised by a person holding such a certificate.

All farmers should be aware of the two statutory approved Codes of Practice that accompany the legislation. These are the Code of Practice for suppliers of pesticides to agriculture, horticulture and forestry; and the Code of Practice for the Safe Use of Pesticides on Farms and Holdings (the 'COSSH Combined Code').

Those farmers whose actions cause a deterioration in water or air quality or cause other statutory nuisances may find themselves held legally responsible under the Water Resources Act 1991 or Environmental Protection Act 1990. Under the former it is a criminal offence to cause or knowingly permit any poisonous, noxious or polluting matter, or any solid waste matter, to enter any controlled waters. In the Control of Pollution (Silage, Slurry and Agricultural Fuel Oil) Regulations 1991 (SI 1991 No 324), minimum standards are set for the construction of silage, slurry and fuel oil installations. These regulations apply to constructions built after 1 March 1991 (see Chapter 25). However, it should be appreciated that the National Rivers Authority (NRA) may give notice to a farmer to achieve the statutory standards on an installation built before 1 March 1991 if its condition presents a 'significant' risk of pollution.

Farmers should always seek to follow the Code of Good Agricultural Practice for the Protection of Water. This code has statutory approval but does not itself create any criminal liability. The Code does not furnish a defence for charges brought under the 1991 Act. This is a major change from the position under the previous legislation (Control of Pollution Act 1974) where a farmer could state by way of defence that he had acted in accordance with the Code.

As a result of regulations made under the Environmental Protection Act, air quality has sought to be protected by the prohibition of straw and stubble burning as from 1993. Further emissions of odours or smoke from farms may create statutory nuisances under the Act as defined by S79(1). If a local authority is satisfied of the existence of such a nuisance, it must serve an abatement notice. Failure to comply with such a notice, without reasonable excuse, constitutes a criminal offence.

Employer's liabilities

A farm enterprise has the same legal obligations towards its employees as any other business. The law on contracts of employment, job security and related issues continues to expand as we adapt our law, largely as a response to our European obligations. The law of employment is of a particularly technical nature, so although its main features are indicated below, farm management will require more detailed information – which must be kept up to date – than is here outlined.

The obligations which a farmer undertakes as an employer – whether towards an employee or other people – will arise only when there is a contract of employment between him and the person employed. When work is done by an independent contractor, the employer will not normally be responsible for wrongs done by the contractor unless on express instructions from the employer. If a third party, such as a road user run down by a negligent tractor driver, suffers injury because of the act of an employee done in the course of employment, then the employer will be vicariously liable for his employee's wrongful act. The employee will be liable personally, but this will usually not help the employer greatly as the injured person will almost certainly proceed against him as he, the employer, is likely to be insured against claims.

In fact, while it is not compulsory for an employer to be insured against claims by non-employees, he is obliged by the Employers Liability (Compulsory Insurance) Act 1969 to insure against liability for injury or disease sustained by his employees in the course of employment in Great Britain. Such an injury will, more often than not, come about because of the negligence of another employee.

A contract of employment, known as a contract of service, is often hard to distinguish from a contract for services such as those rendered by a contractor, but most of the modern employment Acts only apply to the former. A farmer has a duty to take reasonable care in choosing a contractor of repute but there it ends. With 'labour only' contracts (on a relief milking scheme for example), the distinction between employee and contractor may be a fine one; the law, however, looks at the reality of the situation and not at the words used by the parties to describe themselves. This means that a farmer would be likely to be responsible for the tortious acts of, say a 'self-employed' herdsman that affected third parties and arose out of that person's work.

There are statutory restrictions on the recruitment of employees and contract workers, notably the Sex Discrimination Acts 1975 and 1986 as amended by S32 of the Trade Union Reform and Employment Rights Act 1993 (TURERA) which makes it unlawful for employers and managers to discriminate against women or men or married persons when advertising for or engaging employees, and the Race Relations Act 1976 which makes it unlawful to discriminate on grounds of colour, race, or ethnic or national origins. The first Act does not apply to firms where five or fewer persons are employed but the second applies universally. This legislation has to be taken into account also when promotion or redundancy is considered.

On 1 July 1993 the Trade Union Reform and Employment Rights Act (TURERA) received the Royal Assent. This Act is a major non-consolidating piece of legislation dealing with an extremely wide range of employment issues from Trade Union affairs, the abolition of wages councils, the rights to written particulars of employment, maternity rights, the jurisdiction of industrial tribunals and health and safety at work. In the ever-expanding field of employment law this piece of legislation cannot be ignored by any employer.

As is often the case with statutory legislation, different sections have different dates of implementation so while, for example, the new law on the provision of written particulars of employment took effect as from the end of November 1993 those relating to maternity were implemented during October 1994. In particular employers should note that S26 substitutes sections 1–6 of the Employment Protection (Consolidation) Act 1978, so creating a new regime of statutory rights for employees to receive a written statement of their main employment particulars. All those employees whose employment lasts for at least one month are entitled to a written statement specifying the terms and conditions of their employment, within two months from the time they commenced work. The items that must be contained in the statement are specified in the Act, and the extent to which they may be given with reference to other documents is limited. Employers may provide the information by instalments as long as all the information is provided within the time limits and certain items are always provided in the initial instalment known as the 'principle statement'. Employers are now required to give to their employees written notification whenever a change is made to the content matter of the written statement. This notification must be given at the earliest opportunity and in any event not later than one month after the change.

When a contract of employment ends, in the absence of prior arrangements, the common law provides only for summary dismissal of an employee for misconduct or the giving of reasonable notice. Because of the one-sidedness of this position, legislation has been passed to give full-time employees the right to minimum periods of notice, to redundancy payments and to compensation or reinstatement if unfairly dismissed. TURERA 1993 makes dismissal automatically unfair if the reason for dismissal is that the employee has sought to assert one of their statutory employment rights by either:

(1) bringing proceedings against an employer to enforce a statutory right, or
(2) alleging that the employer had infringed one of their statutory rights.

The Employment Protection (Consolidation) Act 1978 as implemented by subsequent Orders and amended by the TURERA 1993 are the major pieces of legislation affecting this aspect of employment law. In effect an employee under retirement age is entitled to notice according to length of service, to redundancy payments in accordance with age and length of service, and to remedies for unfair dismissal when the employer dismisses for an inadequate reason or in an unfair manner, e.g. gives the employee little warning or opportunity to justify his actions. By virtue of the Transfer of Undertakings (Protection of Employment) Regulations 1981 as amended by TURERA 1993, when a farmer takes on with a farm its previous owner's or tenant's workers, he will also take on responsibility for their previous service on that farm so far as notice and redundancy payments are concerned. A farmer who has given recognition to a trade union must consult in advance about redundancies. Following the House of Lords decision in *R* v *Secretary of State for Employment, Ex Parte Equal Opportunities Commission and Another*, TLR 1994, the distinction between full- and part-time workers in the degree to which they are entitled to statutory protection has been demonstrated to be discriminatory and incompatible with European Community law. The practical result of this is that both full- and part-time workers must become eligible to receive statutory protection in the fields of unfair dismissal and redundancy after the same period of employment, at present two years. All disputes arising under the Employment Protection (Consolidation) Act 1978 are heard by Industrial Tribunals. Following TURERA 1993 the jurisdiction of industrial tribunals was extended to include specified breach of contract matters, and tribunal chairmen were empowered to determine disputes while sitting alone. Farmers, as with all employers, are advised to implement well-considered employment policies to which they adhere at all times.

Conclusions

A knowledge of the law as it affects agriculture and the business environment, the mechanisms by which law is created and the direction in which it is proceeding is now a necessity for the agricultural land manager. Those working on the land may feel isolated, but their actions and the results thereof, be it with regard to quantity and quality of production, the effect on the land, landscape or labour force are matters of general public concern. The last decade has shown that, if agriculture will not respond, of its own volition, to the wishes of the society it serves, then laws will be introduced to seek to achieve those ends. But, changes in the law provide new opportunities too; should the proposed revision of the agricultural holdings legislation reach the statute book, the market for let farms should not only be revived but should be able to respond in individual cases to the needs of both landlord and tenant. Likewise the freeing up of the milk market provides choice and opportunity. Knowledge of the law, acquired by education and timely professional advice, should be used in a positive fashion to manage and plan effectively.

References

As a general introduction to the English legal system and the general principles of English law:
Keenan, D. (1992) *Smith and Keenan's English Law*, 10th edn. Pitman, London.

For an overall view of those areas of law relevant to the running of a business:
Savage, N. & Bradgate, R. (1993) *Business Law*, 2nd edn. Butterworths, London.

On the latest employment legislation:
Bowers, J., Brown, D. & Gibbons, S. (1993) *Trade Union Reform and Employment Rights Act 1993. A Practical Guide*. 1st edn. Longman, London.

On most aspects of farm tenancies, land ownership and occupation:
Anon (1992) *Croner's Farm Business Management*. Croner Publications, New Malden.
Lennon, A.A. & Mackay, R.E. (1993) *Agricultural Law, Tax and Finance*. Longman, London.

On farm tenancies:
Rodgers, C.P. (1991) *Agricultural Law*. Butterworths, London.

For planning matters:
Anon (1993) *A Farmer's Guide to the Planning System*. Department of the Environment, Welsh Office, MAFF, London.
Chick, M. & Scrase, A. (1990, continuously updated) *Agricultural Diversification and the Planning System*. Town and Country Planning Department, University of the West of England.
Telling, A.E. & Duxbury, R.M.C. (1993) *Planning Law and Procedure*, 9th edn. Butterworths, London.

For detailed knowledge on access to the countryside:
Riddall, J. & Trevelyan, J. (1992) *Rights of Way: A Guide to Law and Practice*. The Open Spaces Society and the Ramblers' Association, London.

On European law:
Weatherill, S. & Beaumont, P. (1993) *EC Law*. Penguin, London.

On environmental aspects:
Hughes, D. (1992) *Environmental Law*. Butterworths, London.

5

Health and Safety in Agriculture

C. Boswell

Introduction

The earliest legislation affecting the health and safety of workers in agriculture was enacted towards the end of the last century. The Threshing Machines Act 1878 and the Chaff Cutting Machines (Accidents) Act 1897 were extremely limited in scope and only applied in England. They were not repealed, however, until 1961.

Attempts to include provisions for agricultural health and safety during the preparation of the Factories Act 1937 were not successful and it was not until the 1950s that further legislation was introduced in the form of the Agriculture (Poisonous Substances) Act 1952 and the Agriculture (Safety, Health and Welfare Provisions) Act 1956. The former enabled Regulations to be made to protect workers from the risk of poisoning when using certain specified substances. The latter, as the title suggests, had a wider application and some of the Regulations made under it are still in force.

In 1972 the Committee on Safety and Health at Work, under the Chairmanship of Lord Robens, published a report containing far-reaching recommendations for improving the health and safety of persons at work. Many of those recommendations were subsequently embodied in the Health and Safety at Work, etc. Act 1974.

In the meantime the agricultural industry had been making a remarkable transition from a labour intensive, low productivity industry, to one which deploys far fewer people and yet achieves considerably greater output. Increased mechanization, electrification and the extensive use of chemicals have undoubtedly been important factors in achieving high productivity but the growth in their use has also had a considerable effect on the nature and scale of workplace hazards.

An analysis of fatal injuries that occurred in the period 1986/87 to 1991/92 provides an indication of the extent and source of these hazards. Table 5.1 shows the rates of fatal injuries per 100 000 workers. The rates are many times higher than those for most other industries and are particularly high for the self-employed.

Table 5.2 provides a breakdown of the kinds of accidents occurring in the industry.

There is therefore little doubt that the Agricultural Industry is extremely hazardous. Farmers and workers are increasingly engaged in complex tasks requiring more detailed knowledge and training and more careful planning than was ever the case in the past. Moreover the reduction in the size of the labour force creates conditions in which workers are expected to be competent in a wide range of tasks and are often required to work alone and without supervision. In these circumstances knowledge and understanding of health and safety matters and the development of a positive attitude to them are prerequisites to the successful prevention of accidents and disease.

The nature of the industry and the fact that many families actually live at the place of work has led to some unique accident patterns. People of 80 years of age or more feature in accident statistics as do children who work and play on the farms. The law governing the employment of children is generally more permissive for agricultural work than for work in other industries and can lead to very young children working for their parents or guardians on light agricultural and horticultural duties, sometimes with tragic results.

The responsibility for ensuring healthy and safe working conditions in the industry rests primarily on those who create the risks, and legislation has been framed accordingly.

The Health and Safety at Work, etc. Act 1974 provides a legislative framework to promote, stimulate and encourage high standards of health and safety at work.

The Health and Safety at Work, etc. Act 1974 (HSW Act)

The HSW Act is in four parts:

Part 1 – dealing with health, safety and welfare in relation to work.
Part 2 – relating to the Employment Medical Advisory Service.
Part 3 – contains matters relating to Building Regulations.
Part 4 – contains a number of miscellaneous and general provisions.

Part 1 is the most relevant for the purposes of this chapter. The objectives of Part 1 of the Act are:

Table 5.1 Fatal injuries to employees and self-employed people in agriculture, 1986/87 to 1991/92 (rate per 100 000 workers)

	1986/87	*87/88*	*88/89*	*89/90*	*90/91*	*91/92*
Employees:						
Number	27	21	21	23	25	18
Rate	8.6	6.8	7.0	8.1	9.0	6.7
Self-employed						
Number	17	31	25	30	27	32
Rate	6.9	12.7	10.3	12.3	10.9	13.0
Employees and self-employed rate	7.8	9.4	8.5	10.1	9.9	9.7

Courtesy of the Health and Safety Executive

Table 5.2 Fatal injuries to employees and self-employed people in agriculture, 1986/87 to 1991/92

Kind of accident	*Employees*	*Self-employed*	*Total numbers*	*Percentage*
Struck by moving vehicle	27	36	63	21
Contact with machinery or material being machined	22	23	45	15
Fall from a height	19	25	44	15
Trapped by something collapsing or overturning	20	18	38	13
Struck by moving, including flying or falling object	12	26	38	13
Contact with electricity or an electrical discharge	17	10	27	9
Asphyxiation	12	7	19	6
Injury by an animal	3	8	11	4
Other	3	9	12	4
Total	135	162	297	100

Courtesy of the Health and Safety Executive

(1) securing the health, safety and welfare of people at work;
(2) protecting people other than those at work against risks to their health and safety arising out of work activities;
(3) controlling the keeping and use of explosive or highly flammable or otherwise dangerous substances, and generally preventing people from unlawfully having and using substances;
(4) controlling the release into the atmosphere of noxious or offensive substances from premises to be prescribed by Regulations.

The HSW Act applies to employment generally. Thus duties are placed on all people at work, that is employers, employees and the self-employed; manufacturers, suppliers, designers and importers of materials used at work; and people in control of premises. The Act does not distinguish between industries.

The HSW Act is superimposed on earlier related Acts such as the Agriculture (Poisonous Substances) Act 1952 and the Agricultural (Safety, Health and Welfare Provisions) Act 1956. For the time being these earlier Acts and many of the Regulations made under them remain in force and become 'relevant statutory provisions'. Thus, for example, a farmer must ensure he complies both with the general duties contained in the HSW Act and the more specific duties laid down in Regulations.

An objective of the HSW Act is to progressively replace the requirements of the earlier Acts and Regulations by Regulations and Approved Codes of Practice made under the HSW Act. Parts of the two Agricultural Acts and the Regulations made under them have already been repealed.

The HSW Act established two bodies, the Health and Safety Commission and the Health and Safety Executive. The Commission consists of a Chairman and up to nine part-time members, all of whom are appointed by the Secretary of State for Employment after consulting employers' organizations about three members, employees' organizations about three other members and Local Authority and other organizations about the rest. It is important to note that the responsibility for developing policies in the agricultural health and safety field rests with the Commission and not with the Ministry of Agriculture, Fisheries and Food or the Department of Agriculture for Scotland.

The Commission's duties include promoting the objectives of the Act, carrying out and encouraging research and training, providing an Information and Advisory Service and putting forward to Government proposals for Regulations under the Act. The Commission has developed arrangements whereby there is wide consultation on any proposals to either change existing legislation or introduce new Regulations.

The Commission has also established a number of

Advisory Committees, one of which is the Agriculture Industry Advisory Committee (AIAC). The Chairman of the AIAC is HM Chief Agricultural Inspector, and its 12 members are drawn from both sides of the Agricultural Industry. Other people with knowledge and experience of the industry or particular expertise are co-opted as necessary to working parties or are appointed as assessors. The Health and Safety Executive consists of three full-time members who are appointed by the Health and Safety Commission with the approval of the Secretary of State.

The Executive's duties include making arrangements for enforcement of the legislation and carrying out any of the Commission's functions which the Commission asks the Executive to take on. In practice the Executive and its staff are the operating arm of the Commission. They are known collectively as HSE.

It is convenient to divide the work of HSE into three main areas: Policy, Technological, Science and Medical, and Operations.

Policy

Staff are involved in advising the Health and Safety Commission on the need for changes in legislation or standards. They are involved in negotiations on European Community proposals and the development of new or revised legislations. Liaison with national and international organizations is an important feature of the work.

Technological, science and medical

Staff supply technological, scientific and medical support to other parts of HSE and to government on industrial health and safety matters, including the extent and nature of risks, the effect of hazards on individuals or the environment and on appropriate standards.

Operations

Operational staff are mainly grouped in inspectorates, with the majority forming the Field Operations Division which includes HM Agricultural, Factories and Quarries Inspectorates, the Employment Medical Advisory Service and seven Field Consultant Groups.

The Inspectorates interpret and implement new control packages and existing legislation and guidance. They are the main enforcement arm of the Executive and secure compliance with legal requirements and accepted standards, interpreting and applying the latter so far as is possible on a national basis. They do this through inspection, advice, the investigation of accidents and, if necessary, enforcement.

An important aspect of their work is contributing, through practical experience and direct knowledge of the facts of industrial life, both to policy and to the development of standards. They thus advise the Policy Divisions on enforceability and practical acceptability of proposed measures and the need for new approaches.

There is close integration between the Agricultural and Factory Inspectorates, with a shared management chain based on a regional structure. In addition to inspection groups, both inspectorates maintain National Interest Groups (NIGs) as focal points for contact and liaison with industry. There are three such groups in agriculture, based on Stoneleigh (at the National Agricultural Centre), Nottingham and Edinburgh. Each assumes responsibility for particular aspects of agriculture. Thus Edinburgh covers forestry, fish farming, etc; Nottingham covers arable crop production, fruit, horticulture, etc. and includes machinery matters; and Stoneleigh leads on livestock farming, stationary machinery, farm buildings, storage, etc.

All three NIGs are involved in national and international standards work.

Some of the statutory requirements of the HSW Act

The HSW Act imposes duties on everyone concerned with work activities, employers, the self-employed and employees. The duties are imposed on individuals, such as farmers and workers and on companies, partnerships, etc. The scope includes manufacturers, designers, suppliers and importers of machinery and materials for use at work.

The duties in the Act are expressed in general terms with more specific requirements covered by Regulations. Thus there are special Regulations for industries such as agriculture and also Regulations having a more general application.

Some of the duties are qualified by the term '*so far as is reasonably practicable*'. This is an important qualification which is not defined in the Act but has been interpreted by the Courts. It implies an assessment of the risk in comparison with the physical difficulties, time, trouble and expense involved in taking steps to avoid the risk. Thus, if the risks to health and safety are very low and the cost or technical difficulties of taking certain steps to avoid the risks are very high, then it might not be reasonably practicable to take them. However, if the risks are very high, then less weight can be given to the cost of measures needed to avoid them. The comparison does not include the financial standing of those who have the duty of compliance.

A duty which is '*so far as practicable*' without the word '*reasonably*' is stricter and means that the most effective means must be used to comply with the duty, taking into account the conditions and circumstances, the current state of technical knowledge and the financial implications.

In a prosecution alleging failure to comply with those duties, it is up to the accused to show that it was not reasonably practicable or practicable (as appropriate) for him to do more than he had in fact done to comply with the duties.

Some of the duties imposed by the HSW Act are outlined below.

Employers have a general duty to ensure, so far as is reasonably practicable, the health, safety and welfare at work of employees. Five of the most important aspects are specified:

(1) maintaining safe systems of work;
(2) ensuring the safe use, handling, storage and transport of articles and substances;

(3) providing adequate instruction, training and supervision;
(4) maintaining the safe premises and other places of work;
(5) providing a safe working environment and adequate welfare facilities.

For example, a safe system of work involving a tractor might include the choice of an appropriate machine with relevant safety features; maintaining it in a satisfactory condition; considering the terrain and operating conditions; the actual operation of the tractor for specific activities and any special precautions which might need to be taken.

The concept of thinking carefully about the nature of the hazard and the workplace and making arrangements to deal with them is also embodied in a requirement for employers with five or more employees to prepare a written safety policy for the undertaking and to bring it to the notice of employees.

Self-employed persons (as well as employers) are required to conduct their undertakings in such a manner as to ensure, so far as is reasonably practicable, that they do not expose people who are not their employees to risks to their health and safety. This applies both on and off the premises and includes, for example, the public and children. The self-employed are also required to conduct their businesses in such ways as to ensure so far as is reasonably practicable that they do not risk their own health and safety.

Employees are required to take reasonable care of their own health and safety and that of others who may be affected by what they do.

They must also cooperate with their employers and others in meeting statutory requirements and must not interfere with or misuse anything provided to protect their health and safety in compliance with the HSW Act.

Persons who have control to any extent of non-domestic premises are required to take such steps as are reasonably practicable to ensure that there are no risks to health and safety when they are used by persons who are not their employees. The duty is not imposed on employees who might have been put in control at any particular time by the employer.

Designers, manufacturers, importers and suppliers of articles and substances for use at work must ensure, so far as is reasonably practicable, that they are safe and without risk to health in prescribed circumstances. They are required to carry out such tests as may be necessary for the purpose of their duties and to make information available about the uses for which the product and substances have been designed and tested. (*Note*: HSW Act, Section 6 duties were amended by the Consumer Protection Act 1987.)

People in general (i.e. the public) have duties not to interfere intentionally with or misuse anything provided in the interests of safety, health and welfare. This might apply, for example, to fences, warning signs, machinery guards, etc.

Enforcement

Inspectors are appointed under the HSW Act and derive their powers from the Act. They are issued with a warrant which specifies those powers. Essentially these enable the Inspector to carry into effect any of the statutory provisions for which he or she is appointed. Thus, in general terms, there are powers of entry; powers to make examinations and investigations and to direct that premises or parts remain undisturbed for those purposes; to take samples; and to require persons to answer questions.

Inspectors have the power to institute proceedings in England and Wales (different arrangements apply in Scotland) and to issue *Improvement and Prohibition Notices*. The circumstances in which Notices may be issued are circumscribed by detailed requirements in the HSW Act (Sections 21–23) but in summary:

(1) *An improvement notice* may be issued if there is a legal contravention of any of the relevant statutory provisions and requires that the matter be remedied within a specified time.
(2) *A prohibition notice* may have an immediate or deferred effect but requires that an activity shall cease or not be carried out until the matter specified in the notice has been remedied.

Appeals against notices are dealt with by Industrial Tribunals, and details of how to appeal are stated on the notice.

Prosecutions can be tried summarily or on indictment. Fines in the lower Courts are up to £5000 for most offences but breaches of sections 2 to 6 of the HSW Act can attract £20 000 and failure to comply with Improvement and Prohibition Notices £20 000 or six months imprisonment or both. The higher Courts can impose unlimited fines or two years' imprisonment, or both.

Regulations applying to agriculture activities

It is convenient to consider Regulations in two groups, those specifically intended to apply to agriculture and those which are multi-industry or wider in concept.

The distinction is interesting because whilst the former may owe their origin to legislation such as the Agriculture (Safety, Health and Welfare) Act 1956 and contain basic but detailed requirements to suit specific machines, activities, etc., the latter may have been made under the HSW Act and reflect the thrust of developments since the introduction of the Act. They may be fairly general in the regulatory requirements but underpinned by Approved Codes of Practice or guidance documents.

An *Approved Code of Practice* is approved by the Health and Safety Commission and has a particular standing in law. If the code appears to the Courts to be relevant to a case, then it is admissible in evidence. If the guidance in the Approved Code has not been followed, it is up to the defendant to show that he has satisfactorily complied with the requirement in some other way.

Regulations specific to agriculture have been reduced in recent years and particularly since 1993 when the implementation of European Community Directives led to the coming into force of six new sets of Health and Safety Regulations. They are part of the continuing modernization of UK law allowing much outdated law to be repealed.

The new regulations make more explicit some of the duties which were imposed by earlier legislation. They are based on the modern approach of placing emphasis on the importance of managing health and safety and of assessing risk and choosing appropriate preventative and protective measures. It should be noted that they operate alongside the HSW Act and other Regulations made under it.

The six sets of regulations are as follows.

The Management of Health and Safety at Work Regulations 1992

These regulations set out broad duties applying to almost all work activities. Principal provisions relate to risk assessment; arrangements for health and safety; health surveillance etc. They include duties on employees to use equipment etc. in accordance with any training or instructions received, and these duties complement those under Section 7 of the HSW Act. There is an approved code of practice for these regulations which might be considered a cornerstone of the modern approach to health and safety legislation.

Provision and Use of Work Equipment Regulations 1992

These regulations apply across all industrial, commercial and service sectors. Work equipment is defined to include everything from hand tools through to complete machinery installations. Tractors, power harrows and potato grading lines are included but not livestock or substances such as slurry or acids. The regulations deal with suitability of work equipment; maintenance; provision of information, training, controls, etc.

Manual Handling Operations Regulations 1992

These regulations revoke the earlier Agriculture (Lifting of Heavy Weights) Regulations and repeal part of the Agriculture (Safety Health and Welfare) Regulations (S2). The Regulations establish a hierarchy of measures such as avoiding hazardous manual handling operations so far as is reasonably practicable; assessing those which cannot be avoided and reducing the risk of injury so far as is reasonably practicable, including the provision of mechanical assistance.

Workplace (Health Safety and Welfare) Regulations 1992

These regulations deal with such matters as maintenance, ventilation, temperature, lighting, etc. They include provisions for preventing falls from roofs and into dangerous substances. However, workplaces which are in fields, woods or other land forming part of an agricultural or forestry undertaking but which are not inside a building and are situated away from the undertaking's main buildings are exempted from most of the requirements except those relating to sanitary conveniences, washing facilities and drinking water (Regulations 20–22). The HSC has approved a Code of Practice for these Regulations.

Personal Protective Equipment at Work Regulations 1992

These are short regulations dealing with the provision, maintenance and accommodation of personal protective equipment (PPE) and the assessment of its suitability, etc. The provision of information, instruction and training is included. However, it should be noted that PPE is regarded as the last resort in a hierarchy of protection of risks to health and safety. Engineering controls and safe systems of work should always be considered first. Employers' duties in this respect are framed in the Management of Health and Safety at Work Regulations as well as in much of the legislation under the HSW Act 1974.

Health and Safety (Display Screen Equipment) Regulations 1992

Display screens are not a major feature of agricultural operations but the regulations apply with some exemptions for workplaces on agricultural or forestry land away from main buildings. They do not apply to drivers' cabs.

Another modern set of Regulations applying across all industries is the *Control of Substances Hazardous to Health Regulations (COSHH)*. These regulations are based on good occupational health practice involving the identification of hazards; the assessment of risk; the substitution of safer substances, and the introduction of appropriate control measures.

Examples of more general regulations are as follows:

The Health and Safety (First-Aid) Regulations 1981

These place general duties on employers and on self-employed persons so that first-aid provisions are made for employees who are injured or become ill at work and so that the self-employed can render first-aid to themselves if they are injured at work.

The Safety Signs Regulations 1980

These Regulations are based on an EEC Directive and provide that safety signs for persons at work and colours in strips identifying places where there is danger to their health or safety shall comply with BS 5378: Part 1 1980.

The Diving Operations at Work Regulations 1981

These cover all diving operations where the HSW Act applies. This therefore includes fish farming establishments or reservoirs where diving operations may take place.

The Reporting of Injuries, Diseases and Dangerous Occurrence Regulations 1985 (RIDDOR)

These Regulations place responsibilities on employers, the self-employed and certain other responsible persons. The Regulations set out reporting arrangements for fatalities and specified major injuries; certain diseases; specified dangerous occurrences and gas incidents. In the case of death or any of the specified serious injuries, the requirement is to notify the responsible authority (HSE in respect of agricultural activities) by the quickest practicable means and forward a report within seven days. Where persons at work are incapacitated for work for more than three consecutive days, then the report has to be forwarded within seven days of the accident. The form of the report is specified in the Regulations. The Regulations also specify matters which are to be kept in records.

The situation regarding regulations tailored specifically for agriculture is temporarily complicated because many are affected by partial or complete revocations which are set out in detail in the six sets of regulations introduced in 1993 and described earlier.

Those affected in this way are:

The Agriculture (Power Take-off) Regulations 1957
The Agriculture (Stationary Machinery) Regulations 1959
The Agriculture (Threshers and Balers) Regulations 1960
The Agriculture (Field Machinery) Regulations 1962

Regulations remaining intact are:

The Agriculture (Avoidance of Accidents to Children) Regulations 1958
The Agriculture (Tractor Cabs) Regulations 1974 – as amended 1984
The Agriculture (Safeguarding of Work Places) 1959

The Poisonous Sustances in Agriculture Regulations 1984 were revoked by COSHH.

Food and Environment Protection Act 1985

The Food and Environment Protection Act 1985 (FEPA) is an enabling Act in three parts. Parts 1 and 2 deal with the contamination of food and with deposits in the sea. Part 3 deals with pesticides and takes effect by means of Regulations. The detailed requirements have been introduced in The Control of Pesticides Regulations 1986.

These Regulations apply to pesticides (which include products such as herbicides and fungicides) used in agriculture, horticulture and forestry as well as to other uses less directly concerned with the agricultural industry. They apply to animal husbandry but not to pesticides administered directly to farm livestock, for example sheep dips or warble fly sprays which are already controlled by the Medicines Act 1968.

The Regulations prohibit the advertisement, sale, importation, supply, storage and use of a pesticide unless it has been approved. All manufacturers, importers and suppliers of pesticides must obtain approval for each pesticide product.

Anyone who advertises, sells, supplies, stores or uses a pesticide is affected by the regulations. Certificates of Competence are required for anyone who stores approved pesticides for the purpose of sale and by those selling or supplying approved pesticides. Similarly contractors and persons born after 31 December 1964 who apply approved pesticides must have a Certificate of Competence.

Conclusions

The legal framework within which people have to work and which requires them to accept responsibilities for health and safety is now well developed. The Health and Safety Commission, its Advisory Committees and the Health and Safety Executive are deeply committed to the provision of help and guidance to industry, and an extensive system of consultation has been developed so that proposals for change or improvement to the law take account of industry's perception of need.

In some circumstances the law has catered for particularly dangerous situations through the imposition of detailed requirements which effectively prevent someone from making an inappropriate judgement. The compulsory introduction of safety cabs on tractors is an example which has greatly reduced accidents and the risk of noise-induced deafness, while at the same time promoting a better and more productive working environment. Similarly, the detailed requirements for pesticide approval and use should do a great deal to reduce risks to the health of people while at the same time safeguarding the environment.

However, in general, the law does not of itself produce safe and healthy conditions. It is people who create the risks, whether they be manufacturers or users of equipment or substances or engaged in the diverse activities of the industry. They can also create the conditions in which agriculture becomes a safer and healthier industry, for those who work in it or are affected by it.

Advances in technology and farming practices have radically changed the industry and introduced new hazards. In dealing with these hazards the modern farmer and farm worker need to combine knowledge and competency with a commitment to healthier and safer working conditions and practices as an integral part of the efficient and profitable running of the business.

Progress has certainly been made and a great deal of help and guidance is readily available from the Health and Safety Executive and from organizations representing employers and the Trade Unions. Farmers and farm workers are better informed about health and safety issues than they were in the past but there is a continuing need for training to recognize and deal with the risks which are inherent in an industry which is quick to respond to changing conditions.

This brief description of health and safety in agriculture in Great Britain is by way of guidance only. It is not intended to be a comprehensive or authoritative interpretation of the Acts and Regulations, copies of which can be obtained through HMSO or booksellers.

It is also important to note that some aspects of health and safety legislation were reviewed in 1993 but up-to-date information and a wide range of priced and free publications on agricultural health and safety can be obtained by contacting any HSE Office listed in the telephone directory.

References

The Health and Safety at Work, etc. Act 1974, Chapter 37.
The Food and Environment Protection Act 1985, Chapter 48.
The Agriculture (Safety, Health and Welfare Provisions) Act 1956, Chapter 49.
The Agriculture (Avoidance of Accidents to Children) Regulations 1958 SI 1958 No. 366.
The Agriculture (Safeguarding of Workplaces) Regulations 1959 SI 1959 No. 428.

The Agriculture (Threshers and Balers) Regulations 1960 SI 1960 No. 1199.
The Agriculture (Field Machinery) Regulations 1962 SI 1962 No. 1472.
The Agriculture (Tractor Cabs) Regulations 1974 SI 1974 No. 2034.
The Agriculture (Tractor Cabs) (Amendments) Regulations 1980 SI 1980 No. 1036.
The Agriculture (Tractor Cabs) (Amendments) Regulations 1984 SI 1984 No. 605.
The Control of Pesticides Regulations 1986 SI 1986 No. 1510.
The Reporting of Injuries, Diseases and Dangerous Occurrences Regulations 1985 SI 1985 No. 2023.
The Health and Safety (First-Aid) Regulations 1981 SI 1981 No. 917.
The Diving Operations at Work Regulations 1981 SI 1981 No. 399.
The Safety Signs Regulations 1980 SI 1980 No. 1471.
The Control of Substances Hazardous to Health Regulations 1988 SI 1988 No. 1657.

Further reading

The Management of Health and Safety at Work Regulations 1992 – Approved Code of Practice. ISBN 0 11 886330 4.
Provision and Use of Work Equipment Regulations 1992 – Guidance. ISBN 0 11 886332 0.
Manual Handling Operations Regulations 1992 – Guidance. ISBN 0 11 886335 5.
Workplace (Health Safety and Welfare) Regulations 1992 – Approved Code of Practice. ISBN 0 11 886333 9.
Personal Protective Equipment at Work Regulations 1992 – Guidance. ISBN 0 11 886334 7.
Health and Safety (Display Screen Equipment) Regulations 1992 – Guidance. ISBN 0 11 886331 2.
Safe Use of Electricity in Farming and Horticulture – Farm Electric Centre, NAC, Stoneleigh.
Safeguarding Agricultural Machinery – Moving Parts. ISBN 0 11 882051 6.

6

Alternative Enterprises

R.W. Slee

One major component in contemporary discussions about the future of farming in the UK and many other developed countries is the role of alternative enterprises in adjustment strategies. Few now doubt that major changes will take place in the agricultural sector in Europe, and similar pressures for change have been operating in North America and Australasia. But, at a time of policy flux and financial stress, attitudes to alternative enterprises are highly variable. Some look to alternatives as a lifeline; others regard their development as an act of desperation, eccentricity or foolishness.

This chapter provides a brief introduction to alternative enterprises. Alternatives must first be defined. The policy context must be understood for it will strongly influence the viability of conventional and alternative enterprises. The main groups of alternatives will be explored and brief comments made on their likely role in future agricultural systems. The particular challenges of managing alternative enterprises must be recognized, for there are some significant differences from mainstream farm management approaches. (For a fuller discussion of all of these, see Slee, 1989.) Finally, two widely practised forms of diversification will be examined in greater detail in order to illustrate and amplify some of the points raised at a general level.

Around one in four UK farms is diversified according to some very detailed surveys carried out in the late 1980s (McInerney *et al.*, 1989). The figure is lower in the Celtic fringe of Scotland, Northern Ireland and Wales than it is in England, and there are distinct regional variations in the type of diversification (see Table 6.1). It should also be noted that many farms have more than one diversified enterprise.

Defining alternatives

There is no universally accepted notion of what constitutes an alternative enterprise. Any definition begs the question: 'Alternative to what?'. The most reasonable response is to suggest that there is an identifiable group of mainstream agricultural products. Normally, the farm enterprises associated with these products do not go beyond the primary production of food and fibre. Normally, in the EU these mainstream products are supported by the policy mechanisms of the Common Agricultural Policy. Alternatives are alternative to these products but any attempt to define alternatives raises a number of problems.

Firstly, there are examples of policy-supported products (e.g. lupins) which are clearly thought of as alternatives. Furthermore, in the light of new policy initiatives it is likely that a wider range of alternatives will be embraced by Common Agricultural Policy support measures. Over time it is likely that alternatives will be much more fully incorporated into what are regarded as conventional farming systems. Oilseed rape would be an example of an alternative that has already been fully incorporated, and linseed can be seen as an example of an enterprise which can no longer be regarded as alternative.

Secondly, at least some of the discussion about alternatives has focused upon alternative uses of farmland once it has passed out of agricultural ownership. Thus, urbanization is an alternative use but it is rarely one carried on within an agricultural proprietal unit. Some alternative enterprises such as golf courses or forestry may take place under agricultural proprietorship but are more frequently managed independently. In the discussion that follows, the sale of land for alternative use will not be considered further, although it clearly represents an adjustment strategy that should be considered in some situations.

Thirdly, there are problems in separating out farm-based alternatives from other gainful activities which exist quite independently of the farm. Thus viticulture

Table 6.1 Patterns of diversification on diversified farms in England and Wales

Enterprise type	*England* %	*Wales* %
Services	46.7	52.3
Contracting	38.5	39.1
Processing/sales	29.3	21.3
Speciality products	19.8	14.6
Miscellaneous	12.8	13.2

Source: McInerney *et al.*, 1989.
Note: This classification of types of alternatives differs from that used in the next section.

and on-farm wine production are farm-based alternatives. A farmer who imports French wine as a business activity may have another source of income but it cannot be regarded as a farm-based alternative. (See Gasson (1986) and Hill (1989) for a detailed examination of the extent of these other gainful activities.) The significance of this off-farm income should not be underestimated. At no time between 1977 and 1987 was more than two-thirds of the total taxable income of farmers in the UK attributable to farming activities (Hill, 1989). This average figure clearly masks a wide variety of situations. For many farmers, the greater proportion of their income will undoubtedly come from off-farm sources, including both investment income and income from other gainful employment.

Rather than offering a verbal definition, it is more helpful to offer a diagrammatic definition of alternatives (Table 6.2).

It is important to realize that the conventional image of the farmer as independent yeoman is largely outdated. There has been a substantial influx of urban money into the countryside, often for leisure motives, and this can have a significant impact on agricultural systems practised and on the development of on-farm diversification. Off-farm incomes might provide the funds necessary to establish capital-hungry alternatives such as deer farming. Although it is outside the remit of this chapter to consider off-farm income sources, the consequences of them cannot be so readily ignored.

It is misleading to see alternative enterprises as something necessarily novel. The past rural economy was much more diverse, as were the component farming systems. As late as the Second World War there was widespread evidence of alternative enterprises. These included many value-adding enterprises, particularly of milk and meat products as well as farm tourism. Many of these operations were associated with the routine domestic economy of dairy and meat processing, partly for domestic consumption, partly for sale in markets or from the farm gate. Alternatives were subsequently pushed into the background by the development of agricultural policies that concentrated on a streamlined product range and by changes in lifestyles (particularly of women) which led them towards off-farm employment rather than on-farm activities. By supporting only mainstream agricultural products, it was inevitable that these would displace the variety of other activities that had taken place on farms. The rehabilitation of alternative enterprises and the reform of agricultural policy go hand in hand, but the major social changes of recent decades must be borne in mind.

Table 6.2 A diagrammatic representation of alternative forms of diversification

	On-farm	*Off-farm*
Income stream	e.g. snail farming, farm tourism	e.g. farm contracting, unconnected off-farm work
Capital realization	e.g. selling off barns	e.g. selling shares

Source: Derived from Slee, 1989a.

The policy climate

There are many reasons why alternatives are receiving greater attention than at any other time in the last 40 years. Post-war agricultural policy, initially in a domestic context, later in a European one, and most recently at a global level, has both transformed the agricultural industry and created a new set of policy problems. From a time of food shortage and austerity in the years following the Second World War, the countries of western Europe have entered a new era of policy-induced surpluses. In the early 1990s the pressures for policy reform at a European level finally induced action by the EC. The fruits of these reforms have yet to be seen. Agriculturalists have become very sensitive to a barrage of criticism from a variety of directions, but it is important that they should understand the forces that have created this criticism. The same forces have had, and will continue to have, a significant impact on policy.

Economic influences

The contemporary imbalance between supply of and demand for major food products in the European Union is well known. The maintenance of prices above those that would normally prevail in a free market has led to increased production. The failure of the market to absorb this increased production has created the lakes and mountains of surplus produce. This is costly to store and is then frequently 'dumped' on other countries at less than the costs of production. At present, the intervention component consumes a major slice of the CAP budget. Over the last decade well over 50% of the cost of agricultural policy in the UK has been swallowed up by intervention costs, although it is interesting to note that the proportionate share in 1992/3 was estimated to be the lowest share for a decade.

It has long been recognized that in a mature economy the agricultural sector is likely to decline in importance. Put simply, the capacity of people to eat is less than their

Table 6.3 Direct costs of CAP support of UK farming

	Direct support (£ million) (1985/86 values)
1984/85	1815
1985/86	2165
1986/87	1401
1987/88	1513
1988/89	1221
1989/90	1033
1990/91	1433
1991/92	1388
1992/93	1385

Source: MAFF *et al.*, 1992.

ability and desire to acquire additional consumer goods. If incomes are rising, expenditure will be directed towards products associated with affluence. Historically, this has led to a growing demand for animal products including meat and dairy products. However, changes in taste have tended to dampen this process. Such growth as has occurred has been mainly in luxury foods, many of which are imported into the UK.

Until very recently the expansion of domestic food production was still seen as economically desirable. The expansionist ethos still prevailed amongst policy makers well into the 1970s. The high cost of pursuing such a policy is evidenced in the figures found in the Annual Review of Agriculture (Table 6.3). The consequences of that policy will be increasingly borne by farmers, as policy makers endeavour to reorientate agricultural policy away from its dominant productivist focus.

Political influences

The politics of domestic and European agricultural policy making must be recognized as a major influence on the current state of the industry. In the UK the strength of the agricultural lobby has been depleted by the collapse of support from its historic power base in parliament and by the growth of green issues on the political agenda in the country at large. This collapse is scarcely surprising when the majority of the residents of the British countryside no longer have any direct links with the farming industry.

The domestic situation differs markedly from that in the rest of Europe. On the continent, farmers remain a powerful pressure group via the ballot box as they constitute a greater proportion of the population. In addition, political parties in a number of countries woo the agricultural vote and maintain their position on account of it. Consequently, actions which damage the farming communities, however 'necessary' from an economic viewpoint, will rarely be endorsed on mainland Europe lest votes are lost. Expediency and horse trading take the place of rationally determined policy, as was illustrated in the politicking surrounding the General Agreement on Tariffs and Trade (GATT) in 1993. In the UK, the current political climate favours free markets over protectionism, and agriculture is viewed no differently from such industrial dinosaurs as the coal and steel industries.

The consequences of the decline in the political power of the farming industry are far reaching. Whilst at an EU level a major reform of the CAP has taken place, there is still a strong commitment to the social subsidy of rural regions permeating the MacSharry reforms, which contrasts markedly with the hard nosed, free market economic thinking which occupies such a significant place on the UK policy agenda. The consensus of opinion is that the 1992/3 reforms have been much less damaging than was anticipated, not least because of the windfall gains arising from the decline in the value of the £ against the major European currencies. However, few expect the reform process to stop and it is widely argued that the breathing space offered should be used to restructure farm businesses to help them to survive on lower levels of public support in the future.

These general pressures for the reform of agricultural policy are a global phenomenon. Some countries, such as Australia and New Zealand, have made massive reductions in public support to their agricultural industries. In these countries, particularly in New Zealand, there has been a corresponding increase in diversification in the farming community as one component of a much more market-oriented agricultural policy.

Environmental influences

Public interest in environmental issues has grown dramatically in the last 20 years. Marion Shoard's polemic, *The Theft of the Countryside* (1979), angered farmers and raised public awareness of rural environmental change. The debate about environmental issues and the countryside received widespread media coverage, and three main areas of discussion emerged. First, it was felt by at least part of the public that landscape had visually deteriorated under the impact of modern agricultural practices. Hedgerow removal, moorland improvement and land drainage emerged as causes célèbres. These and other side effects of intensification caused increased offence to a growing number of people, especially as the costs of supporting agriculture were visibly growing. Secondly, access was seen as threatened, particularly by field enlargement and the extension of arable systems. This debate looks likely to rumble on in the light of a major Countryside Commission study and a new book from Shoard (1987), and with the beginnings in the early 1990s of landowners receiving payments from government schemes for access provision. Thirdly, habitats have been modified by agricultural practices in ways that have reduced their ecological interest, at the same time as public interest in wildlife conservation has been intensifying.

The fourth area which is receiving increased attention is the production of environmentally friendly products. As questions relating to sustainability are raised, so farmland is considered as a place where renewable resources can be produced, including fibres such as flax for linen and biomass for renewable energy. Such products have become of increased interest to farmers because they offer potentially productive uses for set-aside land.

Wartime food shortages have ceased to occupy an important place in the public's mind. Instead a dominant image of contemporary farming is one of surplus, extravagantly produced at considerable environmental cost. The greening of politics will ensure that environmental issues continue to receive widespread attention and remain high on the policy agenda as the development and expansion of Environmentally Sensitive Areas (ESAs) illustrates.

Food and health influences

Attitudes to food have been transformed in the last decade and these attitude shifts have been reflected in dietary habits. Medical evidence has linked diet with certain major causes of disease and mortality. A number of major investigations have advocated changes in dietary behaviour including reducing saturated fat (and total fat) intake, reducing sugar and salt consumption and increasing fibre consumption. It is immaterial whether the detailed proof of relationships is available. The significant point is that the future diet of almost the whole population is being influenced. The demand for what were in the past seen as growth sectors in the food market (red meat, dairy products) has been, as a result, less than was anticipated.

It is important to separate out food faddism from trends in eating behaviour of a more long-term nature. Short-term health scares with some products may result in temporary crises as was the case with eggs and salmonella. But in the longer term, attitudes to food are changing, influenced by a combination of food gurus, dietitians and journalists and brought into focus by changes in lifestyle.

Coupled with the concern about diet has been an interest in additives in food. Although many of these additives are put into food in manufacture rather than 'in the field', there is also concern about hormones, pesticide residues and certain farm-based 'additives'.

The combined result of these various pressures is to challenge the agricultural status quo and to suggest that policy redirection will be needed. The ALURE package (Alternative Land Use and the Rural Economy) launched in 1987 was the first major manifestation of this change in the policy climate. The succeeding years have seen new initiatives, many of which supported diversification into alternatives. However, diversification can no longer be seen as the new business activities and the eccentric enterprises that have characterized some farms in the past. In addition to these it is necessary to take into account the expanding array of government-sponsored schemes to deliver environmental products. In policy cost terms these represent the greatest share of public support for unconventional activities on farms (Fig. 6.1).

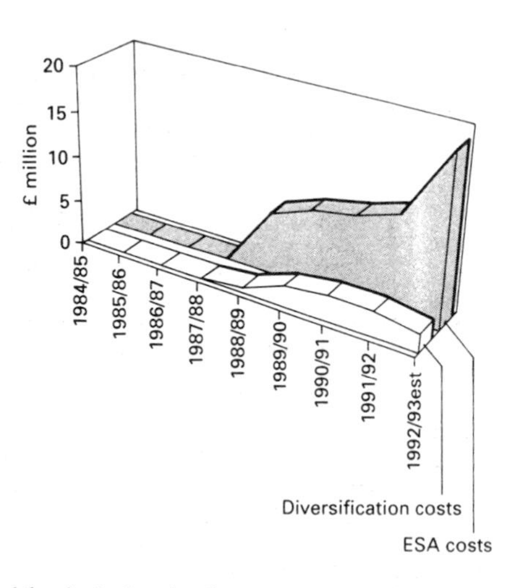

Fig. 6.1 Agricultural policy support costs of diversification (derived from MAFF *et al.*, 1993).

Types of alternative enterprise

Some reviews of alternative enterprises have attempted nothing more than an A to Z of alternatives. There is a case for attempting to classify alternatives in order to identify the characteristics that must be considered in their management. Five main groups of alternatives have been identified (Slee, 1989) and examples are given in Table 6.4.

At times agriculturalists appear to be seeking a set of recipes for production systems which, coupled with healthy gross margin data, will direct them towards a panacea for their income problems. This is wishful thinking. In the turbulent waters of unprotected markets, with unproven enterprises founded on flimsy production data, returns are likely to vary dramatically from location to location. It would be misleading and foolish to advocate simplistic solutions based on diversification. It would be equally foolish, however, to ignore such data as are available in trying to elucidate the qualities and the prospects of the main groups of alternatives. In some cases, as with alternative crops such as the oilseeds, for which production systems are well known and for which there are established support measures, the use of standard estimates derived from farm management handbooks or similar sources is reasonable. However, this is not the case for more individualistic enterprises or enterprises where the returns are more locationally specific.

Tourism and recreation

There is a long tradition of the use of farms and farmland for tourism and recreation. Furthermore, there is evidence of relatively favourable demand trends coupled with a commitment of government to support the development of the leisure industry.

Tourism

Many farm households are involved in farm tourism to a greater or lesser degree. Estimates have been made suggesting that one in ten farmers nationally (DART, 1974) and one in five farmers in Less Favoured Areas (Davies, 1983) are involved with tourist or recreational enterprises. More recent evidence (McInerney *et al.*, 1989) indicates that about 20% of all farmers in England and Wales as a whole are involved in service enterprises (i.e. about one in two diversified farms), but that only half of these are accommodation enterprises. Whilst some of these enterprises create little more than pin money, others provide the major source of income for the business.

An examination of the prospects for farm tourism must recognize both demand and supply influences. The demand for rural tourism is influenced by many factors including incomes, exchange rates, the cost of competing destinations, travel costs, climate and changing tastes. It is extremely difficult to do more than speculate about tourist futures but it is possible to learn lessons from past experience. The 1970s were a boom decade for tourism with increased activity at home and abroad. Numbers of tourists to the countryside grew. Underlying the boom, a restructuring of the tourist sector was taking place, with a growing preference for self-catering rather than serviced accommodation. Tourists were also more likely to tour rather than stay in one place. In spite of the growth of the tourist sector the old resorts were struggling to adapt and the rural sector appeared to be well situated to benefit from these changes. However, the 1980s proved rather more volatile. Upward projections of tourist numbers in the late 1970s had to be adjusted downwards. More distant locations experienced falling visitor numbers, particularly in response to fuel price rises. The early 1990s

Table 6.4 The main groups of alternative enterprises

Tourism and recreation	• Tourism	Bed & Breakfast Caravans
	• Recreation	Farm museums Equestrian enterprises
Value-adding enterprises	• Animal products	Meat processing Skins and hides
	• Crop products	Farm milling Direct marketing
Unconventional enterprises	• Animal products	Sheep milk Fish farming
	• Crop products	Evening primrose Teasels
Ancillary resources	• Woodlands	Fuelwood Craft products
	• Buildings	Accommodation Industrial units
	• Wetland	Fisheries Game
Public goods	• Wildlife	ESA payments
	• Access	Access agreements
	• Landscape	Stewardship payments

Source: Slee, 1989.

show signs of continued volatility. Macro-economic conditions can influence visitor numbers because of the loss of confidence engendered by an enduring recession. Exchange rate changes can also induce significant effects. Political events such as the Gulf War influenced the number of US tourists visiting Britain. Bad summer weather can compound the problem. Optimism about tourist potential is now more qualified.

Rural tourism in general and farm tourism in particular may benefit from a number of favourable trends. First, the general concern and interest of the public in the countryside may induce them to take rural holidays. This may allow the expansion of farm tourism in relatively new tourist areas such as Herefordshire or Staffordshire. Secondly, there is a growing interest in activity holidays, particularly for young people. Farm settings can provide suitable bases for activities ranging from art to archery or traditional rural pursuits such as riding and shooting. Thirdly, the shorter duration second or third holiday may take advantage of an experience different from that of the main holiday. The final trend of significance is the shift towards self-catering. Rural areas do not suffer from the disadvantage of the decaying remnants of the declining serviced sector that characterizes some of the traditional tourist resorts.

The supply side of farm tourism has responded to these changes. There are far more operators than there were a decade ago, and many more people are exploiting the saleability of countryside images in their tourist marketing. Green or soft tourism has been promoted by many public agencies and been given practical expression in a number of projects (Lane, 1989). Farmers have to compete with other providers of country holidays, and a dramatic increase in the supply of bed spaces could easily lead to spare capacity. There is also evidence that the tourism product is becoming more complex and that the farm holiday must have qualities other than cheapness to attract the visitor.

Estimates of the profitability of farm tourist operations show a very wide range of returns (Davies, 1983). However, no explanations of the highly variable returns are offered. It is likely that at least some of the factors influencing profitability, such as occupancy rates, are partly under the influence of the operator's marketing strategies.

Recreation

The demand for recreation in the countryside can be seen to have experienced similar changes to those in the tourist sector. A decade of buoyant growth in the 1970s was followed by a period of instability and appears to have been replaced by a period of rather more selective growth.

However, the demand for rural recreation poses problems for the provider. The landscape is a principal resource and is subjected to largely uncontrolled access on the footpath system and, what might seem to some, totally uncontrolled access off the footpath system. The demand for much rural recreation is not backed up by a transfer of resources. Those who drive through or walk in pleasant countryside pay no price for their pleasure other than transport costs. Consequently, it is in those areas backed up by a willingness to pay that demand must be examined more closely, if farmers are to benefit financially. However, recent changes in countryside policy make it possible for farmers to be rewarded for access provision, both by the provisions of the Countryside Stewardship scheme run by the Countryside Commission (there is, to date (1994), no equivalent scheme in Scotland although discussions are taking place), and by the 1993 agro-environmental proposals which offer payments for access on set-aside land on which there are no extant rights of way.

The demand for traditional countryside sports has been taken up by 'green-wellied yuppies' and showed signs of growth in the early 1980s even when demand for passive recreation in the countryside had slackened. Demand for activity pursuits from rock climbing to hang gliding also appears to be buoyant. But the total demand for these activity pursuits remains relatively small when contrasted with the demand for passive recreation. The supply side has responded. There are many more visitor attractions than a decade ago including rare breeds farms, visitor centres, museums and working farms. Private, public and voluntary sectors have all been major providers, and a growing professionalism in servicing visitor needs will be needed to sustain viability in the future as the effects of competition begin to bite.

It is impossible to speculate on the profitability of these enterprises. Few entrepreneurs will willingly give evidence of their profitability. Some enterprises can produce a very high return on capital in a very short period of time. Others, managed self-indulgently by operators, are little more than disguised hobbies. That there will be increased competition for visitors cannot be doubted. That there are still opportunities for well located and well managed visitor attractions on farmland is equally true. Amateurism must be replaced by a new professionalism, particularly in visitor servicing.

Alternative crops and livestock

The range of alternative crops and livestock is huge as is evidenced in the Centre for Agricultural Strategy Report (Carruthers, 1986). In these pages it is not possible to review the totality of new possibilities. Neither is it possible to explore the alternative uses of existing crops. Instead the main groups of alternative crops and livestock will be identified and brief comments offered on the qualities that should be sought in alternatives. Four main groups are identifiable: ethnic products, health products, luxury products, and craft products. In some cases one particular alternative may fall into more than one category.

Ethnic products

Different ethnic groups exhibit different patterns of demand for food and other products. These ethnic groups may be residents of the UK or of other countries which might or do import UK products. Typical examples of food products sought by specific ethnic groups include goat's meat and milk, feta cheese, Chinese mushrooms and carp.

There is a tendency to think only in terms of domestic ethnic markets but there may be export opportunities in specialist foods which merit exploration. In conventional products the development of niche markets in southern Europe for light lambs is an example of how new markets can sometimes be opened up abroad.

It is also important to recognize the impact of tourism and travel on domestic eating habits and to realize that there may be increasing interest in these exotic foods from the domestic population.

Luxury products

The demand for luxury food is a growth area in the food market. There is evidence, often of a circumstantial nature, to indicate a growing demand for luxury foods, from snails to quails, venison to specialist cheeses. Farm enterprises can either provide the inputs into these luxury foods or, by adding value on the farm, produce the final product.

The principal problem with many luxury products is being able to determine the market with reasonable accuracy. Market research is essential, yet prior to a product launch it can only be indicative and in a free market new entrants could dramatically affect the supply of the product. A major unknown is whether demand will increase, if and when, for example, supermarket chains start stocking a product regularly. If this is the case there may be certain thresholds of supply level which must be crossed in order to achieve the expansion of demand.

The demand for luxury products was, inevitably, adversely affected by the global recession of the early 1990s. At such times buying behaviour may change and the demand for non-essential items may tumble. Equally, an upturn in the state of the economy is likely to be especially beneficial to this sector.

Health products

Few doubt the impact of health-conscious eating on the demand for food. But, instead of bemoaning the loss of markets for traditional products, it is possible to develop new products that offer healthy fare. Low-fat meats such as venison may have advantage over high-fat meats like lamb. Goat's milk may have a growing market with allergy sufferers. Borage and evening primrose both contain gamma-linoleic acid which has a wide range of reputed health benefits. The demand for traditional products such as oat-based cereals has also benefited from health-induced demand.

Organic products fit within this category. They are sought by consumers who want an additive-free diet, and there has been a significant increase in the demand for organic foods. At present significant amounts of domestically consumed organic foods are produced in other countries. There may be scope for an increase of UK producers' market share, although this market has shown a significant slow-down in the rate of growth that was evident in the late 1980s.

Craft products

Craft products encompass a range of products including wood for turning, thatching materials, withies, teasels and specialist animal fibres. In many cases there may be opportunities to add value to these products. Craft products may be sought by working craftsmen such as thatchers or by individuals pursuing craft hobbies. A Shetland sheep's fleece in the south of England sold to a craft worker is worth about four times its value if sold into a conventional wool market. Early retirement and the expansion of leisure time are likely to increase the demand for craft materials.

Value-added enterprises

Value can be added to conventional farm products by processing or by marketing them in different ways. Both avenues merit exploration but both require careful scrutiny. It is important to realize that the objective must be to increase *net* value added. Adding value entails additional equipment and resources, and the capital and running costs must be carefully assessed.

Adding value by marketing

There are a number of conventional marketing channels used by farmers. It is possible to add value by marketing direct to the consumer or by missing certain links in the marketing chain. The simplest examples include pick-your-own (PYO) or farmgate sales. Other examples include sales of eggs or horticultural produce direct to shopkeepers.

In those situations where the customer is expected to come to the farm, location is of prime importance. Where the farmer conducts sales to the public from the back of a van, the precise location of the farm is much less important although proximity to a large market is clearly beneficial.

It should be realized that in many cases the visit to the farm shop or PYO establishment is a recreational visit in addition to being a food shopping trip. Many operators recognize this and cater for the public in a variety of ways. It may also be important to offer particular facilities in order for the farmer to differentiate his establishment from others. There are clearly limits to the potential for PYO, which have probably been reached in certain parts of the country.

Adding value by processing

Value can be added by processing milk into cheese or wheat into flour. Historically, most rural households engaged in various forms of food processing and manufacture. After a period of decline there appears to be a reversal of a trend towards mass-produced foods and a growing demand for farm-produced foods. Following a decline in the quantity of on-farm Cheddar production, the market is now relatively buoyant and in a whole range of products the demand for specialist foods has shown considerable growth (Hill, 1985). Indeed the area of milk processing into new cheeses has been one of the features of the on-farm processing sector in the last decade, and growth in this field looks set to continue. Adding value in this way entails the acquisition of new technical and processing skills as well as business and marketing skills. It cannot be entered upon lightly, but it may offer significant opportunities to an agricultural industry adapting to changing consumer demand.

It is important that on-farm processors should understand the changing structure of food retailing and, in particular, the pre-eminent position of large supermarket multiples. There are a number of examples where farm-produced products have found their way on to the shelves of major supermarkets, but where a supermarket buyer takes up a large proportion of the produce of the enterprise, the producer is inevitably vulnerable.

Ancillary resources

Many farms contain resources which are not used productively for agricultural purposes but which could none the less contribute to farm incomes if developed as an alternative enterprise. Woodland, wetland or redundant buildings are all examples of resources that, if managed differently, may have the capacity to generate income.

Few farmers are aware of the value of timber crops or of the management needs to enhance product value. Governments have shown an increasing willingness to promote farm woodland by providing grant aid. There are now schemes to rehabilitate existing woodlands or plant new woodland. Inevitably with so much talk of surplus land, timber production is being actively considered. There may emerge new markets for timber. The promotion of renewable energy initiatives in the UK is stimulating interest in short-rotation coppice and, in the long run, an expansion of this type of forestry seems inevitable. Detailed survey evidence of the Farm Woodland Scheme in Scotland has shown how wide the range of costs can be. Where the farm workforce has the skill and available time to do much of the work without recourse to outside contractors, the costs of woodland planting can be reduced to virtually nothing (Appleton & Crabtree, 1991). (See Chapter 12 for further details of farm woodland management.)

'Wetlands' is a loose term covering everything from water bodies to wet pasture. Wetlands have alternative value as game areas, for fishing or for conservation and other forms of recreation. For some types of fishing there is no need for large ponds. The potential for fish farming is limited by site characteristics such as water quality, and most of the best sites have now been developed.

Many farm buildings are no longer useful for agricultural enterprises and could be developed for alternative uses. Not infrequently these buildings are converted to tourist use but there are possibilities of developing permanent residences or industrial buildings. Such changes of use of buildings require planning permission which may not always be obtained as local planners have become more fussy about the transformation of vernacular buildings; however, central government circulars have advised that, wherever possible, permission should be given.

Public goods

Public goods are so called because of certain characteristics which make public (i.e. state) support essential if they are to be provided. The distinguishing features of public goods are held to be the fact that non-payers can obtain benefits (non-excludability) and that one person's consumption does not adversely affect another person's welfare (non-rivalry). The environmental products of the countryside are rather like public goods in a number of respects. For example, people cannot be excluded from the public footpath system and an extra person looking at a rare bird such as an osprey does not (usually) diminish the first person's enjoyment.

As a result of the 'public-ness' of these goods, in the absence of public funding no-one can operate commercial enterprises based on them. The use of Management Agreements from the early 1980s to protect wildlife habitats and National Park scenery, and the developing array of agro-environmental measures from the mid 1980s, provide farmers with the opportunity to be paid for providing these beneficial environmental goods. In addition, the Countryside Commission offers payments under the Stewardship Scheme to farmers occupying land in specified habitat types in specified areas.

Most of the payments available for public good provision are not universally available. Eligibility depends on occupying land designated as environmentally valuable under one of the many designations including Environmentally Sensitive Areas (ESAs), National Parks, and Sites of Special Scientific Interest (SSSIs).

There is some debate as to whether farmers might not be able to be rewarded twice for the provision of public goods. Certainly farmers in environmentally attractive areas are likely to benefit from a greater potential to diversify into tourist enterprises than farmers in the middle of areas with few environmentally interesting features. There are, of course, problems in ensuring that the uncompensated owner of an environmental attraction can develop a related revenue-raising enterprise. In some cases this may be impossible. This explains why compensation is offered. Nevertheless, farmers should be aware of the possibilities of creating synergy between environmental (public good) diversification and commercial service-based diversification.

In all the groups of alternatives the farmer must be looking to explore the markets and the production requirements for his particular alternative. There are a number of questions that must be addressed.

- Is the enterprise likely to yield a better return on capital invested than conventional farming?
- Is demand expanding/likely to expand?
- Are there barriers to stop competitors getting involved? That is, does the enterprise possess any distinguishing characteristics (e.g. locational, special recipe, personal attributes of staff, etc.) which yields an advantage over competitors?
- Are there any health/welfare doubts?
- Is the enterprise environmentally acceptable/acceptable to planners?
- Is there any prospect of adverse government intervention?
- Are the business risks acceptable?

In each case the alternative should not be developed until these questions have been answered satisfactorily.

Management of alternative enterprises

The management of alternative enterprises has certain similarities with any farm management problem but there are also differences that must be understood. In many alternative enterprises there is, necessarily, a much greater emphasis on marketing than is the case with most conventional enterprises. Consequently, in addition to concerns with technical efficiency, which have underpinned the farming revolution of the last 40 years, there is a need for a much more highly developed marketing orientation to direct business behaviour at a farm level. As well as the need for a change in the orientation of the business, there are a number of more conventional management concerns which merit attention. Two of these, namely the challenges of establishing an alternative enterprise and the ways in which monitoring and control procedures can be made effective, demand particular attention.

Marketing

Marketing skills have not historically been a major part of farm management training, nor have they been in evidence in the farmer's normal repertoire of business skills. In a supported market and with frequently undifferentiated products this is understandable. Farmers have tended to think about quantities of output and have had a production orientation to their business activity. With alternatives, a different business philosophy should prevail. Markets are unsupported and products often differentiated, and a marketing orientation provides the key to the successful establishment of alternatives. It is not sufficient to think of marketing as an afterthought, born of unsold stocks. Marketing skills and principles should be embodied in a marketing plan which details the business/enterprise strategy with regard to alternatives.

Whilst some alternatives may be founded successfully after an entrepreneurial vision, a safer bet is likely to be market research. Demand should be established from published sources or from market research commissioned or carried out personally. Information on competitors should be collected and it may be possible to visit establishments (preferably in another part of the country) offering a similar product. It is important to consider the market area in which the product will be sold. In the case of farmgate sales it is very local. With a specialist ethnic food it may be international. The marketing plan should be built on this market research. Marketing experts suggest an examination of the four Ps: product, place, price and promotion. Each should be considered in turn (see Chapter 2).

The resource base

Alternative enterprises are founded not just because markets exist but because farmers have resources which can be combined to provide them with a profit (hopefully) and the consumer with a good or service. A thorough appraisal of the resource base of the farm is an essential stage in reviewing options for alternatives. The resource base includes the intrinsic qualities of land, the skills and aptitudes of the whole farm workforce and the capital available for investment.

There is an inevitable tendency for farmers to think about resources in terms of their agricultural value. Land is considered conventionally by its Agricultural Land Classification (ALC) grade, not, for example, by its silvicultural potential. Yet quality of land must be considered in relation to each of the alternative uses. Areas of high tourist and recreational potential may have very low agricultural value. The location of land in the UK, with the exception of outlying islands, has ceased to have a major impact on most types of agricultural production. With respect to alternatives, location is an attribute of land that is of crucial importance in many cases. The importance of location is most marked in enterprises which require the consumer of the diversified product to come to the farm. Location should be considered at at least two levels: the regional level which indicates the general suitability of the area; and the micro level where the exact location of the farm with respect to highways, viewpoints, nearby features of wildlife or archaeological interest, etc., are taken into account.

The skills of the farm household as a whole should be taken into account when considering alternatives, as should those of any hired workers. Retraining may be necessary but there may be skills related to dealing with people or livestock husbandry skills which can be deployed effectively in alternative enterprises. When looking at the labour skills, the seasonal demands that some farming systems create should be considered, for if there is a clash with peak demands for alternatives, the new enterprise is unlikely to be established successfully. There may be an opportunity cost in employing family members in farm-based enterprises where it is more economically rational for them to find off-farm work.

The capital resources consist of the land itself, the buildings associated with the land and any other capital resources in the owner's possession that could be committed to the business and any capital that can be raised from normal sources of borrowing and grant aid. The capital value of land or buildings may be much enhanced by planning permission and neither should be sold to raise capital without first ascertaining development possibilities. Capital can also be locked up in woodlands.

The critical question that should be answered by the resource audit is whether the existing enterprise mix appears to be making the best possible use of the total bundle of resources that comprises the farm and the wider household economy. The audit should yield clues as to the opportunities that should be selected for more detailed appraisal.

Establishing alternatives

The establishment of alternatives should be guided by a marketing orientation and preceded by a thorough resource audit. It is desirable that the fullest use is made of sources of advice. There is unlikely to be a reliable textbook figure to reveal the expected profitability, although attempts have been made (Parker, 1986; Williams, 1988; Williams, 1993). Any figures offered should be used with caution. Consequently, advice from public, private or voluntary agencies should be sought. The pre-eminence of ADAS in conventional agriculture is not matched by a pre-eminence with regard to alternatives. Thus, with a prospective tourist enterprise, the farmer could go to ADAS, the regional Tourist Board or the Rural Development Commission or any number of private sector consultants. Searching out the various agencies can take time, and some agencies, such as NFU Marketing, have developed a useful signposting role.

As well as being uncertain of whom to ask for advice, it may be equally difficult to ascertain where to seek capital or grant aid. Here again signposting agencies are likely to be important. In the late 1980s MAFF embarked on a broadening of the range of grants for diversification and for a while farm diversification grants were given. The demise of these grants in 1993 has left farmers to seek other sources of grant aid. However, in 1994, MAFF launched a Market Development Grant Scheme, which offers discretionary grants of 50% of the cost of approved projects for marketing agricultural produce, up to a maximum of £150 000. Individual businesses or groups of farmers are eligible.

There has been much talk of streamlining and co-ordinating the agencies involved in providing support but this is unlikely to materialize in the short term. However, there have been a number of inter-agency initiatives to create 'one-stop-shop' information and advisory centres, and in the EU-designated Objective 1 and 5b areas inter-agency collaboration may be boosted. In the longer run it seems likely that rural development policy and grant aid will be less specific to farmers than is currently the case. Consequently, it will be vitally important that diversifying farmers maintain an awareness of the total array of business support services, not just the conventional agricultural set that has itself changed dramatically to include environmental grants and farm woodland grants amongst others.

It is not possible to recommend financial evaluation techniques that will be appropriate for every alternative. Much will depend on the level of capital invested, the impact on other farm enterprises and the level of riskiness of the investment. In some cases, where there are only modest investments of capital and there are no knock-on consequences on other enterprises, a simple payback estimation may suffice. In other cases, where the diversification is of a much larger scale, it may be essential to produce detailed business plans including projected cash flows for a number of years ahead.

For a variety of reasons, alternatives are likely to be more risky. As a consequence, it is desirable to conduct sensitivity analyses which assess the outcomes of changes in the key variables determining profitability. Once this sensitivity analysis has been conducted, it should be possible to make a more rational decision (for full details see Slee, 1989).

Running alternatives

Once establishment problems are over, the management challenge of alternatives does not cease. Indeed, many recently established alternatives have arisen out of desperation about the current economic climate confronting farming, rather than a detailed plan for change including diversification. As a result the monitoring and control problems of both new entrants and established practitioners are likely to grow.

The balance between supply and demand for many alternatives has yet to be struck. It is unlikely that it will be struck easily for there is a shortage of accurate supply and demand data. There are no census data for most alternatives to inform us of supply statistics, and market research is often rudimentary. Where the total demand for a product is small compared with conventional agricultural products, oversupply could easily arise. Where markets are unprotected, competition can be harsh, as new entrants into alternatives have sometimes found out to their cost.

Operators can come to terms with competition in a number of ways, both collective and individual. Farm holiday groups producing collective brochures may reduce advertising costs and offer the potential tourist a package of possibilities. A number of Farm Attractions Groups have been established after pioneering developments in Northamptonshire. Food organizations such as Devon Fare have helped individuals to market their products. At an individual level it may be possible to respond by trying to differentiate the product and offer something different or exclusive to the consuming public.

Monitoring and control are important functions of management. They are especially important where there are difficulties of measuring business performance. The performance of conventional enterprises can be contrasted with farm management handbook standards. Many alternatives cannot. Thus, operators must set their own standards, measure performance accurately and explain deviations between actual and anticipated results. Good budgetary control is fundamental to good management. It should provide the information required to enable appropriate adjustments to be made.

One important consideration with alternatives is the determination of the optimal size of enterprise. The 'bigger-is-better' philosophy does not necessarily apply, especially where the initial development of the alternative is based on exploiting a slack resource. Empty bedrooms may create bed and breakfast space and cost little to put in order. Putting up new accommodation by extensions to farmhouses can dramatically increase capital requirements. Direct sales can yield retail price premia. Sales to a wholesaler generate lower prices. As an enterprise grows, so it may have to shift from the retail market to the wholesale one, thus reducing margins. There are, of course, cases where economies of size exist, such as in deer farming or woodland management, but they cannot be taken for granted.

The control and motivation of the workforce can be especially important where enterprises require contact with the public. This arises with tourist and recreational enterprises or in the case of PYO enterprises and farm shops. Social skills are often more important than technical skills. Technical skills cannot, however, be ignored, especially with new crops such as borage where seed-bed preparation and careful harvesting are vital.

In many alternative enterprises the raison d'être is a personal interest of the farmer or a member of the family. This interest may lead to the acquisition of basic knowledge and skills. The same interest may misdirect business behaviour towards self-indulgence. The separation of work and play may be a necessary prerequisite for a healthy balance sheet.

There are certain bureaucratic differences between alternatives and conventional farming. If the enterprise is not a farm enterprise, it is likely to be rated. If a change of use is to take place, planning permission may be required. Opening a farm shop entails satisfying a host of public bodies from planners, to weights and measures inspectors, to public health inspectors. Recent attempts by government in the UK to 'lift the burden' of legislation must be balanced against the desire of the consumer for some kind of protection and by the increasingly onerous requirements of new legislation such as the Food Safety Act.

Two examples of diversified enterprises

Farm tourism

Recent reports in the press have argued that the market for farm tourism is approaching or has reached saturation. Whilst it might be more appropriate to use such arguments against more conventional mainstream farm products, the examination of the market is a necessary starting point of any assessment of the prospects for alternatives. Having examined the market for farm tourism, key management concerns will be identified and the principal causes of variability in profits considered.

The demand for farm tourism must be explored within the framework of tourist demand in general. The general issues have been outlined above. Recent work on farm tourism by Denman & Denman (1993) has yielded a great deal of useful information about the market for farm tourism in England, and the general conclusions are likely to be relevant to other parts of the UK.

It is convenient to break down the market for farm tourism into three main segments: the domestic holiday tourism market; the business tourism market; and the overseas visitor holiday tourism market. Each of these exhibits different demand patterns, both in terms of the product mix sought and the regional variations in the relative size of the market of each segment.

For domestic holidaymakers using serviced accommodation, peace and quiet, an attractive rural setting and value for money were all rated as very important by more than 60% of respondents in the Denmans' survey. Plentiful wholesome breakfasts were also regarded as very important. Less than 20% of those sampled rated a working farm environment around them as very important, and these respondents frequently had young children. This reinforces evidence from studies carried out earlier in the 1980s. However, in spite of the relative unimportance of the farm characteristics, many of those questioned had specifically sought out farm accommodation. Although over 30% of visitors requested en-suite bathroom facilities, the greatest demand, from over 60% of visitors, was for the unsophisticated service of tea/coffee making facilities in the bedrooms. The facilities sought by the average visitor related as much to the surrounding recreational opportunities such as walking or looking at wildlife on the farm as to the formal provision of visitor services. Thus what emerges is a demand from domestic holidaymakers for a rural holiday experience, supported by friendly service. Visitors sought high standards but not standardization, and were concerned that the basic requirements of hygienic, clean, warm accommodation with comfortable beds should be met. It appears that the essence of the experience is being in someone's attractively located home as a guest, rather than an inmate of a tourist institution.

For domestic holidaymakers in self-catering accommodation the same three most important features and the same relative unimportance of the farming activity emerged. In terms of the facilities of the self-catering accommodation, the most important items specified were the provision of linen, washing machines and drying facilities for wet clothes. The quality of the furnishings mattered to the overwhelming majority of respondents. Compared with the serviced sector, the attractiveness of the building was important and there was a distinct preference for traditional cottages on their own, rather than complexes in converted buildings, chalets or caravans. The same kind of surrounding countryside attractions were sought as by those using serviced accommodation.

The business tourist is only interested in serviced accommodation. It is difficult to generalize about this group, but it can be typified by the free-spirited, probably professional individual who dislikes hotels. For business visitors, value for money was the most important characteristic, followed closely by peace and quiet and wholesome breakfasts. Their demands for accommodation facilities are similar to those of holidaymakers, with the exception of their much greater insistence on a television in their bedroom.

It is equally difficult to generalize about the demands of overseas visitors. The principal factors attracting the overseas holidaymakers were peace and quiet, an attractive rural setting, value for money, the character of the place, and interaction with the family. Their demands for facilities were greater than those of UK holidaymakers with a higher proportion seeking en-suite (shower rather than bath) facilities. These visitors were not, in general, interested in farming but a significant proportion were interested in local foods.

These different segments make up different proportions of the serviced farm tourism market in different regions. In parts of south central England less than half of all tourists are domestic holidaymakers, whereas this rises to over 80% in the north of England and the West Country. Overseas tourists are most important in the south and east of England, and comprise over 30% of the market in the south-east of England. Business tourists also comprise over 30% of the market in two tourist board regions, namely the north-west and south-central England.

The majority of farmer households involved in serviced farm tourism had seen an expansion in their enterprise in recent years. Those enterprises which were members of the Farm Holiday Bureau had witnessed more growth than those that were not.

Self-catering accommodation has also experienced generally positive trends in recent years. Occupancy levels were generally good, with Farm Holiday Bureau members exceeding 70%+ occupancy levels from April to October inclusive. Independent operators exhibited a much narrower summer peak level of bookings and thus lower average occupancy levels.

Variations in returns to farm tourist enterprises are considerable, with evidence available to show that whilst perhaps the poorest performing 20% are making a loss on the enterprise, if the unpaid family labour is costed, the top 20% of operators are making very satisfactory returns. One of the tasks of management is to identify those variables over which some degree of control can be exercised. For example, for many farm tourist enterprises location will be an important factor determining overall occupancy levels and will thus influence overall levels of profit. It is also a factor that cannot, under normal circumstances, be altered as the subsidiary nature of the tourist enterprise, and the limited mobility of farm households, will mean that the location of the enterprise is effectively fixed. In contrast, advertising expenditures are controllable variables, with the enterprise manager having to select carefully the advertising mix that will generate the highest returns.

Survey evidence from Scottish farms which have diversified into tourist enterprises (Wilson, 1990) points to high levels of fixed costs, principally depreciation of buildings and fittings, which can adversely affect profits. Clearly, there are very different costs of establishing tourist enterprises from farm to farm. In some cases there are major repair and renewal costs to convert semi-derelict buildings to tourist use. In other situations the accommodation is available in a well maintained family home, with minimal costs needed to create a new enterprise. If budgets hint that depreciation costs and interest charges on any borrowings cannot easily be met by returns received, no development should be countenanced. There may, however, be ulterior motives for establishing the enterprise, such as long-run capital gain, or maintaining the capital stock of buildings.

Gross output of tourist enterprises is determined by the occupancy rate multiplied by the letting price. Occupancy rates tend to be strongly influenced by regional norms but letting prices are much more variable. Critical to any choice of prices should be the tourist's perception of value for money. The promotional mix used should make it absolutely clear what bundle of attributes is being purchased, including whether (in self-catering accommodation) heating is inclusive, etc.

In assessing performance one of the most useful indicators of success is the extent to which new customers are attracted by word of mouth (i.e. personal recommendation). In exceptional situations well over half of all visitors will be recruited from this source. If the figure for personal recommendations begins to fall, this should be taken as a warning that there are serious problems in the enterprise which must be urgently addressed.

A critical decision facing many operators, particularly of serviced accommodation, is whether to try to move their product upmarket, by providing en-suite facilities. Prior to making such a decision, the demands of the clientele should be considered. Is the visitor using the accommodation because it is cheap, because it is on a farm or because he/she is guaranteed a superb breakfast? In areas where there is intense price competition from a large number of operators there may be more cost-effective ways of improving the value-for-money offer and differentiating the product.

Farm tourist enterprises will continue to be the single most important form of farm diversification. Although the market for farm tourism is relatively mature, it still has capacity for growth. The image of the farm as a tourist destination is still associated with overwhelmingly positive images, which are normally reinforced by the use of such accommodation. This strong image should be wrapped up with the multitude of other interesting opportunities that a countryside destination affords, including scenic, historic and wildlife attractions, speciality foods and cultural interests. The image must not only be carefully constructed, but also delivered to the potential consumer in the right form. The pivotal role of the Farm Holiday Bureau as a co-ordinating agency should not be overlooked. In addition there are a number of private agencies that are particularly important in the self-catering sector.

The future for farm tourist operators will be competitive but there are strong grounds for believing that well-managed enterprises have the capacity to contribute significantly to the well-being of farm households.

Horse enterprises

The scope for diversification into horse-related enterprises is considerable. There are over two million horse riders in the UK, many of whom do not own a horse. Of those that do own a horse a significant proportion (40% in a 1993 British Horse Society survey) do not own their own grazing. Consequently, there is a demand for a whole variety of horse-related provision ranging from grazing, at one end of the spectrum, to thoroughbred studs at the other. In this brief analysis the commonly occurring forms of horse-related diversification will be identified and the principal factors affecting their profitability will be explored.

In the first instance a number of general concerns are outlined. First, although there are many horse enterprises that are run commercially, the industry is characterized by many small enterprises for which profit making is not a primary objective. Consequently, this non-profit seeking sector may impact on the profit-seeking sector adversely by pulling down prices to unrealistic levels and by creating competition.

Secondly, there is some evidence that there are significant economies of size in horse-related enterprises. Survey work in 1993 has indicated that many commercial operators have attempted to expand their enterprise size in order to realize these size economies. For example, Cross (1992) asserts that the labour requirements of various horse management duties halve as the number of horses increases from one to twenty. In addition there are other size economies in feed purchasing, marketing, etc.

Thirdly, the horse industry is peopled with many committed and skilled individuals. Those without the requisite skills should rarely contemplate horse-related enterprises, although in exceptional circumstances it may be worth contemplating letting someone else carry the risk of the enterprise by letting out buildings and land for horses, managed by someone else. Also it should not be forgotten that there are at least two types of skill required: equine management and husbandry on one hand and people management skills on the other.

Fourthly, there is abundant scope for product differentiation with horse-related enterprises. The differentiation may take many forms, including the general type of enterprise, the challenge of the mounts, the location of the rides or the ambience of the stables. However, differentiation is only worth pursuing if it can be sustained and should only be pursued if it offers opportunities for competitive advantage.

Fifthly, because of the geographical variations in potential for riding enterprises, with concentrations in suburban areas and holiday areas, there are likely to be situations where competition is very hard as different establishments compete for trade. Given that there are distinct economies of size and the costs of entry into some types of horse-related enterprises is very low (e.g. do-it-yourself livery), such competition is to be expected.

Table 6.5 Approximate livery charges; 1992 values

Type of Livery	*Weekly charge range* £
DIY	17–22
Grass	12–15
Full	45–65

Source: Derived from Cross, 1992.

Livery enterprises can take a variety of forms ranging from year-round care including stabling, feeding, bedding and exercise to do-it-yourself seasonal livery which provides grazing and nothing else. Detailed costings of various types of livery and all horse-related enterprises can be found in the *Horse Business Management Reference Handbook*, published on a regular basis by Warwickshire College (Cross, 1992). Figures used below are all derived from this manual. Table 6.5 shows the prevailing level of charges that can be anticipated.

There are few differences in the fixed or variable costs between average performing and high performing enterprises. The principal difference arises from the ability to charge a premium charge rate because of the quality of service offered.

Riding schools are often complex enterprises to analyse as there is no standard type of enterprise. In some cases there may be working livery horses available for use, when they are not required by their owners. This clearly affects the cost structure of the enterprise. As is the case with livery enterprises, the higher performing riding stables tend to be larger and more intensively run and are able to charge premium fees up to 20% above the average of £9.50 per hour (1992 prices).

Trekking establishments represent the third type of horse-related enterprise. Except on the largest holdings, the opportunities are limited to where there is a good network of bridleways or adjacent open country over which rights of access exist or can be obtained. As with both of the other types of horse-related enterprises, the high performers are those that can charge higher daily rates and still command above average occupancy rates (of the saddle). Indeed the higher performers have higher levels of variable costs but these are more than compensated by the much greater number of trekking days per mount per year.

In recent years returns to riding enterprises in general have tended to fall, although it is not impossible to buck the trend. Indeed, in a recent study of 76 equine enterprises (McInerney and Turner, 1991), the average enterprise made a negative return, if family labour costs were imputed. This tendency for returns to fall is partly owing to the recession-induced drop in demand, and partly owing to new entrants including those who have used farm diversification grants to reduce the costs of establishment. It is a highly competitive market that offers little hope of windfall profits. For the committed and the skilled there may still be opportunities, but there remains a need to be wary about turning time- and money-consuming hobbies into commercial ventures.

Summary

Alternative enterprises are likely to become more important to the farming industry of the 1990s. Pressures of budgetary crisis linked to overproduction in conventional products are so overwhelming that the industry is searching hard for solutions to the problems, and alternatives have a part to play.

The interest and the response on the ground make it very difficult to make generalizations about prospects for individual alternatives. The volatility of the free market for most alternatives exacerbates the difficulties of making long-term appraisals of possibilities.

It is all too easy, though, to resort to hackneyed arguments in dismissing even the thought of diversification into alternatives. Demand may be uncertain; risks may be considerable; markets may not be large. At least some of the dismissals of alternatives may result from the ability of some to take up existing slack in enterprise management. This is not especially difficult in the first instance, but subsequent stages may be more painful and difficult.

At the earliest opportunity a thorough resource audit, guided by a shift towards a marketing orientation, should be carried out. This is not a one-off process but something that should be ongoing as the prices for different products vary in the future. The full range of opportunities embracing both alternatives and conventional farm enterprises should be considered.

Alternatives will not provide salvation for all but they may offer lifelines for some. Those who can adapt their thinking and develop their business skills may well use alternatives as a stepping stone into an uncertain future.

References

Appleton, Z. & Crabtree, B. (1991) *The Farm Woodland Scheme in Scotland: an Economic Appraisal*. Scottish Agricultural College: Economics Report No. 27.

Carruthers, S.P. (Ed.) (1986) *Alternative Enterprises for UK Agriculture*. Centre for Agricultural Strategy, Reading.

Cross, P.L. (1992) *Horse Business Management Reference Handbook*, 4th edn. Warwickshire College, Warwick.

DART Rural Planning Services (1974) *Farm Recreation and Tourism in England and Wales*. Countryside Commission, Cheltenham.

Davies, E.T. (1983) *The Role of Farm Tourism in the Less Favoured Areas of England and Wales*. University of Exeter, Exeter.

Denman, R. & Denman, J. (1993) *The Farm Tourism Market*. The Tourism Consultancy, Ledbury.

Gasson, R. (1986) *Farm Families with Other Gainful Activities*. Wye College, London.

Hill, B. (1989) *Farm Incomes, Wealth and Agricultural Policy*. Avebury, Aldershot.

Hill, S. (1985) *Specialist Foods*. Institute of Grocery Distribution and National Farmers Union, Watford.

Lane, B. (1989) *Will Rural Tourism Succeed?* Paper presented to the Regional Studies Association.

McInerney, J. & Turner, M. (1991) *Patterns, Performance and Prospects in Farm Diversification*. University of Exeter Agricultural Economics Unit, Economic Report No. 236.

McInerney, J., Turner, M. & Hollingham, M. (1989) *Diversification in the Use of Farm Resources*. University of Exeter Agricultural Economics Unit, Economic Report No. 232.

MAFF *et al.* (1993) *Agriculture in the United Kingdom 1992*. HMSO, London.

Parker, R.R. (1986) *Land – New Ways to Profit*. CLA, London.

Shoard, M. (1979) *The Theft of the Countryside*. Temple Smith, London.

Shoard, M. (1987) *This Land is Our Land*. Paladin, London.

Slee, R.W. (1989a) Alternative Enterprises: Their Role in the Adjustment of UK Agriculture. Paper presented to Rural Economy and Society Discussion Group, Annual Conference, Oxford.

Slee, R.W. (1989) *Alternative Farm Enterprises*, 2nd edn. Farming Press, Ipswich.

Williams, J. (1988) *Diversification Guide 1988. New Businesses for Farmers and Landowners*. Cambridge Publications, Swavesey.

Williams, J. (Ed.) (1993) *New Farm Businesses: The Ventures that Work: An Introduction, Case Studies and Selected Directory. Farm Development Review* Special Issue, September 1993.

Wilson, C.J. (1990) *Incomes from Farm Diversification in Scotland*. Scottish Agricultural College: Economics Report No. 25.

Part 2

Crop Production

7

Soil Management

R.J. Parkinson

Introduction

Fertile soil is essential for sustained agricultural production, and has been exploited as a growing medium since animals and plants were domesticated. Cultivation of soils in order to create improved conditions for the sowing of crops can be dated back to at least 6000 BC (Goudie, 1986). Primitive farmers exploited the natural build-up of soil fertility by growing crops on land that was allowed to remain uncultivated for several years after the previous crop was harvested. The intensification that followed the enclosure of land and the adoption of sequential cropping patterns in rotation has ultimately led to greater demands being placed on the soil to supply the needs of the growing crop and to allow machinery access over much of the cropping year.

Today it is possible to grow winter wheat continuously on many soils, with the addition of fertilizers for plant nutrition and agrochemicals to control weeds, pest and diseases. Over the centuries, crop yields have increased dramatically, but our understanding of basic soil properties has not always increased proportionately. During the 1960s and 1970s evidence began to accumulate which indicated that soil fertility could decline under intensive agricultural use, mainly due to modifications of physical properties that had not previously been considered important (e.g. MAFF, 1970). Soil erosion became more common on arable land, particularly under continuous winter cereal cropping. The challenge for farmers in the 1990s is to manage the soil in such a manner as to maintain yields while reducing unnecessary applications of fertilizers and other agrochemicals. This can only be done with a full appreciation of soil physical, chemical and biological properties. In this chapter soil properties are described and the principles of good soil management are discussed.

Soil physical properties

Soil components

Soils are a complex mixture of mineral materials and organic matter that evolved over long periods of time, in some cases thousands of years, as a result of interactions between parent materials, soil organisms and climate. Analysis and understanding of the physical properties of soils and the way they respond when cultivated depend on a detailed knowledge of the basic soil components and how these components are arranged. A fertile soil will contain a mixture of mineral particles, organic matter and void spaces. These voids may be occupied by air or water. In Fig. 7.1 the proportions of these components are given for a well managed topsoil.

The properties of most agricultural soils are dominated by weathered mineral material. The primary components comprise the *texture* of a soil. Some soils, such as peats, are composed mostly of organic matter, in which case a texture description, based on the mineral fraction, is not appropriate. The combination of primary components with organic matter forms the *structure* of soils. In this section the agricultural significance of texture and struc-

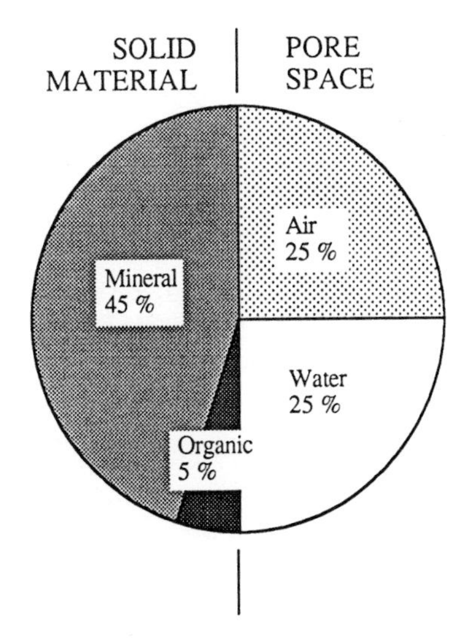

Fig. 7.1 Components of a well managed topsoil, expressed on a volume basis.

ture will be described, followed by other important soil physical properties, most of which are influenced by texture or structure, and may be modified by agricultural practices.

Texture

Mineral particles in soil range widely in size from large stones to minute clay fragments. The proportion of various sized particles has a major impact on soil physical, chemical and biological properties. Hence the description and characterization of soil texture is of primary importance to farmers and growers. Texture determination can be carried out approximately, but rapidly, by hand texturing in the field, or quantitatively by laboratory analysis.

Field determination

Hand texturing is a rapid but relatively crude method of determining soil texture. A soil sample is moistened in the fingers and rubbed in order to break down any natural structures that exist. Coarse sand grains can be seen with the naked eye, and will make the sample feel 'gritty'. Fine sand and silt feel smooth, and can make the sample 'slippery'. Clay binds soils together, and imparts a sticky feel. A description of these different size components is given in Table 7.1. Standard texts such as Simpson (1983) describe the feel and characteristics of texture components in detail. Accurate hand texturing requires technical skill and years of practice. Care must be taken when hand texturing, as some organic matter fractions feel slippery or soapy, similar to silt-sized particles. Mineral soils with high organic matter levels can feel finer in texture than is actually the case. Having described the texture it is possible to determine important soil management characteristics. These are described in the section 'Texture and soil management'.

Table 7.1 Particle size and characteristic properties

Particle	*Diameter, mm*	*Characteristic properties*
Gravel or stones	>2.0	–
Coarse sand	0.2–2.0	Coarse builder's sand or beach sand, particles clearly visible
Fine sand	0.06–0.2	Egg timer sand, just visible with naked eye
Silt	0.002–0.06	Flour, visible with hand lens
Clay	<0.002	Plasticine or putty, visible using electron microscope

Table 7.2 Relationship between particle diameter and surface area

	Surface area $m^2\ g^{-1}$
Fine sand	0.1
Silt	1.0
Kaolinite clay	15–20
Montmorillonite clay	700–800

Laboratory determination

Laboratory determination of texture or particle size distribution by a process known as mechanical analysis is time consuming and is only carried out when detailed information is required. The procedure is based upon sieving and sedimentation in water after thorough disaggregation. MAFF (1986) describe a three phase process. First, the soil sample is passed through a 2-mm sieve to remove stones. The analysis continues on the fine earth fraction that passes through the sieve. Secondly, the soil is dispersed using hydrogen peroxide (which destroys organic matter) and sodium hydroxide (which separates individual particles). Finally, the suspension of soil and water is passed through a series of sieves or a process of controlled sedimentation is carried out, so that the precise proportion of different sized mineral particles can be determined following dry weighing. The equivalent diameter of sand, silt and clay particles is shown in Table 7.1. This is the system employed in the UK (Avery & Bascomb, 1974); in other countries different size classes are used. The results of a particle size analysis can be plotted on a graph which has been subdivided into named texture classes (Fig. 7.2). A plot of percentage clay versus sand includes the silt component by difference, as the total must add up to 100%. For example, a soil containing 40% sand and 30% clay also contains 20% silt, and is described as a clay loam. Having defined texture in a quantitative manner, it is possible to predict more accurately the behaviour of the soil in specific management situations.

Mineral materials

The mineral components of soils derive primarily from the underlying parent material. Hence the character of a soil will depend intimately on the type of rock from which the soil has been formed. In simple terms, rocks composed of coarse particles, such as sandstone and granite, will produce coarse-textured sandy soils. Mudstones and slates will produce fine-textured soils. The chemical character of a soil is also related to the size and origin of the mineral components. Sand and silt-sized particles tend to be dominated by only a few primary rock-forming minerals, of which quartz (silicon dioxide) is the most common. Quartz is chemically unreactive, and therefore plays no part in nutrient retention by soils. In contrast, the clay fraction is dominated by clay minerals that have a varied chemical composition and can be chemically and physically very reactive. Chemical reactivity of soil particles is discussed further in the nutrient-retention section of this chapter. The potential for soil mineral particles to be physically and chemically reactive is partly explained by the increasing surface area with decreasing particle size. Table 7.2 shows the theoretical relationship between particle diameter and surface area. Clay minerals that have an expanding lattice structure, such as montmorillonite, exhibit large surface areas that allow water and nutrients to be retained, while sandy soils possess a small surface area and tend to be chemically inert. Silt is intermediate in behaviour, and does not have the large surface area possessed by clay, nor does it have the often beneficial properties attributed to sand (see next section for more details). In practice, soils are a mixture of all three mineral components, although one may dominate and hence control soil behaviour in the field.

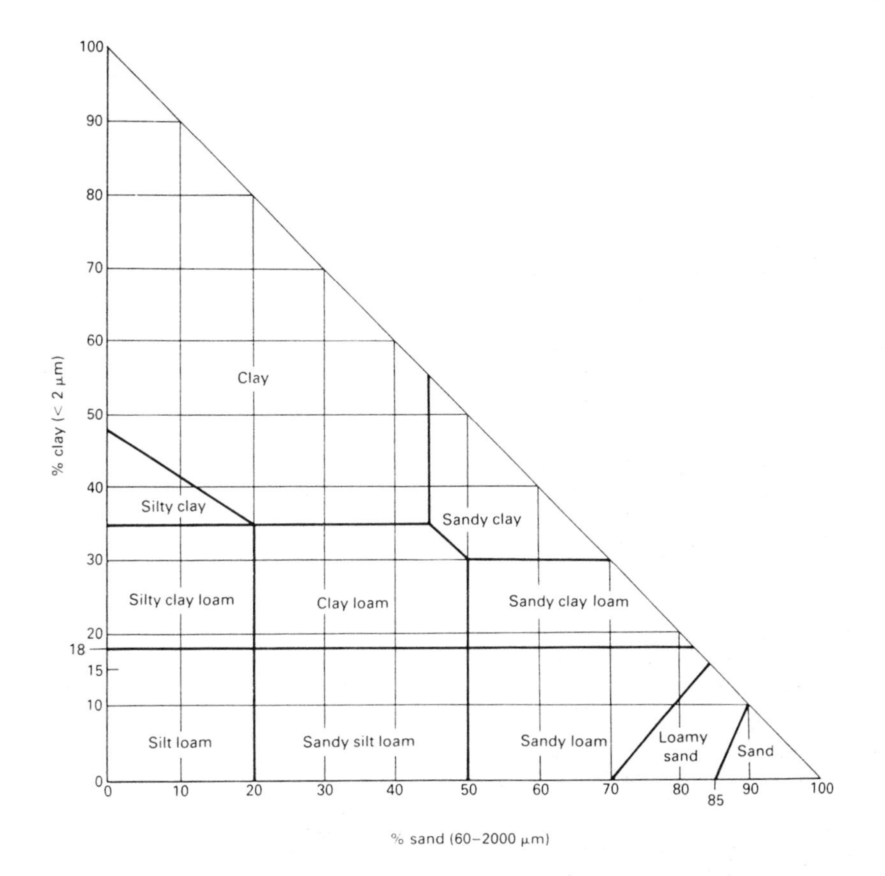

Fig. 7.2 Texture diagram according to the system used in the UK (after Mullins, 1991).

Texture and soil management

Texture exerts a profound influence on soil management. Ultimately the choice and flexibility of cropping as well as potential yields all depend on soil texture. The most important impacts of texture are given below. Many of these relationships are discussed further in other sections of this chapter, and in more detail in Davies *et al.* (1993).

Drainage status

Clay-rich soils retain water against gravity and therefore have high water contents during the winter months. At such a time a clay loam might hold 50–60% water by weight, in comparison with a sandy loam, which might hold only 25–30%. High water contents lead to reduced oxygen levels and poor plant growth (see 'Drainage' for more discussion of the consequences of poor drainage).

Water availability to plants

While finer textured soils retain more water than coarser soils, much of it is held so tightly (due to capillary forces) that crop plants cannot extract the water. Consequently, silt soils tend to have the highest reserves of plant available water, and are the most drought tolerant. Sandy soils tend to be very droughty due to the small volume of stored water.

Workability and trafficability

Access to land in the critical autumn and spring months can be controlled by soil texture. Sandy, well drained soils have few access restrictions, making them suitable for the growth of a wide range of arable crops. Clay textured soils are often difficult to cultivate, being too hard when dry and too soft when wet, and frequently cannot support the weight of agricultural machinery during the winter months (Reid & Parkinson, 1981). Cultivation operations must be timed very carefully in order to minimize the risk of soil damage due to cultivation when too wet.

Nutrient retention

Clay-sized particles retain nutrients very effectively, while sandy soils are often described as 'hungry', that is, fertilizer nutrients need to be added frequently but in small doses in order to supply the requirements of crop plants (see nutrient retention section for further discussion).

Soil texture describes the fine earth fraction (particles <2 mm). Stones or gravel are therefore excluded from this discussion, but quite clearly the presence of significant proportions of stone-sized material can have a major influence on the growth of root crops, and cultivations. Many of the glacially derived soils of northern Britain contain significant quantities of stones which limit the choice of crops. No hard and fast rule can be given as to the amounts of stones which will restrict crop choice or influence cultivations, as stone type and distribution are equally important.

Soil organic matter can influence many of the properties listed above. High levels of organic matter can increase nutrient retention, increase water retention and make soils more workable (see the organic matter section of this chapter).

Structure

Formation of soil structure

As soils develop, mineral particles of sand, silt and clay are mixed with organic matter by soil organisms. This mixing process creates stable aggregates and hence the soil structure. Young soils, such as might be found on a river flood plain or sand dune, tend to show little evidence of structure development, but most soils in agricultural use are well structured. Stable structures develop over long periods of time as soils shrink and swell throughout the year and as organic residues are combined with mineral particles. A visual representation of soil structural components is given in Fig. 7.3. This figure helps to explain the important process of structure formation by the intermixing of soil mineral and organic components. Certain soil characteristics aid structure development, namely organic matter (most important), clay, calcium carbonate and iron oxide (least important). A well structured soil will display fine, equigranular structures in the topsoil, with progressively larger aggregates in the lower horizons of the soil. Figure 7.4 shows several example soil profiles broken down into structure types.

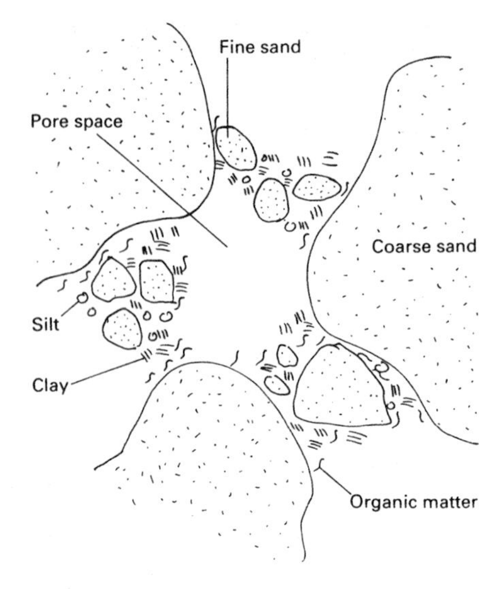

Fig. 7.3 Detailed representation of soil structure formation.

Importance of soil structure

Well structured soils display high porosity, low density, adequate water storage, free drainage and movement of air within the soil profile. Plant roots are able to exploit the whole soil volume, and will display a fine fibrous root system. The fine stable aggregates near the soil surface will be resistant to collapse and therefore will allow free passage of water and air through the surface layers. Roots will extract water and nutrients from within the aggregates, while excess water will drain away through

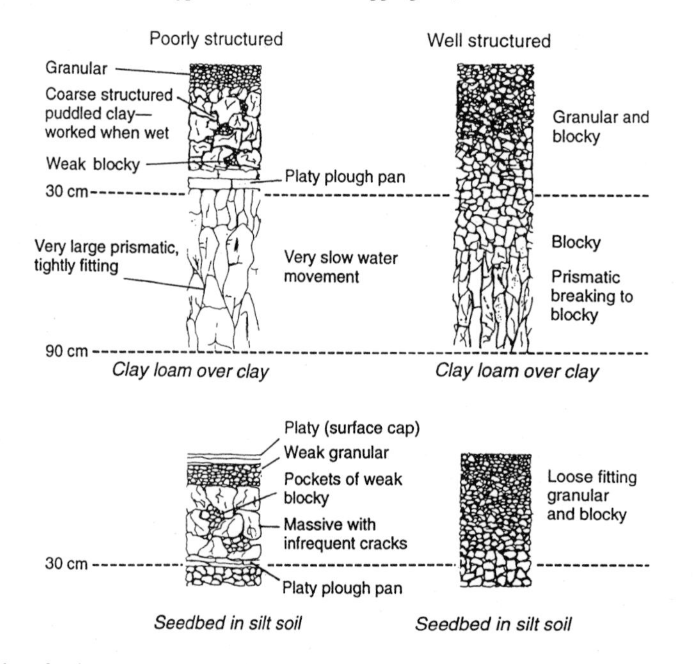

Fig. 7.4 Some examples of soil structure types under good and poor management (after MAFF, 1970).

gaps between the aggregates, resulting in no prolonged periods of waterlogging. The importance of structure cannot be underestimated in agricultural production systems, although assessment of soil structure is a difficult task. Experience can be gained by frequent field soil examination, particularly of soil profile pit faces that have been allowed to weather naturally for a few days, after which time natural structure patterns become more visible.

Modification of soil structure

Many agricultural practices modify soil structure; the creation of a seedbed by ploughing and secondary cultivation, for example, is direct modification of soil structure which is designed to suit the needs of the seed to be sown. Unlike texture, structure is not permanent. Soil aggregate stability can be changed by the cultivation and intensive use of soils. Serious structure breakdown can lead to soil compaction and soil erosion. Many soils in the UK have suffered from problems relating to modification of structure by farming practices. The intensification of agricultural activity in the latter part of the 20th century has led to some deterioration of soil structure. Most notably, arable cropping has caused organic matter levels in some soils to decline, due to the removal of crop products such as grain and straw. This reduction has become critical for many sandy soils which tend to have naturally low organic matter contents. Rates of organic matter decline are variable, but soils with organic matter levels of less than 5% are susceptible to compaction and soil erosion.

Tilth is the term applied to the finely structured surface soil that has been worked down by cultivation implements to create ideal conditions for the germination and growth of crops. Unfortunately the repeated cultivation of soils destroys natural, stable aggregates, resulting in weaker, finer structures that ultimately may collapse, hence leading to a deterioration in the soil physical environment. It is therefore important that soil is not over-cultivated, particularly for autumn-sown crops where soils are exposed to the full force of the winter weather with minimal protection from a growing crop.

Soil structure can be improved by a number of agricultural practices: use of (long) grass leys in rotations, adding manures and other organic materials, adding lime, and the use of deep-rooting green manure crops. All these will help to stabilize the soil and maintain fertility.

Soil density

Figure 7.1 gave a typical breakdown of soil components for a well structured topsoil, which is made up of 50% solids and 50% pore space. As the structural properties of a soil change under agricultural management, so will the density and pore space. As roots need to access water and air held in these pore spaces, an understanding of such changes is important.

Bulk density

Bulk density is defined as the weight of oven-dry soil per unit volume, and depends on the densities of the constituent soil particles and, most importantly, how these constituents are packed together. Bulk density is usually determined by extracting a soil core of known volume, oven drying at 105°C for 24 hours to remove the soil water and then weighing the core. Values of bulk density range widely, as shown in Table 7.3. In general, root access to soil pore spaces becomes difficult above a bulk density of $1.5-1.6\,t\,m^{-3}$. Soil compaction results in an increase in bulk density. In Fig. 7.5 some example bulk density profiles are given for a soil that has suffered surface compaction due to excessive animal grazing (Profile B) and a soil that has been compacted at plough depth due to repeated cultivation when soil conditions were too wet, resulting in smearing and structure destruction (Profile C).

Table 7.3 Bulk density and total pore space values for agricultural soils

Bulk density $t\,m^{-3}$	*Pore space* %	*Description*
0.5–0.8	>70%	Loose, uncompacted topsoils. Peats and organic soils
~1.0	60–65%	Permanent pasture, woodland soils, well structured
~1.5	45%	Compacted, root penetration difficult
~2.0	25%	Dense, no root growth

Under good management, bulk density will tend to reduce until an equilibrium value is reached for a given soil and cropping situation. Bulk density is simply an indication of the general status of the soil in physical terms: to understand the effect on crop plants, it is necessary to describe pore space changes associated with increases in bulk density.

Given bulk density it is possible to calculate the weight of soil in a given area. For example, assuming a bulk density of $1.0\,t\,m^{-3}$, 1 ha of soil down to plough depth (200 mm) weighs 2000 t. This represents a considerable mass of material that is moved every time the soil is ploughed.

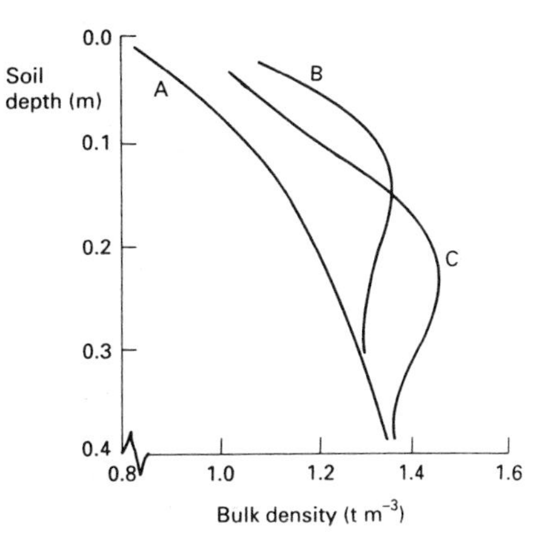

Fig. 7.5 Example bulk density profiles. A: Uncompacted soil; B: Heavily grazed soil showing surface compaction; C: Cultivated soil showing compaction at plough depth.

Particle density

It is necessary to know the density of mineral particles (or when unweathered, solid rock) in order to calculate the pore space in a soil. Particle density is defined as the mass per unit volume of mineral particles in soil, and does not include air spaces or water. Typical values for particle density range from 2.50 to 2.70 $t\,m^{-3}$. For example, a 1-m^3 solid granite block weighs approximately 2.65 t. A value of 2.65 $t\,m^{-3}$ can be assumed for quartz-dominated mineral material in British soils. Organic matter weighs considerably less than mineral material; values of 1.2–1.5 $t\,m^{-3}$ are common for organic material. Given this difference, it is important to state the organic matter content of a mineral soil when describing the pore space of soils.

Pore space

The size, shape and arrangement of soil aggregates controls not only the density but also the total porosity of a soil. When density changes as a result of soil management practices, porosity also changes. Total pore space is defined as the volume of pores expressed as a fraction of the total soil volume, and is usually determined by measuring the bulk density and assuming a particle density of 2.65 $t\,m^{-3}$.

$$\%\ \text{Pore space} = (1 - (\text{bulk density}/\text{particle density})) \times 100$$

For the example given in Fig. 7.1 the total pore space is 50%. By substitution into the equation above, this equates to a bulk density of 1.32 $t\,m^{-3}$. A well structured topsoil under permanent grass with a bulk density of 1.0 $t\,m^{-3}$ would have a total pore space of 62%; i.e. greater than half the soil is pore space. Increases in bulk density that are associated with compaction lead to reductions in pore space. Further examples of increasing bulk density and decreasing pore space are given in Table 7.3.

Total porosity does not provide any direct information about the size of the individual pores, or their function. It is simply an expression of the total volume of a soil that may act as a store of air or water.

Water retention

Water in the soil occupies pore space. The mechanisms by which water is held in soils and then released to plants or allowed to drain out of the profile depend upon the size of the pores. Pores can be classed as having various functions according to their approximate diameter; these are given in Table 7.4. The larger pores (macropores), for example drying cracks and earthworm burrows, allow excess water to drain out of the soil as gravitational forces exceed the low capillary forces in such large pores. These voids are very important during the winter months when heavy textured soils are prone to waterlogging. Macropores tend to be the first to be lost when soils are compacted, hence leading to drainage problems. In addition, as Table 7.4 shows, these pores allow air to enter the soil profile.

The smaller pores (<60 μm diameter) will store water against gravity, due to capillary forces. These forces become stronger the smaller the pore diameter. Eventually a point is reached where the pore is so small that plants cannot overcome the capillary force, and so any water contained in that pore is considered to be unavailable. Figure 7.6 displays the relationship between soil texture and the quantity of water, expressed as a percentage of the soil volume, that is available (held in pores between 0.2 and 60 μm) or unavailable (held in pores <0.2 μm). These differences are of vital importance in cropping systems; sandy soils may have only 5–10% available water, while at the other end of the range silty soils may contain 20% available water. In a dry summer these differences may lead to crop failure or poor yields on sandy soils while heavier textured soils may be able to maintain crop growth throughout a drought period.

As well as soil texture, the depth of a soil exerts a controlling influence over the total amount of water that is available to plants in the soil profile. Table 7.5 gives example calculations of the total available water that might be stored in three example soils. Soil A is coarse textured but deep, while soils B and C are fine textured, although soil C is only 0.5 m deep, below which weathered rock is encountered. The impact that texture and soil depth have on the amount of water available to a plant is clearly seen in this example. The need for irrigation is dependent upon soil texture and hence available water capacity, as well as the climate and weather patterns in a given area. See Chapter 24 for further discussion of irrigation principles and practice.

Table 7.4 Pore size and function

	Diameter μm	*Function*
Macropores	>60	Water transmission, aeration
Mesopores	60–0.2	Water storage – available to plants
Micropores	<0.2	Water storage – not available to plants. Influences mechanical properties of soil

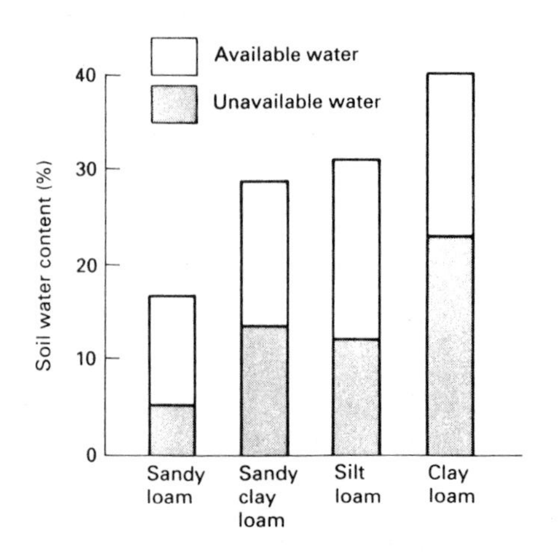

Fig. 7.6 Soil texture and available water capacity.

Table 7.5 Total plant available water for three example soil types

Soil depth mm	*Available water, mm* A. *Sandy loam*	B. *Silt loam*	C. *Silt loam (0.5 m deep)*
0–200	40	80	80
200–500	21	45	45
500–1000	25	50	—
Total	86	175	125

Aeration

Soil air differs from the free atmosphere above the soil surface in being enriched in carbon dioxide and sometimes depleted in oxygen. A steady supply of oxygen is needed in soils for root respiration and to support aerobic bacteria which carry out important functions in the soil (see section on micro-organisms). If the volume of macropores that allow drainage and aeration falls below 10%, it is likely that there will be insufficient air for root respiration. Any practice that increases the volume of air-filled pore space, such as soil loosening, is to be encouraged.

The consequences of poor aeration are that soil conditions change to the detriment of most crop plants, although the susceptibility to poor drainage varies from crop to crop. Breakdown of organic matter in anaerobic conditions can lead to the production of organic acids and ethylene (C_2H_2), both of which inhibit root extension. Nitrate may de denitrified to gaseous nitrous oxide (N_2O) or dinitrogen (N_2), and lost from the soil system.

Soil strength and cultivation

Careful management of soils will lead to continued high productivity and few physical problems. However, it is not always possible to cultivate under optimum conditions, and it is sometimes necessary to keep animals out on the land when high soil water contents will lead to damage occurring. The ability of a soil to resist deformation and damage depends upon the relationship between water content and strength. This relationship is unique for each soil type. In general, high soil water contents lead to low strength, and hence greater susceptibility to potential damage under applied load, for example a tractor tyre or an animal hoof. Figure 7.7 displays this general relationship for a well structured soil. Optimum conditions for cultivation occur when the soil will fail in a brittle, friable manner, as indicated by the asterisk in Fig. 7.7

Various problems can occur if trafficking or animal trampling ('poaching') occurs when the soil is at a high water content such that deformation occurs in a plastic rather than a brittle manner. Some of these problems are described below, and are discussed in more detail in Davies *et al.* (1993).

Cultivation pans
A cultivation pan or plough layer will form following repeated cultivation under conditions that are too wet.

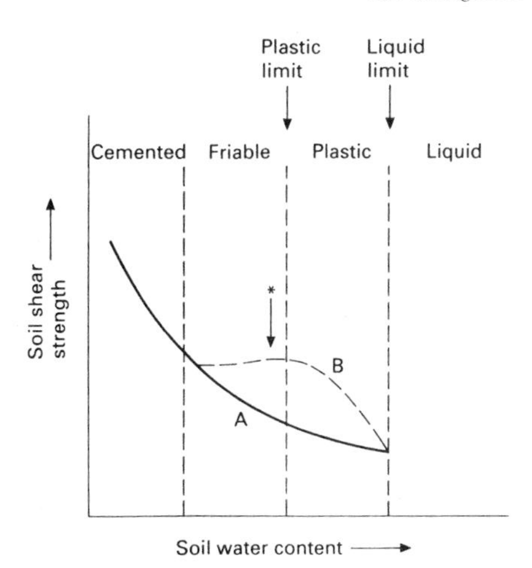

Fig. 7.7 Variation in soil strength with water content. A: Soil aggregate; B: whole soil, if well structured.

Structural aggregates will be destroyed and a compacted smeared layer will be formed (see Fig. 7.5). Such pans can occur in soils of any texture, but they are most common in heavy textured soils that tend to have higher water contents. The presence of a pan will lead to temporary waterlogging, even in better drained soils, as well as higher bulk density, both of which will restrict root development. Alleviation of the problem is usually by soil loosening to below the depth of the pan under dry conditions, when brittle failure will occur.

Surface compaction
Surface compaction may be due to machinery access or livestock grazing of land that is too wet, resulting in plastic deformation, smearing and compaction of the surface layer (see Fig. 7.5). In consequence, aeration will be restricted and surface runoff may occur. Given the opportunity, soils will naturally 'restructure' at the surface, as a result of wetting and drying cycles that occur in all soils during the year. Removing the cause of the initial damage and allowing the soil to recover naturally is often all that is needed in such circumstances.

'Puffy' seedbeds
Overcultivation of soils with high sand and silt contents can lead to structure breakdown with very light, puffy seed-beds. In such cases rolling may be necessary to ensure good soil:seed contact following drilling. Failure to roll may result in uneven germination and poor establishment.

Machinery work days

Autumn and spring are the critical periods of the year when soil damage is possible due to high water contents at times when essential cultivations need to be carried out. The concept of machinery work days describes the potential cultivation opportunities that exist for given soil types. These opportunities are measured during the main autumn and spring period, and describe those occasions when harvesting, tillage and drilling operations can be conducted without risk of structural damage to

Table 7.6 Machinery work days for three example soils in south-west England (after Findlay *et al.*, 1984)

Soil association and texture		*Machinery work days* *Autumn* *(1 Sept–31 Dec)*	*Spring* *(1 Mar–30 Apr)*
Bridgnorth	loamy sand	86	29
Frilsham	sandy clay loam	73	18
Hodnet	silty loam/clay loam	32	3

the soil. Soil Survey regional bulletins (see further reading) contain a detailed description of the system employed in England and Wales. These descriptions form a basis for the discrimination of soils according to their flexibility of management at those critical times of the year when access to the land is necessary in order to be able to establish and then harvest arable crops. Table 7.6 lists the machinery work days for three soil types that occur in south-west England. Texture can be seen to exert a strong influence on the number of machinery work days.

Drainage

Drainage need and benefits

Why drain?

The removal of excess water from soil by an artificial drainage system can reduce soil management problems and increase crop yields. The need for drainage in the UK arises either from heavy textured soils retaining excess winter water, or from a high water table, for example on a river floodplain. Heavy textured soils have an ability to retain nutrients and provide the water needed to satisfy the requirements of demanding crops, and hence are capable of producing high yields, particularly of cereals and grass. The majority of $10\,t\,ha^{-1}$ wheat crops have been obtained from clay loam or clay-textured soils, but only with efficient drainage systems to remove excess winter water from the soil.

Annual patterns of water loss and gain are very variable across the UK. As a result of higher rainfall and less sunshine, the soils in western Britain tend to pose more drainage problems than do those in the east. This does not mean that the arable soils in eastern Britain do not need drainage; the greater demands placed on soils in arable cropping systems has led to the need for increased flexibility of soil management in arable situations, and hence many soils in eastern Britain have been drained prior to being cropped intensively.

The first drainage systems date from Roman times, but it was only after the large-scale enclosure of common lands in the 18th century that drainage became widespread (Trafford, 1970). Techniques used varied widely, with most systems relying on stones, straw or other bulky material to form and stabilize a channel to remove excess water. The mechanical production of clay tiles in the mid-19th century led to dramatic improvements in the quality of drain pipes, and 100-year-old systems can still be found that are working well today. Estimates of areas drained in the 19th and early 20th centuries suggest that up to 50% of the agricultural land in southern England was drained, all by hand. Mechanical drain installation became the norm by the middle part of the 20th century, when 50 000–100 000 ha were drained annually (Trafford, 1970). The current rate of drainage is very much less, in the region of $10\,000\,ha\,a^{-1}$, due to the progressive reduction in grants for drainage schemes and changes in the economics of arable crop production in the 1980s and 1990s. In addition, the conservation value of wetland habitats is now more appreciated than formerly, such that decisions to drain areas of land are no longer based solely on considerations of likely yield increases after drainage. The conservation value of species-rich grasslands, for example, may depend on land remaining undrained. In some cases grants are now available to encourage farmers to choose this option (e.g. under the Countryside Stewardship Scheme). The following discussion refers only to land of little conservation value where drainage is needed to improve soil management and raise yields.

Drainage benefits

Lowering soil water contents by drainage results in several changes in the crop rooting environment, all of which are beneficial to the crop and soil management (Parkinson, 1988). These benefits are summarized below.

Duration of waterlogging

Autumn-sown crops rely on efficient water table control to allow the development of a root system. Failure to control waterlogging in the late autumn/winter/early spring period will result in a stunted root system unable to exploit deeper seated water and nutrient reserves during the following summer.

Soil workability and trafficability

A reduction in water content will increase soil strength. Successful cultivation usually depends on the soil water content being below the lower plastic limit (see Fig. 7.7). Drainage lowers the water content of the topsoil, resulting in increased opportunity for landwork during the critical autumn and spring months. In grazing systems, the benefits from drainage will come from the increased number of days that livestock can graze a sward without risk of damage to the soil structure.

Soil temperature

Drainage reduces soil specific heat capacity and therefore can lead to higher temperatures. Castle *et al.* (1984) noted a 2°C elevation in spring soil temperatures on a clay soil in eastern England when underdrained. Such an increase may lead to the more rapid germination and emergence of spring-sown crops, and may accelerate the development of winter-sown crops. It must be noted, however, that soil temperature does not always rise following drainage. In some field experiments no benefits have been found.

Efficiency of fertilizer use
More aerobic, warmer soil conditions will lead to more efficient use of applied fertilizers, particularly nitrogen top dressings. Growing roots will absorb nutrients more readily and less will be lost by leaching or denitrification (see nutrient availability section).

Arable crop yield benefits
The benefit obtained as a result of installing a drainage system can be most easily measured in terms of crop yield, but the variability of the British climate often produces a wide range of yield benefits from year to year. The yield advantage for most crops when comparing drained and undrained soils is 10–25%. For example, average winter wheat yields can be increased by 1.0 $t ha^{-1}$ (Armstrong, 1978). In wet years, however, efficient drainage can make the difference between crop failure and success.

Grassland/livestock benefits
In the wetter regions of the UK, drainage is essential in order to maximize grass utilization. The benefits can be expressed in terms of dry matter yield or liveweight gain, but it is also important to recognize other advantages. In particular, drainage can alter the composition of the sward (Table 7.7) as well as increasing the response to nitrogen. Table 7.8 shows the benefits of drainage for a clay soil in terms of bullock grazing days and liveweight gain. In addition to weight gain, other benefits in a livestock production system can include a reduction in the incidence of liverfluke and foot problems.

Table 7.7 Effect of drainage on the botanical composition of a grass sward three years after reseeding, on a clay textured soil

	Perennial ryegrass	*Weed grasses (% composition)*	*Other species*	*Bare ground*
Undrained	65	11	14	10
Drained	72	7	17	4

After Castle *et al.*, 1984.

Table 7.8 Influence of drainage and fertilizer application on livestock grazing days and liveweight gain as controlled by drainage of a clay soil in Devon

	Bullock grazing days ($d\,ha^{-1}\,a^{-1}$)	*Bullock LWG* ($kg\,ha^{-1}\,a^{-1}$)
Undrained	239	240
Undrained, fertilized	380	300
Drained, fertilized	451	433
Drained, reseeded, fertilized	433	428

After Trafford, 1970.

Drainage systems

Land can be drained by a system of ditches or pipes laid in the soil, or a combination of both. In the case of pipe drainage, additional short-lived measures, such as mole draining, can be carried out to increase the effectiveness of the drainage system. Both permanent and temporary systems serve specific purposes and must be installed following recommended guidelines (see Castle *et al.*, 1984). The principles and some of the practical points are outlined here.

Open ditch drainage

Most drainage systems find outlets into an open ditch which usually leads to a larger watercourse. These ditches are a vital component of a drainage system, and in some cases may be the sole method of water removal. Careful design, construction and maintenance are therefore very important. Ditches allow direct access for water, have a large capacity to carry storm flows and are easily maintained. However, they can hinder cultivations, they need fencing to exclude livestock and are susceptible to wall collapse and blockage by vegetation.

Ditch specifications
Design standards require that ditches must be of sufficient capacity for the catchment area drained. Theoretical capacity can be calculated given catchment size and design rainfall rate (see for example Farr & Henderson, 1986, p. 131). Ditch width and depth will depend upon soil and geology. The more stable the soil, the steeper the permissible slope. Some examples are given in Table 7.9. Ditches dug in sandy materials must have lower slopes than those in more stable, finer textured soil. Ditch floor gradient will in practice depend upon local topography. The gradient should be uniform and not too steep (leading to channel erosion) or too shallow (leading to silting). A gradient of 0.5–1.0% is generally considered to be adequate. Ditch sidewall collapse can be minimized by guarding against water erosion and livestock damage. Pipe drain outfalls should be fitted with splash plates; ditches should be fenced; and, in areas with highly erodible soils, the sidewalls should be grassed down. Ditches must be piped under roads and farm tracks. The pipes used, normally concrete, must be large enough to carry anticipated peak flows. The design flow can be calculated (see MAFF, 1982), but the minimum recommended pipe diameter is 225 mm. This size of pipe will serve a catchment area of up to approximately 12 ha.

Pipe drainage

Water movement in soil and into pipe drains
Pipe drains remove excess water without reducing the area of land cropped or interfering with field operations.

Table 7.9 Ditch channel side slope ratios and soil type

	*Channel side slope ratio** *Channel <1.3 m deep*	*Channel >1.3 m deep*
Fen peat	Vertical	0.5:1
Heavy clay	0.5:1	.1:1
Clay loam or silt loam	1:1	1.5:1
Sandy loam	1.5:1	2:1
Sand	2:1	3:1

* Horizontal:vertical ratio (h/v).
After Castle *et al.*, 1984.

Installing a permanent system of clay tiles or plastic pipes to carry water below plough depth can solve drainage problems efficiently and can be a worthwhile investment. It is particularly important that the principles of operation, design, installation and maintenance are all understood and followed, as a buried drainage scheme is difficult to inspect and maintain once installed. The successful operation of a drainage pipe system depends upon the flow of water from the soil surface into the drain itself.

Water flow in soil depends upon the hydraulic conductivity. In a uniformly permeable material, such as a sandy textured soil that is underdrained because of a high regional water table (as, for example, would be found on a river floodplain near sea level), water movement takes place through the whole soil (see Fig. 7.8a). In heavy textured soils with low rates of water movement, particularly in the subsoil, water tends to move in the topsoil and into the trench created by the drain-laying operation (see Fig. 7.8b). The bulk of the subsoil plays little part in the disposal of excess water. Rates of water movement depend on soil texture. Table 7.10 gives examples of water flow rates, class descriptions and soil types which are used in drainage design (MAFF, 1982). Water flow rates in coarse-textured soils can be 100 or 1000 times more rapid than in clay soils, hence reinforcing the need for drainage of many heavy soils.

An effective drainage system maintains the water table adjacent to the drains at drain depth. Between drains, the water table will be higher due to the restricted movement of water through the soil. Water enters a pipe system through either the gaps between clay tiles or the slits in plastic pipes (Fig. 7.9). A drain will only conduct water if the water table is at or above drain depth.

Table 7.10 Saturated hydraulic conductivity classification and soil type

Saturated hydraulic conductivity ($m\,d^{-1}$)	*Class*	*Example soil types*
<0.01	Very slow	Clay, silty clay*
0.01–0.1	Slow	Clay loam*
0.1–0.3	Moderately slow	Silty clay†
0.3–1.0	Moderately rapid	Sandy clay loam, clay loam†
1.0–10	Rapid	Loamy sand
>10	Very rapid	Gravel

* Poor structure.
† Good structure.

Materials

Clay or plastic? Clay tiles are normally cylindrical and 250 mm in length. Tiles are supplied in a palletized form, which allows for easy mechanical handling. The most

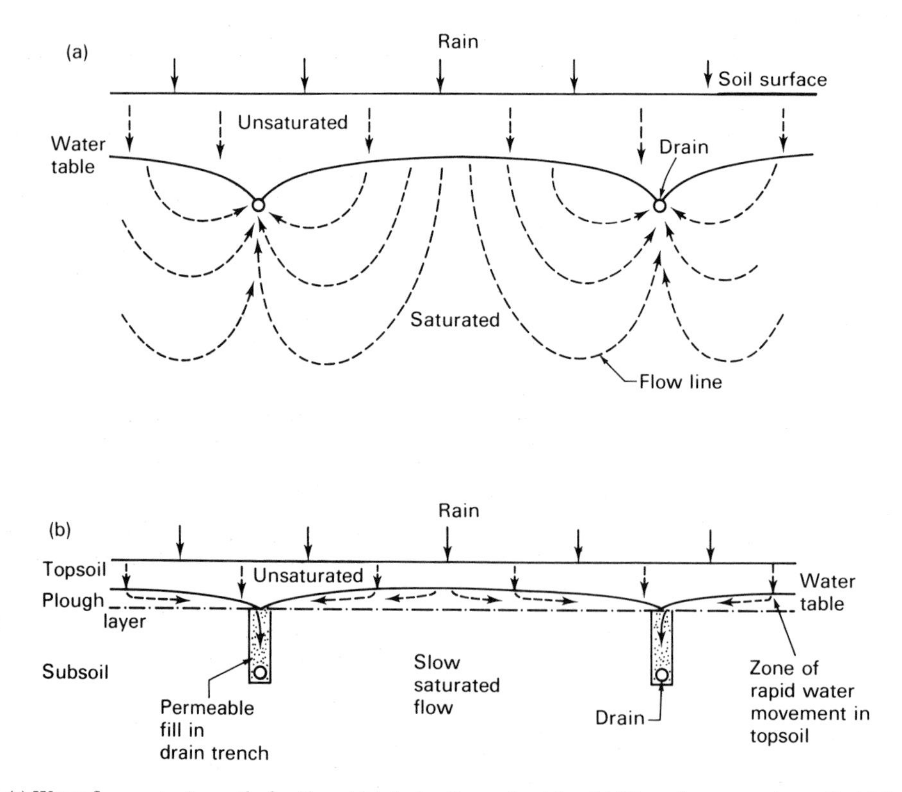

Fig. 7.8 (a) Water flow routes in a soil of uniform high hydraulic conductivity. (b) Water flow routes in a soil with low subsoil hydraulic conductivity.

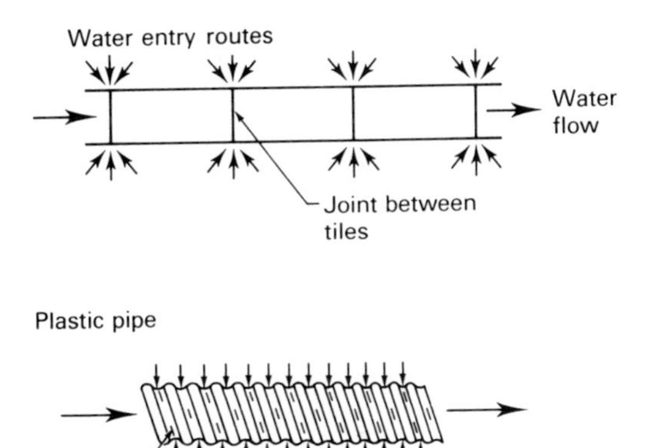

Fig. 7.9 Water entry routes into clay and plastic pipes.

common sizes are 75, 100 or 150 mm ID. Good quality tiles are essential, as the failure of one tile can cause the failure of a complete lateral. The tiles are laid in the trench to butt up to one another, but with a gap of 0.5–2.0 mm resulting from the uneven ends of successive pipes.

Plastic pipes have largely superseded clay tiles. The former have several advantages, being easier and lighter to handle and more suited for mechanical laying. In addition, disjointed drain runs are unlikely. However, the rough interior surface of a plastic pipe results in a lower hydraulic carrying capacity, and the material is inherently weaker. Slots run the length of plastic pipes so water entry is relatively unrestricted. Standard joints and junctions are available for both plastic and clay pipes, so that laterals can be led into main drains. Plastic pipes are available prewrapped with a filter, which may be necessary in silty soils to prevent blockage of the slots by sediment.

Permeable fill Some form of gravel is placed over pipes in about 60% of field drainage systems installed in England and Wales. As permeable fill may account for 50% of total installation costs, it is important to justify its use. Permeable fill acts as a connector to allow water movement from the topsoil to the drain pipe. In addition, backfill forms a permeable surround, to improve water entry to the pipe, and acts as a filter to prevent soil particles entering the drain. Washed gravel with a mean particle diameter of 20–50 mm is the most suitable permeable fill material. The use of permeable fill is not recommended for medium and coarse-textured soils.

Outfalls Where a pipe discharges directly into a ditch, the first 1.5 m of the pipe should be sealed, rigid and frost resistant. The pipe should be able to discharge freely, therefore the invert of the pipe should be at least 150 mm above the normal ditch water level. Where the ditch sides are unstable, the pipe should be supported by a concrete headwall. Failing this, the ditch side should be grassed down.

Installation

Soil and site conditions Drainage work should be carried out when the soil is dry and hence strength is high. Drainage through a crop can be justified in some cases if disruption to farming operations can be kept to a minimum. Under dry conditions compaction of the soil surface and smearing of the drain trench sides are unlikely.

Drain laying Pipes must be laid on a smooth firm bed, shaped to support the pipe. Permanent pipes must be laid at least 600 mm deep to avoid damage by moling or subsoiling operations. Clay tiles must be firmly butted together and located on one side of the drain trench to ensure water flow from one pipe to the next. Care must be taken with plastic pipes when installing them in air temperatures below 5°C, as they become brittle. It is essential to level accurately during the drainage operation so that a continuity of flow occurs. Laser levelling techniques allow grades of 0.1% to be attained, although for practical purposes 0.2% is the limit to which most machines work. Slopes of greater than 4% are likely to lead to erosion problems in the drain trench.

Trenched drainage A trench, normally 150–300 mm wide, is excavated using a vertical endless chain with blades attached to its links. Spoil is brought up by the chain and pushed out to either side of the trench by augers. Figure 7.10 illustrates trenched and trenchless drain installation. Work rates of up to 30 m h^{-1} are possible, although in weakly structured or stony soils progress may be considerably slower. Several important features favour the trenched systems of drain laying:

Advantages	*Disadvantages*
Pipes laid in trench can be inspected before backfilling	In unstable soils, trench wall collapse may occur
Old drainage systems can be found and tied in to the new system	Open trench requires more permeable fill than trenchless
Permeable fill can be placed in the ditch from a hopper to a uniform depth	Slower work rate than trenchless, hence more expensive
Grade of the trench can be easily checked	

Trenchless drainage The drain laying machine features a large plough which cuts a slit to a predetermined depth, and drops a plastic pipe down a narrow box behind the plough. If required, permeable fill is supplied from a hopper mounted above the pipe chute (Fig. 7.10). Currently, only 25% of drain installation is by the trenchless method, which is usually employed to lay close-spaced systems in grassland with small plastic pipes (<75 mm). Surface ridging caused by the passage of the drain laying machine will need to be rolled before normal farming operations can continue.

Secondary drainage

Moling

Heavy textured soils often require, in theory, a drain spacing of 5 m or less in order to dispose of excess winter rainfall. While it may not be economic to lay permanent drains at such a close spacing, temporary channels

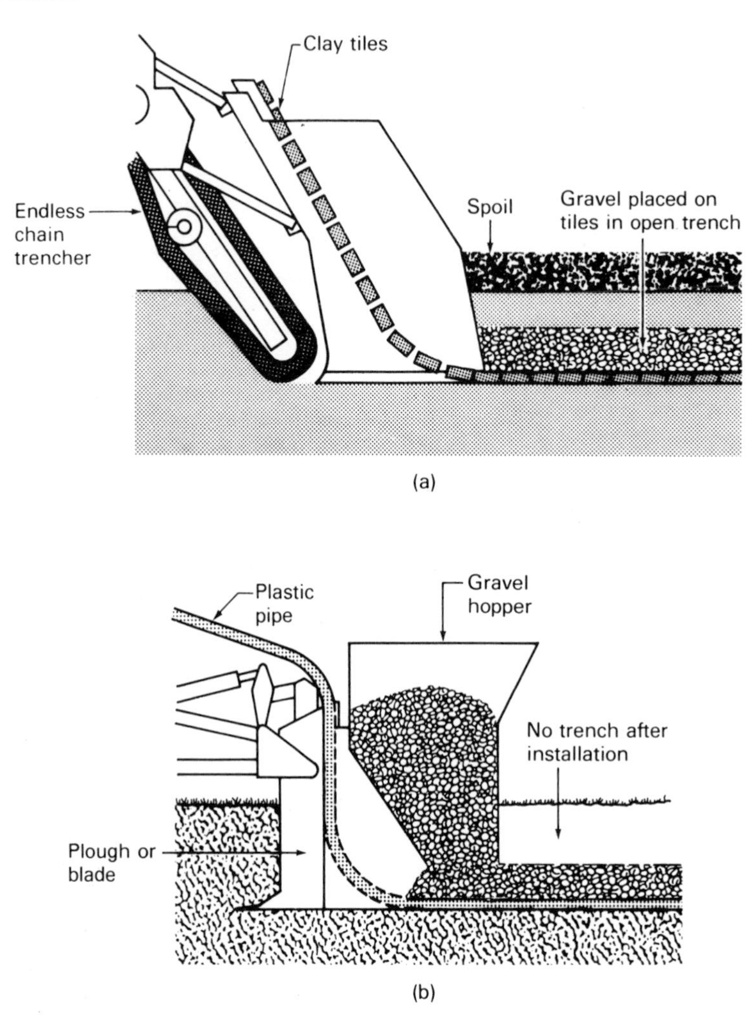

Fig. 7.10 Drain installation methods: (a) trenched; (b) trenchless.

(secondary drainage) may be employed. Mole drainage is a process whereby a system of 'moles' are pulled at 2–3 m spacing across a network of permanent drains that may be spaced at 30–40 m apart. Moling must be carried out under specific conditions and is only appropriate to certain soil types.

A mole plough is essentially a circular bullet followed by an expander which leaves a channel connected to the soil surface by a series of cracks (see Fig. 7.11a). Moling can only be carried out successfully in soils with a substantial clay content. Thirty per cent clay is commonly quoted as the minimum allowable (Castle *et al.*, 1984). Clays which expand and contract in response to changes in water content are more likely to provide stable long-lasting channels than non-expanding clays. Moling should be carried out at a depth of 0.5–0.6 m, at an angle approaching 90° to the field drain (Fig. 7.11b), with a minimum channel slope of 0.5%. These unlined channels are prone to collapse at angles <0.5% due to standing water.

The timing of mole installation is crucial to successful secondary drainage. Figure 7.12 demonstrates the effect of the passage of a mole plough through a soil with a dry surface layer but a moist subsoil. Such conditions will normally occur in early or mid summer in the UK. If a stable channel is formed at depth, it may continue to carry water to the permanent drain for at least five years.

Subsoiling

Subsoiling or soil loosening is not specifically a drainage operation, but can be described as any form of soil cultivation that is intended to shatter the soil beneath normal cultivation depth. When carried out as a drainage operation, the intention is to work the soil when dry so that extensive shattering occurs, thus creating a system of artificial cracks that will conduct water to the drain or at least to lower horizons. Subsoiling is most suitable for those soils which will not hold a mole channel; generally soils with a high silt or stone content. The winged subsoil shoe 'heaves' the soil, allowing it to fail under tension and so produce an extensive network of cracks. This operation is normally carried out after harvest on arable

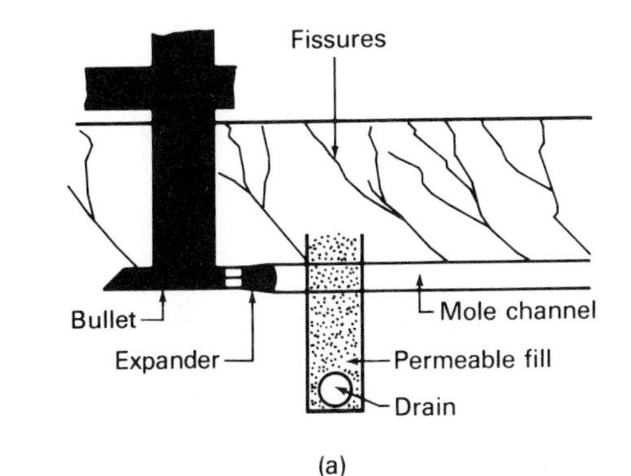

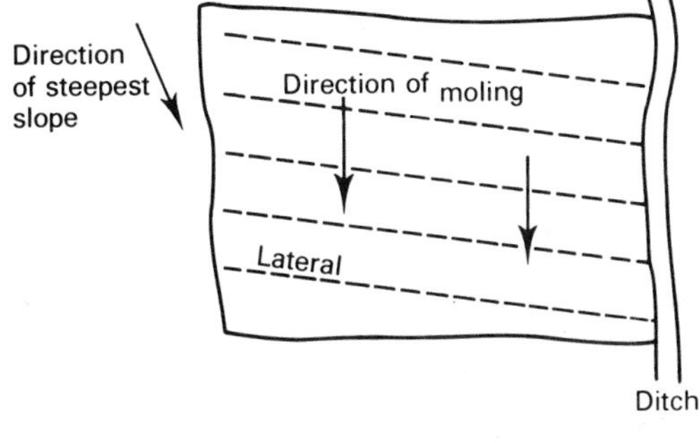

Fig. 7.11 Pulling a mole channel. (a) Section through soil illustrating connection with permeable fill. (b) Field plan of moling direction relative to laterals.

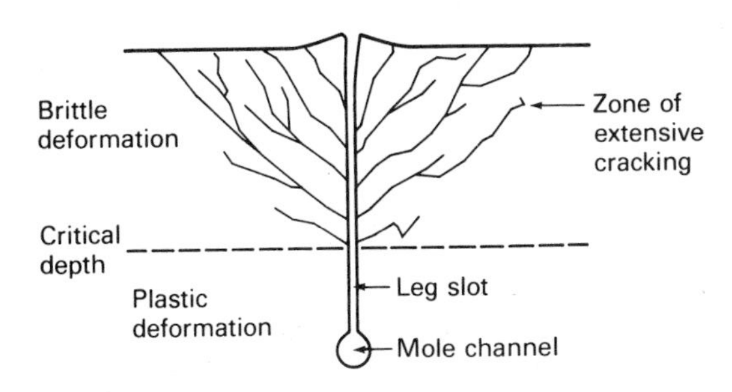

Fig. 7.12 Ideal mole channel configuration.

land. For further details of subsoiler design, see Chapter 24, and for information on the benefits of soil loosening, see Twomlow *et al.* (1991).

Drainage system design

Site investigation

Before carrying out the installation of a single drain or a complete drain network, it is important to identify the causes of the problem. Examination of the soil profile can yield much useful information. The presence of compacted pans, ochreous mottles and grey colours all indicate problems. A drainage contractor or independent drainage advisor will prepare a detailed solution to a particular problem, with a plan, but it is useful for a farmer to be familiar with the initial cause and extent of the problem.

Drainage systems

Regularly spaced systems are used for draining soils with low hydraulic conductivity or uniformly sloping land, with drains spaced according to the rate of water movement through the soil. In heavy textured soils such as clays and clay loams, which can hold a mole, permanent drains may be installed at 30–40 m spacing, with a secondary treatment superimposed upon that system (see Fig. 7.11). A regularly spaced system in a herringbone design can be designed to use the slope of the land to collect water.

Irregular/random systems

Complicated geology or undulating relief may necessitate the design of a random system, which may tap a spring line or drain a wet valley bottom (see Fig. 7.13). The resulting design may leave much of an area undrained, only removing water from natural collecting zones.

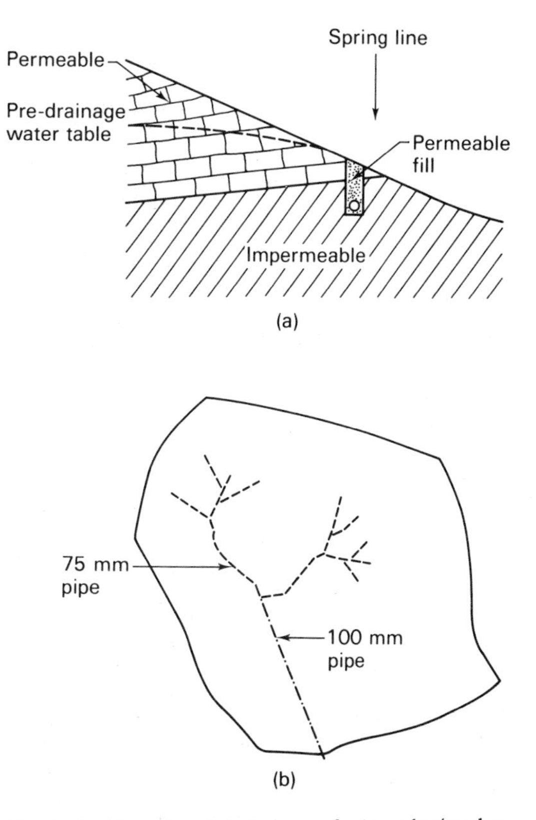

Fig. 7.13 Examples of drain layout for irregular/random systems. (a) Spring-line interception, in section; (b) random layout, in plan.

System design

Recommendations for the design of field drainage pipe systems are given in detail in MAFF (1982). Design criteria are based upon determining the size of pipe needed to conduct water away so that damage to crops is minimized. Pipe sizes for laterals, main drains and ditches can be calculated knowing the characteristics of the site.

- *Land use*: Design rates assume that horticultural crops are more valuable and more susceptible to damage than arable crops, with grass being most tolerant of waterlogging.
- *Soil type*: Soil texture and structure control the relative rate of water movement such that, for example, a well structured clay loam will drain more rapidly than a poorly structured silty clay loam.
- *Rainfall probability*: The probability of a certain amount of rain falling in one day varies across the UK. Drainage design is based upon keeping the water table below a certain depth in the soil so that a given system will be able to dispose of a specific quantity of rainfall in one day.
- *Ground and pipe gradient*: On steeper slopes water disposal is more rapid. However, pipes are often laid across the slope, so it is important to know these gradients when designing a system.

Given this information, and assuming a certain pipe spacing and drainage system to be installed, i.e. 40 m spacing with permeable fill and moles, the pipe size needed can be calculated. This pipe size will allow the system to deal with most (but not all) of the rainfall which can, on balance of probability, be expected in any year.

Drainage and conservation

Wildlife habitats may be affected adversely by land drainage operations. While the principle aim of field drainage is to reduce water contents to enhance crop growth, there are many established methods of reducing the impact of drainage on the flora and fauna of an area. Most important is to plan for conservation so that, for example, an old pond in the corner of a field can remain undisturbed during and after a drainage operation. Ditches can be designed with wildlife in mind, for example with a wide shallow section to carry stormflow. For further details see Castle *et al.* (1984).

Drain maintenance

A drain maintenance programme is essential in order to ensure efficient operation. Annual inspection will reveal any problems that have arisen due to soil instability, vegetation growth or animal disturbance. Ditches should be kept clear of excessive vegetation that will restrict water movement and might lead to silting. Fences should be kept in good order to prevent animal damage of ditch sides. Pipe drain outfalls should be kept clear of obstructions. Blockages can be cleared by rodding if necessary. Moles should be redrawn every five years, while subsoiling should be carried out as conditions and time allow and require.

Economics of drainage

In recent years there has been a steady reduction in the level of government grants available for drainage work, from 65% in the mid-1970s to 15% in 1994. This grant aid is only available when replacing or renewing an existing system, under the Farm and Conservation Grant Scheme. A full plan of the work must be submitted. Higher rates of grant apply in Less Favoured Areas.

Actual costs will depend upon lateral spacing, field size, materials used and soil type. Typical installation costs (1994 prices) for excavating a trench and supplying and laying an 80-mm plastic pipe are £1.05–1.25 m^{-1}. Adding permeable fill over the pipe to within 360 mm of the ground surface adds £1.40–£1.80 m^{-1} to costs (Nix, 1993). An intensive regular-spaced system, say 20 m spaced laterals with permeable fill, may cost between £1300 and £1500 ha^{-1}. Such a system could be expected to have a useful life of at least 25 years. Mole draining costs £45–£60 ha^{-1}. Digging a new ditch (1.8 m top width, 0.9 m depth) costs £1.60–£1.90 m^{-1}. Trenchless systems are cheaper to install, but quality control is not as good so the life of such a system may be less than for a trenched installation.

Current estimates of land drained or redrained each year (10 000 ha a^{-1}) are much less than that needed to be drained in order to replace old drains that have failed (40 000 ha a^{-1}). In the current economic climate, and given recent changes in land use policies and reduction of land values, it is unlikely that the level of drainage activity will increase to match the rate of system decay. As a consequence, soil management problems may become more severe if farmers attempt to continue using land intensively. Given recent moves towards more extensive land use, these problems may not arise.

Soil chemical properties

Crop requirements

Agricultural crops require a wide range of essential nutrients for growth. With the exception of carbon, which is obtained from atmospheric carbon dioxide, these are all stored and released from the soil. Some nutrients are applied commonly from fertilizer sources, such as nitrogen, phosphorus and potassium, while others such as iron, copper and manganese derive from natural soil sources. Full discussion of nutrient uptake processes can be found in Chapter 9, where a detailed list of macro- and micronutrients is given as well as descriptions of nutrient functions in the plant.

Crop requirements for individual nutrients vary widely, as shown in Table 7.11. Nitrogen is taken up by most crop plants in the largest quantities, closely followed by potassium. Crop removals quoted in Table 7.11 for the macronutrients are given in kg ha^{-1}, while removals of micronutrients such as copper or molybdenum are several orders of magnitude smaller. Nevertheless, these micronutrients are essential, and soils must be able to supply them to crop plants. In the absence of fertilizer additions, crops will still grow, indicating that natural soil processes will supply nutrients in the ratios required by plants—albeit at lower levels than the more demanding crops

Table 7.11 Removal of nutrients ($kg\,ha^{-1}\,a^{-1}$) by some crops (after Simpson, 1986)

Crop	*Yield* ($t\,ha^{-1}$)	*Dry matter yield* ($t\,ha^{-1}$)	*N*	P_2O_5	K_2O	*Ca*	*Mg*	*S*
Cereal	6 t grain 3.5 t straw	8	120	50	70	15	10	30
Sugar beet	40 t roots 25 t tops	12	200	45	240	70	25	30
Potatoes (tubers only)	50 t	10	180	50	240	10	15	20
Grass silage	30 t	—	160	40	160	45	15	15
Hay	—	8	100	30	120	30	10	10
Clover	—	5	180	25	120	100	15	15
Kale	50 t	10	200	60	220	250	20	100
Natural grass heath	2 t	0.4	10	3	10	2	1	2

Trace elements (g/ha) removed annually in crops:

Iron (Fe)	600–2000	Zinc (Zn)	100–400
Manganese (Mn)	300–1000	Copper (Cu)	30–100
Boron (B)	50–300	Molybdenum (Mo)	5–20

need to yield well. Nutrient availability depends on soil type, organic matter content and environmental conditions. Matching crop requirements to soil conditions is a difficult task, but one that must be attempted in order to utilize applied fertilizer nutrients efficiently and to minimize losses. Central to good soil nutrient management is the understanding of processes of nutrient retention and release. These processes are complex, and the discussion that follows is grossly simplified. For more detailed discussion, the reader is referred to the further reading section at the end of this chapter.

Nutrient retention – cation and anion exchange

The retention and release of nutrients to plants depends on a variety of soil and environmental factors, of which the chemistry of soil clay and organic matter is the most important. Soils are not inert media: they interact with applied agrochemicals in a very dynamic manner. Before individual nutrients are discussed in detail, the ability of soils to retain and release nutrient ions will be described. These processes not only control the supply of plant-available nutrients, but also act to modify the whole soil chemical environment, with important consequences for crop management.

Charges on soil surfaces

Soil is not an inert material. Most importantly, the clay fraction, iron oxides and organic matter can act in a chemically interactive manner, due to electrical charges at the surfaces of these particles. The origin of these charges depends upon the chemical structure of the material. Charges on clay particles are associated either with defects in clay lattice structures, or with unsatisfied broken bonds at the edges of clay platelets. In both cases the charge imbalance tends to be negative in British soils, resulting in an attraction for ions carrying the opposite, positive charge. Organic matter is likewise negatively charged. Only iron and aluminium oxides show a reversed pattern, tending to attract negatively charged ions due to these components having a positive charge. For a more detailed discussion of this topic, see White (1987).

Ion exchange processes

Nutrient ions are available for root uptake in solution—that is, from water stored in soil pores. This water will be in a quasi-equilibrium with the chemistry of ions held on charged surfaces described in the preceding section. As roots extract certain nutrients, then the equilibrium is upset, resulting in ions leaving the charged surfaces to re-establish the balance. This process is known as ion exchange. Exchange reactions can work in the other direction, so that fertilizer nutrients, for example, can be retained by soils until needed by plants.

Cation exchange

A cation is any ion with a positive electrical charge, such as the hydrogen ion H^+ or the sodium ion Na^+. Both H^+ and Na^+ are monovalent cations, that is, they possess only one charge. A cation may have more than one positive charge, such as magnesium, Mg^{2+} (divalent), or iron, Fe^{3+} (trivalent). The ability of a soil to exchange cations depends primarily on the quantity and type of clay and organic matter present. Exchange reactions in soils can be complex, but they can be simply visualized in terms of 'swops' of equivalent charged ions. Imagine that a clay particle has one calcium ion (Ca^{2+}) attached to its surface (see below). Rainwater percolating down through the soil usually contains dissolved hydrogen ions. This rainwater will wash around the clay particle, so that an exchange reaction may occur, whereby two hydrogen ions will replace one calcium ion. As a result, charge balance is maintained and the calcium ion will be washed further down through the soil profile and may be lost in drainage water. The reaction may go in either direction, as is indicated by the double-headed arrow. The reaction might be expected to move from right to left when lime is added to soil to counteract soil acidity.

$$Ca^{2+}[clay] + 2H^+ \longleftrightarrow H^+[clay]H^+ + Ca^{2+}$$

In acid soils, H^+ and Al^{3+} are the dominant cations, while in calcareous soils Ca^{2+} and Mg^{2+} are dominant. For short periods after fertilizer application, potassium (K^+) or ammonium (NH_4^+) may be dominant.

Soils with a large cation exchange capacity are chemically stable and nutrient retentive. Cation exchange capacities of some example soils are given in Table 7.12. Cation exchange is expressed in a variety of units; those quoted here are in terms of milliequivalents of charge per 100 g of soil. It is the magnitude of the differences between soil types as shown in this table that are important. Clay textured soils have an ability to hold at least five times as many cations on exchange sites as sandy textured soils. In this latter category, the importance of organic matter becomes clear; well humified (decomposed) organic matter can have a cation exchange capacity in excess of 200 meq 100 g^{-1}.

Table 7.12 Cation exchange capacities of mineral and organic soils

	Cation exchange capacity, milliequivalents 100 g^{-1} soil
Sandy loam	5–15
Sandy clay loam	10–20
Clay loam	20–40
Clay	30–50
Lowland peat	150–200
Upland peat	40–60

Anion exchange

Anion exchange occurs in soil to a much more limited extent, and is much less important in terms of any influence on soil chemical reactions. However, for individual nutrient anions such as NO_3^- and $H_2PO_4^-$ it can be locally important. Iron and aluminium oxide surfaces can carry positive charge sites that attract anions, and exchange reactions can occur, but this process is subsidiary to cation exchange.

pH, nutrient availability and liming

The acidity or alkalinity of a soil exerts a strong influence on nutrient availability and uptake, and hence on crop growth. The relationship between soil pH and crop growth is described further in Chapter 9. In this section the causes of this relationship are discussed.

The hydrogen ion concentration in soil solution reflects the soil chemical environment, and hence is used as an indicator of chemical conditions. pH can be defined as follows:

$$pH = -\log[H^+]$$

Expressed in words, pH is the negative logarithm of the hydrogen ion concentration, usually measured in a mixture of 1 part soil to 2.5 parts distilled water. The pH range of some natural and agricultural soils is given in Table 7.13. Most soils in the UK, with the exception of those derived from limestone or calcareous boulder clay, tend to be acid, due to the acidifying effect of rainfall, which is naturally acidic even when unpolluted by oxides

Table 7.13 pH range of some natural and agricultural soils (after Simpson, 1983)

	pH range
Sandy heathland	3.5–5.0
Calcareous brown soil	6.5–8.0
Upland peat	3.5–4.5
Cultivated soil, non-calcareous	5.0–7.0
Cultivated soil, calcareous	7.0–8.0
Permanent pasture, lowland	5.0–6.0
Permanent pasture, upland	4.5–5.5
Lowland peat	4.0–7.0

of sulphur or nitrogen. Soils with a low cation exchange capacity tend to become acidic most rapidly, that is, coarse textured soils. Organic soil such as peats can be either acidic or neutral/alkaline depending upon the origin of the peat.

Nutrient availability

pH affects nutrient availability in a very dramatic manner, as is shown in Fig. 7.14. Although individual crops differ in their response to soil acidity (see Chapter 9) certain general principles hold true regarding nutrient availability to plants. Those nutrients that are stored in organic sources in soils, such as nitrogen, are released when soil organisms break down the organic material, a process known as mineralization. Figure 7.14 shows that the activity of macro-organisms such as earthworms and micro-organisms such as bacteria is pH dependent, with maximum breakdown activity occurring at pH 6–7. Other

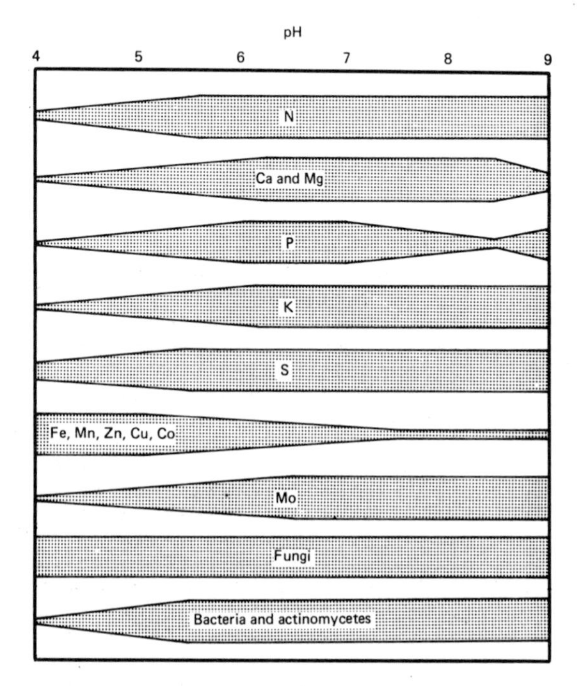

Fig. 7.14 Relationship between pH, nutrient availability and organism activity (Courtesy: Macmillan College Publishing Company 1990 *The Nature and Properties of Soil*, 10th Edn by L. C. Brady).

nutrients such as potassium will be held on exchange sites providing the soil is not swamped with hydrogen ions – hence the restricted availability at low pH, due to the high hydrogen ion concentration. Other nutrients, such as phosphorus, display more complex patterns, for reasons that are discussed in the section on phosphorus availability. With the exception of certain trace elements such as Fe, Cu and Zn, maximum nutrient availability occurs in the range pH 6–7.

Liming

The soil benefits of the addition of lime are well known. In addition to reducing soil acidity, lime will increase soil organism activity (see Fig. 7.14) and enhance structure stability, thus improving physical conditions. These benefits were known to the Romans, but the mode of action of liming materials was only fully appreciated when the principles of soil chemistry, and cation exchange in particular, were elucidated. Soil becomes acid as calcium, magnesium, potassium and sodium are displaced from exchange sites by hydrogen ions and then leached from the profile. Natural weathering processes will replace the leached cations, but only at a slow rate. A recent survey (Church & Skinner, 1986) demonstrated that the proportion of soils with pH values below the recommended levels for arable and grass crops has increased in the last two decades. The reason for soils becoming progressively more acid with time are complex, but the following factors are important:

- acid deposition from polluted rainfall;
- use of acidifying nitrogen fertilizers;
- heavy cropping which removes calcium, magnesium, etc.;
- oxidation of organic matter which generates free H^+ ions in the soil.

Clearly the soil type and parent material are important in supplying the dominant basic, or 'acid resisting' ions, calcium and magnesium. Soils with a large cation exchange capacity will acidify more slowly than sandy, organic-poor soils.

Lime requirement

It is possible to calculate the amount of lime needed to raise the pH of a soil to a value that is deemed suitable for given crop plants. This is known as the lime requirement. Further information on forms of limestone (such as ground limestone, $CaCO_3$) and methods of application in cropping systems is given in detail in Chapter 9. In order to calculate the lime requirement it is necessary to measure or estimate the size of the cation exchange complex, because soil pH measurements only give an indication of 'active acidity' rather than a true measure of the quantity of hydrogen ions that might be released into soil solution over a period of several years. Hence, for any given pH, a soil with a high clay and organic matter content will have a greater lime requirement than a sandy, organic matter deficient soil. Most advisors will use a 'look-up' table, such as is given in Table 7.14.

Having measured the soil pH and assessed the soil textural type, the lime requirement can be calculated using the following equation:

$$\text{Lime requirement (t } CaCO_3\text{ ha}^{-1}) = (\text{Target pH} - \text{measured pH}) \times \text{lime factor}$$

Table 7.14 Lime requirement for different soils and cropping situations (after MAFF/ADAS, 1988)

Cropping	*Soil type*	*Optimum pH*	*Target pH*	*Lime factor*
Arable	Light	6.5	6.7	6.0
	Medium	6.5	6.7	7.0
	Heavy	6.5	6.7	8.0
	Organic	6.2	6.4	10.0
	Peats	5.8	6.0	16.0
Grass	Light	6.0	6.2	4.5
	Medium	6.0	6.2	5.0
	Heavy	6.0	6.2	6.0
	Organic	5.7	5.9	7.5
	Peats	5.3	5.5	12.0

This calculation assumes that the target pH will be slightly higher than optimum pH due to variations in soil pH across any field. The speed of action of a liming material will depend upon when it is applied and the particle size of the material. Lime applied prior to cultivation in the autumn or spring will be rapidly absorbed on to the exchange complex. If a slower release is required, as in an upland situation, then a coarse grade of material can be selected. This will dissolve slowly and therefore counteract any trend towards acidity for a number of years.

Macronutrients

Nutrients are an integral part of the soil : plant system. They 'flow' from one part of the system to another according to a variety of complex processes that interact together. Any attempt to look at one part of the system, for example the effect of applied fertilizers on crop yield, without taking into account other sources of nutrients, must be carried out with care. There are three main stores of nutrients:

- the inorganic store, mainly in the soil;
- the biomass store, mostly living organisms, including plant material;
- the humus store, the dead and decaying remains of plant and animal material.

These stores and the relationships between them are represented diagrammatically in Fig. 7.15. The relative importance of each store varies from nutrient to nutrient; these are discussed individually below. For any given soil:crop situation, the application of fertilizer or manure may only contribute a small proportion of the total soil reserve of a given nutrient. It is therefore important to appreciate that even intensively fertilized soils can supply significant quantities of nutrients from non-fertilizer sources.

Nitrogen

Nitrogen (N) is the most important macronutrient in terms of the quantities taken up by crops, but efficient use of nitrogen in fertilizer and manurial sources is problematic due to the rapid losses of the main plant-available form of N, the nitrate anion (NO_3^-), either by leaching or by conversion to gaseous forms of N. Mini-

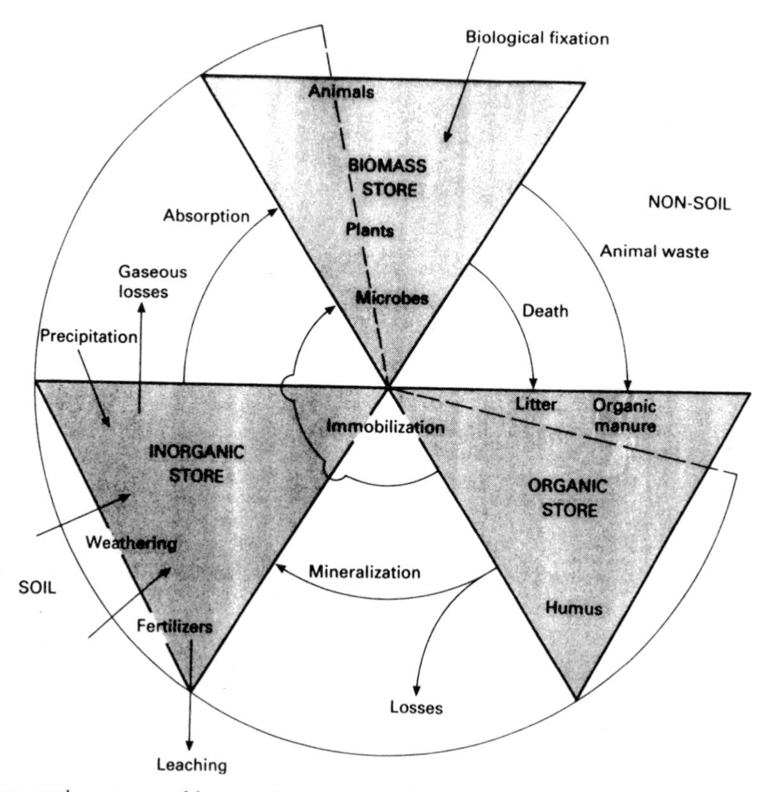

Fig. 7.15 Primary nutrient stores and key transfer processes (after White, 1987).

mizing these losses is best achieved by a full understanding of the N transformation processes in the soil:plant: atmosphere system.

The nitrogen cycle
Nitrogen compounds in soil may be either organic, as components of organic matter, or inorganic, as ammonium (NH_4^+) or nitrate (NO_3^-) ions. The main stores and transfers between those stores are given in Fig. 7.16. It can be seen from this figure that there are no significant rock or parent material sources of nitrogen. All nitrogen in soils is ultimately derived from the atmosphere, which contains 79% N (as the gas dinitrogen, N_2). Rainfall may add $5-20\,kg\,N\,ha^{-1}\,a^{-1}$. These additions are observed to be high where intensive livestock husbandry leads to ammonia volatilization from manure and slurry. During the long process of soil formation, N is fixed into the soil through the action of free-living algae, or bacteria such as *Azotobacter*, or the symbiotic fixation carried out primarily by legumes. These organisms will raise the level of organic N such that a mature soil might contain $2000-5000\,kg\,N\,ha^{-1}$.

Once a substantial pool of organic N has accumulated, the process of mineralization is carried out by bacteria in the soil. Mineralization leads to the production of nitrate, but several steps are involved in the process.

$$\text{Organic N} \longrightarrow NH_4^+$$
$$NH_4^+ \xrightarrow[\textit{Nitrosomonas}]{} NO_2^- \text{ (nitrite)}$$
$$NO_2^- \xrightarrow[\textit{Nitrobacter}]{} NO_3^-$$

All these reactions are controlled by soil and environmental conditions. They can only occur under aerobic conditions, hence waterlogged soils are not a suitable environment for mineralization. However, some water is needed, as mineralization will be restricted in very dry soils. The process is temperature dependent, such that mineralization will be limited below a soil temperature of 4–5°C. Soil pH is also important, with mineralization occurring most rapidly at neutral pH values. Since the production of nitrate is potentially able to proceed quicker than the production of nitrite, the latter is not normally found in significant quantities in soils. The quantity of mineral N present in a soil at any time will be very variable, but might fall within the range $10-50\,kg\,N\,ha^{-1}$.

The products of mineralization, NH_4^+ and NO_3^- (mineral N), react in different ways in the soil. NH_4^+ is attracted to cation exchange sites, and so is less likely to be leached. In contrast, NO_3^- is very mobile in soils, and vulnerable to leaching losses during periods of drainage. The fate of mineral N depends once again on soil and environmental conditions. Plant uptake may occur, mineral N might be leached or converted into gaseous forms (the process of denitrification, which takes place in wet soils), or bacteria might convert the 'available N' into biomass and hence add to the organic reserves. All these processes are summarized in Fig. 7.16.

N additions
Fertilizer N is frequently applied to agricultural soils. Further discussion of the forms and availability of fertilizer N appears in Chapter 9. The common forms of fertilizer N such as ammonium nitrate (NH_4NO_3) are

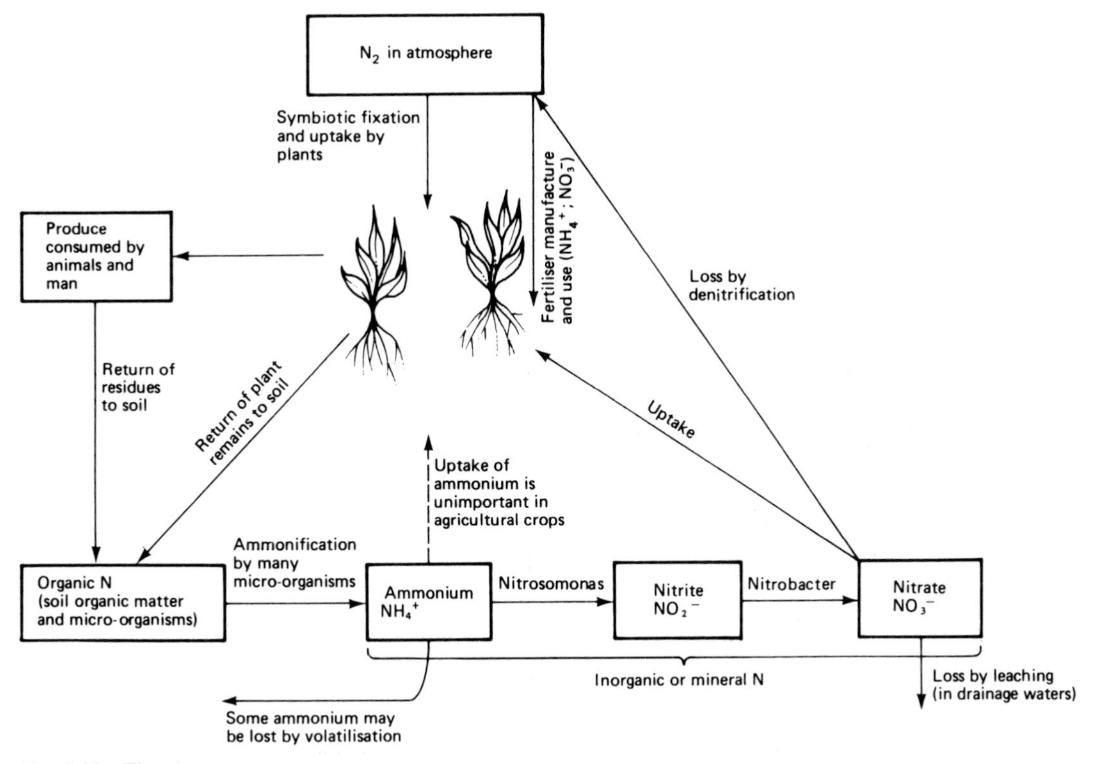

Fig. 7.16 The nitrogen cycle.

readily soluble and plant available. However, experiments have shown that, on average, only half of most fertilizer N is taken up by the crop to which it is applied. The other half may be lost, for example by leaching, or be incorporated into soil organic matter through the action of bacteria. Addiscott *et al.* (1991) give a full description of the fate of fertilizer N.

Organic manures and crop residues are important sources of N, but the availability of the variable quantities of N contained within these materials (see Chapter 9) depends on those processes described above that control the N cycle. Only when bacteria have broken down the crop residues and animal wastes does the N become available to crop plants. Further discussion of fertilizer and manure N utilization can be found in the nutrient management section.

Phosphorus

Sources

Phosphorus (P) is unusual amongst the major nutrient elements in that concentrations in soil solution are frequently extremely low (equivalent to less than 5 kg P ha^{-1} in most situations), but plant requirements (see Table 7.11) can be significant. The main sources of P are from soil parent materials and fertilizer/manure additions. In Fig. 7.17 a simplified P cycle is represented. Release of P from insoluble mineral sources is slow, such that the pool of available P is always small. In mature soils organic matter forms a major reserve of P, which will become available to plants as mineralization occurs. Organic

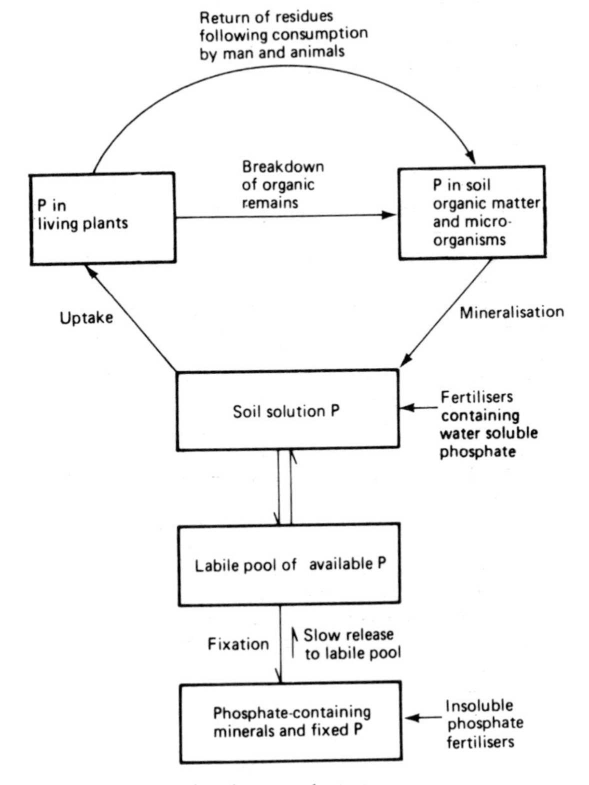

Fig. 7.17 The phosphorus cycle.

matter sources of P become important in sandy soils and peats, which tend to have low levels of mineral P. Most agricultural soils have sufficient reserves of P, built up over many decades. Fertilizer P (as rock phosphate, basic slag, superphosphate or bonemeal) has been added to soils since the mid 19th century. Rainfall additions of P are insignificant. Total P contents range from 500 to 2500 kg ha^{-1}.

Soil reactions and availability
P is very insoluble, so that a large proportion of total P, in excess of 95%, is unavailable to plants at any time. Release of P from insoluble mineral sources is controlled by chemical equilibrium processes. Plant uptake of orthophosphate ($H_2PO_4^-$ and HPO_4^{2-}) removes P from soil solution, leading to more of the labile pool moving into solution (see Fig. 7.17). Plant requirements of P can be up to 3 kg P ha^{-1} d^{-1}, which, given the quantity in solution at any one time, indicates the importance of these chemical reactions. Orthophosphate being an anion does not take part in cation exchange reactions, but may be absorbed on the smaller number of anion exchange sites in soils. *P fixation* is a process whereby soil solution/labile P is rendered insoluble by reactions with calcium (in alkaline soils), iron or aluminium (in acid soils). Fixation leads to added P becoming unavailable to plants over short time periods, and can result in fertilizer P uptake being less than 25% in P-deficient soils. Fixation is least strong at pH 7. In most agricultural soils in the UK there has now been a sufficient history of fertilizer P use and application of manures that the most powerful sites of P fixation have been satisfied, such that newly applied P is fixed less strongly than was the case in previous years. However, it is still important to place fertilizer P close to the developing seed if crop demands are high.

Losses
These tend to be small, due to P insolubility. Leaching through the soil is not a major loss, and there are no gaseous loss processes. Surface soil erosion is the main route of P loss. Experimental evidence (for example, Heathwaite *et al.*, 1990) shows that P losses can be important from overgrazed compacted soils, where surface runoff will carry P bound to sediment and in organic forms down slopes to adjacent watercourses. Arable soil erosion will also lead to P losses. The main 'loss' is the fixation of P, which reduces the plant available supply but adds to total soil reserves. This fixed P may eventually become available to plants over a period of years.

Potassium

Sources
Potassium (K) is derived from the weathering of clay minerals in the parent material, hence any clay-rich soil will be well supplied with K. Many rock-forming minerals, such as feldspars, are rich in K, so that K is released as these break down due to weathering. Organic matter is a poor source of K, as plant material contains K in the fluid phase. Hence most K will be lost at an early stage from decaying plant material. Rainfall may add 1–10 kg K ha^{-1} a^{-1}. Total K contents may be very large (>10 000 kg ha^{-1}), while exchangeable K may range from 500 to 1000 kg ha^{-1}.

Soil reactions and availability
The chemistry of soil K is much less complicated than either N or P. K exists in exchangeable and solution forms as the simple cation K^+ (see Fig. 7.18). Plant uptake of K^+ can be equal to or greater than N (see Table 7.11). Soils with a large cation exchange capacity will be able to retain a large amount of K in a relatively plant available form. Some fixation of available K may occur, but this is a relatively unimportant process in most soils.

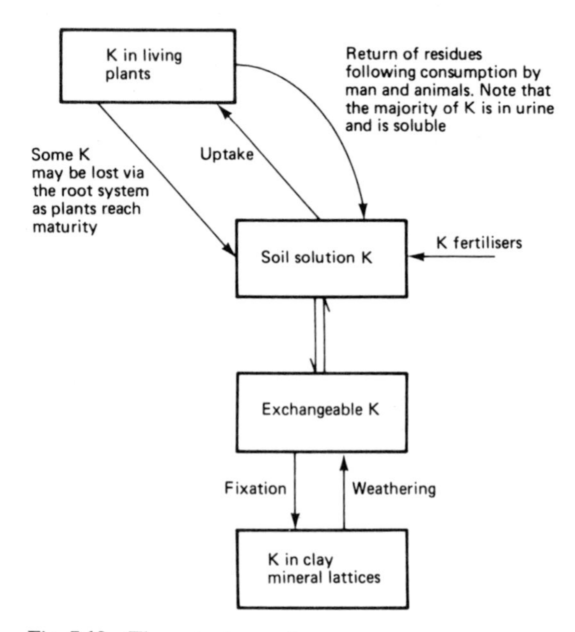

Fig. 7.18 The potassium cycle.

Losses
K is subject to leaching losses, particularly from sandy, organic-poor soils, but this loss is rarely considered to be important in agricultural systems, due to the ease of K availability from soil and fertilizer sources. Ion balance dictates that drainage water must contain an equal proportion of anion and cations. Hence losses of NO_3^-, for example, must be balanced by losses of a cation such as K^+.

Calcium and magnesium

Sources
The weathering of minerals forms the main supply of plant-available calcium (Ca^{2+}) and magnesium (Mg^{2+}). These two macronutrients behave in a similar manner to K in soils. Total soil reserves may be very large, particularly in limestone soils. Exchangeable Ca^{2+} and Mg^{2+} may range between 1000 and 5000 kg ha^{-1}. Soil Mg levels in coastal areas tend to be elevated due to sea salt spray additions.

Soil reactions, availability and losses
As with K, these macronutrients are held on the exchange complex and released into solution as plant uptake occurs. Ca and Mg play an important role in controlling soil pH, as discussed previously. Fertilizer additions accumulate on the exchange complex, and may be

subject to some temporary fixation, but again this process is not of major importance. Large leaching losses of Ca and Mg can occur from agricultural soils (e.g. 10–100 kg Ca $ha^{-1} a^{-1}$), but these are small in relation to total reserves and do not cause a major environmental concern.

Sulphur

Sources

Soil sulphur (S) is derived originally from sulphide minerals, which are oxidized to sulphate (SO_4^{2-}) in weathering. Clays and shales frequently contain large quantities of S. In mature soils, most S is found in the topsoil in the organic form. Additions in rainfall range from 10 to 30 kg S $ha^{-1} a^{-1}$, and are found to be towards the upper end of this range in areas close to industrial sources of atmospheric S such as coal-fired power stations. Sea spray is another source of S. These two sources are sufficient in most cases to maintain adequate S levels in all but the most intensive cropping situations. Total levels may range from 200 to 2000 kg S ha^{-1}.

Soil reactions, availability and losses

Organic S is released by microbial decomposition, to give the plant available form, SO_4^{2-}. In many respects, S behaves like P in soils, but is not subject to the problems of fixation that restrict P availability. Rainfall inputs and dry deposition (dust) from industrial sources account for the majority of crop requirements in many areas of the UK. Leaching losses do occur, but are balanced by natural inputs.

Micronutrients

Micronutrients are essential to plants but are found in small concentrations, often less than 100 mg kg^{-1} dry matter in plant material. A full list of micronutrients is given in Chapter 9. For a detailed discussion of micronutrient availability, the reader is referred to one of the standard texts on soils and plant nutrition, such as Wild (1993) or Mengel & Kirkby (1982). Trace elements include plant micronutrients as well as other elements that are not essential for plant growth, such as selenium (see Chapter 15). These will not be discussed further. Median total micronutrient concentrations of soils in England and Wales are given in Table 7.15. Values range widely according to the element, from <100 mg kg^{-1} dry soil (Zn, Cu, Co) to >10 000 mg kg^{-1} dry soil (Fe).

Deficiencies of micronutrients may occur in agricultural crops, either due to an absolute lack of a particular micronutrient, or due to inadequate availability, which may for example be caused by overliming (see Fig. 7.14).

Table 7.15 Median micronutrient concentrations for all soil, England and Wales (after McGrath & Loveland, 1993)

Element	*mg kg^{-1} dry soil*
Cobalt	10
Copper	18
Iron	26 786
Manganese	577
Sodium	242
Zinc	82

In general, sandy soils and peats are most susceptible to micronutrient deficiencies. Deficiencies have become more common in recent years due to more intensive cropping and fertilizers becoming 'purer', that is containing fewer impurities. Some detailed comments relating to individual micronutrients are made below.

Boron (B)

Boron is present in the rock mineral tourmaline, from which it is weathered to form plant-available borate (BO_3^{3-}). B accumulates in organic matter, and is freely available to plants in all except alkaline soils. Overliming can induce boron deficiency. Boronated fertilizers can be used to overcome these problems when growing B-demanding crops such as sugar beet.

Copper (Cu)

Copper is released from rock minerals by weathering and held on the exchange complex. Cu availability decreases with increasing pH, hence alkaline, sandy or mineral-poor organic soils may be Cu deficient. Cu deficiency may be overcome by a foliar application of copper sulphate.

Iron (Fe)

Iron is very abundant in soils, mainly in rock minerals, hydrated and non-hyrated iron oxides and as plant-available Fe^{2+}, which only occurs in small amounts in soils as this is the reduced form of iron. The oxidized form, Fe^{3+}, is insoluble and immobile. Iron deficiencies are uncommon, but can occur in fruit trees grown on well drained, alkaline soils.

Manganese (Mn)

Manganese is abundant in most soils, being weathered from rock minerals to form the plant available Mn^{2+} ion, which is retained on the exchange complex. Alkaline and freely drained soils can be deficient in plant available Mn, while very acid soils and those that are poorly drained can contain high concentrations of Mn that may be toxic to plants.

Soil:pesticide interactions

A wide variety of pesticides are used in agriculture. Many are short lived and degrade rapidly in the soil whereas others are persistent or may leach out of field soils into adjacent watercourses. Pesticide efficiency may be limited in certain soil types due to adsorption. Under these circumstances it may be necessary to increase application rates to achieve effective control. The environmental consequences of such action may be significant, hence it is important to match pesticide use to soil type as well as to cropping situation.

Adsorption

Adsorption occurs in soils with high organic matter or clay contents. Figure 7.19 displays the general relationship between texture, organic matter and the adsorption of residual herbicides. Reduction in the effectiveness of residual herbicides in direct drilled soils with high surface organic matter and ash contents has been noted in some situations. This adsorption is at its most extreme in peat soils, while on some sandy soils with low organic matter levels, certain soil-acting herbicides cannot be used because of the risk of damage to the crop. The bipyridyl

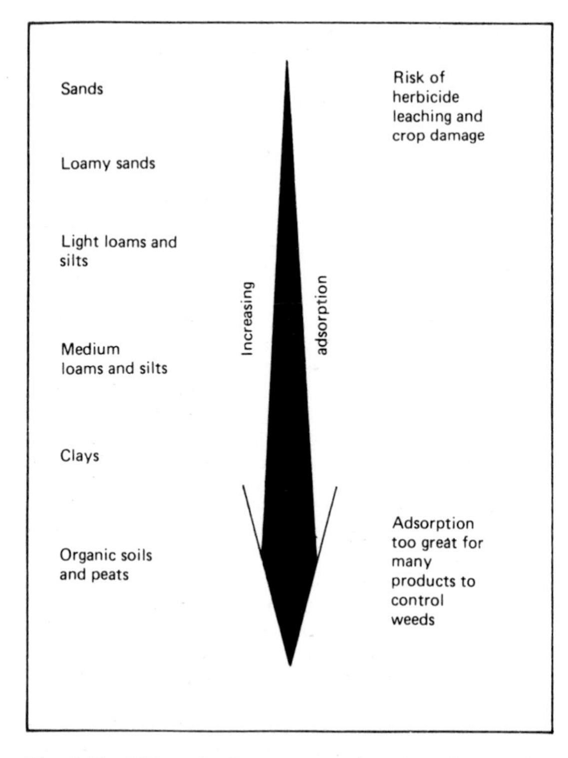

Fig. 7.19 Effect of soil texture on adsorption of residual herbicides.

herbicides paraquat and diquat are strongly and irreversibly adsorbed by clay minerals and hence are rendered inactive on contact with the soil. However, adsorption on to organic matter is reversible, so these herbicides adsorbed by peaty soils can be taken up and may damage crops. Losses of pesticides by leaching may occur from sandy soils which do not exhibit strong adsorption properties. In addition, surface erosion of soil has been found to lead to substantial losses of applied agrochemicals, e.g. atrazine applied to maize crops.

Degradation

The degradation rate, or alternatively the persistence of pesticides, depends upon soil type. Most pesticides in common use are broken down by microbial action or degrade chemically following application to the soil. The rate of breakdown may be rapid, just a matter of days, as shown for example by the synthetic pyrethroids, or may take many months, as in the case of other organophosphorus pesticides. The more rapidly a material breaks down, the smaller the risk of long-term soil organism damage and losses by leaching.

Soil biological properties

A discussion of soil properties would not be complete without a consideration of the role of soil organisms in the maintenance of soil fertility. In many situations, particularly extensive land management, biologically driven processes maintain nutrient supply and optimum soil physical conditions. Biological processes are carried out by organisms which constitute the soil biomass. This living powerhouse of the soil may make up 2–3% of all organic matter, and hence within the topsoil (300 mm) may weigh up to $5\,t\,ha^{-1}$. Once dead, soil organisms become part of the organic matter fraction. When fully decomposed, plant and animal remains are indistinguishable and become part of the humus fraction. All aspects of soil biology are important in agricultural situations.

Micro-organisms

Bacteria

Bacteria are minute soil organisms, often $<5\,\mu m$ in length. They live in water films around soil particles and under ideal conditions can duplicate themselves within 24 hours. Hence the number of these organisms in soil is enormous, but it is difficult to separate bacteria from the rest of the organic matter to calculate total weight or numbers. The commonly accepted range is in the region of 10^7–10^9 organisms g^{-1}, equivalent to a live weight of about 1–$2\,t\,ha^{-1}$. This equates to about 1% of the soil organic matter.

There are many different species of bacteria, which exhibit a wide variety of metabolism and ability to decompose diverse substrates. Detailed descriptions of bacterial and other soil organisms can be found in Jackson & Raw (1966). Most bacteria are classified as heterotrophs, which are organisms that require complex organic molecules for growth. Bacteria secrete enzymes which break down complex molecules into relatively simple, soluble compounds such as glucose, which can be easily absorbed. Some bacteria are autotrophic, that is they can synthesize their cell constituents from simple inorganic molecules given a supply of energy, either from sunlight or from chemical oxidation reactions. In addition to the heterotrophic/autotrophic subdivision, it is also possible to classify bacteria according to their oxygen requirements. Some are aerobic, requiring oxygen for respiration, while others are anaerobic.

Some of the important bacterially moderated processes include the following:

- degradation of organic matter to release many nutrients as by-products: heterotrophic aerobes;
- $NH_4^+ \longrightarrow NO_2^- \longrightarrow NO_3^-$: *Nitrosomonas* and *Nitrobacter*;
- sulphate oxidation: *Thiobacillus*;
- nitrogen fixation: *Azotobacter*, *Rhizobium* spp.

Actinomycetes

These organisms are classified as bacteria, but tend to be larger in size and fewer in number. In morphological and physiological terms they represent a transition between bacteria and fungi. Numbers of actinomycetes may range from 10^5 to $10^8\,g^{-1}$, and most are found in the topsoil. Actinomycetes are important degraders of organic matter, particularly more resistant fractions such as lignin and complex organic molecules. As with bacteria, these organisms do not favour acid or waterlogged soils. *Streptomyces* are responsible for the characteristic smell of freshly turned soil.

Fungi

Fungal colonies tend to be much less numerous than bacteria ($10^5 g^{-1}$), but due to the larger size of fungi these organisms can be equivalent to half the bacterial biomass. They occur in all soils, and are commonly associated with the initial breakdown of organic debris. They can be highly efficient at converting organic matter into fungal tissue, particularly in acidic woodland soils where bacterial numbers may be low. Fungi also tolerate variations in soil water contents better than bacteria, and they are able to degrade resistant materials such as lignin. Mycorrhizal fungi are of particular importance in phosphate deficient soils. The fungus is in a symbiotic association with the host, commonly a coniferous tree. The latter supplies a source of carbon for the fungus which in turn enhances P absorption by the host. The root systems of saplings which are to be planted can be inoculated with mycorrhizal fungi, hence reducing the need to fertilize the soil. Mycorrhizae are not important in most agricultural soils, as the use of fertilizers suppresses their activity.

Macro-organisms

There is a wide range of types and numbers of macro-organisms in agricultural soils. These range from the small arthropods, such as springtails and beetles, to moles and rabbits, all of which may have an important function in disturbing soil and incorporating organic matter.

Only one group of macro-organisms will be discussed here – the earthworms, Darwin's 'natural ploughs of the soil'. Earthworms consume more plant litter material than all the other invertebrates in total. The total earthworm biomass varies widely according to soil conditions, from $<20 \, kg \, ha^{-1}$ under intensive, organic-matter-poor, arable land, to $>500 \, kg \, ha^{-1}$ under neutral pH permanent grassland. Earthworms feed on dead organic matter, and incorporate that material into the soil, at the same time mixing organic with mineral material. Earthworms need calcium for their metabolism, and are therefore not common in acid soils (pH <5.0). Large populations of worms such as *Allolobophora* and *Lumbricus* spp. may ingest and excrete $50-100 \, t \, ha^{-1}$ annually. During that process, large pores will be created which may aid drainage. These organisms are found to be most important in direct drilled soils, where numbers can be much greater than under regular ploughing. It appears that earthworms are sensitive to disturbance, and so many intensively managed soils with low organic matter contents have become depleted in earthworms in recent years. As a result the benefits of litter incorporation, soil mixing and aeration have been lost.

Soil organic matter and the carbon cycle

Soil organic matter consists of plant material and animal remains and excreta. Ultimately all these materials will decompose in the soil to form humus, which contributes to many aspects of soil fertility. Some farming systems, notably the more extensive and the 'organic' systems, rely heavily on this organic matter fraction to supply many of the plant nutrients as well as the physical benefits that derive from high humus levels in the soil. Some intensively farmed soils have in recent years become difficult to manage; undoubtedly falling organic matter levels have played an important role in this decline of fertility.

Organic matter decomposition

The degradation of plant and animal remains is a complex process, carried out by a variety of organisms which have been described above. Initial stages of decomposition are dominated by macro-organisms which incorporate and break down material, and in the process increase the surface area and allow bacterial inoculation of this fresh substrate. The soluble and rapidly decomposed fractions, such as simple sugars and starches, may well break down within a matter of days after incorporation. The more resistant fats, waxes and lignins may take months or even years to decay. Soil conditions are important in this decay process, as described above. Most rapid decomposition will occur in near-neutral, damp, warm soils. The result of these decomposition processes is the production of a suite of complex organic chemicals known in simple terms as humus, an amorphous, dark material in which the structure of the original constituents is not distinguishable. There are various humus fractions, discriminated on the basis of molecular weight and chemistry. For further details, see White (1987). Ultimately, all organic matter in soil will decompose to its constituent components, carbon dioxide, water and minerals.

In natural soils, types and degree of stratification of humus reflect the nature of the decomposition processes. Neutral or weakly acid brown earths rarely show an accumulation of leaf litter at the soil surface. Humus is well decomposed and intimately mixed with the mineral material. Acidic soils are characterized by a build-up of slowly decomposing organic matter at the surface. Poor humification leads to the development of layers of organic material in various stages of decomposition, as would typically be found in a moorland situation. High water contents can also restrict humification, leading to the development of peat. The higher the rainfall and the lower the temperature, the slower the decomposition processes, so that soils in the wetter and cooler parts of the UK tend to have higher organic matter contents.

Role of organic matter

- *Physical benefits* include maintenance of stable soil structure, raising water-holding capacity and creating better conditions for cultivation through reduction of cohesion.
- *Chemical benefits* include retention and release of nutrients and increase in cation exchange capacity.
- *Biological benefits* include the supply of a substrate to various beneficial soil organisms such as bacteria and earthworms.

These benefits are most important for soils with low clay contents. Improvements in fertility of sandy soils following the addition of organic matter such as manure or crop residues can be very marked.

The carbon cycle

The transformation of carbon from gaseous carbon dioxide in the atmosphere to plant, animal and then soil organic matter forms part of the carbon cycle. This cycle is completed when organic matter is released from decomposing humus back to the atmosphere (Fig. 7.20). In agricultural terms the magnitude of the fluxes are important, as carbon accumulates in plant material due to photosynthesis and is lost from soils following organic matter decomposition. In global terms, the influence of industrial activity, notably the burning of fossil fuels, is changing the balance of carbon within the cycle. Fossil fuels represent carbon that has been trapped in rocks by geological processes, and stored for millions of years. Release by combustion is elevating 'natural' levels (Goudie, 1986), and having a major impact on the global carbon cycle. The turnover of carbon in agricultural systems can be measured in years, rather than millions of years. Changes in the input and loss processes within agricultural soils have led to changes in this cycle, which are summarized in the next section.

Impact of agricultural practices on soil organic matter

In general, arable soils have lower organic matter contents than grassland equivalents. Table 7.16 shows the range of organic matter contents observed under selected cropping systems. These values represent the balance between inputs and losses. The major inputs and loss processes are listed below:

- *Inputs*: crop residues (e.g. straw, roots, sugar beet tops), green manures (crops sown and ploughed into the soil, not harvested), animal wastes (manure and slurry).
- *Losses*: frequent cultivation, drainage, liming and fertilizing may all lead to greater losses of organic matter by oxidation.

Intensively managed soils show the greatest changes in terms of organic matter turnover, relative to uncultivated soils. Only permanent pasture and arable soils receiving minimal cultivations escape these changes. Any change in organic matter status reflects the balance between inputs and outputs. Grassland soils often have a stable

Table 7.16 Typical organic matter contents of mineral agricultural soils

	Organic matter %
Arable, straw/residues removed	3–5
Arable, straw/residues incorporated	4–7
Grass/arable rotation	5–10
Permanent grass	10–20
Woodland	15–30

NB: Peats and organic soils (>30% organic matter) are excluded from this table.

organic matter content, while taking a soil out of grass may lead to a decline in organic matter. Figure 7.21 displays two example situations where cropping practice has had an impact on organic matter levels. In both cases the changes occur over a number of years, with the changes being most rapid soon after the change in cropping practice. The fen soils of East Anglia demonstrate the consequences of prolonged intensive cultivation following drainage several centuries ago. Some fen soils have lost more than 0.5 m peaty topsoil, such that the underlying clay is being ploughed up to the surface causing major soil management problems.

Carbon:nitrogen (C:N) ratio

The rate of decomposition of organic materials added to the soil depends on the C:N ratio, which for well decomposed humus is approximately 10–12:1. Bacterial biomass has a C:N ratio of 6:1, while most plant and animal remains contain much more carbon and much less nitrogen. In consequence the rate of breakdown of materials with wide C:N ratios, such as cereal straw, will be slow unless there is a source of nitrogen that the bacteria can use. Following straw incorporation there is often a flush of bacterial activity as decomposition begins. This will lead to nitrate-N being removed from the plant-available pool. This nitrogen is immobilized in bacterial biomass, although it will eventually be released

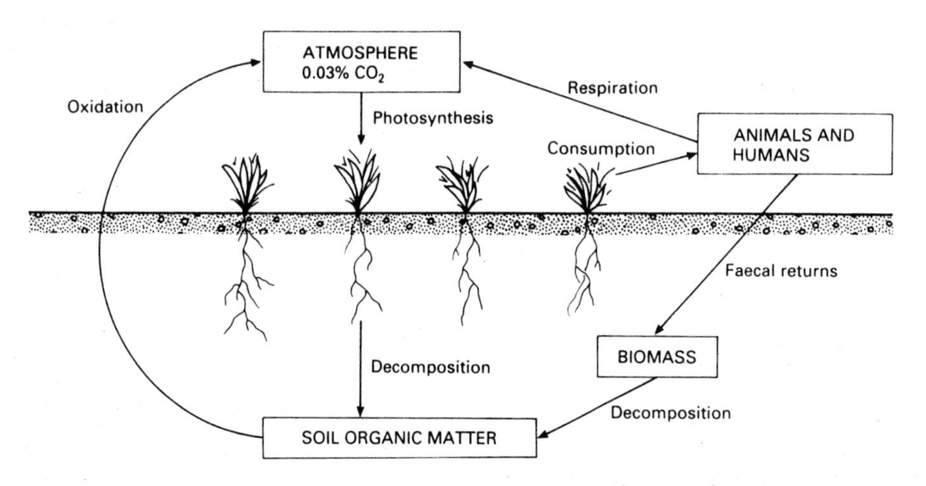

Fig. 7.20 The carbon cycle.

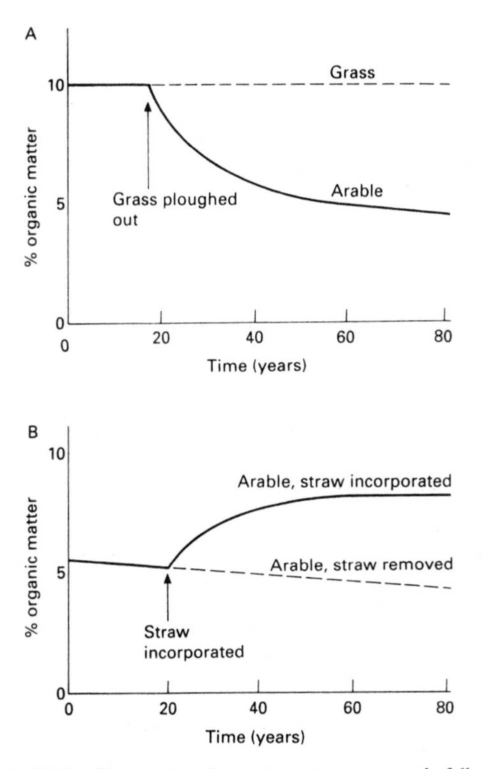

Fig. 7.21 Changes in soil organic matter content. A: following the conversion of grass to arable cropping; B: following the change from straw burning/removal to incorporation.

as the organic matter decays in years to come. Materials with narrow C:N ratios can be decomposed readily, and hence make more suitable forms of organic addition if rapid benefits are needed.

Several general rules should be followed where possible in order to maintain the organic matter status of soils:

- Return as much as the crop residue as possible.
- Minimize disturbance by cultivation (keep shallow, reduce frequency).
- Use green manures.
- Maximize crop cover.
- Use longer grass/clover leys.
- Incorporate manures quickly to minimize oxidation losses.
- Maintain pH levels at optimum for soil organisms.

Adherence to these guidelines will raise soil fertility and reduce the dependence upon fertilizers, while making the soil more easy to manage.

Soil classification and analysis

Soil types and classification

Soil formation

Soils develop as a result of interactions between a number of soil-forming factors, namely parent material, climate, vegetation, organisms, topography and time (Jenny, 1941). As each of these factors varies, so too will the resulting soil characteristics. Detailed discussion of soil formation processes is beyond the scope of this chapter, and can be found in one of the many texts on soil development (e.g. Curtis *et al.* 1976). In agricultural situations, some soil formation processes are modified – for example the relationship between soil and vegetation – while a basic knowledge of the different soil types created by these processes will help to give an insight into reasons for variations in soil fertility between fields and between regions.

The main characteristics of significance to land users derive from the parent material. The geology of an area will be very closely related to the pattern of soil types. In this case geology includes not only the 'hard' rocks, but also the superficial materials, such as river floodplain deposits, from which a soil might be derived. Soils form over long periods of time. Most of the soils in the UK have developed since the end of the last ice age, approximately 15 000–20 000 years ago. Processes of soil evolution are slow whereas accelerated soil erosion from arable fields can be very rapid, leading in extreme cases to a significant reduction in soil depth. Soils deepen by weathering at a rate of between $0.01-0.1\,\text{mm}\,\text{a}^{-1}$. Any losses greater than this (0.1 mm is equivalent to approximately $1\,\text{t}\,\text{ha}^{-1}$) will result in the soil becoming less deep and hence less able to support crop plants.

Soil types

Brown earths

Brown earths are well drained soils of varying texture and of moderately acid to alkaline pH, are warm brown in colour, and are usually well structured with no marked organic matter accumulation at the surface. Such soils are often uninteresting to look at, due to lack of marked horizonation, but this shows that the soil is fertile with a large organism population actively mixing the soil and creating a favourable rooting environment. These soils are capable of supporting most crop types, site and climate permitting.

Gleys

Gleys are poorly drained soils with mottled yellow/grey colours at depth, reflecting waterlogging during some time of the year. Waterlogging may be due to heavy texture or a high regional watertable, i.e. on a floodplain. The grey colours develop when iron is changed from an oxidized (Fe^{3+}, red/brown) to a reduced (Fe^{2+}, grey/blue) state. With drainage and careful management these soils can support some arable crops, notably cereals. These soils possess the advantage of high available water contents, and hence are drought resistant.

Podzols

The acid, sandy soils found on heathlands, moorlands and many coniferous plantations are described as podzols. These soils show marked stratification of the soil profile, with layers of undecomposed organic matter at the surface and possible iron pan at depth. Deep cultivation and the addition of lime and organic matter can raise the fertility of these soils, but they will continue to be droughty and develop patchy acidity under intensive management, for example horticultural cropping.

Peats
Peat soils are unusual in being organic, not mineral based. Peats may accumulate in two environments. Lowland peats (e.g. Somerset Levels, Cambridgeshire Fens, Lancashire Moss) develop over thousands of years in wet, marshy situations at or close to sea level. The soils can exhibit a neutral or alkaline pH, and can be very productive when drained. However, rapid soil erosion and conservation pressures have virtually stopped further drainage and agricultural improvement taking place. Upland peats (e.g. much of the Pennines, Dartmoor, Cambrian Mountains, Southern Uplands) form where high rainfall prevents organic matter decomposition. These soils are invariably acid and incapable of much improvement due to the dominant influence of climate.

Soil classification

In order to understand differences in soil type and management characteristics across the farmed landscape, it is necessary to classify and then map the soils. Soils in Britain have been described and classified at a variety of scales by the Soil Survey and Land Resource Centre (England and Wales) and the Soil Survey of Scotland. Complete mapping of British soils was completed in the early 1980s at a scale of 1:250 000. More detailed maps are available for selected areas, usually at scales of 1:25 000 or 1:63 360. Further details and sources of information are given at the end of this chapter. As the scale of mapping changes, so does the level of detail contained within the map. For example, the National Soil Map of England and Wales, at 1:250 000, provides information which gives a guide to soil types in an area and is therefore useful in general terms, but cannot be used for detailed farm management purposes. Soils in England and Wales are grouped into *associations*, which are combinations of soil series, defined by parent material, texture and mineral composition. Each association is described on the basis of parent material, main soil and site characteristics, cropping and land use.

Land classification and crop suitability maps

The agricultural potential of land, as opposed to just soil, depends on site and climate. The agricultural land classification system employed by MAFF (MAFF, 1988) classifies land into five grades in terms of limitation, typical cropping range and the expected level of consistency of yield. Many of the limitations are soil based, such as drainage status, stone content or soil depth, but can also include other factors, such as ground slope or climatic regime. The description of grades is given in Table 7.17. Agricultural land in England and Wales has been classified using this sytem, at a mapping scale of 1:63 360. This information integrates all those physical factors that might affect crop yield or control choice of cropping, and hence forms a useful basis for the assessment of land potential. However, the grades themselves do not indicate the cause of any restricted potential. Only detailed site and soil analysis can yield that information.

Land can also be classified in terms of potential to support specified crop types, such as arable, grassland or forestry. These crop suitability assessment systems require detailed soil and site information, and can only be compiled from soil survey information. These capability classifications are more robust than the ALC system for that reason. Areas of land can be classified not only in terms of their potential to support given crops, but also in terms of land management practices. For example, classifications of land exist that describe nitrate leaching risk, and suitability for straw incorporation. Further information can be obtained from the Soil Survey Regional Bulletins.

Table 7.17 Agricultural land classification grades (after MAFF, 1988)

Grade 1 – Excellent quality agricultural land
Land with no or very minor limitations to agricultural use. A very wide range of agricultural and horticultural crops can be grown and commonly includes top fruit, soft fruit, salad crops and winter-harvested vegetables. Yields are high and less variable than on land of lower quality.

Grade 2 – Very good quality agricultural land
Land with minor limitations which affect crop yield, cultivations or harvesting. A wide range of agricultural and horticultural crops can usually be grown but on some land in the grade there may be reduced flexibility due to difficulties with the production of the more demanding crops such as winter-harvested vegetables and arable root crops. The level of yield is generally high but may be lower or more variable than Grade 1.

Grade 3 – Good to moderate quality agricultural land
Land with moderate limitations which affect the choice of crops, timing and type of cultivation, harvesting or the level of yield. Where more demanding crops are grown, yields are generally lower or more variable than on land in Grades 1 and 2.

Subgrade 3a – good quality agricultural land: Land capable of consistently producing moderate to high yields of a narrow range of arable crops, especially cereals, or moderate yields of a wide range of crops including cereals, grass, oilseed rape, potatoes, sugar beet and the less demanding horticultural crops.
Subgrade 3b – moderate quality agricultural land: Land capable of producing moderate yields of a narrow range of crops, principally cereals and grass or lower yields of a wider range of crops or high yields of grass which can be grazed or harvested over most of the year.

Grade 4 – Poor quality agricultural land
Land with severe limitations which significantly restrict the range of crops and/or level of yields. It is mainly suited to grass with occasional arable crops (e.g. cereals and forage crops) the yields of which are variable. In moist climates, yields of grass may be moderate to high but there may be difficulties in utilization. The grade also includes very droughty arable land.

Grade 5 – Very poor quality agricultural land
Land with very severe limitations which restrict use to permanent pasture or rough grazing, except for occasional pioneer forage crops.

Soil sampling, analysis and nutrient indices

Soil sampling and description

Soil sampling and description are important procedures which must be carried out with care and precision. Fertilizer recommendations are often based on soil analyses, so it is important to ensure that samples taken are representative of the whole of a given field.

Variability

Soils are naturally occurring materials. Chemical properties can vary widely from point to point within a field, even under 'uniform' agricultural management. Preparations for taking soil samples for analysis include an assessment of the area to be sampled. If there are apparent, systematic variations across the area, such as a break in slope or an obvious poorly drained area, then these areas should be sampled separately.

Soil sampling is usually confined to the topsoil, which may be less in grassland than in arable land. For arable land, the top 15 cm should be sampled, whereas in grassland 10 cm is adequate. Sampling is carried out with a hollow circular corer, so that an uncontaminated sample can be collected. Sampling pattern is important. Fig. 7.22 demonstrates a suggested sampling pattern to collect a number of subsamples that can be bulked together to derive one homogeneous sample representative of the whole field. In this example, 25 subsamples have been collected. In general, the more subsamples the better. Samples should be placed in clean dry bags which can be effectively sealed to prevent contamination.

Soil profile description

In addition to sampling for laboratory analysis, field soil profile description is an important part of good soil management practice. Excavation of a profile face, if possible into the subsoil and ideally down to the soil parent material, will reveal much information regarding root proliferation, stone content, compacted layers, straw and other residue decomposition, drainage problems and so on. This physical evaluation is most important when considering soil loosening or a drainage operation.

Soil analysis

Soil samples are normally air-dried, crushed (to separate out but not break the stones) and passed through a 2 mm sieve prior to analysis, which is carried out on the 'fine-earth' fraction. Analysis is then carried out to determine a variety of parameters, such as plant-available nutrient concentrations, pH, organic matter content and soil texture.

Available or extractable nutrients

As described previously, the total amount of a given plant nutrient may be very large and bear little relation to the quantity that a crop plant has easy access to. Hence a variety of methods have been tried and tested to estimate the quantity of a given nutrient that is likely to become available to a crop during the next year. Methods used have been found by trial and error to predict reasonably closely, for most soil types, the nutrient release and hence the rainfall that is needed from fertilizer sources. Details of methods used by ADAS are given by MAFF (1986).

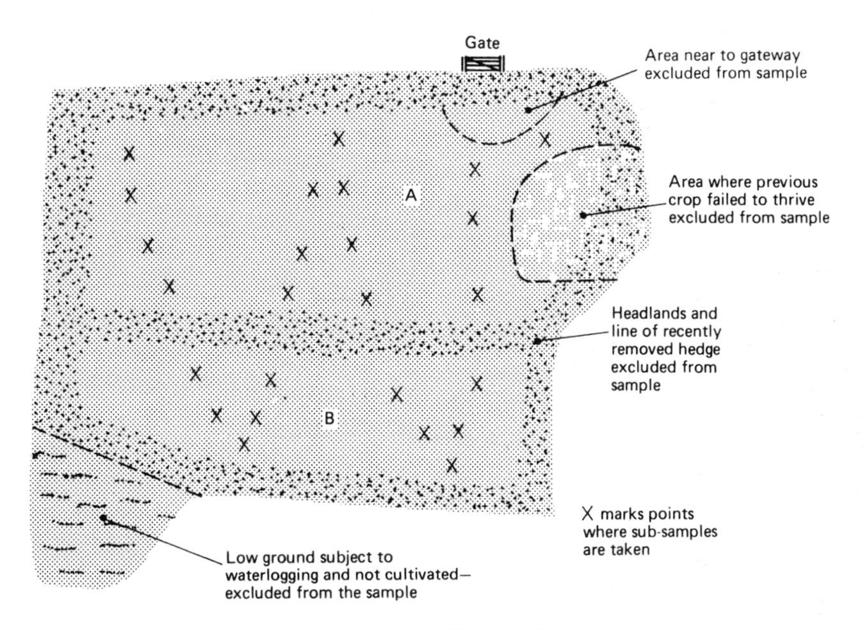

Note – If the soils of parts A and B are known to be different, or the sampler believes them to have been farmed differently in the recent past, two separate samples should be obtained.

Fig. 7.22 Soil sampling pattern.

Nitrogen
Plant-available nitrate is released solely by organic processes which are difficult to predict as they depend upon environmental conditions – rainfall, temperature, etc. – rather than on release from purely chemically mediated processes as is partly the case for P, K and Mg. Nitrate levels in soils can be measured, but the total amount of nitrate that might become available to a crop over a season cannot be so easily determined. Hence the nitrogen index is based on likely organic sources and residues in the soil from previous cropping and manuring.

Total nutrients
Total nutrients are rarely determined for standard agricultural applications. Exceptionally, micronutrient concentrations are determined in soil material to determine deficiency problems. Such analyses are expensive and only warranted when analysing problems in high-value cropping systems.

The nutrient index system

Nitrogen index
The nitrogen index is calculated by ADAS from previous cropping. If lucerne, long leys (three or more years) or permanent pasture have not featured in the last five years' cropping, use Table 7.18 only. If the last crop grown was lucerne, long ley or permanent pasture, use Table 7.19 (first column) only. If one of the above crops has been grown in the first five years but was not the last crop, use both tables and take the higher of the two values obtained.

P, K and Mg index
These are derived from soil analysis. Table 7.20 gives the range of soil nutrient concentrations that correspond to each index value. In general, an index of 3 or above indicates that a soil is well supplied with a particular nutrient, and only a maintenance dressing, to compensate for crop removal, will be required.

Table 7.19 Nitrogen index – following lucerne, long leys and permanent pasture

	1st crop	*2nd crop*	*3rd crop*	*4th crop*	*5th crop*
Lucerne	2	2	1	0	0
Long leys, cut only	1	1	0	0	0
Long leys, grazed or cut and grazed, low N*	1	1	0	0	0
Long leys, grazed or cut and grazed, high N†	2	2	1	0	0
Permanent pasture poor quality, matted	0	0	0	0	0
Permanent pasture, cut only, grazed only or cut and grazed, low N*	2	2	1	1	0
Permanent pasture, grazed or cut and grazed, high N†	2	2	1	1	1

* Low N = less than 250 $kg\,ha^{-1}$ N per year and low clover content.
† High N = more than 250 $kg\,ha^{-1}$ N per year or high clover content.
Source: MAFF, 1994.

Soils and sustainable production

Soil – a renewable resource?

Good soil management is dependent upon a full understanding of the soil's physical, chemical and biological properties described above. This knowledge must be used to appreciate the potential of a given soil to support a particular range of crops under a chosen management regime. The consequences of a failure to identify the limitations of a given soil have been observed many times

Table 7.18 Nitrogen index – based on last crop grown

Nitrogen index 0	*Nitrogen index 1*	*Nitrogen index 2*
Cereals	Peas or beans	Any crop in field receiving large frequent dressings of farmyard manure or slurry
Sugar beet	Potatoes	
Maize	Oilseed rape	
Vegetables receiving less than 200 $kg\,ha^{-1}$ N	Vegetables receiving more than 200 $kg\,ha^{-1}$ N	Lucerne
Forage crops removed	Forage crops grazed	Long leys, grazed or cut and grazed, high N†
Set-aside (rotational)		
Leys (1–2 year), grazed or cut and grazed, low N*	Leys (1–2 year), grazed or cut and grazed, high N†	Permanent pasture, cut only, grazed or cut and grazed
Leys (1–2 year), cut only	Long leys, cut only	
Permanent pasture, poor quality, matted	Long leys, grazed or cut and grazed, low N*	

Source: MAFF, 1994.
* Low N = less than 250 $kg\,ha^{-1}$ N per year or low clover content.
† High N = more than 250 $kg\,ha^{-1}$ N per year or high clover content.

Table 7.20 Soil nutrient concentrations and indices for P, K and Mg

Index	*Phosphorus (mg litre^{-1})*	*Potassium (mg litre^{-1})*	*Magnesium (mg litre^{-1})*
0	0–9	0–60	0–25
1	10–15	61–120	51–100
2	16–25	121–240	51–100
3	26–45	245–400	101–175
4	46–70	405–600	176–250
5	71–100	605–900	255–350
6	101–140	905–1500	355–600
7	141–200	1510–2400	610–1000
8	205–280	2410–3600	1010–1500
9	over 280	over 3600	over 1500

Source: MAFF, 1994.

in agricultural systems. The 'dust bowl' phenomenon of the 1920s and 1930s in the mid-West of the USA was a major environmental catastrophe caused by a combination of agricultural mismanagement and climatic variability (Goudie, 1986). Soil erosion by wind and water lead to declining yields and in some cases to land abandonment.

Accelerated soil erosion following agricultural activity is not new. Historians point to evidence of severe soil erosion being a contributory factor in the decline of the Roman empire following overcropping of vulnerable soils in the Mediterranean region (Seymour & Giradet, 1986). When soil loss exceeds the rate of soil formation, even for a short period of time, the result is declining productivity. In the UK soil erosion is not a common phenomenon, but it has been observed to occur on a wide range of soils in recent years (Speirs & Frost, 1985; Evans, 1990). Losses of soil by erosion are not sustainable in the long term, and measures must be taken to counteract erosive processes on vulnerable soils. The code of good agricultural practice for the protection of soil (MAFF, 1993) details methods for the conservation of the soil fabric, for example:

- stabilize by applying bulky animal manures or other organic wastes;
- maximize plant cover, particularly during the autumn;
- use long grass leys on vulnerable soils;
- cultivate on the contour;
- spread cereal straw on the soil surface.

Efficient nutrient management

Managing nutrients within agricultural situations is a complex and difficult task, due to the interaction between natural and 'artificial' processes. The efficiency of fertilizer utilization is typically 50% (N), 25% (P) and 50% (K), although actual values will vary according to soil, weather and crop management. The nitrogen cycle (Fig. 7.16) demonstrates the complex range of transformations that occur in soils. Fertilizer N is only one source of potential leaching losses: natural processes can also lead to significant removal of N and other nutrients from soils. In recent years concerns have focused on those nutrients that have a major impact on the wider environment,

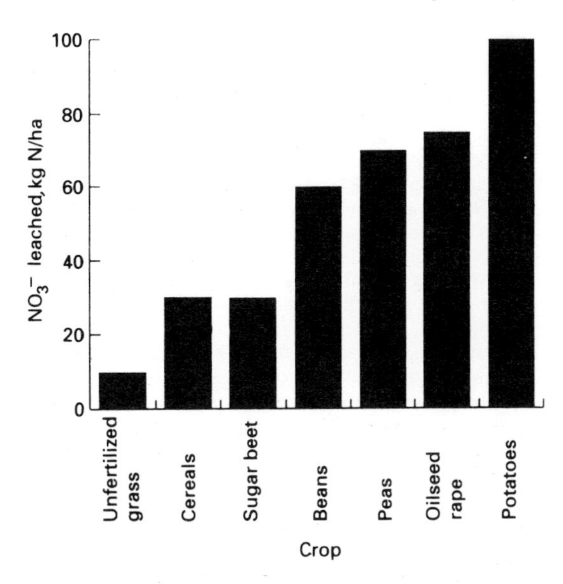

Fig. 7.23 Nitrate leaching loss from selected crops without use of manures (after MAFF, 1993).

notably nitrate and phosphate in water and nitrogen gases in the atmosphere. Efficient nutrient management based on an understanding of soil properties will lead to a conservation of nutrients within the soil, reduced reliance on artificial fertilizers and reduced environmental impact.

Significance nitrate losses from arable and grassland soils have been observed. Figure 7.23, taken from MAFF (1993), plots the losses that typically occur from a range of crops, although individual losses will vary widely according to climate, location, soil type and management regime. Principles of good soil, fertilizer and manure management to reduce these losses have been described in the code of good agricultural practice for the protection of water (MAFF/WOAD, 1991). Adherence to the code will benefit the farmer through efficient nutrient utilization. Some example recommendations are given below:

- Plough up grassland only when necessary, and reseed or drill next crop as quickly as possible.
- Apply organic manures to growing crops and not in the autumn.
- Split fertilizer dressings, and do not apply too early in spring.
- Maximize crop cover in the autumn.
- Incorporate crop residues to fix nitrogen.
- Delay autumn cultivations.
- Reduce grassland grazing intensity in the autumn.

Adherence to these and other recommendations is most important in areas vulnerable to nitrate leaching such as regions where groundwater accumulates in aquifers overlain by permeable soils, e.g. sandy or limestone-derived materials. The designation of nitrate-vulnerable zones where many of the practices described above are mandatory will help to reduce nitrate concentrations in drinking water to levels below 50 mg NO_3 litre^{-1}, the maximum specified by European Community Directive 80/778/EEC. Nutrient enrichment of natural waters, eutrophication, is caused not only by nitrate, but also phosphate, which can be lost from agricultural systems by surface runoff and soil erosion (Heathwaite *et al.*, 1990). Livestock systems have been observed to cause major losses when farm wastes are applied under inappropriate conditions or when surface compaction by animals has enhanced surface water runoff. Again, adherence to the codes of good practice for soil and water protection will minimize such losses.

Organic/sustainable systems

Many of the principles outlined above are embodied in the methods adopted by organic farmers. These are discussed in full by Lampkin (1990). Maintenance of long-term soil productivity can be ensured by practising methods that enhance natural soil fertility and so lead to a reduced reliance on agrochemicals. Reduction in the use of artificial fertilizers and other imported materials is central to the concept of sustainability, as is the aim of minimizing uncontrolled losses from agricultural systems. In practice the wider adoption of such methods is limited by economic constraints, but nevertheless moves towards less intensive systems with more attention paid to the soil and environmental controls on production will lead to greater efficiency of nutrient use and fewer problems of declining soil fertility than have been observed in recent years. Enlightened soil management based on a sound knowledge of soil properties is central to continued productive agriculture.

References

Addiscott, T., Whitmore, A. & Powlson, D. (1991) *Farming, Fertilizers and the Nitrate Problem*. CAB International, Wallingford, Oxon.

Armstrong, A.C. (1978) The effect of drainage treatments on cereal yields: results from experiments on clay lands. *Journal of Agricultural Science, Cambridge*, **91**, 229–235.

Avery, B.W. & Bascomb, C.L. (1974) Soil survey laboratory methods. *Soil Survey Technical Monograph No. 6*. Harpenden.

Brady, L.C. (1990) *The Nature and Properties of Soil*, 10th edn. Macmillan College Publishing Company, Inc., New York.

Castle, D.A., McCunnal, J. & Tring, I.M. (1984) *Field Drainage, Principles and Practices*. Batsford, London.

Church, B.M. & Skinner, R.J. (1986) The pH and nutrient status of agricultural soils in England and Wales 1969–83. *Journal of Agricultural Sciences, Cambridge*, **107**, 21–28.

Curtis, L.F., Courtney, F.M. & Trudgill, S.T. (1976) *Soils in the British Isles*. Longmans, London.

Davies, B.D., Eagle, D.J. & Finney, J.B. (1993) *Soil Management*, 5th edn. Farming Press, Ipswich.

Evans, R. (1990) Water erosion of British farmers' fields – some causes, impacts, and predictions. *Progress in Physical Geography*, **14**, 199–219.

Farr, E. & Henderson, W.C. (1986) *Land Drainage*. Longmans, London.

Findlay, D.C., Colbourne, G.J.N., Cope, D.W., Harrod, T.R., Hogan, D.V. & Staines, S.J. (1984) *Soils and Their Use in South West England*. Soil Survey of England and Wales Bulletin No. 14, Harpenden.

Goudie, A. (1986) *The Human Impact on the Natural Environment*, 2nd edn. Longmans, London.

Heathwaite, A.L., Burt, T.P. & Trudgill, S.T. (1990) The effect of land use on nitrogen, phosphorus and suspended sediment delivery to streams in a small catchment in Southwest England. In: *Vegetation and Erosion*, (ed. Thornes, J.B.), pp. 161–178. Wiley, Chichester.

Jackson, R.M. & Raw, F. (1966) *Life in the Soil*. Edward Arnold, London.

Jenny, H. (1941) *Factors of Soil Formation*. McGraw-Hill, New York.

Lampkin, N. (1990) *Organic Farming*. Farming Press, Ipswich.

McGrath, S.P. & Loveland, P.J. (1993) *The Soil Geochemical Atlas of England and Wales*. Blackie Academic, London.

MAFF (1970) *Modern Farming and the Soil*. HMSO, London.

MAFF (1982) *The Design of Field Drainage Pipe Systems*. Reference Booklet No. 345. HMSO, London.

MAFF (1986) *The Analysis of Agricultural Materials*. Reference Book 427. HMSO, London.

MAFF (1988) *Agricultural Land Classification of England and Wales*. HMSO, London.

MAFF (1993) *Solving the Nitrate Problem*. HMSO, London.

MAFF (1994) *Fertilizer Recommendations for Agricultural and Horticultural Crops*, 6th edn. RB209. HMSO. London.

MAFF/WOAD (1991) *Code of Good Agricultural Practice for the Protection of Water*. MAFF Publications, London.

MAFF/WOAD (1993) *Code of Good Agricultural Practice for the Protection of Soil*. MAFF Publications, London.

Mengel, K. & Kirkby, E.A. (1982) *Principles of Plant Nutrition*, 3rd edn. International Potash Institute, Berne.

Mullins, C. (1991) Physical properties of soils in urban areas. In: *Soils in the Urban Environment*, eds. P. Bullock & P.J. Gregory, pp. 87–118. Blackwell, London.

Nix, J. (1993) *Farm Management Pocketbook*, 24th edn. Wye College, London.

Parkinson, R.J. (1988) Field drainage and the soil environment. *Outlook on Agriculture*, **17**, 140–145.

Reid, I. & Parkinson, R.J. (1981) Too wet, too dry – clay soil problems. *Soil and Water*, **9**, 7–9.

Seymour, J. & Giradet, H. (1986) *Far from Paradise*. Green Print, Basingstoke.

Simpson, K. (1983) *Soil*. Longmans, London.

Simpson, K. (1986) *Fertilizers and Manures*. Longmans, London.

Speirs, R.B. & Frost, C.A. (1985) The increasing incidence of accelerated water erosion on arable land in the east of Scotland. *Research and Development in Agriculture*, **2**, 161–167.

Trafford, B.D. (1970) *Journal of the Royal Agricultural Society of England*, **131**, 129–152.

Twomlow, S.T., Parkinson, R.J. & Reid, I. (1991) *Agricultural Engineer*, **46**, 11–14.

White, R.E. (1987) *Principles and Practice of Soil Science*, 2nd edn. Blackwell, Oxford.

Wild, A. (1989) *Russell's Soil Conditions and Plant Growth*, 11th edn. Longmans, London.

Wild, A. (1993) *Soils and the Environment: An Introduction*. Cambridge University Press, Cambridge.

Further reading and sources of information

In addition to those texts quoted in the reference list, the reader is directed to the sources of information listed below. Soil maps may be obtained from the Soil Survey and Land Resource Centre, Silsoe Campus, Bedford MK 45 4DT (England and Wales), or Soil Survey of Scotland, Macaulay Land Use Research Institute, Craigiebuckler, Aberdeen AB9 2QJ.

Archer, J. (1988) *Crop Nutrition*. Farming Press, Ipswich.

Briggs, D. & Courtney, F. (1985) *Agriculture and Environment*. Longmans, London.

Farr, E. & Henderson, W.C. (1986) *Land Drainage*. Longmans, London.

Hodge, C.A.H., Burton, R.G.O., Corbett, W.M., Evans, R. & Seale, R.S. (1984) *Soils and Their Use in Eastern England*. Soil Survey of England and Wales Bulletin No. 13, Harpenden.

Jarvis, M.G., Allen, R.H., Fordham, S.J., Hazelden, J., Moffat, A.J. & Sturdy, R.G. (1984) *Soils and Their Use in South East England*. Soil Survey of England and Wales Bulletin No. 15, Harpenden.

Jarvis, R.A., Bendelow, V.C., Bradley, R.I., Carroll, D.M., Furness, R.R., Kilgour, I.N.L. & King, S.J. (1984) *Soils and Their Use in Northern England*. Soil Survey of England and Wales Bulletin No. 10, Harpenden.

Ragg, J.M., Beard, G.R., George, H., Heaven, F.W., Hollis, J.M., Jones, R.J.A., Palmer, R.C., Reeve, M.J., Robson, J.D. & Whitfield, W.A.D. (1984) *Soils and their use in Midland and Western England*. Soil Survey of England and Wales Bulletin No. 12, Harpenden.

Rowell, D.L. (1994) *Soil Science – Methods and Applications*. Longmans, London.

Rudeforth, C.C., Hartnup, R., Lea, J.W., Thompson, T.R.E. & Wright, P.S. (1984) *Soils and Their Use in Wales*. Soil Survey of England and Wales Bulletin No. 11, Harpenden.

8

Crop Physiology

M.P. Fuller & A.J. Jellings

Fundamental physiological processes

Analysis of green plant material shows that about 60–90% of fresh weight is made up of water. Of the remainder, the dry weight, about 45% is carbon, 45% oxygen, 5% hydrogen, 2–3% nitrogen, and 2–3% other elements such as potassium, phosphorus, calcium, magnesium and sulphur.

The crop obtains the carbon, oxygen and hydrogen from photosynthesis, and water, nitrogen, potassium and other mineral elements by uptake from the soil. These two processes are fundamental to crop physiology, and thus to efficient crop production.

Water and nutrient uptake

Water enters the roots near the root tips by diffusion across the cell wall. It must reach the xylem vessels embedded in the root by passing through or around the cells of the root to be delivered to the pores in the leaf surface (stomata) via the leaf veins (vascular bundles). It is the evaporation of the water from the stomata (transpiration) which creates the main driving force for water uptake from the soil. The moving column of water from the soil to the atmosphere through the plant is the transpiration stream. The creation of the transpiration stream is largely a physical not a biological process, in which the plant exists as a tube for the flow of water.

The pathway of water from soil to atmosphere thus involves four key steps: (i) through the soil to the roots, (ii) in and out of cells across membranes, (iii) along the xylem, and (iv) out of the leaves into the air. Water moves along the pathway down gradients of water potential (water potential may be defined as the concentration of water, and is thus always a negative value since pure water is said to have a water potential of 0), encountering resistance to flow in the soil, at membranes and cell walls, and at the stomata. Water uptake will be at a maximum when the stomata are fully open in conditions of warm moving air, in full sunlight, with high soil water availability. Water uptake will be reduced by low soil water availability, frozen soil conditions, low soil temperatures, low light levels, high humidity, and low air temperatures. A reduction in water uptake will eventually result in wilting and death.

Of the water taken up from the soil 95% is not used directly by the plant, but moves through and out of the plant (a plant may transpire its own weight of water in 1 h). Approximately 5% of the water taken up is used by the plant to give turgidity to the cells, to provide a medium for metabolism and to participate in the reactions of the cell.

Water uptake is the means by which plants gather mineral ions from the soil (or other root medium). Mineral ions are taken up by the roots in solution and thus the ions travel by the same pathway in the plant to be delivered to the cells of those leaves which are actively transpiring. Mineral ions in excess of the requirement of transpiring tissues are passed into the phloem to be retranslocated to non-transpiring plant tissues.

Root cells can be selective in their uptake of mineral ions, i.e. the relative concentrations of ions in different parts of the plant will be different from each other, and different from that of the soil solution. In this way plants can absorb elements in relative proportions different from that of the soil mineral composition.

Selective uptake operates through selective permeability of cell membranes. Active uptake up the electrochemical gradient requires energy from respiration (some passive uptake occurs by diffusion down the electrochemical gradient), so that a plant under stress or whose roots are in anaerobic conditions will be unable to acquire an adequate mineral supply. In addition mineral uptake is reduced by the same factors which reduce water uptake, e.g. crops often exhibit symptoms of mineral deficiency during a cold winter as soil water has been frozen and mineral uptake prevented.

Photosynthesis and respiration

Ninety-five per cent of the dry matter (DM) of plants is created from the gaseous environment, by the process of photosynthesis.

The molecules of the green pigment of plants, chlorophyll, absorb radiation in the visible light range 400–700 nm (photosynthetically active radiation or

PAR), and by means of the electron transport chain are used to make ATP (adenosine triphosphate) and NADPH (nicotinamide adenine dinucleotide phosphate). The energy thus conserved in these compounds can be used to drive the reactions of the carbon fixation cycle which fixes atmospheric carbon dioxide by converting it to sugars. Thus photosynthesis stores radiant energy by converting it into a chemical form. Agriculture manages the process so that human society can utilize its products as food and fuel.

The summary equation for both parts of photosynthesis is:

$$6CO_2 + 6H_2O \xrightarrow{\text{light}} C_6H_{12}O_6 + 6O_2$$

All the reactions of photosynthesis occur in the chloroplasts. The key enzyme to the fixation of carbon dioxide is ribulose bisphosphate carboxylase/oxygenase or rubisco. This is the most fundamentally important enzyme to life on earth, and it makes up a large part of the world's soluble protein fraction; in crop production it is one of the main beneficiaries of added fertilizer nitrogen.

The rate of photosynthesis depends on both internal and external factors. Internally the amount and activity of the participating enzymes, the amount of chlorophyll, and the anatomy of the leaf all dictate the photosynthetic capacity. The rate achieved at any one time will be affected by several environmental factors, of which the most important are light intensity, temperature and carbon dioxide concentration, the rate always being determined by the factor in shortest supply. Figure 8.1 demonstrates the general relationship between these three main factors.

The sugars produced by photosynthesis are used by the plant to provide energy through the process of respiration which occurs in the mitochondria and is essentially the reverse of photosynthesis. Photosynthesis and respiration are thus the means by which the plant utilizes light energy (which cannot itself be stored) to provide a continuous and translocatable energy source. This means that not all of the carbon fixed is translated into DM increase as a portion will be respired to provide energy and carbon dioxide, which must be refixed or lost.

Thus, growth analysis which measures the DM accumulation estimates net photosynthesis, and to measure gross photosynthesis more sophisticated techniques must be employed, such as gas exchange analysis under varying conditions, or the incorporation of radioactive carbon dioxide. In addition the process of photorespiration, which occurs under conditions of high light and high temperature, also takes up oxygen and releases carbon dioxide which may further disturb the interpretation of gas exchange measurements.

In most common temperate crops at average and above average levels of light intensity, carbon dioxide concentration is the commonest limiting factor for the rate of photosynthesis as light saturation is achieved at less than the light intensity experienced on most UK summer days, and photorespiration can be severe under good summer conditions (above 30°C) making carbon dioxide compensation points high.

Many tropical crops, e.g. sugar cane and sorghum, which use a modified photosynthetic pathway with little or no accompanying photorespiration, are not limited in their photosynthetic rate by carbon dioxide concentration until much higher light intensities. The only UK crop with this tropically adapted metabolism is maize, which, on hot summer days, may gain the advantage of higher photosynthetic rates and faster DM accumulation than its neighbouring temperate crops.

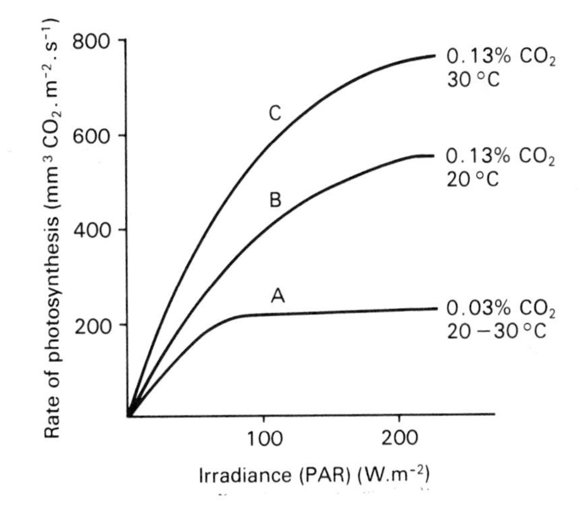

Fig. 8.1 The general relationship of photosynthetic rate with light intensity, carbon dioxide concentration and temperature in leaves of temperate crops. Curve A demonstrates that light saturation occurs at very moderate light levels under atmospheric carbon dioxide concentrations. Curve B shows that, if carbon dioxide concentration can be raised above atmospheric levels, photosynthetic rate can be stimulated, thus carbon dioxide concentration is the limiting factor to photosynthetic rate under normal conditions (curve A). In curve B temperature becomes limiting, as at higher temperature (curve C) there is a further increase in photosynthetic rate; this latter case is only applicable in controlled environments where carbon dioxide concentration may be artificially raised, e.g. glasshouse lettuce production.

Atmospheric carbon dioxide concentration changes very little with time of day or of year, but the internal leaf carbon dioxide concentration can be controlled to some extent by the opening and closing of the stomata. The stomata may open passively in response to changes in the gradient of water potential, but are mostly controlled by light levels, leaf temperature and internal carbon dioxide levels, which cause active, energy-requiring, stomatal opening. Photosynthesis will cease if water stress causes stomatal closure.

The carbon fixed by photosynthesis is moved around the plant as sugars in the phloem. The extent and direction of translocation is very important in crop production since it determines the partitioning of assimilates within the plant. A tissue actively importing sucrose is termed a sink, and one which is exporting sucrose is termed a source. Rate of growth and metabolic demand affect the strength of a sink, and positive feedback between sources and sinks allows increased demand by the sinks to stimulate increased rate of assimilation by the source tissues. Thus, as tubers form within a potato crop the photosynthetic rate of the leaves tends to increase (under standard conditions), but falls again if tubers are removed. The rate of translocation, determined by the size of the phloem vessels, also has some control over assimilate partition.

For temperate crop plants photosynthesis is only about 1% energy efficient. This low efficiency occurs for a number of intrinsic reasons: the conversion of light energy into chemical energy is about 25% efficient; about half of the total radiation is unusable by the plant; about half of the PAR is lost by reflection and transmission, and through respiration and photorespiration. This gives a possible maximum photosynthetic efficiency of about 5% which is further reduced to 1% by imperfect crop production conditions e.g. stress from pathogens, climatic factors, etc.

Growth

Growth is strictly the accumulation of dry matter, though the term is often used loosely to describe any increase in size.

Accumulation of dry matter (DM) occurs fundamentally at the cellular level. DM fixed by photosynthesis is laid down as new compounds within the cell so that individual cells tend to increase their mass with time. At the same time cells may divide to form new cells thus redistributing DM. Thus, when an individual leaf increases in DM, this is through a combination of both an increase in cell size and an increase in cell number. The way in which the cells are organized will determine whether the increase in DM is visible as an increase in leaf area, in leaf thickness or in leaf density.

An overall increase in size will include a certain amount of water accumulation. This is not true growth. In order to analyse crop growth it is usual to determine dry weight and leaf area at known time intervals. These data can be used to obtain the mean crop growth rate (Fig. 8.2) between successive sampling dates (t_x and t_y) and mean leaf area index:

Mean crop growth rate ($\overline{\text{CGR}}$) = $(W_y - W_x) \times 1/P$
(between t_x and t_y)

Mean leaf area index ($\overline{\text{LAI}}$) = $(A_x + A_y)/2P$
(between t_x and t_y)

where W = dry weight,
A = leaf area,
P = cropping area.

From the measurements of dry weight and leaf area an estimation of crop photosynthetic efficiency (net assimilation rate) can be made:

$$\text{Mean net assimilation rate } (\overline{\text{NAR}}) = \frac{(\log_e A_y - \log_e A_x)\,(W_y - W_x)}{(A_y - A_x)}$$

The relationship between these three parameters of growth is:

$$\overline{\text{CGR}} = \overline{\text{LAI}} \times \overline{\text{NAR}}$$

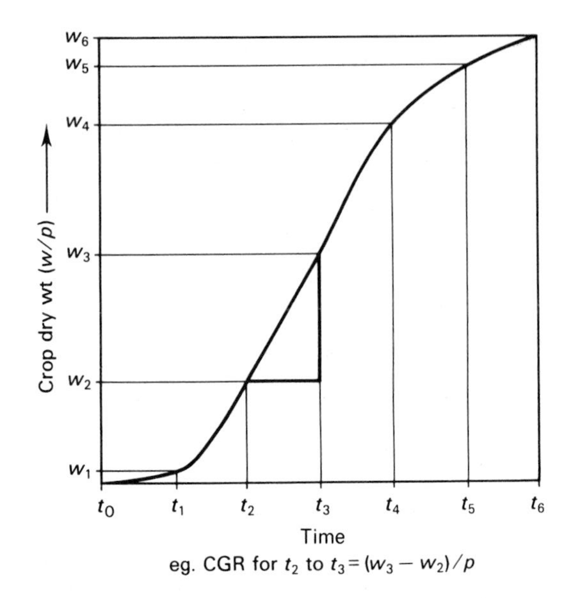

Fig. 8.2 Typical crop dry matter growth curve showing calculation of mean crop growth rate (CGR).

There are a number of other growth analysis formulae which are used to analyse growth in more depth and these have been the foundation of mathematical modelling of crop growth.

Potential yield

The total DM of a plant is termed its biomass. Potential biomass is determined largely by photosynthetic capacity. The proportion of the potential biomass which is economic yield is determined by the extent to which assimilate is partitioned into useful plant parts; this gives rise to the expression Harvest index.

$$\text{Harvest index} = \frac{\text{Economic yield}}{\text{Total biomass}}$$

Both photosynthetic capacity and harvest index are very largely genetically determined so that potential yield for any crop can be theoretically calculated for any given climate.

Many factors may limit photosynthesis to less than capacity, and may influence the partitioning of assimilates to give a less favourable harvest index so that actual yields rarely reach potential yields. This shortfall is termed the yield gap. The practice of efficient crop production tries to close this yield gap by understanding the relationships outlined in this chapter, and acting upon them as described in other chapters of this book.

Crop development cycle

From germination to maturation, crops go through a development cycle which involves an increase in complexity usually accompanied by an increase in biomass. The cycle is a result of the interaction of the genotype (see the 'Plant breeding' section at the end of this chapter) of the plant and the environment it is grown in. The environment provides not only the basic needs of the crop for growth, i.e. light, nutrients, water and heat, but also provides stimuli to trigger developmental sequences.

During its development the crop exists in several 'states' and passes from one state to another in an organized manner via a number of 'processes' (Fig. 8.3).

Germination

The germination of a non-dormant seed will be initiated given the presence of water, an aerobic medium and sufficient warmth. (For most temperate crops the temperature must be above 0–5°C for germination, with an optimum of about 30°C. A few cold-sensitive crops will not germinate below a higher temperature, e.g. maize requires a minimum temperature of 10°C for germination.) Water is imbibed through the testa or through the micropyle and returns the existing cells to full turgor. When a seed enters germination it passes from a quiescent state with extremely low metabolism and a high resistance to drought and frost stress into a state of high metabolic activity and extreme susceptibility to stress. Initially water content of the plant rises from 5–10% to 30–40% as water is taken up by imbibition (the colloidal attraction of water by large molecules in the seed). At this point many plants progress directly into active metabolism, but in those which show dormancy mechanisms germination may cease until dormancy has been broken. Thus, many weed seeds exist in the soil in an imbibed state but ungerminated. In the absence of dormancy or when dormancy has been broken, germination can continue and metabolic rate rises dramatically, with an increase in protein and nucleic acid synthesis. Food reserves are mobilized, often as a result of hormone stimulation of enzyme synthesis. For example, in barley, gibberellic acid stimulates the mobilization of protease and the synthesis of α-amylase, which catalyse the breakdown of starch in the endosperm for translocation as sugars.

The expansion of the embryo and its development into seedling roots and leaves rupture the testa, from which time the seedling ceases total reliance on its stored food reserves and begins to harvest the resources of its environment. Initially, a greater proportion of the reserves are directed to root growth in most species to ensure an adequate supply of water and minerals before the photosynthetic machinery is fully synthesized and operational.

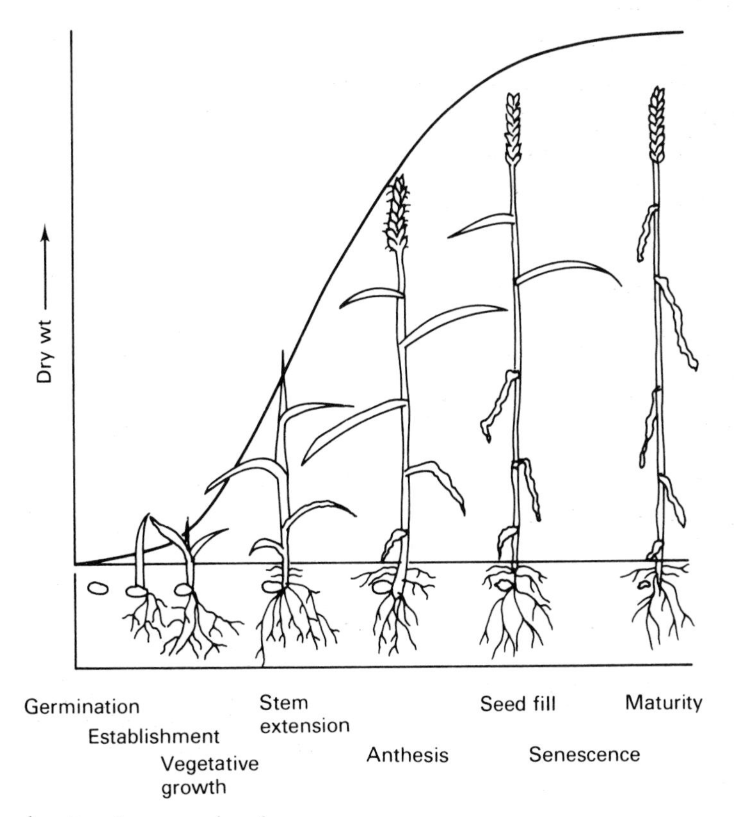

Fig. 8.3 Growth cycle pattern in an annual seed crop.

The size of the seed, which tends to be related to the volume of food reserves, dictates the length of time for which the developing embryo can delay becoming fully independent, e.g. germinating field beans can survive on seed reserves for much longer than germinating oilseed rape. This factor has considerable influence on the optimum depth of sowing for each crop species. In order to achieve early photosynthesis to supplement the seed reserves, the cotyledons are frequently expanded above ground in small-seeded crops.

The germination percentage of a seed lot is an important concept in terms of both the calculation of correct sowing rates, and for complying with the regulations governing the sale of seed.

Vegetative development

During vegetative development, the shoot apex develops primordia which develop into leaves. A leaf first appears as a bump (primordium) on the side of the apex and then undergoes cell division and differentiation. A midrib consisting of protoxylem and protophloem develops and establishes contact with the vascular system of the stem so that it can obtain assimilates and water for its continued development. As the leaf emerges into the light, cell components differentiate, proplastids develop into chloroplasts and the leaf appears green. Assimilation can proceed as soon as the chloroplasts and stomata are functional.

The point of attachment of the leaf on to the stem is important as it is here that the vascular connections are made allowing water to pass to the leaf for transpiration and for assimilates to pass from the leaf for relocation to other growing parts or storage organs. This connection of leaf and stem is called a node, and the stem between two nodes an internode. In many plants, particularly cereals and oilseed rape, there is a stage of development when the internodes are not expanded and the plant takes on a prostrate or rosette appearance. This adaptation keeps the apex of the plant close to the ground where it is more protected from frost and is a useful feature of over-wintered crops. In the spring, such plants demonstrate internode expansion and the apex is pushed up into the air as the crop approaches flowering. Internode expansion is primarily expansion growth of stem cells and is followed by the strengthening of the cell walls to give the stem stiffness. At any one time usually only one internode is expanding rapidly although expansion phases of successive internodes frequently overlap. In most plants, this internode expansion is under the control of gibberellin in the plant, and chemicals which interfere with its synthesis or action are commercially available as stem shorteners.

The final number of leaves that a plant bears (assuming it does not branch) is dependent on the duration of the vegetative phase of the stem apex. At some point the stem apex will receive a stimulus which will trigger a change from producing leaf primordia to producing floral primordia. This change in function of the apex is invisible to the naked eye and usually does not alter the outside appearance of the plant for some considerable time. For example, in winter barley, floral induction often occurs in November or December but the ear does not appear until the following May. Microscopical examinations of stem apices have allowed the development of growth keys which can have a direct bearing on agricultural practices, e.g. nitrogen timing to winter wheat. The stimuli which trigger such changes in the development of the stem apex include photoperiod and low temperature (vernalization) and, in some plants, growth for a set number of days of thermal time.

At the same time as above-ground growth and development are occurring, root growth and development are also progressing. Root apices branch to provide a network of structures specialized for the uptake of water and soil nutrients. Apices are located behind a protective root cap and produce cells which go through a cycle of development. Early in development new external cells have root hairs which increase root surface area and therefore absorptive area for water and nutrients.

Normally in a crop the root system develops in a progressively deeper and more extensive manner, maintaining a balance with the shoot size (root:shoot ratio); however, often as rapid spring growth occurs root growth declines. It becomes important therefore to establish a well rooted crop prior to rapid spring growth.

Reproductive development

Reproductive development starts with the induction of flowering. Externally the plant will continue to expand leaves which have been initiated up to the point of floral induction. By the time the floral structures are visible to the naked eye they are already fully developed with stamens bearing immature pollen and ovaries with developed ovules. The actual process of flowering involves the opening of the flowers, rupture of the stamens and release of pollen (called anthesis), pollination of the stigma, growth of the pollen tube and, finally, fertilization of the ovules. In relation to the whole period of floral development, flowering is only a brief phase.

Some crop plants show remarkable synchrony of flowering, e.g. cereals, with all florets flowering within a few days (determinate flowering). Others, however, have a protracted flowering period lasting weeks (indeterminate flowering), e.g. field beans and oilseed rape. Determinate flowering is a desirable feature in a crop plant since it leads to an evenness in ripening.

Seed development

Following fertilization cell proliferation of embryo and storage cells occurs. The embryo tissues develop into a polarized structure which defines the sites of the stem apex and the root apex. Often the shoot apex undergoes further development and will produce leaf primordia, e.g. in wheat the first three to four leaves are already initiated in the embryo before the seed is shed from the plant. The mass of storage tissue cells becomes filled with storage compounds – starch, lipids, protein.

The duration of seed filling is temperature dependent, lasting longer at cooler temperatures. If the supply of assimilates during this time is limited, the seed may not fill to its maximum capacity and will appear shrivelled or may abort. In a crop plant, therefore, greater seed filling capacity will occur in a cool environment where the length of seedfill will be protracted, provided the crop is not stressed by drought and provided photosynthesis is not impaired. This contributes to the higher yield potential of winter wheat in Scotland compared with southern England. As seed filling progresses, the moisture content of the seed falls. At the end of the seed filling, the process of maturation begins with continued loss of moisture from the seed, the cessation of development of the embryo and frequently the induction of dormancy. Vascular connections between the seed and the mother tissues are discontinued and the seed becomes an independent body. Some further seed development can occur after seed shed but is usually completed when moisture content declines to 5–10%.

Agriculturally, seeds are of major importance as the main food source for animals and man (e.g. wheat, barley, rice maize, beans), as well as providing many beverages (e.g. coffee), medicines (e.g. evening primrose), spices (e.g. mustard), and industrial oils (e.g. linseed).

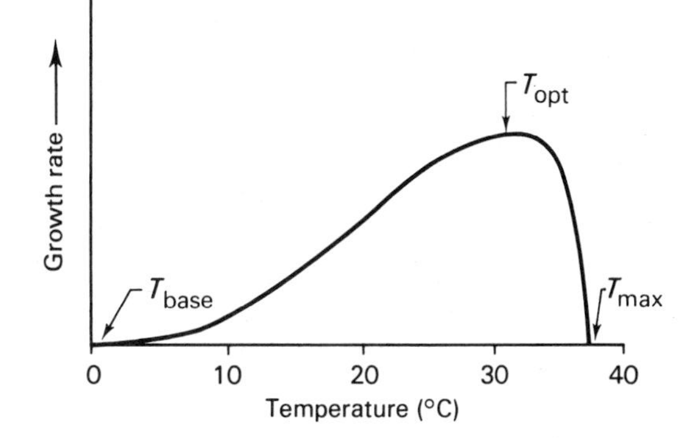

Fig. 8.4 Growth rate response of plants to temperature.

Senescence

Senescence is a two-phase process the first phase of which is an organized breakdown of the tissues where many of the products of the breakdown can be remobilized to other parts of the plant. In this way certain nutrients can be reused for new growth. The initial stages of senescence are not visible to the naked eye but involve a depression of photosynthetic activity followed by a dismantling of the photosynthetic apparatus leading to a breakdown of chlorophyll. The second phase of senescence is the rapid self-destruction of cellular integrity by powerful hydrolytic enzymes.

Following senescence, leaves are often shed from the plant after the formation of an abscission zone in the stem at the point of attachment of the leaf. The cells in this zone become filled with inert substances which effectively act as a barrier to pathogens at the scar left when the leaf is shed. Abscission layer formation is a feature of dicotyledonous plants; monocotyledonous plants such as cereals do not form such layers.

Senescence is occurring throughout the life of a crop plant as old leaves are being replaced by young leaves. However, 'crop senescence' is a term which is usually applied to a crop where the rate of senescence is greater than the rate of growth and the net effect is that the crop is dying. Crop senescence is usually noticeable from about mid-seedfill onwards.

Environmental influences on growth and development

Temperature

Temperature and growth rate

The growth rate of most plants responds to changes in temperature (Fig. 8.4). The maximum temperature for growth (T_{max}), the optimum (T_{opt}) and the base temperature (T_{base}) vary between species but for the UK crops typically all fall between 0 and 35°C. For crops native or well adapted to the UK, e.g. perennial ryegrass and winter wheat, T_{base} is around 0–1°C. However, for crops originating from warmer climates, e.g. maize and sugar beet, T_{base} is around 6–8°C. As a consequence, winter wheat and ryegrass are able to grow early in the spring (March and April) while maize and sugar beet do not commence rapid growth until much later (June/July). In a temperate climate this has important consequences with respect to canopy development (see the section on 'Light capture by crop canopies'), radiation capture and yield potential.

Temperature and development

The progression of a plant through a sequence of development is closely related to the temperature it experiences over a period of time. Temperature varies both diurnally (within a day) and on successive days, and plants integrate their response to temperature over time. Thus, the amount of heat a plant experiences in a day can be called its daily heat unit and a set number of heat units is required between developmental stages.

Some heat unit systems are simply the sum of the average daily air temperatures while others use the daily maximum and minimum air temperatures in a more complex formula.

Heat units may be used in crop production in at least two important ways. First, with reference to past meteorological data, areas can be mapped for suitability for growing particular crops. For example, forage maize can only be grown successfully in the UK in areas which experience over 1250 heat units over 6°C during the months of May to October inclusive (see forage maize section in Chapter 10). Secondly, by using up-to-date meteorological data, the stage of development of a crop in the field can be pinpointed and its subsequent development over the forthcoming week or two can be predicted. Where husbandry inputs are very time responsive, such prediction can be of great benefit in helping to achieve optimum use of inputs. The use of such prediction schemes are available for nitrogen and growth regulator timing to cereals.

Heat units may be modified by photoperiod, which can effectively delay or hasten the thermal response. Temperature can also affect development by acting as a trigger for a change in the development sequence. In particular, low temperature can trigger flowering (vernalization) and dormancy.

Vernalization
Many biennials (e.g. sugar beet), winter annuals (e.g. winter cereals) and perennials (e.g. perennial ryegrass) require a period of vernalization over winter before flowering is possible the next spring. In the UK this is normally achieved quite adequately during the winter but if winter cereals are sown very late (after February–March) then flowering and yield may be seriously impaired. Within winter wheat there exist varietal differences in the vernalization requirement and care should be taken not to sow varieties with low vernalization requirement early in the autumn and, conversely, not to sow varieties with a high requirement too late.

Maincrop sugar beet production exploits only the first year of the biennial cycle of this plant, and flowering in the first year (bolting) reduces crop yield, interferes with machinery and can introduce a weed beet problem. The vernalization requirement of the crop usually prevents flowering in the first year, but if the crop is sown early in the spring or if unusually low temperatures are experienced, flowering of a proportion of the crop may occur. Plant breeders have attempted to produce varieties of high vernalization requirement so that early sowing can be practised.

Dormancy
Low-temperature-induced dormancy can be regarded as a survival strategy in certain plants. True dormancy is common in woody perennials and bulbs but is rare in UK agricultural crops. Perennial ryegrass exhibits a winter quiescence which appears similar to dormancy but upon close study it is found that new leaves are continually being produced at the same time as the death of old leaves occurs. Low temperatures in winter and the lack of net growth are often mistakenly referred to as dormant periods. Close inspection would reveal that development has progressed whereas in true dormancy development ceases.

Dormancy is also apparent in seeds and helps prevent germination in unfavourable environments. Thus weeds which have little or no frost resistance in the vegetative state require a cold shock to the seed (obtained over winter) to break their dormancy before germinating in the following spring. In this way the weed avoids frost which may kill off the species.

Other dormancy-breaking stimuli include extremes of heat, exposure to light, exposure to high concentrations of specific ions, degradation of seed coverings, washing away of inhibitors and the passage of time to allow seeds to fully mature or ripen. Seed dormancy is common in weed species and this, coupled with the prolificity of seed set of many weeds, makes it virtually impossible to eliminate completely a weed species from a field. In contrast, crop species have had dormancy mechanisms bred out of them to ensure rapid and even germination after sowing. Even so, cereals retain some dormancy and it helps prevent sprouting in the ear before harvest which is extremely important both for breadmaking wheat and malting barley.

Temperature-induced stress

Temperature stress to plants can be caused by extreme heat or cold although heat stress is not normally encountered in UK crop production. Cold stress however, particularly frost damage, commonly occurs and can lead to complete plant death or to a degree of damage. Some crop plants have little or no inherent frost tolerance, e.g. maize and potato, and for this reason are restricted in their cropping to the frost-free months (May–October) in lowland areas. Other crop species have varying degrees of frost tolerance and there is frequently varietal variation within a species.

The obvious advantage of frost resistance in a crop species is that it allows the crop to be planted in the autumn which generally raises the yield potential. Species of plants which are capable of withstanding frosts in the winter are not equally frost-hardy throughout their life cycle. During the declining temperatures and daylengths of autumn they undergo a conditioning process known as hardening. This raises the frost-hardiness in 'anticipation' of the ensuing winter. In the following spring as temperatures and daylength increase, this hardiness declines and the crop may become susceptible to damage by late frosts. Hardening temperatures are typically in the range 0–8°C.

Some plants escape damage by frost by a frost-avoidance mechanism. The extreme of this is for the plant to exist in a very resistant form, e.g. a seed or a bulb. Another avoidance mechanism is undercooling (or supercooling) where the tissues attain a temperature below their freezing point without the formation of damaging ice crystals. This may only be effective for mild frosts (−1 to −2°C) of short duration and can be dependent upon the absence of ice nucleators (e.g. certain bacteria). Other plants protect their sensitive tissues with layers of hardy tissue which delay and therefore prevent the penetration of damaging frosts, e.g. the wrapper leaves on winter cauliflowers and the bud scales on woody deciduous perennials. With some high value crops it can be worthwhile investing in frost protection measures such as the use of polythene sheeting, e.g. for early potatoes, or of fogging or misting techniques, e.g. to provide blossom protection in orchards.

Light

Photoperiod

Plants have adapted to use photoperiod in many ways because it is the only true predictable parameter of the environment detectable by the plant. Notable examples occur with flowering and germination which help a species to synchronize critical phases of development. All photoperiodic responses involve the compound phytochrome as the detector compound at the cellular level.

In respect of flowering, plants can either be photoperiodic responsive or day-neutral. Day-neutral plants flower after a set time from germination or dormancy breaking, e.g. annual meadow grass. Photoperiodic responsive plants are either 'long-day' or 'short-day' types requiring exposure to the correct daylength to either start or to accelerate flowering. Temperate plants are frequently long-day types, which helps ensure flowering in the summer months.

Light capture by crop canopies

The crop canopy is defined as the structure of leaves, stems and flowers found above ground. At full development the depth of the canopy can vary from 30 cm (grass) to more than 30 m (agroforestry).

It is important that for a crop to achieve high growth rates and high yield potential it must (i) fully intercept the incoming solar radiation, (ii) make efficient use of that radiation within the canopy, and (iii) produce its canopy at a time when incoming radiation is at its maximum.

It is found that the growth rate and the dry matter production of many crops is linearly related to the quantity of radiation intercepted during the life of the crop. Incoming radiation and therefore cropping potential vary considerably throughout the year in temperate regions with average daily radiation totals of $16\,\mathrm{MJ\,m^2\,d^{-1}}$ in June/July falling to only $2\,\mathrm{MJ\,m^2\,d^{-1}}$ in December/January.

Within a day, actual radiation receipts (the daily insolation) depend on the time of year, which dictates the daylength and maximum radiation intensity, and the degree of cloud cover. Cloud cover may reduce the actual radiation reaching the crop by as much as 75% compared with a clear sunny day. In this way excessive cloud cover can reduce cropping potential whilst in years of high radiation receipt (little cloud) cropping potential is raised. This was certainly a major contributory factor in the record-breaking yields obtained from cereals in 1984.

Measurements of crop canopies have revealed two important expressions which are useful: leaf area index (LAI) and leaf area duration (LAD). LAI is a measure of the surface area of the photosynthetic area of a crop per area of ground occupied whilst LAD is the integral of LAI with time (LAI × time). Thus, on a graph of LAI versus time, LAD is the area under the LAI curve. The value of LAI that has to be achieved by a crop before efficient interception of radiation occurs varies with the crop and the plant spacing but commonly needs to exceed 3–4. Some crops that are well adapted to improving radiation conditions in the spring, e.g. cereals, are able to adapt their canopy structure and raise their LAI to more than 10 and achieve high growth rates and high physiological efficiency.

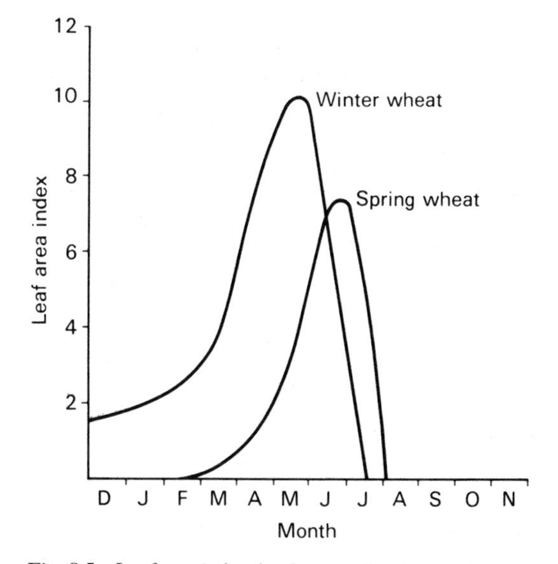

Fig. 8.5 Leaf area index development in winter and spring wheat.

Factors affecting light interception by crops

Early sowing Advancing the sowing date of most crops leads to a better chance of early ground cover and a better timing of the peak LAI closer to mid-summer when radiation levels are higher. This effect is maximized if the crop is frost resistant and sowing can be done the previous autumn, e.g. cereals, oilseed rape. Such early sowing usually also prolongs the growing period allowing more radiation to be intercepted by the improved LAD. This can be illustrated with reference to wheat (Fig. 8.5) where autumn sowing leads to increased yields through improved canopy timing, higher LAI and increased growing period leading to greater light interception over the life of the crop. Many crops are, however, limited in the degree to which the sowing date can be advanced by temperature limitations, e.g. sugar beet, potatoes, maize, peas. Here, the advantages of early canopy development are offset by the risks of frost damage or poor slow germination. In some high value cash crops temperature protection can be supplied by plastic film which acts as a 'mini-greenhouse', e.g. for first early potatoes, early vegetables and sweetcorn.

Seedrate Increasing seedrate can have the advantage of producing an early rise in LAI with ground cover occurring early. This may be of particular importance in crops with short growing periods, e.g. spring and summer sown crops. Indeed, in spring cereals seedrate becomes more and more critical with later and later sowing. For crops with long growing periods compensatory growth (side-branching, tillering) frequently diminishes the advantages of high seedrate. Using high seedrates can produce disadvantages to the crop by leading to excessively high competition between plants. This has the effect of lowering the size of important structures, e.g. grain or root size, and can lead to lodging (canopy collapse).

Nitrogen fertilizers The application of nitrogen fertilizer to most arable crops raises yield potential. This is achieved through a stimulation of cell growth, particularly in the developing leaves, which increases leaf size and thus leaf area index. Also, leaf longevity is improved, i.e. senescence is delayed, which helps to improve LAD. Nitrogen also improves photosynthetic capacity (see the earlier section headed 'Photosynthesis and respiration').

If nitrogen is available in very high amounts, then over-stimulation of growth can occur which can be counter-productive. Leaf area index may be increased to such an extent that lower leaves in the canopy are severely shaded and fall below the compensation point (respiration greater than photosynthesis) and such leaves effectively become 'parasitic' on the plant. Also, excess nitrogen may cause a weakening of stems leading to complete canopy collapse (lodging). Furthermore, stimulation of vegetative growth by nitrogen can be at the expense of the growth of important structures, e.g. tubers in potatoes, roots in sugar beet.

Canopy structure Photosynthesis in the leaves of most crop plants is saturated at radiation intensities of about $300\,\mathrm{Wm^{-1}}$ (or $150\,\mathrm{Wm^{-1}}$ of photosynthetically active radiation). This means that, on many days in spring and

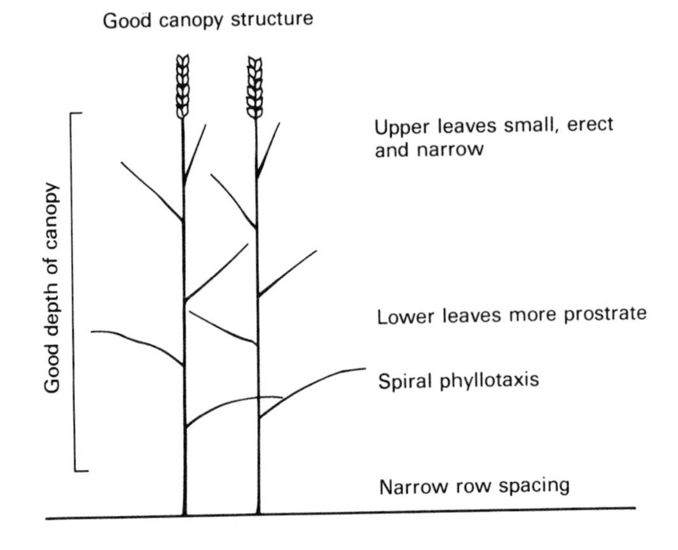

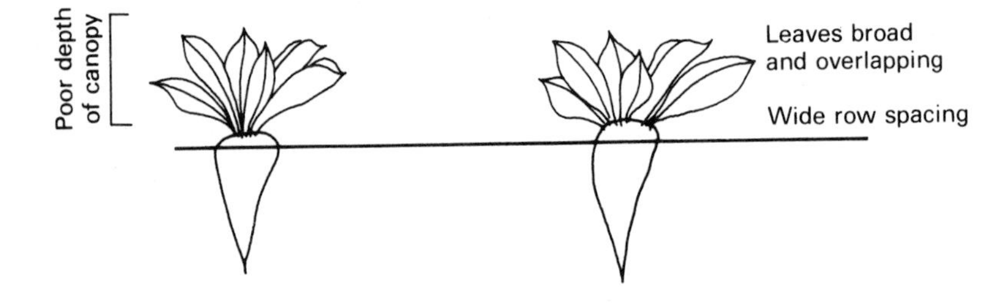

Fig. 8.6 Crop canopy structures.

summer, leaves at the top of a canopy are saturated in radiation for maximum photosynthesis. Unfortunately, such leaves will continue to absorb radiation but will be unable to utilize it and the radiation is therefore wasted because it is not available to the rest of the canopy. There is benefit to crop photosynthesis if the extra radiation can be allowed to penetrate the canopy and be intercepted and used by lower leaves. Canopy structures that allow such penetration have been shown to improve crop growth rate and yield (Fig. 8.6). It is perhaps not surprising to find that crops such as cereals which are well adapted to the UK climate show the best type of canopy structure.

Common problems depressing canopy effectiveness

Lodging The collapse of the canopy caused by wet and/or windy weather conditions and/or the weakening of the stem and/or roots will drastically reduce crop growth rate. Lodging is most serious in deep well-structured canopies with high yield potential, e.g. cereals. As well as depressing yield, lodging may also lead to a spoiling of the economic product by soil contamination or rotting diseases, and can lead to a surge of weed growth up through the crop. Lodging may not be serious if the crop is nearly ripe but will frequently slow down and delay harvest.

Lodging commonly results from weakened stems through high nitrogen use, high seedrates, stem-based diseases or weak-stemmed varieties and from poor supportive root systems occurring in unconsolidated seedbeds.

Foliar diseases Foliar diseases depress crop yield by depressing the effective leaf area and thus lowering the LAI. Many diseases are actually affecting a greater leaf area than is apparent from their visual symptoms, and this means that they have to be controlled at a relatively early stage in their infection. Infection on lower leaves in the canopy is generally insignificant since these leaves are contributing least to the gross production of the crop. Similarly, in autumn-sown crops autumn disease infection is frequently not worth controlling since crop growth rate is slow and protection of leaf area is not as necessary as during the main growth period in spring.

Unfortunately, well structured crop canopies are frequently ideal environments for disease multiplication and spread, and in attempting to achieve high yields from good canopies, disease control will normally be necessary.

Drought Most UK crops can withstand a good deal of water stress without injury. However, even mild stress can provoke a drought response which lowers productivity of the canopy. Stomata close to conserve moisture under drought stress and at the same time restrict carbon dioxide uptake into the leaf and depress photosynthesis. If drought stress is severe or prolonged, it will have an adverse effect on canopy structure since older leaves will senesce and die whilst younger leaves may wilt or roll up. New leaves produced during drought stress will be smaller and less photosynthetically responsive.

Since the risk of drought stress is frequently towards the end of a growth cycle, it is likely to affect the crucial partitioning of assimilates to the economic portions of the crop and can drastically affect economic yield.

Competition

Competition occurs when two or more plants are growing in an environment and the combined demands of the plants exceed the supply of one or more of the limiting factors for growth and development. These factors include water, soil nutrients, soil oxygen, carbon dioxide and light. Space is frequently referred to as a limiting factor but in reality space embraces two or more of the factors already listed.

Within a crop, competition occurs between plants of the same species and is termed intraspecific competition. Between plants of different species it is termed interspecific competition.

Intraspecific competition (plant populations)

In the extreme case of a crop plant growing in complete isolation, i.e. in the absence of competition, its individual yield will give an indication of the maximum yield possible per plant. As the population is increased, competition between plants will increase and individual yield per plant will decrease. However, yield per unit area (crop yield) will increase (Fig. 8.7). This is because, despite each plant not fulfilling its full potential production, a more efficient use of the limiting factors for growth is being made by the community of plants (the crop). This is particularly the case with respect to the efficiency of utilization of light and the ability of the crop canopy to capture more light at higher populations. As population increases, the yield response starts to diminish until a plateau is reached and no further response to plant population can be achieved (asymptotic response curve – Fig. 8.8). In biological terms the optimum plant population in such a case is at the point where the plateau starts whereas in practice the optimum is lower than this when seed costs are taken into account. A dense community of plants also has the advantage of synchronizing and accelerating ripening.

Plant density affects the partitioning of assimilates within the plant with high densities tending to lead to a greater vegetative component and a lower reproductive or storage component per plant. If the adverse effect of partitioning away from an economic portion is not outweighed by the advantage of increased plants per unit area, then economic yield will decline beyond a certain point (parabolic response curve – Fig. 8.8). Often an asymptotic yield response is associated with total biomass production and crops grown for complete utilization, e.g. grass and other fodder crops. In contrast, the parabolic yield response is associated with crops grown for economic yield, e.g. cereals, oilseed rape, potatoes, sugar beet, peas.

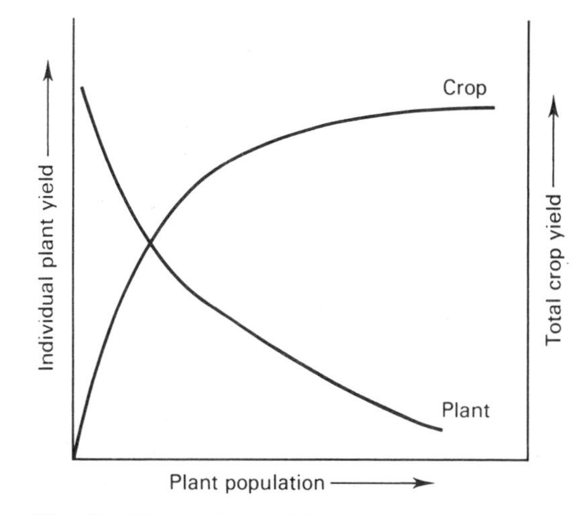

Fig. 8.7 Plant and crop yield in response to increasing plant population.

When considering crops showing a parabolic yield response, obtaining the optimum plant population is very important. However, if a suboptimal population is established, then compensatory growth may occur as individual plants can have a greater share of the limiting resources and attain a larger size. In crops able to show a high degree of compensatory growth, e.g. winter cereals (by way of tillering), the parabolic curve becomes a wide flat-topped response curve.

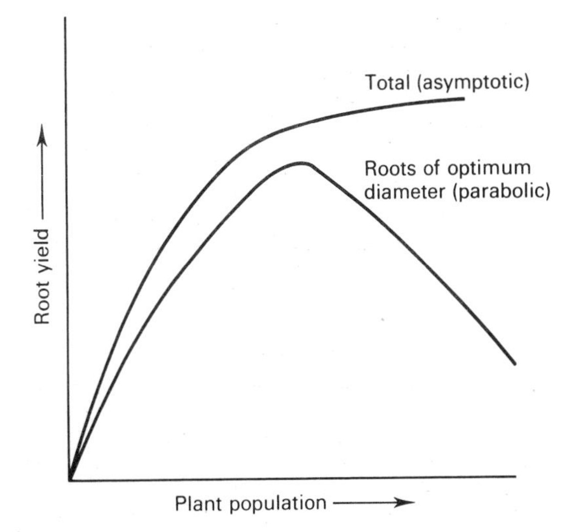

Fig. 8.8 Response of root crops to increasing plant population.

In many crops, especially those which show a high degree of compensatory growth, the unit of population may not always be simple to define. For example, in potatoes it is the stem population and in cereals the tiller population that is of the greatest value. Competitive demands in the crop are not constant but change with time and may lead to alterations in the density of the unit of population (Fig. 8.9). In the case of cereals, the mortality of stems in the spring is a result of competitive demands within the plant for assimilates and can be greatly offset by applying nitrogen to improve leaf area and thus assimilate supply.

Increasing plant populations can also lead to higher mortality rates within a crop (thinning response). Usually, in a highly competitive situation, 'weaker' genotypes or individuals located in an unfavourable position will be disadvantaged and may well die. This could be considered to be a waste of seed but may be a necessary loss to achieve the desired plant population.

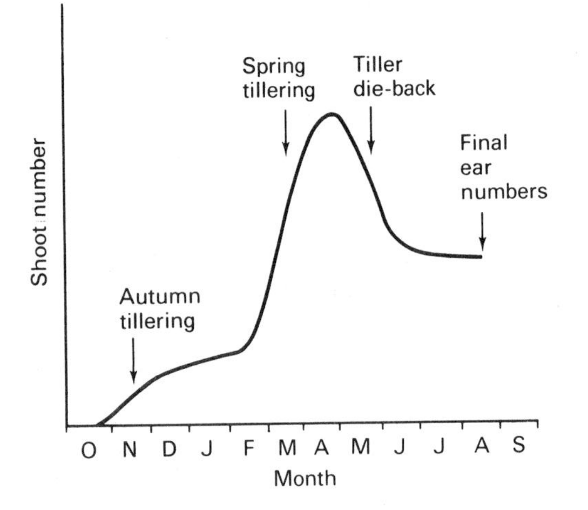

Fig. 8.9 Variation with time of the unit of density in a winter cereal crop.

Interspecific competition

Competition for resources occurs between species in crop mixtures, e.g. grass/clover swards, cereal/legume arable silage crops, where two or more crops are harvested (or grazed) together, and between a crop and its weeds. (Intercropping, where two or more crops are grown together but harvested separately, e.g. maize and beans, is very little used in the UK but is common in other countries. The physiological principles of intercropping are identical to those of mixtures.)

The total biomass produced per unit area of ground when more than one species is present will frequently be greater than that produced by one of the species grown alone at the same density. This is because species of different habit are able to utilize the available resources more efficiently, chiefly by better spatial distribution within both the aerial and soil environments, e.g. by forming a more efficient leaf canopy, or by tapping a greater depth of soil for water and nutrients. In general the greater the difference in habit of the component species the greater will be the yield advantage of the mixture, and conversely the less will be the degree of competitive pressure between the component species. This yield advantage is exploited by crop mixtures, but may be disadvantageous in crop/weed situations, depending on the degree of weed competition.

Quantitative analysis of the degree of competition and its effects is important to determine (i) the optimum mixture proportion, and (ii) the economic threshold for weed control. Many forms of analysis have been devised appropriate to each type of interspecific competition situation. One of the simplest and most widely useful is the replacement series analysis for two-species competition. Yield data are collected from the monoculture of each species, and from a range of mixture proportions (it is usual to employ the same plant density throughout the experiment though this may lead to bias in the case of two species with widely differing single plant sizes), and analysed using a replacement series diagram (Fig. 8.10). This form of analysis can indicate the optimum mixture proportion in yield terms, the extent of the yield advantage, and the behaviour of the component species within the mixture. Three cases are possible:

(1) mutual cooperation, in which the yield of both components is enhanced giving an overall yield advantage;
(2) mutual inhibition, in which the yield of both components is depressed giving an overall yield reduction;
(3) compensation, in which one component is enhanced and the other depressed; this case may lead to a yield advantage, a yield depression or a balanced result, depending on the relative effects on each component.

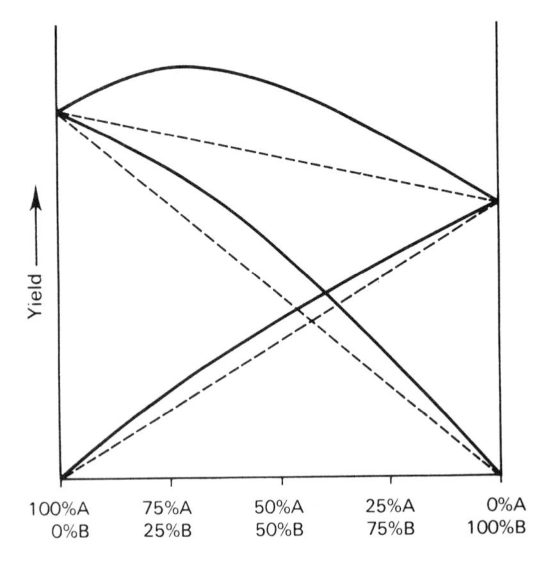

Fig. 8.10 Theoretical replacement series diagram to analyse the performance of a mixture of species A and B. Solid lines indicate observed yields; broken lines indicate expected yields (extrapolated from monoculture yields) if no competition were to occur. Diagram demonstrates mutual cooperation with a maximum mixture yield benefit from 70%A and 30%B.

Analysis of this type often demonstrates that one of the component species has competitive superiority. Competitive superiority is conferred by general spatial superiority, e.g. greater height, a broader leaf canopy, better root distribution, all enabling the species to gain a disproportionately high share of the light, water and nutrients, by shading its neighbours, achieving a higher photosynthetic rate, increasing its transpiration rate and thus its rate of mineral uptake, resulting in higher growth rates and even greater competitive superiority. Competitive advantage may also be achieved through temporal means, e.g. weeds emerging prior to crop emergence will tend to dominate, and vice versa, though this relationship will be mediated by intrinsic, and temperature influenced, growth rate and habit differences. Yield advantage from competition may be enhanced by symbiotic relationships, e.g. legumes donate atmospherically fixed nitrogen from their decaying nodules to their non-legume partners. Competitive advantage is gained by some species through allelopathy, the exudation of toxic substances from their roots.

Spatial and temporal distribution interact and are profoundly influenced both by the environment and the overlying plant density, making physiological interpretation and general prediction of competitive situations extremely difficult, particularly when more than two species are involved.

Modification of crops by growth regulators and breeding

Plant growth regulators

Regulatory compounds known as plant hormones (Table 8.1) influence many of the processes of plant growth and development. These compounds act both independently and interactively at several levels of concentration to achieve control. The internal (endogenous) levels of some of these hormones can be manipulated either by applying further hormone to the plant (exogenous application) or by applying synthetic regulators which depress the endogenous hormone levels.

Table 8.1 The plant hormones and their major functions

Auxins	Cell elongation, cell division, prevent abscission, apical dominance, direct translocation, RNA and protein synthesis
Gibberellins	Cell elongation (particularly stem extension), inhibit organ formation, accelerate germination, enzyme induction
Cytokinins	Release apical dominance, prevent senescence, mobilise nutrients, stimulate organ formation (in association with auxin)
Abscisic acid	Stomatal closure, gibberellin antagonist induces dormancy, control of embryo development (with cytokinins and gibberellins)
Ethylene	Ripening of fruit, senescence, epinasty, abscission, geotropism
Florigen	Flowering stimulus following photoperiod induction (hypothetical hormone)
Vernalin	Flowering stimulus following vernalization (hypothetical hormone)
Anthesin	Ripeness to flower stimulus, needed before vernalin and/or florigen are effective (hypothetical hormone)

Plant growth regulators may be regarded as a short-term answer to a plant breeding problem though they may in effect be a more cost-effective solution.

Discovery and characterization of plant growth regulators can be extremely difficult since effects on plants are frequently transitory and compensated by subsequent growth phases.

Plant breeding

The form of any organism (phenotype) is fundamentally determined by its genes (genotype). The degree to which any character is expressed is modified by the environment.

Crossing of individuals or within populations, followed by a lengthy programme of selection, enable the plant breeder to change the genotype of an organism and so to modify its phenotype, thus creating a new variety.

The objectives of any breeding programme will be numerous but will usually include those leading to an improvement in yield, e.g. improved climatic adaptation, pest and disease resistance, partition of biomass, and those leading to an improvement in quality of product. A considerable proportion of the post-war increase in UK crop production is attributable to the plant breeders. For example, it has been estimated that more than 60% of the yield increase for winter wheat has been due to improved varieties.

Plant breeding has been accelerated in recent years by the advances made in plant biotechnology. In particular, plant tissue culture and molecular biology are improving the speed of variety improvement and breaking down some of the barriers in species hybridization. Transfer of individual genes between unrelated organisms is now possible. Genetic fingerprinting also allows for more precise crossing and selection programmes.

Further reading

Evans, L.T. (1975) *Crop Physiology: Some Case Histories.* London: Cambridge University Press, London.

Forbes, J.C. & Watson, R.D. (1992) *Plants in Agriculture.* Cambridge University Press, Cambridge.

Harper, F. (1983) *The Principles of Arable Crop Production.* Granada Publishing, London.

Hay, R.K.M. & Walker, A.J. (1989) *An Introduction to the Physiology of Crop Yield.* Longmans, Harlow.

Hurd, R.G., Biscoe, P.V. & Dennis, C. (1980) *Opportunities for Increasing Crop Yields.* Pitman Advanced Publishing Programme, London.

Milthorpe, F.L. & Moorby, J. (1979) *An Introduction to Crop Physiology.* Cambridge University Press, London.

9

Crop Nutrition

J.S. Brockman

All the essential nutrients required by green plants are taken up by the plant in inorganic form, unlike animals where some organic compounds are also needed. Indeed, organic materials have to be broken down to inorganic before they can become of value to the plant. There are 16–19 elements known to be essential for plant growth and these are listed in Table 9.1 together with a summary of their main functions in the plant and the form in which they are taken up by the plant. Much further information on the uptake and role of plant nutrients is contained in Mengel and Kirkby (1982).

An essential nutrient is defined as one which is required for the normal growth of the plant, is directly involved in the nutrition of the plant and cannot be substituted by another nutrient. Not all plants have the same essential requirements, e.g. sodium (Na), silicon (Si) and cobalt (Co) are needed by some plant species only. Sometimes essential plant elements are divided into macronutrients (the top 11 elements in Table 9.1) and micronutrients (the remaining eight elements in Table 9.1). This division is based on the probable content of each element in the plant, but plants differ in their nutrient composition so such a distinction is not always meaningful. Also the importance of a nutrient is not related to the quantity contained in the plant tissues, and an insufficient quantity of any nutrient can lead to a restriction in crop growth and/or quality.

Summary of soil-supplied elements

Nitrogen (N)

Nitrogen is an essential constituent of all proteins, so is vital to any living plant.

Broadly, there are three stages in the utilization of N by the plant:

(1) *stage 1*, uptake of NO_3^- and NH_4^+ ions with N present in the plant in inorganic form;
(2) *stage 2*, N present in low molecular weight organic compounds such as amino acids, amides and amines;
(3) *stage 3*, N in high molecular weight organic compounds such as proteins and nucleic acids.

If there is plenty of N available in the soil, the rate of N uptake (stage 1) can exceed the rate of protein synthesis (stage 2–stage 3), particularly if this occurs under conditions of low radiant energy, such as in early spring or autumn. Whilst this is not harmful over a short period because protein production can catch up with a short-term flush of N uptake, if the imbalance continues for a long period there can be:

(1) excessive elongation of vegetative cells of high moisture content, leading to soft, weak growth, e.g. lodging in cereals;
(2) decrease in carbohydrate production; this can lead to unpalatability in grazed grass, fermentation problems in grass cut for silage and poor quality in plants that store carbohydrate in roots or tubers, such as sugar beet and potatoes.

Deficiency symptoms

These first show in older leaves, as young tissues get priority for any limited N supply available to the plant: older leaves become senile prematurely, show a general yellow/brown colour and probably become detached from the stem. Nitrogen-deficient plants become spindly, lack vigour and are dwarfed, often featuring less vegetative development, including tillering, and are lower yielding; also they have a pale colour and a low leaf:stem ratio. Often N-deficient plants ripen earlier than those receiving adequate N nutrition. Nitrogen deficiency in young plants can be corrected rapidly by application of nitrate-containing fertilizers, but over-application of these fertilizers should be avoided as this will lead to the imbalance problems listed above.

Phosphorus (P)

Phosphorus is associated with the transfer and storage of energy in the plant. As such it is vital for living processes within the cell. A supply of P is essential to the plant during the early stages of growth and also during seed formation – the latter because P is found in phytin seed reserves that can supply P to the seedling in the next generation. P is taken up from the soil in inorganic form but can be converted to organic forms

Table 9.1 Essential nutrients, form of uptake and functions in plant

Element	*Form of uptake by plant*	*Main functions in plant*
C	CO_2 from atmosphere (HCO_3 from soil solution)	Major constituent of organic material; can account for 40% of dry weight of plant
H	H_2O from soil water (and from atmosphere if wet)	Linked with C in all organic compounds
O	CO_2 and H_2O; O_2 during respiration	With C essential in carboxylic groups; with H in oxidation–reduction processes
N	NO_3^- and NH_4^+ from soil solution (N_2 from air if fixed by micro-organisms)	Essential constituent of all proteins
S	SO_4^{2-} from soil solution; SO_2 absorbed from atmosphere	Constituent of some essential amino acids, e.g. cysteine and methionine; also forms S–S bridge in enzyme proteins
P	HPO_4^{2-} and $H_2PO_4^-$ from soil solution	Vital constituent of living cells, associated with storage and transfer of energy within the plant
K	K^+ from soil solution	Essential for efficient water control within the plant; formation and translocation of carbohydrate; found in some enzymes
Na*	Na^+ from soil solution	Essential in some plants only (e.g. of marine origin); partially interchangeable with K^+ in most plants
Ca	Ca^{2+} from soil solution	Essential role in cell walls and biological membranes
Mg	Mg^{2+} from soil solution	Essential constituent of chlorophyll, some enzymes and some organic acids
Si*	Probably $Si(OH)_4$ from soil solution	Used in cellulose framework; interacts with P in plant (mechanism uncertain)
B	H_3BO_3 or BO_3^{3-} from soil solution	Assists in carbohydrate synthesis, uptake of Ca^{2+} and absorption of NO_3^-. Big variation in need for B; risk of toxicity
Mn	Mn^{2+} from soil solution (availability to plant affected by soil pH)	Associated with chlorophyll formation and some enzyme systems
Cu	Cu^{2+} or copper chelates in soil solution	Small quantities needed in chloroplasts and for enzyme systems converting NO_3^- to protein
Zn	Zn^{2+} from soil solution	Assists with starch formation and some enzyme systems
Mo	MoO_4^{2-} from soil solution	Small quantity essential for enzymes controlling N nutrition (also for N-fixing bacteria)
Fe	Fe^{2+} or Fe^{3+} or Fe chelates	Essential in chlorophyll and enzyme activities; associated with enzymes in photosynthesis
Cl	Cl^- from soil solution	Involved in evolution of O_2 during photosynthesis (excess Cl more common than a deficiency)
CO*	Co^{2+} or Co chelates† from soil solution	Not essential for most species; essential for N fixation by bacteria and could be used in N nutrition in plant

* These three elements are not essential in all plants species.
† Chelates are organic-type compounds that can maintain the availability of some metal ions over a wide soil pH range.

in a matter of minutes: hence most of the P found in young tissue is in organic compounds. Phosphorus in older tissue is mainly in inorganic form.

Deficiency symptoms

These are not easily detected from appearance. Early stages of deficiency cause a reduced growth rate and limitation in root development, often with leaves looking a rather healthy dark green–bluish green colour as the plant is less able to grow in response to other nutrients such as N. Also there will be poor seed formation and poor vigour from any seed that is subsequently used. It is only when deficiency is severe that the plant will appear stunted, and at this stage the stem can be reddish in colour. Phosphorus deficiency is hard to correct in existing plants once visual symptoms are found, because such symptoms appear at a late stage in the deficiency. Where plants are to be sown in P-deficient soil, it is essential that some water-soluble P, such as superphosphate or ammonium phosphate, is placed near to the seed.

Potassium (K) and sodium (Na)

Potassium is essential for efficient water relationships in the plant, both in controlling water content of cells and movement of water through tissues and in transpiration (by influencing the stomatal guard cells). K is also associated with the formation and translocation of carbohydrate and activation of some enzymes. Plant requirements for K vary from high to very high and many plant species contain more K than any other soil-supplied nutrient. As size of plant increases, so does the need for K; N × K interactions can occur and lack of K will limit full yield increases from N application. The only known form of K uptake by the plant is K^+ and this is true whether the K is applied in inorganic or organic materials.

Some plants, particularly those of marine origin, need Na^+ as an essential nutrient (e.g. sugar beet, fodder beet, mangolds and spinach). Other plants (e.g. maize, rye, soya and perhaps most cereal species) do not require any Na. For other species there is some acceptable replacement of Na^+ for K^+ and vice versa. However, in *all* species K^+ is needed to fulfil the enzyme activities associated with K.

Deficiency symptoms

For K these are seen as a loss of plant turgidity because the water control mechanisms become impaired, leading to reduced growth rate and lower plant metabolic efficiency. Older leaves show yellow/brown spots on leaves and/or whole leaves become brown at the edge.

In potatoes and brassicas initial K deficiency is shown by a blue-green colour, leading to bronzing and marginal necrosis of the leaves. In barley the necrosis of leaves can be spotted as well as on the leaf margins and can be whitish-brown or purplish-brown. In legumes the first symptoms are often whitish spots on leaves followed by marginal browning or bronzing. Generally, application of fertilizers containing muriate or sulphate of potash will correct the deficiency provided this has not reached an advanced stage.

In Na-demanding crops, deficiency of Na shows in reduced plant water movement and an apparent moisture stress. Correction is by the application of an Na-containing material such as Kainit.

Sulphur (S)

Most of the sulphur acquired by plants is taken from the soil as SO_4^{2+} but plants can utilize atmospheric SO_2 absorbed through stomata. S is used in essential amino acids such as cysteine and methionine, and also the presence of disulphide (S−S) bonds is an essential feature of many protein enzymes.

Deficiency symptoms

Sulphur nutrition is closely linked to N, and in most crops receiving fertilizer N a deficiency will show as a failure to respond fully to N. Also a shortage of S will lead to S-demanding proteins and enzymes being in short supply and this can affect crop quality (e.g. insufficient S can lower baking quality of flour).

Crops particularly sensitive to S are brassicas (high content of S-dependent proteins), grass receiving high rates of N (need for S to support action of N) and legumes (high requirement by rhizobial bacteria). Where S is deficient, S-containing fertilizers (superphosphate, ammonium sulphate and potassium sulphate), gypsum or S-coated fertilizers of any type should be used. Normal application rate is 10−50 kg/ha/year of S.

Calcium (Ca) and strontium (Sr)

Calcium is essential for plant growth, being used in cell walls and biological membranes. As soils differ widely in their natural Ca content (and resultant pH) some plants have evolved either a very low need and tolerance for Ca (*calcifuges*) or a high need and tolerance for Ca (*calcicoles*). Apart from this, problems from soil Ca content as such are rare as Ca is the major element controlling soil pH: soil pH is far more important in the overall nutrition of plants than the soil content of Ca. Whilst vegetative growth of plants is rarely restricted by lack of Ca, in some plants poor translocation of Ca within the plant can cause *calcium deficiency* in fruit, e.g. bitter pit in apples and blossom-end rot in tomatoes. Prevention of such Ca deficiency is by spraying calcium nitrate on the leaves when the fruit is setting.

Strontium is chemically related to Ca and is taken up and deposited in cell walls in much the same way as Ca. If excessive Sr uptake becomes a problem, it can be avoided by heavy liming.

Magnesium (Mg)

Some 20% of the magnesium in the plant is found in chlorophyll and much of the rest is in enzyme systems or associated with the movement of anions (mainly phosphates). Mg uptake is rapid in the early stages of growth. Where the plant has access to abundant levels of K^+ and NH_4^+ the uptake of Mg can be reduced, leading to induced Mg deficiency. Sensitive crops include sugar beet, potatoes, tomatoes, grass and many fruit and glasshouse crops. In the past many fertilizers contained Mg as an impurity; now this is not the case and Mg deficiencies are most likely on highly leached acid soils and heavily limed sandy soils.

Deficiency symptoms

These are always seen first in older parts of the plant, as young tissues get priority. Typical deficiency symptoms are intervenal chlorosis, with the leaf veins standing out as dark green. In sugar beet the chlorosis first shows on the margins of older leaves and can be followed by black necrotic areas. In potatoes, mottling on the leaf can be followed by reddish/purple tinting and loss of older leaves.

Low levels of Mg can cause foliar symptoms without reduction in yield, although yield loss has been found in severe deficiency in sugar beet and potatoes. In grass, Mg deficiency does not affect the yield of grass, but it can lead to Mg deficiency in stock eating the grass (hypomagnesaemia).

Where Mg deficiency is caused by a genuine lack of available Mg in the soil, the application of 250 kg/ha of magnesite (45% MgO) or 400 kg/ha of kieserite (27% MgO) will help maintain Mg supplies. For rapid action, foliar application of 25−35 kg/ha of Epsom salts (16% MgO) in 40 litres water is recommended. If Mg deficiency is *induced* by K^+ or NH_4^+, then foliar application is essential for plant deficiency to be corrected, although with grass it is better and more certain to feed an Mg supplement to the stock at risk.

Boron (B)

Boron acts in a similar way to P in hydrolysis reactions and it facilitates sugar translocation. It is an important minor element and its availability is much reduced at high soil pH (e.g. as a result of overliming). Adequate B is essential for good storage quality in root crops.

Deficiency symptoms

They have not been found in UK cereals or grass, but they occur commonly in root crops (particularly swedes and sugar beet), legumes and leafy brassicas. Boron deficiency is characterized by the death of the apical growing point of the main stem and failure of lateral buds to develop shoots. Leaves may become thickened and sometimes they curl. In sugar beet, fodder beet and mangolds, mis-shaped young leaves are the first sign, followed by death of primary leaves and a 'rosette' of secondary leaves covering a scarred root crown. Hollowing may extend from the crown into the root and fungal decomposition can occur. In swedes and turnips plant growth may appear normal, but at harvest the roots may have brown hollow areas known as 'crown rot'. In lucerne, terminal leaves become yellow or red; plants are stunted and adopt a rosette form with death of terminal buds, and fruit formation can be impaired.

Treatment is by use of a boronated fertilizer or borax at a rate equivalent to 20 kg/ha of borax. Alternatively, foliar application can be made using a soluble boron product at 5–10 kg/ha in 250–500 litres/ha of water using a wetting agent.

Toxicity symptoms

Boron is toxic to some plants at a level only slightly above that needed for normal growth. Leaf tips become brown and rapid necrosis of the whole leaf can follow. The problem is worse in dry areas and is exacerbated where irrigation water high in B is used. Sensitive crops include runner beans and grapes; semi-sensitive crops are barley, potatoes, tomatoes and legumes.

Manganese (Mn)

Manganese is involved in chlorophyll formation and enzymatic control of oxidation–reduction processes, and in some ways is similar to Mg. Most UK soils contain adequate available Mn, but some organic soils of high pH are deficient and so are some heavily leached podzols. Manganese deficiency can also be found in badly drained soils and can be induced by overliming or deep ploughing of calcareous soils or by the presence of high levels of Mg.

Deficiency symptoms

They are chlorosis of younger leaves (cf. Mg, where effects are on older leaves). In oats this deficiency is called 'grey speck'; in barley the leaves have brown spots and streaks; in wheat there are intervenal white streaks. In all three of these cereals maturity is delayed and ear emergence reduced with a high incidence of blind ears. In sugar beet the chlorosis is called 'speckled yellows' and is most severe in early stages of growth, accompanied by an upright leaf habit with curling edges. Brassicas show chlorotic marbling; potatoes have stunted leaves with small terminal leaves rolled forward; peas and beans often show brown lesions on the inner surface of cotyledons, the term 'marsh spot' referring both to the condition on the plant and its occurrence on badly drained organic soils. Although Mn deficiency may be controlled by soil application of 125–250 kg/ha of manganese sulphate, this treatment has no long-lasting effect as the Mn is oxidized rapidly. Recommended treatment is foliar application of 6–10 kg/ha of manganese sulphate in 225–1000 litres/ha of water using a wetting agent.

Toxicity symptoms

These are brown spots and uneven chlorophyll on older leaves. Silicon (Si) can minimize the harmful effects of excess Mn.

Copper (Cu)

Only small quantities are needed and the function of Cu in the plant is uncertain, but it is associated with enzymes that convert N to protein and also the redox processes in cells. Cu is a constituent of chloroplasts and aids the stability of chlorophyll.

Deficiency symptoms

They can be found in fruit and some cereal crops, particularly wheat growing on some peats in well-defined areas, on leached acid sandy soils such as heaths and on some shallow, puffy chalks with high organic matter content. In cereals, deficiency symptoms are not seen until the plants are well-developed, at which time the symptoms can change rapidly from yellowing of the tips of youngest leaves to spiralling of leaves, giving a stunted, bushy appearance to the plant; the ears have difficulty emerging, have white tips and are devoid of grain. Wheat grown on copper-deficient chalk soils specifically can show blackening of ears and straw.

Treatment can be by application of 60 kg/ha of copper sulphate to the soil – a treatment that can last two the three years on peaty and sandy soils but not on the deficient chalks: otherwise foliar application of 2–3 kg/ha of copper oxychloride, cuprous oxide or chelated Cu can be effective when applied at a fairly late stage of growth (and this is when symptoms may first show).

Toxicity symptoms

When present in excess, Cu^{2+} replaces other metal ions, particularly Fe^{2+}; root growth is restricted and chlorosis occurs.

Iron (Fe)

Iron is essential for the proper functioning of chlorophyll and related photosynthetic activity. Iron uptake is strongly related to soil pH, and some species, such as sugar beet, brassicas and beans, can show lime-induced Fe deficiency, particularly where soil aeration is poor (e.g. in a badly structured soil).

Deficiency symptoms

These are chlorotic markings on younger leaves (in contrast to Mn deficiency which can affect all leaves irrespective of age). Iron deficiency is found more widely in horticultural crops than agricultural ones, being found particularly in fruit and calcifuge plants such as azalea and rhododendron. The only effective treatment for deficiency is foliar application of an iron chelate (see Table 9.1).

Molybdenum (Mo)

Molybdenum is essential for the enzymes nitrogenase and nitrate reductase. It is also important for N fixation by rhizobial bacteria. Mo deficiency can occur on some acid soils and on soils of high pH. Excessive Mo availability in the soil can induce low Cu uptake, where in grassland it gives rise to 'teart' pastures with Cu deficiency in grazing stock. Excess Mo in herbage can cause toxicity in animals.

Deficiency symptoms

They are found in cauliflowers where hearts fail to form in plants nearing maturity, leading to 'whiptail'. Sometimes liming will alleviate whiptail, otherwise plants should be sprayed early in their life with 0.25–0.5 kg/ha of sodium molybdate in solution.

Zinc (Zn)

Plants need Zn in very small quantities only and there are signs that an increasing number of crops may respond to Zn application. Apart from fruit trees, other sensitive crops are maize, flax and field beans (and perhaps other legumes). Excess Zn can be toxic to plants and induce Fe deficiency. Some sewage sludges contain large amounts of Zn and regular use of such materials on the same area of land can lead to an undesirable build-up of Zn. Although grass is not itself affected by excess Zn, it will raise the concentration in the herbage and this can lead to high levels of Zn in milk. Zn deficiency can be corrected by using 4–10 kg/ha of $ZnSO_4$ as a fertilizer or by using a Zn chelate.

Silicon (Si)

Silicon is the second most abundant element in the lithosphere after O, and so it is not likely to limit plant growth. Si is used by plants in the cellulose framework of cells, and plants deprived of Si under experimental growth conditions have a very limp growth habit. Application of up to 450 kg/ha of sodium silicate has been shown to increase P availability on some soils.

Cobalt (Co)

Cobalt is essential for micro-organisms which fix N, but is probably not essential for higher plants. Excess Co can induce deficiency of both Fe and Mn.

Chlorine (Cl)

Chlorine is essential to plants in small quantities only, where it is used in processes connected with the evolution of O_2 during photosynthesis. As large amounts of Cl are applied with K in muriate of potash, the effects of excess Cl are common, particularly on soils that have been affected by salt. Cl toxicity is seen as burning of leaf tips, bronzing and premature yellowing of leaves. Sugar beet, barley, maize and tomatoes are tolerant, but potatoes, lettuce and many legumes are sensitive to excess Cl. Where there is a risk of a sensitive crop being affected, sulphate of potash should be applied in place of muriate.

Potentially toxic elements

Some elements such as iodine, fluorine, aluminium, nickel, chromium, selenium, lead and cadmium are not essential for plant growth and their presence can lead to toxicity. For more information *see* Mengel and Kirkby (1982).

Organic fertilizers and manures

Organic fertilizers are derived from either plant or animal materials. Not all the nutrients contained in such materials are in organic form and those that are in organic form are not readily or completely available to plants. Complex organic compounds will become part of the soil organic cycle and could perhaps have an eventual nutrient value, depending on the activity of the soil biomass.

Compared with inorganic sources of nutrients, organic sources have the following features:

(1) not immediately soluble in water and so not readily leached;
(2) because they have to break down to become partially soluble, they can act as a slow-release source of plant nutrient;
(3) they can be applied at heavy rates without risk of injury to roots or germinating seeds as they have little ionic activity;
(4) they can stimulate microbial activity;
(5) they are much more costly per unit of plant food (unless by-products);
(6) the recovery of nutrients contained in the materials is low.

Because of the above features, organic fertilizers have little general use in agriculture, but they do have a place in market gardening and horticulture where the slow-release characteristics have application for some of the high-value crops that are grown. In recent years there has been increasing interest in 'organically grown' food: at present there is no generally accepted standard code of practice (but see UKROFS (1992)). Table 9.2 sets out the approved fertilizers recommended by the Soil Association and from this it can be seen that not all organic materials are approved.

Organic nitrogen fertilizers

Hoof and horn meals, hoof meal, horn meal

These contain 12–14% N and can be obtained either coarsely or finely ground: fine materials release N more quickly. They should be worked into the soil before planting or sowing.

Dried blood

This contains 12–13% N. It is very expensve but of great value in glasshouse crops where it is quick-acting.

Shoddy (wool waste)

Analysis varies from 3 to 12% N, depending on the proportion of wool contained amongst other wastes. Shoddy is a very slow-release material, should be worked

Table 9.2 Example of materials permitted by Soil Association (1993)

Fertilizers	
N sources	Blood meal; hoof and horn meal
P sources	Natural rock phosphate; calcined aluminium phosphate rock; basic slag; meat and bone meals
K sources	Rock potash (if it has low Cl content and low water solubility); wood ash
Ca and Mg	Calcareous magnesium rock; gypsum; ground chalk and limestone; calcified seaweed; Epsom salts
Minor/elements	Liquid and dried seaweed; sulphur; sulphate of potash (if K index below 2 and clay content below 20%); other minor elements – based on evidence of deficiency, e.g. soil analysis
Manures and plant wastes	
Straw, FYM, slurry, dirty water, poultry manure, provided they come from organic sources	
Plant waste materials and by-products, provided they come from organic food processing	
Sawdust, shavings and bark, provided they come from untreated timber	
Compost activators based on microbial and plant extracts	
Bio-dynamic preparations	

into the soil before planting and should be analysed before purchase.

Organic phosphate fertilizers

Bone meals

These contain 20–24% P_2O_5 (insoluble in water) together with 3–4% N. The phosphate acts very slowly and is of most value on acid soils.

Steamed bone meals and flours

These contain 26–29% P_2O_5 (insoluble in water) together with about 1% N. These materials are made from bones that are steamed to obtain glue-making substances and a good deal of the N is removed in the process. The bones are generally ground after extraction and the phosphate acts more quickly than in ordinary bone meal.

Meat and bone meal (also meat guano or tankage)

They contain 9–16% P_2O_5 (insoluble in water) together with 3–7% N. These are made from meat and bone wastes and analysis varies; the phosphate has slow availability.

Fish and meals and manures (also fish guano)

These contain 9–16% P_2O_5 (insoluble in water) together with 7–14% N. They are waste products from fish processing.

Organic potash fertilizers

Potassium does not occur in chemically organic form but it does occur in organic materials (such as livestock manures and bird guano). Also potassium occurs in some 'natural' materials (such as wood ash, mica and adularian shale).

Livestock manures

On stock farms large quantities of faeces and excreta are accumulated from housed animals. Quite apart from the environmental and health aspects of the storage and disposal of these products, all animal manures have some value as plant nutrients. The application of these manures to farmland can represent both a relatively safe and a money-saving method of disposal. Before outlining the probable nutrient value of animal manures (including slurry) it must be stressed that all such materials can be extremely variable in composition for two main reasons:

Variation in nutrient content

The main factors affecting the nutrient content of the manure as collected for field distribution are:

(1) the quantity of excreta produced, based on the size and type of animal involved (see Table 9.3);
(2) the composition of the excreta is influenced by the animals' diet;
(3) the method of collection and storage of excreta, involving the degree of dilution by rainwater and/or washing-down water or straw.

Variations in losses during storage

The main factors are:

(1) gaseous loss of N as ammonia; the quantity lost will vary with the conditions of storage and temperature. Loss of 10% N is average but loose-stacked farmyard manure (FYM) that is turned prior to spreading can lose up to 40%. Where slurry is stored for long periods, 10–20% of N can be lost and agitation before removal will aggravate the loss;
(2) leaching losses from FYM stored in the open are in the range 10–20% of N, 5–8% of P_2O_5 and 25–35% of K_2O: the flatter the heap the greater the likely loss;
(3) seepage losses from slurry can result in 20–25% loss of N, a little loss of P_2O_5 and 20–30% loss of K_2O.

The above sources of loss can be additive, so that material that has been stored outdoors for a long period will have a greatly reduced manurial value due to gaseous, leaching and seepage losses.

Table 9.3 sets out the standard ADAS guidelines for estimating the composition and likely nutrient value of livestock manures. Points to note are the relatively high K value associated with cattle manures and the high N value of poultry manure.

Simplified estimation of nutrient value

This can be based on the type of animal and the number of days over which the material was collected. For example, if dairy cows are kept indoors over a 200-day winter, then each cow will produce $200 \times 57 = 11\,400$ litres of undiluted slurry. From Table 9.3 it can be seen that each 1000 litres of cow slurry should be worth

Table 9.3 Guide to composition and nutrient value of manures

Type of stock	*Body weight (kg)*	*Faeces and urine voided (litres/day)*	*Composition*				*Available nutrients**		
			Approx. DM%	*% of fresh weight* N	P_2O_5	K_2O	*N*	P_2O_5	K_2O
Slurry (undiluted)							*(kg/m³ or kg/1000 litres)*		
Dairy cow	550	57							
Two-year bullock	400	27	10	0.5	0.2	0.5	1.5	0.8	3.5
One-year bullock	220	15							
Pig (dry meal fed)	50	4	10	0.6	0.4	0.3	4.0	2.0	2.7
Pig (pipeline fed)	50	7	6–10	0.5	0.2	0.2	2.0–3.5	1.0–2.0	1.5–2.7
Pig (whey fed)	50	14	2–4	0.3	0.2	0.2	0.8–1.6	0.4–0.8	0.8–1.5
1000 laying hens	2000	114	25	1.4	1.1	0.6	9.1	5.5	5.4
Farmyard manure (FYM)							*kg/t of material*		
Cows and cattle			25	0.6	0.3	0.7	1.5	2.0	4.0
Pig			25	0.6	0.6	0.4	1.5	4.0	2.5
Poultry (deep litter)			70	1.7	1.8	1.3	10.0	11.0	10.0
Poultry (broiler litter)			70	2.4	2.2	1.4	14.5	13.0	10.5
Poultry (dried house droppings)			70	4.2	2.8	1.9	25.0	17.0	14.0

* Availability from spring application; for availability of N from autumn and winter application see text. Table based on MAFF (1994)

1.5 kg N, 0.8 kg P_2O_5 and 3.5 kg K_2O. So the following estimates can be made:

	Per cow Over 200 d	Daily rate
N (kg)	17.1	0.086
P_2O_5 (kg)	9.1	0.046
K_2O (kg)	39.9	0.200

If this slurry is spread over a known area, then a quick estimate can be made. For example, if there are 80 cows kept indoors for 200 days and all their slurry is spread over 25 ha of grass, then the potash made available is $(80 \times 39.9)/25 = 128$ kg/ha of K_2O.

Effect of time of application

The nutrient values in Table 9.3 are based on assumed spring application, as this is normally the most effective time for nutrients contained in manures – particularly those liable to leaching over winter. If manures are applied at other times in the year, then the N value should be reduced to the following comparative effectiveness:

if spring application = 100 then
autumn = 0–20
early winter = 30–50
late winter = 60–90

Phosphate and potash values are not affected by leaching and should not be reduced.

Whenever livestock manures are applied to a crop, their nutrient value and cost-saving potential should be recognized and adjustment made in the quantity of inorganic fertilizers used. Standard fertilizer recommendations normally allow for this, for example:

Fertilizer recommendations for maincrop potatoes
N index = 1, P and K index = 2

	kg/ha to apply N	P_2O_5	K_2O	
Total required	160	250	250	(A)
From 50 t/ha of FYM	75	100	200	
From fertilizer	85	150	50	(B)

It should be noted that not only is much more fertilizer needed where no muck is applied (situation A), but also the N:P:K ratio required in the fertilizer is different when the nutrients in muck have been deducted (situation B).

Risks from application of livestock manures

Hypomagnesaemia in grazing stock

Farmyard manure and slurry from ruminant stock are rich in potash and where high rates of such manures are applied to grassland in late winter and spring, there will be an increase in the potash content of the grass. Spring herbage is naturally low in Mg, and as a high plant content of K depresses Mg content, then there is an enhanced risk of hypomagnesaemia in animals grazing grass that has received cattle manures. Where such manures have to be applied to grass that is allocated to spring grazing, then it should be applied well before Christmas.

Animal disease

The main hazard that exists is from the infection of pasture with bacteria of the *Salmonella* group – normally as a result of applying infected cattle or poultry manures and slurries. The following advice is offered to minimize the risk:

(1) As most problems arise where fresh material is applied to grass, store the manure for two to three weeks before application.

(2) Allow rain to wash the manure from the herbage before grazing.
(3) Ensure watercourses are kept free from contamination.

Toxic elements

The main risk is from manure obtained from fattening pigs, as this can contain high levels of Cu and Zn, both derived from feed additives. Cu and Zn build up slowly in the soil and frequent application of pig manure to the same field can lead to potential toxicity problems. Particular risk arises on grazed grassland, where evidence suggests that slurry contamination on the herbage is more at fault than high Cu and Zn levels in the herbage itself: so physical contamination should be allowed to disappear prior to grazing.

Specifically, sheep should *never* be allowed to graze grass contaminated with Cu as extensive liver damage can occur.

Pollution

All animal manures have a high moisture content, and under some storage conditions this can lead to loss of effluent from the store. To minimize the risk of effluent it is vital that additional water (rain water, washing-down water) be kept out of manure and slurry stores.

There are strict legal requirements concerning the discharge of effluent into watercourses, quite apart from the loss of potential plant nutrients that results. The legal aspects of effluent also apply to fields where heavy dressings of manure lead to effluent being lost into field drains or from surface run-off. Consequently it is unwise to:

(1) apply very heavy dressings of manures during autumn and winter;
(2) apply any manure at all in winter to sloping fields near ditches and watercourses.

(See also later section in this chapter, 'Fertilizers and the environment'.)

Management aspects of use of livestock manures

Mixed grass/arable farms

Where suitable arable crops are grown, the most beneficial crops for high application rates are potatoes, maize and cereals.

All grass farms

These produce more manure per total farm hectare than mixed grass/arable farms and have fewer suitable crop situations for its use. If storage does not exist for the whole winter's production, then the following should be considered:

(1) *November/December* apply to next year's spring grazing area;
(2) *January–March* apply to areas for spring conservation cuts;
(3) *May/June* apply thinly to areas cut for conservation in spring.

Other points to note are:

(1) High rates applied to grass can cause physical shading of the grass and loss of tillers, particularly if the grass is not growing rapidly at the time.
(2) Cattle manures are high in potash and its use in spring can predispose hypomagnesaemia.
(3) Most manures cause taint and subsequent refusal problems by grazing stock: some farmers find sheep graze behind cattle slurry better than cattle.
(4) There is a *Salmonella* risk where fresh cattle slurry is applied to grass regrazed by cattle.
(5) Pig slurry is often high in Cu and Zn: grass should be free from slurry contamination before grazing, particularly with sheep.

Pig and poultry units

These are always 'exporters' of manure and this manure requires disposal at regular intervals throughout the year as the stock are nearly always housed. A mixed farm can more easily accommodate and efficiently utilize pig and poultry manure than either an all-grass or all-arable farm. Poultry manure has a higher N content than other manures and is valuable for application to grassland in the March–September period.

Straw contamination

Where manure contains much fresh, unrotted straw, it is possible that the soil bacteria breaking down the straw need more N than is contained in the manure: thus the material can *deplete* N status in the short term. To correct this, additional fertilizer N may be used at the rate of 10 kg of N for every 1 t of fresh straw used: note that this N should be applied to the next crop and *not* to the manure or the soil at the time of manure application.

Other organic manures (see Table 9.4)

Straw can be incorporated into the soil, preferably after it has been chopped, and this will provide some phosphate and potash but, as noted above, it may cause a deficiency of nitrogen that will need correction in the first year or so. However, in the long term, regular annual incorpor-

Table 9.4 Analysis of some organic materials

	*DM %**	*% composition as spread†*		
		N	P_2O_5	K_2O
Straw compost	25	0.40	0.20	0.30
Fresh seaweed	20	0.20	0.10	1.20
Liquid raw sludge	4	0.17	0.15	trace
Liquid digested sludge	4	0.21	0.20	trace
Dewatered digested sludge	50	1.50	1.10	trace–0.50
Dewatered sludge with lime	40	0.90	0.90	trace–0.50

* Sometimes expressed as the synonymous 'per cent total solids'.
† As with all organic manures, composition can vary widely and not all the nutrients are either immediately or eventually available.

ation of straw will stimulate the N cycle in the soil so that 20–30 kg/ha of *extra* available N per year can be anticipated.

Composts can be made from straw alone, straw mixed with another crop waste, or from other wastes alone. As a general rule all such materials suffer from lack of sufficient nitrogen for proper bacterial decomposition and it is recommended that ammonium nitrate is added at a rate of 30 kg of fertilizer/tonne of composting material. Subsequent rotting down may well release N as ammonia unless the mass is kept under anaerobic conditions, and so the resulting compost will still be deficient in N. For this reason good compost is made from very short-cropped material that is well-consolidated and kept in a compact heap. Even so, composts are not highly rated for their nutrient value and it is often easier to apply waste materials to suitable fields at the earliest opportunity as a disposal exercise, plough them in and let decomposition take place within the soil: in this way any available nutrients find their way into the soil/plant complex at minimum cost.

Sewage sludge is often available, either from cesspits on the farm or from urban treatment works. Depending on the treatment process different types are produced: *liquid raw sludge* is material taken straight from sedimentation tanks: this material can be allowed to digest anaerobically to remove oils and reduce pathogens, giving *liquid digested sludge*. The above materials may be *dewatered* and may also be *conditioned with lime* or other chemicals.

Domestic sewage sludge often contains high levels of potentially toxic elements, notably Zn. Sludges arising from industrial plants should be analysed as one or more of some 12 high-risk elements may be present.

Fresh seaweed should not be confused with the liming material calcified seaweed. Seaweed is rich in potash (and sodium) as shown in Table 9.4 and a dressing of 25 t/ha would provide some 300 kg/ha of K_2O. Sometimes seaweed is used to make composts, but like so many organic manures, it is bulky material and expensive to transport. Thus seaweed is used only in close proximity to the coast and is valuable for some horticultural crops and potatoes. Also it is used to provide humus on some light soils.

Inorganic fertilizers

Inorganic fertilizers form an important part of the basis of modern crop production for very few soils are able to supply sufficient quantities of all the nutrients necessary for high yields and quality in crops. However, profitable and environmentally safe use of inorganic fertilizers depends on careful assessment of the total level of nutrients required by the crop on the one hand and the extent of supply from the soil and organic sources on the other.

The three nutrients needed from inorganic fertilizers in greatest quantity are nitrogen, phosphorus and potassium. By law every bag of inorganic fertilizer that contains one or more of these three nutrients must state the content of that nutrient on the bag – in terms of percentage composition of N, P_2O_5 or K_2O. There are strict EU regulations that specify the narrow permitted tolerance between the stated composition on the bag and the actual analysis of the contents, thus it can be assumed that there is great consistency in the composition of different batches of a similar fertilizer.

It is important to realize that fertilizers do not contain N, P_2O_5 or K_2O: these are merely a convenient way of giving a common basis for expressing the composition of all inorganic fertilizers. Nitrogen is usually in the form of nitrates or ammonium compounds, P_2O_5 in the form of phosphates and K_2O in the form of potassium chloride or sulphate. In the case of phosphates, some are water-soluble (and rapidly available in the soil, at least for a short period) and some are insoluble in water. If these latter are to have some agronomic value, they must become available in the soil and sometimes the solubility of these materials in 2% citric acid or neutral or alkaline ammonium citrate is given as a guide to probable value as a fertilizer.

In some specific crop and soil situations other nutrients are needed, such as Mg, B or S, and these can be applied on their own or mixed with one of the normal fertilizers. If a fertilizer is stated as containing any of these additional nutrients, then the % composition *must* be stated.

Nitrogen fertilizers

The major N fertilizers are as follows

	Formula	*% N*
Ammonium nitrate	NH_4NO_3	34.5
Urea	$CO(NH_2)_2$	46
Ammonium sulphate	$(NH_4)_2SO_4$	21
Calcium ammonium nitrate	$NH_4NO_3 + CaCO_3$	21–26
Anhydrous ammonia	NH_3	82
Aqueous ammonia	$NH_3 + H_2O$	25–40
Potassium nitrate	KNO_3	14
Sodium nitrate	$NaNO_3$	16
Calcium cyanamide	$CaCN_2$	21

Ammonium nitrate (AN)

This is an important fertilizer material and is sold as a straight fertilizer or as a component in compound fertilizers. Half the N is in ammonium form and half in nitrate form; in this way it has only half the acidifying effect of ammonium sulphate for any given amount of N applied. Practically pure ammonium nitrate is sold as a straight fertilizer containing 33.5–34.5% N: this material is very soluble in water and also hygroscopic, so it must be stored in a dry place in sealed bags. It is very powerful oxidizing agent and if subject to heat or flame it can explode: thus straight ammonium nitrate and compounds containing a high proportion of ammonium nitrate should not be stored in barns with hay or straw and *never* stored in bulk, as the risk of explosion is thus enhanced.

Being soluble in water, ammonium nitrate is often used as a major N source in liquid fertilizers.

Urea

Urea is the most concentrated solid source of N that is currently available on any wide scale. It is very soluble in water and hydroscopic. Sometimes its granules are softer than those of ammonium nitrate and do not spread so well. When urea is applied to most soils it breaks down rapidly to NH_4^+. However, ammonia (NH_3) can be formed if the soil contains free Ca, or is very dry.

(1) If the ammonia is lost to the atmosphere, then the fertilizer is less effective as a source of N for crop nutrition; this loss of ammonia is most likely when the soil is alkaline and if the urea is applied to the soil surface in dry weather.
(2) In seedbeds the high concentration of ammonia near germinating seeds can cause toxicity and serious loss of plant stand.
(3) On poorly structured soils liable to surface capping the released ammonia can be trapped in the soil and not only kill germinating seedlings but also severely check the growth of roots of established plants.

It follows that urea is best used on acid-to-neutral soils, surface-applied to established crops during periods of frequent rainfall. Thus, in the British Isles, urea has a place on grassland and for top dressing cereals in spring. Compared with AN, the cost per unit of N in urea varies widely. Each tonne of urea contains 33% more N than AN. It follows that even if it is assumed that N in urea is 15% less effective than N in AN, when urea is less than 20% dearer per tonne than AN, urea is a better buy.

Urea is widely used as a major N source in liquid fertilizers, as liquid application overcomes some of the physical problems of urea noted above and when mixed with phosphoric acid to make a compound fertilizer the acid nature of the material further reduces the risk of ammonia loss.

Ammonium sulphate

Ammonium sulphate was once the most important source of inorganic N; it is soluble in water and although all the N is in NH_4^+ form it is quick-acting under field conditions. However, because of its high ammonium ion content it has an acidifying action in the soil and also its N concentration is limited to about 21%; it does, however, contain S. Ammonium sulphate has been replaced by ammonium nitrate in UK and some other temperate countries and by urea in tropical and subtropical countries and some high-rainfall temperate areas.

Calcium ammonium nitrate (CAN)

This is a mixture of ammonium nitrate with calcium carbonate and is sometimes called nitrochalk. It has a higher N content than ammonium sulphate and has little acidifying action in the soil. Also the hygroscopic and explosive properties of ammonium nitrate are nullified by the presence of chalk, making it easy and safe to store, even in bulk. However, CAN tends to be more expensive per unit of N purchased than ammonium nitrate.

Anhydrous ammonia

Anhydrous ammonia is the most concentrated source of N available, being pure ammonia. At normal temperature and pressure this material is a gas, but when stored at high pressure it is a liquid containing the equivalent of 82% N. At this concentration it is very economical to transport on a weight-to-value basis, but the whole process of storage, transport and application needs special equipment, both to maintain up to the point of application the high pressure needed and to prevent hazards from uncontrolled loss of ammonia. Anhydrous ammonia is applied below the surface of the soil using special injection equipment, and because of the high cost of this operation it is economical to apply only fairly high rates of N (at least 100 kg/ha of N) at each application. Also it is essential that loss of ammonia to the atmosphere be kept at a minimum during application by careful sealing of the slits. For these reasons the use of anhydrous ammonia is best restricted to row crops that need high individual N dressings. Although widely used in some countries where suitable crops are grown in stone-free, friable soils, anhydrous ammonia has not established itself in the UK, mainly because it was not found suitable for the potentially major market on grassland due to sward damage caused by injection equipment and loss of ammonia from the slits in the soil.

The very high concentration of ammonia released in the soil in the vicinity of the slits causes partial sterilization of the soil and has the effect of retarding the action of the bacteria which convert NH_4^+ to NO_3^- so that anhydrous ammonia can act as a slow release fertilizer.

Aqueous ammonia

Aqueous ammonia is a solution of ammonia in water, a solution of normal pressure containing about 25% N. Aqueous ammonia has most of the advantages of the anhydrous form except high N concentration, but to offset that it does not need such specialized equipment. In crop situations where a liquid straight N fertilizer is required, then aqueous ammonia has an important place. Its concentration can be increased by:

(1) partial pressurization, where storage and application under only a modest pressure can enable concentration to increase to 40% N;
(2) mixture with urea and/or ammonium nitrate to give 'no pressure' solutions of about 30% N.

To prevent gaseous loss of ammonia all these materials should be either injected into the soil or worked in immediately after application: they should *not* be used on calcareous soils.

Potassium nitrate and sodium nitrate

These are both very quick-acting and highly soluble forms of N which are expensive but valuable in some horticultural crops and in particular for use in foliar feeds.

Calcium cyanamide

This is not important in the UK but is a fertilizer containing N in both amide and cyanide forms. It is soluble in water, and in the soil it is converted to urea: during this conversion some toxic products are released which can kill germinating weeds and slow the rate of nitrification. Breakdown requires water and so the product is not effective in dry conditions.

Slow-release fertilizers

Most of the commonly used and cheaper forms of N are rapidly available to the growing plant and yet many crops would gain greatest benefit from applied N if it were available over a period covering most of the vegetative growth of the plant. This is probably the most convincing reason for considering organic N sources. When inorganic sources are used there are three ways in which this objective can be met:

(1) Apply frequent small dressings of conventional N

fertilizers during vegetative growth, for example as with spring N applications on some winter wheat crops and on grassland.

(2) Use a conventional fertilizer material which has received a coating to reduce its rate of breakdown into plant-available forms, for example sulphur-coated urea.

(3) Application of complex compounds of N which require considerable chemical change in the soil before they are available to the plant.

There is much research to find suitable materials in the last of the above categories. To be successful such a compound must be economic in price and supply available N at a rate required by the crop. The materials on the market tend to be expensive, make available to the crop a relatively low proportion of the total N they contain and have rates of release controlled by soil temperature and moisture so that during humid weather most of the N is released too rapidly for full crop benefit.

Among synthetic slow-release fertilizers are:

- *Urea formaldehyde* (ureaform) 40% N: this is made by reacting urea with formaldehyde and its rate of breakdown to release available N is controlled by soil bacteria.
- *Isobutylidene di-urea* (IBDU) 32% N: rate of release of available N depends on differential particle size and soil moisture status.
- *Sulphur coated urea* (36% N) and *resin coated AN* (26% N): both depend on differential times for the coatings to break down before releasing available N.

Urea and IBDU are used in commercial glasshouse production because not only is the slow rate of N release of particular value to some of the crops grown, but also the rate of breakdown can be controlled to a large extent by adjustment to the management of the house.

Phosphate fertilizers

The major P fertilizers are:

	Formula	*% P_2O_5*
Water soluble		
superphosphate	$Ca(H_2PO_4)_2 + CaSO_4$	18–22
triple superphosphate	$Ca(H_2PO_4)_2$	45–47
mono-ammonium phosphate	$NH_4H_2PO_4$	48–50
di-ammonium phosphate	$(NH_4)_2HPO_4$	54
Water insoluble		
basic (or Thomas) slag	$Ca_3P_2O_8 \cdot CaO + CaO \cdot SiO_2$	10–20
Rhenania (sinter) phosphate	$CaNaPO_4 \cdot Ca_2SiO_4$	25–30
ground rock phosphate	$CaPO_4$ (apatite)	30
Senegal rock phosphate		30

In addition to the above, which can all be used in solid forms, there are a number of liquid products based on *polyphosphates*. The basic component of these solutions is superphosphoric acid, made from orthophosphoric acid and one of a number of polyphosphoric acids. An ammonium compound is used to neutralize the appropriate acid, and NP solutions of ratios in the order 11:37:0 can be obtained. These products are expensive but do provide a highly concentrated source of available P in liquid fertilizers. There can be problems when potassium salts are mixed with polyphosphates, as they can cause crystallization and precipitation.

Superphosphate

Superphosphate is made by treating ground rock phosphate with sulphuric acid and producing water-soluble monocalcium phosphate and calcium sulphate. Usually some rock phosphate remains in the product as the quantity of sulphuric acid is restricted to prevent the final product containing free acid. Superphosphate was once the most widely used form of water-soluble phosphate and formed the basis of most compound fertilizers. Because of its relatively low P_2O_5 content it limited the concentration of phosphate in compounds, but because of the $CaSO_4$ present it did apply useful quantities of sulphur.

Triple superphosphate

This is made by treating rock phosphate with phosphoric acid. This produces mainly water-soluble monocalcium phosphate with no calcium sulphate. Thus, triplesupers contains nearly two-and-a-half times as much phosphate as supers, but no sulphur. Because of its high concentration, triplesupers is used widely in the production of high-analysis compound fertilizers.

Ammonium phosphate

This is made by adding ammonia to phosphoric acid, producing both mono- and di-ammonium phosphates. These compounds are very soluble in water and quick-acting in the soil, and supply both N and P in a highly available chemical combination. They are not used as straights to any extent because of the few crop situations that require a low N:high P ratio, but they do form an important base for many compound fertilizers.

Basic slags

Basic slags are by-products of the steel industry and contain phosphates, lime and trace elements. In the smelting process, P contained in iron ore is held in the furnace bound to CaO, becoming a solid 'slag' as the material cools. The phosphate is in complex chemical combination with calcium and has to be ground before it can be used as a fertilizer. Although the phosphate in slag is not water soluble, it has been found that it can release phosphate slowly over a period of years on soils of pH range 4–7, although the rate of release is greater on acid soils. The trace element content of basic slag is also important, particularly Mg, Fe, Zn, Si and Cu.

Modern steel plants do not produce slags containing much P, although some older plants (mainly outside the EU) can provide slag with 8–10% P_2O_5. None of the 'slag substitutes' on the market seems able to emulate all four of the main attributes of traditional slag, namely:

(1) steady release of available P over several years on both acid, neutral and alkaline soils;

(2) supply of valuable trace elements;
(3) some liming value;
(4) cost per unit of P about half the cost of water-soluble P.

Farmers should check carefully the P content of slag and the cost of that P.

Rock phosphates
These are now being offered to farmers as a source of phosphate where rapid release of P is not essential but soil P status needs to be maintained. Rock phosphates will only release P under acid conditions, although some experiments have shown that soil pH from conventional analysis is not necessarily a true indication of the pH that might exist around root hairs, so that at times rock phosphate has shown some value on neutral soils. Rock phosphates vary considerably in their composition and potential agronomic value. In all cases they must be ground finely to have any value at all, current legislation being that 90% of the material should pass a 100-mesh sieve.

Some of the hard crystalline apatites have very little fertilizer value even when crushed finely and used under acid conditions. On the other hand, soft rock phosphates such as *gafsa* do have value and are regarded as the best of untreated rock phosphates.

Rhenania phosphate is produced by disintegrating rock phosphate with sodium carbonate and silica in a rotary kiln: this 'sintered' product contains much of the P in the form of calcium sodium phosphate and as such the P is rendered a little more available than in the original rock.

Senegal rock phosphate contains aluminium phosphate and when this is heated in a kiln and then allowed to cool in a humid atmosphere, an expanded type of rough granule is produced that greatly enhances the surface area of the material in contact with the soil after application. This calcined Senegal rock is claimed to be an effective P source, even on alkaline soils.

No rock phosphate has a rapid release of available P, so that these materials should not be used in situations where the need of the crop for P is rapid, for example at crop establishment or on crops that have a big demand for P, such as potatoes. Also, the crop recovery of P from rock phosphates is less than the recovery from water-soluble sources, so that the amount of *useful* P in rock phosphate is less than in the water-soluble types: this fact should be remembered when the costs of rival products are considered.

As a guide, the cost of each kg P_2O_5 in a water-insoluble phosphate should be less than half that in a water-soluble product before it becomes a better buy for most uses.

Potash fertilizers

The major K fertilizers are:

	Formula	*% K_2O*
Muriate of potash	KCl	60
Sulphate of potash	K_2SO_4	50

Other potassium-containing fertilizers sometimes used are:

	Formula	*% K_2O*
Potassium nitrate	KNO_3 (13% N)	44
Potassium metaphosphate	KPO_3 (27% P_2O_5)	40
Potassium magnesium sulphate	K_2SO_4, $MgSO_4$ (18% MgO)	22
Magnesium kainite	$MgSO_4$ + KCl + NaCl (6% MgO; 18% Na)	12

Muriate of potash (potassium chloride)
This is mined and is sold in either a powdered form or a fragmented (granular) form. It is the source of K in nearly all compound fertilizers and its use as a straight fertilizer in the UK is very limited. Where crops require substantial amounts of K it should be remembered that equal quantities of Cl are also applied if muriate of potash is used. Hence in a crop such as potatoes it is sometimes advisable to use potassium sulphate instead of muriate.

Sulphate of potash
This is made by treating muriate with sulphuric acid and so is a more expensive source of K. However for Cl-sensitive crops the extra cost of the sulphate salt is recommended. Some manufacturers make a series of compound fertilizers containing potassium sulphate and these are widely used by horticulturists, not only because of the fact that some crops are sensitive but also because of the risk of a build-up of Cl^- ions in the soil where regular heavy manuring is carried out.

Potassium nitrate (saltpetre)
This is expensive and its main fertilizer use is in foliar applications on fruit trees and some horticultural crops.

Potassium metaphosphate
This compound is a water-insoluble source of K (unlike all the other sources mentioned which are soluble in water), and it can be used where it may be necessary to keep ionic concentrations at a low level in the root vicinity.

Potassium magnesium sulphate and magnesium kainite
These are sometimes used where some Mg is required in addition to K; also kainite is often used for sugar beet because of the Na content.

Agricultural salt (mainly NaCl)
This can replace KCl in some situations, for example where sugar beet or mangolds are grown. Sodium chloride should not be applied to poorly structured heavy soils as it may predispose deflocculation, and it should not be applied less than three weeks before sowing lest it affect germination: in fact it is often applied in autumn and ploughed in. Application of Na reduces the need for K.

Forms of fertilizer available

Solids

This is the most common form in which fertilizers are available and normally manufacturers go to great lengths to ensure that the material stores without becoming compacted and spreads evenly and freely at the time of application. Water-insoluble materials, such as some types of phosphate, have to be in solid form, but most other materials could be sold in liquid form if necessary. The main advantages of solid types are:

(1) high concentration of nutrients in the material, reducing transport and storage costs and leading to a high rate of work at spreading;
(2) when sold in plastic bags, very cheap storage is possible and the material will keep in good storage condition for many months;
(3) when sold in bulk, relatively simple modifications to existing buildings will enable successful storage and make possible spreading systems which have low labour requirements;
(4) the farmer can buy large quantities at times when market conditions are favourable;
(5) the farmer can have a range of fertilizer types available, can use only that quantity which is needed at a given time, and can judge the application rate easily by counting the number of bags used.

Where water-soluble ingredients are used in solid fertilizers, it normally takes very little soil moisture to render these ingredients available. Similarly, where the same ingredients are applied to a dry soil, the quantity of water applied with the fertilizer is so minute that it has no irrigation effect. So it can be assumed that 'once brought into contact with the soil, liquids behave in the same way as comparative solid fertilizers, and generally no differences are observed in relation to growth and crop yields' (Mengel and Kirkby, 1982).

Solid materials are sold as powders, granules, prills or fragments.

Powders

These can vary in degree of fineness and for some materials (e.g. water-insoluble phosphates) powders are essential to ensure activity of the material in the soil; powders can be difficult to spread evenly and this is critical if relatively low rates/ha are needed; under these conditions full-width spreaders are advised.

Granules

They are made during manufacture by passing the fertilizer powder through a rotary drum, often with an inert granulating and coating agent; sometimes several fertilizer materials are mixed at the same time, so that each granule contains a mixture of materials. All granules made in this way are designed to break down quickly in the soil and release their nutrients, but sometimes the inert carrier remains visible in the soil for some days. The size and texture of the granules will have a marked effect on the sowing rate and spreading width of fertilizer distributors, and both these granule characteristics are specific to the type of material being granulated and the conditions under which the granulator is working: it follows that granules will vary in spreader performance and each consignment should be checked when being used.

Prills

Prills are made by the rapid cooling of a hot fertilizer liquid, are homogeneous in character and usually have a very smooth surface. Like granules they can vary in size, and great care is necessary at spreading time as prills usually flow more freely than granules and spreader-setting is critical to get correct rates and even spread. Prills normally consist of one fertilizer material only but this may contain more than one nutrient, for example a prilled ammonium phosphate would contain both N and P.

Fragments

Fragments are made by very coarse grinding of a solid material, usually followed by a screening process to produce material of a certain size range, for example some samples of muriate of potash are made in this way. Fragments behave in a similar way to very uneven granules (but note that granules are always spherical in shape because of the way they are made), but fragments and granules should not be mixed and spread together as marked segregation will occur both in the spreader and during the spread itself. Some fragmented material has a very abrasive action on working parts in the spreader and can cause rapid wear.

Most manufacturers take great trouble to ensure compatibility within each fertilizer type, but mixing of different fertilizer types on the farm prior to or during application can lead to poor spreading performance.

Liquids

Some materials must be used in liquid form, either because they do not exist as solids at normal temperature and pressure (such as polyphosphates) or because their intended use precludes solid form (e.g. foliar nutrient application).

The main features of liquid fertilizers are:

(1) easy handling on the farm, as they can be pumped in and out of store and spread by sprayer fitted with non-corrosive working parts and nozzles;
(2) relatively low concentration of nutrients, as some chemicals will crystallize out during storage and at low temperatures if concentration is raised;
(3) storage tanks required, which can be expensive; this can mean that storage capacity is restricted so that an annual requirement for fertilizers cannot be purchased at the lowest price;
(4) the possibility that herbicides can be mixed with fertilizer to further reduce spreading costs.

In the UK liquid fertilizers hold a fairly small share of the total fertilizer market, unlike the situation in some other countries. There is no simple guide as to whether a farmer should use liquids or solids as the decision rests on several aspects of general farm management, such as fertilizer purchasing policy, availability of existing stores for solid fertilizer storage, contacts for contra-trading with merchants selling either solid or liquid fertilizers, types of crop grown and labour and/or machine availability for spreading.

It is likely that the use of liquid fertilizers will continue to increase, particularly as a result of changing farm management practices, such as contracting out fertilizer application as a means of both saving costs and obtaining access to up-to-date technology in terms of application equipment and products.

Gases

As mentioned earlier, anhydrous ammonia can be used as a source of N. However, because of the high pressure needed for storage and the special injection equipment needed, this is a contractor-based service. It is valuable on relatively few crops and soils in the UK.

Compound fertilizers

When fertilizers are sold with a single main chemical ingredient, they are known as *straights*: examples are ammonium nitrate, superphosphate and muriate of potash. Sometimes straights can supply more than one nutrient, for example ammonium phosphate supplies N and P and potassium nitrate supplies N and K. Occasionally straights may be comprised of a mixture of ingredients, but if they still supply only one main nutrient they are still known as straights, for example some phosphatic fertilizers contain a mixture of water-soluble and water-insoluble phosphate, but still supply only P.

Many crops need more than one of the three major nutrients N, P and K, and in most of these situations it is convenient to apply all the nutrients at the same time. For this purpose, manufacturers have produced a wide range of *compound* fertilizers, that is fertilizers which contain more than one ingredient and supply more than one nutrient. Compound fertilizers are more expensive than straights, but many farmers find the extra cost is justified because if several straights were used the cost of application would be increased and it is difficult to mix and effectively spread straights on the farm.

As a guide, the extra cost of nutrients in a compound compared with straights should reach 15% before the compound looks expensive.

Comparing fertilizer costs

The easiest way to compare the relative costs of two compounds is to relate each of them to the cost of straights, using the following steps:

(1) establish the cost of 1 kg of each nutrient in straights;
(2) calculate the nutrient value of the compound, costing all nutrients in the compound as if they were straights;
(3) express the extra price of the nutrients in the compound as a percentage of the cost as straights.

Example

Assume ammonium nitrate (34.5% N) costs £140 per tonne, then 345 kg N cost 14 000 p = 40.6 p/kg.

Triple-superphosphate (45% P_2O_5) costs £180 per tonne, then 450 kg P_2O_5 cost 18 000 p = 40.0 p/kg.

Muriate of potash (60% K_2O) costs £150 per tonne, then 600 kg K_2O cost 15 000 p = 25.0 p/kg.

Compare the value of 20:10:10 @ £164 per tonne
with 22:8:8 @ £168 per tonne

	20:10:10	£	*22:8:8*	£
N	200 × 0.406 =	81.20	220 × 0.406 =	89.32
P_2O_5	100 × 0.400 =	40.00	80 × 0.400 =	32.00
K_2O	100 × 0.250 =	25.00	80 × 0.250 =	20.00
	Value =	146.20	=	141.32
	Cost =	164.00	=	168.00
	Difference =	17.80	=	26.68
	Extra cost =	12.2%	=	18.9%

so 20:10:10 is the better buy.

Fertilizer placement

Band placement

Most fertilizers are spread evenly over the surface of the soil or crop, and are used at the period when the crop is most in need of the nutrients contained. For example, PK fertilizers are worked into a seedbed prior to sowing and N fertilizers are top-dressed on to some growing crops, particularly winter cereals in the spring and on to grassland.

In some row-crops, placing fertilizer in a band near to the zone of the soil in which the majority of the plant roots will feed can both reduce the total quantity of fertilizer needed and check weed growth outside the crop rows. Fertilizer placement in this way tends to reduce the speed of fertilizer application and if carried out at sowing time by *combine-drilling* will also slow down the rate at which drilling can be carried out. Also, if the fertilizer rate is too high or too close to the seed or plant root, the material can cause scorch and this can reduce plant survival and check growth in established plants. Band placement of fertilizer is most advantageous where:

(1) soil nutrient status is very low, for here responses to placed fertilizer can be greater than to very high rates evenly spread over the field;
(2) crops need very high rates of nutrient application and are grown in discrete rows set well apart. Here, placement not only reduces the overall amount of fertilizer needed but also ensures the roots come into contact with sufficient material. A good example of this is the main-crop potato grown on normal mineral soils;
(3) the applied nutrient has limited mobility in the soil: this applies particularly to P which scarcely moves in the soil even when applied in water-soluble form.

Rotational placement

Some soils are reasonably well supplied with nutrients but further fertilizer application is recommended to maintain the adequate level: this situation applies particularly to P and K. It is thus suggested that it may not be necessary to use P and K fertilizers for every crop but adopt a *rotational* fertilizer programme where PK fertilizers are used periodically and used immediately before a PK-sensitive crop is grown. This approach can save money but needs careful monitoring to ensure that soil nutrient levels do not decline: also it should be noted that PK fertilizers will be needed on some fields on the farm each year as the rotation dictates.

Deep placement

In dry weather the growth of crops slows down, mainly due to lack of water for growth and transpiration but partly because the roots cannot absorb the soil nutrients which are located near the surface. Experiments have shown that *deep-placement* of fertilizers can maintain a faster growth rate in crops in dry weather than where the same nutrients are placed on the surface in a conventional way. It should be noted that, for this deep-placement to work, the nutrients should be at least 150 mm below the surface and this is deeper than normal injection of anhydrous or aqueous ammonia.

Fertilizers and the environment

When fertilizers are used, consideration must be given to the potential loss of nutrients from the soil. Phosphate is remarkably immobile in the soil and scarcely any is leached. Some potash is present in the soil in very soluble forms but the K^+ ion is held from leaching by both clay particles and organic matter, so that significant leaching of potash will only occur on sandy soils low in organic matter. Nitrogen is the most likely nutrient to leach, a specific problem being caused by NO_3^- being leached into water used for drinking. Excess nitrate in water can lead to *methaemoglobinaemia* (blue baby disease) in which the action of the red blood corpuscles is impaired. A considerable amount of nitrate is lost from agricultural land in the natural course of events (e.g. when grassland is ploughed or when there is heavy application of farmyard manure or slurry). Use of high rates of fertilizer N, particularly when used at periods in the growth cycle of the plant when N uptake is low, will exacerbate the problem. To minimize loss of fertilizer N by leaching:

- *for grassland* apply no N from mid-September to the expected start of spring growth in February or March;
- *for arable crops* apply no autumn N to seedbeds; make spring applications at the appropriate recommended growth stages in March and April (sometimes in February) and not earlier.

If fertilizer N is applied to bare soil, there is a risk of much of it leaching as nitrate. To minimize leaching, only apply N to a crop that has an active root system that will take up the N. If it is known that the natural soil N value is high, then nitrate can leak out and be leached if there is no crop present: in these circumstances, a catch crop should be grown.

Cultivation disturbs the soil micro-organisms and can lead to nitrate loss, so minimal cultivations should be used where possible.

Optimum fertilizer rates

Fertilizer recommendations (e.g. MAFF, 1994) are based on consideration of both physical response and financial return.

Physical response

This is obtained from field experiments, where the shape of the crop response curve to increasing levels of nutrient input is studied. Figure 9.1 gives a typical crop response curve. Although the exact shape of this curve will vary with crop, soil type and climate, the following principles hold good.

(1) Under field conditions, some yield is obtained where no fertilizer nutrient is used, the yield A depending on the level of available nutrient in the soil. Fertilizer recommendations take this into account by allowing for soil index and making specific recommendations for each crop.
(2) There is a maximum yield (B) which is obtained from a certain level of nutrient input (M): using more nutrient than M will not give a greater yield – it may give the same yield as M or less depending on the nutrient and the crop (shown by the shaded area in Figs 9.1 and 9.2.
(3) The shape of the response curve is such that the yield return per each extra kg of nutrient declines: for example it can be seen from Fig. 9.1 that the first half of the nutrient dressing M gives a much greater yield increase than the second half of the rate concerned. Thus maximum yield is reached at a point where yield return per unit of extra nutrient is so low that it would not pay for the extra nutrient used.

Financial return

Financial return is considered in establishing fertilizer recommendations by giving monetary value to the crop yield and nutrient costs. Figure 9.2 is based on the data shown in Fig. 9.1 and shows a curve for 'profit from fertilizer' which is taken as value of crop minus nutrient cost. The point O is very important as it is the highest rate of nutrient application that will show an increasing financial return. It is called the *optimum rate* and is below the maximum rate (M). Optimum rate depends on the ratio of crop value: nutrient cost, as well as on the shape of the agronomic response curve.

Optimum rates are not very sensitive to changes in nutrient costs or value of crop output. For most crops in the UK there is a large bank of experimental data that gives reliable information on nutrient responses over a wide range of soil and climatic conditions. Thus, ADAS and manufacturers' recommendations are sound guides to probable optimum rates of application. The periodic update of recommendations takes steady trends into account.

Where a new crop situation develops, either by the extension of an existing crop into a new soil type or climate area or by the introduction of a new crop into

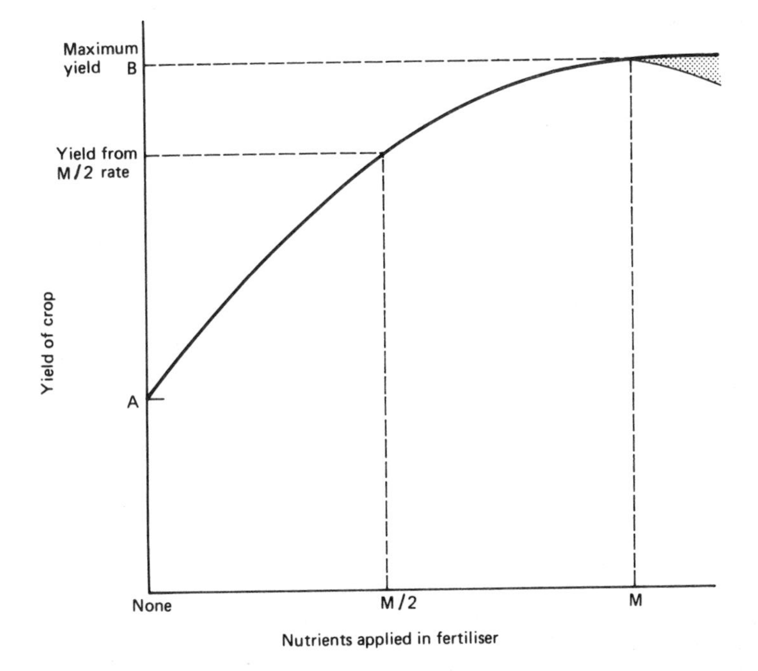

Fig. 9.1 Response of crop yield to increasing rate of fertilizer nutrient.

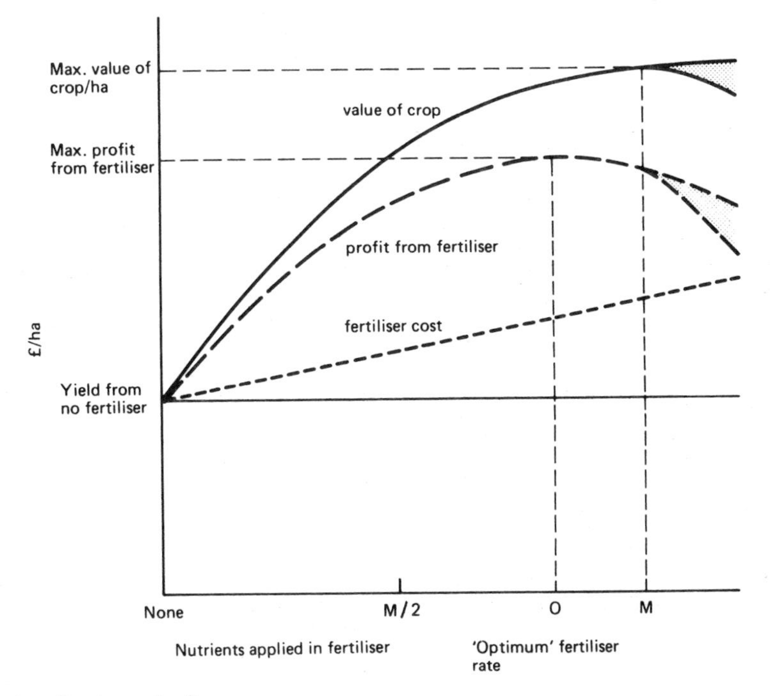

Fig. 9.2 Calculation of optimum fertilizer rate.

the country, then soil scientists and agronomists soon conduct experiments to establish agronomic criteria for economic optima to be calculated. Good examples in recent years have been the work on forage maize and oilseed rape, where recommendations were updated regularly for several years and now the fertilizer recommendations for these crops in the UK are different from those suggested in other countries where soil, climate and cropping systems are not the same as in the UK.

Fertilizer recommendations

Figure 9.3 illustrates the principles used to link knowledge of soil nutrient availability with understanding of crop nutrient need, to reach a fertilizer recommendation.

For all nutrients other than N, soil analysis will give a good indication of the quantity of readily available nutrients in the soil. Assuming a crop needs the same total quantity of nutrient for optimum growth, Fig. 9.3 shows how the need for inorganic fertilizer supplementation will vary with soil index and also how organic manures may or may not give high yields. The same principle is used in deciding N recommendations, except that the soil index is determined by the previous cropping history (see Chapter 7).

Sometimes the concentration of nutrients within the plant is assessed by *tissue analysis*, as there is a good general relationship between the concentration of available nutrients in the soil and the quantity found in the growing plants. However, results of tissue analysis need careful evaluation, as the quantity of any nutrient found in a plant will be affected greatly by:

(1) physiological age of plant when tested;
(2) part(s) of plant taken for analysis;
(3) species (and variety) of plant;
(4) growing conditions prior to testing;
(5) interactions with other nutrients.

Soils differ in their ability to release available nutrients, and crops vary in their nutrient requirements. ADAS and major fertilizer manufacturers publish crop fertilizer recommendations that take into account not only soil differences but also differences in the management of the crop: for example whether grass is cut or grazed, whether cereal yields are consistently above or below average, whether straw is removed or ploughed in. It should be remembered that fertilizer recommendations are based not only on chemical aspects of plant nutrition but also on the results of many field trials and farmer experience: thus recommendations are not clinically

Table 9.5 Example fertilizer recommendations (for winter wheat: normal mineral soil, expected yield approx. 7 t/ha)

kg/ha of nutrients to apply				
In spring		*In autumn*		
Soil index	*N*	*Soil index*	P_2O_5	K_2O*
0	180	0	110	120
1	130	1 and 2	60	90
2	70	3	40	Nil

* 45 less if straw incorporated.

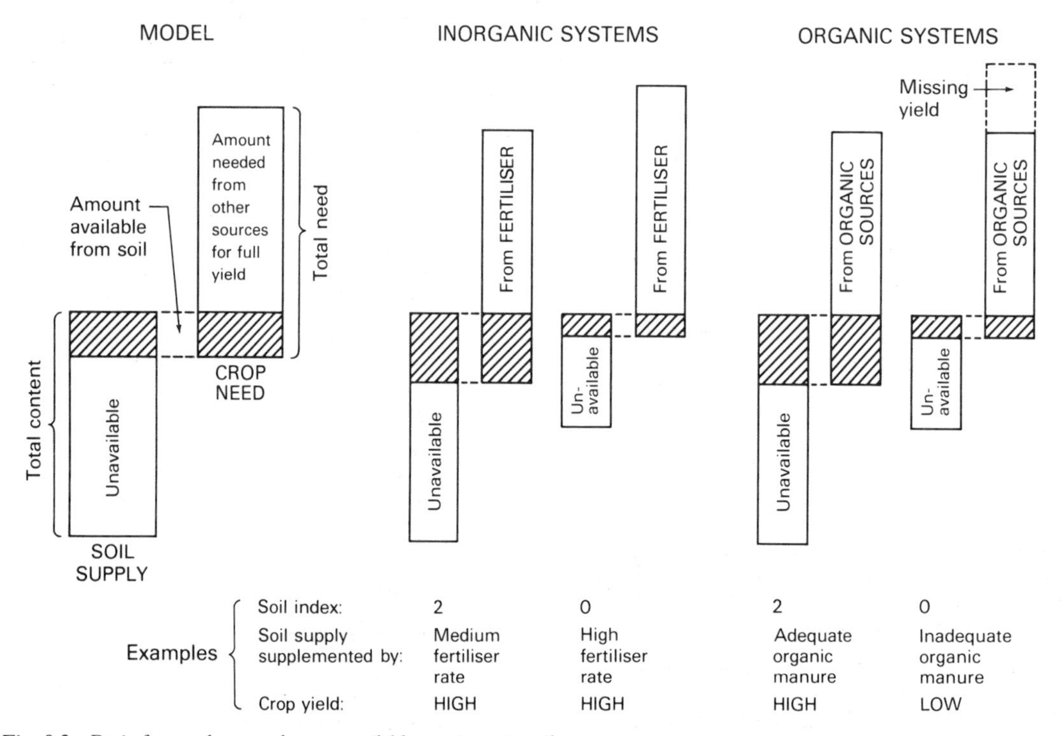

Fig. 9.3 Basis for need to supplement available nutrients in soil.

correct but best approximations which are updated from time to time as new information is obtained.

A typical fertilizer recommendation is given in Table 9.5.

Calculating product use from recommendations

By law every bag or consignment of fertilizer must contain on it a statement of the plant food value of its contents, and under EU regulations the phosphate value must be as P_2O_5 and the potash value as K_2O. Thus a compound fertilizer might be called 20:10:10 and state on the bags that its contents were:

20% N   10% P_2O_5 10% K_2O

If the farmer applied 300 kg/ha of this 20:10:10 fertilizer to a field, then the application of N would be 60 kg/ha (20% of 300), P_2O_5 would be 30 kg/ha (10% of 300) and K_2O would also be 30 kg/ha.

Recommendations are given as the requirements for N, P_2O_5 and K_2O and so the quantity of product to apply needs to be calculated. A useful equation is:

$$P = \frac{R}{A} \times \frac{100}{1}$$

where P = product rate to apply; R = nutrient requirement from recommendations; A = % analysis of nutrient in product.

Example

For first-cut silage it is recommended that a specific crop receives 125 kg/ha of N, 30 kg/ha of P_2O_5 and 30 kg/ha of K_2O, with the phosphate and potash and some of the N in late February and the rest of the N in mid-March. The farmer has in stock a 16:16:16 compound fertilizer and ammonium nitrate (34.5% N). How much of each fertilizer should he apply?

In February:

$$\frac{30}{16} \times \frac{100}{1} = 187.5 \text{ kg/ha of the 16:16:16 fertilizer}$$

In March:

The compound applied 30 kg/ha of each nutrient, including N, so N rate to apply in March is 125 − 30 = 95:

$$\frac{95}{34.5} \times \frac{100}{1} = 275 \text{ kg/ha of ammonium nitrate}$$

Some textbooks give phosphate and potash recommendations in terms of P and K and not P_2O_5 and K_2O. Great care should be taken when reading literature to ensure confusion does not occur. The following conversions can be used:

(1) To convert P to P_2O_5 multiply by 2.29 (P_2O_5 to P multiply by 0.44).
(2) To convert K to K_2O multiply by 1.20 (K_2O to K multiply by 0.83).

Liming

Adequate lime status is essential for the following reasons:

(1) It reduces soil acidity, and most plants will not thrive under acid conditions (Table 9.6 gives pH guidelines for major crops).
(2) It increases the availability of certain plant nutrients in the soil.
(3) Adequate soil pH encourages soil biological activity, thus enhancing the organic matter cycle in the soil, often releasing available N.
(4) It improves the physical condition of the soil by increasing the stability of crumb structures.

As explained in Chapter 7, there is a constant loss of lime from the soil, partly by crop removal and partly through leaching. Thus it is necessary to have a regular liming policy on all soils other than those with a natural Ca content. The frequency of liming will depend on soil type, rate of leaching (soil texture, rainfall, fertilizer N use), and sensitivity of crops grown. Soil analysis gives a good indication of soil pH and lime measurement, and each field should be sampled on a regular basis – say every four years.

On *grassland farms* soil pH should not be allowed to fall below 5.5 and soil should be limed to bring the pH up to 6.0.

On *arable farms* soil pH must be kept above the level required by the most sensitive crop grown. General

Table 9.6 Soil pH and crop tolerance

pH	*Agricultural crops*	*pH*	*Horticultural crops*
6.2	lucerne	6.6	mint
	sainfoin	6.3	celery
6.1	trefoil	6.1	lettuce
6.0	beans	5.9	asparagus
5.9	barley		beetroot
	peas		peas
	sugar beet	5.8	spinach
	vetches	5.7	brussels sprouts
5.8	mangolds		carrot
5.7	alsike clover		onion
5.6	rape	5.6	cauliflower
	white clover	5.5	cucumber
	maize		sweet corn
	wheat	5.4	cabbage
5.4	kale		mustard
	linseed		parsnip
	mustard		rhubarb
	swede		swede
	turnip		turnip
5.3	cocksfoot	5.1	chicory
	oats		parsley
	timothy		tomato
4.9	potato	4.1	hydrangea
	rye		
4.7	wild white clover		
	fescue grasses		
	ryegrass		

At a pH below the indicated level the growth of the crop may be restricted.

advice is to apply lime when pH falls to 6.0 and bring the pH up to 6.5.

Where grass is grown in rotation, it is essential that soil pH is kept up in preparation for succeeding arable crops, as it is difficult to restore rapidly a low pH. If acid grassland is ploughed for arable crops, not only must some lime be worked into the soil, but a sensitive crop such as barley should not be grown for several years.

Liming policy can affect the two diseases – 'club root' of brassicas (*Plasmodiophera brassicae*) and 'common scab' of potato (*Streptomyces scabies*). Club root is less severe on well limed soils whereas common scab is more prevalent where lime has been applied recently. Thus, in a rotation where brassicas and potatoes are grown, lime that may be required should be applied before the brassica crop, leaving at least two years before potatoes are grown.

Purchasing lime

It is a legal requirement that the purchaser of liming materials must be given a warranty by the vendor. Under present regulations there are three main considerations:

(1) fineness of grinding;
(2) neutralizing value;
(3) permissible variation (normally ± 5%).

Fineness of grinding affects the speed of action of the lime and the total quantity of the lime applied that may become useful: the finer the material the more rapid its action and some very coarse materials may never break down sufficiently to be of value. Not only is fineness important, but also the hardness of the material; it is more essential for hard material such as limestone to be ground fine than a softer form such as chalk.

Neutralizing value (NV) is the standard basis for comparing liming materials. NV can be determined by laboratory analysis and is expressed as a percentage of the effect that would be obtained if pure calcium oxide (CaO) had been used. For example, if a sample of ground limestone has NV = 55, then 100 kg of this material would have the same neutralizing value as 55 kg of CaO.

Lime recommendations, based on soil pH and the buffering capacity of the soil, are given in terms of the weight of CaO needed.

Example
A field has a lime requirement of 1.0 tonne/ha of CaO and two materials are available, ground limestone (NV 55) at £24/tonne delivered and hydrated lime (NV 70) at £36/tonne delivered. The two materials can be compared by calculating the cost of NV:

$$\text{ground limestone} = \frac{2400}{55} = 43.6\,\text{p/kg CaO equivalent}$$

$$\text{hydrated lime} = \frac{3600}{70} = 51.4\,\text{p/kg CaO equivalent}$$

Thus the ground limestone is the cheaper buy, except that the hydrated lime may be easier to apply (if not being applied by a contractor) and is quicker acting, factors that may override the price difference in some circumstances.

Some liming materials contain magnesium as well as calcium; where the addition of Mg is considered beneficial, then the use of such material may be justified even though the cost of each unit of NV is greater than a straight Ca lime.

Liming materials

Liming materials occur naturally over a wide area of the UK and they are mostly sedimentary rocks laid down as a result of:

(1) accumulation of shells and skeletal remains of animals and plants;
(2) deposition by marine calcareous algae;
(3) precipitation of Ca (or Mg) carbonates from water;
(4) aggregation of fragments from pre-existing limestones.

Ground limestone

This is a sedimentary rock consisting largely of calcium carbonate and containing not more than 15% of MgO equivalent. One hundred per cent of material must pass through a 5 mm sieve, 95% must pass a 3.35 mm sieve and 40% through a 150 μm sieve. NV is about 50 (maximum for pure $CaCO_3$ is 56). Legal declaration is NV and % through 150 μm sieve.

Screened limestone (limestone dust)

This is a by-product of grading limestone for non-agricultural purposes such as road metal and railway ballast. One hundred per cent must pass 5 mm sieve, 95% must pass 3.35 mm sieve and 30% through a 150 μm sieve. NV is about 48 and legal declaration is NV and % through 150 μm sieve.

Coarse screened limestone (coarse limestone dust)

This is a by-product of grading some road stones and is similar to the above but coarser in that 100% must pass 5 mm sieve, 90% pass 3.35 mm sieve and 15% through a 150 μm sieve. Legal declaration is as above and expected NV is about 48. However, not all the lime in this material is likely to be available: it should be applied at 120% of normal rate and not used where a rapid liming effect is needed.

Ground magnesian limestone

This is similar to the above materials, and subject to the same legal requirements, except that it must contain not less than 15% magnesium expressed as MgO. These materials are useful where there is a long-term need to maintain soil Mg level; however, they do not provide rapid treatment for an existing Mg deficiency.

Ground chalk

This is natural chalk (cretaceous limestone) ground so that at least 98% will pass through a 6.3 mm sieve. The legal declaration is for NV only, which is usually about 50. Some chalks are quite wet when quarried and have to be dried before grinding: the NV value should refer to the material as supplied (i.e. partially dried) and not to a completely dried sample.

Screened chalk

This is produced from chalk outcrops by scarifying the surface with tractor-drawn implements and passing the collected loose material through a screen. This material is produced widely in south and east England. Ninety-eight per cent must pass a 45 mm sieve and the legal declaration is for NV only (usual value is about 45). However, producers often guarantee fineness to pass 25.4, 12.7 or 6.3 mm sieves.

Calcareous sand (shell sand)

This is found on some beaches, particularly in Cornwall and Scotland, and arises because of the large proportion of shell fragments in the sand. In such localities this material is cheap and effective for arable and grass crops. All the material must pass 6.3 mm sieve and the legal declaration is for NV only (variable in the range 25–40).

Calcified seaweed (Lithothamnion calcareum)

These exist in some coastal waters. The material, of coral-like consistency, is dredged, screened to remove shells and either sold as graded material or dried and sold as milled powder. A number of properties have been ascribed to these products from time to time, but they are very effective liming materials, with an NV of 40–50. They tend to be much more expensive than other liming materials.

By-product liming materials

A number of industrial by-products can be used as liming materials; their availability will depend on production in the locality as transport costs can soon cancel out any price/tonne advantage of these materials.

Sugar beet factory sludge

This contains a large quantity of calcium carbonate, but is very wet with about half its weight as water. Nevertheless, this wet material has an NV of 16–22 and contains some N, P and Mg as well: it can be spread by normal farm muck spreaders.

Sometimes this sludge is dried; it is then more expensive and has an NV of 50.

Water works waste

This waste occurs in some areas and consists of a very wet material that contains both calcium carbonate precipitated from water and some calcium hydroxide. Some works partially dry the waste to a sludge containing 25–30% water.

Whiting sand

Whiting sand is a by-product of the manufacture of whitening. It is a pure form of calcium carbonate arising from the process where chalk is ground under water, but normally has a high water content.

Other materials

They include waste from soap works, paper works and bleach works. Many of these materials are wet, but can be spread by contractor and they will dry out to form a fine powder. Some have a slight caustic quality and should be applied to the soil rather than to a crop.

Historically other liming materials have been used. For example when fuel was cheap, it was common practice to burn limestone or chalk in kilns to produce burnt lime (calcium oxide). With an NV of 100 this material was the cheapest form of lime to transport, but it had to be kept free from contact with water. When water is added to burnt lime it forms hydrated (or slaked) lime (calcium hydroxide) and the chemical reaction can evolve considerable heat if allowed to proceed without control. Hydrated lime has an NV of about 70, and although more expensive than the liming materials mentioned earlier, it can be used in cases of emergency as it is extremely fine and quick acting.

Blast furnace slag and basic slag

These materials can have an NV of 35–45 and were widely available until changes in the steel-making processes rendered slags with good NV scarce.

Overliming

Overliming can cause problems and arises when either a regular liming policy is carried on without regard to high soil pH or when the soil is so acid that substantial quantities of lime are needed and mistakenly applied in one heavy dressing. In the latter case, it is essential that some lime is worked into the soil, so that extreme acidity can be alleviated and no part of the soil is overlimed and the remainder of the requirement applied some months later. A good rule is never to apply more than 3 tonnes/ha of CaO equivalent at one dressing.

Overliming can cause the following deficiencies: Mn – causing grey speck in oats, speckled yellows in sugar beet and marsh spot in peas; B – causing heart rot in sugar beet and crown rot in swedes; Cu – causing severe stunting, particularly in cereals; Fe – causing chlorosis in many crops, particularly fruit-bearing species. Overliming also encourages scab in potatoes.

References

MAFF (1994) *Fertiliser Recommendations.* ADAS Reference Book 209. HMSO, London.

Mengel, K. & Kirkby, E.A. (1982) *Principles of Plant Nutrition*, 3rd edn. International Potash Institute, Berne.

Soil Association (1993) *Standards for Organic Food and Farming.* The Soil Association Organic Marketing Company Ltd, Bristol.

UKROFS (1992) *United Kingdom Register of Organic Foods Standards*, MAFF. HMSO, London.

Further reading

Archer, J. (1988) *Crop Nutrition and Fertiliser Use*, 2nd edn. Farming Press, Ipswich.

Cooke, G.W. (1967) *The Control of Soil Fertility.* Crosby Lockwood, London.

Finck, A. (1982) *Fertilisers and Fertilising: Introduction and Practical Guide to Crop Fertilisation.* Verlag Chemie, Weinheim.

Jollans, J.L. (1985) *Fertilisers in UK Farming.* Centre for Agricultural Strategy, University of Reading, Reading.

10

Arable Cropping

A.J. Jellings & M.P. Fuller

All crop production seeks to utilize the natural, free resources of the biosphere (energy as light and heat, oxygen, soil, water, mineral nutrients) to obtain maximum productivity from plants useful to man as food, fuel, fibre, flavourings, and medicines. The biological and physical principles upon which this exploitation is based are outlined in Chapters 7, 8 and 9. These principles underlie crop production everywhere.

Commercial crop production in the UK operates the fundamental principles of productivity from natural resources in a particular economic, political and social context. This context changes with time. Thus, artificial inputs and energy can be added to the available natural resources in order to increase productivity, but only if profitability can be achieved and if compliance with the appropriate policy and legislation framework can be maintained.

This chapter will consider arable cropping in the context of the UK at the present time, by presenting first the principles which are generally applicable, and second the particular husbandry of the major arable crops. The reader should recognize that husbandry practice changes with the introduction of new varieties, improved understanding of biological and agronomic principles, and the development of more efficient techniques and machinery. Cropping practice is also influenced by consumer and societal requirements regarding the quality of food and care for the environment, and by economic and political trends which change the relative value of inputs and outputs.

Arable Crop Planning and Production Practice

Choice of arable crop

The choice of an appropriate arable crop for any given situation depends on many factors. Potential choices must be evaluated against these factors, using sound information about the crop and the situation in which it will be grown. These factors can be grouped around three key questions:

- Can it be grown at this site?
- Can it be sold (or used) profitably?
- What effect will it have on the farm system and business as a whole?

There is no substitute for local knowledge and experience, and observation of the types of crops grown in the locality will give an instant indication of what can be grown. This common-sense method does not indicate what alternatives might be grown, however (or what cannot be grown), as the local producers may be conservative in their choice of crop enterprises. An analysis using basic principles can be very valuable.

Climatic constraints

The climate will impose restrictions on the crop types and varieties that can be grown in a particular location. A good knowledge of the normal weather pattern is essential; accurate long-term data can be found in the MAFF Bulletin *The Agricultural Climate of England and Wales*, or from the local weather station service. Key climatic features are the number of days of frost per year, the average monthly temperatures, and the rainfall pattern. These should be considered in relation to the crop features of: minimum and optimum temperature for germination, base temperature for growth, accumulated temperature (thermal time) requirement from sowing to harvest maturity, sensitivity to frost, sensitivity to drought, and any particular requirements in relation to development, for example for vernalization, pollination, or ripening (see Chapter 8).

The greatest flexibility for cropping with respect to climate is found in areas with mild frost-free winters, relatively even rainfall distribution, and summers which supply more than 750 day degrees above 10°C between May and October.

Soil constraints

The soil will restrict the crop types that can be grown in a particular location. A good knowledge of the soil characteristics, and their interaction with rainfall pattern, for the farm or holding is essential (see Chapter 7). Crops vary in their requirements with respect to soil depth, soil texture and optimum pH, and in their sensitivity to compacted or stony soils, drought, and periods of waterlogging. In general the root crops have the most demanding soil requirements.

The soil and soil water characteristics not only affect the crop directly but also influence the timeliness of mechanized operations, particularly workability at time of establishment, and trafficability at times for major inputs and harvest.

Soils offering the greatest flexibility for cropping are deep and medium textured with few stones in areas with evenly distributed rainfall.

Potential productivity

The potential yield from a particular crop at a particular site is of considerable significance in relation to choice of crop and cropping system. For most crops the value of the output will be strongly correlated with yield per unit area, and this will be one of the determinants of whether it will be profitable to grow the crop. The potential yield will also be a strong determinant of the level of inputs, and thus variable costs, to be devoted to the crop, e.g. recommended nitrogen requirements vary with potential yield (see Chapter 9); in this way potential yield also influences the cropping system employed.

As was seen in Chapter 8, dry matter production is fundamentally determined by the amount of radiation intercepted by the crop, the amount of radiation potentially available is determined by latitude, and the amount potentially intercepted is determined by the genotype of the crop. Thus it is possible to calculate the theoretical maximum dry matter production for any geographical location and any crop. In the UK this theoretical dry matter production is about 35 t/ha for winter wheat; other annual crops have lower theoretical limits due to inferior phenotypes especially with respect to canopy structure. Not all of the dry matter produced becomes yield; much is partitioned into parts of the plant with less commercial value (for example for wheat the theoretical maximum grain yield in the UK is approx. 17 t/ha); canopy structure and partitioning can be improved by plant breeding.

Much more significant to the producer is the effect of environment and management on dry matter production and thus yield. The theoretical dry matter yield is likely to be much reduced by the effects of site and season, and the mangement of the crop within the geographical and genetic constraints will become critical for profitability. Site effects on potential yield must be assessed carefully to inform crop choice decisions, and crop management planning (seasonal weather effects cannot be predicted with accuracy and will only be known as they develop; day-to-day crop management must monitor these effects and adapt accordingly). Site effects on potential yield are related to climate and soil factors and their interaction. For example, a site with low summer rainfall and shallow soil will suffer a yield penalty because dry matter production will be limited by water stress, even with high radiation interception (see Chapter 8). Prediction of potential yield for a particular site is an inexact science. The site class system for grass production (Chapter 11) is one predictive system, and the Agricultural Land

Classification of England and Wales (Chapter 7) can also be used in this way. In addition, experience of a particular site will give a good indication of the potential yield of a 'new' crop. Thus, if cereal yields for a particular site are generally in line with the UK average, then it is likely that, say, oilseed rape yields will also reflect the UK average. However, care should be taken to compare crops with a similar growing season, eg. winter cereal yields will not necessarily give a good indication of spring crop yields.

It should always be remembered that good crop management can only ensure that the potential yield is approached: it can never raise the potential yield. Input decisions must reflect this crucial fact, within an appropriate financial management framework (Chapter 2).

The enterprise planning section of this chapter indicates the range of average commercial yields for each crop. This is not to imply that yields cannot be outside this range in poor or exceptional conditions.

Crop marketing

It is fundamentally important that the output from a crop must be saleable (or useable on farm), and it is vital that the producer is aware of the range of market outlets available, the quality of product required by each outlet, and any special arrangements, such as quotas or contracts, associated with each outlet.

The basic marketing chain is similar for all crop products and is shown below.

Producer/producer group/co-op
↓
Processor/packer
↓
Retailer
↓
Consumer

All crop products require some form of processing and/or packing before retail to the consumer, though the sophistication of this step varies widely from the superficial trimming of the outside leaves of a cauliflower for farmgate sale to the high technology of sugar extraction from sugar beet.

The producer sells either directly to the processor/packer, or to a merchant or wholesaler who sells on to the processor/packer. In some situations it is possible for the producer to take on the roles of processor/packer and retailer and to sell direct to the consumer through a farm shop, farmgate sales, pick-your-own enterprise, or delivery service. Most examples are those of fresh vegetables and potatoes, but such things as straw, rolled oats, flour, bread and herbs can provide marketing opportunities.

The importance of knowing the market options for any crop product and their associated quality requirements cannot be overemphasized. It is crucial for sound strategic planning. Generally, the greater the sophistication of the processing step, the fewer marketing options are available to the producer (e.g. see sugar beet in this chapter), and the more likelihood there is of needing a contract for production of the crop. Where there is a limit to demand, a contract to grow will reduce the risk to an acceptable level (sale of the product is secure provided the quality requirements are met), and may be a legislative requirement where quotas exist, or in relation to crops grown on set-aside land. It is rare for a 'new' producer to obtain a contract for the highly profitable crops requiring sophisticated processing and with a distinct limit to demand (e.g. evening primrose) as existing contract holders are unlikely to go out of the crop. Where limited outlets exist for a crop product, location will be a strong influence on the suitability of a crop for a specific site, as transport costs could reduce financial margins considerably (e.g. sugar beet).

For crops with a range of market outlets it is essential to know the quality requirements of each. To make sensible judgements regarding the relative profitability of contrasting outlets, it is generally true that the more stringent quality requirements have a higher cost of production, though this may be balanced by a premium price, i.e. the value of the output may be higher. Crops grown under contract must meet the quality specifications, or the output may be worthless. Crops grown for on-farm use, e.g. fodder crops, should also be evaluated in terms of quality requirements and cost of production, particularly in comparison with bought-in feed.

The value of the output will depend on the price per tonne and the yield; the price at any one time will vary with quality. Both price and yield are unstable (to a greater or lesser extent depending on the crop): the price will vary over time according to supply and demand, and yield will depend on site and season. Thus it is impossible to predict accurately the exact value of a crop product. However, the potential value can be estimated for financial planning purposes using typical seasonal patterns in conjunction with recent price trends. John Nix's *Farm Management Pocketbook* is useful in this regard, together with weekly prices published in trade journals such as *Farmers Weekly*. The income from the value of the output may be supplemented by an Arable Area Aid Payment for those crops included in the Scheme. Net income will depend on the costs of production (see Chapter 2), which will be influenced by the cropping system.

It is rare for a single crop enterprise to be managed in isolation, and usually potential crops must be evaluated in relation to a profile of enterprises being managed on a single (farm) unit. Labour planning, capital investment in machinery and buildings, and business objectives will also need to be considered as part of the strategic planning process (see Parts 1 and 4), and will in turn influence the cropping system employed.

Choice of cropping system

Political influences

As a consequence of internal (EU) and external (GATT) influences, the CAP has undergone reform (see Chapter 1) which has led to all previous arable crop support payments being replaced by the Arable Aid Payments Scheme. In the short term this will compensate producers for the lower (world) prices for those crops whose production was previously subsidized. This change in the price structure of the industry has led to a greater emphasis than before on maximization of production from natural resources with less dependence on expensive inputs, in order to reduce the variable costs and optimize gross margins.

The increased emphasis on more effective use of freely available natural resources, in combination with consumer pressure against perceived risks to human health and the environment from 'intensive' farming methods, have increased the need to develop and adopt more sustainable cropping systems. Sustainable systems seek to make the best possible use of natural resources whilst minimizing both the use of external inputs and the polluting effects of waste products. Thus efficient cropping systems minimize the need for artificial nutrients and pesticides by utilizing sound rotations, making use of manures and slurries, increasing biodiversity, employing cultural control methods whenever possible, and ensuring that artificial inputs are applied safely and efficiently.

Integrated crop management

Integrated crop management systems (ICM) integrate care for the environment into safe, efficient food production by the use of farming practices which balance the economic production of crops through the application of rotations, cultivations, choice of variety and judicious use of inputs, with measures that preserve and protect the environment. They have been developed as an extension of the well known integrated pest management techniques (see Chapter 13) which rely on a combination of cultural, genetic, biological and chemical control to minimize losses from weeds, diseases and pests. ICM systems appropriate to the UK are being developed and demonstrated by the LEAF (Linking Environment and Farming) organization set up by the Royal Agricultural Society of England; a joint initiative between the NFU and several of the major supermarket retailers is also developing ICM protocols for vegetable production.

Specialized systems

There may be special features of the geographical location which will influence the cropping system employed, for instance if the holding is within a National Park, an Environmentally Sensitive Area, or a Nitrogen Vulnerable Zone. Management agreements may be optional or obligatory in such situations, appropriate advice should be sought from the relevant authority or from MAFF in these cases.

The most well known of the specialist systems of production is organic farming, for which producers conform to particular production 'rules' based on the almost complete elimination of artificial inputs to production, and for which a premium is obtainable for some products, depending on the market demand. For detailed information the reader is referred to Lampkin (1990) and the Soil Association/UKROFS Guidelines.

Crop rotations

The majority of crop production systems depend on an effective and appropriate crop rotation. A crop rotation is a sequence of different crops on the same piece of land. For example, a simple six-course rotation might consist of two courses of winter wheat followed by winter beans, a further course of winter wheat, winter barley, and oilseed rape. On a farm scale the temporal arrangement of the rotation in one field will be reflected spatially each year across fields (Table 10.1).

There are many workable crop rotations, reflecting the diversity of crop enterprises and their combinations within production units across the UK, and it is impossible to give all the possibilities here. On mixed farms a two- or three-year ley will form the backbone of the rotation with forage maize or other fodder crop also in the sequence. (Rotational set-aside may also need to be included, if Arable Area Payments are being claimed.)

Crop rotations are used to:

(1) maintain soil fertility, and obtain maximum benefit from natural nutrient cycling; (particularly important in organic farming systems where artificial fertilizer cannot be used);
(2) give opportunity for cultural control of weeds and/or cheaper chemical control of weeds;
(3) control, or minimize effect of, soil-borne pests and diseases;
(4) encourage biodiversity within the farm system;
(5) make optimum use of natural resources.

Factors to be taken into account when planning rotations include:

- choice of crop enterprises;
- sowing and harvesting dates of each crop;
- work days needed for preparation/sowing and for harvest/clearing;
- soil-borne disease and pest risks associated with each crop;
- weed species present and potential for cultural control and/or cheaper chemical control;
- soil fertility;
- field factors (e.g. soil pH, slope, stoniness, access, previous use of persistent herbicides).

Historically, the chief importance of rotations was in relation to soil fertility. Before the invention of artificial bagged fertilizers, methods for maintaining soil fertility rotations were essential to ensure that yields did not decline as the land became exhausted. There was great reliance on manure and legumes for restoring the fertility after exhaustive crops. Rotating different crops around the same piece of land helped this process.

For example, the Romans often used a three-course rotation of two cereals followed by a fallow year. The fallow was allowed to regenerate naturally with weeds

Table 10.1 Example of a simple crop rotation, shown temporally and spatially

Year	*Field A*	*Field B*	*Field C*	*Field D*	*Field E*	*Field F*
1	wheat	wheat	beans	wheat	barley	OSR
2	wheat	beans	wheat	barley	OSR	wheat
3	beans	wheat	barley	OSR	wheat	wheat
4	wheat	barley	OSR	wheat	wheat	beans
5	barley	OSR	wheat	wheat	beans	wheat
6	OSR	wheat	wheat	beans	wheat	barley

and spilt cereal seed and this was used for grazing. Grazing animals dunged on the regenerating fallow thus replenishing the nutrients to some extent.

The famous Norfolk Four-Course rotation (developed by 'Turnip' Townsend in the 18th century) used wheat, followed by a root crop, followed by barley undersown with clover, followed by grass and clover. The clover added to the nitrogen status, and the ley and the root crop allowed manure to be naturally added back to the soil during grazing.

Although bagged fertilizers reduce our dependence on rotations for the maintenance of soil fertility, good rotational design can reduce the amount of bagged fertilizer needed thus cutting costs, and in some situations can assume considerable importance, e.g. organic systems and in Nitrate Sensitive Areas. In the future, artificial nitrogen restrictions may be imposed. This would increase the importance of rotational design to the maintenance of soil fertility. The effect of each crop on the soil N index (see Chapter 9) should be noted to aid rotational design in this respect.

Rotations ensure that all fields benefit from restorative crops such as leys, clover, beans, applications of farmyard manure and deeper cultivations. Soil structure will benefit and residues from legumes and manure will provide substantial quantities of nitrogen and other elements, and thus cut fertilizer costs.

Many soil- and residue-borne pests and diseases are specific to a crop or group of crops and related weeds, which must be present for the population to multiply. Growing one crop too frequently allows the associated pest/disease to build up to infestation levels, so that subsequent crops of the same type fail completely or produce uneconomic yields. The pest/disease must then be starved out by growing crops which are not susceptible. Unfortunately some pests and diseases survive in the ground without a host plant for many years and starving out becomes a long-term commitment; growing resistant varieties (when available) is of value in some cases. Hence it is cheaper and safer to adopt a sound rotation, in conjunction with other measures such as use of clean seed, destruction of trash and volunteers, and the provision of good crop growth conditions, to prevent pest/disease build-up. See Chapter 13 for further information on the use of rotations in maintaining crop health.

Rotations have to work practically. One of the most basic considerations is that the previous crop must be out of the ground and the soil prepared for sowing the next crop at the optimum time. When the gap between harvest and sowing is too long, potential leaching problems are created and the empty field may be an opportunity lost. High value crops will take precedence, sometimes requiring lower value crops to 'fit' round them. For example, wheat is often sown later than the optimal time when following sugar beet or a vegetable crop. It should also be remembered that few rotations are 'perfect' and fulfil all the ideal requirements.

Green manures and set-aside cover crops

The function of a green manure crop is to occupy ground which would otherwise be fallow, and to take up available nutrients from the soil, primarily to avoid leaching and loss of nutrients. The nutrients thus 'stored' in the plants are returned to the soil prior to planting the true crop by incorporating the green manure into the soil by appropriate means. Any plant may be used as a green manure crop but ideally the plant type should be cheap to establish and tolerant, and should grow quickly thus enabling rapid uptake of available nutrients even in less than ideal climatic and soil conditions. The choice of green manure should be considered carefully in relation to the practised rotation: it is very important that it should not present potential weed, volunteer, disease or pest problems to the rotation. There is some potential for additional benefit by considering the green manure as a short break crop, i.e. in relation to soil-borne pests and diseases, and if leguminous it may contribute additional nitrogen to the system. The use of green manures as 'trap' crops for troublesome pests, e.g. nematodes, requires skill and careful monitoring of pest development stage but can be useful in organic systems. Typical green manures are winter rye, mustard, *Phacelia*, rape and vetches.

Cover crops for set-aside are essentially green manures, having the same function, but with restrictions on choice and specified management guidelines. Currently, acceptable sown covers include grass, mustard, mixtures of at least two crop groups (excluding legumes), or any non-agricultural crop, e.g. *Phacelia*. Legumes are not acceptable unless included at less than 5%. The cover must be cut short between 15 July and 15 August, and must be destroyed by 31 August. The Arable Area Payments Explanatory Guide should be consulted for detailed guidance as the regulations change from time to time.

Enterprise planning

Each enterprise must be planned within the overall strategic business plan in order to predict labour, equipment, and input requirements. This plan must be modified on a day-to-day basis according to the season and other unpredictable events.

Field constraints

Individual fields should be assessed in relation to the proposed cropping. An effective field record system is essential for this. Previous cropping, associated inputs, and historical 'problems', e.g. poorly drained areas and weed infestations, must be taken into account during enterprise planning, particularly in relation to rotational constraints.

Features of the site should also be accounted for, such as access, slope, topography, exposure, footpaths, archaeological sites, and boundary features such as woodland, hedges and field margins. The effects of existing features and past cropping must be incorporated into the enterprise plan for effective crop management; in particular, field margin management should not be considered in isolation from the crop and the cropping system, as it is likely to have a considerable effect on the crop, and on the ease and cost-effectiveness of operations.

Varietal choice

Many varieties are available for each crop type, each variety exhibiting a unique profile of characteristics. The characteristics of importance will vary for each crop type but will usually be related to yield, quality, resistance to disease, time to maturity, and height. Varieties should be carefully chosen to suit the market/end use, the site of production, and the farming system. Recommended lists of varieties are published by the National Institute of Agricultural Botany and other organizations, and these should always be consulted to effect appropriate choices.

Crop establishment

It is false economy to sow anything but seed of high germination and vigour. Only licensed seed merchants can legally sell seed, and are required to state the germination rate. The statutory germination test is carried out under ideal laboratory conditions and may not indicate the field germination rate if the seed is of low vigour, particularly in cold, poor-soil conditions. The vigour of a seed lot is attributable to many factors including age, condition of mother plant, size, weight, mechanical damage, and presence of pathogens. There are various methods of testing vigour. High vigour seed should always be used in less than perfect field conditions. If home-saved seed is to be used, the germination should always be tested, and if suitable for sowing the seed should be cleaned and dressed.

Seed treatments are used to promote good seedling establishment, minimize yield loss, improve quality, and avoid the spread of harmful organisms. Successful practical seed treatments must be consistently effective under varied conditions; be safe to handlers and operators; be safe to wildlife; have a wide safety margin between the dose that controls harmful organisms and the dose that harms the plant; and not produce harmful residues in the plant or soil. The harmful organisms associated with the establishment phase may be seed-borne, soil-borne, or air-borne pathogens that attack the emerging and developing seedling. Seed treatments can be a cheap, easy and effective way to protect crops in the early stages of growth, and to avoid the need for less well placed agrochemical inputs.

Crop seedrate recommendations are of little practical value as varieties, and seed lots of the same variety, can vary significantly in their thousand seed weight (TSW), and allowance must be made for losses due to the pertaining field conditions. It is the final plant population that is important for optimal yield and quality, therefore the appropriate seedrate should be calculated for each situation depending on TSW and field conditions. The following formula should be used:

$$\text{Seedate (kg/ha)} = \frac{\text{Target plant population (No./m}^2\text{)} \times \text{TSW (g)}}{\text{Germination (\%)} \times \text{Field factor}}$$

(The inclusion of a field factor in the seedrate calculation attempts to allow for field losses due to adverse seedbed conditions such as temperature, moisture or condition. Such field losses can only be estimated and may vary from 0.5 (50% establishment of the seeds capable of germinating) for poor conditions to 0.9 (90% establishment) for good conditions.)

Example calculation
Crop with target plant population of 250 plants/m^2, TSW of 50 g, germination of 80%, and estimated establishment of 60%, i.e. field factor of 0.6.

$$\text{Seedrate} = (250 \times 50)/(80 \times 0.6) = 260.42\,\text{kg/ha}$$

Sowing ungerminated seed directly into open field

Seedbed quality is very important to successful establishment, especially for small seeds (e.g. oilseed rape) and sensitive species (e.g. peas). Seed needs to be surrounded by moist aerated soil of adequate temperature for germination. Degree of fineness of tilth should reflect seed size in order to ensure contact between seed and soil moisture. Very fine seedbeds can be liable to capping on some soils, so some small surface clod is desirable in appropriate conditions. Rough seedbeds cause uneven germination, slow establishment, and plants lacking vigour; in addition residual herbicides do not work satisfactorily, and precision drills sow inaccurately. Seedbeds should be neither compacted, nor loose, but reasonably firm.

Depth of sowing depends on seed size and availability of soil moisture. In general sow no deeper than necessary to ensure good coverage as quick emergence ensures a vigorous plant, but recognize that seed needs adequate moisture to germinate and availability will decrease towards soil surface. Smaller seed must be sown shallow because of limited food reserves; larger seed can be sown deeper without risk.

Broadcasting seed is quick and cheap, and can be performed using a suitable fertilizer distributor for cereals, grass, and forage brassicas. Its disadvantages are uneven depth of sowing (seed is often harrowed in) and uneven spatial distribution, making it suitable only in low yield potential situations or as a last resort, e.g. when sowing has been seriously delayed by poor weather. A special case is winter field beans which are best broadcast and ploughed under, often given a yield benefit compared with drilled crops.

Precision drills are used where accurate seed placing and precise plant populations are required. Singling and thinning by hand are now rarely necessary due to drilling to a stand using graded or pelleted seed, with allowance for non-germinating seeds. Success depends on a high quality seedbed, precise drilling, effective protection against seedling pests and diseases, and a successful weed control programme.

Direct drilling uses special drills with zero or minimal cultivations following the use of a total herbicide (paraquat or glyphosate). It can be used for cereals, grass, oilseed rape and forage brassicas, but well structured soils are needed, and if used sequentially certain weeds may become a problem.

Bed systems are used for root and vegetable crops to eliminate fanging, to give quick vigorous establishment, and to reduce reliance on good ground conditions for timely machine operations (Fig. 10.1). Crops are grown in strips or beds which the tractor straddles, thus restricting wheelings and compaction to the areas between the beds. All equipment widths must be matched to the bed width (or multiples of the bed width) to use this system effectively. Bed width is usually determined by harvester design.

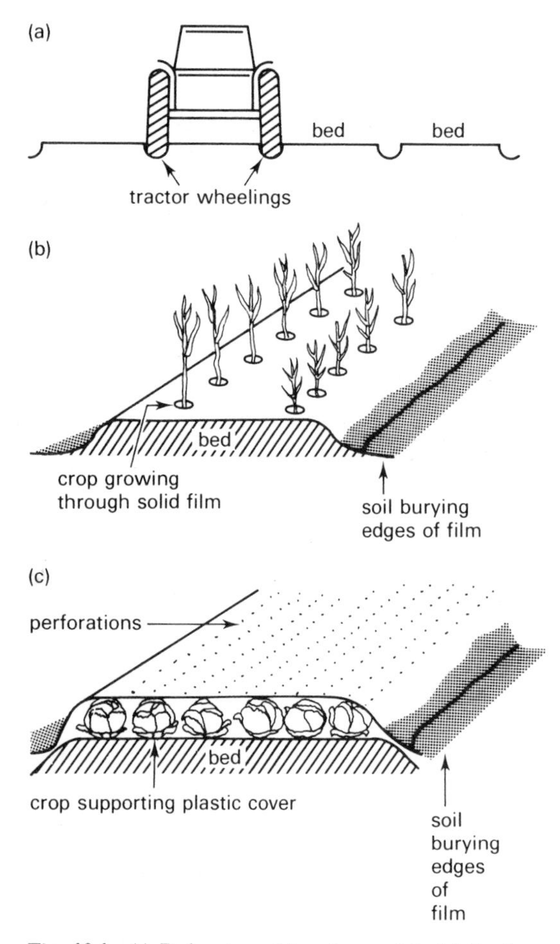

Fig. 10.1 (a) Bed system, (b) surface mulch, (c) floating plastic film.

Plastic covers for protection during establishment in the field

Plastic covers can be used to raise the soil temperature and offer protection from cold weather, in order to achieve earlier emergence and growth and advance maturity in comparison to crops established without protection. The use of covers is only economic for high value crops, particularly field vegetables and early potatoes. Covers may be used as surface mulches or floating films (Fig. 10.1). For surface mulching, solid plastic film is laid in contact with the soil surface, held down by soil at bed edges, and the crop grows through holes pierced at appropriate positions. Photodegradable film is used so that it disintegrates after a certain time (dependent on grade of plastic), from one to four months after laying. Floating films are perforated to allow exchange of gases and water/water vapour as the crop grows beneath the film, giving a 'cloche' effect. Different grades of film are available with different perforation, and therefore ventilation, characteristics. Timing of floating film removal can be critical as plants may not be hardened to exposed conditions, and temperatures below the film can become excessive as ambient temperature increases. It should also be remembered that light levels under the film are less than those 'outside'.

Transplants

The high value and demanding market quality specifications of some vegetable crops can make the increased costs of transplanting (as opposed to sowing seed directly into the field position) very worth while. The use of transplants can give a higher yield of the highest quality product, reduce the need for stringent seedbed preparation, reduced weed control costs, avoid unfavourable field conditions, and release field space for other crops. Transplants may be 'bare rooted' (lifted from raising beds with little soil attached to the roots) or raised in individual blocks, modules or cells in growing medium which is tranferred to the field with minimal root disturbance. Raising transplants is a specialized business requiring careful hygiene and critical management of nutrition and environment during establishment. Most growers purchase transplants from nurseries/suppliers.

Inputs and their timing

Efficient use of inputs and their timing depend on close monitoring of the crop and growing conditions, together with the use of reliable product information. Site and season have very strong effects on crop growth and development and there can be no foolproof 'recipe' for the production of a particular crop. Valuable aids to decision making are the development keys available for the major, and some of the minor, crops, e.g. the Zadoks key for cereals (Table 10.3). Inputs should always be applied according to the state of the crop (or its associated weeds, diseases or pests), not by calendar date.

Arable Crop Enterprises

The annual MAFF census should be consulted for current detailed crop production figures. The 1993 returns give the following production areas for the UK (excluding minor holdings in Scotland and Northern Ireland):

	ha
Cereals	3 030 400
Oilseed rape	374 200
Linseed	149 100
Field beans	163 600
Field peas	81 700
Sugar beet	196 900
Potatoes	166 700
Vegetables	120 800
Fodder	138 900

Combinable crops

Combinable crops are those grown for the grain or seed, harvested at relatively low moisture contents. Grain/seed ripening requires warm dry conditions and thus all the UK combinable crops are grown for harvest in the summer/early autumn period in order to take advantage of the seasonal climate; products are then stored to supply the market all year round. The majority of the UK arable cropping hectarage is used for combinable crops, and thus the cropping system on the majority of UK arable farms is based on the combinable crop season and requirements.

Cereals

Markets

Cereals are grown for animal feed and human consumption both for use in the UK and as exports. The animal feed grain market accepts wheat, barley, oats and triticale and the price obtained reflects the extractable starch or energy content of the grain. Thus wheat commands a higher price than barley and barley a higher price than oats. Oats can be sold into the specialist horse-feed market at higher prices. Quality standards in the UK are appearing in the animal feed grain sector and the price can be reduced if grain is below a target specific weight. Generally, though, the main quality requirement is sound grain of below 15% moisture content with a contaminants content below 5%. The feed price is generally the lowest price in the market.

Grains sold for human consumption include the major markets of wheat for bread and biscuits and barley for brewing. Other minor/specialist markets include wheat for cake flours, barley for other domestic uses, wheat and oats for breakfast cereals and wheat for starch and gluten extraction. The human consumption markets command premiums over feed grains but demand strict quality requirements which are assessed in the laboratories at the intake site. Examples of the quality requirements for bread wheat and malting barley are given in Table 10.2. Export markets have been growing in recent years and specific restrictions apply for each destination country.

Table 10.2 Quality standards for bread wheat and malting barley in the UK

Standards	*Notes*
Bread wheat	
Named varieties	Only HGCA Group 1 varieties are accepted which have been bred as breadmaking types and have the right combination of grain proteins
Hard endosperm	Not essential but preferred by the miller
Hagberg falling number over 250	Assessment of α-amylase activity (high levels ruin the crumb structure of the loaf and are indicated by low HFN)
Protein content over 11%	Indication of high gluten content necessary to give strength to the dough
Specific weight above 76 kg/hl	Indicates well filled grains and good flour extraction rate
Absence of ergots	Poisonous fungal fruiting bodies
Contaminants below 1%	
Good colour and free of smells	
Malting barley	
Named varieties	Varieties bred for malting quality and approved by the Institute of Brewers
Germination above 96%	Indicates good maltose extraction potential
Low nitrogen content	Normally below 1.6%; higher levels can cause cloudiness in beer
Pale lemon colour	
Free from odours	
Contaminants below 3%	

Intervention markets also exist for wheat and barley. This is where the EU buys surplus grain on the market, stores it and resells it on to the market at a later time. In recent years the standards for intervention have risen to dissuade farmers from dumping poor quality grain into intervention. All wheat grain must now meet quality standards similar to those for bread wheat in the UK yet command a price often lower than the feed wheat price.

Growth cycle

Cereals are monocotyledonous plants belonging to the Gramineae family (which includes all the grasses). They are all annual or winter annuals completing their growth cycle in less than one year. Autumn-sown varieties (winter types) differ from spring-sown varieties in that they require a vernalization period (exposure to cold) before they will flower and set seed. Thus winter varieties are in the ground for longer than spring varieties and this gives them a potential yield advantage. For wheat, approximately 97% of the UK crop is winter wheat whereas for barley approximately 55% is winter barley.

Each seed produces a single plant which during its vegetative phase initiates leaves and side shoots (tillers). Each tiller has the ability to produce one ear but normally a plant overproduces tiller numbers and many die off in spring leaving only the strongest to produce ears. The stimulus for the apex of the plant to produce a flowering head (ear) is made early in the growth cycle of the plant, e.g. after only 6 to 8 weeks in a winter barley crop and at the 3- to 5-leaf stage in spring barley but the flower may not be exposed until several months afterwards. Nevertheless it continues to develop inside the sheaths of the developing leaves.

After ear emergence, pollination takes place (anthesis) and then grainfilling. Most of the grain sites of wheat and barley are self-pollinated as the flower opens. The grain is filled with starch produced by photosynthesis of the top two leaves in the canopy during the grainfilling period (80%) and from the stored starch in the stem and photosynthesis in lower leaves (up to 20%). At the end of grainfilling the grain loses moisture and matures in the field before it is ready to combine. Harvesting is best carried out when grain moisture is about 15% but in wet summers harvesting at 20% may be practised and grain artificially dried to less than 15%.

In order to harmonize the chemical supply industry and the growers of cereals a key has been produced for the classification of the development of cereal crops (Table 10.3). All recommendations of agricultural chemicals and fertilizers to cereals are referenced to this growth and development key.

The components of yield of a cereal crop are: Ears/m^2 × grains/ear × seed weight. These components vary with the developmental stages of the crop (Fig. 10.2) but for ears per square metre and grains per ear there is an overproduction and a dying-off phase as the plant adjusts its potential grain sinks to the supply of carbohydrate coming from the leaves (source). The grower has most influence over the ears per square component of yield.

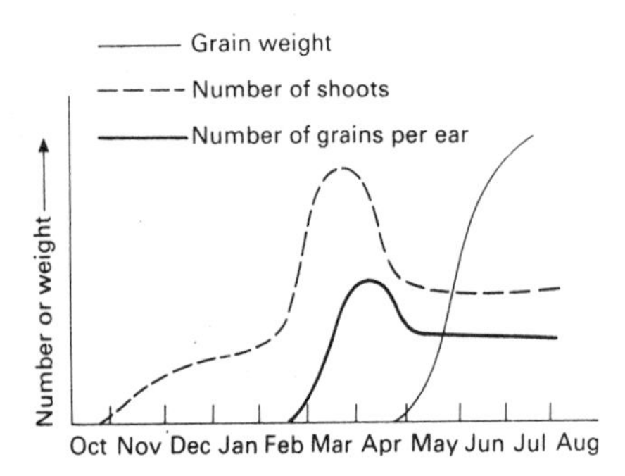

Fig. 10.2 Diagram to show how the components of yield of a winter cereal crop vary with time.

Soil types and rooting

Cereals can be grown on a wide variety of soil types provided a suitable seed bed can be cultivated. The yield potential of the crop (particularly wheat) is however related to the moisture-holding capacity of the soil. This is particularly the case in a dry summer. Cereals yielding about 6 t/ha will transpire about 100–125 mm of water during grainfilling, and whilst some of this will come from rainfall most must come from the soil's storage capacity. Early maturing crops should be grown on soils which are particularly low in available water holding capacity.

The grainfilling periods of the major UK cereals in the south of England are:

Winter barley – June
Winter wheat – July to early August
Spring barley – July to early August
Spring wheat – mid-July to mid-August

Rye and triticale have good tolerance of low soil moisture holding capacity and can outyield wheat on droughty soils in dry years.

Cereals can root down to 2 m but good crops are frequently produced on soil depths of 1 m. Winter-sown crops root more deeply than spring crops because they have longer to expand their roots; root growth practically ceases when rapid shoot growth commences in spring. Cereals have two root systems; the seminal root system which originates from the seed and is the deep penetrating root system, and the nodal or adventitious root system which originates from the base of the crown of the plant and is the root system which proliferates in the topsoil and takes up the majority of the spring applied nitrogen and summer rainfall.

Seedbed

Traditionally, winter and spring seedbeds differ, spring seedbeds being much finer than winter seedbeds. Fine winter seedbeds are sometimes prepared and have been favoured when soil residual herbicides were used. The commonest winter seedbed has a fine tilth at drilling depth (2–5 cm) with a mixture of sizes of clods on top, the biggest being about 5 to 10 cm in diameter. Overly deep seedbeds should be avoided or drilling will be too deep, and loose seedbeds should be avoided or problems with lodging could arise. This does not mean that winter seedbeds should be rolled following drilling as this runs the risk of breaking the surface clods which are necessary to help prevent soil capping and erosion during the heavy winter rains (particularly important on silty soils). If rolling is necessary, then a ring or Cambridge roll should be used. Spring seedbeds can be much finer than winter seedbeds, and rolling helps to consolidate the drying soil and gives better germination.

The use of no-tilth or mini-tilth seedbeds has diminished with the reduction in straw burning and its banning in 1992. More cultivations are needed now to help dispose of unwanted straw and stubble. On suitable soil types (well structured, calcareous loams) direct drilling into stubble can still be beneficial. On heavy soils, reduced cultivations are still necessary to avoid the problems of large sun-baked clods following ploughing.

Plant population and seedrate

An adequate plant population is essential to achieve a good, evenly ripened crop. Given ideal conditions a winter cereal will tiller sufficiently to give an adequate

Table 10.3 Zadok's decimal code for the growth stages of cereals

Code	Stage	
0	GERMINATION	
00	Dry seed	
01	Start of imbibition	
02	–	
03	Imbibition complete	
04	–	
05	Radicle emerged from seed coat	
06	–	
07	Coleoptile emerged from seed coat	
08	–	
09	Leaf just at coleoptile tip	
1	SEEDLING GROWTH	
10	First leaf through coleoptile	
11	First leaf unfolded	
12	2 leaves unfolded	
13	3 leaves unfolded	
14	4 leaves unfolded	
15	5 leaves unfolded	
16	6 leaves unfolded	
17	7 leaves unfolded	
18	8 leaves unfolded	
19	9 or more leaves unfolded	
2	TILLERING	
20	Main shoot only	
21	Main shoot and 1 tiller	
22	Main shoot and 2 tillers	
23	Main shoot and 3 tillers	
24	Main shoot and 4 tillers	
25	Main shoot and 5 tillers	
26	Main shoot and 6 tillers	
27	Main shoot and 7 tillers	
28	Main shoot and 8 tillers	
29	Main shoot and 9 or more tillers	
3	STEM ELONGATION	
30	Pseudostem erection (winter cereals only) or stem elongation	
31	1st node detectable	
32	2nd node detectable	
33	3rd node detectable	
34	4th node detectable	
35	5th node detectable	
36	6th node detectable	
37	Flag leaf just visible	
38	–	
39	Flat leaf ligule just visible	
4	BOOTING	
40	–	
41	Flag leaf sheath extending	
42	–	
43	Boot just visibly swollen	
44	–	
45	Boot swollen	
46	–	
47	Flat leaf sheath opening	
48	–	
49	First awns visible	
5	EAR EMERGENCE	
50	–	
51	First spikelet of ear just visible	
52	–	
53	$\frac{1}{4}$ of ear emerged	
54	–	
55	$\frac{1}{2}$ of ear emerged	
56	–	
57	$\frac{3}{4}$ of ear emerged	
58	–	
59	Emergence of ear completed	
6	FLOWERING	
60	–	
61	Beginning of flowering (not easily detectable in barley)	
62	–	
63	–	
64	–	
65	Flowering half-way	
66	–	
67	–	
68	–	
69	Flowering complete	
7	MILK DEVELOPMENT	
70	–	
71	Seed coat water ripe	
72	–	
73	Early milk	
74	–	
75	Medium milk	
76	–	
77	Late milk	Increase in solids of liquid endosperm visible when crushing the seed between fingers
78	–	
79	–	
8	DOUGH DEVELOPMENT	
80	–	
81	–	
82	–	
83	Early dough	
84	–	
85	Soft dough (Finger-nail impression not held)	
86	–	
87	Hard dough (Finger-nail impression held, head losing chlorophyll)	
88	–	
89	–	
9	RIPENING	
90	–	
91	Seed coat hard (difficult to divide by thumb-nail)	
92	Seed coat hard (can no longer be dented by thumb-nail)	
93	Seed coat loosening in daytime	
94	Over-ripe, straw dead and collapsing	
95	Seed dormant	
96	Viable seed giving 50% germination	
97	Seed not dormant	
98	Secondary dormancy induced	
99	Secondary dormancy lost	

crop density from a plant population of about 100 plants per square metre. However, higher populations of 250 to 300 are recommended because they give a safety margin for losses, they give a more uniform crop and they give greater early weed suppression. Populations above 350 are wasteful of seed and can lead to lodging (commonly seen on headlands where double drilling has occurred). Seed costs are a relatively small component of the variable costs of growing cereals and therefore a high seedrate is frequently considered to be a good insurance policy. Spring cereals have less time to produce strong tillers and as a consequence are established at higher plant populations to compensate (spring barley, 350–400 plants per square metre; spring wheat, 400–500 plants per square metre).

The seedrate for a field is determined from the following:

- the desired plant population;
- the seed weight (thousand grain weight);
- the expected establishment percentage (field factor).

Of these factors the field factor is of course unknown and must be guessed in relation to previous experience. As a guide, good drilling conditions at the optimum drilling date should give an establishment of about 85%; moderate conditions about 70%; poor conditions only about 55%. Conditions, particularly soil temperature, deteriorate quite rapidly after mid-October and it is recommended that seedrate is raised substantially for November sowings. Early spring sowings are frequently made into poor germination conditions (wet and cold) as are late sowings (dry and poor seedbeds). Guideline seedrates are given in Table 10.4. In the mid 1980s most varieties within a species had similar seed weights, and standard seedrates were often used; however, there is more variation in seed size now, especially in winter wheat, and a knowledge of thousand grain weight can influence the seedrate markedly.

Poorly established winter wheat crops should not be ploughed up and planted with spring barley unless there are substantial areas of the field at less than 50 plants per square metre. Stitching in some spring crop into poorly established areas of a winter crop can be effective but will delay the harvest of the field whilst the spring crop ripens.

Table 10.4 Guideline seedrates for cereals in England and Wales

Crop	*Sowing date*	*Normal seedrate (kg/ha)*
Winter wheat	Late Sept–mid Oct	140–160
	Nov–Dec	180–220
Winter barley	Mid Sept–early Oct	140–150
	Mid Oct–end Oct	150–190
Winter oats	Early Oct	150–160
	Late Oct	160–180
Rye	Late Sept–mid Oct	150–180
Triticale	Early Oct–mid Oct	120–150
Durum wheat	Oct–early Nov	180–210
	Feb–mid March	190–220
Spring wheat	Late Feb–early March	190–220
	Mid March	220–230
	Nov–Dec	200–220
Spring barley	Feb–March	125–150
	April	160–170
Spring oats	Feb–early March	170–190
	Late March	190–200

Sowing date

Cereals are remarkably tolerant of a wide variety of sowing dates but perform best over a relatively short window of sowings. Winter wheat, for example, can be sown from early September to March but performs best if sown at the end of September to early October. Winter varieties of course need to experience a period of cold in order to flower and produce grain, but the cold requirement is met even if the crop is sown as late as February/March. Wheat has a greater requirement for cold than does barley which in turn requires more than oats. The windows and optimal sowing dates are as follows:

Crop	*Normal window*	*Optimal*
Winter barley	Early Sept to early Oct	Last 2 weeks of Sept
Winter oats	Late Sept to mid Oct	First week of Oct
Winter wheat	Late Sept to end Oct	First 2 weeks of Oct
Spring wheat	Late Nov to end March	February
Spring barley	Late Feb to end Apr	March
Spring oats	March to end Apr	March

For winter cereals earlier sowing has the advantages of a higher yield potential, higher germination rate and greater tillering capacity but it carries with it a greater risk of disease infection, pest infestation (particularly aphids carrying Barley Yellow Dwarf Virus – BYDV) and greater weed germination. Many growers deliberately delay the sowing of winter cereals in order to reduce the need for autumn pesticides. This is particularly important for growers of organic cereals who will commonly sow winter wheat in November at very high seedrates and avoid the use of herbicides and reduce the damage from BYDV. Growers of large cereal hectarages will, however, need to begin sowing as early as possible in the autumn in order to complete before winter rains bring the soil to field capacity.

The priorities of sowing a large winter acreage are many and varied but good husbandry should take account of the following:

- First wheats can be sown early with little disease carryover risk thus maximizing yield potential (if possible this should be exploited with a milling wheat).
- Fields at risk of take-all (third to fourth cereal after a 2-year break of second cereal after a 1-year break) should be sown towards the end of sowing.
- Heavy soils should receive priority.
- Ploughed-out grassland should always be left at least two to three weeks before sowing to avoid frit fly and aphid carry over.

Varieties of winter wheat show some variability in their ability to tolerate different sowing dates. Generally, varieties which are disease susceptible and/or have a low vernalization requirement are less suited to early sowing.

There is less pressure on spring sowings because there is less area to establish on most farms. However, with the switch to an area support system from the price support system there is a swing towards more spring sowings. Sowing is dictated more by soil conditions than anything else and early sowing is only possible on light, easily drained soils. Germination percentage is likely to be low in cold spring soils and seedrates need to be higher to compensate. With late spring sowings seedrate also needs to be increased as less time is available for tillering.

Sowing depth

Cereals have relatively large seeds with a good carbohydrate store and therefore can germinate from quite deep sowings. It should be remembered though that the seed stores are used for early leaf production as well as for emergence growth. Thus, deeply planted cereals will show low early vigour. Plants which have come from seed sown too deep often have a hooped appearance often referred to as 'hockey sock'; this is common on volunteer plants which have germinated following ploughing. Shallow sowings can suffer from an inadequately developed root system which is susceptible to frost heave over the winter. Frost heave is where ice forms under the crown of the plant and pushes the plant upwards, snapping the roots. The normal depth of sowing is 2–2.5 cm and is adequate for both winter and spring varieties. Under warm germination conditions it can be advantageous to sow slightly deeper than 2.5 cm, especially if the seedbed is dry. Sowing depth is controlled by drill settings and influenced by the level of consolidation of the seedbed. Slow drilling leads to the even placement of the seed and a better control of depth. The practice of broadcasting and cultivating the seed into the seedbed results in the poorest control of sowing depth and inevitably results in greater establishment losses. When broadcasting is used, seedrates are increased by 5–10% to account for this.

Seed dressings

Most cereal seed is dressed to protect it against seedbed pests and against germination and establishment diseases. In the past a routine and cheap double dressing was used based on gamma HCH (pest control) and mercury (disease control) but mercurial seed dressings have now been banned in the EU. Replacement fungicides are now available for seed dressing but are more expensive than before. These fungicides also have a systemic action, which gives some early disease protection to the establishing seedling and can be very effective in spring cereal production where early mildew infection can be significant. The seed dressings are applied to the seed by the seed supplier or by a mobile seed cleaning firm if home-saved seed is used. The technology of seed dressings is advancing rapidly and it may be possible in the near future to obtain cereal seed dressed with a greater range of pesticides and/or germination stimulants such as hydrogen peroxide.

Establishment pests

Cereal crops can fail in some instances following the attack by certain pests during establishment. A more complete description is given in Chapter 13 but a brief consideration of the most important pests is given here.

Slugs

Damage caused by slugs is most serious when they graze on the germinating seed underground causing grain hollowing. Effective control can be obtained by mixing slug pellets with seed before sowing in high risk situations. The post-sowing broadcasting of slug pellets is less effective since the application is frequently made too late, i.e. after damage has been caused.

High risk situations include: heavy soil types; high presence of trash; previous crop favoured slug multiplication in summer, e.g. potatoes, grass, oilseed rape; wet summer and autumn; direct drilling.

Fly larvae

Various fly larvae cause damage to cereals that is known as 'dead heart'. This is where the larvae have penetrated the shoot and eaten away the apical meristem. These larvae include frit fly, wheat bulb fly and yellow cereal fly. Frit fly are common following grass and can be largely controlled by a delay in the sowing of the cereal of at least 14 to 21 days after sward destruction Wheat bulb fly is more localized in its distribution, only being found in the east of England and the East Midlands. The fly lays its eggs on bare soil in July and August and therefore is commonly found after second early potatoes. Post-emergence chemical control of fly larvae is possible after diagnosis of infestation.

Wireworm

This small soil-borne pest is also frequently found after grass. Control by the routine use of a gamma HCH seed dressing is effective for low populations. Where there is a risk of wireworms the seedrate should be increased to compensate for plant losses.

Establishment diseases

Diseases at establishment can be divided into two groups: first, those that cause germination failures; and secondly those that lead to early plant infection which has the potential to cause epidemic disease at a later stage.

Germination failure caused by disease is rare but can occur in poor germination conditions with seed of low vigour. On the whole, certified cereal seed will have a good germination capacity (EU minimum of 85%) and be vigorous enough to establish satisfactorily.

Early plant infection organisms include the seed-borne diseases bunt (*Tilletia caries*), leaf stripe (*Pyrenophora graminea*), *Fusarium* spp, net blotch (*Pyrenophora teres*) and loose smut (*Ustilago nuda*); and the foliar diseases mildew (*Erysiphe graminis*), *Septoria* spp and *Rhynchosporium*. Special mention is warranted of the virus diseases Cereal Yellow Dwarf Virus (commonly known as Barley Yellow Dwarf Virus or BYDV) and Barley Yellow Mosaic Virus (BYMV). BYMV is soil-borne and is serious enough to force some growers out of barley on infected land, but some tolerant varieties are now being developed and used successfully. BYDV is more widespread and has an aphid vector. Aphids carrying the virus can infect a newly established winter cereal either by migrating from ploughed-under grass or grassy stubble and/or by flying in from the airborne population which exists during autumn. The incidence of BYDV is more common in the south of England and in particular in the south west where the aphid population continues to fly later in the autumn thus putting more sowings at risk. Control of the aphids is the key to control of the BYDV. Where the cereal follows a ley or grassy stubble, then it is important to completely destroy the sward and this in turn will destroy the aphids. Chemical destruction of the sward using paraquat leads to a fast sward destruction; sowing can follow two weeks later. If glyphosate is used, the 3- to 4-week delay is necessary as sward destruction is slower. If only ploughing is used, at least 4 weeks' delay is necessary. Flying aphids need to be controlled using insecticides sprayed in the first week of November when flying has ceased. In high risk areas (south west and south) and high risk years (high percentage of infectivity), early sown crops of barley may need to be sprayed twice, once in mid October and again in early November. Only crops which emerge by mid October are at risk of BYDV, which includes most winter barley and early sown winter wheat and winter oats. Spring sowings do not usually suffer yield loss by BYDV since infection occurs late in the life cycle of the crop when aphids begin flying again in summer. Current research has identified varietal resistance to BYDV, and breeding programmes are aimed at producing resistant varieties in the near future.

Weed control

Weed control in cereal production is an integral part of modern growing systems. During the expansion days of cereals in the UK (late 70s and early 80s) profitability from cereals was good and many herbicides were used with up to five different chemicals per crop. In the 1990s the use of herbicides has declined due to a combination of environmental pressures and most importantly a decline in profitability. A summary of weed control is given below but details of herbicide choice are given in Chapter 13.

Weed control options comprise the following:

(1) Pre-sowing – applied to the soil and incorporated before sowing of the crop; only used for soil-acting granules to control wild oats.
(2) Pre-emergence – applied after sowing and before emergence; chemicals are soil acting (residual); good approach for the control of germinating grassy weeds (blackgrass, sterile brome).
(3) Post-emergence – applied after crop and weed emergence; chemicals may be both soil acting and contact or systemic acting:
    (a) Autumn applied – controls seedling weeds germinating with the crop.
    (b) Spring applied – controls larger weeds present in the crop and seedling weeds germinating in the spring.

Winter cereal weed control is also aided by a high seedrate, late sowing and a vigorously growing crop.

Each region and each field has its own particular weed problems but there is a list of common cereal weeds. The competitive ability of these species varies from species to species and from year to year (Table 10.5). The key to cost-effective chemical weed control is the identification of weed species and the ability to assess high risk infestation levels.

Chemical weed control is now subject to the Government regulations contained in Food and Environmental Protection Acts and all sprayer operators must pay due regard to the regulations. For new operators, a competence certificate must be held for the safe use and handling of crop chemicals (see Chapter 5).

Table 10.5 Competitive ability of weeds in cereals

Very competitive	Wild oats Cleavers
Moderately competitive	Blackgrass Poppy Rough stalked meadow grass Chickweed Brome grass Mayweeds Forget-me-not
Poorly competitive	Speedwells Red deadnettle Field pansy

Nutrients

Most soils contain all the necessary micronutrients for continuous cereal production but will frequently be deficient in the three macronutrients nitrogen (N), phosphate (P) and potassium (K). Also, over time, arable cropping leads to the removal of calcium and magnesium. This results in the acidification of the soil, which needs to be corrected by the application of lime.

Phosphate and potassium

The removal of phosphate and potassium by a cereal crop is summarized in Table 10.6. Grain removes virtually equal amounts of P and K whilst straw removes disproportionate levels of K. Application levels of P and K to the soil for a cereal crop therefore depend on the expected grain yield, whether the straw is to be removed from the field or not, and the level of P and K reserves already in the soil. Soil reserves are indicated by the soil index (see Chapter 7) with a high index indicating a good reserve. Good husbandry aims to maintain the P and K indices at about 2 or above but a cereal crop only demands an index of about 1. If the index is a low 1 or a 0, then P and K must be applied to the seedbed of the crop for that crop. If the index is a 1 or low 2, then the P and K are also applied to the seedbed but may be used by the succeeding crop. If the index is a 2 or higher, then the P and K can be applied as a top dressing at any time during the life of the crop since it is only for use by the succeeding crop(s). Some soils, particularly heavy clays, have naturally high levels of K and may not need any application of K to the soil. Recommended levels of application of P and K are given in Table 10.7.

Lime

Details of lime requirement are given in Chapter 9. For cereals, a soil pH of about 6 to 6.5 is required. If the pH drops below 6, then lime is required. Among the cereals, the tolerance of low pH is as follows:

Crop	*Critical pH (pH below which yield is restricted)*
Rye	4.9
Triticale	5.2 (estimated)
Wheat	5.5
Barley	5.8

It can be seen that barley is the most sensitive to acidic soils and liming should take place before the barley crop in a rotation.

Table 10.6 Phosphate and potash removed per hectare by cereal crops

	kg/tonne		*As a 7 t/ha crop*	
	P_2O_5	K_2O	P_2O_5	K_2O
Grain	6	5	42	35
Straw	1	8	7	56
Total	7	13	49	91

Table 10.7 Phosphate and potash requirements of cereals (kg/ha). (m) indicates maintenance dressing (MAFF, 1994)

	Soil index			
	0	*1*	*2*	*3*
Straw incorporated				
P_2O_5 (winter wheat @ 8 t/ha)	110	85	60(m)	60(m)
P_2O_5 (spring wheat; all barley, oats, rye, triticale @ 6 t/ha)	95	70	45(m)	45(m)
K_2O (winter wheat @ 8 t/ha)	95	70	45(m)	0
K_2O (spring wheat; all barley, oats, rye, triticale @ 6 t/ha)	85	60	35(m)	0
Straw removed				
P_2O_5 (winter wheat @ 8 t/ha)	120	95	70(m)	70(m)
P_2O_5 (spring wheat; all barley, oats, rye, triticale @ 6 t/ha)	100	75	50(m)	50(m)
K_2O (winter wheat @ 8 t/ha)	140	115	90(m)	0
K_2O (spring wheat @ 6 t/ha)	120	95	70(m)	0
K_2O (barley, oats, rye, triticale @ 6 t/ha)	135	110	85(m)	0

Nitrogen

Nitrogen fertilizer is the most cost-effective input into a cereal crop. Its present low cost and high yield response mean that the economic optimum is virtually the same as the optimum for maximum yield. Where soil N reserves are low, the response to the application of a kilogram of N can be 30 kg of grain and averages 10–20 kg of grain (Fig. 10.3). With soils which are more fertile the response is lower but still cost effective.

Nitrogen is the only input to the established crop that actually lifts yield potential. All other inputs, fungicides, herbicides, insecticides, growth regulators, merely protect the yield potential that is established.

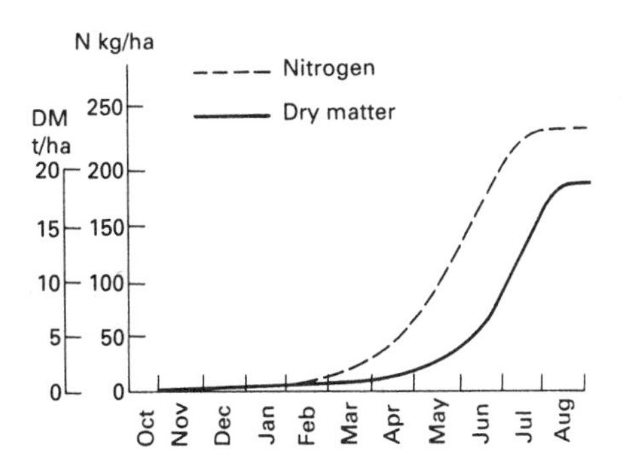

Fig. 10.3 Nitrogen uptake and dry matter accumulation of winter wheat (Lidgate, 1981).

Nitrogen is available to the crop mostly in the form of nitrate although ammonium can also be taken up and can be converted to nitrate in the soil (see Chapter 9). Both are soluble forms of nitrogen and as such can be lost from the soil by leaching if rainfall is high, leading to drainage flow. The majority of nitrogen is taken up by the roots but small amounts can be taken up by the leaves if urea solutions are sprayed on to the foliage. Soil leaching of N is most likely to occur when the soil levels of nitrate are high but the crop demand is low, a situation that frequently occurs in autumn. Leaching can also occur if fertilizer is applied in early spring when the soil is near field capacity and the application is followed by heavy rainfall.

Amounts of nitrogen

The nitrogen taken up by a cereal crop originates from two sources: the soil reserve and that applied by the grower. The soil reserve varies with the fertility of the soil (judged on an index basis by soil type and previous crop – see Chapter 9) and the climatic conditions, particularly temperature. The soil reserve is released from organic matter by microbiological breakdown which is sensitive to both temperature and, in the summer, moisture. Unfortunately it is impossible to forecast just how much of the soil reserve will be made available in any growing season. The applied nitrogen makes up the balance between the crop's requirement and the soil reserve, and since the soil reserve cannot be predicted then neither can the optimum applied levels. Thus although nitrogen is probably the most important input to the cereal crop, it is also the input that is applied with least accuracy. In reality, field trials are conducted every year to assess the response of crops to nitrogen and over time an average response can be gauged and recommended. Current recommendations are given in Table 10.8. Research is continuing into methods of more accurately predicting nitrogen requirement using assessments of both soil nitrate levels and crop 'sap' nitrate levels.

In some regions of the UK where the drinking water is collected in underground reservoirs (aquifers) there is real concern over the leaching of nitrate into this water. Such regions have been designated Nitrogen Sensitive Areas (NSAs), and growers have management agreements to apply reduced levels of nitrogen. Compensation is made to these growers for the reductions in yield achieved.

Timing of nitrogen

The timing of nitrogen application is very important in cereal growing. Uptake precedes the growth response (Fig. 10.3) and it is important to have soluble nitrogen available at the time of uptake. Unfortunately, the release of the soil nitrogen is slow in the spring because the soil temperature is low, and it is important that the grower applies fertilizer to meet the early nitrogen demands of the crop.

The main response timing to applied nitrogen is at the beginning of stem extension or growth stage 31 (first node detectable). For winter barley and winter oats, applying the nitrogen fertilizer at this time will give the full response to that fertilizer. For winter wheat, however, a further 10–20% response will be obtained if the

Table 10.8 Nitrogen requirements of cereals (kg/ha) (MAFF, 1994)

	Soil index		
	0	*1*	*2*
Winter wheat*			
sandy soils	175	140	80
shallow soils	225	190	130
deep silty soils	180	90	0
clay soils	190	110	0
other mineral soils	210	150	70
organic soils	120	60	0
peaty soils	80	20	0
Winter barley for feed/seed†			
sandy soils	160	120	60
shallow soils	200	160	100
other mineral soils	160	120	40
organic soils	100	40	0
peaty soils	60	0	0
Winter barley for malting†			
sandy/shallow soils	120	80	40
other mineral soils	110	60	40
Winter oats, rye, triticale†			
sandy/shallow soils	125	100	50
other mineral soils	110	60	30
organic soils	70	20	0
peaty soils	40	0	0
Spring wheat			
sandy/shallow soils	170	130	70
other mineral soils	170	110	30
organic soils	100	40	0
peaty soils	40	0	0
Spring barley for feed/seed			
sandy soils	125	90	30
shallow soils	150	110	50
other mineral soils	150	90	20
organic soils	80	20	0
peaty soils	40	0	0
Spring barley for malting			
sandy/shallow soils	100	75	20
other mineral soils	90	50	20
Spring oats, rye, triticale			
sandy/shallow soils	125	90	20
other mineral soils	100	60	30
organic soils	70	35	0

* Requirements are given for 8 t/ha (7 t/ha on sandy and peaty soils), adjust by 20 kg/ha nitrogen per t/ha expected yield variation.
† Reduce by 25 kg/ha if lodging risk is high.

nitrogen is applied as a split dressing with 40% at growth stage 30 (pseudostem erect) and the balance of 60% at growth stage 32 (second node detectable). Splitting the nitrogen fertilizer in this way on winter barley leads to early spring greening up which allows the early proliferation of disease, especially *Rhynchosporium*. The increase in disease often counteracts any growth stimulation effect. Owing to the crucial nature of these timings, a revision of Zadok's key has been published to explain in detail the accurate diagnoses of these growth stages (Fig. 10.4). For spring-sown crops the commonest timing of N is in the seedbed but if the crop is sown very early, e.g. in February, then it is recommended that up to 40% of the N is applied in the seedbed and the balance is top-dressed at the 3- to 4-leaf stage (GS 13/14).

Nitrogen is also applied at other timings to assist a poorly developed winter crop which has insufficient root growth. Such crops include late sowings, low plant populations, high winter damage and take-all affected. With these crops, a top-dressing of 25–40 kg/ha of N in February can be beneficial. Seedbed nitrogen is rarely applied to winter cereals except if the crop is direct drilled.

Late nitrogen applied at the flag leaf to ear emerged growth stage (GS 39–59) is commonly practised by growers of winter wheat for milling (bread wheat). This practice can lead to a higher grain protein content but it is not guaranteed. Normally approximately 40 kg/ha of N is applied and in recent years the tendency has been to apply this as a liquid feed direct to the foliage. In this way leaf absorption means a higher uptake of the N and a greater chance of a response, but leaf scorching can occur if the application is made during hot sunny weather.

Disease control

Detailed consideration of cereal diseases and their control is given in Chapter 13. Winter-sown cereals suffer more disease than spring-sown crops, and this is especially true of barley.

The first major consideration for disease control is varietal resistance. Choice of an appropriate variety for the region is of paramount importance since there is regional variation in the serious occurrence of several foliar diseases. For example, yellow rust is not found very often in the south west whereas *Rhynchosporium*, brown rust and *Septoria* are more prevalent in this region.

Unfortunately there are no varieties resistant to all diseases, and any resistance is only temporary since fungal organisms can reproduce many times in a season and have the ability to eventually overcome resistance. Thus, varieties resistant to mildew frequently succumb to a new race of mildew within three years of widespread use. Cereal variety diversification schemes for mildew and yellow rust (NIAB) can help to reduce the effect of epidemic spread of these diseases on the farm. Until long-lasting disease resistance is achieved in varieties, it will be necessary for growers to continue to use fungicides. Problems with resistance also exist with the overuse of certain fungicides, e.g. eyespot resistance to carbendazim. Diversification of fungicides is also a recommended practice. Thus, when a series of fungicides is planned, they should be chosen from different groups (Table 10.9).

Timing of fungicide application is important and a knowledge of the disease and its epidemic spread pattern can help in deciding the timing of application. In the 1970s and 1980s several fungicides were applied to each crop but in the 1990s this is unlikely to be economic and crops may only justify one or two fungicide sprays. For spring cereal only one fungicide is justified and this may be applied as a seed dressing. A summary of the main diseases of winter cereals and the important time of their control is given in Table 10.10.

Plant growth regulators

Plant growth regulators are frequently used in winter cereals to protect against lodging. High risk situations are: weak or long strawed varieties; highly fertile conditions or high N fertilizer applied; poorly consolidated seedbed; high incidence of eyespot likely. The growth

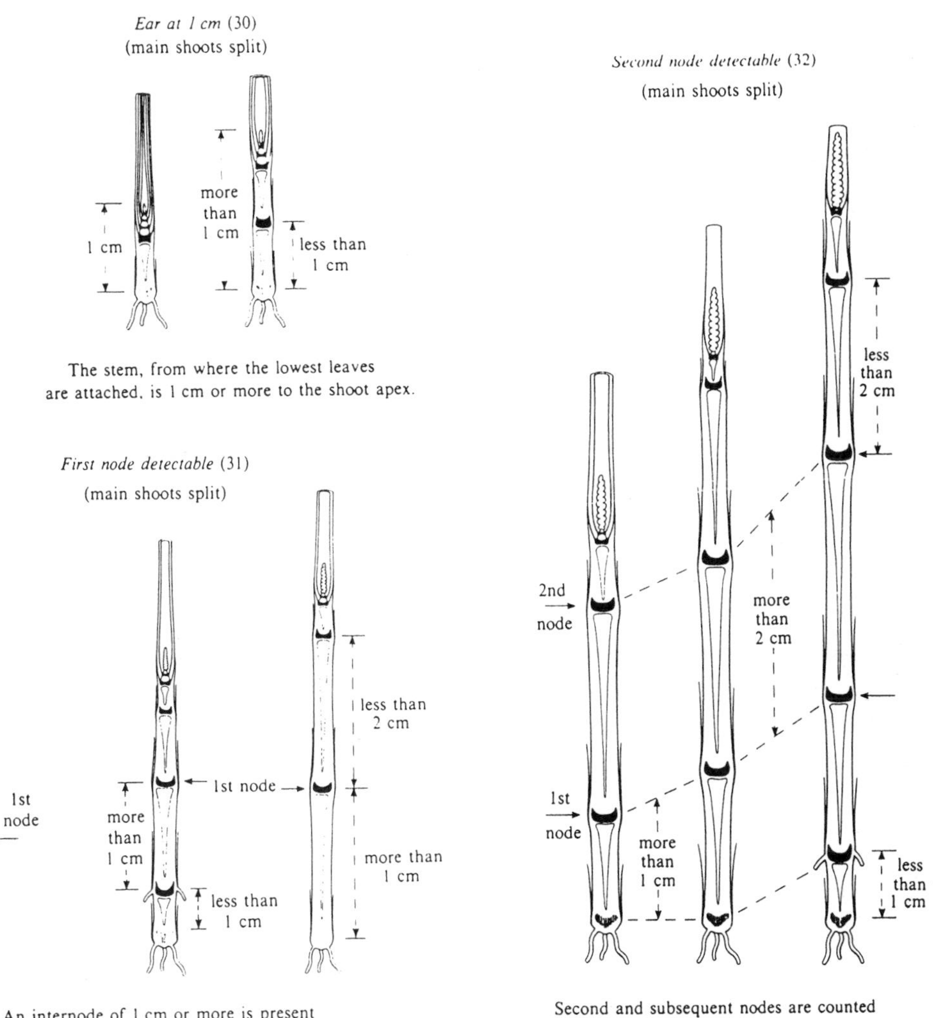

Fig. 10.4 Definition of early jointing growth stages of the decimal growth key of cereals (Tottman, 1987).

regulators shorten and thicken the internodes on the stem, making them resistant to bending and buckling. A summary of the timings of application is given in Table 10.11. Plant growth regulators are not usually used in spring cereal production.

Plant growth regulators have also been used to manipulate tillering although there are no manufacturers' recommendations for this type of use.

The fungicide group the triazoles have some growth regulatory properties and when used as a seed dressing can lead to a shortening of the underground rhizome which is responsible for adjusting the depth of apex in the soil. This could be important for winter oats as the apex of this plant is pushed to the soil surface thus exposing it to freezing winter temperatures and lowering its overwintering ability.

Pest control

The most important pests of cereal in the main growth phase are aphids. Populations can build up on the ear in summer and can lead to small grain size and sooty moulds growing in the honey-dew excreted by the aphids. Control at present is only by means of insecticides sprayed at a threshold average of five aphids per ear. Approved products include those which are selective against aphids and less damaging to beneficial insects. Aphid control is only usually practised on wheat crops.

Desiccation

Following a wet summer and/or lodging of a crop desiccation may be necessary to kill green weeds in the crop prior to harvest. A rapid kill of green material can be obtained with the use of diquat. A slower but more thorough kill is obtained with glyphosate and this chemical kills couch grass rhizomes.

Table 10.9 Cereal fungicide groupings

Fungicide group	*Common name of active ingredient*	*Examples of products (* indicates more than one ingredient)*
Site specific		
(1) Benzimidazoles	benomyl	Benlate
	carbendazim	Bavistin, Bravocarb*, Derosal, Delsene M*
	fuberidazole	Baytan*
	thiobendazole	Ferrax*
	thiophanate-methyl	Cercobin, Mildothane, Compass*
(2) Dimethylation inhibitors		
Triazoles	flutriafol	Impact, Impact Excel* Early Impact*
	propiconazole	Tilt, Radar, Hispor, Tilt Turbo*
	triadimefon	Bayleton
	triadimenol	Bayfidan, Baytan
Imidazoles	prochloraz	Sportak
	imazalil	Fungaflor
Pyrimidines	nuarimol	Triminol, Kapitol*
Piberazines	triforine	Saprol
(3) Morpholines	fenpropimorph	Corbel, Mistral, Sirocco*
	tridemorph	Calixin, Dorin*, Tilt Turbo*
(4) Hydroxy-pyrimidines	ethirimol	Milstem, Ferrax*
(5) Carboxyamides	carboxin	Dual Murganic RPB*
(6) Dicarboxyamides	ipridione	Rovral, Sirocco*, Compass*
(7) Organophosphates	pyrazaphos	Missile
Multi-site fungicides		
(8) Phthalonitriles	chlorothalonil	Bravo 500, Repulse, Bravocarb*, Impact Excel
(9) Dithiocarbamates	maneb	Delsene M*
	mancozeb	Kombat*
(10) Sulphur	Sulphur	Thiovit

Minor cereals

There is always a specialist market in the UK for minor cereals and these can be very profitable but have very small market size.

Durum wheat

This wheat is used to produce semolina flour from which pasta is manufactured. It is a difficult crop to grow reliably in the UK because of quality considerations. Hagberg falling number is frequently low if the crop does not ripen under hot dry conditions, and grain protein content can sometimes be low. It requires no vernalization and can be grown as a winter or spring crop. If sowing in the autumn, then delay until the end of the October. Otherwise treat as a winter wheat.

Triticale

This is a man-made hybrid between rye and durum wheat. It carries the environmental tolerance of rye and the quality characteristics of durum wheat. It can outyield wheat on poor soils but its long straw can lead to problems of lodging. It is relatively disease resistant but more disease is appearing as the crop is being grown more widespread. Only autumn-sown varieties are worth considering and husbandry is the same as for winter wheat. The high protein grain is particularly suitable for pig rations but as yet feed manufacturers are not giving large enough premiums on triticale to make it competitive with wheat.

Grain maize

Generally speaking the UK is at present too cold for the reliable widespread production of grain maize but small areas of production are possible in southern Hampshire and low level areas of Essex and Suffolk. However, the UK imports large quantities of maize and maize products. The crop is grown in the same way as forage maize with a slightly lower plant population (10 plants per square metre). Specialized harvesting machinery is necessary and this is not widely available in the UK.

Oilseeds

Oilseed rape (*Brassica napus*)

Oilseed rape seed is grown for crushing to extract the oil which, depending on variety, can be used for human consumption (cooking oils and margarine), industrial use, or for production of biodiesel. At present the bulk of the UK production is produced for human consumption, only very small hectarages being used for the latter purposes. After crushing, a high protein meal remains which can be valuable for inclusion in animal feed rations. Varieties suitable for oil for human consumption must be low in erucic acid, and for the meal to be suitable for animal feed they must also be low in glucosinolates; such varieties are known as 'double lows'.

Oilseed rape is the most popular UK cereal break crop. It fits well into cereal-based rotations, being broadly

Table 10.10 A summary of winter cereal diseases, their times of appearance and the target diseases for control. (Bold = target diseases)

Growth phase:	Establishment and tillering ⟵⟶	Early stem extension ⟵⟶	Flag leaf emergence ⟵⟶	Ear emergence grain filling ⟵⟶
Decimal growth stage:	11 13,21 15,23–25	30 31 32	37 39 45	51 59 69 71
Approx. time				
W. wheat:	Oct Feb	Mar Apr May	June	July Aug
W. barley:	Sept Oct Feb	Mar Apr	May June	July
W. wheat Diseases present:	*Septoria tritici* Mildew Eyespot Sharp eyespot Fusarium	**Eyespot** *Fusarium* *Septoria tritici* *Septoria nodorum* Mildew	**_Septoria_ sp.** **Mildew** **Yellow rust** **Brown rust**	Sooty moulds
Control:	No control in autumn necessary	Target control against eyespot	Keep ear and top two leaves clean	Only control if milling variety
W. barley Diseases present:	Mildew Net blotch Brown rust *Rhynchosporium*	**Net blotch** ***Rhynchosporium*** **Eyespot** **Mildew** Sharp eyespot *Fusarium*	**Net blotch** ***Rhynchosporium*** **Mildew** **Brown rust**	
Control:	Only control in exceptional circumstances	Control foliar and stem-based diseases	Keep top 2–3 leaves clean	
Winter oats Diseases present:		Mildew	**Mildew** **Crown rust**	
Control:			Control if for human consumption	

Table 10.11 Cereal plant growth regulators and their application timings

Active ingredient(s)	*Application timing*	*Crop*	*Notes*
Chlormequat Chlormequat + choline chloride	GS 30/31	Winter wheat Winter oats	Shortens and stiffens lower internodes; may be followed by second application at GS 31/32/33
Ethephon Ethephon + mepiquat chloride	GS 37	Winter barley	Shortens upper internodes and reduces necking and ear drop
Chlormequat followed by Ethephon	GS 30/31 + GS 33/34	Winter wheat	Combination of the two regulators gives excellent lodging control in high-value bread-wheat crops
Chlormequat Chlormequat + choline chloride	GS 13/15	Winter barley	For crops in Scotland, evens up tiller development going into winter

tolerant of soil conditions and using the same equipment but having different sowing and harvest dates thus easing labour peaks, and providing an opportunity for cheaper grass weed control. As a brassica it does not share any soil-borne pests or diseases with cereals. At present oilseed rape is included in the Arable Area Payments Scheme (see Chapter 1). Yields average 3 t/ha for winter types and 2 t/ha for spring types.

Market

Crushers pay according to the following standards:

40% oil content
35 μmol/g glucosinolate
9% moisture
2% maximum impurities

Variety selection
Selection from the varieties available should consider glucosinolate level, disease resistance, yield potential, height, lodging risk, and time to maturity. Each should be considered in relation to the characteristics of the site and/or the intended system of production. For example, tall varieties will be more suitable for swathing. Glucosinolate level is of great importance in varietal selection. Research to date has given little promise of husbandry control over levels, and site and season have overriding but unpredictable effects.

Soils and site
Well drained soils (pH 6.0–7.0, 6.5–7.0 preferred), free from compaction, are necessary for good yields, medium to sandy loams being ideal for good tap-root development. Soils and sites naturally low in sulphur may be important for low glucosinolate levels, hence light or chalky soils in areas of low atmospheric sulphur are favoured. The latter are found away from the heavy industrial areas of the London, midland and northern conurbations. Rape should be grown on a four-year rotation due to the risk of *Sclerotinia sclerotiorum*; other hosts are peas, spring beans and linseed, and should not be grown in close rotation with other brassicas to avoid build-up of clubroot. Beet nematode can be encouraged in sequences including both rape and sugar beet, and these should therefore be avoided.

Winter types

Seedbed and establishment Rape seed is very small and requires a fine clod-free seedbed. Every effort should be made to conserve moisture. Optimum sowing date is from mid-August to early September (the latter suitable in southern England), when seed should be drilled to a depth of 2–3 cm on 25-cm row widths, followed by rolling. Late sowing is very detrimental to rape yields as growth made before the onset of low winter temperatures (growth ceases below 2°C) determines yield potential; conversely, plants which have advanced too far into stem elongation due to early sowing can suffer badly from frosts, thus sowing date is critical. Ideally plants should have at least five leaves as they enter the coldest part of the year.

Spring plant population should be 80 plants/m^2 and seedrate should be carefully calculated to achieve this because:

- small inaccuracies have big effects because the seed is small;
- varieties vary significantly in their TSW;
- losses prior to spring can be significant as a result of failure to establish, losses to pests, and winter kill (50% losses for late sown crops in poor conditions, 20% for optimal timing in good conditions). Rape responds to low plant density by branching more extensively which compensates for yield but increases the extent of uneven ripening at harvest. Fifty plants/m^2 is generally considered to be the minimum for a worthwhile crop; below this level it may be economic to redrill with a spring crop. Over-thick crops will underyield due to an inefficient canopy structure and increased disease, and may lodge causing difficulties with ripening and harvest.

Establishment threats Weed competition during the establishment/autumn phase will reduce yield, with grass weeds and volunteer cereals having significant effects, particularly in later sown crops. Vigorous crops should compete well with broadleaved weeds, but chickweed and cleavers will cause yield reductions in a high proportion of situations. A weed control programme which deals with grass weeds, volunteers, cleavers and chickweed is recommended (see Chapter 13) to ensure yield losses are minimized; crop development stage must be taken into account when using herbicides as they vary in their 'safe' timings and few are safe once flower buds are present (November in most regions). Additional benefits will be gained from sample purity (cleavers seed is similar in size and shape to rape seed), savings on subsequent cereal crop weed control, and easier harvest.

Pigeons can devastate rape crops, particularly 'gappy' ones as the birds are less likely to land in well established vigorous stands, effective control is difficult and rape may have to be dismissed as an appropriate crop for high risk sites. Slugs, too, can decimate establishing rape and must be controlled in all but very low risk situations (see Chapter 13). Other pests associated with establishment are cabbage stem flea beetle (*Psylliodes chrysocephala*) and rape winter stem weevil (*Ceutorhynchus picitarsis*) which should be controlled on the basis of crop monitoring and threshold guidelines (see Chapter 13). Two diseases should be monitored in the autumn: light leaf spot (*Pyrenopeziza brassicae*) and stem canker (*Phoma lingam*); some varietal resistance is available in both cases but chemical control may be economic in some seasons (see Chapter 13).

Nutrients Nutrient requirements are given in Table 10.12. Timing of nitrogen addition is critical: rape is very responsive to spring nitrogen, and the rapid growth encouraged by rising temperatures should not be checked by tardy application. If conditions permit and growth has begun, nitrogen should be applied from mid February; in appropriate situations (e.g. light soils, heavy applications) the dressing can be split but applications after early April are wasted except where flowering is likely to be delayed when slightly later applications may still be effective. Seedbed nitrogen has no effect on yield: up to 50 kg/ha (in addition to requirements given in Table 10.12) may

Table 10.12 Nutrient requirements of oilseed rape (kg/ha) (MAFF, 1994)

	Soil index *0*	*1*	*2*	*3*
N, autumn sown*				
Mineral soil	160**	100**	40**	0
Organic soil	100	50	0	0
Peat soil	60	0	0	0
N, spring sown				
Mineral soil	120	60	0	0
Peat/organic soil	40	0	0	0
P_2O_5	100	75	50(m)	50(m)
K_2O	90	65	40(m)	0

* 30 kg/ha nitrogen should be applied to the seedbed in addition.
** Nitrogen recommendations for the autumn-sown crop are for 3 t/ha crop, adjust by ±30 kg/ha for each 0.5 t/ha variation from this.

assist canopy development, but due to the risk of loss through leaching this measure should only be used on crops which will use it and will benefit from a boost to establishment to avoid losses.

Rape has a high demand for trace elements during rapid spring growth and in some areas may suffer deficiencies of manganese, magnesium, boron or sulphur (see Chapter 9 for remedies). Care should be taken in the latter case as the response to sulphur is interactive with the response to nitrogen and may boost glucosinolate levels in some situations.

Spring/summer threats Light leaf spot and stem canker may continue to be a problem; also *Botrytis*, *Alternaria* and *Sclerotinia* may be encouraged by warm wet weather. Pollen beetle (*Meligethes* spp), seed weevil (*Ceutorhynchus assimilis*) and pod midge (*Dasyneura brassicae*) are the main insect pest threats. Chemical control is available for these pests and diseases but should be applied according to threshold guidelines, and never when rape is in open flower due to the risk to bees; bees enjoy rape pollen and should be encouraged to forage in rape crops as they contribute to pollination.

Harvest Rape reaches maturity in late July/early August, but ripening is usually uneven due to the structure of the plant: seeds on the terminal (main, central) branch are older than those on the side branches, and maturity varies along each branch. Two methods are commonly used to even up ripening: desiccation with diquat, or swathing into rows. In both cases combining is not possible until 7–14 days later, leaving the crop vulnerable to seed losses from pod shatter or poor weather. Choice of method will depend on availability of a swather or high clearance sprayer, and suitability (e.g. swathing is preferable for exposed sites) as there are no consistent effects on yield or quality. Moisture content at harvest will normally be in the range 10–18%.

Spring types

Seedbed and establishment Good establishment is critical to the success of spring rape but can be difficult in dry springs. Seedbed requirements are as for winter types with no compaction or capping and moisture conserved. Seed can be sown from the beginning of March to the end of April; end of March/early April is ideal in most years, at the same depth, widths etc. as winter types to achieve a plant population of 120 plants/m^2.

Nutrients Nutrient requirements are given in Table 10.12.

Threats Spring rape is a popular crop in suitable areas due to its low input requirements; a quickly established crop will compete well with spring-germinating weeds, and though susceptible to the same diseases as winter rape it rarely suffers unless grown in close proximity to an infected winter variety. Downy mildew (*Peronospora parasitica*) may attack early sown crops. Similarly spring rape is vulnerable to the same insect pests as the winter varieties but usually only pollen beetle requires control. (See Chapter 13 for further advice).

Harvest The spring crop is later to maturity than the winter type, and is harvested in late August to late September. It suffers much less uneven ripening, however, due to its simpler, less branched structure, and can often be combined without any pretreatment. If uneven ripening is a problem, then the procedures used for winter rape are equally applicable.

Linseed (*Linum usitatissimum*)

Linseed is grown for oil extraction from the seeds. The oil has a range of industrial uses, chiefly in the paint and varnish industries, and for the manufacture of linoleum and oilcloths. The cake remaining after extraction is included in ruminant feeds. Whole seed has a small market for horse and bird feed. Straw is potentially useful for paper making, or for high density brickettes as furnace fuel.

At present linseed is included in the Arable Area Payments Scheme (see Chapter 1), and may be grown on set-aside land as an industrial crop subject to certain conditions.

The advantages of linseed are its wide tolerance of conditions throughout the UK and its relative ease of production. It fits well into cereal-based systems; however, establishment success is critical and harvest can be problematic.

Linseed can be grown throughout the UK, but harvesting problems will increase in later and/or wetter regions. Establishment can be difficult so sites with poor soil structure should be avoided. The main rotational threats are *Sclerotinia sclerotiorum* (other hosts are oilseed rape, peas and spring beans) and wilt (*Fusarium oxysporum*), requiring a 4-year break, and fields infested with perennial grass weeds are a problem due to the very poor competitiveness of the linseed plant. Key characters on which to make varietal choice are regional suitability, time to maturity, and stem height and strength. Short-strawed varieties are less suitable for light or drought-prone soils.

A wide range of soil types are suitable but heavy clays should be avoided because of the difficulty of obtaining a fine seedbed for good establishment. Once established, linseed is reasonably tolerant of drought provided its tap root is able to penetrate down to available water. The flowering period is the most vulnerable to dry conditions. Optimum pH is 6.0–7.0.

Good seedbed conditions are essential for successful establishment. A fine tilth with no compaction and sufficient moisture should be provided by autumn ploughing, minimal spring cultivations and a roll prior to drilling. There is no advantage from early sowing into a cold, poor seedbed; appropriate seedbed conditions for quick establishment are of paramount importance. Optimum sowing period is mid-March to mid-April but acceptable yields can be achieved from sowings up to early May (late April in northern areas). Seed should be sown at 1.5–2.5 cm depth (3 cm may be necessary for dry seedbeds) on row spacings of 8–18 cm (narrower spacing gives better plant distribution). Even distribution of seed is difficult due to the 'flat' shape and the slipperiness of the seed. A plant population of 400–500 plants/m^2 is optimal for yield; a lower density offers less weed competition and encourages the plants to branch leading to more uneven ripening; higher populations give weak individual plants with a lower yield potential, a propensity to lodge and vulnerability to disease. Seedrate should take account of seed size, seedbed conditions, soil type and row spacing, establishment ranges typically from 50% to 80%.

There is little response to P and K, which should be applied in the seedbed. Nitrogen should be applied in the seedbed or at time of unfolding of the first pair of true leaves, according to soil characteristics, sowing date and amount needed (Table 10.13).

The structure of the linseed plant renders it a poor competitor against weeds, especially during the establishment phase, so good weed control is essential to prevent yield reduction and harvest difficulties. Preparatory control, e.g. for perennial weeds, and post-emergence chemical control are favoured; MCPA products can stress linseed crops and should be avoided. (See Chapter 13 for further weed control advice.)

Linseed diseases are increasing as the hectarage grows, and some chemical control may be necessary; however the avoidance of thick, dense canopies and lodging will lessen the risk. The main threats are grey mould (*Botrytis*) and blight (*Alternaria linicola*), both at flowering, particularly in wet weather. (See Chapter 13 for further disease control advice.)

The crop matures in late August/September but can be later in northern areas or in cool years. Desiccation with diquat is usually advantageous, primarily to dry out the green fibrous stems, but also to overcome uneven ripening and achieve timely harvest. Desiccation should be carried out when the seeds rattle inside golden-brown capsules; there is little risk of seed loss from pod shatter following desiccation and the crop can stand until harvest is convenient. At least 10–20 days will be necessary to dry the stems, shrivel the leaves, and ripen the seed to dark brown with a moisture content of about 10–15%. The fibrous stems are the biggest problem at harvest as they are tough, need sharp knives, and have a tendency to wrap around the reel; the slippery thin seed flows easily and will be lost through any leaks in the combine and trailers.

The shape of the seed makes efficient, early drying vital as the seed packs down in store and has a high resistance to airflow. For storage the aim is to achieve 8% moisture from a maximum drying temperature of 65°C (higher temperatures can reduce oil quality); the recommended maximum storage depth is 80 cm (see Chapter 14 for further drying and storage advice).

Average yields are approx. 2 t/ha. Payment is based on a standard of 9% moisture, 38% oil content and 2% impurities. Plant breeding effort continues to achieve higher yields and greater ease of production by means of a more efficient canopy structure and better partitioning from fibre to seed, though true dual purpose varieties (seed and fibre) are also sought. Greatest effort is towards varieties with specialized oil profiles, including those suitable for human consumption.

Table 10.13 Nutrient requirements for linseed (kg/ha) (m) indicates maintenance dressing (MAFF, 1994)

	Soil index *0*	*1*	*2*	*3*
N	80*	40*	0	
P_2O_5	100	75	50 (m)	50 (m)
K_2O	90	65	40 (m)	0

* Reduce to zero on fen peats and organic soils.

Legumes

Arable legumes are relatively large seeded combinable dicotyledonous crops grown for their protein-rich seeds. They share the ability of all legumes to fix atmospheric nitrogen and add 'no-cost' nitrogen to the soil system. They are thus classed as restorative crops, usually leaving the soil with a nitrogen index of 1. In reality the amount of N fixed and available for the following crop can vary widely according to soil conditions and the health of the crop. Arable legumes are a valuable contributor to sustainable crop rotations. Field beans and peas are the important UK arable legume crops currently; others such as lupins, navy beans, soya beans, chick peas and lentils require plant breeding effort to provide varieties consistently suitable and economically viable for the UK climate.

Field beans (*Vicia faba*)

Market
All field beans are grown for harvesting 'dry' and are sold as animal feed or used on farm through mill and mix units (though similar varieties are grown as broad beans for harvesting as a green vegetable). Their value as an animal feed lies chiefly in their crude protein content (18–26%) (see Chapter 17). Some smaller seeded varieties enjoy a premium niche market as pigeon feed or as export for human consumption; a contract is required for these markets. Many varieties have high tannins in the seed coat which restrict their inclusion in diets; some tannin-free varieties are available and are more attractive as a feedstuff though they tend to be lower yielding.

The Common Agricultural Policy has supported field bean production since 1978 in order to encourage the use of home-grown feed in place of imported soya. At present beans are included in the Arable Area Payments Scheme (see Chapter 1).

Advantages
Field beans are grown widely in the UK. They have a reasonable gross margin potential, are relatively tolerant and easy to grow, and fit well into cereal-based systems using the same equipment and spreading sowing and harvesting dates. They are an effective cereal break crop as they provide an opportunity for grass weed control and share none of the cereal-associated soil-borne pests and diseases. As a legume they leave some residual nitrogen to benefit the following crop (usually restoring the site to N index 1).

Disadvantages
The major disadvantages of field beans are their sensitivity to drought, their unstable yields and their relatively late harvest. Winter beans can form extremely tall dense canopies in wet years leading to poor pollination, disease control problems and harvesting difficulties.

Types
There are both winter and spring types of field bean available with contrasting agronomic characteristics, and suitable for different situations; yield and harvest date differences are not large. Winter beans are more frost hardy and form a much taller, thicker leaf canopy than spring types; they can be ploughed in rather than drilled. Traditionally beans were also categorized by seed size

with those smaller than a TSW of 530 g being tick (or tic) beans, and those greater than a TSW of 530 g being horse beans. Smaller beans are rounded and similar to a slightly elongated pea; larger beans are flattened and kidney shaped and are more difficult to handle, for example through augers. All winter beans are of the larger seeded horse type.

Varietal choice

Key characters on which to make varietal choice are end-use potential, maturity date, straw height, and disease resistance. Resistance to downy mildew is available in some spring varieties, but there is little difference in resistance to bean rust and chocolate spot between varieties at present. Shorter varieties should be chosen for soils not likely to suffer from moisture stress, and taller varieties for drier sites. It should be noted that shorter varieties tend to suffer more from bean rust (*Uromyces viciae-fabae*). Later maturing varieties should be avoided in cooler, wetter areas of the country.

Suitable sites

Beans were traditionally grown on heavy soils but a wide range of soil types are suitable where drought is not likely to be a problem. Generally spring beans are likely to be preferable to winter types in areas where spring sowing is appropriate and feasible as there are more varieties available with better marketing and husbandry characteristics. Winter beans are preferable in areas where spring sowing is problematic, summer droughts are more likely, and the yield differential will be pronounced, or where earlier harvest is important. Diseases of winter beans will easily cross-infect to spring beans, and close proximity of crop, trash or volunteers should be avoided.

Sites should not have grown legumes in the previous 4 years in order to control footrot (*Fusarium* spp), downy mildew (*Peronospor viciae*), and stem nematode (*Ditylenchus dipsaci*) though if the site is known to be infected with nematode an 8-year break should be practised (other hosts are legumes, oats, sugar beet, onions). Spring beans should be separated from oilseed rape and linseed by 4 years to control *Sclerotinia sclerotiorum*.

Waterlogged soils should be avoided as bean seeds are sensitive to poor aeration; compacted soil is detrimental to beans but is less likely to be a problem for winter types. Light soils are unsuitable due to the risk of drought. Soil pH should be 6.5–7.5, ideally 6.8–7.0.

Establishment

Seed should be treated to help prevent damping off (*Pythium* spp), and should be tested to avoid introducing stem nematode (*Ditylenchus dipsaci*).

Winter beans should be sown four weeks before frost is likely to occur to encourage good quick establishment without too much above-ground development; depending on the location this might be late October or November, or early December in mild areas. Ploughing in winter bean seed to a depth of about 15 cm after broadcasting is a quick, cheap method of sowing and has often been shown to give a yield advantage over drilling as it allows the use of soil residual herbicide, protects from pest damage (birds, rodents), and leaves a rough, free-draining surface which offers protection from cold winds. Winter beans are only hardy down to −12°C and the young shoots can be severely damaged by frost. This can lead to the production of many weak branches, which if excessive can lower yields. Optimum plant population is 25 plants/m^2, rising to 40 plants/m^2 for later sowings which are less likely to branch thus reducing yield potential (optimum number of flowering stems is 40 stems/m^2. Accurate seed rates must be calculated to take account of the often large variations in seed size between varieties and seed lots (see this chapter).

Spring varieties should be sown as soon as soil conditions permit (minimum temperature for germination approx. 5°C, and moist and aerated) in late February/early March; yield reductions are likely from sowings after mid-March, and the risk from drought will be exacerbated. Optimum plant population is 50–60 plants/m^2 but, depending on seed costs, the economic optimum is likely to be 40–45 plants/m^2. Spring varieties should be drilled on 15–35 cm row widths (the narrower widths should be used for shorter varieties) at a depth of 6–8 cm, especially if soil residual herbicide is used. The seedbed should be well aerated.

Nutrients

Requirements are given in Table 10.14; no nitrogen is needed.

Table 10.14 Nutrient requirements of field beans and field peas (kg/ha). (m) indicates maintenance dressing (MAFF, 1994)

	Soil index			
	0	*1*	*2*	*3*
P_2O_5	75	50	30(m)	30(m)
K_2O	120	50	30(m)	0
MgO	150	100	0	0

Threats

Winter beans are uncompetitive to weeds in the establishment phase due to the low plant density and relatively slow development of a canopy. Some chemical control may be necessary, particularly to control grass weeds for the benefit of the rotation, though late sown crops will encounter few weed problems for cultural reasons. The only major disease threat is from chocolate spot (*Botrytis fabae*), and possibly seed-borne leaf/pod spot (*Ascochyta fabae*). It is rare for winter beans to require insect pest control. (See Chapter 13 for detailed advice.)

Good early weed control is important for spring beans as they are uncompetitive in the establishment phase and tall/long weeds continue to compete and cause harvesting difficulties e.g. volunteer rape, fat-hen, cleavers. The most serious disease threat is bean rust (*Uromyces viciae-fabae*), also chocolate spot and leaf/pod spot as for winter beans. Black bean aphid (*Aphis fabae*) is a common pest which damages flowers and pods as well as encouraging chocolate spot. Stem nematode (*Ditylenchus dipsaci*) and bean weevil (*Sitona lineatus*) may also affect spring bean crops. (See Chapter 13 for detailed advice.)

Harvest

The usual harvest date is September for both types, when pods are black and brittle and moisture content is below 20%, though up to 25% is combinable. Moist conditions should be chosen, if possible, to minimize pod shatter. Yield averages 3.5–4.5 t/ha.

Field peas (*Pisum sativum*)

Market

Also referred to as combinable peas, field peas are grown for harvesting 'dry' (and should be distinguished from vining peas grown for harvesting as a green vegetable). The majority of the UK crop is used for animal feed by the compounders, or directly on-farm through mill and mix units. A small but significant proportion of the national crop is used for human consumption as canned ('processed', 'marrowfat' and 'mushy') peas or dried peas, with potential for increase in the 'snack' pea market (comparable to peanuts). Their value as a foodstuff lies chiefly in their crude protein content (18–26%) (see Chapter 17), though some current research effort is directed to their potential utilization as an oilseed crop. Peas destined for human consumption have additional quality requirements (colour and freedom from defects such as staining and splitting) compared with those for animal feed, but can be sold at a premium of up to 50% of the feed price. Contracts are usually necessary for reliable sale for human consumption.

The Common Agricultural Policy has supported field pea production since 1978 in order to encourage the use of home-grown feed in place of imported soya. At present peas are included in the Arable Area Payments Scheme (see Chapter 1).

Advantages

Field peas are grown widely in the UK. They have a reasonable gross margin potential, are relatively tolerant and easy to grow, and fit well into cereal-based systems using the same equipment and spreading sowing and harvesting dates. They are an effective cereal break crop as they provide an opportunity for grass weed control and share none of the cereal associated soil-borne pests and diseases. As a legume they leave some residual nitrogen to benefit the following crop (usually restoring the site to N index 1), and their early summer harvest date provides good entry opportunities.

Disadvantages

The major disadvantages of field peas are their sensitivity to poor soil/soil water conditions, their unstable yields, and their extreme susceptibility to lodging which frequently leads to difficult harvest operations. Pea yields can be highly variable from year to year and between sites. This yield instability is a result of extreme sensitivity to environmental conditions (for example waterlogging at the pre-flowering and flowering stages), inter-plant competition (see Chapter 8) leading to wide variation in individual plant size, and strong intra-plant competition arising from an indeterminate flowering habit leading to flower, pod and seed abortion. 'Wild type' peas are climbers with long stems and many tendrils to anchor the plant, and with little need for a strong supporting stem; consequently peas rarely stand without falling over or lodging, even when short stemmed. Plant breeding effort, using the vast *Pisum* gene bank, continues to seek varieties with more efficient and productive plant and canopy structures, resistance to common pests and diseases, and seed components appropriate for a wider range of end uses.

Types

There are several categories of pea variety depending on their seed type (marrowfat, white, large blue, and small blue) and on their leaf type (normal, tare leaved, or semi-leafless, Fig. 10.5). All are suitable for animal feed but only some varieties are suitable for human food use, and these are mostly of the marrowfat type. Leaf type can be an indicator of potential for lodging, normal-leaved types having a heavier canopy and thus a greater propensity to lodge, particularly in fertile conditions. However, some normal-leaved types have been selected to lodge early and thereafter to hold their pods clear of the ground, thus avoiding harvest difficulties. Tare-leaved and semi-leafless types were bred to reduce the weight of the canopy and to increase the number of tendrils to create self-supporting canopies with less tendency to lodge.

Varieties

Key characters on which to make varietal choice are end-use potential, lodging and combining risk, resistance to pea wilt (*Fusarium oxysporum* f. sp. *pisi*) and downy mildew (*Peronospora viciae*), haulm length, and maturity date. Shorter varieties shoul be chosen for moist fertile soils, and taller varieties for dry poorer soils, to avoid combining problems. Late maturing varieties should be avoided in cooler, wetter areas of the country. Varieties suitable for human consumption tend to be lower yielding and should therefore be avoided if the intention is to sell or use for feed.

Suitable sites

Peas are traditionally grown in lower rainfall areas as they are not compatible with wet conditions particularly during the later stages of development when the canopy is vulnerable to disease and infection will reduce both yield and quality, and during the harvest period when rainfall adds further difficulty.

Peas should not be grown more frequently than 1 in 5 years on the same site. The main soil-borne threats which can be prevented by sound crop rotations are *Sclerotinia sclerotiorum* (other hosts are oilseed rape, linseed, spring beans and most other brassicas and legumes), downy mildew *Peronospora viciae*, the footrot complex mainly *Fusarium* spp. (other hosts are legumes), pea wilt *Fusarium oxysporum* f. sp. *pisi* (other hosts are beans), and pea cyst nematode *Heterodera goettingiana* (other hosts are legumes).

A wide range of soil types are suitable but very heavy soils should be avoided as peas are very sensitive to waterlogging, particularly at emergence and flowering (for example, 5 days' waterlogging at flowering can reduce yield by 75%) and compaction; very light drought-prone soils should also be avoided. Lighter free-draining soils are ideal. Optimum soil pH is 5.9–6.5; higher pH is acceptable but is likely to lead to manganese deficiency and a foliar manganese spray may be required (see Chapter 9). Stony soils can add to harvesting difficulties.

Establishment

A good seedbed is essential for successful pea establishment. Compaction must be avoided and peas should not be sown in wet conditions or 'forced' seedbeds. A good tilth should be obtained with the minimum of spring cultivations; ground should be ploughed in the previous autumn.

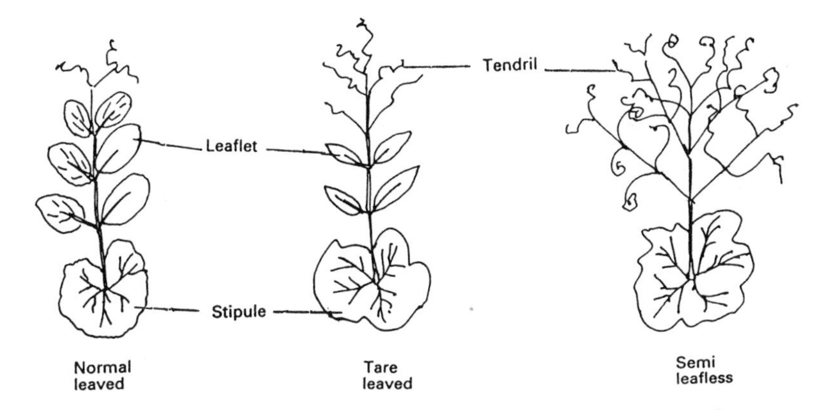

Fig. 10.5 Field pea leaf types (diagrammatic representation of the structure at a single node).

Seed should be treated to help prevent damping-off (*Pythium* spp.), downy mildew (*Peronospora viciae*), leaf and pod spot (*Ascocyta pisi*) and *Mycosphaerella pinoides*.

Peas should be sown as soon as soil conditions are favourable (minimum temperature for germination approx. 5°C, and moist and aerated) from mid-February; yield potential drops by approx. 125 kg/ha/week after the first week in March, with sowing after mid-April unlikely to be profitable. (Recent trials of a high-yielding autumn-sown variety suggest that winter field peas may become more common in the future.) Seed is drilled on 10–20 cm rows at a depth of 4–5 cm, with care taken to achieve even spacing in order to reduce inter-plant competition. Guideline optimum plant populations are:

Marrowfats	65 plants/m^2
Whites and large blues	70 plants/m^2
Small blues	95 plants/m^2

However, as the seed cost is a high proportion of the variable costs of production (about 60%), the economic optimum plant population can vary from year to year, depending on seed costs, and from site to site, depending on likely output value. Accurate seed rates must be calculated to take account of variations in seed size (see this chapter), and likely field losses. Field losses can be high in peas due to the sensitivity of the seed and young seedling to soil conditions; seed lots should be tested for their vigour and only high vigour seeds used for early sowing or sowing in wet conditions. Expected losses for marrowfats are 10% when planted in March and 15% when planted in February, and for small blues 15% in March and 20% in February. Other types show intermediate losses. An additional 5% expected loss should be included in the calculation for heavier, wetter soils.

The seedbed should be rolled after drilling to conserve moisture and to provide a flat surface for effective pre-emergence weed control and ease of combining; rolling can also reduce loss of germinating seeds to bird pests.

Nutrients
No nitrogen is required and the response to phosphate is small, but potash is important: rates are shown in Table 10.14. Germinating pea seed is sensitive to high concentrations of nutrient solution so fertilizer is best incorporated in the autumn to avoid seed scorch; no more than 50 kg/ha potash should be added by combine drilling. Maintenance dressings are better applied at other courses in the rotation.

Weed control
The pea crop is not very competitive against weeds (particularly the semi-leafless types) and good weed control is therefore essential as infestation will reduce yields, increase disease risk and add to harvesting difficulties. Pre-emergence herbicides are cheaper and effective against a broad range of weeds and are preferable unless soil type (organic or sandy) or conditions (dry or cloddy) prevent their use. Post-emergence application may be necessary for some troublesome weeds (oilseed rape volunteers, broad-leaved perennials, and well established cleavers), or when a pre-emergence product is inappropriate or ineffective. It should be noted that varieties vary in their tolerance of some post-emergence herbicides, and that all varieties can be sensitive to post-emergence herbicides if their leaves are poorly waxed. This can result from harsh weather conditions, disease or previous spray applications. If control has been unsuccessful, or if perennial grass weeds need controlling, a desiccant can be used pre-harvest. (See Chapter 13 for further weed control advice.)

Diseases
Most pea diseases are soil-borne or seed-borne, and are prevented by utilizing sound rotations (footrot, downy mildew, pea wilt), resistant varieties (downy mildew, pea wilt), healthy seed (leaf and pod spot, *Mycosphaerella pinoides*, bacterial blight), and seed treatment following appropriate testing (damping-off, downy mildew, leaf and pod spot *M. pinoides*. Most are encouraged by wet weather. Pea wilt and downy mildew are the greatest threats to yield. Botrytis or grey mould (*Botrytis* spp) does not arise from soil or seed: it is encouraged by wet weather from flowering onwards, particularly in crops with a thick leaf canopy, and can be controlled with a protectant fungicide programme. (See Chapter 13 for further disease control advice.)

Pests
The pea aphid (*Acrythosiphon pisum*) is the most likely pest of the pea crop. It appears before flowering onwards and should be controlled if one plant in five is infested.

Other potential pests are pea/bean weevil (*Sitona lineatus*), thrips (*Thrips angusticeps*), pea moth (*Cydia nigricana*) (all controllable with insecticide); and pea cyst nematode (*Heterodera goettingiana*). The latter should not be a problem if sound rotations are practised. Pea moth affects quality rather than yield, and control is only necessary in crops for human consumption. (See Chapter 13 for further pest control advice.)

Harvesting

Earliest varieties mature in July in warmer areas, though in the north harvesting can be in late August/early September. Ideally harvest should take place at 16–20% moisture content, but harvesting at up to 25% followed by drying may be necessary to avoid a reduction in quality during wet weather or to reduce splitting in crops for human consumption. If the crop is weedy or shows uneven ripening, it may be desiccated with diquat when the lower pods are dry and brown and the seed dry, the middle pods are yellow and wrinkled and the seed rubbery, and the upper pods are green and wrinkled. The crop can be combined 7–10 days later.

Shedding losses can be high during harvest. To reduce losses, harvest into the 'lay' of the lodging, use lifters and a slow drum speed with maximum fan, open up the concave, and use correct sieve sizes (see Chapter 24). For crops for human consumption avoid harvesting when the pods and haulm show surface moisture as this leads to the peas becoming coated with dust which lowers the quality.

Wet peas should be dried immediately to avoid spoilage from moulds; splitting can result from incorrect drying and renders peas unfit for the human food market. Peas will store for 4 weeks at 17% moisture, for 6 months at 15% moisture, and long term at 14% or less (see Chapter 14).

Average yields are 3.5 t/ha. Payment for peas is based on a standard of 14% moisture and 3% impurities, price is variable and can be very low for poor samples. Quality samples for human consumption or seed attract a premium.

Root and vegetable crops

Sugar beet (*Beta vulgaris*)

Market

Sugar beet is grown exclusively for the sugar trade and is controlled in the UK by British Sugar plc (BSC) by a system of contracts with growers. The closely related crop of fodder beet is not controlled and is grown mostly by dairy farmers in the south and south west of the UK. Sugar beet needs to be processed to extract the sugar, and growers are usually clustered around processing factories although some transportation supplements may be paid so as to assist growers far from the factories. A by-product of processing is sugar beet pulp which is sold to the feed supply industry mostly for cattle feed.

Contracts with British Sugar guarantee a market but only for a fixed tonnage of beet at the top price (A quota). Excess production will be paid at a lower price (B quota) and excess to this may be rejected or paid at an even lower price. Payment is based on the weight of washed and correctly topped beet delivered to the factory at a sugar content of 16% (additions/deductions for deviations of 0.1%) modified by delivery and transport allowances. The payment is levied for research and development. The processing 'campaign' begins in September with a higher price being paid for early delivery where yield is being sacrificed. Full details of the contract system can be obtained from British Sugar plc, Peterborough, or from the NFU.

Refined beet sugar (Silverspoon) competes on the UK market with Cane Sugar (Tate & Lyle) which is imported under agreements with former UK colonies and developing countries. Thus the UK is only 50% self sufficient in sugar whereas the EU is 135% self-sufficient.

Growth cycle

Sugar beet originates from the Mediterranean coast and has a salt requirement and shows sensitivity to cold temperatures. This is manifested by a slow germination rate at soil temperatures lower than about 8°C and a slow growth rate in early spring. Furthermore the crop is a biennial requiring exposure to cold temperatures (vernalization) to induce flowering. It is the first year's vegetative growth that is exploited as the sugar crop when the tap root swells and fills with sucrose. However, the vernalization requirement can be fulfilled in the young seedling stage and flower production can occur in the first year of growth (bolting). For this reason the crop is sown in the spring and bolting risk is increased with early sowing or in cold springs. The crop does not meet in the rows until July and this lowers the yield potential in the UK. It keeps growing well into October and can be harvested as late as November although to avoid undue soil damage many growers lift earlier and temporarily store the roots in a clamp.

Varieties differ in their bolting and downy mildew resistance and to some degree in their virus yellows tolerance. Furthermore, small canopied varieties are available for fertile sites and on peat soils. Varieties are reviewed annually by the NIAB and the BSC.

Soil types

Sugar beet must be grown on good, deep, well drained, stone-free soils free of compaction and pans to allow a good development of the tap root. Poor soil conditions lead to fangy roots which impede lifting and result in a crop which is difficult to clean and has a lower sugar content. Medium textured soils frequently give the best crops although crops are grown on a wide range of soil types. Soils must have a pH above 6.5.

Seedbed

The seedbed must be prepared with care for sugar beet to avoid compaction. Medium and heavier soils benefit from winter ploughing whilst light soils can be ploughed and furrow pressed in spring. Winter cultivations should avoid working in very wet conditions when smearing and pans can be formed, and double or cage wheels are recommended on cultivation tractors. Spring work should only be carried out on dry soils and the number of passes should be kept to a minimum to avoid wheelings. The final seedbed preparations should produce a fine and level seedbed and should be worked down on the day of drilling to facilitate moisture retention. Too fine a seedbed can result in capping of the soil following heavy rain and this can lead to poor crop emergence.

Sowing
In order to achieve high yields early sowing is recommended. However, this carries with it a bolting risk. Early drilling and cold soils should be sown with bolt-resistant varieties. Normally drilling will commence as soon as possible after 20 March and be finished by 10 April. Best crops are achieved with an evenly distributed plant population of about 75 000 plants/ha. Populations below 62 000 give serious yield decreases whilst above 100 000 harvesting is impeded. Irregular stands interfere with harvesting equipment and can lead to a large wastage of unharvested roots and irregular topping. Row spacing is dictated by harvesting machinery but yield is decreased if row widths exceed 500 mm.

Beet seed is a fused fruit containing several seeds and in the past had to be thinned after emergence or rubbed and graded before sowing. All varieties used today are monogerm (single viable seed) and this enables the crop to be precision sown to a stand. An establishment rate of over 70% is expected and seed is generally sown at an interseed spacing of about 175 mm. For early sowings and on soils and fields known to give germination problems an interseed spacing down to 155 mm can be used.

Nutrient requirements
Sugar beet has a high requirement for potassium and a requirement for sodium (salt) because of its maritime origin, for boron to counteract heartrot, and for magnesium especially on light soils. It does not require much phosphorus and its nitrogen supply must be regulated to avoid an overstimulation of the leaf canopy which can reduce root yield and lead to a high level of nitrogenous compounds in the root. Recommended fertilizer rates are given in Table 10.15.

Soil pH should be 6.5 to 7.0 on mineral soils although a lower pH is tolerated on sandy soils (6.0 to 6.5) and peat soils (6.0). Use of magnesian limestone helps to provide adequate magnesium. Lime should ideally be applied to

Table 10.15 Nutrient requirements of sugar beet (kg/ha). (m) indicates maintenance dressing (MAFF, 1994)

	Soil index			
	0	*1*	*2*	*3*
N				
Sandy/shallow soils	125	100	75	
Deep silty soils	80	50	25	
Other mineral soils	100	75	50	
Organic soils	75	50	25	
Peaty soils	50	25	0	
P_2O_5	100	75	50(m)	50(m)
K_2O*	200	100	75	75
MgO	165	85(m)	0	0

* Increase by 150 kg/ha if tops are carted off. For all soils except organic, peat or fen silts, also apply 150 kg/ha sodium (as 400 kg/ha salt) or increase potash by 100 kg/ha.

the previous crop in the rotation but if this is impossible apply as two dressings and incorporate well to avoid acid layers which will lead to fanging.

Sugar beet germination can be impeded by fertilizers applied just prior to sowing. Most fertilizers (P, K, Mg, Na) are best applied in the autumn or winter prior to ploughing or on light soils just prior to seedbed preparation. Nitrogen is normally applied in spring to avoid winter leaching, and 40 kg/ha is applied just post-drilling and the balance at the two to three-leaf stage.

Weed control
Sugar beet is a slow-growing crop in April and is very sensitive to weed competition. Many herbicides do not give an adequate length of protection and it is common to spray three or more herbicides during establishment. Other methods of weed control include inter-row hoeing which is frequently used in conjunction with band spraying (spraying herbicide only on the crop rows). Inter-row weeds are sometimes left to grow quite large before being controlled to hold the soil together and to prevent soil erosion (wind-blow) which can seriously damage the young beet. Another approach to this problem is to 'plant' chopped straw between the rows or even to sow cereals. Once the crop meets in the rows it suppresses further weed growth.

An increasing problem in sugar beet crops has been the appearance of weed beet. This weed is a beet plant which flowers in its first year rather than producing a harvestable root. This should not be confused with bolters which flower because of a cold spring or early sowing. Weed beet if allowed to set seed can cause a problem in the field even when the rotation is long as a proportion of the seeds remain dormant in the soil. Weed beet cannot be controlled by spraying a herbicide but can be killed by using a weedwiper and a total translocated herbicide.

Pests and diseases
Sugar beet is susceptible to beet cyst nematode (*Heterodera schachtii*) which can only be controlled satisfactorily by using a sound rotation. Use of nematicides gives some suppression in infected soils.

Most pests of the crop are only damaging during establishment where damage leads either to plant death or to subsequent plant distortion. Seedling pests include millipedes, symphylids, springtails and pygmy mangel beetle. Wireworm can also be a problem if grass is in the same rotation. Blanket treatment for these seedling pests using an insecticide is not usually carried out unless a problem is foreseen. Fieldmice also devour seed and are encouraged by spilled seed and shallow drilling. The most significant pests are peach potato aphid and green aphid which carry Beet Virus Yellows (BVY) and Beet Mild Yellowing Virus (BMYV) causing severe disease especially if infection occurs early. Crop losses can be as high as 40–50%. Control of aphid infestation begins with controlling the overwintering sites of the pest, i.e. in clamps or ground keepers, on spilt plants at the loading bays and in gardens and neighbouring allotments. Severe winter frosts reduce aphid numbers in early spring. Early aphid attack can be prevented using soil-applied granular insecticides but the expense of this form of control limits their use only to areas where severe attacks are likely (e.g. East Anglia). Foliar applications of insecticides are applied when one infected plant in four is recorded or when the British Sugar Corporation gives warnings. Usually one spray is sufficient but in bad years a second may be advisable. There is evidence that some aphids are building up resistance to organophosphate insecticides and the use of alternatives, e.g. carbamates, is recommended.

A serious root-infesting disease of sugar beet, rhizomania, has appeared in the UK in recent years. This disease is found on the continent of Europe but until recently was not present in the UK. The exact source of the infection is not known but infected fields have been quarantined and the growers banned from growing sugar beet or moving their equipment on to uninfected ground in an attempt to contain the spread. At present there is no known cure for the disease with the exception of ceasing to grow sugar beet. It is not known how long the organism can survive in the soil in the absence of the crop and therefore rotational control is difficult. The disease causes a proliferation of lateral roots and stops the correct development of the tap root thereby reducing sugar yields drastically. Because of the risk of rhizomania the BSC is seeking fresh ground for sugar beet, and contracts for delivery to the Kidderminster factory have even been placed in Cornwall.

Irrigation

The crop is sensitive to drought during June when the tap root is extending and the leaf canopy is developing, and during July and August when the canopy is in full sucrose production for storage in the tap root and the drought leads to stomata closing which slows photosynthetic rate. The extension of the tap root (down to 1.8 m) provides adequate moisture for crop development during September. Irrigation water (up to 25 mm) should be applied in June when the soil moisture deficit approaches 50 mm. During July and August the soil moisture deficit should not exceed 50 mm and irrigation can be applied to bring the soil to field capacity if required.

Harvesting

There are several types of sugar beet harvester available and all perform three functions. First they cut off the green leaf (topping), then they dig the swollen part of the tap root out of the ground and elevate it into a hopper or trailer. The sugar beet 'campaign' begins in September with early lifted crops delivered direct to the factory. Later liftings can be temporarily stored on concrete mats where they can be washed. As the soil conditions deteriorate in October and November, later delivered crops are lifted and clamped under bales of hay to protect them from frost. Frosting severely damages the tap roots and leads to rejection at the factory. Clamps must be managed to keep the temperature below 10°C by ventilation, or respiration will lead to sugar losses. The sugar beet crop yields about 40–50 t/ha of roots with a sugar content of 16%, giving 6–7 t/ha of sugar.

By-products

The leaves and crowns of sugar beet have feeding value to livestock but must be wilted for a few days to diminish high levels of oxalic acid in the leaves. If the tops are free of dirt they can be ensiled. The tops from a 50 t/ha root crop will have a feeding value equivalent of about 0.5 ha of marrowstem kale. Tops can be fed to both ewes (1 ha of tops lasting 250 ewes one week) and dairy cows (limit to a maximum of 19 kg/d per cow). The pulp of the processed beet roots also has feeding value to livestock. It is provided in several forms, including dry powder, nuts or molasses. Feed value is equivalent to that of oats.

Potatoes (*Solanum tuberosum*)

Markets

There are several types of market for potato tubers, which require different varieties to be grown. Growers tend to target only one market. Production is regulated by land quota which is obtained from the Potato Marketing Board; growers are not allowed to produce potatoes for sale unless they possess potato quota.

General market requirements are for smooth even tubers with shallow eyes for minimal wastage and reduced labour in peeling. The tubers need to be free of greening, growth cracks, disease (blight, scab, silver scurf, gangrene), pest damage (wireworm, cutworm, slugs) and free of mechanical damage. Cooking quality is related to variety, growth and storage conditions. Early harvest of the maincrop in autumn tends to give the best quality tubers.

Maincrop – ware potatoes

This is the largest volume market requiring heavy yields of saleable medium- to large-sized potatoes. It is now becoming divided into medium-sized tubers for general domestic use and larger tubers for oven baking (bakers). More and more of the crop is now being retailed by supermarkets who require the crop to be washed, graded and bagged in small plastic bags. Bakers are sometimes sold in packs of 2 potatoes and must be very clean and blemish free. The traditional unwashed crop bagged in 25-kg paper sacks is still available, especially from farm shops. The crop is harvested when fully mature, beginning in early October. Some crop is stored in ventilated cold stores and some is left in the field until required. There have been considerable advances in storage technology over the last 10 years in response to the market requirements for good blemish-free potatoes.

Earlies – new potatoes

This is a high value sector of the market and is traditionally the province of the frost-free areas of the UK particularly west Cornwall, south Pembrokeshire, east Kent and Jersey. The major advance of the last 10 years has been the widespread use of plastic film to advance the crop and give an early harvest thus achieving a higher price. However, the traditional crop is being undermined by production from the Mediterranean (Cyprus, Greece and Egypt). Prices are at their highest in April (up to £2000/t) and drop as more and more of the crop comes on to the market reaching a base in July (about £200/t). The potatoes are lifted in an immature state and have a waxy texture after cooking. Later plantings and second early varieties give a continuity of fresh potatoes throughout summer until the maincrop arrives. Out-of-season early potatoes are produced by some growers in plastic tunnels where the potato is often fitted into a fallow period in lettuce production.

Second earlies

The second early potato crop fits between the earlies and maincrop. These varieties achieve higher yields than earlies but harvest earlier than maincrops, in July and August.

Processed potatoes
Potatoes sold for processing constitute an expanding sector of the market, now taking over 20% of the total UK potato production particularly for frozen chips and potato crisps. Only a very small amount is canned and this is usually early crop.

Processing in the potato crisp industry requires specific named varieties with high dry matter content (over 20%), low reducing sugar content (below 0.25%) with shallow eyes and regular shape.

For frozen chips the potato must have high dry matter content (over 20%) and low reducing sugar content (below 0.4%), and be regular with shallow eyes and a white creamy flesh which does not discolour upon cooking or suffer any taints.

Seed potatoes
This is a specialist market confined generally to regions where a cool summer means a lower incidence of aphids and consequently less chance of aphid-transmitted virus diseases, e.g. Scotland and Northern Ireland. The objective of seed tuber production is to produce a high yield of tubers in the range of 32–60 mm using clean certified seed so as to give seed tubers which are true to type and as free of disease as possible. Some growers of early potatoes produce their own seed tubers, and many maincrop growers keep the small tubers from a certified maincrop to provide 'once grown' seed tubers for the subsequent year.

A specialist method of seed production is now appearing in commercial practice with the availability of minitubers produced through a combination of micropropagation followed by intensive glasshouse raising at high plant populations.

Varieties
Growers should refer to the NIAB list for varieties of potatoes but need to determine their market before choosing a variety. Processors will often specify only one or two varieties. Early varieties are best chosen with experience of performance within a region and soil type.

Seed certification
Registered seed producers comply with MAFF seed production regulations and can sell seed that is relatively virus free and is given a certificate of health with one or more of the following codes:

(1) Seed for multiplication purposes:
 - VTSC – Virus tested stem cuttings: the highest grade of seed tubers that have been raised in aphid-proof glasshouses
 - SE – Super Elite seed: derived from VTSC of SE stocks
 - E – Elite seed
 - 1, 2 or 3 – Number of years of multiplication since VTSC

(2) Seed for commercial crops:
 - CC – Certified Seed: not suitable for further seed multiplication
 - NI – Indicates that the variety is not immune to wart disease.

In areas of intensive maincrop production it is advisable to purchase certified seed each year as disease spread through tubers once grown is likely. In less intensive regions it may be possible to keep some tubers once grown for the following year's crop.

Seed size
The plant population of potatoes is a function of both seed size and tubers per unit area. As seed size increases, so does the number of potential shoots per tuber and logically these can be sown at lower seedrate. However, there is a complex competition between shoots per tuber and most growers prefer to plant smaller seed tubers. Seed of size 30–60 mm is preferred for most crops, with the lower range for earlies especially. Seed at the top end of the range or above tends to produce numerous small tubers in maincrop production and in Scotland these are retained for seed crop production.

Seedrate
Seedrate is determined by the average weight of the seed tubers and the number of tubers planted per hectare. The variety also influences the seedrate with some varieties tending to produce either large or small tubers. Therefore seedrate has to be adjusted to force the variety to produce the required tuber size. Seedrate is also influenced by the target market. Seed tubers are very expensive and this can influence seedrate too. Lastly, the number of stems produced per tuber can be influenced by the pre-planting storage conditions, i.e. chitting, which can stimulate one or several shoots per tuber. Table 10.16 gives recommendations on seedrate for some common varieties. Seedrates are commonly 2.5–4.5 t/ha for varieties producing large tubers, e.g. Desirée and Pentland Crown, and 1.8–2.4 for varieties such as King Edward and Record which tend to produce undersized tubers. Seedrates for earlies, seed production and canning are given in Table 10.17.

Row width
Row width often varies depending on tractor tyre width and harvester width. On soils which tend to give green tubers, wider rows tend to allow more soil per row which counteracts greening. Maincrops tend to be grown in 710 mm rows although 760, 800 or even 900 mm is possible. Earlies tend to be grown on narrower rows because they are harvested before maturity, and using narrower rows tends to raise yields because of the better spatial distribution of the plants. Row widths of 540–600 mm are common. Seed and canning potatoes are likewise grown on slightly narrower rows of 660–760 mm.

Spacing of tubers within the rows is determined by the row width and the seedrate, and is given in Table 10.18.

Seed treatments (sprouting or chitting)
The process of sprouting leads to quicker emergence after planting and an earlier leaf canopy. This can be vital in obtaining an early harvesting crop of earlies and even for maincrops can lift yields by 3 to 5 t/ha. Well sprouted seed tubers must be handled with care at planting and may even still be planted by hand in very early regions such as Jersey. For seedcrops sprouting is desirable since it leads to earlier bulking and maturity of the crop allowing an early desiccation and avoidance of aphid risk, thus keeping the level of virus low in the seed tubers. Sprouting can

Table 10.16 Optimum seed tuber populations and seedrates for some maincrop potato varieties

Variety and grade (mm)	*No. of sets/50 kg*	*Assuming ratio cost of seed; value of ware = 2:1*	
		Optimum population (10^3/ha)	*Optimum seedrate (t/ha)*
Varieties tending to produce undersized tubers:			
King Edward			
30–50	850	37.5	2.2
30–60	700	33.8	2.4
Maris Piper			
30–50	850	36.2	2.1
30–60	700	32.3	2.3
Record			
30–50	750	38.7	2.6
30–60	600	34.8	2.9
Varieties tending to produce over-sized tubers			
Desirée			
30–50	750	48.8	3.3
30–60	600	42.6	3.5
Pentland Crown			
30–50	750	48.8	3.3
30–60	600	42.6	3.5
Pentland Squire			
30–50	750	57.5	3.8
30–60	550	47.0	4.3

ADAS figures.

advance development in the field by 10 to 14 days. Sprouting is, however, expensive and requires the availability of frost-free chitting houses or glasshouses. Large maincrop producers will usually sprout only a part of the crop to spread out the harvest.

Sprouting is carried out in glasshouses or more usually in heated barns. Tubers are placed upright in special chitting trays and stacked up to 5 metres high. Natural daylight or supplementary lighting (warm white fluorescent tubes) with a daylength of 8 to 12 hours is necessary for sprout development. Temperature must be controlled above a set minimum for a variety:

Below 4°C	Sprout development is inhibited
5.5°C	Only fast sprouting varieties grow
7°C	All varieties commence growth, and fast sprouting varieties grow quickly
10°C	All varieties grow rapidly, fast sprouting varieties grow excessively

For most varieties, a low-temperature thermostat and additional heating are usually sufficient to keep the temperature above 7°C. Fast sprouting varieties can be more troublesome and may need to be cooled by ventilating at night to keep the temperature down.

Ideally, sprout length should be about 19 to 25 mm. If sprout length is advancing too quickly, the store should be cooled; conversely if sprout length is too short, the temperature should be increased. Sprouts are very sensitive to frost, and during freezing weather the store must be kept frost free. The store should also be kept well ventilated to avoid high humidity which forms condensation at lower temperatures and can cause rots. A careful watch for aphid should also be kept and the store fumigated if necessary; this is a particular problem for earlies where the time in store can be very long.

Single-sprouted tubers are preferred by many growers and this can be achieved by commencing sprouting in autumn at 16°C when a single growth dominates and suppresses further sprout development (apical dominance). Multi-sprouted tubers can be achieved either by delaying the sprouting process until January or by rubbing off the main sprout and releasing the apical dominance.

Table 10.17 Guide to seedrates for potatoes

Type of production	*Seedrate (kg/ha)*	*Remarks*
Earlies	3200–4500	Single sprouted seed required for earliest crops; multiple sprouted seed satisfactory for later lifting and second earlies. First early seed should be tightly graded and planted 200–360 mm apart according to size
Canning	Optimum 7500 Economic range 2500–5000	High stem density 160–220/m^2 required. Use multisprouted seed spaced 150–230 mm apart according to size. Size grade seed. Optimum seedrate depends on cost of seed and value of produce
Seed production Home produced seed, e.g. Scotland	5000–7500	High seed rates are fully justified where cheap seed is available; home produced ware size, which is of lower value than seed, is planted
Purchased seed	3800–5000	High seed rates are very expensive if seed is bought at high price, e.g. early growers producing once grown seed. Use lower rate and multisprouted seed

NB When growing new variety check on suitable seedrate – there are large differences.

Table 10.18 Spacing of potatoes according to seed rate

	Required seed rate (t/ha)												
	17.5	*20.0*	*22.5*	*25.0*	*27.5*	*30.0*	*32.5*	*35.0*	*37.5*	*40.0*	*45.0*	*50.0*	*75.0*
No. of sets/50 kg	*Spacing between sets (mm × 10) in 760 mm rows*												
400	94	84	74	66	61	56	51	48	43	41	37	33	21
500	76	66	58	53	48	43	41	38	35	33	29	26	17
600	63	56	48	43	41	38	33	30	30	28	24	21	15
700	53	48	43	38	35	30	28	28	25	23	21	19	12
800	48	41	38	33	30	28	25	23	23	20	19	16	11
900	43	38	33	30	28	25	23	20	20	18	16	15	10
1000	38	33	30	28	25	23	20	18	18	18	15	14	9
1100	35	30	28	25	23	20	18	18	15	15	14	12	7
1200	30	28	25	23	20	18	18	15	15	13	12	11	6

Single-sprouted seed is essential for earlies whilst for seed crops and canning multi-sprouts are preferred.

Many growers do not like to use fully sprouted seed because the care needed at planting does not fit in with the mechanization of large planting machines. An alternative method is to use mini-chitted seed with sprouts only 2 to 3 mm in length. With this method seed is first 'cured' at 13–16°C for about 10 days upon arrival at the farm, then cold stored at 3°C to suppress sprout development. Then, 3 weeks before planting, the temperature is raised to 7–10°C to begin the sprout development. With this method, seed does not need to be placed in special trays but can be kept in bags or in 500-kg bulk containers.

Sowing date

Sowing date varies with region, soil type and crop type. The potato foliage is very sensitive to frost and so should be sown at about the time of the last frosts so that it emerges in frost-free weather.

Early crops in the frost-free growing regions are sown as soon as soil conditions allow. The earliest are sown in Jersey on the steep south-facing hillsides (cotees) in January and the remainder follow in February in Cornwall, Pembrokeshire and Kent. Second earlies follow earlies in March.

Maincrop varieties begin planting in late March in the south and early April in the north. Unsprouted seed may often be planted first as this will take the longest time to emerge. The aim is to complete planting by the end of April.

Plastic film

One of the biggest developments in potato production in the last 10 years has been the widespread use of plastic film on potatoes. Various types of film are available and are graded on the number of holes per square metre. Ventilation is essential or early blight may attack the crop; the holes also let rainfall and carbon dioxide into the developing crop. Some early growers also use spun fibre floating film to good effect. The cost of the covers makes it only economic to use them on the most valuable crops, i.e. earlies and possibly second earlies. Covers can advance lifting of earlies by 10 to 14 days and their widespread use has undermined the natural advantage of the extremely early regions. For example, the Isles of Scilly do not now grow early potatoes! Best responses to covers tend to occur in poor years when the greenhouse effect created by the cover is maximized.

Crops should be covered as soon as possible after sowing following the use of a soil residual herbicide, and the crop should be covered within 14 days of planting. Timing of removal of the cover depends on weather factors and crop development. Optimum timing appears to be no longer than 4 weeks after 95% crop emergence.

Covers give some frost protection to the crop by virtue of raising the soil temperature by a couple of degrees, but well developed crop touching the plastic will be easily scorched since the dew on the underside of the plastic frequently freezes during a radiation frost.

Soils

Early crops are grown on light soils which can be easily cultivated in early spring. Maincrops prefer deep, moisture-retentive soils. Soils should preferably be well prepared and free from compaction and stones and clods which damage the crop when it is harvested. Many growers now use a stone and clod separator before planting a field as it has been shown that this results in a better quality crop. Various machines are available. Some remove the stones and clods to a trailer for dumping and others bury the stones and clods under the furrow. Nearly all potatoes are grown in ridges to facilitate harvesting but there is some interest in bed systems. The ridges may be drawn up twice in order to give some weed control by cultivation but most growers rely on soil residual herbicides for weed control. The residual herbicide may be mixed with a low rate of paraquat and applied just before emergence to kill germinated seedlings and to prolong the effect of the residual chemical.

Potato fields are usually left to run slightly acid (pH 5.8–6.0) as this helps prevent potato scab. In a rotation, routine liming should always be applied to a field after a potato crop, never before.

Nutrient requirements

Potatoes respond well to applications of well rotted manure provided they are applied well in advance of planting in the previous autumn. Maincrop potatoes yield best in fertile conditions and, as such, fertilizer applications are high. The crop does not remove all of this fertilizer and the residual amounts left in the soil benefit following crops such as winter wheat.

Early potatoes do not require as much nitrogen or potassium as maincrops because they are harvested before maturity and high nitrogen will delay tuberization and thereby harvest date. Excess nitrogen reduces dry matter of the tubers thus reducing cooking quality, and delays tuber initiation and haulm maturity thereby harvest date. It also predisposes the leaf to potato blight attack. Nitrogen should be reduced in fertile conditions and after heavy dressings with manure. Phosphorus encourages early tuber growth which is important for all crops especially earlies. It helps give good skin strength and counteracts blight susceptibility. Only soluble forms of phosphorus should be used. Early crops grown on acid soils in the West Country give good responses to phosphorus.

Potassium helps to produce a tuber which has a low dry matter content and is less likely to bruise; it also reduces the incidence of after-cooking blackening. Best responses to potassium are obtained with maincrops. Nutrient recommendations are given in Table 10.19.

Irrigation

Potatoes respond to irrigation, especially maincrops, and yields can be doubled in dry years. Irrigation water should not be applied until after marble-sized tubers are present in the crop, otherwise too many tubers will be stimulated and an undersized crop produced. The late application of irrigation to a droughted crop can result in growth cracks and deformities to the tubers leading to rejections. This also occurs in unirrigated crops which experience heavy late summer rainfall following a drought. Irrigation also helps reduce the incidence of scab but may lower dry matter percentage of the tubers.

Diseases and pests

The single most important disease of potatoes is potato blight (*Phytophthora infestans*). This disease is so devastating to the crop that growers must routinely spray fungicides to keep the disease in check every 10 days during periods of risk. Recent fungicide resistance has made the job of complete protection more difficult in recent years. Most crops of earlies are harvested before conditions are favourable for blight spread and therefore escape the use of fungicides. The most important pest of potatoes is potato cyst nematode. This persistent soil-borne pest can only be controlled satisfactorily by crop rotation, 1 year in 6 cropping. However, many growers want to crop more frequently than this and use soil-applied granular nematicides to control the pest. A soil screening service is offered to growers by ADAS. A second pest is aphid carrying virus. Where insecticide granules are used, some protection against aphid is given.

Harvest and storage

Earlies are mostly harvested without destroying the haulm whereas with maincrops routine desiccation is commonplace. Destruction of the haulm is important to prevent blight spreading into the tubers, and it also starts the maturity phase of the tubers which commences with skin development. All potatoes need time to develop a skin which will protect them during subsequent handling. For early crops this occurs within a day or two after careful lifting. For other crops it will occur 10 days after destruction of the haulm or as much as 21 days if the haulm was particularly vigorous.

Many potatoes are now contract lifted or grown by large growers with sophisticated lifting machinery which is carefully set up to lift the crop without causing damage to the crop. Early harvest of maincrops in mid-September to mid-October tends to give the least harvesting troubles with a cleaner crop, less bruising, better healing, less dirt and better cooking quality.

Most storage is indoors in a wide variety of farm buildings, but best storage with lowest wastage is obtained from temperature-controlled stores. Indoor stores must be regularly checked for hot spots or sites of rotting which can ruin a crop. Potatoes are frequently treated with sprout suppressants as they go into store to improve their keeping quality beyond February. However, supermarkets are becoming increasingly dissatisfied with the use of sprout suppressants and are forcing large growers to use more temperature-controlled storage.

Table 10.19 Nutrient requirements of potatoes (kg/ha) (MAFF, 1994)

	Soil index				
	0	*1*	*2*	*3*	*4*
N (earlies/canning/seed)					
Sandy/shallow soil	180	130	80		
Other mineral soils	160	110	70		
N (maincrop/second earlies)					
Sandy/shallow soils	240	200	130		
Other mineral soils	220	160	100		
Organic soils	180	130	80		
Peat soils	130	90	50		
P_2O_5 (earlies/canning)	350	300	250	200	200
P_2O_5 (seed)	350	300	250	150	100
P_2O_5 (maincrop/second earlies)	350	300	250	200	100
K_2O (earlies/canning)	180	150	120	60	60
K_2O (maincrop/second earlies/seed)	350	300	250	150	100
				0	
MgO	165	85	0		0

Crop yields Maincrop potatoes yield 30–40 t/ha rising to 50 t/ha or more for good crops, especially with the use of irrigation.

Vegetable crops

Quality is paramount for vegetable crop products. Quality requirements vary for each type of product and according to the market outlet, even between different supermarkets and between different presentations for the same supermarket. General quality requirements include size, uniformity, shape, colour, freedom from damage from handling or pathogens and pests, and freedom from chemical residues. Fresh vegetables are perishable and many may only be stored for very limited periods, thus all-year-round production is necessary using varieties and techniques to overcome climatic constraints. Vegetables are grown under contract to processors or to packhouses who co-ordinate production and undertake grading and packing to supply the major supermarkets, or are sold to wholesalers who supply the smaller retail outlets and the catering trade. Local opportunities exist for pick-your-own, farm shop sales, and community marketing, direct to the public. Prices can very widely through the year according to supply and demand, with out-of-season production generally attracting higher prices. Integrated crop management (ICM) systems are becoming increasingly important in vegetable production for the major supermarkets.

Root vegetables

Carrots (Daucus carota)
Carrots are the most widely consumed vegetable in the UK and are grown all year round for the fresh market and for canning, freezing and processing. Clean, straight, undamaged roots with good colour are required, of appropriate size depending on end use. Varieties are grouped into several types which vary in their shape (e.g. Nantes type are cylindrical, Chantenay are conical), and maturity period (e.g. Amsterdam Forcing are early, Autumn King are late).

Best carrot soils are well drained, deep, stone-free sands or light, loamy peats with good water-holding capacity to allow easy root penetration and unrestricted root expansion. This ensures good shape and easy harvesting of clean roots. Sites should be rotated with a 5-year break between carrot crops to prevent build-up of carrot cyst eelworm (*Heterodera carotae*) and violet root rot (*Helicobasidium purpurea*).

A fine, firm, level, clod-free seedbed should be achieved with minimum working. Bed systems are preferable to eliminate compaction effects on root development. Dressed seed should be drilled as shallowly as possible into soil moisture at even depth. 'Out of season' production uses floating film covers for protection during establishment and/or overwintering as appropriate. Sowings are made between March and June, and for overwintering in October. An effective weed control programme is essential through the life of the crop as carrots are poor competitors (see Chapter 13). Chief pests are carrot fly (*Psila rosae*) and carrot willow aphid (*Cavariella aegopodii*), and both should be controlled according to timing of their life cycles and populations in relation to crop timing (see Chapter 13); pest monitoring systems are extremely useful for control within ICM principles.

Nutrient requirements are given in Table 10.20. Response to P and K is good, to N minimal. On sandy soils 400 kg/ha salt should be ploughed in before drilling. Ideal pH is 5.8 on peats and 6.5 on sands; the crop is sensitive to overliming which can cause deficiencies of manganese and copper (and boron on sands).

Root size is influenced by plant population, sowing date, time to harvest, soil fertility and water supply; complete uniformity is unachievable, but, as each market requires different root size, grading of harvested crop can ensure a high proportion of saleable roots. Seedrate should be carefully calculated to take account of variety, situation and required grade. Target populations will vary widely, for example a cylindrical type grown for small carrots (20–25 mm shoulder diameter) will require about 170 plants/m^2, but the same yield (40 t/ha) of large carrots (45–50 mm shoulder) will be obtained from 16 plants/m^2. Field factors must be incorporated into calculations of seedrate: for cold soil/poor tilth 0.5; average conditions 0.6; good conditions 0.7. Spacing will depend on system, soil and grade required; single, double, triple and multi-row systems are all used as appropriate.

Harvesting uses top lifting harvesters or elevator digger harvesters. Average yields are 40 t/ha, but exceptional crops can produce 100 t/ha. Field storage can be achieved by earthing over or clamping.

Swedes (Brassica napus)
Most of the UK culinary swede production is from Devon or from Scotland, supplying the market all year round from sowings made between March and July (or March transplants from January sowing).

Soil should be well drained but moisture retentive, of pH 6.0–6.5, and on a rotation of six years to prevent

Table 10.20 Nutrient requirements of carrots, swedes and parsnips (kg/ha) (MAFF, 1994)

	Soil index				
	0	*1*	*2*	*3*	*4*
Early bunching carrots					
N	60*	25*	0		
P_2O_5	300	250	175	100	75
K_2O	250	150	125	50	0
Maincrop carrots					
N	60*	25*	0		
P_2O_5	250	150	125	50	25
K_2O	250†	150†	125†	50†	0
Swedes					
N	100‡	40	0		
P_2O_5	150	100	50	50	25
K_2O	250	200	150	75	0
Parsnips					
N	100‡	40	0		
P_2O_5	175	100	75	60	30
K_2O	225	150	150	75	0
All					
MgO	150	100	0	0	0

* Reduce to zero on fen peats.
† On sands and light loams 150 kg/ha sodium (400 kg/ha salt) can be helpful; reduce potash by 60 kg/ha.
‡ Reduce to 60 kg/ha on fen peats.

clubroot and *Phoma*. Dressed seed is drilled into moisture 10–20 mm deep to give approx. 10–15 plants/m^2 (a lower density will give an earlier harvest date), on 40–50 cm row widths.

Nutrient requirements are given in Table 10.20. Main threats are clubroot, powdery mildew, *Phoma* and cabbage root fly (see Chapter 13 for further advice on control). The yield is approx. 30–40 t/ha.

Parsnips (Pastinaca sativa)

Parsnips are marketed all year round though demand is strongest in the winter months. The roots have good frost resistance, and flavour and sweetness are improved by frosting. The market demands clean, 'white', straight roots without fangs, bruising or other damage. Size depends on outlet, from 35–65 mm shoulder diameter for supermarket presentation packs, to 150 mm shoulders for processing. As for carrots, varieties show different shape characteristics from bulbous to long and tapering. Varieties also show variation in their susceptibility to bruising, and their resistance to cankers: orange brown canker (*Itersonilia* spp) causes rot in the field; black canker (*Phoma* and *Mycocentrospora* spp) causes rot at sites of damage.

For best quality crops, soils should be deep, sandy, well drained and stone free. Heavy soils cause root fanging and make harvesting of clean roots very difficult. It is also difficult to obtain clean roots from peaty soils as the peat clings to the 'wrinkled' surface. A wide rotation with carrots (not less than five years) should be practised. A fine seedbed with no compaction should be prepared from minimum passes; bed systems are preferable. The main drilling period is from January to May, depending on site and target harvest date. Natural or pelleted seed is precision drilled up to 15 mm deep to give a plant population of 40–60 plants/m^2 for smaller roots, or 20–30 plants/m^2 for larger roots, on 40–50 cm rows depending on system.

Carrot fly and canker are likely to be most serious problems (see Chapter 13). Nutrient requirements are given in Table 10.20. Organic soils may encourage manganese deficiency; a foliar spray of manganese sulphate will be economic if deficiency symptoms are noted. Soil pH should be at least 6.5 on mineral soils and 5.8 on peats.

Top lifter harvesters can be used if top still present, but after they have died back digger elevator harvesters must be used. Parsnips are very susceptible to bruising and damage, and handling procedures should take account of this. Yields range from 15 to 30 t/ha.

Alliums

Dry bulb onions (Allium cepa)

Grown for the fresh market and for processing and pickling, bulbs must be globe shaped with good appearance, no staining, appropriate colour, and thin dry necks. Many varieties are available with different colour, shape, maturity date, and keeping qualities. Varieties are either 'autumn sown' to be harvested in June/July, or 'spring sown' to be harvested in August/September.

Most onions are grown in the Eastern counties as warm sunny conditions are necessary for quality and ease of harvesting. Soils should be well drained, deep, moisture retentive, stone free and free working, with a pH of 6.3–7.0 (5.5–7.0 on peats). Silts, brick earths, medium loams, sandy soils and peats are most suitable. Onions and leeks should not be grown on a rotation of less than six years to prevent build-up of white rot (*Sclerotium cepivorum*); crops are also susceptible to stem eelworm (*Ditylenchus dipsaci*) whose host range also includes legumes, oats, carrots, parsnips and sugar beet.

Seedbeds must be fine, firm, level and clod free with no compaction; bed systems are preferable. Overwintered crops should be sown in the second half of August (early August in the north of England); earlier sowing creates risk of bolting. Spring crops are sown in February/March. (Improved yield and quality can be obtained from module-raised transplants sown in January/February and planted out in March/April.) Seed is drilled into moisture 12–25 mm deep, on row widths to fit the system but usually 30–45 cm. Plant population will influence the size of bulbs: 65–85 plants/m^2 for ware production (over 25 mm diameter), 320 plants/m^2 for picklers (under 25 mm diameter). Field factor for seedrate calculation is 0.5 for cold soil, 0.7 for good conditions.

Onions are highly responsive to P and K but not to N. Keeping quality is improved by K but reduced by N (for nutrient requirements see Table 10.21). Neck rot (*Botrytis allii*) must be controlled to prevent heavy losses in store; downy mildew (*Peronospora destructor*) and leaf spot (*Botrytis* spp) should also be controlled. White rot and stem eelworm are controlled by rotation; onion fly (*Delia antiqua*) is controlled by seed dressing.

The crop is mature when tops go down. The drying procedure is crucial for quality. Traditionally onions were undercut, windrowed and field dried for 7–10 days. Sophisticated drying/store rooms now allow the crop to be lifted immediately (after flailing tops), followed by a controlled staged drying process in store; quality is more secure by this method than by field drying. Yields range from 30 to 45 t/ha.

Leeks (Allium ameloprasum)

A wide range of types and varieties, and the use of transplanting techniques allow leeks to be harvested almost year round from July to May. The market requires clean, long, straight shafts with a good length of blanch; size depends on the presentation required.

Table 10.21 Nutrient requirements of onions and leeks (kg/ha) (MAFF, 1994)

	Soil index				
	0	*1*	*2*	*3*	*4*
Bulb onions					
N (spring sown)	90*	60*	30*		
N (autumn sown)	100*	70*	40*		
P_2O_5	300	250	150	50	25
K_2O	180	150	0	0	0
Leeks					
N‡	150*	90*	30*		
P_2O_5	300	250	150	50	25
K_2O	275	150	125	50	0
All					
MgO	150	100	0	0	0

* Reduce by 60 kg/ha on fen peats.
† Applied as spring top dressing, except at Index 0 when up to 50 kg/ha can be applied to seedbed.
‡ No more than 100 kg/ha should be applied in the seedbed. Up to 100 kg/ha top dressing may be required in addition.

Leeks grow on a wide range of well drained but water retentive soils, with pH 6.5–7.0. Deep loams and peats are preferred; coarse sands make cleaning difficult as particles catch in the leaf sheaths, giving a gritty, unpleasant texture. Rotation is as for onions.

Direct sowing of seed into the field position is possible for March/April sowing, when seed should be drilled 12 mm deep in beds on 25–30 cm rows or on 45–50 cm rows to allow earthing up. Transplanting of bed-raised or module-raised plants (sown January to April) is used for most leeks, especially early sown, as quality is improved (uniformity and long blanch). Planting out (April to July) uses row widths of 25–60 cm depending on the bed system used. Final plant population is 20–50 plants/m^2 depending on the size of leek required.

The main threats are white rot (*Sclerotium cepivorum*), rust (*Puccinia allii*), onion fly, stem eelworm, and onion thrip (*Thrips tabaci*); see Chapter 13 for further advice. Leeks respond well to a heavy dressing of farmyard manure (60–70 t/ha) ploughed in. Nutrient requirements are given in Table 10.21. For every 10 t FYM applied; reduce fertilizer by 15 kg/ha N, 20 kg/ha P_2O_5, 40 kg/ha K_2O, 8 kg/ha Mg.

Leeks are harvested by hand after undercutting, or by elevator diggers, or by complete harvesters. Yield is 20–30 t/ha.

Brassica vegetables

Cauliflowers (Brassica oleracea – var. botrytis)
Cauliflowers are produced all the year round in the UK. The main types of cauliflower are:

- summer – July to September
- autumn – September to December
- winter-heading (known as 'broccoli' in Cornwall) – December to May
- overwintered – May to June

The majority of the crop is produced on the fen silt soils of the Wash in south Lincolnshire but other significant areas of production are west Cornwall, south Pembrokeshire, east Kent (Isle of Thanet), Evesham, Lancashire and Jersey. When in head the crop is sensitive to frost which causes blemishes on the curd. For this reason winter-heading types are mostly found in west Cornwall and Jersey where the climate is milder. Winter-heading types are also produced in large numbers in north-west Brittany, France, and exported to the UK. Heading date is a function of vernalization requirement of the variety, which is controlled genetically and hence can be selected for.

The traditional varieties were open pollinated types which showed variable performance and a wide range of heading dates within a variety such that fields had to be trafficked and cut over a number of weeks. These are rapidly being superseded by hybrid varieties with more uniformity and vigour leading to a higher number of heads cut per field of class 1 curds. Hybrids are now available in all of the cauliflower groups but are yet to be proven commercially in significant numbers for the winter-heading group. Growers are referred to the NIAB vegetable growers leaflet number 1 'List of Cauliflowers' and to seed merchants' catalogues for new varieties.

The different groups of cauliflowers have differing husbandry requirements with regard to plant spacing, plant raising, fertilizer requirements and crop protection.

Rotational position Cauliflowers should be rotated with non-brassica crops in a frequency of cropping of 1 year in 5 because of the risk of clubroot (*Plasmodiophora brassicae*), a soil-borne disease. On high pH soils cropping can be more frequent since this discourages clubroot build-up. Another problem with long term cropping is stem canker (*Phoma lingam*). Cauliflowers, like all brassicas, can carry over beet cyst nematode and should not be mixed in rotation with sugar beet.

Cauliflowers are grown in rotation with early potatoes in Cornwall, Jersey, Thanet and Pembrokeshire, but in Lincolnshire they are often the most important crop in the rotation. It is possible to transplant winter-heading cauliflower after an early harvested winter barley crop in the south west.

Plant production The entire cauliflower crop is produced from plantlets transplanted into the field and is never direct sown. Two sources of plants are used: grown in seedbeds and lifted as bare-rooted plants (peg plants), and grown in modular trays (modules). The move has been towards more and more module-grown plants, but peg plants are still preferred for winter-heading varieties. Since hybrid seed is much more expensive than open-pollinated seed, these varieties are more commonly grown in modules. Specialist plant raisers exist within the production zones, and growers often buy direct from these rather than buying seed.

The timing of sowing and transplanting is very important for all cauliflower types with only narrow windows of opportunity. Module-raised plants need to be transplanted earlier than peg plants since they are generally smaller and slower to come away. Only very few varieties show tolerance of a wider window of sowing and transplanting, and these are typically summer types with no or a very low vernalization requirement. These varieties can be used as summer- or early autumn-heading varieties. Sowing and transplanting dates are given in Table 10.22.

Plant beds (for bare root transplants) A well cultivated plant bed must be prepared on clubroot-free land. Drill seeds in beds about 250 mm apart at a depth of 20 mm, ensuring that seed is placed into soil moisture. Sow about 30 g per metre run to give about 36 plants per metre run. Seed is placed 18 mm apart with a precision drill. Treat the seedbed against cabbage root fly if sown after April. If necessary, protect seedlings against cabbage aphid and caterpillars using an insecticidal spray.

When selecting plants for transplanting reject small and weak plants and any with damage or aphid. Transplants should be protected from drying out too much during transplanting.

Module-grown plants are usually raised by specialists but large growers may raise their own. Early crops are often sown in modules with a large cell size, with later transplanted crops being grown in small cells (9–15 ml volume of compost per cell). Plants are routinely fed and watered, and treated with insecticides, and are usually dipped in an insecticide prior to planting to give cabbage root fly protection.

Table 10.22 Sowing and transplanting dates for the production of cauliflowers

Heading date	*Harvest months*	*Sowing dates*	*Transplant dates*
Early summer	July–Aug	Feb–early March	March–April
Mid summer	Sept	Mid March	Mid May
Early autumn	Oct–early Dec	Late April	Mid–late June
Late autumn	Nov–Dec	Mid May	Mid Jun–early July
Winter-heading	Dec–May	Early May	Mid–late July
Overwintered (spring-heading)	May–June	Late May–early June	July

Soil types Choose free-draining soils with a pH of 6.5–7.0. Cauliflowers thrive in good soils with good heart and high organic matter. The use of leys to achieve this is good practice, otherwise use farmyard manure. Alluvial soils with good water-holding capacity are necessary to grow summer types succesfully, and crops should not receive any checks in the growing season or buttoning and poor curd quality can occur.

Cultivations Cultivate land well and deeply plough for summer types and work to a plant bed quickly to retain moisture. Land should be prepared well in advance of planting; a final pass just before planting gives some weed control and ensures an aerated bed for successful establishment. Ridging up of rows is practised in windy areas and in overwintered crops to divert rain away from the stem. Ridging gives some weed control also but should be carried out once the plants are well established but completed before the crop meets in the rows to avoid damaging roots.

Plant population The grower of cauliflowers is concerned with achieving the maximum number of good quality heads per hectare. Plant population influences both curd size and its quality. If population is too high, then only small curds will be produced and with winter-heading types the quality will be markedly affected. Some supermarkets are now marketing 'baby' curds and these are produced at very high populations. Supermarkets are now demanding smaller curds even for maincrop cauliflowers. It may be possible to increase seedrate slightly when using F1 hybrid seed since the increase in vigour may allow a higher degree of plant competition without detriment to curd quality. Higher plant populations are possible with summer-heading types, and the winter-heading types require the most space. Plant populations vary depending on the type of cauliflowers being grown (Table 10.23) and the harvesting machines being employed. Whatever harvesting machinery is being used, the row width must be an exact multiple of the distance between the wheels. The trend through the last 10 years has been towards specialist harvesting rigs, either tractor mounted or purpose built.

Nutrient requirements Cauliflowers require fertile land to grow to best quality; nutrient requirements are given in Table 10.24. High levels of potash are required but heavy dressings should be applied well before transplanting to avoid scorch. Likewise, nitrogen applied at over 150 kg/ha should be split and half applied one month after transplanting. Winter-heading and overwintered types are not given as much nitrogen at transplanting but are top dressed later to encourage growth before heading. Thus winter-heading types are top dressed in late autumn and overwintered types in early spring (Table 10.25). Boron is sometimes applied to avoid dead heart or stem rot.

Table 10.23 Plant populations and spacings for cauliflowers

Crop	*Plant population/ha*	*Spacing (mm)* *Rows*	*Within rows*
Early summer	30 300–37 000	600	450–550
Late summer and early autumn	27 700	600	600
Late autumn	23 800	600	700
Winter-heading	20 000	750	750
Overwintered	23 500	650	600

Table 10.24 Nutrient requirements of cauliflowers (kg/ha) (MAFF, 1994)

	Soil index *0*	*1*	*2*	*3*	*4*
N					
Summer/autumn	250	190	130		
Winter/Roscoff	75*	40*	0*		
P_2O_5	175	125	75	50	25
K_2O	300	200	125	60	0
MgO	150	100	0	0	0

* In addition, add up to 200 kg/ha as top dressing 6–8 weeks before harvest.

Irrigation Summer and autumn crops need moist soil to ensure good quality, and need irrigating on demand with about 10 mm water at each irrigation. Other crops do not need irrigation although some growers trickle irrigate the plants at transplanting in July to aid establishment. Bare root transplants with their larger stem size appear to resist transplanting drought stress more than module plants. Their greater height also allows for deeper transplanting than modules, thus putting the roots into moister soil.

Harvesting and quality The UK market is now dominated by the multiple supermarkets with packhouses, cooperatives and individual large growers having specific contracts with supermarket buyers. The wholesale market also still exists, e.g. Covent Garden.

Typical market requirements include:

good shape – well rounded, undercurved base, not 'blown' or breaking up;
good colour – as white as possible, not yellow or discoloured;
specific size – by weight, e.g. 700 g min; or by diameter, e.g. 12 cm min;
free of blemishes – no rot, frost damage, bruises, caterpillar droppings;
free of bracts – small rudimentary leaves growing through the curd;
not ricey – small white rice-like protrusions from the curd;
not too many wrapper leaves – well trimmed.

Most crops are cut to Class 1 standard. A Class 2 exists but commands a lower price and may be rejected totally by a supermarket buyer. Crops are not selected on nutritional or taste criteria.

Packaging varies according to the market. In extreme cases the curds may be over-wrapped with clingfilm but the majority are now appearing in facepacks or are still in the more traditional crates. Packaging is carried out either on the harvesting rig or in the packhouse. Most cauliflowers are packed in set size boxes or crates into which a set number of heads are packed with the set number determining the head size, eg. 6, 8, 11 or 12 per box or crate.

Cauliflowers are supported by an EU Intervention Board which can help to put the bottom into the market, but this is only really taken advantage of during peak production times in the summer and early autumn months. Price fluctuates widely from week to week depending on supply which is very weather dependent. Growers often cooperate with a packhouse to try to even out the supply problem, particularly where contracts are made with supermarkets who demand year-round supply.

Calabrese (Brassica oleracea – var. italica)
Also known as broccoli and a 'tight-headed' version of sprouting broccoli. The edible portion is a bright green succulent stem terminating in a tight head of green flower buds known as a spear. The most valuable portion of the plant is the primary spear but secondary spears can sometimes be produced and if the price is right may be worth harvesting. Calabrese has risen in popularity in the last 5 years at the expense of cauliflower and as a consequence the area grown has increased dramatically. This rise in popularity has also meant that there has been an improvement in the quality and number of varieties available.

Varieties All commercial varieties are now F1 hybrids. Varieties do not have much vernalization requirement and mature some 75 to 90 days after sowing. Thus they behave as summer cauliflowers and produce spears from late May to late November in the UK. Early crops are obtained using module-raised transplants sown under glass in February or March, whereas later crops are direct drilled to a stand in the open field. The spears are not protected by wrapper leaves and are easily damaged by hail during late summer storms.

There is plant breeding interest in producing winter-heading varieties for UK production since year-round supply to the supermarkets can only be achieved with imports of crop from Spain and Italy during the winter months.

Soil types The crop must be grown without any growth checks but a wide range of soil types are suitable. Early crops are transplanted on to sandy soils which are accessible early in the spring but these crops may require irrigation. Moisture-retentive soils in good heart are preferred for direct-sown crops, and soils not prone to soil capping are preferred. Soil pH should be 6.5 to 7 minimum since the crop is highly susceptible to clubroot.

Establishment Early crops raised in modules must be transplanted at an early development stage (two true leaves), and large transplants should be avoided since these suffer too much of a check at transplanting, reducing yied. Prior to transplanting, modules should be soaked in a solution containing high nitrogen which helps establishment and reduces the checking effect. If the soil is dry at transplanting, then establishment will be improved if trickle irrigation is applied with the transplants. Transplants must be protected against cabbage root fly.

Direct-sown crops are usually drilled to a stand using a precision drill. A well prepared level seedbed is essential, and some growers prefer to grow in beds where compaction is avoided. A plant population of 8.5 to 10 plants/m^2 is desired, with the low population giving larger spears. Normal spacing is 500-mm rows and 230 mm within the row or in beds 1.68 to 1.80 m wide containing four rows 400 mm apart and plants 200 mm apart within the row. For the lower populations and larger spears the within-row spacing is increased to 250 to 300 mm.

Continuity of production Transplanted crops often take longer to mature than direct-sown crops. Choice of variety is important, especially for the early crops with varieties being designated early varieties. The following programme gives continuity of production:

Transplanted crop: Sowing date under glass	*Transplant date*	*Harvest period*
Mid February	Early April	Late May
Mid March	Mid April	Early June
Late March	Mid April	Mid June
Direct-sown crop: Sowing date in field		
Mid March	–	Late June
Late March	–	Early July
Mid April	–	Mid July
Late April	–	Late July
Mid May	–	Early August
w/b 28 May	–	Mid August
w/b 4 June	–	Late August
w/b 10 June	–	Early September
w/b 17 June	–	Mid September
Late June	–	Late September
Mid July	–	Early October
Late July	–	Mid October
Early August	–	Late October–Early November
Mid August	–	Mid–late November

Successful production of late maturing crops can only be carried out successfully in the milder south and south west of the UK and on Jersey.

The AFRC Horticultural Research International and ADAS have now produced a computer model of calabrese growth and maturity prediction which is available to growers and grower groups to help plan production and supply.

Nutrient requirements Like most brassicas calabrese prefers fertile soils and requires adequate levels of nitrogen and potassium. Recommended fertilizer rates are given in Table 10.25. Only apply up to 100 kg/ha of N in the seedbed and top dress the balance after emergence. If applying more than 75 kg/ha of K, then apply well before drilling and work into the soil or germination and establishment may be affected. Calabrese has a magnesium requirement, and compound fertilizers with added magnesium can be beneficial.

Table 10.25 Nutrient requirements of calabrese (kg/ha) (MAFF, 1994)

	Soil index				
	0	*1*	*2*	*3*	*4*
N	250	190	130		
P_2O_5	150	75	60	60	30
K_2O	150	100	0	0	0
MgO	150	100	0	0	0

Irrigation Irrigation is necessary to ensure quality spear production. Throughout the growing season 25 mm of irrigation can be applied whenever the soil moisture deficit reaches 25 mm. An application of 50 mm of irrigation 20 days before expected harvest helps to ensure quality spear production. Very dry soils at drilling should be irrigated before drilling, allowed to drain and then sown in order to ensure high germination.

Crop protection The crop is susceptible to most brassica diseases and pests, and needs to be protected accordingly. Herbicide choices are limited.

Harvesting and marketing Spears must be harvested with a sharp knife whilst the heads are tight with no yellow petal showing. The stem must be succulent with no trace of woodiness. The preferred size is a head diameter of 60–80 mm with a maximum of 100 mm. The spear should be 140–150 mm in length and the stem diameter 15–20 mm with a maximum of 30 mm. Most crops need cutting over two or three times since maturity within a crop is slightly variable.

Harvesting early in the morning is preferable, especially in summer to preserve high water contents in the spear. The harvested crop should be cooled immediately using a wet air cooler or a vacuum cooler. Conventional cool stores do not always give a satisfactory cooling rate and can lead to loss of water and flaccid spears.

Marketing through a cool chain is necessary to preserve quality. Some supermarkets demand overwrapping with cling film. Presentation will be dictated by the buyer but is commonly in single or small groups of spears. It is essential to check the buyer's requirements for presentation before cutting and preparing.

The crop yields from 2 to 12 t/ha with an average of 6 t/ha.

Spring cabbage (Brassica oleracea)
This is the unhearted or semi-hearted 'spring greens' type cabbage, now produced all year round from different areas (December–March: south-west England; February–April: Kent; summer/autumn: Lincolnshire). Stemminess and bolting must be avoided for quality; yield is very variable.

Seed should be sown 2 cm deep into a fine, firm seedbed on free-draining soil of light to medium texture with adequate moisture and a pH of 6.5–7.0. A bed system is preferable to avoid compaction and achieve uniformity. Plant population will influence size of heads and time to maturity; aim for 30–40 plants/m^2 for spring greens, and 10 plants/m^2 for a fully hearted crop, on approx. 35–40 cm rows. Nutrient requirements are given in Table 10.27. Top dressing depends on heaviness of crop; timing should ensure a steady supply of nitrogen throughout the growing period.

Summer and autumn cabbage (Brassica oleracea – var. capitata)
The range of varieties provides hearted cabbage for marketing from May to October (earlies are pointed 'Hispi' types; later are round). February and March sowings for early harvest are made under protection for transplanting; April and May sowings can be made outdoors, either for transplanting or direct into field position. Yields are variable but can reach 25 t/ha.

Soil should be of pH 6.5 (6.0 on peats), permit good root development and have good drainage; loams and peats are preferable for early crops; water-retentive medium to heavy soils are better for later crops. Beds for transplanting should be fine and level with no compaction; very good seedbeds are required if sowing direct. Plant population will influence the size of heads and time to maturity. For summer harvest aim for nine plants/m^2 on 30–40 cm spacings, for autumn harvest four plants/m^2 on 40–60 cm spacings. Nutrient requirements are given in Table 10.26.

Winter cabbage (Brassica oleracea)
A range of types is available including winter whites, January King, Savoy, and red cabbage for fresh sales, storage or processing. Harvest is from September to March, depending on varietal characteristics and frost hardiness.

Table 10.26 Nutrient requirements of cabbage (kg/ha) (MAFF, 1994)

	Soil index				
	0	*1*	*2*	*3*	*4*
N					
Spring cabbage*	75	50	25		
Summer/autumn/winter (pre-Xmas) cabbage†	300	250	200		
Winter (post-Xmas) cabbage	150	125	100		
Winter cabbage for storage†	250	190	130		
P_2O_5	200	125	75	50	25
K_2O	300	250	175	75	0
MgO	150	100	0	0	0

* In addition, add up to 250 kg/ha as top dressing 6–8 weeks before harvest.
† Apply 100 kg/ha at planting, and remainder when fully established.

Free-draining, deep soils are essential. Both transplanting (seed sown in April/May and transplanted into field in June/July) and direct sowing into the field (May/June) are suitable. In either case moisture stress must be avoided and irrigation is often necessary. Plant population will influence size of heads and time to maturity; for supermarket quality aim for four to six plants/m^2; for larger heads for processing reduce to two and a half to three plants/m^2. Average yield is 20–30 t/ha. Nutrient requirements are given in Table 10.27; excess nitrogen will impair frost hardiness and potential for storage.

Legume vegetables

The main legume vegetables are broad beans, runner beans, french beans and peas, all harvested as green, immature seeds in the pod. Precise timing of harvest is vital for the necessary quality characteristics of sweetness and tenderness. The products are highly perishable as all have high respiration rates and quickly deteriorate; careful handling and cool chain transport are therefore essential. All are grown for freezing, canning or selling fresh.

Beans and peas for processing are confined to the Eastern counties close to the processing plants, and are grown under contract only in order to programme production to provide an even, continuous supply over the growing season. At the optimum stage of maturity mobile viners and harvesters work round the clock to harvest the product for rapid transport and processing. Varieties and husbandry are dictated by the processor.

Beans and peas to be sold fresh are generally hand harvested, picking over the crop several times to maximize yield of the correct quality, thus labour costs are high. Variety lists and technical advice are published by the Processors and Growers Research Organization (PGRO), Peterborough. See Table 10.27 for nutrient requirements and combining pea and bean sections of this chapter for general considerations. Precise technical details vary widely according to site, type, variety, system of establishment and target harvest date. Effective weed, disease and pest control is vital to achieve the required quality, and irrigation is often beneficial.

Table 10.27 Nutrient requirements of vegetable legumes (kg/ha) (MAFF, 1994)

	Soil index			
	0	*1*	*2*	*3*
Broad beans				
N	60	25	0	
P_2O_5	250	200	150	50
K_2O	250	150	100	50
French/runner beans				
N	150*	100*	75	
P_2O_5	250	200	150	50
K_2O	275	175	100	50
Vining peas				
N	0†	0†	0†	
P_2O_5	75	50	25	25
K_2O	120‡	50	25	0
All				
MgO	150	100	0	0

* Reduce if using *Rhizobium* inoculum.
† Up to 25 kg/ha may be used for early fresh peas if spring rainfall was heavy and available N is likely to have leached.
‡ No more than 50 kg/ha should be combine drilled.

Fodder and forage crops

Forage maize (*Zea mays*)

Forage maize is restricted in its distribution in the UK by temperature. It is not frost tolerant and requires temperatures above 6°C to germinate and grow. The crop is therefore a summer crop with a 5-month growth cycle (May to Septemer). In recent years plant breeders have made considerable advances with varieties which can mature in less time (early maturing varieties) although the yield of these is slightly less. These advances have improved the reliability of the crop in the UK and pushed the potential area for growing the crop further north. The crop is favoured by dairy farmers since it produces a high quality, high-yield silage with the minimum of arable expertise, and the availability of contractors for drilling and harvesting the crop means only a low investment in machinery is needed. Maize thrives on fertile soils and can be grown without the use of fertilizers on land that has had large amounts of slurry or farmyard manure applied. The crop commonly produces a fresh weight yield of some 30–50 t/ha (7.5–12.5 t/ha dry matter yield) with a dry matter content of 25–30%.

Growth cycle

Maize is a monocotyledonous crop with a growth cycle similar to that of other cereals. It has a vegetative phase when only leaves are produced, a stem extension phase, pollen shed (anthesis) and grain filling. It differs from other cereal species in that the male and female parts are carried in separate structures. The pollen is borne on the terminal tassel of the stem and is the last structure to emerge (tasselling).

Suitable areas

Suitable areas for growing forage maize in the UK are mapped using long-term meteorological records and accumulating the average daily temperature from 1 May to 30 September after first subtracting 6°C (mean accumulated heat units above 6°C, AHU). For growing extra early maturing varieties at least 1250 AHU are needed; for second earlies 1300 AHU, for mid-season 1350 AHU, and for late varieties 1450 AHU (see Fig. 10.6). Growers also need to take account of altitude when considering which variety to choose, with approximately 25 AHU lost for every 100 ft (30 m) above sea level. Furthermore, 25 AHU are lost if the seedbed is cold and wet and 75 lost if the slope of the field is north facing. Very early sowing can also stress the crop if low temperatures are experienced, losing up to 50 AHU if the crop is set back.

Once the average daily temperature is above 10°C the crop grows well, and at temperatures above 15°C the C4 photosynthetic pathway of maize becomes a distinct advantage and growth rate is very rapid.

The crop is quite tolerant of low soil moisture but will yield better if moisture is available. For this reason yields are frequently higher in the west of the UK where summer rainfall is higher. In drier climates, e.g. central

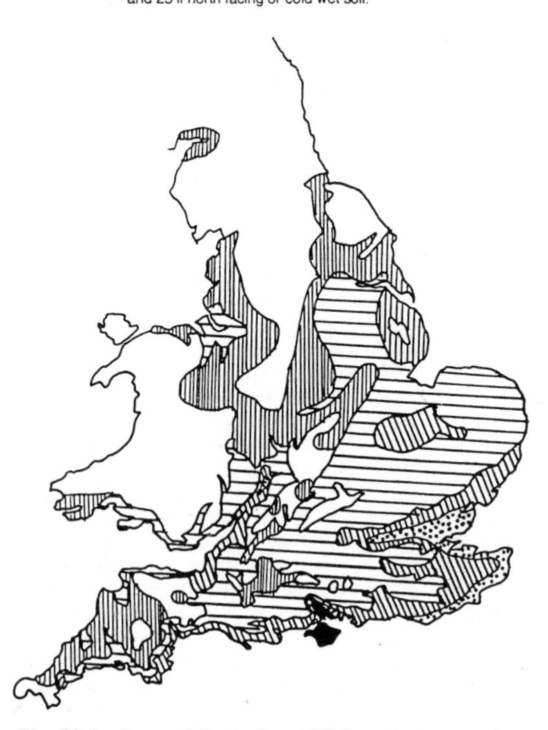

Fig. 10.6 Areas of England and Wales suited to growing forage maize based on accumulated heat units (AHU) over 6°C from 1 May to 30 September. (Adapted and redrawn from *Forage Maize growing guide*, Semences Coop de Pau.)

USA and Africa, the crop is frequently irrigated but in the UK it is not economically feasible to irrigate. Summer droughts lead to reduced seed set as the critical pollen shed becomes out of phase with the receptivity of the silks (styles).

Seedbed
Forage maize can be sown on any soil type on which a good, reasonably fine tilth can be produced in May. This excludes the heavier soils which are cold and slow to warm up, thus slowing down germination, and which can be difficult in a late harvest. Available summer moisture often excludes the lighter and shallower soils.

Ploughed land should be quickly worked down in the spring to avoid 'cobbly' seedbeds which impair both germination and soil residual herbicide action. Seedbeds need to be reasonably fine, but firm and ring rolling after drilling is often desirable.

Rotation
Maize fits well into any rotation and does not carry over any pests, diseases or serious weeds of arable crops. It can be grown sequentially. It can suffer serious damage from frit fly and should not follow a late ploughing out of a sward without at least four weeks between the crops. Since forage maize can be late to be harvested (October or even November) it can be difficult to get a winter cereal in as a succeeding crop. The crop is frequently sown on to a sacrifice field which has been used to dispose of slurry stores. This has been traditionally carried out through the winter but it is becoming recognized that this is an environmentally unsound practice. Winter spreading leads to losses of nitrogen through volatilization of ammonia, and runoff of slurry into watercourses is a major pollutant. Slurry should be emptied on to the field in spring to minimize these problems.

Sowing
Sowing date depends on prevailing temperatures in spring. A good guide is to wait until the soil temperature is above 10°C. Occasionally this is late April in the south but more usually early to mid-May. Late sowings result in crops that have poorly filled cobs and are late harvesting and result in poor quality silage.

The crop must be drilled at a depth of 50 mm using a precision drill usually on 75-cm row widths. Plant populations are relatively low: 11 plants per m^2 and a seedrate of 12 seeds per m^2 is used. Higher seedrates can improve overall yield but depress seed set and lower the quality of the silage. Since precision drilling is used, seed is graded and often sold in packs by seed number. Growers and contractors need to adjust drills to the seed size of the particular variety and seed lot.

Seed is normally dressed with a fungicide and insecticide, and a bird repellant can be added. This is to deter rooks which can devastate a field during germination. Harrowing or ring rolling after sowing removes the drill lines, which also helps to deter rooks.

Pests and diseases
The crop is particularly sensitive to frit flies which eat out the apex of the plant. Since forage maize varieties do not tiller freely, a damaged plant is a virtual write-off. If bad frit fly attacks are common or expected, then insecticidal granules can be broadcast or trickle applied at drilling.

Other pests include rooks and badgers. Badgers can be important in dry summers when they knock plants over and feed on the grain as a protein supplement when slugs and snails are in short supply.

Several foliar diseases attack maize including *Septoria* leaf blotch, but it is not economic to apply fungicides to forage maize in the UK.

Nutrient requirements
Most forage maize is grown on land which has been well fertilized with farm wastes, particularly slurry. If applications are heavy, no potash or phosphate fertilizers are needed but crops should be given a further 40 kg/ha of nitrogen in the seedbed or just prior to emergence. In the absence of heavy slurry applications, fertilizers should be applied as given in Table 10.28. The levels of potash recommended are high because the crop removes large quantities when silaged; however, there is not a large yield response to the application of either P or K. There

Table 10.28 Nutrient requirements of forage maize (kg/ha). (m) indicates maintenance dressing (MAFF, 1994)

	Soil index 0	1	2	3
N	60	40	30	0
P_2O_5	80	60	40(m)	0
K_2O	180	150	120(m)	

is a trend towards higher yields from forage maize, and crops are now planted earlier in late April in line with developments made in northern France. In these situations the crop can respond to a 'starter' fertilizer applied in the drill rows of about 10 kg/ha of N, P and K. With the high yielding crops nitrogen may be increased to 120 kg/ha in the absence of slurry applications.

Harvesting

The first frosts of autumn mark the end of the growing season and accelerate maturity but crops should be harvested within 10 to 14 days of frosting or quality begins to deteriorate. The crop is harvested when it has dropped below 75% moisture content (25% DM) and the grain is at least at the hard doughy stage; harvesting at below 70% moisture content when the grain is hard and flinty can improve ensilaging and leads to less effluent. Generally the leaves of the crop are dry and papery at harvest. Cool summers delay harvesting. Harvesting begins in the earliest areas with early/mid maturing varieties in late September, and can last to the end of October in late areas and with late varieties. In exceptionally late years harvesting may not be complete until early November.

Many crops are harvested using multi-row harvesters, and it is important that the crop is well chopped and the grain well cracked as this will aid digestion by livestock. Many crops are contract harvested. Ensiling is relatively easy because of the high level of soluble carbohydrates located in the stems at harvest, and additives are rarely necessary. The crop has an average D value of about 70 at harvest (see Chapter 11) with a relatively low crude protein content of 9%.

Yield

Good crops of forage maize yield 50 to 70 t/ha of fresh matter (12.5 to 18.75 t/ha dry matter) which gives approximately 45 to 60 t of silage per hectare.

Lucerne (*Medicago sativa*)

This legume is grown in various parts of the world and is also known as alfalfa. Providing it is grown on a deep, well-drained soil it can develop a very deep taproot and this enables the plant to exhibit a high degree of drought resistance. Growing stems develop from a crown just above ground level and the plant becomes dormant over winter. During this dormant period the plant must not be waterlogged, otherwise it will rot. Also, lucerne is very intolerant of acidity throughout the full depth of its root system: this means that it should not be grown on acid soils, as even if these were limed, it is unlikely the lime could correct the acidity throughout the whole rooting zone.

Lucerne is a perennial. It is slow to establish and it gives a low yield in its seeding year, but in subsequent years it should yield about 12 t/ha of DM on deep calcareous soils in southern England and continue this level of production for at least four full harvest years. Provided plant density is maintained, there is no reason why the crop should not remain productive, yielding, say, 8–10 t/ha of DM for up to ten years. Normally lucerne should be cut rather than grazed, as cutting ensures that the whole crop is defoliated at the optimum stage of growth and also prevents the hooves of grazing animals having a deleterious effect on plant density (and longevity) by knocking stem buds from the crowns.

In NIAB trials using the simulated silage-cutting management, the DM yield of lucerne, achieved without the use of any fertilizer N, has been similar to that obtained from grass swards receiving about 350 kg/ha of N. Also, the life expectancy of well-managed lucerne is greater than that of grass that is regularly cut, so that in terms of cost per tonne of DM, lucerne can be cheaper than grass. However, care must be taken in the conservation of lucerne, whether it be for hay or silage. If it is made into *hay*, the highly nutritious leaves dry much more quickly than the stems, and so the leaves tend to shatter and be lost in the final stages of field drying and baling. Barn drying of lucerne is better, and in some areas lucerne is grown as a very satisfactory crop for high-temperature drying. When making lucerne into *silage*, it must be remembered that the crop is difficult to compact sufficiently to exclude air unless it is chopped into short lengths: also there can be fermentation problems as it is a plant with a high N but low water soluble carbohydrate content. An additive that enhances the acidity of the material is well advised (e.g. molasses or an acid additive).

One important aspect of the successful management of lucerne is based on the appreciation that the very rapid growth of stem and leaf, both in spring and throughout the season following defoliation, depletes the plant's nutritional reserves. These must be replaced before the next regrowth period, otherwise the plant will be weakened and eventually die. Depending on growing conditions it takes 4–6 weeks for the plant to replace its reserves and so lucerne should never be defoliated at a greater frequency than this. There is a specific danger period in the autumn (often cited as September), because if the reserves are used to provide growth at this time, insufficient photosynthesis will occur to replace reserves; then with the onset of winter the plant will become dormant and die back. In this circumstance, it will have little in reserve to stimulate succeeding spring growth and many plants will die.

Lucerne can be grown as a pure stand or with a companion grass. It has been said that if lucerne is to be grazed, the presence of some grass is desirable as it presents the animal with a better balance of carbohydrate/protein in its diet and also it reduces the trampling damage caused by the animals' feet. However, the presence of grass can be highly competitive and act as a weed to the lucerne. On balance, it is preferable to grow lucerne pure and mainly cut it.

Lucerne can be susceptible to *Verticillium* and bacterial wilts, and also to stem nematode (*Ditylenchus* spp). There are varietal differences in resistance to these, although no present varieties are immune to any of these diseases. NIAB (1994) gives details of disease resistance and an indication of the yield expectancy from the varieties available.

For lucerne to grow satisfactorily its root nodules must be inhabited by *Rhizobium meliloti*. This is a different species from the bacteria needed by the other commonly grown legumes in the UK, and so, unless lucerne has been grown successfully in the field within the last ten years, the seed should be inoculated with the appropriate bacterium. This is best done by mixing the seed with a vacuum-packed, peat-based inoculum immediately prior to sowing. The efficacy of the inoculum decreases rapidly on exposure to air, and the use of more convenient methods, such as pelleted seed, is not as satisfactory.

The following highlights the main cultural aspects of successful lucerne production.

Establishment

As the plant is slow to germinate and develop, and must grow sufficiently to store reserves to support spring growth, sowing should take place before mid-July. In areas where there is likely to be moisture stress at this time, sowing needs to be earlier. Some successful growers, realizing that the first-year yields will be low and the risk of weed competition in a slowly growing crop will be high, sow lucerne in April with 100 kg/ha of barley or wheat that is taken off as arable silage.

The seedbed must be as weed-free as possible, very fine and level, and 17 kg/ha of seed is usually drilled in rows 100–150 mm apart and no more than 12 mm deep. Unless the crop is sown with a companion, weed problems can be anticipated and the use of an appropriate herbicide (depending on weed species present) is nearly always required.

Defoliation management

Bearing in mind that it is better to cut than graze, and also that there should be a minimum of 6 weeks between cuts, the first cut should be taken when the flower buds start to appear, often in mid-May. Subsequent cuts can normally be made again on the appearance of flowers and it is likely that two more cuts will be taken before September, when the crop should be left until well into the autumn. As a rule of thumb, the first cut is likely to provide 40% of the total annual yield and the next two cuts another 50% between them, so that the yield in autumn, also a difficult time for cutting silage, represents about 10% of yield. Provided the ground is not wet, this growth can be grazed. If this or cutting proves difficult, it can be left to die back: this will represent a small loss of yield but should have no harmful effect on the following year's production (in contrast to the consequence of allowing a grass sward to carry appreciable growth into the winter and then die back).

Fertilizers

Some authorities recommend 25 kg/ha of N in the seedbed if the soil N index is 0. Apart from this, no fertilizer N should be used at any time on this crop.

PK requirements are slightly greater than would be applied to a high-N grass sward regularly cut for silage. Table 10.29 sets out PK recommendations.

The annual P requirement can be made in one or several applications, but the K fertilizers should be applied in *at least* two dressings, one in spring and the other after the second cut. It follows that a convenient compound fertilizer is 0:1:1 in spring, followed by straight K later.

Table 10.29 Nutrient requirement of lucerne (kg/ha)

	Soil index			
	0	*1*	*2*	*3*
Seedbed				
P_2O_5	100	80	50	30
K_2O	100	80	50	nil
Conservation (per year)				
P_2O_5	120	100	80	40
K_2O*	240	200	160	80

* On K-rich boulder clays these rates can be reduced by 80 kg/ha.

Weed control

Apart from weed problems at establishment, the only other major weed problem associated with a healthy, dense stand of lucerne is likely to be from competitive plants such as grasses or chickweed, that develop over winter when the crop is dormant. These can be conveniently and cheaply controlled by spraying with a non-selective herbicide such as paraquat just before the stem buds begin to develop in the spring. (Timing is crucial here, as if the spraying is left too late, damage will be done to the crop.)

Sainfoin (*Onobrychis viciifolia*)

This is a forage legume with a deep taproot and, like lucerne, it demands a free-draining calcareous soil. It is rarely grown in UK, as for normal farming purposes its average yield of 7–8 t/ha of DM is well below that of lucerne and it may not be as persistent. However, it is less stemmy than lucerne, is easier to make into good hay and has a higher feed value. In situations where the high feed value carries a premium, then this may more than compensate for the low yield. For example, some racehorse trainers are prepared to pay very high prices for sainfoin hay, and crops are grown near to these markets. Also the aftermath growth is very good for fattening lambs.

Seed is available in husk (unmilled) or milled and may be sown in either form: seedrates should be 125 kg/ha for unmilled and 65 kg/ha for milled. Seedbed conditions and sowing times should be as for lucerne. Once established, no N is required and PK rates should be about half of those recommended for lucerne.

Roots

Fodder beet and mangels (*Beta vulgaris*)

Fodder beet is a hybrid of mangels and sugar beet, and the three crops share many characteristics. Both fodder beet and mangels are capable of giving high yields of digestible dry matter suitable for feeding (after lifting) to ruminants; the roots store well thus providing winter feed. Mangels provide 10–15% dry matter, and fodder beet 15–20% dry matter, depending on variety, from yields typically up to 120 t/ha. Thus up to 12 t DM/ha can be expected from these roots, with fodder beets also supplying up to 5 t DM/ha from the tops.

All types require the warm, sunny conditions more typically found in southern England, and well drained, highly fertile soils. Heavy soils make establishment and harvesting difficult. The use of monogerm seed eliminates the need for hand thinning; all fodder beet varieties are available in this form. A fine firm seedbed should be prepared to drill in late April or as soon as conditions are suitable; variety susceptibility to bolting should be taken into account with respect to sowing date. Seed should be sown at 12–18 mm into moisture, at a row spacing of 50–60 cm (adapted to suit harvesting machinery), to achieve a plant population of 7–10 plants/m^2. Precision drilling will give the even stand needed to reach yield potential and ease machine harvesting. Once established they are relatively drought tolerant. Flea beetle can cause serious damage to young seedlings, but beet cyst eelworm can be avoided by the use of sound rotations. Aphids should be controlled, particularly where there is risk of transmission of virus yellows. Varietal resistance should be used to counter powdery mildew (*Erysiphe betae*) and rust (*Uromyces betae*). Nutrient requirements are given in Table 10.30.

Lifting is usually in October or November (before severe frosts), when the outer leaves begin to wither, using modified swede or sugar beet harvester. However, hand harvesting is possible for most types except those whose roots sit well down in the ground (usually high dry matter fodder beet varieties). Only fully mature roots will store well, and damage should be avoided during lifting and handling; low dry matter varieties are more susceptible to damage. Depending on conditions some cleaning may be necessary as dirt will cause feeding problems. Some varieties are more prone to fanging and thus a higher potential dirt tare. Roots can be stored with protection from frost, e.g. in earthed-up clamps, for several months. Fodder beet types produce a useful amount of tops, which should be kept clean during lifting, and wilted or ensiled before feeding.

Swedes (*Brassica napus*)

Swedes are a useful frost-hardy fodder crop yielding storable roots of 8–12% dry matter (depending on type), from a fresh yield of up to 60 t/ha. Green-skinned varieties have higher dry matter content and can be stored for longer than the purple- and bronze-skinned types; the former are more popular in the north and Scotland. Some varieties are dual-purpose, being suitable for fodder and the culinary market (see under 'vegetable crops' in this chapter for swede production for human consumption).

Table 10.30 Nutrient requirements of fodder beet, mangels and swedes (kg/ha). (m) indicates maintenance dressing (MAFF, 1994)

	Soil index			
	0	*1*	*2*	*3*
Fodder beet and mangels				
N	125	100	75	
P_2O_5	100	75	50(m)	50(m)
K_2O*	200	100	75	75
Swedes				
N	100	50	25	
P_2O_5	150	100	50	50
K_2O	150	125	100	60

* Increase by 150 kg/ha if tops are carted off. For all soils except peat or fen silts, also apply 150 kg/ha sodium (as 400 kg/ha salt) well before drilling or increase potash by 100 kg/ha.

Crops are tolerant of a wide range of soils but they should be well drained and moisture retentive with a pH of 6.0–6.5. There is some varietal resistance to clubroot and sound rotations should complement this. They are best suited to cooler, moister parts of the UK. Seed is drilled from April to June for utilization from November until March, either *in situ* or after lifting and storing. Sowing date and utilization will be strongly influenced by the particular farm system, but swedes tend to be sown earlier in the north as a break crop and later in the south as a catch crop (to avoid powdery mildew and cabbage root fly attack). Dressed seed should be drilled 10–20 mm deep into a fine firm seedbed at a row spacing of approx. 50 cm to give a population of 8–15 plant/m^2. A higher density gives smaller roots which are easier to feed and store. A residual herbicide is recommended, and flea beetles may need control. Powdery mildew (*Erysiphe cruciferarum*) can be a problem in dry conditions, and control may be economic if young plants are infected. It is not generally economic to control pests after establishment in swedes for fodder; seed dressings should be used to prevent problems as far as possible. Swedes are responsive to P and K; excess N reduces root yield by encouraging top growth, and roots are more tender and store less well. Nutrient requirements are given in Table 10.30.

Leafy brassicas

Leafy brassica forage crops should not be included in rotations with brassica cash crops such as oilseed rape or vegetables due to the potential for build-up of soil-borne diseases, notably clubroot. Powdery mildew (*Erysiphe cruciferarum) and Alternaria brassicae* are common to them all, the former particularly in dry weather.

Kale (*Brassica oleracea*)

Kale is a suitable crop for all but exposed sites, being tolerant of soil type and typically yielding up to 7 t DM/ha at 15% dry matter content. It can be sown from April to June depending on type and target date for utilization (*in situ* or zero grazed): marrowstem types are sown early for utilization before Christmas (not frost hardy) when the thick fleshy stems are palatable and provide the bulk of the yield; thousand-head types (tall and dwarf types available) are sown late for utilization of the leaves and sideshoots from Christmas to March (frost hardy); hybrid types have the advantages of both with thick high-yielding stems that remain palatable, and most varieties are frost hardy, allowing utilization from successional sowings from August to March (however, crops become less hardy with maturity and heavy N use). The crop is tolerant of soil type but heavy or poorly drained land may create utilization problems. It may be direct drilled or sown into a fine moist seedbed at 2.5–5.0 kg/ha broadcast (1.5 kg/ha precision drilled). Nutrient requirements are given in Table 10.31.

Forage rape (*Brassica napus*)

Forage rape is a useful green forage catch crop suited to western and northern moister areas, providing feed until heavy frosts are experienced. It may be sown from mid-April to August, into a fine firm seedbed. Optimum soil pH is 6.0–6.5, at 3.5–6.5 kg/ha (10 kg/ha broadcast). It is usually fed *in situ* to sheep from about 10 weeks after

Table 10.31 Nutrient requirements of kale, forage rape and quick-growing turnips (kg/ha). (m) indicates maintenance dressing (MAFF, 1994)

	Soil index 0	1	2	3
Kale				
N	125	100	75	
P_2O_5	100	75	25(m)	0
K_2O	100	75	50(m)	50(m)
Forage rape and stubble turnips				
N	100	75	50	
P_2O_5	75	50	25(m)	0
K_2O	100	75	50(m)	50(m)

sowing, during the period from July until December, and is excellent for fattening lambs (care should be taken to introduce gradually, with grass or other roughage available to avoid gorging). Forage rape can yield up to 4.5 t DM/ha at about 12% dry matter. Nutrient requirements are given in Table 10.31. (Hungry Gap kale and rape kale are also rapes but with much greater frost hardiness than forage rape varieties, thus they are sown into cereal stubbles for utilization in March/April.)

Quick-growing 'continental/stubble' turnips (*Brassica campestris*)

These are quick-growing (ready for grazing 8–12 weeks after sowing) white-fleshed turnips grown primarily for the leaf as a forage crop, and should be distinguished from the longer season white- and yellow-fleshed turnips grown for their roots. Seed can be sown between April and August, for summer or autumn grazing, at 3–6 kg/ha (6–8 kg/ha broadcast). Yield will be up to 3.5 t DM/ha at 7–10% dry matter. Nutrient requirements are given in Table 10.31.

Fibre, fuel and other non-food crops

Non-food crops offer arable diversification possibilities with good potential for expanding markets in contrast to the food crops where, generally, demand is inelastic and market size is unlikely to grow within the EU. Many can also take advantage of being grown on set-aside land as industrial crops though the financial benefit is not necessarily greater as a result of this as market values for most non-food crops are low relative to the currently supported food crops. The future is likely to see an expansion of non-food crop area, and varieties will be 'designed' to give unique composition profiles for a wide range of non-food uses from fuel to medicines. Wide-scale viability awaits the appropriate infrastructure and plant breeding advances.

Fibre crops

The fibre-producing crops hemp and flax are supported by a CAP area payment (independent of the Arable Area Payments Scheme), or may be grown as an 'industrial' crop on set-aside land. The fibre extracted from these crops is used as the raw material for a range of paper and textile products. Contracts are essential for the successful marketing of both crops.

Hemp (*Cannabis sativa*)

Hemp was grown widely in Britain for centuries but production ceased in the 1950s when its use as a drug became illegal. Low THC varieties are now available and their production is permitted under Home Office licence. Outlets are limited at present but are likely to expand.

The crop is easy to grow and requires very few inputs, but areas with low rainfall at harvest (late summer/autumn) are favoured due to the importance of drying to prevent deterioration during storage.

Approved variety seed is sown at a rate of 80–90 kg/ha and a depth of 3 cm in spring as soon as soil conditions allow and there is no risk of frost. A firm fine seedbed is preferable for good establishment. Appropriate weed control may be necessary until a full canopy is produced. The crop can reach a height of 2–3 m and should be swathed, dried and baled in warm dry conditions. Deterioration of the fibre will occur if the bales are not dry. Yields of approx. 25 t/ha can be expected, but may be lower in dull years or for late-sown crops. Good drying, bulk handling, and storage facilities are vital.

Flax (*Linum usitatissimum*)

Flax is the same species as linseed and shares many of its characteristics though each crop has varieties with appropriate features. For example, flax varieties are taller with fewer capsules and seeds than linseed varieties. Seeds from flax plants grown for fibre are crushed for oil and contribute significantly to the EU crush, and 'dual purpose' varieties also exit (fibre can be obtained from linseed stems but is of poor quality, particularly with respect to fibre length).

Flax is more demanding than linseed due largely to the higher plant density needed (up to 1800 plants/m^2) to produce the long fine stems for good fibre quality. The higher density leads to a greater sensitivity to low soil moisture (due to reduced root development), and a greater vulnerability to disease (particularly stem diseases); very low rates of nitrogen are used to avoid the production of coarse fibre, typically 20 kg/ha applied in the seedbed.

Energy crops

The production of crops for energy has the advantage of creating a renewable fuel source which could use land surplus to requirements for food production, as well as providing a greater diversity of crop enterprises; the chief disadvantage of crops as a renewable energy source is their relative bulkiness per energy unit, making it necessary for processing to be carried out near the site of production if the energy yield is not to be used up for transport purposes. Such regional localized processing facilities are not currently in place in the UK.

In theory any crop can be a source of energy for heating or electricity/power generation since all crop dry matter is stored energy. However, an economically viable energy crop needs the following characteristics: to be an efficient converter of solar radiation into dry matter, and to yield close to the theoretical dry matter maximum for the UK; to be harvested at low moisture content; to be hazard free; and to have a low relative cost of production.

Perennial crops capable of efficient radiation interception year round and easily harvested could theoretically provide 50–60 t/ha dry matter (equivalent to about 30 kW/ha/year). In practice such crops do not yet exist but both short-rotation coppice (willow, poplar) and *Miscanthus* have shown potential in trials. Annual arable crops cannot yield so highly but are more flexible for the arable producer. At present only oilseed rape is grown in the UK in this way, for the production of biodiesel (rape methyl ester), by a small number of producers under contract. Bioethanol can be produced from cereals, sugar beet etc. and trials are underway in other EU countries. The expansion of the market for energy crops in the UK awaits political commitment, industrial investment and crop research and development.

Medicine, flavouring, and other minor crops

The production area needed to supply the market from such crops as evening primrose (*Oenothera biennis*) and borage (*Borago officinalis*) – both grown for the extraction of gamma-linolenic acid – and condiment mustard (*Brassica juncea*, *Sinapsis alba*) is small, and the associated lucrative contracts are filled by a few favoured growers. However, there are potential niche markets for a wide range of crops in this general category for the entrepreneurial producer to develop; in the future many producers may become niche producers as genetic transformation technology enables the plant breeders to develop new varieties with characteristics tailored to very specific end uses, for example antibodies.

Acknowledgements

The authors gratefully acknowledge the help and assistance given by Mr R.D. Toosey in the form of Chapter 5 on Arable Crops in the previous edition of *The Agricultural Notebook*.

References

Lampkin, N. (1990) *Organic Farming*. Farming Press, Ipswich.
Lidgate, H.J. (1981) Matching nitrogen with cereal needs. Marketable yield of cereals. Course Papers 1981.
MAFF (1994) *Fertiliser Recommendations* (6th edn). Reference Book 209. HMSO, London.
NIAB (1994) *Recommended List of Grasses and Herbage Legumes*. National Institute of Agricultural Botany, Cambridge.
Tottman, D.R. (1987) The decimal code for the growth stages of cereals, with illustrations. *Annals of Applied Biology*, **110**, 441–454.

Bibliography and further reading

The publications marked with an asterisk are published annually.

ADAS (1982) *Cauliflowers*. MAFF Reference Book **131**. HMSO, London.
*AFRC *Institute of Arable Crops Research Annual Report*. AFRC, London.
Anon. (1993) *NFU/Retailer ICM Protocol – Fresh Cauliflower*. NFU, London.
Anon. (1994) *NFU/Retailer ICM Protocol – Fresh Carrots*. NFU, London.
BASF (1989) *The Guide to Maximising Profits, 1989/90. Winter Crops*. BASF, Hadleigh, Suffolk.
BASF (1990) *The Guide to Maximising Profits, 1990/91. Spring Crops*. BASF, Hadleigh, Suffolk.
BASF (1991) *The Guide to Maximising Profits, 1991/92. Winter Crops*. BASF, Hadleigh, Suffolk.
BASF (1992) *The Guide to Maximising Profits, 1992/93 Spring Crops*. BASF, Hadleigh, Suffolk.
*BCPC *Brighton Conference Proceedings*. BCPC, Farnham, Surrey.
Boatman, N. (1994) *Field Margins: Integrating Agriculture and Conservation*. BCPC Monograph 58.
Carr, S. & Bell, M. (1991) *Practical Conservation: Boundary Habitats*. Hodder & Stoughton, London.
Cooke, D.A. & Scott, R.K. (1993) *The Sugar Beet Crop*. Chapman & Hall, London.
Gill, N.T., Vear, K.C. & Barnard, D.J. (1980) *Agricultural Botany*. Duckworth, London.
Harris, P.M. (1992) *The Potato Crop*. Chapman & Hall, London.
Hebblethwaite, P.D. (1983) *The Faba Bean*. Butterworths, London.
Hebblethwaite, P.D., Heath, M.C. & Dawkins, T.C.K. (1985) *The Pea Crop*. Butterworths, London.
*Ivens, G.W. *The UK Pesticide Guide*. CAB/BCPC, Farnham, Surrey.
LEAF (1994) *A Practical Guide to Integrated Crop Management*. RASE, Stoneleigh.
*MAFF *Agriculture in the United Kingdom*. HMSO, London.
*MAFF *Arable Area Payments. Explanatory Guide*. HMSO, London.
MAFF (1984) *The Agricultural Climate of England and Wales*. HMSO, London.
Milford, G.F.J. *et al.* (1990) *Cereal Quality II. Aspects of Applied Biology* **25**.
Milford, G.F.J., Evans, E.J. (1991) Factors causing variation in glucosinolates in oilseed rape. *Outlook on Agriculture* **20**, 31–37.
NAC (1984) *Marketable Yield of Cereals. Course Papers 1984*. RASE, Stoneleigh, Warcs.
NAC (1985) *Marketable Yield of Cereals. Course Papers 1985*. RASE, Stoneleigh, Warcs.
NIAB *Recommended Variety Lists*. National Institute of Agricultural Botany, Cambridge.
*Nix, J. *Farm Management Pocketbook*. Wye College, Ashford, Kent.
Scarisbrick, D.H., Atkinson, L. & Asare, E. (1989) Oilseed rape. *Outlook on Agriculture* **18**, 152–159.
Sheldrick, R., Thompson, D. & Newman, G. (1987) *Legumes for Milk and Meat*. Chalcombe Publications, Marlow.
Turner, J. (1987) *Linseed Law*. BASF, Hadleigh, Suffolk.
Wibberley, E.J. (1989) *Cereal Husbandry*. Farming Press, Ipswich.

11

Grassland

J.S. Brockman

Grassland occupies a major part of the agricultural land of the UK, as shown in Table 11.1.

Rough grazing

Rough grazing can be defined as uncultivated grassland found as unenclosed or relatively large enclosures on hills, uplands, moorland, heaths and downlands. Thirty-three per cent of all agricultural land is in rough grazing, and nearly two-thirds of this area is in Scotland. As shown in Table 11.2, the contribution of the rough grazing area to national ruminant nutrition is only 5% and this low figure indicates the low level of production associated with rough grazing. The natural vegetation has low yield and nutritional characteristics: the period of summer growth and the total yield are restricted by a combination of altitude, poor soil fertility, high rainfall, difficult access and the consequent poor control of defoliation management.

The main types of rough grazing are as follows.

Mountain grazing

This is land over 615 m in altitude, where high rainfall, low soil temperatures and extreme acidity restrict the plant species that will grow and limit their growth. Winters are very cold in these areas, farms are isolated and the stock-carrying capacity of the land is very low.

Table 11.2 Estimated contribution of grass and forage to total ME requirements of UK ruminants

	Overall %	*Estimated for:* *dairying*	*beef*	*sheep*
From grassland	60			
From rough grazing	5	65	80	90
From forage crops	5			
From other foods	30	35	20	10

Moors and heaths

Moors and heaths are found at lower altitudes and are often dominated by heathers (*Calluna* spp) in relatively dry areas and purple moor grass (*Molinia caerulea*) or moor mat grass (*Nardus stricta*) in wet areas. Heather can be utilized by sheep and these areas can be improved by liming, selective burning and application of phosphates; grass will be encouraged, often the indigenous sheep's fescue (*Festuca ovina*).

Stock will eat moor mat grass in spring, but as the season progresses it becomes unpalatable, especially to sheep. Most natural species occur in patches and clumps, their occurrence reflecting the environment and the selective grazing of animals. Normally cattle will graze moor mat grass and purple moor grass more readily than sheep, and some improvement in overall production can be encouraged by regular grazing together with the

Table 11.1 Grassland in the UK, 1989 (10^6 ha)

	England and Wales	*Scotland*	*N. Ireland*	*Total UK*	%		
Grass							
rough grazing	1.7	4.0	0.2	5.9	33		71
over five years	4.0	0.7	0.6	5.3	30	38	
under five years	0.9	0.4	0.2	1.5	8		
Cereals	3.3	0.5	0.05	3.9	22		
All other crops	1.2	0.1	0.02	1.3	7		
Total	11.1	5.7	1.07	17.9	100		

use of lime and phosphate. Under these circumstances, encouragement will be given to bents (*Agrostis* spp), sheep's fescue and red fescue (*Festuca rubra*). Although these three grass species have a low production potential compared with grass species that are sown in lowlands, they have a greater value than *Molinia* and *Nardus*.

Some rough grazings occur on lime-rich soils, such as the downs in the south of England where the soil is too shallow or too steep for cultivations: here sheep's fescue and red fescue dominate the sward.

Table 11.3 Classification of permanent grass on botanical composition

	% of fields surveyed containing (*Baker, 1962*)	(*Green, 1974*)
15% or more of PRG	20	26
Mainly *Agrostis*	59	56
Fescue dominant	7	10
Others	14	8

Permanent grassland

Permanent grassland can be defined as grassland in fields or relatively small enclosures and not in an arable rotation; 30% of all agricultural land is in permanent grassland and four-fifths of this area is in England and Wales. The 5.3 million ha of permanent grassland represent many different types of grassland, and, as the above definition suggests, much of the area is on land not suited to arable cropping.

It has been estimated that 30% of permanent grass in the UK suffers from serious physical limitations that impede use of machines: the main factors are rough and steep terrain, stones and boulders on the surface and very poor drainage. Farmers can improve drainage, and such a step can lead to a marked increase in output from this type of land. Much useful information on permanent grass is given by Hopkins & Green (1978).

Several surveys have classified permanent grass on the basis of its botanical composition, an account of the classification being given by Davies (1960). In simple terms, swards can be placed in order of probable agricultural value as follows (bearing in mind that grassland is rarely homogeneous in composition, and most natural species occur in irregular patches in the field).

Rough grazing

poor quality

- heather (*Molinia, Nardus*);
- areas with red fescue;
- *Agrostis* with red fescue (with rushes and sedges on wet soils).

Permanent grass

- mainly *Agrostis*;
- *Agrostis* with a little perennial ryegrass (PRG) (less than 15% PRG, often with rough-stalked meadow grass, meadow foxtail, cocksfoot and timothy);
- second-grade PRG (15–30% PRG with *Agrostis*) (often with crested dogstail, sheep's fescue, cocksfoot and tufted hair grass);
- first-grade PRG (over 30% PRG) (often with white clover, rough-stalked meadow grass, *Agrostis*, cocksfoot, sheep's fescue and timothy).

↓ good quality

Table 11.3 gives a summary of surveys conducted by Baker (1962) and Green (1974). It shows that *Agrostis* dominates the permanent grasslands of the UK, although the proportion of ryegrass swards increased slightly from 1962 to 1973; for further information see Hopkins (1979). From 1971 to 1980 a special Permanent Pasture Group studied permanent grassland (Forbes *et al.*, 1980). Subsequently, work on this subject has been based at the Institute for Grassland and Environmental Research, North Wyke, Okehampton, Devon.

Rotational grass

Rotational grass (temporary grass) can be defined as grass within an arable rotation. Eight per cent of all agricultural land is in rotational grass and it has one or more definite functions, quite apart from its role in supporting an animal enterprise. These functions include:

- being an essential part of rotation to
  - break disease and/or pest cycle
  - improve fertility
  - break weed cycle
  - improve soil structure.
- enabling the growth of grass species best suited to animal enterprise, giving
  - better fertilizer response than indigenous grass
  - high yields of grass and animal product
  - good seasonal distribution of grass production.

It is estimated that about 600 000 ha of grassland are sown annually, of which 450 000 ha are in rotational systems and 150 000 ha are sown after grass as a probable method of grassland improvement.

Distribution and purpose of grassland in the UK

Grassland is not distributed evenly over the UK, although every county within the UK contains a good proportion of farmed grass. For example, predominantly arable counties such as Norfolk and Kent have 20% and 40% respectively of their agricultural land in grass. Most grassland occurs in the western half of Great Britain and in Northern Ireland, where rainfall is high and the soils are heavier, and so less suited to arable cropping, than in the east. (For more details of crop distribution see Coppock, 1976.)

In some mainly arable counties grass is grown either as a break from regular arable cropping or on land that is unsuited for other crops. In the truly grassland areas, grass is grown as a highly specialized crop, often on farms growing little but grass.

Most of the grass grown in the UK is eaten by the ruminant animals, dairy cows, beef cattle and sheep, and the proportion of total individual farm income derived from these three animal enterprises varies considerably, ranging from 100% down to a very low proportion indeed. It follows that there is a great diversity in the way in which grass is farmed. This chapter outlines general principles that underlie economic grassland production, irrespective of the type of grassland farming being practised.

Use of grass

Grass as such is rarely a cash crop, unless sold as hay, dried grass, or big-bale silage. On nearly every farm, grass is processed by ruminant animals which are adapted to digesting the cellulose found in grass and also synthesizing proteins from simple nitrogen compounds (see Chapter 15). Whilst it is possible to measure grassland production as weight of grass produced (and research results are often expressed in this way), it is important to realize that growing grass is only a part of the production process. Apart from producing grass, the farmer is concerned about the proportion of the grass that is grown which is eaten by stock, i.e. its degree of utilization. Also, ruminants are fed material of non-grass origin, e.g. cereals. Figure 11.1 illustrates the essential steps.

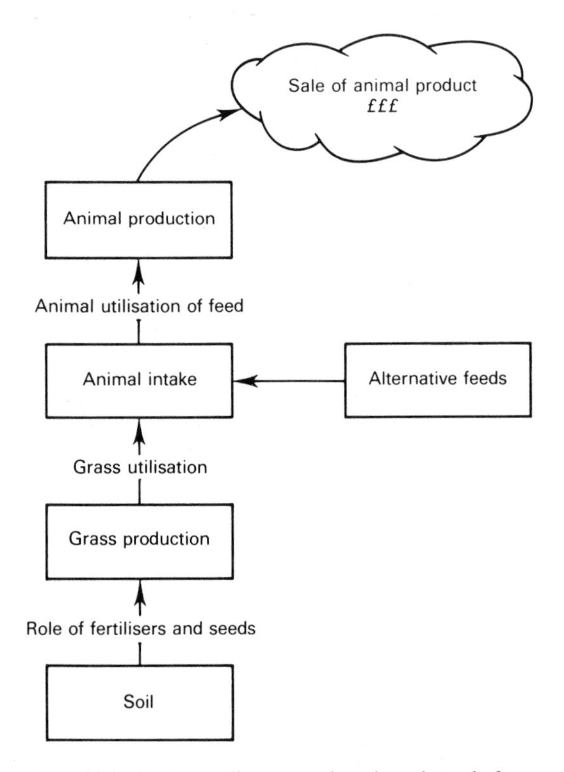

Fig. 11.1 Factors affecting cash value derived from grassland.

Although the ruminant can extract most of the energy and protein it requires from grass, it has been estimated that in the UK some 30% of the energy used by ruminants comes from cereal feeds, as shown in Table 11.2. In pastoral countries, such as New Zealand, where animal products have a low cash value, virtually all ruminant production is derived from grass: this illustrates that the relative proportion of grass and other feeds in the ruminant diet is related to the economic value of the animal product. Thus it is possible that grass could contribute more significantly to ruminant nutrition in the UK in the future; such a move would depend on improvements in grass production and utilization, or other feeds becoming relatively more expensive. On the other hand it would be possible for grass to become even less significant than now if alternative feeds became more attractive – either in monetary terms or from the point of view of convenience to grow and feed.

Grassland improvement

The botanical composition of permanent grass and older rotational grass is related to the nature of the environmental factors operating there. The main constraints are as follows.

Within farmers' control
Drainage
Compaction
Acidity
Low fertility
Weed problems
Low stocking rate
Poor grazing control

Outside farmers' control
Steepness
Altitude
Climate
Rough or stony terrain
Aspect

If grassland improvement is being considered, the above factors can be used as a checklist. In fields where drainage, acidity and low fertility are not major problems, grassland production can often be improved by better defoliation management. The following section gives a brief account of basic grass physiology and shows how the farmer's management of grass can have a marked influence on the type of sward produced.

Nature of grass growth

Figure 11.2 illustrates a plant of perennial ryegrass (PRG) in a vegetative state. All the while the plant is not producing seed heads, the region of active plant growth (or stem apex) remains close to the ground and below cutting or grazing height. Growth of the plant continues in two possible ways:

- Increase in size of existing tillers. As each leaf grows (Fig. 11.3), cells close to the stem apex divide and

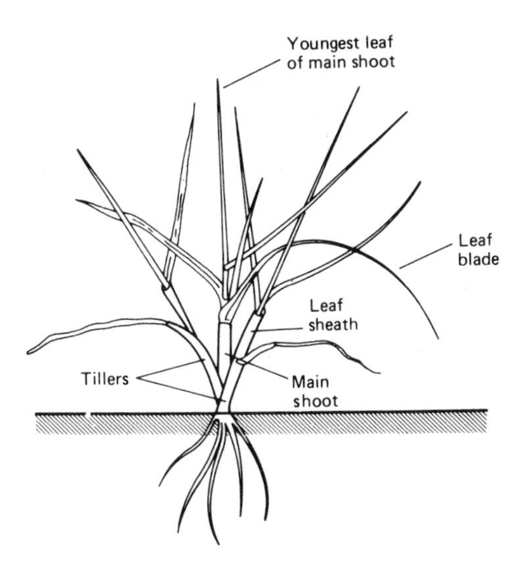

Fig. 11.2 Vegetative development in grasses.

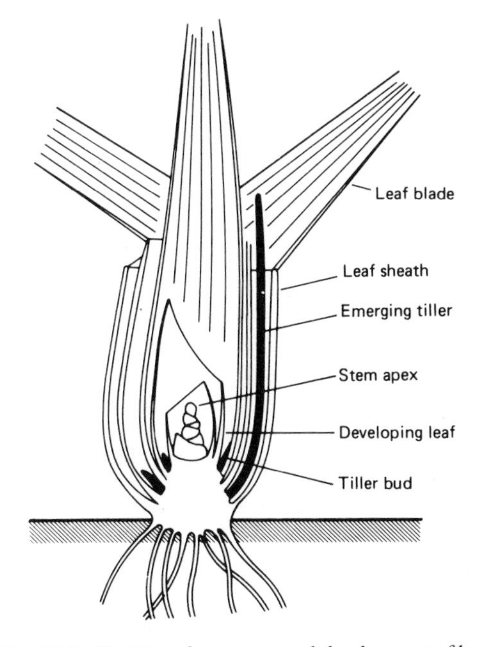

Fig. 11.3 Position of stem apex and development of leaves and tillers in grass.

elongate so that the youngest part of the leaf is at its base and the oldest at the tip. As each leaf develops, its base surrounds the apex, so the stem apex is continually sheated by developing leaves. This method of vegetative growth maintains the stem apex at ground level and protects it from damage caused when the oldest part of the leaves may be removed during cutting or grazing.

- Tiller production. Each tiller bud contains a replica stem apex and further tillers will develop as the plant grows, giving rise to vegetative reproduction. These young tillers are also well protected from damage and rapidly develop a root system and an identity of their own – to further continue the process of vegetative growth and development.

When grasses have the ability to tiller freely and are encouraged to do so, they can spread laterally over the ground surface and form a thick or 'dense' sward, characterized by a large number of grass tillers per unit area.

During inflorescence development (sexual reproduction) the pattern of grass growth is quite different in two important ways. When inflorescence development is triggered, by a mechanism based on day length (and slightly modified by temperature), two changes alter the normal vegetative growth:

- New tiller development is suppressed and existing tiller buds start to form the inflorescence.
- The stems of these inflorescence buds elongate rapidly, to carry the developing ear well above the ground.

When ear-bearing stems are defoliated, they are incapable of regrowth as the inflorescence was their apex, and plant regrowth can only occur from further developing tiller buds at the base of the plant.

When a grass plant bears visible ears, much of its impetus for vegetative growth is lost, and the longer the plant is left bearing ears, the slower will be the subsequent development of vegetative organs. Thus removal of a heavy, tall grass crop with ears emerged will leave a thin or 'lax' sward, with few tillers per unit area and a slow propensity to regrow. However, if regular frequent defoliation is then practised, the sward will become more dense, particularly if there is a high population of PRG, as this species tillers freely in the vegetative state. The effect of cutting date is well illustrated in Table 11.4, where a delay in taking the first silage cut gave a high silage yield but seriously delayed speed of regrowth.

Tillers and plants are in a state of continual change, the life of an undefoliated leaf in PRG being about 35 d in summer and up to 75 d in winter. If grass is left undefoliated, there is a rapid burst of vegetative (tiller) development in spring which is suppressed in May/June by tall inflorescence-bearing stems that restrict further vegetative development until autumn, followed by drastic tiller restriction during winter. Frequent defoliation

Table 11.4 Effect of first cut date on regrowth (S23/Melle)

Date cut	*DM (t/ha)*	*Delay in regrowth (d)*	*Days to grazing stage (2.5 t/ha DM)*	*Total yield up to 20 July (t/ha)*
20 May	4.2	8	32	9.7
30 May	5.4	11	35	9.5
10 June	6.6	15	39	9.2
20 June	7.8	20	44	8.9

Based on Wolton (1980).

in spring will restrict ear development, maintain tiller formation and lead to a dense sward. On soils of low to medium fertility, application of fertilizer N will further encourage overall plant growth and tiller production.

The timing and frequency of defoliations will have a marked influence on the seasonal distribution of grass growth, as shown in Fig. 11.4. It is important to realize that there is always an uneven pattern of grass growth during the growing season, but this pattern is exaggerated when spring defoliations are infrequent.

Most varieties of PRG can tiller freely compared with less desirable 'weed' species of grass, so that regular frequent defoliation enables PRG to compete favourably for space in the sward and become dominant. Also most varieties of PRG are comparatively prostrate in growth habit compared with weed grasses, and so defoliation close to the ground will further encourage development of a PRG dominant sward.

The practical importance of the basic physiology described above was demonstrated 60 years ago in classic experiments by Jones (1933). He showed on a relatively fertile lowland site that, provided there was at least 15% PRG in the sward, an *Agrostis* dominant sward could become PRG dominant in two years if subjected to regular, tight grazing. Similarly, he showed that a newly sown PRG sward would rapidly degenerate in composition if subjected to irregular high defoliation.

Some other plants, such as white clover and creeping bent (*Agrostis stolonifera*), can propagate vegetatively by stolons: under some circumstances close defoliation will favour these plants as well.

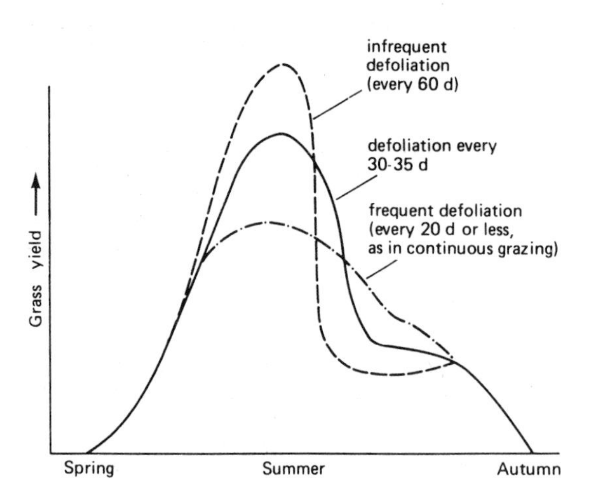

Fig. 11.4 Effect of defoliation frequency on seasonal distribution.

Why improve grassland?

Any method used to improve the botanical composition of grassland will cost something, and clearly it is of no economic value to improve for cosmetic reasons; improvement must lead to an increased profit. On most farms it has been found that a combination of some of the following factors are most likely to lead to profitable improvement;

- increased output (usually from a higher stocking rate being carried *or* higher yields of silage or hay);
- improved herbage quality giving better animal production;
- longer growing season;
- quicker regrowth;
- bigger response to fertilizer N;
- more palatable grass;
- repair of areas damaged by poaching or drought.

Methods of improvement

The farmer has three improvement strategies available:

- Reseed – plough or chemically destroy the previous crop and reseed.
- Renovate – introduce new species into the old sward, with or without partial chemical destruction.
- Retain – improved defoliation technique, often with attention to drainage, pH and fertilizers.

Which improvement method?

	Reseed	*Renovate*	*Retain*
Speed of action	very rapid	quick	slow
Cost	expensive	moderate	very cheap
Stability	can revert easily	adjustable to acceptable level	improvement based on farmer's ability
Special considerations	other limitations must be corrected	topography may limit	no real physical limits

Methods of grassland establishment

Grassland is established by sowing seeds either as part of a replacement/renovation procedure where grass follows grass or as part of a rotation involving other crops. Time of sowing will vary, but there are two six-week periods in the year when sowing should take place. These are: mid-March–April *and* August–mid-September.

Sowing outside these periods is risky and can be unsuccessful due to summer drought or winter cold. The timing and method of sowing the seeds will be influenced by the purpose of the crop, soil texture and structure, climate and farm type. Where good white clover establishment is desired, then sowing should be either in spring or in early August, as sowing later will limit seriously the vigour of the clover plants and their ability to compete with grass. Whatever method of sowing is used, it must be remembered that the grass seed has a relatively small food reserve and so must have an adequate supply of available nutrients near the surface of the soil. The biggest single demand is for phosphate, and except on high phosphate soils a fertilizer supplying 80 kg/ha of P_2O_5 should be applied at seeding time. Also grass seeds should be sown no deeper than 25 mm below the soil surface; where sowing is preceded by cultivations, care should be taken to ensure that the seedbed is level and fine. After sowing, grass establishment is enhanced greatly if the seedbed is well consolidated; the firmer the better, provided soil structure is not impaired or surface percolation rate reduced.

Undersowing

This can be a very successful technique provided a few simple rules are observed. The most important rule is to prevent the 'nurse' crop, usually spring cereals, from overshadowing the establishing grasses and thus depleting the density of the sward before the nurse crop is removed. Where cereals are undersown, an early-maturing, stiff-strawed variety should be used. Whilst farmers successfully undersow wheat, barley and oats, and the choice often depends on rotational considerations, barley is the most suitable, having short straw and an early harvest. Seedrate should not exceed 125 kg/ha, and to avoid lodging fertilizer N rate should be about two-thirds that which would have been used had the crop not been undersown. The seeds mixture may be broadcast or drilled, with drilling always being preferred if conditions are likely to be dry. The grasses may be sown immediately after the spring cereals are sown and the ground then harrowed and rolled: alternatively, some farmers prefer to wait until the cereal is established before sowing the grass seeds. If the cereals are sown in February, it is a good idea to delay sowing the grasses until March. When the cereals are taken through to harvest, the straw must be removed as soon as possible and often the grasses will benefit from a dressing of complete fertilizer, say 400 kg/ha of a 15:15:15 compound, followed by grazing three to four weeks later to encourage tillering.

Some farmers do not take the cereal to harvest, but cut the crop for silage when the grain is still soft: others grow cereal/forage pea mixtures for use in this manner. Removal of the crop at this earlier stage benefits grass establishment and is good practice, provided there is a use for the arable cereal silage that is made.

Whenever crops are undersown, it must be remembered that herbicide use on the cover crop may be restricted, particularly if the seeds mixture contains clover.

Seeding without a cover crop

Here the seed mixture is sown as a crop in its own right, the nature and timing of operations being adjusted to the needs of the grass crop. In the past this technique has been known as *direct seeding* but this term is not encouraged now as it can be confused with *direct drilling*.

Where grass is sown without a cover crop, establishment must be rapid both to minimize weed competition and reduce the period that land is unproductive. Thus adequate fertilizer must be applied to the seedbed; 80 kg/ha of both P_2O_5 and K_2O and 60–80 kg/ha of N if no clover is sown. If clover is in the mixture, it is advisable to use no fertilizer N in the seedbed otherwise clover establishment will be seriously impaired. An early grazing will benefit the establishment by encouraging tillering, but a September sowing should not be grazed before winter as it will drastically reduce the sward's ability to build up root reserves during the winter.

Cocksfoot and meadow fescue establish more slowly than ryegrass, with tall fescue the slowest of all. Seeding without a cover crop is best suited to ryegrass mixtures, particularly straight Italian and hybrid mixtures with no clover.

Direct drilling

Grass and clover seeds can be direct-drilled successfully provided:

- the drilling date is within the periods given in the preceding section;
- the previous crop has been thoroughly killed off using a herbicide such as paraquat (if annual weeds are present) or glyphosate (if perennial weeds exist). If there was a dense mat of previous vegetation and paraquat is used, then two half-rate sprays at three- to four-week intervals should be made. After spraying with either chemical at least two weeks should elapse before sowing;
- the seeds are placed 10–20 mm deep in mineral soil. Where there has been a previous mat of vegetation and/or where the ground is rough from poaching by grazing animals, it is not easy to ensure that all the seeds are sown where they are required. Direct-drilling grass into grass can result in failure, principal reasons being insufficient time between spraying, and sowing seeds into the surface mat and not into the soil proper.

Normal direct-drilling practice includes application of fertilizer at similar rates to the preceding section and use of 8 kg/ha of mini slug bait. Autumn sown crops are particularly susceptible to damge from frit fly and use of an insecticide is advised.

Slit-seeding

This technique is used for sward renovation, the principle being that at least some of the original sward is allowed to survive.

Three basic rules are:

- Ensure aggressive weeds in the old sward are killed (particularly stoloniferous species), otherwise they will 'strangle' the newly emerging young plants.
- Use the narrowest width of drill feasible and cross-drill if possible, to get the most rapid knitting together of the sown species.
- Sow at time of year to give seedlings ample time to grow.

If the old sward contains a reasonable proportion of perennial ryegrass (PRG) but is lax and beginning to fill with annual meadow grass (*Poa annua*), then slit-seeding may be the quickest and cheapest way of renovating the sward and boosting production. Also some farmers use this technique to introduce Italian ryegrass into old swards in an attempt to extend the grazing season and increase production. Furthermore, this method can be used to introduce white clover into an existing grass sward. Where there are no real problems from either grassy or broad-leaved weeds, then no herbicide treatment is needed. Otherwise, paraquat or glyphosate will have to be band-sprayed ahead of the slit-seeding of the grass.

There are several types of slit-seeder available, but the basic technique is to drill rows of grass seed into an existing sward with no previous cultivation. Some machines can apply a band of herbicide ahead of the drill; some cultivate the narrow band of the drill before the seed is sown; some can apply fertilizer and slug bait at the time of sowing.

The fertilizer rates and sowing periods are similar to the preceding sections. In some cases, slit-seeding may be tried on very old swards and on fields not suited to normal cultivations. It is essential to check soil pH and also ensure that bad drainage will not cancel out any good that may arise from sowing new grasses.

Scatter and tread

On an opportunist basis it is possible to renovate grassland by simply scattering the seeds on the soil surface under moist conditions. This should be followed immediately by turning in some grazing stock (preferably sheep) for 2–3 days to tread in the seeds. This technique is a gamble as its success depends on a thorough, quick tread and then some two weeks of continuing moist, but not wet, weather to allow the seeds to establish. The technique is cheap, and can be used to patch up worn areas in grazed paddocks. Also, some farmers report successful grass establishment following mixing of grass seeds in a slurry tanker and application with the slurry.

Steps to increase the white clover content of existing swards

Many swards contain little clover, and the following procedure, developed by research workers at the Institute of Grassland and Environmental Research (IGER), North Wyke, has been found to be an inexpensive way of increasing white clover content on suitable soils.

(1) Check soil suitability, particularly drainage and compaction; have soil analysed.
(2) Take silage (or hay) cut in June/July, having used a moderate N rate of up to 80 kg/ha.
(3) Unless sward is very open, spray paraquat at 1.5 litres/ha to kill 'bottom' grasses.
(4) If little clover is present, oversow or slot-seed white clover seed at 2 kg/ha; also apply slug pellets and Dursban as a precaution against frit-fly damage.
(5) If PK indices are under 3, apply 60 kg/ha of nutrient(s) in short supply. If pH is below 6, apply lime.
(6) Rotationally graze with cattle at approximately monthly intervals (if no cattle, use sheep). *Do not*:
    - use any fertilizer N;
    - apply any herbicides (including any 'clover-safe' ones);
    - set-stock.
(7) Next spring *apply no fertilizer N*. Ideally, take a low-yielding silage (or hay) cut. If this cannot be done, continue management (6).
(8) By August of year 2, the clover should occupy at least 40% of the surface area when viewed from above. If below 40%, check out what is wrong. If 40% or over, start to manage it as a *grass/clover sward*.

Characteristics of agricultural grasses

In this section, the important characteristics of the commonly used grass species are reviewed, as a precursor to information on the formulation of seeds mixtures. All the information on the performance of grass species and on seeds mixtures is based on information from NIAB and ADAS. An essential publication is NIAB Farmers' Leaflet No. 4 (revised annually).

Perennial ryegrass (PRG)

Perennial ryegrass is very successful under British conditions; not only do PRG varieties account for over half the grass seeds sown, but also the best permanent pastures have an appreciable PRG content. Provided they are defoliated regularly and are grown under conditions of medium-to-high fertility, they will give high yields of digestible herbage over a long season. PRG has the following valuable characteristics:

- tillers well and so forms dense swards under good management;
- fairly prostrate and ideal for grazing (but must be protected from prolonged overshadowing by taller species on cut swards);
- recovers well from defoliation, both because of high tiller numbers and good rate of replacement of root reserves;
- persists well and is suited to long-term grassland.

PRG is classified into three main groups, based on date of ear emergence in spring. The following heading dates are for central-southern England:

Very early and early group	Up to 20 May
Intermediate group	21 May–4 June
Late and very late group	5 June onwards

These heading dates are approximately 12 d later in northern Scotland: if a ruler is placed from Swindon to Aberdeen, then each twelfth of the distance represents one day later in heading.

In general, when early heading varieties of PRG are compared with late heading varieties, the early heading varieties:

- grow earlier in spring, but not as much earlier as the difference in heading date: for example a three week earlier heading date would give 7–10 d earlier spring growth;
- are more erect;
- tiller less freely;
- are easier to cut for conservation, both because of erect growth and good stem elongation at heading;
- do not grow well in mid-season.

Compared with early heading varieties, late heading varieties:

- are more prostrate;
- tiller well and are more persistent;
- give good mid-season growth (June–August).

PRG has a limited temperature tolerance compared with other grasses, being unable to withstand very hot summers where temperatures exceed 35°C (not found in the UK but can occur in some temperate Continental areas) or very cold winters where temperatures persist below about −10°C. Lack of winter hardiness is a problem with PRGs in the north of Scotland, where only the most winter-hardy varieties are recommended. Winter survival in PRGs is enhanced if little or no autumn nitrogen is applied and the plants are defoliated to at least 100 mm before the onset of winter.

Breeders have produced *tetraploid* varieties of PRG. Compared with diploid varieties of the *same heading date*, tetraploid varieties:

- appear higher yielding and may have a greater yield above, say, 80 mm, but total DM yields are generally the same. Cattle may be able to graze tetraploids more easily under some conditions;
- have a higher soluble carbohydrate content which may enhance palatability when grazed;
- are more erect, with leaf of higher moisture content; the higher moisture content can render them less suitable for conservation;
- tiller less freely and are less persistent;
- more winter hardy and more drought resistant;
- have larger seeds and so require a higher seed rate.

Italian ryegrass (IRG)

These are grasses with an expected duration of two to three years and heading dates in the period 10–30 May (in southern England). They are very erect, and although capable of tillering they rarely tiller enough to prevent a steady decline in sward density from establishment onwards. IRG will grow for a long season, starting growth in spring before PRG and continuing well into the autumn under conditions of high fertility. However, the bulk of growth from IRG is in the April–June period, making them particularly useful for early spring grazing and a heavy conservation cut. During the summer they tend to run into seedhead every 35–40 d and this can lower their feeding value at this time. Most varieties of IRG are even less winter hardy than PRG. The spring heading date of IRG coincides with the early group of PRG. IRG is specially useful as a rotation grass, where its two- to three-year duration is not a disadvantage. *Tetraploid* varieties are important, as tetraploidy emphasizes the natural advantages of IRG in terms of erectness, leaf size and soluble carbohydrate content.

Hybrid ryegrass (HRG)

During recent years plant breeders have combined the vigorous pattern of growth found in IRG with the greater tiller density and persistence found in PRG. The New Zealand hybrid Grassland Manawa (formerly known as H1) was available for many years, but was only suited to mild areas. Now there are a range of bred hybrids on the market, nearly all of which are tetraploid. They are best regarded as slightly more persistent types of IRG that are also rather more leafy in summer than IRG and can head in early June rather than May.

It is likely that other hybrid grasses, not necessarily involving a ryegrass parent, will become available as breeders perfect new techniques for developing useful crosses between grass species.

Westerwolds ryegrass

These are annuals. When sown in the spring or summer they flower in the year of sowing and do not persist over winter. They can be sown in autumn in mild areas and can then provide very early spring growth. They are only of real value where high yields of grass are required for a short period, and they are best regarded as catch crops, otherwise IRG should be used, as it will give equivalent yields and longer duration for a similar or lower seed cost.

Timothy

Timothy is a persistent and winter-hardy grass that will grow well in cooler, wetter parts of Britain. Also, it has survived well in drought years in other areas. Varieties of timothy head 1–25 June – as late as and later than the latest heading of the PRG varieties. Timothy is very palatable and grows well in summer and autumn, but when grown alone it does not give a dense sward or very high yields. It is an ideal companion grass in seed mixtures and historically was included with meadow fescue and clover. Today it is used with PRG and is found in most mixtures sown in cool, wet parts of the country. Timothy seed is some three times lighter than ryegrass seed, so its weight in a mixture is an underestimate of its real importance. Timothy is very winter hardy and is widely used in Scotland, where sometimes it is grown alone for heavy hay cuts followed by sheep grazing.

Cocksfoot

Cocksfoot is a native grass that is very responsive to nitrogen and, with a very well developed rooting system, is useful on light soils in low rainfall areas, where it shows drought resistance. It grows and regrows rapidly, is erect in habit and has relatively broad leaves for a grass; the leaves are also coarse and hairy. Cocksfoot heads in early to mid May and has an aggressive growth habit that makes it competitive in swards that are not closely defoliated during the main growing season. Over recent years cocksfoot has declined in popularity as it was found that its digestibility was below that of PRG and IRG. Breeders are producing new varieties that will correct this deficiency, and also are seeking to breed varieties that are less coarse and hairy, thus increasing its palatability.

Meadow fescue

This is a very adaptable grass that tolerates a wider range of climatic conditions than ryegrass. It heads mid to late May. It does not tiller as freely as PRG and so forms an open sward that is more prone to weed invasion, particularly as it is not aggressive. Meadow fescue makes an ideal companion grass in mixtures, for example with timothy and/or cocksfoot: also, it remains leafy during summer under conditions of fairly low fertility, unlike PRG which demands higher fertility if it is to remain leafy. As fertilizer use on grass has increased over the last 30 years, the importance of PRG has risen and the popularity of meadow fescue has declined, except where the winters are too severe for PRG – for example in Scandinavia. Historically, meadow fescue and timothy were the favoured grass companions to white clover. Meadow fescue could become more important again if there is a resurgence in the use of white clover, where timothy/meadow fescue/white clover swards might become very important.

Tall fescue

This is one of the earliest grasses to grow in spring. It is winter hardy, drought resistant and very persistent under regular cutting (which PRG is not). It is very slow to establish, taking up to one season to become fully productive. Also it has a very rough, stiff leaf that is extremely unpalatable to grazing stock, although this unpalatability does not apply to silage made from the grass. Tall fescue's main role is on grass drying farms in eastern England, although a few farmers on light land have used tall fescue in areas they cut regularly for silage.

Forage bromes

When regularly cut under some dry soil conditions, forage bromes can yield and persist well. Forage bromes should not be grown on heavy and badly drained soils and should not be grazed regularly. They do not tiller very well and are prone to weed invasion. *Bromus willdenowii* has outyielded IRG when cut for silage. *Bromus sitchensis* gives lower yields, but is leafier in summer and is to be preferred if some grazing is contemplated.

Herbage legumes

Legumes have root nodules that can be inhabited by rhizobial bacteria which can fix atmospheric nitrogen (N) and make it available to the host plant. Herbage legumes are grown either on their own ('straight') or mixed with grasses where their ability to acquired fixed N can also benefit associated grasses.

The important herbage legumes used in UK are as follows.

White clover (WC)

White clover is scarcely ever grown straight in the UK: even white clover seed is produced from suitable defoliation management on a grass/clover sward. Unlike red clover, white clover is persistent under grazing, often developing well when associated grass competition for light is reduced by constant defoliation. White clovers are classified according to their leaf size:

(1) *Small leaved* These are very prostrate and have been selected from cultivars existing in old pastures (e.g. Kent Wild White). They are very hardy, but always remain small even when under favourable growing conditions; thus they do not stand up to grass competition when fertilizer N is used or when a conservation cut is taken. Their real value is on sheep-grazed swards, conditions of low fertility and at high altitudes.
(2) *Medium leaved* Leaf size is not a consistent characteristic in the field: if a medium–large clover is used in a sward that is regularly and tightly grazed, its leaf size will be quite small. Also, there is a general tendency for persistence to decrease as leaf size increases.
(3) *Large leaved* In recent years breeders have produced bigger types of WC that can persist better under high N usage and in conserved swards. Even so, large-leaved WC does not persist for as many years as medium or small leaved WC at lower levels of N use.

It is important to select the most appropriate type of white clover for the expected management conditions: often it is advisable to sow a range of types in the mixture.

Red clover (RC)

There are three main types of red clover:

(1) *Single cut* Traditionally this was used for one hay cut in rotational farming where the clover break lasted one year. It is slow to grow in spring and lower yielding than other types.
(2) *Early red* (also Double-cut or Broad Red) These give early growth and high yields over two cuts and the possibility of autumn grazing. It is not very persistent, particularly when grazed regularly, and is best grown for two years only, either straight or in mixtures with erect high-yielding grasses such as IRG for cutting.
(3) *Late red* These flower two to three weeks later than Early Red and are more persistent, being suited to medium-term leys where a red clover constituent is required. Such leys should be cut periodically as red clover does not persist well if regularly defoliated more frequently than every 35 d.

Single cut clover is scarcely used now and NIAB do not recommend varieties. In both other groups there are recommendations for both diploid and tetraploid varieties. Tetraploid red clovers do not necessarily give a higher yield than diploid counterparts, and choice of variety can depend on selection for disease resistance. Red clover can suffer severely from two diseases:

- clover rot (*Sclerotinia trifoliorum*), and
- clover stem eelworm (*Ditylenchus dipsaci*).

Both diseases can be problems in areas where red clover has been commonly grown, e.g. in the arable areas of East Anglia. Where either disease is present, resistant varieties must be grown. Tetraploids give low seed yields and have bigger seeds than diploids, so tetraploid seed is expensive.

Alsike

This is rather more persistent than red clover under

grazing and is also more tolerant of wet acid conditions and more winter hardy and disease tolerant. It is often used with late red clover in mixtures.

Trifolium (crimson clover)

An annual, its main use is to provide a heavy crop of green material in May following late summer sowing. It can be sown into cereal stubbles, either straight, or with rye (corn) or IRG. Seed rates are 25 kg/ha if alone, or 10 kg/ha with 125 kg/ha of rye, or 15 kg/ha with 10 kg/ha IRG.

Trefoil (black medick)

Very useful for heavy yields of catch crop on thin calcareous soils. A mixture of 6 kg/ha with 20 kg/ha IRG sown in August will give excellent production the following spring for early grazing conservation.

Herbage seed production

The successful production of herbage seeds is very specialized and the majority of herbage seeds produced in UK are grown on large arable farms, on land that is relatively free from 'weed' grass species and with adequate seed harvesting and handling equipment. Grass and clover seeds are small, and yields low (see Table 11.5), so that attention to detail is important. Also, the management of the crop prior to harvest should be adjusted so that the number and size of fertile seed heads is maximized and the presence of lush, leafy growth is minimized.

Depending on the species and variety being grown, the date of harvest can vary from end-of-June (for early cocksfoot) to mid-September (for late timothy and some white clover crops). Thus the defoliation management in spring and early summer will depend on the date of expected harvest, the principal objective being to minimize leaf and maximize stem elongation and seed heads in the growth leading up to the cut.

Most seed crops are established specifically in the year prior to harvest. A firm, fine seedbed is essential and seeds are usually drilled. Weed control is very important and it is essential that no grasses develop other than the variety being grown. As there is a premium on a clean, vigorously growing crop, regular use of herbicides and application of high rates of seedbed P and K are advised. Also, spring-applied N rates of 100–150 kg/ha are often used, to encourage tillering and leaf growth, prior to defoliation and subsequent development of the seed-bearing stems.

Provided they remain free from other species and the yield potential is still high, seed crops may be harvested for at least two years (except IRG, HRG and red clover). In the years of seed production, there will be some grazing available from the field, and the threshed straw has a feed value equivalent to poor hay. Once seed production has ended, there is the possibility that the sward can continue as a normal grassland field.

It must be emphasized that herbage seed production is a very specialized business. Existing growers have built up their expertise for specific varieties and usually continue to grow these as long as there is a market for them. Normally, selection of new varieties by existing seed producers and the introduction of new seed growers is done through seed houses and merchants, who have the necessary seed stock, the market outlets and the necessary inspection procedures.

Basis for seeds mixtures

Seeds mixtures should be made up from specific varieties that are recommended for their individual merits. There are very many seeds mixtures available to farmers in the UK and the following outlines a way in which the suitability of mixtures can be examined and appropriate mixtures formulated. A golden rule is to keep the mixture as simple as possible commensurate with its function, also ensuring that sufficient of each ingredient is present to confer its expected characteristic.

Table 11.5 Some basic facts on herbage seed production

Species	*Type*	*Harvest period*	*Probable yield range (kg/ha)*
IRG		July–Aug	800–2000
HRG		July–Aug	800–2000
PRG	early	early July	750–1000
	intermediate	late July	800–1250
	late	up to mid Aug	600–1000
Timothy	early	mid Aug	375–650
	late	late Aug–mid Sept	375–650
Cocksfoot	early	late June–mid July	450–1000
	late	mid–July	450–1000
White clover	large	late July–mid Sept	350–450
	small	late July–mid Sept	150–350
Red clover	early	Aug	400–500
	late	Sept	350–500

Purpose of mixtures

Spread growth pattern

A single variety will exaggerate the spring peak in production, and use of several varieties with varying heading dates can spread this peak to some extent, e.g. a mixture of early, intermediate and late PRG.

Speed up regrowth

When grasses are defoliated for conservation, regrowth is slow from grass plants that were fully headed. If a mixture is used, it is possible that some grass plants are relatively vegetative and will give rapid growth from tillers when the crop is cut. However, the date of cutting such mixtures is critical, as if cutting is delayed, not only will the later heading varieties start to head, but also the feeding value of the early heading varieties will fall rapidly.

As a general rule:

(1) *for cutting areas*, have grasses with well-matched heading dates;
(2) *for grazing areas*, a greater spread of heading dates can give a greater spread of growth over the season than a single cultivar.

Easier grazing management

With a mixture the timing of grazing is less critical than with single-variety swards. Also, if a range of varietal types is present, there is a good chance that varieties well suited to the particular management will develop. However, there is always a danger of incursion of weed species if this management is too lax, or if some of the varieties in the mixture are too short-lived for the expected duration of the ley.

Cheaper seed cost per hectare

This is possible as merchants can buy seed under contract or in bulk. The individual components of a mixture should be checked, as sometimes mixtures include non-recommended and unsatisfactory varieties.

Lessen disease risk

Grasses can suffer from a number of diseases and pests such as crown rust and ryegrass mosaic virus, but so far research has not shown the economic merit of controlling these. It is likely that some varieties are less susceptible to pests and diseases than others, and a mixture spreads the risk.

Types of mixture

Mixtures can be classified into 12 possible groups, based on expected duration and intended use, as follows:

Duration		Use
very short term (1–2 years)	×	mainly for cutting
short term (2–3 years)		mainly for grazing
medium term (3–5 years)		general purpose
long term (over 5 years)		

Table 11.6 gives an outline of the basis for formulating seeds mixtures in ten of these groups and suggests the other two areas are not appropriate.

Seedrates

The correct seedrate will vary to some extent with the climatic and soil conditions at sowing and expected during the establishment period. The most common single problem is that seeds are sown too deep, either because of bad drilling practice or when broadcast on to a rough seedbed. Under ideal sowing conditions, experiments have shown that a PRG seedrate of 4 kg/ha will establish a good sward. Farmers always use much higher rates than this because field conditions are not ideal, but even so it should be remembered that although a very high seedrate may give a rapid green cover, it will not lead to higher yields of grass as plant competition kills out many of the young plants within three months of sowing.

The following rates are guidelines for seedrates of straight species under good seedbed conditions:

IRG (diploid)	35 kg/ha	PRG (diploid)	25 kg/ha
IRG (tetraploid)	40 kg/ha	PRG (tetraploid)	30 kg/ha
Meadow fescue	25 kg/ha	Cocksfoot	25 kg/ha
Timothy	15 kg/ha*	Tall fescue	40 kg/ha
White clover	2 (+ grass)	Red clover	10 (with IRG) 4 (in leys)

* But rarely sown alone.

For mixtures, the appropriate mean can be taken, for example:

20% of mixture tetraploid IRG,	20% of 40 =	8 IRG
60% of mixture diploid PRG,	60% of 25 =	15 PRG
20% of mixture timothy,	20% of 15 =	3 timothy
		26 kg/ha

For difficult seedbeds, add 20% to all the above seedrates.

Check-list for studying mixtures

The main points to note are:

(1) rate/ha of individual components, at least 4 kg/ha for grass (2 kg/ha for timothy), at least 2 kg/ha for clover;
(2) use of recommended variety; if not recommended, is there a reason? e.g. suited to locality, cheapness, personal preference;
(3) diploid or tetraploid, bearing in mind characteristics of each;
(4) heading dates, generally little spread for spring cutting swards;
(5) date at which the grass is at 67-D (note the silage made from this grass would have a D-value of 64–65);
(6) mid-seaon D-value;
(7) score for level of water-soluble carbohydrate in first growth;
(8) score for ground cover (related to persistence).

Table 11.7 illustrates the way in which a seeds mixture can be analysed. Points to note are:

Table 11.6 Basis for seeds mixtures

Main use	*Very short term (1–2 years)*	*Short term (2–3 years)*	*Medium term (3–5 years)*	*Long term (over 5 years)*
Cutting	35 of 1a (diploid) or 40 of 1a/b (tet) or 40 of Westerwolds or 25 of 1a/b + 8 of 8a	40 of 1a/b or 20 of 1a/b + 15 of 2a	10 of 1b + 10 of 2a + 10 of 2b or 10 of 2a + 10 of 2b + 5 of 2c + } dip only 2 of 7c*	Not applicable except 35 of 6a for grass drying
Grazing	40 of 1a/b (tet) or 20 of 1a/b + 15 of 2a (tet)	20 of 1a/b + 15 of 2a/b or 15 of 2a (tet) + 15 of 2b + 2 of 7c*	8 of 2a + 8 of 2b + 8 of 2c + } dip and tet 4 of 3a/b + 2 of 7b/c* or 20 of 5a + 5 of 3a/b + 2 of 7b/c*	12 of 2b + 12 of 2c + 4 of 3b + 2 of 7a/b* or 24 of 2c + 4 of 3b + 2 of 7b (some 7a for sheep)*
General purpose	Not appropriate	Not appropriate	8 of 2a + 8 of 2b + 8 of 2c + 4 of 3b + 2 of 7b/c* or 8 of 2a/b + 4 of 3b + 8 of 5a + (4 of 4a +) 2 of 7b/c + 2 of 8b	10 of 2b + 10 of 2c + 4 of 3b + 2 of 7a/b* or 10 of 2b/c + 4 of 3b + 4 of 5a + (4 of 4b +) 2 of 7b + 2 of 8b

All rates are in kg/ha: several varieties may be used within a species group, but no variety should be used at less than 4 kg/ha (2 kg/ha for white clover, red clover and timothy varieties).
* If good white clover content is required, double rate to 4 kg/ha.

Key to species groups:

1a = Italian ryegrass	3a = Early timothy	7a = Small white clover
1b = Hybrid ryegrass	3b = Late timothy	7b = Medium white clover
2a = Early PRG	4a = Early cocksfoot	7c = Large white clover
2b = Intermediate PRG	4b = Late cocksfoot	8a = Early red clover
2c = Late PRG	5a = Meadow fescue	8b = Late red clover
	6a = Tall fescue	

Table 11.7 Example of analysis of a seeds mixture, sold as 4–5-year ley, mainly for dairy cow grazing

kg/ha	*Species*	*Variety*	*Ploidy*	*NIAB rec*	*HD*	*Date 67-D*	*Mid-S D*	*WS carbo. score*	*Ground cover score*
5	IRG	RVP	dip	O	57	57	64.3	7.5	3
7	PRG	Frances	dip	G	44	51	65.8	6	6.5
2	PRG	Bastion	tet	G	44	55	66.0	7	5.5
5	PRG	Melle	dip	O	77	65	68.0	6	7.5
2	PRG	Condesa	tet	G	72	68	67.5	6	7
2	Tim	Motim	dip	G	79	60			
2	WC	Huia		X					

- The presence of IRG in a mixture sown to last over three years is undesirable: as the IRG dies out there is a serious risk that weeds will take its place.
- Although the use of tetraploid PRG can be an asset under cow grazing, both are present in insufficient quantity.
- There is a reasonable spread of HD in the PRG varieties.
- The mixture uses two outclassed varieties and one unrecommended variety.
- The mixture could be cut about 25 May for grass at 67-D.
- The total seedrate is for ideal conditions and should be raised to 30 kg/ha for other conditions.

Fertilizers for grassland

As with any crop, the soil is able to supply some or all of the essential nutrients needed for growth. Fertilizers are used to make up the difference between those which the soil can supply and the quantity required by the particular crop. In grassland the nitrogen (N) supply is so important that it has the effect of controlling the level of grassland yield, and the requirement for other nutrients is related to the yield of grass produced and hence the N supply. Because grass has to be utilized through an animal system to achieve monetary value, not all farmers seek the highest possible grass yields, but rather to balance the level and seasonality of grass yield with the requirements of their particular stock enterprises. The skill of economic grassland production lies in balancing production with requirement very carefully, thus ensuring high utilization.

It follows that farmers are justified in operating at a wide range of N inputs on their grassland, depending on the yields needed. As the next section shows, the quantity of N available from non-fertilizer sources can vary, and so the amounts of fertilizer N used to 'top-up' N supply will range from nil up to more than 400 kg/ha of N (equivalent to 1160 kg/ha of ammonium nitrate fertilizer). It is important for both economic and environmental reasons that the rate of fertilizer N applied is adjusted for the level of production required and the other sources of available N.

Sources of available nitrogen

Soil nitrogen

Very large quantities of N are present in the organic matter in the soil, for example a value of 10 000–12 000 kg/ha can be expected under old grassland, but most of this N is present either in living matter or complex organic compounds and is not available to plant roots. However, biochemical activity does release some N in available forms each year, the quantity depending on:

(1) organic matter (OM) content of the soil.
(2) biochemical activity of that organic matter,
(3) length of time in grass.

There can be a very wide range in available N supply from the soil, from 20 kg/ha on soils low in OM (e.g. old arable and light-textured soils) to 150 kg/ha on soils with large quantities of active OM (e.g. old grassland). In the UK there is no accepted method of soil analysis for predicting available soil N under grass, and estimates have to be based on the above actors.

For example, assume the following annual values:

30 kg/ha for a young ley;
100 kg/ha for a fairly productive grazed sward over three years old;
150 kg/ha for a highly productive permanent pasture.

Animal-recirculated N

Grazing animals retain only some 5–10% of the N they eat in herbage, voiding about 70% of returned N in urine and 30% in dung.

Potentially, grazing animals can recirculate a high proportion of herbage N, but both dung and urine occur in relatively small patches in the field, affecting only a small proportion of the sward in one grazing season. Recirculation of N by animals is not very effective in the first two years of grazing and after that the effect is related to stocking rate (SR). For grazed swards over two years old, assume the following values for animal recirculated N:

nil if grazing season SR less than 2 'livestock units' (LSU)/ha;
20 kg/ha if grazing season SR 2–2.5;
40 kg/ha if grazing season SR 2.5–3.5;
60 kg/ha if grazing season SR 3.5–5.0;
90 kg/ha if grazing season SR over 5 LSU/ha.

Where slurry (and farmyard manure) is applied to grassland, there is an effective recirculation of animal N, the value depending on the rate of application and the N-value of the material (see Chapter 7).

Clover nitrogen

Rhizobial bacteria in clover root nodules with fix atmospheric N and this N is used by the clover plant. As the clover plants grow, old roots die and some fixed N in these roots becomes available to grass roots: this is termed *underground N transfer*. Also clover foliage contains fixed N and if the clover is grazed, then some of this N is returned to the soil in dung and urine: this is termed *overground transfer*.

The value of clover N depends on the quantity of clover in the sward. Where there is a very large amount of clover, likely to average 30% or more of the weight of herbage grown in the year, then clover N can be equivalent to 150 kg/ha of fertilizer N (and even 200 kg/ha under some favourable conditions). At the present time it is rare to find a grass/white clover sward that contains enough clover to average 30% over the year. Figure 11.5 shows the differing seasonal growth patterns of grass and white clover. The best time to assess clover content is in July/August, when clover content must exceed 60% if the sward is to average 30% over the year. Scoring clover content in summer can give a guide to the value of clover

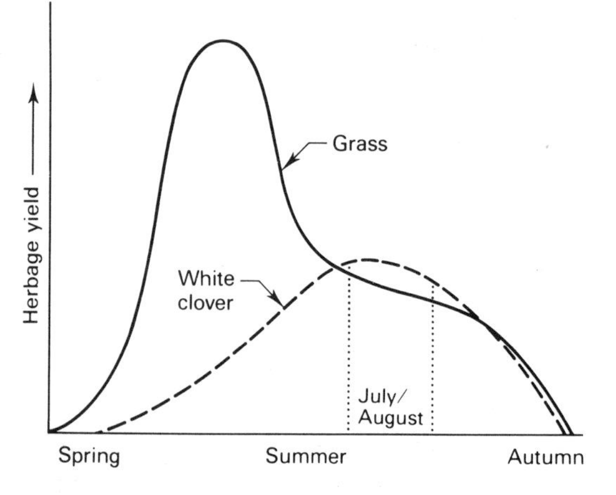

Fig. 11.5 Relative seasonal production of grass and white clover in grass/clover swards.

in grazed swards. If 60% clover, assume 150 kg/ha N, 30% clover assume 75 kg/ha, and so on.

Even with skilful management the application of fertilizer N will decrease the clover content, so that fertilizer N and clover N are not additive. As a rule of thumb, application of every 2 kg/ha of fertilizer N will decrease clover N by 1 kg/ha. This means that if 300 kg/ha of fertilizer N is applied over the season to a grass/clover sward, then clover N contribution will be nil (and pro rata for lower fertilizer N rates).

Fertilizer nitrogen

This should be applied to make up the difference between the total available N supply from other sources and the required N level (see Table 11.10).

Efficient use of fertilizer N

The seasonal use of fertilizer N can have a marked influence on the quantity of grass produced and the economic value obtained from the fertilizer.

As shown in Fig. 11.6, grass takes up N much more rapidly than it grows in response to that N. Not only is N uptake always more rapid than growth response, but also N uptake can occur when temperature and moisture limit growth. The greenness of grass is associated with N content (i.e. N uptake) and not grass growth, and it does not follow that very green grass following N application is growing rapidly. Research has shown a definite relationship between the quantity of N grass can use for growth and the number of days of active growth available to the grass. From this work, a *maximum* N rate of 2.5 kg/ha for every day of active growth is suggested. Thus *maximum* individual dressings of fertilizer would be:

for 21 d paddock grazing, $21 \times 2.5 = 52.5$ kg/ha
monthly application in continuous grazing, 30×2.5 = 75 kg/ha
for silage grown 40–50 d, $40–50 \times 2.5$ = 100–125 kg/ha

Because grass takes up N so rapidly and completely, there is very little residual N left following defoliation: where high annual rates of fertilizer are needed, some N should be used for each growth period rather than relying on infrequent heavy dressings, and a season of 200 d of active growth would justify a maximum of $200 \times 2.5 = 500$ kg/ha of N.

Grass grows most rapidly in spring (May and June) and it gives the biggest responses to N in this period. However, very high rates of N application in the spring may deplete tiller production and limit valuable mid-season growth. It is suggested that in grazing systems the maximum daily N rate should never exceed 2.5 kg/ha, even in spring. Where silage cuts are taken, a lack of big N responses in summer should be recognized. A guide is 125 kg/ha of N for spring silage, 100 for July cuts and 80 for September cuts. From September onwards the day length is decreasing rapidly and grass growth slows appreciably: it is recommended that no N dressing should exceed 40 kg/ha after mid-September.

N rate and site class

As a result of an analysis of the National Manuring Trials, it was found that the major factor determining grass response to fertilizer N was soil available water content (AWC) (Morrison *et al.*, 1980). In situations where AWC was reasonably good throughout the growing

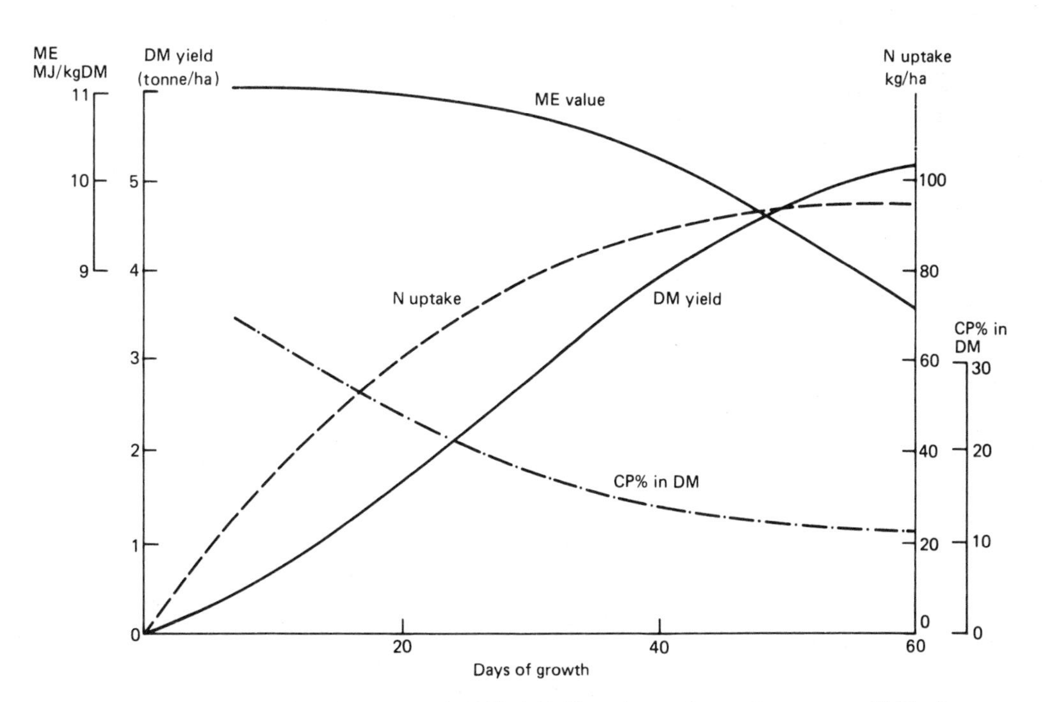

Fig. 11.6 Relationship between days of grass growth, DM yield, N uptake, crude protein content and ME value.

season, the growth of grass could be maintained by adequate N application. Where AWC fell to low levels during the summer, grass growth was restricted by lack of water and failed to respond fully to N application.

AWC can be estimated from two easily assessed factors:

- water-holding capacity of the soil, based on the combination of soil texture and rooting depth;
- summer rainfall.

Table 11.8 (Thomas *et al.*, 1991) shows how the use of the two factors can lead to an appropriate *site class number* being given to a farm (or part of a farm).

It was also found that altitude had some effect on grass growth (mainly via temperature) and so it is recommended that 1 is added to the value derived from Table 11.8 if the altitude is over 330 m above sea level.

It follows that fields with a high site class number have a more limited yield potential than low site class sites: thus it is logical that the *maximum* fertilizer N rate and total annual DM yield will vary with site class (see Table 11.9).

Table 11.9 gives a refinement to the '2.5 kg/ha/d of growth' rule for N outlined earlier. Two hundred days of growth (giving a maximum N rate of 500 kg/ha) would be found only on Class 1 sites, and all other sites, with less growing days, would justify less N over the season. Normally, the reduction in fertilizer N use in these other sites should come from decreasing the number of fertilizer applications and *not* by diluting the 2.5 kg/d rule in the first part of the season.

Phosphate and potash

The soil can supply some or all of the phosphate and potash needed by grass, fertilizer being required to bridge the gap between supply and demand. As a guide, total phosphate demand is 60–80 kg/ha of P_2O_5 and that for potash is 150–300 K_2O, depending on the level of grass production (and hence available N supply). For both phosphate and potash, soil analysis can be a useful guide to the quantity of fertilizer needed (see Table 11.10).

Grazing animals

These return a substantial proportion of both the phosphate and potash eaten, virtually all the phosphate being

Table 11.8 Basis for estimation of site class

Soil type	*April–September rainfall (mm)* Over 500	425–500	350–425	Under 350
All soils except those listed below	1	2	2	3
Shallow soils over chalk and rock; gravelly and coarse sandy soils	2	3	4	5

If field is over 330 m above sea level, add 1 to above score.

Table 11.9 Probable grass DM yields (t/ha) at range of fertilizer N rates and maximum N rate to apply*

Site class	*Fertilizer N (kg/ha)* 0	50	100	150	200	250	300	350	400	450	*Maximum to apply*
6	1.4	2.5	3.8	4.9	6.1	7.2	7.5 at				270
5	1.6	2.9	4.2	5.4	6.6	7.7	8.4 at				300
4	2.0	3.3	4.6	5.9	7.2	8.2	9.1	9.5 at			330
3	2.4	3.8	5.1	6.4	7.7	8.7	9.6	10.3	10.5 at		370
2	2.8	4.2	5.6	7.0	8.3	9.4	10.3	11.0	11.5	11.6 at	410
1	3.2	4.7	6.1	7.5	8.9	10.0	10.9	11.6	12.2	12.7 at	450

* Averaged over grazing and cutting managements.

Table 11.10 Basis for fertilizer recommendations on grassland

Annual stocking rate (LSU/ha)	*Amount of nutrient to apply (kg/ha)* N		P_2O_5 *soil index*				K_2O *soil index*				
	High clover and/or soil N	*Low clover and/or soil N*	0	1	2	3†	*Utilization*	0	1	2	3
1.0–1.3	none	50–100	60	40	20	0†	cut	100	80	40	0
1.3–1.7	40–100	100–160					grazed	80	40	20	0
1.7–2.2	80–150*	160–300	80	60	30	0†	cut	200	120	90	45
over 2.2	150–300*	over 300					grazed	80	60	30	0

* High clover N unlikely at these stocking rates.
† Some may be needed every three to four years to maintain level.

in dung and most of the potash in urine. Dung is very unevenly distributed over the field and broken down very slowly to release the phosphate: as a result animal recirculation of phosphate is not assumed. Urine covers a much greater area of the sward than dung, and all the potash in urine is in a readily available state: thus allowance is made for potash recirculation on grazed swards.

Plant uptake of phosphate is steady and is well related to plant needs: it follows that the annual fertilizer requirement for phosphate can be applied in one annual application if desired, without causing nutrient imbalance. However, uptake of potash by the plant is related to the quantity of potash available to the roots and *not* to the plant's needs: thus grass can take up in a single growth period as much as twice the potash it needs for adequate growth. This excessive uptake is called 'luxury uptake' and can have two harmful consequences:

- Mg content is inversely related to K content in grass, so the unnecessarily high level of K leads to very low Mg content and a greater risk of hypomagnesaemia in stock eating the herbage, particularly in spring and autumn when herbage Mg contents in grass are naturally low.
- Growth following defoliation of this herbage may be restricted by lack of available potash.

Where high annual rates of potash are needed, some potash should be applied for each growth period, except none should be applied in spring to swards that will be grazed unless the soil K index is 0.

Fertilizers for grassland establishment

Young developing grass and clover plants have a high demand for phosphate and potash, particularly phosphate. As the young plant has a very small root system, it is necessary to ensure that fertilizer supplies large quantities of available phosphate and potash.

For undersowing use normal cereal recommendation for phosphate and potash at sowing: if soil is index 0 or 1, apply 50 kg/ha of P_2O_5 and/or K_2O immediately after harvest.

For other methods of establisment apply the following before or at time of sowing:

	kg/ha of P_2O_5				*kg/ha of K_2O*			
Soil index	0	1	2	3+	0	1	2	3+
	100	80	60	40	100	80	40	nil

Young grass also benefits from some fertilizer N, particularly when sown in spring. However, little or no seedbed N should be used if good clover establishment is expected.

	kg/ha of N in seedbed	
	spring sowing	*autumn sowing*
IRG, no clover	80	60
PRG types, no clover	60	50
PRG types, with clover	0–30 (max)	0

Sulphur

Application of fertilizer N to grass increases the synthesis of S-containing proteins, so the S requirement of grass is related to N use. At low N levels the S requirements are modest but high N grass needs the relatively large amount of about 40 kg/ha of S over a growing season. Few fertilizers contain much S and in recent years most available S has come from atmospheric pollution. As the pollution is reduced, more S is needed from other sources to meet the requirements of high N grass.

Table 11.11 gives results from 23 sites of an ADAS experiment studying grass grown for silage. The results show two important aspects.

- S response interacted with N, with the biggest yield response to added S at the highest N rate;
- major S responses occur after utilization of the first growth in the season, as S deposition over winter is often adequate to support spring growth.

Table 11.11 Grass response to sulphur under regular cutting (t/ha of DM)

	N rate (kg/ha)			
	0	*200*	*400*	*600*
Nil sulphur	5.0	10.3	12.2	12.5
50 kg/ha of S	5.2	10.6	13.2	13.6
	% increase in DM yield from each cut due to S			
	(mean of N rates 400 and 600 only)			
	cut 1	0		
	cut 2	10		
	cut 3	20		
	cut 4	18		

Calcium

Calcium is applied in lime, and maintenance of a pH of 5.8 or above will ensure that acidity does not limit the growth of grasses and clovers *and* that adequate calcium is present in the soil.

Magnesium

Magnesium deficiency can occur in grazing animals as hypomagnesaemia. This is a problem that cannot be cured reliably by applying magnesium to the soil; the best prevention of the disease is to place magnesium compounds in the animals' food or water. However, where soil magnesium is low, it is sound practice to use a magnesium-containing form of lime when lime is applied.

Other nutrients

Grassland in the UK suffers rarely if ever from other nutrient deficiencies, although occasionally grazing stock may suffer from lack of minerals such as sodium or copper. The safest rule is to supply the deficient element directly to the animal rather than through the herbage: this not only ensures that the animals concerned receive a correct dosage but also ensures that areas of grassland with adequate levels of minerals are not enhanced to reach toxic levels.

Patterns of grassland production

The basic pattern of grassland production is shown in Fig. 11.7 and shows that grass can be available for grazing over a long period of the year, with conservation as hay or silage removing surplus grass and making this grass available for winter feeding. As conserved grass has a lower feed-value than fresh grass, other feeds have to be used with hay and silage, normally a cereal-based concentrate.

Generally over 50% of total annual grass production has occurred by the end of May and some management practices can emphasize the spring peak even more than this. For example, heavy use of N during spring with much lower rates later in the year can result in 70% of annual production by the end of May: where one very heavy conservation cut is taken in early June, this cut can account for as much as 85% of annual production.

Spring peak in production will be emphasized by:

- very high N use in April–May, particularly as this can limit July–September production below that which would otherwise be obtained;
- taking one very heavy cut in spring, particularly as this will deplete tiller numbers and subsequent regrowth;
- use of a single grass variety, particularly an early-heading one;
- low summer rainfall reducing production from June onwards.

Spring peak will be minimized by:

- restricting N use in April–May, although this will lower annual grass yield as grass is most responsive to N during this period;
- very frequent defoliation in the May–June period (Fig. 11.4);
- using a mixture of grasses with different spring growth patterns.

Effect of site class

Figure 11.8 shows the effect of site class on the seasonal growth pattern of grass receiving optimum N supply throughout the season, and indicates that there are two parts to the growing season:

(1) up to the end of May, when site class has little effect on yield;
(2) from June onwards, when differences in site class can give a 100% variation in expected grass yields.

Farmers have to take into account such differences in grass growth when planning their grassland management strategy.

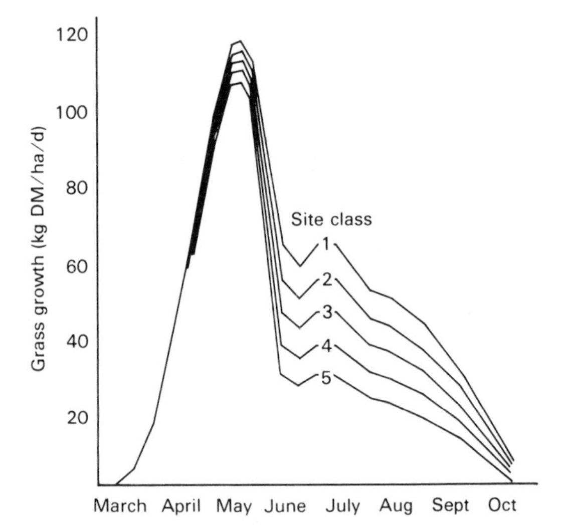

Fig. 11.8 Seasonal pattern of DM production from a perennial ryegrass sward at five site classes.

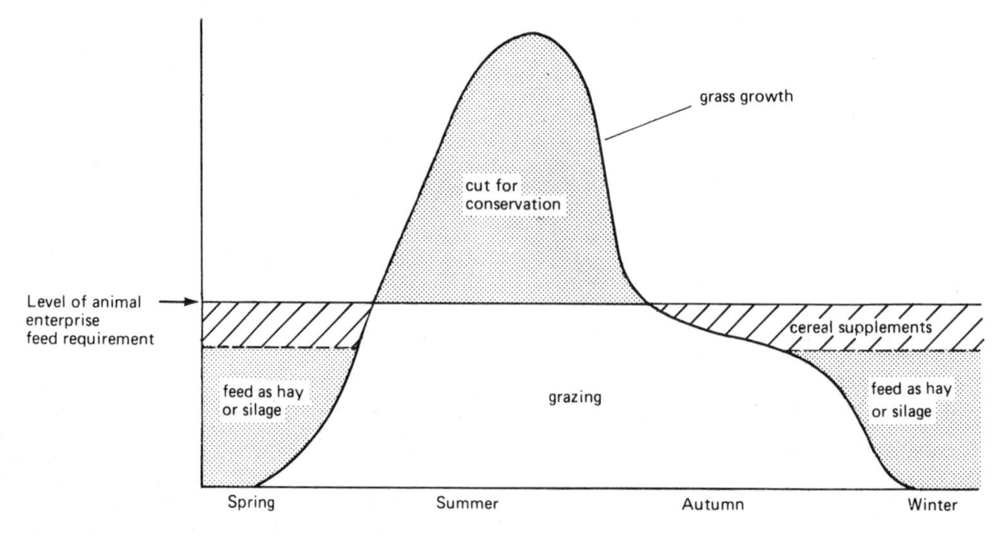

Fig. 11.7 Pattern of grass production and use.

For example, if it is assumed that grazing animals require the same amount of grass every day throughout the season and the accumulated 'surplus' can be cut periodically for silage, then it might be possible to take two to three cuts on site class 1 or 2 farms. On site class 5, only one spring cut will be possible and often the overall stocking rate will be lower as well.

Estimation of site class is based in part on average rainfall. If, in a particular year, rainfall is *above* average, the grass growth will be better than expected and surplus grass can be ensiled as an effective way of maintaining good grass utilization. In years when rainfall is *below* average, less silage will be made and on site class 4 or 5 the growth of grass may be below even that required for grazing: here other forage may be needed, e.g. 'buffer feeding' of silage.

Expression of grasland output on the farm

Farmers seldom weigh grass and so have little idea of grassland output as 't/ha'. Also, it is not the yield of grass *grown* that is important but the quantity of grass *utilized* by the livestock systems. There are two general ways in which the output of grassland can be assessed. There is a 'quick' way based on types of stock and overall stocking rate, and a more detailed method based on utilized metabolizable energy (UME).

Stocking rate method

The forage intake of ruminant stock is closely related to their liveweight and it is convenient to take a standard 'livestock unit' (LSU) of 550 kg and express all other stock on the basis of this LSU (sometimes called 'cow equivalent': see Table 11.12). For example, at a particular time in the season, a farm has 250 ewes of average weight 60 kg plus 360 lambs of average weight 20 kg plus 40 heifers of average 200 kg. The total LSU would be

$$\left(250 \times \frac{60}{550}\right) + \left(360 \times \frac{20}{550}\right) + \left(40 \times \frac{200}{550}\right) = 54.9 \text{ LSU}$$

Growing animals change weight over the year, and some animals are sold and others bought, so that the annual expression of stocking rate as LSU depends on:

- conversion of all stock to LSU.
- allowance for changing stock weight and/or numbers.

This latter point is best assessed on a monthly basis.

If, in the previous example, the ewes were kept on the total grass area of 35 ha for the whole year, the lambs grew from 10 to 40 kg liveweight over seven months and the heifers increased in weight over the year from 160 to 350 kg, then the annual LSU carried on the area is:

ewes 250, weighing 60 kg, for whole year $= 250 \times \frac{60}{550} = 27.3$

lambs 360, average weight 25 kg for 7 months

$$= 360 \times \frac{7}{12} \times \frac{25}{550} = 9.5$$

heifers 40, average weight 255 kg, for whole year

$$= 40 \times \frac{255}{550} = 18.5$$

$$\text{total} = 55.3$$

So stocking rate $= \frac{55.3}{35} = 1.58$ LSU/ha.

An estimate of forage intake can be obtained from the equation

$$y = 0.025x + 0.1z$$

where y = DM intake (kg/d), x = animal liveweight (kg) and, for dairy cows only, z = milk yield (kg/d).

For example, a 550-kg non-dairy LSU would have an appetite of 13.75 kg/d of DM. A dairy cow of this weight, and yielding 20 kg/d of milk, would have an appetite of 15.75 kg/d of DM.

Table 11.12 Cow equivalents of other ruminant stock

Liveweight (kg)	*Proportion of LSU*		*Liveweight (kg)*	*Proportion of LSU*	
10	0.04	growing sheep			
20	0.06	growing sheep			
30	0.08	growing sheep			
40	0.10	growing sheep	40	0.08	ewes
60	0.13	growing sheep; growing cattle	60	0.10	ewes
80	0.17	growing sheep; growing cattle	80	0.13	ewes
100	0.20	growing cattle			
150	0.29	growing cattle			
200	0.38	growing cattle; heifers			
250	0.47	heifers			
300	0.56	heifers			
350	0.64	Channel Island cows			
400	0.73	Channel Island cows	400	0.56	beef cows
450	0.82	Channel Island cows	450	0.61	beef cows
500	0.91	Friesian dairy cows	500	0.68	beef cows
550	1.00	Friesian dairy cows	550	0.75	beef cows
600	1.09	Friesian dairy cows			
650	1.18	Friesian dairy cows			

From Forbes *et al.* (1980).

Utilized metabolizable energy (UME)

The total energy value of the food eaten by the animal is its *gross energy* (GE). Some of this energy is passed out in faeces and the remainder is *digestible energy* (DE). Some of the DE is lost as methane from the rumen and some is passed out in urine; that remaining is termed *metabolizable energy* (ME). Some ME is lost as heat and the remainder, *net energy* (NE), is used for maintenance and production. Whilst NE is the nearest assessment to the production potential of a food, it has been found that ME is well-correlated with production and is much easier to assess on a routine basis than NE.

Energy is expressed as megajoules (MJ) or gigajoules (GJ), where 1000 MJ = 1 GJ.

Animals extract useful (metabolizable) energy from their feed to

- maintain their body functions;
- increase body weight;
- lactate (where appropriate);
- provide nutrition for any fetus carried.

The energy required to fulfil the above functions in all classes of stock is known and so it is possible to estimate the total quantity of ME utilized by animals (this can be for an individual animal or a livestock unit or a whole farm, depending on the most convenient way to calculate). The ME value of all commonly used supplementary feeds is also known, so that, where grass is used along with other feeds, it is possible to produce a balance sheet that leads to an estimate of UME from grass, as shown in the example below. Note this example illustrates the UME method of estimating output from grass; the exact UME standards for size and type of animal, its weight change and ME value of milk of specific quality can be obtained from published tables (and some organizations offer computerized methods).

Example 100-cow Friesian herd, averaging 5500 litres of milk and using 1.3 t/cow of concentrates per year in addition to grazing and silage made from 50 ha of grassland.

On average over a year each cow utilizes the following ME:

Maintenance	365 d at 60 MJ/d	=	21 900 MJ
Weight change over year, none			
1 calf/cow		=	2 400 MJ
5500 litres milk × 5.3 MJ/litre		=	29 150 MJ
Energy utilized per cow over year		=	53 450 MJ
Less energy from concentrates @ 85% DM			
Say 13 MJ/kg of DM	1300 × 0.85 × 13	=	14 365 MJ
Difference is ME/cow utilized from grassland		=	39 085 MJ
	or		39.08 GJ
Stocking rate is 2 cows/ha, so			
UME/ha	2 × 39.08		78.2 GJ

Walsh (1982) found that cows rarely obtain more than 45 GJ per year from forage; good herds average 40–45 and moderate herds average 35–40. If the value falls below 35, then the herd is not utilizing forage very well.

Efficiency of grassland use

The UME figure gives a value for energy used; if the total quantity of ME available were known, then the % utilization of the energy could be calculated. It is very difficult to measure grass yield on a farm, but some organizations use a prediction based on information similar to that given in Table 11.9.

Example The farm above has a site class of 2 and average fertilizer N use is 300 kg/ha. It is assumed that the average energy value of grass (grazing and silage) over the year is 10.5 MJ/kg DM.

Grass growth	estimated at	=	10 300 kg DM/ha
ME produced	10 300 × 10.5		1081 50 MJ/ha
		or	108.15 GJ/ha

Utilization of ME $= \frac{78.2}{108.15} \times \frac{100}{1} = 72\%$

But note all such calculations are based on assumed yields and so are only a guide.

Grass digestibility

In the 1950s workers at the Grassland Research Institute found that the *digestibility* of grass (i.e. the percentage of grass eaten that was digested by the animal) was well correlated with the animal intake of grass and with the subsequent animal production from it. Since that time much advice and literature refers to digestibility. By convention, the term 'D-value' (i.e. the use of the capital letter D) should refer specifically to *the digestible organic matter as a percentage of the dry matter.* There is an approximate relationship between D-value and ME in grass, such that between D-values of 60–70, ME (MJ/kg of DM) = 0.16 D.

Figure 11.6 shows the effect of increasing grass growth period (maturity) on yield, N uptake, CP content and ME value. As grass matures it increases in yield but decreases in % CP and ME value: it is important to define the quality of grass required and defoliate it at the appropriate stage of growth. Generally, high grass quality is associated with low yields and vice versa (see Fig. 11.9 and Table 11.13). Not only does a feed low in ME provide a low feed value, but also the animals eat less of such a feed. Table 11.14 shows the double action of low ME value on intake and production (Brockman & Gwynn, 1961).

Clearly the target ME value of grass must depend on the productive potential of the animals concerned and on whether the grass is grazed or conserved: the latter point being influenced by the inevitable loss of feed value when grass is conserved and the need for at least a moderate yield to justify the costs of conservation. Guidelines for ME levels for good production from grass are:

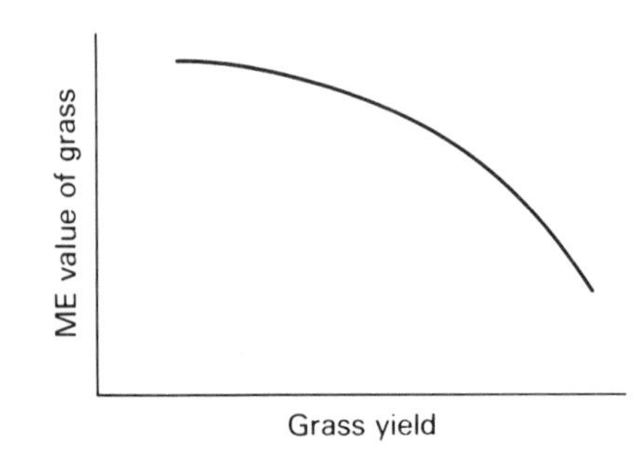

Fig. 11.9 General relationship between yield and quality in grass.

Table 11.13 Effect of cutting frequency on grass yield and quality

	Three cut system		Two cut system		
Date	*DM yield (t/ha)*	*ME (MJ/kg DM)*	*Date*	*DM yield (t/ha)*	*ME (MJ/kg DM)*
(a) *Field data*					
late May	4.6	10.6	early June	7.8	9.6
early July	3.2	9.8	mid-August	3.7	9.0
mid-August	1.8	9.6			
Total	9.6	Mean 10.0	Total	11.5	Mean 9.2
(b) *Feeding data*					
concentrate feeding	low	high		low	high
concentrate intake (kg/d)	5.0	9.3		5.0	9.3
silage intake (kg/d DM)	11.2	9.0		10.3	8.4
milk yield (litres/d)	19.9	20.8		17.3	19.7
land required for 180-d winter (ha/cow)	0.25	0.21		0.20	0.16

Source: Leaver and Moisey (1980).

Table 11.14 Effect of grass ME value on intake and production in dairy cows

Grass ME value (MJ/kg DM)	*Grass DM intake (kg/d)*	*ME intake (MJ/d)*	*Milk yield (litres/d)*
11.3	12.8	145	14.6
9.5	12.0	114	11.8
8.4	10.6	89	10.0

Utilization	*Minimum ME (MJ/kg DM)*	*Optimum ME*
Dairy cow grazing	10.5	11.5
Grazing other growing stock	10.0	11.0
Grazing dry cows	10.0	10.5
Grazing dry ewes	9.5	9.5
Silage and barn-dried hay	10.0	10.8
Field hay	9.0	10.0

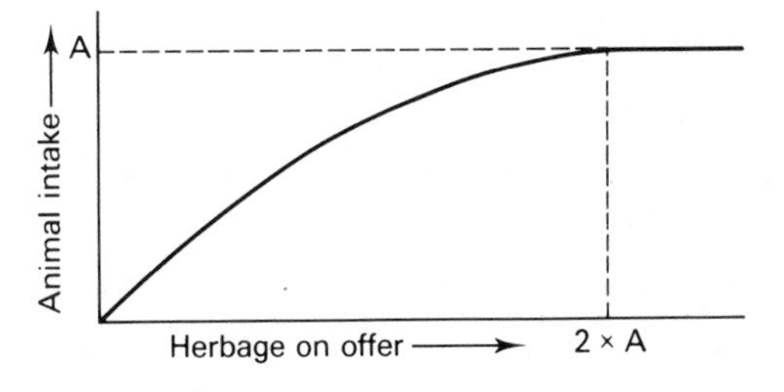

Fig. 11.10 Relationship between herbage on offer at grazing and animal intake.

Output from grazing animals

Grazing experiments with dairy cows, beef cattle and sheep have all shown that if the quantity of herbage available to the grazing animal is reduced below a certain value, then intake per animal falls below appetite, and production per animal is below potential.

Figure 11.10 summarizes the data and shows that if herbage on offer is less than double that required at any grazing occasion, then intake will decline. This means that during a specific grazing period herbage utilization should not exceed 50%. From this conclusion, two major practical points arise:

- Can utilization in grazing systems exceed 50% without loss of production?
- Is it possible to monitor grazed swards to ensure animals have sufficient grass to reach appetite and thus potential production?

The remainder of this section deals with these two questions.

Utilization

Over a season a grazed field is grazed on a number of occasions (from, say, 5–6 in a rotational system to 80–100 in a 'continuous' system). Whilst on no single occasion the utilization should be allowed to exceed 50%, the key to good overall utilization is to ensure that most of the unutilized herbage on one grazing occasion is available and utilized at a subsequent one.

In this way, it is feasible to achieve 70% (or even 80%) utilization over a grazing season. Indeed, the successful use of any of the grazing methods outlined in the next section is to ensure that:

- animals are not forced to eat too tightly;
- grass uneaten early in the season is available for utilization later.

Monitoring herbage available

Many experiments have shown that sward height is a useful practical guide to the availability of herbage.

For example, work with spring calving dairy cows has given the relationship shown in Fig. 11.11. Similar relationships have been found for other classes of stock and critical sward heights have been established, which are:

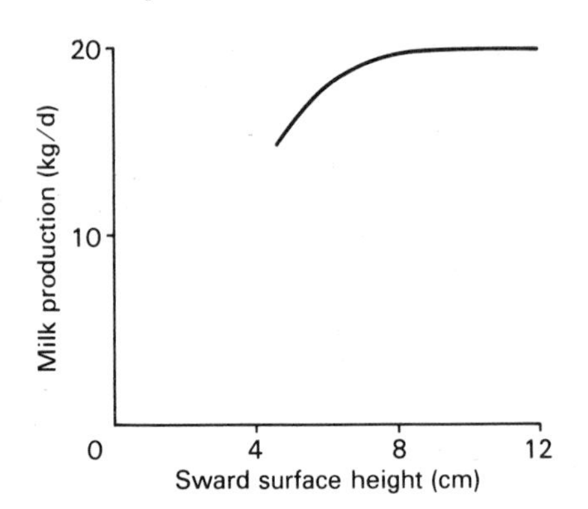

Fig. 11.11 The influence of sward surface height on milk yield: average grazing season yields for spring-calving cows.

Type of stock	*Minimum sward height (mm)*	
Sheep	40	If very high output is required* from grazing, add 20 mm to these values
Yearling cattle	60	
Dairy cows	80	

* Such as flushing ewes, finishing lambs, high milk yield from freshly calved cows.

Note that as animals become larger, so the critical height is higher.

Measuring sward surface height

Place a ruler (or similar marker) vertically in the sward, with the lower edge just touching the ground. Find the height at which a finger descending the ruler touches a green leaf. Points to watch are:

- Ensure the ruler is vertical and is not pressed into the soil or standing in a depression.
- 30–40 readings are taken in the field, avoiding bias and including a fair proportion of grazed and ungrazed areas.
- Contact with stems or seed heads is ignored.

In *rotational systems* the assessment of sward surface height is a guide for moving the stock.

In *continuous grazing* it is suggested that readings be taken every one to two weeks, so that adjustments can be made to stock numbers and/or grazing area.

(For more on the practical application of sward height, see Hodgson *et al.*, 1986.)

Grazing systems

Many areas of grass, particularly those for beef and sheep, are grazed on a very extensive basis that is not systematic in nature. On the other hand most dairy farmers have a very definite basis to their grazing policy, and an increasing number of the more successful beef and sheep farmers are adopting grazing strategies similar in principle to dairy systems. This section will deal with the major methods used to graze dairy cows; some of the methods outlined are appropriate for other stock.

Whatever the grazing method or stock used, the principles outlined in the previous section apply.

Two sward system

Here one area is regularly cut for conservation and the other regularly grazed. It is difficult to accommodate the seasonal growth pattern in the grazed area, but the system has advantages where part of the grass area is inaccessible to grazing (e.g. on a split farm) or where the cutting grass is in an arable rotation and does not justify fencing and a water supply.

Otherwise the integration of cutting with grazing is a powerful tool in smoothing out the availability of grass for grazing, because whereas only one-third of the total area may be needed for grazing in May, two-thirds may be needed in June–July and the whole area from August onwards.

For	*Against*
Useful if part of area cannot be grazed	Risk of surplus grass in grazed area in spring and insufficient later

Set-stocking

This implies a given number of stock on a fixed area for a long period, often the whole season. The term *set-stocking* is sometimes used erroneously for *continuous (or full) grazing* (see below). Set-stocking is advised under very extensive conditions only at low stocking rates, as there is a tendency for undergrazing in spring, leading to poor quality mature herbage, followed by overgrazing in late summer and autumn resulting in low animal performance. One advantage is that provided the perimeter of the area is stockproof, fencing costs and water supply problems are minimal.

For	*Against*
Simple and cheap	Does not match grass growth
Good for large groups	Possible disease build-up

Continuous grazing (or full grazing)

This refers to systems where the stock are allowed to graze over a large area for a fairly long period, say two to three months. Many farmers use this system for dairy cows and associate it with a high stocking rate, high and regular fertilizer N usage and a willingness to bring in other feeds if grass supplies become inadequate. As rate of grass growth slackens during the season, farmers can both reduce stock numbers as cows dry off and increase the grazing area by adding some silage aftermaths. Thus part of the area may be grazed for the whole season, but neither the total area nor the stock numbers are fixed, and other feed is used as a buffer against periods of low grass availability. This system depends on grass growing as rapidly as possible over the whole season and is best associated with high N application, with monthly applications of about 70 kg/ha of N from March to August and site class 1 or 2.

Some farmers have a 'rapid rotation', where the area is divided into three or four blocks and each is grazed for about one day at a time. In other situations there is one area for day grazing and another for night grazing. Because of the frequent defoliation that occurs, these are variants of continuous grazing and are not types of rotational or block grazing.

Overall there should be 1 ha of available grazing for every three cows: at turnout in spring the whole area may be needed but as growth picks up the area for spring grazing will be about 1 ha per five cows; following spring silage cuts the area may need extension to 1 h per four cows, with the whole area in use from August onwards, depending on the season.

For	*Against*
Reduces poaching as stock dispersed	No visible assessment of grass growth (but sward height can be measured)
Saves cost in fencing and water	Regular check on performance essential
Increases sward density and clover	Cow collection can take time
Eliminates daily decision on grazing area allocation	Willingness to supplement feed at grass essential

Three field system

This is one that has much to commend it if the whole grass area can be grazed conveniently (this is often difficult with cows that have to walk to and from milking and for this reason this method is more often used in beef and sheep systems). Basically this method formalizes a continuous grazing system into three periods in the season and in a very simple way adjusts to the expected pattern of grass prodcution. The grassland area should be split by stockproof fencing into two areas, one area being approximately double the size of the other. In spring the stock graze the smaller area and the larger area is cut, preferably for silage as this will provide more reliable aftermaths than where hay is taken. As soon as the aftermaths are available, the stock are switched and the smaller area is taken for a conservation cut. Once the smaller area has available aftermaths, the whole area is grazed.

For	*Against*
Simple self-adusting to seasonal growth	Inflexible
Suitable for large animal groups	Based on *expected*, not *actual* growth
Good parasite control; 'clean' grass	Often unsuited to dairy grazing

Block grazing systems

These are systems in which the grazing area is divided into a number of fairly large blocks with the aim of grazing on a rotational basis. Thus one block of grass is grazed with a large number of animals for a short period, usually 1–7 d, and then they are moved to other blocks whilst the grazed block regrows. After about 20–28 d, the block can be grazed again. Blocks that are surplus to grazing are cut for conservation. This basic principle is used for all classes of grazing stock, with the following common amendments.

For dairy cows Farmers often like cows to have a fresh allocation of herbage each day, so an electric fence can be used to ration the grass. This was formalized in the *Wye College* system where cows were given one-seventh of a block each day and one block lasted one week.

For ewes and lambs The fence separating adjacent blocks can have spaces wide enough for the lambs to pass but not the ewes. Thus lambs can obtain the pick of the next block to be grazed by the ewes (*forward creep grazing*) or creep into an area never grazed by ewes (*sideways creep grazing*).

For cattle rearing Young stock may be grazed one paddock ahead of older cattle so that the younger ones obtain the pick of the grass and the older cattle clean up, ensuring good utilization. This system is called *leader follower grazing* and has been used particularly in heifer rearing.

For dairy cows Occasionally farmers split their herd and do a leader/follower with the cows, putting the high yielders in the leader group, or putting dry and nearly-dry cows as followers behind the main herd.

For	*Against*
Efficient grass utilization	Good fencing round each paddock essential
Good parasite control for young stock	Regular stock-moving decisions needed
Areas large enough for conservation	Many water points needed

Paddock grazing

This represents a very formal method of rotational grazing, where the grazing area is divided into some 21–28 permanently fenced and equal-sized paddocks. The aim is that one paddock is grazed each day and the rotation around the area is completed as soon as the first paddock is ready for grazing again. Surplus paddocks can be taken out for conservation but usually the operation is restricted by the small size of the paddocks. Paddock size must be related to the number of grazing animals, as the stock density must be sufficient to ensure that the grass is efficiently utilized during the short grazing period. A guide is to allow 100–125 cows/ha daily, depending on stocking rate.

Thus, for a highly stocked herd of 100 cows and where 25 paddocks are selected, each paddock should be about 1 ha in size. Every paddock must have a water point and independent access on to a track to avoid poaching.

For	*Against*
Easiest rotational system to manage	High fencing and water costs
Gives objectivity to grazing plan	Small areas for conservation
	Some wasted land in trackways

Strip grazing

This involves the use of an electric fence to give a fresh strip of herbage once or twice daily. Ideally, stock should be confined to the daily strip only by use of a regularly moved back fence to prevent the regrazing of young regrowth, but this is done rarely because of problems with animal access and water supply. Although this is the most sensitive system to allow adjustment for fluctuations in grass growth, it is not common because:

- It requires daily labour to decide on area allocation and to move the fence.
- There is a tendency for grass to become over-mature ahead of grazing in large fields.
- There is a risk of serious poaching along the fence line in wet weather.

Many of the best features of strip grazing are achieved more easily in block grazing.

Zero grazing

This implies that grass at grazing stage is cut and transported to stock that are either housed or kept in a 'sacrifice' area. Zero grazing is practised widely in some countries where grass is inaccessible for grazing (e.g. in 'strip field' systems) or where a range of crops needs to be grown to ensure continuity of supplies (e.g. in semi-arid areas). In temperate climates there is research evidence (Marsh, 1975) that stocking rates can be higher under zero grazing because:

- The sward does not suffer the deleterious physical effects of treading and selective grazing.
- Herbage can be harvested at the ideal stage for the stock concerned.
- Utilization can be high as there is no field refusal of grass (although this implies that the stock eat most of the grass carted to them).

Both milk and meat production per hectare have been shown to be greater from zero grazing than from other grazing systems. However, zero grazing is not used commonly in the UK except for specific opportunist reasons such as to:

- reduce poaching in early spring and autumn;
- provide 'grazing' from distant or inaccessible fields;
- Use other crops during crisis periods, as in drought.

The reasons for the lack of popularity of zero grazing despite its technical excellence for high production per hectare are:

- high labour and machinery costs for feeding *and* for removing slurry;
- complete dependence on machinery, with risk of breakdowns;
- output per animal tends to be depressed (but high SR gives high output per hectare);
- problems of refused grass at feeding face, particularly as wet grass heats rapidly when heaped.

Buffer grazing and feeding

This term does not refer to a grazing system but to the provision of additional feed if and when that available from the grazing area is insufficient. The buffer (or additional) material can come from.

- another (often adjacent) grass area;
- a specifically grown forage crop;
- a bulk forage such as silage, hay or straw;
- a dry feed such as cereals, concentrates or dried grass.

Conservation of grass

Conserved grass is used as the basis of ruminant feeding when grass for grazing is not available. Grass for conservation is either grown specifically for this purpose or taken as a surplus in a grazed area: often both situations apply in any one year on an individual farm. Where grass is grown for conservation, it is possible to tailor the crop to fit the particular requirements of the unit, in terms of herbage varieties, yield and feeding quality. Where grass arises as a grazing surplus, an over-riding factor is the rapid removal of the crop in order to allow regrowth for a further grazing (see Table 11.4).

When green crops are cut, biochemical changes occur and, if these are allowed to continue unchecked, degradation of the material will take place, releasing heat and effluent (water that contains much of the soluble cell contents). There are four methods by which this progressive degradation can be restrained, but at present only the first two methods are used in commercial practice and only these two methods will be discussed in detail.

Acidification

Degradation ceases when the pH of the material falls below 4.2–5.0 (depending on the moisture content) provided the material is anaerobic (oxygen-free). This material is *silage*, and can be made by heaping herbage in as near oxygen-free conditions as possible and allowing the natural process of biochemical change to produce acids which can effectively 'pickle' the material. Additives can be used to both accelerate the process and give a more reliable end-product.

Dehydration

Degradation ceases when the material reaches 85% DM (grass in the field averages about 20% DM). This means the production of each tonne of hay at 85% DM necessitates the loss of 3.25 t of water.

Dehydration can take place under natural conditions in the field and is *haymaking*. Sometimes hay that is almost dry is placed in a building and subjected to forced-draught ventilation to complete dehydration; this is *barn hay drying*. On some specialized units grass at or near field moisture is carted to a high-temperature drier and dehydrated very rapidly; this is *green crop drying*.

Preservation

Some chemicals can prevent the process of degradation and so preserve the material in almost the same chemical state as at cutting.

At present research is seeking suitable chemicals that will preserve herbage economically and yet not restrict animal intake, affect rumen fermentation or be carried into the animal products.

Freezing

Freezing will inhibit chemical breakdown almost completely in the material.

For many years research institutes have been using frozen grass in feeding experiments as a normal means of preserving fresh grass and all its intrinsic value. At present the cost of freezing and storing herbage on a farm scale is prohibitive.

Silage making

The process of ensilage consists of preserving green forage crops under acid conditions in a succulent state. When such green material is heaped, it respires until all the oxygen in the matrix is exhausted: during respiration carbohydrates are oxidized to carbon dioxide with evolution of heat. Continued availability of oxygen, as in a small outside heap of grass, will lead to enhanced oxidation with resultant decomposition and overheating.

Assuming the oxygen supply is restricted, bacteria can control the fermentation process. These bacteria, present on the crop, the machinery and in soil contamination, fall into two categories, desirable and undesirable.

The desirable bacteria are ones which can convert carbohydrate into lactic acid and are mainly *Lactobacillus* and *Streptococcus* species. These are anaerobic bacteria and their even distribution and activity throughout the grass in the silo is encouraged by mechanical chopping of the grass, rapid consolidation and exclusion of air. Lactic acid is a relatively strong organic acid and its rapid production within the ensiled grass leads to a low pH and conditions which inhibit the lactic acid producing bacteria and *all* other bacteria as well. The pH at which this 'pickling' occurs depends on the moisture content of the grass: the wetter the grass the lower the pH needed and the greater the quantity of lactic acid that has to be produced.

Silage DM%	*pH for stable silage*
18	3.8
20	4.0
22	4.2
26	4.4
30	4.6
35	4.8

Silage with a good lactic acid content is light brown in colour, has a sharp taste and little smell: it is very stable and can be kept for years if necessary provided nothing is done to permit oxygen to enter the material.

The undesirable bacteria are:

- aerobic species of bacteria that can oxidize carbohydrate to carbon dioxide and water;
- obligate anaerobes of the *Clostridium* species that can ferment carbohydrate and lactic acid to form butyric acid; *Clostridium sporogenes* can break down amino acids to ammonia and amines, and some of the latter are toxic to stock.

An indicator of clostridial activity is the ammonium-N content of silage, as the ammonia produced by these bacteria is retained in the silage.

	Fermentation quality		
	Good	*Moderate*	*Bad*
NH_4-N as % of total N	0–10	10–15	over 15

Butyric silage is olive-green in colour, has a rancid smell and is unpalatable to stock. Also it has a higher pH than lactic silage, is unstable and will not keep for more than a few months.

Clostridial activity in silage can be inhibited by:

- reducing moisture content of the grass, as this will lessen the quantity of acid needed to prevent decomposition;
- ensuring adequate carbohydrate is present for lactic acid bacteria, or applying an acid to assist in lowering pH;
- adding inoculants of live lactobacilli;
- avoiding contamination from soil or animal manure, both of which contain large quantities of clostridia.

Intrusion of air *during* fermentation will delay or even prevent the achievement of a stable pH and will lead to an excessive amount of carbohydrate being used, thus lowering the nutritional value of the silage. Intrusion of air *after* the silage has reached a stable condition will lead to secondary respiration and a further progressive loss of carbohydrate. Secondary respiration can shorten the storage life of well made silage and can occur when the silo is opened for use if the exposed feeding face is too large for the rate of silage removal.

Silage making can result in considerable loss of material during the wilting, fermentation and feeding periods. Field losses can range from 0 to 10% of the DM yield depending on degree of wilting in the field: fermentation losses are inevitable and commonly range from 10 to 20% of the DM yield: effluent can give a loss of up to 5%: not all of the material present in the silo is suitable for feeding, due mainly to side and top waste, and in clamp systems this 'visible' loss can be 5–15% of the DM yield. Thus, in a really good silage system, only 80% of the weight of grass cut will be available for feeding as silage and often the figure is as low as 65–70%.

For the best fermentation the crop should have a high carbohydrate content (to provide ample substrate for fermentation) and a low moisture content (to reduce the volume of material requiring acidification). For these reasons emphasis is given to cutting crops when their carbohydrate (soluble sugar) content is high and when they can be wilted quickly. ADAS advice is that sugar content should be at least 3% at time of cutting and grass should be wilted from its normal moisture content of about 80% down to 70–75%. Often these two desirable conditions cannot be met and farmers can use additives to help alleviate the problem.

Additives can be used to:

- *Increase water soluble carbohydrate (WSC)* By adding extra WSC (e.g. molasses) the lactobacilli are better able to produce lactic acid. (Note also that wilting decreases the quantity of lactic acid needed for effective pickling.) Also *enzymes* such as cellulase and hemicellulase can be used to convert some of the non-soluble material in the plant into WSC. Note there is no starch in grass and so amylase is not effective.
- *Increase effective bacteria* Inoculants containing *Lactobacillus* and *Streptococcus* can be used to ensure lack of suitable bacteria does not limit the rate of lactic acid production.
- *Add acid* In some countries inorganic acids are used at high rates in the silo at time of filling to create a pickled effect. In the UK acids such as formic and sulphuric are applied at 3–5 litres/t of grass as the grass is picked up from the field. These acids reduce the quantity of lactic acid needed to reach a stable pH and consequently reduce the time taken for stability to be reached in the silo. Although effective additives, these products are dangerous chemicals: strict precautions are needed when they are used.
- *Use preservatives* Some acid additives also contain chemicals that should suppress unwanted biochemical reactions: examples of such chemicals are formalin and sodium metabisulphite. Table 11.15 summarizes the Liscombe Star system that can be used to warn of possible fermentation problems that might be alleviated by additive use.

Storage

Storage of silage is in bulk silos (clamp or tower) or big bales (contained in sealed polythene bags).

Clamps

Clamps are found in a variety of forms, e.g. walled or unwalled, roofed or open, on the surface or in pits.

Towers

Towers are made of concrete or galvanized–vitreous enamelled steel: material for ensiling is always added to the top but, depending on type, silage is removed from either the top or the bottom.

Big bales

Big bales can be stored on level sites around the farm. They must be protected from wind which can damage the polythene, and from rodents who can eat holes in the bags.

Table 11.15 The Liscombe star system[1]

Grass variety	Timothy/meadow fescue	*
	Perennial ryegrass	**
	Italian ryegrass	***
Growth stage	Leafy silage	○
	Stemmy mature	*
Fertilizer Nitrogen	Heavy (125 kg/ha +)	– *
	Average (40–125 kg/ha)	○
	Light (below 40 kg/ha)	*
Weather conditions (over several days)	Dull, wet (less than 2% sugar)	– *
	Dry, clear (2.5% sugar)	○
	Brilliant, sunny (3% sugar or more)	*
Wilting	None (15% DM)	– *
	Light (20% DM)	○
	Good (25% DM)	*
	Heavy (30% DM)	**
Chopping and/or bruising	Flail harvester or forage wagon	*
	Double chop	**
	Meter/twin chop	***

[1] A total of 5 stars needed for a good fermentation. Consider each factor, such as variety, growth stage etc, and add up the stars (*).

How the star system works

For example a perennial ryegrass sward (**), in leafy silage growth stage (○) and heavily manured (– *), being ensiled in dry weather (○) and only lightly wilted (○) with pick-up double chop (**), gives a total scores of *** and will show a benefit from additive use.

In comparison an Italian ryegrass sward (***), in a leafy stage (○), which is heavily manured (– *), in average weather (○) but well wilted, 25% DM (*) and meter chopped (***) will give a total score of ****** and not require additive.

5 Stars – no additive needed.
4 Stars – use additive at recommended rate.
3 Stars – use additive at recommended rate.
2 Stars – use additive at higher recommended rate.
1 Star – use additive at higher recommended rate.
0 Stars – unsuitable conditions for making silage.

Making silage in clamps

Assuming the material to be ensiled is either high in sugars and low in moisture or having an additive applied, the main principle during the filling process is to eliminate as much air from the matrix of herbage as possible and keep the material airtight. Polythene sheeting is an essential feature of silage making and is available for this purpose in 300 or 500 gauge and in widths up to 10 m. The use of polythene sheets can be taken to the ultimate in the production of *vacuum silage*, where the crop is stacked on a sheet laid on the floor of the silo: then another sheet is placed over the heap and the two sheets are joined together at ground level by a plastic 'strip-seal', after which the whole mass is evacuated by a vacuum pump. Such complete removal of oxygen leads to excellent fermentation, but the process is laborious and difficult to carry out on a large scale. Polythene sheeting is easily punctured and on some farms polythene in any form is nibbled by rodents.

Farmers have found that many of the advantages of vacuum and similar techniques can be obtained more simply by making silage in a walled pit using a *wedge-filling* principle (sometimes called *Dorset wedge*).

On the first day of filling, the cut crop is stacked at one end of the silo, against an end wall that is either solid or has a polythene sheet lining. If the side walls are not solid, they too must have a polythene lining. The material is normally put into the silo using a push-off buckrake, and the buckraking tractor maintains the slope at the steepest reasonable angle. When the material has reached the maximum intended height, the slope is progressed forwards, leaving a fixed height of material. One principal objective of the wedge system is to prevent warm air rising out of the silo, as this will encourage oxygen-rich cold air to come in at the bottom and sides. A polythene sheet is therefore placed over the grass each night, and when one section of filling is complete, the sheet is left in place so that the silo is gradually wrapped in polythene sheets. If the technique is carried out correctly, the oxygen in the air in the silo is soon used up, lactic acid fermentation proceeds, and the crop consolidates under its own weight, often resulting in a drop in crop height to two-thirds to three-quarters of the original. Because of this shrinkage the polythene sheet should not be fixed rigidly, but rather covered with a flexible and convenient material such as old tyres, sand bags or even a net. If the sheet is not held down tightly, it will flap in the wind, allowing in more air and eventually tearing the sheet.

It is essential to fill a silo rapidly: a rate of 100–200 t/d, depending on silo width, should be the minimum.

Long material can be made into good silage in clamps, but chopped material is easier to handle and consolidates better, with less oxygen trapped in the matrix. Fairly wet grass can be placed in a clamp, but it will produce effluent. ADAS figures show that grass ensiled at 20% DM will release an average of 200 litres of effluent per tonne of grass ensiled, and the quantity of effluent decreases progressively until material ensiled at 28% DM should give no effluent.

Silage effluent

This is a real problem if it enters a watercourse, as it has a high Biological Oxygen Demand (BOD), and will kill many oxygen-demanding organisms in the water, including fish. Many farmers wilt their grass simply to avoid or minimize the effluent problem. As a precaution it is essential to construct an effluent tank adjacent to the silo, taking care to ensure that *only* silage effluent can enter it (i.e. no surface or rainwater) (see Chapter 25). Silage effluent is very acid, and all effluent conducting channels and ducts must be coated with acid-resisting material. Silage effluent contains some plant nutrients (say 2, 1 and 1.5 kg/1000 litres of N, P_2O_5 and K_2O respectively) but as it is very acid it is very phytotoxic. It can be applied to arable land by tanker or a slurry irrigation system, but care is needed to ensure it does not get into land drains and hence into a watercourse.

Silage effluent can be given to stock, normally either cows or pigs. Care must be taken in collecting and storing the material to ensure there is no seepage. Also effluent deteriorates on storage and becomes unacceptable to stock, so a preservative such as formalin should be added.

Making silage in towers

The crop has to be blown into the top of the tower and at feeding time the silage is removed by mechanical means: thus it is vital that the ensiled material is well chopped and sufficiently dry to remain friable after compaction in the tower during storage. For this reason, grass going into a tower must be above 35% DM and many tower operators prefer 45–55% DM (sometimes called *haylage*). Very wet material must *never* be placed in a tower as the effluent from it will cause very serious corrosion to the tower structure leading to possible collapse of the tower. Because the material going into a tower must be well chopped, and because a tower is almost airtight, compaction in a tower is good, and excellent fermentation is assured. Towers have a deserved high reputation for producing good silage with minimum in-silo losses of 5–10%. (But field losses during the necessary prolonged wilting period will be 10–15%). Some farmers have shown that application of 'tower' techniques in terms of wilting and chopping the grass but then placing the material in a polythene-lined clamp produce silage equivalent to the tower: even so the tower is a first-class starting point for mechanized feeding and on some farms is justified for this reason alone.

Making big-bale silage

Precut grass can be picked up by special big balers. Unwrapped bales can be stacked very close together and tightly covered by a polythene sheet. Generally, they are either placed in individual polythene bags after baling or wrapped in polythene. Whatever system is used, it is essential that the minimum volume of air is trapped with the grass and that no further air gets near the silage until feeding time. If this practice is followed, fermentation will proceed rapidly and well, giving a well-preserved and very edible silage. This rapid fermentation may produce enough CO_2 to cause well-sealed bags to swell up: it is important to ensure that bags do not flap loosely in the wind once this distension of the bags has passed.

Effluent is very undesirable in big bags and so normally grass is wilted to at least 28% DM and some farmers prefer 35% DM. Compared with clamp silage, big bags have the following features:

Advantages	*Disadvantages*
Effective way to deal with small quantities of grass	Rupture of bag can lead to spoilage of whole contents
When baled by contractor, very low capital expenditure	Bags can be damaged by wind, rodents and birds
Bales can be stored in field and moved when needed for feeding	Feeding systems need careful planning
Can have very efficient fermentation (low losses)	
Excellent when small amounts of silage needed (e.g. for buffer feeding)	

Feeding clamp silage

Self-feeding

This takes place when the animals are allowed to help themselves to the silage. The settled height of the silage must not be higher than the animals can reach and usually the animals' access to the face is restricted by some physical means, such as an electric fence, so that they cannot climb on the silage or pull out big lumps of silage and waste it by treading it under foot. Where animals are given 24 h access to the face, then some 20 cm of face width should be allowed per cow equivalent. If access time is limited, the width per beast must be greater. The rate of feeding can be controlled by the distance the electric fence is moved each day, and by this means the silage can be rationed to last the whole winter. Self-feeding is a very cheap and effective method of feeding silage and is very popular with farmers. The only real justification for using other feeding techniques for clamp silage is where several groups of animals are to be fed from one silo face, or where a balanced feeding programme is planned in which silage must be premixed with other feeds.

Mechanical removal

Mechanical removal of silage from clamps can be done by fore-loader, grab, block cutter or a cylindrical silage cutter. Where the silage is greatly loosened during removal, it must be eaten within 12 h or it will deteriorate from re-fermentation: in this respect the block cutter is good as it removes blocks weighing some 0.6–0.7 t without losing much of the original density of the silage.

Feeding big-bale silage

Normally big bales are moved using a single protruding tine fixed to a fore-end loader, preferably with a levelling control and push-off device. Big bales are heavy and care is needed to ensure safe movement in this way. ADAS state that the weight of a big bale can be estimated by:

wt of bale (kg) = 725 − (7 × DM% of grass)

e.g. if 30% DM, then guide weight = 515 kg

With each bale holding about 0.5 t of silage, the feeding method must be based on mechanical placement of the bale. Where big-bale silage is used to supplement other forage, then ring feeders placed in the animals' exercise area can be very effective in allowing easy access for moving bales. If the big-bale silage is the major feed, and if this must be fed in a building, then there must be easy access for the tractor carrying the bale and proper control of feeding to prevent animals climbing on to the bale.

Crops for ensilage

Many green crops can be ensiled and also some arable by-products such as sugar beet tops and pea haulms. By far the most common crop is grass (including grass/clover herbage), which should be ensiled when its digestibility is at least 65, and, for high quality silage, the D-value at cutting must be 67–70.

The crop will lose about 2 units of D during the ensilage period even where there is a good fermentation, and as much as 5 units of D can be lost if fermentation is poor.

Whole-crop cereals

These can be made into silage, the cereals being cut about two weeks after full ear emergence. Yields of 8–10 t/ha of DM can be obtained from this crop, but digestibility is often low at around 62–63D, so that animal intake and performance are fairly low: also it is possible for whole-crop barley to have an intake below expectation due to physical dislike of the barley awns in the silage.

Special arable mixtures

These are sometimes grown for arable silage, usually based on the traditional oats–legume combination: this has the advantage of combining the high protein value of the legume with the good carbohydrate content of the cereal. Also this type of mixture requires less fertilizer N than straight cereals or grasses. Examples of such mixtures are:

	125 kg/ha oats	125 kg/ha oats
	35 kg/ha vetches	45 kg/ha forage peas
or	35 kg/ha beans	

Legume silage

This, usually based on red clover or lucerne, can have a high protein content, but it is more difficult to get a good fermentation than with grass, cereals or maize because of the low sugar content in legumes: an additive is very often essential. Also legumes tend to have fibrous stems when cut for silage and the material compacts much better in the silo if a precision-chop harvester is used.

Maize

Maize can make excellent silage provided it is well chopped before ensiling. A crop of maize with grain at the 'pasty' stage has a high carbohydrate and low protein content, a D-value of 63–65 and a moisture content of about 70% as it stands in the field. In the silo, this crop ferments well without additives and it is very well suited to mechanical handling. Provided its low protein value is recognized at feeding time, it is excellent material for silage and a good contributor to stock nutrition. The main snag with maize is the relatively low yield obtained in the UK coupled with the late harvesting in October or even November. A good crop of maize should produce 40 t/ha of made 30% DM silage in mild, sunny parts of southern England, but in practice yields of only about 30 t/ha often occur.

Silage – facts and figures

High yields of crop are essential for silage to justify the machinery, labour and fuel costs involved. Table 11.16 is a guide to the fertilizer rates and expected yields of grass silage cut at three different times in the season.

Density of silage

clamp silo	unchopped	20–25% DM = 720–800 kg/m^3
	chopped	20–25% DM = 800–850 kg/m^3
tower	chopped	35–45% DM = 500 kg/m^3

Feeding values (Table 11.17)

Haymaking

Haymaking is still a popular method of conservation, involving the reduction of moisture from fresh grass at 80% to about 20% when the product can be stored. Successful haymaking should follow the following principles.

- Grass should be at the correct stage of growth when cut. As grass is allowed to mature its total yield increases and its moisture content falls, so it may be tempting to allow a very heavy, mature crop to develop before cutting. *But* digestibility falls at a rate of one-third to one-half of a D-value unit/d once the seed-heads have formed; mature hay has low feed value and low intake characteristics, even though it may be well made.
- Losses should be kept to a minimum. Losses can arise in the following ways:

Table 11.16 Good yields from grass cut for silage (t/ha)

Cutting period	*Fertilizer to apply (kg/ha)*	*Site class 1*			*Site class 3*			*Site class 5*		
		DM cut	*Grass cut @ 25% DM*	*Silage @ 25% DM*	*DM cut*	*Grass cut @ 25% DM*	*Silage @ 25% DM*	*DM cut*	*Grass cut @ 25% DM*	*Silage @ 25% DM*
May	125 N (+ 30 P_2O_5 and 50 K_2O on index 0 and 1 soils)	6	24	20	6	24	20	6	24	20
July	100 N (+ at least 50 K_2O on soils below index 3)	4	16	12	3	12	10	second cut likely in wet year only		
Aug/Sept	80 N (+ at least 40 K_2O on soils below index 3)	$2\frac{1}{2}$	10	8	third cut likely in wet years only					
Total over season		$12\frac{1}{2}$	50	40	9	36	30	6	24	20
Expected number of cuts			3			2			1	

Table 11.17 Typical values for a range of silages

	DM%	*ME (Mj/kg DM)*	*D-value*	*Crude protein (% in DM)*	*DCP (g/kg)*
Grass silage					
excellent	25	11.2	70	18	120
good	25	10.7	67	16	105
average	25	10.2	64	14	95
poor	25	9.8	61	12	75
Whole-crop cereals					
barley	30	9.3	62	10	50
oats	30	8.6	57	9	60
wheat	30	8.4	55	8	35
Oats and vetches	27	9.6	60	16	95
Lucerne	27	9.4	62	20	160
Forage peas	22	9.4	62	20	150
Maize	24	10.6	65	10	70

- Respiration losses will occur in the field as the herbage continues to respire after cutting. Rapid drying will minimize these losses, which can amount to 1–5% of the DM yield.
- Mechanical losses occur if herbage is fragmented by machinery into pieces too small to be picked when the hay is collected for storage. Mechanical losses tend to become greater as the herbage becomes drier and when the action of the machines is abrasive. A mechanical loss of 10% is acceptable.
- Leaching of nutrients can take place when the cut material is exposed to rain. Ideally hay is made and removed from the field without rainfall, when this loss is zero, but long periods of heavy rain can lead to a DM loss of up to 15% and a soluble nutrient loss far in excess of this.
- Some of the hay may itself be inedible due to dust and mould. Both these factors arise when hay is made under adverse conditions: the loss can be up to 15% of the DM and it emphasizes the importance of making hay under good climatic conditions.

When grass is cut and a wide area of swath is exposed to wind and sun, there is a rapid initial loss of moisture, as external moisture is lost and water from the outer cells of the leaves and stems. At this stage drying will take place without much sun provided the atmosphere has a low relative humidity. Also at this stage of drying the material can be treated quite roughly by machines. As drying proceeds it becomes progressively more difficult to remove water and the rate of moisture loss declines, together with an increasing risk of mechanical damage. For continued drying sun is necessary to provide heat, and wind is valuable for removing water vapour from within the swath. The rate of drying can be speeded up if the grass is cut with a flail mower or passed through a crimper/conditioner immediately after cutting: however, grass treated in this way suffers more in bad weather, and flail-mown material can suffer mechanical losses as high as 50% if subjected to prolonged wet weather.

Field drying rate is maximized if the cutting machine leaves the largest possible leaf area exposed to sun and wind, and not tight swaths. If bad weather threatens during the drying process, the material should be windrowed to present the smallest area to the rain. Also it is sound practice to swath-up the grass at night to minimize the effect of dew, ensuring the ground and surface of the swath are dry before the material is again spread next morning.

The rate of drying depends on the weather, proper use of machinery, rates of N used, varieties in the sward, stage of maturity at cutting, bulk of grass present, time of year and desired moisture content at transporting from the field. Most hay is baled. It should not be baled until it is fit for storage, i.e. with a moisture content of about 20%. Frequently hay is baled at a higher moisture content than this, and then the bales are left in the field to 'cure'. Very little further moisture loss can occur in the field from the bale, and yet the bales will acquire considerable moisture if rain falls. As a general rule, baling should be regarded as the first step in the process of transport into storage, and grass should not be baled until the herbage is dry and transport into storage is organized.

Barn hay drying

This is a very useful technique for reducing the risk of bad weather, minimizing losses and making hay from younger material (at a higher feed value) or where higher N rates have been used. The material is cured as far as possible in the field, certainly down to at least 30% moisture. It can then be baled, taking care not to overcompress in the bale, and the bales carefully placed over a grid or ducts through which air can be blown. There must be no gaps between the bales, otherwise air will take the line of least resistance and fail to pass through the bales. If the outside humidity is low and only some 5–8% of moisture needs to be removed, then cold air blown for up to a week will suffice, often ceasing to blow at night when humidity might be high. During periods of sustained high humidity or when considerable moisture must be removed, the air must be heated. The air can be heated by either a flame or electric heaters, but in any event the heating of air is *very expensive*. Also, barn-drying installations cannot deal with a sudden large volume of material to cure. Barn drying is best regarded as a means of producing a limited quantity of quality hay.

Success in barn drying depends on:

- justifying the extra costs involved, rather by the higher quality of the hay than by a salvage operation on a mediocre crop;
- allowing moisture content to fall to below 30% (and certainly 35%) before baling;
- using bales at a low–moderate packed density;
- stacking the bales carefully to avoid cracks through which the majority of the air can pass;
- having sufficient fan capacity to obtain a good flow of air, with heating available if humidity is high or bales are wet;
- having a dry secondary store, so that dried bales can be moved to allow more bales to be dried.

Hay additives

Hay additives are based on either propionic acid or ammonium bispropionate. Both chemicals have a strong antifungal activity and will reduce the rate of decomposition of hay (and control heating) when the material is still too wet for immediate storage. However there are four snags to hay additives:

- It is very difficult to apply these chemicals evenly to the herbage. Attempts to apply within the bale chamber or spray on the swaths before baling have both failed to obtain the necessary even application (in contrast to the addition of silage additives in a forage harvester).
- Although both the chemicals mentioned give good initial control of decomposition, they are broken down gradually and give protection for some two to three weeks only. Thus the hay must be dehydrated to a proper storage moisture content soon after baling and storage.
- Under warm conditions at baling some 50–75% of the additive may be lost by volatilization.
- Rate of application should depend on the moisture

content of the hay – and this will vary from swath to swath and throughout the day, for example:

% moisture	*kg of additive/t of hay*
28	7
30	9
32	11
34	13

Development of a successful hay additive is proving difficult and at present it should not be regarded as a means of storing wet hay but rather as an aid to delaying the need to get to a safe storage moisture content.

Hay facts and figures

Quality in hay can be judged by its colour, which should be bright green/yellow, by its sweet smell and an absence of dust. Also the feeding value of hay can be determined by its analysis (Table 11.18).

Yields
Yields of hay vary considerably, and high N rates should not be used to grow a hay crop because it aggravates the curing problem: a maximum N rate of 80 kg/ha for the growth period is recommended.

Light crop = 2–3 t/ha of made hay
Medium crop = 4.5 t/ha of made hay
Heavy crop = 5 t/ha and over

A standard bale of hay measures approximately 0.9 m × 0.45 m × 0.35 m and weighs 20–30 kg depending on type of material and density (33–50 bales/t). A big round bale of hay measures about 1.5 m high and 1.8 m in diameter and contains some 25 times more material than a standard bale, weighing 500–600 kg. A big rectangular bale is approximately 1.5 m × 1.5 m × 2.4 m with the same weight as the round bale.

Approximate storage volumes

	m^3/t
Loose medium-length hay	12–15
Standard bales	7–10
Big rectangular bales	10

Green crop drying

When green crops are passed through an efficient high temperature drier, they are rapidly reduced to a stable moisture content with little or no loss of nutrient value. Also artificially dried green crops are very palatable to stock. In the UK the majority of green crops dried are either grass or lucerne; both are traded under the general term 'dried grass'. Less than 1% of the grassland area of the UK is used for dried grass production, and most of the drying is done in very specialized large units where the size of operation justifies the use of the large, very efficient high-temperature triple pass drier, the smallest of which can produce some 3000 t of product per year, needing about 400 ha of adjacent land to provide material. Most green crop driers are members of the British Association of Green Crop Driers, who supply specialist information to members on a variety of topics related to the industry. Although dried grass has a very high reputation as a supplement to silage in cattle and cow rations, it is unlikely that green crop drying will expand in the foreseeable future because it demands a high input of fossil fuel – equivalent to about 200 litres of oil/t of material produced, quite apart from the large equipment needed to cut and transport the grass to the drier and the power needed by the mill and cuber to package the material after drying.

In an efficient grass-drying unit, dry-matter losses are only about 3–5%, and in this respect the technique is far superior to hay and silage making (Table 11.19).

Grassland farming and the environment

As grassland is the most widely occurring crop in the UK it is inevitable that grassland farming has a broad interface with many aspects of the environment.

Main points to consider are as follows:

Pollution of watercourses and drinking water

Animal wastes and manures should never be allowed to run directly into drains and watercourses: a barrier ditch

Table 11.18 Typical values for a range of types of hay

	DM%	*ME (MJ/kg DM)*	*D-value*	*Crude protein (% in DM)*	*DCP (g/kg)*
Grass hay					
excellent*	85	10.7	67	13	100
good	85	9.6	64	11	60
average	85	9.0	60	10	50
poor	85	8.4	56	9	40
Grass/clover hay	85	10.0	64	15	110
Lucerne	85	8.8	55	20	140
Sainfoin	85	9.3	58	16	120

* Barn-dried samples can have higher values for ME, D and protein.

Table 11.19 Feed analysis figures for dried-grass and lucerne

	DM%	*ME (MJ/kg DM)*	*D-value*	*Crude protein (% in DM)*	*DCP (g/kg)*
Dried grass					
leafy grass	90	11.2	70	16	110
average	90	9.6	64	15	95
lucerne	90	9.6	60	24	170

system may help to alleviate the problem but most of the material is best applied to the land, where the maximum recommended rate is 25 000 litres/ha of material in any one year.

Silage effluent is a very potent pollution hazard. The best single measure of the harmful effect of such materials is the amount of oxygen taken out of the water, which is measured as BOD (Biological Oxygen Demand). As the following figures show, silage effluent is 168 times more demanding than domestic sewage and only milk is worse, being twice as demanding as silage effluent.

	BOD (mg/litre)
Raw domestic sewage	400
Dairy washings	2 000
Cow slurry	35 000
Silage effluent	67 000
Milk	140 000

Other farm wastes can cause problems, notably used chemical containers (from dairy hygiene, animal health and crop protection materials). Guidelines for disposal area:

(1) Empty the container completely at time of use for its intended purpose (do not wash out and let water down drain).
(2) Destroy containers as soon as possible and do not use them for any other purpose.
(3) *Combustible containers* should be burnt well away from habitations, stock and edible crops.
(4) *Metal containers* should be flattened and buried at least 450 mm deep well away from drains. The spot should be marked and recorded for future reference.
(5) *Glass containers* should be broken in a sack and buried as (4).

Nitrate losses

Fertilizer N does not leach rapidly from grassland if it is applied in the February–September period. There is substantial leaching of nitrate from all grassland in the autumn and this leaching will be greater from high N grassland than from low N grass or grass/clover swards. Ploughing of grassland gives very substantial losses of nitrate over the next few months.

Effect on botanical composition of swards and soil

As shown by Jones (1933), the present composition of a farmed area of grassland is a reflection of its immediate past management. If the yield and speed of regrowth of desirable species is encouraged, then other species in the sward will suffer from enhanced competition. Thus, productive grassland will contain a narrower range of plant species than non-productive land.

Grassland encourages the build-up of good soil structure and an increase in soil organisms. If an area of permanent grass is productive enough to support 2 LSU/ha, it has been estimated that the total weight of organisms below ground is equivalent to 20 LSU/ha.

Table 11.20 Earthworm biomass (kg/ha)

	Drained land	*Undrained land*
Permanent pasture		
200 kg/ha N	165	100
400 kg/ha N	285	125
Reseed		
400 kg/ha N	60	105

Table 11.20 (based on Garwood & Gilbey, 1985) shows the interactive effect of fertilizer N and drainage on earthworms and the harmful effect of reseeding.

Other factors

Good grassland farming should not give grounds for serious environmental concern. Care should be taken to avoid offence from annoying odour caused by animal units and farm waste. Also, the appeal of the countryside can be marred by such items as scattered polythene sheets, untidy fencing and restricted access to public footpaths.

Glossary of grassland terms

Based on Hodgson (1979) and Thomas (1980), with additions.

Anthesis: flower opening.
Biomass: weight of living plant and/or animal material.
Browsing: the defoliation by animals of the above-ground parts of shrubs and trees.
Buffer feeding (or grazing): provision of additional forage when quantity in grazing area is insufficient.
Canopy: the sward canopy as it intercepts or absorbs light.
Canopy structure: the distribution and arrangement of the components of the canopy.
Closed canopy: a canopy which either has achieved complete cover or intercepts 95% of visible light.
Cover: the proportion of the ground area covered by the canopy when viewed vertically.

Cow equivalent (CE): an aggregated liveweight of animals equivalent to a standard cow of 550 kg (synonymous with LSU).
Crop growth rate (CGR): the rate of increase in dry weight per unit area of all or part of a sward.
Crop (standing): the herbage growing in the field before it is harvested.
Crown: the top of the tap-root bearing buds from which the basal leaf rosette and shoots arise (appropriate to clovers but *not* grasses).
Culm: the extended stem of a grass tiller bearing the inflorescence.
D-value: digestibility of organic matter in the DM eaten (as %).
Date of heading (or date of ear (inflorescence) emergence): for a sward, the date on which 50% of the ears in fertile tillers have emerged.
Defoliation: the severing and removal of part or all of the herbage by grazing animals or cutting machines.
Density: the number of items (e.g. plants or tillers) per unit area.
Flag leaf: in grass this is the final leaf produced on the flowering stem.
Foliage: a collective term for the leaves of a plant or community.
Forage: any plant material, except in concentrated feeds, used as a food for domestic herbivores.
Forage feeding: the practice of cutting herbage from a sward (or foliage from other crops) for feeding fresh to animals.
Grassland: the type of plant community, natural or sown, dominated by herbaceous species such as grasses and clovers.
Grazing: defoliation by animals.
Grazing cycle: the length of time between the beginning of one grazing period and the beginning of the next.
Grazing period: the length of time for which a particular area of land is grazed.
Grazing pressure: the number of animals of a specified class per unit weight of herbage at a point of time.
Grazing systems:
Continuous stocking: the practice of allowing animals unrestricted access to an area of land for the whole or a substantial part of a grazing season.
Creep grazing: the practice of allowing young animals (lambs or calves) to graze an area which their dams cannot reach.
Mixed grazing: the use of cattle and sheep in a common grazing system whether or not the two species graze the same area of land at the same time.
Rotational grazing: the practice of imposing a regular sequence of grazing and rest from grazing upon a series of grazing areas.
Set stocking: the practice of allowing a fixed number of animals unrestricted access to a fixed area of land for a substantial part of a grazing season.
Harvesting: defoliation by machines.
Harvest year: first full harvest year is the calendar year following the seeding year.
Herbage: the above-ground parts of a sward viewed as an accumulation of plant material with characteristics of mass and nutritive value.
Herbage allowance: the weight of herbage per unit of live weight at a point in time.
Herbage consumed: the herbage mass once it has been removed by grazing animals.
Herbage cut: the stratum of material above cutting height.
Herbage growth: the increase in weight of herbage per unit area over a given time interval due to the production of new material.
Herbage mass: the weight per unit area of the standing crop of herbage above the defoliation height.
Herbage residual: the herbage remaining after defoliation.
Inflorescence emergence: the first appearance of the tip of a grass inflorescence at the mouth of the sheath of the flag-leaf.
Leaf: in grasses = lamina + ligule + sheath; in clovers = lamina + petiole + stipule. Note leaf is *not* lamina.
Leaf area index (LAI): the area of green leaf (one side) per unit area of ground. *Critical LAI* is LAI at which 95% visible light is intercepted; *maximum LAI* is the greatest LAI produced by a sward during a growth period; *optimum LAI* is the LAI at which maximum crop growth rate is achieved.
Leaf area ratio: total lamina area divided by total plant weight: *specific leaf area* is the lamina area divided by lamina weight.
Leaf burn (or scorch): damage to leaves caused by severe weather conditions, herbicides, etc. This contrasts with *leaf senescence* which is a genetically predetermined process where leaves age and die, usually involving degradation of the chlorophyll in the leaves.
Leaf emergence: a leaf in grass is fully emerged when its ligule is visible or when the lamina adopts an angle to the sheath. In clovers the leaf is emerged when the leaflets have unfolded along the midrib and are almost flat.
Leaf length to weight ratio: lamina weight divided by lamina length.
Livestock Unit (LSU): see CE.
Net assimilation rate (NAR): defined as $1/A{\cdot}\mathrm{d}W/\mathrm{d}t$ where A = total leaf area, W = total plant weight, t = time, units = $(\mathrm{g\,m^2})/\mathrm{d}$.
Palatable: pleasant to taste.
Plastochron: the interval between the initiation of successive leaf primordia on a stem axis.
Preference: the discrimination exerted by animals between areas of a sward or components of a sward canopy, or between species in cut herbage. *Preference ranking* is based on the relative intake of herbage samples when the animals have a completely free choice of the materials.
Pseudostem: the concentric leaf sheaths of a grass tiller which perform the supporting function of a stem.
Regrowth: the production of new material above the height of defoliation after defoliation, often with initial regrowth at the expense of reserves stored in the stubble.
Rest period: the length of time between the end of one grazing and the start of the next on a particular area.
Seed ripeness: the stage at which the seed can be harvested successfully.
Seeding year: the calendar year in which the seed is sown.
Seedling emergence: a seedling has emerged when the shoot first appears above the ground.
Selection: (by animals) the removal of some components of a sward rather than others.

Shoot bases: the part of the sward below the anticipated height of defoliation. This becomes *stubble* after defoliation.
Simulated sward: an assemblage of plants intended to represent in convenient form a 'normal' sward.
Sod: a piece of turf lifted from the sward, either by machine or grazing animals.
Spaced plant: a plant grown in a row so that its canopy does not touch or overlap that of any other plant.
Standing crop: the herbage growing in the field before it is harvested.
Stem: the main axis of a shoot, bearing leaves.
Stocking density: the number of animals of a specified class per unit area of land actually being grazed at a point in time.
Stocking rate: number or weight of animals kept on a unit area of land for a long period (preferably for 12 months, so including grazed area *and* conserved area).
Sward: an area of grassland with a short continuous foliage cover, including both above- and below-ground parts, but not any woody plants.
Sward establishment: the growth and development of a sward in the seeding year. *A primary sward* is one that has never been defoliated. *A mixed sward* is one that contains more than one variety or species. *A pure sward* is one that contains a single stated variety or species.
Sward height: height above ground at which a descending finger first touches vegetative material (ignoring stems and seed heads).
Tiller: an aerial shoot of a grass plant, arising from a leaf axil, normally at the base of an older tiller. *An aerial tiller* is one that develops from a node of an extended stem.
Tiller appearance rate (TAR): the rate at which tillers become apparent to the eye without dissection of the plant.
Tiller base: the part of a growing tiller below the height of defoliation.
Tiller stub: the part of a tiller left after defoliation.
Turf: the part of the sward which comprises the shoot system plus the uppermost layer of roots and soil.
Utilized metabolizable energy (UME): the quantity of ME accounted for in animal production. *If UME of grass*, then UME after deduction of ME supplied from non-grassland sources.
Winter burn: leaf burn in winter.
Winter hardiness: the general ability of a variety or species to withstand the winter.
Winter kill: death of plants in winter.

References

Baker, H.K. (1962) *Proceedings of the Sixth Weed Control Conference, Brighton* **1**, 23–30.
Baker, R.D. (1964) *Journal of the British Grassland Society* **19**, 149–155.
Brockman, J.S. & Gwynn, P.E.J. (1961) *Journal of the British Grassland Society* **16**, 201–202.
Coppock, J.T. (1976) *Agricultural Atlas of England and Wales*. Faber & Faber, London.
Davies, W. (1960) *The Grass Crop*. Spon, London.
Forbes, T.J., Dibb, C., Green, J.O., Hopkins, A. & Peel, S. (1980) *Factors Affecting the Production of Permanent Grass*. Grassland Research Institute, Hurley.
Garwood, E.A. & Gilbey, J. (1985) Grassland Manuring Occasional Symposium No 20, 94–96. British Grassland Society, Hurley.
Green, J.O. (1974) *Preliminary Report on a Sample Survey of Grassland in England and Wales. Report 310.* Grassland Research Institute, Hurley.
Hodgson, J. (1979) *Grass and Forage Science* **34**, 11–18.
Hodgson, J., Mackie, C.K. & Parker, J.W.G. (1986) British Grassland Society: Grass Farmer No. 24, 5–10.
Hopkins, A. (1979) *Journal of the Royal Agricultural Society of England*, 140–150.
Hopkins, A. & Green, J.O. (1978) Changes in Sward Composition and Productivity, Occasional Symposium No. 10, 115–129. British Grassland Society, Hurley.
Jewiss, O.R. (1979) British Grassland Society: Grass Farmer No. 4, 3–7.
Jones, M.G. (1933) *Empire Journal of Experimental Agriculture* **1**, 43–47; 122–128; 361–367.
Leaver, J.D. & Moisey, F.R. (1980) British Grassland Society: Grass Farmer No. 7, 9–11.
Marsh, R. (1975) *Pasture Utilisation and the Grazing Animal, Occasional Symposium No. 8*, 119–128. British Grassland Society, Hurley.
Morrison, J. Jackson, M.V. & Sparrow, P.E. (1980) *Report of Joint GRI/ADAS Grassland Manuring Trial.* Report No. 27, Grassland Research Institute, Hurley.
Thomas, H. (1980) *Grass and Forage Science*, **35**, 13–23.
Thomas, C., Reeve, A. & Fisher, G.E.J. (1991) *Milk from Grass*, 2nd edn. Scottish Agricultural Colleges, Perth.
Walsh, A. (1982) *The Contribution of Grass to Profitable Milk Production* (Rex Paterson, Memorial Study). British Grassland Society, Hurley.
Wolton, K.M. (1980) AgTec, Summer 1980. Fisons Fertilizer Division, Felixstowe.

Further reading

Frame, J. (1992) *Improved Grassland Management*. Farming Press, Ipswich.
Holmes, W. (1989) *Grass: Its Production and Use*, 2nd edn. Blackwell, Oxford.
Jones, M.B. & Lazenby, A. (1988) *The Grass Crop*. Chapman & Hall, London.
Minson, D.J. (1990) *Forage in Ruminant Nutrition*. Academic Press, London.
NIAB Farmers' Leaflet Recommended Varieties of Grasses and Clovers. *National Institute of Agricultural Botany*, Cambridge.
Sheldrick, R., Thompson, D. & Newman, G. (1987) *Legumes for Milk and Meat*, Chalcombe Publications, Marlow.
Wilkinson, J.M. (1984) *Milk and Meat from Grass*. Granada Publishing, London.

See also
British Grassland Society Symposium Proceedings published regularly on specific grass and forage topics, published by British Grassland Society.

12

Farm Woodland Management

A.D. Carter

Woodland covers 2 300 000 ha or about 10% of the total land area of Britain, of which farm woodlands make up about 300 000 ha (Hart, 1991). Less productive than many larger commercial forests, farm woodlands typically occur in small fragmented parcels and comprise a significant proportion of mixed broadleaved woods, especially in the lowlands of Britain, where they form important features in the landscape and may have considerable value for wildlife and game. All too often, however, the poor condition of the growing stock, difficult access, small size and neglected or inappropriate management all conspire to reduce the productivity and quality of timber. Fortunately, both site conditions and the quality of the trees can be improved by the application of appropriate silvicultural techniques to create a high quality timber resource with multiple benefits for the landscape, amenity, wildlife and game. New farm woodland planting is also expanding, as surplus agricultural land becomes available and new silvicultural systems such as agroforestry and short-rotation coppice become more economically viable.

Despite this clear potential, a number of perceived obstacles deter landowners from either managing existing woodlands or planting new ones, including a lack of knowledge of silviculture or how to market small volumes of timber. The most enduring financial obstacle is the long gestation period involved in forest investments, requiring a wait of at least twenty years for most crops before thinnings even begin to offset the early establishment costs. This 'income gap' can be partially offset by some of the following options:

- Revenue from harvesting an existing crop can be used to offset the cost of establishment.
- Grants are available to compensate landowners for planting trees and other forms of woodland management.
- Development of shorter rotation crops (e.g. coppice).
- Agroforestry systems allow the spaces between the trees in the early years to be utilized for reduced-level grazing or other agricultural use.
- Integration of forestry with other revenue-generating activities in the early life of the crop, such as game management, recreation or growing Christmas trees.

Setting management objectives

While woodlands can be successfully managed for multiple benefits, competing objectives must be prioritized so that an appropriate silvicultural management plan can be formulated. For the majority of farm woodlands, maximum timber production may have to be compromised to a certain extent to incorporate secondary objectives, such as game and wildlife habitat or to meet the requirements of landscape designations. Once the overall objectives are set, the specific silvicultural operations most appropriate will depend on an assessment of the site conditions and the growing crop, together with a knowledge of the external environment, such as the market for timber produce, the wider policy framework and grant arrangements.

Forest products and markets

Timber is normally sold on a price per unit volume basis, which is set according to the volume available, the state and location of the timber (standing, stacked in the forest or at the roadside), ease of access and the method of sale (auction, tender, or private sale). Each species will have recognized characteristics which must be met for particular markets, such as length, diameter (Fig. 12.1), straightness and freedom from defects. Some markets are much more exacting than others (e.g. veneers) demanding higher prices, so that the harvested timber may be graded and separated into different price bands according to quality. For most broadleaves, the difference between the market price for low and high quality is enormous. For example, oak veneers may fetch up to £300/m^3, while poor quality oak may only be worth £10/m^3 as firewood.

Smaller farm woodlands cannot compete with the large upland forests, which supply vast quantities of timber to bulk markets, such as pulp processors and particle board manufacturers. The farm woodland owner is best advised to produce high quality timber, to take advantage of local specialized markets and to add value wherever possible by processing on the farm. For further

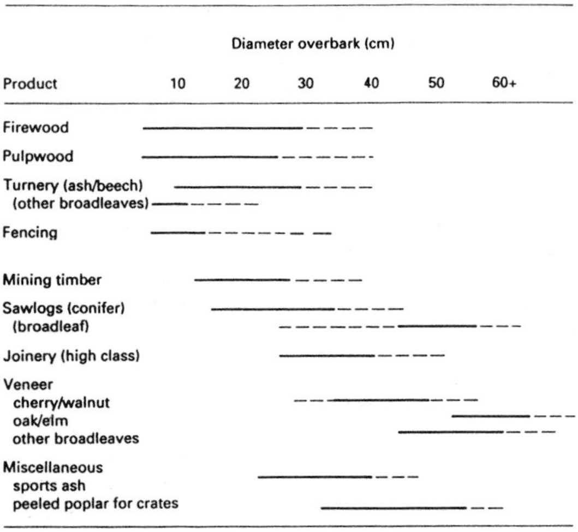

Diameter overbark (cm)

Product 10 20 30 40 50 60+

Firewood

Pulpwood

Turnery (ash/beech)
(other broadleaves)

Fencing

Mining timber

Sawlogs (conifer)
(broadleaf)

Joinery (high class)

Veneer
cherry/walnut
oak/elm
other broadleaves

Miscellaneous
sports ash
peeled poplar for crates

Fig. 12.1 Diameter specifications for various end uses (after Hibberd, 1988, by permission of Forestry Authority, 1994).

information on timber properties of farm woodland trees, see Brazier (1990).

Silvicultural options for existing woodlands

Clear felling

Clear felling involves the harvesting and removal of the whole forest crop at once. As the system of choice for maximum timber production it enables the creation of a uniform and even-aged restocking offering simplified forest treatment and associated economies of scale. The use of transplants enables the forest manager to take advantage of superior species, provenances or cultivars better adapted to the site.

The open nature of the clear fell creates forest conditions substantially different from those under the woodland canopy which have a number of important environmental implications. The increased levels of light, rapid breakdown of organic matter and the exposure of mineral soil encourage the vigorous growth of weed species. The exposed soil may be prone to erosion and increased run-off, especially on sloping ground. The brash left from felling may provide a home for weevils and beetles which may damage future crops. The edges of neighbouring stands will be exposed to increased levels of light and risk of wind damage. Thin-barked species, such as beech, may suffer from sun scorch, and oak trees may develop epicormic growth. On some sites, the large-scale clear felling of mature trees may be inadvisable due to amenity or landscape considerations.

Natural regeneration

The majority of existing woodlands have the potential to regenerate from seed that has fallen on the forest floor, given enough time and the right environmental conditions. For the forest manager, this 'free gift' can provide a low cost means of restocking woodland, as well as important conservation advantages.

Successful natural regeneration relies on interventions to enhance the natural process of regeneration in the forest. Where timber production is an objective of management, the parent crop must be of good quality, well suited to the site and capable of producing large quantities of viable seed. The most prolific in this respect are ash, sycamore, birch, alder, Norway maple, Scots pine, sweet chestnut, cherry and rowan. Some species produce more seed in mast years (cycles of 3–5 years), such as oak and beech, although some seed may be produced every year. Common practice is to carry out a series of heavy late thinnings to encourage seed production from the best trees, by removing 40–60% of the crop, or overstorey. Ground preparation may be required to break up weed growth, incorporate humus layers and expose the mineral soil, with the aim of producing a moist, friable soil surface free from competing vegetation. Rotovation, disk ploughing or the wallowing of pigs have proved successful in this respect. An assessment of the restocking of seedlings should be undertaken in the year following seeding, with gaps larger than 7 × 7 m planted up with transplants, if full stocking is required. The regeneration will require protection from grazing by either fencing (which can prove expensive for small areas) or individual protection by tree shelters. The retention of the over-storey allows rank weed growth to be controlled, as well as giving protection from frost and sun scorch. As the young trees become established, the overstorey is gradually removed, increasing the amount of available light to the growing crop. Subsequent operations include the re-spacing of the regeneration to at least 1500 trees/ha.

The creation of a fully stocked woodland through natural regeneration can prove a complex and consequently costly exercise. The key drawbacks for timber production include the reliance on seeding years rather than markets, the need for skilled management, the possibility of adverse ground conditions (including brambles or rhododendron), or parent trees of inferior quality.

In a review of broadleaved natural regeneration, Evans (1988) suggests that natural regeneration could be used for infilling gaps, adding to woodland, stocking inaccessible pockets of ground (especially in the uplands), improving wildlife habitats, increasing amenity value and for regeneration operations when a heavy seed year coincides with harvesting.

A number of silvicultural systems take advantage of the potential for natural regeneration. In the *shelterwood system*, a partial overstorey is retained for 2–30 years, depending on the species, creating an uneven-aged woodland structure for at least part of the silvicultural cycle. This can be advantageous where the complete removal of the forest crop may be too drastic for landscape or amenity reasons. In *group systems*, the size of the forest unit is reduced in size to a small felling coupe (30–50 m diameter), rather than the whole stand, normally centred on existing gaps or regeneration. Larger gaps (0.1–0.5 ha) are suitable for light-demanding species and easier to manage, but have more weed growth and a higher risk of windthrow, and natural regeneration may not be able to penetrate the centre of the gap. The orientation affects the microclimate in the gap: north–south of elliptical shape provides a good compromise between wind and sun. Other drawbacks of the group system include the opportunity costs foregone by the small and hence uneconomic size of gaps and the opening of the canopy to wind.

Coppice

Coppicing is a silvicultural system suitable for mainly broadleved species, which are capable of regenerating from the stumps, or *stools*, of a previous crop cut near to ground level. The coppice shoots arise from dormant or adventitious buds on the stool, at or below the cut surface. This vegetative regeneration has the advantage over seedlings in that ample supplies of carbohydrates are available from the parent stool and its root system. The vigorous early growth of coppice shoots means that weeding is not normally required, although the stools may need some protection from grazing damage in the first few years. Stools are normally cut in the dormant season, although the best time is early spring as the buds begin to swell (March to early April) and the roots have their maximum carbohydrate stores. Autumn and winter cutting leads to some loss of stools from bark separation and late spring frost damage. Regularly coppiced stools are known to survive many rotations, with some lime stools living to many hundreds of years old, although the ability to coppice generally decreases with the age of the coppice regrowth to 40 years or about 40 cm diameter. Dormant buds are increasingly likely to have their connections with the pith severed and development made incapable, especially as the bark at the base of the trunk also becomes thickened and more difficult to break through.

The most vigorous coppice species are willow, sweet chestnut, hornbeam, lime, pedunculate and sessile oak, ash and hazel. Some species readily produce suckers, such as wild cherry and some poplars, and beech coppices only very weakly. The vast majority of conifers do not coppice, with the exception of Coast redwood *Sequoia sempervirens*.

Traditional coppicing arose by the regular cutting of natural mixed woodland. Stools which died were replaced by cuttings, seedlings or by a process of *layering* involving bending over a remaining coppice shoot from an existing healthy stool, pegging it to the ground and covering with earth. Coppice is one the oldest forms of forest management, supplying fuelwood and roundwood for traditional rural crafts producing hurdles, thatching spars and tool handles. The length of rotation is set to the type of product required, short rotations of 5–7 years for small diameter pliable material and longer rotations of 20–40 years for material to be turned, cleaved and squared. During the 1800s, coppice expanded to supply charcoal for the iron and glass industry and oak bark for tanning. Since the middle of the last century, however, the substitution of coal for fuelwood and the decline in demand for coppice products has led to much former coppice falling into disuse. The only significant market remains for sweet chestnut coppice for split fence palings and stakes, although local markets exist for fuelwood, pulp and specialized rural crafts.

For suitable species, coppice offers rapid early growth and assured establishment; however, the burst of growth is not sustained over longer rotations. In terms of volume production, coppice is not as productive as high forest systems, with yields of 2–4 m^3/ha/year being normal.

Coppice with standards

Coppice with standards is a two-storey system comprising an understorey (simple even-aged coppice) and the overstorey of uneven-aged standards, usually of seedling origin. Standards characteristically have short boles and a greater proportion of branches than high forest trees, especially when the coppice rotations are short. The proportion of coppice to standards in a compromise: ideally there are between 75–100 standards of all sizes per ha. Oak is traditionally the most common standard, but may not be the most suitable, due to its branching habit. Trees with a more marked apical dominance, lighter canopy, and less tendency to produce epicormics (e.g. ash, lime and cherry) may be more suitable. Thin-barked trees such as beech and sycamore tend to suffer from sun scorch when released and are unsuitable.

Common understorey species include oak, ash, hornbeam, beech, sweet chestnut, sycamore, field maple, lime, hazel, alder, birch, cherry, sallow, elm and aspen. Where the density of standards is high, shade-tolerant species are favoured (e.g. lime, sycamore).

Stored coppice

Coppice can be converted to high forest by reducing the number of coppice shoots to one for every stool. The faster growth in the first few years does not continue through the rotation, so that volume out-turn equals that of maiden trees. Stored coppice typically produces poor quality timber, as a result of basal sweep, shake and a tendency to decay.

Short-rotation coppice for firewood and biomass

Firewood is traditionally a secondary product from the branchwood and low quality timber of broadleaved woodland, especially coppice. As a fuel, only air-dried wood should be burnt, since wood with a high moisture content gives off little heat and causes tar deposits in stoves and flues. The best timbers for firewood include ash, beech, birch, hazel, holly, oak and old fruit trees, which may be cut and split or chipped by machine.

High volume production can be secured over short rotations of coppice (5–10 years) from selected broadleaved hybrids planted at high densities for energy generation, firewood or industrial cellulose. Experimental trials of willow (*Salix* × *viminalis*, Bowles Hybrid) and poplar hybrids from Belgium promise yields of up to 30 m^3/ha/year, equivalent to 13–15 tonnes of dry wood/ha/year or the energy produced from 7 tonnes of coal. To prove economically viable as an alternative to agricultural production, intensive production methods are required, including the use of high-yielding disease-resistant hybrids, fertilizer and herbicide applications and mechanical harvesting. Further expansion of the biomass market will depend on the relative cost of fossil fuels and advances in energy generation technology.

Christmas trees

The production of Christmas trees can be successfully incorporated into woodland management by utilizing dead ground under electricity pylons or other temporary spaces in newly established mixtures and along forest rides. The most commonly grown species is Norway spruce; although more attractive and needle-retentive species, such as Noble fir (*Abies procera*) and Caucasian fir (*Abies nordmanniana*) are gaining ground. The objective is to produce healthy, well shaped and bushy trees by planting at a close spacing of 1 × 1 m (10 000/ha) with careful attention to weeding and inspection for the presence of insect damage. Shearing may be required in the fourth year to restrict the length of the leading shoot and to create a better shaped tree. The trees are normally harvested in a range of sizes to suit local markets, with the main crop being harvested between 4 and 10 years.

Improvement of neglected woodland

The poor quality of many farm woodlands generally owes more to inadequate or inconsistent past forest management than to any inherent poor site fertility. Where inspection reveals the presence of a potential final crop of defect-free marketable species, complete replacement is rarely advisable. The three broad options below consist of increasingly drastic actions, depending on the level of stocking.

Improvement (even stocking, 150–300 stems/ha final crop)

Improvement involves the identification of final crop trees and carrying out silvicultural treatments to release those chosen stems. From the pole stage onwards (10 m top height), all competing trees and climbers are removed from around the final crop trees to allow complete crown freedom. Pruning is advisable to produce clean stems and to remove epicormic growth. The felling of the poorer quality matrix will often offset the cost of these silvicultural treatments.

Enrichment (uneven stocking, <150 stems/ha final crop)

Enrichment involves the planting of additional trees in a stand to increase the stocking of utilizable stems. The degree of enrichment required depends on the quality of the existing crop. Where some utilizable trees remain, the focus of enrichment is usually existing gaps or bare ground. The width of the planting gaps should be at least 1.5 times the height of the tallest neighbouring trees. If the existing crop does not contain sufficient utilizable stems, the previous crop can be partly cleared in swathes or bands at least 4 m wide. Enrichment species are chosen for rapid early growth to beat weed competition and coppice regrowth, and include Norway maple, cherry, southern beech, sycamore, western hemlock and western red cedar. The individual establishment cost per tree is quite high and the use of treeshelters at 2–3 metre spacing has proved most successful, although attention is needed to ensure that the growing trees are not shaded out by crown expansion of the remaining overstorey.

Replacement (no potential final crop)

Replacement involves the partial of full removal of the existing crop and its replacement by a new superior crop, where the present stand is unlikely to produce a potential final crop. The system is effectively a clear fell, although most of the trees will not have reached conventional harvestable size. Regrowth from cut stumps can be killed by the use of a foliar herbicide (e.g. glyphosate) or ammonium sulphate on larger stumps. The incidence of honey fungus (*Armillaria* spp) may preclude the use of some species for replanting. Many former broad-leaved woodlands have been converted to more productive coniferous plantations by this method since the 1940s, with mixed success and enormous losses to wildlife and landscape value. The cost of clearance nowadays does not generally justify the replacement of woodlands to conifers and may be inadvisable on the basis of current grants geared to broadleaved woodland. A compromise may be to clear and replant a woodland in several stages by clearfelling only part of the woodland at any one time.

Forest measurement

Forest measurement is usually required to provide an estimate of the quantity of timber for sale. In addition, forest inventory can provide the woodland owner with useful information for planning and forecasting future production. A periodic inventory will show the current state of the stand, its performance over time and any consequent deviations from the expected which may require treatment (e.g. restricted drainage). The amount of time invested in forest measurement is a balance between the degree of precision required and the costs involved. Generally more precise measures are used for more valuable timber in larger quantities.

Forest measurement and inventory planning are fully described in two Forestry Commission publications, Hamilton (1985) and Edwards (1983). Timber is normally measured as solid volume, although other measures exist for specific purposes (green weight, dry weight, number of pieces). A number of conventions exist, as follows.

Diameter

Diameter (cm) is normally measured using a girthing tape around the circumference of the tree or log, rounded down to the nearest whole centimetre. *Diameter at breast height (dbh)* is measured at 1.3 m above ground level.

Length

Length (m) is measured by tape following the curvature of the log, rounded down to 0.1 m for lengths up to 10 m and 1 m for lengths over 10 m.

Height

Height (m) can be measured by an observer standing a known distance (at least 1.5 times the height of the tree) away from the base of the tree by either trigonometry or from direct readings from specifically designed instruments, such as the clinometer, altimeter or hypsometer. *Total height* is the straight-line distance from the base to the uppermost tip of the standing tree. *Timber height* is the height to the lowest point on the main stem where the diameter is 7 cm overbark. In broadleaves, timber height is the lowest point at which no main stem is distinguishable. The *top height* of the stand is the average total height of the 100 trees of largest diameter (dbh) per hectare and is used to estimate the yield class. A more practical method of estimating top height is given below for single-species even-aged plantations or for the dominant species in mixtures.

(1) Lay down random plots through the stand, the number of samples required depending on the size and variability of the crop.

Area of stand (ha)	*Uniform crop*	*Variable crop*
0.5–2.0	6	8
2.0–10.0	8	12
Over 10.0	10	16

(2) Record the total height of the single tree of largest dbh within 5.6 m of the plot centre.
(3) Calculate the average height of the samples taken to give an estimate of the top height of the stand.

Basal area

Basal area (m^2) is the cross-sectional area of a tree at 1.3 m (dbh).

$$\text{Basal area (individual tree)} = \frac{\pi \times \text{dbh}^2}{40\,000} \text{ m}^2$$

where dbh is measured in cm.
The basal area of a stand (m^2/ha) is the sum of the basal areas of the trees in the stand. The basal area of a stand can be measured by laying down sample plots or using a relascope (or angle gauge) to take direct readings (see Hamilton, 1985). Basal area is a useful value as it gives a better measure of the level of stocking than tree density alone.

Volume

Volume (m^3) is measured overbark to 7 cm diameter, or where no main stem is distinguishable, not including the volume of branchwood. The measurement of timber volume is always an estimate, due to the variable nature of trees and logs; in particular, the volume must be corrected to account for the taper of different species of tree. A number of useful tables have been produced in the aforementioned Forestry Commission booklets, which allow volume to be estimated by measuring height and diameter of standing or felled timber. The following procedures are drawn from those publications.

Estimating the volume of a standing tree

The volume of a standing tree can be estimated by correcting the total height (or timber height for broadleaves) by the use of a 'form factor' to account for the taper. The degree of taper increases in widely spaced stands and among open-grown broadleaves. Appropriate form factors range from 0.5 for mature plantation conifers, 0.4 for open-grown conifers or close-spaced broadleaves to 0.35 for younger trees. To calculate the volume using this method, first find the basal area at breast height (in m^2) and multiply this by the total height (m) and the appropriate form factor.

Example
A mature plantation-grown Douglas fir with a total height of 22 m and dbh of 46 cm:

$$\text{Basal area} = \frac{\pi \times 46^2}{40\,000}$$
$$= 0.166$$
$$\text{Volume} = 0.166 \times 22 \times 0.5$$
$$= 1.83 \text{ m}^3$$

An alternative method in Hamilton (1985) is based on the use of single tree tariff charts, by measuring dbh and total height.

Estimating the volume of a stand

The method of estimating stand volume will depend on the value of the timber and the method of sale. For moderate to high value crops the *tariff system* is appropriate. This system is based on the relationship between tree diameter and volume, which will vary depending on the tree species, age, and site conditions. Tables have been produced for a wide range of conditions in British forestry, where this relationship is classified by tariff numbers from 1 to 60. By calculating the appropriate tariff number for any particular stand of trees, the volume of the stand will be found by consulting the tariff tables in Hamilton (1985).

Measuring the volume of felled timber

The volume of felled timber can be reasonably accurately calculated by measuring the diameter of the log at the midpoint along its length (midpoint diameter) and using Huber's formula below:

$$\underset{(\text{m}^3)}{\text{Volume}} = \frac{\pi \times (\text{mid point diameter})^2}{40\,000} \times \text{length of log}$$

where diameter is measured in cm and the length is measured in m. Logs over 15 m should be measured in two or more sections. Volumes can also be calculated using the midpoint diameter tables in Edwards (1983) and Hamilton (1985). In addition, tables have been produced for sawlog and roundwood volumes on the basis of top diameter (normally underbark) and length. This assumes a standard taper of 1:120 and gives a quicker but less accurate measure of log volumes.

Other measures

An estimation of timber in *stacks* is sometimes useful, although the volume of the stack must be converted to a solid volume by the use of conversion factors, which take account of log diameter, length, taper, straightness, and method of stacking. Typical conversion factors range between 0.55–0.65 for broadleaves and 0.65–0.75 for conifers, although for short, straight billets the figure may be as high as 0.85.

The yield class system

The annual increase in timber volume for even-aged stands of trees is termed the *current annual increment* (CAI, m^3/ha/year), which rises rapidly in the early life of the plantation and falls gradually as growth rates decline in old age. The mean annual increment (MAI, m^3/ha/year) is the average annual yield up to any particular age, which reaches a maximum at the ideal rotation age for maximum volume production. The maximum MAI also defines the *general yield class* (GYC, m^3/ha/yr) of the stand, which is a measure of the productivity of the trees under those site conditions (Fig. 12.2).

The GYC of an even-aged plantation can be estimated without detailed volume measurements, since a good correlation exists between top height, species, age and yield class. By estimating the top height of the stand, the yield class for that species can be read from the appropriate table in Edwards & Christie (1981) or Rollinson (1988). It should be remembered that these tables give estimated future volume production on the basis of normal silvicultural treatments and full stocking. A reduction of 15% from the total area is normally made for forest tracks, rides and stacking areas, although this should be increased where other unstocked ground exists.

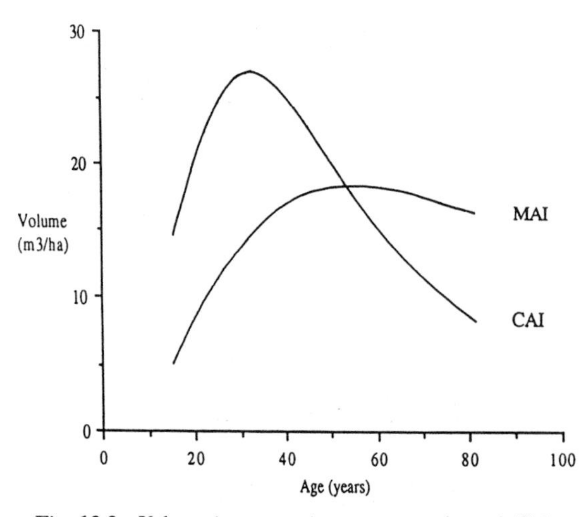

Fig. 12.2 Volume increment in an even-aged stand (Sitka spruce GYC 18) (after Edwards & Christie, 1981, by permission of Forestry Authority, 1995).

Thinning

The spaces between trees in a young plantation quickly close, as the crowns of individual plants coalesce. Trees compete for light, moisture and nutrients and eventually some trees become increasingly suppressed or overtopped by their more dominant neighbours. The growth of these suppressed trees is arrested and they become increasingly susceptible to disease and decay. This natural process of self-thinning reduces the numbers of stems in a woodland by some 50–70% of the original planting density at harvest.

The removal of some trees artificially does not increase the total yield from woodland, although it does allow the forest manager to remove trees of poor quality and to selectively release the better trees. The effective utilizable yield and stand quality is therefore increased. Where the products of thinning are marketable, thinning yields a useful financial return. This may be essential for broadleaved trees to offset the long wait for a return from the final harvest.

The major determinants of timber quality are:

- *Sufficient length of clean stem or bole.* This is achieved by a high initial planting density, with a consequent rapid canopy closure, which reduces to a minimum the size of the juvenile core. The competition from neighbouring trees suppresses the development of side branches and promotes apical growth (especially in broadleaves). In more open-grown conditions, high pruning of selected final-crop trees (to 5 m) may be necessary.
- *Large girth.* This is achieved by reducing competition from neighbouring trees to allow for the full development of the canopy of the selected trees.
- *Evenness of growth.* For most conifers, the faster the rate of growth, the lower the timber density and strength, while for broadleaves, evenness of growth may be more important for some markets, such as for veneers. Oak, by contrast, is stronger the faster the rate of growth (greater proportion of denser latewood).

The date of first thinning depends on timber species and the growth rates reached on the site, normally within the range 15–25 years old or about 8 m in height. Faster growing trees on good sites are thinned earliest.

While the overall objective may be to maximize the financial return from the woodland, normally the first thinning in conifers is likely to yield very little utilizable timber, although its effect on the subsequent crop value may be considerable. Over the life of the crop, 80–90% of the trees could be removed as thinnings, without affecting the productive capacity of the stand. The amount removed at each thinning and the method of selection are determined by the thinning intensity, thinning cycle and thinning regime.

Thinning intensity

This is the rate at which volume is removed from the stand, often termed the *annual thinning yield*. The higher the intensity, the greater the amount of timber removed and consequently the larger the amount of growing space is left around the remaining trees. Higher intensities also produce larger diameter trees, and consequently a greater financial return. The maximum intensity which can be maintained without a loss in overall volume production is termed the *marginal thinning intensity*. For a wide range of species and conditions, this point approximates to an annual rate of removal of 70% of the maximum mean annual increment (MAI). For a stand of yield class 20, the annual thinning yield at this point would be 14 m^3/ha, although this rate of removal only holds for a stand within the normal thinning period, which may vary according to species. See Rollinson (1988) for further information.

Thinning cycle

This is the number of years between each thinning, which is determined by the silvicultural requirements of the timber species and the thinning volume generated. Longer cycles allow for a greater volume out-turn, but may reduce volume production and increse the risk from windblow. Where small amounts of thinnings are used on the farm, it may be preferable to carry out light thinnings at more frequent intervals, with distinct silvicultural advantages since the trees are released more gradually. Typical thinning cycles range between 3 and 6 years for conifers and up to 10 years for some slow-growing broadleaves.

Thinning yield

This is the actual volume removed at any one thinning, that is, the thinning intensity multiplied by the thinning cycle. For most commercial operations about 50–60 m^3/ha is the minimum amount of timber that can profitably be extracted. For example, a stand of Douglas fir (yield class 24) might have a thinning cycle of 3 years, which enables about 50 m^3/ha to be removed at each thinning (annual thinning yield 16.8 m^3/ha).

For even-aged stands of trees that are fully stocked it is possible to determine when thinning is required according to published thinning tables. Where stocking or management differs from normal (i.e. many farm woodlands), the point of thinning may have to be judged by visual inspection. A more accurate method is to estimate the basal area of the stand and to compare this with threshold basal area tables for fully stocked stands (see Table 12.1). If the basal area is equal to or more than the figure in the table, the stand is ready for thinning. The amount of timber removed is also best controlled by the volume or basal area of the thinnings, on the basis on marked plots (see Rollinson, 1988).

Thinning type (selective thinning)

This is the type of tree in the stand which is removed by thinning. Figure 12.3 classifies the types of tree found in a stand. It is normal practice to identify the final crop trees at an early stage, although it may be prudent to choose a larger number than required to allow for further selection at a later date. Thinning should aim to remove trees which are forked, spiral barked, coarse branched or leaning and any that look unhealthy or spindly. While marking for thinning, canopy and crowns must be continually observed so that excessive gaps are not produced after thinning. A *low thinning* removes trees which are currently in the lower levels of the canopy (sub-dominants and suppressed trees), while a *crown thinning* is directed to releasing the crowns of selected final crop trees, which may mean felling competing dominants. An *intermediate thinning* is the commonest form of thinning, and comprises a low thinning to remove suppressed and sub-dominant trees, together with opening up of the canopy to release better dominant trees and create a more uniform stand.

Systematic thinning

The first thinning differs from most other thinnings in that very little, if any, financial return is likely, due to the

Table 12.1 Before-thinning basal areas for fully stocked stands (basal areas in m^2/ha)

Species	*Top height (m)*										
	10	*12*	*14*	*16*	*18*	*20*	*22*	*24*	*26*	*28*	*30*
Scots pine	26	26	27	30	32	35	38	40	43	43	—
Corsican pine	34	34	33	33	33	34	35	36	37	39	—
Lodgepole pine	33	31	31	30	30	31	31	32	33	34	—
Sitka spruce	33	34	34	35	35	36	37	38	39	40	42
Norway spruce	33	33	34	35	36	38	40	42	44	46	49
European larch	23	22	22	22	23	24	25	27	28	30	—
Japanese/hybrid larch	22	22	23	23	24	24	25	27	28	29	—
Douglas fir	28	28	28	29	30	31	32	34	35	37	40
Western hemlock	32	34	35	36	36	36	37	38	38	39	40
Western red cedar	—	49	50	51	53	55	57	60	63	66	70
Grand fir	—	39	39	39	39	39	39	40	41	43	45
Noble fir	—	45	46	46	47	48	49	51	52	54	—
Oak	24	26	25	24	24	25	25	25	25	—	—
Beech and sweet chestnut	22	24	-25 18	25	27	29	31	33	36	36	37
Sycamore, ash, alder and birch	—	17		20	22	25	28	33	—	—	—

After Rollinson (1988), by permission of Forestry Authority, 1994.

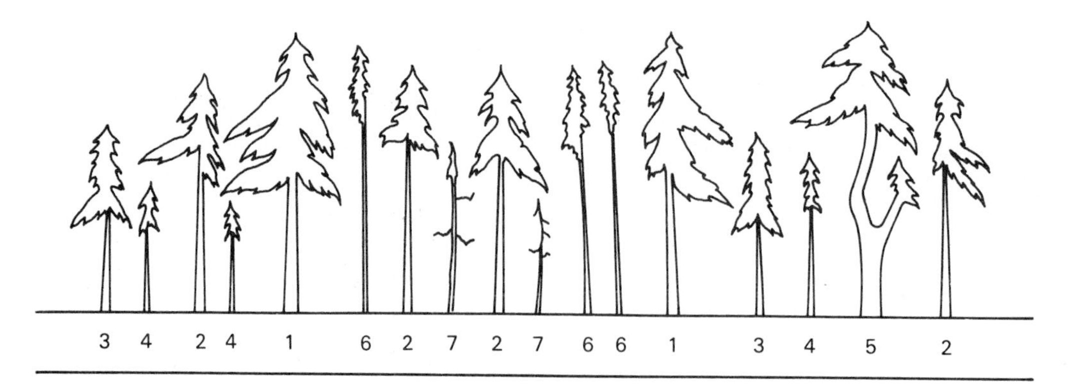

Fig. 12.3 Classification of types of tree likely to be found in the crop. 1) dominants, 2) co-dominants, 3) sub-dominants, 4) suppressed trees, 5) wolf trees, 6) whips, and 7) leaning, dead or dying trees.

small size of trees and low volumes produced. In order to reduce the cost of this operation while opening up the canopy, a number of systematic thinning systems are used. This involves the removal of trees according to a predetermined pattern, irrespective of an individual tree's merits. Extraction and marking costs may be reduced greatly by the adoption of systematic methods of thinning, particularly in early thinnings as trees felled in selectively thinned stands are easily caught up in the dense thicket. *Line thinning* is a systematic form of thinning where trees are removed in lines along the planting rows or in more sophisticated systems as a series of interconnected lines (e.g. chevron thinning). The most common systems in practice are based on removing one planting row in four or five, although for close spacings (less than 2 m), two adjacent rows may be removed to allow for the passage of extraction machinery. The benefit of line thinning is restricted to those trees neighbouring the line that was removed; for 1:5 or 1:6 row thinning it may be advisable to carry out an additional selective thinning in between the rows. The creation of rows wider than 5 m should not be contemplated, due to the loss in volume production and especially on sites prone to windthrow.

No-thin systems and delayed thinning

Where the danger of windthrow is particularly high (Windthrow Hazard Classes 5 and 6), in the extreme north and west of Britain, a no-thinning system with a reduced rotation age is preferable. Although total volume and production may be reduced, the volume of sawlogs is not significantly lower than for thinned stands. Other recent developments include the use of chemical line thinning and self-thinning mixtures, which appear to be more cost effective than harvesting and have a lower risk of windthrow. Elsewhere, the decline in the real value of small roundwood in recent years has led to a reappraisal of the economic advantages of thinning on large commercial forest estates. On some poorer, windfirm sites (GYC < 10), the time of first thinning may be delayed by up to ten years, allowing for a greater out-turn of volume at a larger average diameter. For small farm woodlands, however, thinning may always be advisable on silvicultural grounds, in addition to the advantages of a more open canopy for wildlife, game and amenity.

Where previous thinning has been neglected, this may lead to a large proportion of weak or spindly trees. It is important at this stage not to correct the past neglect by a single heavy thinning, as this may open the canopy to a greater risk of wind or snow damage, epicormic growth or sun scorch. Some species will recover better than others: for example, crown expansion is most rapid in oak, although ash will recover only weakly. In some older stands a decision has to be taken whether or not sufficient trees will respond to thinning, especially if there are already signs of windthrow or crown dieback. Any thinning which may be carried out should be restricted to 5% of the total basal area.

Harvesting

In the life of the woodland, a time comes for the harvesting of the mature final crop trees. The age of the trees at this point or *rotation length* is set according to the management objectives of the owner. For some markets, the rotation length is relatively fixed, according to specific size and diameter requirements, such as for sawlogs or veneers. The point of maximum volume production (max MAI) can be a useful indicator of the ideal rotation length, although it takes no account of the different prices obtained for different sizes of timber. Many commercial growers prefer the concept of *financial rotation*, whereby the maximum income from timber is the overriding objective for the site. In this respect the length of rotation can be drastically affected by the discount rate chosen, which in practice forecasts a harvesting date some 5–10 years before the point of maximum volume production. These concepts are useful in forest planning, although the exact date of felling will often be determined by external factors such as the local market conditions.

In recent years the efficiency of harvesting has been greatly advanced by the use of specialized felling, processing and extraction machinery for use in large commercial forests. The high capital expense of this machinery largely precludes its use in smaller farm woodlands, so that harvesting is likely to involve the use of chainsaws and smaller extraction machinery. The harvesting operation itself requires careful planning, to determine the course and direction of felling and the location of extraction routes and stacking areas. Chainsaws and other forest machinery are extremely dangerous, requiring specialized training in their safe operation. The job is therefore best left to experienced contractors. The choice of harvesting system is a choice for the forest manager, and is determined by the nature of the terrain, the silvicultural system, the product specification, the ease of access, degree of mechanization available and landscape considerations. Where simple product mixes are required (e.g. sawlogs and pulp), the trees are normally cut to size in the wood and extracted separately (the *shortwood system*). For a more complex range of sizes and lengths, trees are extracted whole from the forest to a cleared area, where the timber is cut to size (the *tree length system*). The cost of felling and clearing will depend on the terrain and size of the timber cut, although it should be indicated whether the price includes the treatment of stumps with urea (against *Heterobasidion annosum*) or the burning of lop and top.

Control of tree felling

The woodland owner or tenant, if entitled, must obtain a *felling licence* from the Forestry Commission (now Forestry Authority) before felling growing trees. Exceptions to this requirement include a small entitlement of up to 5 m^3 per quarter for own use, small diameter trees (under 8 cm dbh) and for dead or dangerous trees. Special permission may also be required for sites within areas designated as Conservation Areas or Sites of Special Scientific Interest (SSSI) or where the trees are protected by a Tree Preservation Order (TPO) imposed by the local planning authority.

New farm woodlands

The establishment of new farm woodlands gives the landowner a greater degree of control than for existing woodlands over the size, location and most appropriate type of silvicultural system. New planting tends to be concentrated on areas of lower productivity or agriculturally marginal sites, such as steep slopes or awkward corners of fields. Other sites include extensions to existing woodland, shelterbelts or for amenity around farm buildings. Many farmers are now incorporating woodland planting with general conservation or landscape work around the farm.

Site and species selection

For new farm woodlands, the most important decisions for the forest manager are the correct choice of tree species and site for planting. Subsequent forest management will not be able to make up the lost ground as a result of a poorly adapted species in the wrong conditions. The location of new farm woodlands and species choice should be dictated with reference to the following site factors.

Climate and topography

While the mild climate of the UK is generally favourable for the growth of a wide range of trees, the choice of species for any particular area will be influenced by the altitude, exposure, rainfall, length of growing season and the incidence of drought or frost. With increasing altitude, yield is reduced as a result of lower ambient temperature and shorter growing season, giving an effective economic tree line of about 450 m above sea level. For the same altitude, more exposed sites will have lower overall productivity and trees on the edge of woodlands will be shorter and more branchy. These sites will also be prone to a greater risk from windthrow (see section on wind). The amount of precipitation differs across the country, so that species with higher water requirements (e.g. Douglas fir, larch, spruce, western hemlock, and western red cedar) thrive in the wetter north and west of Britain (1000–1500 mm/year), while the drier eastern counties are most suited to less thirsty species (e.g. Corsican pine, Scots pine). Summer warmth is required for some species, particularly sweet chestnut and walnut, which restricts economic growth of these species to the southern counties. The incidence of seasonal frosts, often exacerbated by local topography, may cause injury to young trees, especially at the beginning or end of the growing season. The risk of frost damage is greatest in hollows or valleys where cold air from higher ground is able to collect. Only frost-hardy species should therefore be planted in these locations.

Soils and ground vegetation

While healthy forest growth is attainable on a wide range of soil types, tree planting tends to be concentrated on those soils less suitable for agriculture, including less fertile mineral soils, gleys, podzols and deep peats. Adequate root development is the essential prerequisite for sustained growth through anchorage and supply of water and nutrients. Where root growth is constricted through impeded drainage, an iron pan, shallow or compacted soils, there is a greater risk of subsequent windthrow or moisture stress in drought periods, requiring further cultivations or drainage work to be carried out. The owner is well advised to dig a series of soil inspection pits to check for soil depth, fertility, soil texture, drainage and the existence of hard pans. Further information can be obtained through soil maps and land capability maps for forestry, available from the Soil Survey and Land Research Centre (SSLRC) for England and Wales, or the Forestry Commission's survey of land capability classification for forestry (Bibby *et al.*, 1988) for Scotland.

Ground vegetation can be used as an indication of soil and site conditions as a guide to tree species selection (Table 12.2), although care is needed since the type and pattern of vegetation present can be greatly modified by previous land management. The systems most commonly in use are based on the presence of particular indicator species, which denote site fertility, soil pH, and drainage conditions.

Species selection (Table 12.3)

The nature of the woodland site will impose certain restrictions on the choice of tree species for planting, such that less demanding sites will enjoy a far longer list of possible tree species than an inhospitable one. Even in the most hostile locations, however, acceptable forest growth can be achieved by planting trees most adapted to

Table 12.2 Assessment of site quality from vegetation present

Site quality	*Indicator species*
Good	Ash, beech, hazel, hornbeam, field maple, oak, bluebell (*Hyacinthoides non scriptus*), dog rose (*Rosa* spp), primrose (*Primula vulgaris*), wild garlic (*Allium ursinum*), wild raspberry (*Rubus idaeus*), Dog's mercury (*Mercurialis perennis*).
Moderate	Alder, bracken (*Pteridium aquilinum*), honeysuckle (*Lonicera periclymenum*), horsetails (*Equisetum* spp), rhododendron (*Rhododendron ponticum*), soft rush (*Juncus effusus*)
Poor	Bilberry (*Vaccinium* spp), cotton grass (*Eriophorum* spp), deer grass (*Trichophorum caespitosum*), cross-leaved heath (*Erica tetralix*), heather (*Calluna vulgaris*), moor mat grass (*Nardus stricta*), purple moor grass (*Molinia caerulea*), sphagnum moss (*Sphagnum* spp)

those conditions. Following the site assessment, therefore, a list of suitable trees should be drawn up by relating the site conditions to their silvicultural characteristics. Once this short list of trees has been prepared, the final selection can only then be chosen on the basis of desired products, yield and other management objectives.

Tree species differ in their silvicultural characteristics according to their natural position within the woodland ecosystem. An important distinction relates to how tree species react to different light conditions. *Light demanders* tend to have thin crowns, light foliage and less dense timber and will grow rapidly in reasonably open situations, although in shade they are quickly suppressed and die. *Shade bearers* tend to have more layered crowns and heavier timber and will only grow slowly at first, although they are able to survive moderate shade. Where maximum volume production is a major objective, light-demanding species are clear winners. Other important silvicultural characteristics include climatic and soil requirements, timber quality, response to silvicultural treatment and sensitivity to drought, frost, wind, exposure, insect attack, browsing and disease.

In addition to appropriate species choice, consideration should also be given to the origin or *provenance* of the planting stock, since, for the same species, the silvicultural

Table 12.3 Species characteristics and site requirements

Species	*Recommended sites*	*Unsuitable conditions*	*Silvicultural characteristics*	*Timber quality*
Scots pine (*Pinus sylvestris*)	Low rainfall areas on heather, poor gravel or sandy soils. Frost hardy.	Wet or soft ground, chalk, limestone, high rainfall moorlands.	Light demander, although slow growth rate and volume production. Useful as a nurse for hardwoods.	General purpose softwood timber, good strength, takes preservative well.
Corsican pine (*Pinus nigra* var. *maritima*)	Low rainfall areas and elevations on sand and clay soils especially near the sea. Plant only in southern and eastern England.	High elevations, wet moorlands (increased risk of dieback from *Gremmeniella abietina*).	Higher volume production and better form than Scots pine, although more difficult to establish. Seedlings in Japanese paper pots give good results.	Timber similar to Scots pine but coarser and a little weaker.
Lodgepole pine (*Pinus contorta*)	Pioneer species on poor heaths, deep peats and sand dunes.	All but the poorest sites where no other tree will grow.	Coastal provenances give higher volume production but poorer form. Vulnerable to pine beauty moth in North Scotland.	Timber similar to Scots pine.
European larch (*Larix decidua*)	High rainfall areas on moist, free-draining loams.	Dry, poorly drained and very exposed sites or frost hollows. Avoid areas with rainfall under 750 mm.	Deciduous conifer. Good nurse for hardwoods, although runs out of top growth quickly so thin from age 15–18. Susceptible to butt rot (*Heterobasidion annosum*).	Heavy and generally strong timber, best quality used for boat building. Heartwood is naturally durable, and makes a good farm timber for fencing, gates and other estate uses.
Japanese larch (*Larix kaempferi*)	Mild and wet regions, less exacting than European larch. Pioneer in uplands on heather and grass.	Dry, poorly drained and very exposed sites or frost hollows. Avoid areas with rainfall under 750 mm.	Corkscrews on very fertile sites.	
Hybrid larch (*Larix* × *eurolepis*)	Hybrid larch is slightly hardier than European larch.	As for Japanese larch.	Higher yielding than Japanese larch. Only use first-generation seedlings.	
Douglas fir (*Pseudotsuga menziesii*)	Sheltered valley slopes on well drained and moderately fertile soils. Grows well in the wetter regions.	Exposed or frosty sites, heather ground, wet, soft or shallow soils.	A high yielding species although susceptible to windblow on shallow soils.	Strong construction timber with a high weight:strength ratio.

Continued

Table 12.3 *Continued*

Species	*Recommended sites*	*Unsuitable conditions*	*Silvicultural characteristics*	*Timber quality*
Norway spruce (*Picea abies*)	Moist grassy or rushy sites, most reasonably fertile soils and fairly heavy clays.	All dry, exposed sites and frost hollows.	Often grown on old woodland sites.	Good general purpose timber, works and nails well. Stable in changing humidity conditions so suitable for building. Not suitable for preservative treatment or outdoor use.
Sitka spruce (*Picea sitchensis*)	Wet exposed uplands in the north and west of Britain. Thrives on peats and grasslands in high rainfall areas.	All sites liable to dry out and rainfall areas under 1000 mm.	Will withstand very exposed conditions, although avoid previous scrub land where there is risk of honey fungus (*Armillaria* spp).	Timber superior to Norway spruce, but too coarse for joinery. Good pulpwood.
Western hemlock (*Tsuga heterophylla*)	Tolerant to high and low rainfall on acid mineral soils and better peats.	Slow to establish on heather and open ground without shelter.	Strong shade bearer, often grows best in mixture. Susceptible to butt rot (*Heterobasidion annosum*) and honey fungus on previous conifer sites.	General purpose building timber and pulpwood.
Western red cedar (*Thuja plicata*)	High rainfall areas on moderately fertile soils on sheltered sites.	Exposed, poor and dry sites.	Shade bearing and narrow crown, useful in mixtures or nurse species.	Light coloured heartwood which is very durable. Cladding used outside for greenhouses and sheds.
Lawson's cypress (*Chamaecyparis lawsoniana*)	Requirements not exacting but best on deep fertile soils with moderate to high rainfall.	Dry infertile sites and heather ground. Avoid frosty, exposed and waterlogged sites.	Slow-growing shade bearer for use in underplanting or in mixtures (e.g. with oak).	Heartwood reasonably durable for general purpose uses. Small sizes used for fencing.
Grand fir (*Abies grandis*)	Well drained moist deep soils.	All poor soils, dry, frosty or exposed sites.	High volume producer, useful for underplanting, as moderately shade bearing.	Soft white timber of only moderate quality.
Noble fir (*Abies procera*)	Well drained deep moist soils, tolerates acidity.	Poor dry soils.	Withstands exposure well. Tolerates drier sites than Sitka spruce. A useful shelterbelt tree in Scotland.	
Pedunculate oak (*Quercus robur*)	Deep clay loams with ample moisture, although will tolerate heavy clays.	Shallow, infertile or poorly drained soils on exposed sites.	Oak is a strong light demander and will not grow under shade. For top quality timber, close planting (at least 2500/ha) is required for suppression of side branches. Bole shading by underplanting or pruning will prevent problems from epicormics. Oak grown on lighter soils may suffer from 'shake'. Windfirm.	Oak has strong, naturally durable heartwood, although the sapwood does require preservative treatment if used outside. Timber quality (and price) varies tremendously. Prime quality oak is scarce and is used for veneers, planking and furniture. Lower grades for beams, fencing, gates and other estate purposes. The branchwood and lowest grades may only be fit for firewood or pulp.
Sessile oak (*Quercus petraea*)	Best on deep porous brown earths, although will tolerate clay soils.	As above and wet clays or fluctuating water tables (risk of shake).		

Continued

Table 12.3 *Continued*

Species	*Recommended sites*	*Unsuitable conditions*	*Silvicultural characteristics*	*Timber quality*
Beech (*Fagus sylvatica*)	Light well drained soils on chalk or limestone, deep sands and acid brown earths.	Cold, wet, poorly drained soils, dry infertile sands.	Shade bearing, although may require a nurse (e.g. Scots pine) on exposed sites. An excellent amenity tree and suitable for underplanting. May suffer from severe squirrel damage, requiring close spacing for adequate selection.	Timber strong with a fine even texture which takes a stain well. Used for a wide range of interior uses: furniture, kitchenware, turnery and flooring. Lower grades used for firewood or pulpwood. Not suitable for exterior uses.
Ash (*Fraxinus excelsior*)	Only thrives on moist, well drained fertile soils, ideally deep calcareous loams in valley bottoms. Wild garlic or dog's mercury are good indicators.	Dry, shallow, heavy clay or badly drained soils, heaths and moorlands. Avoid frost hollows and exposed sites.	Ash is usually grown in mixture with beech, oak, cherry or larch. The stand needs to be thinned regularly to encourage large crowns.	Spring timber with high shock resistance. Top grade ash (annual rings 1.5–6 mm) is used for sports goods, tool handles and furniture. Lower grade timber makes excellent firewood.
Sweet chestnut (*Castanea sativa*)	Deep fertile soil in warm climate, ideally in southern England. Warm sunny acid sandy loam banks are ideal (pH 4.0–5.0).	Cold, wet, badly drained, exposed, frosty or infertile sites. Chalk, limestone or alkaline soils.	Chestnut often grown productively as coppice for estate products. Timber trees over 80 years are prone to shake on free draining soils.	Sawn timber makes a strong substitute for oak. Coppice material used for cleft fencing, poles and stakes.
Sycamore (*Acer pseudo-platanus*)	Moderately fertile free draining soil. Grows on exposed upland sites and resistant to frost.	Dry, shallow, ill-drained or heavy clay soils.	Moderate shade bearer and regenerates freely. Performs well on poor upland sites as windfirm. Prone to heavy squirrel damage.	White timber used in turnery and where in contact with food. Figured sycamore valuable for veneer and furniture.
Wild cherry (gean) (*Prunus avium*)	Deep fertile well drained soil, especially over chalk.	Heavy soils, depressions and all infertile sites.	Grows best in small groups in mixture with oak, ash or beech. Not attacked by squirrels, but may be severely affected by bacterial canker and deer browsing. Regular thinning and pruning in summer for quality timber.	Decorative timber with rich reddish brown heartwood used for veneers, furniture and panelling.
Common walnut (*Juglans regia*); black walnut (*Juglans nigra*)	Sheltered south-facing slopes on deep, fertile, well drained, medium-texture soils. Optimum pH 6.0–7.0. Chalk or limestone with more than 600 mm of overlying soil. Require warm summers (ideally south and central England).	Only planted on carefully selected sites which meet requirements, avoiding frost hollows.	Best grown in open situation or in small groups, although pruning will be needed to improve form in July/August.	Valuable decorative dark brown timber, often dug out and taken to the sawmill with main roots as 'figuring' occurs in the base. Branches for firewood.
Common alder (*Alnus glutinosa*)	River and stream banks and wet marshy places in lowlands and uplands. Frost hardy.	All dry sites. For very acid sites prefer grey alder.	Good shelterbelt tree (windfirm) for use on nutrient poor sites or in coastal locations, as resistant to salt spray. Mixes with birch and ash and is a good nurse for oak on wet clay soils. Fixes atmospheric nitrogen in root nodules. Grows rapidly for the first 25 years and coppices vigorously.	Timber dries red brown, good for turnery, staining and polishing.

Continued

Table 12.3 *Continued*

Species	*Recommended sites*	*Unsuitable conditions*	*Silvicultural characteristics*	*Timber quality*
Grey alder (*Alnus incana*)	Requires drier situation than common alder, but tolerates poorer sites.	Dry infertile situations, heathlands, shallow and thin chalky soils.	Widely planted for wind breaks, shelter and as a pioneer species for reclaiming derelict land. Extremely hardy.	Timber worthless.
Italian alder (*Alnus cordata*)	Drier chalk and limestone soils of southern England.	Avoid dry thin acid or infertile soils.	Pioneer species used for single-row windbreaks.	Timber worthless.
Black poplar (*Populus robusta*); black hybrids (*Populus* × *euramericana* var. Eugenii, Gelrica and Heidemiz T-78)	Base-rich loamy soils in sheltered position with water table 1–1.5 m below the surface in summer.	Exposed acid dry or infertile sites.	Poplars often used for perimeter horticultural windbreaks.	Light white timber used for pallets, chip baskets and pulpwood. The demand for match timber is much reduced.
Balsam poplar (*Populus trichocarpa* var. Fritzi Pauley and Scott Pauley, *Populus tacamahaca* × *trichocarpa* var. 32)	Tolerates more acid soils than black hybrids and more suited to the cooler wetter parts of Britain.	As for black poplar.	Balsam poplars subject to bacterial canker; plant only resistant clones. Preferred to black hybrids as they are fastigiate and trim more easily. Hybrid *Populus txt* Clone 32 most commonly used. Other poplars include white and grey spp.	
Cricket bat willow (*Salix alba* var. *coerulea*)	Only margins of flowing streams or rivers with highly fertile soils.	Useless elsewhere.	Planted as sets 10–12 m apart, with all sides shoots removed on the bottom 3 m of the stem for knot-free timber.	Cricket bats.
Silver birch (*Betula pendula*)	Brown earths, podzols, sands and gravels. Withstands frost and exposure.		Pioneer species with light crown. Regenerates easily on mineral soils, although not widely planted.	Grown in Scandinavia with improved cultivars for veneer. Low-grade timber makes good firewood.
Southern beech (*Nothofagus procera*; *Nothofagus obliqua*)	Wide range of soils from heavy clay to deep sand.	Badly drained exposed or frosty sites.	Both species are fast growing. Prefer *N. procera* in the wetter west of the country, *N. obliqua* in the drier east. Light-demanding species, but start under a thin canopy; thin early in life. Frost damage leads to stem cankers.	Timber of both species similar to native beech, but with 20% less bending strength.
Norway maple (*Acer platanoides*)	Moist, deep, free-rooting soils with high base status.	Will not thrive on infertile soils, although will survive on thin chalk soils. Avoid frost hollows and exposed sites.	Good amenity tree, suited for screens and mixed shelterbelts. Heavily damaged by squirrels.	Timber as for sycamore; flooring, furniture, turnery and veneer. Good firewood.
Hornbeam (*Carpinus betulus*)	Moist damp clays, chalk, limestone and acid brown earths. Frost hardy.	Thin, infertile, dry and very acid sites.	A substitute for beech on clay soils and where high frost resistance is required. Shade bearing and slow growing, therefore valuable as coppice understorey for shelterbelts or under oak to control epicormics.	Hard, heavy and tough timber giving a very smooth finish ('white beech' on the continent) used for carving and turnery. Makes very good firewood.

characteristics of the trees may be slightly different for seed drawn from different geographical sub-populations. Many commercially grown exotic conifers, for example, are native to the western seaboard of North America and seed from the northern populations (e.g. Alaska) will be more resistant to winter cold, although slower growing than seed drawn from more southerly populations (e.g. California). Since the first introductions of these exotic conifers, seed may now be obtained from seed orchards (or provenances) in Britain, even though the origin remains unchanged. Both seed origin and provenance are recorded in tree planting catalogues which must be matched with the environmental conditions of the planting site for optimum growth. Further information is available from Lines (1987).

Pure crops and mixtures

Pure crops are easier and simpler to manage than mixtures, although mixtures do offer other benefits if managed skilfully. Broadleaves are often grown in mixture with conifers to provide an early economic return from the coniferous thinnings to offset the delayed returns from the longer rotation of slower growing broadleaves. Appropriate mixtures may also be able to more fully utilize the site through differences in rooting depth and are far preferable on landscape, game and conservation grounds. Mixtures are often used where one species or *nurse* offers benefit to the main tree species, either in the form of shelter against frost or cold, smothering weed growth, nutrition, support or suppression of side branches. Timely thinning is often required to prevent the nurse from outgrowing and suppressing the main tree species, so that care is needed in the choice of appropriate mixture. As a general rule, for broadleaved/conifer mixtures, the expected growth rate of the conifer should not be more than double that for the broadleaved species. Broadleaves should be established as blocks (9 or 25 to each block) within a conifer matrix, at an appropriate spacing to provide a final broadleaved crop. Alternatively line mixtures can be used, although these should be lines of at least three of each species to reduce the risk of suppression. Normal practice is to remove the adjacent lines of conifers to the broadleaves at the first thinning followed by selective thinning thereafter. Greater care is required on hillside plantings to avoid regimented or geometric patterns of mixtures which are highly visible in the landscape.

Suitably robust mixtures include:

- oak, ash and cherry on moist, deep, fertile soils or clay over chalk;
- oak and European larch on free-draining and lighter soils;
- oak and Norway spruce on heavy acid clays;
- oak and western red cedar on free-draining soils;
- beech with western red cedar, Scots pine or Corsican pine;
- sweet chestnut and European larch;
- sitka spruce and European larch.

Establishment

The effective establishment of trees is an important management operation in the life of the woodland. Neglected or poor establishment practice, in an effort to reduce costs, will lead to lower initial stocking and a consequent deterioration in the quality of the remaining trees which will be difficult to remedy later. On the other hand, careful planning and organization of site preparation, planting, weeding and tending will more than recoup the initial outlay in the improved quality of the woodland.

Site preparation

The aim of ground preparation is to create suitable planting conditions for rapid establishment. Where planting is to take place on the site of a recently harvested woodland, the brash from the former crop must be cleared as part of the harvesting operation. Larger material can be cut up for firewood and the lighter lop and top either burnt, chipped or pushed into heaps between the planting lines. The whole clearance operation can often prove expensive. Additional problems with restocking include the rewetting of the ground as soil moisture increases after felling, and compaction caused by heavy harvesting machinery.

On upland sites, the principal method of ground preparation is by *ploughing*, using either a single furrow or double mouldboard, set at the appropriate planting distance. Ploughing creates a raised ridge of cultivated mineral soil, through which the trees are planted. This can almost eliminate the need to weed during the first growing season, as the vegetation is completely buried under the inverted ridge. The main advantage, however, is through improved root development, as a result of increased soil aeration, drainage, soil temperature and the release of nutrients from the breakdown of organic matter. On heavy, wet or organic soils, an even larger mound may be required, using a machine to dig out *dollops* of soil or peat and deposit them on the planting position. This technique is also suitable for restock sites where the presence of large stumps precludes ploughing. An alternative to ploughing where drainage is not required (mainly on lowland sites) is *scarifying*, which involves the removal of surface vegetation by mechanical scraper. Scarifiers and mounders provide improved planting conditions without major site disturbance, so reducing the potential for soil erosion and its consequent effects on reducing water quality.

Site preparation on former agricultural land

The release of agricultural land for planting offers the potential of more fertile land than is normally the case for commercial upland forestry, although slightly amended establishment practice is required. For further information consult Williamson (1992).

The initial preparation of the planting site will depend on the previous land use and soil conditions. The ploughing of former grassland or arable stubble can be counterproductive, as the disturbance of the soil both exposes viable weed seeds and creates a bare seed bed for the rapid invasion of arable weeds which are difficult and expensive to control. Conversely, undisturbed stubble will harbour herbicide residues which will reduce the quantity of seed in the surface layers of the soil and delay invasion by broadleaved annual weeds. Any straw can be chopped and left as a mulch to retain moisture and control the growth of volunteers and other grass weeds. A grass sward is the most controllable of ground conditions, and can be achieved on arable sites by underplanting a previous crop with a suitable grass mixture. Heavy soils present particular problems due to shrinkage and waterlogging. In a dry summer, heavy land will shrink, leading to a network of tiny cracks. On recently planted land, this shrinkage is often concentrated in the bare strips along planting lines or around the circumference of trees which are spot weeded, leading to excessive cracking and loss of plants due to moisture loss from exposed roots. The poor structure and texture of heavy soils can pose problems at planting too, which should be delayed until the soil has a suitable moisture content in the autumn. If the soil is too hard and dry, it will be difficult to firm in the roots, while if the soil is too wet, the sides of the planting hole will be smeared, which may crack and expose the roots in a dry summer. Some sites may suffer from compaction as a result of poaching or heavy machinery, and the presence of a soil pan at some depth could impede the growth of trees or increase the risk from windthrow. Ripping will improve soil structure, although the ripped lines should not be subsequently planted due to the risk of soil cracking.

Drainage

The prime function of forest drains is to remove surface water and to prevent waterlogging, which over prolonged periods will severely impede the growth of trees and may increase the risk of windthrow as root development is restricted. In the uplands, substantial areas of planting land suffer from excessive soil water, particularly on gleyed clays, peaty gleys and deep peat where the lateral movement of water is severely limited, requiring extensive drainage systems. Open drains (60–90 cm deep) are cut to intercept surface water from above the plantation at a slight angle to the contour (no more than 2°), together with smaller collecting drains at 30–40 m spacing through the planting site. The environmental effects of increased levels of sediments in upland streams leaving afforested land have caused some concern in recent years, due to the impact on aquatic life. The Forestry Commission have produced a useful guide on silvicultural practices which can mitigate some of the worst of these effects (Forestry Commission, 1988). In particular, drains should end 5–10 m from smaller streams and 15–30 m from main watercourses to reduce sediment loading.

Once fully established, the trees themselves draw water from the soil through transpiration which adds to the efficiency of the drainage system. The drainage system should be monitored regularly, although maintenance operations should be restricted to removing serious blockages and overflowing ditches. At harvesting, substantial damage to the drainage system can be avoided by working during drier periods and installing temporary culverts to remove surface water. Inevitably, some reinstatement work will be necessary, although the root channels formed by the previous crop will improve water movement, especially where these have broken through any surface pan or compacted layers.

Drainage work can be an expensive and time-consuming operation, normally involving the use of contractors with specialized excavators. Drainage systems should be planned on the basis of local soil and climatic conditions with observations being made regularly through the year and drains marked on scale maps. Where high production is not an overriding objective, the owner may be better advised to plant more tolerant tree species on wetter sites, such as alder or willow.

Fencing

New plantations will require protection from grazing animals (rabbits, hares, deer and farm stock) while the young trees become established. The choice of protection will depend primarily on the size and shape of the plantation and the density of planting. For sites smaller than 2–3 ha, individual protection by tree shelters is normally the cheapest method, while for larger areas, fencing becomes increasingly cost effective. Fencing may also be required on woodland sites bordering stock fields, roads or public places. Irregularly shaped and narrow belts of trees are the most expensive to fence, having a greater perimeter than square plantations.

The specification of the fence will depend on the type of animal present and the durability required. Rabbit fences should be at least 0.75 m high with 31 mm mesh wire netting, while for deer and stock a stronger fence is required to a height of 1.8–2.0 m high. Spring steel wire is commonly chosen as a support for various grades of wire netting, since it will return to normal if accidentally stretched by an animal or fallen trees. If the fence is required for longer than five years, only treated wooden posts and struts should be used. The line of the fence should be chosen to avoid wet or shallow soils and snow hollows, which may allow entry of animals in winter. Where the fence cuts across badger runs, two-way gates should be provided. Further details of fence specifications are available in Pepper & Tee (1986).

Plant origin

The production of high quality timber depends on the selection of healthy planting stock grown from selected parents of superior quality from the correct choice of provenance. Under the Forest Reproductive Material Regulations, cuttings and plants of listed species (oak, beech, poplar and most conifers) intended for the production of timber may not be sold unless obtained from sources approved and registered by the Forestry Authority in Great Britain or other approved authorities elsewhere in the European Union. For unlisted species, seed from recommended sources is preferable to 'unknown' sources.

Planting stock

Although trees are available in a variety of sizes (BS 3936), smaller plants offer the advantages of cheapness, ease of handling, faster early growth and less chance of dieback or drought stress than larger whips or standards. Plants should be ordered well in advance from a forest tree nursery to ensure delivery and quantity required (names and addresses can be obtained from the Royal Forestry Society of England, Wales and Northern Ireland). Only healthy plants should be accepted, which are stout and well balanced with plenty of fibrous moist roots. Check any suspect trees by nicking the bark; live plants will have a greenish white appearance under the bark. The following types of planting stock are commonly available.

- *Bare rooted transplants* are widely available as forest trees, being reasonably priced and less bulky than containerized trees, although greater care is needed in planting and handling. Transplants are raised in a seedbed for one or two seasons and then lifted and planted out in transplant beds for a further year or two, which encourages the growth of a vigorous root system. Nursery catalogues may refer to these plants in a coded system (e.g. a 1 + 1 plant has spent one year in the seedbed and one year in the transplant bed). In some nursery systems, plants remain in the same seedbed but are sown at a lower initial density and are undercut by a reciprocating bar passed through the seedbed, to promote bushy root growth, e.g. $\frac{1}{2}$ u $\frac{1}{2}$. ($\frac{1}{2}$ u $\frac{1}{2}$ indicates a one-year-old plant which has been undercut midway through the growing season.)

- *Container-grown plants* retain a growing medium around the roots of the tree at planting, allowing planting outside the growing season and a lower likelihood of mortality through poor handling. Plants are available in a range of containers and sizes, from reusable stiff plastic pots to degradable paper cells (Japanese paper pots) or flexible root trainers which produce small plants as 'plugs'. The plants are normally raised in polytunnels and hardened off outside prior to planting. Always be careful not to accept pot-bound trees.
- *Cuttings and rooted sets* are used for poplars and willows which will grow from unrooted cuttings (20–25 cm by 1–2 cm diameter).

Time of planting

The planting season for bare-rooted transplants runs from the end of September until the end of March, the exact date determined by the onset of dormancy, which is later in a mild season, together with the ground conditions. Planting is most successful when coupled with higher soil temperatures, although delaying planting until after budburst renders trees susceptible to decay. Budburst can be delayed, while extending the planting season, by protecting the trees in cold storage before delivery. Broadleaves are best planted by mid November to allow the roots to become settled in over the winter and better able to withstand drought in the following season. Conifers are normally planted during early spring to avoid winter frost damage.

Plant handling

Although disease, frost or drought may account for some losses of trees in the first year of planting, a major cause of death is the result of poor handling between the nursery and the planting site. Particular problems include the following.

- *Root desiccation* is a significant cause of tree mortality from the exposure of bare roots to drying winds, particularly while the planting hole is being prepared. Bare-rooted transplants should ideally be kept in co-extruded opaque plastic bags, which are white on the outside to reflect heat and black on the inside to keep the roots cool. Fertilizer bags are not suitable due to the presence of chemical residues which may scorch the plants. Trees are traditionally stored prior to planting by 'heeling in', whereby the trees are lined out in a shallow, slanted trench (30 cm deep) to cover the roots. This practice should be avoided where storage is only required for a short period, as disturbance of the roots is likely. Where there is a danger of frost, straw or bracken should be placed over the trees.
- *Overheating* of trees in black or translucent plastic bags can lead to excessive moisture loss and eventually death. Trees should always be stored out of bright sunlight.
- *Physical damage* can be caused by poor handling between the nursery and the planting site, leading to broken, bruised or damaged shoots, roots and buds. Plants should never be dropped off the back of a trailer or treated roughly.

Planting

Before the planting operation, all site preparations should be completed and the plants and equipment safely delivered to the site. Ideal planting weather is mild and wet, and trees should not be planted in severe frost or while the ground is waterlogged. The first planting line is laid out with sticks or twine, from which subsequent lines are measured. Experienced planters will judge the distance between plants by stepping or by the length of the planting tool.

Bare rooted trees are normally planted by the *notch method*, which involves the cutting of a T or L shape slit in the soil. Although a garden spade can be used, a special planting spade with a strengthened handle and straight blade is preferable. The first slit is levered open with the spade, allowing the tree to be inserted into the notch until the old ground level mark (*soil collar*) of the tree coincides with the surface of the disturbed ground. The roots should be evenly distributed and the soil firmed down around the tree, which should remain vertical. For larger trees or standards, *pit planting* is the preferred technique. This involves the digging of a hole large enough to take the spread-out roots of the tree, the bottom of which may be broken up to aid drainage. Well rotted organic matter or a sprinkle of bone meal may be added, and for large specimens a stake should be driven in before inserting the tree. The tree should be planted at the same depth as it grew in the nursery (note the position of the soil collar), and for pot-grown trees some of the fine roots can be gently teased out. The soil should be firmed in around the roots with the foot, tied securely to the stake with appropriate expandable ties and watered in well.

On level and workable sites (free from large stones, stumps and ditches) mechanized planting by a tractor-mounted machine may be possible, especially on new farm woodland sites. This technique offers the advantages of quicker planting and improved survival from the reduced handling time.

Spacing (Tables 12.4 and 12.5)

The choice of spacing between the trees will reflect the management objectives for the woodland. If good quality timber is required, it is important to obtain a high initial stocking density, which will lead to straighter and less branchy trees, better returns from thinning and a greater number from which to select the final crop trees. Where the objective is for amenity or game, wider spacings will be more cost effective, although pruning of selected final crop trees may be required. On exposed sites prone to windthrow, wider spacings are advisable to allow individual trees to develop more extensive, windfirm root systems.

Table 12.4 Spacing of trees

Species and situation	*Approximate spacing (m)*
Christmas trees	0.75–1.00
Oak/beech	1.50–1.75
Slower growing conifers (Pine and Norway spruce on poor sites)	1.50–1.75
Conifers/mixed broadleaves	2.0
Amenity mixed broadleaves in tree shelters	3.0
Poplar	8.0

Table 12.5 Tree requirements (number of trees required/ha planted on the square)

Spacing (m)	*No. of trees/ha*	*Spacing (m)*	*No. of trees/ha*
0.75	17 778	2.75	1 322
1.00	10 000	3.00	1 111
1.25	6 400	4.00	625
1.50	4 444	4.50	494
1.75	3 625	5.00	400
2.00	2 500	6.00	278
2.25	1 975	8.00	156
2.50	1 600	10.00	100

Tree shelters

These translucent open-ended tubes, placed over each tree, have revolutionized the establishment of broadleaves on sheltered lowland sites and for protecting naturally regenerated seedlings or small plantings (e.g. in hedgerows or enrichment schemes). The shelters act as 'tree greenhouses' providing the tree inside with a warm, moist microclimate. The results are very rapid initial growth and better root establishment, which increase tree survival, especially in dry summers. In addition, the shelters give protection against animals and allow for the easier application of foliar herbicides. Height growth can be doubled for some species (notably oak), although this effect is reduced once the tree emerges from the top of the shelter. The advantage of treeshelters is therefore to speed up the expensive process of tree establishment by faster growth and increased survival over the first 3–4 years. Following this, the shelter continues to give support to the tree until it disintegrates between 5 and 10 years. Shelters are now available in a variety of different shapes, sizes and colours: 0.6 m for protection against rabbits, 0.75 m for hares, 1.2 m for roe and muntjac deer, 1.5 m for sheep, 1.8 m for red, fallow and sika deer, and 2 m for cattle. Treeshelters should not be used on exposed or waterlogged sites and any existing vegetation on the planting position should be screefed away. The base of the shelter will need to be pushed into the ground at least 5 cm to prevent air circulating up through the shelter, and secured using a stout and preferably treated stake together with a special nylon tie. Different trees vary in their response to treeshelters (Table 12.6), which should only be used with appropriate species and using good transplants with a root collar of at least 6 mm. It is important to note that the use of treeshelters does not mean that weed control can be avoided (see section on weed control).

Table 12.6 Height increment improvement with treeshelters compared with mesh guards 3 years after planting

>100% increase	Oak, beech, walnut, lime, sycamore, field maple, birch, hawthorn.
50–100% increase	Douglas fir, grand fir, Sitka spruce, Norway spruce, Corsican pine, Japanese larch, Norway maple, alder, sweet chestnut, ash, holly, southern beech
<50% increase	Western hemlock, western red cedar, wild cherry, black walnut, hornbeam, horse chestnut, rowan, whitebeam

After Potter (1991), by permission of Forestry Authority, 1994.

Treeshelters have been the cause of great debate among foresters over the last 10 years. Concerns have been raised over their stability in exposed conditions, the growth of trees after the shelter is removed, weeds in the shelter and not least the unattractive appearance of lines of treeshelters in the landscape. Some of these problems have been overcome by recent research, such as the use of different coloured treeshelters to blend in with the ground vegetation and management advice on their correct removal by splitting the treeshelter at the base with a sharp knife. The longer term implications of treeshelters for growth performance over the life of the crop are still the subject of research, although early indications are that their correct use will not jeopardize future growth potential.

Beating up

Beating up involves the replacement of dead trees which are lost during the first year. As it is not normally economic to replace every tree, up to 20% of trees may be lost without requiring beating up. Beating up is advisable where failures occur in groups, and at wider spacings where losses are more noticeable. It must be carried out in the year following planting to be fully effective, using good quality plants, although it should be borne in mind that these trees will be a year behind in growth and more likely to become suppressed by their neighbours.

Fertilizer application

Trees respond only weakly to nutrient supplements compared with agricultural crops, so that the marginal improvement in growth from fertilizer applications is unlikely to be economic for farm woodlands on fertile mineral soils. In commercial upland plantations, fertilizer use is generally restricted to the establishment phase to improve initial growth rates of conifers until canopy closure, the decision to fertilize being based on soil or foliar analysis. After canopy closure, further applications are not normally necessary due to the shading of competing vegetation, improved nutrient cycling and the capture of atmospheric nutrients by the crowns of the trees. On restocked sites, the increased availability of nutrients from the breakdown of brash and litter normally precludes the use of fertilizers except for very poor heathland or moorland soils.

Phosphate, potassium and nitrogen are normally applied as top dressings, using ground rock phosphate (50–75 kg P/ha), muriate of potash (75–100 kg K/ha) and prilled urea or ammonium nitrate (80–120 kg N/ha). The use of fertilizer in farm woodland planting should be restricted to the following specific conditions:

- *Heathland mineral soils* (iron pans, podzols, gley soils) may suffer from phosphate deficiency. Sitka spruce planted on heather-dominated heathland or moorland suffers from a well documented 'heather check' which restricts growth through the limited availability of nitrogen and may delay canopy closure by up to 10 years. Although heather competition may be reduced by herbicide applications, additional inputs of nitrogen may be required. Alternatively, Sitka spruce may be planted in mixture with a suitable provenance of Scots pine, lodgepole pine or Japanese larch, which appears to suppress the competing heather and increases the availability of nitrogen.
- *Peaty gleys and deep peats* are likely to suffer from phosphate and potassium deficiency where the peat depth exceeds 30 cm.
- *Chalk downland* may suffer from nitrogen and potassium deficiency on shallow soils.
- *Derelict and restored soils* often suffer from limited nitrogen availability, in addition to specific nutrient deficiencies or toxicity. Further soil analysis will normally be necessary.

Aftercare

Weeding

The most significant threat to the successful establishment of forest plantations is competition from weed growth, especially on drought-prone sites. Grasses and broadleaved weeds compete effectively with the growing trees for water, light and nutrients, reducing height growth and potentially increasing mortality. On fertile sites the growth of rank weed vegetation may also cause physical damage by smothering or collapsing over the trees in the autumn. The cutting back of these weeds and grasses will reduce light competition and physical damage, although root competition will not be affected. In fact, mown grass exerts an even greater competition for soil moisture than unmown grass. Recent research and field trials into the control of weed growth has clearly shown that a weed-free zone of at least 1 m diameter significantly improves the chances of successful establishment.

The weed control treatment most appropriate should be determined by regular inspection of the site prior to planting to reveal the type and abundance of weed species present, together with an assessment of the soil conditions and potential for weed invasion. The total removal of weeds from the site is rarely justified either by cost or on silvicultural grounds: in fact, there is some evidence that the retention of some weed growth may benefit the trees by providing shelter from frost or sun scorch. A totally weed-free plantation may also be prone to excessive soil erosion and has a substantially reduced wildlife value. Weed control therefore normally involves the selective spot treatment of competing vegetation around the trees or in strips along the planting lines, retaining weed growth in the central aisle. The timing of treatments will depend on the site, although it should be borne in mind that the time of greatest moisture competition is during March–April when tree roots are actively growing. Control should therefore aim to anticipate future weed growth before this becomes obvious on the ground. Current practice involves pre-planting treatments to establish a weed-free planting position, together with additional treatment during the growing season if necessary.

The following forms of treatment are available.

Herbicides

Herbicides provide the most cost-effective form of weed control available, and when used in accordance with the manufacturer's recommendations the environmental risks are minimized. A wide range of chemicals are available to enable the control of forest weeds, although most will only affect a limited number of species or may only be active at certain times of the year in particular weather conditions. It is therefore important to match the weed control spectrum of the herbicide with the weeds found on the site. Herbicides can be divided into two main groups, foliar-acting and residual.

(1) *Foliar-acting herbicides* are applied to the aerial parts of actively growing weeds, from where they are taken up through the stem and leaves. While some can be applied over the trees without harming them, few can be used during the growing period (May–July) or during bright sunlight without scorching the foliage.
(2) *Residual herbicides* are applied where the soil is moist and friable for maximum effect. The active chemical works through the soil and is taken up by the weed's roots. These herbicides are particularly useful for broadleaves susceptible to foliar herbicides, often being applied in the dormant winter season to control weeds during the spring and early summer.

The need to reduce the cost of herbicide applications has led to the development of low-volume sprayers (weed wipers, spot guns or controlled drop application sprayers), which also reduce the environmental impact of these chemicals. Where larger woody weeds are present (e.g. gorse, laurel, rhododendron), the shrubs should be cut back by mechanical means followed by herbicide treatment of the regrowth from the stumps.

Current farm woodland practice involves the use of residual herbicides, such as propyzamide in granular form in the winter (slowly volatilizes in cold soil) followed by the use of foliar sprays in the summer (e.g. glyphosate) to cope with localized weed problems. For further instruction on appropriate herbicides, storage and application, see Williamson & Lane (1989).

Mulching

Mulching involves the prevention of weed germination and growth by covering the ground around the tree with either organic material or an artificial sheeting. Mulches also retain soil moisture close to the surface by reducing evaporation losses, which promotes root growth and nutrient cycling. Artificial mulches include specially formed black polythene sheets (125 μm thickness) with ultraviolet inhibitors to slow degradation. Stones or earth should be used to weigh down the corners of the mats to prevent them from being displaced by the wind and to deter voles from burrowing underneath. The mats should last for 3–4 years, although they increasingly suffer from damage by perennial weeds and animals (foxes will rip the sheeting to catch voles). Successful organic mulches include straw, wood chips or crushed bark, although they may reduce tree growth by locking up available nitrogen as the mulch breaks down and are prone to being blown away by the wind. Where mechanization allows, a cover crop (e.g. rye grass) can be sown between the planting rows and then sprayed off with herbicide before tree planting to provide a more stable mulch than loose organic material.

Inter-row vegetation management

The inter-row vegetation between the weed-free spots or strips will grow unchecked if no weed control is carried out. This may lead to invasion by persistent perennial species or tall growing weeds which fall over and smother the trees in the autumn. The development of rampant inter-row vegetation may also impede access by farm machinery. This naturally occurring vegetation will need to be controlled by regular mowing to prevent further problems and seeding into neighbouring fields. As an alternative, the ground between the weed-free areas may be sown with a cover crop which is much more controllable or has uses other than weed suppression (e.g. game cover or conservation). Inter-row crops should be chosen for low productivity, which do not compete for moisture with the trees and are easily controlled by herbicide in the weed-free zone (e.g. fescues). Kale has been successfully used as a game cover crop (the instant spinney); it provides a sheltered habitat for game birds for about 3 years, in addition to the physical support it lends to tree shelters. A hardy variety should be chosen, sown in rows 50–60 cm apart, in June–July following tree planting. Fencing should be chosen to exclude farm stock but to allow game birds through holes in the netting. Where the new woodland is part of a project to increase the value of wildlife habitats on the farm, the inter-row area may be sown with a low-productivity grass and wild flower mixture, which should be chosen to match the underlying soil type. This type of mixture produces an attractive meadow habitat which is most appropriate for the edges of the woodland and areas adjacent to footpaths and other public spaces.

Cleaning

This refers to the removal of woody plant growth which is either in competition with the tree crop or physically damaging. The operation normally involves the cutting back of unwanted broadleaved trees or regrowth from hardwood stumps, and the severing of harmful climbers such as honeysuckle (*Lonicera periclymenum*) and old man's beard (*Clematis vitalba*). Methods of cleaning depend on the nature of the site, although the work generally involves the use of clearing saws to cut or girdle shoots. Chemical treatment is a cost-effective operation, using glyphosate, applied to either foliage or cut surfaces. Where individuals or groups of crop trees have failed, naturally regenerated broadleaves or regrowth from stumps should be retained in these areas to suppress the development of side branches on neighbouring crop trees.

Inspection racks

As the new plantation reaches canopy closure, it becomes necessary to cut rackways through the thicket of trees in order to allow assessment of its silvicultural condition. These are normally made by cutting off the lower branches of two adjacent rows with a pruning saw, cut flush to the trunk.

Brashing

This involves the removal of dead and dying branches from crop trees up to a height of about 2 m, normally undertaken during the thicket or pole stage prior to first thinning, to improve access for marking and extraction. Brashing may also be carried out where access for game beaters and amenity is required or to release broadleaves when grown in mixture with conifers. The persistence and difficulty of removal of the lower branches varies with species and crop spacing. Shade-bearing trees tend to have the most long-lived branches (e.g. western hemlock), while larch tends to self-prune well. Spruce tends to have quite tough branches, which are difficult to remove; larch branches break off easily with a stick; and pine is intermediate between the two. The use of a special curved brashing or pruning saw is preferable to a billhook, especially with larger branches where extra care is needed to reduce the risk of damage. Complete brashing of plantations is now rarely economic, so that only sufficient trees are now brashed to allow for inspection and marking, concentrating on the final crop trees. Where plantations are line thinned, the rows to be removed are left unbrashed.

Pruning

Pruning involves the removal of side branches above the height of normal brashing to produce longer lengths of clean knot-free timber. The operation is restricted to selected final crop trees, in order to achieve the standards of quality required for high grade timber (e.g. planking, joinery, beams and veneers). Pruning should be carried out before the side branches become larger than 5 cm, for ease of working and to reduce the incidence of disease entry. A sharp pruning saw or chisel is used with an extendable handle for high pruning. The branches are sawn not quite flush with the trunk, but slightly proud to avoid bark damage. High pruning will inevitably remove some 'live' branches, although there are no great harmful effects if the amount of crown removed ranges between 25 and 45%. Pruning regimes differ for conifers and broadleaves.

Conifers

Initial pruning to 4 m is carried out once the trees reach about 10 cm diameter or 20 years old. Further pruning to 5–6 m height is carried out once the trunk at this point reaches 10 cm diameter. The operation may be carried out at any time of year, although preferably during March to May for quicker healing. Pruning of conifers is not widely practised in larger commercial estates, although for smaller farm woodlands, if labour is available, it may be justified to increase the capital value of the woodland. An expanding market exists for pruned pine (Scots and Corsican), Douglas fir and European larch for 'boatskin' quality.

Broadleaves
Pruning begins early for broadleaves (5–10 years) to create a single main stem through the removal of competing shoots (*formative pruning*) and later by the removal of lower side branches from the trunk to about 5 m. The operation is particularly suitable for selected final crop trees grown for decorative veneers (e.g. oak, walnut, cherry and maple). Pruning should never been undertaken during flushing (March–May), since resistance to infection is lower at this time. For most species winter is the optimum time for pruning, although cherry should be pruned between June and August to reduce the risk from canker and silver leaf disease (*Chondrostereum purpureum*).

Epicormic branches
Epicormic branches arise from adventitious buds around pruning scars or from dormant buds on the stem, which are triggered into sprouting by increased levels of light (e.g. following thinning). These epicormic shoots remain semi-moribund in normal woodland conditions, although a further increase in light levels will enable the shoots to develop into larger branches, which will severely affect the quality of the timber. Only small knots will be tolerated for high quality sawnwood and may have to be completely absent for decorative veneers. Pedunculate oak, sweet chestnut, poplar and cricket bat willow are the worst affected. The growth of epicormics can be restricted by avoiding sudden changes in the environment of the stand (e.g. through heavy thinning) or by pruning. If carried out annually, epicormics can be removed by rubbing. Classical silvicultural treatment of oak stands to reduce the incidence of epicormics involved growing an understorey of beech, hornbeam, Norway spruce or western red cedar to keep the boles of the oak trees shaded.

Forest protection

Fire

The chances of fire damage depend on the nature, size and location of the woodland. A key factor is the amount of inflammable material in the forest, either dead wood and brashings or dry undergrowth in young plantations before canopy closure. March to May is the most dangerous period, when a large amount of dead vegetation remains from the previous year, especially when combined with low humidity, high temperatures and windy conditions. Woodlands close to main roads or residential areas also tend to have a higher risk of fire outbreak. The risk of fires spreading can be reduced by cutting fire breaks, strips of at least 10 m width which are kept free of inflammable material by mowing or cultivation. Japanese larch or alder do not easily catch fire and can be used as fire belts, 10–20 m wide, between compartments. Forest tracks and roads also act as internal fire barriers. Other precautions which should be taken include:

- Inform the local fire brigade of the location of woodland access points and fire fighting equipment
- Ensure adequate access to woodlands by removing fallen trees and repairing broken culverts.
- Maintain fire fighting equipment at strategic points, to include spades, buckets and axes, and provide birch brooms and beaters at entrance points to woodlands.
- Assess the proximity to water and if necessary dig additional ponds.
- For high-risk sites (along footpaths and roads) clear inflammable material and brash young conifers early.
- Ensure that all staff and contractors observe the fire precautions and are trained in the fire drill.
- Insure against fire.

Wind

Prolonged exposure to high winds significantly reduces tree growth rates and in extreme conditions will lead to stem damage from abrasion by neighbouring trees. Where root development is restricted by waterlogging or an iron pan, woodlands are more susceptible to windthrow during autumn and winter gales since the shallow root plate gives less anchorage. In persistently windy regions, the risk of windthrow can be reduced by shortening crop rotation lengths and either avoiding line thinning or not thinning at all. Site drainage and cultivation can improve tree anchorage and root development. Some species are more susceptible than others; Douglas fir and larches are particularly prone to windthrow on shallow clay soils. Wind-thrown trees can be economically harvested, although care is required in extracting the timber as the opening up of the woodland could lead to further windthrow along exposed unstable edges. Windthrown pockets should always be cleared back to windfirm edges or existing compartment boundaries for this reason.

In order to predict the likely effect of wind on the woodland, the Forestry Commission has devised a Windthrow Hazard Classification based on windiness, elevation, degree of exposure and soil conditions. The classification gives an indication of the height to which a tree crop can be grown before windthrow will become a limiting factor. For further information see Miller (1985).

Insects

A large number of insects are dependent on trees for some part of their life cycle, either for food, by siphoning off plant sap, eating leaves, needles, shoots, bark or sapwood, or for reproduction, using the trunk, stump or shoots for laying eggs or building brood chambers. The vast majority of these associations cause insufficient damage to be of concern, and some insect–tree relationships are mutually beneficial. The most damaging insects are associated with particular stages in the life of the tree crop, together with site and climatic conditions favouring a rapid increase in insect population size. There is some evidence in this respect that trees already under some environmental stress (e.g. from incorrect choice of species for site) are more prone to serious damage. The most susceptible stage in the life of the woodland is at establishment, particularly where this is adjacent to neighbouring mature trees of the same species or where stumps from the previous crop are still present, which may provide the initial source for an insect outbreak. While some insect pests are fairly general in their tastes, others are specific to particular trees or groups, conifers tending to be more susceptible than broadleaves. The most important insects associated with particular stages in the life of the forest are as follows.

Establishment of all species on restock sites may be prone to attack from large pine weevils (*Hylobius abietis*) and black pine beetles (*Hylastes* spp), which will breed and multiply in the stumps and roots of the previous crop to emerge and feed on the newly planted trees, causing severe losses. The trees should be dipped in insecticide prior to planting, which will give protection for the first year, or alternatively a top spray can be applied after planting.

From the *thicket stage onwards*, many tree species suffer varying degrees of defoliation from leaf and needle feeders. Most trees will recover from even quite severe defoliation, and control measures are normally only economic in large commercial woodlands. Specific examples of damage include pine looper moth (*Bupalus piniaria*) on Scots pine, pine beauty moth (*Panolis flammea*) on lodgepole pine and the oak leaf roller moth (*Tortrix viridana*).

On *windblown sites* or where *felled timber* is available, a number of secondary pests (weevils and bark beetles) will breed and multiply under the bark of damaged, dead or felled timber. Where subsequent populations are high, the surrounding healthy crop may be damaged; spruce, larch and pine being the most severely affected. The pine shoot beetle (*Tomicus piniperda*) is the most important pest, requiring good forest hygiene to control outbreaks in susceptible crops. Felled timber should be either removed or debarked within six weeks, during the period March to August, to prevent broods of insects from being raised.

The identification and control of insect pests is a specialized activity, so, if insect damage is suspected, further advice should be sought from the Forestry Authority. Further information is available in Bevan (1987).

Disease

There are a number of fungal and bacterial diseases of trees, which can lead to deterioration of the main stem, destruction of roots, shoots or cambium and ultimately death or windblow. As with insect damage, the severity of the disease infection is often exacerbated by both climatic conditions and environmental stress; wounding of trees from extraction damage or pruning being examples. Control measures are rarely justified, so correct planting choice using more resistant tree species is to be advised in susceptible conditions. The most important diseases to be aware of are as follows.

Honey fungus (*Armillaria* spp) can be a problem in old broadleaved woodlands, affecting both broadleaves and conifers. It spreads through the soil from infested wood, especially old broadleaved stumps, via a network of black strands (*rhizomorphs*) to infect young trees. Losses are sporadic and rarely affect a whole wood, although it may be wise to restock with a more resistant tree, such as Douglas fir or a broadleaved species.

The stem and root rot *Heterobasidion annosum* is the most serious disease of coniferous woodland, leading to deterioration of the lower trunk and roots. Larch, spruce, western hemlock, western red cedar and pines (on former agricultural land or alkaline soils) are most susceptible, especially on the sites of previous coniferous crops. The risk of disease can be reduced by treating the freshly cut stump surfaces of felled trees with a concentrated coloured solution of urea within 15 minutes of felling. More resistant species include Douglas fir, grand fir and Corsican pine.

Larch canker is caused by the fungus *Lachnellula willkommii* on European larch, favouring its replacement by hybrid larch. Group dying of conifers caused by the fungus *Rhizina undulata* spreads through litter from fire sites, therefore fires should be excluded from the forest interior. Beech bark disease develops from the association of the scale insect *Cryptococcus fagisuga* and the fungus *Nectria coccinea*, forming a matt of white greasy wool on the main stem and branches. Infected trees should be removed during thinning operations.

Squirrels

Grey squirrels damage pole-size trees by stripping bark from susceptible species (beech, sycamore, ash and pine) between May and July, causing deformities in the main stem and increased forking. The extent of damage can be reduced by controlling squirrels from April to July, by live cage trapping or using hoppers with warfarin-poisoned bait. Warfarin is potentially hazardous to both people and wildlife, requiring training in its safe application. The use and dimensions of the hoppers are dictated by the Grey Squirrels (Warfarin) Order 1973, so that animals larger than the grey squirrel are prevented from access to the hopper. A recent modification has been to incorporate a transparent flap door with a securing magnet into the entrance tunnel of the hopper, which prevents access by smaller, less inquisitive animals (voles, mice and small birds), reducing the rate of bait removal by 30–50%. For maximum control the hoppers should be placed under large trees clear of vegetation at 200 m centres or 3–5/ha. Further information is available from the Forestry Authority.

Landscape design for farm woodlands

A woodland which is designed and managed using the principles of good landscape practice will enhance the surrounding landscape, together with benefits for wildlife and public enjoyment. The Forestry Authority and other agencies (e.g. the National Parks and local authorities) are increasingly encouraging the adoption of sympathetic forest design and management through control of felling licences, forest strategies and grant schemes, and there is also evidence that landscaped forests command greater prices in the market.

On level ground the landscape sensitivity of the forest is less than for highly visible forests on steep slopes at higher elevations, or in designated areas of great landscape value. Good landscape practice therefore begins by assessing the nature of the surrounding landscape, which will dictate the most appropriate design solutions.

For *hedgerow landscapes* in the lowlands, the design of woodlands should mirror the often geometric shape of the surrounding hedgerow pattern, the shape of the woodland interlocking harmoniously with the surrounding fields.

For *open and upland landscapes*, the design of woodlands needs more careful consideration. In these areas the topography or landform will be the dominant force in the landscape so that any new woodland planting must reflect the scale and shape of the surroundings. Small woodlands may look out of scale and should be located close to existing woodlands or on lower slopes where a hedgerow pattern is more evident. The shape of the forest should follow natural boundaries and vegetation, rather than the contour or the edge of the fence. As a general rule the edges of the forest should rise up the valleys and hollows and fall down the shoulders of hills and ridges.

The landscape of all forests can be improved by attention to the following measures:

- *Diversity* creates visual interest and enhances the landscape value of the forest. Different textures and colours can be created by using different species, such as broadleaves along watercourses or compartments of larch, which change from fresh green in summer to brown in winter. A change of species can reduce the scale of the forest and allow more light into recreational areas, although the main species should make up two-thirds of the composition for aesthetic reasons. Diversity can also be increased by the provision of open ground, exposing outcrops of rock and creating a mosaic of different ages of trees.
- *Forest boundaries* are particularly important at the skyline, where the forest should either completely cover the skyline or should cross it at the lowest point, cutting diagonally across the main view or else curving gently over the skyline. The edges to the forest should be as natural as possible, varied in scale with the landscape by the use of irregular groups, different species or plant spacing and detailed shaping.
- *Forest operations* should be carried out in sympathy with the landscape. At planting, ploughing should be organized to create open areas which follow the landform, and forest tracks should cut diagonally across the contour rather than along it. The harvesting of the forest provides an excellent opportunity to improve the appearance of a previously poorly designed forest by introducing different species and a greater diversity of age structure and by reshaping the edges of compartments. Any such work will need to be carefully phased to avoid immediate drastic changes.

For larger woodlands, especially in sensitive locations, it is advisable to undertake a full landscape appraisal using photographs taken from key viewpoints surrounding the woodland. Acetates and overlays can be produced to show the effect of different operations, the phasing of harvesting operations and the sites for new planting. For further information see Forestry Commission (1989).

Forest management for non-timber uses

Shelterbelts

The provision of correctly designed and located shelterbelts within arable field crops or grassland will have the effect of reducing wind speed and moisture loss from the crop, leading to earlier harvests and greater productivity. Solid or impermeable windbreaks should be avoided, as they cause an upward deflection of the airstream, producing an area of low pressure to leeward of the barrier which results in intense turbulence more damaging than the original wind. More porous windbreaks (ideally 50% permeability) not only reduce turbulence but also reduce windspeed for a distance up to 30 times the height of the barrier. On sloping ground the location of the windbreak should be carefully considered as a trap is created for frosty air flowing down from higher up the hill. Single or double row windbreaks are normally used on horticultural and productive agricultural land, as wider windbreaks give no additional reduction in windspeed and may harbour pests such as pigeons or bullfinches and prevent normal drying, favouring fungal diseases such as grey mould (*Botrytis cinerea*). Shelterbelts may be inappropriate in some situations as the trees will shade neighbouring crops, or on heavy wet soils where the drying effect of the wind is reduced.

In upland livestock areas, shelterbelts afford protection to stock from the cold driving winds, reducing heat loss and hence improving survival and cutting food consumption. Dense shelterbelts are more suitable, as animals tend to pack in tightly to the lee of the wood during winter gales. Narrow belts of trees become draughty around the lower trunk with age and pose problems at maturity as restocking will generally mean clear felling, leading to a temporary loss of shelter. Wider shelterbelts should therefore be used (at least 20 m and 45 m if possible), which allow for the use of a mixture of shrub and tree species to create a more densely graded profile. Wider shelterbelts allow restocking to take place progressively without loss of protection, and have a greater proportion of utilizable timber together with other benefits for wildlife and game. Livestock can be allowed access to the woodland in extreme weather for shelter and feeding on ground vegetation, although this should only be a temporary measure as long-term entry by livestock will lead to root damage, poaching and browsing of natural regeneration.

Good shelterbelt species for field crop protection are balsam poplar hybrids (*Populus tacamahaca* × *trichocarpa*), grey alder, or Italian alder on drier soils, all at 1.0–2.0 m spacings. For wider livestock shelterbelts a mixture of species is advisable, especially beech, pine, sycamore, cherry and rowan, with Sitka spruce on exposed ground in the west.

Pheasants

Farm woodland management for pheasants need not be in conflict with sound silvicultural practice. The value of woodlands can be substantially increased by incorporating some of the following specific measures to increase its potential for pheasants.

- Pheasants do not thrive in cold draughty conditions, so the woodland should be designed to give maximum protection to ground level using a mixture of taller trees and shrubs. The perimeter of the woodland may be planted with hedging species or Christmas trees. Roosting trees should be retained on restocked sites and at the intersection of rides.
- The holding capacity of the wood can be increased

by establishing ground cover shrubs, which provide additional shelter and food for the birds. Suitable species include butcher's broom (*Ruscus aculeatus*), rose of Sharon (*Hypericum calycinum*), dogwood (*Cornus alba*), elder (*Sambucus nigra*), flowering nutmeg (*Leycesteria formosa*), privet (*ligustrum vulgare*), raspberry (*Rubus idaeus*) and snowberry (*Symphoricarpos albus*).
- A rich ground layer should be maintained by avoiding planting trees which cast a dense shade (beech, sycamore and close-spaced conifers), favouring trees with a lighter canopy (oak, ash, birch, cherry, rowan, larch and pine) and thinning at the appropriate time to increase the amount of light reaching the forest floor. Coppice is particularly suitable as it maintains a higher proportion of ground vegetation and a diversity of ages in the stand.
- Pheasants are woodland-edge birds, using the forest for winter cover and roosting. By planting a series of smaller and longer woodlands, the proportion of edge can be increased.
- In larger woodlands, open spaces and rides (30–50 m wide) should be provided to allow birds room to fly up and over the guns.

Conservation

The following section gives general guidance on the improvement of woodland for wildlife, although some woodlands will already have important conservation status or designation (SSSI, NNR, ancient semi-natural woodland) and will require a more specific management prescription normally involving expert advice from either local or national wildlife organizations.

The following actions are designed to increase the diversity of habitats in the woodland.

- Native trees support large numbers of species (especially invertebrates) and by planting trees native to the area the local genetic stock is preserved. Natural regeneration of native trees should be encouraged wherever possible.
- In commercial coniferous woodlands, some broadleaves should be planted (minimum 5% for Woodland Grant Scheme), especially along watercourses and woodland edges.
- Open space is an important woodland habitat with characteristic ground vegetation and associated wildlife. These areas can often be incorporated at planting by leaving bare ground around rock outcrops, stream sides and wetlands. Woodland glades and rides (at least 5–10 m wide) can be managed by cutting back vegetation and mowing to create a graded profile.
- The ground vegetation which builds up following tree establishment supports a thriving population of small mammals and their predators. Following canopy closure, much of this vegetation is lost under the dense shade of the thicket of trees, especially conifers. By thinning earlier (up to 5 years) and slightly heavier (up to 10% more volume removed) ground vegetation is more likely to survive.
- Structural diversity can be achieved by planting an understorey of native shrubs such as hazel, hawthorn, holly or juniper, particularly along woodland rides and edges.
- Harvesting creates the opportunity to increase the age diversity of uniform woodlands by clearing smaller areas (0.2–0.5 ha) phased in over a longer period.
- Some mature trees should be retained beyond the normal rotation until physical maturity (at least 1%), to provide old growth habitats and habitat continuity.
- Dead wood habitats are scarce in most commercial woodlands, so that stacks of dead branchwood should be left to rot down (where forest hygiene permits) and standing dead trees retained for hole-nesting birds, specialized invertebrates and epiphytes. In addition, nest boxes can be provided in a variety of designs.
- Forest operations should be carried out with respect to wildlife. The passage of felling and extraction machinery should be organized to avoid disturbance to breeding sites and other sensitive areas such as streams. Where a fencing line passes over a badger track, a swing gate should be provided. Herbicide use should be carefully controlled to avoid contamination of watercourses.

The landowner may find it useful to incorporate some of these improvements into a conservation management plan for the woodland. Following an appraisal of the wildlife value of the site, appropriate management prescriptions should be proposed to reflect the overall aims for the woodland.

Forest investment

In financial terms the forest plantation represents a long-term investment, where the owner faces a patient wait before revenue from the later thinnings and final harvest offset the early establishment expenditure. In order to compare forest investment with other alternatives, a discounted cashflow may be calculated, using an appropriate interest rate (net of inflation), to give a *net present value* (NPV) for the proposed woodland. The return from different rotations and species can be compared easily, since all future revenues and expenditure are discounted back to year 0. The *internal rate of return* (IRR) is calculated as the interest rate at which NPV = 0. The choice of discount rate is critical to the effect on NPV, since the higher the discount rate chosen, the lower the value of future returns, especially for longer rotation crops such as oak. A figure of 3–5% is normally appropriate for most commercial woodlands, although lower rates may be acceptable where other less tangible benefits such as landscape and wildlife are considered important.

Integration of agriculture and forestry

The integration of farming and forestry may have mutual economic advantages as a result of efficiencies of labour deployment and the utilization of marginal land for tree growth. Recent interest has focused on the potential for *agroforestry systems*, which involve the establishment of widely spaced trees with grazing or arable crops in between the rows. The productivity of the agricultural component will depend on the tree spacing and resulting point of canopy closure. The timber objective is to produce a main stem of at least 40–50 cm dbh over a short rotation of 40–50 years, combined with pruning to remove side branches. Suitable tree species include ash, cherry, walnut, poplar, Douglas fir and hybrid larch. Economic models have shown that such systems may prove financially viable, particularly in marginal areas, although further field testing is required. Further information can be obtained from your local ADAS or Forestry Authority office.

Grants

The Woodland Grant Scheme (Forestry Authority)

The Woodland Grant Scheme (WGS) aims to encourage the management of existing woodland together with the expansion of private forestry in a way which achieves a reasonable balance between the needs of the environment, increasing timber production, providing rural employment and enhancing the landscape, amenity and wildlife conservation. Grants are available for restocking, new planting or natural regeneration on areas larger than 0.25 ha or 15 m wide, although smaller areas may be accepted by agreement with the Forestry Authority.

A number of other provisions must normally be met, including the maintenance of a broadleaved component of the woodland, protection of statutory designations (e.g. ancient monuments, SSSIs, rights of way), appropriate management of ancient woodland and compliance with guidelines on drainage and fire protection. Planting may only be undertaken once approval for the scheme is confirmed by the Forestry Authority. An *establishment grant* is payable in two instalments following successful planting or natural regeneration, with higher rates of grant for broadleaves and areas under 10 ha.

Additional supplements may also be available for arable fields or improved grassland (*Better Land Supplement*), woodlands within five miles of urban areas (*Community Woodland Supplement*) and for professional advice in the preparation of management plans. For applicants wishing to maintain or enhance the environmental value of the woodland or to provide for public access, an *annual management grant* is payable. In addition, a *livestock exclusion annual premium* may also be available for woodland in Less Favoured Areas of high environmental value. Information packs about the grant scheme are available from the Forestry Authority.

The Farm Woodland Premium Scheme (MAFF)

The Farm Woodland Premium Scheme aims to encourage the planting of woodland on farmland by compensating for the loss of revenue during the early years of the new woodland. The scheme runs in conjunction with the Woodland Grant Scheme and covers arable and improved grassland which qualifies for the Better Land Supplement:

- Farmland which has been under arable cultivation for at least 3 years prior to the WGS application.
- Grassland which has been improved for at least 3 years prior to the WGS application, the sward comprising at least 50% ryegrass, cocksfoot, timothy or white clover.

Within Less Favoured Areas the criteria for improved grassland status are less stringent and unimproved land may also qualify for the grant, where this has been in agricultural use for at least three years prior to the application.

The annual payments are higher in disadvantaged areas and are based on a stocking of at least 1100/ha for broadleaves and 2250/ha for conifers. Plantations at wider spacings, coppice establishment and agroforestry projects are paid on a pro rata basis. Up to 20% of the total area may be left unplanted for woodland glades, forest margins or along watercourses, and up to 10% non-timber trees and shrubs may be planted for conservation or game.

Other sources of grant assistance

Grants are available for approved *amenity planting* of areas under 0.25 ha financed by the Countryside Commission, Countryside Council for Wales or Scottish Natural Heritage operated through the relevant local authority. For woodlands of high wildlife conservation value, assistance may be available through English Nature (or equivalent organization), National Park Authorities or through charitable organizations such as the Woodland Trust. Additional help may also be available for sites within other schemes such as Countryside Stewardship or Environmentally Sensitive Areas.

Tree planting may form an important component of an application for the Farm and Conservation Grant Scheme (MAFF), especially for small amenity areas, hedgerow trees, shelterbelts and the enclosure of broadleaved woodlands to promote natural regeneration. Further information on these and other local schemes can be obtained from the regional office of the Forestry Authority, your local Farming and Wildlife Advisory Group (FWAG) officer or the Woodlands Advisor of the local authority.

References and further reading

Bevan, D. (1987) *Forest Insects*. FC Handbook 1.
Bibby, J.S. *et al.* (1988) *Land Capability Classification for Forestry in Britain*. Mccauley Land Use Research Institute, Aberdeen.
Blyth, J. *et al.* (1987) *Farm Woodland Management*. Farming Press, Ipswich.
Brazier, J.D. (1990) *The Timbers of Farm Woodland Trees*. FC Bulletin 90.
Crockford, K.J., Spilsbury, M.J. & Savill, P.S. (1987) The relative economics of woodland management systems. Occasional Paper 35, Commonwealth Forestry Institute, University of Oxford.
Davies, R.J. (1987) *Trees and Weeds*. FC Handbook 2.
Edwards, P.N. (1983) *Timber Measurement: A field guide*. FC Booklet 49.
Edwards, P.N. & Christie, J.M. (1981) *Yield Models for*

Forest Management. FC Booklet 48.
Evans, J. (1984) *Silviculture of Broadleaved Woodland*. FC Bulletin 62.
Evans, J. (1988) *Natural Regeneration of Broadleaves*. FC Bulletin 78.
Forestry Commission (1988) *Forests and Water Guidelines*.
Forestry Commission (1989) *Forest Landscape Design Guidelines*.
Forestry Commission (1990) *Forest Nature Conservation Guidelines*.
Hamilton, G.J. (1985) *Forest Mensuration Handbook*. FC Booklet 39.
Harding, T. (1988) *British Softwoods: Properties and Uses*. FC Bulletin 77.
Hart, C.E. (1991) *Practical Forestry for the Agent and Surveyor*. Alan Sutton, Stroud.
Hibberd, B.G. (1988) *Farm Woodland Practice*. FC Handbook 3.
Hibberd, B.G. (ed.) (1991) *Forestry Practice*. FC Handbook 6.
Insley, H. (ed.) (1988) *Farm Woodland Planning*. FC Bulletin 80.
James, N.D.G. (1989) *The Forester's Companion*, 4th edn. Blackwell, Oxford.
Lines, A. (1987) *Choice of Seed Origins for the Main Forest Species in Britain*. Forestry Commission Bulletin 66.
Lucas, O.W.R. (1991) *The Design of Forest Landscapes*. Oxford University Press, Oxford.
McCall, I. (1988) *Woodlands for Pheasants*. Game Conservancy, Fordingbridge.
Miller, K.F. (1985) *Windthrow Hazard Classification*. FC Leaflet 85.
Pepper, H.W. & Tee, L.A. (1986) *Forest Fencing*. FC Leaflet 87.
Peterken, G. (1993) *Woodland Conservation and Management*. Chapman & Hall, London.
Potter, M.J. (1991) *Treeshelters*. FC Handbook 7.
Price, C. (1989) *The Theory and Application of Forest Economics*. Blackwell, Oxford.
Rollinson, T.J.D. (1988) *Thinning Control*. FC Fieldbook 2.
Savill, P.S. & Evans, J. (1986) *Plantation Silviculture in Temperate Regions*. Clarendon Press, Oxford.
Williamson, D.R. (1992) *Establishing Farm Woodlands*. FC Handbook 8.
Williamson, D.R. & Lane, P.B. (1989) *The Use of Herbicides in the Forest*. FC Fieldbook 8.

Useful addresses and sources of advice

The Forestry Commission has split into the Forestry Authority (responsible for advice, grants, etc.) and the Forestry Enterprise (responsible for the management of the state-owned forests), although reference may still be found to the former name.

Forestry Commission – main headquarters
231 Corstorphine Road, Edinburgh EH12 7AT. Tel: 0131 334 0303.
Forestry Commission – publications and research
Forest Research Station, Alice Holt Lodge, Wrecclesham, Farnham, Surrey GU10 4LH. Tel: 01420 22255.
Arboricultural Association
Ampfield House, Ampfield, Nr Romsey, Hampshire SO51 9PA. Tel: 01794 687177.
Association of Professional Foresters
7–9 West Street, Belford, Northumberland NE70 7QA. Tel: 01668 213937
British Christmas Tree Growers' Association
12 Lauriston Road, Wimbledon, London SW19 4TQ. Tel: 0181 946 2695.
Coed Cymru
Ladywell House, Newtown, Powys SY16 1RD. Tel: 01686 26799.
Game Conservancy
Fordingbridge, Hampshire SP6 1EF. Tel: 01425 52381.
Institute of Chartered Foresters
7a St. Colme Street, Edinburgh EH3 6AA. Tel: 0131 225 2705.
National Small Woods Association
Hall Farm House, Preston Capes, Northants NN11 3TA. Tel: 01327 361387.
Royal Forestry Society of England, Wales and Northern Ireland
102 High Street, Tring, Hertfordshire HP23 4AH. Tel: 01442 822028.
Royal Scottish Forestry Society
11 Atholl Crescent, Edinburgh EH3 8HE. Tel: 0131 229 8851.
Timber Growers United Kingdom
5 Dublin Street Lane South, Edinburgh EH1 3PX. Tel: 0131 5387111.
Tree Council
35 Belgrave Square, London SW1X 8NQ Tel: 0171 235 8854.
Woodland Trust
Autumn Park, Dysart Road, Grantham, Lincolnshire NG31 6LL. Tel: 01476 74297.

13

Crop Health

G. Moule

Weed Control

Weeds are injurious or harmful to growing crops, and if not controlled reduced yields from crops and grass result, harvesting and other operations are hindered, produce contaminated or taints imparted and the product rendered unfit for sale. A number of weeds are poisonous.

Most weeds of significance are annuals or perennials. Annuals complete their life cycle within a year and reproduce only from seed, of which they set an abundance. Growth is rapid, some species producing two or more generations in a year. Perennials also produce seed but do not die after seeding and, as they also reproduce from vegetative storage organs, serious infestations can build up rapidly if not controlled. Biennials only live two years. They make vegetative growth in the first and set abundant seed in the second, after which they die; they reproduce only from seed. Few are important, e.g. ragwort (*Senecio jacobea*) and spear thistle (*Cirsium lanceolatus*), which infest badly managed grassland. For convenience of weed control, weeds may also be classified as grasses, broadleaves, rushes and sedges and woody weeds.

The problem of pesticide resistance has long been more widespread and serious with continued and persistent use of certain groups of insecticides (OPs – organophosphorus compounds and carbamates) and fungicides (EBIs – ergosterol biosynthesis inhibitors, and MBCs – methyl benzimidazole carbamate) against their respective and specific targets, e.g. peach–potato aphid (*Myzus persicae*) on sugar beet and potatoes; powdery mildew (*Erysiphe graminis*) and eyespot (*Pseudocercosporella herpotrichoides*) on cereals.

This has in the main been because of the highly mobile nature of aphids and fungal spores and of the short generation times from mother to daughter aphids and spore to spore. This often gives rise to several generations per season or per year. This extremely rapid multiplication can give rise to populations shifting from mostly susceptible to largely partially resistant types in a few seasons. This has been the case with insecticide and fungicide use in the 1980s in the UK and the rest of western Europe and many other parts of the world.

Life cycles with weeds are generally much larger and normally annual or perennial. Thus herbicide resistance problems have taken longer to materialize and, because seeds are far less mobile than airborne aphids or fungal spores, have been much slower in their spread from site to site. The triazine herbicides, atrazine and simazine, when continually used for weed control in maize and fruit and for total weed control, have selected out resistant types of weeds but this has been largely confined to the subtropical and maize-growing regions of the world.

In the early 1980s (Moss & Cussans, 1985) blackgrass (*Alopecurus myosuroides*) populations showing variable resistance to the main cereal herbicides used for grassy weed control had become evident in several areas of Eastern and Southern England. In 1991 (Clark & Moss, 1991) the problem had been identified on 46 farms on 19 counties of England. The resistance is variable and complex and to such a wide range of differing herbicides that only fully integrated control programmes over the whole length of the rotation are likely to be successful. An example of a programme for blackgrass control is given in Table 13.1. Other programmes for different weed problems can be based on principles similar to those that apply to blackgrass. The ban on straw burning since 1992 has not helped the situation but changes in set-aside could aid future programmes. Pesticide resistance problems are in general likely to increase with continued intensification of agriculture during the 1990s.

Occurrence of weed problems

Grassy weeds

Annuals

Annuals are serious problem in cereals, especially wild oats (*Avena* spp.), blackgrass (*Alopecurus myosuroides*) and sterile brome (*Bromus sterilis*), but can be controlled chemically; not usually a serious problem in broadleaved crops. Only annual meadow grass (*Poa annua*) occurs in grassland. The correct use and integration of set-aside to help control herbicide-resistant blackgrass and other arable weed problems has not been fully explored yet.

Perennials

Perennials are a widespread problem in grassland and occur on poorly managed arable land, especially cereals,

Table 13.1 Control strategies in blackgrass (and most autumn-germinating weeds)

Control factor	*Best strategy*	*Worst strategy*
Rotations	Non-cereal crops Continuous spring-sown crops	Continuous winter cereals
Stubble hygiene	Complete weed kill during long moist intercrop period	Incomplete weed kill in a short dry intercrop period
Cultivations	Plough	Direct drill or reduced cultivations
Drilling date	Late autumn using stale seedbed technique	Early autumn with immediate sowing
Herbicide	Use effective herbicides on all crops in rotation	Non-effective herbicides
	Use non-selective herbicides in all intercrop periods	No herbicide
	Use specific graminicides on broadleaved break crops	No herbicide
	Correct early post-emergence use in cereals under good conditions	No herbicide/ poor timing and conditions
Crop competition	High seed rate	Low seed rate
	High rate of N fertilizer	Low rate of N fertilizer
	Vigorous cereal variety	Weak-growing cereal variety
	Thick barley crop	Thin wheat crop
	Well drained fields	Poorly drained fields

After Clark & Moss, 1991.

where they cannot be controlled chemically while the crop is growing actively; stubble treatment or spraying shortly before harvest is usually necessary. Good chemical control in young trees, no problem later.

Broadleaves (non-woody)

Annuals

Except for mouse-eared chickweed (*Cerastium vulgatum*) these occur almost entirely on arable land, where there are good chemical controls in most crops. Elsewhere they are only a problem following soil disturbance, i.e. on new leys or poached grassland.

Perennials

Sprayed cereals are an excellent cleaning crop for these, where they are readily controlled by translocated herbicides. They cannot usually be controlled chemically in broadleaved arable crops. Some are difficult to control chemically in grassland, e.g. docks (*Rumex* spp) unless ploughed prior to treatment. Adequate controls in young trees.

Rushes and sedges (perennial)

These are found on poorly drained and managed grassland, and are a sign of low fertility. They should not be an arable weed.

Woody weeds

Woody weeds are mainly a problem in young woodland. They can be readily controlled by chemicals in establishing conifers, but may establish in undergrazed grassland, e.g. gorse (*Ulex* spp), brambles (*Rubus* spp), etc.

General control measures – applicable to all systems

Prevention of spread of weeds by seed

Preventive sanitation is necessary to stop the spread of seed of a wide range of weeds, e.g. wild oats (*Avena* spp), docks (*Rumex* spp) and gorse (*Ulex* spp). Sow clean seed, burn straw infested with weeds and avoid buying weedy hay or straw or transporting on to land destined for seed production. Allow farmyard manure to heat up in a neat pile before spreading. Prevent spread of weed seeds from hedgerows or banks, e.g. sterile brome (*Bromus sterilis*) and waste places, e.g. docks (*Rumex* spp), ragwort (*Senecio jacobea*), and seeds and vegetative organs from headlands.

Cultural and rotational methods

See also major cropping systems – Arable Crops (Chapter 10), Grassland (Chapter 11) and Woodland (Chapter 12).

Principles of chemical (herbicidal) weed control

The effect of any herbicide or chemical weedkiller depends on the manner and circumstances in which it is used, namely the herbicidal treatment. If recommendations are to be successful, they must be given in relation to the whole treatment. The three basic components of herbicidal treatment are the crop, type of weed, and type and dosage rate of herbicide. Herbicidal treatments can be:

- Total or non-selective. Applied with the object of killing all vegetation, either before planting the crop or in non-crop situations, e.g. paths, around buildings and waste land;
- Selective. Designed to suppress weeds without damaging the crop. However, all treatments become total if a large excess of herbicide is applied.

Herbicides may be applied either to the foliage (foliar application) or to the soil (residual application).

Foliar application

Foliar-applied herbicides may be of two types.

Contact types
These kill only the parts of the plant they contact. Single applications are suitable only for use against annual weeds.

Translocated types
These are transported within the plant to parts remote from the point of application. They can be used for perennial weeds with well protected storage organs.

Applications may be made at these stages of crop growth:

(1) pre-sowing or pre-planting before the crop is sown or planted;
(2) pre-emergence, after sowing but before emergence;
(3) post-emergence, after crop emergence.

Each may consist of contact, translocated or residual applications or a combination of two or more of these.

Herbicides may be applied *overall*, when the whole crop area or soil is covered. Most spraying for fallows, cereals or grassland is so treated.

Alternatively the application may be *directed*, where only part of the area is sprayed. Special equipment is needed, hence the acreage must justify the capital cost. With *band spraying*, where a band of chemical about 180 mm wide is applied along the rows, say for sugar beet (pre- or post-emergence), some two-thirds of the cost of herbicide is saved. Weeds between the rows are controlled by rowcrop tackle. Chemical hoeing, where the herbicide is applied to weeds or soil between the rows of an emerged crop, the crop being protected by metal guards on either side of the row, is used only on a limited scale. Directed applications may also be made from *tractor-mounted spraylances*, *knapsack sprayers* or *granule applicators*. Directed applications are widely used in woodlands and forests.

Selective treatments

Contact applications
These are used almost entirely for controlling annual weeds but kill only emerged weeds, being applied to the seedling or young plant. Contact herbicides usually have little or no residual effect and are thus ideal for application immediately before sowing. A high degree of operator efficiency is necessary to achieve adequate spray penetration and coverage. Higher pressures and smaller droplets are required.

Uses
(1) *Pre-sowing or pre-planting.* For (a) control of emerged annual weed seedlings, a 'stale seedbed' technique; (b) sward desiccation prior to direct drilling or ploughing (paraquat); (c) stubble cleaning, desiccation of stubble for burning and couch control (paraquat). Can be combined with a residual herbicide to control weed seedlings not emerged at the time of application. Several residual herbicides have some contact action.
(2) *Post-emergence.* Specific herbicides used on brassicas, sugar beet, mangolds and similar crops. Generally used in mixture with a translocated herbicide for cereals.

Translocated applications
These are used for controlling annual and perennial weeds. These usually act more slowly than contact herbicides but a much lower degree of cover is necessary, provided there are no misses. Damage to susceptible crops from drift is a serious hazard. High pressures and small droplets should not be used. Usually applied in low or low–medium volume. Many translocated herbicides need 6 h of fine weather but thereafter are unaffected by rain; others need longer.

Uses
(1) *Pre-sowing.* This is for (a) control of couch and other grassy perennials in stubbles or fallows; (b) control of late starting perennials in stubbles and seedlings germinating after a first spraying; (c) control of perennials for direct drilling, provided chemical has little residual effect.
(2) *Post-emergence or post-planting.* By far the greatest proportion of land sprayed is treated in this way, including most of the cereals treated for broadleaved weeds and all permanent and temporary grassland where selective weed control is required. Also used on clovers, lucerne and peas and in forestry and some other crops.

Soil or residual applications

These remain active in the soil for a variable period, depending on rate of application and speed with which herbicide is dissipated from the soil by rainfall and microbial activity. Cropping restrictions exist according to the individual herbicide and dosage rate and the time necessary between application and sowing a susceptible crop. Degree of activity, effectiveness and requisite dosage rate of residual herbicides depend on soil type. Some are of limited value in certain soils, e.g. organic soils and peats.

Uses
(1) *Pre-sowing.* These herbicides are usually incorporated into the soil and those with high volatility need immediate incorporation. The ideal tool is a rotary cultivator and some are fitted with spray bars. Alternatively, two passes of heavy harrows at right angles to each other may be used. Some herbicides are also formulated as granules, obviating need for incorporation. They control difficult perennial weeds such as couch, or, if applied shortly before sowing, annual broadleaved weeds, wild oats or blackgrass in a wide range of annual crops, cereals, peas, brassicas and sugar beet.
(2) *Pre-emergence.* Treatment is highly dependent on adequate moisture and a smooth, high-quality tilth, and is almost useless in dry conditions and rough seedbeds. Some herbicides in this group are mixed with a contact herbicide to give immediate control

of annual weed seedlings. Several have some contact properties of their own used to control annual grassy and broadleaved weeds in a wide range of broadleaved and cereal crops.

(3) *Post-emergence.* Largely used for directed application to perennial crops, orchards, soft fruit, and forestry; of less importance in farm situations.

A number of herbicides are used for crop desiccation prior to harvesting, e.g. diquat, glufosinate-ammonium and glyphosate for the desiccation of leafy red clover seed crops, potato haulm and green material in a cereal crop ready for harvesting. They may be used for root and shoot destruction on potato and mangold clamps or old clamp sites.

Choice of herbicide

Choice of herbicide is governed by the following factors.

Crop

Only crops and varieties for which a herbicide is specifically recommended should be treated; others are liable to be severely damaged or even destroyed. Even where several herbicides are recommended for a particular crop, they may vary in their degree of crop safety.

Weeds to be controlled

The occurrence of weed species and the nature of weed problems vary greatly between different types of crop and from farm to farm; species which are very troublesome in one situation may not even be present nearby. In some cases a weed infestation may consist of a single species, which may be dealt with by a simple herbicide, but the problem is usually much more complex, involving a wide spectrum of weeds, some of which are resistant to the simple herbicide. Successful weed control, at reasonable cost, involves a combination of the following approaches.

Broad spectrum

These treatments generally contain a mixture of two or more (sometimes four) herbicides given as a single application. These can be very successful for the control of a wide range of weeds, which have reached simultaneously an appropriate growth stage at the time of application, especially for the late spring foliar treatment of broadleaved weeds in cereals. Herbicides must be miscible. The same principle is used also for residual applications.

Herbicidal programmes

Single applications, even when using broad-spectrum mixtures, are only partially effective on mixed weed floras which develop over a relatively long period, e.g. winter cereals, where the crop may first be infested by autumn germinating annual grassy and broadleaved weeds, and then by spring-germinating annual grass and by a wide range of broadleaved weeds. An autumn application, including a residual herbicide, may be highly effective for much of the winter but may lack persistency to give adequate control of spring-germinating wild oats and broadleaved weeds. Conversely, if application is delayed until the spring, the autumn weed germination may well have reached infestation proportions. A *sequence* of applications, designed to deal specifically with the weeds being encountered, is therefore given at appropriate times in the life of a particular crop and is referred to as a herbicidal programme. The principle is now used in a very wide range of crops – sugar beet, root crops, vegetables, cereals, perennial crops and forestry.

Integration of herbicide use with those used on other crops

Crops differ greatly in the ease with which certain weed types may be controlled selectively during the life of the crop and in the post-harvest opportunities which they offer for chemical or mechanical cleaning. Cereals are an excellent crop for cleaning broadleaved weeds with herbicides: a sequence of two or three cereal crops sprayed with translocated herbicides will control most, if not all, perennial broadleaved weeds.

A winter cereal stubble allows ample time for cleaning grassy perennials before a spring-sown crop. Conversely, it is not often possible to control selectively broadleaved perennials in broadleaved crops such as sugar beet and potatoes; these weeds usually present serious problems if not controlled in the preceding cereal crop. Grassy annuals, which are frequently expensive to control in cereals – especially early sown winter varieties, can be readily controlled in spring-sown broadleaved crops, where they should not present any problem. Integration of herbicidal treatments over a run of crops pays dividends.

Stage of growth of crop and weed

Treatment at the wrong stage of crop growth is likely to result in damage to the crop, which can be very severe or even lethal, while application of a herbicide to weeds at an unsuitable stage results in unsatisfactory control. Foliar applied contact herbicides used to control annual weeds need very early application while translocated types, especially for control of perennials, need later treatment, when there is an adequate leaf area to absorb enough herbicide. A wide degree of crop or weed tolerance is especially valuable when climatic or weather conditions restrict the period of application (e.g. late spring applications to winter cereals) when a more expensive herbicide or one with a narrower spectrum may be justified. Avoid, where possible, herbicides with very narrow tolerances of the stages of growth of crop or weed between which application must be made.

Farming system and cropping programme

Use of sophisticated herbicidal treatments requires considerable expertise and equipment, both generally available on intensive arable farms. On many livestock farms, however, neither the equipment nor the experience exist, and simple treatments, often applied by a contractor, are preferable. Herbicidal treatments should be planned in conjunction with the cropping programme and must take probable residual effects into account.

Cost of treatment

This must be considered in relation to type and degree of weed infestation controlled. The value of crop and saving in labour and cultivations must also be taken into account. Heavy infestations of wild oats, blackgrass or couch grass in wet conditions justify expensive herbicidal treatment. Light infestations or conditions where cultivations are cheap and effective seldom do. Herbicides which result in complete mechanization of row crops or permit use of narrow rows giving higher yields, as with carrots, are well worth while. High value crops like sugar beet, potatoes or carrots permit the use of sophisticated herbicidal treatment. Less valuable fodder crops seldom justify heavy expenditure on herbicides. Where labour is freed by the use of herbicides, it must be saved or used for other profitable enterprises. When labour is in short supply, use of herbicidal treatment makes the difference between growing a crop or not growing it.

Spray application

Success or failure of herbicidal treatment depends largely on efficient application. A high standard of maintenance is required and sprayers should be overhauled and worn parts replaced before winter storage. Sprayers should be calibrated before use and output checked periodically. Use correct pressure volume rate and nozzle type for each job. Good marking and correct height of spray bar are essential; fans or cones of spray should meet a few centimetres above the top of the crop being sprayed.

The manufacturer's latest instructions on dosage rate, mixing, application, and use of protective clothing must be followed. Use clean tap water; dirty water blocks filters. Establish a proper spraying routine using gloves, face shield and protective clothing when handling dangerous concentrates. Empty drums must be rinsed out into a spray tank, the empty drum being disposed of properly or burned to avoid contaminating water supplies or streams. Wash out the sprayer thoroughly, using a synthetic detergent, after each day's work or when changing types of chemical or crop, on waste land away from watercourses. *Even small traces of herbicide can severely damage susceptible crops.*

Field conditions and stage of crop and weed growth must be right, for spraying at the wrong stages gives poor control or severely damages the crop. Best results are obtained when weeds are growing vigorously, not in cold weather or droughts. Spraying immediately before heavy rain is expected should be avoided. *Drift* is a major hazard and does irreparable damage to susceptible crops. Even slight traces of growth regulator herbicides ruin glasshouse, fruit and many market garden crops. Spray only in calm weather, never up-wind with susceptible crops, and avoid unnecessarily high pressure or small droplets.

Always possess adequate insurance cover against third-party claims from damage by spray drift. Cover needs to be substantial because glasshouse crops involve very large sums of money.

Nearly all herbicides are applied as a spray, water being the normal diluent. The volume of water used may be:

Volume	*litre/ha*
very low	<90
low	90–200
medium	201–700
high (rarely used)	>700

In ULV (ultra low volume) water is *not* used. Special formulations required (mainly forest application) 5–20 litres/ha.

Low or low–medium volume applications of 280 litres/ha or under require less labour and a larger acreage can be sprayed in a given time.

Tank mixes

Chemicals are mixed in the sprayer tank as opposed to buying a ready made mixture. Ingredients may include herbicides, fungicides, insecticides, aphicides, trace elements, wetting agent and growth-regulating substances. The practice is common. NB Tank mixes *must only be applied within label recommendations or in accordance with Pesticides 1995/Pesticides Register or subsequent revision of every product in the mix.* It is also essential to ascertain (especially where mixing additional herbicides to broaden the weed control spectrum) that varietal, stage of growth of crop and weed and any other restrictions do not differ from those that apply to the main or initial herbicide or other ingredient.

The method and order of mixing *must* conform to mixing instructions provided and must only be done in that way.

Rope wick applicators consist of a rope wick soaked in a strong solution of total herbicide (glyphosate) which is wiped over the leaf surface of the weed to be killed. Application relies for selectivity on weed being taller than crop to be treated, e.g. bolters and flowering weed beet in a sugar beet crop or developed rushes in grazed pasture. Applicators vary from wide boom-type machines for field use to small hand-held units where the wick is mounted on a cross piece (cf. broomhead) and the herbicide is contained in a hollow handle. The operator then wipes the target foliage. The latter models are suitable for small areas such as spot treatment, parks, gardens and in forest work.

Food and environment protection act 1985 The Control of Pesticides Regulations 1986 came into operation on 6 October 1986. The term pesticide includes chemical substances and certain micro-organisms prepared or used to destroy pests, which include plants as well as animals. Herbicides are therefore pesticides and their advertisement, storage, sale or supply and application are controlled. Only herbicides or mixtures of herbicides registered for each product by the manufacturer and currently listed in Pesticides 1995 (or appropriate subsequent revision) Part A 1.1. Herbicides (Reference book 500, Ministry of Agriculture, Fisheries and Food (MAFF), Health and Safety Executive), may be applied to cropped or uncropped land; this book is obtainable from MAFF Publications, Lion House, Alnwick, Northumberland NE66 2PF, UK.

Purchase and use of herbicides

Before purchasing a herbicide or herbicidal product, it is essential to ensure that it is entirely suitable for the purpose for which it is intended. Before mixing and application, it is equally essential to *read the manufacturer's instructions carefully and to follow them exactly*, taking all recommended precautions for the safety of operator, livestock, wildlife and beneficial insects and to avoid drift and pollution of streams or watercourses. The herbicidal product should *only* be used for purposes and applied only by methods and in circumstances which the manufacturers specifically recommend.

Herbicide recommendations for a wide range of field and vegetable crops are published in booklet form by ADAS and revised regularly; obtainable (priced) from MAFF Publications.

The purpose of this chapter is to outline the main principles and methods of cultural, rotational and chemical weed control. Individual herbicides listed are given as examples; *their inclusion does not constitute a field recommendation for use* nor does the omission of any herbicide necessarily suggest it to be inferior in any way. While every reasonable precaution has been taken to ensure the correctness of this chapter, neither the author, editor nor the publishers will be held responsible for any losses arising from its use.

Arable crops and seedling leys

Once established, a heavy leaf canopy suppresses further weed growth but individual crops differ in density and duration of the canopy. Crops such as kale produce a heavy, lasting canopy but cereals and potatoes may 'open up' late in the season allowing weed growth to re-start. The problem of weed control is thus twofold:

(1) To establish a dense leaf canopy as quickly as possible and maintain it throughout the life of the crop. All factors which promote rapid growth and increase yield, such as free drainage, good soil structure, adequate soil moisture, ample fertilizer, high quality seed and freedom from pests and diseases, allow crops to form a thick, well maintained leaf canopy. Conversely, drought or poor plant population favour weed growth.

(2) Control measures are needed to suppress weeds until a complete leaf canopy is formed or where the crop canopy is poor. These measures include preventive sanitation, rotations, cultivations and chemical weed control.

Rotations

A well balanced rotation prevents build-up of certain weeds by changing the environment from one year to another. If one crop is grown continuously, management and timing of cultivations remain constant and particular types of weed increase, e.g. blackgrass in winter cereals, wild oats in all cereals. Rotations including three-year leys and roots give much less trouble. The alternation of cereals with broadleaved crops also allows herbicides to be rotated and used much more efficiently. Build-up of couch and other perennial grasses can be prevented by routine stubble cultivations.

The rotation may also be adapted to deal with other weed situations. Wild oats die out if left undisturbed under a long ley (seven to eight years), while two- to three-year perennial ryegrass leys, if *grazed hard* and not mown, usually eradicate couch.

Cultivations

Annuals

(1) *Preparation of 'false seedbed' prior to sowing crop*. The seedbed is prepared early, but not too fine, to encourage weed seeds to germinate. They are then killed by harrowing, and another crop of weeds is allowed to grow and then killed; the process may be repeated several times. The method results in a very clean seedbed and is suitable where time allows, e.g. summer seedbeds for lucerne, direct sown leys, swedes or kale, for general annual weed control. Only suitable for the control of weeds whose seed will germinate at the time of year cultivations are carried out.

(2) Thus, the similar technique, *stubble cleaning* or early autumn cultivation of cereal stubbles for annual weed control, is highly effective against autumn-germinating weeds, e.g. blackgrass, (*Alopecurus myosuroides*) or sterile brome (*Bromus sterilis*), but is useless against weeds which germinate mainly in spring, e.g. spring wild oat (*Avena fatua*). The latter may be killed by delaying the sowing of spring crops to allow harrowing in March and April.

(3) *Delayed sowing of crop*, combined with repeated harrowing, is an effective method of controlling some weeds, e.g. delaying sowing of winter cereal until 5 November gives good control of blackgrass and seedling grasses but usually reduces yield; also loss of yield from delaying spring cereal sowings.

(4) *Inter-row cultivations for crops grown in wide rows*, e.g. sugar beet. Hand hoeing obsolete. Weeds within rows most effectively controlled by herbicides but overall application expensive. Use of band sprays (directed band of spray 99 mm each side of row), with inter-row cultivations saves two-thirds of chemical cost at each application.

(5) *Green smother crops*, e.g. cereal silage may be grown to control wild oats, which ensile satisfactorily and which prevents seeding. May be followed with a catch crop or undersown if not planted too thickly.

Perennials

Cultural methods for controlling perennial weeds are nowadays restricted to grassy species, couch (*Agropyron repens* and *Agrostis gigantea*) and others, in cereal stubbles in hot dry weather. Broadleaved perennials are most economically controlled in a run of sprayed cereal crops.

Fallows may be used but where possible stubble cleaning is cheaper. Costs are minimal where routine stubble cleaning is employed as a *preventive* measure. Regular attention should be given to the *headlands* where infestations frequently start.

There are two approaches to controlling infestations:

Desiccation

This process relies on hot dry weather, when it is cheapest and very effective. The ground is loosened by chisel plough or heavy cultivator, the couch rhizomes worked to the surface and desiccated in the sun by frequent shallow cultivations. The rhizomes should not be ploughed in during autumn and the seedbed for the following cereal should be prepared by shallow cultivations.

Exhaustion
Normal growing weather is ideal. Soil and rhizomes are broken up, usually with a rotary cultivator, to encourage fresh growth. This is then destroyed by further cultivations when shoots are 50 mm long or have an average of one and a half to two leaves each. The process is repeated several times to exhaust the rhizomes and buds. A *bare fallow*, which covers a full year, is rarely used. The *bastard fallow*, the land being broken up with heavy cultivators or ploughed in late summer and then baked in the clod, is excellent for dirty, worn-out leys or old pastures prior to a winter cereal.

General chemical weed control

Pre-sowing sward desiccation

Used prior to direct drilling, minimal cultivations or to kill grass swards before ploughing to give grass-free seedbed. Only herbicides with minimal residual effect suitable. Use (1) either glufosinate-ammonium or paraquat (foliar applied contacts) for general purpose where couch grass (*Elymus repens*) and rhizomatous grasses and perennial broadleaved weeds are absent, or (2) glyphosate (foliar applied translocated) for couch and other rhizomatous grasses and perennial broadleaved weeds.

Pre-sowing control of grassy perennials

Suitable herbicides are listed in Table 13.2. Herbicides are more reliable than cultural methods for heavy infestations in wet weather. Routine annual treatments around headlands should be considered.

Weed control in cereals

Cereals not undersown

Winter cereals

These occupy the land for some ten months of the year and when they follow another winter cereal little time is available for cleaning. Crops are now sown earlier to increase yield (September and early October) and are then more exposed to infestations of autumn germinating annual and other weed grasses, including blackgrass (*Alopecurus myosuroides*), sterile brome (*Bromus sterilis*), annual and rough stalked meadow grasses (*Poa* spp) and winter wild oat (*Avena ludoviciana*).

Cultural and rotational controls include: prevention of seeding from banks and hedgerows and control on headlands where infestations of sterile brome (*B. sterilis*) often start, stubble cultivations and delaying sowing until 5 November, which usually reduces yield somewhat, or changing to a spring sown crop (incomplete control of *A. ludoviciana*).

Herbicidal treatments
Those for autumn application are shown in Table 13.3. If needed, failure to treat at this stage results in a heavy infestation by spring. Some pre-emergence herbicides used for grass seedlings also control several species of broadleaved annual weeds, but may not control wild oats and do not persist long enough to control spring germinating broadleaves: post-emergence applications for the latter two situations are frequently necessary. A full programme for an early sown winter cereal crop may thus include as many as four applications:

(1) stubble application for perennial grasses (Table 13.2);
(2) autumn applied residual for annual grasses and broadleaved weeds (Table 13.3);
(3) spring post-emergence application for wild oats (Table 13.3);
(4) spring post-emergence application for broadleaved weeds (Table 13.4).

Table 13.2 Presowing control of perennial weeds – stubble cleaning and elsewhere

Herbicide	*Type of activity*	*Species controlled*	*Limitations of herbicide*
Amitrole	Foliar: translocated	Couch (*E. repens*) and other grassy and broadleaved weeds	Specified time interval before sowing crop, 125 mm regrowth required before spraying
Glufosinate-ammonium	Foliar: contact	Annual and stoloniferous perennial grasses. Couch only if repeated applications are given. *Not* a single application	No residual effect. Heavily grazed fields should show active regrowth
Glyphosate	Foliar: translocated	Couch, most grasses and perennial broadleaved weeds	No residual effect. 125 mm regrowth required before spraying
Paraquat	Foliar: contact	Annual and stoloniferous perennial grasses. Couch only if repeated applications are given. *Not* a single application	No residual effect except on trash; then allow 10 days for photochemical degradation. 75 mm regrowth required

Table 13.3 Selective weed control in cereals: annual/seedling grasses

Herbicide	*Weeds controlled**						*Suitable crop*[†]	*Remarks*	*Activity against herbicide-resistant clones of blackgrass*[‡]
	Type of activity	*Barren brome (B. sterilis)*	*Blackgrass (A. myosuroides)*	*Meadow grasses (Poa spp)*	*Wild oat (Avena spp)*	*Some annual broadleaved weeds*			
Chlorotoluron	CoR		Yes	Yes	Yes	Yes	Winter wheat and winter barley – named varieties only		Poor/variable
Clodinafop-propargyl	Tr		Yes	Yes	Yes		Wheat, durum wheat, rye and triticale		
Cyanazine	CoR	Yes	Yes	Yes		Yes	Winter wheat and barley		Variable
Diclofop-methyl	Tr		Yes	Yes	Yes		Barley, wheat, rye and triticale	Also controls awned canary grass, ryegrasses	Variable
Diclofop-methyl + fenoxaprop-P-ethyl	Tr		Yes	Yes	Yes		Barley		Variable
Difenzoquat	Tr				Yes		Barley, wheat, winter rye	Also crops undersown with ryegrass/clover	
Fenoxaprop-P-ethyl	Tr		Yes	Yes	Yes		Wheat	Also controls awned canary grass	Variable
Flamprop-M-isoprophyl	Tr				Yes		Barley, wheat, rye and triticale	Also crops undersown with ryegrass/clover	
Imazamenthabenz-methyl	CoR		Yes		Yes	Yes	Winter wheat and winter barley	Also onion couch, volunteer oilseed rape and charlock	Poor/variable
Isoproturon	TrR	(Yes)	Yes	Yes	Yes	Yes		Barren brome controlled only after tri-allate	Poor/variable
Methabenzthiazuron	CoR		Yes	Yes		Yes	Winter barley, winter oats, winter wheat		Poor/variable
Metoxuron	CoR	(Yes)	Yes	Yes		Yes	Winter barley, winter wheat – named varieties only	Barren brome controlled by double application or after triallate, also controls mayweed	Poor/variable

Continued

Table 13.3 *Continued*

Herbicide	*Weeds controlled**						*Suitable crop*†	*Remarks*	*Activity against herbicide-resistant clones of blackgrass*‡
	Type of activity	*Barren brome (B. sterilis)*	*Blackgrass (A. myosuroides)*	*Meadow grasses (Poa spp)*	*Wild oat (Avena spp)*	*Some annual broadleaved weeds*			
Pendimethalin	R		Yes	Yes	Yes	Yes	Barley, winter rye, triticale, winter and durum wheat		Poor/variable
Terbutryn	R		Yes	Yes		Yes	Winter wheat, winter barley, winter oats, rye and triticale	Also controls ryegrasses	Variable
Tralkoxydim	Tr		Yes		Yes		Wheat, barley, durum wheat, rye, triticale		Poor/variable
Triallate	R		Yes	Yes	Yes		Barley, wheat		Variable
Trifluralin	R		Yes			Yes	Winter barley, winter wheat	Main purpose broadleaved weed control and annual grasses/ seedlings	Good

Key
Type of application: Co = contact foliar; Fo = foliar; R = residual or soil acting; Tr = translocated foliar.
* Weeds controlled depend on mixture (where applicable).
† Suitability to crop depends on mixture (where applicable).
‡ Use of carbetamide, ethofumesate, trifluralin and propyzamide in non-cereal broadleaved crops within the rotation can give good control of herbicide-resistant blackgrass.

Table 13.4 Selective control of annual broadleaved weeds in cereals *not* undersown

Herbicide or predominant herbicide in mixtures (M = mixtures only suitable/available)	*Type of activity*	*General weed control*	*Combinations difficult to control with simple herbicide**							*Suitable crops*†	*Remarks*
			Black bindweed (P. convolvulus) Redshank (P. persicaria)	*Common chickweed (S. media)*	*Cleavers (G. aparine)*	*Corn marigold (C. segetum)*	*Knotgrass (P. aviculare)*	*Mayweed spp (Anthemis spp)*	*Speedwell spp (Veronica spp)*		
Amidosulfuron	CoR	Yes			Yes					Wheat, barley, oats, rye, triticale	Also charlock, shepherd's purse, forget-me-not
Bentazone M	Co	Yes		Yes	Yes	Yes	Yes	Yes	Yes	Barley, winter wheat	Also field pansy
Bifenox M	R	Yes			Yes				Yes		Also field pansy
Bromoxynil M	Co	Yes	Yes			Yes	Yes	Yes	Yes	All cereals	
Clopyralid M	Tr	Yes	Yes			Yes		Yes		Barley, oats, wheat	
Cyanazine M	FoR	Yes	Yes	Yes			Yes	Yes	Yes	Barley, oats, wheat	

Continued

Table 13.4 *Continued*

Herbicide or predominant herbicide in mixtures (M = mixtures only suitable/available)	*Type of activity*	*General weed control*	*Combinations difficult to control with simple herbicide**							*Suitable crops*†	*Remarks*
			Black bindweed (P. convolvulus) Redshank (P. persicaria)	*Common chickweed (S. media)*	*Cleavers (G. aparine)*	*Corn marigold (C. segetum)*	*Knotgrass (P. aviculare)*	*Mayweed spp (Anthemis spp)*	*Speedwell spp (Veronica spp)*		
2,4-D	Tr		Yes							All cereals exc. spring oats‡	
Dicamba M	Tr	Yes	Yes	Yes	Yes		Yes	Yes		All cereals	
Dichlorprop and mixtures	Tr	Yes	Yes	Yes						Barley, oats, wheat	Not rye
Diflufenican M	R	Yes	Yes	Yes	Yes	Yes	Yes	Yes	Yes	Winter wheat and winter barley	
Fluoroglycofen-ethyl	Tr				Yes				Yes	Winter wheat and winter barley	Also field pansy
Fluoroxypyr and mixtures	Tr	Yes	Yes	Yes	Yes		Yes			Wheat, barley, durum wheat, rye, triticale	Also hemp nettle, docks and forget-me-not
Ioxynil M	Co	Yes	Yes	Yes	Yes	Yes		Yes	Yes	All cereals	
Isoproturon M	TrR	Yes		Yes		Yes		Yes		Winter barley, winter wheat, rye	Also most grasses
Linuron	CoR	Yes	Yes	Yes		Yes		Yes	Yes	Spring cereals	
Linuron + trifluralin	CoR	Yes	Yes	Yes			Yes		Yes	Winter barley, winter wheat	May require spring treatment for perrenials
MCPA	Tr	Yes								All cereals†	
Mecoprop and mixtures	Tr	Yes	Yes	Yes	Yes					All cereals exc. rye‡	
Metsulfuron-methyl	CoR	Yes		Yes	Yes			Yes	Yes	Wheat, barley, oats, durum wheat, triticale	Also field pansy
Pendimethalin	R	Yes			Yes				Yes	Barley, winter wheat, triticale	Also most grasses
Pyridate	Co	Yes			Yes	Yes			Yes	Wheat, barley, oats, rye, triticale, durum wheat	Also dead nettle
Thifensulfuron-methyl M	Tr		Yes		Yes		Yes			Wheat and barley	
Triasulfuron	CoR	Yes	Yes	Yes	Yes		Yes	Yes	Yes	Wheat, barley, oats, durum wheat, rye, triticale	Also charlock
Tribenuron-methyl	CoR	Yes		Yes				Yes		Winter wheat, winter and spring barley, rye, durum wheat, triticale	Also charlode

Key to Table 13.4
Type of activity: Co = contact foliar; Tr = translocated foliar; R = residual or soil acting; Fo = foliar acting.
* Weeds controlled depends on mixture (where applicable); † Suitability to crop depends on mixture (where applicable).
‡ Cheap general-purpose translocated herbicide for controlling broadleaved perennial weeds and alternating with broad-spectrum herbicides (as required).

One of the problems of spring applications of broadleaved herbicides to winter cereals is to complete the job before the cereals have passed out of a safe growth stage. Select a herbicide with a *wide application tolerance* – several broad spectrum herbicides are now available for winter cereals which can be applied up to the two-node stage (Zadok scale 32, see Table 10.3). Autumn-germinating grassy weeds are unlikely in late sown winter cereals, but these crops are subject to spring germinating wild oats and broadleaved weeds.

Spring cereals

Spring cereals (sown some six months after winter crops) allow ample time for stubble cleaning; residual effects from herbicides applied in the previous season are much less likely. The weed situation is much simpler and, in the absence of wild oats, only a single application for broadleaves is normally necessary.

Herbicides for broadleaved weeds

Perennials require treatment with translocated herbicides: contact types are useless. The 'simple' herbicides, e.g. mecoprop, MCPA and 24-D, give satisfactory control in cereals provided adequate leaf area is allowed to develop before spraying. Do not spray too early: wait as long as possible. Control is enhanced by a series of sprayed cereal crops. More complex mixtures containing translocated herbicides are only necessary where the presence of other difficult annual weeds requires them.

A wide range of annuals is controlled by simple herbicides but some species are resistant, and more complex herbicides are necessary (Table 13.4). Choose according to crop and weeds present. Check stages of crop and weed growth, etc. suitability from manufacturers' literature.

The more expensive 'broad spectrum' herbicides are not usually necessary every year: cheaper simple herbicides or those with a more limited weed spectrum may be considered for intervening years, but in no circumstances is it advisable to omit the annual cereal spray for *broadleaved* weeds – even if only MCPA is used – otherwise, docks and thistles soon develop.

Undersown cereals and seedling reseeds containing clover

Use only herbicides to which clover is tolerant. General purpose herbicides such as MCPB or MCPB + MCPA control many weeds but a number are resistant (see Table 13.5). Selection should be made on this basis. Sprays which control chickweed should always be chosen where the weed is known to be a problem, as it can rapidly suffocate young leys. *All* undersown and seedling leys should be sprayed, as dock seedlings, which are usually present and inconspicuous, are highly susceptible in the seedling stage to all translocated herbicides – adult docks are much more difficult to control.

Root crops, row crops and other broadleaved crops

The principles of weed control are similar in all these crops – the main differences lie in the herbicides used.

Weed problems

Unlike the situation in cereals, perennial weeds are generally difficult or impossible to control selectively during the growing period of these crops. Grassy perennials (couch *Elymus repens* and others) are best controlled in cereal stubbles or pre-sowing by the herbicides listed in Table 13.2. Exceptions include couch in maize – use atrazine (high rate) but crop must be grown for two successive years; and cycloxydim, fluazifop-P-butyl, propaquizafop, quizalofop-ethyl and sethoxydim in many broadleaved crops.

Broadleaved perennials

These should be controlled by translocated herbicides applied to preceding cereal crops wherever possible or, if feasible, by pre-sowing application of glyphosate. Exceptions include creeping thistles (*Cirsium arvense*) in peas – apply MCPB post-emergence (*not* beans or MCPB + MCPA) and sugar beet – clopyralid.

Grassy annuals

Grassy annuals (e.g. blackgrass *Alopecurus myosuroides* and wild oats *Avena* spp) may be controlled by herbicides listed in Table 13.6.

Broadleaved annuals

For broadleaved annuals consult Table 13.6. Selection should be made according to the weeds present or expected to be present and with residual herbicides on soil type; on peaty soils the range is very limited.

Herbicidal programmes

These are necessary in broadleaved crops which germinate slowly and take a long time to form a complete leaf canopy. Such crops are extremely vulnerable to weed competition when there are successive germinations of different weed species. Sugar beet is an excellent example, where programmes, which are constantly changing, have become sophisticated and costly.

Weeds from shed crop seed or groundkeepers

Problems arise from weed beet, shed oilseed rape, potato and beet groundkeepers and, when oilseed rape or stubble brassicas are to follow, shed cereal grains. Wherever possible prevention is better than cure; methods include preventive sanitation, adequate length of rotation, cultural controls as for annual weeds and herbicides (see Table 13.7)

Weed beet

Weed beet is an annual form of beet which produces no harvestable root and sheds large quantities of seed in the year of germination. Although indistinguishable from sown beet in the seedling stage, weed beet may be detected as they occur in patches. Any beet out of place between or within sown rows must be treated as weed beet; they grow from seed shed by bolters in beet crops, which may come from varieties with low bolting resistance and weed beet or groundkeepers elsewhere. Bolting must be prevented as sugar beet varieties are genetically

Table 13.5 Selective control of annual broadleaved weeds in undersown cereals; seedling leys (grass and clover, lucerne and sainfoin) (a) Established legumes and seed crops

Herbicide or predominant herbicide in mixtures and type of action. (M = mixtures only suitable or available)	*General weed control*	*Combinations difficult to control with simple herbicides*							*Suitable crops*	*Limitations of herbicide*
		Black bindweed (P. convolvulus) and redshank (P. persicaria)	*Common chickweed (S. media)*	*Cleavers (G. aparine)*	*Corn marigold (C. segetum)*	*Knotgrass (P. aviculare)*	*Mayweed spp (Anthemis, etc.)*	*Speedwell spp (Veronica spp)*		
Benazolin M Tr	Yes	Yes	Yes	Yes		Yes			Undersown cereals, direct sown leys; *not* lucerne or sainfoin	
Bentazone M Co	Yes		Yes	Yes	Yes		Yes		Undersown cereals; *not* undersown lucerne	
Bromoxynil M Co	Yes	Yes			Yes	Yes	Yes		Undersown cereals	Annual weeds only
2,4-DB Tr	Yes	Yes				Yes			Lucerne – undersown cereals, direct sown leys	
2,4-DB + MCPA Tr	Yes	Yes				Yes			Seedling red and white clover, undersown cereals, direct sown leys not lucerne	Not established clovers
2,4-DB + Bromoxynil + Ioxynil Co Tr	Yes	Yes				Yes			Undersown cereals, seedling leys not lucerne	
MCPB Tr Herbicides used only in MCPB + MCPA above mixtures:	Yes								Sainfoin, seed clovers undersown leys, *not* lucerne	Not alsike for seed or yellow trefoil
2,4-D Ioxynil MCPA	Yes								Sainfoin, leys and undersown cereals, *not* lucerne	

For control of wild oats (*Avena* spp) in some cereals undersown with grass and clover *see* difenzoquat, flamprop-M-isopropyl in Table 13.3.
Key: Type A activity; Co = contact foliar; M = mixtures only; R = residual or soil acting; Tr = translocated foliar.
Italics denote herbicide also or solely used by itself.

Table 13.5 (b) Selective weed control in established legumes and herbage seed crops

Herbicide and type of action	Purpose	Suitable crops	Limitations of herbicide
Legumes			
Carbetamide Fo R	General broadleaved weeds, grasses and volunteer cereals	Lucerne, sainfoin, clovers	Winter period control mainly
Propyzamide Fo R	General broadleaved weeds, grasses and volunteer cereals	Lucerne, clover, seed crops	Winter period control mainly
Diclofop-methyl Tr	Grasses and volunteer cereals	Lucerne	
Fluazifop-P-butyl Tr	Grasses and volunteer cereals	Lucerne, sainfoin, clovers, trefoil, red fescue seed crops	
Sethoxydim Tr	Grasses and volunteer cereals	Lucerne, sainfoin, clovers, trefoil, red fescue seed crops	
Grass for seed/pure grass stands			
Ethofumesate Co R with or without loxynil and bromoxynil	General weed control	Most grasses	Winter period control mainly
Methabenzthiazuron Co R	General weed control	Ryegrasses	Winter period control mainly
2,4-D, dicamba MCPA and mecoprop mixtures Tr	General broadleaved control – not grasses	Pure grass swards	
Isoxaben R	General broadleaved control – not grasses	Pure grasses and grass seed crops	
Difenzoquat Tr	Winter oats	Ryegrasses	Very limited weed spectrum

Key:
Type of activity; Co = contact foliar; R = residual or soil acting; Tr = translocated foliar.
Desiccation of clover seed crops: diquat.

Table 13.6 General selective herbicidal control in the major broadleaved field crops

Field crop	*Grass weed and volunteer cereal control*	*General broadleaved and grass weed control*
Field beans	Alloxydim-sodium	Carbetamide
Broad beans	Cycloxydim Diclofop-methyl (not volunteer wheat/barley) Fluazifop-P-butyl Sethoxydim Propaquizafop Triallate (not cereals)	Cyanazine Chlorpropham and fenuron Diquat (pre-harvest desiccation) Glufosinate-ammonium (pre-harvest dessication) Prometryn ± pendimethalin or terbutryn Propyzamide Simazine ± trietazine Terbutryn ± trietazine or terbuthylazine Trifluralin
Brassicas, OSR	Cycloxydim Diclofop-methyl (not volunteer wheat/barley) Fluazifop-P-butyl Propaquizafop Quizalofop-ethyl Sethoxydim	Benazolin + clopyralid (no grass control) Carbetamide Chlorthal-dimethyl Clopyralid (no grass control) Cyanazine Glufosinate-ammonium (pre-harvest dessication) Metazochlor Napropamide Propachlor Propyzamide Pyridate (no grass control) Tebutam Trifluralin
Fodder brassicas (radish, rape, swedes, turnips, kale)	Fluazifop-P-butyl	Clopyralid (no grass control) Desmetryn (not swedes, turnips, radishes) Propachlor Sodium monochloroacetate Tebutam Trifluralin
Vegetable brassicas (broccoli, Brussels sprouts, cabbage, calabrese, cauliflower)	Cycloxydim Diclofop-methyl (not volunteer wheat/barley) Fluazifop-P-butyl	Aziprotryne (Brussels, sprouts and cabbages) Carbetamide (cabbages only) Chlorthal-dimethyl Clopyralid (no grass control) Desmetryn (not broccoli/ cauliflowers) Metazachlor Pendimethalin (transplanted brassicas) Propachlor Tebutam Trifluralin
Carrot	Alloxydim-sodium Diclofop-methyl (not volunteer wheat/barley) Fluazifop-P-butyl Propaquizafop Sethoxydim	Chlopropham Isoxaben Linuron Metoxuron Pendimethalin Pentanochlor Prometryn Trifluralin

Continued

Table 13.6 *Continued*

Field crop	*Grass weed and volunteer cereal control*	*General broadleaved and grass weed control*
Leeks/onions	Alloxydim-sodium Diclofop-methyl (not leeks) Fluazifop-P-butyl (not leeks) Propaquizafop (not leeks)	Aziprotryne Bentazone (not leeks) Chlorbutfam Chloridazon Chlorpropham Chlorthal-dimethyl Clopyralid (not leeks) Cyanazine (not leeks) Fluoroxypyr (not leeks) Ioxynil Monolinuron (not onions) Pendimethalin Prometryn (not onions) Propachlor Sodium monochloroacetate
Linseed	Diclofop-methyl (not volunteer wheat/barley) Propaquizafop Sethoxydim	Bentazone Clopyralid ± bromoxynil or cyanazine
Maize and sweet corn	Triallate	Atrazine Clopyralid (not grasses) Cyanazine Pendimethalin Pyridate (not grasses) Simazine
Peas	Alloxydim-sodium Cycloxydim Diclofop-methyl (not volunteer cereals) Fluazifop-P-butyl Propaquizafop Sethoxydim Triallate	Aziprotryne Bentazone ± MCPB Chlorpropham Cyanazine Glufosinate-ammonium (pre-harvest desiccation) MCPB ± MCPA Pendimethalin Prometryn Simazine Terbutryn Trietazine Trifluralin
Potatoes	Alloxydim-sodium Cycloxydim Diclofop-methyl (not volunteer cereals)	Bentazone Cyanazine Diquat ± paraquat Glufosinate-ammonium (pre-emergence and pre-harvest) Linuron ± terbutryn Metribuzin Monolinuron ± paraquat or glufosinate-ammonium Pendimethalin ± prometryn Prometryn ± terbutryn Terbutryn ± terbuthylazine or trietazine
Sugar beet, fodder beet, mangels	Alloxydim-sodium Cycloxydim Diclofop-methyl (not volunteer wheat/barley) Fluazifop-P-butyl Propaquizafop Quizalofop-ethyl	Carbetamide Chloridazon Chlorpropham Clopyralid (not grasses) Ethofumesate Glufosinate-ammonium (pre-emergence)

Continued

Table 13.6 *Continued*

Field crop	*Grass weed and volunteer cereal control*	*General broadleaved and grass weed control*
	Triallate	Glyphosate (post-drilling) Lenacil Metamitron Paraquat (pre-emergence and inter-row directed) Phenmedipham Propyzamide Trifluralin

Table 13.7 Control of volunteer crops as weeds in the following crops

Volunteer crop	*Affected crop*	*Herbicide*	*Limitations*
Selective controls			
Field beans	Cereals	Clopyralid Ioxynil Mecoprop	Post-emergence treatments
Weed, sugar beet	Cereals	Bromoxynil Isoxaben Metsulfuron-methyl	
	Sugar beet, fodder beet, mangels	Glyphosate	Rope wick application only
Cereals (also oats and most grasses)	Most broadleaved field and vegetable crops	Alloxydim-sodium	
		Cycloxydim	Not carrots
		Diclofop-methyl	Does not control volunteer wheat or barley
		Fluazifop-P-butyl	Not beans or potatoes
		Propaquizafop	
		Quizalofop-ethyl	Not peas, beans, potatoes
		Sethoxydim	Not potatoes
		Carbetamide (also general broadleaved weed control)	Not beets, peas, potatoes
		Propyzamide (also general broadleaved weed control)	Not beets, peas, potatoes
	Grass crops	Ethofumesate (also general broadleaved weed control)	Winter period control mainly
Oilseed rape	Cereals	Bromoxynil Imazamethabenz-methyl Isoproturon Isoxaben Mecoprop Metsulfuron-methyl Terbutryn	
Potatoes	Cereals and grasses Linseed	Fluoroxypyr Metsulfuron-methyl	
Non-selective controls			
Beans, beets, cereals and oilseed rape	Pre-sowing and pre-harvest desiccation in many crops	Glufosinate-ammonium Glyphosate Paraquat ± diquat	
Potatoes	Pre-sowing	Glyphosate	Autumn stubble treatment
	Pre-sowing of barley	Amitrole	Autumn stubble treatment with specified time interval before sowing crop

unstable when allowed to cross-pollinate freely in an uncontrolled environment and soon degenerate into the annual weed beet form.

A single bolter is capable to producing up to 2000 viable seeds in a single season; if not removed it can soon build up an infestation.

Control

Good isolation and tight control of cross-pollination by seed growers is essential. Root growers should walk their fields before row cropping, looking for patches of seedlings outside rows or in cereal crops and identify the problem. *Prevention of seeding is better than cure.*

Measures

Pull *all* bolters before flowering (i.e. yellow pollen visible); if pulled after flowering remove from field and destroy to prevent shedding of viable seed. Destroy all weed beet at cleaner loader sites or on wasteland to prevent cross-pollination with bolters in root crops. If hand labour is not available control bolters by mechanical cutting at least twice in July. Obtain advice from the British Sugar field staff as to correct stage for cutting. Drill only after 20 March to avoid bolting; use only *non-bolting varieties* (NIAB Farmers' Leaflet No. 5).

Additional measures (*once weed beet are identified*): lengthen rotation with non-beet crops and use effective herbicides in other crops (e.g. bromoxynil mixtures in cereals). Spread beet over whole farm. Tractor hoe frequently, close to rows to remove about 90% of weed beet seedings, and use trifluralin for late inter-row weed control. Alternatively use rope wick with glyphosate when beet is above crop. If seed of weed beet has been shed, leave it on the surface; avoid ploughing in, which preserves it until brought to the surface by later ploughings. Use direct drilling or minimal cultivations for cereals on suitable soils. Only relatively few fields are as yet seriously affected but weed beet poses a serious menace to the whole British home-grown sugar industry; it could equally become a major problem with fodder beet.

Weed control in established grassland

The presence of infestations of weeds or weed grasses is generally an indication that all is not well with management and growing conditions. Weed infestations can also arise in leys, where the species originally sown lacked persistency and faded out, leaving the field to be colonized by weeds.

The first step is to decide whether the sward justifies retention and subsequent improvement or whether to kill all vegetation and reseed or grow a run of one or more arable crops before reseeding. Retention is often best on moist fertile soils or in wetter situations as a more poaching-resistant sward is maintained; costs are lower and no grass production is lost. Conversely, where a temporary sward is worn out, in dry situations or where a poor sward is infested with perennial weeds which are difficult to control by herbicides while the field remains down to grass, destruction of the sward has much to commend it. A herbicide which gives a total kill if properly used (e.g. glyphosate) is a valuable first step. The majority of grassland weeds will not withstand well managed cultivations and arable crops for long, especially where the arable break starts with cereal crops sprayed with translocated herbicides.

The main weeds of grassland and their appropriate controls are listed and summarized in Table 13.8. Where weedy grassland is to be retained, improvements in management, growing conditions and fertility levels are essential. Herbicides are often an essential part of the programme but cannot maintain any improvement by themselves.

Table 13.8 Control of weeds in established grassland

Weed	*Situations where epsecially troublesome*	*Preventive cultural and management controls*	*Most effective herbicidal treatments and herbicides*
Bent, creeping (*Agrostis stolonifera*)	Poorly managed lowland pastures, overgrazed in winter and spring, undergrazed in summer	G, E, F, G	Asulam used for docks gives some useful effect
Bracken (*Pteridium aquilinum*)	Heavily understocked sites, upland, moor and hill lands, poisonous	B, C, I, K, L	Asulam
Buttercup, bulbous (*Ranunculus bulbosus*)	Common in most permanent grassland. Poisonous except in hay. *R. bulbosus* occurs in drier conditions	If very serious N otherwise CE	QR (2,4-D*) (MCPA*)
Buttercup, creeping (*R. repens*)		A, C, E	P2,4-D* MCPA* MCPB and mixtures P MCPA*
Buttercup, tall (*R. acris*)		A, C, E	MCPB and mixtures
Chickweed (*Stellaria media*)	Poached grassland along with other annual weeds	D	*Benazolin mixtures* mecoprop[†]
Daisy (*Bellis perennis*)	Overgrazed poor permanent grass and lawns	C, E R if necessary	2, 4-D* MCPA*
Dandelion (*Taraxacum officinale*)	Common in poorly managed grassland	C, E, F	2,4-D* MCPA*

Continued

Table 13.8 *Continued*

Docks, broad and curly leaved (*Rumex obtusifolius* and *R. crispus*)	Universal on land stocked only with cattle; also treated with slurry and mown	J, M, N	O *Asulam Dicamba mixtures** or *Triclopyr** and mixtures
Gorse (*Ulex* spp)	Grossly undergrazed acid permanent grass and waste places	B, C, D, E, G, H, J, M	*Triclopyr* and mixtures*
Horsetails (*Equisetum* spp)	Soils with wet subsoil; remains poisonous in hay and silage	A, C, E, F, G	RT(2,4-D)* (MCPA*) (MCPB mixtures)
Meadow grass, annual (*Poa annua*)	Very common in reseeds and poor, open or poached grassland. May suddenly appear/disappear	D, E	Ethofumesate* *if* cost is justified
Mouse-eared chickweed (*Cerastium holosteoides*)	Universal in grassland	C, D	(2,4-D*) (MCPA*) (MCPB mixtures)
Ragwort (*Senecio jacobea*)	Neglected and overgrazed old pastures and wasteland, hedgerows. Remains poisonous in hay and silage	C, D, E, J, M Undersow new leys	2,4-D* MCPA* Once sprayed, keep stock out until ragwort has disappeared
Rush, hard (*Juncus inflexus*) Rush, soft (*Juncus effusus*)	Badly drained wet land of low fertility; establish from seed in poached or open sward	A, C, D, E, G, I Reseed new leys *without* cover crop A, B, C, D, G, E, I Reseed new leys *without* cover crop	Resistant to selective herbicides P, 2,4-D* MCPA*. Cut 4 weeks after spraying
Sorrel, common (*Rumex acetosa*)	Poor permanent grass, usually damp	B, C, G	(2,4-D*) (MCPA*)
Sorrel, sheep's (*Rumex acetosella*)	Indicates acidity on poor permanent grass	B, C, G	(2,4-D*) (MCPA*)
Stinging nettle, great (*Urtica dioica*)	Mainly on grazed grassland, often with loose structure; encouraged by surface litter	E, H, J, K N if serious	PQRS, *Triclopyr* and mixtures*
Thistle, creeping (*Cirsium arvense*)	Universal in undergrazed grassland – rarely seeds	C, E, F, G N if serious	PQR 2,4-D* MCPA* *MCPB mixtures*, Clopyralid*
Thistle, spear (*C. vulgars*)	Universal. Establishes readily from blown seed	M	O 2,4-D* MCPA* MCPB mixtures
Tussock grass (*Deschampsia caespitosa*)	Badly managed wet grazings, establishes from seed	A, C, D, E, H, M	No selective chemical control
Yorkshire fog (*Holcus lanatus*)	Wet, acid or badly managed lowland grassland	A, B, C, E, F, G	Asulam used for docks gives useful effect

* Clovers killed or severely checked. Prefer 2,4-D to MCPA if suitable.
† Mecoprop is effective against chickweed but kills clover. Use only on grass seed crops, pure grass swards and in emergency, when chickweed (*S. media*) is likely to smother sward or oversowing.
( ) denote limited effect.
Italics denote first choice herbicide.

Key to Table 13.8

A Improve drainage.
B Apply lime.
C Increase fertility. Apply N, P and K as required; a base dressing of 190 kg/ha insoluble P_2O_5 given as basic slag or ground mineral phosphate on phosphate-deficient pastures in high rainfall areas or wet situations.
D Avoid poaching; keep cattle off grassland in winter.
E Avoid overgrazing in winter, spring or early summer.
F Avoid undergrazing in summer.
G Increase stocking rate, mow or top over after grazing.
H Change management to include close cutting, using forage harvester to cut silage.
I Cut twice annually for two or three years.
J Heavy grazing with sheep reduces infestation.
K Heavy trampling with cattle but unstock cattle on bracken to avoid bracken poisoning.
L On ploughable land plough bracken deeply with digger plough when fronds are three-quarters open in spring or early summer. Cut up rhizomes with several passes of disc harrows and consolidate. Grow pioneer crop, e.g. rape or turnips, graze in autumn, disc harrow in December, do not plough. Follow with re-seeded ley or potatoes; re-seed must be intensively stocked and managed or bracken returns.
M Prevent flowering and seed production by spudding, cutting, flail or swipe mower.
N Several perennial weeds are difficult to control with inexpensive herbicides in established grassland but are easily controlled by such herbicides in the presence of cultivations. Where practicable, ploughing and cultivations, followed by either (a) re-seeding with Italian ryegrass and spraying with MCPA, 2,4-D or mecoprop as soon as possible, followed by two further sprayings during the summer, or (b) growing a run of two or more sprayed cereal crops.
O Undersown leys and reseeds must be sprayed with an MCPB or 2,4-DB mixture (see Table 13.5) to kill perennial broadleaved weeds in the susceptible seedling stage. Once established these weeds become much more resistant to herbicidal treatment.
P Spray in spring or early summer before flowering.
Q Spray in autumn.
R Spray in at least two successive years.
S Spot treatment – wet clumps to run-off point.
T Spray when plants have made maximum growth.
Note. When pasture contains a very large amount of weed and little or no valuable grass or clover, reseeding should be considered, especially if intensive stocking is intended.

Weed control in forest and woodlands

Adequate weed control is essential for the satisfactory establishment of young trees and their subsequent growth. The nature of the problem and weed spectrum is much more variable than with normal agricultural crops where much of the land is reasonably level and the weed spectrum is confined largely to weed grasses and broad-leaved herbaceous species. In woodland a much wider weed spectrum occurs and terrain is often difficult. The situation varies from sites where little or no weeding is necessary to those where weed growth is vigorous and rapid and may cause total failure of the young trees; the commonest cause of smothering is the collapse of tall vegetation on top of them in the winter. Frequently the control of one type of weed allows another to develop, e.g. grasses often follow cutting down of brambles (*Rubus* spp) while ploughing temporarily controls some grass species but may result in heavy growth of thistles (*Cirsium* spp) or gorse (*Ulex* spp). It is necessary to be prepared for these changes. Total eradication of weeds over the whole site is unlikely to repay the cost; generally the suppression of weed growth in the immediate vicinity of the young tree is adequate, provided that tall woody broadleaves are not allowed to develop. Considerable time and money can be saved by knowing when to intervene and when intervention is unnecessary.

Weed types

For practical purposes weeds may be grouped as follows:

(1) 'soft' or 'fine' grasses: bents (*Agrostis* spp), fine leaved fescues (*Festuca* spp), Yorkshire fog and creeping soft grass (*Holcus lanatus* and *H. mollis*) and meadow grasses (*Poa* spp);
(2) coarse grasses and rushes: oat grasses (*Arrhenatherum* spp), cocksfoot (*Dactylis glomerata*), purple moor grass (*Molinia caerulea*) and rushes (*Juncus* spp);
(3) mixed herbaceous broadleaved weeds and grasses;
(4) bracken (*Pteridium aquilinum*);
(5) heaths (*Erica* spp) and heather or ling (*Calluna vulgaris*);
(6) low growing woody weeds, e.g. gorse, furze or whin (*Ulex* spp);
(7) taller woody broadleaved weeds. These include:
    (i) trees and large shrubs or bushes: birches (*Betula* spp), sallow and goat willows (*Salix* spp), rhododendron (*Rhododendron ponticum*), laurel (*Prunus* spp), blackthorn (*Prunus spinosa*) *et al.*;
    (ii) coppice regrowth – most broadleaves; sweet chestnut (*Castanea sativa*) is especially vigorous;
    (iii) trailers: blackberry (*Rubus* spp), dog rose (*Rosa* spp);
    (iv) climbers: honeysuckle (*Lonicera periclymenum*) and ivy (*Hedera helix*).

Weeding and cleaning

Low growing weeds, i.e. groups 1–6, suppress the growth of young trees and may even kill them unless controlled by weeding. This is usually carried out during the first three to four years after planting, when the young trees should have made enough growth (about 2 m high) to suppress any further low weed growth. Grass, even if kept tightly cut, suppresses growth badly, as its all-pervading root system robs the young trees of water and nitrogen (N). If not controlled, grass and bracken can also collapse on top of the young trees in winter and smother them. The best methods of controlling grass in the immediate vicinity of young trees are (i) ploughing before planting, or (ii) the application of a grass killing herbicide (glyphosate or paraquat) pre-planting in strips 1 m wide along the proposed planting lines, or (iii) spraying around the transplants with a knapsack sprayer with its lance fitted with a tree guard.

Cleaning, i.e. the removal of unwanted growth, is carried out in the pre-thicket stage, usually 6–12 years after planting according to the state and type of the growth present; if growth is substantial, give an early 'low cleaning' with a second 'high cleaning' later.

Unwanted types include: fast growing broadleaved 'weed' trees, especially birches (*Betula* spp), sallow or goat willow (*Salix caprea*), which can seriously damage or even suppress a young plantation of conifers as well as broadleaved crops. These can be removed by companion saws or application of herbicide.

Harmful climbers, e.g. honeysuckle (*L. periclymenum*) must be removed from broadleaves as they wind round the trunk and produce serious distortion. Unwanted coppice, laurel or rhododendron regrowth is best prevented by treating the old stumps with herbicide *immediately* after cutting (see Table 13.9). Where retained to draw up broadleaved crop stems, e.g. between tree shelters, regrowth is cut out as and when it starts to compete with the crop.

Methods of control

Control measures must be carefully integrated to suit the weed flora, topography and general conditions of the site. Each operation should be carried out as part of the overall establishment plan and not in isolation. Methods can be grouped into the following categories: (a) hand, (b) mechanical, and (c) herbicidal.

Handwork

Weeding

Weeding with hand tools, e.g. grass hooks, is slow, laborious and expensive, especially if two cuts are wanted in a single season; it has largely been superseded by ploughing, mechanical aids and herbicides. On some ploughed upland sites no weeding may be required for two to three years after planting, sometimes not at all. The effect of handwork is only temporary and may not last a whole season; it does not prevent grass from reducing the growth rate of young trees. Handwork also creates a summer labour peak when on the farm there are heavy labour demands.

The traditional method is to walk down a line of plants, trimming away vegetation around each plant for a sufficient distance to prevent smothering, using a curved grass or reaping hook. *Do not make any cuts until the plant is located.* A light stick is carried to push weed growth away and if of the same length as the planting distance within the row, location of small trees is much quicker.

Table 13.9 Selective weed control in forests, woodland, shelterbelts and other tree situations (except fruit and ornamentals)

Weed situation	*Herbicide and type of activity*	*Method of application*	*Apply pre- and/or post-planting*	*Months of application*	*Remarks*
Predominantly grasses	Atrazine RTr	FS	2	*Feb–Apr* (May). Conifers only	Not organic peaty soils. *Molinia*, *Calamagrostis* and *Deschampsia* spp resistant
	Glufosinate-ammonium	FS	1 or 2	Feb–May and May–Oct but any other time to green foliage	All spp susceptible and must be protected. Not rhizomatous grasses
	Glyphosate Tr	FS	1 or 2	*July–Sept* but any period of active foliage growth	Post-planting; protect actively growing trees; some conifers can be sprayed overall when dormant
	Hexazinone RTr	FS	1 or 2	*Mar–May* (June–Aug)	Limited range of spp. Spruces and pines only
	Imazapyr Tr R	FS	1 or 2	July–Oct	Conifers (Sitka spruce, lodgepole pine, Corsican pine) only. Site preparation 5 months before planting
	Paraquat ± diquat C	FS	2	*Feb–May* and *Aug–Oct* but any other time to green foliage	All spp susceptible and must be protected. Not rhizomatous grasses
	Propyzamide R	Granules or FS	1 or 2	*Oct–Dec* only exc. Jan in upland Britain	Reduced activity on organic soils; max. peat depth 10 cm *Dactylis*, *H. mollis* and *Calamagrostis* show some resistance
	Dalapon + dichlobenil Tr R	Granules	1 or 2	Dec–March	Apply 1-m-wide band 2 months before planting. Conifers and broadleaved trees pre- and post-planting
Grasses with substantial proportion of broadleaved herbs	Glyphosate Tr R (± simazine)	FS	1 or 2	*July–Sept* but any period of active foliage growth	Post-planting: protect actively growing trees: some conifers can be sprayed overall when dormant
Bracken (*Pteridium aquilinum*)	Asulam Tr	FS	1 or 2	*Late June–Aug* Best effects finds unfurled and canopy almost complete	Best control but only controls bracken. Conifer tolerance is high, except western hemlock (*T. heterophylla*)

Table 13.9 Continued

Weed situation	*Herbicide and type of activity*	*Method of application*	*Apply pre- and/or post-planting*	*Months of application*	*Remarks*
	Dicamba Tr	FS	1 or 2	March–May	Concentrated solution applied as narrow band midway betwen tree rows
	Glyphosate Tr	FS	1 or 2	*July–Aug* but before senescence starts	Post-planting: see above. Wider weed spectrum than asulam
Heather (predominantly ling, *Calluna vulgaris*)	2,4-D Ester Tr	FS	2	*June–Aug* before mid July only as directed spray with knapsack sprayer	Never on broadleaves; larches and lodgepole pine sensitive. Some crop damage usually occurs. Toxicity to crop increases in hot weather. Water catchment area limitations
	Glyphosate Tr	FS	1 or 2	*Late Aug–end Sept* after conifer growth has hardened; timing varies with season	Spruces and pines tolerant but avoid leaders
Gorse (*Ulex* spp) and broom (*Sarothamnus scoparius*)	Triclopyr Tr	FS	1 or 2	*July–Sept*	Post-planting: larch and broadleaves very susceptible to damage. Spruces will tolerate overall sprays if leader growth has hardened. Pines more sensitive. Do *not* apply VLV or ULV. Wear face shield
Rhododendron (*Rhododendron ponticum*)	Ammonium sulfamate Tr R	FS	1 only	Any time of year dry	Allow 3 months between application; all species can be heavily damaged
	Glyphosate Tr	FS	1–2 only if trees are protected	*June–Sept*	All species heavily damaged if spray comes in contact
	Triclopyr Tr	FS	1–2 only if trees are protected	*June–Sept*	Not VLV or UVL. Wear face shield. All spp can be heavily damaged

NB All Rhododendron treatments are best given pre-planting

Woody weeds including brambles (*Rubus* spp)	Fosamine Ammonium Tr	FS	1 or 2	*July–Sept* prior to leaf senescence	Only deciduous broadleaved woody spp controlled. Rhododendron, gorse and broom resistant. Protect wanted broadleaved species. Conifer sensitivity varies
2,4-D + dicamba + triclopyr Tr		FS	1 or 2	*July–Nov*	Conifers only. 1–3 months before planting and as a directed spray between rows and plants
Foliar applications	Glyphosate Tr	FS	1 or 2	*June–Aug* when weeds are growing actively	Post-planting: protect all wanted broadleaves. Only spray suitable conifers overall after leader growth has hardened. Not larches
Tree shelters (Protected broadleaf crop) Control of all unwanted spp.	Glyphosate Tr	FS	2	*June–Aug* when weeds are growing actively	Use only glyphosate. Direct spray at vegetation in ring round each shelter. Minimum diameter 1.2 m for 2 years after planting
Woody weeds Stem treatments	Ammonium sulfamate Tr R (Water soluble crystals)	Frill girdle or notch	1 preferable	Any time of year in dry weather	Corrosive: only apply by recommended plastic applicator. Allow minimum 3 months before planting. Especially suitable for rhododendron
	2,4-D Amine Tr	Tree injection only	1	Any time of year	Allow at least 1 month before planting
	Glyphosate Tr	Notch	1 or 2	Any time of year except sap rise period Mar-late April	Unwanted individual stems can be safely treated in any spp. Cut 2 notches for larger trees
Woody weeds Cut stump treatment	Ammonium sulfamate Tr R (Water soluble crystals)	Dry crystals or spray	1 only	Any time of year on a dry day	All spp can be heavily damaged by direct contact or root uptake. Allow *at least* 3 months before planting
	Glyphosate Tr	Knapsack sprayer, paint brush or clearing saw with spray attachment	1 or 2	*Oct–Apr* before spring sap flow	Use 20% solution in water with red dye added. Planting can occur immediately after treatment. Post-planting; safe in any spp but avoid foliage contact

Key for Table 13.9
Type of activity; C = contact (foliar); R = residual (soil); Tr = translocated (foliar). *Pre- or post-planting*: 1 pre-planting; 2 post-planting.
Method of application; FS = foliar spray.
Note: Detailed recommendations for all suitable herbicides and treatments are given in Forestry Commission (1986) and subsequent revisions; consult this before application.

Work is much easier where rows are quite straight and trees are evenly spaced.

Cleaning
If no mechanical facilities are available, handwork is suitable for dealing with light infestations or small patches, e.g. laurel or rhododendron. It is essential for dealing with injurious climbers on broadleaves unless a hand-held rope wick herbicide applicator is available.

Mechanical weeding and cleaning

Machines can be tractor mounted, pedestrian controlled or hand held. Tractor-mounted machines with a suitable attachment, e.g. rotating flails or chains, can give a high output at low cost on soft weed or deal with considerable woody growth, but cannot operate on very rough or steep terrain. A good scarifier can deal with lop and top if not too large and produce a surface tilth for planting. The output of hand machines is low, but they are very useful on steep land, in confined spaces, for small patches and light infestations. String trimmers ('strimmers') are suitable for soft growth but it is important to avoid damaging the bark of young trees with the plastic 'string'. Companion saws are invaluable for most hand cleaning work, except injurious climbers. Machines are useful on very mixed weed floras, where the vegetation is out of hand or where it is impracticable to use herbicides, e.g. around reservoir catchment areas.

Herbicidal treatments

These have now displaced handwork on large areas of forest as they are normally cheaper, last for a whole season or more and save labour. A wide range of well tried herbicides for most weed situations is now available. Some methods require no diluent (e.g. ULV and granules) and are well suited to steep or rough sites or where water supply is difficult. The main problems lie in persuading some woodland owners to use herbicides. Under no circumstances may the use of herbicides be entrusted to anyone but fully trained and properly instructed and supervised operators. (Consult The Control of Pesticides Regulations 1986 SI 1986/1510 and the Consents to Sale, Supply, Storage and Use of Pesticides, also the MAFF leaflet: *Pesticides: Guide to the New Controls*, UL 79.)

No single herbicidal treatment can be relied on to control the very broad spectrum of weeds that may develop in the early years of a plantation: some weeds are resistant and different herbicides may be needed later. Some treatments have a short period of use in a growing crop, while timing in relation to growth stages of trees and weeds is important; if wrong, weed control may fail or the crop may be damaged. Careful forward planning is therefore essential to integrate the herbicidal programme with all other operations during the establishment period – from the initial preparation of the site until the trees have formed a dense thicket.

Herbicidal application

Equipment

Steep or rough terrain and tree growth largely preclude the use of normal farm equipment for woodland. The use of lances or tractor-mounted sprayers may be feasible in some situations. The following are suitable, according to herbicide, species and type of application.

Knapsack MV sprayers
Directed applications around base of young trees – some herbicides require use of *tree guards* (i.e. to cover tree or spray jet). NB glyphosate, paraquat. Also stump and basal bark treatments.

Mistblowers LV
Knapsack or tractor mounted for overall foliar applications. The latter machine only of limited value in lowland forest.

Ultralow volume (ULV)
Overall foliar application by incremental spraying.

Distributors for granular herbicides
Airflow type distributors are essential for accurate granule distribution. Handwork is too laborious, inaccurate and wasteful of expensive herbicides; overdosing can occur. Suitable for overall or strips 1–2 m wide; flow can be cut off with wide tree spacings for spot application. ULV and granules require no diluent and are ideal for areas where water supply or carriage is difficult.

Tree injectors
These are for injecting undiluted herbicide into standing trees at (a) breast/waist height – cut made with axe, or (b) base of tree with chisel bit for penetration.

Rope wick application
Handheld applicators are suitable as a replacement for handwork on climbing or tall growing weeds or climbers.

Types of treatment

Pre-planting
These treatments are given to control unwanted vegetation prior to planting, either overall or in strips, in the centre of which trees are subsequently planted; selectivity at this stage is not needed but some herbicides require an appropriate interval between application and planting. Avoid drift on to adjacent susceptible crops.

Post-planting
Post-planting treatments if given overall *must* be selective in the species concerned, as determined by age, size and growth stage of the trees, time of year and herbicidal activity; otherwise *directed* applications must be given, using tree guards where necessary (e.g. glyphosate or paraquat applications in susceptible species). Foliar applications are dependent for success on the active growth of weeds at the time of application, e.g. with broadleaved woody weeds apply before leaf senescence. Selectivity is often obtained by application when the crop is not making rapid growth or is dormant. Timing is all important for control of weeds and crop safety, as application periods are frequently *very limited*. Surface or soil applications, granular or liquid, depend on adequate rainfall for activity. Restrictions exist on timing of applications.

Treatment of individual large trees

Stump treatment

This treatment is given to freshly cut stumps (within 24 h of felling) of broadleaved trees or woody weeds to prevent coppice regrowth. Give as a *pre-planting* treatment. Use triclopyr, saturating cut surface and remaining bark. Use paraffin or diesel oil as diluent. Hawthorn (*Crataegus* spp), laurel (*Prunus laurocerasus*) and rhododendron (*Rhododendron ponticum*) require high dosage rates.

Unwanted standing trees

Basal bark spray
Saturate bottom 300 mm (thin bark) or 450 mm (thick bark) of whole circumference of trunk to run off; use triclopyr in paraffin or diesel oil. Application in conifer crops should be delayed until growth extension has ceased and resting buds formed.

Frill girdling
A ring of overlapping downward cuts encircling the trunk close to ground level is first made with a light axe or bill hook, to penetrate the cambium and preferably the sapwood; herbicide is then sprayed on to bark just above cuts and runs into them. Use triclopyr. Also suitable for ammonium sulphamate (AMS) for trees less than 150–200 mm diameter breast height (1.3 m about ground level).

Table 13.10 Weed control in non-crop situations

Herbicide, type of activity and persistency of initial dose in soil	*Main species/types affected when used for initial knockdown of foliage*	*Availability in mixtures with very persistent residual herbicides; also other leaf applied herbicide. Yes/No. Remarks*
Foliage absorbed herbicides		
Amitrole, Tr, LP	Rhizomatous grasses	Yes. Also affect broadleaved herbaceous weeds
Dalapon Tr, MP[1]	Rhizomatous grasses	Yes. Broadleaves resistant to straight herbicide
Glufosinate-ammonium C None	All green foliage	No. Straight herbicide only burns tops of perennials
Glyphosate Tr None	All green foliage	Excellent control of all species
Paraquat C None	All green foliage	Yes. Straight herbicide only burns tops of perennials
Triclopyr Tr LP	Woody spp	No. Also docks (*Rumex* spp) and nettles (*Urtica* spp)
Foliage and/or root absorbed or vaporizing herbicides		
Dichlobenil VP	Bracken (*P. aquilinum*) Docks (*Rumex* spp) grasses and general weed control	No + dalapon only — Granules only. Do not use near glasshouses, hops or susceptible crops
Imazapyr Tr P	Bracken, grass and general weed control	No. Conifers only. Do not use near broadleaved trees or shrubs of value
Picloram Tr VP	Ragwort (*S. jacobea*) General weed control	Yes. Straight herbicide general broadleaved weeds
Sodium chlorate Tr		Yes. Straight herbicide highly inflammable. Fire risk.
MP[2]		Do not apply near roots of trees or shrubs of value
Root-absorbed herbicides (residual)		
Bromacil VP	Used for long-term residual weed control	Yes. Do not use near trees or shrubs of value
Diuron VP	Used for long-term residual weed control	Yes. Do not use near trees or shrubs of value
Tebuthiuron VP		No. Do not use near trees or shrubs of value

Key to Table 13.10
Type of activity: C = contact; Tr = translocated.
Persistency: Persistency of chemicals is determined to a large extent by the dosage rate (increased dosage rate = longer persistency), soil type and rainfall. The following times are *only approximate* and apply to the *straight herbicide only*. Additives may increase persistency greatly: LP (Low persistency, up to 8 weeks); MP[1] (moderately persistent, 3 months to a season); MP[2] (rather less persistent than MP[1] 3 to 4 months); None (no persistency; sowing can follow soon after treatment if required); P (persistent a season); VP (very persistent, a season or longer).
NB There is a need for great care when using herbicides in the proximity of glasshouses, orchards, ornamentals and any susceptible crops. Avoid drift at all times. Many of the above herbicides are quite unsuited for amateur use.

Notching
Notching is for applying solid AMS crystals to larger trees; cut a ring of steps with an axe, maximum 10 cm edge to edge apart at base of tree, floors of steps sloping slightly inward to retain AMS crystals, 15 g per step.

Tree injection
Tree injection requiring a special tool for injecting undiluted herbicides into the translocation systems of a tree may be carried out safely at any time of the year.

Choice of herbicidal treatment

A sequence of treatments is likely to be necessary in the early life of the crop; the treatments for various weed situations are summarized in Table 13.9. Particular attention should be paid to the requirements and limitations of each herbicidal treatment.

Weed control (unselective) in non-crop situations

The object is usually to keep the ground free from vegetation for a prolonged period, as with yards, drives, paths and roadways. If temporary control only is needed and cropping is to follow, select only herbicides with short residual effects.

It is not possible to keep land permanently free from vegetation by herbicidal application unless there is an 'ongoing' programme. There are three distinct aspects to the problem: (a) The '*initial knockdown*' to kill existing vegetation. A translocated foliar-acting herbicide is usually preferred, e.g. glyphosate. Contact types alone, e.g. paraquat, only kill annuals; they simply defoliate perennials. (b) *Prevention* of immediate re-establishment with a very persistent residual herbicide, e.g. bromacil, diuron. Established perennials require a heavy dose for a complete kill. (a) and (b) may either be given as separate applications or, where a particular mixture is suitable, are given as a single combined application, e.g. glyphosate + simazine: paraquat + diuron. (c) *Maintain freedom from weeds*. If (a) and (b) are fully effective, only a low rate, low cost application will be required in the following spring for maintenance; if (a) and (b) are not fully effective, a translocated (foliar) application will be required later.

Persistency of herbicides is temporary and depends on (i) *herbicide*: in some, e.g. paraquat and glyphosate, there is little persistency; others, e.g. diuron and bromacil are very persistent; (ii) *dosage rate*: persistency and cost of residual herbicides increases with dosage rate; (iii) *rate of leaching or decomposition* of herbicide is accelerated by high rainfall on soil types which leach readily (e.g. sand) and with increasing soil temperature, resulting in higher microbial activity. Adsorptive surfaces, e.g. peat, clay or ashes require higher dosage rates.

When choosing a herbicide, particular reference must be made to surrounding crops, gardens and glasshouses. Spray drift dangers apart, trees can be killed by uptake of herbicides through their roots, especially with sodium chlorate; glasshouses are very vulnerable to vapour given off by chlorthiamid and dichlobenil. Always ensure that drift, run-off or leaching will not affect adjacent or local sites. Obtain advice before treating slopes or hard impenetrable surfaces where run-off may affect watercourses, streams or wanted vegetation. A selection of suitable herbicides is given in Table 13.10.

References

Clarke, J.H. & Moss, S.R. (1991) The occurrence of herbicide resistant *Alopecurus myosuroides* (blackgrass) in the United Kingdom and strategies for its control. *Brighton Crop Protection Conference – Weeds*, 3, 1041–1048.

Forestry Commission (1986) *The Use of Herbicides in the Forest.* Booklet No. 51 Forestry Commission, Edinburgh.

Ivens, G.W. (1995) *The UK Pesticide Guide.* British Crop Protection Council and CAB International, Oxford and Farnham.

Ministry of Agriculture, Fisheries and Food (1993, superseded by subsequent editions). *Pesticides 1993.* Pesticides approved under the control of Pesticides Regulations. HMSO, London.

Moss, S.R. & Cussans, G.W. (1985) Variability in the susceptibility of *Alopecurus myosuroides* (blackgrass) to chlorotoluron and isoproturon. *Aspects of Applied Biology*, 9, 91–98.

Roberts, H.A. (1982) *Weed Control Handbook – Principles*, 7th edn. British Crop Protection Council. Blackwell Scientific Publications, Oxford.

Further reading

British Sugar plc publish herbicide recommendations (for herbicidal programmes) annually: 'Herbicides recommendations for post emergence low volume sprays for annual broadleaved weed control'.

Ministry of Agriculture, Fisheries and Food (ADAS) publish a range of priced booklets on weed control in crops. Obtainable from MAFF Publications, Lion House, Willowburn Estate, Alnwick, Northumberland NE66 2PF.

Processors and Growers Research Organisation (PGRO), Thornhaugh, Peterborough PE8 6HJ, publish leaflets (free to members) on choice of herbicides in peas and varietal susceptibility to herbicides; also associated leguminous crops.

Pests of Crops

Farm crops are subject throughout their growth to attack by pests belonging to various animal groups, the most important being insects, mites, nematodes, slugs and mammals. Different pest species vary in their mobility, host specificity, period of peak abundance, regularity of occurrence and response to climatic conditions.

The rapid expansion is areas devoted to 'new' crops such as oilseed rape, other oilseed crops and field grown vegetables, and the development of trends in crop production such as the early sowing of winter cereals, have created new pest problems, some of which have been only partially solved.

It is well known that problems such as the development of resistance in pests, the killing of beneficial organisms and the entry of persistent chemicals into food chains have resulted from the use of pesticides. Because of these problems, and the increasing cost of chemical control, a great deal of effort is now being directed towards the development of techniques aimed at improving the effectiveness of pesticide usage so that smaller quantities may be used in a more selective manner. Examples of such techniques are:

- establishing economic thresholds so that growers can be advised more precisely on pest population levels at which control is justified;
- forecasting the abundance of pests so that growers can be advised whether or not preventive control measures are necessary;
- employing biological and ecological information so that other methods of control can be used to replace or enhance chemical control;
- employing equipment which is capable of applying pesticides accurately in smaller quantities.

A growing body of knowledge related to these is now available and is incorporated in the text where appropriate.

There may be legal constraints or recommended codes of practice associated with the control of a pest or the use of a pesticide. All the factors mentioned above should be considered when contemplating prevention or control of a pest and additional information should be sought, if necessary, from appropriate sources such as ADAS, growers' organizations, technical representatives and other professional advisers. The way in which such information is obtained and disseminated is changing rapidly and much advice which was recently freely available through ADAS and other sources may now have to be paid for. Information technology is rapidly being applied to pest control, and systems such as Prestel-Farmlink and Agviser point the way to further developments in the near future.

In this chapter the pests are dealt with under 'host- crop' headings but it must be remembered that many of the general feeders are associated with a wide range of crops. It is not possible to include every pest which might injure a crop and those covered have been selected somewhat arbitrarily on the basis of their actual or potential economic importance or on their frequency of occurrence.

Cereal pests

Many pests whose normal hosts are grasses infest cereals either directly via mobile winged adults or indirectly when the feeding stages migrate from ploughed or desiccated grass to a following cereal crop. Direct drilled cereals are especially prone to the latter.

Wheat bulb fly (*Delia coarctata*)

A serious pest of winter wheat in eastern Britain; winter barley, triticale, rye and early sown spring wheat or barley may also be attacked. Adult fly lays eggs in cracks and crevices in bare soil in July–September, eggs hatch and larvae invade susceptible cereal plants in following January–March; hatching delayed by cold conditions. Larvae feed inside base of main shoot or tillers and migrate, as they grow, to infest more shoots.

Identification and high-risk situations

Patches of dead or damaged plants in January–March, expanding in April–May. Damage shows as 'deadhearts': centre leaf dies and turns yellow, outer leaves remain green. The larva, a typical legless fly maggot blunt at posterior end, is revealed on peeling outer leaves away from base of shoot. Wheat after set-aside land is likely to be at risk.

Host crops drilled in fields which were fallow or sparsely covered in previous July, e.g. roots. Late sown winter wheat, untillered, is especially susceptible. ADAS forecasts, based on egg counts of soil from sample fields, are issued.

Cultural and organic control
Sow winter wheat before end of October at a shallow depth. Sow spring wheat and barley after mid-March. Avoid wheat after potatoes, sugar beet, row crops and set-aside land.

Chemical control
(1) Seed treatment of chlorfenvinphos or fonofos. These are unlikely to be beneficial on wheat sown before late October and must not be used after 31 December because of the risk to wild birds. Control less effective on deep drilled seed.
(2) Fonofos granules or spray of chlorfenvinphos, chlorpyrifos or fonofos mixed into top 50 mm of soil at, or immediately before, drilling give better protection but cost more. Use when high risk is forecast.
(3) Sprays of chlorpyrifos, chlorfenvinphos, or pirimiphos methyl at egg hatch where soil conditions permit; advice on timing from local ADAS.
(4) Where earlier treatments have been omitted, or have failed to stem an attack, spray with dimethoate when damage is observed; not later than mid-March.

Note. Some of the above treatments are approved only for use on winter wheat.

Yellow cereal fly (*Opomyza florum*)

A pest of early sown winter wheat and occasionally winter barley; spring wheat and barley unlikely to suffer. Eggs laid in soil near base of host plants in October and November. Start of egg hatch at any time between late January and March, actual time varying from year to year. Newly hatched larvae climb the shoots, enter them between the outer leaves and burrow downwards to the growing points which they destroy.

Identification and high-risk situations
Untillered plants and tillers developing 'deadhearts' and then dying; distinguish from wheat bulb fly damage by distinct thin brown line circling or spiralling down the central shoot. Larvae small, pointed at both ends, complete their development inside one shoot.

Wheat drilled before mid-October in districts where this species has caused damage previously is especially at risk.

Cultural and organic control
Early sown crops, well advanced at egg hatch, compensate well for tiller damage, and chemical control is unlikely to give an economic return.

Chemical control
Probably justified only on slow growing winter wheat crops in high risk situations, which have not already been treated with a synthetic pyrethroid insecticide against BYDV. Sprays, including chlorfenvinphos and various synthetic pyrethroids such as alpha-cypermethrin, bifenthrin, cypermethrin and deltamethrin, should be applied at egg hatch (timing from ADAS) or following ADAS warnings.

Frit fly (*Oscinella frit*)

Grasses, especially ryegrass, are the natural host plants of the frit fly whose larvae feed inside the bases of the shoots. There are three generations (sometimes four in southern counties) of flies in a year and larvae overwinter in the shoots of their host plants. Cereals may be attacked in two different ways.

Migration from sward

Identification and high-risk situations
Slight angular bend above coleoptile where larva has penetrated at single-shoot stage of growth, followed by 'deadheart', i.e. central leaf dies and turns yellow. Legless, transparent larva 2–5 mm long, inside base of shoot. Young plants usually killed.

Autumn-sown wheat, barley, oats and rye drilled over ploughed grass or grassy stubbles, or directly drilled into desiccated swards may be invaded by larvae migrating from the decomposing grass.

Cultural and organic control
Leave an interval of 4–6 weeks between ploughing/desiccating and drilling. Late sown crops suffer less because lower soil temperatures reduce migration.

Chemical control
Spray with chlorpyrifos, fonofos, pirimiphos methyl or triazophos

(1) at drilling in high risk situations on farms where damage has occurred frequently;
(2) at crop emergence following local ADAS warnings;
(3) when centre leaves of 10% of plants at one-leaf to two-leaf stage detach easily on being gently pulled, and show a small brown feeding scar at the base.

To be effective the treatment should be given before obvious damage is seen.

Direct oviposition
Female flies lay eggs directly on spring oat and maize plants, the eggs hatch after a few days and the larvae enter the shoots.

Identification and high-risk situations
First-generation attack results in 'deadhearts' and death of young oat and maize plants or excessive tillering of older plants. Second-generation flies lay eggs in spring oat spikelets, the larvae feeding producing blindness or shrivelled and damaged grain.

Later sowings of spring oats are at greater risk of attack by first generation flies because the latter are attracted only to plants or tillers with fewer than five leaves. Maize is often at risk because of its late sowing date but attacks are unpredictable.

Cultural and organic control
Sow spring oats before mid-March in south, late March in Wales and north Britain.

Chemical control
For spring oats, spray with chlorpyrifos if crop has not reached the four-leaf stage by first half of May in south, late May in Wales and north. Or apply triazophos immediately damage is observed. Economic yield response likely to be small unless attack is severe. For maize, granules of bendiocarb, phorate and carbofuran in the seed row as a precaution. Or spray with chlorpyrifos, fenitrothion or triazophos at crop emergence.

Wireworms (*Agriotes* spp)

The larvae of click-beetles; natural habitat is the soil under permanent grass where they feed on the underground parts of plants.

Identification and high-risk situations
Main damage occurs in autumn and spring when wireworms feed on underground stems and hypocotyls of cereals. Seedlings may be completely severed and turn yellow and die while still remaining upright in the ground. Several plants in a row may show progressive symptoms. Damage often occurs in patches where other factors are contributing to poor growth, e.g. disease, poor drainage. Careful excavation of soil around injured plants reveals stiff, smooth, yellow larvae up to 20 mm long.

Only likely to be a problem in the first three years after permanent grass. All cereals are affected but wheat and oats are most susceptible.

Cultural and organic control
Increase seed rate in high-risk situations. Assist recovery by rolling and nitrogen top dressing if soil conditions are suitable.

Chemical control
Seed treatments based on gamma HCH are cheap and are recommended when wireworm population is low. To reduce the risk of plant injury the seed should be sown soon after treatment; moisture content should not exceed 16% and the dressing should be applied uniformly at the prescribed rate. For higher populations apply gamma HCH* as a spray to the soil when the seedbed is being prepared. Fonofos treatment for wheat bulb fly control will also provide some degree of protection against wireworms.

* Gamma HCH should not be used in this way if potatoes or carrots are to be planted within 18 months because of risk of taint.

Leatherjackets (*Tipula* spp)

The larvae of the daddy longlegs or craneflies. The flies lay eggs in grassland mainly in September. The eggs hatch in 10–14 d and the leatherjackets feed in the soil on the underground parts of plants but will come to the surface in dull, moist and relatively warm conditions to feed at ground level. Activity is reduced in cold and dry conditions.

Identification and high-risk situations
Young plants may be chewed below or at ground level and spring cereals in particular may be completely severed or roughly grazed down. Leaves with ragged holes. Damage often in patches coinciding with wet areas of fields. Confirm by examining soil in vicinity of affected plants for presence of legless, grey-brown, fleshly grubs up to 50 mm long, with tough, wrinkled skins and a number of small, pointed protuberances at tail end.

Cereals following grass or grassy stubbles are at risk especially when preceding September–October has been wet. All cereals suffer; winter cereals may be attacked in early winter but the greatest damage normally occurs on spring cereals in April and May. ADAS monitors populations in most regions and forecasts are issued or are available on request. Rooks are attracted to cultivated fields with large numbers of leatherjackets and may be seen searching for them by turning over surface clods.

Cultural and organic control
Plough or desiccate grassland before September; kill grassy stubbles immediately after harvest. Rolling and top dressing with nitrogen when soil conditions permit will encourage the crop to grow away from an attack.

Chemical control
Examine crop daily for signs of attack in risk situations because damage can occur very rapidly and insecticide must be applied immediately to avoid excessive loss. Chemical control is justified when a *total* of 15 or more leatherjackets are found on examining ten separate 30-cm lengths of row selected at random diagonally across the field. The figure applies when the rows are 17.5 cm apart; at narrower spacings a total of ten or more is the critical level. The soil within 3–4 cm of the plants should be carefully searched.

Insecticides may be applied in these circumstances as sprays, granules or poison baits, the latter normally giving best results. In all cases the application is best made when the leatherjackets are active on the surface, i.e. in the evening in damp, warm weather.

Sprays Chlorpyrifos, fenitrothion gamma HCH*, quinalphos and triazophos. These may also be applied during the preparation of the seedbed when leatherjacket populations are known to be high.

Granules Apply chlorpyrifos using an applicator capable of distributing small quantities/ha.

Poison baits Mix fenitrothion or gamma HCH* with bran moistened to a crumbly consistency. Thorough mixing to ensure even distribution is essential, and the bait should then be distributed as evenly as possible over the field. Bran plus gamma HCH pellets are available commercially.

Methiocarb bran pellets used pre-/post-planting for slug control will give useful reductions in leatherjacket populations. Both pests commonly occur together and dual control is often obtained.

Field slug (*Deroceras reticulatum*)

This is the most important of several species of slugs which may feed on cereals, and is the only one which regularly feeds above or close to the soil surface. Slug numbers vary considerably according to soil type, previous cropping and climate; their activity is also affected by temperature and humidity but, in general, their numbers are greatest in autumn and late spring.

Identification and high-risk situations
Winter wheat, rye and barley may be severely damaged by slugs feeding on the germ of the seed shortly after drilling, giving the impression of total or partial seed failure. Young shoots of all cereals may be eaten as they germinate, or grazed at ground level and their growing points destroyed. The leaves of older plants may be shredded longitudinally but this damage, though conspicuous, is not normally of great importance.

Autumn sown cereals, especially wheat, are most likely to suffer economic loss; damage may be seen on spring cereals but this is not normally important. Populations are highest in undisturbed soil, especially if it has good moisture-retaining properties and where there is dense ground cover. Direct drilled cereals or cereals following grass, oilseed rape, peas, beans and other crops with bulky residues on silt or clay soils are most at risk, especially late-sown crops in wet autumns.

---

* Gamma HCH should not be used in this way if potatoes or carrots are to planted within 18 months.

Cultural and organic control
Prepare a fine, well consolidated seed bed which restricts movement of slugs through soil. Avoid late-sown crops.

Chemical control
Treat with metaldehyde, methiocarb or thiodicarb baits when seedbed conditions and cropping history indicate a high risk. Test baiting is no longer recommended for forecasting slug damage but is useful in indicating when slugs are active on the surface and therefore amenable to chemical control. Best results are obtained with baits applied to the surface before drilling, preferably following rain, and left undisturbed for 3–4 d. Failing this, post-drilling applications or mixing and drilling slug pellets with the seed may be carried out.

Cereal cyst nematode (*Heterodera avenae*)

Mainly a problem on spring oats. Winter oats, maize, wheat, barley and rye are progressively more resistant to damage. Present in small numbers in grassland which increase rapidly when cereals are grown frequently. Small, less than 1 mm long, lemon-shaped cysts, the bodies of dead females, containing approximately 400 eggs when young, may persist in the soil for many years their contents declining slowly in the absence of suitable hosts. Minute larvae emerge from the cysts in spring to response to substances exuding from growing cereal roots and invade and feed inside the rootlets. Females swell and form young cysts in July which may be seen attached to roots of affected plants, white at first and darkening as they age.

Identification and high-risk situations
Damage normally shows up as patches of pale, stunted plants whose roots are short and much branched in comparison with those of healthy plants.

Mainly confined to chalky and light soils in southern England. Oats following several successive crops of wheat and/or barley are most at risk but other cereals may suffer when populations are high. Suspect fields can be sampled to determine population levels.

Cultural and organic control
Reduce high populations by sowing a ley. Avoid growing oats after several successive crops of wheat or barley. Resistant varieties may be grown where cyst counts are high and are useful in reducing nematode populations; the spring barley Brewster and the spring oats Keeper and Rollo all exhibit resistance to the two principal pathotypes (races) of cereal cyst nematode. Use NIAB Farmers Leaflet No. 8 for variety choice.

Chemical control
Chemical control is not economically justified at present.

Cereal aphids

Three species are important; like all aphids they feed by sucking plant sap causing direct damage to their host plants; they may also transmit cereal virus diseases. Correct identification, and careful observation of population trends wherever possible, is important because different species may require different treatment. Special care must be taken when deciding whether or not control measures are justified because spraying is costly in both economic and ecological terms on a crop which occupies such large areas of land.

Grain aphid (*Sitobion avenae*)

Identification and high-risk situations
Largish aphids 2–3 mm long, colour varying between reddish brown and green, two black tubes (siphunculi) projecting from upper surface at rear of body. Early arrivals feed on leaves but transfer to flowering heads as latter emerge and feed on rachilla and developing grain; resulting grain is light and shrivelled.

Mainly a problem on winter wheat causing direct damage in summer, but may occur on all cereals. Winged females migrate from grasses in late May–early June in southern Britain, later in the north. Numbers build up in hot, humid conditions. Losses greatest when heavy infestations occur when flowering heads are developing. Also transmits BYDV from grasses to cereals and then within the cereal crop, especially in the north.

Cultural and organic control
Direct damage in summer: none recommended. BYDV transmission: see under Bird cherry aphid.

Chemical control
Direct damage in summer: spray winter wheat when, at the beginning of flowering (GS 61), an average of five or more aphids per ear is found on at least 50 ears selected at random from all parts of the field except the headlands,

and when weather is conducive to aphid build-up. Alpha-cypermethrin, chlorpyrifos, cypermethrin, deltamethrin, demeton-S-methyl, dimethoate, esfenvalerate, fenvalerate, heptenophos, oxydemeton-methyl, pirimicarb and thiometon are all effective. Local beekeepers should be warned of impending spraying and, where there is a risk to bees, pirimicarb should be the insecticide of choice.

BYDV transmission: see under Bird cherry aphid.

Rose grain aphid (*Metapolophium dirhodum*)

Identification and high-risk situations

Light green aphids, 2.25–3.0 mm long, with darker green stripe down middle of back. Normally on undersurfaces of lower leaves but will colonize upper leaves and surfaces and stems as numbers build up. Only likely to cause economic loss when flag leaves are heavily infested.

Winged females disperse from brambles and wild roses to cereals and grasses late May–early June. Rapid build up in warm, humid conditions.

Cultural and organic control

No cultural control recommended.

Chemical control

In winter wheat and spring barley spray when average number of aphids per flag leaf exceeds 30 at any time between flowering and milky ripe stage (GS 75). Insecticides and precautions as for grain aphid.

Bird cherry aphid (*Rhopalosiphum padi*)

Mainly important as a transmitter of BYDV from grasses to cereals and then within the cereal crop. Overwinters naturally as eggs on bird cherry but, in areas with a mild winter climate, can also overwinter as small colonies of wingless adults and juveniles on grasses and early sown winter cereals; may then build up and spread rapidly in the spring.

Identification and high-risk situations

Small, 1.5–2.3 mm long, brown to greenish-brown aphids with rust red patches at rear of body; found on all parts of the plants. Cereals infected in the autumn do not normally show BYDV symptoms until the following spring when patches of affected plants can be seen. Infected plants in spring-sown crops normally become conspicuous at ear emergence.

Main risk of virus transmission is in southern England and Wales on early sown, September–early October, winter cereals. Late sown spring cereals may also become infected by aphids migrating in May. Cereals following grass, in close proximity to grassland, and in fields with high hedges are more likely to suffer.

Cultural and organic control

Direct drilled crops least, ploughed crops most severely affected due to poorer survival of populations of aphid's natural enemies.

Table 13.11 Spraying to control BYDV in winter cereals

BYDV	*Sowing date:*		
	Before mid-September	*Mid–late September*	*Early–mid-October*
Severe in past	A	C	D
Little previous history	B	D	D

A, Spray in mid-October.
B, Spray in mid-October only if aphids present in crop or following ADAS warnings.
C, Spray in late October or early November.
D, Spray only if aphids present in crop or following ADAS warnings.

Chemical control

(1) BYDV may be introduced into winter cereals up to mid-October by winged aphids migrating from infected grasses, the risk increasing in proportion to the earliness of drilling, the severity of previous infections and the percentage of migrating aphids which are carrying the virus (Infectivity Index). General recommendations on spraying are summarized in Table 13.11.
Sprays include bifenthrin, cyfluthrin, cypermethrin, deltamethrin and lambda-cyhalothrin.
(2) BYDV may also be transmitted by aphids which survive when cereals follow grass. This is best prevented by killing the sward with paraquat 7–10 d prior to cultivation and 14 d before sowing.

Potato pests

Pests of potatoes fall into two main categories, those that reduce the productivity of the plants by direct feeding damage or by transmitting virus diseases, and those that disfigure the tubers thereby reducing their market value. In both cases the most effective control measures are normally applied before or at the time of planting.

Peach potato aphid (*Myzus persicae*)

Overwinters as eggs on peach trees and as small colonies of adult and juvenile females on a wide range of host plants especially in mild winters and in protected situations such as glass houses and chitting houses. Winged females arrive on potato plants in spring and feed by sucking the sap. Colonies of wingless aphids build up rapidly in hot dry conditions. Movement within the crop is responsible for transmission of virus from infected plants. Mainly important as a vector of leaf roll virus and severe mosaic (virus Y) into and within home-grown 'seed' crops and in the chitting house. Early migration from overwintered colonies and rapid build-up in early summer may create a virus problem in main-crop potatoes in some years.

Identification and high-risk situations
Small scattered colonies of green to pinkish aphids, relatively inactive even when disturbed.

More likely to be a problem following a mild winter and early spring and when calm, warm weather prevails in summer.

Cultural and organic control
Destroy volunteers, discards, clamp site debris and other potential sources of potato viruses. Plant certified healthy 'seed'. Grow 'seed' crops in isolated sites and rogue virus infected plants as soon as symptoms appear. Burn off foliage early.

Use NIAB *Potato Variety Handbook* for variety choice.

Chemical control
Routine control is recommended only for home-grown 'seed' production and is more effective against leaf roll virus which, in contrast to virus Y, is not transmitted instantly by aphids. Granular applications of aldicarb, disulfoton and phorate to the seedbed or planting furrow will generally protect until mid-June; thereafter one or more sprays should be applied until haulm destruction. Alternatively, spray at 80% plant emergence using deltamethrin, demeton-S-methyl, dimethoate, heptenophos, nicotine, pirimicarb or thiometon and repeat at 14 d intervals.

On ware potatoes, direct damage may be caused by this and other aphids including the potato aphid (*Macrosiphum euphorbiae*), a larger species which may form dense colonies in hot, dry summers. Spraying is justified only when an average of 3–5 aphids per true leaf from a sample of equal numbers of upper, middle and lower leaves is found in July. Do not spray if aphid predators (ladybird and hoverfly larvae) and parasites (indicated by swollen, brown, 'mummified' aphids) are present. In chitting houses and seed stores, pirimicarb smokes used as required from the end of November will kill aphids on the sprouts.

Chemical control is complicated by the increasing prevalence of strains of aphids which are resistant in varying degrees to some of the insecticides listed above. This pattern of resistance is constantly changing and growers are advised to obtain local advice on the best choice of insecticide.

Potato cyst nematodes (*Globodera rostochiensis* and *G. pallida*)

The most important pests of potatoes in the UK. *Globodera rostochiensis*, which forms yellow cysts on the roots, is present in all potato-growing districts; *G. pallida*, with white cysts, is more restricted in its known distribution but is probably present at low population densities even in areas where it is not considered to be a threat. Life cycles are similar to that of the cereal cyst nematode except that potatoes are the only important field hosts; tomatoes are also susceptible.

Identification and high-risk situations
Infestations normally show up first as patches of stunted plants which wilt readily. Their root systems are shortened and much branched and small spherical cysts may be seen attached to the rootlets.

Yields are reduced when potatoes are grown in soils containing viable cysts, and cyst numbers increase rapidly when potatoes are grown frequently in infected fields. The number of eggs/g of soil can be estimated by advisers and the information should then be used to plan a cropping and control programme in consultation with specialist advisers.

Cultural and organic control
The main objective of control is to contain the cyst/egg population to an acceptable level. It is best achieved by combining lengthened rotations, resistant varieties and nematicide treatment. EC regulations require that potatoes sold as 'seed' can only be grown in fields which prove to be eelworm free following a statutory test. Where initial egg counts are high, potatoes should not be grown for

Chemical control
Incorporate granules of aldicarb, carbofuran ethoprophos or oxamyl in the seedbed. Rates of application may vary according to soil type and whether an early or maincrop variety is to be grown. Effectiveness depends on thorough mixing with the soil to the correct depth. Autumn application of dazomet or dichloropropene may also be used; special applicators may be required.

several years; thereafter the rotation can be shortened. Destroy volunteer plants.

The effectiveness of resistant cultivars is lessened by the presence in the UK of one pathotype of *G. rostochiensis* (RO1) and four of *G. pallida* (PA1, PA2, PA3 and PA4). RO1 predominates in field populations in eastern and southern England, Scotland and Ireland. PA3 predominates in the Humber basin and the Channel Islands. In other regions these two main pathotypes are randomly intermingled and field infestations of PA1, PA2 and PA4 are also present. The first early varities Accent, Minerva, Pentland Javelin, Premiere and Rocket, and the maincrop varieties Cara, Fianna, Lady Rosetta, Kingston, Maris Piper, Nadine, Sante, Saxon and Stemster, are resistant to pathotype RO1 only. They tolerate invasion by the nematodes and alter the sex ratio so that few new cysts are formed. It is unfortunate that where RO1 resistant varieties are grown frequently, populations of *G. pallida* will almost certainly increase. The new varieties Rocket and Sante, combining RO1 resistance with partial resistance to *G. pallida*, may lead the way to more effective control.

Use NIAB *Potato Variety Handbook* for variety choice.

Free-living nematodes (*Trichodorus* and *Paratrichodorus* spp)

Minute, less than 1.00 mm long, soil-inhabiting nematodes. Migrate through the soil and feed externally, sucking the contents of surface cells of tubers and roots.

Identification and high-risk situations
Loss of yield negligible. Important as vectors of tobacco rattle virus (TRV) causing spraing which renders tubers unmarketable (see under Potato disease).

Associated with coarse, sandy soils.

Cultural and organic control
The first early variety Premiere and the main-crops Record and Romano are resistant to spraing.

Use NIAB *Potato Variety Handbook* for variety choice.

Chemical control
Aldicarb, 1,3-dichloropropene or oxamyl incorporated pre-planting. Phorate granules used primarily against wireworms also bring about some control.

Garden slug (*Arion hortensis*)

The principal slug species affecting potatoes, it lives almost entirely underground only coming to the surface in numbers during rainy weather in July and August. The keeled slug, *Milax budapestensis*, may also contribute to tuber damage.

Identification and high-risk situations
Slugs penetrate tubers through relatively small, rounded entrance holes and excavate wider tunnels and cavities thereby reducing the marketability of the crop. The garden slug is black with a yellow sole. Millipedes may be present as secondary feeders.

Numerous in wet, heavy soils, especially those with high organic matter content. Damage likely to be greatest during a wet autumn following a mild, wet summer; also in irrigated crops.

Cultural and organic control
Improve drainage. Harvest as soon as possible. Cara, Estima, Fianna, Kingston, Marfona, Maris Peer, Maris Piper, Saxon and Stemster are highly susceptible and should not be grown in slug risk situations. Lady Rosetta, Pentland Dell and Romano are among the least susceptible varieties.

Use NIAB *Potato Variety Handbook* for variety choice.

Chemical control
Surface baits containing metaldehyde or methiocarb can be useful when applied in wet conditions in late July and August when the slugs come to the surface. A number of marked plants of Maris Piper within the crop can be examined regularly to provide an early warning of an attack.

Wireworms (*Agriotes* spp)

See wireworms in cereals. Relatively low populations can cause serious economic loss.

Identification and high-risk situations
Small round holes in the tuber leading into tunnels of the same diameter; wireworms may be found on cutting tubers open. Entry of slugs, millipedes and other soil pests is encouraged. Loss in yield is negligible but crop may be rejected or market value greatly reduced.

Numbers are likely to be high in the two years following permanent grass but damaging populations, above 75 000/ha, may persist even in arable land. Numbers are estimated by soil sampling. Wireworm activity increases in autumn. Unlikely to be a problem on early lifted varieties.

Cultural and organic control
Lift maincrops as soon as possible after maturation. Avoid potatoes after permanent grass.

Chemical control
Work ethoprophos into the soil prior to planting when population levels merit chemical control; higher rates are needed on peaty soils. Alternatively, apply phorate granules in the furrow at planting. This will also reduce spraing and early aphid attack. On no account should gamma HCH be used.

Cutworms

The caterpillars of a number of related species of nocturnal moths, which damage a wide range of crops. The most important species in recent years has been the turnip moth (*Agrotis segetum*). Eggs are laid on foliage and stems of many crop and weed plants from May to July. The small caterpillars hatch and feed on the leaves for two to three weeks, during which time they are highly susceptible to wet conditions, and then descend into the soil feeding just below ground level although they may come to the surface at night.

Identification and high-risk situations
The fleshy, smooth, greenish-brown caterpillars, up to 5 cm long when fully grown, excavate irregular shallow pits in the surface of the tubers, especially those close to the surface.

Prevalent when weather has been warm and dry during early feeding stages, especially on light, well drained soils and weedy fields. Irrigated crops are unlikely to suffer.

Cultural and organic control
Control weeds.

Chemical control
Spray with alpha-cypermethrin chlorpyrifos, triazophos or cypermethrin when the newly hatched caterpillars are feeding on foliage prior to descending into the soil.

Sugar beet pests

The seedling is the most vulnerable growth stage of any plant, a fact of particular relevance to the sugar beet crop which is now almost entirely sown to a stand with pelleted monogerm seed. This results in a low seedling population, and soil pests feeding on seed or seedlings can damage or destroy a high proportion of the plants very quickly even though they themselves are present in relatively small numbers. Effective weed control increases the problem in the case of general feeders. It is essential, therefore, that protection should be applied in advance against predictable soil pests and that the crop should be examined every day from sowing until it is well established so that control measures can be applied in time to be effective.

Seed furrow treatment with granular insecticides, extensively used against aphids and free-living nematodes, will also control or reduce the injurious effects of most soil pests. In general, they are preferable on ecological grounds to overall seedbed or post-emergent treatments against soil pests which may kill important predators and parasites of aphids.

Sugar beet pests also affect fodder feet and mangels but the cost of control on these crops is only justified in exceptional circumstances.

Soil pest complex

This includes millipedes, symphylids and springtails and it is convenient to deal with them as a group. Main damage is to the roots and underground stems of seedlings which may result in death or permanent stunting of growth. Tolerance to attack normally develops between the 2- and 4-true-leaf stage of development.

Millipedes
Body composed of many similar segments. Examination with a hand lens reveals the presence of two pairs of legs per segment. The spotted snake millipede (*Blaniulus guttulatus*), cylindrical in section with a line of reddish-brown spots along each side, is the most injurious but flat millipedes including *Polydesmus angustus* may also contribute to the damage.

Seedling roots and stems tunnelled or rasped causing collapse of plants; large numbers may congregate around individual seedlings in May, extending the damage caused by other soil pests. Damage often patchy.

Symphylids (*Scutigerella immaculata*)

White, up to 15 mm long, 12 pairs of legs when fully grown. Become active in May and June; may be seen running rapidly over surface of disturbed soil.

Springtails (*Onychiurus* spp)

Minute, up to 2 mm long, white insects with three pairs of legs. Active in the soil at lower temperatures than other soil pests.

High-risk situations

Damaging populations likely only in soils with a good structure containing many minute fissures and macropores which the pests can move through to congregate rapidly around seedlings. Silts and chalky soils generally most heavily infested. Damage by symphylids is normally confined to crops in silty soils, whereas millipedes and springtails are present in a wider range of soil types. Sandy soils carry the lowest populations. Field populations show enormous variation even between similar neighbouring fields, and predictions of damage have to be on a field basis.

Cultural and organic control

None.

Chemical control

At present, there are no accurate guidelines as to when control is justified, and decisions on preventive treatment are based on soil type and especially on the previous history of damage in a field. To this end it is desirable that small portions of a field should be left untreated to help in assessing the effect and value of any control measure. Investigations have shown that an assessment of springtail population, which can be made at lower temperatures prior to drilling, will give an indication of the overall risk of attack; the use of baited traps for this purpose is under development. The main objective of any control measure is to bring about a uniform stand of between 75 000 and 100 000 plants/ha. Current recommendations are:

(1) Seed treatments with imidacloprid or tefluthrin. These are environmentally more beneficial and imidacloprid also gives excellent control of aphids, virus yellows and mangold fly;
(2) Seed furrow applications of granular formulations of aldicarb + HCH, bendiocarb, benfuracarb, carbofuran, carbosulfan, or oxamyl;
(3) Spray soil with gamma HCH and work into the seedbed (not on some peat soils) – but see introductory remarks.

Woodmouse (*Apodemus sylvaticus*)

The mice live in burrows in the open field, hence their alternative name of long-tailed field mice. They dig up and break open the husks and feed on the ungerminated seed, working their way along rows. Each individual has a large foraging area. Populations vary enormously from year to year during the relatively short time that seed is at risk.

Identification and high-risk situations

Soil disturbed along rows. Seed pellets fragmented.

Damage is greatest on early sowings in dry soils and where some seed is exposed on the surface. BS field staff monitor populations during the spring.

Cultural and organic control

Avoid spillage and exposure of seed.

Chemical control

Place proprietary mouse baits or breakback traps at a density of three to four/ha in the open fields away from hedgerows as soon as damage is observed. Baits and traps should be placed in pipes or other suitable containers to protect non-target mammals and birds. Traps must be reset and baits replenished until the seed germinates.

Pygmy beetle (*Atomaria linearis*)

Small, 2 mm long beetles which disperse, by flying when temperatures exceed 15°C and wind speeds are below 7 km/h, during mid-April to June from previous year's beet fields into fields carrying the current year's crop where they feed on the seedlings. Populations build up during the year but little harm is caused to plants beyond the seedling stage. Adults overwinter in soil and under clods and plant debris on the surface. Numerous in intensive beet growing areas. Blackened feeding pits in the hypocotyl below or just above the soil surface which may bring about complete severance and death. At a slightly later stage beetles feed in the young heart leaves which become very distorted and misshapen as they unfurl and grow.

Cultural control
Avoid close rotations and close proximity to previous beet crops.

Chemical control
Imidacloprid, methiocarb or tefluthrin, normally present in the seed pellet, gives some protection against early attack. In areas where damage occurs regularly, control as for soil pests; chlorpyrifos used as a foliar spray is also effective.

Wireworms (*Agriotes* spp)

See under Cereal pests. On sugar beet, relatively low populations can lead to poor establishment.

Identification and high-risk situations
Hypocotyl chewed below ground level; seedlings die.

Most likely in the second or third year after permanent grass, but damaging populations may persist longer. Damage more serious on early drilled crops and chalky soils.

Cultural and organic control
Avoid growing beet in grass-dominated rotations.

Chemical control
Seed furrow treatments of bendiocarb, benfuracarb, carbosulfan and carbofuran used against other pests give reasonable control. Otherwise, work gamma HCH* into the seedbed, using higher rates on soils with a high organic content.

Beet flea beetle (*Chaetocnema concinna*)

Small, 2.5–3.5 mm long, shiny bronze beetles active in dry sunny weather. Adults overwinter in hedgerows and other sheltered situations close to the fields in which they have been feeding. Emerge on first warm days of spring and disperse to fields of seedling beet to feed on cotyledons and leaves. Seedlings beyond the cotyledon stage are normally able to withstand an attack.

Identification and high-risk situations
Small circular pits, later developing into holes on the cotyledons – 'shothole' damage. In bad attacks the cotyledons are completely consumed and the seedlings killed. Beetles may be seen on the cotyledons, jumping when disturbed.

Outbreaks sporadic, more likely in dry areas with abundant shelter and in cold dry springs when beetles are active but seedling growth is slow.

Cultural and organic control
Early sown crops are more likely to be past the susceptible stage at the time of attack.

Chemical control
Seed treatment with imidacloprid gives good control when used against the soil pest complex mentioned earlier.

Seed furrow treatment with granular applications of benfuracarb, carbofuran, carbosulfan or foliar sprays of deltamethrin, gamma HCH or trichlorfon give good control. These should only be used when absolutely necessary, because of danger to predators and parasites of aphids.

* Do not plant potatoes or carrots within 18 months following this treatment.

Leatherjackets (*Tipula* spp)

See under Cereal pests.

Identification and high-risk situations
Leatherjackets feed in spring, cutting off the seedlings at or just below ground level. Grubs in soil near damaged plants.

A problem in wetter, western beet-growing areas and on fields with a high water table; also when beet is grown after grass or weedy stubbles. Much lower populations are injurious to beet than to cereals.

Cultural and organic control
Plough grassland before September. Avoid growing beet in grass-dominated rotations.

Chemical control
Apply gamma HCH or chlorpyrifos as a spray as soon as damage is seen. If slugs are also likely to be a problem, use methiocarb as a surface bait.

Cutworms

See under Potato pests. Sporadic attacks by a number of species occur on sugar beet. In the fens the caterpillars of the garden dart moth (*Euxoa nigricans*) cause occasional damage to beet seedlings, grazing and severing them at ground level between April and mid-June. Control, when damage is observed, by band sprays of alpha-cypermethrin, cypermethrin, gamma HCH or triazophos applied in large volumes of water in the late afternoon in moist conditions when cutworms are more likely to come to the surface. Effectiveness is increased if the insecticide is then hoed into the surface soil alongside the seedlings. Other species may feed later in the season on the tap roots but control is not normally justified.

Peach potato aphid (*Myzus persicae*)

See also under Potato pests. Sugar beet is one of the numerous host plants of this species which is of major importance as a vector of beet yellows virus (BYV) and beet mild yellowing virus (BMYV): see section on sugar beet diseases.

Viruses are not transmitted via the true seed of sugar beet, so every plant starts life free from infection but may become infected by aphids migrating into the crop after having fed on other infected sources. Further spread then occurs within the crop as aphid populations increase and disperse.

Identification and high-risk situations
Green aphids may be seen on the plants.

Early migration, and therefore early virus infection, follow a mild winter. Crops in which a lot of bare soil is exposed are more attractive to aphids. The Virus Yellows Warning Scheme provides early warning of the likelihood of infection in time for preventive seed furrow treatments to be carried out, and updates the information as the season progresses. The highest risk is normally in the south of the beet-growing area served by the Bury St Edmunds and Ipswich factories.

Cultural and organic control
Sow early and encourage rapid early growth. Break the cycle of virus transmission by eliminating overwintering sources of virus; plough in plant residues on previous year's beet fields; destroy debris on cleaner-loader and mangel and fodder beet clamp sites before aphid migration gets under way in April. The seed crop (now rarely grown in the UK) may be an important overwintering 'bridge' for virus and aphids. Stecklings should be grown in isolation and/or under cover crops. Avoid close proximity to overwintered brassica crops. Severe and prolonged frosty weather in January/February can dramatically reduce the incidence of aphid/virus yellows infection.

Chemical control
The development of resistance in peach potato aphid populations to many widely used insecticides, especially in areas where virus yellows is a recurrent problem, is a major threat to the beet industry. It is therefore very important that insecticides are used only when absolutely necessary in order to prolong their active field life.

Seed treatment with imidacloprid is one of the most environmentally beneficial methods of aphid/virus yellows control.

Preventive seed furrow treatments, specifically directed against virus transmission, should only be used in high risk 'Yellows' areas, and following receipt of early 'Yellows' warnings. Materials include aldicarb, disulfoton, oxamyl or phorate granules which should protect up to the 6- to 8-leaf stage. Carbofuran, carbosulfan and benfuracarb, used against soil pests, provide shorter lived protection.

Sprays containing deltamethrin + heptenophos, demeton-S-methyl, dimethoate, pirimicarb, or thiometon may be used on crops which are otherwise unprotected following later warnings of aphid migration or when an average of one wingless green aphid per four plants is observed. Sprays may have to be repeated in years when aphid migrations persist, but not after the plants have reached the 15–20-leaf stage of growth.

Black bean aphid (*Aphis fabae*)

This aphid can seriously reduce the yield and quality of beet crops as a result of direct feeding damage. It can transmit BYV but is relatively unimportant in this respect. Overwintering occurs almost entirely as eggs on spindle trees (*Euonymus europaeus*) from which winged females migrate to summer hosts, including beet, in May and June on which they do not normally become numerous until July. Dense colonies of wingless females grow rapidly in hot weather and more plants become infested as the season progresses. Numbers decline in September as a result of emigration and build-up of predators and parasites.

Identification and high-risk situations
Conspicuous black aphids often co-existing with green peach potato aphids. Damage most important on heart leaves which become distorted and brown.

Injurious in hot dry seasons especially on late sown crops and when plant growth is retarded by drought. Loss in yield occurs when average number of aphids per leaf exceeds two.

Cultural and organic control
Early sowing with reduce severity of attack.

Chemical control
Normally controlled by sprays used against peach potato aphid. Where specific action is required, high volume applications (450 litres/ha) give better coverage of the larger plants. Pirimicarb, because of its selective action, enables aphid predators and parasites to persist and 'mop up' surviving aphids.

Beet cyst nematode (*Heterodera schachtii*)

The life cycle is similar to that of the potato and cereal cyst nematode, dealt with under their respective headings. There are, however, some special features which are significant when considering control measures. The development time from the emergence of the larvae from the cysts to the formation of the next generation of cysts is short and up to three generations may be completed in a single growing season, resulting in a much greater increase in cyst population when a host crop is grown. This species also has a wide range of hosts including all beets, various weeds and most brassicas including oilseed rape; the latter, although efficient hosts producing many cysts, are relatively unaffected by the invading larvae.

The Beet Cyst Nematode Order of 1977 gives MAFF the power to prevent host crops being grown on infected land, but these powers have not yet been used. The clause governing length of rotation in British Sugar contracts was removed in 1983 and since then growers have been able to follow rotations with no formal constraints. In recent years, rotations in the main beet areas have tended to become shorter and surveys have shown that numbers of infected fields and levels of infection have risen, especially in the Fens. Significant yield reductions may be expected if this trend continues.

Identification and high-risk situations
Causes 'beet sickness' in fields where no rotational control is carried out. Patches of stunted, unhealthy plants whose outer leaves wilt readily in dry conditions, and turn yellow and die prematurely. Tap roots poorly developed with numerous fibrous side roots. Lemon-shaped cysts visible on roots.

Injurious populations are more likely to occur in light sandy or peaty soils. Frequent cropping with sugar beet, fodder beet, red beet, spinach, mangel and brassicas (except radish and fodder radish) will lead to a rapid increase of cyst populations in fields which are infected.

Cultural and organic control
The only reliable control is to adopt long rotations covering all host crops, combined with weed control and the prompt destruction of all host crop residues. Length

Chemical control
Too costly and unreliable at present. Soil fumigation may be justified on some soils where potato cyst nematodes are also a problem.

of rotation is influenced by the likelihood of damage and the soil type, and growers should carefully examine roots from suspect crops for the presence of cysts, or have the soil sampled to ensure early detection of infestations. The following guidelines are suggested:

Field category	*Maximum host crop frequency**
(1) No known infection	1 year in 3
(2) Infected, but no heavy yield loss	1 year in 4
(3) Yields affected	Rest 4 years then follow a long rotation

Free-living nematodes (*Docking disorder*) (*Trichodorus* and *Longidorus* spp)

Similar to the species which attack potatoes.

Identification and high-risk situations
Patchy damage in crop, associated with sandier areas. Individual healthy plants often conspicuous among surrounding stunted plants. Attacked roots distorted, fangy or with lateral extensions.

A problem on light, sandy soils only, particularly when plants are under stress from other causes.

Cultural and organic control
Provide good growing conditions.

Chemical control
Seed furrow applications of aldicarb, benfuracarb, carbofuran, or oxamyl in high-risk situations and where there is a history of infection.

Oilseed rape pests

This relatively new crop has raised a number of new pest problems and as the area and intensity of cropping increases it is probable that more will arise. Pests which affect the developing pods, flowers and flowering stems are of particular importance and in this connection it is necessary to distinguish between winter and spring sown crops because their susceptible growth stages occur at different times and, since the pests normally have relatively fixed times of maximum abundance, they are subject to a different range of pests.

Bees are strongly attracted to rape flowers and, although they make little impact on seed set in this self-pollinating crop, sprays should not be applied to crops in flower. This is not normally necessary, but if it should be, e.g. when flowering is uneven, the sprays should be applied in the early morning or the evening using pesticides which are *relatively* safe to bees. Local beekeepers should be given adequate warning of any intention to spray.

Spray booms should be set high to promote even distribution when advanced crops are being treated.

Routine insecticide treatments are not recommended for any pest; populations should be carefully monitored and sprays applied only when threshold levels, if these have been established, are reached.

Field slug (*Deroceras reticulatum*)

See also under Cereal pests.

Identification and high-risk situations
Leaves shredded, seedlings grazed.

A problem mainly in the autumn and early winter on winter rape directly drilled into cereal stubbles on wet soils and in cool conditions which retard germination and plant growth.

Cultural and organic control
Sow heavy soils first, light soils last in autumn.

Chemical control
Drill slug pellets containing metaldehyde, methiocarb or thiodicarb with the seed *or* apply pellets as soon as damage is seen.

* 1 year longer on mineral or high organic soils.

Brassica flea beetles (*Phyllotreta* spp)

Small, 2.5 mm long, dark, shiny beetles often with a yellow stripe running down each wing case. Life-cycle and injurious effects similar to those of beet flea beetle. Mainly a problem in spring sowings. Most seed is dressed with gamma HCH as a protection but it may be necessary to band spray the seedling rows with gamma HCH in the face of sustained attacks when seedling growth is slow. Potatoes and carrots should not be grown for 18 months after such treatment.

Cabbage stem flea beetle (*Psylliodes chrysocephala*)

A pest of winter rape at present mainly in parts of the East Midlands, East Anglia and south-eastern England, but is widely distributed and is a potential pest in other areas. Adults migrate in August or September from fields of the current year's crops of rape and other brassicas into fields where autumn-sown seedlings are emerging. The adults cause shothole damage on the seedling but this is normally of little importance. Females lay eggs in the soil in October–November; these hatch at temperatures above 5°C. The larvae burrow into the stems of the plants via the petioles, mainly between October and January, and accumulate there as hatching continues. When feeding is completed, they emerge to pupate in the soil in April and May giving rise to adults in May and June.

Identification and high-risk situations

Attacked plants collapse as their stems are hollowed. Small, white grubs with dark heads and six small legs may be seen on opening affected stems. Distinguish from cabbage leaf miner larvae, found widely in petioles, which is a typical fly maggot with no legs and no obvious head.

The presence of large numbers of the small, 4 mm, dark metallic blue beetles with thickened hind legs in the current year's crop and/or shothole damage in autumn seedlings indicate that damage is likely to occur during the winter until early spring. Crops growing next to previously infected crops are especially at risk.

Cultural and organic control

Isolate new crops from previous crops as far as possible.

Chemical control

Only when, in autumn or early winter, an average of more than five larvae per plant are found in the petioles on a random sample of at least 20 plants. Later treatment of larger plants is not likely to give an economic response. In general, the last time for treatment is in November or early December, or following official warnings. Treatments include granules of carbofuran, phorate or sprays of alpha-cypermethrin, bifenthrin, lambda-cyhalothrin, cyfluthrin, cypermethrin, deltamethrin, fenvalerate, gamma HCH or pirimiphos-methyl. Synthetic pyrethroid sprays may also be applied earlier if large numbers of adults are seen; the crop should still be monitored later for larval damage.

Rape winter stem weevil (*Ceutorhynchus picitarsis*)

A pest of winter oilseed rape; a rare species until 1981 since when localized outbreaks leading to economic damage have occurred in eastern England south of the Humber. Adults assemble in oilseed rape fields late September–October. Eggs laid, usually in batches of up to 12 in punctures and small cracks at base of leaf stalks. These hatch during the winter and the larvae burrow into the plant and move towards the crown, which may be hollowed out; and leave to pupate in the soil in March. Young plants may be killed outright or die later in the winter; surviving plants develop secondary shoots and are stunted with a rosette-like habit of growth leading to uneven ripening and reduced yield.

Identification and high-risk situations

Adults 3–4 mm long, black with minute yellow scales. Typical weevil snout, long and curved. Larvae creamy white, up to 6 mm long, distinct brown head, no legs, body curved and wrinkled.

No obvious pattern as yet; attacks likely to become more frequent in intensive rape-growing areas.

Cultural and organic control

Isolate new crops from previous crops as far as possible.

Chemical control

No thresholds established. Control as for cabbage stem flea beetle with bifenthrin, carbofuran, cyfluthrin, cypermethrin, deltamethrin or phorate.

Pigeon (*Columba palumbus*)

Winter crops are attractive to flocks of pigeons which can cause extensive damage as a result of their grazing activities, especially on late-sown crops of small plants with a high proportion of bare soil. Control by sowing early and encouraging rapid germination and growth. The deterrent effect of scaring devices is usually short lived and their use is best restricted to the period mid-January to mid-March when most of the serious damage is caused. A combination of several devices, moved around if necessary, is likely to give the best results. These include ground-based mechanical scarecrows and the inflatable man, and aerial devices such as hawk-shaped silhouettes suspended from poles, and kites and balloons. The latter must not be used within 5 km of an airfield and must not be flown more than 60 m above ground level.

Cabbage stem weevil (*Ceutorhynchus quadridens*)

Affects spring crops. Adults migrate from fields of previous year's crops, feed on seedling leaves and lay eggs. Larvae hatch and tunnel in petioles and stems in May and June causing parts of the plants above where they are feeding to wilt and collapse. Larvae white with darker heads and legless. Control is not normally directed specifically against this pest; the sprays applied to control pollen beetles and seed weevils in spring will control it.

Cabbage aphid (*Brevicoryne brassicae*)

Occasionally a problem, especially on spring crops in hot, dry seasons when dense colonies of the mealy grey aphids form on the flowering stems. Specific control is necessary only when an appreciable proportion of the stems is colonized, care must be taken to avoid killing bees if insecticides are applied. Pirimicarb is the least harmful to bees and natural enemies of aphids.

Pollen beetles (*Meligethes* spp)

Also known as blossom beetles. A problem mainly on spring crops but may affect backward winter crops. The adult beetles hibernate in sheltered situations such as hedges and copses and fly to rape crops, assembling in increasing numbers during hot, sunny periods in April and May. Eggs are laid in the early green flower buds, and the larvae, which hatch after a few days, feed on and destroy the reproductive parts of the flowers. Feeding by adults and larvae continues as the flowers open. Large numbers of beetles cause a significant reduction in the number of pods formed.

Identification and high-risk situations
Look for small, 1.5 mm long, oval, shiny black beetles with conspicuous knobs on the ends of their antennae. Fly readily in summer weather when disturbed.

Common in most districts especially where there is an abundance of winter shelter. Most damage occurs when peak numbers coincide with the green bud stage of development.

Cultural and organic control
None.

Chemical control
Because of the damage to bees and other beneficial insects, and the potential waste of money, it is essential that sprays are applied only before the late yellow bud stage and only when pest population levels exceed the threshold value. It is well worth while to make careful counts at the appropriate stage of crop growth. This is done by selecting 20 plants at random throughout the field, avoiding headlands, on a sunny day when beetles are on upper parts of plants; move carefully to avoid disturbance. Carefully bend and shake the flower heads over a tray or stiff card approximately 30 × 25 cm and count the beetles dislodged.

(1) On spring rape this must be done at the early green bud stage and treatments given if the *average* population exceeds three per plant. Counting and treatment may have to be repeated in the event of re-infestation but only up to the yellow bud stage.
(2) On winter rape, count during green to yellow bud stage and treat when average per plant reaches 15–20.

Sprays include alpha-cypermethrin, deltamethrin, endosulfan, esfenvalerate, fenvalerate, gamma HCH* and malathion.

---

* Do not plant potatoes or carrots until at least 18 months after application.

Cabbage seed weevil (*Ceutorhynchus assimilis*)

A major pest in both winter and spring crops. The adult weevils overwinter in hedgerows and other sheltered situations and migrate in spring to rape crops. They fly actively in hot, calm weather and rapidly colonize the fields. Some time is spent feeding on the leaves and other parts of the plants but this causes little or no loss. Eggs are laid only in the soft young green pods shortly after flowering. Normally only one egg is laid per pod and the damage is caused by the larvae feeding on and destroying the young seeds. The larvae emerge when fully fed and pupate in the soil, giving rise to a second generation of adults which move to other brassica crops in the area in July–August.

This pest is often associated with the brassica pod midge which exploits the holes made in the pods by the weevil as egg-laying sites.

Identification and high-risk situations
Small, 2–3 mm long, grey beetles with long thin snouts are seen on the plants, flying actively in sunny weather, often in association with pollen beetles. White, legless grubs with brown heads in the seed pods, normally only one per pod.

Highest populations develop in areas of intensive oilseed rape production where the crop has been grown for many years. High levels of parasitism occur, especially in unsprayed crops, and many larvae do not complete development.

Cultural and organic control
None.

Chemical control
Because of the danger to bees and other beneficial insects, sprays must never be applied when crop is in flower.

(1) On spring crop maximum populations usually assemble on the plants before flowering and sprays such as alpha-cypermethrin, fenvalerate, gamma HCH, phosalone and triazophos should be applied at the late yellow bud stage when an average of one or more adult weevils is found per plant. Use the same counting technique as for pollen beetles.
(2) On winter crops, which are usually more advanced, peak populations often coincide with flowering. In this case it is necessary to make the counts when the crop is in flower and then wait until after petal fall, when the crop presents an overall green appearance, before spraying if the threshold was exceeded at any time during flowering. Phosalone should be used if late flowers are persisting in the crop. Local beekeepers should be warned at least 48 h before spraying and the application should be made preferably in the evening after bees have ceased flying actively.

Brassica pod midge (*Dasyneura brassicae*)

Sometimes known as the bladder pod midge because of the swellings characteristic of pods which may contain up to 40 minute white maggots feeding on and weakening the inner walls of the pods. Similar, but pink, maggots are found sometimes but these are fungus feeders and can be disregarded. Affected pods ripen and shed seed prematurely. The midge is associated with the seed weevil because the adults, which are fragile and weak, make use of the weevil's oviposition punctures and exit holes to penetrate the pod for egg laying. Control of seed weevil usually protects the crop against the midge but when exceptionally large populations occur a specific treatment using alpha-cypermethrin, endosulfan, fenvalerate, phosalone or triazophos, on winter rape only, may be required. In many instances, because of the weak flight of the midge, only the headlands need treatment.

On spring rape the midge is controlled by sprays directed against seed weevils pre-flowering.

Kale pests

The main pests of kale, grown as animal fodder, are those which affect the establishment of the seedlings.

Brassica flea beetles (*Phyllotreta* spp)

Life cycle and injurious effects similar to beet flea beetles.

Identification and high-risk situations
'Shothole' damage to cotyledons and stems, leading to complete loss in severe attacks. Small, 2–3 mm, beetles, some with a yellow stripe down each wing case, jumping actively when disturbed. Mainly a problem in hot, dry conditions in April and May on spring sowings when the seedlings are in the cotyledon stage and growth is slow. Plants are much more tolerant of injury after the first true leaves develop.

Cultural control
Sow early and encourage rapid early growth by providing a fine, well fertilized seedbed.

Chemical control
Dress seed with gamma HCH, usually combined with a fungicide. This may be replaced by granules of carbofuran or a spray of deltamethrin in the face of sustained attacks. Potatoes and carrots should not be planted within 18 months of the treatments using gamma HCH.

Cabbage stem flea beetle (*Psylliodes chrysocephala*)

See under Oilseed rape pests. Mainly a problem on late sown overwintering kale crops. Early sown, well grown crops are more tolerant of injury. Chemical control as on oilseed rape.

Pea and bean pests

Apart from the general pests which disrupt the normal growth of the crops, growers of peas in particular have to contend with a number of pests which feed within the pods and affect the appearance of the peas. A relatively small percentage of damaged peas can reduce the market value of the crop and may even lead to rejection by the processor.

Pea and bean weevil (*Sitona lineatus*)

Hibernating adults become active from mid-April onwards and move into pea and bean fields, where they feed on the leaves. In May and June they lay eggs in the soil and the larvae burrow and feed on the root nodules.

Identification and high-risk situations
Characteristic U-shaped notches in the leaf edges or on the folds of unopened leaves.

Leaf injury is only likely to be important when the leading shoots of young plants are heavily attacked at a time when plant growth is slow. Extensive leaf-notching, however, indicates that more serious nodule damage is likely to occur later.

Cultural and organic control
Encourage rapid early growth.

Chemical control

(1) *On peas* Foliar sprays containing cyfluthrin, cypermethrin, deltamethrin, esfenvalerate, fenitrothion, fenvalerate, lambda-cyhalothrin or triazophos, when significant leaf damage is seen are more likely to lead to an economic yield response. Oxamyl or phorate granules applied pre-drilling or at sowing against other pests will give some control.

(2) *On beans* Spring-sown crops are most at risk and treatment with foliar sprays containing cyfluthrin, cypermethrin, deltamethrin, esfenvalerate, lambda-cyhalothrin or triazophos when significant leaf notching occurs are more likely to be economic. Phorate granular treatments as with peas will give some control.

Pea cyst nematode (*Heterodera goettingiana*)

Life cycle similar to cyst nematodes of cereals, potatoes and sugar beet but with peas, beans and vetches as hosts. Peas are seriously damaged but beans, although efficient hosts, are little affected by larval invasion. Important in parts of eastern England where peas are grown intensively; occurs locally in other regions.

Identification and high-risk situations
Patches of stunted, prematurely senile pea plants are normally the first indication of trouble. Lemon-shaped cysts can be seen on the short, fibrous roots which bear few root nodules.

Cyst numbers build up rapidly in fields cropped too frequently with peas and beans.

Cultural and organic control
Lengthen the interval between host crops. Cysts are persistent, and expert advice from ADAS/PGRO on the length of the interval should be sought.

Chemical control
Reduce damage by mixing oxamyl granules in the seedbed before drilling. This must be considered to be a supplementary treatment and is not an acceptable alternative to lengthening the rotation.

Pea aphid (*Acyrthosiphum pisum*)

Winged females migrate from a wide range of leguminous plants into pea crops during the summer months. These found colonies of wingless aphids and further spread occurs within the crop. Seldom a problem on beans.

Identification and high-risk situations
Colonies of large, long-legged, green aphids on the growing points and spreading to the pods when numerous. Growing points yellow, leaves and pods distorted. Plants may be covered with sticky honeydew.

Colonies build up rapidly in hot, settled weather reaching a maximum in June–July.

Cultural and organic control
Early sowing and early vigorous growth will reduce severity of attack.

Chemical control
Foliar sprays applied at high volume to assist penetration into the crop on peas growth for human consumption when colonies are easily seen and numbers are increasing. On peas for animal consumption, spray when 50% of plants are infected between flowering and four trusses set. The choice of aphicide may be limited if the crop is to be harvested shortly after spraying. Active ingredients include: bifenthrin, cyfluthrin, fatty acids, cypermethrin*, deltamethrin*, demeton-S-methyl, dimethoate, fenitrothion, fenvalerate*, heptenophos*, malathion, pirimicarb*, thiometon and triazophos.

Granular treatments with disulfoton and phosate at sowing time may also be beneficial on late sown crops.

Pea moth (*Cydia nigricana*)

The moths emerge from the previous year's pea fields and congregate in the current year's crop, the flight period lasting from early June to mid-August with a peak round about mid-July. Eggs are laid on the leaves and the minute caterpillars crawl to and penetrate the soft young pods and feed inside for three weeks. They emerge after this and overwinter in the soil.

Identification and high-risk situations
Peas burrowed into and fouled by the excrement of the small, pale caterpillars with black heads.

Common in all pea-growing areas and particularly prevalent in East Anglia and Kent. Especially important on peas for processing because even a light infestation may result in rejection. All peas which flower in the main flight period are at risk.

Cultural and organic control
Early picking peas which flower before the main flight period escape serious attack. Early sowing recommended.

Chemical control
Commercially available pheromone traps set up in individual fields should be used in all fresh pea and vining crops to indicate the arrival of the adult moths. The crop must be sprayed if any moths are caught up to full flowering; treatment may have to be repeated after 10 d. In peas to be combined for human consumption, pheromone traps may also be used to determine the need to spray and the timing of the treatment. Peas grown for animal feed need no treatment. The following insecticides can be used: bifenthrin, cyfluthrin, cypermethrin, deltamethrin, esfenvalerate, fenitrothion, fenvalerate, lambda-cyhalothrin and triazophos.

* Short minimum harvest interval: see manufacturers' labels.

Pea midge (*Contarinia pisi*)

Mainly a pest of vining peas. Midges migrate from previous year's pea fields and lay eggs near the developing flower heads; the small pale, jumping maggots feed in the developing buds, on the leaves and sometimes inside the pods. Attacks localized and sporadic.

Identification and high-risk situations
Flower heads sterile; terminal parts of flowering stems shortened and leaves distorted to form 'nettle-heads'.
Rain stimulates emergence from previous year's pea fields. Small, fragile, yellowish flies concealed within the terminal clusters may be found by careful examination when the midges assemble about a week before flowering.

Cultural and organic control
Isolate new crops from previous crops as far as possible.

Chemical control
In vining peas in areas where there is a regular occurrence of damage, crops should be sprayed when the oldest green buds are about 6 mm long, about 7 d before flowering. In slow growing crops a second application may be required 6–10 d later. In other crops the leading shoots should be examined as they become susceptible, and sprayed if adult midges are found in about 15% of the plants. ADAS or PGRO may issue spray warnings. Suitable insecticides are: demeton-S-methyl, dimethoate, disulfoton, fenitrothion and triazophos.

Black bean aphid (*Aphis fabae*)

See also under Sugar beet pests. Mainly a problem on spring-sown field beans.

Identification and high-risk situations
Dense colonies of black aphids on stems and growing points. Pods distorted, poorly filled, contaminated by honeydew and moulds. Bean virus diseases transmitted.
Serious infestations more likely on late sown spring crops and in southern England. Colonies build up rapidly in hot, dry weather.

Cultural and organic control
Sow early and encourage rapid early growth.

Chemical control
Beans are bee pollinated and aphicides should be applied just before flowering when forecasts issued by ADAS in conjunction with Imperial College suggest that this is necessary. In areas where the risk is lower, treatment may be restricted to the headlands.
Foliar applications of granules of disulfoton or phorate, or sprays of demeton-S-methyl, dimethoate, fatty acids, heptenophos, malathion, nicotine, pirimicarb, resmethrin or thiometon are recommended. Granular formulations and pirimicarb sprays are the least harmful to bees and should be used if it becomes absolutely essential to treat field beans which are being worked by bees. If this is the case the application should be made in the evening.

Field vegetable pests

Uniformity of appearance and freedom from blemish are the main requirements of vegetables grown for human consumption, and very low pest populations, which have a negligible effect on yield, may therefore be a cause of considerable financial loss. Standards of control must be exceptionally high and often involve routine and intensive use of insecticides. This has created problems of increasing resistance to many of the chemicals which have been successfully used in the past. The choice of insecticide may be further limited by the short interval between final treatment and harvest, and other factors such as soil type.
Anyone growing such crops is recommended to take full advantage of forecasts, spray warning and advice provided by the local ADAS and commercial organizations. The principal pests of the main vegetable crops are:

(1) brassicas – flea beetles, cabbage root fly, cabbage caterpillars (various species), cabbage aphid;
(2) broad and french beans – black bean aphid, bean seed flies;
(3) carrots – carrot fly, willow-carrot aphid;
(4) onions – onion fly, stem and bulb nematode.

Schemes, regulations and acts relating to the use of insecticides

Virtually all the insecticides in use are potentially hazardous, some more than others, to users, consumers, wildlife and beneficial organisms, and there are many schemes and legal requirements governing their purchase, use, application, storage and disposal. The principal schemes and Acts are as follows:

(1) The Food and Environment Protection Act 1985 – and the Control of Pesticides Regulations 1986 (made under the Act).
(2) The Control of Pollution Act 1974.
(3) The Health and Safety at Work Act 1974 – and the Poisonous Substances in Agriculture Regulations 1984 (made under the Act).
(4) The Poisons Act 1972 and the related Poisons Rules.
(5) Wildlife and Countryside Act 1981.

References and further reading

Anon. *British Sugar Beet Review*. Quarterly. SBREC. British Sugar plc, Peterborough.

Biddle, A.J., Knott, G.M. & Gent, G.P. (1988) *Pea and Bean Growing Handbook* Vols 1 and 2, 6th edn. PGRO, Peterborough.

Gair, R., Jenkins, J.E.E. & Lester, E. (1987) *Cereal Pests and Diseases*. Farming Press, Ipswich.

Gratwick, M. (1992) *Crop Pests in the UK*. ADAS/MAFF, Chapman & Hall, London.

Greig-Smith, P., Frampton, G. & Hardy, T. (1992) *Pesticides, Cereal Farming and the Environment – The Boxworth Project*. MAFF, HMSO, London.

Gunn, J.S. (1990) *Crop Protection Handbook – Potatoes*. BCPC, Farnham.

Jaggard, K.W. (1989) *Sugar Beet – A Grower's Guide*, 4th edn. SBREC, Bury St. Edmunds.

Jones, F.G.W. & Jones M.G. (1984) *Pests of Field Crops*, 3rd edn. Edward Arnold, London.

McKinley, R.G. (1992) *Vegetable Crop Pests*. Macmillan, London.

NIAB (1995) *Variety Handbooks and Recommended Variety Leaflets*. NIAB, Cambridge.

Scopes, N. & Ledieu, M. (1983) *Pest and Disease Control Handbook*, 2nd edn. BCPC, Farnham.

Diseases of Crops

Recent changes in farming practice have resulted in changes in the disease problems of crops. The increase and intensification of the cereals and oilseed rape cropping in recent years, coupled with the trend towards increased autumn sowings, has resulted in the widespread incidence of many air-borne foliar diseases. Many seed-borne diseases, however, are now quite rare due to the general use of the various seed treatments available. Similarly, disease problems in the future are only likely to change rather than disappear totally. Set-aside, for example, will create new problems as well as being beneficial in other respects.

The cost-effectiveness of any crop protection treatment is dependent upon many factors which may vary widely with the region, season, the chemical used and the individual crop and variety concerned. In the short term the cost of the treatment has to be set against the increased value of the crop, and there are likely to be large differences between the treatment of high value crops such as field-scale vegetables compared with fodder crops for livestock. In the long term it is likely to be preferable to undertake uneconomic measures with an initial infection of a persistent problem (e.g. long-lived soil-borne disease or pest) to prevent or reduce recurring losses in future crops. Many problems of this nature are best controlled by strict preventative measures in the first instance (e.g. potato wart and sugar beet rhizomania) and are invariably subject to strict legislation as notifiable diseases.

Crop diseases can be divided into the following general groups:

- air-borne,
- soil-borne,
- seed-borne,
- vector-borne.

Many diseases have two or more distinct phases of attack and may belong to more than one of these groups, e.g. canker (*Leptosphaeria maculans*) of brassicas may be seed- and soil-borne as well as having a very distinct and destructive air-borne phase. However, each disease has been allocated to the most appropriate group for control measures.

Air-borne diseases

These quickly establish in the crop and mostly attack the foliage and stems producing an overall blanket-type field infection, i.e. virtually all plants are affected to the same degree. They are all of fungal or bacterial origin and produce spores which are easily dispersed by wind and air currents. Spore production may be vast and rapid during suitable environmental conditions, and long-distance spread can occur in a short period of time. Many are associated with wet weather conditions, particularly when the air temperatures are above 10°C, i.e. April to September. Important diseases included in this group are potato blight (*Phytophthora infestans*), the yellow rusts (*Puccinia striiformis*) of wheat and barley, *Septoria* spp of wheat, *Rhynchosporium* leaf blotch of barley, rye and ryegrasses, crown rust (*Puccinia coronata*) of oats, canker (*Leptosphaeria maculans*) of brassicas and various leaf and pod spots (*Ascochyta* and *Botrytis* spp) of peas and beans. Their activity is often curtailed by hot dry weather. Others tend to be more prevalent during the warm dry spells of summer and early autumn, e.g. powdery mildews (*Erysiphe* spp) and brown rust *Puccinia* spp) of various crops. The powdery mildews, however, are more independent of specific climatic conditions for spread than most other diseases.

Many air-borne diseases are very host specific and produce large numbers of asexual spores giving rise to specific races of the fungus with the ability to attack only a very few species or even several varieties (cultivars) of one crop. This extreme specificity can be utilized to advantage when choosing varieties of certain cereals (see National Institute of Agricultural Botany Diversification Schemes).

Control

Disease epidemics are only likely to result if

- a susceptible host is sown,
- a virulent race of the pathogen occurs, and
- the environment is favourable for disease attack and spread.

If any of these three factors is limiting in any way, crop damage is not likely to be severe. Wherever possible all forms of control measures should be used to give a fully integrated control programme (Fig. 13.1). If, however, a resistant variety is used or climatic conditions have not been favourable for disease spread, chemical control can safely be reduced or omitted on economic grounds.

Cultural and organic control

Crop rotations are not very useful in controlling air-borne diseases as they cannot prevent wind-blown spores coming in from neighbouring infected areas. Similarly, efficient stubble clearing after harvest drastically reduces carryover of spores on to the following crops but at best, like crop rotation, is only likely to delay subsequent re-infection. Both are very much more useful for soil-borne diseases and some types of weed control. Burning of stubble followed by deep ploughing to bury the remainder is preferable to the use of heavy cultivators with their poor burial ability. The date of sowing may affect the

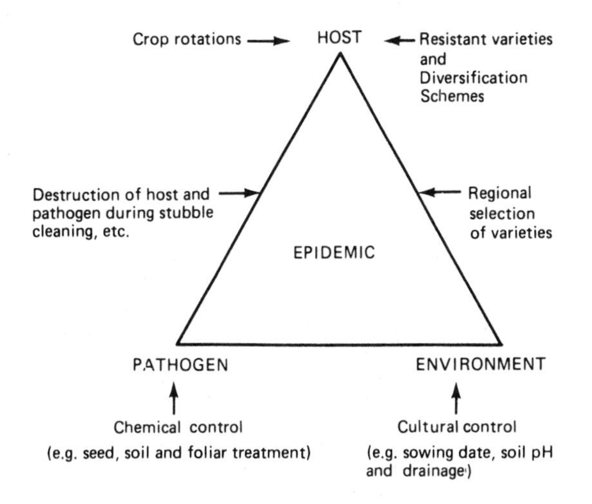

Fig. 13.1 A hypothetical epidemic and its integrated control.

length of the intercrop period considerably and the subsequent disease carryover. Many diseases are very common on late tillers and volunteers, consequently early sown autumn crops are particularly prone to disease carryover from the previous crop. The seedlings of these crops often become greatly infected with various seed-borne diseases from the later germinating shed corn after stubble cleaning (Fig. 13.2).

September sowings of winter cereals are often very high yielding but seldom give as high returns as October-sown crops due to the increased costs incurred for the early autumn disease, pest and weed control required. For these reasons spring-sown crops may be useful as cleaning crops in a sequence of autumn-sown crops, although returns tend to be considerably lower.

Air-borne diseases are best controlled by the use of resistant varieties. Where these do not exist or break down due to new pathogen races, chemical control should be used. All variety choice should be made after consultation with the National Institute of Agricultural Botany (NIAB) recommended leaflets, paying particular attention to those diseases common in the region. Where such schemes exist, use should also be made of the NIAB Variety Diversification Schemes to help reduce the spread and severity of those air-borne diseases which exist in many physiological races.

Farmers should grow, preferably, several varieties chosen from different diversification groups and not just one or two popular varieties that may often belong to the same diversification group and thus the same pattern of susceptibility. Diversification in time is also useful, e.g. where wheat follows wheat it is advisable if the second wheat crop variety is chosen from a different diversification group from the first wheat variety.

Chemical control

Seed treatments

These are useful in controlling many seed- and soil-borne diseases and give good control during the first half of the plant's life. Few are persistent enough to give effective control of most air-borne diseases of later critical stages of growth such as flag-leaf emergence and heading of cereals. Those that are persistent are often more expensive than similar foliar treatments and, in years of relatively low disease levels, are unnecessary. Seed treatments for foliar diseases are best restricted to very susceptible varieties, only as they may also exacerbate fungicide resistance problems.

Foliar treatment

This treatment is considerably more flexible than seed treatment particularly with regard to the number and choice of chemicals available. The correct chemical(s)

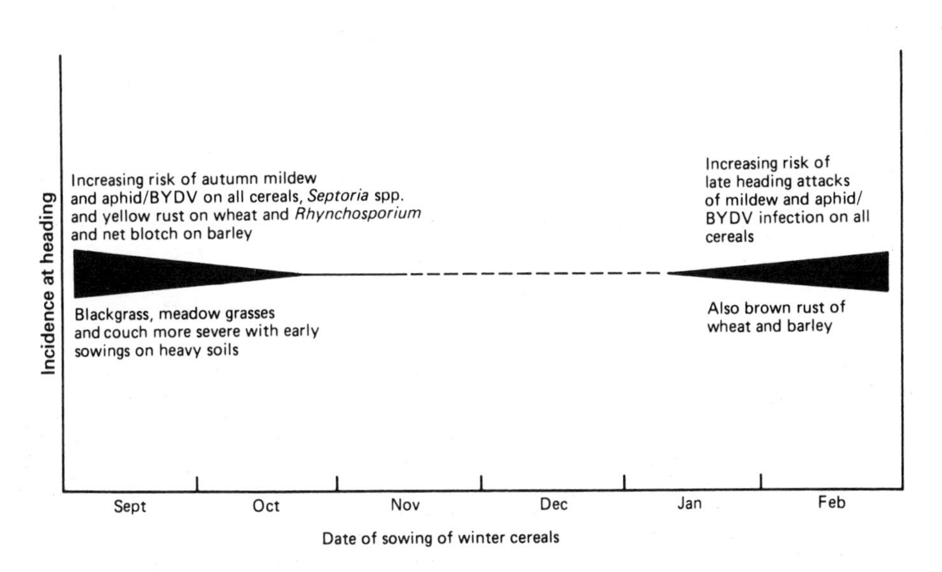

Fig. 13.2 The relationship between the date of sowing of winter cereals and associated disease, pest and weed problems (BYDV, barley yellow dwarf virus).

can be chosen accurately for the specific disease(s) as and when they occur. Protectant fungicides with little or no eradicant action must always be applied before disease build-up. As a general rule even when using systemic fungicides with good eradicant activity the best economic responses are obtained when applications are made at the first sign of disease build-up, particularly if this coincides with weather conditions that favour disease development (Fig. 13.3).

It is normally more cost effective to use a fungicide in high risk situations, for example

(1) susceptible varities,
(2) high disease levels in nearby crops,
(3) suitable weather conditions for disease spread,
(4) predisposing factor(s) operative.

Broad-spectrum fungicides generally give better responses than specific types particularly if the variety is susceptible to more than one disease and/or two or more diseases are likely to attack the crop. With any crops there are often critical growth stages that benefit from good disease protection, e.g. flag-leaf/ear emergence in cereals and rapid bulking up of potatoes in August. Generally crops should not be sprayed with any crop protection chemical if they are suffering from stress. Fungicides used during periods of frost, waterlogging and drought often give rise to quite significant yield reductions.

Fungicide resistance
In recent years, as a result of the overuse of fungicides on certain crops, e.g. eyespot and powdery mildew control on cereals, *Botrytis* grey mould control on many horticultural and agricultural crops, insensitive/resistant strains of the fungal population have been artificially selected out and now dominate the population. This can result in a lack of effective disease control. To minimize the problem fungicides with different modes of action are mixed, alternated and/or rotated to prevent selection and build-up of resistant strains. The ADAS fungicide grouping scheme for cereal disease control contained in Booklet 2257 'Use of Fungicides and Insecticides on Cereals' should be consulted for further information.

Soil-borne diseases

The main symptoms usually occur on the roots and stem bases which often give rise to wilting and stunted plants. Affected plants occur in patches of varying sizes in the

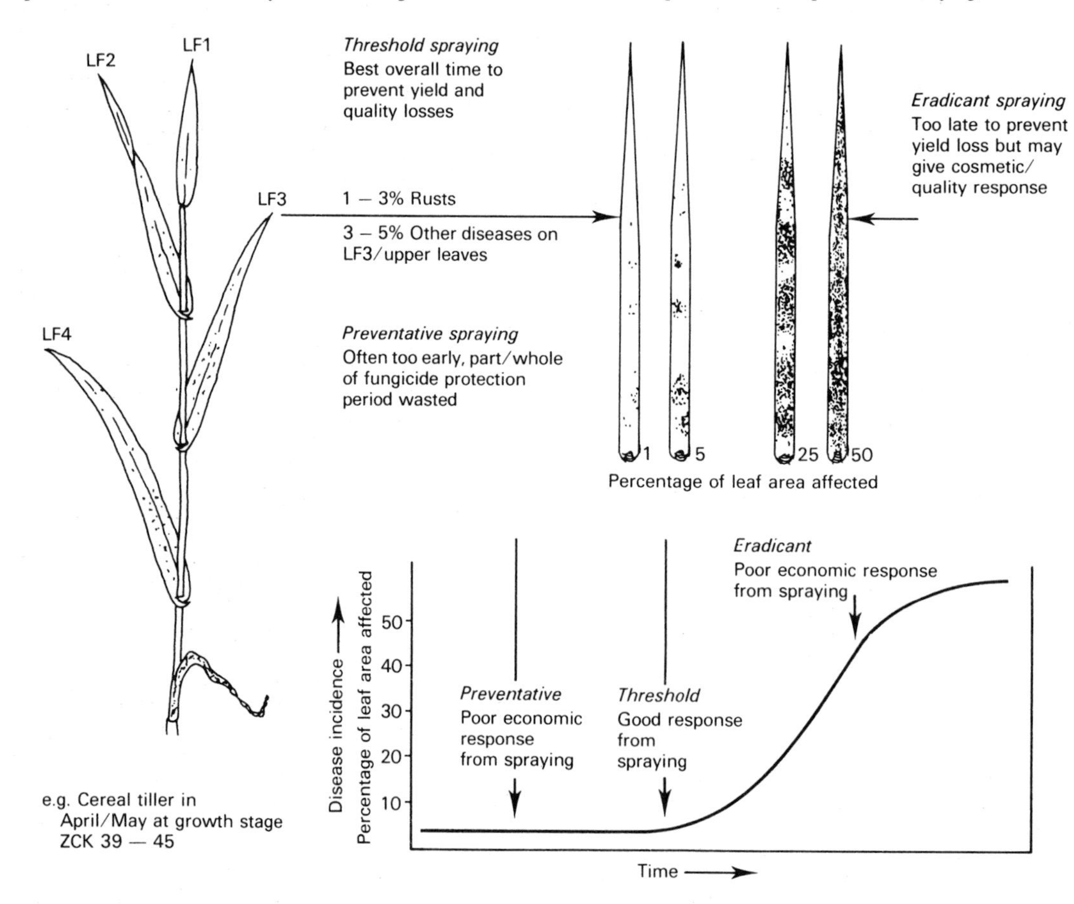

Fig. 13.3 The effect of time of spraying on the likely yield response.

field and may be confused with poor drainage, shallow soil depth and various mineral deficiencies. These diseases generally have very limited powers of mobility and are commonly dispersed in contaminated soil on machinery wheels, animal's feet, clothing and footwear. Severe wind and water erosion of soil may be significant in certain areas also.

The organisms generally exist in a limited number of races and either attack a wide range of crops and are short lived without a host, e.g. 'take-all' (*Gaeumannomyces graminis*) of cereals and grass, or possess a narrower host range coupled with long-term survival in the soil as a resting spore, e.g. potato wart (*Synchitrium endobioticum*), onion white rot (*Sclerotium cepivorum*) and pea wilt (*Fusarium oxysporum*). Club root (*Plasmodiophora brassicae*) of brassicas is a particularly difficult disease as it has a wide host range including most agricultural/horticultural brassicas and many cruciferous weeds such as charlock (*Sinapis arvensis*), shepherd's purse (*Capsella bursa-pastoris*) and field pennycress (*Thlaspi arvense*). Its soil-borne resting spore is also capable of surviving for at least eight years between susceptible hosts. Once established, eradication of these diseases is extremely difficult and often quite costly, therefore the main form of control measure should be preventative using a sensible cropping sequence. Overcropping should be avoided and extreme care should be taken with contaminated fields to prevent further spread. Sugar beet rhizomania is similar but more persistent, and thus even more difficult to control.

Often the most severe effects of the disease can be alleviated by various cultural measures such as generous fertilizer application, drainage and liming where appropriate. Some specific crops can be sown on infected soils provided resistant varieties are used. As a result of legislation (The Wart Disease of Potatoes (Great Britain) Order 1973) requiring the use of immune varieties to the UK races of potato wart since the 1920s this particular disease has been virtually eliminated. Similar MAFF legislation is being used to contain and hopefully eradicate the recent outbreaks of sugar beet rhizomania in East Anglia since their discovery in 1987. Chemical control is considerably more difficult, expensive and generally less effective than with air-borne diseases. Often it is only worthwhile with high value arable crops.

Seed-borne and inflorescence diseases

Symptoms are most likely to be seen in the seedling and young plant stages followed later by further major attack at the end of the crop's life on the flowers, seed pods and seeds in general. They may often cause a systemic infection of the plant causing no visible external symptoms during the vegetative phase and then appear quite dramatically on the ears, e.g. loose and covered smuts (*Ustilago* spp) of cereals, bunt (*Tilletia caries*) of wheat, and maize smut (*Ustilago maydis*). Others cause severe attack on the seedlings and foliage during the early stages of growth and relatively little damage until the heading stage, e.g. various seedling blights of cereals caused by *Fusarium*, *Pyrenophora* and *Septoria* spp. This latter group is often associated with untreated seed such as shed corn as a result of poor stubble cleaning. The diseases develop on the volunteer plants, then infect the newly sown autumn crops. This is particularly common with early sown winter cereals.

Seed-borne diseases are potentially very serious indeed and can cause considerable yield reduction. In the UK most of them are controlled by the general use of healthy seed produced in various Seed Certification Schemes and such fungicidal seed dressings as thiabendazole, carboxin, iprodione and thiram. Seed treatments are an inexpensive form of disease control and should always be used where the health of a seed sample is in doubt. Bunt of wheat and the covered smuts of barley and oats are now exceedingly rare as a result of the standard organo-mercury seed dressings used in the past. The recent trend towards the use of untreated home-sown seed to reduce costs in many combinable crops has resulted in a resurgence of some seed-borne diseases, e.g. bunt of wheat.

Vector-borne diseases

The important diseases in this group are all viruses and the main vectors are either aphids above ground or nematodes and fungi below ground. Aphid-borne diseases may show a superficially similar blanket pattern of distribution to air-borne diseases in the field. However, not all plants are infected and the appearance and severity of symptoms is more variable. Crop attack and appearance of symptoms are directly dependent upon aphid activity initially and therefore seldom occur much before May and continue until October. Yield reductions are related to the age of the plant at the time of infection. Seedling infection, as often occurs with early sown autumn and late sown spring crops, can cause severe reductions in yield in individual plants whereas late attacks on mature plants may cause negligible loss. The yield loss is therefore directly related to the numbers of individual plants affected and their age at infection. Most true seed is naturally virus-free at planting and will give rise to healthy seed despite being infected during vegetative growth. However, where vegetative 'seed' is used, e.g. potato seed tubers, once infected these will give rise to infected seed. Consequently to maintain virus-free stock for commercial growers, potato seed tubers are produced in areas such as Scotland and Northern Ireland with low aphid populations and sold subject to statutory restrictions (Prevention of Spread of Pests (Seed Potatoes) (Great Britain) Order 1974 and Seed Potatoes Regulations 1978) on purity and health. Similar schemes also exist in horticulture for production and sale of virus-free stocks of strawberries, raspberries and top fruit that are typically reproduced by vegetative means.

Nematode- and fungal-borne viruses occur in patches in the field similar to most other soil-borne pathogens. Attack by free-living types of nematodes and subsequent virus infection is much more common on lighter sandy soils where nematode populations are naturally higher. Many crops and weeds may be attacked by these nematodes but potatoes, sugar beet and raspberries are at greatest risk. The fungal vectors of sugar beet rhizomania and potato powdery scab are most successful in transmission of their respective viruses when both soil water tables and temperatures are high, particularly during periods of summer irrigation. Further spread of the diseases often occurs in drainage water as well as in contaminated soil on beet roots, potato tubers, other root crops and machinery, boots, livestock feet, etc.

Once infected with viruses, plant yields can be severely reduced and chemical control is not possible. Control should be aimed at preventing or delaying infection for as long as possible by the use of healthy seed, isolation from infected sources of material and then chemical control against the vectors concerned. The control of aphid vectors tends to be generally less expensive than for nematodes but prevention of virus infection is variable and dependent upon virus type.

Lastly it should not be forgotten that plants are often attacked by several organisms simultaneously and virus infection may occur alongside or be confused with fungal diseases and in particular mineral deficiencies.

Diseases of wheat

Leaf and stem diseases

Powdery mildew (*Erysiphe graminis*)

Superficial grey-white fungal pustules on leaves, stems and ears, particularly at heading. Pustules darken with age. Attacks wheat, barley, oats, rye and grasses but cross-infection unlikely. Common in May/June (ears) and Oct/Nov (seedlings) and favoured by warm dry conditions. Air-borne disease surviving on stubble, late tillers, volunteers and early sown winter wheat.

Cultural and organic control

(1) Use NIAB Cereal Variety Handbook for variety choice and Diversification Schemes.
(2) Destroy stubble and preferably plough soon after harvest.
(3) Avoid early autumn and late spring grown crops.
(4) Avoid close proximity of winter and spring crops and wheat after wheat.

Chemical control

Foliar treatment Use of one of the following fungicides alone or in mixtures at first sign of disease threshold in spring and up to full ear emergence (ZCK 59) especially on NIAB rated susceptible varieties.

cyproconazole
fenpropidin
fenpropimorph
flusilazole
flutriafol
mancozeb
nuarimol
prochloraz
propiconazole
sulphur
tebuconazole
triadimefon
triadimenol
tridemorph

Seldom economic to apply fungicides in autumn or more than one well timed application at start of epidemic in spring/summer months especially on NIAB rated resistant varieties. The numbers of applications best minimized to one or two per season to prevent fungicide resistance.

Yellow rust (*Puccinia striiformis*)

Orange-yellow pustules occurring in stripes on mature leaves or groups on leaves of young plants. Air-borne diseases favoured by cool moist conditions in May/June and wet summers in general. Survives on volunteers, late tillers and early sown winter wheat. Occurs also on barley and rye but cross-infection from barley is unlikely. Very common in Eastern England.

Cultural and organic control

(1) Use NIAB Cereal Variety Handbook for variety choice and Diversification Schemes.
(2) Destroy stubble and preferably plough soon after harvest.
(3) Avoid early sowing of winter wheat.
(4) Avoid close proximity of winter and spring crops and wheat after wheat.

Chemical control

Foliar treatment Use of one of the following fungicides alone or in mixtures at first sign of disease threshold in spring and up to full ear emergence (ZCK 59) especially on NIAB rated susceptible varities

cyproconazole
fenpropidin
fenopropimorph
flusilazole
flutriafol
mancozeb
nuarimol
oxycarboxin
prochloraz
propiconazole
sulphur
triadimefon
triadimenol
tridemorph
tebuconazole

Brown rust (*Puccinia recondita*)
Orange-brown pustules randomly scattered or grouped in patches on leaves. Air-borne disease favoured by hot dry conditions in June and July. Seldom severe until after ear emergence, therefore less common or as damaging to yield as yellow rust. Survives on stubble, late tillers, volunteers and early sown winter wheat in mild winters. Also occurs on rye and barley but a different species is involved in barley and therefore cross-infection is not possible. Generally of infrequent occurrence.

Cultural and organic control
(1) Use NIAB Cereal Variety Handbook for variety choice.
(2) Destroy stubble and preferably plough soon after harvest.
(3) Avoid late sown and late maturing crops.
(4) Avoid close proximity of winter and spring crops and wheat and spring crops and wheat after wheat.

Chemical control

Foliar treatment Use of one of the following fungicides alone or in mixtures at first sign(s) of disease threshold in spring and up to full ear emergence (ZCK 59) especially on NIAB rated susceptible varieties

cyproconazole	propiconazole
fenpropimorph	tebuconazole
flusilazole	triadimefon
flutriafol	triadimenol
mancozeb	tridemorph

Septoria diseases (*Leptosphaeria nodorum* syn. *Septoria nodorum*) and leaf spot (*Mycosphaerella graminicola* syn. *Septoria tritici*)
Septoria spp cause brown, often irregularly shaped lesions on leaves and purple-brown glume blotch phase on glumes at heading. Difficult to diagnose unless leaves and head are still green. Associated with high rainfall areas, wet summers and high humidity at heading. Rain splash/air-borne disease surviving on infected stubble and seed.

Cultural and organic control
(1) Use NIAB Cereal Variety Handbook for variety choice.
(2) Destroy stubble and preferably plough soon after harvest.
(3) Avoid early sown winter wheat and wheat after wheat.

Chemical control

Seed treatments Those containing the following will give partial control of *Septoria* spp.
fuberidazole
guazatine

Foliar treatment Most cost effective when applied to protect flat leaf and ear (ZCK 39–59) at first sign of disease threshold especially on NIAB rated susceptible varieties.

carbendazim (± maneb. manocozeb)

chlorothalonil	iprodione
cyproconazole	nuarimol
fenpropidin	prochloraz
flusilazole	propiconazole
flutriafol	tebuconazole
	triadimenol

The number of applications is best minimized to one or two per season to prevent fungicide resistance.

Barley yellow dwarf virus (*vectors – cereal aphids*)
Seedling infection can result in stunted plants of wheat, barley, oats, rye and grasses. Infection in early summer results in purple-red leaves in wheat, bright canary yellow in barley and pink leaves in oats. Spread by two main cereal aphids, bird-cherry (*Rhopalosiphum padi*) and grain aphid (*Macrosiphum avenae*), during warm dry weather in autumn and May/June. Most common in early sown winter and late sown spring crops. Survives in late tillers, volunteers and grasses.

Cultural and organic control
(1) Winter wheat most tolerant, spring barley most susceptible to yield reductions.
(2) Destroy stubble and preferably cultivate soon after harvest.
(3) Avoid early sown winter and late sown spring crops.
(4) Avoid close proximity of cereals and grasses.
(5) Direct drilled crops suffer least and ploughed the most infection.

Chemical control

Foliar treatment

(1) Delay infection to as late in the plant's life as possible by avoiding high aphid populations.
(2) September and May sown cereal crops are high risk for aphid/BYDV and mildew infection. Dual low cost treatment of both may be worthwhile.

Root and stem base diseases

Eyespot (*Pseudocercosporella herpotrichoides*)
Dull brown indefinite eyespot at base of stem eventually causing lodging. Attacks wheat, barley, oats and rye. Common in intensive cereals on heavy damp soils. Soil- and stubble-borne surviving two to four years between susceptible crops. Favoured by long wet cold periods in winter and spring.

Cultural and organic control
(1) Use NIAB Cereal Variety Handbook for variety choice.
(2) Deeply plough stubble soon after harvest.
(3) Avoid three or more years of continuous wheat and/or barley especially on heavy land.
(4) Use three-year break from wheat/barley for disease reduction.
(5) Avoid early sown lush crops.

Chemical control

Foliar treatment Use of one of the following fungicides alone or in mixtures at late tillering stage (ZCK 30–31).
benomyl
carbendazim
cyproconazole (especially if MBC fungicide resistance suspected)
flusilazole
prochloraz (especially if MBC fungicide resistance suspected)
propiconazole

One application only per season required.

Take-all (*Gaeumannomyces graminis*)

Dead, stunted and thin patches of varying sizes in fields of wheat and barley. Black fungus kills roots causing empty bleached 'wheat heads' and premature ripening. 'White heads' often attacked by sooty moulds at harvest. Oats and rye are usually resistant to wheat/barley race of fungus. Disease is favoured by light alkaline soils and above average rainfall in winter and spring. Survives on couch (*Elymus repens*) and stubble for two years. Common in intensive cereals. More severe on early sown crops.

Cultural and organic control
(1) Use two-year break from wheat and/or barley for eradication.
(2) Prevent build-up by avoiding more than three years of continuous wheat or barley.
(3) Avoid wheat after barley.
(4) Direct drilled crops often less affected than traditional sown crops.
(5) Oats, maize and grasses are effectively resistant in most commercial situations.
(6) No resistant varieties of wheat or barley.

Chemical control
(1) No economical chemical control.
(2) 40 kg/ha N extra top dressing at first sign of disease in spring may alleviate most damage.

Sharp eyespot (*Pellicularia filamentosa* syn. *Rhizoctonia cerealis*)

Attacks stems and stem bases causing numerous and clearly defined 'eyespot' lesions. Soil-borne disease with wide host range (including barley, oats and rye), and therefore not easily controlled by rotation. Most common on light sharp soils of neutral/acid nature during cold dry conditions. More prevalent in rotation with grass leys, peas and root crops. Less prevalent in intensive cereals.

Cultural and organic control
(1) No resistant varieties, oats and rye most susceptible, barley least so.
(2) Avoid late sown winter and early sown spring crops.

Chemical control
No economic chemical control but prochloraz ± carbendazim has given some disease suppression.

Brown foot rot and ear blight (*Fusarium* spp)

Ill-defined brown rotting of stem base, roots and seedlings especially on cold, heavy, poorly drained acid soils in wet autumns and winters. Also occurs on other cereals.

Cultural and organic control
(1) No varietal resistance known at present.
(2) Avoid late sown winter and early sown spring crops in cold soils.

Chemical control

Seed treatment Use of a seed dressing containing one of the following will give partial control.
fenpiclonil
fuberidazole
guazatine
imazalil
thiabendazole

Foliar treatment Use of one of the following alone or in mixtures may give good disease suppression when applied at late tillering (ZCK 30–31).
benomyl
carbendazim
prochloraz
tebuconazole

Ear diseases

Loose smut (*Ustilago nuda*)

Ears only visibly affected. All grains destroyed and replaced by black spores. Conspicuous in early June, at the start of heading. Not serious in UK. Seed-borne disease.

Cultural and organic control
(1) Use NIAB Cereal Variety Handbook for variety choice.
(2) Use clean certified seed.

Chemical control

Seed treatment Use of seed dressing containing one of the following may be necessary for many seed certification schemes.

carboxin	thiabendazole
flutriafol	triadimenol + fuberidazole

Bunt (*Tilletia caries*)
Ears small and stunted. Internal contents of grain replaced by black fishy-smelling spores but seed coat intact initially. Now very rare in UK. Seed-borne disease mainly, but can be soil-borne on rare occasions.

Cultural and organic control
(1) Use clean certified seed.
(2) Avoid untreated home-saved seed.

Chemical control

Seed treatment Use of a seed dressing containing one of the following

carboxin	guazatine
fenpiclonil	triadimenol + fuberidazole
flutriafol	thiabendazole

Black/sooty moulds (*Cladosporium herbarum/Alternaria* spp)
Black sooty appearance on overripe/late harvested crops in wet seasons and/or on prematurely ripened grains as a result of other disease attacks, e.g. take-all, foot rots. Cosmetic damage only.

Cultural and organic control
(1) Avoid late harvesting of crops where possible.

Chemical control

Foliar treatment Chemicals used for *Septoria* glume blotch control at heading will give incidental control of sooty moulds. Otherwise not economic to spray.

Diseases of barley

Leaf and stem diseases

Powdery mildew (*Erysiphe graminis*)
Superficial grey-white fungal pustules on leaves and stems particularly during stem elongation. Pustules darken with age. Attacks wheat, barley, oats, rye and grasses but cross-infection unlikely. Common in May/June (adult plants) and Oct/Nov (seedling and shed corn). Favoured by warm dry conditions. Air-borne disease surviving on stubble, late tillers, volunteers and early sown winter barley. Very common and serious on spring barley.

Cultural and organic control
(1) Use NIAB Cereal Variety Handbook for variety choice and Diversification Schemes.
(2) Destroy stubble and preferably plough soon after harvest.
(3) Avoid early sown winter and late sown spring barley crops.
(4) Avoid close proximity of winter spring crops and barley after barley.

Chemical control

Seed treatment Use of a seed dressing containing one of the following especially on NIAB rated susceptible varieties.

ethirimol	triadimenol
flutriafol	

Foliar treatment Use of one of the following alone or in mixtures at first sign of disease threshold in spring and up to full ear emergence (ZCK 59) on susceptible varieties.

cyproconazole	propiconazole
fenpropidin	sulphur
fenpropimorph	tebuconazole
flusilazole	triadimefon
flutriafol	triadimenol
mancozeb	tridemorph
nuarimol	triforine
prochloraz	

Seldom economic to apply fungicides in autumn or more than one well-timed application at start of epidemic in spring/summer months especially on NIAB rated resistant varieties. The number of applications also best minimized to prevent fungicide resistance.

Yellow rust (*Puccinia striiformis*)
Distinct form from that on wheat and cross-infection unlikely. Very similar in all other respects and full account given under 'Wheat'. Control similar to that on wheat with addition of triadimenol and flutriafol seed dressing mixtures which may control infections on young plants. A later foliar application is also likely to be needed with these particular seed dressings.

Brown rust (*Puccinia hordei*)

Distinct species from that on wheat and cross-infection is not possible. Very much more common than wheat brown rust and likely to cause greater crop damage. Yields and quality are likely to be reduced. Most common in low rainfall areas of Southern and Central England during July at heading time, especially in hot, dry seasons. Control is similar to that of brown rust of wheat with addition of triadimenol and flutriafol seed dressing mixtures which may give control on young plants although these early infections tend to be of irregular occurrence. Seed dressing most useful and economic on late sown spring barley only.

Leaf blotch (*Rhynchosporium secalis*)

The disease causes blotches with purple-brown borders on the leaves and stems of barley, IRG, PRG and rye. Blotches initially pale grey-green, often diamond shaped or at base of leaf blade. Air-borne foliar disease often very severe in high rainfall areas, coastal regions and generally widespread in wet summers. Yield and quality reduction can be very severe when weather is cool and wet in May/June particularly. Survives on stubble, late tillers, volunteers and early sown winter barley crops. Also sometimes seed-borne.

Cultural and organic control

(1) Use NIAB Cereal Variety Handbook for variety choice.
(2) Destroy stubble and preferably plough soon after harvest.
(3) Avoid early sowings of both winter and spring barley.
(4) Avoid close proximity of winter/spring crops and infected rye grasses.
(5) Avoid barley after barley.

Chemical control

Seed treatment Seed dressings may give control of early infections.

flutriafol + ethirimol + thiabendazole triadimenol + fuberidazole

Foliar treatment At first sign of disease threshold in spring and up to full ear emergence (ZCK 59).

benomyl
carbendazim
chlorothalonil
cyproconazole
fenpropidin
fenpropimorph
flusilazole
flutriafol
mancozeb
nuarimol
prochloraz
propiconazole
tebuconazole
triademifon
tridimenol

Barley yellow dwarf virus (*vectors – cereal aphids*)

Extremely damaging on yield where infection occurs in seedling stage on early sown winter or late sown spring barley. Later infection at heading time in May/June has minor yield effect. Barley and oats are much less tolerant of attack than wheat. Control measures are similar for those on wheat. Aphicidal control is likely to be more cost-effective on barley particularly when combined with powdery mildew control which often accompanies BYDV infection. Aphicide control after symptoms appear on plants will only reduce further losses and will not affect initial infection damage. A full account is given under 'Wheat'.

Net blotch (*Pyrenophora teres*)

The disease causes dark brown net-like blotches on seedlings of volunteers in particular. Symptoms spread to young crops in autumn and attack mature plants in June. Favoured by warm wet conditions in Sept/Oct and May/June. Disease may be seed-borne or from crop debris and volunteers left in field.

Cultural and organic control

(1) Avoid untreated seed.
(2) Avoid early sown winter barley.
(3) Avoid winter barley after spring barley.
(4) Destroy stubble and preferably plough soon after harvest.

Chemical control

Seed treatment (partial control only)

carboxin + thiabendazole + imazalil
flutriafol + ethirimol + thiabendazole
guazatine + imazalil
triadimenol + fuberiadazole
(± imazalil)

Foliar treatment Spray to protect flag leaf in May/June with one of the following:

cyproconazole
fenpropidin
fenpropimorph
flusilazole
flutriafol
iprodione
mancozeb
prochloraz
propiconazole
tebuconazole
triforine

Leaf stripe (*Pyrenophora graminea*)

Long brown stripes on leaves usually running entire length. Leaves often split and shred later. Most common on untreated seed in cool wet weather.

Cultural and organic control

(1) Avoid untreated seed.
(2) Avoid early sowings.
(3) Avoid barley after barley.

Chemical control

Seed treatment

fenpiclonil
imazalil

Halo spot (*Pseudoseptoria stomaticola* syn. *Selenophoma donacis*)

Small angular spots with dark margins and pale centres. Most common on top leaves and lawns in May/June during cool moist conditions. Seldom damaging in UK. Seed- and stubble-borne but may also survive on wheat, rye, cocksfoot and timothy.

Cultural and organic control
(1) Avoid untreated seed.
(2) Avoid barley after barley.

Chemical control

Foliar treatment If symptoms are severe, spraying with one of the following may be beneficial:
flutriafol
prochloraz
propiconazole

Root and stem base diseases

Eyespot (*Pseudocercosporella herpotrichoides*)

Similar but generally less important than that on wheat. Cross-infection is likely and taken into account for crop planning. Chemical control measures are unlikely to be as necessary or as economic compared with disease control on wheat. A full account is given under 'Wheat'.

Take-all whiteheads (*Gaeumannomyces graminis*)

Similar but generally less important than that on wheat. Cross-infection occurs but barley is more tolerant of attack and there is less time for the disease to spread in the soil with spring-sown barley. Spring barley usually shows a lower incidence of attack than winter barley due to the longer intercrop period. A full account is given under 'Wheat'.

Sharp eyespot (*Pellicularia filamentosa*) syn. *Rhizoctonia cerealis*

Rarely economic on barley. See under 'Wheat'.

Brown foot rot and ear blight (*Fusarium* spp)

Seldom recorded on barley probably due to the majority of barley being spring sown and on lighter free-draining soils which discourage disease development. See under 'Wheat'.

Ear diseases

Loose smut (*Ustilago nuda*)

This disease is a distinct form from that on wheat and cross-infection does not occur. Otherwise very similar in all other respects and for further information see under 'Wheat' plus the addition of seed treatments containing flutriafol.

Covered smut (*Ustilago hordei*)

Very similar to bunt of wheat: for control see under this disease plus the addition of seed treatments containing flutriafol.

Black/sooty moulds (*Cladosporium herbarum*/*Alternaria* spp)

Identical to those on wheat causing cosmetic damage to diseased and/or late harvested crops. Chemicals used for *Septoria* glume blotch control on wheat are likely to be uneconomic for sooty mould control solely. Many are, however, used for control of various leaf diseases of barley also.

Diseases of oat

Leaf and stem diseases

Powdery mildew (*Erysiphe graminis*)

Grey-white superficial fungal pustules on leaves and stems. Pustules darken with age. Often very severe on late sown spring oats during warm dry weather in May/June and on winter oat seedlings in Oct/Nov. Also occurs on wheat, barley, rye and various grasses but cross-infection unlikely.

Cultural and organic control
(1) Use NIAB Cereal Variety Handbook for variety choice.
(2) Avoid early sown winter and late sown spring oat crops.
(3) Destroy stubble and preferably plough soon after harvest.
(4) Avoid close proximity of winter/spring oat crops.
(5) Avoid oats after oats.

Chemical control

Seed treatment
fuberidazole + triadimenol

Foliar treatment Use of one of the following at first sign of disease threshold in spring and up to full ear emergence (ZCK 59).
fenpropimorph
nuarimol
propiconazole
sulphur
triadimefon
triadimenol
tridemorph

Crown rust (*Puccinia coronata*)

Orange-brown pustules on leaves and sometimes stems. Attack cultivated and wild oat species as well as several important grasses (PRG and IRG). Cross-infection from grasses is unlikely. Air-borne disease favoured by cool wet conditions especially during May/June and Oct/Nov. Survives on stubble, late tillers, volunteers and early sown winter oats. Most severe in high rainfall areas.

Cultural and organic control

(1) Use NIAB Cereal Variety Handbook for variety choice.
(2) Avoid early sown winter and late sown spring oat crops.
(3) Destroy stubble and preferably plough soon after harvest.
(4) Avoid close proximity of winter/spring oats.
(5) Avoid oats after oats.

Chemical control

Seed treatment
triadimenol + fuberidazole

Foliar treatment Use of one of the following at first sign of disease threshold in spring and up to full ear emergence (ZCK 59).
triadimenol + tridemorph

Leaf spot and seedling blight (*Pyrenophora avenae*)

Seedlings discoloured and stunted, adult lower leaves with short purple-brown stripes and upper leaves with spots. Seed- and stubble-borne disease; most damage occurs on seedlings during cold wet weather.

Cultural and organic control

(1) Avoid untreated seed.
(2) Destroy stubble and preferably plough soon after harvest.
(3) Avoid very early autumn sowings of winter oats.
(4) Avoid sowing in cold wet soils.
(5) Avoid oats after oats.

Chemical control

Seed treatment
carboxin + thiabendazole + imazalil
guazatine + imazalil
triadimenol + fuberidazole

Foliar treatment Very seldom necessary to use fungicides on adult infections.

Barley yellow dwarf virus (*or red leaf*)

Similar to virus infection on other cereals. For further details see under 'Wheat'.

Oat mosaic virus (*fungal vector*)

Plants stunted in patches in field. Few tillers formed and leaves have dark green mosaic appearance. Fungus vector on *Polymyxa graminis* present in infected soil causes virus infection. Most common in winter oats during cool wet springs and summers. Infested soils remain so for long periods and thus crop rotations are not very effective. No chemical treatment economic on a farm scale.

Root diseases

Take-all (*Gaeumannomyces graminis*)

Almost immune to 'take-all' that affects wheat and barley and often used as a break crop where appropriate. However, it is susceptible to a distinct oat strain occurring in several western and northern parts of the UK. Control measures for oat strain of take-all are similar to those used for wheat and barley.

Ear diseases

Oat smuts (*Ustilago hordei* and *Ustilago avenae*)

Similar to those on wheat and barley. Rarely economic on the oat crop. For more details see under 'Wheat'. Most seed treatments are effective including guazatine mixtures which are likely to be the cheapest available.

Diseases of rye/triticale

See under 'Wheat diseases' for the following diseases of rye: powdery mildew, yellow and brown rust, eyespot, sharp eyespot, take-all, brown foot rot, bunt and sooty moulds.

See under 'Barley diseases' for leaf blotch of rye and barley.

Diseases of forage maize and sweet corn

Stalk rot (*Gibberella/Fusarium* spp)

Base of stalks rot, plant wilts and often lodge especially in July/Aug. Some strains may also attack wheat. Control includes the use of resistant varieties found in NIAB Forage Maize Variety Handbook and prevention of lodging by avoiding windy exposed sites and late harvesting. Thiram seed treatments may give partial control.

Take-all (*Gaeumannomyces graminis*)

Roots attacked mainly but seldom damaging in UK. May act as an alternative host of the disease for following crops of wheat or barley. No recommendations for control of the disease on maize. For further details see under 'Wheat'.

Maize smut (*Ustilago maydis*)

Typically causes black galls on the cob or elsewhere. Spectacular in appearance but of little economic importance at present in the UK. Soil- as well as seed-borne survival. No chemical control measures can be recommended. Growers should avoid maize growing more than one year in four and avoid known infected sites.

Damping-off (*Pythium* spp)

Poor emergence and slow early growth especially under cold wet conditions. Control measures include the use of treated seed and avoidance of cold early sowing especially in poorly drained and/or low lying fields. Chemical seed treatment with thiram is recommended.

Diseases of potato

Leaf and stem diseases

Potato blight (*Phytophthora infestans*)

Grey-brown leaf and stem lesions spreading rapidly to kill haulm and infect tubers. Very common and serious on main crop potatoes after long periods of wet windy weather in July, August and September. Very common in high rainfall areas and the main potato growing regions of UK. Yield and quality losses with first and second earlies may be slight but the plants may act as a source of infection for nearby maincrops. Infected tubers may cause very high storage losses as a result of secondary bacterial infection. Fungicide resistance problems are common and widespread.

Cultural and organic control

(1) Use NIAB Potato Variety Handbook for variety choice.
(2) Use clean pre-sprouted seed tubers.
(3) Avoid late sowings.
(4) Avoid close proximity of main crops with earlies or previous potato fields.
(5) Maintain stable and well earthed-up ridges to reduce tuber blight.
(6) Destroy groundkeepers and potatoes on dumps and in clamps.
(7) Isolate potato dumps from growing areas.
(8) Defoliate and harvest early.

Chemical control

Foliar treatment Protectant fungicides.

(1) Start spraying maincrops approx. every 10 days just before haulm meets in rows or at first sign of disease if earlier.
(2) Vary spraying interval rate and chemical used depending upon incidence of wet weather and age of crop.
(3) Vary fungicide to reduce incidence of resistance problems.

Young plants

Non-systemic:
- chlorothalonil
- copper oxychloride
- mancozeb
- maneb
- zineb

Systemic:
Max. five applications per season until mid August
- benalaxyl + mancozeb
- cymoxanil + mancozeb
- dimethomorph + mancozeb (max. eight applications per crop)
- fluazinam (max. ten applications per crop)
- metalaxyl + mancozeb
- ofurace + bisdithiocarbamate
- oxadixyl + mancozeb

Mature plants and tuber blight control
- chlorothalonil
- dimethomorph + mancozeb
- fentin acetate + maneb
- fentin hydroxide
- fluazinam

Restrictions on systemic mixtures designed to prevent fungicide resistance. If an acceptable yield has been formed, stop spraying and kill haulm where 5% disease level occurs on varieties susceptible to tuber blight. Haulm destruction may be delayed in those varieties showing good tuber resistance. If the crop is required for long-term storage, do not lift for at least 10 days after the haulm is completely dead.

Potato leaf roll virus (main aphid vector, peach potato aphid *Myzus persicae*)
Symptoms vary with variety of potato, strain of virus and time of infection. Plants stunted, leaflets with margins rolled upwards and inwards often with purple tinges. Foliage is hard and leathery; tubers are smaller, fewer in number and yield is much reduced. Mother tuber seldom rots during season and remains hard till harvest. Infection after flowering produces no visible symptoms till after planting next year. Very common in southern England with home-saved seed.

Cultural and organic control
(1) Use certified virus-free VTSC seed.
(2) Use NIAB Potato Variety Handbook for variety choice.
(3) Isolate new seed from home-saved seed.
(4) Isolate potato dumps and clamps from growing fields.
(5) Destroy groundkeepers and potatoes on dumps.
(6) Isolate growing fields from allotments, market gardens and housing areas.
(7) Rogue virus-infected plants early in season.

Chemical control (aphid vectors)
Prevent aphid infestations at all times in chitting houses, stores, potato fields and dumps using appropriate aphicides. Spread in field can be drastically reduced by efficient use of one of the following:

aldicarb	nicotine
deltamethrin + heptenophos	oxamyl
demeton-S-methyl	phorate
dimethoate	pirimicarb
disulfoton	thiometon
malathion	

Potato virus Y (severe mosaic and leaf drop streak, main aphid vector *Myzus persicae*)
Symptoms vary with variety of potato, strain of virus and the time of infection. Plants stunted, leaves small, crinkled, showing mosaic, mottling and necrosis. Some varieties show leaf drop streak symptoms with certain strains of Y. Mosaic symptoms usually more severe than Virus X.

Cultural and organic control
As for leaf roll virus.

Chemical control
Prevent aphid infestations at all times in chitting houses, stores, potato fields and clamps. Spread in the field is *not* easily reduced by aphicides due to non-persistent nature of the virus in the aphid.

Potato virus X (mild mosaic)
Typically symptomless or a very mild mosaic. No reduction in size of leaflets and yield losses usually slight but may accentuate effects of other viruses. No vector involved, mechanically transmitted by machinery, footwear, clothing and general physical contact of contaminated with healthy plants. Physical method of spread coupled with general lack of symptoms makes roguing difficult and spread can be rapid in susceptible crops. No visible symptoms in tubers. NB Several other viruses of generally minor importance can also cause mild mosaic symptoms. Best controlled by the use of certified virus-free VTSC seed.

Cultural and organic control
(1) Use certified virus free VTSC.
(2) Use NIAB Potato Variety Handbook for variety choice.
(3) Isolate new seed stocks from home-saved seed.
(4) Destroy groundkeepers.

Chemical control
None

Black leg (Erwinia carotovora var. atroseptica)
Soft black bacterial decay on lower parts of stem and heel end of tuber. Plants may wilt and die and stems are easily pulled out. Storage losses greater than in field. More common under wet conditions in field and store. Mainly seed-borne.

Cultural and organic control
(1) Certified virus-free VTSC seed is generally safer than home-saved seed.
(2) Avoid cutting or damaging seed at planting.
(3) Avoid poorly drained fields.
(4) Avoid putting wet tubers in store.
(5) Lift early under clean dry conditions thus reducing tuber damage.

Chemical control
Storage fumigation using a bactericide such as dichlorophen or an organo-iodine complex may reduce infection in seed tubers.

Spraing (tobacco rattle, vectors nematodes)
Leaves show yellow mottle, lines or rings with distorted leaf margins. Locally common on light dry sandy soils with high populations of free living nematodes. Tubers unmarketable due to brown concentric rings or arcs internally. Symptoms not usually visible till tuber cut. Yield reductions negligible compared with loss of tuber quality. Most of the progeny from spraing-infected tubers are virus free.

Cultural and organic control
(1) Use NIAB Potato Variety Handbook for variety choice.
(2) Efficient weed control in preceding crops may be beneficial in lowering nematode populations.
(3) Avoid home-saved seed in virus prone areas.

Chemical control
Use of one of the following nematicides may be useful on known infected sites.
aldicarb
ethroprophos
carbofuran
oxamyl
1–3 dichloropropene
Also phorate when used for wireworm control.

Spraing (potato mop top virus, vector powder scab fungus)
Symptoms and control similar to spraing caused by tobacco rattle virus. The powdery scab fungus vector (*Spongospora subterranea*) is most common in the wetter west and northern parts of the UK. Most mop top spraing is not associated with any particular soil unlike tobacco rattle spraing. Mop top virus is more readily transmitted in infected tubers than tobacco rattle so they should not be used for seed purposes.

Tuber and storage diseases

Gangrene (*Phoma exigua* var. *foveata*)
An increasingly troublesome disease of storage maincrop and seed potatoes especially from northern Scotland and Northern Ireland. Dark coloured round or oval shallow depressions in tubers one to two months after lifting. Soil-borne fungus enters damaged areas at lifting especially under cold late conditions. Often secondary invasion by dry rot and black leg organisms occur also. Large dry hollow cavities usually form in the centre of the tuber in store. Well sprouted but infected tubers usually produce normal plants.

Cultural and organic control
(1) Use of NIAB Potato Variety Handbook for variety choice.
(2) Early lifting of maincrops (i.e. before 10 October).
(3) Avoid damage at lifting/riddling time.
(4) Curing period of 10 d at 13–16°C after lifting and riddling will help wound healing.

Chemical control
(1) Fumigation with 2-aminobutane 14–21 d after lifting controls gangrene and skin spot (seed potatoes only).
(2) Thiabendazole dips, dusts and ULV mists at harvest or within three weeks after also controls silver scurf and skin spot.
(3) Carbendazim + tecnazene (pre-storage).
(4) Imazalil (pre-storage and pre-planting seed tubers only).
(5) Iodophor + thiabendazole.

Dry rot (*Fusarium solani* – f.sp. *caeruleum*)
Soil- and seed-borne disease causing wrinkled concentric rings with pink, white or blue fungal pustules on tubers within one to two months after storage. Eventually tuber dries out, shrink and mummifies. Bacterial wet rots may also invade under damp storage conditions. Less important than gangrene.

Cultural and organic control
(1) Use of NIAB Potato Variety Handbook for variety choice.
(2) Early lifting of maincrops (i.e. before 10 October).
(3) Avoid damage at lifting and riddling time.
(4) Curing period as for gangrene.

Chemical control
(1) Use of sprout suppressant tecnazene (TCNB) dust at lifting.
(2) Thiabendazole treatments as under gangrene.
(3) Carbendazim + tecnazene.
(4) Imazalil (pre-storage and pre-planting seed tubers only)

Common scab (*Streptomyces scabies*)
Very common superficial skin blemishing disease especially on light/gravelly alkaline soils low in organic matter. Seed/soil-borne disease affecting selling quality but not yield.

Cultural and organic control
(1) Use of NIAB Potato Variety Handbook for variety choice.
(2) Avoid liming potato or preceding crop.
(3) Avoid low organic matter soils.
(4) Irrigate crop especially during June/July.

Chemical control
No chemical control available.

Powdery scab (*Spongospora subterranea*)
Very uncommon except on wetter soils of the west and north. Most common in wet seasons and low-lying areas of field. Long-lived soil- and seed-borne disease.

Cultural and organic control
(1) Use NIAB Potato Variety Handbook for variety choice.
(2) Avoid use of known infected sites for at least five years after last potato crop.

Chemical control
maneb + zinc oxide (seed potatoes only)

Silver scurf (*Helminthosporium solani*)
Very common superficial skin blemishing disease affecting appearance but not yield. Silvery grey lesions develop during storage under high temperature and humidity conditions, causing loss of fresh weight in storage. Mostly seed-borne.

Cultural control
(1) Avoid planting infected tubers.
(2) Lift early as soon as crop is mature.
(3) Avoid high temperatures during storage.

Chemical control
As per gangrene control.

Skin spot (*Polyscythalum pustulans*)
Superficial skin blemishing disease common and important. Purple spots appear during storage and can affect crop emergence if used for seed. Most common on cold dry or heavy loam soils.

Cultural and organic control
(1) Presprout seed before planting and discard affected tubers. Use disease-free or chemically disinfected seed.
(2) Avoid late cold wet harvesting.
(3) Store tubers in boxes under dry ventilated conditions.
(4) Allow an adequate curing period of 14 days at 12–15°C.
(5) Avoid low temperatures during storage (below 7°C).

Chemical control
As per gangrene control (on seed potatoes only).

Pink rot and watery wound rot (*Phytophthora* and *Pythium* spp)
Soil-borne diseases causing wilting in field and pink rot and/or rapid rot in store. Very sporadic occurrence. Locally common in hot summers on heavy, badly drained soils. Commonly spreads in store.

Cultural and organic control
(1) Avoid damaging tubers at lifting/riddling time especially under damp soil conditions.
(2) Extend period between potato crops to eight years or more and prevent infection of clean fields.
(3) Improve drainage.

Chemical control
No chemical control available.

Wart disease (*Synchytrium endobioticum*)
Large external warts and deformities on tubers and stolons. Now very rare due to government legislation and use of field immune varieties. Common in Europe. Soil-borne disease. More common in wet north and west.

Cultural and organic control
(1) Use NIAB Potato Variety Handbook for selection of immune varieties.
(2) Avoid spread on infected implements and dung fed with infected tubers.
(3) Rest field for 30 years. Outbreaks must be reported to MAFF (notifiable disease).

Chemical control
Not permitted.

Black scurf and stem canker (*Thanatephorus cucumeris*)
Superficial black scurfy skin blemishes and brown or white girdling of stem bases. Black patches easily removed from skin. Young sprouts destroyed on seed tubers causing delayed emergence. Soft leaf rolling and wilting may occur (cf. virus leaf roll). Control includes the use of well-sprouted healthy seed. Avoid early/deep plantings in cold dry conditions. It is generally less severe if the crop is harvested early.

Chemical control
Apply to seed potatoes preplanting.

imazalil + pencycuron
imazalil + thiabendazole
iprodione
pencycuron
thiabendazole
tolclofos-methyl

Diseases of brassicas

Leaf and stem diseases

Powdery mildew (*Erysiphe cruciferarum*)
Silvery white patches on most brassicas causing eventual defoliation. Very common on swedes, turnips and brussels sprouts in dry summers. Air-borne disease.

Cultural and organic control
(1) Use of NIAB Oilseeds Variety Handbook for variety choice.
(2) Delayed sowing of spring and summer sown brassicas may be beneficial.

Chemical control

Foliar treatment At first sign of disease use one of the following
copper sulphate + sulphur
triadimefon
triadimenol

Canker (*Leptosphaeria maculans/Phoma lingam*)
Beige leaf spotting in autumn followed by stem cankers and lodging in spring and eventually pod and seed infection. Very important disease of oilseed rape, and all brassica seed crops may be affected. Seed-, stubble- and debris-borne disease favoured by wet weather.

Cultural and organic control
(1) Use of NIAB Oilseeds Variety Handbook for variety choice.
(2) Isolate crop from other brassica crops and previous oilseed rape fields.
(3) Use treated seed.
(4) Chop/burn and deeply plough oilseed rape stubble soon after harvest.
(5) Use at least four-year break between brassica crops.

Chemical control

Seed treatment Used seed treated with one of the following
carboxin + thiram
thiabendazole + thiram
thiram + fenpropimorph

Foliar applications If infection occurs in autumn or spring on foliage, use of one of the following may be beneficial. Maximum of three applications per crop to prevent fungicide resistance.
carbendazim
prochloraz

Light leaf spot (*Pyrenopeziza brassicae/Cylindrosporium concentricum*)
Pale green/bleached areas on leaves and passing on to stems and inflorescences especially during wet weather in May/June. Common on most brassicas especially seed crops. Seed- and debris-borne.

Cultural and organic control
(1) Isolate crops from other brassica crops and previous stubble.
(2) Use treated seed.
(3) Chop/burn and deeply plough stubble soon after harvest.
(4) Use at least four-year break between brassica crops.

Chemical control
At 20% of the leaf area affected use of one of the following may be beneficial
benomyl
carbendazim
chlorothalonil
fenpropimorph (not OSR)
iprodione + thiophanate methyl
mancozeb
prochloraz
propiconazole
triadimenol (not OSR)
tebuconazole

Dark leaf spot (*Alternaria* spp)
Small dark leaf spots on foliage and seed pods especially during wet weather. Very common on all brassica seed crops and rape and stubble turnips in general. Seed- and debris-borne.

Cultural and organic control
(1) Isolate crop from other brassica crops and previous stubble.
(2) Use treated seed.
(3) Chop/burn and deeply plough stubble soon after harvest.
(4) Avoid early sowing (before 20 August).

Chemical control

Seed treatment
fenpropimorph + thiram
iprodione

Foliar treatment (*Botrytis* spp control also) From early flowering onwards one or two applications to prevent pod infection.
carbendazim
chlorothalonil
fenpropimorph
iprodione (± thiophanate-methyl)
maneb + zinc
prochloraz
propiconazole
tebuconazole (OSR only)
triadimenol (not OSR)
vinclozolin (OSR only)

Downy mildew (*Peronospora parasitica*)
Yellowing of lower leaves with white fungal growth on undersurfaces especially on autumn-sown seedlings and young plants during winter months, and wet weather. Not usually important on mature plants or during summer period. Air-borne disease.

Cultural and organic control
(1) Isolate crop from other brassica crops and previous stubble.
(2) Chop/burn and deeply plough stubble soon after harvest.
(3) Avoid very thick plant populations and low-lying cold wet fields.

Chemical control
Not normally necessary but foliar applications on seedlings and young plants with one of the following may be beneficial at times.
benalaxyl + mancozeb
chlorothalonil (± metaloxyl)
copper oxychloride + metalaxyl
dichlofluanid (protected crops only)
fosetyl-aluminium
mancozeb
maneb + zinc
propamocarb hydrochloride

Cabbage ringspot (*Mycosphaerella brassicola*)

Brown/black concentric ringspots on foliage of brassicas especially cauliflowers, brussels sprouts, kale and cabbages in south-west England during winter months. Seed- and stubble-borne.

Cultural and organic control

(1) Isolate crop from all other brassica crops and previous stubble.
(2) Chop and deeply plough stubble soon after harvest.

Chemical control

Seed infection is deep-seeded, requiring hot water treatment and thiram dressing.

Foliar treatment One of the following at first sign of infection during Nov–Feb:

benomyl
chlorothalonil
fenpropimorph
tebuconazole (OSR only)
triadimenol

Cauliflower mosaic virus (aphid vectors *Myzus persicae* and *Brevicoryne brassicae*)

Mottling and vein clearing of leaves, plants stunted and yields severely reduced. Very common on all brassicas especially cauliflowers where curd production is affected.

Cultural and organic control

(1) Isolate crops from all other brassica crops and previous stubble.

Chemical control

Control aphid populations at all times and delay virus infection to late in plant's life.

Foliar treatment Use of one of the following aphicides at first sign of aphid attack will reduce spread:

aldicarb
chlorpyrifos
cyfluthrin
cypermethrin
deltamethrin
demeton-S-methyl
dimethoate
disulfoton
fatty acids
fenvalerate
heptenophos
malathion
nicotine
permethrin
phorate
pirimicarb
quinalphos
resmethrin
thiometon

Clubroot (*Plasmodiophora brassicae*)

Stunting of plants as a result of tumour growths on and later rotting of crop roots. All brassica crops may be affected but particularly severe on summer crops though rarely damaging on kale. Swedes and turnips on cold wet acid soils may be very severely affected. Soil-borne disease occurring in patches in fields. Resistant spores may survive for more than eight years between susceptible crops in affected soils. Potentially very serious on oilseed rape crop.

Cultural and organic control

(1) Use of NIAB Recommended Variety Leaflets for variety choice.
(2) Avoid overcropping of brassicas (not more than one year in five as a preventative measure).
(3) Avoid transport of infected soil on boots, etc.
(4) Avoid feeding of infected roots on fields planned for future brassica production.
(5) Liming to at least pH 6.5 will reduce severity of attack.
(6) Improved drainage may reduce severity of attack.
(7) Use a break of at least eight years from brassicas after infection.

Chemical control

Soil treatment Sterilization of seedbed for high value brassica crops with dazomet in summer months above 10°C or cresylic acid as a soil drench in autumn when crop removed.

Sclerotinia stemrot (*Sclerotinia sclerotiorum*)

Bleached areas on the stem from May onwards with hard block sclerotia within cavity of affected stem area resulting in lodging. *Sclerotinia* affects a wide range of plants – beans, brassicas, beets, carrots, celery, peas, potatoes and many weeds. Sclerotia contaminate seed, persist in soil for eight years or more and may produce air-borne spores in May to infect through wet petals shed on to leaves and stems. Disease favoured by wet weather during flowering and petal fall.

Cultural and organic control

(1) Avoid over-cropping of brassicas and/or other host crops (not more than one year in five as a preventative measure).
(2) Use *Sclerotinia*-free seed and clean equipment.
(3) Chop or burn *Sclerotinia*-infected stems and stubble followed by deep ploughing of remaining debris.

Chemical control

An application at first petal fall with one of the following will control *Sclerotinia*, *Botrytis* spp and *Aternaria* spp infections.

iprodione (± thiophanate-methyl)
prochloraz (± carbendazim)
tebuconazole
vinclozolin

Diseases of sugar beet, fodder beet, mangold

Rhizomania (beet necrotic yellow vein virus, fungal vector *Polymyxa betae*)

One of the most important diseases of sugar beet, can affect fodder beet and spinach also. Notifiable disease: contact MAFF, Plant Health Inspectorate, ADAS and British Sugar plc if infection suspected. Common in Europe and USA; first UK outbreak in East Anglia in 1987. A few sporadic cases reported since. Patches of poor growth, slightly yellowed leaves with abnormal and heavy proliferation of rootlets to produce root bearding or 'rhizomania'. Long-lived soil-borne vector and virus. Dissemination and new infection occur by soil transport from infected fields and by sugar beet factory effluent. Control is very difficult at present, and up to date advice must be obtained from MAFF and British Sugar plc.

Virus yellows (aphid vectors peach-potato (*Myzus persicae*), black bean aphid (*Aphis fabae*)

Leaves turn yellow and may be more susceptible to fungal attack. Yield and quality reduced depending upon earliness of infection. Aphids spread the two viruses concerned after feeding on infected plants. Beet yellows virus (BYV) carried for only a few hours in aphids but beet mild yellowing virus (BMYV) carried by aphids for most of its life. Reduce spread in spring to young crops from clamps, seed crops and any over-wintered beet, plants or remnants left in field. Many aphicides now less effective due to aphid resistance in south and east.

Cultural and organic control

(1) Use of NIAB Recommended Variety Leaflets for variety choice.
(2) Sow early.
(3) Avoid very low plant populations.
(4) Use aphicides and delay virus infection to as late as possible in the season.
(5) Avoid close proximity of root crops to steckling beds and previous crops.
(6) Inspect crops regularly and spray if one plant in four has aphids.

Chemical control (aphid vectors)

Seed treatment
imidacloprid

Seed-furrow treatment Use one of the following:
aldicarb, oxamyl
carbosulfan, phorate
disulfoton
followed by:

Foliar treatments Use any one of:
demeton-S-methyl, pirimicarb
dimethoate, thiometon
disulfoton

Downy mildew (*Peronospora farinosa*)

Lower leaves yellow with white mycelium underneath especially during winter six months on seed crops and early sown root crops in spring. Root yields and juice quality can be seriously affected in wetter seasons and low lying areas.

Cultural and organic control

(1) Use of NIAB Recommended Variety Leaflets for variety choice.
(2) Avoid close proximity of seed and root crops.
(3) Avoid early sowing of root crops on heavy soils.

Chemical control
None approved.

Powdery mildew (*Erysiphe* spp)

Powdery white mycelium on foliage especially during dry weather in late summer. Sugar yield may be significantly reduced in south and east. Spray foliage with wettable sulphur, propiconazole, triadimefon or triadimenol at first sign of disease in July/August. Less cost effective after 1 September.

Blackleg (*Pleospora bjoerlingii*) and damping-off diseases

Important fungus causing damping off and poor seedling growth leading to low plant populations. Seed-borne disease more important in crops drilled to a stand. Use seed treated with thiram + hymexazol.

Beet rust (*Uromyces betae*)

Orange-brown pustules randomly scattered or grouped in patches on leaves. Air-borne disease favoured by hot dry conditions July to September. Spraying at first sign of disease with propiconazole or triadimenol will give good disease control in July/August. Less cost effective after 1 September.

Leaf spots (*Ramularia beticola*, *Phoma betae*)

Spray *seed crops only* at first sign of disease(s) during summer months. There should be a maximum of three applications at 14–21 d intervals with fentin hydroxide. Propiconazole will give good suppression of *Ramularia* on non seed crops.

Diseases of peas and beans (field, broad/dwarf, French)

Damping off, foot rot and seed-borne diseases (*Pythium*, *Phytophthora*, *Fusarium*, *Ascochyta* and *Colletotrichum* spp but not halo blight)

Poor germination and early growth.

Cultural and organic control
(1) Use disease-free or treated seed.
(2) Avoid sowing into cold wet soils.
(3) Chop/burn and preferably plough stubble soon after harvest.
(4) Avoid poorly drained or low lying fields.
(5) Use appropriate fungicides on growing crops.

Chemical control

Seed treatment
Peas
benomyl
fosetyl-Al
metalaxyl + thiabendazole + thiram
thiabendazole + thiram
thiram
Field/broad beans
benomyl
metalaxyl + thiabendazole thiram
thiabendazole + thiram
thiram
Dwarf beans
metalaxyl + thiabendazole + thiram
thiram

Leaf, stem and pod spots (*Ascochyta pisi* – peas, *Ascochyta fabae* – field and broad beans, *Botrytis fabae* – chocolate spot of field and broad beans, *Colletotrichum lindemuthianum* – anthracnose of dwarf/navy beans)

Various types and sizes of brown/grey coloured spots of leaves, stems and pods. Cause partial defoliation, collapse of stem and discoloration of pods and seeds especially during wet periods in May/June, July. Chocolate spot is most severe on winter beans; seed- and stubble-borne. Use disease-free seed wherever possible.

Cultural and organic control
As above for seed-borne diseases.

Chemical control

Seed treatment
Peas
metalaxyl + thiabendazole + thiram
thiabendazole + thiram
Field/broad beans
benomyl
thiabendazole + thiram
Dwarf/French/runner
metalaxyl + thiabendazole + thiram

Foliar treatment
Peas
benomyl
carbendazim + chlorothalonil
fosetyl-aluminium
iprodione (± thiophanate-methyl)
vinclozolin
Field/broad beans
benomyl
carbendazim
chlorothalonil
iprodione
thiophanate-methyl
vinclozolin
Dwarf/French/runner
benomyl
carbendazim
iprodione
thiophanate-methyl
vinclozolin

Pea wilt (*Fusarium oxysporum*)

Rapid wilting of plants within patches in fields during late May and June. Foliage turns grey then finally goes yellow starting at base of plant then upwards. Very persistent soil-borne disease severely affecting yields in the affected patches. May be seed-borne. Mostly controlled by rotation and resistant varieties.

Cultural and organic control
(1) Use disease-free or treated seed.
(2) Use NIAB Recommended Variety Leaflets for variety choice.
(3) Use at least a four-year break after a *healthy* crop of peas *or* beans. Longer if unhealthy.

Chemical control
No chemical control available.

Downy mildew (*Peronospora viciae*) of peas and field/broad beans
Lower leaves yellow with white mycelium on undersurface especially during wet periods/seasons. Seed and stubble-borne; plough deeply after infection. Use NIAB Peas and Beans Recommended Variety Leaflets for variety choice.

Chemical control

Seed treatment
metalaxyl + thiabendazole + thiram

Foliar treatment
carbendazim + chlorothalonil (peas only)
chlorothalonil + metalaxyl (beans only)
fosetyl-aluminium

Grey mould/pod rot of peas/beans (*Botrytis* spp)
Lesions with grey fungal growth on pods especially those damaged or in contact with soil. Very common and severe in wet seasons causing serious reduction in quality.

Cultural and organic control
(1) Chop/burn and deeply plough stubble soon after harvest.
(2) Avoid poorly drained or low-lying soils.

Chemical control

Foliar treatment Spray at flowering with one of the following to protect pods:
benomyl
carbendazim
chlorothalonil
iprodione
thiophanate-methyl
vinclozolin

Broad bean rust (*Uromyces fabae*) of field/broad beans
Orange-brown pustules randomly scattered or grouped in patches on leaves. Air-borne foliar disease favoured by hot dry conditions July/August. Spraying at first sign of disease with fenpropimorph will give good disease control.

Diseases of linseed

Damping-off, seedling blight and root rots (*Pythium*, *fusarium*, *Ascochyta* and *Colletotrichum* spp)
Various seed- and soil-borne fungi causing rotting of seeds and seedlings, and root rots of young plants.

Cultural and organic control
(1) Avoid early sowing in cold wet soils.
(2) Avoid acid or waterlogged soils.
(3) Avoid soil compaction and surface capping.
(4) Extend normal 5-year rotational break to 7 years with *Fusarium* spp infection.

Chemical control

Seed treatment
prochloraz
thiabendazole + thiram

Alternaria spp
Cause poor germination, damping off, seedling blight and leaf blight during cool, wet seasons especially on wet soils and in lush crops. Seed- and stubble-borne.

Cultural and organic control
(1) Use 5-year rotational interval.
(2) Chop and deeply plough debris soon after harvest.
(3) Use disease-free or treated seed.
(4) Avoid high plant populations.
(5) Restrict cropping to well drained sites and low-rainfall areas if possible.

Chemical control

Seed treatment
iprodione
thiabendazole + thiram

Foliar treatment
iprodione (off label only)

Diseases of carrots

Damping-off and leaf blight (*Alternaria dauci*)
Causes damping-off and leaf blight on wet soils and during wet seasons. Seed- and soil-borne.

Cultural and organic control
(1) Use disease-free or treated seed.
(2) Avoid poorly drained or low-lying cold fields.
(3) Avoid early sowings.

Chemical control

Seed treatment
iprodione + metalaxyl + thiabendazole
metalaxyl ± thiabendazole
thiabendazole + thiram

Violet root rot (*Helicobasidium brebissonii*)

Roots covered with purple fungal growth at lifting time especially on cold poorly drained soils. Wide host range including sugar beet, beetroot, parsnips, potatoes and weeds such as docks and dandelions. Brassicas are resistant. Soil- and stubble-borne disease. Yield reduced in store and quality markedly reduced.

Cultural and organic control

(1) Avoid susceptible root crops being grown more than one year in five.
(2) Chop and deeply plough crop debris soon after harvest.
(3) Practise good weed control.

Chemical control

No chemical control available.

Black rot (*Stemphylium radicinum*)

Large black sunken lesions on mature roots and rotting in store. Also considerable seed losses can occur in seed crops and damping-off of seedlings in the field. Seed-borne disease mainly.

Cultural and organic control

(1) Use disease-free or treated seed.
(2) Chop and deeply plough crop debris soon after harvest.

Chemical control

Seed treatment
iprodione + metalaxyl + thiabendazole
thiabendazole + thiram

Carrot motley dwarf virus (vector, willow carrot aphid)

Yellow mottling of leaves and stunting. Also affects parsnips and celery.

Cultural and organic control

(1) Isolate carrot field from other aphid hosts and previous cropped fields.
(2) Large fields are less prone to overall attack than small fields.
(3) Reduce aphid populations at all times.

Chemical control

Seedbed treatment Aldicarb carbofuran, carbosulfan disulfoton or phorate granules at crop emergence.

Foliar application At first sign of aphids use one of the following

disulfoton
demeton-S-methyl
dimethoate
malathion
phorate
pirimicarb
thiometon

Storage rots (*Sclerotinia sclerotiorum, Botrytis cinerea*)

Grey and white fungal mycelium on roots in store especially when damaged and stored under damp conditions.

Cultural and organic control

(1) Avoid late lifting under cold wet conditions.
(2) Avoid damaging roots at harvesting.
(3) Provide adequate ventilation to keep roots dry and cool in storage.

Chemical control

None approved.

Diseases of onions and leeks

Downy/mildew (*Peronospora destructor*)

Pale oval lesions on leaves and dieback on tips. Spreads extensively within field under cool wet conditions. Fungus overwinters in bulbs and in soil for many years, causing further infection. Common on autumn-sown onions during winter months. Attacks all onions and shallots.

Cultural control

(1) Isolate crops from other onion crops and previous onion stubble.
(2) Avoid using known infected fields for at least five years.
(3) Avoid low-lying cold or poorly drained sites.
(4) Practise good weed control to help air circulation within crop.

Chemical control

Foliar treatment Apply at first sign of infection and repeat every 14 d to a maximum of six applications.
chlorothalonil + metalaxyl
mancozeb + metalaxyl (off label)

Smut (*Urocystis cepulae*)

Leaves blister and rot to release black powdery fungal spore masses. Can be very common and serious on seedlings and young plants of onions, leeks, shallots, chives and garlic. Fungus penetrates in seedling stage only. Therefore, if raised in disease-free seedbed, transplants cannot become infected later. Fungal spores survive in soil for at least 10 years.

Cultural and organic control

(1) Avoid contaminating soil with infected debris, or soil on machinery, boots, wheels, etc.
(2) Burn all infected plants immediately. Do *not* bury in soil.
(3) Encourage fast germination and early growth. Avoid sowing early in cold wet soils.
(4) Avoid using infected fields for at least 10 years.

Chemical control

Seed treatment
thiabendazole + thiram

White/rot (*Sclerotium cepivorum*)

Plants yellow, stunted/wilted with rotten base and covered with white fungal growth. Soil-borne fungus surviving in soil for many years.

Cultural and organic control

(1) Avoid contaminating soil with infected debris, or soil on machinery, boots, wheels, etc.
(2) Burn all infected plants immediately. Do *not* bury in soil.
(3) Avoid any infected soils for at least eight years.

Chemical control
None approved.

Neck and storage rots (*Botrytis allii*)

Onions soften and rot internally while in store. Discoloration and rotting of neck occurs after several weeks in store. Very common in December and January. Mostly seed-borne but crop debris and onion dumps can be an important source of infection. Infected seedlings and plants appear healthy in field.

Cultural and organic control

(1) Use disease-free or treated seed. Care is needed in using home-saved seed.
(2) Deeply bury crop debris soon after harvest.
(3) Completely cover old onion dumps with soil.
(4) Avoid damage at harvest and ensure adequate curing/drying occurs.

Chemical control

Seed treatment
benomyl
benomyl + iodofenphos + metalaxyl
iprodione
thiabendazole + thiram

Foliar application Several applications of one of the following to the plant foliage at three- to four-week intervals may be beneficial particularly to machine-harvested crops during wet season.

benomyl | chlorothalonil
carbendazim | iprodione

Leek rust (*Puccinia allii*)

Orange-brown pustules randomly scattered or grouped in pustules on leaves. Air-borne foliar disease favoured by hot dry conditions August/September. Spraying at first sign of disease with chlorothalonil, fenpropimorph or triadimefon will give good disease control.

Diseases of grasses and herbage legumes

Crown rust of rye grasses (*Puccinia coronata*)

Crown rust causes orange fungal pustules on the leaves of rye grasses, fescues and cultivated oats, but cross-infection is unlikely. Most common on late summer/autumn silage cuts of rye grasses (especially IRG). Associated with hot weather and cool dewy nights coupled with low nitrogen application. Reduces yield, palatability and digestibility of forage. Disease is more severe in a pure stand of rye grass cultivars especially early heading types. Very common in south and west England. Less frequent in north. Air-borne disease surviving on establish leys.

Cultural and organic control

(1) Use NIAB Recommended Variety Leaflet for variety choice.
(2) Increase defoliation by cutting or grazing more frequently.
(3) Change management to all grazing in late summer through to early winter.
(4) Increase nitrogen levels to 250 kg/ha or more.

Chemical control
No chemical treatment economically feasible at present but the following will give good control at low disease levels.
propiconazole
triadimefon

Leaf blotch of rye grasses (*Rhynchosporium secalis* and *Rhynchosporium orthosporum*)

Leaf blotch causes dark brown blotches with light centres on barley, cocksfoot, couch, timothy and rye grasses (especially IRG). Cross-infection is possible but generally restricted. Most common under cool moist conditions on spring IRG silage crops. Yield, palatability and digestibility are reduced. Air-borne disease surviving on established plants.

Cultural and organic control

(1) Use NIAB Recommended Variety Leaflet for variety choice.
(2) Increase defoliation by cutting or grazing more frequently.
(3) Change management to all grazing in early spring and summer months.

Chemical control

No chemical control economically feasible at present but the following chemicals will give good control at low disease levels.

propiconazole
triadimefon

Powdery mildew (*Erysiphe graminis*)

Grey/white superficial fungal growth on leaf surface. Attacks a wide range of grasses/cereals but cross-infection unlikely. Common during and after dry periods. More conspicuous under high soil nitrogen condtions. Reduces yield, palatability and digestibility but generally less damaging than crown rust and leaf blotch. Air-borne disease surviving on established plants.

Control

Use NIAB Farmers Leaflet No. 4 for variety choice. Increase defoliation by cutting or grazing more frequently. No chemical control economically feasible at present but propiconazole or triadimefon will give good control at low disease levels.

References

Ivens, G.W. (1995) *The UK Pesticide Guide*. CAB International and British Crop Protection Council, Oxford and Farnham.

MAFF/HSE (1992) *Pesticides*, Reference Book 500. HMSO, London.

NIAB (1995) *Variety Handbooks* and *Recommended Variety Leaflets*. NIAB, Cambridge.

Further reading

Dixon, G.R. (1981) *Vegetable Crop Diseases*. Macmillan, London.

Smith, I.M. *et al.* (1988) *European Handbook of Plant Diseases*. Blackwell Science, Oxford.

14

Grain Preservation and Storage

P.H. Bomford

Grain is stored after harvest so that it can be marketed, or used, in an orderly manner. Prices are generally at their lowest at harvest time, and can be expected to rise with length of storage, until the next harvest approaches. If livestock are to be fed with home-produced grain, careful storage can ensure a year-round supply of high quality feed.

A successful storage system will preserve those qualities of the product which are important for its proposed end use. These qualities may include a high germination percentage, good baking or malting properties, ease of extraction of constituents such as starch, oils or sugars, freedom from impurities, discoloration or taint, the nutritive value of the grain to livestock, or the conditions of moisture content and temperature necessary to meet Intervention standards.

Grain is generally preserved in good condition by controlling its moisture content and temperature so that the organisms which cause deterioration cannot develop. Grain stored in this way can be sold into any appropriate market, or fed to livestock on the farm.

Conditions for the safe storage of living grain

The main agencies which cause deterioration in stored grain are moulds, insects and mites. Some of these organisms occur naturally on cereal crops, and so are likely to be brought into the store with the crop. Residual populations of insects and mites may be present in crevices or crop residues in the 'empty' grain store, but thorough cleaning and insecticide/miticide treatment of the store before it is filled can greatly reduce this hazard.

The grain is protected from attack by a tough outer skin or pericarp and impermeable seed coat or testa. Unfortunately, the severe threshing treatment to which the crop is subjected at harvest can damage these layers, allowing easy access by mould organisms to the starchy endosperm or the embryo. In one study between one-fifth and one-third of all wheat grains were found to have threshing injuries to the skin of the embryo area. The embryo contains sugars, fats and proteins and is thus an excellent source of nutrients for the moulds. Damage to the embryo will impair the viability of the seed.

The relationship between the possibility of grain deterioration (over a 35-week period) and storage conditions is illustrated clearly in the classic diagram (Fig. 14.1) which has been produced by research workers at the MAFF Slough Laboratory.

In the figure, line A defines the conditions under which mites will breed, and therefore infestations can build up. Breeding can be prevented by keeping the temperature below 2.5°C, or the grain moisture content below 12%. In fact, protection from most major mite species is achieved if grain is dried below 14% moisture content (MC) (Wilkin, 1975). Careful cleaning and acaricide treatment of the store when empty will minimize the risk of mite populations persisting from one season to the next.

Line B covers the risk of loss of germination due to embryo damage, and also the deterioration of baking (wheat) and malting (barley) qualities. Grain stored under conditions to the left of the line may be expected to retain all these qualities and thus be suitable for sale into premium markets.

Line C defines the conditions under which moulds will grow on and subsequently in the grain. In addition to the 'invisible' forms of damage mentioned in the previous paragraph, mould growth can lead to a loss of crop weight as the mould organisms feed on the grain, discoloration due to visible mould growth, and the unpleasant health hazards of myocotoxins (poisons produced by the mould organism which can kill or debilitate livestock or even humans who eat contaminated foodstuffs) and mould spores (a very fine dust which can cause farmers' lung, a pneumonia-like affliction, when inhaled). It is essential to wear an adequate *organic dust respirator* if working with grain (or hay) which is suspected of being mouldy.

Moulds, like other living organisms, produce heat as they break down food with oxygen, liberating water and carbon dioxide. Thus, any large concentration of moulds (or insects or mites) will make the grain around it warmer and more moist. As can be seen from the diagram, this will have the effect of making conditions even more favourable for their own development, and may also encourage the development of other pests or even allow the grain to germinate, which will happen if the local moisture content increases to 25–30%.

Severe mould infestation, if not detected and con-

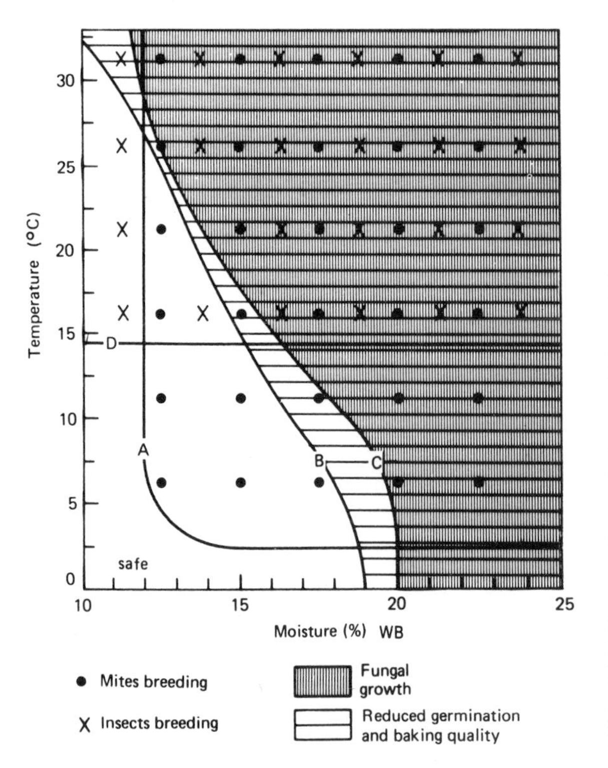

Fig. 14.1 Combined effects of temperature and moisture content on grain condition and grain pests over a 35-week storage period. (Crown copyright. Reproduced with permission). After Burges & Burrell, 1964.

trolled, will result in the grain being caked together, particularly at or just below the surface and against the walls of the store, and sprouted at the surface.

This can reduce or eliminate the market value of the corn, and may also render it hazardous to livestock and to the farmer.

Line D shows that insects such as grain weevils and beetles can generally be prevented from becoming active and breeding if the temperature of the grain is kept below 14.5°C. In fact, the cooler the grain, the less active the insects and the greater the margin of safety, even down to freezing point. Reducing grain moisture content has no effect on controlling insect activity.

Insect activity in the grain will lead to loss of dry matter and reduced germination, contamination with insect remains and faeces, and as above, to an increase in temperature and moisture levels which can lead to moulding and even sprouting of the grain.

Where an insect problem has been found to exist and it is not possible to control it by cooling the grain, the insects may be controlled by fumigation of the grain store with substances such as methyl bromide or phosphine. These substances are toxic and their successful use involves use involves sealing the grain store and the deployment of specialized control and safety equipment. This should be left to specialist pest control contractors.

The grain may also be treated with an approved insecticide as it is conveyed into store if insect problems are expected.

As is the case with mites, the best defence against insect infestations is 'good housekeeping' in cleaning out dust pockets and crop residues in which populations can survive in the empty store, and treatment with an approved insecticide. During storage, the extent of insect populations should be assessed by trapping.

It can be seen from Fig. 14.1 that (ignoring mites) the preservation of living grain in top condition for a 35-week storage period may be achieved by keeping conditions within the area below line D and to the left of line B. If the grain is intended for use or sale as livestock feed, then the the right-hand boundary moves out to line C, allowing a higher moisture content to be accepted.

Generally, the crop is safest when temperature and moisture content are low, and at increasing risk as either factor rises.

The time factor in grain storage

As may be expected, insect, mould or mite infestations take time to build up. This means that grain, too moist or too warm to be stored safety for seven months without drying or cooling, can be held for several weeks with little or no harm. The length of time such a parcel of grain can be held without loss of quality decreases with increasing temperatures or moisture content. Table 14.1 shows the maximum recommended moisture content for bulk grain stored at various temperatures and for various periods. (It must be remembered that storage for any extended period at temperatures over 14.5°C entails the risk of insect build-up *however dry the grain.*)

Table 14.1 Estimated maximum number of weeks of mould-free storage of barley at a range of moisture contents and temperatures (ADAS, 1985)

MC	*Temperature °C*				
(%)	*5*	*10*	*15*	*20*	*25*
16	120	50	30	12	5
17	80	28	10	4	2
18	22	9	4	2	1
19	14	6	3	1.5	0.5
20	10	4.5	2	1	0.5
22	7	3	2	1	0.5
24	4	2	1	0.5	0.2
26	2.5	1	0.5	0.2	0.2

Grain moisture content and its measurement

Grain moisture content is expressed as wet-base (WB) percentage, or the ratio of the weight of water contained in a sample of the grain, to the total weight of the sample, including the water.

In a laboratory the moisture content of a sample of grain is found by drying the weighed sample in an oven until no water remains, and then re-weighing to find the weight of the water which has been driven off. The moisture content (wet base) is found from the expression:

$$\text{MC\% (WB)} = \frac{\text{Loss of weight on drying}}{\text{Original weight of sample}} \times 100$$

This is the only completely accurate method of measuring the moisture content of a sample of grain, and is always used in scientific research or when moisture meters are to be calibrated. However, the complete oven-drying of even a milled sample of grain will take $1\frac{1}{2}$ to 4 h. Faster moisture measurement systems have been produced to satisfy the need for an 'immediate' evaluation of grain moisture content.

Most grain moisture meters measure an electrical property of the grain such as its resistance or capacitance, which have been found to vary with moisture content. A good quality instrument, well maintained, can be expected to give a reading which is correct to within $\pm 1\%$ of the true moisture content of the sample, and this reading can be available within minutes of the sample being taken.

Some instruments require the grain to be ground, which increases the time needed to make a measurement. However, a whole-kernel instrument may give false results with grain that has just passed through a drier, or with very cold grain since in these cases the condition of the exterior of the grain may not reflect the internal situation.

When measuring the moisture content of a sample of grain, it is most important to ensure that the sample is truly representative of the bulk of grain from which it has been drawn. This can be best achieved by systematically extracting several small samples from different points and depths in the load or store, mixing them together and then taking the final sample from this mixture.

Without great care in the taking of samples, even the best of moisture meters and the most careful of operators will fail to produce satisfactory results.

Drying grain

Where incoming grain is at too high a moisture content for long-term storage, it must be dried at the beginning of the storage period. In the process of drying, heat is provided to evaporate the moisture from the grain and a stream of air picks up the water vapour and carries it away into the atmosphere.

The more a batch of grain has to be dried, due either to a high initial moisture content or to a low final moisture content, the greater will be the cost both in terms of energy usage and drying equipment operating time.

When grain is dried the resulting product will weigh less than it did before drying, by the amount of water which has been removed. It is often important to know what this weight loss will be, either for the purpose of estimating drying costs or for converting harvested grain weights to equivalent weights of dry grain. The following formulae are helpful.

$$X = \frac{W_1\,(M_1 - M_2)}{100 - M_2}\ X\ \frac{W_2\,(M_1 - M_2)}{100 - M_1}$$

where

X is the weight loss on drying,

M_1 and M_2 are initial and final moisture contents (% WB)

W_1 and W_2 are weights of grain before and after drying.

In addition to the weight loss on drying, a further minimum weight loss of 1–1.5% can be expected over the storage period, due to respiration of the grain itself and to the activities of storage pest organisms.

Drying systems range from those which are capable of drying grain within 2 or 3 h of its arrival from the field to those where complete drying can take two weeks or more.

Fig. 14.2 Timely harvesting of a clean crop is the first stage in its successful preservation and storage. Courtesy John Deere Ltd.

Because of the air temperatures which are used, high-speed drying systems may be classified as 'high-temperature driers', while the slowest are known as 'low-temperature' or 'storage' driers, the latter term because drying takes place after the grain has been put into the store and the store does double duty as a drier.

High-temperature driers

High-temperature driers can extract moisture from grain very rapidly because a high temperature allows moisture to diffuse rapidly out from the interior of each grain, and the very hot air which is blown through the grain has a very great attraction for water vapour and a very large water vapour carrying capacity.

As the hot air passes through the grain, it gives up some of its heat to evaporate water and is thus cooled as it picks up water vapour. The distance that the air travels through moist grain must be limited so that there is no risk of it being cooled to the point where it becomes saturated (i.e. can no longer carry all its water, and water is deposited back on the grain by condensation). In high-temperature driers the thickness of the grain layer rarely exceeds 250 mm, while in low-temperature storage driers it may be as much as 3 m.

The higher the temperature at which a given drier is operated, the faster is the rate of moisture extraction, the higher is the grain througput of the drier, and in many cases the lower is the energy consumption per unit of water removed. Hence, operating a drier at a higher temperature can reduce drying costs in terms of both overhead and fixed costs per tonne dried.

If the temperature of the drying air is too high, grain temperature will rise to a level high enough to cause damage to germination, baking or malting properties, extraction of oils or starch or even to the livestock feeding value of the grain. Table 14.2 shows the recommended maximum air temperatures at which to dry again.

Unless the grain is completely dry, grain temperature will always be lower than air temperature due to the cooling effect of evaporating water; the greatest cooling effect will be where the grain is moist and the evaporation rate is high. Even at the end of the drying process, average grain temperatures will still be 10–20°C lower than drying air temperature. If in doubt, drier manufacturers' recommendations should be followed.

Table 14.2 Recommended maximum air temperatures for high-temperature grain driers

	Moisture content (% WB)	*Drying air temperature (°C)*
Malting barley and seed grain	≤24	49
Miling wheat	>24	43
Grain for stockfeed	≤25	66
	>25	60
		82–104

From Blakeman, 1982.
These recommendations are based on research carried out in the 1930s. New recommendations are likely to be based on a grain-temperature/exposure time factor rather than the temperature of the drying air.

High-temperature driers be classified as 'continuous' where there is a constant flow of grain through the machine and into store, or 'batch', where a quantity of grain is loaded into the machine where it remains while it is dried, and is then unloaded into store.

Continuous flow driers

Traditional continuous grain driers are of the crossflow type, where the grain moves in a layer 200–300 mm thick while the drying, and later cooling, air is directed through the grain at right angles to its direction of travel. In order to give an equal drying effect to all grains, the grain should be turned or mixed several times as it passes through the drier.

The crossflow principle is popular due to its simplicity, but other airflow systems can offer advantages and a few have appeared on the market. In a counterflow drier system, drying air flows in the opposite direction to the grain so that the hottest and driest air meets the driest corn. This gives faster and more efficient drying, but there is a greater risk of damaging the grain by over-heating. This method is in fact only used in the cooling section of some driers where the temperature gradient is in the reverse direction and this system gives a more gentle temperature reduction and the advantage of maximum heat transfer can be fully utilized.

In a concurrent flow drier, grain and drying air move in the same direction although the air moves at a much faster rate. Thus, the hottest air always meets the wettest corn, where the cooling effect of evaporation will be greatest. By the time the air reaches the driest corn, its temperature will have been much reduced, minimizing the temperature stress on the grain. This may permit air temperatures above the recommended maximum, without excess grain temperatures. There is a faster througput from a given drier size, and more efficient use of heat (Brooker *et al.*, 1974).

The mixed-flow drier combines features of all three types. Heat transfer is efficient, and high air temperatures can be used without risk of overheating the grain. At a plenum temperature of 120°C and when drying grain from 20 to 15% MC, grain temperature only reached 50°C (Bruce and Nellist, 1986).

Not all the drying is done in the drying section of the drier; a final 1 or 1.5 percentage points of moisture are removed while the grain is being cooled, utilizing the heat energy remaining in the grain from its passage through the drying section. The cooling section occupies one-third to one-quarter of the drier, and is generally capable of cooling grain to near ambient temperature before it is discharged from the drier. The greatest demand on the cooling section is when only a small amount of moisture needs to be removed from the grain, and drier throughput is higher than rated.

If there is cooling available in the store, drier capacity can be increased since the whole of the drier can be used for drying the grain, which is cooled in the bulk store later. A refinement of this system, where the grain is held, still hot, in a sealed bin for a 12-h period of 'tempering' before being cooled by low-volume aeration, is known as 'dryeration'. 'Tempering' will improve efficiency, as better use is made of the residual heat in the grain (Bakker-Arkema, 1984).

The drying capacity, or 'throughput' of a continuous flow drier is expressed as the number of tonnes/h of grain at 20% MC which can be dried to 15% at a specified drying-air temperature (often 65°C). Output of the drier will be increased if a higher air temperature can be (safely) used, or if the grain requires drying by less than five percentage units, and vice versa.

The amount of drying done to each grain depends on the length of time for which that grain is exposed to the stream of hot air in the drying section of the drier, and this is controlled by the rate at which grain moves through the machine. This rate is adjustable, and adjustments are made according to the moisture content of the incoming grain and the resulting final moisture content of the dried product. Both of these values should be checked regularly. Several manufacturers offer automatic drier control units, which can monitor final grain moisture content (or grain temperature at the end of the drying section which is a related value) and regulate the flow of grain accordingly.

It should be remembered that grain can take 3 h or more to pass right through a continuous-flow drier, so that any adjustment made to the machine will take this long to come into full effect.

Energy consumption by continuous driers

Continuous-flow and other high-temperature driers, which evaporate water at a very rapid rate, consequently consume energy at a rapid rate (Table 14.3).

Such a high heat requirement clearly rules out many sources, and at present the available types are fired by diesel (gas oil) or LP gas (propane). Potential alternative fuels are natural gas (if a piped supply is available) or coal.

To convert 1 kg of water into vapour at 20°C, 2.45 megajoules (MJ) of heat energy are required. Tests have shown an average energy use of 6.7 MJ/kg of water evaporated, for a range of traditional crossflow continuous driers. Clearly, significant amounts of energy are used up in other ways. Apart from the energy used for the operation of fans and conveyors, and the proportion which leaves the drier in the form of latent heat locked up in water vapour, most of the 'wasted' energy is lost in the form of heated, unsaturated air either through the lower part of the drying section or from the cooling of the drier. Recycling of this exhaust air to exploit the energy it still carries can give energy savings of up to 40% (Bakker-Arkema, 1984).

Several designs of mixed flow driers can achieve specific energy figures of less than 4.0 MJ/kg under test conditions, although they are not often attained under farm conditions.

In operating the drier to save energy, the most important factor is to *avoid over-drying*. Every kilogram of water that is removed from the grain adds to the cost of the operation and slows output. Reducing the drying air temperature will *not* save energy, and will certainly increase the cost of drying a tonne of grain. Automatic control of continuous driers can produce a uniform product and minimize energy use (Bruce & McFarlane, 1993).

On farms the major component in drying costs is the overhead cost of the equipment. Any cost-saving exercise should address this aspect first, before moving on to matters of energy consumption.

Table 14.3 Furnace capacity of continuous driers

t/h of wet corn to be dried from 20–15% MC at 65°C air temperature	*Drying capacity* 2.5	5.0	10.0	20	30	60
Evaporation rate (kg water/h)	150	300	600	1200	1800	3600
Furnace capacity (kW)	220	450	900	1860	2700	5300

High-temperature batch driers

The high-temperature batch drier carries out the same functions as the continuous drier, but it operates in a series of separate steps instead of a flow process. The sequence of events can be controlled manually or automatically so that the drier can be left to run unattended. Each batch spends 2–4 h in the drier.

Batch driers can be installed on a permanent base, or some models can be moved on their own wheels from one site or one farm to another. Portable driers can be easily re-sold. The unit may be rated by the amount of grain in a batch, or by the number of tonnes of grain that can be dried by five percentage points in 24 h operation. Capacities range from 38–135 t/24 h. The capacity of a drier in 24 h should be equal to the daily harvesting capacity.

Storage, or low-temperature, driers

Where grain is to be stored on the farm after it has been dried, it is possible to reduce costs by combining the functions of drying and storage. Tables 14.4 and 14.5 refer to the storage volumes of various grain crops. Capital cost is generally less than for a separate drier and a store, and typical energy consumption is 3.5 MJ/kg of water evaporated. Against this, the system gives a slow drying rate, 0.5%/d or less, and its successful operation demands a high level of management.

Table 14.4 Crop weights and volumes

	kg/m^3	m^3/t
Wheat	785	1.28
Barley	689	1.42
Oats	513	1.95
Beans	817	1.20

Table 14.5 Round bin capacities (wheat)

Bin diameter (m)	*Volume (m^3 m)*	*Weight (t/m)*
2.4	4.5	3.5
3	7.1	5.5
3.7	10.8	8.4
4.3	14.5	11.3
4.9	18.9	14.8

Measurement and control of air relative humidity

The operation of any low-temperature drier relies on the fact that there is an equilibrium relationship between the moisture content of grain and the relative humidity (RH) of the air which surrounds it (Table 14.6). Grain is brought to a safe moisture content by ventilating it for two weeks or more with air at the corresponding equilibrium relative humidity. Any deviation in air quality will affect both the rate of drying and the final moisture content, so it is clear that the accurate measurement and control of relative humidity must be a prime requirement for successful drier management.

Relative humidity is measured with a wet and dry bulb hygrometer, which consists of a pair of identical thermometers. The bulb of one thermometer is covered with a wet cloth sleeve. After exposure to the air stream until readings stabilize, the two thermometers are read and the relative humidity is found from a slide rule or chart supplied with the instrument. All other instruments for measuring or controlling relative humidity must be calibrated regularly against the basic, and inexpensive, wet and dry bulb instrument.

If the relative humidity of the air is too high, it may be reduced by heating the air. Equipment which will raise the air temperature by 6°C is adequate to deal with any weather situation. Electrical heater banks are commonly used, because of their ease of installation, operation and control, but oil- or gas-fired heaters are available, or a stove may be fuelled with wood or straw. Diesel-engine-driven fans supply the waste heat of the engine, whch can raise the air temperature by up to 4.5°C. A heat input of 1.1 kW is needed to raise the temperature of 1 m^3/s of air by 1°C, and thus reduce the relative humidity by 4.5 units.

Heater output must be adjusted with every change in ambient conditions if the maximum drying rate is to be maintained without unnecessary use of heat. An automatic humidity control (humidistat) which will regulate the output of a series of electric heaters to maintain the drying air at the desired relative humidity will soon pay for itself in energy savings and reduced overdrying.

An alternative method of controlling air relative humidity is by means of a heat pump, which cools the air to condense moisture and then re-heats the dehumidified air to further reduce RH. A unit of adequate size can give a satisfactory performance at lower energy cost than heaters, but at a very much greater capital cost (Brook, 1986).

Table 14.6 Equilibrium moisture contents of a range of seeds at 25°C (% WB)

Species	*Air relative humidity (%)* 40	50	60	70	75	80	90
Wheat	10.7	12.0	13.7	15.6	16.6	17.6	23.0
Barley	9.8	11.5	13.2	15.1	16.1	17.3	23.1
Oats	9.5	11.0	12.5	14.3	15.4	16.7	22.0
Oilseed rape	5.7	6.6	7.4	8.8	10.2	11.9	17.4
Linseed	6.3	7.3	8.5	10.0	11.0	12.5	18.2
Field peas	9.5	11.2	13.0	15.8	18.0	20.9	28.6
Perennial ryegrass	9.2	10.6	11.9	13.9	15.3	17.4	24.4

After McLean, 1989

Note: As temperature rises, the equilibrium shifts so that RH increases in relation to MC and the air gains moisture from the grain. The reverse happens if temperature falls (Pixton & Warburton, 1971).

The low-temperature drying process

As the drying air moves upwards through the grain mass it picks up water and is cooled. Grain at the bottom of the store dries first, and a 'drying front' then moves slowly upward through the grain until all the gain is dry after two weeks or longer. During drying, the grain above the drying front remains as wet (or wetter) than when it was harvested. This portion of the grain is preserved before drying by the cooling effect of the air which has dried the lower layers of grain.

If grain at a moisture content above 22% is to be dried, it must not be loaded to a depth of more than 1 m. This reduces the drying time, increases drying efficiency and also prevents the soft grains being squashed by the weight of a full depth of corn. If grains at the bottom were crushed, this would reduce the air spaces between them and restrict the flow of drying air.

Drying rate and efficiency are improved for grain of any moisture content if the grain store is filled in layers as drying proceeds, rather than piling the grain in one portion of the store first (Sharp, 1984).

Heat should not be used for drying grain above 18% MC. Initial drying should be carried out with unheated air, with heat being applied only in the final stages of drying. Use of heat at an earlier stage will result in re-condensation of moisture in the upper layers of grain.

Progress of the drying front towards the grain surface can be followed by extracting samples from different depths with a sampling spear. The depth of the transition between dry and wet grain indicates the position of the drying front.

Air supply and distribution

The drying air is distributed over the base of the grain mass by means of a system of ducts, or through a fully perforated floor. An adequate quantity of air must be provided to make the drying period as short as possible. An airflow rate of 0.1 (m^3s)/m^2 of floor area, or 0.05 (m^3s)/m^2 of grain, whichever is greater, is recommended (ADAS, 1982). Since not all the grain in the store is being dried at once, the recommendations are applied to one-half to three-quarters of the store according to geographical region. Adequate fan capacity is a cornerstone of successful crop drying.

Fans for grain stores

A fan moves air by creating a pressure behind it. The air output of the fan decreases as the back pressure or static pressure against which it is working increases. Back pressure builds up as the air is forced along ducts, round bends and through the grain. The greater the air speed, the higher the back pressure (Cory, 1991).

The pressure generated by a fan is measured with a water manometer (WG = water gauge), and fan output can be determined from this reading by reference to manufacturers' performance curves. Working pressures for crop drying fans are 60–150 mm WG.

A fan is 40–85% efficient. The remaining 60–15% of its power consumption is lost as heat, which warms the air by up to 2.5°C (Sharp, 1982). This may be enough heat to dry a crop, but it is a disadvantage when the fan is used for cooling. Cooling systems sometimes operate under suction with the fan at the outlet; this way, the fan's heat does not enter the crop.

Some fans absorb more power as back pressure drops and air output increases, and the motor may become overloaded and overheat. Only non-overloading fans are suitable for crop drying.

Centrifugal fans

Centrifugal fans develop pressure by accelerating the air in the rotor and then converting its kinetic energy into pressure energy in the diverging housing or volute. They are quiet in operation and bulky. A form of centrifugal fan with backward-curved (b-c) blades on the rotor is non-overloading, and is frequently used in grain stores. Pressures up to 300 mm WG are possible, according to rotor diameter and speed.

Axial-flow fans

Pressure is created by the rotation of an airscrew-like impeller in a close-fitting circular housing. Air passes in a straight line through the fan, giving high peak efficiencies. The fan alone can only produce a pressure of 80–120 mm WG, but stationary intake guide-vanes can increase this by 20–60%. Two fans contra-rotating in the same tubular housing can increase the pressure up to three times. The fan is non-overloading, and more compact than a centrifugal fan of the same output. It is noisy, particularly when guide-vanes are fitted. Silencers are available, or the fan can be shielded with bales to absorb the noise.

Table 14.7 shows the output of various crop-drying fans against a range of static pressures.

Table 14.7 The output of various crop drying fans against a range of static pressures (manufacturers' figures)

Type and power of fan	*Air output (m^3/S)*					
	Static pressure (mm WG)					
	50	*75*	*100*	*125*	*150*	*175*
3.8 kW b-c	4.6	3.9	3.0	–	–	–
7.5 kW b-c	8.2	7.7	7.0	5.9	4.4	1.8
18.7 kW b-c	13.7	13.1	12.4	11.5	10.5	9.4
30 kW axial	10.0	8.8	7.1	4.7	1.8	–
52 kW axial	18.8	17.7	15.9	14.1	11.8	9.4

Ducts and drying floors

After being pressurized by the fan and heated if necessary, air reaches the grain via a main duct and a series of lateral ducts spaced at 1–1.2 m. To limit the build-up of back pressure, ducts should have a minimum cross-sectional area of 1 m^2 for every 10 m^3/s of air that passes. This will give a maximum air speed of 10 m/s in the ducts. Similarly, the size of the openings through which the air passes from the ducts into the grain should ensure that air speed at this point does not exceed 0.15 m/s. Lateral ducts do not normally exceed 10 m in length because the output of air along the duct becomes uneven beyond this length. A better distribution of air to the grain lying between the ducts is achieved when air escapes from the sides rather than the top of the duct.

Above-floor ducts convert any sound, level concrete floor into a drying floor, and can be removed if the building is needed for some other use. However, the operations of filling and emptying the store are made more difficult by the presence of above-floor ducts. Many stores are equipped with permanent below-floor ducts which give a flat floor and are strong enough to support the weight of trailers or bulk lorries. Although the flexibility of use of the store is curtailed and cost is increased, moving the grain in and out of store is greatly facilitated.

Completely even distribution of drying air over the floor area can be achieved if the entire floor is perforated, and air passes up from a plenum chamber below the floor, through the perforations and into the grain. This system is used in some grain bins. A further refinement to this type of floor is an arrangement of louvres which direct jets of air to 'sweep' the last few tonnes of grain out through an unloading chute using the power of the drying fan. This eliminates the dusty task of sweeping or shovelling the last corn out of a flat-bottomed bin.

Uniform air flow throughout the grain mass also demands that the grain itself be uniform. Any differences in depth or compaction, or any local concentrations of rubbish, will affect uniformity of air flow. Pre-cleaning the grain before it enters the store will remove most of the rubbish. Even layers of grain should be built up over the whole floor area by the careful use of conveyors, spreaders or grain throwers, to prevent rubbish, or a particular batch of wet grain, from being concentrated in one spot.

If air flow to any part of the store is restricted, the grain in that part of the store will dry slowly or not at all. Air flow through the grain can be measured at many points on the surface using an inverted funnel type of airflow meter, and any shortfall can be corrected. The instrument costs only the price of a few tonnes of corn, and is simple and quick to use.

To ensure that moisture-laden air can escape easily from the roof space above the grain store, exhaust openings of at least 1 $m^2/(2.5\, m^3 s)$ of air flow must be provided at eaves, ridge or gable ends.

Grain cleaners

It is desirable to remove chaff, dust and impurities by pre-cleaning grain before it is dried or put into store. This avoids the expense of drying valueless material, and reduces the risk of fire in high-temperature driers. The presence of impurities in a bulk store can block the spaces between the grains, and inhibit air flow. Concentrations of such impurities can build up under the discharge of stationary conveyors.

Grain may also be cleaned or graded when it is removed from store prior to sale; the ability to do this is particularly valuable where the grain is to be sold at a premium for seed, malting or baking, or in a situation of over supply where it is necessary to produce a good-looking sample.

Dual-purpose grain cleaners are available from many sources. For pre-cleaning, machines can accept 4–40 t/h; this output is halved for cleaning after storage when more time is available and a more thorough job is required. Most machines combine one or two aspirations (separation by airblast) with a series of vibrating sieves. Airblast and sieve sizes are adjusted to suit the crop being cleaned. Inexpensive aspirator pre-cleaners are available with outputs to 12 t/h.

Cleaners with reciprocating sieves should be mounted on a rigid base so that no damage will result from their constant vibration.

Management of grain in store

Monitoring grain temperature

After the grain has been dried and placed in store, careful management and monitoring of its condition is essential if the full value of the crop is to be retained. Grain temperature is normally monitored, since heating is a symptom of most of the problems which affect stored grain. Any local increase in grain temperature must be detected and the cause corrected before there is time for grain damage to occur.

Many manufacturers offer permanently installed systems of grain temperature sensors, with a central monitoring station where all the measurement points can be checked at least weekly. Some systems will themselves perform the checking, and sound an alarm if a pre-set temperature is exceeded. At the other end of the scale, a temperature sensing spear can be pushed into the grain and the temperature read off in a number of positions and depths in the grain store to build up a composite picture of grain temperature.

Aeration

Any grain store more than 20 t in capacity should be fitted with aeration equipment. Storage driers will have this equipment anyway, but where the grain has been dried elsewhere and then put into store, small fans and air distribution ducts must be provided. The required airflow is 0.5–0.7 m^3/min for each m^2 of floor area, or 10 m^3/ht of grain (ADAS, 1985). A 0.375 kW fan unit can aerate up to 50 t of grain at a time. To give even air distribution, ducts should not be spaced more widely apart than twice the depth of the grain.

Aeration helps to maintain grain quality in the following three ways.

General cooling

The cooler the grain even down to freezing point, the better it will keep (see Fig. 14.1). Without refrigeration, grain can only be cooled to ambient temperature. The ability to force cold air into the grain as the season cools, and to take advantage of cold nights in order to reduce the temperature of the grain, is a valuable aid to the management of the store.

Where aeration equipment is available in the store, the output of a continuous drier can be uprated by converting its cooling section into additional drying space, and cooling the dry grain in the store.

In addition to the conditioning of dry grain, aeration has been used to preserve undried grain of up to 21% MC which is to be used on the farm. The success of this system is dependent on the season; cold nights during harvest, and a cool dry autumn are ideal; a mild, moist autumn might not permit the grain to be cooled quickly enough to prevent mould development. As with other low-cost storage systems, careful management and a thorough understanding of the principles involved are essential for success.

Elimination of hot spots

When localized heating of the grain is detected, the traditional solution is to 'turn' the grain to break up the hot spot and allow it to cool down by exposure to the air. When aeration equipment is fitted, cool air can be blown through the grain to reduce spot temperatures without the need to move the grain.

Maintaining an even temperature throughout the grain

When the grain goes into store, it is all at approximately the same temperature as the outside air. As the season progresses, the layers of grain in contact with the walls of the store will cool, while the grain in the centre of the store will remain at its original temperature. Sunshine will heat the grain at the south side of a store. Any temperature inequalities of this sort will set up convection currents in the air which fills the spaces between the grains. This will result in moisture removal from the warmer grains and condensation on the cooler grains, particularly just below the surface at the centre of the store. This process can result in mould development, which may be followed by heating, insect activity and even sprouting.

If the grain is cooled in line with the cooling of the season, temperature will be uniform throughout the grain mass and moisture movement will not occur. This eliminates the need for costly over-drying which gives a 'safety margin' to allow for the effects of moisture movement. As well as minimizing any losses due to deterioration of the grain, the correct use of aeration equipment will permit drying costs to be reduced.

Handling grain (Fig. 14.3)

Most grain storage installations rely on conveyors for moving the grain. The characteristics of the common types of handling equipment are summarized below.

Auger conveyors

The grain is moved by a screw, or 'flight', which rotates inside a steel tube. Grain is drawn in at one end, and is normally discharged at the other end although intermediate discharge points are possible.

Augers are made in diameters of 90, 115, 150, 200 and 300 mm. Outputs of dry wheat, when the conveyor is inclined at 45 degrees to the vertical are as follows: 114 mm–25 t/h; 150 mm–35 t/h; 200 mm–58 t/h; 300 mm–120 t/h. Output is reduced with other grains, or if the grain is damp, or if the angle is steeper, and is increased if the angle is flatter. Portable augers are made in lengths up to 15 m: large units have a wheeled tripod stand.

The auger is a versatile conveyor which will operate in many positions without adjustment. There is considerable churning and abrasion of the conveyed material, particularly if the auger is only partially full. The churning effect may be used to advantage in mixing a preservative with grain, but augers must not be used for conveying fragile materials such as rolled barley or pelleted feeds, because of the risk of damage.

Fig. 14.3 Eighteen tonne per hour mixed-flow drier with associated grain handling equipment installed at a farm grain store. Courtesy: Law-Denis Engineers Ltd.

An auger may be used to meter the flow of, for example, an ingredient into a mixer by regulating its speed of rotation. This gives very accurate control of output.

The flight of the auger projects beyond the tube at the intake end. This portion must be guarded at all times with a welded-mesh screen to avoid the risk of serious personal injury.

Chain and flight conveyors

The grain is moved along a smooth wooden or steel trough by an endless chain fitted with horizontal scraper or 'flights'. The grain can be fed into the trough at any point, and is discharged at the end of the conveyor or at intermediate discharge points if these are fitted. Some models are reversible.

Chain and flight conveyors are normally horizontal, although some can work at small slopes. They are quiet in operation, long-lasting, economical in power, and do little damage to the grain. Lengths up to 60 m are available; outputs range from 15 t/h to more than 100 t/h. They are widely used as top conveyors in bin-type grain stores. 'Double-flow' models can supply grain to a continuous drier, and also take excess grain back to store.

Belt conveyors

The grain lies on an endless fabric-reinforced rubber belt which slides along a flat bed. Grain can be loaded onto the conveyor at any point, through a hopper which concentrates the grain in the centre of the belt. Discharge is over the end of the belt, or at a movable discharge point where the belt passes in an S-shape over a pair of rollers. The grain shoots over the upper roller and is deflected into a side discharge trough to right or left.

A motorized discharge can move slowly back and forth along a conveyor distributing the grain evenly along the whole length. In conjunction with a grain thrower to project the grain across the store, a level layer of grain can be built up over the complete area.

The belt conveyor is quiet, and very gentle with the material being conveyed. It is found as top and bottom conveyors in bin-type grain stores. Conveyors with cleated belts can operate at steep angles, but the typical grain version has a smooth or textured surface without cleats, and operates horizontally.

Belt conveyors are available in lengths up to 45 m and capacities from 20 to more than 50 t/h. Power requirement is greater than for the chain and flight machine.

The bucket (or belt) elevator

This is a device for raising grain vertically. A series of steel or plastic scoops or buckets are bolted to a flat endless belt which moves vertically between a pair of pulleys. The motor and drive pulley are at the top, while the belt passes round a tensioning idler pulley at the bottom. Grain is fed into the conveyor through a regulating orifice up to 1 m above its lowest point, and is discharged from a spout up to 1 m below its highest point. For this reason, the elevator usually stands in a hole at least 1 m deeper than the bottom of the receiving pit, and is often subject to flooding in winter when the water table is high. The top of the conveyor is generally accommodated in a small extension of the grain store roof where access for maintenance may be less than good.

No damage can occur to the grain while it is in the buckets, but careful operation of the intake slide and correct design of intake and discharge points are necessary if damage is to be avoided entirely.

Bucket elevators are available in capacities from 5 to 120 t/h, and heights to 25 m. 'Double-leg' models combine two separate conveyors with a single drive unit. The two legs may be used separately; for example, one leg may load wet grain into a continuous drier while the other takes dry grain from the drier and raises it to the top conveyor of the store. When a high throughput is required, for example when loading a lorry, both legs can be used to raise grain from the bottom conveyor into the reversed top conveyor which carries it to a discharge chute outside the store.

Multiple valving devices can be used to direct the grain from the conveyor's discharge to alternative destinations. Controls at ground level simplify the operation of such valves.

Pneumatic conveyors

The grain is carried along a closed tube 130–210 mm in diameter by a stream of air moving at 70–100 km/h. The air stream is produced by a narrow, large-diameter fan driven by electric motor or tractor power take-off.

Intake may be by a flexible suction spout, which is very convenient for emptying flat-bottomed bins or clearing floor stores. Alternatively, the grain may be injected into the air stream through a metering hopper. Discharge is from the far end of the system, which may be up to 200 m from the intake if sufficient power is applied, at outputs of 5–40 t/h. The grain does not pass through the fan but considerable abrasion can occur as it moves along the pipes and round bends; this shows itself in the large amount of dust which is generally produced at the discharge end.

This is a very versatile conveying system, as the pipework can be joined together to accommodate any storage arrangement, and the power unit is generally portable and can be used at several locations during its season of work. Against this must be set a high initial cost, and the fact that the power consumption of 1–2.5 kW/(th), is high, particularly for those models with a suction intake. Fragile materials such as rolled grain or pelleted diets should not be handled in this way.

The grain thrower

A floor store can be filled in a series of flat layers if a grain thrower is available. The grain is metered onto a fast moving conveyor belt or a rotating paddle-wheel which throws it up to 12 m horizontally, and to a height of up to 5 m. Larger models stand on the floor; smaller units may be mounted at the discharge of other types of conveyors. Outputs range from 5 to 120 t/h, at a power consumption of 1 kW per 4–7 t/h. There is little damage to the grain, although some dust is produced.

A further advantage of loading a floor store in this way is that any rubbish present in the grain is distributed evenly throughout the store and will not have any local effect on the drying process.

Grain weighers

A very important instrument in the grain producer's armoury is the weigher. It permits him to determine yields at harvest time, to control his stock of grain, and to weigh the grain as it comes out of store for sale or for livestock feed on the farm.

The weigher consists of a counterbalanced hopper which is positioned in the flow path of the grain. When the hopper has received a pre-set weight of grain, the flow is cut off, one increment of weight is recorded on a counter, the hopper is discharged and the cycle is repeated. Very accurate weighing is possible at rates from 6 to more than 100 t/h according to model.

Loading lorries

If the crop is to be sold in bulk, it may have to be discharged into lorries of 38 t or more. If conveying equipment of 20 t/h is used, which is adequate for most other functions in the grain store, the lorry will be tied up for almost 2 hours and this may make the product less attractive to a potential customer. High capacity conveyors are available, but would only be fully utilized for a few hours per year.

To overcome this problem some grain stores are equipped with covered V-bottomed hoppers of 40 t capacity with several large discharge ports mounted high enough for a bulk lorry to be driven underneath. The hopper can be filled slowly with the store's existing conveyors, allowing plenty of time for the grain to be passed through cleaning equipment if required, and yet the lorry can be filled in 10 minutes or so with the minimum of effort or dust. In the likely event of there being a buyer's market for grain, it is this type of detail, plus a quality product, that will ensure a sale in a competitive situation.

Storage of grain in sealed containers

Where grain is to be processed into livestock feed on the farm, there are many advantages to storing it without drying. Apart from the fact that there is no drying cost, the grain can be handled into store at a very fast rate so there are no harvest bottlenecks. Moist grain can be rolled easily, and produces little dust when ground; this can improve palatability to stock and improve working conditions for the stockman.

When stored under airtight conditions, the oxygen concentration is quickly reduced to less than 1% by the respiration of the grain and other aerobic organisms such as moulds. The grain is killed and most destructive organisms die or become dormant (Hyde, 1965). Some fermentation may take place if the moisture content is above 24% and this will taint the grain.

As long as the container remains sealed, the condition of the stored grain is stable. When grain is removed from the store, care must be taken to allow only the minimum of air to enter, and to reseal as soon as possible.

Once exposed to the air, the moist dead grain will deteriorate rapidly especially if the weather is warm. Normally, only enough for 1 or 2 d use should be withdrawn at a time.

The container for airtight storage may be a plastic or butyl bag, or the more durable enamelled steel tower silo. Capacities are from 15 to 800 t/silo. Unloading is by means of augers, a sweep arm which can traverse the circular bottom of the silo, and a fixed arm which brings the grain from the centre of the silo to discharge outside through a resealable spout.

If grain above 22% MC is to be stored in a tower silo, the lower layers are likely to be compressed by the weight above and may bridge and be difficult to unload. A few tonnes of dry corn at the bottom of the silo will facilitate this initial stage of unloading, as will the use of a gentle method of filling such as a bucket elevator rather than the more convenient but damaging blower.

Use of chemical preservatives to store grain

An alternative method of controlling moulds and pests in undried grain is to treat the grain with a chemical preservative such as propionic acid, at a rate which increases with the moisture content of the grain.

After treatment, no special storage structure is required; the grain must be kept dry to avoid dilution or leaching of the preservative so floor, walls and roof must be waterproof. Since propionic acid is highly corrosive, even to galvanized steel, the materials of the store must resist corrosion or be coated with plastic or a chlorinated rubber paint for their protection. Any equipment which handles either the preservative, or grain which has been freshly treated, must be thoroughly washed out with water after use to minimize corrosion.

The preservative liquid or its fumes can cause irritation or injury to the skin, eyes, mouth or nose. Protective garments should be worn, and a supply of water kept close by in order to wash off splashes. The operator should stand upwind of working application machines or piles of recently treated grain.

The applicator generally consists of a pump, flow regulator and nozzle which draw the chemical from a container and spray it on to the crop at the intake of a short (1.5 m) length of auger. The tumbling effect of the auger is very effective in distributing the chemical over every grain, an important requirement of this sytem. Charts are provided so that the chemical flow rate can be adjusted to match the grain moisture content (up to 30% and more) and the (measured) throughput of the auger.

The cost of the chemical is relatively high, so in order to be viable the handling and storage parts of the system must be inexpensive.

Since the grain is killed and tainted by the preservative, there is no market for the grain other than for stockfeed. Weed seeds are also killed, which may be an advantage.

Because the chemical is absorbed into the grain, protection continues after the grain is removed from store, so quite large batches of feed can be prepared at a time, even in warm weather, without risk of deterioration.

References

ADAS (1982) Booklet 2416. *Bulk Grain Driers*. MAFF Publications, Alnwick.

ADAS (1985) Booklet 2497. *Low-rate Aeration of Grain*. MAFF Publications, Alnwick.

Bakker-Arkema, F.W. (1984) Selected aspects of crop processing and storage: a review. *J. Agric. Engng. Res.* **30**, 1–22.

Blakeman, R. (1982) Fuel saving in grain drying. *Agr. Eng.* **37**, 60–67.

Brook, R.C. (1986) Heat pumps for near ambient grain drying: a performance and feasibility study. *Agr. Engr.* **41**, 52–57.

Brooker, D.B., Bakker-Arkema, F.W. & Hall, C.W. (1974) *Drying Cereal Grains*. Avi Publishing Co., Westport, CT.

Bruce, D.M. & McFarlane, N.J.B. (1993) Control of mixed-flow grain driers: an improved feedback plus feedforward algorithm. *J. Agric. Engng. Res.* **56**, 225–238.

Bruce, D.M. & Nellist, M.E. (1986) Simulation of continuous-flow drying. *Proc. BSRAE Association Members' Day*, 26–27 February.

Burges, H.D. & Burrell, N.J. (1964) Cooling bulk grain in the British climate to control storage insects and to improve keeping quality. *J. Sci. Fd. Agric.* **15**, 32–50.

Cory, W.T. (1991) Fans for today's agriculture. *Agr. Engr.* **46**, 2–8, 34–38, 66–70.

Hyde, M.B. (1965) Principles of wet grain conservation. *Agr. Engr.* **21**, 75–82.

McLean, K.A. (1989) *Drying and Storing Combinable Crops*, 2nd edn. Farming Press, Ipswich.

Pixton, S.W. & Warburton, S. (1971) Moisture content/relative humidity equilibrium of some cereal grains at different temperatures. *J. Stored Prod. Res.* **6**, 283–293.

Sharp, J.R. (1982) A review of low-temperature drying simulation models. *J. Agric. Engng. Res.* **27**, 169–190.

Sharp, J.R. (1984) The design and management of low-temperature grain driers in England – a simulation study. *J. Agric. Engng. Res.* **29**, 123–131.

Wilkin, D.R. (1975) The effects of mechanical handling and the admixture of acaricides on mites in farm-stored barley. *J. Stored Prod. Res.* **11**, 87–95.

Part 3

Animal Production

15

Animal Physiology and Nutrition

R.M. Orr & J.A. Kirk

Animal production involves the conversion of feedstuffs, mostly of vegetable origin, to animal products in the form of meat, milk, eggs and wool. In this chapter those physiological and metabolic processes which are especially relevant to this conversion are considered. The chapter is divided into eight sections:

Regulation of body function
Chemical composition of feedstuffs and animals
Digestion
Metabolism
Voluntary food intake
Reproduction
Lactation
Growth
Environmental physiology and behaviour

Since animal production involves various species of farm animal, an attempt is made throughout to include reference to the comparative aspects of the subject. In addition, since animal productionists are concerned with the efficiency of the conversion of feedstuffs into products, reference is likewise made where appropriate to this important aspect of livestock science.

Regulation of body function

In order that animals can survive their environment and changes in that environment, they must possess the necessary coordinated mechanisms to maintain their own internal environment in a steady state. These physiological reactions that maintain the steady states of the body have been designated 'homeostatic' and are achieved by the combined action of the various organ systems of the body. This coordinated action is carried out by the nervous and endocrine systems. These systems must initiate the necessary adjustments that enable animals to respond to external environmental changes such as the availability of food, temperature and light.

The nervous system

The nervous system in mammals is an extremely complex collection of nerve cells which plays an essential part in the functioning and behaviour of the animal. The basic units of the nervous system are the nerve cells or neurons and the associated receptors.

The ability of animals to maintain homeostasis is dependent on the nervous system receiving and responding to information from specialized receptor cells. To this end mammals have evolved receptors that are sensitive to a wide range of physical (e.g. temperature), chemical (e.g. specific components of the blood) and mechanical (e.g. muscle stretch) stimuli. When any response to information received at receptors is necessary, the information is carried from receptors to the cell or tissue (effector) where appropriate adjustments are made. This is the function of the neurons.

There are three different types of neurons:

(1) sensory or afferent neurons which carry information from receptors to the spinal cord and brain;
(2) motor or efferent neurons which carry information from the spinal cord and brain to the effector organ; and
(3) interneurons or association neurons which make connections between sensory and motor neurons within the spinal cord and brain.

Neurons consist essentially of a cell body with an elongated projection, the axon, and numerous shorter projections, the dendrites (Fig. 15.1). For our purposes the cell body can be thought of as performing two functions. First, it receives information from the axons of other cells both directly on to its surface membrane and also via the dendrites and, secondly, it acts to sum all the effects of inputs, which may be either excitatory or inhibitory, that it receives. Should this sum of inputs exceed a critical level, then an impulse is triggered which is then carried along the axon to the synaptic region which in turn passes the neuronal signal to the next stage in the nervous system, either another neuron or, alternatively, a muscle or gland. Propagation of signals along an axon is dependent on changes in the electrical potential between the inside and the outside of the axon membrane which results from changes in the permeability of the membrane to sodium, potassium and chlorine ions. Passage of neuronal signals at synaptic regions involves the release of a chemical transmitter substance into the small

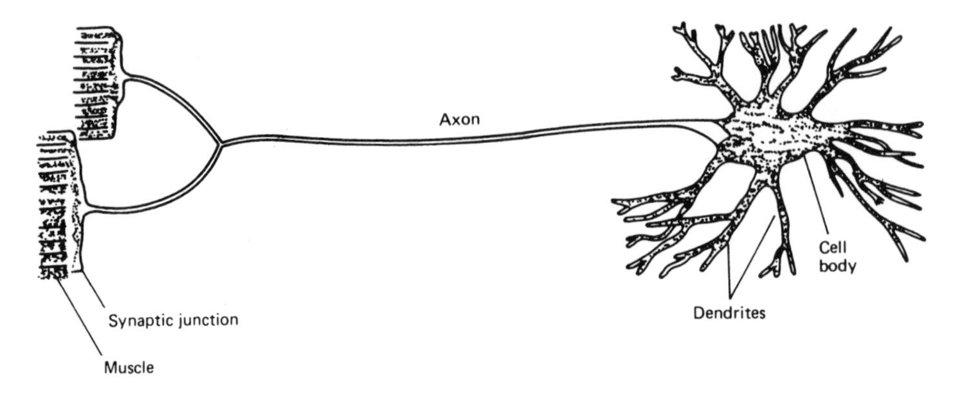

Fig. 15.1 A nerve cell.

gap between the synaptic region and the membrane of the next neuron. The nature of the transmitter substance determines whether the influence on the next neuron is excitatory (e.g. acetylcholine and noradrenaline) or inhibitory (e.g. γ-amino butyric acid). These transmitter substances released by the synapse are received by specialized sites on the dendrites of the next neuron which have the effect of changing the electrical potential of the neuron.

It is the nerve cells – the basic building blocks – from which the various components of the nervous system of mammals are built (Fig. 15.2).

Division of the nervous system between central and peripheral is made simply on the basis of whether the fibres or cell bodies lie outside the spinal cord. The peripheral nervous system can be further subdivided into the somatic nervous system and the autonomic nervous system. The former is involved in the control of the skeletal muscles in the body whereas the latter controls specific target organs in the body such as the heart, lungs, blood vessels and intestines. The autonomic nervous system has two subdivisions, the sympathetic and the parasympathetic. In many cases these two divisions have opposite effects on the target organs and glands. The sympathetic nervous system acts to increase the level of activity or secretion in an organ whereas stimulation of the parasympathetic will have the opposite effect.

The central nervous system consists of the spinal cord and brain. The spinal cord is contained in a continuous

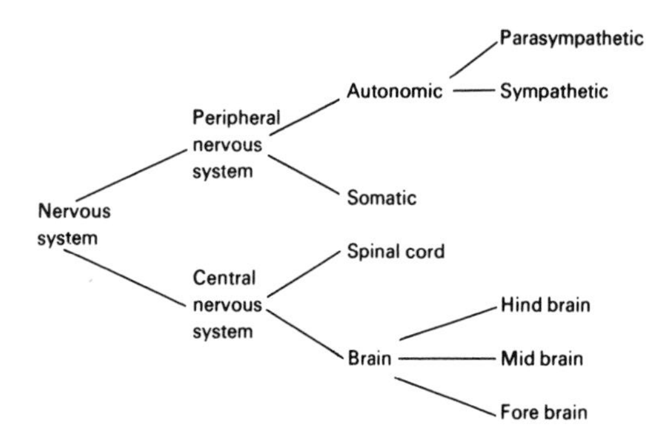

Fig. 15.2 The nervous system of mammals.

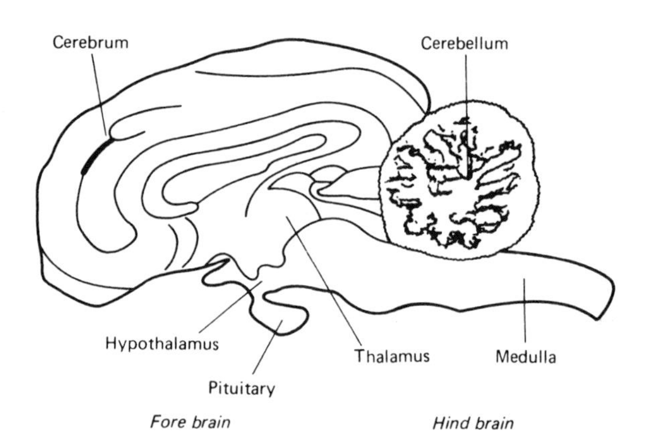

Fig. 15.3 Midline section of the brain showing the location of major structures.

channel within the vertebrae, which form the backbone. At each vertebra there are two openings in the base through which nerves can pass in and out of the spinal cord. Sensory nerves enter through openings on the dorsal surface, motor nerves leave through openings on the ventral surface. The brain itself is a vastly complex organ consisting of billions of nerve cells, interconnected by neurons. However, certain areas of the brain (Fig. 15.3) have been found to have specific functions.

(1) The hind brain: this consists of the medulla and the cerebellum. The medulla is involved in the regulation of the heart, breathing, blood flow and posture. The cerebellum is involved in the coordination of muscle activity.
(2) The fore brain: this consists of the thalamus, the hypothalamus and the cerebrum. The thalamus plays an important role in the analysis and transmission of sensory information between the spinal cord and the cerebral cortex. Lying directly beneath the thalamus is the hypothalamus which, despite its small size, has been shown to be involved in the control of basic behaviours such as hunger, thirst and sexual behaviour. The hypothalamus also has close connections with the pituitary – a hormone-

Table 15.2 The composition of the animal and feedstuffs (%)

	Water	*Carbohydrate*	*Protein*	*Fat*	*Ash*
Pig, 30 kg	60	<1	13	24	2.5
Adult cow	58	<1	16	20	4
Hen	56	<1	21	19	3
Grass – leafy	80	12	5	1	2
Wheat straw	10	75	3	2	9
Wheat grain	13	72	12	2	2
Turnips	90	7	<1	1	1
Soyabean meal	12	35	45	2	6

Table 15.3 Major amino acids occurring in proteins

Essential	*Non-essential*
Arginine	Alanine
Histidine	Aspartic acid
Isoleucine	Citrulline
Leucine	Cystine
Lysine	Glutamic acid*
Methionine	Glycine
Phenylalanine	Hydroxyproline
Threonine	Proline*
Tryptophan	Serine
Valine	Tyrosine

* Amino acids also essential to the chick.

the conversion of inputs that are mainly in the form of carbohydrates with lesser quantities of protein into outputs that are mainly composed of fat and proteins. The minerals (ash) and vitamins are present in relatively small proportions and are collectively termed micronutrients.

Proteins

One of the most characteristic chemical features of all organisms is their content of proteins. In both plants and animals, proteins perform the important function of *enzymes* which are responsible for metabolic processes. In the animal they act as the structural proteins of the skin (collagen and keratin), connective tissues, tendons and bones (collagen) and muscle (actin and myosin), as hormones (e.g. insulin), as *antibodies* and as substances performing transport and osmotic functions in the blood (e.g. haemoglobin, albumin).

The fundamental units from which all proteins are constructed are amino acids. All amino acids contain the elements carbon, hydrogen, oxygen and nitrogen. The nitrogen content of most proteins varies between 15–17% with an average of 16%. Measurement of protein content of foods is based on estimation of nitrogen content. Traditionally protein content is expressed as crude protein (CP) where $CP = N \times (100/16)$. Three amino acids, cystine, cysteine and methionine also contain the element sulphur. Thus the sulphur content of proteins is a function of the proportion of sulphur-containing amino acids present, e.g. wool has a high sulphur content reflecting its high cystine content. Proteins are synthesized from a pool of 20 different amino acids (Table 15.3). These conform to the general formula

$$\begin{array}{c} R \\ | \\ H_2N - C - COOH \\ | \\ H \end{array}$$

in which a central carbon atom has attached to it an amino group ($-NH_2$), a carboxylic acid group ($-COOH$), a hydrogen atom and a variable side chain designated here by the letter R. The condensation of the amino group of one amino acid with the carboxyl group of the next, forming peptide bonds, is the mechanism by which proteins are produced. Proteins generally contain several hundred amino acid residues and are thus large molecules with molecular weights ranging from 35 000 up to several hundred thousand. Proteins are synthesized in plant and animal cells where the cell nucleus contains genetic material which determines the nature of the newly synthesized protein. The physical and chemical characteristics of proteins are altered by the different proportions of amino acids, the sequence in which they occur in the protein, the degree of cross-linking that occurs between different parts of the molecule and by the presence of other compounds in their structure. For example, some proteins contain lipids (lipoproteins), carbohydrates (glycoproteins) or mineral elements (haemoglobin, casein). From the viewpoint of the nutritionist the proportion of the different amino acids present in a protein and those characteristics of the protein which may influence their availability to the animal are of primary importance.

Whereas plants and many micro-organisms are capable of synthesizing all of the different amino acids found in proteins provided they have an adequate supply of inorganic nitrogen and organic compounds capable of supplying the other elemental components, higher animals are not capable of synthesizing all amino acids required by the various tissues. Thus, some amino acids are required in the diet of most animals and are referred to as *essential* amino acids. Those not specifically required in the diet are called non-essential amino acids (see Table 15.3).

Thus, the ability of a food protein to furnish the animal with essential amino acids is a parameter that particularly influences its value to the animal. In practical animal nutrition the amino acids most likely to be limiting in their supply are lysine, methionine and tryptophan. Ruminants do not require dietary amino acids to the same extent as monogastric species such as pigs and poultry. This is because the microflora in their digestive tracts are capable of synthesizing amino acids (both essential and non-essential) from the simple organic compounds found in the rumen. This supply of amino acids is, however, inadequate to meet the demand of high producing animals.

The ease with which food proteins may be hydrolysed during digestion may influence their nutritional value. This is influenced by the solubility (characteristics of the protein. Feed processing techniques (e.g. heat or formaldehyde treatment) which bring about denaturation of the physical structure of a protein may influence its solubility characteristics.

In ruminant nutrition the proportion of dietary protein

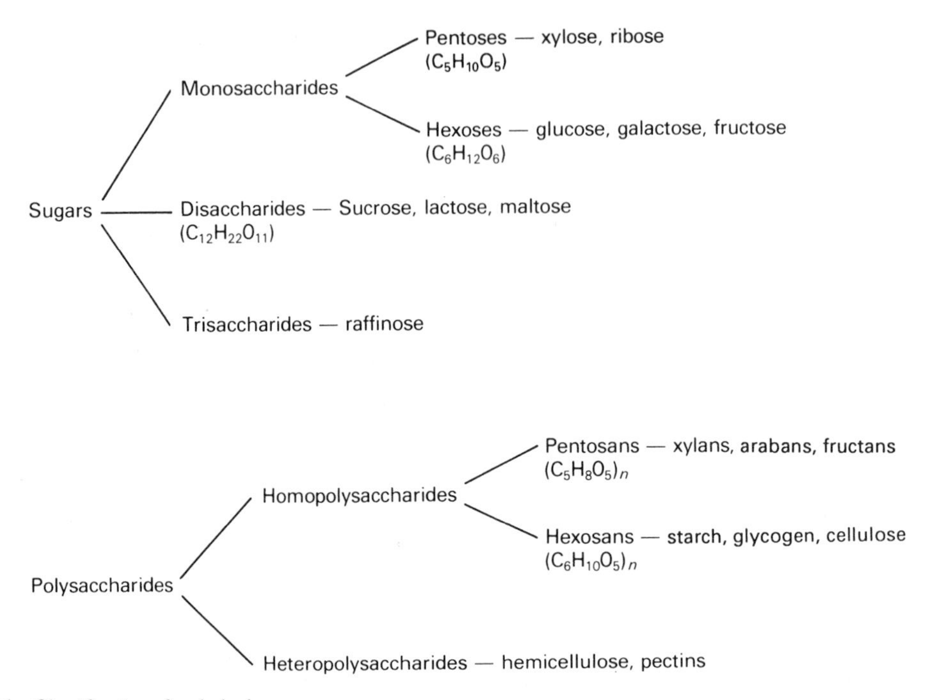

Fig. 15.6 Classification of carbohydrates.

hydrolysed by the rumen microbial population is referred to as its *degradability*. Protein hydrolysed by rumen micro-organisms is termed rumen degradable protein (RDP) and that which is resistant to hydrolysis is undegradable protein (UDP).

A comparison of the crude protein, rumen degradability parameters and essential amino acid content of animal feeds is given in Table 17.3.

Carbohydrates

Carbohydrates are the main product of photosynthetic activity in plants and contribute up to 80% of the dry matter in many feedstuffs. The nutritive value of these carbohydrates to the animal is variable and dependent on both the chemical structure of the carbohydrate and the digestive capacity of the animal. Generally speaking the variety of carbohydrate types in plants is far wider than in animals where it accounts for only a very small proportion of the body composition.

Carbohydrates are composed of the elements carbon, hydrogen and oxygen. The general formula $C_n(H_2O)_n$ may be used to represent any carbohydrate and indicates that the ratio of hydrogen to oxygen atoms is always 2:1, as in water. A classification of commonly occurring carbohydrates is given in Fig. 15.6 and the carbohydrate content of a range of plant tissues is shown in Table 15.4.

From the point of view of the nutritionist a distinction may be made between those carbohydrates that occur within plant cells, either as simple sugars or storage reserve compounds (e.g. starch, sucrose and fructans) and those that occur in the cell wall performing a structural role (e.g. pectins, cellulose, hemicellulose). Such a distinction is based on the digestibility characteristics of carbohydrates, the former being readily hydrolysed

Table 15.4 Carbohydrate content of plant tissues, % of DM

Component	*Tropical grasses*	*Temperate grasses*	*Cereal seeds*	*Alfalfa*	*Green vegetables*
Sugars	5	10	negligible	5–15	20
Fructans	0	1–25	0	0	–
Starch	1–5	0	80	1–7	low
Pectin	1–2	1–2	negligible	5–10	10–20
Cellulose	30–40	20–40	2–5	20–35	20
Hemicellulose	30–40	15–25	7–15	8–10	low

From Van Soest, P.J. (1982) *Nutritional Ecology of the Ruminant*, O & B Books Inc., Corvallis, with permission.

whereas the latter are relatively resistant, their degradation being largely a function of microbial activity in the digestive tract.

The non-structural carbohydrates of plant material can be further categorized in terms of their cold water solubility. The term 'water-soluble carbohydrates' refers to the monosaccharides, oligosaccharides and some polysaccharides, principally fructans, and distinguishes them from the starches that are the principal component of most seeds. Cereal grains are the major source of starch to farm animals.

Although small amounts of various free monosaccharides may be detected in feeds (e.g. glucose, fructose), most are of sufficiently low concentrations to be of little importance in nutrition. Sucrose is the main sugar in the sap of plants and in the case of root crops serves as the primary form of energy storage. Many plants convert sucrose into other forms for storage. Temperate grasses store fructans in leaves and stems and starch in the seed. There are two general types of fructans, the levans that occur in grasses, and inulins that are characteristic of the Compositae, e.g. the Jerusalem artichoke. Fructan content may account for up to 25% of dry matter in perennial ryegrasses, especially in cool growing seasons.

The water soluble carbohydrate of forages is especially important in the ensilage process (see Chapter 11) and may also influence their palatability. The water soluble carbohydrate content is chiefly influenced by the physiological conditions of growth. High light intensity and photosynthetic rate increase water soluble carbohydrate content. Hence marked diurnal variation in the water soluble carbohydrate content occurs in the living plant. Respiration of cut and drying forage may markedly reduce sugar content.

Starch is the most important storage carbohydrate in plants. Two types of starch exist, amylose and amylopectin. Both have crystalline structures which are disrupted by heating. The temperature at which this occurs is called the gelatinization temperature and these changes in the starch structures upon moist heating are partly responsible for the improvement in utilization resulting from steaming, flaking, micronizing and pelleting. Physical processing of the gelatinized starch is often required to prevent recrystallization (retrogradation).

Excessive heat treatment may cause caramelization of the carbohydrates which has a detrimental effect on utilization. The conflicting results that have been obtained in feeding studies of processed grains are most likely due to these interacting effects of gelatinization, retrogradation and caramelization.

Of the structural carbohydrates the most abundant are the celluloses along with lesser quantities of hemicelluloses, principally xylans. The former are polymers of glucose, the latter polymers of xylose. The proportion of these carbohydrates in the plant increases as it matures. Their nutritional availability to the animal varies from total indigestibility to complete digestibility and depends on the animal to which it is fed and on the degree of lignification. Differences between animals in their ability to utilize these carbohydrates are largely a function of microbial activity in their digestive tracts – ruminants and to a lesser extent other herbivores (e.g. horses, rabbits) having a greater capacity than monogastric species.

Lignin is a complex substance, a phenylpropanoid polymer of high molecular weight which associates with the structural carbohydrates in the cell wall to form an amorphous matrix. It is particularly resistant to degradation and is the main factor limiting digestibility of forages. The lignin content of plant cell walls increases with maturity and accounts for the reduction in digestibility of herbage as it matures. Cereal straws are examples of highly lignified feeds. Chemical treatment of cereal straws with alkalis such as sodium and ammonium hydroxide breaks the mainly ester bonding between lignin and the structural carbohydrates in the cell wall and increases the susceptibility of the cell wall to degradation in the digestive tract of animals. The efficiency of the treatment is dependent upon the proportion of lignin–carbohydrate bonds that are broken.

In animal tissues the main carbohydrates represented are glucose, which features in the energy metabolism of animals, and glycogen which is synthesized in muscle and liver from glucose and acts as a readily available form of energy storage. The amount of glycogen and its rate of metabolism are of particular relevance to changes which occur in the muscles of meat animals after slaughter. Lactose is a disaccharide synthesized in the cells of the mammary gland, representing over 95% of the total carbohydrate present in milk.

Lipids

The lipids are a diverse group of substances which share the common property of being relatively insoluble in water and readily soluble in organic solvents such as ether or chloroform. Most animal feeds contain up to 5% of lipid in the dry matter whilst the animal body may contain up to 40% of lipids. A variety of different types of lipids are found in plant and animal tissues performing a range of important biochemical or physiological functions. A chemical classification of lipids and some of their functions are given in Fig. 15.7.

Fats and oils are constituents of both plants and animals and account for about 98% of all naturally occurring lipids. They have the same chemical structure and properties, differing in that oils occur in liquid form in plants and fats chiefly in the solid form in animal tissues. Structurally, they are mainly triglycerides (neutral fats) in which three fatty acids are joined by ester linkages to glycerol. The fatty acids have the general formula $CH_3(CH_2)_nCOOH$. As well as differing in chain length – naturally occurring fatty acids are mainly of chain length 4–18 carbons – they also differ in their degree of unsaturation (Table 15.5).

In most triglycerides there is a mixture of fatty acids, and in both fats and oils there is a mixture of triglyceride types. Animal body fats are characterized by a high proportion of saturated and mono-unsaturated fatty acids – particularly stearic and oleic – plant and fish oils by a high proportion of poly-unsaturated fatty acids (Table 15.6). Three of the fatty acids, linoleic, linolenic and arachidonic acids, are known as the essential fatty acids since they cannot be synthesized by the animal body and must be provided in the diet from vegetable sources. However, arachidonic acid can be formed from dietary linoleic acid. The essential fatty acids are precursors of the prostaglandins.

The energy content of triglycerides is considerably

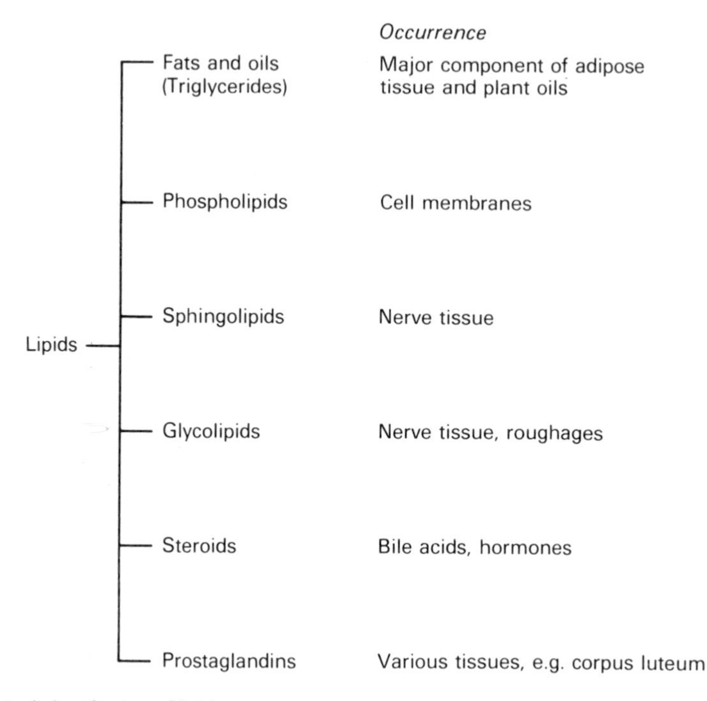

Fig. 15.7 Chemical classification of lipids.

greater than that of other nutrients – whereas 1 g of carbohydrate has a gross energy content of about 16 kJ, 1 g of triglyceride has a gross energy content of about 40 kJ. Thus fat may be included in animal diets as a means of increasing their energy density. Triglyceride, unlike the glycogen of liver and muscles, can be laid down in virtually unlimited amounts in adipose tissue (white fat) and serves as a more economical means of energy storage. Much of the white adipose tissue of the body is deposited under the skin (subcutaneous) with smaller proportions of the body's fat to be found around the kidneys, between muscles (intermuscular fat), and between and within muscle fibres. Species and strain of animals are important factors determining the distribution of fat depots. For example, 'beef' breeds of cattle have lower proportions of internal fat depots than 'dairy' breeds of cattle. The composition of the body fat in simple-stomached species may be influenced by the type and amount of dietary fat consumed.

Table 15.5 Commonly occurring fatty acids

	*Abbreviated designation**
Saturated fatty acids	
Acetic	C2:0
Propionic	C3:0
Butyric	C4:0
Caproic	C6:0
Caprylic	C8:0
Capric	C10:0
Lauric	C12:0
Myristic	C14:0
Palmitic	C16:0
Stearic	C18:0
Arachidic	C20:0
Unsaturated fatty acids	
Palmitoleic	C16:1
Oleic	C18:1
Linoleic	C18:2
Linolenic	C18:3
Arachidonic	C20:4

* The first number after C indicates the number of carbon atoms and the second number indicates the number of double bonds present.

The membranes of cells and intracellular organelles in animals are largely comprised of a group of lipids termed phospholipids characterized by their phosphoric acid component. The most commonly occurring are the lecithins which also contain the water soluble vitamin choline. The phospholipids are higher in unsaturated fatty acids than the triglycerides of adipose tissue. Arachidonic acid is especially prominent and acts as a depot of this metabolically important fatty acid. The phospholipids are, along with another group of compound lipids, the sphingolipids, major components of nervous tissue. As substances with emulsifying properties phospholipids fulfil important functions in lipid transport in the blood in which they combine with simple fats, cholesterol and proteins to form lipoproteins. Blood lipoproteins are classified on the basis of their density which is, in turn, a reflection of the proportions of their balance of lipid and protein.

Whereas triglycerides are the major lipid components of concentrate feeds and seeds, the lipids in roughages are mainly in the form of glycolipids which in addition to containing fatty acids, chiefly linoleic and linolenic, have

Table 15.6 Fatty acid composition (%) of some common fats and oils

Fatty acid	*Palm oil*	*Soyabean oil*	*Rapeseed oil*	*Butter*	*Beef tallow*	*Lard*	*Fish oil*
C14 and less	1	1	—	15	2	1	5
C16:0	45	12	5	23	35	32	13
C18:0	4	4	2	9	16	8	4
C18:1	40	27	56	35	44	48	22
C18:2	10	50	25	3	2	10	6
C18:3	—	6	9	—	0.4	1	4
C20 and above	—	—	3	—	—	—	37

a carbohydrate component in the form of galactose.

The steroids are a large group of physiologically important compounds in plants and animals. All are derivatives of cyclic alcohols, the parent compound being the sterol nucleus.

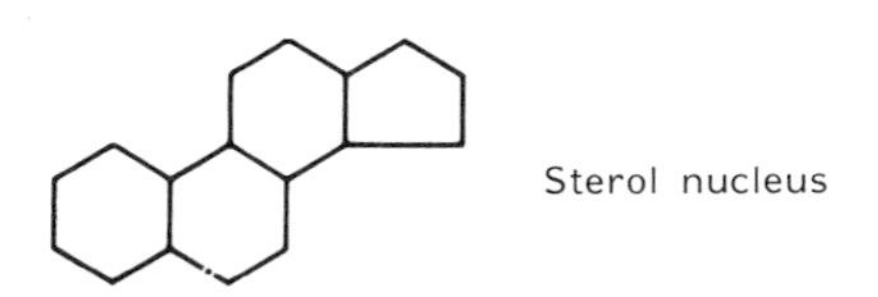

In animal tissues the sterol cholesterol is the most common, occurring in cell membranes, nervous tissue and the blood. It is important as a precursor of such substances as the bile acids, sex hormones (oestrogens, androgens and progestins), hormones of the adrenal cortex (cortisol, corticosterone, aldosterone) and vitamin D. In recent years cholesterol has received much attention in the context of levels in the blood being associated with the condition of coronary heart disease in humans.

Micronutrients

Micronutrients are dietary components that need to be present in only relatively small quantities compared with carbohydrates, lipids and proteins. There are two broad groups of micronutrients, the *minerals*, which are elements other than carbon, hydrogen, oxygen and nitrogen required by the animal, and the *vitamins*, which are organic compounds required for particular body functions. Details of the occurrence of individual micronutrients in feeds and their role in the animal are given later (see 'Mineral/vitamin nutrition and metabolism').

Water

Water is contained in all plant and animal cells. In foods the water content varies widely from over 90% in certain roots to 10–15% in dried feeds such as cereal grains and hay. Such variation in the water content of feeds makes it essential that comparisons of the nutritive value are made on a dry matter basis. The water content of the animal body is also highly variable. Since muscle tissue contains some 75% of its weight as water whereas fat contains only 12–15%, the water content of the whole animal varies with the proportions of these two tissues, being highest in young lean animals and least in mature, fat animals.

Water is obtained by drinking and from food. Additionally, water is produced by metabolic processes during the oxidation of nutrients within the body. This is termed metabolic water. The oxidation of 1 kg of carbohydrate yields 0.6 kg water whilst for every kg of fat oxidized 1.1 kg water is produced. Metabolic water production is usually of the order of about 5–10% of the total water consumed. Animals lose water through faeces and urine, water vapour in expired air and through skin. Milk represents a major loss in lactating animals, whilst the pregnant animal deposits large quantities of water within the uterus particularly in late pregnancy.

Within the animal, body water is the major component of all body fluids and internal secretions. In the ruminant especially large amounts of water are secreted into the digestive tract through saliva and digestive juices. Although the bulk of this water is reabsorbed from the digestive tract, faecal water losses are considerably higher in ruminants than in other species. The high specific heat and latent heat of evaporation of water aids in the regulation of body temperature, whilst chemically water takes part in hydrolytic reactions and in the absorption of digested nutrients.

Water intake and requirements are influenced by dietary and environmental factors. Thus water intake is related to dry matter intake and to the composition of the dry matter, especially in relation to the mineral salt content. The levels of sodium chloride and to a lesser extent the protein content of the diet influence urinary excretion and therefore requirements for water. Increases in the environmental temperature which increase water losses through respiration and sweating result in greater water consumption. All stock need adequate supplies of water at all times; insufficient water means reduced DM intake with resultant depression in productivity.

Digestion

The process of digestion breaks down complex food materials in the diet to simple products which can be readily absorbed from the digestive tract into the blood and lymph.

Anatomy of the digestive tract

Farm mammals may be divided into either simple-stomached (monogastric) animals such as the pig, or ruminants which possess a compound stomach of four regions, cattle and sheep being the main examples.

The monogastric digestive tract

The arrangement of the monogastric digestive tract is shown in Fig. 15.8. Food is digested by physical and chemical means while in the digestive system. The process begins in the mouth with the physical breakdown of food by the teeth. Saliva from the salivary glands helps the mastication and swallowing of the food. Movement of food through the digestive tract is brought about by waves of alternate contraction and relaxation of the muscular layers of the digestive tract wall, an action which is termed peristalsis.

Chemical digestion in the stomach begins with the secretion of acidic gastric juices. The digesta passes from here into the duodenum and the rest of the small intestine, where it is digested further by the actions of bile, pancreatic juice and intestinal juice. A more detailed description of the chemical processes involved is given later in this section. Digestive products are absorbed mainly in the small intestine. The large intestine is chiefly concerned with absorption of large quantities of water from the digestive waste, or faeces, before it is voided from the body (defaecation).

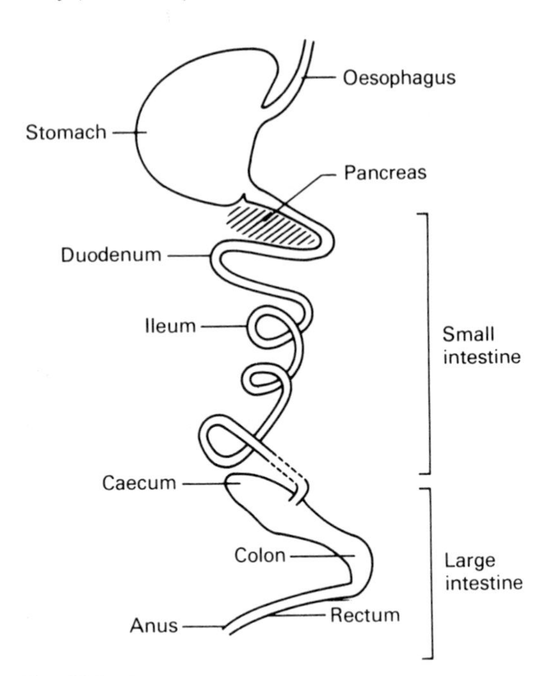

Fig. 15.8 Diagrammatic representation of a monogastric digestive tract.

The ruminant digestive tract

The ruminant stomach, unlike that of the monogastric animal, is divided into four compartments: rumen, reticulum, omasum and abomasum. The first three of these are known as the forestomachs and as such have no digestive glands. The rumen is the largest of these compartments, comprising more than 60% of total stomach capacity. The fourth stomach, the abomasum, closely resembles the structure and function of the stomach of the monogastric animal. The passage of food through the ruminant stomachs is somewhat complicated (see Fig. 15.9). After swallowing, food enters the rumen to be mechanically mixed and subjected to fermentation by the host population of microbes. Food may spend up to 30 h in the rumen. From time to time, coarse material is regurgitated into the mouth to be rechewed (chewing the cud) and then returned to the rumen for further digestion. The more fibrous the diet, the greater the rumination time, which may be as much as 8 h a day. In total, a cow may regurgitate 50 kg of food per day. Eventually the finer material in the rumen passes into the omasum, where water is extracted, and then on to the abomasum. Here any bacterial activity within the digesta is inhibited by the acidicity of the gastric secretions. Microbial protoplasm is digested by the proteolytic enzymes present. From this stage onward the ruminant digestion resembles that of the pig.

The pre-ruminant digestive tract

At birth the rumen and reticulum are very much underdeveloped compared with the adult, thus the animal is said to be at a pre-ruminant stage. Consequently, little or no rumen fermentation takes place in newborn animals. Instead the liquid milk diet bypasses the rumen-reticulum and goes directly to the abomasum, where milk proteins are clotted and partially digested. The rumen bypass is made possible by the presence of an oesophageal groove running from the oesopahgus to the omasum. Suckling or the drinking of a liquid diet stimulates the groove to close over into a muscular tube so that the contents are prevented from entering the rumen or reticulum. As the young animal is slowly introduced to a more solid diet, the reflex closure mechanism gradually diminishes, thus allowing solid material to enter the rumen so that fermentation can commence (see description of calf rearing in Chapter 18).

Biochemistry of digestion

Few of the dietary nutrients present in an animal's food can be directly absorbed by the animal and must therefore be modified to make them available. This is accomplished in the digestive tract by enzymes either produced and secreted into the digestive tract by the host animal or of microbial origin, the micro-organisms, living in symbiotic association with the animal. The former is most characteristic of monogastric species whereas the latter is characteristic of the ruminant. Neither monogastrics nor ruminants are, however, solely reliant on either host-produced or microbial enzymes for nutrient breakdown.

Carbohydrate digestion

Monogastrics

Only non-structural plant carbohydrate such as starch and the various water-soluble carbohydrates are subject to hydrolysis by digestive secretions. There are no enzymes present in digestive secretions capable of hydrolysing cellulose and hemicellulose.

Saliva contains significant quantities of only one enzyme, α-amylase, and like pancreatic amylase this enzyme initiates the breakdown of starch into a mixture of maltose and dextrins. With an optimum pH of 7, salivary amylase activity is short-lived since as the food

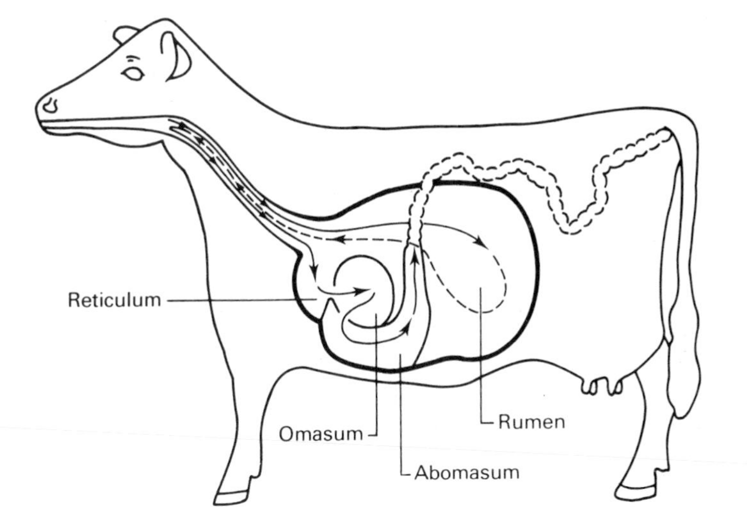

Fig. 15.9 Diagrammatic representation of the pathway of food through the ruminant digestive system of the cow.

reaches the stomach the pH falls to a value of around 2. In the duodenum the food mixes with pancreatic and duodenal secretions and bile, the pH rises and conditions become suitable for the action of pancreatic amylase which continues the breakdown of starch principally to maltose.

The mucosal cells of the small intestine produce a number of carbohydrases. These are located on the brush border of the mature cells and include maltases, lactase and sucrase. Lactase activity is maximal in the suckling animal and declines as milk makes a declining contribution to the diet. These carbohydrases can hydrolyse the appropriate simple sugars to their monosaccharide constituents which are then readily absorbed from the small intestine, mainly by 'active transport' mechanisms. Thus the main end product of carbohydrate digestion is glucose.

Cell wall carbohydrates pass through the small intestine of pigs and poultry and reach the caecum where they are substrates for the microbial population that exists in this region. This microbial population can ferment cellulose and hemicellulose (and any starch evading breakdown in the small intestine) to volatile fatty acids in a similar fashion to ruminal fermentation (see next section) but it is quantitatively limited. These volatile fatty acids are absorbed from the rear gut and have in the case of the pig been estimated to supply between 10 and 20% of the absorbed energy.

Ruminants

All the dietary carbohydrates are subject to some fermentative degradation in the rumen. The major end-products of this fermentation are volatile fatty acids (VFA), mainly in the form of acetic, propionic and butyric acids, along with the gases carbon dioxide and methane. VFA production represents up to three-quarters of the effective energy value of the diet. The rate at which dietary carbohydrates are fermented to these end products is dependent on type, soluble carbohydrate being fermented more rapidly than starches which in turn are fermented more rapidly than cell wall carbohydrates.

The concentration of VFA in the rumen is a function of their rates of production and rates of absorption. Absorption of VFA is in the free form without active transport. The acids produced during fermentation are partially neutralized by the buffers present in the saliva. Where diets containing high quantities of readily fermentable carbohydrates are fed, acid conditions may result. Such lowered pH conditions can interfere with rumen fermentation and may lead to acidosis in the host (see the section headed 'Rumen-related metabolic disorders'). Where such diets are being fed, violent fluctuations in rumen pH may be prevented by more frequent feeding, the inclusion of certain minimum amounts of long roughage to induce greater production of saliva, and the inclusion of agents such as sodium bicarbonate in the diet.

The proportion of the various types of VFA produced by the fermentation of carbohydrate is dependent on diet. Typical values for a range of diets are given in Table 15.7. The acetic + butyric/propionic (non-glucogenic VFA/glucogenic VFA) ratio is of particular relevance to the efficiency of dietary energy utilization, and efforts to manipulate this ratio are a feature of ruminant production. For example, the inclusion of ionophore-type and other antibiotics such as monensin and avoparcin to feeds can increase the proportion of propionic and thus narrow the ratio, whereas increasing the frequency of feeding promotes an increase in the proportion of acetic and thus a wider ratio. The former change is of benefit to fattening animals, the latter of benefit to lactating animals. Differences in the proportions of VFA end products are a function of the type of microbial population present in the rumen in that different microbial species use different nutrients as their principal substrates. Thus, for example, whereas *B. amylophilus*, utilizes starch, *B. succinogenes* is a principal utilizer of cellulose.

Table 15.7 The effect of diet on the molar proportions of VFA in rumen liquor

Diet	*Molar % of VFA*			
	Acetic	*Propionic*	*Butyric*	*Higher**
Grass silage	76	13	7	4
Fresh grass	63	20	13	4
60% hay: 40% concentrates	60	25	12	3
40% hay: 60% concentrates	55	30	10	5
20% hay: 80% concentrates	51	36	10	3
10% hay: 90% concentrates	45	39	11	5

* Higher VFA include valeric, isovaleric and caproic acids.

Under some circumstances a portion of the fermentable carbohydrates may escape rumen fermentation. This is mainly in the form of structural carbohydrate which is sufficiently lignified to prevent microbial degradation. Further fermentation of this fraction may take place in the caecum, leading to a further source of VFA to the animal, but in practice the extent of this is very limited. Under some circumstances dietary starch may also escape rumen fermentation and reach the lower gut. This is a possibility when high cereal diets processed in particular ways are fed (e.g. finely ground maize). This starch can be digested in the small intestine in the same way as occurs in monogastric species with glucose as the principal end-product. Although this starch is used more efficiently, the small intestine of the ruminant is more limited in its carbohydrase activity than the monogastric.

Gas production resulting from the fermentation of carbohydrates is lost by eructation and in energy terms accounts for about 7% of the food energy going into the rumen. This is in the form of methane and results from the reduction of carbon dioxide by hydrogen. A wide range of compounds have been shown to be capable of depressing methane production with the aim of increasing the efficiency of the fermentation. None of these are at present being used commercially.

Protein digestion

Monogastrics

The enzymes concerned with protein digestion may be considered either as exopeptidases (enzymes which hydrolyse peptide bonds at the ends of polypeptides) or endopeptidases (those which hydrolyse peptide bonds within a polypeptide). The exopeptidases are either carboxypeptidases or aminopeptidases; the endopeptidases include pepsin, trypsin, chymotrypsin and dipeptidases. With the exception of pepsin, which operates in the acid environment of the stomach, the protein-digesting enzymes are to be found in the small intestine. Trypsin and chymotrypsin are secreted in the pancreatic juices; the other enzymes are secreted from cells within the mucosa of the small intestine. A feature of these proteolytic enzymes is their specificity for peptide bonds involving particular amino acid types. For example, chymotrypsin has a preference for peptide bonds involving the carboxyl group of aromatic amino acids.

The end result of such proteolytic activity is the hydrolysis of dietary protein to free amino acids. These amino acids are absorbed by cells lining the small intestine and subsequently enter the portal blood and are transported to the liver. Small quantities of short peptides may also enter the absorptive cells of the small intestine where they are hydrolysed to allow free amino acids to enter the blood.

During the passage of food through the gut, considerable quantities of endogenous protein are added in the form of digestive secretions and desquamated mucosal cells. This protein is itself digested and absorbed like dietary protein and complicates the assessment of dietary protein requirements.

Ruminants

In the ruminant animal consideration must include not only the digestive utilization of dietary protein but also the utilization of other sources of dietary nitrogen, termed non-protein nitrogen (NPN) (Fig. 15.10). Although it is technically more correct to refer to both the true protein and NPN in terms of 'nitrogen', feed manufacturers and farmers tend to use the concept of crude protein (N in feed × 6.25).

Within the reticulo-rumen, microbial activity results in the degradation of all the NPN and large and variable amounts of true protein. Ammonia is produced from these degradative processes. Collectively the NPN and true protein degraded in the rumen are termed the *rumen degradable crude protein* (RDP) fraction of the feed. Rumen degradable crude protein can be further partitioned into a *quickly degraded crude protein* (QDP) fraction and a *slowly degraded crude protein* (SDP) fraction. Quickly degraded crude protein includes the NPN and all the true protein that is quickly and completely degraded to ammonia. This is essentially the feed nitrogen that is readily soluble in cold water. Slowly degraded crude protein describes the fraction which releases ammonia at a slower rate and is essentially the total RDP in the feed minus the QDP fraction. It should be noted that this SDP fraction has a variable rate of degradation and is quantitatively influenced by the effect of level of feeding on its rate of outflow from the rumen. A further component of dietary crude protein, that fraction which is resistant to degradation by microbial activity in the rumen, is termed *undegradable crude protein* (UDP). The ratios of RDP/UDP and of QDP/SDP in rations are variable and dependent on the sources of crude protein and the effect of any processing procedures such as ensilage, heat or formaldehyde treatment.

The true protein within the RDP fraction is initially hydrolysed by bacterial proteases to yield free amino acids. These may be subsequently utilized by micro-organisms to synthesize *microbial crude protein* (MCP), but the bulk of these free amino acids are further degraded to produce ammonia, organic acids and carbon dioxide. This ammonia is the major source of nitrogen for the synthesis of MCP. Non-protein nitrogen within the RDP fraction is likewise degraded to release ammonia; in the case of the most commonly used commercial source, urea, it is rapidly hydrolysed by bacterial ureases. The rate of MCP synthesis and the efficiency with which the RDP is converted to MCP is dependent on the rates of ammonia release and energy availability in the rumen. The availability of energy to microbes in the

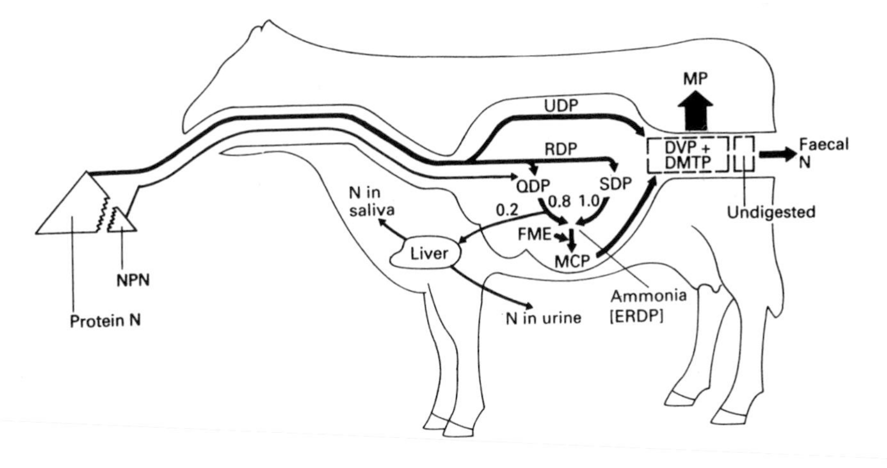

Fig. 15.10 Summary of nitrogen utilization by the ruminant.

rumen is dependent on the dietary intake of *fermentable metabolizable energy* (FME). Fermentable metabolizable energy consists of the metabolizable energy (ME) which rumen microbes can utilize, that is the ME in the diet minus the ME in dietary fat and dietary volatile organic acids such as a those found in silages (see Chapter 17, section on metabolizable energy). The maximal efficiencies of conversion of QDP and SDP to MCP are 0.8 and 1.0 respectively. This gives us the expression *effective rumen degradable crude protein* (ERDP = 0.8 QDP + 1.0 SDP). The rumen microbes' ability to synthesize MCP in relation to FME intake is recognized to be affected by level of feeding and to vary between 9 g MCP/MJ of FME (maintenance) and 11 g MCP/MJ of FME (3 × maintenance).

In situations where there is an excess of ERDP, especially QDP, in relation to FME, ammonia concentrations build up in the rumen and significant absorption of ammonia into the blood may occur. On reaching the liver it is converted to urea, an energy-demanding process, and is mainly excreted in the urine, although a small quantity (≈20%) is recycled to the rumen via the saliva. In cases of extensive ammonia absorption from the rumen, ammonia toxicity may occur (see the section headed 'Rumen-related metabolic disorders'). However, protein rationing systems aim to minimize ammonia losses through providing an optimal ERDP/FME ratio in the diet.

In addition to ERDP and FME, the rumen microorganisms also require sulphur and phosphorus for protein synthesis. Sulphur is required for the synthesis of the sulphur-containing amino acids. A ratio of N:S of 10:1 is considered optimal, and phosphorus is a component of the nucleic acids which are involved in protein biosynthesis.

The MCP resulting from ruminal biosynthesis reaches the abomasum and small intestine in the form of microorganisms carried by fluid out of the rumen once they lose their attachments to food particles. They pass into the abomasum and small intestine together with any dietary UDP. The processes of digestion of these proteins in the ruminant abomasum and small intestine are little different from those occurring in the monogastric animal.

Thus the mixture of amino acids available for absorption from the small intestine is comprised of those that make up MCP and those present in UDP. It is thus markedly different in composition from dietary protein. The sum of these two forms of protein absorbed as amino acids from the small intestine make up the *metabolizable protein* (MP) supply of the animal. In the metabolizable protein rationing system (see Chapter 17 under 'Ruminants – metabolizable protein') it is assumed that MCP contains 75% of its total nitrogen as amino acids and has a true digestibility of 0.85 to give the digestible microbial true protein fraction (DMTP) underlined in italics. The true digestibility of UDP, the digestible undegraded protein (DUP), is related to the crude protein contained in the acid-detergent fibre (ADIP) according to the equation, DUP = 0.9 (UDP−ADIP).

The metabolizable protein system allows nutritionists to formulate rations that will provide animals with the appropriate amounts and types of dietary protein to meet their requirements. It does not, however, take into account the ability of that dietary supply and its subsequent digestive utilization to furnish the animal with the correct amounts of individual essential amino acids. Although this is less important than in the monogastric because a major advantage of the rumen microbes is their ability to synthesize essential amino acids in MCP, the future development of even more precise systems will necessitate this information. For example, it is currently known that in silage-based diets methionine is the limiting amino acid for milk protein synthesis. By protecting supplementary pure methionine from rumen decomposition inside a coat of hydrogenated lipid, it is possible to supply this and perhaps other limiting amino acids.

A feature of the nitrogen content of micro-organisms reaching the small intestine is that a proportion (25%) of it is in the form of nucleic acids. The pancreatic juices of ruminants contain high activity of nucleases which break down the nucleic acids. The phosphorus released in this process is absorbed and recycled to the rumen via the saliva but the other end products are of no value to the animal and represent a further source of nitrogen loss in the digestive utilization of dietary nitrogen by the ruminant.

Lipid digestion

The digestion of lipids requires that they become miscible in water before they can be absorbed through the villi of the intestine. In this respect fat digestion and absorption differ from those of carbohydrate and protein.

Monogastrics

The small intestine is the site of fat digestion and absorption. As it enters the small intestine, fat becomes mixed with bile salts from the gall bladder in the form of taurocholic and glycocholic acids which have emulsifying properties reducing the size of fat particles and giving an increased surface area for digestive hydrolysis.

The major enzymes with lipolytic activity are to be found in the duodenum and are secreted in the pancreatic fluids. Pancreatic lipase hydrolyses triglycerides to a mixture of mono-, di- and triglycerides. In the presence of bile, these contribute to the production of dietary fat forming micelles which after disruption on contact with the microvilli can enter the mucosal cells. Other lipid components in the form of fat-soluble vitamins and sterols are likewise components of these micelles. Within the mucosal cells the various lipid fragments absorbed are resynthesized into triglycerides and phospholipids and along with sterols and protein combine to form particles called chylomicrons. These pass into the lymph and then into the general circulation.

Differences in the efficiency with which different fats are digested and absorbed exist. Unsaturated fatty acids are digested better than saturated ones and digestibility decreases with chain length. Synergistic effects between fatty acids exist such that absorption of saturated fatty acids is greater in the presence of unsaturated fatty acids.

Ruminants

Ruminant diets normally contain only some 3–5% lipid either as triglyceride in concentrate feeds or galactolipids in forages but may contain up to 10%. The fatty acid composition can vary widely but in forages and cereals high proportions of linoleic and linolenic acid are present. In the rumen, microbial lipases hydrolyse a high proportion of these fats releasing fatty acids. Of the polyunsaturated fatty acids some 80–90% are rapidly hydrogenated to saturated and mono-unsaturated fatty acids. In addition this may also result in the production of fatty acids in the *trans* configuration. Ruminal micro-organisms also synthesize long-chain fatty acids many of which are of the odd-numbered and branched chain type. These processes account for the presence of these unusual fatty acids in both ruminant body and milk fat.

Unlike VFA the long chain fatty acids resulting from dietary and bacterial fat are not absorbed until the digesta reaches the small intestine where mechanisms of absorption are similar to those of the monogastric.

Although the modification of fat in the rumen is restricted to hydrolysis and hydrogenation, dietary fats themselves may inhibit fermentation with consequent effects on the extent of digestion in the rumen. The digestion of cellulose is particularly affected and may result in a 'high propionate type' fermentation. This occurs when the fat content of the diet is around 7–10% but is dependent on the type of fat and manner of incorporation in the diet. For example, unsaturated fat types have a greater effect on the fermentation than saturated fats, and free fat a more deleterious effect than fat present in 'whole' oilseed meals.

In recent years systems of 'protecting' fat by coating the fat with undegradable protein have been developed. The inclusion of additional fat is of greatest relevance to dairy cows where there is often the need for more energy dense diets, although it may also be used to manipulate the amount and composition of milk fat. Appropriate processing of whole oilseeds may produce a similar 'protected-fat' system.

Metabolism

Metabolism is the name given to the sequence of chemical processes that take place in the tissues and organs of the animal. Some of these processes involve the breakdown of compounds (catabolism), others involve the synthesis of substances (anabolism). Catabolic processes are frequently oxidative in character and are primarily concerned with generating energy for mechanical work and for the chemical work of synthetic processes. Thus the anabolic processes of carbohydrate, protein and fat synthesis in the body are inextricably linked to the catabolic processes. The common currency of energy production and energy utilization in catabolic and anabolic processes is the substance adenosine triphosphate (ATP).

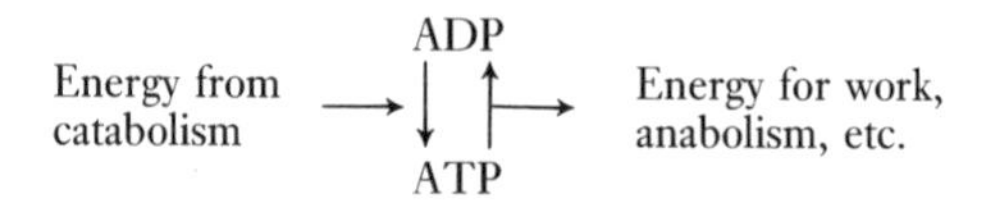

The starting point of metabolism may be looked upon as the substances absorbed from the digestive tract.

Dietary nutrient	*Major end-product of digestion*	
	Ruminant	*Monogastric*
Carbohydrate	VFA-acetic, propionic, butyric acids	Glucose
Proteins	Amino acids	Amino acids
Fats	Mono-, di- and triglycerides, fatty acids	Mono-, di- and triglycerides, fatty acids

As can be seen, these end products are different in ruminant and non-ruminant species and give rise to differences in metabolism between these animal groupings.

In the coverage of metabolism given here only a general outline of the processes involved is given. Further detail of particular pathways may be obtained by referring to appropriate textbooks.

Glucose metabolism

The major pathways by which glucose is catabolized to yield energy as ATP involves firstly the glycolytic pathway which results in the production of pyruvate, and secondly the tricarboxylic acid cycle which results in the complete oxidation of pyruvate to carbon dioxide and water (Fig. 15.11).

Glucose for these pathways is mainly obtained in the monogastric through absorption from the digestive tract. Glucose that is absorbed but not immediately catabolized

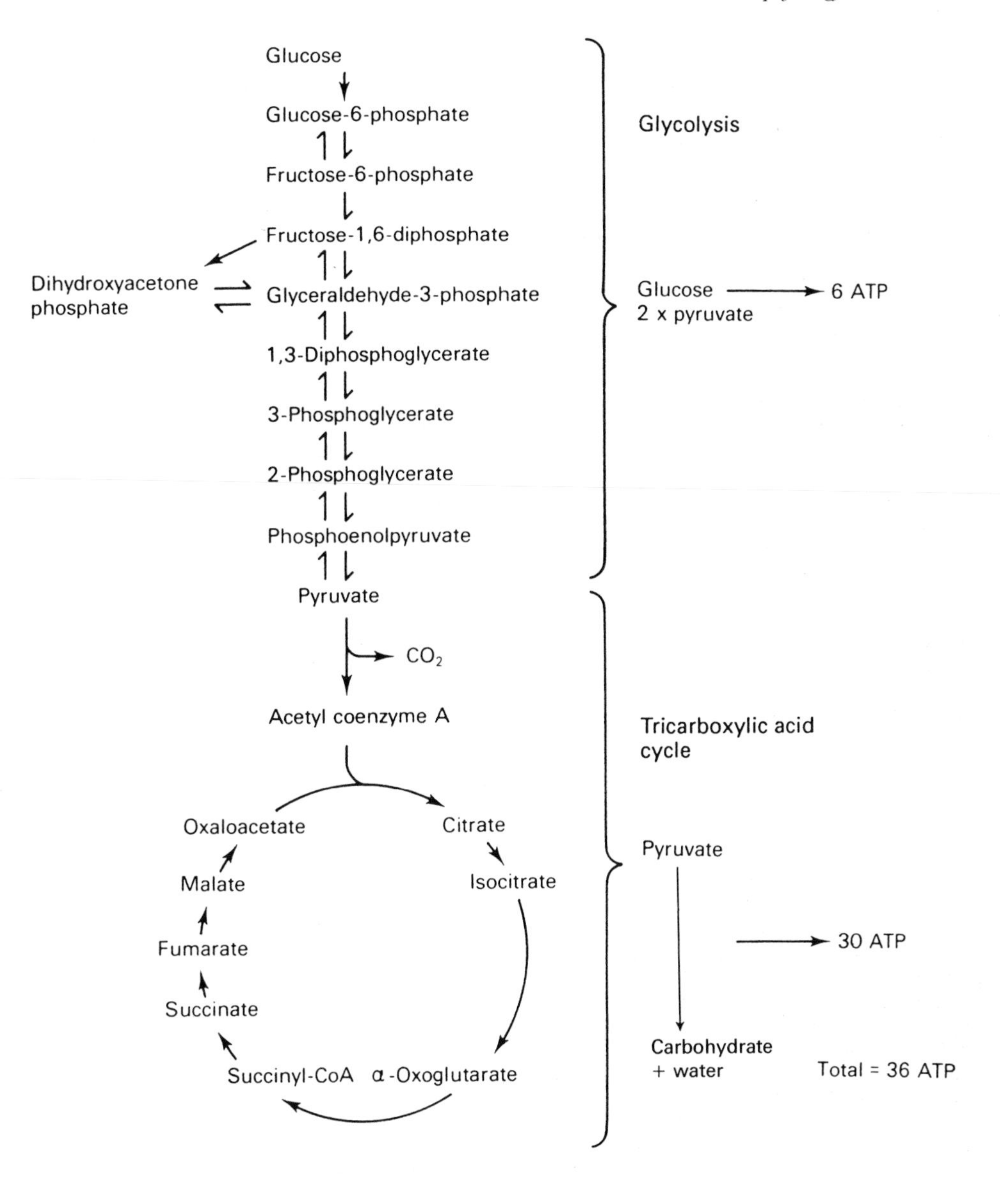

Fig. 15.11 The oxidation of glucose via glycolytic and tricarboxylic acid pathways.

may be stored as glycogen in muscle and liver cells. The amount of glucose that can be stored as glycogen is relatively limited and the bulk of glucose that is not catabolized is stored in the animal's adipose tissue. The concentration of glucose in the general circulation is kept within fairly narrow limits through the action of a variety of hormones but especially insulin and glucagon. Thus, although the rates of absorption of glucose from the digestive tract fluctuate with meal patterns, the levels of glucose in the blood are kept relatively constant. The storage of surplus glucose as glycogen and the ability to reconvert glycogen, especially that in the liver, back to glucose are the means by which blood glucose levels are regulated. Liver cells are also able to synthesize glucose from certain metabolites (e.g. amino acids, oxaloacetic acid) in order to maintain blood glucose levels. This is termed gluconeogenesis. Body fat reserves may also be called upon as an energy source in order to spare glucose catabolism when rates of absorption are lower than rates of utilization.

In addition to acting as a principal energy source for the organs and tissues of the animal, glucose is also used as the precursor for lactose synthesis in the mammary gland of lactating animals and is the principal energy-supplying substrate to cross the placenta and be used by the growing fetus.

For the complete oxidation of glucose the presence of oxygen in cells is necessary. Under anaerobic conditions the pyruvate produced by glycolysis is converted to lactic acid. This is of relevence in the muscle cells of the animal post-slaughter where the catabolism of cell glycogen to lactate produces a fall in the pH of the muscle. The amount of lactic acid produced and the resulting pH can thus be influenced by the reserves of glycogen at slaughter. When animals have low muscle glycogen reserves at slaughter, only a relatively small fall in pH results. The meat from such animals (most commonly bulls) is usually characterized as being dark in colour and firm and dry in texture (DFD meat), both of which are undesirable characteristics from the viewpoint of the consumer. Under some circumstances in animals that show excessive adrenalin release at slaughter, most commonly pigs with particular genetic characteristics, glycogen is quickly metabolized to lactic acid giving a rapid fall in muscle pH while the carcass is still warm. This tends to give rise to meat which is of pale appearance and soft and exudative in texture (PSE). Again these are undesirable characteristics to the consumer.

VFA metabolism

The volatile fatty acids are absorbed across the rumen wall down a concentration gradient. Rates of absorption are thus mainly dependent on rates of production in the rumen although other factors such as rumen pH may have an influence. Although fluctuations in the rates of absorption exist and are dependent on feeding regime and ease with which dietary carbohydrates are fermented, the diurnal variation in absorption of energy-supplying metabolites is considerably less in the ruminant than the fluctuations in glucose absorption in the monogastric.

Of the three main VFA produced in the rumen, only acetic appears in the peripheral circulation. This is because propionic is rapidly converted to glucose by the liver and butyric to 3-hydroxybutyric acid during passage across the rumen wall.

Acetic acid is the major VFA absorbed and as such is catabolized via the tricarboxylic acid cycle by a variety of tissues to provide energy. When acetic acid is absorbed in amounts surplus to immediate energy needs, the surplus is synthesized into long chain fatty acids and stored in the adipose tissue. Although the acetic acid acts as the precursor for fat synthesis, the energy required for this anabolic process is supplied by glucose. Likewise in the lactating ruminant acetic acid is the precursor for milk fat synthesis within the mammary gland.

Although little glucose is absorbed from the ruminant digestive tract, levels of glucose in ruminant peripheral circulation are only slightly lower than those found in the monogastric. A major source of this glucose is propionic acid although gluconeogenesis from amino acids also makes a significant contribution. Whilst glucose is not used as a principal energy source in the ruminant, it is utilized in many ways as in the monogastric. Thus, for example, it is the main source of energy for the nervous tissue and the fetus, for glycogen and lactose synthesis and as an energy supplier for fat and protein biosynthesis.

Although usually absorbed in somewhat lesser quantities than acetic acid and propionic acid, butyric acid can be used as an energy source by a variety of tissues after being converted to 3-hydroxybutyrate in its passage through the rumen wall. Like acetic it may also serve as a precursor for both body and milk fat synthesis.

The proportions of the different VFA absorbed into the peripheral circulation can have an influence on the endocrine balance within the animal such that the partition of nutrients into different body processes may be affected. Thus high concentrate, low forage diets which favour the production of propionate in the rumen depress milk fat secretion in the lactating ruminant and promote lipogenesis in adipose tissues. It is thought that the stimulation of insulin secretion by propionate is the primary factor influencing partition.

Fat metabolism

Body fat metabolism is used as a means of regulating the energy metabolism of the animal. Thus dietary energy consumed in excess of immediate requirements is deposited as fat in adipose tissue and is released from adipose tissue as free fatty acids when the dietary supply of energy is inadequate. Such fluxes in fat reserves are characteristic of lactating and pregnant animals which have to call on fat as an energy source when food intake is inadequate to meet requirements and replace this fat at other times in their production cycle. Even on a diurnal basis, particularly in monogastric animals being fed discrete meals, a cycle of deposition and mobilization of fat occurs. Fat is a considerably better energy store than either glycogen or glucose because not only does it have more than twice the energy content per unit weight than these nutrients, it also has very little water associated with it (~12% in fat) whereas glycogen and protein are heavily hydrated (~75%). Thus the energy stored per kg of tissue is about 32 MJ for fat and 4 MJ for glycogen and protein. Thus fat can supply a starved animal with energy for several months whereas the energy reserves in glycogen are sufficient for only about a day.

A principal site of fat metabolism is the liver. Both catabolism and anabolism of fatty acids are located in this organ. Fatty acids may be completely oxidized by the processes of β-oxidation and the tricarboxylic acid cycle to carbon dioxide and water and release energy. Such catabolism also takes place in muscle. Glucose and acetate are the major precursors for fatty acid synthesis. The relative importance of these two substances depends on species, the former being more important in monogastrics, the latter more important in ruminants. Adipose tissue can also synthesize fatty acids and is the major site of fatty acid synthesis in some species, including pigs and ruminants.

Fat circulates in the blood in a variety of forms. Fat resulting from the digestion and absorption of dietary fat is in the form of particles called chylomicrons. These particles consist mainly of triglyceride with small amounts of protein. Mostly they are processed in the liver but may release their fatty acids to adipose tissue or mammary tissue in the lactating animal, and thus contribute to body fat and milk fat synthesis. It is in this way that the fatty acid composition of dietary fat may influence that of the body and milk fat. Fatty acids that have been synthesized in the liver are transferred for storage in adipose tissue in the form of lipoproteins. It is also in this form that liver-synthesized fatty acids are transferred to the mammary gland for direct inclusion into milk fat. The other form in which fat circulates in the plasma is as free fatty acids (FFA), also termed non-esterified fatty acids (NEFA). These are mostly transported in the plasma bound to albumin. Although the amount of NEFA in the plasma is very small, this fraction is important metabolically because lipid is released in this form from adipose tissue and transported to the liver or other tissue to supply energy in times of need. The levels of NEFA are typically elevated in the undernourished animals. These NEFA may also be taken up by the lactating cell and be incorporated into milk fat.

Amino acid metabolism

Amino acids resulting from the digestion of dietary protein are absorbed from the small intestine into the portal blood. These amino acids contribute to the metabolic 'pool' of amino acids in the blood and tissues. In the body amino acids are also constantly being liberated into the blood from the breakdown of tissue proteins. This pool serves as a source of amino acids, some of which are used to build nitrogenous substances such as muscle or milk proteins or certain hormones whilst the vast majority are catabolized, their nitrogen being excreted as urea (Fig. 15.12). The process of tissue protein synthesis and degradation is referred to as the 'turnover' of body proteins. Clearly in the growing animal synthesis exceeds degradation.

Muscle accounts for up to 60% of body protein. Thus the growth of animals involves a substantial uptake by muscle tissue of amino acids from the pool for protein synthesis. The efficiency with which amino acids in the pool are utilized is dependent on the proportions of the different amino acids that make up the pool. The proportions of amino acids in the pool are mainly influenced by the profile of amino acids absorbed from the intestine. The idealized profile is where the amount of all essential and non-essential amino acids absorbed from the intestine is the same as that required for tissue protein synthesis. Only the total of non-essential amino acids supplied need be considered since there exists in the liver a capacity to synthesize non-essential amino acids that are in deficit from others that are in surplus by the process of transamination. Such an idealized profile does not exist in practice and some imbalance in the supply of amino acids exists (Fig. 15.13). In this example lysine is the amino acid in greatest shortage, termed the 'limiting' amino acid, and is responsible for the restriction in protein synthesis.

In monogastric species the profile of amino acids absorbed from the intestine is a reflection of the amino acid composition of dietary protein. In pig diets the amino acid most likely to be limiting is lysine, levels of which are especially low in cereals. Hence considerable attention is given to ensuring adequate lysine levels in practical diets through the inclusion of protein feeds high in lysine or of synthetic lysine.

In ruminant species the profile of absorbed amino acids making up the metabolizable protein is influenced by the composition of the microbial protein produced from dietary RDP and any DUP present in the diet. In many production situations the supply of amino acids from microbial protein is sufficient to meet the animal's requirements. In certain situations of high productivity, however, in particular the high yielding cow, the supply of particular essential amino acids may be limiting. In grass silage based diets, methionine is most likely to be limiting whereas, in corn silage based diets, lysine may be limiting. Through the inclusion of DUP of appropriate amino acid composition or synthetic amino acids pro-

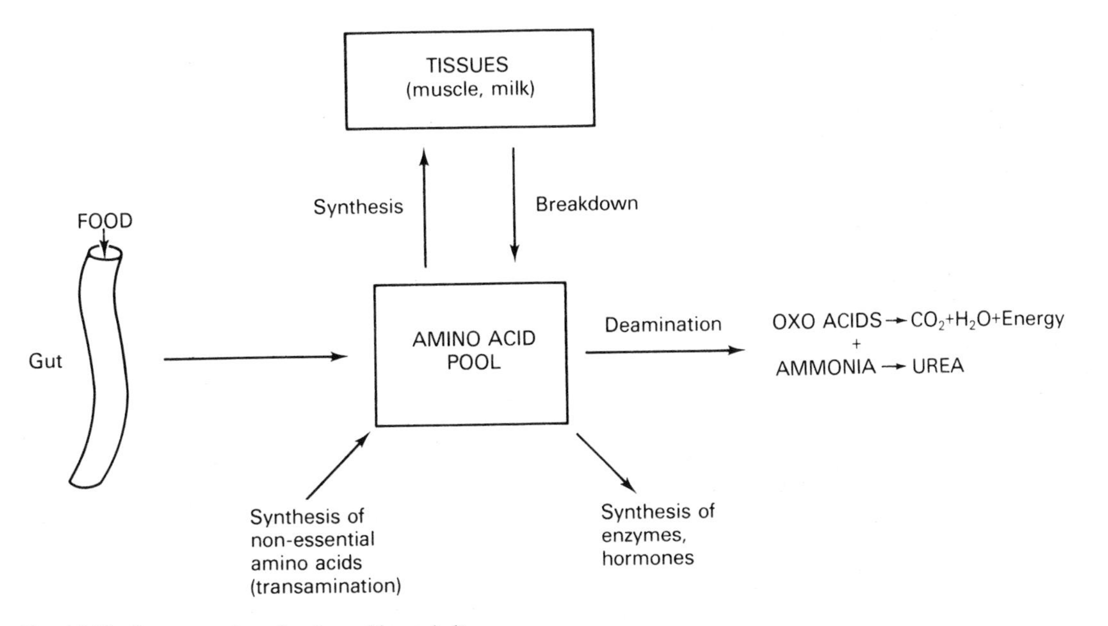

Fig. 15.12 Representation of amino acid metabolism.

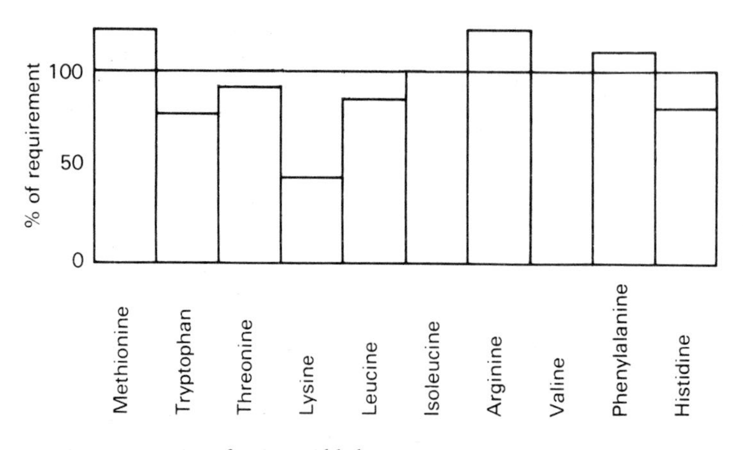

Fig. 15.13 Diagrammatic representation of amino acid balance.

tected from rumen degradation, an appropriate balance may be obtained.

Amino acids which are surplus to requirements for protein synthesis are converted by the liver in a process known as deamination to oxo acids and ammonia. The ammonia is rapidly synthesized to urea which is subsequently excreted in the urine. The oxo acids may be used as energy sources mostly by way of the tricarboxylic acid cycle. Surpluses of amino acids occur when an excess of protein is fed and/or where the profile of absorbed amino acids differs significantly from requirements. A further situation where the extent of deamination may be elevated is in starvation where tissue protein may be catabolized to supply energy and amino acids used to maintain blood glucose levels.

Mineral nutrition and metabolism

The essential mineral elements are designated as either macro- (or major) or micro- (or trace) elements depending on their concentration within the animal body (Table 15.8). Although animal tissues and feeds contain about 45 mineral elements, only about half have been shown to have an essential function. Three general functions may be identified for minerals:

- a structural function in bones, e.g. Ca, P, Mg and fluorine;
- as electrolytes in body fluids maintaining acid/base balance, osmotic pressure and inducing excitation of nerves and muscles, e.g. Na, K, Cl and Ca;
- as integral components of enzymes and other biologically important compounds. Trace elements especially function in this capacity.

Animals obtain most of their mineral requirements from the feed they consume. Animal feeds vary widely in their mineral composition, and individual feeds may vary in their mineral content according to the soil, fertilizer and other environmental influences on its production. The trace mineral composition of feeds tends to vary more widely than that of the major elements. As well as the absolute amount of individual mineral present in feed, the availability of the element in terms of the proportion absorbed and utilized is also important. The availability of individual mineral elements may be influenced by other dietary components. Thus, for example, a surplus of phosphate may reduce the availability of calcium through the formation of insoluble salts. An excess of molybdate may likewise reduce the availability of copper. In some instances, availability of certain ions may be increased through the formation of soluble chelates. Such interactions present difficulties in the determination of requirements and in some instances the identification of particular deficiencies and imbalances.

Table 15.8 The essential mineral elements

Major	*Trace*
Calcium	Iron
Phosphorus	Zinc
Magnesium	Copper
Sodium	Cobalt
Potassium	Manganese
Chlorine	Molybdenum
Sulphur	Selenium
	Iodine

In situations where feedstuffs fail to provide the animal with sufficient amounts of particular minerals to meet requirements, mineral supplementation may be given in concentrated forms such as finely divided powders that may be efficiently distributed through feeds; suitable licks containing the deficient elements; or through 'bullets' and 'needles' that lodge in the reticulum and slowly release the mineral throughout the animal's lifetime. The latter are particularly useful for the supply of trace elements and as a means of giving supplementary minerals to animals kept under extensive systems of production.

Calcium, phosphorus and magnesium

These three elements are found together in the animal's skeletal tissue. The calcium and phosphorus exist mainly in the form of crystalline hydroxyapatite [$Ca_{10}(PO_4)_6(OH)_2$] which mineralizes the organic matrix. About one-third of the magnesium in bones is bound to phosphate, the remainder being absorbed on the surface of the mineral structures. The main component of the organic matrix is protein in the form of collagen. Some 98% of the body calcium, 80% of the body phosphorus and 65% of the body magnesium are present in bone tissue. The remaining amounts of these elements are present in the blood either as free ions or in complexed forms and in the soft tissues. In these tissues they have diverse functions. Thus calcium participates in neuromuscular activity and has a role in blood clotting; phosphorus functions in energy metabolism as a component of ATP and is present in the phospholipids of cell membranes and certain proteins; whilst magnesium, like calcium, has an involvement in the excitation of nerves and muscles, and acts as an enzyme cofactor in metabolism.

The overall metabolism of these elements is illustrated in Fig. 15.14. An important element of this metabolism is the maintenance of constant levels of calcium and phosphorus in the blood. This is achieved through the interaction of the two hormones, parathyroid hormone and calcitonin and the active metabolite of vitamin D_3, 1,25-dihydroxycholecalciferol(1,25$(OH)_2D_3$) which control the absorption from the digestive tract, influence the balance of resorption and deposition in bone and influence the extent of excretion in faeces and urine. Parathyroid hormone (PTH) is secreted in response to a fall in blood calcium and its action in raising blood calcium through increasing absorption efficiency from the gut, increasing resorption relative to deposition in bone and depressing urinary excretion, is mediated through (1,25 $(OH)_2D_3$). The effects of calcitonin are antagonistic to PTH but do not involve 1,25$(OH)_2D_3$. PTH and 1,25 $(OH)_2D_3$ are likewise involved in the regulation of plasma phosphate concentration. The amounts of ionized magnesium in blood serum are normally in the range of 20–40 mg/litre and although there is constant exchange of magnesium ions between serum and bone surfaces there do not appear to be the same type of homeostatic regulatory mechanisms of the type that exist for calcium.

Simple deficiencies of calcium, phosphorus or vitamin D result in bone abnormalities such as rickets in the young and osteomalacia in the mature animal. The metabolic disease milk fever, most commonly found in dairy cows just after parturition, is the main problem associated with a breakdown in the regulation of calcium metabolism (see later section on 'Metabolic disorders in livestock'). Because of the animal's lesser ability to mobilize bone phosphorus, low levels of blood phosphate may readily arise when intakes of phosphorus are inadequate. Low levels of blood phosphate have been associated with poor fertility in animals. Low levels of serum magnesium (hypomagnesaemia) may arise through dietary insufficiency, but especially in ruminants low coefficients of absorption may be a contributory factor. Absorption which occurs from the reticulo-rumen may be adversely affected by high levels of ammonia, potassium and phosphates. Outbreaks of either chronic or acute hypomagnesaemia (grass staggers) occur in ruminant livestock as discussed later.

The amounts of calcium and phosphorus found in different feeds vary widely whereas most of the commonly fed diets for farm animals contain sufficient magnesium

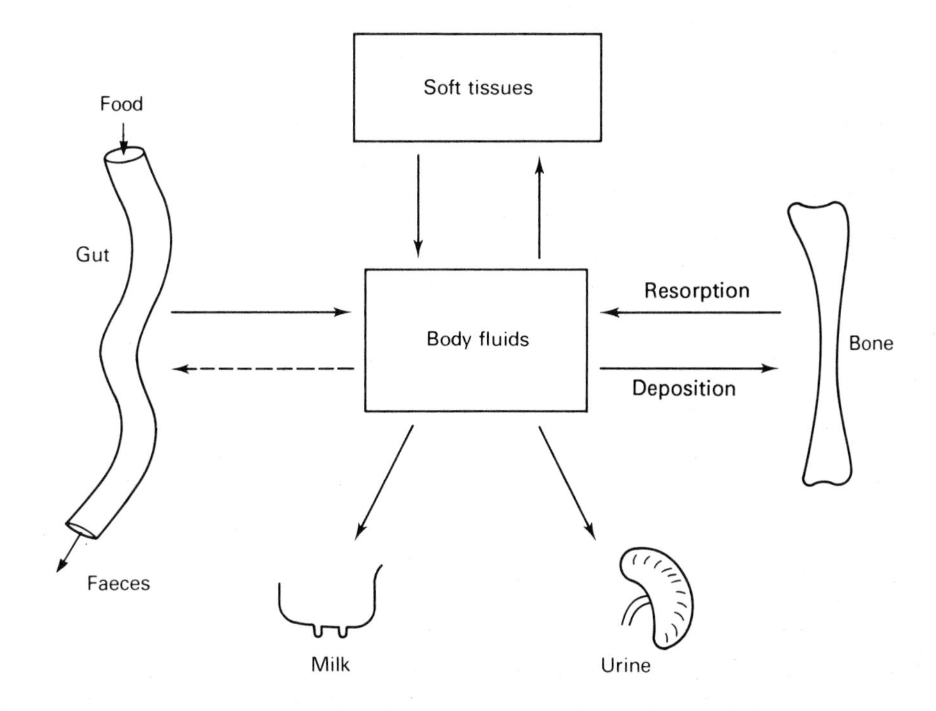

Fig. 15.14 Diagrammatic representation of the calcium, phosphorus and magnesium metabolism.

to meet body needs. As a generalization, green forages are relatively high in calcium but low in phosphorus whilst animals on diets high in concentrates (cereals and oilseeds) are likely to be receiving insufficient calcium but sufficient phosphorus. A number of supplementary sources of both minerals are available to correct problems of either insufficiency or imbalance in the dietary supply of calcium and phosphorus, e.g. ground limesetone, mono-, di- and tricalcium phosphates, sodium phosphate and bone flour. These supplementary sources supply different amounts of the two elements and thus particular supplements may be used in given situations. Phosphates used as supplements must be defluorinated before feeding since fluorine is toxic.

Sodium, potassium and chloride

These three minerals are found mainly in the soft tissues and body fluids and together are largely responsible for the maintenance of osmotic pressure, fluid balance and acid–base balance and also have, along with calcium and magnesium, an important function in neuromuscular activity. Sodium is the major cation in body fluids, almost exclusively in the extracellular fluid and blood, whereas most of the potassium is found intracellularly. Chlorine is the major anionic constituent of body fluids. It is a constituent of hydrochloric acid secreted in the gastric juice.

Deficiency of these elements is unlikely since the potassium content of most foods is high and, although sodium and, to a lesser extent, chlorine are not always present in sufficient amounts, the practice of including common salt or sodium bicarbonate in the diet means that these elements are supplied in sufficient amounts. The intake of potassium is frequently much higher than that of sodium but a balance between the two is maintained by the hormone aldosterone which ensures the excretion of excess potassium and reabsorption of sodium by the kidney. Excessive intake of potassium may, however, limit the absorption of magnesium in the ruminant.

In situations of inadequate water intake or where either an excess or imbalance of sodium and chloride in the diet exists, animals may be unable to regulate osmotic and acid–base balance allowing the development of either an alkalosis or acidosis. This is more likely in monogastric species and in young animals. There is now evidence that, since these elements are all involved in the maintenance of acid–base and water balance, an optimum balance must be achieved between them and that imbalances, even in the absence of deficiency or toxicity, can adversely affect production. For chickens it has been suggested that the sum of $Na + K - Cl$ should be between 250 and 300 mmol/kg of feed.

Sulphur

Sulphur present in the body and food is largely present in the form of protein since sulphur is a constituent of the amino acids cysteine, cystine and methionine. Wool is particularly rich in cystine and contains about 4% of sulphur. Some sulphur is contained in the vitamins biotin and thiamine. In monogastrics sulphur deficiency is reflected in a shortage of these essential sulphur-containing amino acids which are often limiting in tissue protein synthesis. In the ruminant the micro-organisms utilize elemental sulphur or sulphate to synthesize these sulphur-containing amino acids. Deficiency in the ruminant is unlikely in most situations but there is evidence that animals receiving a high proportion of their nitrogen from urea, or animals grazing pastures growing on soils low in sulphur, may benefit from supplementation.

Iron

More than half the iron (60–70%) of the animal body is present as a constituent of the protein haemoglobin found in red blood cells and functioning in the transport of oxygen from the lungs to the tissues. Small amounts are to be found in myoglobin which serves as an oxygen store in muscle and in certain enzymes but the remainder of body iron is present in the form of the storage compounds ferritin and haemosiderin. These are to be found in the liver, spleen, kidneys and bone marrow.

Although iron is poorly absorbed from the digestive tract – a coefficient of absorption of about 5–10% pertains in the adult – that which is absorbed is efficiently retained. Thus iron liberated from the breakdown of red blood cells and haemoglobin is efficiently recycled in the resynthesis of haemoglobin in the bone marrow and only small amounts are lost through excretion in the faeces (as components of bile) and in urine.

Iron requirements in the adult are generally low and easily met by dietary sources but young suckling animals are at risk due to the low level of iron in milk and the poor reserves of the newborn. The pale colour of the meat in veal calves is due to the low levels of iron-containing pigments. Suckling piglets are especially vulnerable to deficiency and should be routinely given supplementary sources either orally or by intramuscular injection. Piglets reared out of doors acquire sufficient iron by rooting in the soil. Supplementing pregnant females with iron compounds may have the effect of increasing the iron reserves of the newborn, but dietary intake of iron has no influence on the level of iron in the milk of lactating animals.

The main consequence of iron deficiency is anaemia, characterized by low blood haemoglobin, subnormal growth and diarrhoea. Low levels of haemoglobin may arise for reasons other than iron deficiency. For example parasitic infection, insufficient dietary protein and the presence of the unusual amino acid S-methyl-cysteine sulphoxide, which is present in brassicas, may contribute to the development of anaemias.

Cobalt

The chief role of cobalt is as a constituent of the vitamin B_{12} molecule. In the ruminant micro-organisms synthesize this vitamin provided they have a source of dietary cobalt. Vitamin B_{12} is an essential cofactor in the metabolism of propionic acid in the liver.

Cobalt deficiency occurs in grazing areas where the levels and uptake of soil cobalt are low. The condition is given various names, for example pine, and is characterized by a general unthriftiness, poor appetite and anaemia. Prevention of deficiency depends on oral administration of cobalt either as salts in the diet or as 'bullets'. The latter are small pellets that contain cobalt oxide which remain in the reticulum and release cobalt slowly.

Copper

Copper is widely distributed in the body and functions as a cofactor in several enzyme systems. It is involved in the formation of haemoglobin, is present in certain pigments found in the hair and has an involvement in the production of the characteristic physical properties (crimp) of wool. The liver has an ability to concentrate copper and thus acts as a storage organ.

Various conditions arise, especially in ruminant stock, when there is a shortage of copper. These range from anaemia resulting from copper being necessary for the utilization of iron in haemoglobin synthesis, bone disorders resulting from copper being required to produce the structural integrity of the organic matrix, pigmentation failure, diarrhoea and impaired fertility. A condition in lambs known as swayback is associated with copper metabolism. Affected animals are unable to walk properly, suffering incoordination of the muscles and rear limbs. Postmortem examination reveals damage to the spinal cord which arises through inadequate synthesis of the myelin sheath in which copper has a role.

Such conditions may arise from a simple deficiency due to grazing areas with low levels of soil copper but frequently are due to poor utilization of ingested copper. The absorbability of copper is influenced by a number of other dietary components. The presence of molybdenum and sulphate may, for example, make copper unavailable through the formation of insoluble copper thiomolybdate in the rumen. Such interactions influence the requirement of the animal for copper.

Whilst copper is an essential element it may also be highly toxic when consumed in quantity. In this context there is considerable species variation. Pigs can tolerate high levels of copper (it is routinely included at up to 150 ppm to stimulate growth), whereas sheep and to a lesser extent cattle are particularly susceptible to copper toxicity. This toxicity is due to the accumulation of copper in the liver and its subsequent liberation into the blood where it causes haemolysis.

Molybdenum

This is an element required in traces for certain enzyme systems in both the animal and ruminal micro-organisms but which is toxic at higher levels. In practice, deficiency is rarely encountered. Toxicity is related to its interaction with copper (see above) and may occur when animals graze herbage with a high molybdenum content. Molybdenum uptake by herbage is influenced by soil pH being greater at higher pH, such that molybdeosis can follow overliming of pastures. Control is through oral administration of copper sulphate.

Zinc

Zinc is distributed widely in the animal body with higher concentrations in bones, skin, hair and wool and the testes. Its major role within the animal is as an integral constituent of certain enzyme systems. Most diets contain sufficient zinc for farm animals but insufficiency may arise through inadequate absorption from the small intestine. Excesses of calcium and copper in the diet may depress zinc absorption as may also the phytic acid present in cereals and oilseeds. Deficiency initially results in skin lesions, termed parakeratosis in growing pigs, a condition readily alleviated through the addition of zinc to the diet.

Manganese

This element occurs in most tissues but is present in higher concentrations in bone, liver, kidney and pancreas. It is a component of enzymes involved in the synthesis of the mucopolysaccharides that occur in bone. In pyruvate carboxylase, an enzyme central to fat and carbohydrate metabolism, it has a role along with the vitamin biotin, and it is a cofactor for enzymes involved in the synthesis of cholesterol and urea. Deficiency is rare in farm livestock but when it occurs it can cause retarded growth, skeletal deformity and reduced fertility in breeding animals. The element is widely distributed in feeds but only some 5–10% is absorbed. The presence of excessive calcium and phosphorus may further depress this coefficient of absorption.

Manganese tends to be concentrated in the exterior layers of cereal grains such that bran is a good dietary source.

Iodine

The sole role of iodine in the body is as a constituent of thyroxine, a hormone secreted by the thyroid gland and unique in having an inorganic constituent. Thyroxine is primarily involved in the regulation of energy metabolism, but may also influence development and fertility. The secretion of thyroxine is regulated by hypothalamic (thyroid releasing factor, TRF) and pituitary (thyroid stimulating hormone, TSH) factors which are subject themselves to negative feedback. A deficiency of iodine leads to the production of insufficient thyroxine to prevent TSH from continually stimulating the thyroid leading to its enlargement, a condition known as goitre.

Iodine deficiency is generally a problem in regions with soils containing low levels of iodine. Where such situations exist, supplementation with iodized salts is necessary taking care to avoid excessive levels. Some iodine deficiencies can arise in situations of apparently adequate intake due to the presence of dietary constituents termed goitrogens. These occur most commonly in brassicas in the form of thiocyanates and thiocarbamides, and interfere with thyroxine synthesis. Whereas the goitrogenic effects of the thiocyanates can be overcome through the addition of extra iodine to the diet, the effects of the thiocarbamides can only be partially overcome by this measure.

Selenium

The role of selenium as an essential nutrient is closely related to that of vitamin E in that both micronutrients are involved in the prevention of oxidation of tissue lipids. As a component of the enzyme glutathione peroxidase, selenium is involved in the destruction of peroxides produced by the oxidation of unsaturated fatty acids before they have an adverse effect on cell membranes. It is also necessary for the production of pancreatic lipase.

A deficiency of selenium causes a variety of syndromes in farm animals ranging from the nutritional muscular dystrophy (white muscle disease) found most commonly in calves and lambs and characterized by degeneration of skeletal muscle; mulberry heart disease found in piglets in which the heart muscle is affected and sudden death may occur; and the exudative diathesis found in chicks in which damage to the cell walls leads to leakage of fluids from cells and oedema of the breast. As well as being cured by inclusion of trace amounts of selenium to the diet these conditions also respond to vitamin E supplementation.

The margin between sufficiency and toxicity levels of

selenium in the diet of farm animals is a relatively fine one and considerable care is required in providing supplementary sources. Selenium in foods exists as either the inorganic form of selenites or in organic forms bound to sulphur-containing amino acids, e.g. selenomethionine. The organic forms are more available than the inorganic forms. The amounts of these different forms in feeds can vary widely and feeds originating from areas with low soil selenium may be deficient.

Vitamins in metabolism and nutrition

Vitamins are organic compounds required in extremely small quantities for the normal function, health and productivity of animals. As a generalization animal cells are unable to synthesize these substances and must therefore obtain them from exogenous sources, either the diet or in some instances through microbial synthesis within the digestive tract. Individual vitamins are required for specific metabolic roles, frequently as integral parts of various enzyme systems. Thus deficiencies reveal a variety of disorders and symptoms that are frequently non-specific especially in the marginal deficiencies that are more likely to pertain in farm livestock.

Vitamins encompass a variety of chemical structures but may be readily classified according to their solubility characteristics into fat-soluble and water-soluble vitamins (Table 15.9). All the vitamins have chemical names but many continue to be known by letters of the alphabet by which they were designated prior to knowledge of their chemical identity.

The solubility characteristics of vitamins have implications for their absorption and storage. Fat-soluble vitamins are absorbed along with fats in micelles and may be stored in fat-containing tissues whereas water-soluble vitamins are absorbed mainly by passive diffusion and there is little or no capacity for body storage necessitating frequent intake in the diet.

Most foodstuffs contain some vitamins or in some cases the precursors from which the animal derives vitamins (provitamins) but the amount of individual vitamins in feeds varies widely. Monogastric animals and young ruminants must obtain most of their vitamin requirements from feedstuffs but in the case of the ruminant the microbial population in the rumen synthesizes the B vitamins and also vitamin K which subsequently become available to the animal. In most instances the supply of these vitamins is sufficient to meet the ruminant's requirement for them but in situations of high productivity there may also be a need for dietary supplementation. Microbial synthesis of these vitamins also occurs in the rear gut of animals, both ruminants and monogastrics, but very little is absorbed and is only of real value to animals which practise coprophagy.

Table 15.9 Vitamins of importance in animal nutrition

Vitamin	*Chemical name*
Fat-soluble vitamins	
A	Retinol
D_2	Ergocalciferol
D_3	Cholecalciferol
E	Tocopherol
K	Phylloquinone
Water-soluble vitamins	
B_1	Thiamine
B_2	Riboflavin
—	Nicotinamide
B_6	Pyridoxine
—	Pantothenic acid
—	Biotin
—	Folic acid
—	Choline
B_{12}	Cyanocobalamin
C	Ascorbic acid

Vitamin A

All animals have a dietary requirement for vitamin A. In considering the supply of this vitamin two groups of compounds are of interest. One group are the carotenoids, the most important of which is β-carotene. Carotenoids are principally found in the leaf tissue of plants, to a lesser extent in seeds, and are precursors of vitamin A. The other group of compounds are forms of vitamin A itself which are only found in animal products and thus, from the viewpoint of farm animals, are present as supplementary rather than naturally occurring components of the feed.

The utilization of dietary carotenes differs between species. In the pig, sheep and goat they are largely converted to vitamin A in the intestinal mucosa prior to absorption whereas in cattle and poultry some carotene escapes conversion and appears in the blood. Such absorbed carotenoids can be converted to the active vitamin in the liver and kidney but are also evident in the pigmentation of egg yolks, poultry carcasses and the milk and fat of cattle. The efficiency of conversion of carotene to vitamin A varies from almost 100% in poultry to less than 30% in ruminants. In all animals the conversion efficiency declines with increasing intake of either carotenes or vitamin A. Surpluses of vitamin A in the body may be stored in the liver and protect the animal during periods of vitamin A insufficiency.

Vitamin A performs a variety of functions in the body but many of these relate to the maintenance of the integrity of epithelial tissues. Thus in vitamin A deficiency, keratinization of epithelia in the respiratory tract, genitourinary tract, alimentary tract and cornea occurs. Such damage to the membranes in these tissues not only leads to poor absorption from the digestive tract and also respiratory and reproductive problems but also allows ready entrance of bacteria such that secondary infections may arise.

Vitamin A has an important role in the formation of bones, especially in the formative stages. Deficiency causes retardation of growth, bones being shorter and thickened. The synthesis of certain glycoproteins, a major constituent of the organic matrix, is depressed.

The animal normally receives its vitamin A from carotene in plant materials. The carotene content is high in green crops, especially in young leafy material, whilst roots and cereals are generally poor sources. The carotene content of feeds can be influenced by harvesting and storage processes.

Both enzymic and non-enzymic oxidation may lead in some instances, e.g. haymaking and storage, to up to 80% destruction of carotene in feeds. When vitamin A supplements are added to animal feeds it is usually protected in a gelatin–carbohydrate coating containing antioxidant to prevent loss in activity during feed storage.

Vitamin D

Two forms of vitamin D are of practical importance, namely D_2 and D_3. These active forms of the vitamin, chemically known as ergocalciferol and cholecalciferol, are formed from the effects of ultraviolet irradiation on the steroids ergosterol and 7-dehydrocholesterol which are the respective provitamins. Ergosterol occurs commonly in plants and is transformed during the sun-curing of forages to the active form of the vitamin. The provitamin of D_3 is synthesized in animal tissues, especially skin. Vitamins D_2 and D_3 have similar effectiveness for mammals but D_2 is virtually inactive in poultry.

The role of vitamin D is chiefly concerned with calcium and phosphorus metabolism and the mineralization of bone (see the earlier section headed 'Calcium, phosphorus and magnesium'). It is now clear that vitamin D undergoes metabolic change in the liver (to 25-hydroxyvitamin D_3) and the kidney (to 1,25-dihydroxyvitamin D_3) before it can perform its role of influencing the intestinal absorption, bone mobilization and urinary excretion of calcium and phosphorus. The mode of action of $1{,}25(OH)_2$ D_3 is similar to that of steroid hormones through inducing the production of messenger RNA.

Intensively kept livestock that do not receive direct access to sunlight and which receive little of the vitamin from cereal based diets require supplementation whereas grazing animals receive adequate amounts from irradiation. Synthetic metabolites of vitamin D have been used in the prevention of milk fever.

Vitamin E

Vitamin E refers to a group of substances called tocopherols, the most important of which is α-tocopherol. Although first identified to have a role in reproduction it is now known to function primarily as an antioxidant preventing peroxide damage to cell membranes resulting from the oxidation of polyunsaturated fatty acids. In this role it is supported by the selenium-containing enzyme glutathione peroxidase. The conditions that result from selenium deficiency (see 'Selenium') may also arise if vitamin E is deficient.

Green foods and cereal grains are good sources of the vitamin although some of the activity may be lost during storage, e.g. high moisture storage of cereal grains. Where supplementation is required, it is usually provided as α-tocopherol acetate. The need for supplementation is increased if the diet has increased levels of polyunsaturated fatty acids as, for example, when calves are turned out to grass in spring.

Vitamin K

Vitamin K is required for normal blood clotting through its involvement in the synthesis of prothrombin. Deficiency of vitamin K is rare in mammals since it is synthesized by the micro-organisms of the rumen in ruminants. In pigs hind-gut synthesis combined with a degree of coprophagy is usually sufficient to meet requirements, although young piglets housed on slatted floors require supplementation. Likewise, poultry are routinely supplemented with vitamin K concentrates or with feeds such as lucerne meal that are high in vitamin K. The dietary requirement is increased during periods of treatment with antibiotics, which reduce intestinal synthesis.

A relative vitamin K deficiency can occur when its action is blocked through antagonists such as dicoumarol which is found in mouldy clover hay. It can be prevented by giving extra vitamin K. Agents which similarly block the action of vitamin K, e.g. warfarin, have found use as rat poisons and in the treatment of thrombosis in humans.

B vitamins

The group of vitamins referred to as B vitamins function primarily as components of enzyme systems involved in carbohydrate, fat and protein metabolism. Ruminant animals usually obtain their requirements from microbial synthesis in the digestive tract and since the B vitamins are present in significant quantities in a wide variety of feeds it was formerly thought that only in particular situations did pigs and poultry require supplementation. The need for supplementation, however, is constantly being increased through the improved performances demanded of animals such that not only monogastrics but even high producing ruminants may now benefit.

Some of the metabolic roles and symptoms arising in deficiency are given in Table 15.10. Identification of problems resulting from B vitamin insufficiency is frequently difficult because they are often associated with non-specific symptoms such as reduced appetite, poor performance, diarrhoea and general poor condition.

In meeting the requirements for individual vitamins not only is the vitamin content of feedstuffs relevant but also the availability and stability of the vitamin. Thus, for example, thiamine deficiency in ruminants results not from a lack of thiamine in the digestive tract but from a combination of thiamine destruction by thiaminases and reduction in availability through the action of antagonists. Antivitamin factors have likewise been associated with biotin (streptavidine), folic acid (sulphonamide), niacin and pyridoxine. As far as stability of vitamins in feeds is concerned, the effect of heat and light during processing and storage can lead to loss of thiamine, pyridoxine and riboflavin activity. In the case of biotin, levels in feed do not relate to activity of the vitamin. Thus, for example, in wheat and barley, although biotin is present in significant amounts, it is almost totally unavailable whereas in feeds such as soya and fish meal all of the biotin is available.

Dietary supplementation using commercial forms of vitamins is primarily dependent on the ability of dietary raw materials to supply the vitamin but also on the cost of the vitamin. Thus since excess of water-soluble vitamins is not harmful, supplementation may in some instances (e.g. riboflavin) be carried out merely as a precautionary measure in cases where the vitamin is inexpensive, but in the case of a highly expensive vitamin such as biotin there is greater need to justify and be precise in supplementation. A further important parameter in deciding whether supplementation is necessary is the degree of confidence that can be placed in tables of composition and requirements.

Vitamin C

It is often considered that since farm animals can synthesize this vitamin there is no need for its consideration in the diet. Some evidence exists, however, that extra dietary vitamin C may help animals deal with stress

Table 15.10 The metabolic roles and deficiency symptoms associated with the B vitamins

Vitamin	*Metabolic role*	*Deficiency symptoms*
Thiamine (B_1)	Decarboxylation in carbohydrate metabolism	Nervous disorders – cerebrocortical necrosis (CCN), polyneuritis
Riboflavin (B_2), Niacin	Hydrogen transfer in energy metabolism	Non-specific; curled toe paralysis (chick)
Pantothenic acid	Part of coenzyme A	Non-specific; 'goose-stepping' (pigs)
Pyridoxine (B_6)	Amino acid metabolism	Non-specific; skin lesions, anaemia
Biotin	Fatty acid metabolism	Skin and hoof lesions, reduced fertility
Choline	Phospholipids, methionine metabolism	Fatty liver, perosis (birds)
Folic acid	Nucleic acid synthesis	Non-specific; anaemia, reproductive problems
Cobalamin (B_{12})	Nucleic acid synthesis Propionate metabolism	Non-specific; dermatitis (pigs), poor feathering (poultry)

resulting from adverse environmental conditions such as high temperature and ill health. Such an effect of vitamin C may result from its role in the synthesis of corticosteroids. Beneficial effects of supplementation have been reported for poultry kept at high temperatures and for early weaned piglets.

Vitamin C is also known to function in collagen synthesis, as an antioxidant and as an enhancer of iron absorption from the gut.

Metabolic disorders in livestock

A number of disorders occur in farm animals which appear to be related to an imbalance in the dietary supply of and demand for nutrients. These disorders tend to occur in animals with a high demand for particular nutrients and are sometimes referred to as 'production diseases'. They occur where the classic physiological mechanisms which normally deal with fluctuations in supply and demand and partition of nutrients are unable to cope or adapt quickly enough, thus leading to metabolic imbalances.

Ketosis

This condition occurs in both the cow and the ewe. Typically it occurs in cows (bovine ketosis/acetonaemia) in early lactation when milk yield is at a peak, and in ewes (pregnancy toxaemia/twin lamb disease) in late pregnancy when carrying two or more lambs. The metabolic origins of the condition are an imbalance in the animal's energy supply and demand. The energy intake in these conditions is limited in the case of the pregnant ewe by the growing fetus restricting rumen volume whilst in the dairy cow maximum voluntary food intake is not achieved until some time after peak yield. In both ewes and cows, animals that are overfat tend to have more restricted food intake.

Glucose is the energy supplying metabolite particularly in demand in the pregnant animal for the metabolism of the fetus and in the lactating cow for milk synthesis. Thus a characteristic of the condition is low blood glucose levels. Further to this the imbalance in energy supply and demand leads to the metabolism of body fat reserves. The metabolic consequence of this is that in its attempts to maintain glucose supplies through gluconeogenesis the mobilized body fat is incompletely oxidized and leads to the formation of the ketone bodies that are characteristic of the condition. The accumulation of ketones can be recognized through blood analysis, the use of the Rothera test on milk samples, or the presence of the sweet odour of acetone on the animal's breath. As well as ketone bodies being produced, fatty acids mobilized from adipose tissue may lead to fat deposition in the liver ('fatty liver' syndrome). This restricts gluconeogenesis and further exacerbates the problem. The low availability of glucose eventually leads to nervousness, inappetance and in the cow a drop in milk yield. The drop in milk yield in the cow can lead to spontaneous recovery as the supply and demand for energy gets back in balance but in sheep the condition is often fatal unless the fetuses are aborted.

Prevention of the condition involves taking measures which allow the animal to keep energy metabolism in balance. Avoidance of overfatness, maintenance of an appropriate energy intake through ration formulation and avoidance of sudden changes in diet that may affect the rumen function are essential. The inclusion of appropriate amounts of protein, particularly UDP, has also been shown to be an important preventative measure.

Milk fever (parturient paresis)

Milk fever is a metabolic disturbance affecting high producing dairy cows normally within two to four days of calving. It is less frequently found in sheep in late pregnancy.

The condition arises from an imbalance in the animal's calcium metabolism which leads to a dramatic fall in blood calcium from the normal of 2.5 mmol/litre to less than 1.5 mmol/litre. The clinical signs include initial excitement and change in muscle tone developing into muscle tremor, stiffness of gait and eventual recumbency and coma. The condition arises through the inability of the mechanisms which normally regulate calcium metabolism to adjust to the sudden increase in demand for calcium for milk synthesis at the onset of lactation. Under normal circumstances any imbalance between the supply of calcium in the diet and demand for calcium is taken care of through the mobilization and resorption of calcium from and into skeletal reserves. Such regulation is mainly brought about through the actions of parathyroid hormone, calcitonin and vitamin D. Higher yielding and older cows are more susceptible, the latter being due to the lesser ability to mobilize calcium from the skeleton.

Preventative measures suggested include the feeding of low calcium diets in late pregnancy to stimulate the regulatory mechanisms into action. In practice, it is difficult to feed diets with sufficiently low levels of calcium. Alternatively, treatment with vitamin D_3 and its analogues at appropriate times pre-calving can lead to the mobilization of bone calcium but the timing of such treatment is critical. Therapy involves the intravenous administration of calcium borogluconate solution. Other compounds such as calcium, hypophosphate and magnesium sulphate are necessary in certain cases. Recovery from therapeutic treatment is usually rapid although relapses may occur.

Grass staggers (hypomagnesaemia tetany)

This is a disorder of cattle and sheep and is characterized by low levels of blood magnesium resulting from an inadequate absorption of magnesium from the digestive tract. In its acute form, which most commonly occurs in lactating cows shortly after 'turn-out' in spring and less commonly during periods of rapid growth in autumn, animals show progressively hyperirritability leading to staggering gait, tetany, violent convulsions and death. A chronic form is found particularly in beef cattle being fed poor quality winter feeds and in outwintered animals.

The inadequate absorption of magnesium from the gut may be due to a restricted food intake, low magnesium content in the diet or low absorption efficiency. Very often a combination of these factors is involved. Unlike the situation with calcium metabolism the animal has little ability to buffer fluctuations in supply and demand for magnesium since there is little skeletal reserve and this is largely immobile. A steady supply of magnesium in the diet is therefore essential. Amongst a variety of factors that have been implicated in influencing the efficiency of magnesium absorption are rates of nitrogen and potassium fertilizer application to grass, rumen pH and the stress that sudden adverse weather conditions can cause.

It has been well demonstrated that daily administration of magnesium oxide (calcined magnesite) in the concentrate part of the feed will prevent grass tetany during seasons when it is prevalent. Dosage levels of 50 g/d for cattle and 7 g/d for sheep are adequate. The use of magnesium bullets which lodge in the rumen and slowly release magnesium are a useful alternative in more extensively managed stock.

Rumen-related metabolic disorders

The health of ruminant animals may be adversely affected by imbalances in rumen metabolism. Such imbalance may occur through the excessive or sudden ingestion of concentrates leading to *acidosis* and when excessive non-protein nitrogen and quickly degradable protein is fed leading to *ammonia toxicity*.

The problem of acidosis results from the production of very high levels of lactic acid and low rumen pH. This may influence gut function causing rumen stasis and epithelial damage, or alternatively the rapid entry of lactic acid into the blood may so upset the animal's acid–base balance that hypotension and respiratory failure ensue.

Sudden ingestion of concentrates brings about the acute form of the condition but chronic forms may be seen in cattle given very high concentrate diets with little or no forage. Such diets are associated with characteristic changes in the microbial population of the rumen, Gram-negative organisms being replaced by Gram-positive organisms such as lactobacilli.

Ammonia concentration in the rumen is a function of its rate of production from dietary RDP and NPN and its rate of utilization for microbial protein synthesis. The latter is mainly dependent on energy availability in the rumen. An imbalance in the production and utilization of ammonia in the rumen leads to accumulation of ammonia in rumen liquor. This eventually spills over into the blood upsetting the animal's acid–base balance and producing signs of hypertension and in extreme cases respiratory failure.

Prevention of both these conditions is dependent on care in ration formulation and feeding régimes. Observation of appropriate ratios of forage to concentrate and of degradable protein and NPN to fermentable metabolizable energy are essential. The mixing of ingredients and frequency of feeding need also to be considered.

Blood chemistry and nutritional status

Under normal circumstances the concentration of blood metabolites is kept within well defined limits and it is only in circumstances of metabolic imbalance that deviations from the 'normal' occur. Identifying deviations from normal in the composition of blood allows the diagnosis and correction of subclinical conditions which may adversely affect production. The analytical details of levels of metabolites present in a blood sample taken from an animal are referred to as its 'metabolic profile'.

Metabolic profile testing has to date been mainly used to monitor the nutritional status of dairy herds. Animals considered to be representative of high yielding, medium yielding and non-lactating groups in the herd are sampled and the blood analysed for glucose, 3-hydroxybutyrate, free fatty acids (indicators of energy status), albumin, globulin, urea and haemoglobin (all indicators of protein status), packed cell volume (an indicator of blood dilution) and various minerals (e.g. Ca, P, Mg). After analysis of samples the results are processed by computer and presented as a histogram allowing comparison of levels of individual metabolites with 'normal' values. Normal values are considered to be the mean for the population ± 2 s.d. from the mean.

Full interpretation of the data requires information on both the feeding regimes and performance of animals. Only then may recommendations on changing the nutritional inputs be made.

Voluntary food intake

Voluntary food intake may be defined as the weight of food eaten by an animal per unit of time when given free access to food. Its importance as a parameter influencing both the biological and economic efficiency of production has been increasingly realized in recent years. The level of intake dictates the rate of production, the proportion of the intake going towards production and thus the efficiency of food conversion. The level of intake may also influence the products of production. For example, when intake is excessive, then excessive fat deposition may occur in growing and lactating animals. Thus an aim in production is to match intake to the required level and type of production.

An ability to predict and manipulate the voluntary food intake of farm animals requires a knowledge of those factors that influence it. Like other animals, farm animals appear to control their food intake primarily by monitoring their consumption of available energy (energostasis). Thus, if animals are fed a range of diets of differing energy concentration, they will alter their dry matter consumption through altering either meal frequency or meal size in order to achieve energy balance. The precise mechanisms by which such regulation occurs is still unknown but involves centres in the hypothalamus of the brain that receive neural and endocrine information pertaining to energy balance and thus regulate the 'drive' to eat.

Mechanisms of energostasis

The mechanisms by which animals are capable of regulating their energy balance are imprecisely understood. Many of the early theories on the control of food intake have proposed single factors. These include such classic ideas as the *chemostatic* theory in which it is proposed that levels of primary blood metabolites, such as glucose in the monogastric and volatile fatty acids in the ruminant, provide the feedback information; the *thermostatic theory*, which proposes appetite to be linked to the animal's thermoregulatory mechanisms; and the *lipostatic control theory*, which proposes that the feeding drive is in some way controlled by the body's fat reserves. It is unlikely, however, that food intake is regulated by any single mechanism and that whilst some aspects of these theories are involved, the centres in the brain concerned with the control of feeding are more likely to receive a variety of feedback signals which they integrate to determine feeding behaviour. As well as the involvement of primary blood metabolites it is likely that metabolic hormones, such as growth hormone, insulin and glucogen, and gut peptides, such as cholecystokinin, have a part to play in energostasis.

Limits to energostasis

Although the underlying mechanism by which animals regulate their intake is related to energy balance, a number of factors may limit the capacity of animals to achieve 'energostasis'.

The ultimate limiting factor to the intake of food must clearly be the physical capacity of the digestive tract, the rates of digestion and absorption and the rate of passage of food through the digestive tract. Thus, although animals may seek to control their energy consumption through physiological mechanisms, they may fail to achieve this when fed bulky and/or indigestible feedstuffs. In practice the extent of dietary dilution necessary to produce limitations to gastric distension is unlikely to pertain in the monogastric and is more likely to arise in the ruminant animal. Evidence for the physical limitation of intake comes from observations relating intake to the available energy concentration of the diet. This is illustrated in Fig. 15.15. The range of energy concentrations over which energostasis may be accomplished or physical limitations to intake prevail is dependent upon the physiological status of the animal and characteristics of the feed other than simply energy concentration. Thus, for example, in pregnant or overfat animals where the developing uterus or internal fat deposits restrict rumen volume, physical limitation occurs at a lower dietary energy concentration. As far as feed characteristics are concerned these features which influence the rate of passage of food through the gut are relevant. Thus digestibility and rate of digestion and those features of a feed such as chemical composition, level of processing and water content which may influence these parameters are influential. The extent of gut fill is monitored by stretch receptors in the gut wall and this information is relayed to the brain via the nervous system where it influences feeding behaviour.

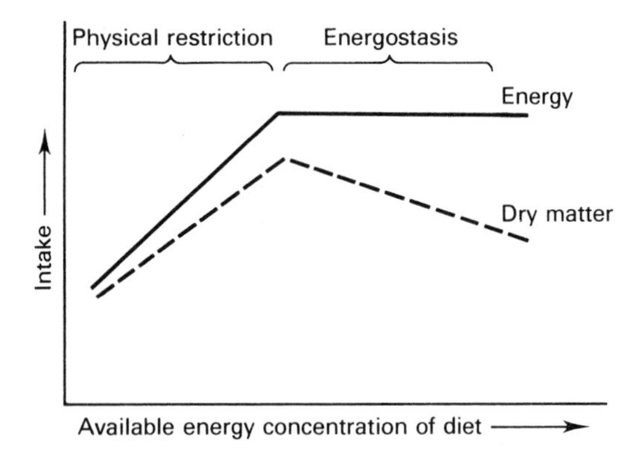

Fig. 15.15 Representation of the relationship between the energy concentration of the diet and energy and dry matter intake.

The prediction of intake

Since a knowledge of the dry matter intake which an animal will eat is essential for the precise and economic formulation of diets, the ability to predict intake is of the utmost importance. For ruminants, particularly dairy cows, many relationships have been proposed. The simplest is that proposed by MAFF and widely used in advisory work:

$$\text{Dry matter intake (kg/d)} = 0.025\,W \times 0.1\,Y$$

where W is body weight in kg and Y is daily milk yield in kg.

Such a simplistic prediction equation has obvious drawbacks in that characteristics of the food, such as physical form, palatability, acidity and chemical position and characteristics of the animal other than bodyweight and milk yield (e.g. pregnancy, degree of fatness, liveweight change and stage of lactation), all of which are known to influence intake, are not taken into account. An example of a more complex equation, predicting the dry matter intake of lactating dairy cows, is:

$$\text{DMI (kg/d)} = 0.076 + 0.404c + 0.013W - 0.129n + 4.121 \log_{10}(n) + 0.14Y$$

where DMI is dry matter intake, c is kg of concentrate dry matter, n is the week of lactation, W is body weight (kg) and Y is milk yield (kg).

Prediction of intake for pigs has received less attention since the practice in pig feeding has been to restrict intake in order to achieve appropriate carcass composition at slaughter. In prediction equations derived from poultry a parameter with a considerable influence is environmental temperature.

Manipulation of intake

The objectives in the manipulation of intake are dependent on the species involved. In pigs there has been some manipulation of appetite by genetic selection. In the past, selection for leanness has also led to a reduction in intake. For the future it would appear that, if genetic increases in potential lean tissue growth are to be realized, then there is a need to breed for increased appetite.

With ruminant animals production is frequently limited by physical restraints on the intake of forage and the objective is to increase total dry matter intake, especially the proportion of forage in the diet.

The greatest scope for manipulating intake at present lies in altering the characteristics associated with the feed which affect intake. Thus the processing of foodstuffs can have a major effect on the intake of feeds. The grinding and pelleting of forage, the chemical treatment of forages with alkali, and the processing of cereal grains either physically or through techniques such as micronization that gelatinize the starch may all influence intake through their effects on the rate of passage and/or digestibility of feeds. The method of feeding may likewise influence total daily intake. The mixing of the various dietary components to produce a 'complete diet' may increase intake as may the method of feeding employed in feeding systems involving the separate allocation of compound and roughage. Thus the more frequent feeding of compound through the use of 'out-of-parlour' feeders and the mechanized feeding as opposed to self-feeding of silage may increase intake.

In the grazing animal there is now a conceptual basis for understanding the influence of plant morphology and sward structure on herbage intake in terms of the effect of such parameters as bulk density, sward height and shearing strength. Quantification of such relationships so that sward variables may be objectively manipulated is now required.

The facility to manipulate the animal in ways which will influence intake is more limited. Since body fatness has a clear influence on intake, attaining appropriate body condition targets in the production cycle of pregnant and lactating animals is important. Social interaction can influence feeding behaviour such that group size, pecking order and social facilitation of feeding are characteristics with potential for manipulation. Feeding behaviour and intake are affected by photoperiod, and although the mechanisms involved differ in different species it may be possible to manipulate this parameter to advantage. The observation that Brahman cattle consume considerably more forage dry matter in relation to body weight when compared with Friesians has led to the suggestion that selection for the characteristic of rumen volume and/or giving animals early experience of very bulky, low quality diets may be worthwhile.

Reproduction

Reproduction is of crucial importance to all animal populations, both for herd or flock replacement and as a means of genetic improvement. Reproductive physiology is the science of reproductive functions within the animal. As such it holds the key to increased reproductive efficiency. It may also be able to provide explanations when things go wrong – such as subfertility or infertility in breeding stock. The manipulation of reproduction through the application of reproductive technology is now standard practice in most modern livestock enterprises. The following section seeks to provide a basic outline of the physiology of reproductive processes and to review the more recent advances in the technology of reproduction.

The female reproductive system

Anatomy

In all mammals the reproductive system of both sexes is bilaterally symmetrical, which means that organs consist of right and left paired structures. The female reproductive system is comprised of two ovaries, two oviducts or fallopian tubes, two uterine horns, a uterus, a cervix, a vagina and an external opening, the vulva (see Fig. 15.16). Although the female reproductive anatomy is essentially similar between farm species, there exists a distinction in the degree of fusion of the uterine horns and their relative length with respect to the body of the uterus. This relationship depends on the litter size of the species. Thus the pig has long uterine horns compared to the body of the uterus since the many embryos of the sow develop within these horns. In contrast, the cow normally has a single offspring which develops in the uterine body, therefore the uterine horns are relatively much shorter. The ewe reproductive tracts falls somewhere between these two extremes.

Sex determination

The sex of an offspring is determined by its genetic makeup or chromosomes which are donated by its parents. A female mammal has a pair of identical sex chromosomes, designated XX, whilst those of the male are non-identical, XY. The primary reproductive organ,

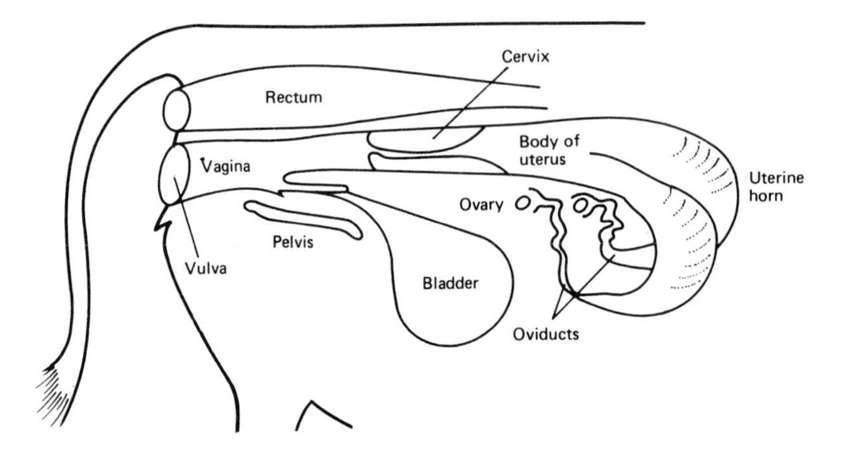

Fig. 15.16 Reproductive tract of cow.

the ovary, produces female gametes or eggs which carry only half the chromosome complement of the nuclei of the parent cells. Therefore each egg cell can carry only one sex chromosome and this must be an X chromosome. Male gametes or sperm cells on the other hand can be either X- or Y-bearing. Thus, at fertilization, the sex of the resulting embryo is determined by the 'sex' of the fertilizing sperm cell. Within ejaculated semen there are approximately equal numbers of X- and Y-bearing sperm, thus the chances of a fertilized egg being male or female is 50:50. This is known as the primary sex ratio. The secondary sex ratio is the number of male:female offspring at birth.

The genetic sex of the female is determined at fertilization. The phenotypic expression of that sex, or more simply, the appearance of 'femaleness' is largely determined during embryonic development. Of primary importance in this respect are the embryonic ovaries which secrete the female sex hormone, oestrogen. This hormone determines that the embryo reproductive tract shall develop into a female one rather than a male reproductive system.

Impairment of normal embryonic sexual development sometimes occurs, as for instance in the case of the freemartin heifer. In the small proportion of twins that occur in cattle, when a male and female embryo share the uterus, the production of male hormone testosterone, from the testes of the male embryo, causes the reproductive system of the female embryo to be effectively 'masculinized'. The result is a female calf, or freemartin, which is usually infertile.

Sexual development

The ovaries are the site of production of the female gametes or oocytes. Their production is known as oogenesis. At birth there are something like 200 000 primary oocytes already present in the ovaries, far in excess of those needed for the animal's entire reproductive life. During their development into mature oocytes, female gametes go through a stage of meiotic division which results in a halving of the chromosome number of each egg cell. Oocytes are surrounded by layers of ovarian cells and the whole structure is termed a follicle (see Fig. 15.17). Follicle development is controlled by the gonadotrophins from the anterior pituitary, chiefly FSH. Maturing follicles produce increasing quantities of the steroid hormone oestrogen, which influences both growth of the reproductive tract and the sexual behaviour of the animal.

Follicle development does not reach completion until puberty occurs.

Puberty

This signals the beginning of the female's active reproductive life when the animal commences breeding activity. Thus puberty is indicated by the first heat or oestrus during which the female is receptive to the male. The age at which puberty is reached is dependent on the animal's body maturity which is related to its bodyweight and liveweight gain. Onset of puberty coincides with the point of inflexion on the animal's growth curve (see Fig. 15.34). Breed size influences the onset of puberty; smaller, faster maturing breeds reach puberty before larger breeds. Other factors are nutrition, health, season of birth and the presence or absence of a mature male. The introduction of a mature boar to late prepubertal gilts advances their onset of puberty. Data on the age at puberty for farm species are given in Table 15.11.

Oestrus cycles

All females of farm species, once puberty has been attained, exhibit sequences of reproductive activity known as oestrous cycles. They may be continuous throughout the year as in cattle and pigs or may be restricted to a specific breeding season as in the case of sheep, goats and deer, all of which are autumn breeders. The non-breeding time of the year is known as anoestrus. In sheep seasonal anoestrus is much longer for temperate or northern breeds, such as Scottish Blackface, than it is for breeds of more equatorial origin such as Merinos, which tend to breed for most of the year. For seasonal breeders, daylength or photoperiod is the cue to onset of breeding activity. Photoperiodic changes influence the activity of the pineal gland, via the optic pathway. The pineal regulates seasonal changes in reproduction through its control of gonadotrophin release from the anterior

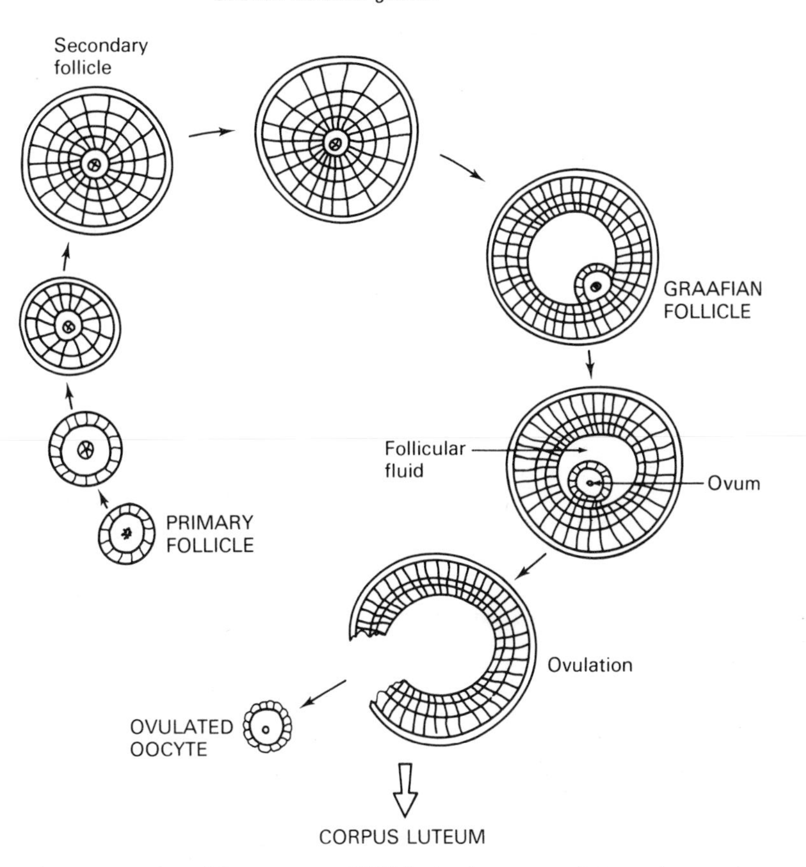

Fig. 15.17 Diagrammatic representation of the sequence of follicle development, ovulation and corpus luteum formation during the ovarian cycle.

Table 15.11 Some data on normal female reproduction

(a)

Animal	*Onset of puberty*	*Age at first service*	*Length of cycle*	*Duration of oestrus*
Cow	8–12 months	15–18 months	21 d (18–24 d)	18 h (14–26 h)
Ewe	4–12 months (first autumn)	First autumn or second autumn	17 d (14–20 d)	36 h (24–48 h)
Sow	$4\frac{1}{2}$–10 months	7–10 months	21 d (18–24 d)	48 h (24–72 h)

(b)

Animal	*Time of ovulation*	*No. of ova shed*	*Length of gestation*	*Optimum time for service*
Cow	10–15 h after end of oestrus	1	282 d (277–290 d)	Mid to end of oestrus
Ewe	Near end of oestrus	1–4	149 d (144–152 d)	16–24 h after onset of oestrus
Sow	Middle of oestrus	10–20	114 d (111–116 d)	15–30 h after onset of oestrus

pituitary. Melatonin, a hormone secreted by the pineal, is produced in increased amounts during darkness. This hormone has an important role as a mediator between environmental light patterns and seasonal differences in gonadotrophin secretion (see later section on melatonin and out-of-season breeding).

The sequence of events in the oestrous cycle of the sow, ewe and cow are essentially similar, although their precise timing and duration do differ. (For a more detailed comparison see Table 15.11). The oestrous cycle may be divided into four main stages: oestrus, met-oestrus, di-oestrus and pro-oestrus.

Oestrus

This is the stage of the oestrous cycle when the female comes into heat and is characterized by physiological and behavioural changes, such as reddening of the vulva; increased mucus secretion of the cervix and vagina; increased irritability and vocalization; mutual grooming activity; and most importantly, a willingness to stand for the male. In cattle in particular, in the absence of a male the oestrous female will allow mounting by other females, who themselves are often in oestrus or approaching it. For a more detailed coverage of behaviour and detection in oestrus, see chapters on cattle, sheep and pigs (18, 19 and 20).

Duration of oestrus and length of the oestrous cycle differ between individual animals and may be influenced by such factors as age of animal, its standing in the social hierarchy, stage of the breeding season, time of year, climate, nutritional status, health and stress.

In pigs and sheep, egg release (ovulation) occurs at the end of oestrus. In contrast, ovulation in the cow occurs some 10–12 h after the end of behavioural oestrus. There is evidence that the interval between onset of heat and time of ovulation is significantly shortened when females are naturally mated compared with those artificially inseminated.

Met-oestrus

This is the stage immediately post-oestrus when the corpus luteum forms in the ovary from the cells of the post-ovulatory Graafian follicle (see Fig. 15.17).

Di-oestrus

This is the longest stage of the oestrous cycle and coincides with the maturation of the corpus luteum. It is otherwise referred to as the luteal phase.

Pro-oestrus

This is of relatively short duration when one or more follicles undergo the final stage of maturation. Animals in pro-oestrus may show a reddening of the vulva and increased sexual activity, although they will not stand to be mounted.

Hormonal control

The oestrous cycle is regulated by pituitary and ovarian hormones. Generally speaking the concentrations of these hormones in the blood reflect the physiological changes occurring in the animal (see Figs 15.18a and b). The endocrine events in the oestrous cycle of the cow have been assumed as typical of all farm mammals.

If the beginning of oestrus is day 0, then this is the stage at which Graafian follicles reach maturity under the influence of the gonadotrophins FSH and LH. In the cow, ovulation occurs shortly after the end of oestrus as a result of the increases in the levels of the luteinizing hormone, LH. Thus a mature egg is released into the oviduct and the follicle cells within the ovary form the corpus luteum. The corpus luteum gradually matures during di-oestrus, secreting progesterone, whose function is to prolong the luteal phase of the oestrous cycle until the fate of the ovulated egg has been decided. Progesterone, through its effects on the pituitary, inhibits the release of FSH and thus prevents maturation of further follicles. Progesterone, in conjunction with the low levels of circulating oestrogen at this time, helps prepare the uterus lining or endometrium for the possible arrival of a fertilized egg. Sometimes animals become acyclic due to the presence of a persistent corpus luteum, or luteal cyst as it is generally termed. Thus a return to oestrus is indefinitely delayed.

Normally, however, in the absence of fertilization, the corpus luteum naturally regresses at about day 16 or 17 of the cycle. Luteal regression is accompanied by a fall in progesterone. The cause of these events is the release by the uterus of a hormone known as prostaglandin.

With the end of the luteal phase, follicle development accelerates. Maturing follicles are the major source of oestrogen production, the chief oestrogen in the cow being oestradiol. During this pro-oestrus stage, blood concentrations of oestrogen are therefore greatly elevated. Oestrogen is the hormone responsible for the physiological and behavioural changes taking place during pro-oestrus and oestrus.

The number of follicles that mature and are eventually released by the ovaries is termed the ovulation rate. Although influenced by gonadotrophins, ultimately ovulation rate is genetically determined. Thus some species and breeds have a greater ovulatory capacity than others. Other factors influencing ovulation rate are age, stage of the breeding season and nutritional status.

The practice of flushing, where the level of nutrition is increased shortly prior to mating, usually results in an improved ovulation rate.

Synchronization of oestrus

Synchronization of oestrus, or controlled breeding, is possible in farm species through the manipulation of oestrus cycles by means of hormone administration. This can be achieved in two ways:

- by premature regression of the corpus luteum (luteolysis) using prostaglandins;
- by prolongation of corpus luteum activity beyond the normal length of the luteal phase, using progesterone; followed by sudden withdrawal of treatment.

In both cases animals return to heat shortly after treatment, in a synchronized fashion. See Fig. 15.19 for a summary of hormonal changes associated with oestrus synchronization.

Prostaglandins

Prostaglandin $F_{2\alpha}$ or its synthetic analogues, when administered to a breeding female with an active corpus luteum, causes an effective shortening of the oestrous cycle. For instance, in cattle, the injection of prostaglandin causes the corpus luteum to immediately regress and

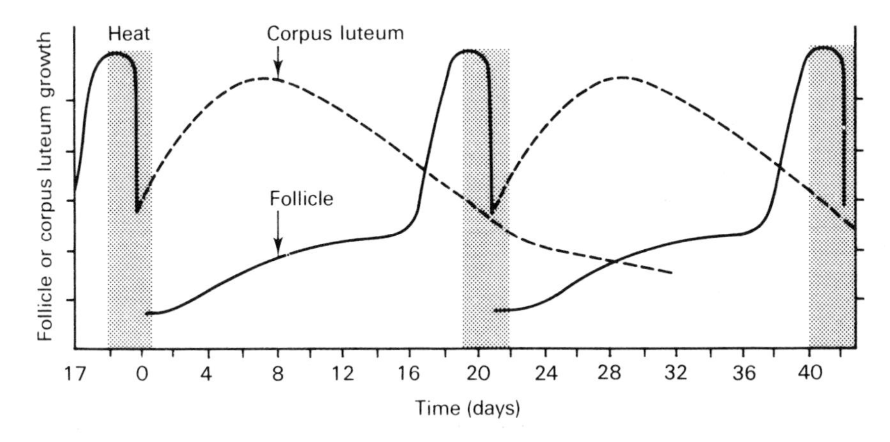

Fig. 15.18(a) Follicle and corpus luteum development during the oestrous cycle.

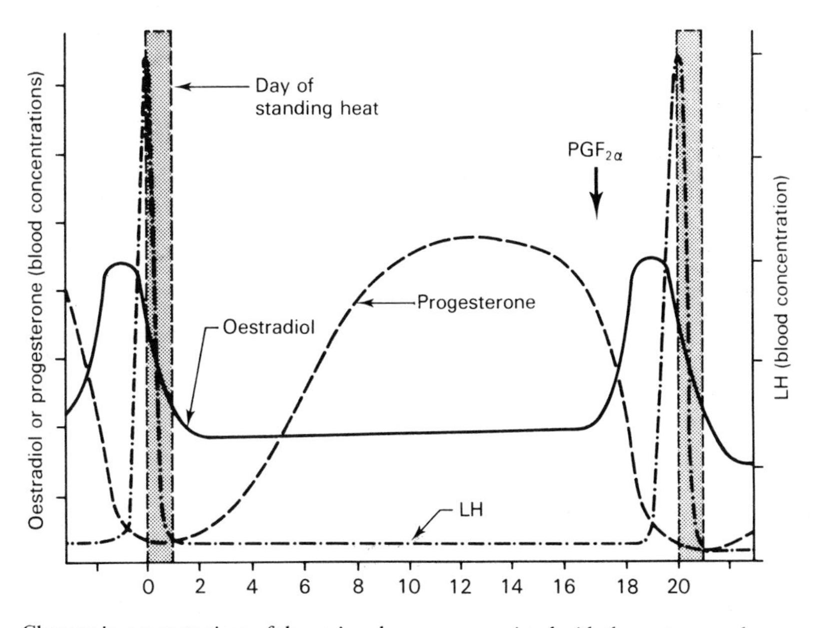

Fig. 15.18(b) Changes in concentrations of the various hormones associated with the oestrous cycle.

oestrus normally follows two to three days later. Insemination can then be carried out three to four days after prostaglandin treatment. However, not all cows respond to prostaglandin in the same manner. Animals between days 1–5 and days 17–21 of the oestrous cycle are unaffected by prostaglandin administration. Between days 1–5, the corpus luteum is immature and therefore unresponsive to prostaglandin and cows between days 17–21 have no corpus luteum present. The solution to this problem has been either to use two injections 11 days apart (when all animals should then be between days 5–17 of the cycle) or to use one injection and combine this with oestrus detection of unaffected animals. The use of prostaglandins to synchronize oestrus in pigs is largely ineffectual since the corpora lutea of the sow will only regress if treated between days 12–15 of the 21-day cycle.

Progesterone

Use of progesterone or its synthetic analogues (collectively referred to as progestagens) simulates the activity of a corpus luteum thereby artificially prolonging the luteal phase of the oestrous cycle. Thus once progestagen treatment is ceased, the animal returns to oestrus shortly afterwards. Greater synchrony of oestrus is achieved by either combining the progestagen with oestrogen or by $PGF_{2\alpha}$ injection following progestagen withdrawal.

In cattle, progestagens are administered intravaginally with oestradiol, in the form of a coil known as a PRID (progesterone-releasing intravaginal device). PRIDs have an advantage over prostaglandins in that their effectiveness is not dependent on the presence of a corpus luteum and unlike $PGF_{2\alpha}$ inadvertent use on pregnant animals is not likely to cause abortion. In sheep, progestagen-impregnated vaginal sponges are used in a similar fashion

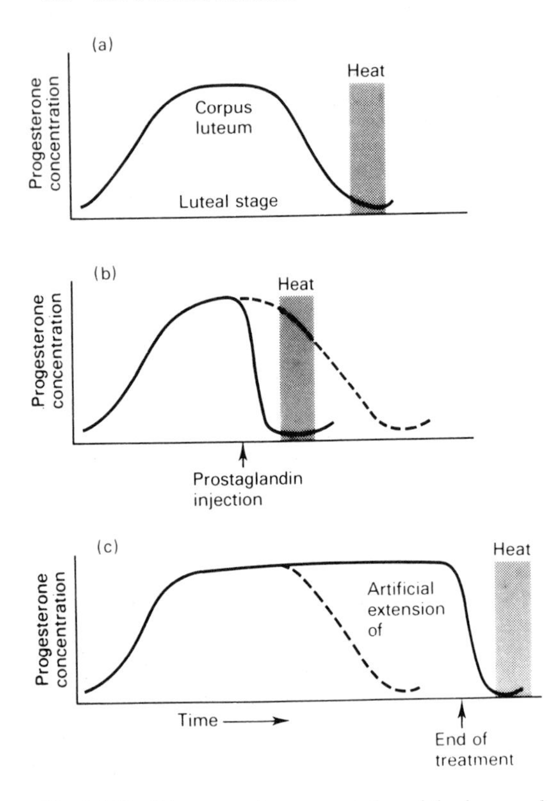

Fig. 15.19 Diagrammatic representation of the hormonal events of the oestrous cycle in (a) a normal cycle, (b) after prostaglandin injection, and (c) after progestagen treatment.

to PRIDs. However, immediately following sponge removal, PMSG may be injected to stimulate ovulation rate when synchronizing sheep at the beginning of their breeding season. (For further details of oestrus synchronization procedures in sheep see Chapter 19.)

The success of the above techniques depends on the degree of oestrus synchrony that can be achieved. Unfortunately, there is an inevitable variability in animals' responses to treatment, which means that in practice synchronization is never perfect. Pregnancy rates following synchronization are generally similar to those of untreated animals.

The male reproductive system

Anatomy

The male reproductive system is comprised of paired primary and secondary sex organs, together with accessory sex glands (see Fig. 15.20). The primary sex organ is the testis whose function is to produce sperm and male sex hormones. The testes are situated outside the abdominal cavity in the scrotal sac. The secondary sex organs consist of the reproductive tract, extending from the testis to the urethra. The chief role of the reproductive tract is to transport sperm. The urethra, in addition, carries urine from the bladder. Accessory sex organs (the seminal vesicles, the prostate gland and the bulbo-urethral glands) are situated at the base of the penis and

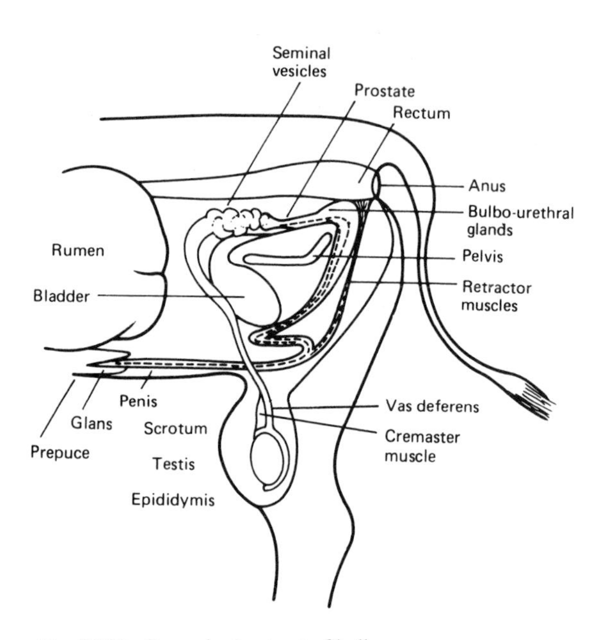

Fig. 15.20 Reproductive tract of bull.

are responsible for seminal fluid production (see later section).

Sexual development

The possession of a Y chromosome by the male fetus causes the undifferentiated gonad to develop into a testis rather than an ovary. The production of male hormone, testosterone, by the fetal testis regulates the subsequent development of the male reproductive tract. The entire reproductive system is thus fully differentiated prior to birth and by then the testes have normally descended into the scrotal sacs via an opening in the ventral abdominal wall known as the inguinal canal. In a small proportion of males, the testes may remain within the abdomen, a condition known as cryptorchidism. Since sperm production depends on the testes being maintained at 2–4°C lower than body temperature, such animals are invariably infertile.

After birth, significant further development is limited until puberty.

Puberty

Puberty in the male is characterized by both physical and behavioural changes. The marked increases in testosterone output at puberty stimulate sperm production and the development of secondary sex characteristics associated with the mature male. These characteristics include further growth of the reproductive tract, an increase in muscle formation, development of a masculine voice and body odour, together with an increase in sexual and aggressive behaviour. The exact onset of puberty, in contrast to the female, cannot easily be determined. Thus puberty in the male may be said to have occurred when enough sperm can be produced to successfully impregnate a female. For example, in the bull, the ejaculate should contain a minimum of 5×10^6 spermatozoa with at least 10% motility. Such a definition, however, takes no account of sexual behaviour of the animal. Sexual activity may reach full intensity considerably later than puberty. Thus puberty should not be confused with full sexual maturity. For bulls, puberty usually occurs at between seven and nine months of age, whilst sexual maturity may not be attained until four to five years of age.

Male sexual behaviour has two components: libido (or sex drive) and the ability to copulate (mating behaviour). Whilst libido is largely genetically determined, mating behaviour may depend on the social conditions in which the male is reared.

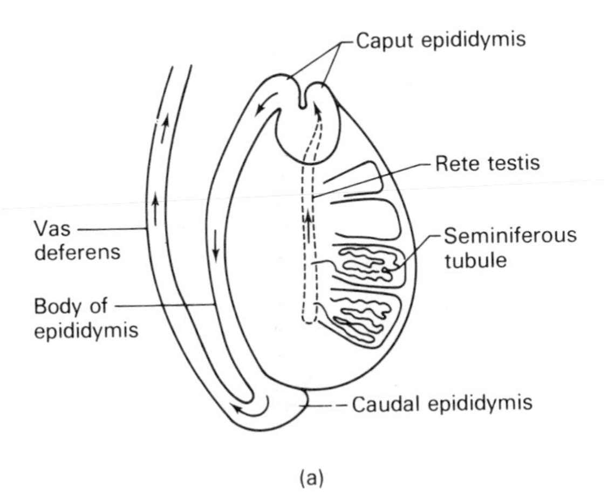

Fig. 15.21(a) Diagrammatic representation of a longitudinal section of the testis of the bull (→ represents passage of spermatozoa).

Castration, which is the removal or destruction of the testes, results in a loss of both sperm production and libido. Males castrated soon after birth fail to develop the characteristic male appearance. Growth is also affected by castration (see section on growth). It is important not to confuse castration with vasectomy which is the sectioning of the vas deferens to render the male infertile. Unlike castration, vasectomy has no apparent effect on sexual behaviour or the animal's ability to sexually arouse the female.

Sperm production

Sperm is produced by the seminiferous tubules of the testis (see Figs 15.21a and b). The germinal epithelium only begins significant sperm output at puberty. Spermatogenesis is temperature-dependent, the temperature of the testes being lower than that of the body. This is achieved by countercurrent heat exchange between the blood of the spermatic artery and vein supplying the testes. Maintenance of testes temperature is also helped by the action of the cremaster muscle which in cold conditions draws the testes closer to the body and in hot conditions moves them further away.

Sperm production is seasonal in wild species. In farm species, seasonal breeding in the male is not so clearly defined. However, the ram shows marked changes in sperm production and testis size depending on the season of the year. There are, however, breed differences and this should be borne in mind when choosing the ram for out-of-season breeding of ewes. Sperm production, as well as being daylength dependent, is also affected by environmental temperatures. For example, hot weather may seriously affect sperm production capacity in the boar.

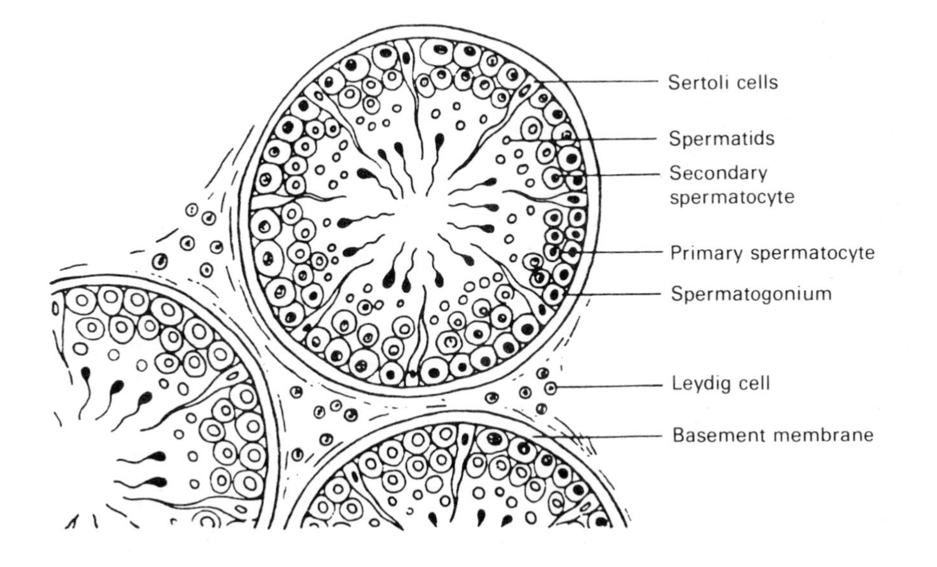

Fig. 15.21(b) Cross-section of the seminiferous tubules of the testis.

Spermatogenesis
Sperm production by the seminiferous tubules is controlled by FSH and LH. The Leydig cells are stimulated by LH to produce testosterone, which in turn stimulates spermatogenesis. The process of spermatogenesis is outlined in Fig. 15.22.

Spermatogenesis is characterized by a massive cell proliferation accompanied by the eventual halving of the chromosome complement (meiosis) of each spermatozoon. By the time spermatozoa have collected in the lumen of the seminiferous tubule, they have attained their distinctive appearance, as shown in Fig. 15.23. They do not, however, become fully motile until they have been transported to, and resided in, the caudal epididymis. The complete sequence of spermatogenesis takes between 40 and 60 days in farm species.

Testis diameter is a useful criterion for sperm production. The testis of the ram noticeably increases in size as the breeding season approaches.

Seminal fluid is added to spermatozoa prior to ejaculation by the prostate, seminal vesicles and bulbo-urethral glands. A comparison of sperm and semen output in farm species is given in Table 15.12.

Artificial insemination (AI)

The technique of artificial insemination permits the semen from a genetically superior male to inseminate a large number of females. The number of inseminations that can be performed using the semen from a single ejaculate depends on the extent to which the semen can be diluted without impairing its fertilizing capacity. As many as 1000 cows can be inseminated with the diluted semen from one ejaculate of a bull. This figure is far less for the boar and the ram (see Table 15.12). Another factor limiting the widespread use of AI in pigs and sheep is the difficulty of long-term preservation of semen. Unlike cattle semen, which can be stored deep-frozen, boar and ram semen do not survive the process without some reduction in their fertilizing capacities. Thus AI in pigs and sheep has largely been confined to the use of fresh semen.

Artificial insemination involves a number of distinct stages: semen collection; examination and evaluation of

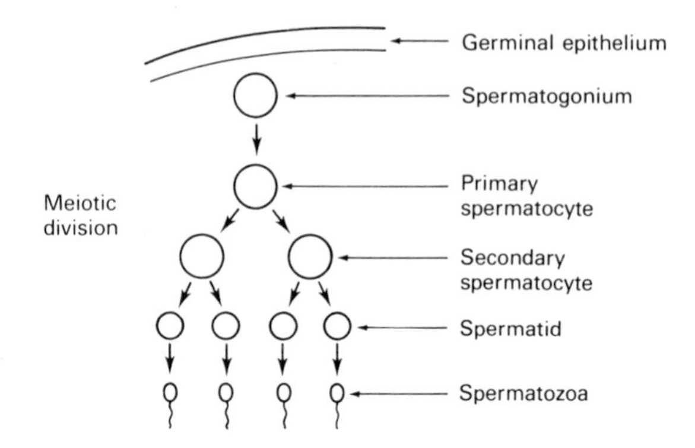

Fig. 15.22 Spermatogenesis from Peters & Ball, 1995.

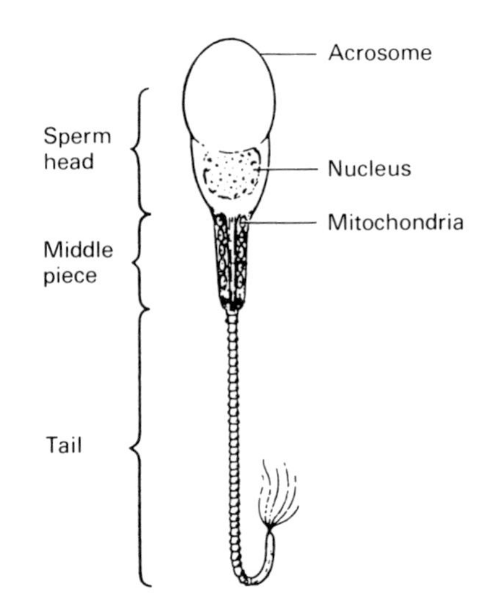

Fig. 15.23 Structure of a spermatozoon (or sperm cell) from Peters & Ball, 1995. *Reproduction in Cattle.*

Table 15.12 Data on sperm and semen production in farm animals

Species	*Volume of ejaculate (ml)*	*Expected sperm concentration* ($\times 10^6$/ml)	*Total number of sperm per ejaculation* ($\times 10^9$)	*Number of collections per week*	*Site of semen deposition at:* mating	AI	*Potential number of inseminations per ejaculate*	*Volume of inseminate after dilution (ml)*	*Number of motile sperm* ($\times 10^6$)
Bull	4–8	1200–1800	4–14	3–4	Anterior vagina	Cervix and/or uterus	400	0.25–1.0	5–15
Ram	0.8–1.2	2000–3000	2–4	12–20	Anterior vagina	External cervical os	40–60	0.05–0.2	50
Boar	150–500*	200–300	40–50	2–4	Uterus†	Uterus†	15–30	50–100	2000
Stallion	30–>150	100–250	3–15	2–6	Uterus†	Uterus†	5	20–50	1500‡

* Includes gelatinous secretion of the bulbo-urethral glands.
† The ejaculate makes passing contact with the cervical canal.
‡ Total number deposited in three inseminations during the prolonged period of oestrus.
Source: courtesy of Hunter, R.H.F. (1980) *Physiology and Technology of Reproduction in Female Domestic Animals.* Academic Press, London.

semen; semen dilution and storage; and, lastly, techniques of insemination.

Semen collection
One commonly used method in cattle is the use of a teaser female or a dummy, with the ejaculate collected into an artificial vagina. With pigs, a dummy mounting stool is used and the penis is grasped manually. A technique which is especially successful in the ram is that of electro-ejaculation, where a rectal probe provides a low voltage stimulation of the musculature of the reproductive tract.

Examination and evaluation of semen
During collection and examination every effort is made to avoid temperature shock to the semen, which might result in a drastic reduction in sperm viability. Semen examination involves an evaluation of a number of aspects of sperm quality: sperm numbers; morphological appearance of the spermatozoa; and sperm motility. The sperm count is important in that it determines the extent to which the semen can be diluted, which in turn dictates the number of inseminations possible from a single collection. Morphological examination assess the quantities of abnormal or immature spermatozoa present. For instance, in the case of a young male or a sire that has been over-used, a high proportion of immature spermatozoa may be present. In these instances the semen would not be used for commercial inseminations. Sperm with poor motility would also be rejected on the basis of its probable low fertilizing capacity.

Semen dilution and storage
Semen is diluted with chemical 'extenders' so chosen because of their ability to preserve the life of the sperm and at the same time have a negligible effect on the fertilizing ability of spermatozoa. Such diluents contain nutrients, a buffer to protect against pH changes, and, in the case of deep-frozen semen, a chemical such as glycerol to protect against the effects of chilling. Antibiotics are also included to prevent bacterial contamination. In pigs and sheep, where fresh semen is used, such chemical extenders may prolong its effectiveness for up to three days. Once diluted, bull semen is stored in liquid nitrogen at − 196°C in plastic straws. Each straw contains sufficient diluted semen for a single insemination.

Insemination techniques
Insemination is performed using specially designed catheters or insemination 'guns'. Such devices in the case of cattle and sheep permit the semen to be deposited further up the female reproductive tract than would be achieved by natural insemination. This allows a more dilute semen to be used without adverse effects on conception rates. It is important to note that with any artificial insemination procedure, stressful circumstances surrounding the event may have detrimental effects on its successful outcome. The timing of insemination is also vital relative to the stage of oestrus of the female, and this is dealt with more fully in Chapter 18.

Factors affecting male fertility

In summary there are a number of factors which affect optimum fertility in the male:

- Genetics – there are considerable individual and breed differences in male fertility.
- Age of the animal – performance increases with age from puberty.
- Nutritional status – e.g. rams should be in good body condition at mating to sustain performance throughout the mating season.
- Environment – daylength and ambient temperature affect level of sperm production.
- Frequency of use – both overuse and underuse affect fertilizing capacity of the male (a mature boar may be used 2–3 times per week and a bull 3–4 times per week).
- Health – e.g. physical damage to the penis may cause sufficient pain to affect libido. Damaged legs or feet may hinder mating ability.

Mating

Mating, the process which ensures that spermatozoa are deposited in the reproductive tract of the oestrus female, represents the culmination of a complex sequence of behavioural and physiological events. Prior to actual mating the male and female seek each other out by a combination of visual and olfactory signals. For example, the bull and the ram both display the characteristic Flehmen response which involves curling the upper lip and sniffing the air. This enables the male to detect the sexual pheromones given off by the oestrous female. In a similar way, the female may be sexually aroused by the odours of the mature male. Pheromones are believed to play a part in individual animal recognition. That a male may show a clear preference for some females over others is important with regard to mating management of herds or flocks. For example, running several rams with a flock or using males in rotation takes account of this fact.

Mating itself consists of mounting the female, thrusting, intromission and ejaculation. The exact site of deposition of sperm in the female reproductive tract depends on the species. In the cow and the ewe, sperm is deposited in the vagina, whilst in the sow, sperm is placed beyond the cervix into the uterus. Once in the female reproductive tract, sperm need to make their way to the site of fertilization, the oviduct. This may take as little as 2 h in the case of the pig but a much longer time may be necessary in cattle and sheep. Only a few hundred of the billions of sperm in the ejaculate manage to reach the oviduct. Fertilization of the egg is still not possible until the spermatozoa have completed the process of sperm capacitation. This term refers to a minimum period of 'acclimatization' to the uterine conditions which is necessary before sperm have the capacity to successfully fertilize an egg.

Spermatozoa and eggs undergo the effects of ageing while in the female reproductive tract. For sperm the effective fertilizable lifespan is 1–2 d and for the egg it is 10–12 h from time of ovulation. Given the rapid ageing of an ovulated egg and the delay in time between insemination and when spermatozoa are able to effect fertilization, it is clear that mating or artificial insemination must precede ovulation by an appropriate amount of time if conception rates are to be optimized (see Chapter 18 for advice on timing of AI in cattle).

Fertilization

Once spermatozoa have become fully capacitated, the acrosomal portion of the sperm head releases enzymes which enable the sperm to penetrate the protective membrane, the zona pellucida, which envelops the egg. There then follows a fusion of nuclear contents of the egg and sperm cells. The zona pellucida undergoes a subsequent chemical change which prevents entry of further sperm. The fused egg and sperm are termed an embryo or zygote.

Pregnancy

Embryonic development

Whilst in the oviduct the embryo begins to divide mitotically thus doubling its cell mass at each successive division. The embryo, by the time it has reached the uterus after about 3–4 d, has developed into a hollow ball of cells.

Implantation

It is generally considered that pregnancy only becomes established once the embryo has successfully attached itself to the lining of the uterus wall or endometrium. This process is known as implantation.

Fetal development

An embryo may be regarded as having attained the status of a fetus once recognizable organ development is apparent; heart, limb buds, liver and spinal cord are all present at an early stage of pregnancy. The majority of growth taking place in the first half of pregnancy, however, is mainly that of the fetal membranes and the placenta. Most of the growth of the fetus itself is confined to the latter stages of pregnancy (see Fig. 19.7).

A summary of the sequence of the major events of pregnancy is shown in Fig. 15.24.

Maintenance of pregnancy

Hormones, especially progesterone, play an important

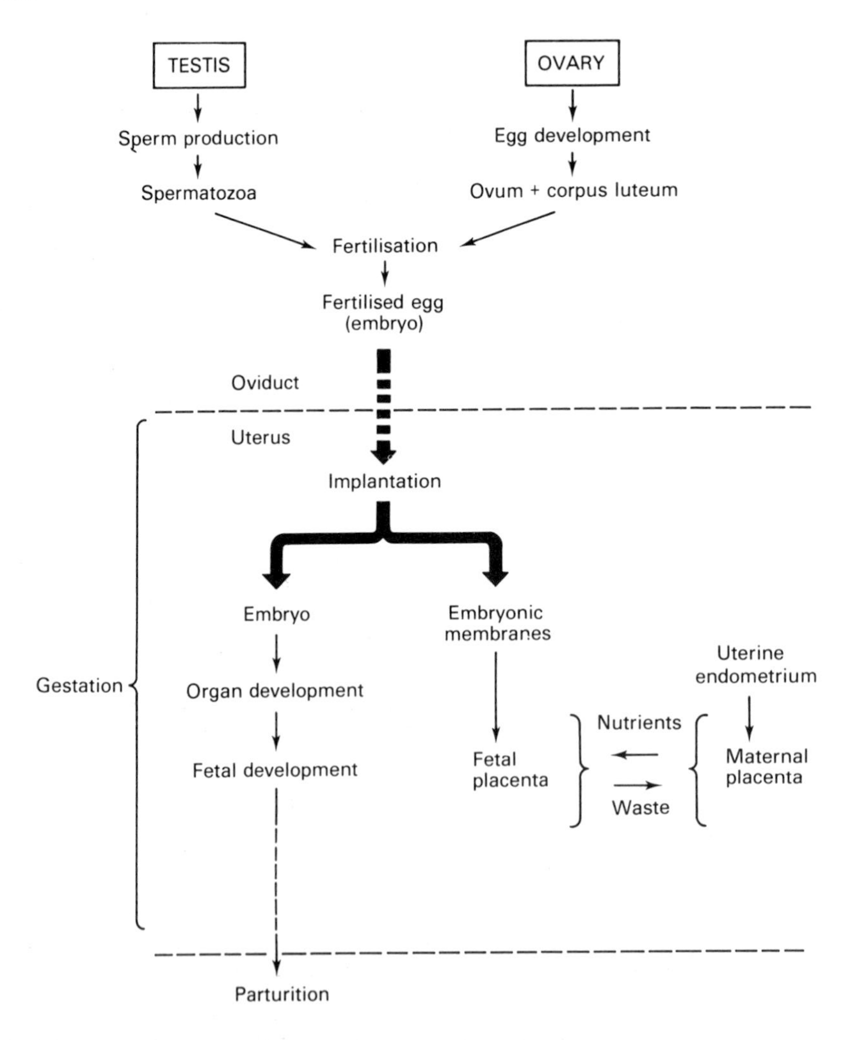

Fig. 15.24 Schematic representation of the principal events of gestation.

part in the establishment and maintenance of pregnancy following mating and fertilization. In the cow, for example (see Fig. 15.25), following ovulation and service the blood and milk progesterone levels remain identical up to day 17 or 18 whether the egg has been fertilized or not. After this time, in the non-pregnant animal, progesterone levels fall, as described earlier (Fig. 15.25). In the pregnant animal, however, progesterone levels remain elevated due to the continued presence of a corpus luteum. It is thought that the embryo itself plays a major role in preventing the prostaglandin $F_{2\alpha}$-induced regression of the corpus luteum that occurs if fertilization has not taken place. Thus the continuing production of progesterone after day 18 in the pregnant animal both ensures the maintenance of a uterine endometrium to receive the fertilized egg and also acts on the hypothalamus/anterior pituitary to inhibit gonadotrophin release. Progesterone therefore prevents further oestrous cycles. The high circulating levels of progesterone characteristic of pregnancy eventually fall at the end of pregnancy, shortly before parturition.

The source of progesterone through pregnancy differs in farm species. In the sow, the corpus luteum persists throughout the whole of pregnancy and is the main source of progesterone. In the cow and the ewe, in contrast, the responsibility for progesterone production switches from the corpus luteum to the placenta approximately halfway through pregnancy. There are obvious practical implications for prostaglandin treatment. Administration of prostaglandin to pregnant animals with an active corpus luteum would result in an abortion.

Embryonic and fetal mortality

It is estimated that 30–40% losses in potential offspring occur between fertilization and parturition in farm species. In cattle, where normally only one egg is ovulated, the death of an embryo or fetus means a failed pregnancy. In contrast, in pigs, where there is a much higher ovulation rate, success or failure may be calculated in terms of the proportion of ovulated eggs (as measured by the total number of corpora lutea on both ovaries) surviving to term.

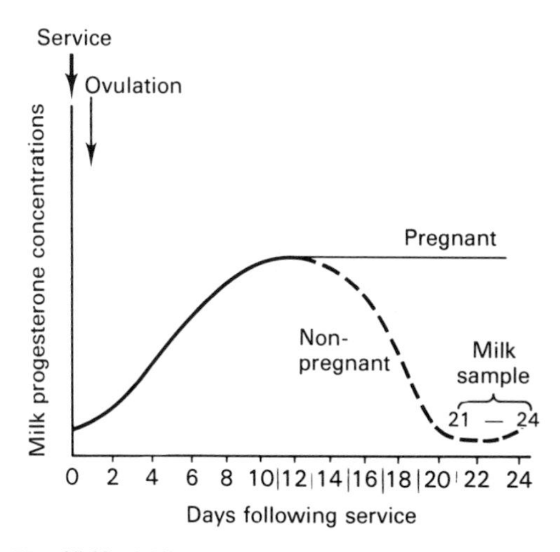

Fig. 15.25 Milk progesterone profiles in a pregnant and non-pregnant cow following service and ovulation.

Some two-thirds of all pre-natal deaths can be put down to embryonic losses. Much of this loss tends to occur either in the first few days following fertilization or at around the time of implantation. Genetic abnormalities, hormonal insufficiency or failure due to an aged egg are thought to be among the chief causes. Any nutritional problems or imbalance occurring at this stage will only serve to exacerbate such losses. If embryo mortality is largely attributed to failure of the fertilized egg, then fetal losses are much more likely to be the result of some deficiency on the maternal side, such as placental inadequacy or uterine infection. A dead fetus may result in an abortion or in some cases it may be retained within the uterus in a mummified condition, where the fetal fluids have become reabsorbed by the uterus lining.

Pregnancy diagnosis

An accurate diagnosis of pregnancy is vital to the reproductive efficiency of breeding livestock. Any pregnancy diagnosis service should also be early, fast and reliable so that animals not successfully inseminated may be re-bred as soon as possible. In addition, as in the case of sheep, a determination of the number of fetuses present allows the farmer to feed the mother accordingly. Techniques which can offer early diagnosis suffer the disadvantage of the high rates of early embryo mortality and are therefore less able to give an accurate estimate of viable offspring at the end of pregnancy. Some of the more commonly employed pregnancy diagnosis techniques are outlined below.

Hormone assay

This involves the measurement of hormones found in the blood or milk of pregnant animals, by means of a highly sensitive, laboratory-based procedure known as radio-immunoassay. Hormone assay techniques have been confined commercially to dairy cattle. There are two types of hormone diagnosis available: an early progesterone test and a later oestrone sulphate test.

Progesterone test

This method allows pregnancy diagnosis to be made as early as 21–24 d after ovulation. Samples of blood or, more conveniently, milk, are taken at this time and the levels of progesterone present are determined.

Pregnant animals will have high progesterone values, while low values will indicate that the animal has returned to oestrus (see Fig. 15.25). Although the laboratory procedure is highly accurate, the technique does produce a proportion of false positive diagnoses, which may be due either to cows which have abnormal length oestrous cycles or to insemination at the wrong time of the cycle. There is a further drawback implicit in such an early dignosis, in that an animal correctly diagnosed as pregnant may subsequently lose the embryo.

Oestrone sulphate test

This is a much later test (at around 15 weeks) which can confirm pregnancy with almost 100% reliability. This is because the method, unlike the progesterone test, measures a hormone, oestrone sulphate, which is derived from the fetus itself. The test is of limited practical use because of the lateness of the diagnosis.

Rectal palpation
The technique involves the insertion of an arm into the rectum of an animal in order to manually detect (palpate) the presence of a fetus within the underlying uterine horns. This procedure is largely limited to cattle and needs to be performed by a trained operator. The fetus can be palpated from around day 40 up to the fifth month of pregnancy. After this time palpation becomes impossible until after month seven because the uterine horns become greatly enlarged and 'disappear' below the pelvic brim.

Ultrasonic techniques
Such procedures consist of ultrasonic detection of fetal or placental structures. A variety of techniques have been developed, all of which involve the same basic principle. A probe is applied to the surface of the animal's abdomen which transmits and receives an ultrasonic beam. The reflected signal, which is able to detect the presence of the fetus and associated tissues, is then converted into an auditory or visual output. Recently, real-time ultrasound scanners, used routinely in human medicine, have been developed for farm use. This device allows an image of the uterine contents, including the fetus, to be displayed on a screen. This method has been commercially used in sheep to detect fetal lambs between 50 and 100 d of pregnancy (see Chapter 19 on Sheep for a fuller description). With all the available ultrasonic methods, a skilled operator is needed to correctly interpret the results obtained. Commercially, ultrasonic pregnancy diagnosis has been restricted to pigs and sheep.

Parturition

Gestation length
The onset of parturition or birth is determined by the length of gestation. Gestation length is largely determined by the fetus. Whilst it is the mother that decides the hour of birth, it is the fetus that decides the month, week and day. Other factors which influence gestation length include species, breed, litter size, sex of the fetus, age of the dam and the genotype of the sire.

Initiation of parturition
As stated above, it is the fetus that is chiefly responsible for initiating birth. Recent evidence has shown that once the fetus achieves its maximum growth, hormonal changes cause progesterone production to be 'switched off'. Thus the withdrawal from circulation of the hormone responsible for maintaining pregnancy signals the beginning of parturition. Increases in oestrogen and prostaglandin $F_{2\alpha}$ follow, with a resultant release of oxytocin, which is responsible for causing uterine contractions (i.e. labour). A diagrammatic summary of events in the sheep is shown in Fig. 15.26.

Stages of labour
The process of parturition may be divided into three stages:

The first stage – the preparatory stage involving dilatation of the cervix and relaxation of the pelvic ligaments.
The second stage – the expulsion of the fetus through the pelvic canal.

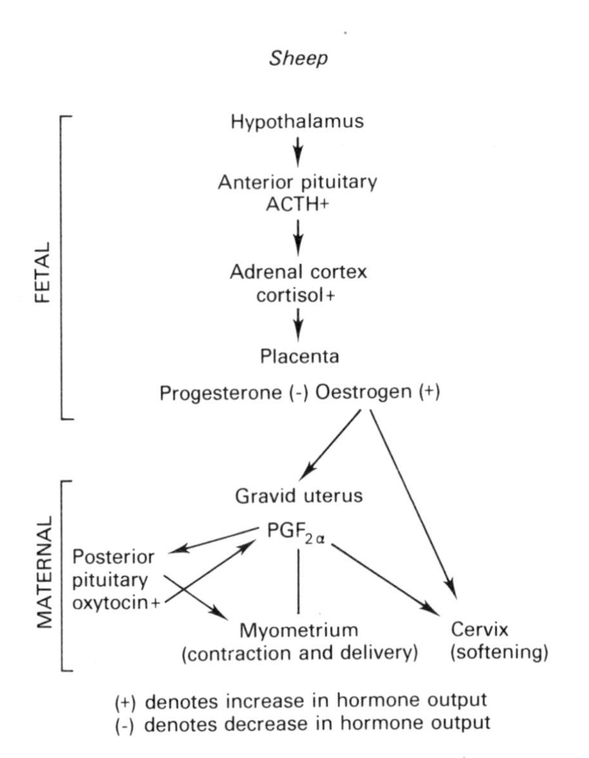

Fig. 15.26 A postulated sequence of hormonal events controlling the onset of parturition in the sheep (courtesy of First, N.L. (1979) Mechanisms controlling parturition in farm animals. In *Animal Production*, pp. 215–257. Ed. by H. Hawk. Allanheld Osmun, Monclair, New Jersey.

The final stage – the expulsion of fetal membranes and the initiation of uterine involution.

Any difficulty or prolongation of the birth process is termed dystocia.

Induction of parturition
Simulation of the events which initiate parturition (as outlined in Fig. 15.26) can be brought about by exogenous hormone treatment of late pregnant animals. Corticosteroids, prostaglandin $F_{2\alpha}$ and their synthetic analogues, when administered to animals nearing the end of their natural gestation, will induce parturition within two to three days of treatment. The method is not without its problems however. For example, in cattle such inductions of parturition have been associated with increased calf mortality and greater incidence of retained placentas. Used commercially the procedure allows a more exact timing of birth to fit in with management and labour requirements. (For a detailed description of induction in lambing see Chapter 19).

Lactational anoestrus

This describes the situation, following parturition, when active lactation is accompanied by minimum ovarian activity. Behavioural oestrus and complete resumption of breeding cycles are not observed for several weeks in cattle, sheep and pigs, depending on such factors as milk yield, nutritional status and suckling intensity.

Recent advances in reproductive technology

This section will attempt to review some of the more recent technological developments not already mentioned earlier in this chapter.

Superovulation

This process involves the injection of hormones such as FSH and pregnant mare serum gonadotrophin (PMSG) to increase the number of ovulatory follicles and hence the number of eggs released by the ovaries. Superovulation has been used in association with embryo transfer to allow the genetic potential of superior breeding females to be more fully exploited.

Immunization against ovarian steroids

Immunization against ovarian steroids is a newly developed technique which improves the ovulation rate without the risk of excessively large litters than sometimes results from superovulation. The female is injected with ovarian steroid and the immune system is stimulated to produce antibodies against the hormone. Normally, within the animal ovarian steroids regulate, by a process of negative feedback, the gonadotrophins that are responsible for follicle development and ovulation. Immunization is believed to interrupt this negative feedback and thereby produces an increase in ovulation rate. The ovarian steroid androstenedione, commercially available as Fecundin, has been successful in improving lambing percentages. Two injections are required, at eight and four weeks prior to mating.

Melatonin and out-of-season breeding

The pineal gland and its secretion, melatonin, have a central role in the photoperiodic control of reproduction in seasonal breeders. Melatonin, when fed in the afternoon to out-of-season sheep, has the effect of creating a chemical 'darkness' which convinces the animal that it is experiencing a shortened daylength. This results in an advance of the breeding season. Melatonin is licensed for use in sheep as a subcutaneous implant marketed under the trade name Regulin.

Embryo transfer

The commercial application of the technique of embryo transplantation has been largely confined to cattle. A number of eggs are produced by superovulation in a donor cow and, after fertilization, non-surgically collected and transplanted into the uteri of recipient cows. In this way a greater number of calves can be produced from a genetically valuable female than would be achieved normally during its reproductive lifetime. It is possible to freeze embryos for long-term storage and still achieve acceptable pregnancy rates with the thawed embryos. This facility therefore offers a useful extension to the scope of embryo transfer.

Sex determination

This can be achieved either by sexing of embryos or by attempting to separate the X- and Y-bearing spermatozoa. Although laboratory techniques exist to do this, no commercial method is yet available for use in animal production. The Y chromosome of the male produces a protein known as the HY antigen, which it is possible to detect by immunological means. Alternatively, it is possible to karyotype the cells of the embryo by a procedure which examines and identifies the sex chromosomes. A successful method of sex determination would have enormous benefits for the AI industry, given that half the number of dairy calves produced consist of males that may be unwanted.

A rapid milk progesterone test

A rapid, highly sensitive progesterone assay is now available commercially, employing ELISA procedures (enzyme-linked immunosorbent assay), which can be used 'on-farm' by untrained operators. A bulk milk sample is obtained between 21 and 24 d after service in the same way as for the laboratory-based radioimmunoassay. The ELISA method shares the same disadvantage as its more established counterpart in the number of false positives that are produced (see section on pregnancy diagnosis).

Lactation

The growth and development of the mammary gland and the subsequent manufacture and secretion of milk represent an important phase in the reproductive cycle of mammals. This section describes the structure and development of mammary tissue, the processes involved in the synthesis, secretion and release of milk, together with the underlying hormonal control of these events.

Anatomy of the mammary gland

The structure of the adult mammary gland of the cow is shown in Fig. 15.27. The majority of the gland consists of closely packed lobules, each made up of numerous clusters of alveoli. It is the epithelial cells lining the lumen of each alveolus, the secretory cells, which are the site of milk synthesis. The secreted milk is stored in the alveolar lumen prior to passing into the small ducts leading from each alveolus. The arrangement of the alveoli and ducts is rather similar to a large bunch of grapes, with the alveoli being the grapes and the ducts the stem branches. The ducts merge into ever larger ducts until the largest of these, the collecting ducts, empty into the gland cistern. In the cow prior to milking, 40–50% of the milk is held in the gland cistern and larger ducts, while the rest is retained within the small ducts and alveolar lumen. In contrast, in the sow approximately 90% of the milk is held in these latter structures.

Continuous with the gland cistern is the teat cistern and the streak canal through which the milk is eventually secreted. The streak canal is surrounded by a sphincter composed of circular smooth muscle fibres. In cows which tend to leak milk, this sphincter is not tight enough, whilst the opposite is the case with slow milkers. The mammary gland of the sow differs in that each teat is served by two separate streak canals, each leading to the teat cistern and gland cistern.

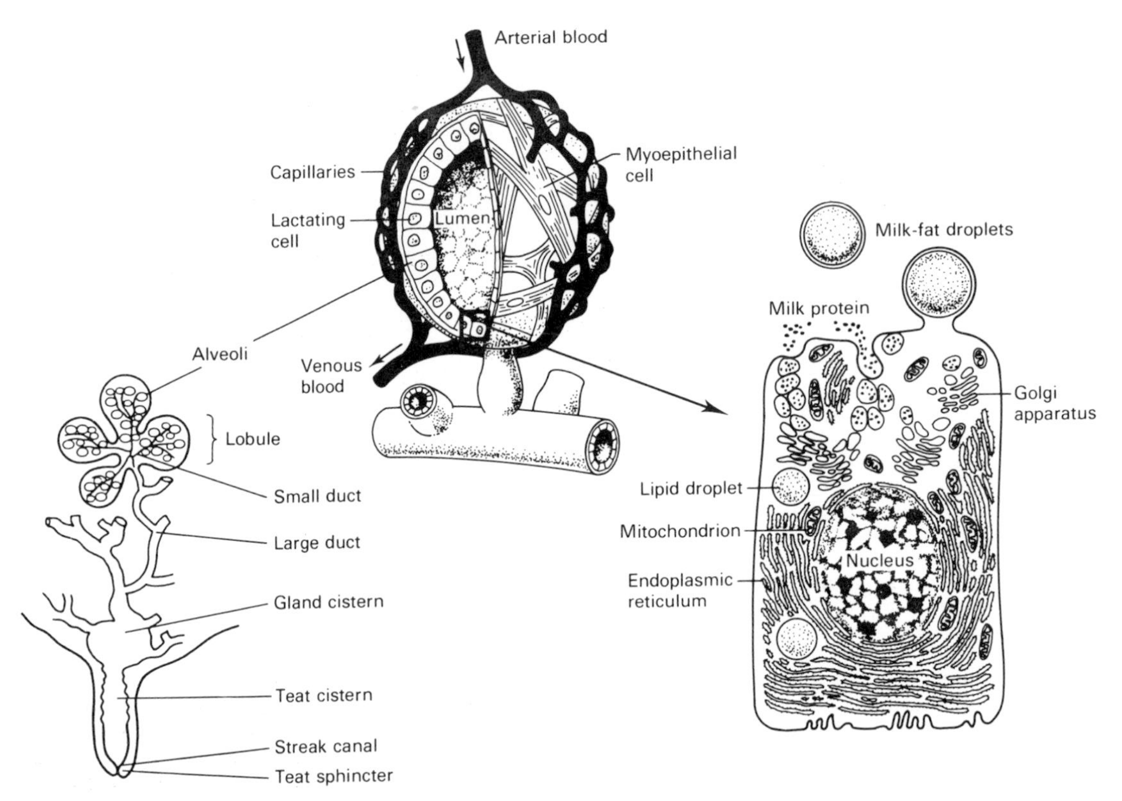

Fig. 15.27 Diagram showing structure of the mammary gland of the cow.

The sow normally has seven pairs of glands distributed bilaterally on the ventral surface of the body from thorax to inguinal region, a single teat serving each gland. The cow has two pairs of glands, found in the inguinal region, the rear two accounting for approximately 60% of total milk production. The ewe has a single pair of glands, inguinally situated.

The udder of the cow may weigh as much as 15–30 kg and as much again when full of milk. Support for the udder is therefore very important. This is achieved by means of suspensory ligaments which are attached to the body wall and pelvic girdle by tendons (see Fig. 15.28). The lateral suspensory ligament is fibrous and inelastic while the median suspensory ligament is more elastic; thus, when the udder fills with milk, the median ligament stretches to allow the udder to expand.

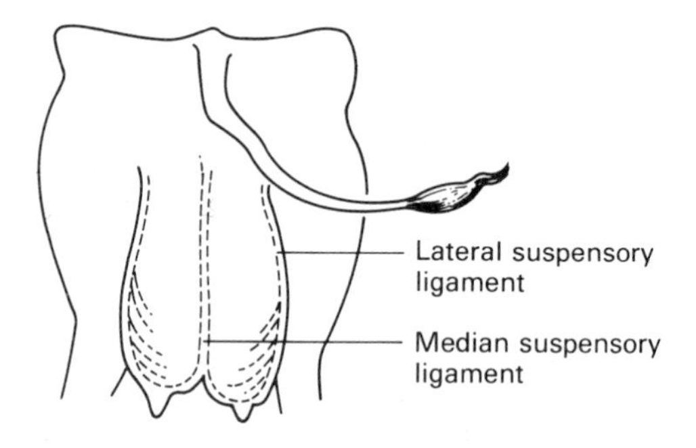

Fig. 15.28 Support structures for the udder of the cow.

Mammary development

At birth there is little mammary development present apart from teats, connective tissue and a rudimentary duct system. There is no secretory tissue, its place being taken by fat. At puberty there is an increase in ductal development, but it is only when the animal is pregnant that significant development of the duct and lobule-alveolar system commences.

Mammary gland differentiation is under the control of a complex battery of hormones from the anterior pituitary, ovary and adrenal gland. A summary of their effects on mammary development is given in Fig. 15.29. Lactogenesis, which is the initiation of milk secretion, occurs towards the end of pregnancy and at parturition. It is thought that the trigger for this event may be the decline in progesterone concentrations at parturition, although prostaglandin $F_{2\alpha}$ may also be involved. Following parturition there is a proliferation in the secretory cell population which, together with an increased output of each secretory cell, is responsible for the dramatic mammary secretion at this time. Suckling activity, which is itself related to the number of offspring, encourages

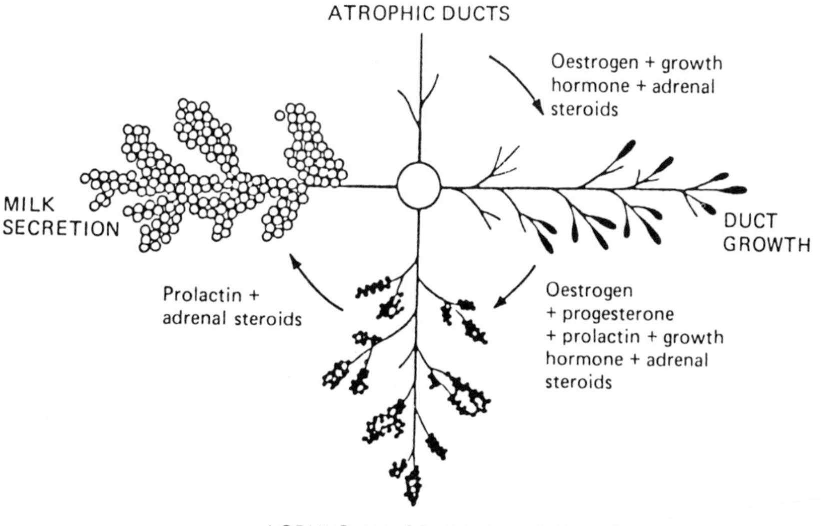

Fig. 15.29 Summary of hormonal control of mammary developments from Peters & Ball, 1995.

greater secretory response. Conversely, a decline in suckling or milking leads to a diminished milk output. At weaning or 'drying off', the mammary gland regresses or involutes and this is marked by a dramatic reduction in the secretory cell population.

Milk biosynthesis

Whilst the composition (Table 15.13) and quantity of milk secreted by the dams of the different domestic species vary considerably, the milk from all species has the same basic qualitative composition. Milk essentially consists of two phases; an aqueous phase, in which are partitioned proteins in colloidal suspension and lactose, minerals and water-soluble vitamins in solution; and a lipid phase containing triglycerides, phospholipids, sterols and fat-soluble vitamins. The lipid in milk is about 99% triglyceride and exists in the form of globules ranging from 3 to 5 μm in diameter surrounded by a complex membrane comprised of the other lipid components consisting of phospholipids, cholesterol and fat-soluble vitamins and which is acquired at the time of secretion from the lactating cell.

Table 15.13 Approximate average composition of milk of different species (% by weight)

	Total solids	*Fat*	*Protein*	*Lactose*	*Minerals/ water soluble vitamins*
Cow	12.5	3.7	3.4	4.7	0.7
Sow	20.0	8.0	6.0	5.2	0.8
Ewe	17.0	5.5	6.2	4.5	0.8
Rabbit	35.0	18.3	14.0	2.0	0.7
Human	12.4	3.8	1.0	7.0	0.6

The protein fraction in milk consists of two major protein groups, the caseins, which exist in the form of colloidal particles called micelles and of which there are four types – α_s, β, γ and κ-caseins – and the serum (whey) proteins consisting of principally β-lactoglobulin and α-lactalbumin.

Because of the commercial importance of cow's milk as a food, more is known about it than about the milk of other mammals and the mechanisms of synthesis of milk have been much investigated. In this section, therefore, synthesis will be described with reference to synthesis in the ruminant and attention drawn where appropriate to differences in the monogastric.

Site of biosynthesis

The site of biosynthesis is the vast number of epithelial cells which line the alveoli, each cell in an alveolus discharging its milk into the lumen or hollow part of the structure (see Fig. 15.27). A feature of the secretory cells is the large number of mitochondria present, being indicative of the high energy demand of biosynthesis and also the marked hypertrophy of the Golgi apparatus and endoplasmic reticulum which occurs at the onset of lactation, these being the sites of synthesis of milk constituents. The alveoli of the mammary gland are well supplied with blood capillaries and it is from the arterial supply that the metabolites used in milk synthesis are drawn. Arterio-venous (AV) difference studies have shown that acetate, 3-hydroxybutyrate, free fatty acids, glucose, amino acids and proteins are withdrawn from the blood as it passes through the mammary tissue and act as precursors in milk synthesis. The origins of these precursors are illustrated in Fig. 15.30.

Milk protein synthesis

A-V difference studies have shown that the total free amino acids withdrawn from blood are apparently sufficient for synthesis but since the balance of absorbed

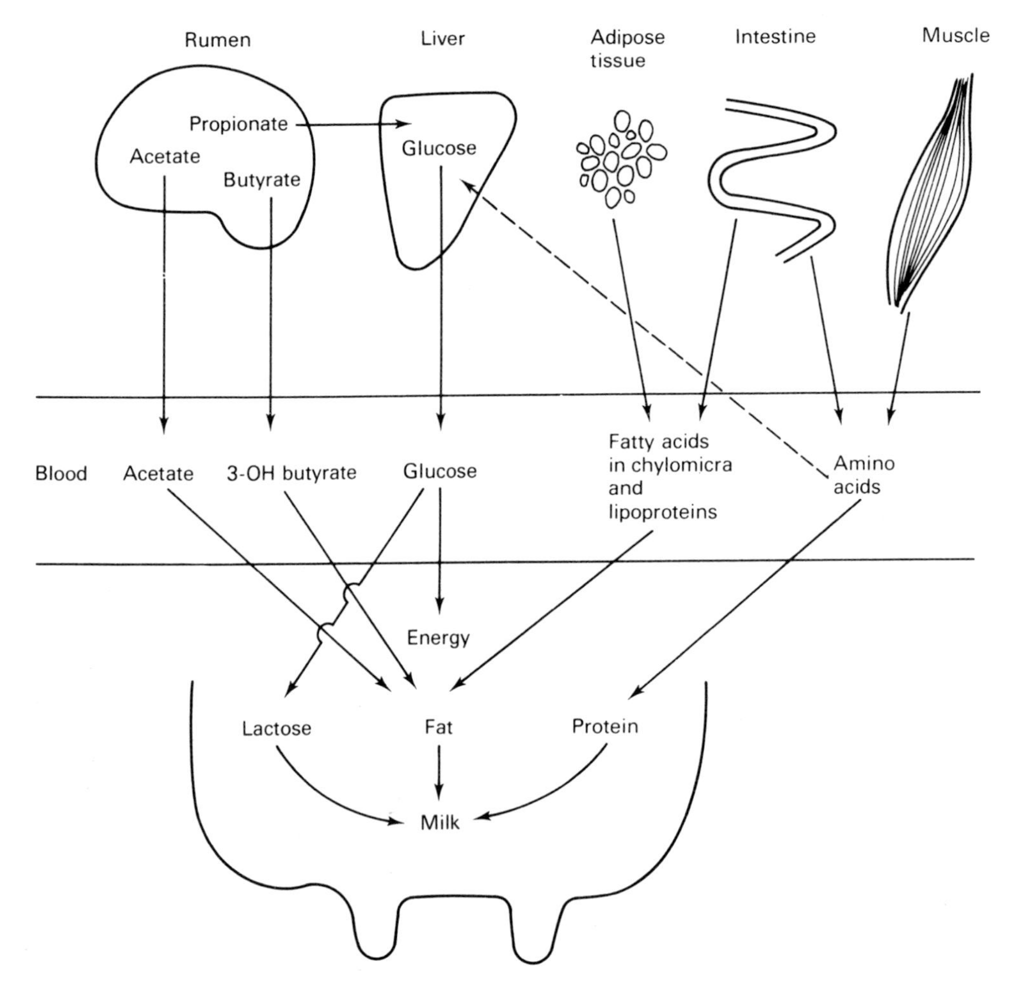

Fig. 15.30 Diagram showing precursors for milk synthesis in the ruminant.

amino acids is not the same as that present in the milk proteins a certain amount of amino acid synthesis must occur. Of the amino acids incorporated into milk protein the provision of the amino acids essential for protein synthesis is of greatest concern since the provision of non-essential amino acids can be achieved by synthesis within the lactating cell. A number of studies have indicated that under the dietary circumstances encountered by dairy cows in the UK, methionine is likely to be the first limiting amino acid for milk production and that there may be value in giving 'protected' methionine supplements. Protein synthesis involves the assembly of amino acids into a polypeptide chain through peptide bonds. The sequence of the amino acids in the various milk proteins is dictated by the usual method of transcription by ribonucleic acid on the ribosome of the cell.

Following the release of the protein from the endoplasmic reticulum they are 'packaged' in secretory vesicles derived from the Golgi apparatus and transported to the apical membrane of the cell before being secreted by exocytosis into the lumen of the alveolus.

Milk fat synthesis

The fatty acids present in the triglycerides in milk fat range from C_4 to C_{18} in chain length and in comparison with other fat sources are characterized by having a relatively high proportion of short chain fatty acids. The origins of the different fatty acid groups vary (Fig. 15.31). Thus short chain fatty acids and most of the medium chain fatty acids are synthesized within the lactating cell using acetic acid and to a lesser extent 3-hydroxybutyrate as precursors whereas the long chain fatty acids and some of the medium chain fatty acids are derived directly from circulating lipoproteins in the plasma. Fatty acids are removed from circulating lipoproteins by the enzyme lipoprotein lipase present in the capillary membrane. Whereas acetic and 3-hydroxybutyric acids are mainly derived through absorption from the rumen, the fatty acids present in circulating lipoproteins may be derived from the mobilization of adipose tissue or absorption of dietary or microbial lipid from the small intestine. The presence of odd-numbered and branched chain fatty acids in milk fat is due to the latter source.

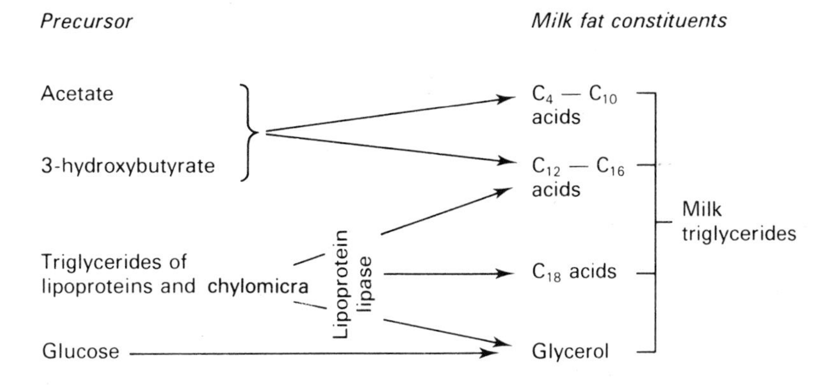

Fig. 15.31 The origins of milk fat triglycerides.

Non-ruminant animals appear to use glucose for the synthesis of milk fat. In all species the metabolism of glucose by the lactating cell is the precursor of the glycerol moiety for triglyceride synthesis and in addition is the main energy source for both fat and also protein synthesis in the lactating cell.

In the secretion of fat from the lactating cell, droplets of fat coalesce and migrate towards the apical membrane of the cell where they are enveloped in membrane as they push their way into the lumen of the alveoli. The true nature of the process remains obscure.

Lactose synthesis

Of the major milk constituents lactose is chemically the simplest. Synthesis takes place from glucose withdrawn from the blood as it passes through the mammary tissue. In ruminant animals this utilization of glucose for lactose synthesis represents a major drain on the animal's glucose reserves. Lactose secretion would appear to share the same secretory route, via vesicles, with the milk proteins, together with water, ions and other water-soluble materials.

Nutritional influences on milk biosynthesis

The rate of synthesis of milk components by the lactating cells is dependent on the availability of milk precursors and the ability of the animal to partition these precursors into milk production. The latter appears to be mainly a function of the genetic make-up of the animal but it is also influenced by diet.

The ability of animals to partition nutrients into milk production as opposed to body gain is controlled by a variety of hormones. It appears that the most important are growth hormone (GH), insulin and glucagon. Research has shown that GH, also termed somatotropin, has a major role in partitioning nutrients away from deposition in body tissues and towards milk production. Research using bovine somatotropin (BST) derived from recombinant-DNA (rDNA) technology has shown that the injection of BST can increase yields in cows by between 10 and 40%. The BST is thought to operate through supplying the mammary gland with more nutrients through increasing the rate of body fat mobilization and hence the levels of plasma free fatty acids, acetoacetic acid, 3-hydroxybutyric acid and lipoproteins available to lactating cells and perhaps also through increased uptake of glucose and amino acids by the mammary gland.

The greatest effects of diet on the secretion rates of milk constituents are caused by the form of dietary carbohydrate and the fat content of the diet. Dietary protein, within normal limits, has much less effect.

It has long been recognized that the ratio of acetic + butyric:propionic acid in the rumen, which is largely a function of type of dietary carbohydrate (see the section headed 'Carbohydrate digestion'), has an influence on milk fat content. Thus increasing the roughage: concentrate ratio increases the yield of milk fat and vice versa. This is due to the fact that propionic acid derived from the fermentation of concentrates in the rumen is gluconeogenic and elevates plasma insulin. As a consequence there is reduced mobilization of fatty acids from body fat, and milk fat secretion falls due to a reduced supply of precursors for milk fat synthesis. In contrast to the effect that propionic acid has on milk fat, increased levels of propionate absorbed from the rumen increase milk protein synthesis.

The effects of dietary fat on milk composition are complex and are mediated through direct effects on the availability of diet derived fatty acids in the plasma and the indirect effects of dietary fat on rumen fermentation (see 'Carbohydrate digestion'), and on the *de novo* synthesis of short chain fatty acids in the rumen (Fig. 15.32). Thus the yield of milk fat depends on a balance of these effects. Not only total yield but also the composition of the milk fat can be influenced through the inclusion of dietary fat. The use of 'protected' fat systems can allow the incorporation of dietary fat without some of the indirect effects of dietary fat being incurred.

The milk ejection mechanisms

As a result of secretion, milk accumulates in the alveoli and ducts of the mammary gland. This milk is ejected from the alveoli and ducts by contraction of the smooth muscle fibres (myoepithelial cells) which surround the alveoli. This process of milk 'let down' or, more accurately, milk ejection, is the result of a neurohormonal reflex (see Fig. 15.33). The reflex is made up of a neural sensory input and a hormonal output. Tactile stimulation of the teats (through the butting and suckling of the offspring) provides the stimulus which evokes the release

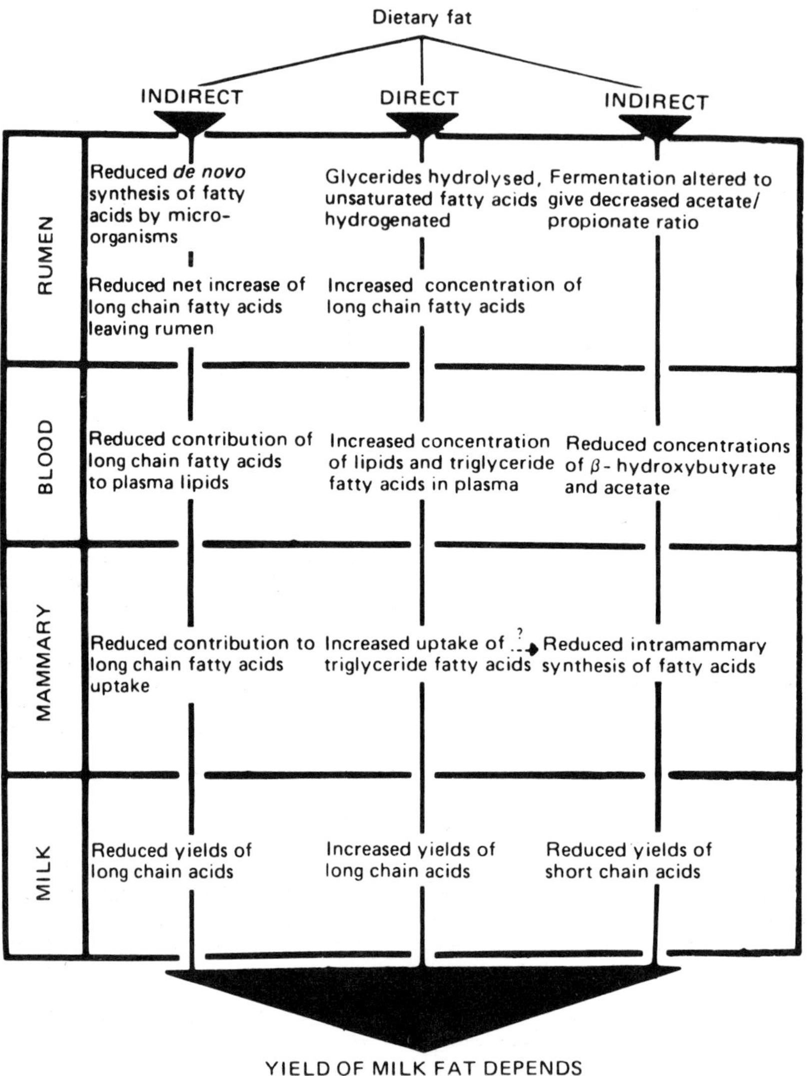

Fig. 15.32 Effects of dietary lipid on aspects of metabolism related to milk fat synthesis (courtesy of Storry, J.E. (1981) *Recent Advances in Animal Nutrition*. Butterworths, London).

of oxytocin from the posterior pituitary. Oxytocin then acts on the myoepithelial cells to cause their contraction, and hence milk ejection. The time from initial stimulation to the start of alveolar contraction is less than 1 min; in the cow the effective duration of milk ejection is 2–5 min, and in the sow it is about 10–30 s.

Because a proportion of milk is stored in the gland cistern and larger ducts, initially milk is passively withdrawn from the gland, to be followed by the active milk-ejection phase. The completeness with which milk is removed from the gland depends on the amount of oxytocin released from the posterior pituitary, which itself depends on the strength of the initial stimulation. Natural suckling is more efficient in this respect than either hand or machine milking. Once the pituitary store of oxytocin has been depleted, there is a substantial delay while more milk is synthesized and the store of oxytocin is replenished.

In machine-milked dairy animals the normal routine prior to milking is sufficient to trigger a milk ejection reflex. This is a good example of what is termed a conditioned reflex. The milk ejection reflex can be inhibited by emotional disturbance. It is likely that this is due to a release of adrenaline which, by its vasoconstrictive action on the blood supply to the udder, reduces oxytocin availability to the gland.

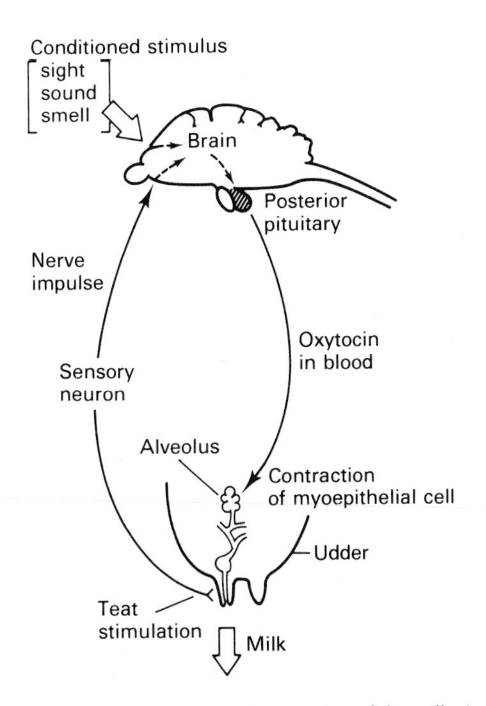

Fig. 15.33 Diagrammatic representation of the milk ejection reflex.

Growth

A knowledge of the processes involved in animal growth and an ability to manipulate these processes may be used to influence the efficiency of production. In the production of animals for meat a knowledge of mechanisms involved in the control of muscle, fat and bone deposition is relevant such that we may efficiently produce animals with a high proportion of muscle and an optimal amount of fat at market weight. Growth is also a characteristic of animals that is related to reproductive performance (see earlier section).

The nature of growth

Growth has two aspects. The first is the increase in mass (weight) per unit time of the whole animal or part of the animal. The second involves changes in form and composition resulting from differential growth of the component parts of the body. In studies of meat animals we are primarily concerned with the growth of the major tissues of the carcass which are muscle, fat and bone and with the proportions of these three major tissues in the carcass.

A growth curve for weight or size in the whole animal follows the pattern shown in Fig. 15.34. From birth the animal will typically grow along a sigmoidal curve showing acceleration about puberty and slowing down as maturity is approached. Animals are usually slaughtered around the pubertal phase of their growth curve at between one-half and two-thirds of mature body size.

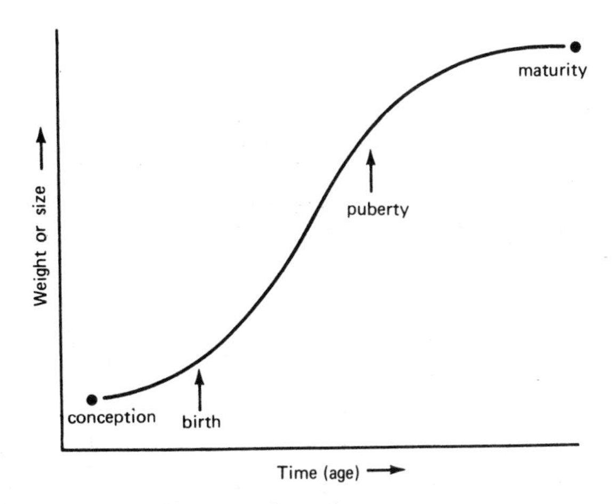

Fig. 15.34 The sigmoid growth curve.

The liveweight growth of farm animals is the gross expression of combined changes of the carcass tissues, organs and viscera and gut fill. The animal productionist is primarily interested in the growth of the carcass tissues comprising the muscle, bone and fat. The carcass weight of the animal is comprised of the live weight less the viscera, gut fill, blood, hide, head and feet and may be influenced by a variety of biological factors, particularly level of fatness and diet type and by carcass handling procedures.

The carcass is an extremely variable commodity reflecting the temporal growth of its component tissues. Although each of the carcass components follows a sigmoidal growth curve similar to that of liveweight, the different growth curves are not in phase one with another. This is due mainly to the fact that the different tissues vary in priority for the available nutrients. The bone tissue reaches maximal growth rate prior to maximal muscle growth with adipose tissue being the latest of the body tissues to attain peak growth intensity (Fig. 15.35). Thus, carcass composition in terms of the proportions of muscle, fat and bone changes as an animal grows (Fig. 15.36).

The chemical composition of the body changes with age in a way which clearly substantiates the phasic development of the tissues. The most marked changes with age are the decrease in the proportion of water in the body and the increase in the proportion of lipid. The almost inverse relationship between the water and fat content of the body reflects the lower water content of fat (10%) compared with that of muscle tissue (75%). The percentages of protein and mineral matter decline only slowly as growth proceeds, mainly as a result of changes in lipid content; in fact if chemical composition is expressed on a fat free basis the chemical composition of the body remains remarkably constant during growth reflecting that the bones and muscles of the limbs and trunk remain in proportion to one another.

Tissue growth

The growth of individual tissues is a reflection of the

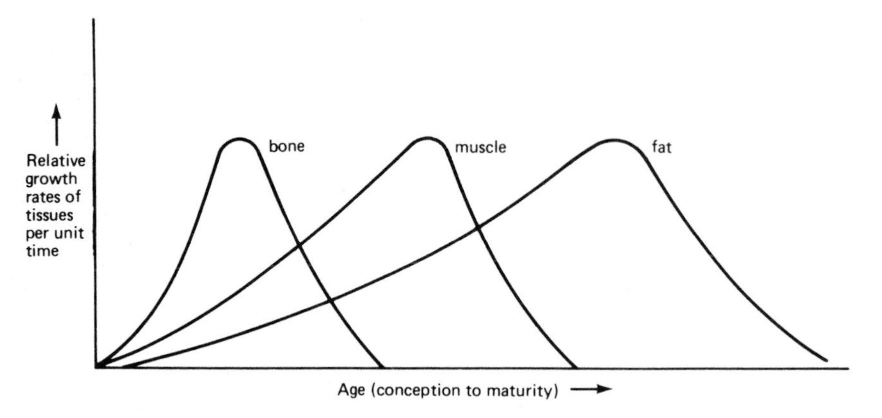

Fig. 15.35 Diagram indicating relative growth rates of body tissues from conception to maturity.

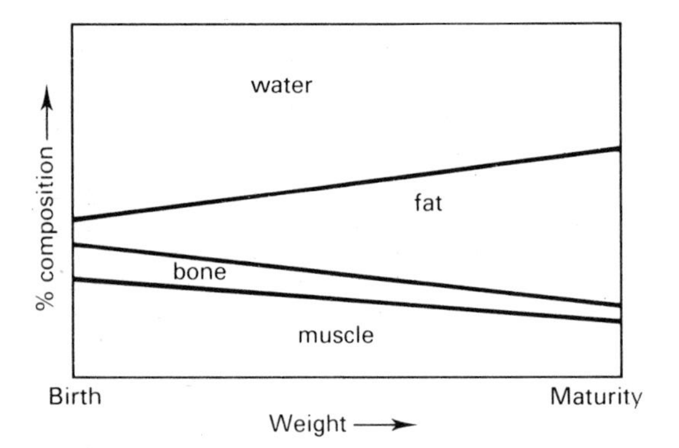

Fig. 15.36 Diagrammatic representation of changes in carcass composition.

increase in both the size and number of cells. An increase in number of cells is called *hyperplasia*, and an increase in cell size is termed *hypertrophy*. Hyperplasia and hypertrophy of all body cells occur during embryonic life.

Bones grow during both prenatal and postnatal periods through the interaction of three different cell types: chondrocytes, osteoblasts and osteoclasts. *Chondrocytes* are cells that produce cartilage, *osteoblasts* produce bone collagen and other bone components, and *osteoclasts* break down bone during the process of resorption. Bone grows in length by ossification of the epiphysial cartilage. Once this is complete the bone will stop growing in length. In mature animals bone is composed of approximately 50% mineral [$Ca_{10}(PO_4)_6(OH)_2$] and 50% organic matter and water, on a weight basis.

In muscle tissue development the number of muscle cells is fixed just prior to birth. Thus the growth of muscle that follows birth is due to hypertrophy of cells through protein deposition. Deposition of protein in muscle represents the net effect of two processes, protein synthesis and protein degradation. Control of protein deposition may be exerted through either or both of these processes although growth clearly involves synthesis exceeding degradation. The supply of nutrients, especially amino acids and energy, and the hormonal status in the animal, particularly the balance of insulin, growth hormone, the glucocorticoids and the sex steroids, are influential.

Fat consists of fat cells (adipocytes) and supporting connective tissue. The period during which hyperplasia of adipocytes occurs is unclear but some postnatal as well as prenatal development does occur; in pigs up to five months of age; in sheep up to 11 months of age; and in cattle up to 14 months of age. Hypertrophy of adipose tissue occurs through the net deposition of fatty acids in adipocytes. Adipocyte size may vary considerably according to the animal's energy balance, reducing in size in times of negative energy balance and increasing in times of energy surplus. There are two types of adipose tissue, which are referred to as *white fat* and *brown fat*. Most body fat is white and it functions as a depot of stored energy. Brown fat is very active metabolically and can be used to maintain body temperature, which is especially important in neonatal animals (see the section headed 'Environmental temperature and animal production').

Factors influencing growth

The patterns of tissue growth and consequent changes in chemical composition of the body are influenced by several interrelated environmental and genetic factors.

Genetic effects

Genetic differences occur in that animals of the same species vary in their mature size and weight, a feature which is reflected in differences in carcass composition. Generally animals which reach the asymptote of the growth curve at an early age have lower mature body weights. The converse also applies. This is exemplified in the Angus and Friesian breeds of cattle and largely reflects differences in the timing of adipose tissue growth. Differences in mature weight are, however, probably not solely responsible for differences in composition between breeds. Thus, for example, Soay sheep have been shown consistently to contain less fat than Down breeds at the same stage of maturity. Although it is often implied that animals which have large mature body weights are more efficient because of their lower maintenance energy requirement per unit of body weight, there is little evidence that this is the case and the overriding factor governing efficiency within a species is the choice of slaughter weight.

In addition to these differences in total body size and body composition there are genetic influences on the proportions of muscle and bone tissue. These are most noticeable in the comparisons of dairy and beef breeds of cattle, the latter generally having more muscle relative to bone when comparisons are made at equal levels of fatness and stage of maturity. In cattle and sheep, although not pigs, there are clear breed differences in the distribution of body fat, especially between the carcass (subcutaneous and intramuscular) and intra-abdominal (perinephric fat, omental fat and mesenteric fat) depots. Thus 'dairy' breeds of cattle and sheep breeds such as the Finnish Landrace tend to deposit a higher proportion of fat intra-abdominally than 'beef' breeds of cattle and sheep such as Suffolks and Hampshires.

Differences in the growth and composition of body tissues occur between the sexes. In cattle and sheep, if comparisons of body composition are made at equal weights or ages, entire males have less fat than castrates and castrates less fat than females. The situation differs slightly in pigs where gilts have less fat than castrates at equal body weight. Differences in mature weight between males and females explain some of the differences in composition attributable to sex. The rest is most likely due to differences in the levels of testosterone found in males, females and castrates. The effect of androgens is specifically shown in the more pronounced development of the forequarter muscles of the bull and boar compared with the female or castrate. The influence of sex on muscle to bone ratio is unclear but at the same level of fatness it is usual that entire males have higher muscle: bone ratios than females illustrating further that the impetus for fattening supersedes the impetus for muscle growth at lighter weights in females than in males.

Nutritional effects

Nutrition is generally the dominant factor influencing the expression of growth potential in farm animals. Both quantitative (plane of nutrition) and qualitative (diet composition) variation in nutrition influence growth. The general influence of plane of nutrition on growth develops from the priorities that exist for available nutrients as an animal grows (Fig. 15.35). On the high plane of nutrition the growth curves are telescoped together, whereas on a low plane the sequence is extended. The most profound effects are on the deposition of fat. This is illustrated in Fig. 15.37 and shows that not only does a plane of nutrition above maintenance influence the proportion of the different tissues synthesized but also that a plane of nutrition below maintenance can lead to 'negative' growth, especially of fat tissue in the first instance.

The main feature of diet composition which can influence growth is the energy and protein content of the diet and the ratio of energy:protein in the diet. Fat growth relative to muscle and bone growth is dependent on the level of energy intake. Increasing the level of energy intake at a specified protein intake level increases the deposition of fat relative to muscle. Additionally with any specified level of energy intake animals increase in fatness if their muscle growth is restricted by the amount, and in the case of non-ruminants the quality, of protein in the diet. Thus a balance of energy and protein is necessary for the production of carcasses with the desirable level of fatness. The plane of nutrition can be used to adjust the growth rate in different stages of the growing/fattening period. Thus, for example, by restricting the energy intake of pigs the amount of fat in the carcass at slaughter may be limited and the time to slaughter in cattle may be manipulated through adjusting the plane of nutrition in order to make best use of price fluctuations.

A further nutritional influence that requires mention here is that of 'compensatory growth'. Animals whose liveweight growth has been retarded by restriction of feeding exhibit this phenomenon which is characterized by more rapid than normal growth when introduced to a high plane of nutrition. Compensatory growth is characterized by rapid deposition of cellular protein so that the maximum protein/DNA ratio in the cell is achieved. Although in biological terms the issues involved in compensatory growth are complex, in economic terms the phenomenon can be put to good use. Under UK conditions the growth rate of cattle may be restricted during the winter (stored period) with 'compensatory growth' being achieved on summer grass.

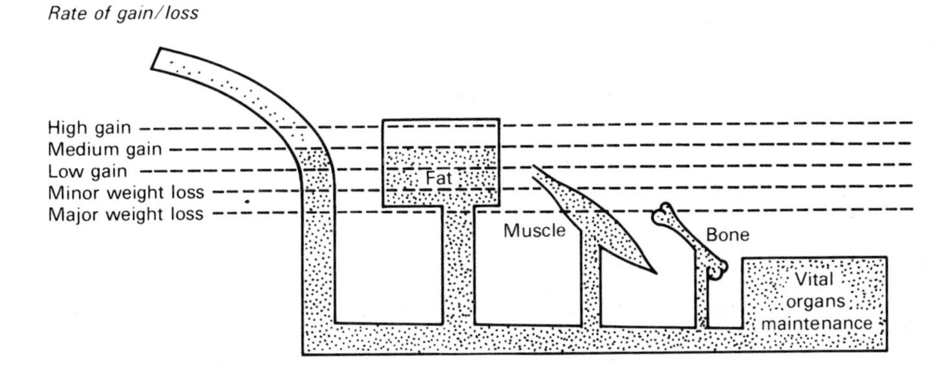

Fig. 15.37 A model of priorities for nutrients during growth (after Berg, R.T. & Butterfield, R.M. (1976) *New Concepts in Cattle Growth*. Sydney University Press, Sydney).

The hormonal control and manipulation of growth

The endocrine system is a major regulator of animal metabolism and probably all hormones either directly or indirectly influence animal growth. Those considered to have specific effects on growth are the hormones of the anterior pituitary, the pancreas, thyroid, adrenals and gonads. These hormones affect body growth mostly through their effects on nitrogen retention and protein deposition. The precise modes of action of the individual hormones is only partly understood and is complicated by the fact that in some cases at least their effects depend on the sex, species of animal and balance of other hormones. When workers have attempted to correlate the concentration of hormones in the blood and growth, no clear relationships have emerged.

Although there is little evidence of an association between blood growth hormone (GH) levels and growth in livestock, some workers have shown there to be greater growth hormone secretory activity in genetically superior cattle and pigs. More recently it has been shown that exogenous growth hormone may produce an anabolic response in farm livestock. The anabolic response of GH appears to be mediated through a group of peptide hormones termed somatomedins. Not only do these increase growth rate but they also produce leaner carcasses. The secretion of GH from the pituitary is controlled by the balance of an inhibitory hypothalamic peptide, termed somatostatin, and a stimulatory GH releasing factor (GHRF) which is also a hypothalamic peptide. Much recent research has looked at the possibilities of either removing or neutralizing the effects of somatostatin by active immunization techniques and at supplementing the effect of GHRF as a means of manipulating growth.

It is now generally accepted that the natural steroid hormones – oestrogens, androgens, progestins and glucocorticoids – exert an effect on growth. The difference in growth rate and mature body size of male, female and castrate animals suggests that the sex hormones play an important role in the control of growth. There can be no doubt that these differences are in the main androgen-dependent. Deprived of the anabolic effect of androgens, castrates divert energy intake into the synthesis of fat rather than muscle and are thus less efficient in food conversion efficiency terms. Castration can only be justified on the basis of its effects on sexual and aggressive behaviour and in some situations the possible risk of carcass taint. Such characteristics only reveal themselves in the period close to slaughter and the technique of late castration using immunological techniques is at present being considered by research workers as a means by which the maximum anabolic advantage of the testes may be obtained. Immunological castration involves the active immunization of animals against gonadotrophin releasing hormone (GnRH). This has the effect of inhibiting the secretion of follicle-stimulating hormone (FSH) and luteinizing hormone (LH) such that spermatogenesis ceases and testosterone secretion declines sharply.

Several synthetic compounds have been used to promote growth and in many cases increase the protein content of the carcass. Classically, androgenic compounds would be expected to increase protein synthesis and hence protein deposition in muscle. However, the synthetic androgen trenbolone acetate has been shown to decrease protein synthesis and exerts its growth promoting effect through decreasing protein degradation in muscle. This it appears to do through depressing the catabolic effects of the glucocorticoids. As with trenbolone acetate, the oestrogenic anabolic agent zeranol, which has been widely used commercially, would also appear to promote growth through reducing the rate of protein turnover rather than through stimulating muscle protein synthesis. It has also been suggested that oestrogens stimulate circulating GH concentration and change thyroid hormone status, and that this accounts in part for their growth stimulatory effects.

Exogenous anabolic agents to stimulate the growth of animals, especially cattle, were widely used in production practice until 1986 when EEC legislation banned their use on the grounds of their possible risk to human health. The main substances used prior to the ban were trenbolone acetate (a synthetic androgen) and the oestrogenic substances hexoestrol and zeranol. The ability of these agents to stimulate growth is dependent on their producing an optimum balance of androgens and oestrogens in the animal. Thus response to implants is determined by the sex of the animal; best responses have been obtained by the exogenous administration of androgens to females, oestrogens to entire males and a combination of androgenic and oestrogenic agents to castrates.

One of the major concerns of the meat producing industry in recent years has been the excessive amount of fat deposited in carcasses. The origins of this concern have been primarily due to public concern about the health risks associated with the consumption of animal fats although excessive fat production also represents a source of inefficiency in livestock production. Although leaner carcasses may be achieved in the short term by slaughtering at lighter body weights and in the long term by appropriate selection programmes, an alternative presently being investigated is the feeding of a group of substances termed β-agonists. Agents such as clenbutarol and cimeterol which are substituted catecholamines have been shown to reduce fat deposition and increase the protein content of the carcass. Their mode of action is not fully understood but influences on fat deposition would appear to be due to a reduction in lipogenesis and an increase in lipolysis whilst the effects on protein content appear to be by a combination of reduced protein degradation in muscle and an increase in the rates of protein synthesis. As yet, however, β-agonists have not been approved for commercial use.

Environmental physiology

Throughout their evolution animals have been subjected to environmental influences, and farm animals, like their ancestors, have to employ a range of structural and functional strategies to adjust to changes in their environment. The term environmental physiology describes those physiological mechanisms involved in the process of homeostasis, and the external stimuli, such as temperature, daylength and stocking rate, that necessitate such responses.

The nature and importance of photoperiod in regulating seasonal activities such as reproduction have already been mentioned in this chapter and are also discussed in Chapter 19. Perhaps the most important environmental influence, particularly with regard to animal productivity, is that of temperature. This governs the amount of heat energy lost from the animal, which in turn determines the amount of dietary energy retained by the animal.

Maintaining thermal balance

Farm animals maintain a high (37.5–39°C) and relatively constant body temperature despite large fluctuations in environmental temperature. The combination of physiological and behavioural changes which regulate body temperature are controlled by the hypothalamus.

The maintenance of a constant body temperature requires that there be a balance of heat production within the body and heat loss from the body. Heat is produced as a by-product of the metabolism of ingested nutrients used for maintenance and the various productive processes. Heat transfer between the body and the environment occurs by means of conduction, convection, radiation and evaporation. Since generally the animal body is at a higher temperature than the environment, heat may be lost by all these methods. The partition of losses between the various routes of heat exchange varies according to the air temperature, with evaporative losses being high at high air temperature, whereas radiation and convection are the main avenues of loss at low and intermediate temperatures. In addition to air temperature, other interacting environmental factors – wind, precipitation, humidity and sunshine – are determinants of rates of heat transfer. Wind, precipitation and humidity particularly affect evaporative losses; wind also affects convection losses by increasing air movements around the animal's body, thus destroying the insulating layer of air trapped in the animal's hair or wool, and sunshine especially affects the net radiation exchange between the animal and its environment in animals kept outdoors.

The mechanisms by which animals can regulate body temperature fall into two categories: alteration in the rate of heat production and the rate of heat loss. The animal has a number of mechanisms by which the rate of heat loss can be altered: by varying the degree of vasodilation, sweating, piloerection (erection of body hair), and numerous behavioural responses such as change in posture, activity and food intake. It is these mechanisms which come into play first of all in the regulation of body temperature. The range of temperature within which body temperature can be regulated without changes in metabolic heat production is termed the thermoneutral zone. The air temperature at which these mechanisms are no longer successful in dissipating sufficient heat to prevent a rise in body temperature is termed the upper critical temperature; and the temperature below which the animal must increase its rate of heat production is termed the lower critical temperature (see Fig. 15.38). These critical temperatures are not fixed and are especially dependent on the animal's metabolic body size and level of production.

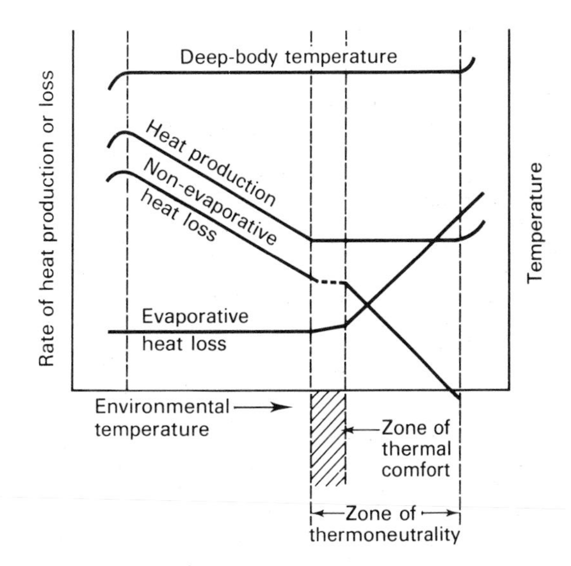

Fig. 15.38 Graph of thermoregulatory responses to a range of environmental temperatures.

Environmental temperature and animal production

To date much of the precise work on the effects of the environment on the animal, in temperate regions, has been concerned with intensively kept pigs and poultry and has primarily studied their interaction with the nutrition of the animal. Energy is the main nutrient affected, higher environmental temperatures releasing more food energy for productive purposes but having secondary effects on protein metabolism, food intake and carcass composition.

Of all the farm species the pig is perhaps the most vulnerable to the effects of environmental temperature. It has little body hair for insulation and few sweat glands to help with evaporative cooling. However, pigs do make certain behavioural adaptations to temperature, such as huddling together in cold weather and wallowing in hot conditions. The lower critical temperature of pigs changes markedly with age and body size. Small pigs with a larger surface:body mass ratio will lose relatively more heat from their skin surfaces than will their older and larger counterparts. Thus a newborn pig is particularly vulnerable to a cold environment. Figure 15.39 shows how the lower critical temperature of a pig improves with increase in body weight. Apart from the level of dietary energy, lower critical temperature is also affected by factors such as ventilation rate, wetness and type of flooring, and provision of straw.

In the light of these effects it is clear that where quantitative data on environmental effects are available, they may be taken into account in the planning of feeding programmes, the design of buildings, and optimizing of investment in insulation and supplementary heating.

In the outdoor situation experienced in the UK, most interest in environmental effects centres on the combined effects of extreme cold, wind and rain. The effects

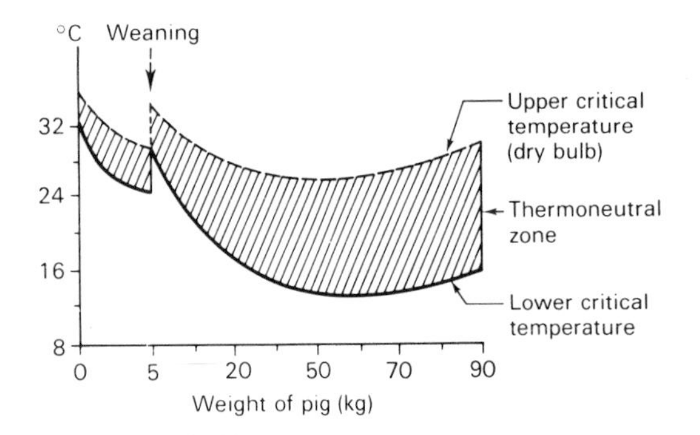

Fig. 15.39 Graph showing the relationship between upper and lower critical temperatures and bodyweight in a growing pig (courtesy of ADAS (1982) *Pig Environment*, Booklet 2410).

on adult animals are primarily discomfort and loss of production, but in the young and newborn such weather conditions are a threat to life. Surveys have shown that perinatal lamb mortality can vary from 5 to 45% on hill farms and that, at the higher mortality rates, extreme cold is the main contributor. In extreme conditions, shivering thermogenesis is often insufficient to maintain body temperature in the neonate. These animals may increase heat production by non-shivering thermogenesis, a process mediated by the effects of cold on the sympathethic nervous system and particularly by the action of noradrenaline on brown adipose tissue. The breakdown of this special type of adipose tissue results in much higher levels of heat production than are produced from other catabolic processes, and serves to help prevent hypothermia in the neonates of many species. It appears likely that variations in the amounts of brown adipose tissue present in the body at birth may explain in part the genetic variation in the ability of the newborn lambs to survive adverse conditions.

Environmental physiology and animal welfare

In recent years the welfare of farm livestock has come under increasing scrutiny, particularly the more intensive methods of animal production. In such systems, an animal may be housed in a largely artificial environment. The extent to which animals are able to adapt to their surroundings will depend mainly on the design of the accommodation and the husbandry skills of the stockman.

Failure to interpret correctly the animal's physiological and behavioural needs may have serious consequences for both animal productivity and animal welfare. Animals that are physiologically or behaviourally deprived may suffer stress, a response which in the long term can result in an increased incidence of disease through a lowered resistance to infection and reduced growth rates. Even off the farm it has been shown that pre-slaughter stress results in a loss of carcass quality and weight, as in the case of dark-cutting meat of beef animals and the pale, soft, exudative meat of pigs stressed immediately prior to slaughter. Thus attention to the conditions during transport, auction and lairage and at slaughter may have benefits for animal productivity as well as animal welfare.

The Farm Animal Welfare Council (FAWC) in their codes of welfare recommendations offer guidance to the farmer on the physiological and behavioural needs of farm animals that will help safeguard their welfare.

Further reading

Church, D.C. (1988) *The Ruminant Animal:Digestive Physiology and Nutrition*. Prentice Hall, New Jersey.

Cole, D.J.A., Haresign, W. & Garnsworthy, P.C. (1993) *Recent Developments in Pig Nutrition 2*. Nottingham University Press, Nottingham.

Frandson, R.D. (1986) *Anatomy and Physiology of Farm Animals*. Lea & Febiger, Philadelphia.

Hafez, E.S.E. (1987) *Reproduction in Farm Animals*. Lea & Febiger, Philadelphia.

Haresign, W. & Cole, D.J.A. (1988) *Recent Developments in Ruminant Nutrition 2*. Butterworths, London.

Hunter, R.H.F. (1982) *Reproduction in Farm Animals*. Longmans, London.

McDonald, P., Edwards, R.A. & Greenhalgh, J.F.D. (1994) *Animal Nutrition*. Longmans, London.

Peters, A.R. & Ball, P.J.H. (1995) *Reproduction in Cattle*, 2nd edn. Blackwell Science, Oxford.

Rook, J.A.F. & Thomas, P. (1983) *Nutritional Physiology of Farm Animals*. Longmans, London.

Whittemore, C. (1993) *The Science and Practice of Pig Production*. Longmans, London.

16

Animal Welfare

J.C. Eddison

The welfare of animals under the care of humans has been of concern for many years. Most livestock producers both recognize the economic benefit of managing their animals in a humane manner, and would also be able to identify essential elements common to all high welfare husbandry systems. These would include:

- housing designed both for the animals and the producer;
- stocking levels appropriate to production and welfare;
- consistency of management practices;
- high quality stockmanship;
- an awareness of the needs of the animal.

The aim of this chapter is to provide an introduction to the key principles that underpin practical animal welfare.

Background

The momentum for the current debate regarding the welfare of agricultural livestock in particular originated in the 1960s. The concerns expressed at that time were, perhaps, most famously voiced in a book entitled *Animal Machines* (Harrison, 1964). In this book, Ruth Harrison described aspects of intensive livestock production that many people found abhorrent and which she described by the term *factory farming.* As a result of the adverse publicity caused by the contents of both this book and other publications, the UK government established a Technical Committee (commonly referred to as the Brambell Committee after its chairman, Prof. Brambell) to review the welfare of animals kept under intensive livestock husbandry systems (HMSO, 1965). One of the many proposals of that committee was the establishment of a standing committee to advise government on farm animal welfare. This group is now known as the Farm Animal Welfare Council and its functions include both the review of scientific and other evidence and the provision of advice to government on legislative changes that may be necessary.

Public awareness and debate concerning welfare issues are not confined to the United Kingdom: many countries throughout the world have enacted welfare legislation and so has the European Union. In fact, one author (Ryder, 1989) has argued that the United Kingdom is a nation '. . . strong on ideas and weak on action . . .'. This particular view would be disputed by livestock producers, especially since the UK led the rest of the European Community in banning individual stalls and tethers for pregnant sows (HMSO, 1991).

Whatever differences may exist between nations with regard to their attitudes towards welfare legislation, we have to accept that the issue of animal welfare is now important in agriculture throughout western Europe as well as in several other countries throughout the world. Not only is there a greater public awareness of welfare issues but, as will be outlined later, the interests of welfare and productivity have a great deal in common. As a consequence, all those individuals with livestock in their charge need to have a basic understanding of animal welfare for two very important reasons. First, so that they can manage their stock in such a way that the animals are maintained in an environment that is as humane and stress-free as possible. And, secondly, so that the animal welfare debate can be conducted in an informed and reasoned manner with contributions from all those involved with or with an interest in livestock production.

This chapter is divided into three sections. The first introduces and defines some important terminology and reviews some of the biological principles that underlie welfare. The second section outlines methods by which welfare can be assessed. Finally, the chapter concludes with a discussion of the factors that influence how we determine acceptable levels of welfare.

What is welfare?

Many emotive words are used during discussions of welfare and so, before any reasoned welfare debate can take place, we must have a clear understanding of what is meant not only by the term welfare, but also by terms such as *pain*, *suffering* and *stress*.

Welfare

Several authors have offered definitions of welfare (HMSO, 1965; Hughes, 1976; Kilgour & Dalton, 1984; Broom, 1986) from which emerge some general principles.

The Brambell Committee (HMSO, 1965) stated that the meaning of the word welfare embraced '. . . both the physical and mental well-being of the animal'. Equally as important as defining the meaning of welfare, the members of that committee identified the need to state what welfare is not. They were quite adamant that an animal should not be judged to be in a state of good welfare merely because it is growing or producing some non-meat product such as milk, eggs or wool. In this context, they viewed production as merely indicative of an animal being fed adequately. Furthermore, by including both the mental and physical states of the individual in their definition of welfare, they clearly did not simply equate welfare with health (in the layman's sense of the term), although they did recognize that good health is a necessary prerequisite of good welfare.

Hughes (1976), in his examination of how behaviour can be used to assess welfare, proposed the following definition: '. . . a state of complete mental and physical health, where the animal is in harmony with its environment'. As it stands, this definition is very similar to that of the Brambell Committee, but Hughes went further and discussed the difficulties of interpreting evidence concerning the welfare of animals. For example, an animal may modify its behaviour when it is introduced to a new environment. The difficulty for anyone who wishes to judge the welfare of an animal in a novel environment lies in distinguishing between behavioural changes that are part of an adaptive response to the novel surroundings, and those that are indicative of impaired welfare due to the inadequacies of the new environment.

Broom (1986) developed the themes explored by Hughes and defined the welfare of an individual as '. . . its state as regards its attempts to cope with its environment'. He argued that, in some environments, an individual would have to exert very little effort to cope, while under other conditions much more energy might be required to survive and in some circumstances individuals might not be able cope at all. These latter individuals would be regarded as in a poor state of welfare.

A difficulty raised by Broom is that it is far easier to recognize individuals that are not coping (and whose welfare is poor) than to discriminate between those individuals who are in a state of either average or good welfare. Moreover, one cannot regard the absence of indicators of poor welfare as positive evidence of good welfare. To be able to identify those animals in a state of good welfare, and thereby isolate those factors that promote a state of well-being, would enable us to take action to improve the conditions of other, less-advantaged individuals. This difficulty in recognizing good welfare has been echoed by the Farm Animal Welfare Council (1993) which identified as a research priority the development of new measures of welfare and also the search for measures of good welfare in particular. The difficult task of determining *how* to measure welfare is discussed in a later section.

There is a great deal of commonality between the various definitions of welfare, and all authors are in total agreement that it is judged at the level of the individual. This can be illustrated by a very simple example. In a cubicle house for dairy cattle, most individuals might be coping with their environment very well, but there will be some variation around the average. At the bottom of the scale there may be weaker or subordinate individuals that might not be able to lie in the best locations and may, perhaps, regularly have to lie in damp, dirty or draughty areas. Those individuals would have to expend more energy coping with their environment than the others and may be exposed to a greater disease risk. They may be able to cope with this less than ideal situation, but they would be said to be in a poorer state of welfare than the cows that do not have to lie in draughts or in dampness. This simple example demonstrates that welfare is a continuum with poor at one extreme, and good welfare at the other.

A further point to make in this respect is that individuals can respond to a given stimulus in a diversity of ways, and therefore it is extremely important to assess the welfare of an individual using a variety of measures rather than relying upon a single index. A classic example demonstrating this need was reported by Duncan & Filshie (1979). They compared two strains of hen: one flighty and another regarded as placid. They observed the behaviour and monitored heart rate of both strains when the hens were confronted with an approaching human. The flighty strain exhibited a stronger behavioural response to the stimulus than the placid strain. However, the heart rate of the placid strain was elevated for far longer than the flighty strain. The two measures, behavioural and physiological, provided a more detailed description of how the hens had coped with the approach of the human. Independently, neither measure accurately described the responses.

The ability of an animal to cope with its environment and the very act of switching on its coping mechanisms depend to a great extent on the individual's perception of its situation. If an animal perceives its environment to be satisfactory, then it will not have any stimuli to which it has to respond: it does not perceive itself to have diminished welfare. Conversely, if an individual is provided with all its needs but, at the same time, it *wants* some additional commodity or change to its environment (e.g. a pen mate), then that animal will perceive itself to have decreased welfare and, in the context of the definitions of welfare by both the Brambell Committee and Hughes, it is not in a state of physical and *mental* harmony with its environment. These cognitive aspects of welfare have become a key area of research in recent years (for review, see Lawrence & Rushen, 1993).

Taking into consideration the fact that many homeostatic mechanisms exist (e.g. physiological, behavioural, neurological and cognitive) and that individuals vary in the way they respond to similar stimuli, it is important to employ a combination of measures (including both physiological and behavioural indices) to assess welfare.

Pain

Although various definitions of pain have been proposed, there is a consensus about its meaning. Fraser & Broom (1990) offered the following definition: '. . . pain is a sensory stimulus which is itself aversive'. However, the real difficulty with pain as a concept is not what is meant by the term itself, but that it is a very personal experience and therefore there are great problems in assessing pain levels or in comparing the intensity of pain between individuals.

There are some points in Fraser & Broom's definition that are extremely important to the animal welfare debate. First, the definition correctly emphasizes the importance of the sensory or nervous system in pain detection. This is particularly important since much time has been devoted to assessing whether the experience of pain is similar across all vertebrates. The dangers of attributing human-like feelings (anthropomorphism) to other groups of animals have hindered discussions of the effect of specific surgical practices such as beak trimming of poultry and de-horning of cattle. In fact, pain receptors (or *nociceptors*) are found throughout bird and mammal species (which include all traditional farm animals) and are also found in many other groups of animals. Not only are there structural similarities between species in this respect, but evidence also exists to show that similar physiological processes in different groups occur after traumatic surgery. For example, Gentle (1986) showed that neurological processes occurred in chickens after beak trimming that directly paralleled those that occurred in human amputees and which, in the latter case, are associated with extreme pain. Whilst such evidence is not conclusive proof, a cautious approach would be to assume that animals do experience pain and that beak trimming, for example, is painful to the individual.

The second point of importance to emerge from the definition of Fraser & Broom is that the sensation of pain itself is aversive; this is quite distinct from any effect on the individual that may result directly from the specific cause of the pain.

Clearly, there is an adaptive advantage to be gained from responding to pain. For example, removing a paw from a hot surface will minimize injury. But different groups of animals vary in the way in which they respond to pain or try to avoid it. Some, such as sheep, remain silent when subjected to a painful experience whilst dogs, in contrast, scream loudly. Since the neural anatomy of mammals is fairly similar, it is unlikely that this variation in response is because one species feels pain and the other does not.

There are a number of reasons that could explain variations in response to pain. One possibility is that evolutionary pressures have led to the suppression of the visible expression of pain in prey species such as sheep. An injured sheep that announced itself as such would attract the attention of predators which could lead to rather dire consequences.

Mechanisms other than behavioural modification (which is the easiest to detect for a human observer) exist that enable animals to cope with pain. Natural analgesics (β-endorphin and enkephalins) have been found to be secreted within the brain. These neuropeptides, which are in fact natural opiates, will reduce the sensation of pain.

The variations in the way animals cope with pain make the task of assessing it very difficult. However, it is important to have some methods through which assessment can be made. A very useful checklist was proposed by Morton & Griffiths (1985) which provides a starting point (Table 16.1). However, irrespective of the tools that may be available, only attentive stockmanship will ensure that an animal that is in pain will be identified quickly. In fact, the key to the achievement of high levels of welfare is high quality stockmanship.

Suffering

The relationship between pain and suffering was explained in the Brambell Report (HMSO, 1965), suffering being used as a collective noun that includes not only pain, but also fear, frustration and exhaustion. Dawkins (1980) defined suffering as '. . . a wide range of unpleasant emotional states' and added that the loss of a companion (bereavement), anxiety, and conflict should also be included. While there may, of course, be other forms of suffering, Dawkins' definition seems to be perfectly adequate.

An important aspect of the concept of suffering to note is that, while pain may be extremely intense and perhaps even transient, other forms of suffering can be much less intense but, because of their much longer duration, can be equally distressing to the individual animal. This is illustrated by the chronic frustration experienced by predators that are kept in confined enclosures in some zoos. This frustration is manifested in stereotypic pacing along well worn paths within their enclosures.

Whilst the terms welfare and suffering are closely related, they do not share the same meaning. For example, an animal that has suffered a physical injury or that is diseased would be in a poor state of welfare but, if it is anaesthetized or has received painkillers, then it will not be suffering: it will not be perceiving its poor welfare. In essence, if an animal is suffering, then it is in a state of poor welfare but the converse is not necessarily true.

Stress

Of all the terms associated with animal welfare, stress has probably been the subject of the greatest number of interpretations. Its use in discussions of animal welfare has been made even more difficult because it is used very widely, and precisely, in many scientific disciplines and is also in common usage in non-scientific conversation. As an example of the confusion, physicists and engineers talk of the stress *applied* to a structure, i.e. an external pressure imposed upon a body but, at the same time, they sometimes use the term to mean the *effect* experienced by the body caused by that external pressure. To have a single term referring to both cause and effect is bound to lead to some confusion.

In many situations, livestock have to cope with adverse conditions. Examples would include: vibrations during transport; the disruption of dominance hierarchies when new individuals are introduced into established groups; and the frustration experienced by farrowing sows and laying hens in the absence of suitable nesting material. Most people who acknowledge that animals can suffer (and there are some who do not share this view) would agree that some or all of these circumstances can be stressful to the animals that are involved. However, if we are to engage in objective discussions of stress, we need to have a clear understanding of what is meant by the term.

Much of the way in which biologists view the term

Table 16.1 A simple checklist of assessing pain and distress that illustrates the relationship between signs and degree of pain (after Morton & Griffiths, 1985)

	Normal *(0)*	*Mild* *(1)*	*Moderate* *(2)*	*Severe* *(3/4)*
Appearance		Coat loses sheen, hair loss, starey – harsh		
		Failure to groom, soiled perineum		
		Discharge from eyes and nose		
		Eyelids partly closed		
			Eyes sunken and glazed	
		Hunched up look		
		Respiration laboured, abnormal panting		
			Grunting before expiration; grating teeth	
Food/water intake		Reduced		Zero (prolonged)
		Faecal/urine output reduced		Zero
Behaviour		Away from cage mates, isolated; or bullying from mates	Unaware of extraneous activities	
		Self-mutilation		
		Restlessness, reluctant to move, recumbent		
		Change in temperament		
		Squealing, howling, etc, especially when provoked		
Clinical signs Cardiovascular	Strong pulse		Weak pulse	
		Cardiac rate increased or decreased		
		Abnormal peripheral circulation		
		Pneumonia, pleurisy		
Digestive		Altered faecal volume, colour, consistency		
		Abnormal salivation		
		Vomiting (high frequency)		
			Boarded abdomen as in peritonitis	
Nervous (musculoskeletal)			Lameness and arthritis	
		Twitching		Convulsions

stress results from the work of Selye who identified a consistency of response to adverse conditions across several species (Selye, 1956). He formulated the notion of a general adaptive syndrome (GAS) to describe the way an individual adapts to a set of adverse conditions. He divided this response into three stages: alarm, resistance and exhaustion. A number of authors (for review, see Broom & Johnson, 1993), with the benefit of techniques unavailable to Selye, have demonstrated that the universal stress response that he proposed is no longer acceptable. There is now evidence of considerable differences in the way different species and also individuals within the same species respond to stressful situations. However, his three-stage framework still provides a very useful structure within which to discuss stress, and it will be used here.

The *alarm stage* is a relatively short period during which an animal becomes aware of the noxious stimulus and takes short-term compensatory action. An individual may be able to cope with a novel environment or adversity by physiological or behavioural modification. In such cases we would say that the individual has *adapted* to its conditions (Hughes, 1976). This may require the rapid provision of energy which can be made available in the short-term by the mobilization of glycogen stores in the liver. The glycogen is converted to energy through the action of the hormones adrenaline and noradrenaline which are produced in the adrenal medulla under stimulation of the sympathetic nervous system. These have been referred to as *flight* or *fight* hormones because their activity is often associated with agonistic interactions or extreme defensive action. But their action is not solely confined to aggression: they provide short-term energy for any purpose.

The release of adrenaline will also cause the constriction of peripheral blood capillaries and increased blood supply to skeletal muscles. It has been suggested that this is the causal mechanism underlying the interruption of milk flow in frightened cows: the blood supply to the udder is reduced leading to a reduction in oxytocin and a halt to milk release.

If the animal's response to adversity is longer term, then other mechanisms have to be activated (Selye's *resistance stage*) in order to maintain homeostasis. The adrenal cortex, stimulated by *adrenocorticotrophic hormone* (ACTH) released by the hypothalamus, produces cortisol, hydrocortisone and corticosterone (glucocorticoids). These have the effect of releasing energy from non-carbohydrate sources, such as protein, over a period of several hours.

Neither the alarm nor resistance stages are physiologically damaging to an individual: the various response or adaptation mechanisms have evolved to assist the animal to cope with adverse conditions. For example, increased energy provided to increase body temperature in cold conditions would ensure survival. Moreover, it has been suggested that brief activation of adrenal cortex activity early in life can have positive benefits (Fraser & Broom, 1990).

The control mechanisms underlying the way in which an individual adapts to its environment can be divided into two distinct groups. The first of these is termed *negative feedback control* where the animal responds to change, and the second is *feed forward control* where action is taken in anticipation of change. These terms are borrowed from control engineering.

It is relatively easy to understand how an animal can respond to a stimulus through negative feedback control. Sensory input is continually being received and integrated in the brain which, in turn, initiates appropriate actions by such as muscle, endocrine and physiological systems. For example, in cold weather, sows will lie closer together on straw in order to minimize heat loss. In addition to centrally processed actions, peripheral reflex actions act in order to counteract stimuli such as the vibration of lorries during transport. Under those circumstances, the tension of several muscles in an individual is adjusted in order to maintain its balance in the face of the vibration of the vehicle and the jostling of the other animals being transported.

Modification in physiological states or behaviour that occur in anticipation of environmental change may occur through a variety of routes. Environmental cues, such as diurnal changes in light and seasonal changes in daylength, are strongly correlated with endocrinological and behavioural changes in many species (e.g. the reproductive and migratory cycles of birds and several mammals). Predictable components of the management of a husbandry system are learned quickly by the animals. Pigs, for example, will soon learn the timing of feeding and modify their behaviour during the time immediately prior to the arrival of food. Anticipation of aversive experiences will stimulate avoidance or escape behaviour: sheep that have been poorly handled will be reluctant to move along a race. While predictable events that are aversive will be stressful, it has been shown that the occurrence of similar stimuli that are unpredictable will be much more stressful (Weiss, 1971).

The final (*exhaustion*) stage of Selye's general adaptive syndrome occurs when the reserves of the animal have been exhausted and it is no longer able to cope with the conditions and, unless rapid action is taken, it will die. Whilst the short-term secretion of glucocorticoids can be beneficial, chronic activation of the adrenal cortex results in many problems. These include reductions in: food consumption, protein synthesis, growth, gonadal activity (ACTH being secreted in preference to GH and other gonadotrophins in the pituitary), lymphocyte count and immune response. It also leads to changes in cardiovascular activity, gastric ulcers and increased disease susceptibility, and has been proposed as the cause of sudden deaths, particularly in pigs and poultry (Fraser *et al*, 1975). This is the final stage of GAS, when the body's defence mechanisms are exhausted resulting in a reduction of the animal's biological fitness, i.e. its reproductive potential or chances of survival decline. It is only this stage that should be referred to as stress. Accordingly Broom (1983) defined stress as: 'The process by which environmental factors overtax control systems in an individual, thus activating responses whose effects are prolonged and ultimately detrimental to that individual', and equated the term stress with the exhaustion stage of GAS. Stress is a detrimental process; confusion would arise if a beneficial process were to be described by the same term. Therefore, it is important that activation of the adrenal cortex early in life should not be referred to as stress but as stimulation.

There is also a genetic component in stress: breeds and strains of both pigs and poultry have been shown to inherit a susceptibility to stress (Mills *et al.*, 1988). For example, the halothane gene in pigs is beneficial for growth and body conformation but, when present as the homozygous genotype, makes individuals possessing it more susceptible to stress.

Stress terminology can be summarized as follows: *stress* is a detrimental process to which an individual may be subjected by a *stressor*, the result of which is that the individual is said to be *distressed*.

How do we assess welfare?

Various authors (Fraser *et al.*, 1975; Broom, 1983) have emphasized the importance of using several indicators to assess welfare. However, what should we measure? There is no simple SI unit for welfare and we have also seen that there is not even a standard definition to which all scientists adhere. Therefore, in order to perform such an assessment, we need to refer to the various definitions for guidance on what we should measure.

The view taken by the Brambell Committee (HMSO, 1965) and Hughes (1976) was that welfare was concerned with the *well-being* of an individual. So, in order to gauge welfare using this definition, we need to ask questions like: what is good or bad for a particular animal? This logically leads on to the problem of *who* should decide what is good or bad for an individual. It is also important to note that this problem is quite independent from another important question of how bad conditions can become before we say that the welfare of an animal is deemed to be unacceptable: this is discussed in the final section.

There is a great danger of ascribing human feelings to animals without evidence to show that there is such a similarity of feeling. Therefore, wherever possible, the animal should be the appropriate decision-maker regarding its need or wants. Duncan & Poole (1990) reviewed various methodologies used to assess the needs of animals. These needs might be environmental (e.g. preferred floor types for laying hens) or behavioural (e.g. the opportunity to build nests). They concluded that simple preference tests were flawed; for example, individuals may choose to satisfy short-term needs to the detriment of their longer-term welfare. The use of operant conditioning techniques, where an individual works for a reward, offers a better guide to an animal's motivation for a particular environmental feature or opportunity. The amount of work that an individual is prepared to perform in order to obtain a reward is regarded as a measure of its motivation for the reward. Some difficulties with this technique have been identified (Dawkins & Beardsley, 1986), but they are not insurmountable, and so such methods are useful as long as some care is taken when designing experiments. These methods offer scientists the opportunity to gain an insight into the animal's perception of its environment which, given the definitions of welfare discussed earlier, is extremely important.

In order to summarize the needs of farm livestock (i.e. what is good for them), the Farm Animal Welfare Council (1979, 1992) adapted ideas from Brambell's report (HMSO, 1965), and established a pragmatic framework within which welfare can be discussed and on which new developments can be based. This framework has become known as the *Five Freedoms*, which incorporate the basic pre-requisites for good welfare:

(1) freedom from thirst, hunger or malnutrition;
(2) appropriate comfort and shelter;
(3) prevention, or rapid diagnosis and treatment, of injury and disease;
(4) freedom to display most normal patterns of behaviour;
(5) freedom from fear.

It is important to note, though, that satisfaction of these criteria does not automatically ensure that good welfare will necessarily follow: they provide a guide to the basic requirements of a husbandry system. These Freedoms have been incorporated into the Codes for Recommendations for the Welfare of Livestock (MAFF, 1983) and are implicit in all UK welfare legislation.

An assessment of welfare that incorporates use of the Five Freedoms measures the *provision* of the animal's environment with respect to welfare. Broom's (1986) definition, in contrast to those of Brambell (HMSO, 1965) and Hughes (1976), is couched in terms of the animal's ability to *cope* with its environment: by measuring how well an individual is coping we would be measuring welfare and, therefore, would measure the functional capability of the individual. The two approaches complement each other.

Mason & Mendl (1993) highlighted the difficulties of measuring welfare and identified points that had to be considered when designing experiments. For example, different individuals may cope with adversities in their environment in various ways (e.g. Duncan & Filshie, 1979), thereby making the choice of indicators quite difficult. Although Duncan & Poole (1990) correctly identified behavioural indicators as useful because no disturbance is caused when they are measured, they did recommend the use of a multitude of different types of measure for assessing welfare. For convenience these have been grouped under the following headings: physiological and biochemical, production and behavioural.

Physiological and biochemical indicators

Several physiological and biochemical correlates of welfare have been identified. The example of increased heart rate in fearful poultry has already been described. Similar increases in cardiovascular activity have also been recorded when sheep were introduced to a flock, during loading into a vehicle and during transportation (Baldock & Sibly, 1990). However, care must be taken when interpreting such results since perfectly normal activities, such as play, running, courtship and mating, will also elicit similar responses.

Under chronically poor welfare conditions, the plasma concentration of glucocorticoid hormones will increase. However, simple assays of hormone levels can be misleading for several reasons: diurnal variation occurs in many species; there is considerable inter-individual variability in hormone activity; and the duration of the intervals between stimulus, response and the time when samples are taken can have a profound effect on the magnitude of observed response. All of these factors will have a profound effect upon the magnitude of the measurement. Furthermore, like heart rate, the secretion of glucocorticoids is not confined to times of adversity: elevated levels will also be observed, for example, during courtship.

One methodology that does seem to be useful is ACTH challenge, in which an injection of ACTH is administered and the adrenal response is measured. An individual that has been subject to chronic stress will have enlarged adrenal glands which will produce more corticosteroids than smaller adrenals in an animal that has not been stressed. This method seems to overcome the problems associated with taking simple measures of glucocorticoid activity, and has been used to demonstrate the stress associated with mixing and space limits in dairy cattle as well as the effects of stocking density in pigs.

In addition to the enlargement of adrenal glands, the development of gastric ulcers is also a symptom of chronic stress.

Biochemical changes in skeletal muscles post-mortem can occur in response to adverse conditions during the period prior to slaughter. Pale, soft and exudative muscle (PSE) and dark, firm and dry meat (DFD) are both indicators of earlier poor welfare.

Rapid glycolysis (the breakdown of glycogen stores) immediately prior to slaughter is a problem particularly associated with pigs and can result from long or arduous journeys. It causes the production of high levels of lactic acid and a dramatic decrease in muscle pH after slaughter which, in turn, leads to a decrease in the water-binding properties of the muscle. Meat derived from pigs stressed in this way is pale grey and soft and exudes water: it is of low value.

Long-term depletion of glycogen stores leads to an unpopular and, therefore, low value meat for different reasons. With low levels of glycogen, there is little or no production of lactic acid in the muscles after slaughter. The pH therefore remains high, giving rise to a dark-cutting meat that is firm and dry. This particular problem is found in cattle and pigs and can be caused by a variety of factors including fighting during the mixing of unfamiliar individuals during transport or in lairage at the slaughterhouse (Guise & Penny, 1989; Warriss *et al.*, 1990).

Production indicators

It has already been stated that normal growth and reproduction are indicative of adequate food provision but not necessarily of high welfare (HMSO, 1965). It is important to remember that animal species utilized by man to provide meat or other products have been genetically selected for their muscle growth and reproductive performance. As a consequence, growth and reproductive performance will generally be affected only when individuals are subjected to extreme or chronic adversity. If individuals do not grow at the normal rate, fail to come into oestrus, or have reduced conception rates or litter sizes, we can probably conclude that they are not in a state of good welfare. They need to be examined swiftly in order to identify the underlying reasons and to take appropriate action.

The immediate (or proximate) causes of reduced growth, loss of weight, or diminished reproductive performance are generally either illness, nutritional deficiencies, digestive abnormalities or behavioural factors. Obviously, inadequate food allowance, rations that are deficient in particular nutrients (e.g. protein deficiencies) or trace element deficiencies in pasture will limit growth or cause metabolic diseases (e.g. swayback in sheep due to copper deficiency). Disease may have a direct effect on growth and it may also depress appetite. Behavioural observations of many farm species have demonstrated that subordinate individuals may not display normal reproductive behaviour in the presence of more dominant conspecifics (members of the same species) thus leading to reduced productivity.

Ultimately, however, such factors may really be symptoms of inadequate or inappropriate housing, or poor management. For example, if insufficient bedding material is available for group-housed cattle or pigs, then some individuals may be forced to lie on cold or wet flooring thereby diverting energy to thermoregulatory activity that would otherwise be used for reproduction or growth. Such conditions will also provide sites for disease organisms to multiply. Limited space will prevent subordinate individuals from avoiding more dominant animals. Lack of protection when feeding has been shown to result in reduced access to food by subordinate animals (Albright, 1969; Bouissou, 1970). Several solutions to such problems are available. Sufficient feeding stalls for all individuals in a group can be provided (e.g. Morris & Hurnik, 1990). Alternatively, transponder-based systems with a smaller number of feeders have been designed for cattle and pigs in which every individual has its own food allowance and is protected from the rest of the group while it feeds. Such systems prevent food stealing and therefore do not reinforce this behaviour in the more dominant individuals.

Advances in technology applied to animal production can be used to provide information to stockpersons concerning animals that are potentially at risk. Daily action lists generated by computerized feeding systems for cattle and pigs indicate which individuals have not consumed all of their food allowance and thereby give the stockperson an early warning of possible problems.

Production indicators are not all confined to growth or production deficiencies. Laying hens that are stressed may retain the egg in the oviduct longer than normal, which can lead to the deposition of extraneous calcium carbonate in the eggshell (Mills *et al.*, 1987).

Welfare is a concept that is based upon the individual, therefore herd averages of production or food utilization do not provide an adequate index of welfare. However, if group averages do fall, they may well be indicative of diminished welfare or of a potential decrease in welfare which will require rapid action.

Behavioural indicators

The behaviour of an animal can provide an insight into many aspects of its welfare and, because animals are not disturbed while their behaviour is being observed, behavioural indicators can be very useful indeed. Abnormal behaviour can indicate illness or disease, inadequacies of the physical or social environment, and even nutritional problems. For example, if a sow has difficulty in walking or lying down, it may be suffering from a lameness disorder, it may be fatigued due to disease, the stocking density might be too high, or its pen may be too restrictive for easy movement.

One of the major reasons why behavioural indicators are so fundamental to welfare assessment is that animals are highly motivated to perform certain behaviours (e.g. nest-building by pre-parturient sows or pecking for food by hens). When frustrated, individuals may perform vacuum activities such as displaying nest-building behaviour without a suitable substrate such as straw. Alternatively, their behaviour may be redirected towards themselves or conspecifics such as is the case with feather-pecking in battery hens or belly-nosing penmates in newly weaned piglets who would still suckle their mothers if they had the opportunity. On occasions, frustration may lead to stereotypic behaviour (e.g. repeated pacing in battery hens or bar-bitting and sham-chewing in tethered sows). The importance of such motivation was recognized by the Farm Animal Welfare Council and incorporated into the Five Freedoms: the freedom to display most normal patterns of behaviour.

The diversity of abnormal behaviour is clearly illustrated by the categorization used by Fraser & Broom (1990) in which they catalogued four main groups: stereotypic or repetitive, non-functional behaviour; behaviour directed towards self or inanimate objects; behaviour directed towards other individuals; and abnormal function. While space limitations preclude a detailed discussion here of the diversity of behavioural indicators of welfare, it is important to emphasize some key principles. Fortunately, there are several substantive reviews of this topic (Kiley-Worthington, 1977; Fraser & Broom, 1990; Broom & Johnson, 1993). Furthermore, stereotypic behaviour in particular has recently been the subject of an extensive review by Lawrence & Rushen (1993).

A behaviour may be regarded as abnormal because its performance is observed only rarely. More frequently, however, it is either the inappropriate context in which the behaviour is performed or the excessive repetition without obvious function that signifies its abnormality. Moreover, the fact that a behaviour is performed frequently by a large number of individuals in a group does not necessarily mean that such behaviour is normal: outbreaks of tail-biting amongst piglets certainly cannot be viewed as normal!

The performance of such behaviour is indicative of a problem (as has been illustrated by the earlier examples), and so the task of the stockperson is threefold:

- to be aware that the behaviour is abnormal;
- to identify the causal factors;
- to take appropriate action to effect a remedy.

The key to solving welfare problems is contextual, and it is important to remember that the source of a problem may be related not just to the current physical and social environment of the animal, but may also have an experiential or evolutionary component.

Evolutionary factors were used earlier to explain the differences between sheep and dogs in their expression of pain. The importance of early experience was illustrated by individuals that may be reluctant to move along a race or up a ramp into a vehicle because they have experienced aversive stimuli in the past (e.g. rough handling). Poor stockmanship in one context can have profound effects on later behaviour.

The social structure of a herd, particularly its stability, is extremely important. The introduction of new individuals to a group can be especially traumatic and lead to fighting and serious injury.

What is an acceptable level of welfare?

The preceding sections have reviewed both the various definitions of welfare and also the diversity of scientific indices of welfare. The role of the welfare scientist is twofold: to determine *how* to assess welfare and also to carry out actual assessments. On the other hand, the task of determining what is an acceptable *level* of welfare, and of identifying the cut-off point between acceptable and unacceptable welfare, is not for the scientist to perform, but is one for which society in general must take responsibility.

The minimum level of welfare that society demands is a function, and inevitably a compromise, of many factors. In order to understand how acceptable levels of welfare are determined (and how they might change in the future), we need to examine more closely the relationship between production and welfare. Furthermore, we need to identify those other factors that contribute to the attitude of society towards welfare.

Ultimately, the need for an adequate supply of affordable food and other animal-based products (e.g. wool) drives animal production. So, in addition to deciding on levels of welfare, society also determines the required level of animal production. In post-war Britain, the need for cheap food generated the pressure for livestock producers to turn towards more intensive methods. Frequently, debates concerning animal welfare in agriculture have been portrayed as a conflict between two quite distinct groups: producers satisfying the demand for food and their need to make a living, and the welfare lobby who are protecting the interests of livestock. However, this is a very simplistic view of the situation and is inevitably a gross misrepresentation for two simple reasons. First, in many ways the interests of food production are served by high welfare. And, secondly, there are many pressures other than production itself that determine husbandry methods.

The close correlation between the interests of production and welfare have been very clearly illustrated by earlier examples. Stress can reduce meat quality, disrupt egg production and depress milk yield. Moreover, responses to chronic stress also include the diversion of resources from growth to energy production.

Many factors influence the way livestock are managed. Individually, each may be quite laudable, but their interaction with production or with each other may lead to associated welfare problems. This point is well illustrated by an example relating to food hygiene and human health.

The very natural anxiety about food hygiene particularly at time of slaughter has led to legislation (EC, 1991, 1992) that lay down very stringent requirements for hygiene in slaughterhouses in the EU. The cost of upgrading abattoirs to the required standards has led to many closures. As a consequence, animals have to travel further and for longer on their way to slaughter: their welfare suffers. In addition to the welfare consequences, meat quality will, on occasions, also be affected detrimentally. In this situation, production and welfare are certainly not in opposition.

Welfare issues are clearly not always simple: they can rarely be reduced to a choice between two simple options. Furthermore, in the case of farrowing crates, the welfare of one group of individuals (the piglets) is advantaged at the expense of another (the sow).

Figure 16.1 illustrates some of the factors that influence the way we manage our livestock. The minimum level of welfare that society will accept is a function of the relative importance of these and other factors. Their relative strengths will fluctuate over time and so will acceptable levels of minimum welfare. Welfare is measured along a continuum, and the level of welfare that society deems to be acceptable is a moving point which, during the past 30 years, has been progressing further up the scale requiring from farmers a higher level of welfare for livestock. Given that welfare is only one of the constraints that is placed on livestock production methods, it is important to ensure that, when any legislation or recommendations concerning livestock are prepared, the full welfare implications are taken fully into consideration.

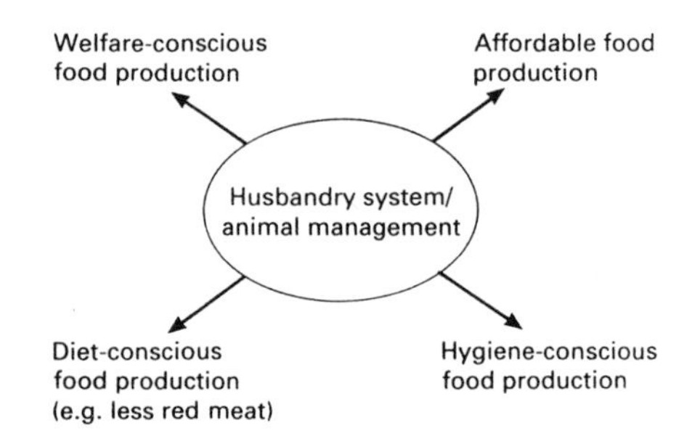

Fig. 16.1 Some of the pressures that determine the husbandry methods in which livestock are reared.

References

Albright, J.L. (1969) Social environment and growth. In: *Animal Growth and Nutrition* (eds E. S. E. Hafez & I. A. Dyers). Lea and Febiger, Philadelphia.

Baldock, N.M. & Sibly, R.M. (1990) Effects of handling and transportation on heart rate and behaviour in sheep. *Applied Animal Behaviour Science*, **28**, 15–39.

Bouissou, M.-F. (1970) Rôle du contact physique dans la manifestation des relations hierarchiques chez les bovins: consequences pratiques. *Annales de Zootechnie*, **19**, 279.

Broom, D.M. (1983) The stress concept and ways of assessing the effects of stress in farm animals. *Applied Animal Ethology*, **11**, 79 (Abstract).

Broom, D.M. (1986) Indicators of poor welfare. *British Veterinary Journal*, **142**, 524–526.

Broom, D.M. & Johnson, K.G. (1993) *Stress and Animal*

Welfare. Chapman & Hall, London.
Dawkins, M.S. (1980) *Animal Suffering: The Science of Animal Welfare*. Chapman & Hall, London.
Dawkins, M.S. & Beardsley, T.M. (1986) Reinforcing properties of access to litter in hens. *Applied Animal Behaviour Science*, **15**, 351–364.
Duncan, I.J.H. & Filshie, J.H. (1979) The use of radio telemetry devices to measure temperature and heart rate in domestic fowl. In: *A Handbook on Biotelemetry and Radio Tracking* (eds C.J. Amlaner & D.W. MacDonald). Pergamon, Oxford, pp. 579–588.
Duncan, I.J.H. & Poole, T.B. (1990) Animal welfare. In: *Managing the Behaviour of Animals* (eds P. Monaghan & D.G.M. Wood-Gush). Chapman & Hall, London, pp. 193–232.
EC (1991) Council Directive amending and consolidating Directive 64/433/EEC on health problems affecting intra-Community trade in fresh meat to extend to the production and marketing of fresh meat. OJ L268/69.
EC (1992) Council Directive amending and updating Directive 77/99/EEC on health problems affecting intra-Community trade in meat products and amending Directive 64/433/EEC. OJ L57/1.
Farm Animal Welfare Council (1979) *Farm Animal Welfare Council* (Press Release: FAWC/1). Farm Animal Welfare Council, Surbiton.
Farm Animal Welfare Council (1992) *The Five Freedoms* (Press Release: 92/7). Farm Animal Welfare Council, Surbiton.
Farm Animal Welfare Council (1993) *Report on Priorities for Animal Welfare Research and Development*. Farm Animal Welfare Council, Surbiton.
Fraser, A.F. & Broom, D.M. (1990) *Farm Animal Behaviour and Welfare*, 3rd edn. Baillière Tindall, London.
Fraser, D., Ritchie, J.S.D. & Fraser, A.F. (1975) The term 'stress' in a veterinary context. *British Veterinary Journal*, **131**, 653–662.
Gentle, M.J. (1986) Neuroma formation following partial beak amputation (beak trimming) in the chicken. *Research in Veterinary Science*, **41**, 383–385.
Guise, H.J. & Penny, R.H.C. (1989) Factors influencing the welfare and carcass and meat quality of pigs 1. Mixing unfamiliar pigs. *Animal Production*, **49**, 517–521.
Harrison, R. (1964) *Animal Machines: The New Factory Farming Industry*. Vincent Stuart, London.
HMSO (1965) Report of the technical committee to enquire into the welfare of animals kept under intensive livestock husbandry systems. Command: 2386.
HMSO (1991) The Welfare of Pigs Regulations, 1991. Command: 1477.
Hughes, B.O. (1976) Behaviour as an index of welfare. In: *Proceedings of the 5th European Conference on Poultry Welfare*, Malta, pp. 1005–1018.
Kiley-Worthington, M. (1977) *Behavioural Problems of Farm Animals*. Oriel Press, Stocksfield.
Kilgour, R. & Dalton, D.C. (1984) *Livestock Behaviour*. Granada Publishing, London.
Lawrence, A.B. & Rushen, J. (1993) *Stereotypic Animal Behaviour. Fundamentals and Applications to Welfare*. CAB International, Wallingford.
MAFF (1983) Codes of recommendations for the welfare of livestock. MAFF Publications. (Individual codes for each species of farm livestock.)
Mason, G. & Mendl, M. (1993) Why is there no simple way of measuring animal welfare? *Animal Welfare*, **2**, 301–319.
Mills, A.D., Marche, M. & Faure, J.-M. (1987) Extraneous egg shell calcification as a measure of stress in poultry. *British Poultry Science*, **28**, 177–181.
Mills, A.D., Faure, J.-M. & Lagadic, H. (1988) Génétique du stress. *Recueil de Médecine Vétérinaire*, **164**, 793–800.
Morris, J.R. & Hurnik, J.F. (1990) An alternative housing system for sows. *Canadian Journal of Animal Science*, **70**, 957–961.
Morton, D.B. & Griffiths, P.H.M. (1985) Guidelines on the recognition of pain, distress and discomfort in experimental animals and an hypothesis for assessment. *Veterinary Record*, **116**, 431–436.
Ryder, R.D. (1989) *Animal Revolution: Changing Attitudes towards Speciesism*. Basil Blackwell, Oxford.
Selye, H. (1956) *The Stress of Life*. Longmans, Green & Co., London.
Warriss, P.D., Brown, S.N., Bevis, E.A. & Kestin, S.C. (1990) The influence of pre-slaughter transport and lairage on meat quality in pigs of two genotypes. *Animal Production*, **50**, 165–172.
Weiss, J.M. (1971) Effects of coping behaviour in different warning signal conditions on stress pathology in rats. *Journal of Comparative Physiology and Psychology*, **77**, 1–13.

17

Livestock Feeds and Feeding

R.M. Orr

Feed represents the major cost to animal production. Thus, the efficiency of its use can have a considerable impact on the performance of an enterprise.

In general terms efficient utilization of feeds is dependent on the following elements:

(1) a knowledge of the nutrient value of the range of feedstuffs available;
(2) a knowledge of the nutrient requirements of animals;
(3) an ability to formulate diets and rations from a mixture of feedstuffs which will meet nutrient requirements within the potential voluntary food intake of animals. Clearly, an important element of this process of formulation is that of meeting nutrient requirements most economically.

In this chapter the procedures used in determining the first two elements are given attention as are characteristics of the major feeds, in terms of feeding value and limitations to their use, that are used in diet formulation. The aspect of feeding particular classes of livestock is dealt with in the chapters on the individual species.

Nutrient evaluation of feeds

The value of a feed is dependent on how much of particular nutrients in the feed the animal is able to use to meet the requirements of the various body processes. In the utilization of feed there are inevitable losses of nutrients due to inefficiencies in digestion, absorption and metabolism. Thus, the available nutrients in a food are the difference between the potential value of a feed as measured by chemical analysis and these inevitable losses of nutrients. Losses through incomplete digestion are a major factor in determining feeding values, and digestibility measurements are in some instances used to give an indication of feeding value – this is particularly true in the evaluation of roughages. For the purpose of diet formulation, however, the nutritive value of foods is determined in terms of their ability to furnish the animal primarily with energy and protein and, to a lesser extent, with minerals and vitamins.

Digestibility measurements

The digestibility of a food is defined as the proportion of the food which does not appear in the faeces and is assumed to be absorbed. This is strictly termed 'apparent' digestibility because a very small fraction of faecal excretion is in the form of fragments of gut mucosa and therefore not of dietary origin, but in practice this is not taken into account. The apparent digestibility of a feedstuff may be determined in animals restrained in metabolism crates which allow the precise measurement of food intake and of faecal excretion over a specified period of time. From this and the analysis of feed and faeces, digestibility coefficients for the food dry matter and also for individual food nutrients may be calculated, i.e. organic matter, protein, energy.

The percentage of digestible organic matter (OM) in the dry matter of a feed is termed the D-value. The D-value is widely used to describe the nutritive value (especially energy potential) of roughage feeds, particularly hay and silage.

Since digestibility assessments using live animals are tedious to conduct and unsuitable for routine estimations of large numbers of samples, alternative laboratory techniques (*in vitro* methods) have been developed. The earliest *in vitro* methods were developed for roughages and aim to simulate ruminant digestive processes. A two-stage process is carried out. The first stage involves incubation of a sample with rumen liquor to simulate rumen fermentation, the residue from this being further incubated with an acid pepsin solution to simulate the digestion of protein in the lower gut. More recently, methods based on the use of fungal cellulases have been used successfully to predict the digestibility of forages and other feeds. Following boiling with neutral detergent solution which hydrolyses the cell contents, the remaining cell wall or neutral detergent fibre (NDF) fraction is incubated with a cellulase solution, the residue from which can be ashed to determine the digestibility of the organic matter in the dry matter. This is referred to as *neutral detergent cellulase digestible organic matter* (NCD). For compounded feeds, additional steps involving a preliminary treatment of the feed with amylase to digest starch and the inclusion of gammanase to the cellulase digestion phase to break down any mannans that may be present are involved. This is termed the *neutral detergent cellulase + gammanase digestibility* (NCDG).

The apparent digestibility of feedstuffs may vary between animals due to such factors as type of animal (e.g. ruminant versus non-ruminant; sheep versus cattle). The main characteristics of food that affect digestibility are its chemical composition, particularly the fibre, protein and lignin content. Thus, for foodstuffs such as cereal grains and protein concentrates that are fairly constant in chemical composition, digestibility values are fairly constant, but for forages where composition varies widely according to growth stage wide variation in digestibility values may be found.

The other main characteristic that can have an effect is the method of feed preparation and processing involved. The effects of processing on the digestibility of cereal grains are complicated by different responses in the different species of farm animal, but such techniques as pelleting, rolling, milling and cooking through micronization and steaming can be of use in particular situations. Likewise, chemical processing through alkali treatment can be used to effect improvements in digestibility of both grains and forages. Physical processing of forages, especially grinding, decreases digestibility through increasing rate of passage. The decreases in digestibility of foods observed when the level of intake is increased are likewise largely due to rate of passage effects. Because the level of intake affects digestibility, measurements are usually made at the maintenance level of feeding.

Measurement of feed energy

Animals use energy for a variety of body functions ranging from energy required for essential muscular activity and maintenance of body temperature to energy required for the synthesis of body tissues. It is primarily energy intake which determines the level of performance that an animal can sustain.

The energy content of feeds may be expressed in the following ways: gross energy (GE), digestible energy (DE), metabolizable energy (ME) and net energy (NE). To understand the meaning of the terms involves a recognition of the ways in which food energy is utilized by the animal.

The large number of chemical bonds that make up the organic compounds present in food contain the energy present in the food. This chemical energy, termed gross energy or heat of combustion, is determined by the technique of bomb calorimetry in which a known weight of food is completely oxidized and the heat liberated measured. Of this gross energy in a food only a portion is eventually available to provide for the animal's energy requirements. In the utilization of the gross energy there are several routes of energy loss, namely:

- losses of indigestible energy in the faeces;
- energy losses in the urine, mainly in the form of nitrogen-containing substances such as urea;
- losses of energy as methane produced during the fermentation of dietary energy in the rumen and to a much lesser extent in the large intestine;
- heat produced from inefficiencies in metabolic processes of the animal and dissipated from the body surfaces, and to a lesser extent heat produced from inefficiencies in the metabolism of microbial cells in the digestive tract (heat of fermentation). The sum of these forms of heat is termed the heat increment of metabolism.

The energy remaining after these losses are accounted for represents the energy the animal utilizes for maintenance and production. This may be represented by Fig. 17.1.

The unit of measurement used in the assessment of food energy is the joule (J). More typically it is customary to use the kilojoule (kJ) or the megajoule (MJ).

$$1\,\mathrm{MJ} = 1000\,\mathrm{kJ} = 1\,000\,000\,\mathrm{J}$$

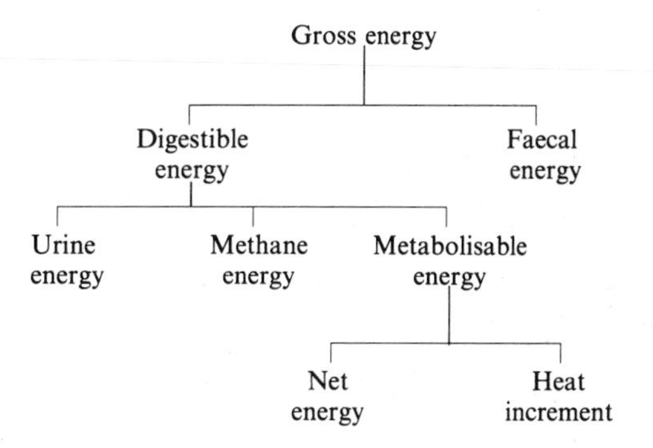

Fig. 17.1 Routes of energy loss.

Gross energy value

This is measured by burning a known weight of the food and recording the total heat produced. The gross energy value of a food is dependent on the proportions of carbohydrate, protein, fat and ash that are present. The approximate energy values of these components are: carbohydrate 17.5 MJ/kg DM, protein 23.5 MJ/kg DM, fat 39 MJ/kg DM and ash 0 MJ/kg DM. Thus the fat and the ash content of a food have potentially the greatest influence on gross energy values. In practice, most animal foods have a low and fairly constant amount of these components, mainly consisting of carbohydrate with lesser amounts of protein, such that the gross energy of a wide range of foods is fairly consistently around 18 MJ/kg dry matter. Although gross energy tells us the potential energy in a feed, it is the least satisfactory assessment of energy value since no account is taken of the availability of that energy to the animal.

Digestible energy

Digestible energy (DE) is a measure of the gross energy of the food minus that part of the food energy lost in the faeces. Strictly speaking this is the 'apparent' DE of a food and is determined in digestibility trials in which the intake and faecal excretion of gross energy is measured. The DE value of foods is dependent on the gross energy value and the factors that influence the digestibility of foods.

Metabolizable energy and fermentable metabolizable energy

The metabolizable energy (ME) of a food is the digestible energy less the energy lost in the urine and as methane. Typically about 8–10% of the digestible energy is lost in urine. About 7–12% of the digestible energy is lost as methane in ruminants whereas in the pig losses are negligible. Thus, for pigs metabolizable energy values are approximately 0.9 of DE values and for ruminants are approximately 0.81–0.86 of DE values.

Metabolizable energy values are determined by feeding trials in which animals are kept in metabolism cages within respiratory chambers that allow the accurate measurement of intake, faecal and urinary excretion and methane production. ME values are primarily influenced by the digestibility of the feed. Although the proportion of digestible energy lost in urine and methane is relatively consistent, urine losses are affected by the protein content of the diet.

In relation to ruminant feeding the measure *fermentable metabolizable energy* (FME) content of a feed or diet is required in order to use the metabolizable protein (MP) system. The FME is defined as the ME minus the ME present as oils or fat in the feed (ME_{fat}) and the ME contribution of fermentation acids (ME_{ferm}) present in silages.

Thus:

$$[FME]\ (MJ/kg\,DM) = [ME] - [ME_{fat}] - [ME_{ferm}]$$

It describes the amount of ME in a feed that is available to the microbes in the rumen for use in the synthesis of microbial protein. Dietary fat (ME = 35 MJ/kg DM) does not supply energy to the rumen microbes such that with a fat content of only 4% the FME is reduced by 1.4 MJ/kg DM. Fermentation acids, mainly lactic and acetic in well preserved silages, cannot be further fermented to provide energy in the rumen. For a silage with 100 g lactic acid/kg silage dry matter, a deduction of 1.51 MJ from the estimated ME must be made. Where lactic acid values are not known, ME_{ferm} may be taken as 0.1 ME.

Net energy value (NE)

The metabolizable energy supplied by the diet may be looked upon as the absorbed nutrients available for the various metabolic processes in the tissues. Thus, in the ruminant, volatile fatty acids, and, in the monogastric, glucose, can be looked upon as some of the main components of ME. The metabolic pathways into which these and other metabolites are directed so that forms of energy (e.g. ATP) that are useful to the animal may be created are not 100% efficient, some of the energy being converted into heat which is termed the heat increment (HI). Some heat is also produced from the work of digestion whilst in the ruminant heat arises from the inefficiencies in the metabolism of rumen microbes (heat of fermentation). The HI of a food is measured by the technique of animal calorimetry, and when subtracted from the ME gives the NE value. Typically the HI is greater in the ruminant (35–65% of ME for mixed diets) than in the monogastric (10–40% of ME for mixed diets.)

In the monogastric considerably less HI results from the metabolism of fat (9%) than of carbohydrates (≃17%) or proteins (≃25%) but in contrast to the ruminant the efficiency with which ME is used for the different body processes, such as maintenance and growth, is very similar. The NE value of a food to the monogastric is therefore a function of its ME value and the nutrients comprising the ME, especially the proportion of fat present.

In the ruminant the extent of heat production in the metabolism of ME and therefore the NE value of a feed are dependent on the purpose for which the ME is utilized. This is because the products of digestion (the constituents that make up the energy of ME) are used at different levels of efficiency for maintenance or conversion to the various productive functions of lactation, growth and reproduction. The situation is further complicated by the fact that the principal energy-containing end-products of digestion, the VFA – acetic, propionic, butyric – are used with differing efficiencies. Thus, diet composition, which influences the molar proportions of VFA, as well as physiological function, have an effect on the efficiency of utilization of ME. The feature of feeds which correlates most closely to this effect of diet composition is its *metabolizability* (qm) which is defined as the proportion of [ME] in the [*Ge*] of the feed:

$$qm = [ME]/[Ge]$$

A mean value for the [GE] of ruminant diets that may be used in this calculation is 18.8 MJ/kg DM. This is illustrated in Fig. 17.2 and means that individual feeds cannot be assigned single NE values.

Using the linear equations given on Fig. 17.2 the NE values of feeds when used for maintenance (NE_m), lactation (NE_l) and growth (NE_g) can be calculated as follows:

$$NE_m = ME \times k_m$$
$$NE_l = ME \times k_l$$
$$NE_g = ME \times k_g$$

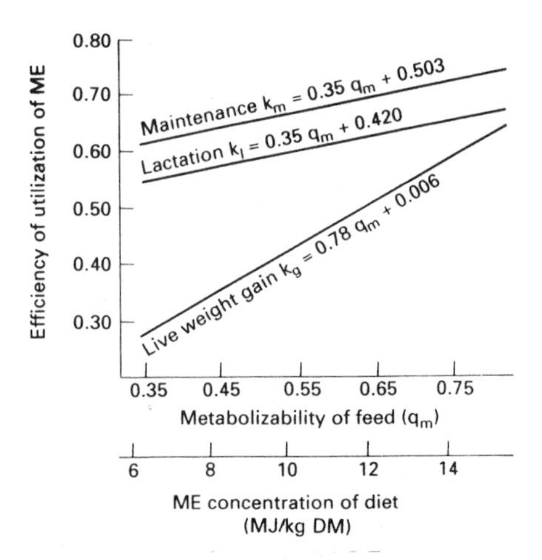

Fig. 17.2 Effect of diet and production on the efficiency of utilization of ME.

Thus, for example, for two diets (A and B) with ME values of 10 and 13 MJ/kg DM respectively and taking an average GE of 18.8 MJ/kg DM for both diets the NE values will be as follows:

MJ/kg DM	*Diet A*	*Diet B*	*Ratio A/B*
ME	10.0	13.0	0.77
NE_m	6.89 (0.69)	9.68 (0.74)	0.71
NE_l	6.06 (0.61)	8.60 (0.66)	0.70
NE_g	4.21 (0.42)	7.08 (0.54)	0.59

This illustrates how the efficiency of utilization of ME (values in parentheses) and NE values differ according to the ME value of the diet and the purpose to which the ME is put.

Protein evaluation of feeds

Animals require protein to provide them with amino acids. These amino acids are used for the synthesis of body tissue proteins and proteins in products such as milk and wool which leave the body. Simple-stomached animals obtain their supply of amino acids from the digestion and absorption of dietary protein whereas in the ruminant the situation is complicated by the presence of the rumen which means that the supply of amino acids comes from a combination of microbial protein produced from dietary rumen degradable protein (RDP) or non-protein nitrogen (NPN) and dietary undegradable protein (UDP). Thus, different approaches to the evaluation of foods as protein sources are necessary for the monogastric and the ruminant.

An estimate of the crude protein in feeds is the starting point of all protein evaluation systems, subsequent measures of protein quality being dependent on its digestive and metabolic use by the animal.

Crude protein (CP)

Proteins in food are the main nitrogen-containing components. Thus, by measuring the nitrogen content of a food using the Kjeldahl technique, it is possible to derive its protein content. Two assumptions are made in calculating crude protein content from nitrogen content. First, it is assumed that all nitrogen in a food is present in protein. In reality this is incorrect in that some nitrogen is present in such compounds as amides. For most feeds this nitrogen is less than 5% of the total but for root crops it may be considerable. Secondly, it assumes that all food protein contains 16% nitrogen although, in practice, a range of 15–18% for different protein types exists. Because of the inaccuracy of these assumptions the protein content of feeds is expressed as crude protein thus:

$$\%CP = \%N \times (100/16)$$

or more commonly

$$\%CP = \%N \times 6.25$$

The crude protein content gives no indication of how efficiently the protein is utilized.

Measures of protein quality

Monogastrics

The usefulness of proteins in the diet of monogastrics is dependent on their digestibility and absorption and the subsequent retention of amino acids in the animal's tissues.

Most simply the *digestible crude protein* content of a diet is determined from the results of a digestibility trial where it is assumed that the difference between the nitrogen in the feed and faeces represents the quantity of nitrogen absorbed. As this does not take into account the fact that some of the faecal nitrogen arises from nitrogen secreted back into the digestive tract in the form of digestive secretions, it is more accurately known as the '*faecal apparent digestibility coefficient*'. More recently, for pigs it has been shown that as no useful amino acids are absorbed in the rear gut a better measure of protein digestibility is *ileal apparent digestibility* (D_{il}). This is determined in cannulated animals through the measurement of nitrogen disappearance prior to the food reaching the distal ileum and is usually about 8% lower than faecal digestibility values although for some feeds it may be as much as 14% lower. The technique can also be used to obtain apparent ileal digestibility coefficients for individual amino acids in a feed. This is useful because individual amino acids behave differently in the gut and frequently have digestibility coefficients different from that of the whole protein. Further data in this area are required.

Although it might be thought that, once absorbed, amino acids are available for incorporation into tissue proteins, utilization efficiency is very much dependent on the balance of amino acids absorbed. There are a number of methods of assessing this aspect of protein quality (e.g. apparent biological value and growth assays), but in practice the efficiency with which digested protein is utilized is largely dependent on the supply and balance of certain essential amino acids, e.g. lysine, threonine, tryptophan and methionine; the amino acid that is available in least supply in relation to requirement determines the extent of protein synthesis (the limiting amino acid).

In mixed diets the term *protein value* (V) can be derived from a comparison of the ileal-digested and utilizable amino acid composition of the diet with the profile of utilizable ileal digested amino acids in an *ideal protein*. That amino acid in the comparison giving the lowest ratio determines the protein value of the diet. In cereal-based diets this is frequently lysine, although in some situations threonine and methionine may be limiting. The use of pure synthetic amino acids can do much to boost the protein value of pig diets. The ideal protein is broadly similar in terms of essential amino acid composition to that of pig tissue and milk protein.

Using the above parameters and knowing the efficiency of utilization of absorbed ideal protein, the term *ideal protein supply* (IP) from a diet may be calculated as

$$IP = \text{Feed intake} \times CP \times D_{il} \times V \times v$$

where CP is crude protein concentration, D_{il} is ileal digestibility of the CP, V is protein value and v is the efficiency of use of an ideal protein. The efficiency of use of ileally digested ideal protein (v) varies from 0.70 to 0.95.

Ruminants – metabolizable protein
Metabolizable protein (MP) is defined as the total digestible true protein (amino acids) available for metabolism after digestion and absorption of the feed in the animal's digestive tract. Metabolizable protein has two components:

- *the digestible microbial true protein* (DMTP) which is a measure of the amino acids being made available for the animal's metabolism from microbial crude protein synthesis in the rumen;
- *the digestible undegraded feed protein* (DUP) which is that fraction of a feed that has not been degraded during passage through the rumen, the *undegradable protein* (UDP), but which is digested in the small intestine.

For the calculation of MP a number of parameters describing the extent and rates of degradability of feed proteins as they pass through the digestive tract are required. They include the following:

- *Total dietary crude protein* (CP) which is the total N × 6.25 and includes nitrogen from true protein and NPN sources. This CP contains a fraction with a variable rate of degradability in the rumen, the rumen degradable crude protein (RDP), and a fraction resistant to degradation in the rumen, the undegradable crude protein (UDP).
- *Quickly degradable crude protein* (QDP) which is determined by placing the dried, ground feed in a Dacron bag and washing it in cold water. The crude protein extracted is the QDP, also referred to as the *a* fraction. Thus,
  $$[QDP]\ (g/kg\ DM) = a \times [CP]\ (g/kg\ DM)$$
- *Slowly degradable crude protein* (SDP) is the fraction of feed protein (termed the *b* fraction) expressed as a proportion of the total CP in the feed which is degraded in the rumen but does not form part of the QDP fraction.

The precise SDP value for a feed is dependent on the rate of degradation of dietary CP in the rumen and the fractional outflow rates of feed from the rumen. Fractional rates of degradability per hour spent in the rumen (designated *c*) vary for different feed protein sources, ranging from 0.02 (fishmeal) to 0.50 (wheat). Fractional outflow rates from the rumen (designated *r*) are highly correlated with level of feeding in the animal. In relation to level of feeding the following fractional outflow rates have been derived:

Maintenance	0.02/hour
2 × maintenance	0.05/hour
3 × maintenance	0.08/hour

Thus:

$$[SDP]\ (g/kg\ DM) = \{(b \times c)/(c + r)\} \times [CP]\ g/kg\ DM$$

- *Undegradable crude protein* (UDP) in a feed is given by

$$[UDP] = [CP] - \{[QDP] + [SDP]\}$$

- *Digestible undegradable crude protein* (DUP) is determined from the UDP and the digestibility of the UDP fraction in the small intestine. This latter factor is related to the amount of crude protein in the acid detergent fibre fraction of the feed, the *acid detergent insoluble crude protein* (ADIP). DUP may be calculated according to the equation:

$$[DUP]\ (g/kg\ DM] = 0.9\ \{[UDP] - [ADIP]\}$$

- *Effective rumen degradable protein* (ERDP) describes the amount of crude protein synthesized by rumen micro-organisms from the QDP and SDP fractions. The efficiencies of nitrogen capture being 0.8 and 1.0 respectively, ERDP may be calculated according to the equation:

$$[ERDP]\ (g/kg\ DM) = 0.8\ [QDP] + [SDP]$$

Thus the [ERDP] value for any feedstuff is dependent on the CP of the feed; its *a* fraction from which QDP content can be determined, and its *b* fraction, the fractional rates of degradability (*c* value) and fractional outflow rate of feed from the rumen (*r* value) from which SDP content can be calculated.

Example For a feedstuff such as ground barley (CP = 129 g/kg DM) which has CP degradability factors of:

$a = 0.25$ $\quad r = 0.05$
$b = 0.70$ $\quad$ [ADIP] (g/kg DM) = 2.5
$c = 0.35$

$$
\begin{aligned}
[QDP]\ (g/kg\ DM) &= 129 \times 0.25 = 32.3 \\
[SDP]\ (g/kg\ DM) &= [CP] \times \{(b \times c)/(c + r)\} \\
&= 129 \times \{(0.7 \times 0.35)/(0.35 + 0.05)\} \\
&= 79 \\
[ERDP]\ (g/kg\ DM) &= 0.8\ [QDP] + [SDP] \\
&= 105 \\
[UDP]\ (g/kg\ DM) &= [CP] - \{[QDP] + [SDP]\} \\
&= 18 \\
[DUP]\ (g/kg\ DM) &= 0.9\ \{[UDP] - [ADIP]\} \\
&= 14
\end{aligned}
$$

It is from these values that values for the synthesis of *microbial crude protein* (MCP), *microbial true protein* (MTP) and finally metabolizable protein (MP) supplied by a diet can be calculated. The synthesis of MCP in the rumen is dependent on the amount of fermentable metabolizable energy (FME) in the diet and the level of feeding. Thus the values for MCP synthesis per MJ FME (value *y*) at different levels of feeding (L) relative to the ME requirement for maintenance are described by the equation:

$$y\ (g\ MCP/MJ\ FME) = 7.0 + 6.0\ \{1 - e^{(-0.35L)}\}$$

Thus when (L = 1) y = 9 g MCP/MJ FME
(L = 2) y = 10 g MCP/MJ FME
(L = 3) y = 11 g MCP/MJ FME

In diet formulation the objective is for ERDP supply to be equal to the potential MCP supply from the available FME. If ERDP supply exceeds its requirement for MCP synthesis, then FME is limiting and the excess ERDP will be wasted and result in elevated levels of blood ammonia and urea. If ERDP supply is less than ERDP requirement, then the diet is ERDP limited and

$$MCP\ (g/d) = ERDP\ (g/d)$$

Of the MCP produced by the rumen microbes about 0.25 is present as nucleic acids, which cannot contribute to the synthesis of body tissue or milk protein. The MTP content of a diet is therefore 0.75 of the MCP. MTP is 0.85 digestible in the small intestine so that digestible microbial true protein (DMTP) in a diet is:

$$DMTP = 0.75 \times 0.85 \times MCP = 0.6375 \times MCP$$

To this source of true protein absorbed from the small intestine we must add the DUP to obtain the MP supplied by a ration

$$MP = 0.6375\ MCP + DUP$$

Micronutrients

The mineral composition of feeds is usually described in terms of the total quantity of the individual elements present. Since availability of minerals in food varies greatly and is influenced by a number of factors there is no attempt to include this parameter in evaluation.

Vitamin concentration of food is usually expressed as milligrams or micrograms per kilogram. However, the levels of fat-soluble vitamins are normally expressed as International Units (IU).

1 IU vitamin A = 0.3 mg crystalline vitamin A alcohol

1 IU vitamin D = 0.025 mg crystalline vitamin D_3

1 IU vitamin E = 1 mg α-tocopherol acetate

Chemical analysis and the prediction of feed value

The most widely used routine methods of evaluating the nutrient content of feedstuffs are based on chemical procedures. Such analyses are only of value, however, if the feed analysed is representative of the entire batch. To this end it is important that appropriate sampling procedures be followed. Using appropriate sampling equipment a number of core samples should be taken from different areas of the feedstuff concerned, these samples thoroughly mixed together and a suitable size of subsample taken for analysis. It is important that samples for analysis are appropriately packaged, labelled and stored prior to analysis. For example, silage samples not analysed immediately should be frozen.

Current methods of analysis of concentrate feeds frequently rely on the Weende or proximate method of analysis whilst the analysis of forage feeds is largely based on the methods developed from Van Soest. From elements of these analytical procedures a number of equations have been developed to predict feeding values that may be used in the processes of ration formulation.

Concentrate and compound feeds

Current methods of analysis are still partly based on the Weende or proximate system in which feeds are broken down into six fractions (Table 17.1). The system has many criticisms, particularly with respect to its failure to completely separate the carbohydrate fraction into fibrous and non-fibrous carbohydrates and its overestimation of true protein. It is likely that the use of certain components of the proximate analysis will decline in the future as new techniques are used more widely, but the contents of crude protein, oil, crude fibre and ash are still declared on bags of both compounds and straight feedingstuffs.

Since the proximate analysis fails to adequately describe the carbohydrate fraction of feeds, several methods have been derived to measure the starch and sugar content. However, the precision of these methods requires improvement. Although not required by law a statement of starch and sugar level is occasionally given on bags of feedingstuffs. A further development is that the fibre fraction in certain raw materials may be characterized according to the Van Soest scheme of analysis (Fig. 17.3) into neutral detergent fibre (NDF) which corresponds to the sum of cellulose, hemicellulose and lignin, and acid detergent fibre (ADF) which consists of cellulose and lignin.

The use of crude protein overestimates the protein content of feeds because of some of the nitrogen being derived from non-protein sources but where NPN sources such as urea are used in the formulation of

Table 17.1 The fractions of proximate analysis

Fraction	*Components*	*Procedure*
(1) Moisture	Water and any volatile compounds	Dry sample to constant weight in oven
(2) Ash	Mineral elements	Burn at 500°C
(3) Crude protein (= N × 6.25)	Proteins, amino acids, non-protein nitrogen	Determine nitrogen by Kjeldahl method
(4) Ether extract (oil)	Fats, oils, waxes	Extraction with petroleum spirit
(5) Crude fibre	Cellulose, hemicellulose, lignin	Residue after boiling with weak acid and weak alkali
(6) Nitrogen-free extractives	Starch, sugars and some cellulose, hemicellulose and lignin	Calculation (100 minus sum of other fractions)

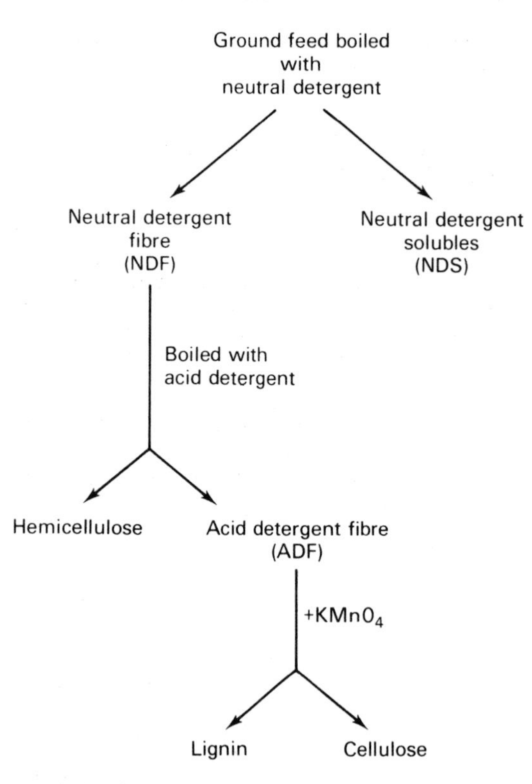

Fig. 17.3 The Van Soest scheme of forage analysis.

ruminant compound feeds a statutory statement of the protein equivalent of the feed derived from NPN sources must be given. Such a statement must also accompany the declaration of protein content of compound feeds for non-ruminants if greater than 1% of protein equivalent is derived from NPN sources. For ruminant feeds a further description of the protein that can be given is the proportion of rumen degradable (RDP) and digestible undegradable protein (DUP) present, and the relevant degradability parameters.

In recent years a considerable amount of research effort has been put into the development of equations which relate some of the above aspects of chemical composition to the energy value of feeds. This is especially useful for compound feeds. Although a vast amount of work has gone into such equations, it is important at the present time to realize their limitations as far as precision is concerned.

For ruminant compound feeds an equation derived by the ADAS Feed Evaluation Unit is:

$$\text{ME (MJ/kg DM)} = 0.0140\ (\text{NCDG}) + 0.025\ \text{oil}$$

where NCDG, the neutral detergent cellulase + gammanase in a feed (see the section headed 'Digestibility measurements'), and oil are expressed as g/kg DM.

An alternative equation which relates proximate analysis to ME but which has a lower degree of precision is:

$$\text{ME (MJ/kg DM)} = 11.78 + 0.0654\,\text{CP}\,\% + 0.0665\ \text{oil}\,\%^2 - (0.0414\ \text{oil}\,\% \times \text{CF}\,\%) - 0.118\ \text{ash}\,\%$$

For compounded pig diets an equation derived by the Edinburgh School of Agriculture is:

$$\text{DE (MJ/kg DM)} = 17.49 + 0.157\ \text{oil}\,\% + 0.078\,\text{CP}\,\% - 0.325\ \text{ash}\,\% - 0.149\,\text{NDF}\,\%$$

or alternatively

$$\text{ME (MJ/kg DM)} = 0.016\,\text{NFE}\,\% + 0.032\ \text{oil}\,\% + 0.018\,\text{CP}\,\% + 0.015\,\text{CF}\,\%$$

and for poultry an equation derived by the Poultry Research Centre to predict the metabolizable energy is:

$$\text{ME (MJ/kg DM)} = 5.39 - 0.113\,\text{CP}\,\% + 0.281\ \text{oil}\,\% + 0.113\ \text{starch}\,\% - 0.136\,\text{CF}\%$$

Analytical descriptions of forage feeds

Forage feeds are characteristically much more variable in their nutritive value than concentrates and must be routinely evaluated prior to their use in ration formulation. The methods of forage analysis developed by Van Soest in the USA (see Fig. 17.3) have served as the basis of forage evaluation in recent years. From such measurements he devised the following regression equation for estimating digestibility of forages:

$$\text{Digestibility} = 0.98\,\text{NDS} + [1.473 - 0.789\log_{10}\text{lignin}]\,\text{NDF}$$

where NDS and NDF are percentages of the forage dry matter and lignin is expressed as a percentage of the ADF fraction.

Such analysis according to the van Soest method is too costly and time consuming for routine analysis, but modification of the acid detergent fibre has served for many years as the basis of routine analysis. Modified acid detergent fibre (MADF) is obtained by refluxing the dried ground forage with an acid detergent solution (cetyl trimethyl-ammonium bromide) for 2 h and recovering the residue that is resistant to hydrolysis. The MADF value can then be used in appropriate regression equations to predict the ME values of various classes of feed. The use of MADF in such equations is currently being replaced by either cellulase enzyme techniques (see under 'Digestibility measurements') or else by the Near Infrared Reflectance Spectroscopy (NIR) methods that are in widespread use in the industry and based on reference samples of known *in vivo* digestibility and ME values. A selection of the equations used to predict the ME value of forage feeds is shown below:

Grass silage

$$\text{ME (MJ/kg DM)} = 15.0 - (0.0140 \times \text{MADF})$$
$$= 5.45 + (0.0085 \times \text{NCD})$$
$$= 2.91 + (0.0120 \times \text{IVD})$$

Maize silage

$$ME\ (MJ/kg\,DM) = 13.38 - (0.0113 \times MADF)$$
$$= 3.62 + (0.0100 \times NCD)$$

Grass hay

$$ME\ (MJ/kg\,DM) = 15.86 - (0.0185 \times MADF)$$
$$= 4.22 + (0.0086 \times NCD)$$
$$= 2.63 + (0.0109 \times IVD)$$

where

MADF = modified acid detergent fibre (g/kg DM)
NCD = neutral detergent cellulase (DOMD) (g/kg DM) (see under 'Digestibility measurements')
IVD = *in vitro* digestibility (DOMD) (g/kg DM) (see under 'Digestibility measurements')

Fermentable metabolizable energy (FME) values for forages are mainly of relevance to silages where some of the energy is present as fermentation acids. As the extent of fermentation is related to the dry matter of the ensiled crop, a prediction of FME can be made using an equation developed by the ADAS Feed Evaluation Unit

$$FME\ (MJ/kg\,DM) = ME \times \{(0.467 + (0.00136 \times (ODM) - (0.00000115 \times ODM^2)\}$$

where ODM is oven dry matter (g/kg) of the silage.

Further routine information on forage composition and quality is also given by the following chemical procedures:

- dry matter or moisture percentage,
- crude protein,
- pH and organic acids (e.g. lactic) in silages,
- ammonia-N, soluble-N and amino-N in silages.

The latter two measurements are indicative of the type of fermentation (see Chapter 11).

Near infrared reflectance analysis

Near infrared reflectance analysis is the most up to date, rapid and precise routine method of analysis currently available. Illumination of a feedstuff in the near infrared region of the electromagnetic spectrum (1100–2500 nm) causes certain molecular changes to occur which can be quantified such that the amount of light absorbed at a particular wavelength is proportional to the concentration of particular organic components. By selecting particular wavelengths and using computer interpretation of the spectra, rapid measurement of constituents such as moisture, starch, protein, oil and fibre can be made in straight concentrate, forage and compounded feeds.

From such information an assessment of the energy status of the feed may be made. Although this method of analysis is only as precise as the traditional 'wet' chemical analysis of materials that are used to calibrate the NIR analysis equipment, it allows analysis of feeds in as little as 10 minutes compared with the 24–48 h period required for 'wet' chemical analysis. For feed compounders it provides an opportunity to analyse routinely each consignment of feed entering the mill and in the light of this information to update feed formulation rapidly. In addition, it allows a check on the analysis of compounded feeds before they leave the mill, giving an opportunity to prevent release to farms if they do not conform to statutory or company specifications.

Raw materials for diet and ration formulation

A great variety of feedstuffs is used for the formulation of animal diets and rations. Whilst pig and poultry diets consists almost entirely of concentrated feedstuffs, the requirements of ruminant livestock are met by a combination of roughages and concentrates, the contribution of concentrates to total requirements being greatest in dairy cow feeding and least in the feeding of sheep and suckler cows.

In considering feedstuffs it is useful to distinguish between the feedstuffs which are relatively high in fibre and/or water content (bulky feeds or roughages) and those which are low in fibre and water (concentrate feeds). Figure 17.4 indicates into which of these major groupings different feedstuffs fall.

The following terms that are frequently used in animal feedingstuffs terminology are also worth defining at this stage.

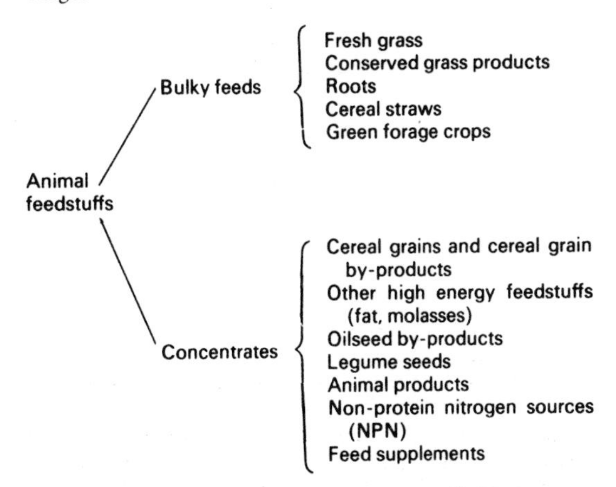

Fig. 17.4 A classification of animal feeds.

Compound feeds A number of different ingredients (including minerals, vitamins, additives) mixed and blended in appropriate proportions, to provide a properly balanced complete diet or, in the case of ruminants, a technically designed supplement to natural foods (e.g. grass or roughages). An indication of the composition of compounds manufactured in Britain for the different classes of livestock in 1993 is given in Table 17.2.

Straights Single feedingstuffs which may or may not have undergone some form of processing before purchase, e.g. wheat, barley, flaked maize, soyabean meal, fish meal.

Protein concentrates Products specially designed for further mixing before feeding, at an inclusion rate of

Table 17.2 Compound feed formulations Great Britain – 1993, June; % of feeds

	Cattle		*Pigs*			*Poultry*		
	Standard dairy	*Summer grazing*	*Pregnant/ lactating sow*	*Rearing*	*Finishing*	*Broiler starter*	*Broiler finisher*	*Layer*
Cereals								
Wheat	4	1	18	36	25	51	55	40
Barley	1	–	8	14	8	–	1	8
Maize	–	–	–		–	2	1	1
Other*	–	1	–		–	–	–	–
Total	5	2	26	50	33	54	56	49
Cereal by-products								
Wheat offals	15	12	24	8	18	–	–	8
Extracted rice bran	3	16	1	–	–	–	–	–
Dried grains	3	6	–	–	–	–	–	–
Maize gluten	18	10	5	1	2	–	–	3
Other†	6	10	5	2	3	1	1	6
Total	46	54	35	11	24	1	1	17
Animal/vegetable proteins								
Soya	4	–	7	22	18	22	16	11
Rape	6	5	4	1	5	6	3	–
Palm kernel	6	11	–	–	–	–	–	–
Fish meals	–	–	1	4	1	4	1	–
Meat and bone	–	–	2	–	1	4	5	4
Other‡	15	11	6	3	4	3	10	6
Total	32	27	22	30	30	39	35	21
Miscellaneous§	16	18	17	8	13	6	7	13

Source: MAFF statistics.
* Oats, rye, sorghum, etc.
† Nutritionally improved straw, biscuit meal, malt culms, oatfeed meal
‡ Sunflower meal, peas, beans, olive residue, cottonseed, citrus pulp.
§ Molasses, sugar beet pulp, oils and fats, minerals and vitamins, manioc, limestone (especially in layers rations).

<5%, with planned proportions of cereals and other feedingstuffs either 'on farm' or by a feed compounder. These usually contain protein-rich ingredients such as fish meal, meat meal and soya which may be supplemented with minerals and vitamins.

Additives Substances added to a compound or a protein concentrate for some specific purpose other than as a direct source of nutrient, e.g. coccidiostats, antibiotics, flavourings.

Supplements Products used at <5% of the total ration and designed to supply specific amounts of trace minerals, vitamins and non-nutrient additives.

In animal feeding a basic consideration is the nutritional value of available feeds. This information has been accumulated over many years and is being continually updated as more precise measures of nutritive values are obtained. Details on composition and nutritive values of the more widely used feedstuffs for the various classes of livestock are given in Table 17.3. More comprehensive information may be obtained from appropriate publications (see Further reading/information sources).

The value of a feedstuff is not solely described by its compositional analysis or nutritive value. Limits to the use of a feed may be imposed by the presence of anti-nutritional or toxic factors or by particular organoleptic or physical properties. In the consideration of individual feeds the possibility of such factors operating will be given attention.

Grazed herbage

Grazed herbage provides a major source of nutrients for ruminant animals. As a feedstuff, however, it has the disadvantage of being variable in nutritive value and in addition its efficiency of utilization is difficult to control. Aspects of this are covered in Chapter 11, the salient points being that whilst a number of factors such as climate, soil fertility and botanical composition of the sward may affect nutritive value, the most important factor is stage of growth (Fig. 17.5).

Early in the growing season grass has a high water, organic acid and protein content and a low content of structural carbohydrates and lignin, making it highly digestible to the animal. As the plant matures and the yield of forage increases, there is an increase in the proportion of stem tissue and a decrease in leaf tissue. This is reflected in increased levels of structural carbohydrates and lignin, decreased protein and thus declining

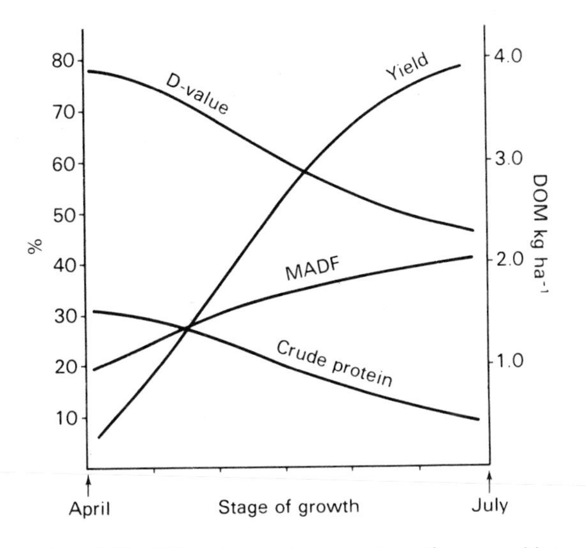

Fig. 17.5 Effect of increasing maturity on the composition and yield of digestible organic matter (DOM) of a typical ryegrass sward.

digestibility and metabolizable energy values. A further consequence of these changes in nutritive value that occur is that the animal's voluntary intake of herbage declines as it matures. Thus, systems of utilization of grassland must take into account those variables together with animal production targets and other management constraints, e.g. amount of herbage required for winter use. A relatively high degree of control of these factors may be exerted on lowland pastures and has led to the development of systems which allow the measurement of the yield and efficiency of utilization of ME from grass (see Chapter 11). In the case of hill pastures the restricted growing season and difficulties in controlling herbage utilization lead to a very cyclical pattern of nutrition in grazing animals. Although during the growing season (May–October) grass in hill pastures will be in the range of 65–55 D, during the winter months D-values may be as low as 40 with little or no protein. Such low levels of protein restrict the rate of herbage breakdown in the rumen and consequently adversely affect the voluntary intake of the animal.

The presence of clover in pasture increases the protein supply to the animal and since the effects of stage of growth on the feeding value of clover are less obvious, due to the lower degree of lignification of stem tissue, the inclusion of clover in a sward results in a reduced rate of decline of nutritive value with advancing maturity. In the grazing of pastures with a high clover component a degree of care must be taken to guard against 'bloat' which may occur due to the stimulation of foam production in the rumen by certain cytoplasmic proteins.

Conserved grass

A feature of all conserved grass is that the nutritive value is dependent on the quality of grass used and the efficiency of the conservation process. Thus, the aim of all conservation processes is to minimize the losses of nutrients from the grass utilized and to produce a feedstuff that fulfills the production objectives of the livestock to which it is to be fed.

The main methods of conserving grass are as hay, silage and artificially dried grass. Most grass in the UK is now conserved as silage. Because of the costs of production artificially dried grass is produced on a relatively small scale for specialized purposes. Details of these processes and factors affecting the efficiency of the processes are given in Chapter 11, Grassland.

As the nutritive value of conserved hays and silages can vary considerably, it is advisable that appropriate samples be routinely analysed so that most prudent use of both conserved forage and supplementary concentrate feeds may be made. Since grass is generally cut at an earlier growth stage for silage than for hay, the nutritive value is generally higher. Thus when compared with the best quality hays good quality silages have approximately 10% more ME and up to 50% more CP. Retention of the vitamins and minerals in grass during hay and silage making processes is variable but in general the mineral content of both materials reflects levels in the crop unless effluent losses or leaching losses are high; silages are good sources of carotene but contain negligible amounts of vitamin D whereas the opposite situation pertains with sun-cured hays.

During haymaking, losses of nutrients through oxidation of water soluble carbohydrates, protein breakdown, leaf shatter and the possibility of leaching losses when the crop is subjected to continuous wetting and redrying combine to increase the proportion of cell wall material and thus reduce the digestibility when compared with the original grass. The calcium and phosphorus content of hay is frequently only half of values for grass. Absence of green colour is a good indication that carotene has been destroyed.

Some reductions in nutrient losses may be achieved through baling at lower dry matters (50–60%) and resorting to some form of barn drying. Failure to achieve dry matters of 85% in store may result in fungal spoilage which represents a hazard both to human health, since it causes farmers' lung, and to animals due to the fungal toxins being a cause of abortion in cattle.

The feeding value of silage whilst primarily dependent on the stage of growth of grass used may be further influenced by the extent of nutrient losses incurred during the ensilage process. The digestibility of silages are, however, usually very similar to those of the original grass except where there have been high losses through continued respiration either during wilting in the field or prior to anaerobic conditions being achieved in the clamp. Restricting the extent of the fermentation in the clamp through the use of either acidifying or inhibitor type additives results in greater retention of water-soluble carbohydrates (WSC). These give the silage a higher energy value through their more efficient fermentation in the rumen compared with the fermentation of the organic acid end products that result from their fermentation in the clamp. Although nutrients such as WSC, proteins and minerals may be lost in the effluent of silages made from unwilted grass, experimental evidence indicates that greater milk production may be obtained from high moisture silages. This is probably the result of an influence on rumen fermentation altering the partition of nutrients in the animal.

Table 17.3 Guide to the chemical composition and nutritive values of some commonly used feedstuffs

Feedstuff name and number	*Dry matter content (%)*	*Chemical composition*					*Nutritive value – energy*		
		CP (*% of feedstuff DM*)	*EE*	*CF*	*NFE*	*Ash*	*Metabolizable energy*[†]		*Digestible energy*
							Ruminants (MJ/kg DM)	*Poultry (MJ/kg A–F)**	*Pigs (MJ/kg A–F)*
(1) *Pasture grass* (average)[†‡]	20.6	13.9	2.2	24.2	52.1	7.6	11.2 (10.5)	–	–
(2) *Conserved grass products*[‡]									
(2.1) Silage (average)	28.9	15.9	3.7	27.1	44.6	8.7	11.0 (8.1)	–	–
(2.2) Hay (average)	85.8	12.2	1.8	30.1	48.1	7.8	9.2 (8.6)	–	–
(2.3) Dried grass (average)	89.0	19.7	3.8	22.4	42.7	11.4	10.7 (9.4)	–	8.90
(2.4) Maize silage	30.4	8.8	3.0	17.2	66.1	4.9	11.3 (9.0)	–	–
(3) *Roots*									
(3.1) Cassava (dehydr. ground)	88.0	2.8	0.3	8.7	84.8	3.4	12.8 (–)	14.43	15.3
(3.2) Fodder beet	18.0	6.8	0.3	5.9	79.1	7.7	11.9 (11.8)	–	2.9
(3.3) Mangels	10.0	9.2	0.8	6.2	76.9	6.9	12.4 (–)	–	1.97
(3.4) Swedes	10.5	10.8	1.7	10.0	71.7	5.8	13.9 (–)	–	1.7
(3.5) Turnips	10.0	12.2	2.2	11.1	66.6	7.8	13.0 (12.9)	–	1.4
(3.6) Sugar beet pulp (dried)	90.0	9.9	0.7	20.3	65.7	3.4	12.8 (12.6)	–	12.9
(4) *Cereal straws*									
(4.1) Barley	86.0	3.7	1.6	48.8	39.2	6.6	6.4 (5.9)	–	–
(4.2) Oat	86.0	3.4	2.2	39.4	49.3	5.7	7.2 (6.7)	–	–
(4.3) Wheat	86.0	2.4	1.5	42.6	47.3	6.2	6.1 (5.7)	–	–
(5) *Green forage crops*									
(5.1) Kale-marrowstem	14.0	15.7	3.6	17.9	49.3	13.6	11.8 (11.1)	–	–
(5.2) Rape	14.0	20.0	5.7	25.0	40.0	9.3	9.5 (–)	–	–

(6) *Cereal grains and cereal by-products*									
(6.1) Barley	86.0	10.8	1.7	5.3	79.5	2.6	13.3 (12.7)	11.13	13.2
(6.2) Wheat	86.0	12.4	1.9	2.6	81.0	2.1	13.7 (13.1)	12.18	14.3
(6.3) Oats	86.0	10.9	4.9	12.1	68.8	3.3	12.1 (10.7)	11.05	11.4
(6.4) Maize	86.0	9.8	4.2	2.4	82.3	1.3	13.8 (12.4)	13.22	14.9
(6.5) Maize (flaked)	90.0	11.0	4.9	1.7	81.4	1.0	15.0 (—)	—	15.3
(6.6) Sorghum	86.0	10.8	4.3	2.1	80.1	2.7	13.4 (—)	12.98	14.3
(6.7) Brewer's grains (dried)	90.0	20.4	7.1	16.9	51.2	4.3	11.5 (9.3)	7.91	7.6
(6.8) Distillers' grains (dried)	90.0	30.1	12.6	11.0	44.3	2.0	12.4 (10.5)	9.37	—
(6.9) Bran	88.0	17.0	4.5	11.4	60.3	6.7	10.1 (—)	8.45	8.9
(6.10) Fine wheat middlings	88.0	17.6	4.1	8.6	65.0	4.7	11.9 (10.4)	11.80	11.8
(6.11) Rice bran	90.0	14.5	2.0	15.0	55.0	13.5	7.1 (6.9)	—	8.0
(6.12) Maize gluten feed	90.0	21.0	3.8	3.9	68.5	2.8	11.8 (10.3)	—	11.8
(7) *Other high energy feeds*									
(7.1) Fat and oil (FGAF)	98.0	—	100	—	—	—	33.0 (0)	35.5	33.0
(7.2) Molasses	75.0	4.5	0	0	84.1	6.9	13.1 (13.0)	7.91	—
(8) *Oilseed by-products*									
(8.1) Ex soyabean	90.0	50.3	1.7	5.8	36.0	6.2	13.3 (12.7)	10.67	13.6
(8.2) Ex rape seed	90.0	41.3	3.4	10.4	36.6	8.2	13.3 (12.1)	7.36	12.2
(8.3) Ex groundnut (decort.)	90.0	50.4	2.1	27.3	31.6	4.7	13.7 (11.3)	11.4	13.7
(8.4) Ex sunflower	90.0	42.3	1.1	18.1	31.2	7.2	9.6 (8.8)	7.5	12.6
(9) *Legume seeds*									
(9.1) Field beans	86.0	26.5	1.5	9.0	59.1	4.0	13.1 (12.7)	10.00	12.2
(9.2) Peas	86.9	26.1	1.4	6.0	66.5	3.2	13.5 (13.0)	10.86	14.34
(10) *Animal products*									
(10.1) Fish meal (white)	90.0	70.1	4.0	0	1.8	24.1	14.2 (11.0)	11.46	11.8
(10.2) Herring meal	90.0	76.2	9.1	0	4.4	10.2	— (—)	13.35	13.5
(10.3) Meat and bone meal (medium protein)	90.0	52.7	4.4	0	1.7	41.2	— (—)	—	9.6
(10.4) Dried skim milk	95.0	37.2	1.1	0	53.2	8.5	14.1 (—)	12.26	15.6

* A–F = as fed.
† Fermentable metabolizable energy values of ruminant feeds in brackets.
‡ Values highly variable and samples require to be routinely analysed.

Table 17.3 *Continued*

* Feedstuff number	Nutritive value, protein							Amino acid composition						Mineral composition	
	Parameters for MP system, ruminants				Digestibility coefficients, pigs		Digestible crude protein Poultry	Glycine + serine	Lysine	Methionine + cysteine	Isoleucine	Threonine	Tryptophan	Calcium	Phorphorus
	a	*b*	*c*	ADIP	Faecal	Ileal									
	(decimal proportions)			g/kg DM			(% A–F)	(% of total feed A–F)						(% of total feed DM)	
(1)	0.34	0.57	0.09	7.5	—	—	—	—	—	—	—	—	—	0.50	0.29
(2)															
(2.1)	0.63	0.26	0.14	8.1	—	—	—	—	—	—	—	—	—	0.62	0.30
(2.2)	0.22	0.60	0.08	7.5	—	—	—	—	—	—	—	—	—	0.55	0.28
(2.3)	0.37	0.63	0.04	14.4	0.36	0.30	—	—	0.75	0.44	0.68	0.68	—	0.92	0.33
(2.4)	0.66	0.19	0.20	—	—	—	—	—	—	—	—	—	—	0.39	0.18
(3)															
(3.1)	—	—	—	—	0.40	0.25	—	—	0.09	0.03	0.05	0.05	0.01	—	0.03
(3.2)	0.25	0.65	0.44	—	—	—	—	—	0.035	0.02	0.024	0.036	0.02	0.22	0.22
(3.3)	—	—	—	—	—	—	—	—	—	—	—	—	—	0.22	0.22
(3.4)	—	—	—	—	—	—	—	—	—	—	—	—	—	0.42	0.33
(3.5)	0.25	0.65	0.34	—	—	—	—	—	—	—	—	—	—	0.44	0.33
(3.6)	0.24	0.70	0.05	—	0.50	0.45	—	—	0.69	—	0.37	0.37	0.09	0.96	0.09
(4)															
(4.1)	0.30	0.50	0.12	6.3	—	—	—	—	—	—	—	—	—	0.39	0.08
(4.2)	0.30	0.51	0.11	3.8	—	—	—	—	—	—	—	—	—	0.24	0.09
(4.3)	0.30	0.50	0.12	5.0	—	—	—	—	—	—	—	—	—	—	—
(5)															
(5.1)	0.25	0.65	0.27	—	—	—	—	—	—	—	—	—	—	2.14	0.31
(5.2)	—	—	—	—	—	—	—	—	—	—	—	—	—	0.93	0.42

(6)															
(6.1)	0.25	0.70	0.35	2.5	0.76	0.72	9.0	0.81	0.35	0.34	0.35	0.30	0.15	0.05	0.38
(6.2)	0.39	0.57	0.13	3.8	0.82	0.75	8.8	1.00	0.31	0.39	0.40	0.31	0.12	0.03	0.40
(6.3)	0.72	0.23	0.40	—	0.75	0.70	8.5	1.00	0.38	0.34	0.36	0.34	0.12	0.09	0.37
(6.4)	0.26	0.69	0.01	—	0.78	0.73	6.7	0.78	0.26	0.29	0.32	0.32	0.08	0.02	0.27
(6.5)	—	—	—	—	—	—	—	—	0.29	0.33	0.36	0.36	0.10	—	0.29
(6.6)	—	—	—	—	0.75	0.70	8.4	0.65	0.22	0.31	0.38	0.30	0.08	0.03	0.68
(6.7)	0.19	0.67	0.07	—	0.70	0.60	14.1	1.80	0.52	0.55	0.94	0.55	0.21	0.32	0.78
(6.8)	0.84	0.12	0.11	81.3	0.60	0.55	—	—	—	—	—	—	—	0.31	0.33
(6.9)	—	—	—	—	0.60	0.50	11.1	1.50	0.60	0.50	0.55	0.45	0.25	0.16	0.84
(6.10)	0.34	0.57	0.11	2.5	0.70	0.60	15.0	1.43	0.64	0.40	0.49	0.49	0.22	0.13	0.91
(6.11)	0.04	0.78	0.06	—	0.50	0.40	—	—	0.50	0.34	0.45	0.34	0.44	0.19	1.10
(6.12)	0.38	0.61	0.12	5.0	0.65	0.55	22.3	5.51	0.55	0.80	0.55	0.70	0.13	0.28	0.80
(7)															
(7.1)	—	—	—	—	—	—	—	—	—	—	—	—	—	—	—
(7.2)	1.0	0.00	—	—	0.30	0.25	—	—	—	—	—	—	—	—	—
(8)															
(8.1)	0.08	0.92	0.08	14.4	0.87	0.80	42.8	3.94	2.94	1.35	2.52	1.81	0.60	0.23	1.02
(8.2)	0.32	0.61	0.16	22.5	0.70	0.60	26.1	3.32	2.04	1.26	1.36	1.58	0.44	0.59	0.94
(8.3)	0.22	0.77	0.09	—	—	—	40.8	—	1.67	1.25	1.67	1.36	0.52	0.12	0.51
(8.4)	0.24	0.71	0.35	16.3	0.75	0.65	24.8	2.91	1.60	1.49	1.71	1.56	0.50	0.41	1.33
(9)															
(9.1)	0.33	0.72	0.09	3.1	0.75	0.68	21.1	2.08	1.50	0.36	0.93	0.84	0.21	0.19	0.67
(9.2)	0.56	0.44	0.09	—	0.75	0.68	20.6	1.05	1.6	0.59	0.97	0.87	0.20	0.09	0.83
(10)															
(10.1)	0.30	0.63	0.02	—	0.92	0.80	59.0	7.12	4.40	1.26	2.46	2.46	0.61	7.93	4.37
(10.2)	—	—	—	—	0.92	0.80	66.6	—	5.62	1.62	3.10	2.91	0.58	3.26	2.22
(10.3)	—	—	—	—	0.65	0.58	41.2	8.00	2.61	1.48	1.26	1.55	0.29	11.44	6.00
(10.4)	—	—	—	—	0.95	0.90	27.5	2.82	2.63	1.26	2.42	1.58	0.47	1.17	0.96

* For Feedstuff Names refer to ***Feedstuff name and number*** column in **Table 17.3** on pages 404-405.

A further factor that may influence the feeding value of silages is the extent of nutrient losses that are incurred due to the exposure of silage to air prior to its consumption. Oxidation of any residual WSC and organic acids by micro-organisms is especially likely and the management of the clamp face and the method of feeding should aim to restrict these losses.

The ensilage of whole crop cereals, particularly forage maize, gives a product of higher energy value than grass silage, but lower levels of CP. Such silages may be usefully supplemented with NPN compounds.

Whilst the nutritive properties of conserved forages may be defined to an extent by chemical analysis, one of the main problems in feeding roughages when they are a major component of the diet is to predict how much animals will consume. This is more of a problem with silages where factors such as digestibility, dry matter content, chop length and the fermentation pattern are operating. Intakes tend to be enhanced by high dry matter, high digestibility and reduced chop length. Whilst intakes of silages with high pH and high ammonia levels resulting from poor fermentation are low, at the other end of the scale high levels of acidity, especially in maize silages, may also restrict intake. At present there is insufficient information to predict accurately roughage intake by different classes of livestock.

Typical artificially dried grass is a product which, by virtue of its relatively low fibre content, occupies an intermediate position between the bulky feeds and the concentrates. Due to the rapid removal of water at high temperatures dried grass has a similar feeding value to the original grass when expressed on a dry matter basis. Since artificial drying is an expensive process, only young leafy grass of high D-value and high crude protein content is used. An effect of high temperature drying is to decrease the degradability of protein. Intakes of dried grass do not present a problem since not only is it highly digestible but also its cost dictates that its use is limited to a supplemental role.

Cereal straw

The gross energy content of cereal straw is similar to that of grains but because of its low digestibility even to ruminants much of this energy is lost in the faeces and makes no contribution to the animal. A further constraint to the use of straw is that a slow rate of digestion and therefore rate of passage through the gut limits the intake.

Since the levels and digestibility of the protein in straw are negligible and since there are very few vitamins and minerals present, the only nutrient straw can supply is energy. The availability of this energy is limited by the very high levels of lignin protecting the fibrous carbohydrates from microbial breakdown. In addition when straw is a major component of the diet the low levels of nitrogen available to the rumen micro-organisms may be a further factor limiting its digestibility. It has, however, been observed that considerable differences in nutritive values exist not only between the different cereal types, oat straw being superior to barley, which in turn is of higher quality than wheat, but also between varieties. Typically, spring varieties give higher nutritive values than winter varieties but even comparisons of spring barley varieties have shown that digestibility values can vary from as low as 40 for varieties such as Delta and Golden Promise to nearly 60 for varieties such as Doublet and Corgi. Likewise for wheat straws digestibilities ranging from 30 to 50 have been found for different varieties. It should be noted that growing conditions may also influence nutritive value although the influences of fertilizer treatment, season and daylight are largely unknown.

In recent years a great deal of research has been carried out with the objective of improving the nutritional value of straw. Chemical treatments with alkali in the form of sodium and ammonium hydroxide lead to improvements in nutritive value and dry matter intake. The extent of this improvement is dependent on the quality of the original straw, greater improvements are obtained with poor quality materials, and the method of treatment chosen. A number of 'on-farm' methods of treatment have been devised ranging from the low-cost stack methods to high capital investment oven-treatment methods. In the production of nutritionally improved straw (NIS) by feed compounders the straw is additionally ground and pelleted. As with the feeding of untreated straw, however, it is important that the remainder of the ration supplements the low levels of protein and nitrogen present. The use of ammonium hydroxide as the alkali helps overcome this problem. Due to their increased digestibility the intake by animals of treated straw is much higher than that of untreated straw allowing much wider inclusion in the diets of ruminant animals.

Both treated and untreated straw may be used to effect in devising feeding strategies. Thus, in the feeding of dairy cows it may be used in the case of dry cows to save silage; in the case of freshly calved cows alkali-treated straw may be used as a chemical 'buffer' to increase appetite; as a buffer feed for cows at grass as a means of maintaining milk fat content; or as a means of ensuring adequate rumination so preventing displaced abomasum in animals being fed very high density diets. Traditionally straw has formed an important part of the ration for suckler cows and 'store' animals but for situations where higher levels of performance are required from growing beef animals the level of inclusion of straw in the ration and therefore the potential benefit from treatment are severely limited. Some farmers have used both treated and untreated straw to effect as an alternative roughage source in the feeding of ewes.

The economic advantage from using treated straw is difficult to assess but is largely dependent on the cost of the original straw and the price of alternative energy sources such as barley.

Roots and tubers

The root crops, turnips, swedes, mangolds and fodder and sugar beet, are characterized by having a high water content (75–90%) and low crude fibre (5–10% of DM). The main component of the dry matter is sugar. Thus roots are highly digestible and the principal nutrient they supply is energy.

The crude protein content is in the range of 50–90 g/kg DM, a fairly high proportion of this being in the form of NPN. As suppliers of vitamins and minerals (potassium excepted) roots are poor and rations containing a high proportion of roots require appropriate supplementation.

Whilst roots can be fed and are palatable to pigs their high water content and bulkiness limit levels of incorporation for young growing pigs although diets for sows may contain up to 50% of their DM as roots. Roots are more commonly used in the diets of ruminant animals either being fed *in situ* to sheep or carted and fed to housed animals. Although roots are capable of producing very high yields of energy per hectare, the costs and difficulty of harvesting represent a considerable problem as far as the feeding of housed animals is concerned. A further problem is that where large quantities of roots are fed to housed animals, recognition should be made of the large quantity of urine that is excreted.

Sugar beet is not primarily grown for animal consumption but sugar beet pulp, the residue left after sugar has been extracted at the factory, is widely used in ruminant diets where in energy terms it can substitute for barley. The higher fibre content compared with barley makes it a useful component of the concentrate diet when cows go out to grass in spring but account must be taken of its low protein value when substituted for cereals. This product is sold either directly to farmers as fresh 'pressed pulp' with about 28% DM and is therefore a perishable commodity or as dried pulp which may be readily stored for many months. Dried pulp is available in shredded form, or as pellets or nuts. Its feeding value is frequently modified prior to sale by the addition of molasses, magnesium, minerals and vitamins.

The feedstuff variously termed tapioca, manioc or cassava is a raw material which acts as one of the main cereal substitutes. It is extracted from the tropical root crop cassava and became widely used by compounders in the 1970s and early 1980s when it often made up around 10–20% of pig, ruminant and pig compound feeds. More recently the reduction in price advantage has led to lower rates of inclusion. Manioc is principally comprised of starch with low levels of protein (2.5%) and oil (0.5%). The energy content is similar to that of wheat and barley but when used as a replacer for these cereals requires greater protein supplementation. Raw manioc contains cyanogenic glucosides which can be enzymically hydrolysed to hydrocyanic acid. Processing must eliminate these glucosides since the hydrocyanic acid has the effect of depleting the body reserves of the essential sulphur-containing amino acids methionine and cystine. A further quality control characteristic that must be assessed is the ash content since on occasion it may acquire excessive levels of silica from the soils on which it is grown. This reduces the energy value. As ground manioc is very dusty it is normally included in pelleted feeds rather than meals.

Green forage crops

The main green forage crops fed are kale, forage rape and stubble turnips. These brassica crops can provide a useful source of succulent feed during autumn and winter. Composition varies between the different brassica species and varieties, being particularly affected by the leafiness. As a generalization they compare favourably with the feeding value of good quality silage in energy terms and are rich in protein, minerals and vitamins.

These crops are generally fed *in situ*, kale principally to dairy cows, and rape and stubble turnips mainly to fattening lambs. For the latter they may be fed without cereal supplementation, but cereals can usefully complement the high protein supply of these crops to give increased liveweight gain.

In feeding these brassica crops account has to be taken of certain agents that may prove harmful when brassicas are fed in excess. The presence of goitrogens which affect iodine utilization (see Chapter 15) and of the chemical S-methyl-cysteine sulphoxide (SMCO) which can produce haemolysis and consequently anaemia mean that their use in diets should be limited. The latter problem is minimized if the crop is grazed before it is too mature and before secondary leaf growth occurs. For dairy cows it is recommended that intake of these brassica forage crops be limited to a maximum of 30% of the total dry matter intake and particular attention paid to the health of animals. Additionally, it should be remembered that the high levels of calcium present in brassicas may upset the Ca:P ratio in the diet, necessitating phosphorus supplementation.

A further problem is that rape can cause a skin condition termed yellows is in white-faced lambs. In this condition the face and ears become sensitive to light.

Cereal grain and cereal grain by-products

Cereal grains and their by-products are the main ingredient of rations for pigs and poultry and provide the major source of energy in compound feeds fed to ruminant animals. The main cereal grains used in the UK are barley, wheat and oats which are mainly homegrown along with imported maize and sorghum.

All cereal grains are rich in carbohydrate which is mainly in the form of starch. Starch accounts for about 70% of the seed, varying between grain types. In raw material terms the crude protein of cereals is rather low and is the most variable item ranging from 6 to 14%. This crude protein is of relatively poor quality being low in the essential amino acids lysine, methionine and tryptophan and containing 10–15% of the nitrogenous compounds in the seed as NPN.

The crude fibre levels in cereals vary with species, being lowest in maize and highest in oats. Its level has a direct bearing on the digestibility and therefore energy value of the whole grain. The oil content also varies with species; oats and maize having higher levels than other cereals. Cereal oils are unsaturated, the main acids being linoleic and oleic, which leads them to become rancid fairly quickly after processing.

Cereals are relatively good sources of phosphorus but much of this is present as phytates which adversely affect its availability to livestock. In general, cereals are deficient in calcium and contain varying levels of other trace minerals. Of the vitamins most cereals are good sources of vitamin E, although under moist storage conditions much of this may be destroyed. Except for yellow maize, cereals are low in carotene and vitamin A and are deficient in vitamin D and most of the B vitamins, thiamine excepted.

The feeding value of cereal grains is relatively constant during prolonged storage but occasionally poor harvest and storage conditions can lead to the presence of fungal toxins such as aflatoxin, zearalenone, vomitoxin and ochratoxin. *Aspergillus*, *Penicillium* and *Fusarium* moulds are particularly responsible for the production of these toxins. Affected grains are usually thin and shrivelled with a pinkish colour. In the case of vomitoxin maximum concentrations in feed should be <1 ppm for pigs and dairy cows and <5 ppm for poultry and other cattle. Cocktails of mycotoxin in contaminated feed are, however, potent at lower levels than are individual mycotoxins.

Maize

Of the commonly used cereals yellow maize has the highest metabolizable energy content. With a metabolizable energy value of 15 MJ/kg DM it is especially useful for broilers where yellow carcass pigmentation resulting from its high carotene content is desirable. For laying hens it not only supplies xanthophylls that enhance the colour of egg yolks but it also supplies linoleic acid which is a necessary dietary component for birds to produce eggs of a satisfactory size. The high proportion of unsaturated fatty acids means that its inclusion in pig diets must be restricted to 35% of the diet otherwise there is the likelihood of soft fat depots high in polyunsaturated fatty acids. The yellow pigmentation in body fat resulting from the feeding of yellow maize is considered undesirable in pig and ruminant carcasses. White maize has all the attributes of yellow maize without the problems of carcass fat pigmentation.

Wheat

Although not appropriate for the production of yellow broilers, wheat is suitable for all other feeding situations. It frequently forms up to 70% of diets for poultry, can be included at high levels for pigs provided that care is taken to control the fineness of grinding, whilst for ruminants inclusion levels of up to 30–40% in compound feeds may be used provided the remainder of the diet contains sufficient fibre. The gluten content of wheat can have a beneficial effect on the quality of heat processed extruded cubes or pellets.

Barley

Traditionally barley is considered to be particularly suitable for pig feeding having an appropriate amount of both fibre and also of oil which is associated with the production of saturated carcass fat. The higher level of fibre than is found in wheat or maize means that the upper inclusion rate for poultry must be limited to about 30% in many cases. High levels of barley in broiler diets have been associated with wet droppings.

Oats

Oats are normally only used for ruminants and horses. Their low energy and high fibre means that they are seldom used for pigs or poultry. In recent years varieties of 'naked' oats have received some attention as being especially suitable for young piglets, since without the husk the grain is of high energy value due to its comparatively high oil content and the protein quality is somewhat better than in other cereals.

Sorghum

From a nutritional viewpoint the grain sorghums (milo, kaffir and hybrids) resemble wheat. They have a low fibre content and in comparison with maize contain more protein and less oil. Certain varieties have a high content of phenolic compounds including tannins which not only influence palatability but also lower protein digestibility and reduce energy values for pigs and poultry.

With respect to other cereals rye can only be tolerated at low inclusion rates because of its content of B-glucans and phenolic compounds which reduce performance and result in wet droppings in poultry and are toxic in quantity. Rice and millet may be used as an alternative to wheat once the husk has been removed. Triticale, a cross between durum wheat and rye, is similar to wheat but has a higher protein content.

Cereal preparation and processing

The aim in all cereal processing methods is primarily to increase the efficiency of utilization of the nutrients. Such improvement may result simply from an increased nutrient availability but other factors such as changes in palatability and nutrient density may also contribute towards improvements in performance.

Most grain processing methods have as their main objective improvement in the availability of the starch present. Some of the techniques involve solely physical change, others chemical and some a combination of both physical and chemical; in addition, some processes are carried out 'wet', others 'dry'; some involving heat treatment, others under cold conditions. Mechanical alterations of the grain are the most widely used and involve physical disruption of the grain such that the starch is made more available. Grinding, rolling and crushing are the most common processes employed. It is worth noting, however, that such treatments have little or no effect on the nutritive value of barley or wheat offered to sheep.

Processing procedures which bring about chemical changes through the gelatinization of the starch include such techniques as steam flaking, micronizing, popping and pelleting. The 'flaking' process has long been applied to produce 'flaked maize' in a process involving steaming of the grain either at atmospheric pressure or in a pressure chamber followed by rolling. Micronizing and popping are both 'dry heat' processes. In the former grain is passed under gas-fired ceramic tiles, the radiant heat from which produces rapid heating within the grain causing it to soften and swell. It is then crushed in a roller mill which prevents reversal of the gelatinization process. Popping is the exploding of grain through the rapid application of dry heat. Popped grain is usually rolled or ground prior to feeding. Popping has been shown to be particularly effective in processing sorghum grain.

When feeding concentrate mixtures to certain livestock classes it is common practice to produce it in the form of pellets. From a management point of view it reduces wastage, prevents selection of ingredients and makes for easier storage and handling. Additionally, there are in some instances nutritional advantages to pelleting, this being due in the main to improvements in the available energy content.

Recent studies involving the treatment of grain for ruminant consumption with alkali appear to indicate potential as a means of increasing the availability of energy.

Cereal by-products

When cereal grains are processed for human consumption a number of by-products are produced which are used extensively as livestock feeds. The main sources of by-products are the flour milling industry, the brewing and distilling industries and, imported from the USA, by-products from maize processing.

Wheat by-products
When wheat is milled to produce flour for human consumption about 28% of the grain becomes available as by-products of the process. Wheat millfeeds are usually classified and named on the basis of decreasing fibre as bran, coarse middlings and fine middlings or shorts. These arise from the removal of the outer layers of the kernel. Wheat bran is comprised of the coarser fraction (about 50% of wheatfeed) and is usually fed to ruminant species. The finer fractions are lower in fibre and widely used in formulating diets for pigs and poultry. Since the protein in wheat grain is concentrated mainly in the outer layers, crude protein levels in these residues are higher than the whole grain. Likewise wheatfeeds are relatively good sources of most of the water-soluble vitamins, except niacin.

Brewing/distillery by-products
The main by-products arising from the brewing and distilling industries are brewers' grains and distillers' grains (draff), which are essentially the part of the grain that remains after the starch has been removed in the malting and mashing processes. They may be purchased without being dried and fed either fresh or after ensiling or alternatively may be purchased after being dried. Since both feeds have a relatively high fibre content their use is limited to ruminant rations. As well as the fibre fraction being more concentrated by the loss of the starch, so also are the crude protein and oil fractions. In the case of distillers' grains the relatively high lipid content (80–90 g/kg DM) is known to interfere with the cellulolytic action of the rumen microflora but this can be overcome to an extent by the addition of suitable amounts of calcium which results in the formation of insoluble calcium soaps.

Other by-products of these industries, namely malt culms, dried brewers' yeast and dried distillers' solubles, may also be fed to livestock.

Rice bran
In recent years rice bran has become increasingly important in animal feeds and up to 10% may be included in compound feeds for ruminants. It is also used in sow diets. Rice bran consists mainly of the bran and outer part of the grain and is obtained as a by-product when rice is 'polished' for human consumption. It tends to be a variable product but good quality rice bran contains very little of the less nutritious hull fragments and is rich in oil (14–15%) making it susceptible to rancidity. This oil is usually extracted either by solvent, leaving less than 1% oil in the product, or by expeller press in which case up to 10% oil may remain. Solvent extracted brans have ME values of the order 6.5–7.5 MJ/kg DM whereas those with higher oil content can have ME values of 11 MJ/kg DM. Such differences emphasize the need for users to ascertain the composition of rice bran before purchase or use in a formulation.

Maize by-product feeds
The wet milling of maize is used in the USA for the production of starch, sugar and syrup. A number of by-products result from this process including bran, germ meal and gluten. The latter is the most important, and large amounts of maize gluten are exported from the USA into the EU.

Maize gluten is a good source of energy and protein for both dairy and beef rations. Energy levels are similar to those of barley with crude protein of the order of 20%. Up to 20% of the total DM may be included in ruminant rations and it may also be included on a more limited scale in pig and poultry diets. The quality of maize gluten from different sources can very considerably such that it is essential to monitor the quality closely.

Other high energy feeds

Whilst feed grains are the main energy-supplying concentrate feeds, other feeds are routinely used to supply energy to livestock.

Fats and oils

These are of particular value in increasing the energy density of the ration since its energy value is more than twice that of digestible carbohydrate. Fats may also improve rations by reducing dustiness and increasing palatability. The inclusion of fats in animal diets has found greatest application in milk replacers for suckling animals and in the diets of pigs and poultry where energy density is a factor controlling total energy intake. In recent years fats have also been included in dairy cow and sow diets. Energy costs can represent up to 75% of total formulation costs for animal feeds, and fat is frequently the cheapest source.

A major source of fat is feed grade animal fat (FGAF), often referred to erroneously as tallow. This is the fat rendered from meat and bone meal and in fatty acid compositional terms is typical of animal fats being high in palmitic, stearic and oleic acids. In recent years FGAF has been increasingly blended with vegetable oil by-products to produce blended fats that have fatty acid profiles appropriate to the class of livestock to which they are fed. The vegetable oil by-products come mainly from edible oil refining and consist of a mixture of neutral oil and free fatty acid. A further source of vegetable oil is oil recovered from such processes as potato crisp manufature. The blending of oils to produce an appropriate fatty acid profile is of relevance in that this parameter can influence the digestion, absorption and consequently metabolizable energy value. Synergistic effects between fats frequently exist as evidenced by mixtures of tallow and soyabean oil having higher metabolizable energy values than either of the individual sources. In nutritional terms the most important aspect is the ratio of saturated to unsaturated fat and the specific requirements for polyunsaturated fatty acids such as linoleic acid for pigs and poultry.

Current opinion is that hard fats based on palm or FGAF are most suited to ruminants, and high levels of free fatty acids which are unacceptable to monogastrics are thought to be beneficial to energy values. The use of 'protected fat' systems (Chapter 15), either through the mixing of fat with a carrier such as vermiculite or with formaldehyde-treated protein sources, allows the incorporation of extra fat into ruminant diets without the adverse effects on cellulose digestion normally associated with supplementary raw fat. It also presents an opportunity to manipulate milk fat output, and through the incorporation of polyunsaturated fatty acids the fatty acid composition of milk fat and carcass fat. These protected fats are usually free-flowing powders which are easily incorporated into the diet without special equipment.

For poultry diets a typical profile for supplementary fat is 35% saturated fatty acids, 20% linoleic acid and less than 50% of the total fatty acids as free fatty acids. Relatively soft fats are also recommended for pig rations provided it is within the limitation of producing soft carcass fat. It is frequently observed that the utilization of the non-fat components of the non-ruminant diet are improved by adding fat to the ration. This is probably due to an effect of additional fat on rate of passage of food through the gut.

A further source of dietary fat that has received attention in recent years is the inclusion of full fat oilseeds in the diet. Most attention has been given to full fat soyabeans. When used in the diet of non-ruminants there is the need for severe physical processing in order to make the oil fully available. Extrusion techniques and to a lesser extent micronization are most effective. Appropriate processing of whole oilseeds also provides a means of supplying additional fat to dairy cows within a 'protected fat' system. Extensive heat treatment, considered to be overheating in the preparation of whole oilseeds for monograstrics, is recommended in the processing of oilseeds for dairy cows.

Molasses

This is a by-product of sugar refining. It is very low in protein, the main constituent being sugar, giving it an energy value of about 85% of the value of cereal grain. It is mainly used in beef and dairy rations and also sow diets where it may be of particular value as a pellet binder or as a component of feeds which include NPN sources. The limiting factor to its inclusion, usually 5–10%, is the difficulty of mixing it into concentrated feeds. This requires specialized equipment making it very difficult to use 'on-farm'.

Citrus pulp

Dried citrus pulp is prepared from the residue resulting from the manufacture of citrus juices and consists of a mixture of pulp, peel, seeds and cull fruits. Since a variety of fruits go into it, it has variable composition and requires routine analysis before use. It is similar in feed value to dried sugar beet pulp with a lower protein content (5–8%). It is mainly used for dairy and beef rations and quantities up to 50% of the total DM in the diet may be used if desired. At such levels it may produce taint in milk. This product is not very palatable to monogastrics although up to 10% may be included in sow diets.

Oilseed residues

Several oil-bearing seeds are grown to produce vegetable oils for human consumption and industrial processes. The residues that remain after the extraction of the oil are rich in protein and are of great value as livestock feeds.

Among such high protein feeds are soyabean meal, rape seed meal, sunflower seed meal, cottonseed meal, groundnut meal, palm kernel meal and sesame meal.

Oil is extracted from these seeds by hydraulic pressure (expelled) or solvent (extracted). Most oilseeds are now subjected to the latter treatment, the efficiency of oil extraction being much higher, leaving little oil (<1%) in the residue whereas expeller methods leave up to 6% of the oil in the residue. The amount of oil left in the residue affects the energy value of the feed.

Fibre levels will also affect the energy value of the feed and in some cases, e.g. groundnuts, fibre levels will vary according to whether the seeds have been decorticated (removal of husk) prior to processing.

Of those oilseeds used widely, soya has the best quality protein followed closely by sunflower and rapeseed. Soya is slightly deficient in methionone, whereas sunflower and rapeseed, whilst being slightly higher in methionine, are deficient in lysine. Groundnut meal, although high in crude protein content, has very low levels of methionine and is low in both lysine and tryptophan.

In addition to nutritive values a number of anti-nutritional factors have to be taken into account when feeding certain oilseed residues.

In rapeseed meal the presence of glucosinolates, which under the action of myrosinase produce compounds (isothiocyanates and oxazolidinethione) that are goitrogenic (see Chapter 15), can limit its inclusion in animal feeds. In recent years newer varieties with lower levels of glucosides have been produced. These newer varieties contain less than 1% w/w of glucosinolate and apart from that grown for industrial purposes now account for almost all the rapeseed grown. Further problems with rapeseed meal can occur when it is fed in excess of 5% to laying poultry in that it can increase the incidence of haemorrhagic fatty liver and through the presence of sinapine promotes the accumulation of trimethylamine which can cause a fishy taint in eggs. In some instances the presence of anti-nutritional factors can be nullified by heat during their processing. This is the case with the trypsin and urease inhibitors present in raw soya and with the yellow pigment, gossypol, which is present in cottonseed meal. Such heat processing reduces the solubility of the protein and in addition reduces the quality of the protein through involving amino acids such as lysine in browning reactions with carbohydrates. A particular problem associated with groundnut meals is that they are prone to infestation by the mould *Aspergillus flavus* which results in their subsequent contamination with the mycotoxin called aflatoxin. Current regulations prohibit the importation of groundnut meal containing >0.05 ppm of aflatoxin B_1.

Legume seeds

The seeds of legumes such as peas, beans and lupins are useful sources of both energy and protein although it is for the latter nutrient that they are primarily included in rations.

The protein in field beans is particularly rich in lysine but the low levels of methionine restrict the extent of its inclusion in the diets of pigs and poultry. Varietal differences in the digestibility of the protein present exist and are related to the amounts of tannins, vicin and convicin present. Similar varietal differences in the digestibility of peas exist. As a consequence of their only moderate protein content the extent of their use tends to be restricted, especially in the diets of young growing stock.

Lupins have a significantly higher protein content than peas and beans but they tend to be deficient in lysine, the sulphur containing amino acids and tryptophan. For pigs and poultry the poorer protein quality restricts the amount that may be included as an alternative to soyabean meal to about 10% of the diet. Further limiting factors to their use at present are the levels of the alkaloids, lupenine and sparteine, which by imparting a bitter taste depress feed intake, and also the presence of α-galactoside sugars, which are fermented in the rear gut of monogastrics causing flatulence.

Animal protein supplements

Feedstuffs of animal origin other than rendered fats are principally included in diets as supplemental sources of high quality protein which can remedy deficiencies in the essential amino acid composition of the rest of the diet. Although this role has long been recognized in the feeding of pigs and poultry, current thinking on the protein nutrition of ruminants indicates that high quality protein, which is resistant to degradation in the rumen, may be of value.

Protein supplements of animal origin are derived from slaughterhouse wastes as meat meal, meat and bone meal, blood meal and feather meal; from milk and processed milk chiefly in the form of skimmed milk powders and wheys; and from fish and processed fish as fish meals. Although the bulk of these by-products are simply available as dried ground meals of the raw materials a recent innovation is the production of protein hydrolysate of these waste products. These are produced through the action of proteolytic enzymes and have the advantage of being virtually odour free and readily produced in a variety of formats to meet market demand, e.g. in extruded admixtures with grain or grain by-products. These protein hydrolysates are especially useful for inclusion in poultry and pig starter rations and with their high degree of solubility may also be included as an ingredient of low antigenicity in milk replacer formulae for calves. A further attraction of protein hydrolysis is that it allows less biodegradable materials such as poultry feathers to become useful protein sources.

A problem with the major animal meals is that their composition tends to be variable according to the composition of the raw materials used. This is especially true of their oil and ash content which is dependent in the case of slaughterhouse waste on the efficiency of rendering and on the proportion of bone present, and in the case of fish meals, on the species of fish used. A further variable factor is that during processing overheating of these products can markedly reduce their digestibility to monogastrics and also reduce the available lysine content. Variations in processing temperatures can also cause variability in the RDP:UDP ratios in these products.

Meat and bone meal

It is available in two types: as a low fat meal (~6% oil, 48–50% protein) or as a high fat meal (12–14% oil, 45% protein). The cost of the latter is often attractive because of its high energy value but the high oil content can present handling difficulties. The ash content within meat and bone meals constitutes a valuable source of available phosphorus as well as calcium.

Meat meal

This is produced from slaughterhouse waste from which all bone, hoof and horn have been excluded and consequently has a higher protein level (50–55% CP). As with meat and bone meal it is relatively low in methionine and tryptophan when compared with fish meals.

An important aspect of the production of meals from slaughterhouse wastes is that adequate sterilization must occur. The bovine spongiform encephalitis (BSE) epidemic in cattle has led to regulations banning the inclusion of both meat and meat and bone meals in ruminant diets.

Fish meals

These are excellent sources of protein being especially high in lysine and methionine and also of minerals. In the UK over 60% of fish meals go into poultry rations, 20% into pigs with the remainder going into ruminant and fish rations. The composition of fish meals is dependent on the type of fish used, demersal species, such as cod, producing meals with a low oil content (2–6%) whilst pelagic species, such as herring, produce meals with fairly high oil levels (7–13%). Such variation in oil contents influences the metabolizable energy values but the use of high oil content meals is more limited because of the possibility of their imparting fishy taints or 'soft' fat in the product. As a component of the diets of early weaned pigs fish meal has certain advantages over a protein source such as soyabean meal in that it appears to have a relatively lower antigenicity whilst for ruminants it is a particularly useful source of DUP rich in lysine and methionine. The latter amino acid is usually first limiting in milk protein synthesis, especially in cows on silage-based diets.

Milk by-products

The main product that can be used in compounded animal feedstuffs is skimmed milk powder. The extent of its use is largely confined to inclusion in high quality milk replacers for young stock. Such powders have been extensively denatured such that protein digestibility to monogastrics is reduced. If included in pelleted feeds the level of inclusion needs to be limited to avoid pellets that are too hard. Liquid milk by-products, such as wheys and ultrafiltrates from cheese-making operations, are frequently used in liquid feeding systems for pigs but levels of inclusion have to be monitored because of their high lactose and mineral contents.

Non-protein nitrogen sources (NPN)

Feedstuffs which contain nitrogen in a form other than protein are termed non-protein nitrogen. Such compounds can serve as useful components of ruminant diets in that they provide a source of nitrogen for rumen micro-organisms to synthesize microbial protein. Although a wide variety of compounds can be used as NPN sources the market is dominated by urea. When urea is fed it is initially broken down to ammonia and carbon dioxide by microbial urease. This ammonia may then be utilized along with appropriate oxo-acids in the synthesis of microbial protein. The efficiency of urea utilization is particularly dependent on the availability of oxo-acids and of energy to meet the needs of protein synthesis. These are affected by the amount and type of dietary carbohydrate. Starch appears to be the best source. Failure of micro-organisms to incorporate ammonia rapidly into microbial protein leads to a loss of nitrogen through urinary excretion; in extreme cases of ammonia production outstripping utilization, toxic levels of ammonia in the blood may result. A variety of factors must be considered in utilizing urea in feeds. These may be summarized as follows.

- The diet to which urea is being added must be suitable in terms of its energy, protein and mineral status to allow efficient use of NPN.
- Diets containing urea must be introduced gradually to allow rumen micro-organisms to adapt.
- Urea should not be used in pre-ruminant diets or where the level of NPN in the diet of adult ruminants is already fairly high, e.g. silage.
- Levels of inclusion should be appropriate to the class of stock being fed. For example, levels exceeding 1.25% in the concentrate ration will affect production in dairy cows, particularly in early lactation, whilst for beef cattle and suckler cows levels should not exceed 2.5% with restricted feeding or 3% with *ad libitum* feeding.

Feed supplements (nutrients)

In formulating a ration the primary aim is to fulfil the animals' requirements for energy and protein and sometimes fibre. Should the ration prove to be lacking in micronutrients, then additions of the appropriate minerals, vitamins or amino acids may be made. Nutrient supplements are commercially available which allow the addition of small amounts of specific nutrients without changing the general make-up of the initial formulation.

Supplementation of rations with specific amino acids is mainly of concern to non-ruminants where cereal-based diets may be sub-optimal in such amino acids as lysine, methionine and tryptophan. In the feeding of high yielding dairy cows there is evidence that supplementation of silage-based diets with 'protected' methionine may give economic responses.

Mineral supplementation may be provided in the form of licks or feeding blocks which allow the animal free access to a suitable combination of minerals or alternatively those specific minerals in deficit in a ration may be added to the ration in the form of a powder. In using mineral supplements the interrelationships among minerals must be recognized since excessive amounts of one mineral can cause a deficiency of another. Many proprietary supplements containing mixtures of macro- and trace minerals appropriate to particular production situations are available commercially.

As with minerals, any vitamins in deficit in a formulation must be made good by supplementation. For ruminants fat-soluble vitamins are the major consideration whilst for pigs and poultry both the water and fat-soluble vitamin content of the diet may require supplementation. A variety of balanced vitamin premixes formulated for particular circumstances and usually containing the vitamins in a chemically pure form such that only very small amounts are required are available commercially. In assessing the need for vitamin supplementation it is important to recognize the variability in the vitamin content of feedstuffs and also that vitamins are easily destroyed by agents such as heat, light and oxidation.

Nutrient requirements

The requirements of animals for nutrients are initially derived in net terms and subsequently converted to and expressed in the same terms and units of measurement that are used to describe the nutrient content of foods in the rationing process. Thus, for example, the energy requirements for maintenance of a ruminant are determined in net energy terms, then converted to and expressed in metabolizable energy terms. Nutrient requirements are a measure of what the average animal requires for a particular function. Tabulated data on requirements generally include a safety margin over and above what the average animal requires in order that animals with requirements higher than the average are adequately fed. Such values are referred to as recommended nutrient allowances.

Recommended nutrient allowances may be expressed in two ways, either as a quantity or as a proportion of the diet. Thus the lysine allowances of a growing pig may be expressed either as 12 g/d or 8 g/kg of the diet, on the assumption that the pig is consuming 1.5 kg/d of diet. In practice, the allowances for ruminants are expressed in quantitative terms on a daily basis whereas those for pigs and poultry are expressed as a dietary concentration of the nutrient in the feed.

The remainder of this chapter is devoted to a survey of how requirements are derived and of the factors influencing the requirements of animals. The total requirement for particular nutrients of animals is arrived at factorially by adding together the requirements for maintenance and the various production functions of lactation, growth and reproduction. For ruminants the nutrient allowances for maintenance and the productive functions are usually given separately whilst for pigs and poultry nutrient allowances for maintenance are usually combined with the appropriate production function. Tabulated data on nutrient requirements and allowances are contained in the following publications: *The Nutrient Requirements of Farm Livestock. No. 1 Poultry* (1975) ARC, London; *Energy and Protein Requirements of Ruminants* (1993) AFRC, CAB International, Wallingfod; *The Nutrient Requirements of Pigs* (1981) Commonwealth Agricultural Bureaux, Slough; *Nutrient Allowances for Pigs* (1982) MAFF/ADAS, Booklet 2089; *Advisory booklet Nutrient Requirements of Sows and Boars* (1990) AFRC, HGM Publications, Bakewell.

In this section energy and protein requirements will be considered for each of the body processes. Micronutrient requirements will be considered at the end of the section.

Energy and protein requirements for maintenance

The maintenance requirements of an animal refer to the amounts of nutrients an animal requires to keep its body composition and weight constant when it is in a non-productive state.

Energy

Since the energy used for maintenance must leave the body through heat, the *net energy requirement* for maintenance may be determined by measuring the heat produced by an animal when kept in the fasted state (fasting metabolism, FM). Such measurements may be made in animal calorimeters and show that energy requirements for maintenance are roughly proportional to the surface area of the animal or more closely, and of more practical value, related *to body weight* (W), or a function of body weight, namely $W^{0.73}$ or $W^{0.67}$. This is termed the metabolic weight of the animal.

A number of factors other than body weight may influence energy requirements for maintenance. Those with greatest influence are the activity of the animal and environmental conditions although lesser effects related to breed, sex and age have also been identified. Clearly animals which have to forage for food expend more energy than stall-fed animals. Likewise animals kept under adverse environmental conditions expend additional energy. These additional energy expenditures are debited to the maintenance requirement.

Environmental effects are chiefly mediated through the interacting effects of temperature, wind speed and rain. Within certain limits (the thermoneutral zone) the animal can maintain thermoregulation by limiting heat loss/gain through altering peripheral blood circulation and behavioural mechanisms. Under more extreme conditions, however, the animal may have to use energy to maintain thermoregulation either to produce heat or to dissipate heat. The environmental temperature at which the animal is forced to produce heat to maintain thermoregulation is referred to as the *lower critical temperature*, and the environmental temperature at which it is required to use energy to dissipate heat is the *upper critical temperature*. Neither of these values is constant, being influenced by wind speed, rain and the effective insulation of the animal as influenced by coat or fleece and subcutaneous fat.

To convert net energy requirements for maintenance to the same units of measurements used in the energy evaluation of feeds, the efficiency with which food is used to meet net energy requirements, the addition of allowances for activity and a safety factor (+5%) must be taken into account.

Thus, for example, the maintenance ME requirements of cattle, sheep and goats, M_m, are given by:

M_m (MJ/d) = $(F + A)/k_m$
where F = fasting metabolism
A = activity allowance as defined below, and
k_m = efficiency of utilization of ME for maintenance (see section headed 'Net energy value')

For lactating cows fasting metabolism (F) is given by:

$$F\ (\mathrm{MJ/d}) = 0.53\ (W/1.08)^{0.67}$$

where the factor 1.08 converts liveweight (W) to fasted body weight, and activity allowance (A) assuming 500 metres walked, 14 hours standing and nine position changes to be:

$$A\ (\mathrm{MJ/d}) = 0.0095\ W$$

Similar equations exist for other cattle, sheep and goats, also incorporating, where appropriate, the effects of housing and terrain on the maintenance requirement.

For pigs, energy for maintenance (E_m), expressed as ME, can be estimated from:

$$E_m\ (\mathrm{MJ/d}) = 0.440\ W^{0.75}$$

and effects of ambient temperatures (T) less than the lower critical temperature (T_c) on additional energy requirements for maintenance (E_{h1}) to be 0.012 MJ ME per kg $W^{0.75}$ for each °C below the lower critical temperature ($T_c - T$). Thus

$$E_h\ (\mathrm{MJ/d}) = 0.012\ W^{0.75}\ (T_c - T)$$

It should be noted that T_c is not a fixed value and dependent on such factors as overall metabolic activity, as influenced by body size, feed intake and rates of growth or milk production, and also the environment in which the animal is kept.

Protein

Proteins of body tissues are constantly being renewed. During this *turnover* of body tissues the amino acids resulting from breakdown are not re-utilized with 100% efficiency for synthesis into new protein. Thus the diet must provide the animal with sufficient protein to remain in nitrogen balance. This *net requirement* for protein for maintenance can be determined by measuring the faecal and urinary protein losses (N × 6.25) from the animal when it is fed a protein-free diet, termed the *basal net protein* requirement (NP_b), and the dermal losses of protein as scurf and hair (NP_d). These combined losses are more closely related to metabolic body weight ($W^{0.75}$) than body weight.

In ruminants these net protein requirements are converted into a *metabolizable protein* (MP) requirement taking into account the efficiency (k_n) with which MP is utilized. The efficiency of utilization of MP for maintenance (k_{nm}) is 1.00.

Thus in cattle, for example, the MP requirement for maintenance (MP_m) is:

MP_m (g/d) = $(NP_b + NP_d)/1.00 = 2.30\ W^{0.75}$
where NP_b (g/d) = $2.1875\ W^{0.75}$
and NP_d (g/d) = $0.1125\ W^{0.75}$

Similar equations are available for sheep and goats.

In expressing the protein requirements of pigs the concept of *ideal protein* (see section on Measures of protein quality – Monogastrics) *requirement* is used. The ideal protein requirement for maintenance (IP_m) can be calculated from

$$IP_m(\mathrm{kg/d}) = 0.0013\ W^{0.75}$$

Energy and protein requirements for lactation

The nutrient requirements of the animal for milk production are dependent on the yield of milk being produced and the composition of that milk.

Energy

The energy value of milk, which is a measure of the net energy required for milk production, is dependent on its compositional analysis. It may be determined by measurement of its gross energy using a bomb calorimeter or more usually by prediction equation using information on its compositional analysis. The energy yielding components of milk, i.e. the fat, protein and lactose, are considered to have energy concentrations of 38.5, 24.5 and 16.5 MJ/kg respectively. However, on a routine basis the constituents of cow's milk commonly determined are fat and protein. With a knowledge of these values, fairly accurate predictions of milk energy value can be made using the equation:

$$EV_l\ (MJ/kg) = (0.376 \times \%\ fat) + (0.209 \times \%\ protein) + 0.946$$

Taking into account the efficiency with which metabolizable energy is used for milk production (k_l) (see section headed 'Net energy value') the ME requirement for lactation (M_l) is calculated for dairy cows from

$$M_l\ (MJ/d) = Y \times EV_l/k_l$$

where Y is the milk yield in kg/d.

Provision of dietary metabolizable energy to lactating dairy cows is complicated by the relationships between milk yield, the partition of dietary energy and the voluntary food intake of the animal through the lactation. The situation is illustrated in Fig. 17.6.

In the early part of lactation it is usually the case that cows are unable to consume sufficient energy to meet energy requirements for milk production and call on body reserves, chiefly fat, resulting in a loss of body weight.

The amount of dietary ME for milk production to which this weight loss is equivalent may be calculated as follows:

NE value of liveweight change = 19 MJ/kg
Efficiency of use of mobilized body reserves for milk synthesis = 0.84

Thus the ME from liveweight loss available for milk synthesis in the lactating cow = 19 × 0.84 = 16 MJ/kg. The mobilized nutrients are then utilized with the same efficiency (k_l) as absorbed nutrients such that the dietary ME equivalent for milk production of liveweight loss is $16/k_l$ MJ/kg.

The tissue that is mobilized in early lactation must be replaced in later lactation through the inclusion of energy in the ration over and above that required for maintenance and production. The amount of dietary ME required for body gain may be calculated from the NE value of liveweight change and the efficiency of utilization of ME for body gain (k_g) to be $19/k_g$ MJ/kg.

Whilst these calculations are appropriate for rationing purposes it is worth pointing out that the partitioning of dietary energy between milk production and body tissue is an individual characteristic of the cow, related in physiological terms to the hormonal – especially the growth hormone/insulin – balance of the animal. In addition the composition of the diet may affect partition, diets with a high energy concentration encouraging the deposition of energy in body tissue at the expense of milk production.

As with the cow the energy requirements for milk production in the sow may be obtained from food or from body fat. The amount of energy being obtained from the latter source depends on the feed intake of the sow, a characteristic which itself may be influenced by level of body fatness; sows fed well in pregnancy and carrying more fat at the start of lactation frequently eat less food and lose weight faster than sows fed less well during pregnancy. Given the energy value of average sow's milk to be 5.4 MJ/kg and the efficiency of use of dietary ME into milk production (k_l) to be 0.70, the ME required per kg milk produced is 5.4/0.7 = 7.7 MJ/kg. When the energy from mobilized body fat is used for milk synthesis, the efficiency factor is 0.85; for catabolized body protein it is much lower at 0.5. Thus assuming 1 kg body fat to contain 39.3 MJ, 1 kg of body fat can convert into 33.4 MJ of milk energy (39.3 × 0.85) or 6.2 kg of milk (33.4 ÷ 5.4) or the equivalent of 47.7 MJ of dietary ME (6.2 × 7.7). When food ME is converted to energy in body fat, the efficiency factor is 0.75, i.e. 52.4 MJ (39.3/0.75) is required per kg.

Protein

The net protein requirements for milk production are dependent on the protein content of the milk. Protein levels in cow's milk are normally reported as CP (% per litre) of which 0.95 is *true protein*. Given that milk has a mean density of 1.03 the true protein content can be calculated as:

$$\text{True protein (g/kg)} = CP\% \times 0.95 \times 10/1.03 = 9.22\,CP\%.$$

Using an efficiency of utilization of absorbed amino acids for milk protein synthesis (k_{nl}) of 0.68 the metabolizable protein requirement for milk production (MP_l) can be calculated from

$$MP_l\ (g/kg\ milk) = (\text{True protein content of milk})/0.68 = 9.22\,CP\%/0.68 = 13.56\,CP\%$$

Similar equations, taking into account the differences in the protein content of their milk, are available for sheep and goats.

In the sow, since the supply of protein is already expressed in terms of available absorbed *ideal protein* (see section headed 'Measures of protein quality – monogastrics') which takes into account all the inef-

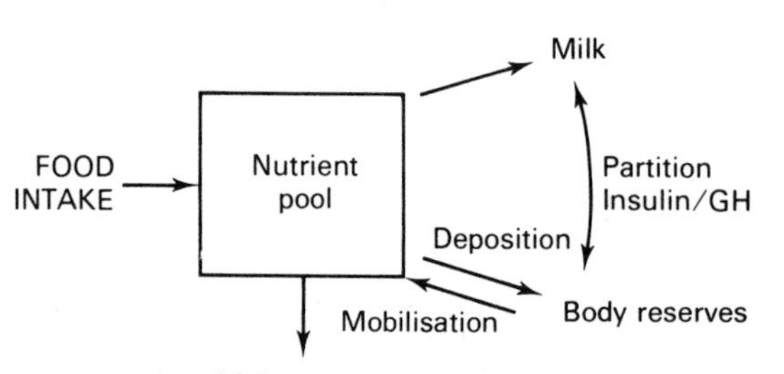

Fig. 17.6 The partition of nutrients in the cow.

ficiencies in the utilization of dietary protein, the protein requirement for milk production expressed as an *ideal protein requirement* is the same as the amount of protein the sow produces in her milk, i.e. protein content (g/kg) × yield (kg).

When animals change weight during a lactation, there are extra requirements to produce liveweight gain and contributions of protein to milk production when animals mobilize body tissue. In cows and ewes the MP requirements for liveweight gain, calculated on the basis of the net protein content of gain and an efficiency of use factor (k_{ng}) of 0.59, are 233 g/kg and 140 g/kg respectively. When ruminants mobilize tissue protein, it is assumed that the mobilized protein is used with an efficiency of 1.0 and thus adds an equivalent amount of protein to the MP supply of the animal. The supply of MP per kg weight loss is estimated to be 138 g/kg and 119 g/kg for cows and ewes respectively. The efficiency with which this source of MP can be used for milk synthesis is exactly the same as all other sources of MP ($k_{nl} = 0.68$). In the sow it is estimated that the conversion of body tissue protein into milk protein is around 0.8–0.9.

Energy and protein requirements for reproduction

The nutrient requirements for reproduction may be considered from the following viewpoints:

- how the nutrient intake of animals may influence reproductive potential in terms of its effect on such parameters as age at puberty, ovulation rate, conception and re-breeding; and
- the determination of the nutrient requirements of the pregnant female animal for fetal and extra-uterine growth in preparation for lactation.

In general the requirements of pregnancy increase exponentially from insignificant levels in early pregnancy and it is only in the last trimester of pregnancy that it becomes necessary to make special dietary provision (Fig. 17.7).

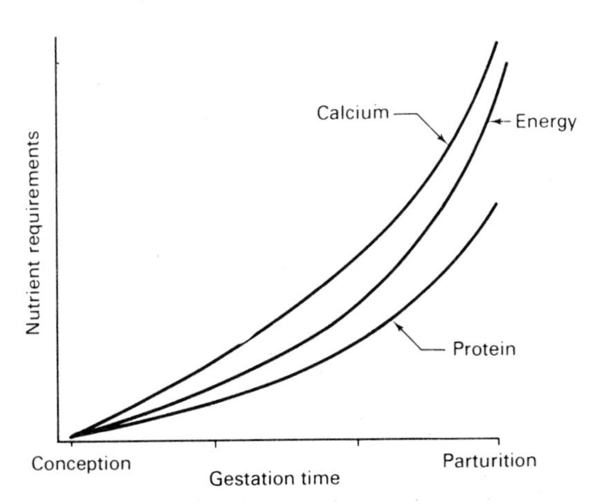

Fig. 17.7 Effect of stage of gestation on the relative requirements for nutrients during pregnancy.

Energy

The plane of energy nutrition during the *rearing* period can influence the age at which puberty is achieved in cattle and sheep. Whilst earlier puberty may be achieved by increased levels of feeding, caution must be taken to ensure that excessive fat deposition does not occur. In practice the level of energy fed and its effect on age at puberty are determined by the requirements of the production system chosen. In pigs, plane of nutrition would appear not to have such a marked effect on age at puberty.

The *period before mating* is considered important since ovulation rate may be influenced by the condition of the animal. In ewes, improving the body condition can have a particularly marked effect on the fertility of hill breeds of sheep but more prolific breeds and crosses of these with hill breeds respond less dramatically. The more short-term practice of 'flushing' ewes prior to mating would also appear to improve fertility especially of ewes with lower condition scores. There is also reason to suppose that a similar relationship operates in the sow. From the viewpoint of feeding, the most practical advice is to ration energy according to body condition taking account not only of those animals that require an improvement in body condition but also of overfat animals which tend to have reduced fertility. In the dairy cow mating must occur during early lactation at which time body weight loss is frequently occurring. Excessive weight loss is associated with infertility and the aim in feeding over this period is to restrict the extent of weight loss.

The period immediately *post-conception*, when the developing embryo is nourished by direct absorption from its fluid environment, is critical in that faulty or inadequate nutrition at this time can jeopardize the successful implantation of the fertilized ova in the uterine wall. Adequate nutrition must therefore be continued over this period although subsequently there is little extra requirement for energy until the last trimester of pregnancy when accelerating fetal growth increases energy demand. Although severe energy undernutrition in early pregnancy can affect placental growth, some loss in body condition can occur in ewes without adverse effects on fetal development.

Just as an awareness of body condition governs feeding in the premating period, so also it is an important parameter to be monitored as pregnancy progresses. Whilst over-thin sows and ewes will produce weak, underweight young with reduced survival rate, the development of over-fatness, especially in mid-pregnancy, can lead to pregnancy toxaemia and problems at parturition. Over-conditioning in late pregnancy may also lead to reduced voluntary food intake in the subsequent lactation. In the feeding of ewes during the last six weeks of pregnancy it is important to take into account the likely number of fetuses being carried. In this context the use of pregnancy scanning techniques has greatly improved the precision of rationing.

With dairy cows it is practice to terminate a lactation some eight weeks before the birth of the next calf. Restoration of body tissue converted to milk in early lactation should take place prior to 'drying off' since the conversion of feed energy into liveweight gain is more efficient in the lactating than the non-lactating state. Provision of extra energy during the dry period is not generally recommended because not only is it less efficiently used but there is the risk of excessive fattening and it has little influence on the birth weight of the calf.

Equations which may be used to calculate the net energy (NE) retention in the gravid uterus are:

For cattle:

$$\text{NE retention (MJ/d)} = 0.025\, W_c\, (E_t \times 0.0201\, e^{-0.0000576t})$$

where E_t is total energy retention at time t (days from conception) and can be calculated from:

$$\log_{10} (E_t) = 151.665 - 151.64\, e^{-0.0000576t}$$

and W_c is the calf birthweight in kg and can be calculated from

$$W_c\ (\text{kg}) = (W_m^{0.73} - 28.89)/2.064$$

where W_m is the mature body weight of the dam.

For sheep:

$$\text{NE retention (MJ/d)} = 0.25\, W_0\, (E_t \times 0.07372\, e^{-0.00643t})$$

where W_0 is the total predicted weight of lambs at birth (kg), E_t is total energy retention at time t (days from conception) and can be calculated from

$$\log_{10} (E_t) = 3.322 - 4.979\, e^{0.00643t}$$

For sows:
Net energy deposition in uterus and mammary tissue (MJ/d)

$$= 0.107\, e^{0.027t} + 0.115\, e^{0.016t}$$

where t is days of gestation.
Metabolizable energy requirements for concepta growth can be derived from these net energy values using an efficiency of ME use for growth of the concepta (k_c) of 0.133 for the ruminant and 0.5 for the sow.

Additional energy requirements need to be satisfied in pregnant animals that are themselves still growing. Sows, for example, do not reach their mature body weight until their fourth parity whilst cow's mature weight is not reached until their third lactation. There is still considerable uncertainty about the quantitative aspects of requirements for this aspect of pregnancy, because of the impact of pregnancy itself on energy utilization. There is some evidence that the maintenance requirements of the pregnant animal are reduced in early pregnancy whilst later in pregnancy the so-called 'heat of gestation' increases energy needs for maintenance.

Protein

In the pregnant animal the net protein requirement is represented by not only the products of conception, fetal and uterine growth, mammary tissue growth and protein turnover losses, but also lean tissue growth in the maternal body. Clearly these various requirements are not easily separated out one from another, and, further to this, precise estimation of requirements is complicated by changing efficiency of utilization during pregnancy. Especially towards the end of pregnancy there appears to be more efficient utilization of protein, termed pregnancy anabolism, probably resulting from the animal's changed hormonal status. Requirements are clearly at their greatest during the latter part of pregnancy when uterine and fetal growth predominate.

Estimation of the metabolizable protein requirement for pregnancy (MP_c) in ruminants can be obtained from the tissue protein (TP) deposition in the products of conception – mainly influenced by stage of pregnancy and birthweight of the calf or lambs – and an efficiency of utilization of MP for pregnancy (k_{nc}) of 0.85. Equations developed to incorporate these parameters are:

$$\text{For cows } MP_c\ (\text{g/d}) = 1.01\, W_c\, (TP_t \times e^{-0.00262t})$$
$$\text{For ewes } MP_c\ (\text{g/d}) = 0.25\, W_c\, (0.079TP_t \times e^{-0.00601t})$$

where t is number of days from conception, W_c is predicted birthweight of the calf/lambs (kg), and TP_t (kg) can be derived from:

$$\log_{10} (TP_t) = 3.707 - 5.698\, e^{-0.00262t} \text{ in cattle}$$
$$\log_{10} (TP_t) = 4.928 - 4.873\, e^{-0.00601t} \text{ in sheep}$$

For sows, ideal protein requirement for uterine and mammary tissue (g/d)

$$= 0.0036\, e^{0.026t} + 0.000038\, e^{0.59t}$$

where t is number of days from conception.

Energy and protein requirements for growth

The commercial measure of growth in farm animals is liveweight gain. In determining the nutrient requirements for liveweight gain the major problem to be faced is that the composition of liveweight gain is not constant. The major influences on the composition of liveweight gain are:

- the physiological age or weight of the animal, and
- the rate of liveweight gain.

Detailed coverage of these influences on the composition of liveweight gain is given in Chapter 15.

Energy

The net energy content of the LWG made by an animal is dependent on the proportions of protein, fat, water and ash present. Thus the energy content is represented by the protein and fat fractions. In general the gain made by young animals has a low energy content since it has a high proportion of water (in lean) and ash (in bone) and little fat. As animals grow and mature the energy content of gain increases, this being due to the declining proportions of water present and the increasing proportions of fat.

The rate of LWG influences the net energy content of the gain since the greater the rate of gain the higher the proportion of fat produced.

In specifying energy requirements for growth it is necessary, therefore, to indicate the rate of growth anticipated. This is especially the case in beef production where it is desirable to achieve target LWG at each stage of the particular production system (see Chapter 18). In some situations the target LWG is the maximum growth potential for some or all of the production cycle. This is normally achieved in practice by allowing *ad libitum* intake of high energy feeds. Some restriction of energy intake is frequently desirable as animals approach slaughter weight as a means of reducing the fat content of the carcass, a principle applied in the production of pigs for bacon where overfatness of carcasses is penalized.

In the rationing of growing ruminants a particular problem arises in that energy requirements are not affected simply by the energy content of the gain but also by the metabolizable energy concentration (total ME in diet/total DM in diet, M/D) of the diet which has a marked effect on the efficiency of utilization of metabolizable energy. What it means in practice is that a higher proportion of the metabolizable energy in a largely concentrate-based diet is laid down as body tissue than in a largely roughage-based diet. The origins of such differences lie in the types of rumen fermentation and particularly the VFA end products that different diets produce. Thus increasing the proportion of propionate decreases the heat increment of metabolism. This extra dimension to assessing the energy requirements for growth has led to the development of a Net Energy System, sometimes referred to as the Variable Net Energy System, which takes account not only of the effects of the weight of the animal and rate of gain but also the M/D of the diet. Details of the Variable Net Energy System are given in Chapter 18.

Whilst energy requirements for gain are mainly influenced by the factors considered above, the breed and sex of the animal also have an influence. Thus for cattle later maturing breeds require less energy per unit of gain at equal body weights than early maturing breeds; the energy content of gain at a particular weight is greater in heifers than in steers which in turn is greater than in bulls.

An equation that may be used to predict the energy value of weight gains (EV_g) in cattle which takes account of these factors is:

$$EV_g\ (MJ/kg) = \frac{C_2\ (4.1 + 0.0332W - 0.000009W^2)}{(1 - C_3 \times 0.1475\Delta W)}$$

where $C_3 = 1$ when plane of nutrition > maintenance and 0 when < maintenance; C_2 corrects for mature body size (±15%) and sex (±15%) of the animal, ranging from 0.70 for late maturing bulls to 1.30 for early maturing heifers; and ΔW is liveweight gain (kg/d), such that the energy retained in the animal's body per day (E_g) is given by:

$$E_g\ (MJ/d) = \Delta W \times EV_g$$

and the ME required to support this gain is given by:

$$ME\ requirement\ (MJ/d) = E_g/k_g$$

where k_g is the efficiency of utilization of ME for growth (see section headed 'Net energy value'). Similar equations are available for growing sheep and goats.

An approach used in pigs is to consider the stoichiometry of protein and lipid synthesis in liveweight gain. The lack of information on the extent of variation in tissue turnover, especially of protein, with body size and rate of growth make it difficult to predict with any certainty the energy costs of retention. Values of the order of 50–70 MJ/kg of protein retention and 50–55 MJ/kg fat retention, at efficiencies of 0.44 for protein and 0.75 for fat, have been suggested. In terms of what this means as far as energy requirements for body tissue gain are concerned, the differences in water associated with protein and fat must be taken into account.

Protein

Growth rate in animals is primarily determined by the energy intake. The same principles which govern the net energy requirements for gain also affect the net protein requirements for gain. In reality the requirements are not so much for protein but more specifically for the various amino acids that contribute to growth.

In cattle the net protein requirement for gain (NP_f) is influenced by sex and breed, which affect mature body weight and also rate of gain. For castrates of medium mature body size,

$$NP_f\ (g/d) = \Delta W\ \{168.07 - 0.16869W + 0.0001633W^2\} \times \{1.12 - 0.1223\Delta W\}$$

where ΔW is liveweight gain as kg/d.

Taking into account the efficiency of use of MP for body gain (k_{ng}) of 0.59 allows calculation of MP requirement for gain (MP_f) as:

$$MP_f\ (g/d) = NP_f/0.59$$

Correction factors of ± 10% are used to take into account each of the effects of sex and breed size, giving, for example, a correction factor of 1.20 for late maturing bulls and 0.80 for early maturing heifers. Similar equations can be used for prediction of MP_f in sheep and goats.

For pigs, the protein requirement expressed in terms of ideal protein (see section headed 'Measures of protein quality – Monogastrics') is precisely equal to the amount of protein deposited in the tissue. Information on the precise rates of protein retention in relation to body weight and the influences of sex and genetic make-up on these rates of retention are required if the potential rates of retention are to be maximized but not exceeded.

Micronutrient requirements

Micronutrients have a vast number of roles in the animal body and failure to meet the animal's requirements for particular micronutrients can at least impair the performance and at most produce clinical conditions. The precision of tabulated data on allowances for micronutrients is limited by the methodology of determination and by the fact that numerous interactions occur between individual micronutrients. Such interaction can influence availability in terms of absorption and their functional activity in a metabolic pathway. In addition, a variety of mechanisms exist in the animal to provide a relatively constant micronutrient milieu for metabolic activity under circumstances of variable dietary intake, thus ensuring the normal physiological functioning of the animal. Nevertheless, such homeostatic mechanisms can break down under more extreme dietary supply situations, leading to impaired physiological function either through an inadequacy or an accumulation of micronutrients.

Allowances for individual mineral elements have been determined either through factorial methods or by feeding trial techniques. In the former method the net requirement for the element is determined by adding together the endogenous losses of the element that occur from the animal when fed a diet free of the element – equivalent to maintenance requirement – to the amount of element that is present in the product of the animal. For a product such as milk this is relatively easy, but for estimation of the mineral composition of liveweight gain it is a somewhat more laborious procedure. To convert the net requirement to a dietary requirement the availability of the particular element in terms of absorbability has to be taken into account. In the performance or feeding trial method of estimation, diets containing different amounts of the element are fed and their influence on performance monitored. The main difficulty is in setting the parameters that relate to the optimum intake of the element. For example, in the case of an element such as calcium, maximum growth rate may be obtained at a level of intake that is inadequate in terms of bone strength. The ability of the animal to store some mineral elements further complicates the situation.

Estimates of requirements for vitamins are obtained from feeding trials. The same criticisms made for the use of this method for estimating mineral requirements also apply to vitamins.

In practice, levels of supplementary inclusion in particular situations are arrived at empirically, taking into account any conditions such as feed composition and health status of the herd which may influence requirements.

Further reading/information sources

AFRC (1993) *Energy and Protein Requirements of Ruminants*. CAB International, Wallingford.

Bolton, W. & Blair, R. (1977) *Poultry Nutrition*, 4th edn. MAFF Bulletin No. 174. HMSO, London.

Church, D.C. (1991) *Livestock Feeds and Feeding*. Prentice-Hall, New York.

Fisher, C. & Boorman, K.N. (eds) (1986) *Nutritional Requirements of Poultry and Nutritional Research*. Butterworths, London.

MAFF (1990) *UK Tables of Nutritive Value and Chemical Composition of Feedingstuffs*. Givens, D.I. & Moss, A.R., eds. Rowett Research Services, Aberdeen.

MAFF (1992) *Feed Composition – UK Tables of Feed Composition and Nutritive Value for Ruminants*, 2nd edn, Chalcombe Publications.

McDonald, P., Edwards, R.A. & Greenhalgh, J.F.D. (1988) *Animal Nutrition*. Longmans, London.

Thatcher, P.A. & Kirkwood, R.N. (1990) *Non Traditional Feed Sources for Use in Swine Production*. Butterworths, London.

Whittemore, C. (1993) *The Science and Practice of Pig Production*. Longmans, London.

Williams, D.R. (1987) *Animal Feedingstuffs Legislation of the UK. A Concise Guide*. Butterworths, London.

Wiseman, J. (ed.) (1987) *Feeding of Non-ruminant Livestock*. Butterworths, London.

Wiseman, J. & Cole, C.J.A. (1990) *Feedstuff Evaluation*. Butterworths, London.

18

Cattle

J.A. Kirk

Domesticated cattle are nearly all descended from two major species: *Bos taurus*, which includes the European types, and *Bos indicus* to which the Zebu cattle belong. Selection from these has led to the development of a number of well defined breeds. These breeds vary from types used primarily for milk production to those developed for beef production.

In some breeds an attempt has been made to combine the desirable qualities of both dairy and beef cattle to provide dual-purpose cattle.

Since the 1950s a number of exotic breeds have been imported into the UK, mainly from Europe. The rationale behind this importation of breeding stock has been to improve beef production although some are dual-purpose breeds in their country of origin. Most of these immigrants have been large bodied, fast growing cattle producing lean carcasses (i.e. Charolais, Limousin, Simmental), characteristics particularly valuable for satisfying consumer demand for lean meat. Similarly, the dairy sector has benefited from the importation of Holstein cattle from Canada, the USA, Denmark and New Zealand.

Fat cattle, calves and milk are an important sector of total UK agricultural output (Table 18.1).

The last two decades have seen a dramatic change in the UK dairy herd structure. The number of registered milk producers has fallen by over 60%, whilst dairy cow numbers have fallen by about 20%. This has resulted in a rapid increase in the average size of dairy herd (Table 18.2).

Milk yield per cow has also risen. This coupled with a decline in liquid milk consumption has resulted in more milk being used for manufacture (Table 18.3).

Milk sold for processing commands a lower price than that in the liquid milk market. Milk price as paid to the producers has to take into account the quantity and lower price of milk sold for manufacture (Fig. 18.1).

A similar pattern is seen when looking at milk utilization between the liquid consumption and manufacturing sectors within the EEC (Table 18.4).

Milk product surpluses arose because of the increased amount of milk being produced and reduced consumption. In 1986 it was estimated that accumulated intervention stocks stood at: butter 1 500 000 t and milk powder 1 100 000 t compared with beef at 620 000 t. Associated with this was the massive cost of the Common Agricultural Policy for the initial purchase of these and other products and their storage. It was for this reason that in March 1984 the EEC Agricultural Ministers agreed on a policy for curbing the increase in EEC milk production by the introduction of a system of quotas. The agreement allowed for a superlevy to be charged on all milk produced above a specific quantity or 'quota'.

In the UK there is a close relationship between milk production and beef production. Dairy herds commonly use beef bulls for crossing, and the crossbred calves, together with pure-bred bull calves from dairy herds, are fattened for beef. A number of heifers from the dairy herd are reared as replacements for suckler beef herds (Fig. 18.2).

Consumption of beef is increasing in all EEC member states except Portugal. Table 18.5 shows the pattern of meat consumption in the UK.

Table 18.1 UK agricultural output, 1984 and 1992

	1984		*1992*	
	£ million	%	*£ million*	%
Livestock				
Finished cattle and calves	1 922	16.1	2 034	14.5
Finished sheep and lambs	579	4.8	1 034	7.4
Finished pigs	1 000	8.4	1 091	7.8
Poultry	664	5.6	919	6.6
Other livestock	94	0.8	1 058	7.5
Total livestock	4 259	35.7	5 217	37.2
Livestock products				
Milk	2 293	19.2	2 930	20.9
Eggs	537	4.5	437	3.1
Wool	37	0.3	41	0.3
Others	26	0.2	33	0.2
Total livestock products	2 893	24.2	3 441	24.6
Total crops	3 541	29.7	3 534	25.2
Total horticulture	1 241	10.4	1 817	13.0
Total agricultural output	11 934		14 009	

Table 18.2 Cattle numbers in the UK (×10)

Year	*No. of producers*	*No. of dairy cows*	*Average herd size*	*Average milk yield (litres)*
1985	48 827	3 147 000	64	4770
1986	47 927	3 135 000	64	4880
1987	46 740	3 039 000	65	4945
1988	43 952	2 908 000	65	4870
1989	42 306	2 861 000	65	4895
1990	41 248	2 884 000	66	5050
1991	39 891	2 768 000	67	5080
1992	38 400	2 683 000	68	5135

After MMB (1992) *UK Dairy Facts and Figures.*

Table 18.3 Utilization of milk in the UK (litres per head per week)

	1985	*1986–87*		*1989–90*		*1991–92*	
Liquid	48	6619	44%	6582	48%	6643	49%
Manufacture	52	8515	56%	7245	52%	6853	51%
Av. consumption	2.36	2.30		2.27		2.24	

After MMB (1992) *UK Dairy Facts and Figures.*

Table 18.4 Milk utilization in the EEC (%), 1990

	Liquid	*Cream*	*Butter*	*Cheese*	*Milk powder*	*Total utilization '000 tonnes*
Belgium	15	11	50	8	6	3 825
Denmark	6	8	37	33	11	4 742
France	9	5	43	28	6	27 591
Germany	14	14	33	18	3	24 161
Greece	29	3	2	54	–	2 057
Irish Republic	11	4	62	13	2	5 577
Italy	24	6	15	46	–	12 931
Luxemburg	15	11	35	7	–	285
The Netherlands	6	4	36	33	11	11 476
UK	47	4	20	20	3	15 224
Total ten	18	7	33	27	5	107 869

After MMB (1992) *EEC Dairy Facts and Figures.*

Table 18.5 UK meat consumption per head (kg)

	1982	*1987*	*1991*
Beef and veal	18.3	19.9	17.6
Mutton and lamb	7.3	6.3	7.4
Pork	12.9	13.9	13.9
Bacon and ham	8.4	7.9	7.5
Poultry	14.4	17.8	20.2
Total	61.3	65.8	66.6

After MLC (1991) *Beef Yearbook.*

Intervention purchases of beef are declining but sales from intervention are also small. By the end of 1985 intervention stocks were 633 000 t. Imports into the EEC, in particular imports of fresh or chilled beef into Italy from Yugoslavia and Austria and live cattle from Eastern Europe into Italy are significant. However, the EEC retained its position of a major net exporter of beef and veal. Self-sufficiency in beef and veal has declined from 110.1% in 1984 (Table 18.6).

Definitions of common cattle terminology

At birth
- male – bull calf, bullock calf if castrated
- female – heifer calf, cow calf

First year
- male – yearling, year-old bull
- female – yearling heifer

Second year
- male – two-year-old bull, steer, ox, bullock
- female – two-year-old heifer

Third year
- male – three-year-old bull, steer, ox, bullock
- female – three-year-old heifer, becomes cow on bearing calf

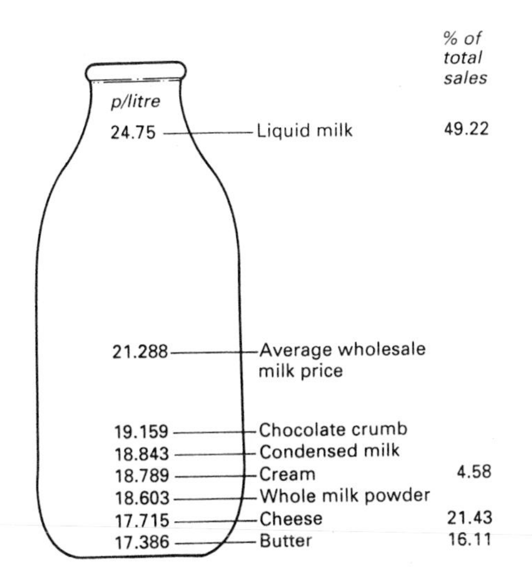

Fig. 18.1 Average return from milk for each product – United Kingdom, 1992. (After MMB (1992) *UK Dairy Facts and Figures*.)

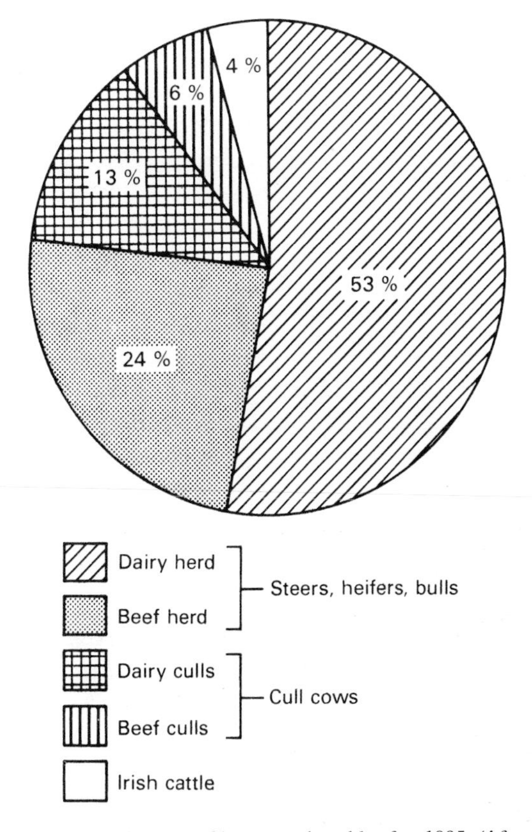

Fig. 18.2 Sources of home produced beef – 1985. (After MLC (1986) *Beef Yearbook*, reproduced with permission.)

Heifer – usually applied to a female over one year old which has not calved. An unmated animal is known as a maiden heifer and a pregnant one as an in-calf heifer. In some areas the term first-calf heifer is used until after the birth of a second calf. A barren cow is either barren, cild or farrow and when a cow stops milking she is said to be yeld or dry.

Stirk – limited to males and females under two years in Scotland. It is usually applied to females only in England, the males being steers.

Store cattle – stores, are animals kept usually on a low level of growth for fattening later.

Veal – the flesh from calves reared especially for this purpose and normally slaughtered at about 16 weeks of age.

Bobby veal, slink veal – flesh from calves slaughtered at an early age, often only a week or two old. These animals tend to be of extreme dairy type, thus making them undesirable for rearing as beef cattle.

Cow beef – beef from unwanted cows, often used in processing.

Bull beef – beef from entire male animals, the majority produced as an end product but some as a by-product from redundant breeding stock.

Table 18.6 Cattle population, 1991 ('000 head)

Country	*Total cattle*	*Total cows*	*Dairy cows*	*Other cows*	*% Dairy cows in total cow population*
Belgium	3 109	1 162	797	365	68.6
Denmark	2 222	849	746	103	87.9
France	20 970	8 740	4 968	3772	56.8
Germany	17 134	6 011	5 632	379	93.7
Greece	631	307	214	93	69.7
Irish Republic	6 073	2 094	1 364	730	65.1
Italy	8 094	3 296	2 751	545	83.8
Luxemburg	205	78	52	26	66.7
The Netherlands	4 876	1 941	1 881	60	96.9
Portugal	1 381	626	394	232	62.9
Spain	4 924	2 657	1 519	1138	57.2
UK	11 623	4 441	2 779	1162	62.6
Total twelve	81 242	32 202	23 097	9105	71.7

After MMB (1992) *EEC Dairy Facts and Figures*.

Ageing

The development of the incisor teeth can be used as an indication of the age of cattle.

The dental formula for a full mouth is:

Permanent molars	*Temporary molars*	*Incisors*	*Temporary molars*	*Permanent molars*	
3	3	00	3	3	
–	–	–	–	–	=32
3	3	44	3	3	

There are four pairs of incisor teeth. These are found in the lower jaw; the upper jaw has no incisors but is a hard mass of fibrous tissue known as the dental pad. Starting in the middle of the jaw the pairs are known as centrals, medials, laterals and corners. The times at which the temporary incisors are shed and replaced by the permanent incisors are important (Fig. 18.3).

Individual animals will vary from these figures. Breed, management and feeding all have an influence.

Calf rearing

The foundations of the future health and well being of the calves are laid by good feeding and management throughout the first three months of age.

Temporary incisors

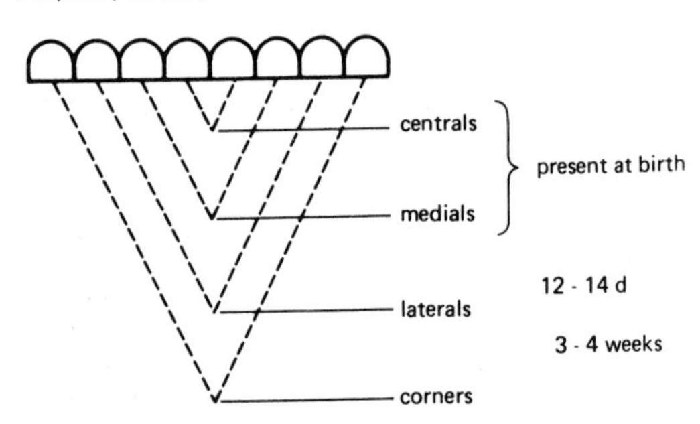

Permanent incisors

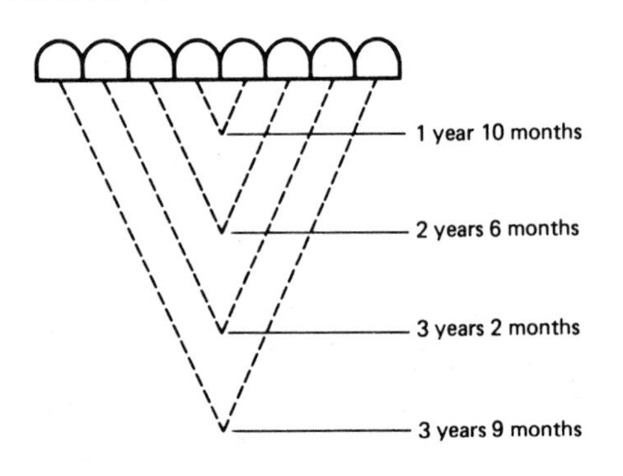

Fig. 18.3 The development of the incisor teeth as an indication of the age of cattle.

The rectal temperature of a healthy calf is 38.5–39.5°C. Common to all the systems of rearing is the need for an adequate supply of colostrum (the secretion drawn from the udder at the time of parturition). At birth the calf is virtually free of bacteria but quickly becomes infected with organisms from its surroundings. The blood of the newborn calf contains no antibodies until the calf has received colostrum.

As can be seen in Table 18.7, colostrum contains a high percentage of proteins especially immunoglobulins with their attendant antibodies, minerals, vitamins (especially the fat soluble A, D and E) as well as carotene which is the cause of the yellow coloration.

It is vitally important that the calf receives colostrum during the first 24 h of life. The calf's ability to absorb the antibody protein is greatest during the first 6–12 h of life. This is due to the calf's stomach wall changing and becoming impermeable to the immunoglobulins (Fig. 18.4). Hence, to obtain the maximum protection, the calf should be fed colostrum during the first 6 h of life.

Colostrum feeding should continue for 4 d. Some farmers achieve this by leaving the calf to suckle; others remove the calf at birth. If the calf is removed at birth, it is easier to teach it to drink from a bucket. Frequently, a greater quantity of colostrum is produced by the cow than is needed by the newborn calf. Any excess colostrum can be diluted with water at the rate of two parts of colostrum to one of water and fed to older calves in place of milk or milk substitute. Alternatively colostrum may be frozen and kept in case of an emergency when it may be fed to newborn calves. The composition of colostrum changes to milk during the first 4 d milking. If pre-calving milking is practised, this change may take place before the calf is born.

The zinc sulphate turbidity test (ZST) enables the concentration of circulating antibodies in the calf's blood to be determined. There is a high correlation between the results of this test and calf health, showing that it is essential that calves get adequate amounts of colostrum. This is undoubtedly one of the major factors involved in

Table 18.7 Composition of colostrum (first 24 h after calving) and of milk

	Colostrum	*Milk*
Fat (%)	3.6	3.5
Non-fatty solids (%)	18.5	8.6
Protein (%)	14.3	3.25
Casein (%)	5.2	2.6
Albumin (%)	1.5	0.47
Immunoglobulin (%)	6.0	0.09
Ash (%)	0.97	0.75
Calcium (%)	0.26	0.13
Magnesium (%)	0.04	0.01
Phosphorus (%)	0.24	0.11
Iron (%)	0.20	0.04
Carotenoids (μg/g fat)	25–45	7.0
Vitamin A (μg/g fat)	42–48	8.0
D (μg/g fat)	23–45	15.0
E (μg/g fat)	100–150	20.0
B (μg/g fat)	10–50	5.0

After Roy (1980).

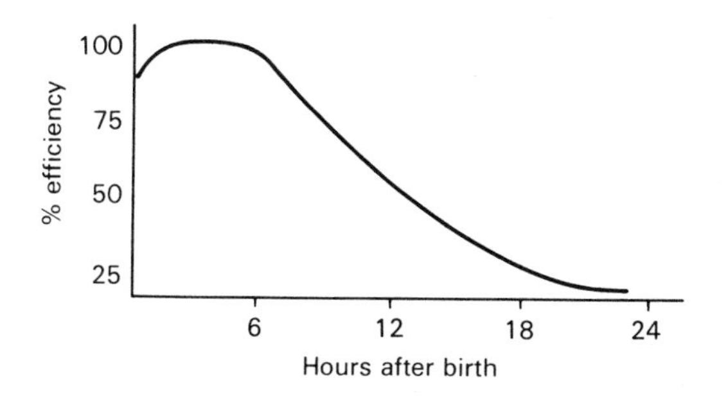

Fig. 18.4 Effective uptake of antibodies from colostrum in the young calf.

ensuring that disease and ill health are kept under control (Table 18.8).

Although the newborn calf has the same four stomach compartments, rumen, reticulum, omasum and abomasum, as that of the adult, they vary in size (Table 18.9) and development.

The same relative proportions would also apply to the volume of the various compartments.

The stomach of the newborn calf is similar to that of a monogastric animal. The oesophageal groove conveys liquids or semiliquids directly to the omasum and from there they pass to the abomasum. It is in the abomasum where true gastric digestion takes place.

The age at which transition to ruminant digestion takes place depends on the diet the young calf is fed. The longer a calf is fed milk, the less is the urge to consume other feeds and the later will be the development of the rumen.

Ideally calves should be kept for at least 7 d on the farm on which they were born. Newly purchased calves should be allowed to rest after a journey. If the calves have only had a short journey and were recently fed, additional carbohydrate, e.g. glucose, may predispose the animals to diarrhoea. Generally calves arriving before noon should be left to rest in a dry, draught-free, well bedded area and then fed in the evening with a drink of an electrolyte/glucose solution. Calves arriving mid to late afternoon should be left until the following morning before being fed the same solution. Proprietary electrolyte solutions are available or a solution of glucose (22 g) with the addition of sodium chloride, common salt (4.5 g), per litre of water can be used.

Table 18.8 Relationship of colostral status to calf performance at five weeks

	Colostrum rating		
	Low	*Medium*	*High*
No. of calves	87	158	182
% Treated for illness			
No treatment	17.2	37.3	48.9
Treated once	27.6	41.2	30.2
Treated twice	33.3	12.0	13.8
Treated more than twice	21.9	9.5	7.1

Source: Thickett *et al.* (1979).

Table 18.9 Proportion of tissues by weight of a calf's

	Age (weeks)			
Stomach compartment	*0*	*4*	*12*	*20–26*
Rumen-reticulum	38.0	52.0	64.0	64.0
Omasum	13.0	12.0	14.0	22.0
Abomasum	49.0	36.0	22.0	14.0

Source: Warner & Flatt (1965).

First drink	1.5 litres electrolyte/glucose solution
Second drink	1.5 litres electrolyte/glucose solution
Third drink	50:50 milk powder/electrolyte solution
Fourth drink	milk powder alone at standard concentration.

Numerically the most common method of rearing calves is 'by hand'. Artificial milk replacers are used instead of cow's milk because the cost per unit of milk replacer is less than the equivalent amount of milk. Calves destined as replacements in the dairy herd are almost always reared by hand. This method of rearing requires specialist accommodation and is more exacting in terms of labour but enables economies in food costs to be achieved. Milk substitutes basically consist of skim milk powder and added fat, protein vitamins and minerals; whey powder is sometimes included. Most artificially reared calves are 'early' weaned.

Early weaning

In this method the calf is weaned onto a diet of dry food by five weeks of age. The early introduction of concentrates, hay and water by day 7 after birth encourages the development of the rumen. To achieve the desired intake of concentrates it is essential that they be palatable and fresh. Milk replacer is restricted, again to encourage the consumption of hay and concentrates. Calves should be eating 0.75–1.00 kg of concentrates daily with a liveweight of 65 kg by weaning at five weeks. The advantages of early weaning are that concentrates are much cheaper per kg of gain that milk replacer.

'Milk' feeding can be practised in several forms. Once- or twice-daily feeding of milk substitutes is the common practice on farms.

Ad libitum feeding of milk substitute was first made possible with the advent of high fat milk substitutes and automatic dispensing machines. The introduction of 'acid' milk replacers has enabled *ad libitum* feeding of cold milk to be practised without the need for sophisticated machines. Acid milk replacers are classified according to two types:

(1) Medium acid based on skimmed milk. These have a pH of 5.5–5.8 and a protein content of 24–26% with 17–18% fat.
(2) High acid usually based on whey from cheese manufacture. These have a pH of 4.4–5.8 with 19–20% protein and 12–15% fat.

The acidity has a positive effect in helping to reduce

the incidence of digestive upsets. The pH of the abomasum prior to feeding is 2.0–2.8 but after feeding conventional milk replacer of pH 6.2–6.5 the abomasum pH rises to between 4.5 and 6.2. It then declines to pre-feeding levels after 3–5 h. Satisfactory digestion depends on enzyme action and the optimum pH for this to occur is between 3 and 4. The conclusion therefore is that by feeding high acid milk replacer the abomasal pH is kept near optimum levels for enzyme activity and below the pH at which most organisms can survive.

Milk replacers may be prepared for 3 d feed supply at a time and the formulation aids the normal digestive process of the calf.

The suckling action by the calf ensures correct closure of the oesophageal groove allowing milk to enter the abomasum without spilling into the developing rumen, where it can ferment and possibly cause digestive troubles. The argument that cold feeding is bad for calves probably grew out of the bucket feeding system where a calf may suffer a physiological shock when consuming large amounts of cold milk in a short period of time. With an *ad libitum* feeding system the 'little and often' effect of food consumption ensures that the small amount of milk taken in at any one time is rapidly warmed up to body temperature.

There are three important elements if the use of *ad libitum* feeding of acidified milk is to be applied correctly.:

(1) Ensuring that calves are consuming enough milk. One of the easiest ways of achieving this is to feed the calves individually for the first 5–10 d by means of a bucket and teat. Any initial difficulty with calves not drinking may be overcome if the milk is fed warm; later the calves readily consume the milk replacer if it is fed at progressively cooler temperatures until cold feeding is practised.
(2) Calves should remain on cold acid milk *ad libitum* for three weeks. At no time should the supply of milk be allowed to run dry as excess consumption may take place when replenished and this may lead to scours.
(3) After three weeks the intake of milk may be reduced by substituting cold water for replacer during the night. This encourages the consumption of concentrate food. Calves must have access to palatable concentrates at all times. These should preferably be sited near the teats where milk is drunk, thus encouraging consumption.

Criticism of *ad libitum* feeding is usually made on the grounds of the over-consumption of milk, linked to the cost of feeding greater quantities of milk replacer. Commercial farmers claim that although it may cost more to rear a calf, the animal experiences less stress and fewer health problems and as a result is a better calf. Added to this is a reduction in labour coupled with a more flexible work routine.

When reared in groups and machine-fed, care should be taken to ensure that individuals know how to suck and that there is no bullying so all have an opportunity to feed. Evenly matched batches and close observation for the first few days are essential.

A pen 1.8 m long and 1 m wide will accommodate an individual calf up to eight weeks old. Less pen space is needed by calves reared in groups; an area of 1.1 m^2 per calf should be sufficient up to eight weeks, this being increased to 1.5 m^2 per calf by 12 weeks.

Whatever system of housing is used, warmth, a dry bed and particularly, prevention of draughts are important to the well being of young calves.

The three most commonly practised systems of natural rearing are single suckling, double suckling and multiple suckling.

Single suckling

This is by far the most popular and is carried out mainly on hill and marginal farms, by pedigree beef breeders, on lowland farms with inaccessible grass and on some arable farms utilizing grass as a break crop. In this method the cow rears her own calf with milk being the main source of nutrients. The calf remains with the dam until weaning at approximately six months of age. Cows are normally calved in either autumn or spring. The advantage of autumn calving is that calves are old enough to make full use of the grass in spring and are heavier at weaning in the following autumn. The disadvantage is that cows and calves often have to be housed during some of the winter period. Spring calving has the advantage that cows can be overwintered outdoors; the disadvantages are that the young calf cannot make as good use of spring grass, and spring grass may cause a flush of milk in the cows resulting in the calves scouring. Concentrates should be fed before weaning so eliminating any loss in condition later.

Double suckling

The objective is for each cow to rear two calves together. After calving a second calf should be introduced, and allowed to suckle with the cow's own calf. Particular care should be practised at first until the cow willingly accepts the new calf. Success of the system depends on a supply of newborn calves as required, cows having an adequate supply of milk (often a cull dairy cow or a Friesian cross cow) and good management.

Multiple suckling

This necessitates nurse cows yielding a suitable quantity of milk, say 4000–5000 litres, and a supply of suitable calves. Each suckling period usually lasts about ten weeks. In the first ten weeks of lactation four calves are suckled, in the second ten weeks three new calves are suckled, in the third period another three and for the last period two calves, giving a total of 12 calves reared during the lactation. Cows are often removed from the calves between feeds.

Veal production

For veal production calves capable of high rates of liveweight gain are required, Friesian bull calves normally being used. These were traditionally reared on an all-milk diet and slaughtered between 140–160 kg liveweight at 14–18 weeks of age. It should be noted that the recently published welfare codes recommend that calves have access to roughage feed. Friesian bull calves will have a killing-out percentage of 55–60%.

Correct feeding is critical if an adequate return on capital is to be achieved. Milk replacers of the high fat type with at least 15% and up to 25% fat are required for maximum gains. The aim is for a daily liveweight gain to slaughter of 1 kg/d or more. Good husbandry with particular attention being paid to hygiene and observation of animals for ill health is of paramount importance.

Calves were normally housed in buildings with a controlled environment and kept in individual pens on slatted floors. New systems of rearing veal calves involve loose housing in barns with Yorkshire boarding sides, floors bedded in straw and the animals fed *ad libitum* replacer from machines.

Management of breeding stock replacements

Rearing policy from weaning depends on two main factors, the season in which the replacement is born (autumn or spring) and the age at which it is to be calved. Age at calving is important as it is related to conception, dystokia, milk yield in first lactation, overall lifetime milk production and the herd calving pattern.

Conception is largely influenced by liveweight, oestrus being associated with weight. Table 18.10 gives the target liveweights for various breeds.

Dystokia problems require particular consideration and help to determine the appropriate weight and age at first calving. Calving problems are greater in younger heifers, especially if calved before 22 months of age. The size of both the calf and the heifer are important, and both are influenced by the feeding of the heifer. Condition scoring is a valuable management aid and heifers should score between 3 and 3.5, six to eight weeks prior to parturition. If the condition of heifers is correct at this time, restricted feeding during the last weeks of pregnancy will not affect heifer size but will help to reduce calving difficulties by minimizing the growth of the calf.

Choice of bull affects calving difficulties, but the choice is also influenced by the values of the calves as herd replacements. Sires from large beef breeds cause the greatest problems in Friesian heifers. As Friesian bulls tend to cause more problems than either of the two smaller beef breeds, Aberdeen Angus and Hereford, their use on heifers is not recommended unless there is a need for a large number of dairy herd replacements.

Table 18.10 Target liveweights (kg)

Breed	*Weight at service*	*Calving weight*
Jersey	230	340
Guernsey	260	390
Ayrshire	280	420
Friesian	330	500
Hereford × Friesian	320	500
Aberdeen Angus × Friesian	290	430

Another advantage in favour of the Hereford and the Angus is that both colour-mark their calves thus adding value. Whatever the choice of breed, individual variation within breeds has a great effect on the incidence of difficult calvings.

First lactation yield is lower in heifers calved early, milk yield being closely related to liveweight at calving (Table 18.11). In subsequent locations, differences in milk yield are minimal. Evidence suggests that because early calved heifers (two years) are kept in the herd to the same age as later calved (three years) they average one lactation more. Thus their lifetime yield is increased, as well as providing the extra calf.

Heifer rearing should be planned so that replacements enter the herd to fit the intended calving pattern. This is one of the main determinants in maintaining a system of block calving. Once the age and the month at which the heifer is to calve have been decided, then growth rates to achieve the necessary target liveweights at service and calving can be calculated.

Differences between growth rates for autumn- and spring-born heifers are largely due to the higher weight gains achieved at grass (Table 18.12). Maximum use of grass means economical rearing and the advantage of compensating growth.

Table 18.11 Milk production according to age at first calving

	Age at first calving (months)				
	23–25	*26–28*	*29–31*	*32–34*	*35–37*
Herd life (lactations)	4.00	4.03	3.84	3.81	3.78
Lifetime yield (kg)	18 747	18 730	17 964	17 991	17 657

Source: Wood (1972)

Table 18.12 Target weights for two-year calving

	Weight (kg)
Autumn-born heifer	
Birth	35
Turnout (6 months)	150–170
Yarding (6 months)	275–300
Service (15 months)	330
Turnout (18 months)	375–400
Calving (24 months)	500–520
Spring-born heifer	
Birth	35
Turnout (3 months)	80–100
Yarding (6 months)	140–160
Turnout (12 months)	250–280
Service (15 months)	330
Yarding (18 months)	400–420
Calving (24 months)	500–520

Source: MLC/MMB joint publication *Rearing Replacements for Beef and Dairy Herds* (reproduced with permission).

Autumn-born heifer calving at two years old

Autumn-born calves should be weaned at five weeks of age when consuming at least 0.75 kg of concentrates. The concentrate should be palatable and contain 17% crude protein. Between five and 12 weeks calves should be fed concentrates containing 15% crude protein *ad libitum* up to a maximum of 2 kg/d and hay to appetite. Silage as a partial substitute for hay can be fed from six weeks of age. From three months to turnout in the spring the concentrates fed can be cheapened by reducing the crude protein to 14%. Hay or silage should be fed *ad libitum*. By six months each calf will have consumed 50 kg of early weaning concentrates, about 300 kg of rearing concentrates and 330 kg of hay.

To achieve the desired gains during the first grazing season (6–12 months of age) a continuous supply of good quality grass and the control of parasitic worms are essential. Calves should be vaccinated against husk (lungworm) before turnout unless clean pastures are available (clean pastures being those that have been free of cattle since the previous midsummer). Supplementary feeding of concentrates (1–2 kg/d) after turnout prevents a check in growth which might otherwise occur. A change to clean silage aftermath after dosing against stomach worms in midsummer is generally recommended. At this time cereal feeding (9% protein plus vitamins and minerals) may be introduced when grass becomes scarce or very wet and lush.

The grazing system should be integrated with the conservation area; 0.25 ha/animal can be divided into three sections. One section is grazed until the end of June, whilst the other two are cut. After this time the two conserved areas are grazed and the other one cut; finally all three areas are grazed.

The second winter is best subdivided into two halves: from yarding until service at 15 months and from service until turnout at 18 months. Cattle benefit from dosing against internal parasites at yarding and from the use of an insecticide against warble fly. A daily liveweight gain of 0.6 kg/d is important to ensure a target service weight of 330 kg and a good conception rate. Conserved forages form the basis of the ration (25 kg silage or 6 kg hay/d) and this is supplemented by 2.5 kg of concentrates (12% protein). The concentrates can be based on barley and the protein content and quantity of concentrate adjusted according to the quality of roughage.

Identification of bulling heifers is often found to be a problem. Careful observation for oestrus, and heat detection devices may prove to be valuable aids. After service the liveweight gain may be reduced to 0.5 kg/d; this allows the concentrate level to be reduced. The overall concentrate use during 12–18 months should be about 250 kg of concentrates and 4–5 t of silage or 1.25 t of hay.

During the second grazing season target weight gain should be about 0.7 kg/d. With good grassland management no concentrates are necessary until steaming-up (generous feeding of cow pre-calving) takes place in late summer. Excess steaming-up should be avoided to prevent overstocking of the udder prior to calving. The best guide to the level of feeding during late summer is body condition score. This should be between 3 and 3.5 six weeks before calving.

Spring-born heifer calving at two years old

The management and feeding of the newborn calf until weaning is the same as for autumn-born calves. As young calves are too small to make efficient use of grass, concentrate (16% crude protein) feeding should be continued after turnout. Hay should also be available during this time (Table 18.13). If growth rate from grass alone falls below 0.5 kg/d, concentrate feeding should be restarted. The target stocking rated should be about ten calves/ha. This may be achieved by a similar system of grassland management as for autumn-born calves. It is important to ensure that grazing is clean, as very young calves are extremely susceptible to parasites. At yarding calves should weigh between 140 and 160 kg and have consumed 100 kg of concentrates and 125 kg of hay from weaning.

From yarding at six months to turnout in the following spring at 12 months, a growth rate of 0.6 kg/d should be maintained. At yarding animals should be dosed against internal parasites and may be dressed against warble fly during the winter. The basis of the ration will be either silage or hay, the amounts fed being 18 kg or 5 kg/d, respectively. The level of concentrate feeding depends on the quality and quantity of the conserved forages, a guide being 2–3 kg/d of 16% crude protein. If the growth rates are not maintained, the target service weight of 330 kg at 15 months of age in the spring will not be achieved and conception rates will be poor.

The second grazing season growth rates from turnout to service will be 0.8 kg/d. After service, target growth rates can be reduced to 0.6 kg/d. Supplementation with mineralized cereals (barley) should commence in early autumn.

During the second winter the aim should be to produce a Friesian heifer calving down at about 500 kg. This can be achieved by feeding 2 kg/d of concentrates and up to 30 kg of silage. This will take approximately 4.5 t of silage or 1.25 t of hay and 360 kg of concentrates. Steaming-up prior to calving can be practised and the heifer's body condition score should be between 3 and 3.5 six weeks prior to calving.

Heifer replacement rearing enterprises compete for resources with milk-producing animals. The two major economies that can be made in the resources required in the production of replacement heifers are the reduction in the number required, and reduction of the age at which they calve. Both these enable considerable economies to be made in land, labour and capital invested in livestock and buildings. However, young heifers grown well enough to calve at two years of age need a high plane of nutrition. This necessitates the feeding of greater quantities of concentrates, thereby increasing the cost of concentrate feed for animals calving at two rather than

Table 18.13 Approximate feed quantities used

	Autumn born	*Spring born*
Milk substitute (kg)	13	13
Concentrates (kg)	700	920
Silage (t)	5.5	7.25
and hay (kg)	100	
or hay (t)	1.9	2.0

three years of age. As far as total feed cost is concerned, heifers calving around two years of age cost nearly as much as those calving a year older. Thus intensification of heifer rearing with a greater reliance on concentrates has important repercussions in that the land and labour saved can be made available for milk production or other more profitable enterprises.

In addition to these direct savings in resources the heifer calving at the younger age will have a longer herd life with a greater total lifetime milk production (Table 18.14).

These higher lifetime yields make up for a slightly lower first lactation yield; 4300 kg for a two-year-old heifer compared with 4500 kg for a three-year-old.

As far as total feed is concerned it appears that heifers calving at two years of age cost nearly as much as those calving a year older. The financial saving in grazing and forages for the younger calving animals is substantially eroded by the need for a higher concentrate input to maintain growth rates. This is of especial importance during the winter periods unless the diet comprises ample good quality forage.

Even if there is no great saving in the cost of producing younger calving heifers there is a saving in capital investment in young stock because of the quicker turnover in animals and a reduction in the number of followers kept. Indirect benefits may result from the successful adoption of such a system of rearing, the intensification required having repercussions throughout the whole dairy enterpruse, especially with respect to better grass production, conservation and utilisation. The effects that the more rigorous discipline involved in rearing heifers to calve at two years imposes are likely to extend right across the rearing enterprise and benefit the whole farm economy.

In beef suckler herds calves out of heifers have a slightly slower growth rate than calves out of five-year-old cows (Table 18.14). These slower calf growth rates are largely due to the early calved beef heifers having a slightly lower milk yield. Over their lifetime beef heifers calved first at two years of age produce more calves over their lifetime than those calving at three years.

Feeding dairy and beef cattle

Details of the characteristics of the main feedstuffs used by cattle may be found in Chapter 17, and discussion of the theory of animal nutrition in Chapter 15. This section outlines the compilation of rations for dairy and beef cattle.

Table 18.14 Effect of age at calving on calf growth (Percentage difference in 400 d weights compared with calf out of a five-year-old cow)

	Percentage difference
Heifer	
First calving at 2 years	−8
$2\frac{1}{2}$ years	−5
Cow	
Second calf	−3
Third calf	−2
Fourth calf	0

Source: MLC/MMB joint publication *Rearing Replacements for Beef and Dairy Herds* (reproduced with permission).

Breeding stock

One of the main constraints involved in the rationing of dairy cows is dry matter intake (DMI). A simple yet relatively accurate estimate of DMI is given by the equation:

$$\mathrm{DMI} = 0.025\,\mathrm{W} + 0.1\ \mathrm{Yield}$$

where W = bodyweight in kg.

For cows in the first six weeks of lactation this prediction should be reduced by 2–3 kg. A more realistic prediction is given by:

$$\mathrm{DMI} = [135\,\mathrm{g/kgW}^{0.75} + (0.2 \times [\mathrm{Y} - 16.4)] \times \mathrm{CF}$$

CF is a conversion factor which takes account of the variation in appetite as lactation progresses, and is taken as approximately 0.78, 0.88, 0.93, 0.96 and 0.98 for weeks 2, 4, 6, 8 and 10 of lactation. Intake will be increased by up to 5% by feeding a variety of feeds, and by feeding the concentrate component on a 'little and often' basis, and will be decreased by 5% by silage with an ammonia-N content above 10%, and by 10% if ammonia-N is above 15%.

Energy requirements depend on liveweight, milk yield, milk quality and stage of pregnancy. Requirement for maintenance is given by the formula $ME_m = 8.3 + (0.091\,W)$ and for production of 1 kg milk by $ME_l = [(0.376\,BF) + (0.209\ Prot.) + 0.946] \times 1.69$.

A summary of requirements is given in Tables 18.15 and 18.16. In early lactation it is often impossible to meet total energy requirements within the constraints imposed by limited DMI. In such circumstances backfat may be utilized. Each 1 kg of fat used replaces 28 MJ of dietary energy. Provided the cow is in good body condition at calving (condition score 3–3.5), the loss of up to 30 kg backfat over this period is acceptable. Utilization of backfat in this way allows an increase in the forage: concentrate ratio of the diet and may well be economically desirable. At some stage it will be necessary to replace any backfat utilized. This is best done during late lactation rather than in the dry period and will require an input of 34 MJ for each 1 kg backfat gained (see Chapter 17). Fetal growth increases dramatically during the later stages

Table 18.15 Computed maintenance allowances for dairy and beef cattle (MJ/d)

Liveweight (kg)	*MJ*
100	17.4
200	26.5
300	35.6
400	44.7
500	53.8
600	62.9
700	72.0

(Based on 8.3 + 0.091 W.)

Table 18.16 Computed requirements (MJ) for 1 litre of milk of varying composition

Protein (g/kg)	*Fat content (g/kg)*						
	32	*34*	*36*	*38*	*40*	*42*	*44*
28	4.61	4.74	4.87	5.00	5.13	5.26	5.39
30	4.68	4.81	4.94	5.07	5.20	5.33	5.46
32	4.75	4.88	5.01	5.14	5.27	5.40	5.53
34	4.82	4.95	5.08	5.21	5.34	5.47	5.60
36	4.89	5.02	5.15	5.28	5.41	5.54	5.67
38	4.96	5.09	5.22	5.35	5.48	5.61	5.74

(Based on 1.69 (0.376 BF + 1.209 Prot. + 0.946).

of pregnancy: requirement for fetal growth is calculated as $1.13e^{0.0106t}$ (where t = number of days pregnant and e = 2.718). In practice, little accuracy is lost if additions of 10, 15 and 20 MJ/d are made for months, 7, 8 and 9 respectively.

Protein requirements for breeding females may be expressed either in terms of rumen degradable and undegradable protein (RDP and UDP) or of metabolizable protein (MP).

Calculation of requirements for RDP and UDP is straightforward. RDP requirement will be 7.8 × ME of the diet. This is the maximum amount of RDP which the micro-organisms of the rumen can capture. Any RDP or non-protein nitrogen (NPN) in excess of this amount will be lost, passing into the bloodstream in the form of ammonia and possibly causing metabolic disturbance. An excess of more than 250 g/d RDP should be avoided. All animals have a requirement for tissue nitrogen (TN). This may be met either from microbial sources (i.e. from captured RDP) or from dietary UDP. Requirements for TN average 12 g/d for maintenance and 4.8 g/litre of milk (5.7 litre for Jerseys). Where backfat is being utilized, 16 g TN/kg fat are saved: to replace fat increases demand by 24 g TN/kg. Total TN requirement is compared with the ability of microbes to meet that requirement. Microbial nitrogen to the tissues (TMN) is given as 0.53 × ME of diet; where TN < TMN there will be no demand for dietary UDP. Where TN > TMN there will be a demand for dietary UDP, calculated as UDP = [(1.91 TN) – ME] × 6.25.

From the above it is evident that there is a significant interaction between energy input and protein requirement. Where, for example, backfat is being utilized as an energy source in early lactation, there is a reduced input of dietary energy. This reduces the rumen microflora's ability to capture RDP and increases demand for dietary UDP. Flat-rate feeding of concentrates, especially at low levels of input, also makes a response to dietary UDP more likely.

In terms of metabolizable protein (MP) the cow's requirement for maintenance may be calculated as 2.3 g $W^{0.75}$ and that for lactation as 1.4 CP (where CP = protein in milk). Backfat contributes 138 g MP for every kg utilized and requires 233 g when it is replaced.

Having established the requirements of the animal, and knowing the analyses of all available foodstuffs, a suitable ration may now be formulated. Amounts of ingredients to be included will depend on availability, cost, palatability and form, but most formulations will aim at maximizing forage input and minimizing use of purchased foods. The overriding constraint in this respect is the relationship between the animal's potential DMI and its dietary energy needs. This relationship is known as the minimum energy density of the diet (minimum M/D = total dietary energy [ME]/total DMI).

As minimum M/D goes up, so the maximum amount of forage possible in the total diet goes down. In a simple, two-ingredient situation the maximum forage DM input is given as:

$$\text{Max. forage DM} = \frac{\text{DMI (ME of concentrate} - \text{M/D)}}{\text{ME of concentrate} - \text{ME of forage}}$$

It is obvious from this equation that increases in potential DMI, or decreases in dietary energy demand through backfat mobilization, can significantly increase forage utilization. Conversely, if concentrate input is restricted and total appetite is limited, for example in the first few weeks of lactation, backfat loss is inevitable. In this regard, few cows are able to eat more than about 12 kg forage DM/d, irrespective of amount of concentrate fed.

Once a diet has been formulated to meet ME requirements, within DMI constraints, it may then be evaluated in terms of RDP/UDP or MP (Table 18.17.) (See Chapter 15). Where discrepancies between provision and requirement exist, they may be corrected by manipulation of the proportion or quality of the compound component(s) of the diet.

Growing/finishing cattle

The metabolizable energy system is usable in formulating rations for mature/lactating cattle because the efficiencies with which such cattle use energy (0.72 for maintenance; 0.62 for production of milk) are constant irrespective of ration. In beef cattle efficiency of energy use for maintenance is constant (0.72), but that of production (kg) varies with diet according to the formula:

$$kg = 0.0435\,\text{M/D}$$

where M/D = energy concentration of the diet.

In situations where the diet is known, therefore, the ME system may be used to predict the likely performance of animals on that diet. Formulation of a ration requires the use of a different approach, namely the net energy (NE) system.

Table 18.17 Daily MP requirements for lactating cows (600 kg liveweight; backfat 40 g/kg, protein 32 g/kg)

	Yield (kg)				
ΔW (kg/d)	*40*	*30*	*20*	*10*	*0*
–1.0	2020	1552	1084	–	–
–0.5	2093	1625	1156	688	–
0	2165	1697	1229	761	420
+0.5	–	1819	1351	883	542

Source: Cooper, R.A. (pers. comm.).

Predicting performance from a known ration

Once the energy (ME) provided by a ration is known, predicting performance depends on calculating:

(1) the energy needed for maintenance (M_m);
(2) the energy used for gain (E_g);
(3) the energy required for 1 kg of gain (EV_g).

The gain predicted (d) will then be the energy used (E_g) divided by the energy required for 1 kg gain (EV_g), namely

$$M_m = 8.3 + 0.091\,W$$

$$E_g = \frac{(\text{Total ME} - M_m) \times K}{1.05^*}$$

$$EV_g = 6.28 + 0.3\,E_g + 0.0188\,W$$

$$DLWG = E_g/EV_g$$

* A 5% allowance for safety in prediction may be added.

The diet must then be checked to ensure that it contains sufficient protein to sustain this rate of gain (Table 18.18). This prediction takes no account of breed or sex effects, and should be taken only as a guide to performance.

Table 18.18 Computed valued for $E_m + E_g$ (MJ NE)

Gain required (kg/d)	*Liveweight (kg)*				
	100	*200*	*300*	*400*	*500*
0	12.4	18.8	25.2	31.6	38.0
0.25	14.7	21.6	28.6	35.5	42.5
0.50	17.4	25.0	32.6	40.1	47.7
0.75	20.7	29.0	37.3	45.6	53.9
1.00	24.6	33.9	43.1	52.3	61.5

Formulating a diet to meet target daily gains

Formulating a ration to meet required daily gains necessitates the use of net energy values. This does not demand knowledge of the efficiency with which the animal uses each feed for production but does require estimation of the average efficiency of utilization of a feedstuff for maintenance and gain (k_{mg}). This estimate is derived from a value known as animal production level (APL), which expresses total energy requirement ($E_m + E_g$) proportional to the requirement for maintenance (E_m). As APL increases (i.e. as the proportion of the energy used for gain increases), so k_{mg} will fall. Since the energy value of feedstuffs is expressed in terms of ME, it is necessary to calculate the net energy (NE_{mg}) value of each feedstuff separately for each diet and each animal group, and to use NE values in compiling the ration. The equations used in this process are as follows:

$$E_m = 5.67 + 0.061\,W$$

$$E_g = 1.05 \times \left[\frac{DLWG\,(6.28 + 0.0188\,W)}{(1 - 0.3\,DLWG)}\right]$$

$$APL = \frac{E_m + E_g}{E_m}$$

$$k_{mg} = \frac{\text{ME of feed} \times APL}{1.39\ \text{ME of feed} + 23\,(APL - 1)}$$

$$NE = \text{ME of feed} \times k_{mg}$$

Having calculated all these values it is then possible to compile a ration, using a modification of the equation used for compiling dairy cow rations:

$$\text{Max. forage DM} = \frac{\text{DMI (NE compound} - \text{N/D)}}{\text{NE compound} - \text{NE forage}}$$

Appetite (DMI) in growing beef cattle is generally lower than that of mature cows, falling linearly from 0.029 W at 100 kg liveweight to 0.021 W at 500 kg liveweight. Some values from the equations above are given in Tables 18.19 and 18.20.

Once again these equations do not take account of breed and sex effects, and modification to diets may be necessary in the light of these factors. Separate consideration must also be given to requirements for protein, vitamins and minerals.

Metabolizable protein requirements for beef cattle vary according to sex and breed class (early, medium or late maturing). Thus there are nine possible combinations. Table 18.19 gives requirements for medium-maturing castrates. Corrections, by ± 10%, may be used for heifers/bulls or for early/late maturity.

As with dairy cows, once the diet has been formulated for DMI and ME its ability to meet MP requirements may be calculated. Where the diet is based upon grass silage it is likely that a DUP source will also be necessary.

Appetite (DMI) in growing beef cattle is generally lower than that of mature cows, falling linearly from 0.029 W at 100 kg liveweight to 0.021 W at 500 kg liveweight. Some values from the equations above are given in Tables 18.20 and 18.21.

Table 18.19 MP requirements of medium maturing steers (castrates)

	Liveweight (kg)				
ΔW	100	200	300	400	500
0.5	220	261	299	335	373
0.75	286	322	355	390	426
1.0	348	379	409	441	477

Table 18.20 Computed values for APL ($E_m + E_g/E_m$)

Gain required (kg/d)	*Liveweight (kg)*				
	100	*200*	*300*	*400*	*500*
0	1.00	1.00	1.00	1.00	1.00
0.25	1.19	1.15	1.13	1.12	1.12
0.50	1.41	1.33	1.29	1.27	1.26
0.75	1.67	1.54	1.48	1.44	1.42
1.00	1.99	1.80	1.71	1.66	1.62

Table 18.21 Computed NE values (NE_{mp}) of feedstuffs

APL	*ME of feed (MJ)*					
	8.0	*9.0*	*10.0*	*11.0*	*12.0*	*13.0*
1	5.8	6.5	7.2	7.9	8.6	9.4
1.2	4.9	5.7	6.5	7.3	8.1	8.9
1.4	4.4	5.2	6.1	6.9	7.8	8.7
1.6	4.1	4.9	5.8	6.7	7.6	8.5
1.8	3.9	4.7	5.6	6.5	7.4	8.3
2.0	3.8	4.6	5.4	6.3	7.3	8.2

(Based on MEF × k_{mg})

$$\text{where } k_{mg} = \frac{\text{MEF} \times \text{APL}}{1.39\ \text{MEF} - 23\ (\text{APL} -)}$$

Beef production

The EEC beef industry is dependent on dairy farming. Agricultural policy restricting milk output will have a profound effect on the beef sector in both the short and the long term. In the short term dairy farmers will react to milk quotas by reducing herd numbers, and as more cows are culled this will increase the supply of cow beef. In addition less young stock will be retained for breeding. In the long term the number of calves available either as surplus bull calves or dairy cross beef calves will be reduced. Since the introduction of milk quotas in 1984, the UK dairy herd has declined by 602 000 head, i.e. by approximately 20%.

Recent developments in beef production systems, using imported breeds, better grassland management and more efficient use of concentrate feeds, have resulted in greater liveweight gains and a move from grass to yard finishing.

Baker (1975) suggests four factors that need to be considered when choosing a system for beef production.

- Financial resources. These include cash flow requirements and the capital availability.
- Physical resources, e.g. the area and quality of grassland available, field structure and the availability of water.
- Date of birth of calves – autumn or spring.
- Type of cattle – pure dairy, dual purpose or beef, or crosses.

Growth is usually measured by the change in liveweight of the animal. As this includes changes in the weight of feet, head, hide and the internal organs, including the content of the gut, besides carcass tissue, it is not always a reliable indicator of the final amount of saleable beef. Liveweight gain follows a characteristic sigmoid curve (see Chapter 15), the rate of growth being influenced by nutrition, breed and sex.

An important phenomenon in some beef systems is compensatory growth. When animals have been fed on low quantities and/or low quality feed, growth will slow down or cease. This is known as a store period. When full levels of feeding are resumed, these animals will eventually catch up animals that have been kept on full feeding levels; this is compensatory growth. It is exploited in beef systems so winter feeding can be kept at as low cost as possible and when spring feeding is started compensatory growth takes place, taking advantage of relatively cheap grazed grass.

The development of the animal varies as different tissues mature at different rates; nervous tissue first, followed by bone, muscle and finally fat (for a fuller explanation see Chapter 15). If at any time the energy intake of the animal is in excess of that needed for the growth of earlier maturing tissues, bone and muscle, then fat will be deposited. In time the fattening phase is reached when fat grows fastest. Fat is deposited in the body in a certain order: subperitoneal fat (KKCF, kidney knob and channel fat) first, then intermuscular fat, subcutaneous fat and, finally, intramuscular fat (marbling fat).

In the later maturing breeds, which have characteristically high growth rates, the fat deposition phase occurs at a greater age and weight than in the early maturing breeds fed on similar diets. Sex effects are also important: bulls grow faster than steers and these in turn grow faster than heifers. In parallel, bulls are later maturing than steers and steers later than heifers. As a result heifers of early maturing breeds quickly reach an age when they are depositing fat and are slaughtered at a younger age. Bulls of the late maturing breeds need high levels of feeding before fattening commences (Fig. 18.5). In practice, different types of cattle are managed differently and are slaughtered at different levels of carcass fat cover to suit the requirements of different sections of the meat trade. An understanding of the relationship between production system and cattle type is fundamental in planning production for a particular market.

The amount of fat in the animal influences killing-out percentage (sometimes called the dressing percentage), i.e. the yield of carcass from a given liveweight. As the animal gets heavier and fatter, the killing-out percentage increases. Systems of describing carcasses have been developed in many countries. The purpose of beef carcass classification is to describe carcasses by their commercially

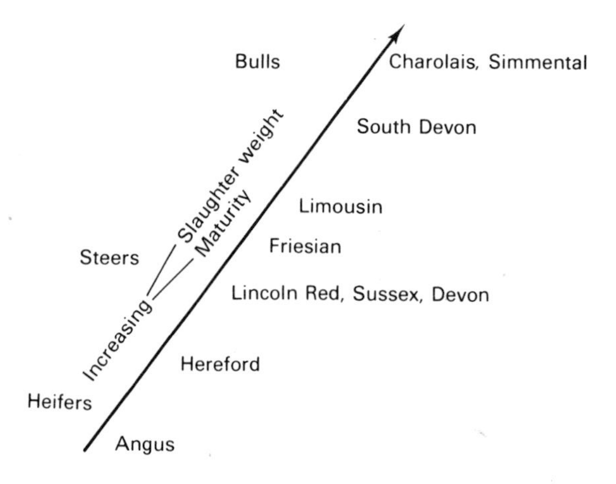

Fig. 18.5 Ranking of cattle of different breeds and types in terms of slaughter weight and rate of maturity. (From MLC (1982), reproduced with permission.)

important characteristics. The development of the classification system sought to improve the efficiency of marketing throughout the whole industry. The description enables wholesalers and retailers to define their requirements. Producers should be able to obtain higher returns by producing animals to match these market requirements. The result is a flow of information from the consumer through the retailer to the farmer. Producers can then assess the economics of providing one type of carcass from others and plan their systems of management accordingly (Fig. 18.6).

The scheme is based on a grid system which describes fatness on the horizontal scale and conformation on the vertical scale (Table 18.22). Table 18.23 gives the average composition of carcasses in each of the EC fat classes.

Fatness and conformation are determined by visual appraisal. Conformation relates to shape and takes into account carcass thickness, blockiness and the fullness of the round. Fatness is always referred to before conformation, e.g. a carcass falling in fat class 2 and conformation R would be described as 2R. The scheme also includes information on weight, sex type and sometimes age. The MLC undertakes many demonstrations of the system throughout the UK every year and anyone interested should contact the nearest MLC Regional Fatstock Offices.

Selecting cattle for slaughter

The two most important aids to the selection of cattle for slaughter are weight and fat cover (finish). knowledge of the weight of animals and their previous performance means that decisions can be based on finish. Once animals reach their minimum weight, the decision to sell should be based on finish. Excess fat is wasteful and has to be trimmed off carcasses besides which feed turned into unwanted fat is wasted and is very expensive. The best way of assessing carcass fat cover is to handle the cattle.

The diagram (Fig. 18.7) shows the five key points for handling:

(1) over the ribs nearest to the hindquarters;
(2) the transverse processes of the spine;
(3) over the pin bones on either side of the tailhead and the tailhead itself;
(4) the shoulder blade;
(5) the cod in a steer or the udder in a heifer.

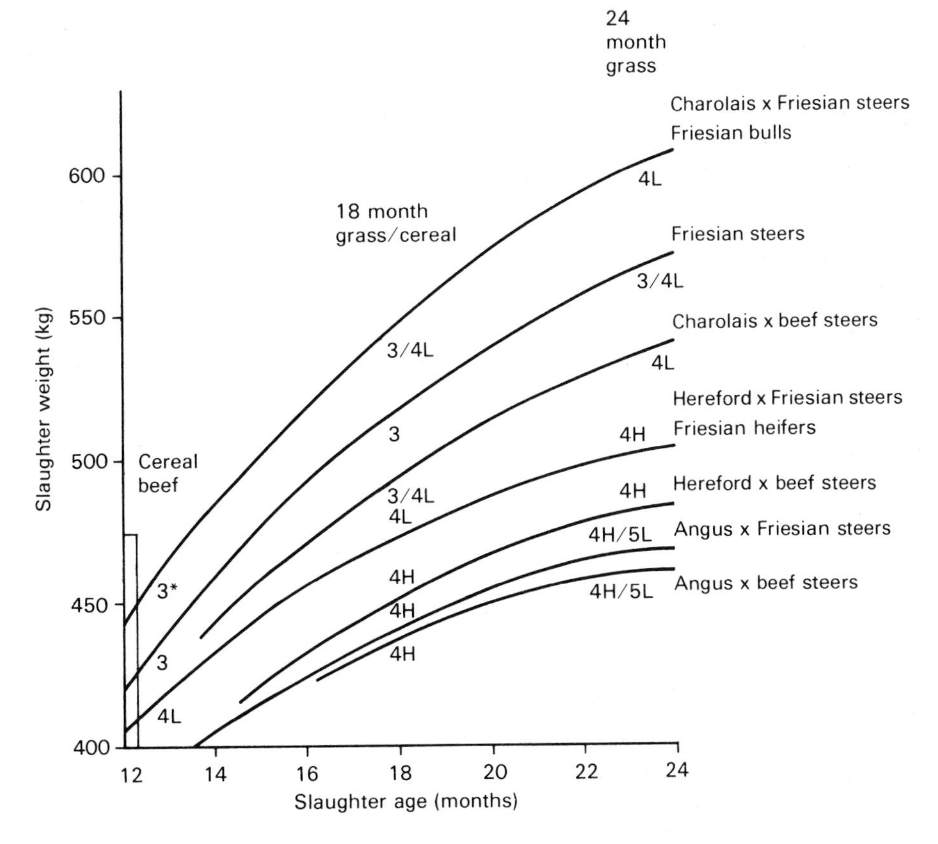

Fig. 18.6 Average liveweight, slaughter age and carcass fat for different breeds and types of cattle. (After MLC (1982), reproduced with permission.).

Table 18.22 Percentage distribution of Great Britain clean beef carcasses in the EEC classification grid (Jan–Dec 1985)

		Fat class							
		Leanest 1	2	3	4L	4H	*Fattest* 5L	5H	*Overall*
Very good	E			0.1	0.1	0.1			0.3
C O	U+		0.1	0.7	1.4	0.9	0.3		3.4
N F	–U		0.5	2.5	5.3	3.6	0.9	0.1	12.9
O R	R		1.0	7.4	17.2	10.7	2.2	0.3	38.8
M A	O+	0.1	1.3	7.9	15.2	8.1	1.6	0.3	34.5
T I	–O	0.1	0.8	2.8	3.4	1.3	0.3	0.1	8.8
O N	P+	0.1	0.2	0.3	0.2	0.1			0.9
Very poor	–P	0.1	0.1	0.1	0.1				0.4
overall		0.4	4.0	21.8	42.9	24.8	5.3	0.8	

Source: MLC (1986) *Beef Yearbook* (reproduced with permission).

Table 18.23 Average composition and saleable meat yield of *R* conformation* carcasses by fat class

Carcass composition	Fat class				
	1 and 2	*3*	*4L*	*4H*	*5L*
	Percentage of carcass weight				
Lean	69	64	61	58	53
Fat	12	18	22	25	31
Bone	19	18	17	17	16
Saleable meat					
Meat yield	74	72	71	70	69
Fat and lean trim	7	10	12	13	15

*Most common conformation class; it accounts for about 40% of classified carcasses.
Source: MLC (1982) *Selecting Cattle for Slaughter* (reproduced with permission).

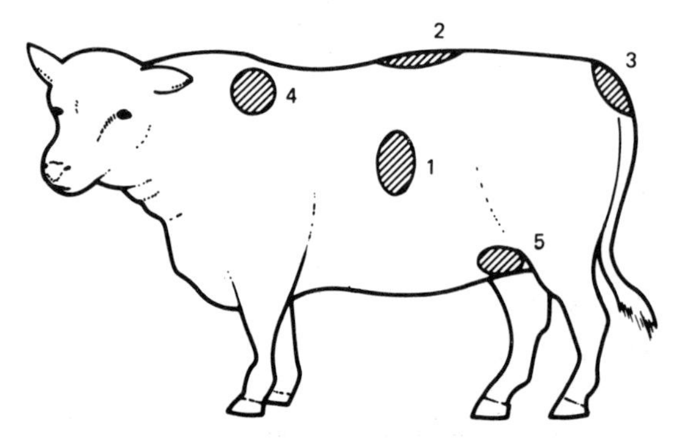

Fig. 18.7 Key handling points.

If producers are doubtful about the fat levels in animals they have selected for slaughter, they should check their classification reports or follow carcasses through the abattoir.

An EEC directive banning the sale of carcasses from animals in which hormone growth promoters were used commenced 1 January 1988. The British Government implemented a ban on the use of hormone growth promoters from 1 December 1986. This ensured that Britain was in line with other member states of the European Community.

At the time of writing the use of feed additive growth promoters, e.g. Romensin and Avotan, is still permitted.

Beef production systems

Cereal beef

Cereal beef, more commonly known as barley beef, is so named because of the system of production. Dairy-bred animals are fed on an all-concentrate ration, usually based on barley. This encourages rapid liveweight gain and animals are slaughtered at 10–12 months of age weighing 430–470 kg. Killing-out percentage is usually 54–56%, thus yielding a carcass of 230–265 kg.

Calves are reared on an early weaning system (see section earlier in this chapter headed 'Early weaning'). Weaning takes place at five weeks of age on to a concentrate ration containing 16% crude protein. At 10–12 weeks of age a diet usually based on rolled barley supplemented with protein, vitamins and minerals giving a mix of 14.5% crude protein is fed *ad libitum*. From six to seven months the protein level of the mix can be reduced to 12% thus reducing costs (Table 18.24).

The system is best suited to late maturing animals as early maturing animals become overfat at light weights. The Friesian is most commonly used because of its ready supply from the dairy herd. As bulls are later maturing than steers and have better food conversion efficiencies (FCE), there has been an increasing number of entires kept on this system. Keeping bulls is facilitated by animals being housed throughout. Their faster and leaner growth, compared with steers, means that bulls can be slaughtered at the same age (10–12 months) weighing 470 kg. For a comparison of bulls and steers see Table 18.25.

Practical problems that may be encountered include respiratory diseases and bloat. With animals housed throughout, building design is of particular importance both to the handling and physical management of stock and to reducing respiratory diseases. Good ventilation and draught-free buildings significantly reduce the incidence of pneumonia. Bloat or rumen tympany is a

Table 18.24 Liveweight and protein levels (% CP) of diet

Period	*Liveweight (kg)*	*% CP in diet*
5–12 weeks	100	16
3–6 months	250	14
6 months-slaughter	430–470	12

Table 18.25 Cereal beef targets (Friesians)

Period	*Gain (kg/d)*		*Economy of gain kg feed: kg gain*	
	Bulls	*Steers*	*Bulls*	*Steers*
0–5 weeks	0.45	0.45		
6–12 weeks	1.0	0.9	2.7	2.8
3–6 months	1.3	1.2	4.0	4.3
6– slaughter	1.4	1.3	6.1	6.6
Overall	1.2	1.1	4.8	5.5
Slaughter age (d)	338	345		
Slaughter weight (kg)	445	404		
Carcass weight (kg)	231	211		
Feed inputs (kg)				
Milk powder	13			
Calf concentrate	155			
Protein supplement	225			
Barley	1500			

Source: Allen & Kilkenny (1980).

greater problem when animals are kept on slats rather than in bedded pens where they can consume roughage. If 1 kg/d of hay or barley straw is fed, this normally prevents bloat.

Cereal beef production is sensitive to the relative prices of calves, barley and beef. Its main advantages are that it makes no direct use of land as all feeds can be bought-in, and it is not seasonal, hence an even cash flow can be established once a regular throughput of animals is established.

Maize silage beef (Tables 18.26 and 18.27).

Continental producers have developed this system of fattening dairy-bred calves and it is now a well established system for bull beef production. British interest has been aroused as maize silage systems are fully mechanized and the crop provides a useful arable break crop.

Calves are reared to 12 weeks in the same way as for cereal beef. Maize silage is then introduced and fed *ad libitum*. A protein supplement must be fed to bring the overall crude protein of the diet to 16%, the main problem with maize silage being its low protein content.

Table 18.26 Maize silage beef production (Friesians)

	Bulls	*Steers*
Gain (kg/d)	1.0	0.9
Slaughter weight (kg)	490	445
Slaughter age (months)	14	14
kg feed DM/kg gain	5.6	6.0
Protein concentrate (kg)	500	500
Maize silage DM (t)	1.75	1.65
Stocking rate (cattle/ha assuming 10 t DM/ha)	5.7	6.1

Source: MLC (1978) *Beef Improvement Services* (reproduced with permission).

Table 18.27 Slaughter weights for maize silage beef

	Slaughter weight (kg)	
	Bulls	*Steers*
Friesian	480–500	430–450
Charolais, Simmental, Blonde d'Aquitaine, Limousin, South Devon × Friesian	510–540	460–480
Devon, Lincoln Red, and Sussex × Friesian	480–500	430–450
Hereford × Friesian	450–480	400–430
Aberdeen Angus × Friesian	420–450	380–410

Source: Allen & Kilkenny (1980).

As with the cereal beef the system favours animals of late maturing type.

Maize silage can also be used for finishing suckled calves. It is important to ensure that the silage has a high dry matter content (i.e. 25% DM). Silages with low dry matter levels will result in lower feed intake and lower liveweight gains.

18-month grass/cereal beef (Tables 18.28 and 18.29)

Autumn-born calves are reared on a conventional early weaning system.

Friesian and Hereford × Friesian calves are commonly used. The autumn-born steer should weigh 180–190 kg when turned out in the spring. Calves born during the winter and early spring weigh less and have weight gains at grass. The daily gain of heifer calves is 20% poorer than steers and they finish at lighter weights. Bulls grow more rapidly and produce carcasses about 10% heavier than steers (Baker, 1975).

During the grazing season the aim is to achieve a daily liveweight gain of 0.7–0.8 kg. To achieve this, grass must be managed to provide a continuous supply of high quality herbage. Supplementary feeding with rolled barley may be practised to maintain the target growth rate; this

Table 18.28 Targets for 18-month grass/cereal beef

	Performance		*Feeds*	
	Weight (kg)	*Gain (kg/d)*	*Concentrates (kg)*	*Silage (t)*
At start	50		406	1.0
		0.75		
At turnout	197		76	–
		0.77		
At yarding	336		456	4.1
		0.95		
At slaughter	494			
Stocking				
N fertilizer (kg/ha)		264		
Stocking rate (cattle/ha)		4.32		
Grazing gain (kg/ha)		1098		

After MLC (1986) *Beef Yearbook* (reproduced with permission).

Table 18.29 Comparison of Friesian steers, Hereford × Friesian steers and Hereford × Friesian heifers in an 18-month beef system

	Friesian steers	*Hereford × Friesian*	
		Steers	*Heifers*
Weight (kg)			
At start	50	50	41
At turnout	208	212	213
At yarding	357	352	330
At slaughter	522	484	410
Daily gain (kg)			
First winter	0.78	0.69	0.66
Summer grazing	0.82	0.77	0.66
Second winter	0.97	1.09	0.65
Feeds			
Concentrates (kg)			
First winter	400	380	310
Summer grazing	80	72	50
Second winter	625	340	190
Silage (t)			
First winter	1.1	1.5	1.3
Second winter	4.5	3.7	2.1

After MLC (1986) *Beef Yearbook (reproduced with permission).*

will usually be necessary for the first few weeks after turnout in the spring and after late August when herbage quality and quantity begins to decline.

The animals should weigh about 320–350 kg at yarding. The target rate of liveweight gain is 0.9–1.0 kg/d. Winter feeding is usually based on silage which is supplemented with mineralized rolled barley. The amount of barley fed will depend on the quality of the silage and the desired weight gains. Cattle are slaughtered at 15–20 months of age weighing 400–520 kg.

Grass silage beef

Grass silage beef is also known as The Rosemaund Beef System or Storage Beef. The system normally uses dairy bred calves – Hereford × Friesian or Friesian bull calves – fed on grass silage to appetite with rationed compound feeds and housing the animal throughout. Animals are slaughtered at 11–15 months of age.

Calves are reared on an early weaning system, and are gradually changed from the early weaning ration to a rearing compound. High quality silage is introduced at this time. At 12 weeks the rearing compound is restricted to 2 kg/animal/d and this level of compound is fed until slaughter. If silage quality is less good the level of compound may be incresed to 4 kg/animal/d (see Tabie 18.30 for further details).

The system depends on the ability to obtain high growth rates from young animals; it is therefore important that the quality and quantity of silage is adequate. Rosemaund results show that when grass is cut at a (D-value of 70 or over satisfactory liveweight gains can be achieved.

A comparison of results from units using Hereford × Friesian and Friesian calves is given in Table 18.31.

Stocking rates are dependent on silage yields and tonnes of silage used per animal. High stocking rates lead to a high working capital requirement and hence high interest charges.

Table 18.30 Feeding bulls from weaning to sale (storage beef)

Time from calf arrival (kg)	*Liveweight (kg)*	*Feeding*
5–6	65–75	Weaned: *ad libitum* early weaning concentrate
6–8	75–85	Calf concentrate replaced by rearing compound fed to appetite
8–12	85–105	Silage first offered to appetite
12–16	105–130	Compound gradually reduced to 2 kg/d
16–sale	130–slaughter	Silage offered to appetite plus 2 kg compound/d rising to 4 kg if needed

Source: Hardy & Meadowcroft (1986).

Table 18.31 Performance of Hereford × Friesian bulls, Friesian bulls and Friesian steers on grass silage beef system

	Hereford × Friesian bulls	*Friesian bulls*	*Friesian steers*
Performance			
Days to slaughter	393	364	395
Mortality	3	3	2
Weight at start (kg)	105	110	111
Weight at slaughter (kg)	489	483	498
Daily gain from 12 weeks (kg)	0.98	1.05	0.98
Feeds			
Concentrates (kg)	824	954	976
Silage (t)	5.4	5.7	5.5
Stocking rate (head/ha)	7.87	7.44	6.88

After MLC (1986) *Beef Yearbook* (reproduced with permission).

Suckled calf production

Suckled calf production is practised under a variety of environmental conditions, from hill land to lowland. Beef suckler cows vary from the larger and milkier types to the smaller and hardier types.

The number of purebred beef calves has declined, crossbreds becoming more prevalent, and this introduces the heterosis effects of greater fertility and calf viability.

The choice of season of calving is affected by the availability of buildings and winter grazing, the bulk of calvings taking place in either the autumn or late winter/early spring. A short calving season enables the cows to be fed as one group without over- or under-feeding of individuals and also gives a more uniform batch of calves to be sold or fattened.

The use of condition scoring has added some precision to the management of sucker herds. The body condition of breeding cows at service and at calving is particularly important in reducing the incidence of barren cows.

Herds that have a condition score of above 2, with the optimum being 2.5–3, have the better calving intervals and the greatest number of reared (Tables 18.32 and 18.33).

Table 18.32 Relationship between body condition and reproductive performance (beef cows)

Cow scores	*Calving interval (d)*	*Herd average*	*Calves weaned per 100 cows served*
1–2	418	1–2	78
2	382	2	85
2–3	364	2–3	95
3+	358	3+	93

Scores on the scale 1 (very thin)–5 (very fat)

Source: MLC (1978) *Beef Improvement Services* (reproduced with permission).

Table 18.33 Body condition score targets

Stage of production	*Target score*	
	Autumn calving	*Spring calving*
Mating	2.5	2.5
Mid pregnancy	2.0	3.0
Calving	3.0	2.5

Suckler calf production usually produces calves for sale post weaning as stores; this is especially true of hill and upland producers where supplies of winter feed are scarce. These will then be fattened by lowland farmers. Lowland suckler herds are more likely to carry their calves through to slaughter (Table 18.34).

Season of calving tends to determine the production system. Autumn calving is most popular as calves sold in the autumn store sales are older and heavier. However, autumn calving involves more buildings and a greater requirement for winter feed.

Table 18.34 Average performance of suckler herds, 1992

	Lowland	*Upland*	*Hill*
Cow performance/100 mated			
Calving spread (weeks)	13	13	12
Calves born live	92	92	95
Calves purchased	2	3	2
Calf mortality	2	2	2
Calves reared	92	93	95
Calves			
Age at sale/transfer (days)	238	278	259
Weight at sale/transfer (kg)	295	301	283
Daily gain (kg)	1.05	0.93	0.93
Feeds			
Cow concentrates (kg)	136	145	223
Calf concentrates (kg)	58	96	142
Silage (t)	2.6	4.1	4.3
Stocking			1.10
Stocking rate (cows/ha)	2.23	1.63	43
N fertilizer (kg/ha)	147	121	

After MLC (1993) *Beef Yearbook* (reproduced with permission).

Grass finishing of stores

Suckler bred stores are purchased in the autumn sales and may be overwintered before being finished off grass the following summer (Table 18.35). Alternatively the stores may be finished over winter for sale the following spring (Table 18.36).

Dairying

Milk production is the largest enterprise in UK agriculture with an annual net sum received by producers of some £2293 million (Table 18.1). This accounts for about 19% of the total value of all agricultural output.

Table 18.35 Average results for units overwintering and grass finishing stores, 1987–88

Performance	
Grazing period (days)	131
Weight (kg)	
At purchase	275
At turnout	365
At slaughter	481
Daily gain (kg)	
Over winter	0.56
Over grazing	0.89
Grazing gain (kg/ha)	567
Feeds	
Concentrates (kg)	
Over winter	235
At grazing	42
Silage (t)	3.6
N Fertilizer (kg/ha)	156
Stocking rate (cattle/ha)	3.62

After MLC (1989) *Beef Yearbook* (reproduced with permission).

Table 18.36 Average results for winter finishing suckled calves, 1988–89

Performance	
Feeding period (days)	157
Weight at start (kg)	344
Weight at slaughter (kg)	499
Daily gain (kg)	0.99
Feeds	
Concentrates (kg)	527
Silage (t)	2.6
Prices	
Suckled calf cost (p/kg liveweight)	127
Sale price (p/kg liveweight)	120

After MLC (1989) *Beef Yearbook* (reproduced with permission).

Lactation curves

If the milk yield of cows is plotted against time, a graph of a lactation curve is produced. The standard lactation is 305 d and the annual cycle can be conveniently split into four distinct parts of early, mid and late lactation and the dry period (Fig. 18.8). After calving, milk yield will rise for a period of four to ten weeks when peak milk yield will be achieved. The time taken to reach peak yield varies with breed, individual, nutrition and yield.

Once a cow has reached peak yield the subsequent decline is approximately 2.5% per week or 10% per month. Many producers are now beating this performance with milk declining by 2% per week, post peak yield. The decline in heifers is about 7–8% per month. The daily peak yield of cows is approximately 1/200 th of total 305 d yield. Thus a cow giving 30 kg at its peak will have a total yield of approximately (30 × 200) 6000 kg.

The heifers' peak yield will be approximately 1/220 th of total 305 d yield. Thus a heifer giving 18 kg at peak will yield approximately (18 × 220) 3960 kg. A useful 'rule of thumb' guide is that two-thirds of the total yield will be produced in the first half of lactation.

It is clear that the height of the curve at peak milk yield has a great influence on the total lactation yield. The factor that is most likely to limit the level at which cows reach their peak lactation is nutrition.

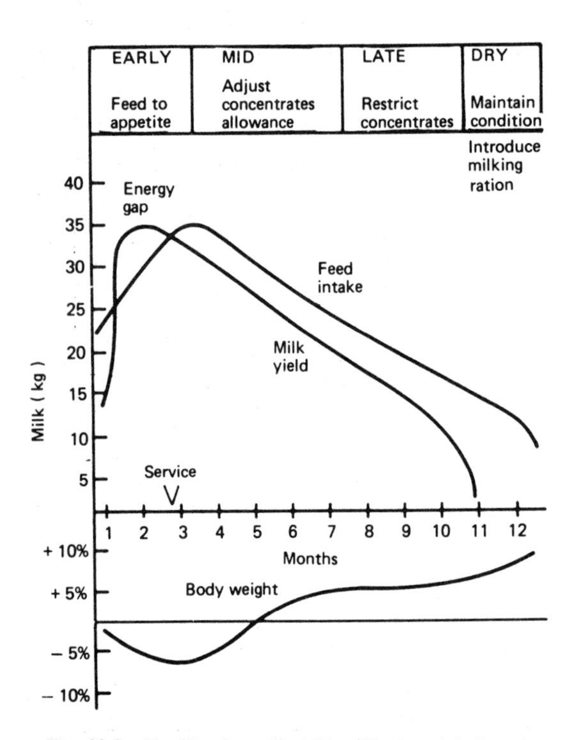

Fig. 18.8 Feed intake, milk yield and body weight lactation relationships for a dairy cow producing 35 kg milk daily at peak. (From the Scottish Agricultural Colleges (1979), reproduced with permission.)

Feeding dairy cattle

The nutrition of cattle can be divided into two distinct requirements: first, energy and nutrients to provide maintenance (maintenance requirement); secondly, to provide nutrients for growth, the development of the unborn calf and milk (production requirement).

Details of rationing are given earlier in this chapter.

Accurate feeding of dairy cows involves long-term planning as the requirements and allowances at any one time can be influenced by previous nutrition.

The practice of feeding prior to calving is a good example of the nutrition in one period affecting the production in another. During the last two months of pregnancy – the latter part of the dry period – the unborn calf grows rapidly and there is extensive growth and renewal of mammary tissue. The practice of steaming-up (the generous feeding of a cow pre-calving) is designed to fulfil these two needs, as well as providing reserves of body tissue which the cow can catabolize during early lactation and accustoming the rumen to consuming increasing quantities of concentrates.

Steaming-up is usually started some six to eight weeks prior to calving, beginning with a small quantity of concentrates which is then increased each week. Traditionally, in the week before parturition the level of concentrates fed would be about one-half to three-quarters of the amount that it is expected will be needed at peak lactation. The amount of concentrates fed will depend on the body condition of the animals, the expected level of milk yield at peak lactation and the quality of bulk fodders. For efficient and economic feeding, bulk fodders should be analysed for their nutrient value. Heifers are not usually given more than 3 kg of concentrates/d unless their body condition is very poor.

As the date of calving becomes close the animal may have a temporary loss of appetite. This should been regained within 3–4 d after calving. As the appetite increases, the cow is commonly fed for the quantity of milk being produced and an extra 1 kg/d of concentrates in an attempt to increase future production: this is usually termed 'lead feeding'. However, too rapid an increase in concentrate feeding can lead to digestive upsets. It is important to realize that feeding dairy cows should not be viewed entirely in terms of concentrate nutrients as the bulk food part of the ration can contribute a substantial part of the total nutrients. This is increasingly the case since the introduction of milk quotas and at a time of falling returns to milk producers.

Correct feeding pre- and post-calving is essential to prevent excess weight loss in the early part of the lactation. Some weight loss is inevitable as the cow's appetite will not reach its maximum until after peak lactation. This difference in time taken for an animal to reach peak lactation and peak DMI means that a nutrient gap occurs. This deficiency in nutrients is made up by the cow catabolizing fat, and a daily liveweight loss of at least 0.5 kg can be expected. Too great a loss in weight can be conducive to a higher incidence of ketosis (acetonaemia). The manipulation of body weight and condition in dairy cows has become necessary to sustain high milk yields. The pattern in Table 18.37 is commonly suggested.

Frood & Croxton (1978) showed that the ability of a cow to reach a predetermined level of milk was closely related to its condition at calving. Cows whose condition

Table 18.37 Liveweight change during lactation of dairy cows

Week no.	*Liveweight change (kg/d)*	*Change during 10 weeks (kg)*	*Net effect on liveweight (kg)*
0–10	−0.5	−35	−35
10–20	0.0	0	−35
20–30	+0.5	+35	0
30–40	+0.5	+35	+35
40–52	+0.75	+63	+98

Source: HMSO (1974) Crown copyright (reproduced with permission).

score was below 2 at calving did not achieve their predicted milk yield, and those whose score was above 2.5 yielded more than their predicted yields. Animals with a high condition score at calving, i.e. having ample body reserves, gave a higher earlier peak milk yield than animals in poor condition whose body reserves could not furnish enough nutrients during the time of low appetite and high energy demands of the cows.

The condition at calving and subsequent weight loss can have repercussions on conception rates at service. To ensure good conception rates cows should have a condition score of at least 2.5 at service. Below this score fertility is adversely affected. Cattle that are 'milking off their backs' must have their diet supplemented with protein and minerals additional to those normally included in the diet.

Once peak milk yield has been achieved and production starts to decline, concentrate use can be reduced and intake of bulk foods increased. The level of concentrates fed will be determined to some extent by cow condition. Animals that have lost much weight will need liberal feeding until after they are served and are in calf. Concentrate levels should still be fed slightly in excess of milk production in an attempt to prevent the decline in yield. Liveweight during the period of mid-lactation should be stable.

During late lactation cows have a maximum DMI relative to milk yield. The aim is to maintain milk production and restore the body condition of the cow. Maximum use of good quality forage with little or no concentrates should achieve the aims.

Suggested condition scores for dairy cows are as follows:

Calving	3.5
Service	2.5
Drying off	3

It should be stressed that the above deals with obtaining the maximum amount of milk from a cow. Farmers coping with quotas by reducing the yield per cow will usually rely on feeding less concentrates and rely on a greater proportion of milk from forage.

Flat-rate feeding

With the traditional method of feeding, differing amounts of concentrates are fed according to the stage of lactation and yield. On a flat-rate feeding system all lactating cows are fed the same daily amount of concentrates throughout the winter regardless of their individual yields. Bulk food is usually silage because of its better feed value; this must be of good quality (at least 10 MJ/kg DM) and must be fed *ad libitum*. If there is not enough silage available, other good quality bulk foods can be fed in its place. On most farms the optimum level of concentrates per cow will be between 6 and 10 kg/d; this depends on the yield of the cows and the quality of the forage fraction of the diet. The system works as the amount of concentrates fed rather than their pattern of feeding is the important factor.

Flat-rate feeding of concentrates is most effective and easier to manage if the herd has a tight calving pattern.

Complete diets

In this system all the dietary ingredients are mixed in such a manner that individual ingredients cannot be selected out from the rest and the mixture is offered *ad libitum*. The whole system can be mechanized and does away with the need for equipment designed to feed concentrates both in and out of the parlour. It is more common with larger dairy herds where cows can be split into at least three groups according to the stage of lactation and appropriate rations can be fed to each group.

Self-fed silage

This system became popular after the Second World War with the advent of loose housing. The silage depth should not exceed 2.25 m for Friesians or they will not be able to reach the highest part of the clamp. If the depth of silage is greater than 2.25 m, the upper part will have to be cut and thrown down, otherwise there is a tendency for cows to tunnel into the clamp and animals can be trapped when the overhang collapses. This is especially dangerous if tombstone barriers are used as the cows can have their necks trapped between the uprights of the barriers. The width of the feeding face should be 150 mm/cow. It is unwise to have a greater width than this or the silage will not be consumed quickly enough resulting in spoilage and secondary fermentation.

The commonest barrier to keep the cows from the silage face is either an electric wire or an electrified pipe; the latter being less likely to break than wire and providing a better electrical contact. Consumption can be controlled by the distance between the barrier and the feed face. The layout of buildings is important as cows need ready access to the silage. The width of the face is relative to the number of animals that will feed from it and one of the problems encountered with such a feed system is that significant herd expansion is difficult because of the inability to increase the width of the silo face.

Milk quality

Milk is a major source of protein, fats, carbohydrates, vitamins and minerals in the human diet. All these are present in milk in an easily digested form. The greater part of milk is water. The remainder is solid which can be divided into two main groups, the milk fat (butterfat) and the solids-not-fats (SNF) (see Fig. 18.9).

The total solids content of milk varies. Milk fat is a mixture of triglycerides and contains both saturated and

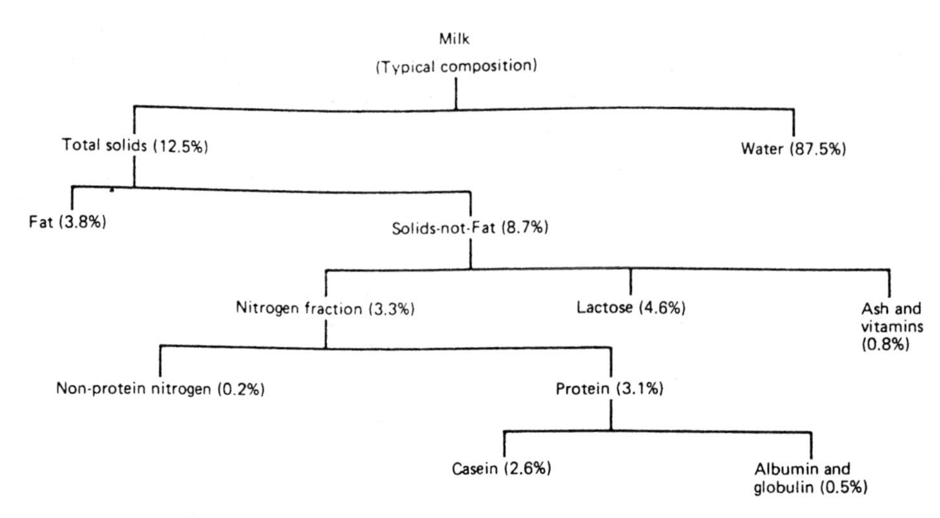

Fig. 18.9 Typical composition of milk from a Friesian dairy cow.

unsaturated fatty acids. Casein is the main protein contained in milk.

There is generally an inverse relationship between milk yield and the milk fat and protein percentages. The higher the yield, the lower the percentage composition of these components.

The breed of cow is an important factor affecting milk composition. Friesians, one of the heaviest milk yielding breeds, have some of the lowest fat and protein contents. Jerseys on the other hand have low yields but high fat and protein content. Individual variation within a breed can also cause a large effect on milk composition. Old cows tend to produce milk with a lower total solids content because yields tend to increase up to the fourth lactation and udder troubles increase with age. These factors have a depressing effect on fat and solids-not-fats. A regular intake of heifers into the herd will help to maintain milk quality, because of the age structure of the herd.

Day to day variations in milk quality can be caused by incomplete stripping of milk from the udder. The first milk to be drawn from the udder at milking contains low levels of fat when compared with the last milk to be drawn off; solids-not-fats change very little during the milking period. Fat percentage is usually lower after the long period between milkings. Part of this is due to the higher udder pressure which causes lower fat secretion.

The fat content of milk often drops when cows are turned out to lush grass in the spring. This is a function of the corresponding increase in yield and because of low fibre levels. The problem can be mitigated to some extent by feeding 2 kg of hay, straw or long roughage before the cows go out to grass and by restricting the grass so that the more fibrous stem fraction is eaten as well as the leaf. Long fibre is necessary for the rumen fermentation to produce acetic acid, acetate being the main precursor of milk fat.

Progressive underfeeding and poor body condition are responsible for cows producing milk with low protein levels. If energy supply is deficient either during the winter or as a result of grass shortage during the summer, then solids-not-fats will drop. Subclinical mastitis can be the cause of a reduction in protein levels of milk.

If the above factors are not the cause and there is still a milk quality problem, then the answer might be to change the breed or to use a progressive breeding programme with bulls selected with milk quality characteristics.

The milk pricing system is based on:

(1) compositional quality payments – based on the fat, protein and lactose content of the milk produced;
(2) contemporary payments – this means compositional quality payments are based on test results obtained in the month being paid for;
(3) seasonal adjustments.

Current values for each of the above can be found in *Milk Producer* or *Farmers Weekly*.

As milk is valued by the value of its component parts, any factors that affect the monthly quality of milk, such as low fat at turnout in spring, can have a marked effect on producers' returns. Because of the seasonality adjustments, milk price is reduced in the spring and prices are increased in autumn to encourage production in August and September. Producers attempting to alter the balance amongst milk constituents, fat, protein and lactose are restricted by seasonal, lactation and breed factors. Variation in milk constituents with stage of lactation and with season are shown in Figs 18.10 and 18.11.

Oestrus detection

Oestrus (bulling or heat) is the time during which the cow will stand to be mounted by the bull. This period lasts on average 15 ± 4 h. During the winter the period may be greatly reduced; a shorter period of 8 h is also common with heifers. The average interval between oestrus is 20–21 d.

Failure to detect animals on heat can be a major problem, especially with the increasing size of herds. Poor conception rates are often blamed on poor artificial insemination techniques, bull fertility and disease, but the actual problem may be poor management.

The first management factor necessary to improve

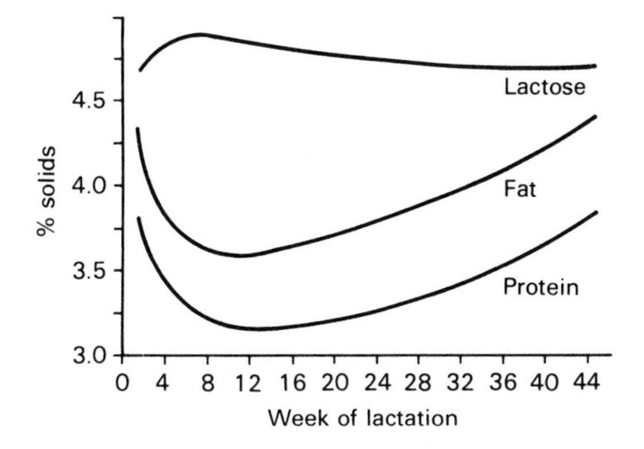

Fig. 18.10 Variation in milk constituents with stage of lactation. (Source: MMB (1983) *The New Pricing Package*.)

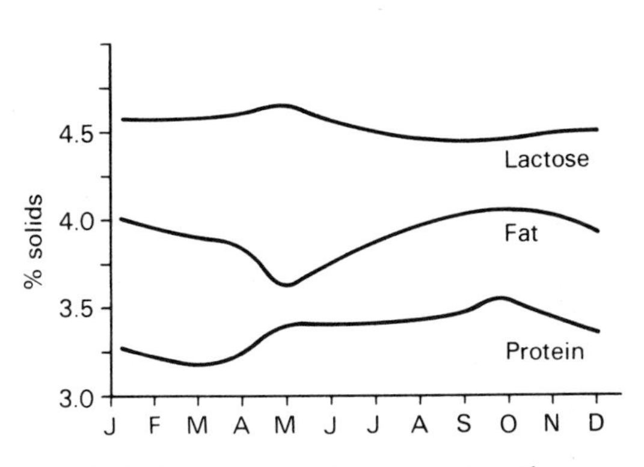

Fig. 18.11 Variation in milk constituents with season. (Source: MMB (1983) *The New Pricing Package*.)

oestrus is good cow identification. There are a number of methods by which cows can by identified clearly, but freeze branding is probably the best. In conjunction with identification should be the keeping of good records. It is essential that simple, accurate and complete records are available whilst cows are being observed for signs of oestrus. Dectecting cows in oestrus is part of good stockmanship as cow behaviour is the best indication of heat.

Signs indicating oestrus are:

- mounting other cows;
- sore, scuffed tail-head and hip bones;
- mud on flanks;
- mucus from vulva;
- restlessness;
- steaming animals.

Stockmen must be allowed adequate time for observation of the herd. The most productive times are after the animals are settled after milking and feeding. Mid morning and late evening are particularly productive. Observations should ideally take place three or four times a day and the herd should be watched for at least half an hour at a time. Any sign of oestrus should be written down and not left to memory. The target should be to detect 80% of all cows in oestrus. Various aids may be used, including vasectomized bulls, pads or paint placed on the cows' rumps, all of which have proved useful.

The timing of service is important. If cows are served too early or too late in the heat period, poor conception rates may result. Ideally, cows should be served in the middle of the heat period between 9 and 20 h after the start of heat to obtain the best results (Fig. 18.12).

The aim should be for a calving interval of 365 d. If the cow is to have a lactation of 305 d and a dry period of 60 d, with pregnancy lasting between 278 and 283 d,

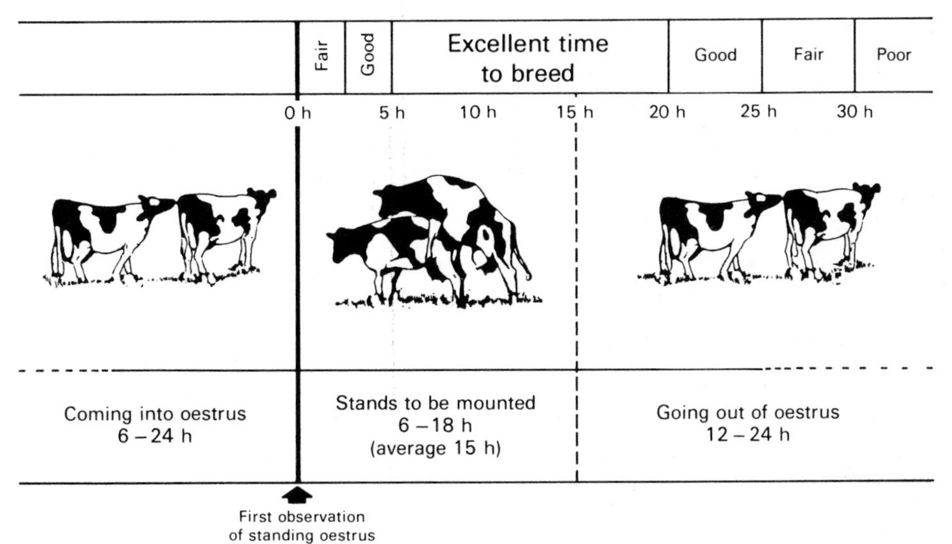

Fig. 18.12 Optimum time for insemination (from HMSO (1984); reproduced with permission).

cows should conceive 81–86 d after calving. If service is delayed until the first heat after 60 d – a common practice – the result will be a slightly higher conception rate but a longer calving interval.

Oestrus can now be controlled and synchronized by the use of prostaglandins. The technique is useful where a group of heifers needs to be inseminated. The animals will come into heat at the same time and the result will be a tighter calving pattern. In older cows the technique can be used on individual animals rather than the herd as a whole. The technique is not a cure for infertility and will only bring an animal on heat that is cycling normally (see section on Prostaglandins in Chapter 15).

The practice of calving the herd over a period of about ten weeks is not as common as would be expected despite the potential simplification of management routine. All the main tasks, calving, oestrus detection, service and drying off can occur during a specific time period for the entire herd. Coupled to this is the fact that the feeding management of the herd can be simplified by having uniform groups of animals at similar stages of lactation. Heifer rearing becomes simpler as the animals have a smaller age range. Not all farmers are attracted to such a routine but the regime can create condition more conductive to the achievement of better results.

Milk production

Milk production and the premises under which milk is produced are controlled by the Milk and Dairy Regulations 1959, a requirement being that all dairy farms are registered by the Ministry of Agriculture. Premises, stock and methods of production are open for inspection at any time.

The milking process

Milk ejection or 'let-down' is controlled by the hormone oxytocin. Milk let-down is a conditioned reflex in response to a stimulus. Quiet handling of animals prior to milking is essential; if cows are nervous or frightened, milk let-down will be inhibited. The natural stimulus for cattle is the calf suckling; in dairy herds a substitute stimulus in the form of feeding and/or udder washing is used. Besides providing this stimulus, udder washing removes dirt which may contaminate the milk as it leaves the teats. Methods of udder washing include sprays or buckets of water; the warm water often includes an antibacterial agent and cloths or paper towels are used to clean and dry the udder and teats. Disposable paper towels are preferable to cloths as there is less chance of infection being passed from cow to cow.

Before or after udder washing, fore-milk (the first milk from the udder) is removed by hand into a strip cup to reveal any signs of mastitis in the form of clots, flakes or watery milk. This is a useful indication of clinical mastitis.

Teat cups should be applied as soon after washing as possible. Maximum rate of milk flow is reached after about 1 min; later the flow rate declines quite rapidly. Once milking has started it should be accomplished as rapidly as possible. Overmilking should be avoided as this may damage the udder, thus predisposing mastitis. Stripping – the removal of the last milk – can be achieved by applying downwards pressure to the teat cup with one hand and the udder massaged downward with the other, the forequarters being done first. In parlours the stripping of milk and automatic removal of teat cups, when the milk flow slows to a predetermined rate, is becoming more common.

Teat disinfection by dipping teats in a cup containing an approved iodophor or hypochlorite solution helps prevent the spread of bacteria causing mastitis.

Efficient milking is largely dependent upon a good milking routine. This should be simple and consistent providing a 'let-down' stimulus for the cow and a routine series of operations which the milker has to perform on each animal.

After milking the milk will be at an ideal temperature (37°C) for the growth and multiplication of most bacteria. As the milk is collected once a day by bulk tankers some must be kept overnight. It is important to cool the milk quickly to prevent its deterioration. Cooling takes place in a refrigerated bulk tank and the Milk Marketing Board requires milk to be cooled to below 4.5°C 30 min after milking. Before milk is allowed into the bulk tank it is passed through a filter.

Milk is routinely tested for keeping quality by direct or indirect tests and penalties are imposed for failure to meet the required standards. Stringent penalties are also incurred if antibiotics are detected in milk; these usually result from the intra-mammary treatment of mastitis.

Poor keeping quality can be caused by inadequate cleaning of the milking equipment giving a build-up of residue. Hand washing is normally used for bucket milking plants. A cold water rinse of approximately 10 litres of water can be drawn through the clusters of each unit, this removes the film of residue left after milking. The equipment is then dismantled and washed once in either detergent or a detergent and sterilant solution at 50°C, then finally rinsed in clean water also containing a sterilizing agent.

With pipeline systems where it is impractical to dismantle the equipment, cleaning is done *in situ* by circulation cleaning. Here the success of the operation relies on the properties of the chemicals and heat for the disinfectant effect. Again the basic process starts with a rinse of cold water; this is followed by the circulation of a detergent solution. The initial temperature needs to be fairly high (80–85°C) as the solution will be cooled during the first cycle round the plant. A final rinse with cold water, which may contain sodium hypochlorite for sterilization, is circulated.

The commonest sterilizing and disinfecting agents are sodium hypochlorite, hypochlorite, bromates and iodophors. They should be used in accordance with the manufacturers' recommendations.

Another method of cleaning equipment *in situ* is by the use of acidified boiling water. Here hot water (96°C) is flushed through the plant for 5–6 min and allowed to run to waste; this pre-rinses and warms up the plant. Nitric or sulphamic acid is mixed with the water; this has no disinfecting effect but removes the milk deposits from the equipment. The aim is to heat the plant to 77°C for at least 2 min to achieve disinfection.

The circulation cleaning method involves the use of expensive chemicals and takes more time (15 min) whereas the acidified boiling water system needs more water (13–18 litres/unit) at a much higher temperature thus using more energy, but only takes 5–6 min.

To prevent a build-up of scale on equipment it may be necessary to use a milkstone remover once per month.

Bulk tanks may be cleaned by hand, using long-handled brushes, or by a mechanical spray. As tanks are cooling mechanisms and often contain an ice bank at the base, cold system of cleaning is normally used. This uses an iodophor or bromate cleaning agent at mains water temperature. The tank is rinsed with cold water by hand, immediately after emptying, and the cleaning agents applied. In automatic systems the rinse is sprinkled into the tank; this is then followed by a solution containing the chemicals. The inside is then rinsed before milking.

Particular attention should be paid to the outlet, paddle, dipstick and underneath the lid and bridge of the tank.

Cattle breeding

The aim of selecting breeding animals is to produce a future generation with improved performance. Improvement is brought about by increasing the frequency of desirable genes and by creating favourable gene combinations. The genotype of an animal is its genetic constitution. The phenotype of an animal is the sum of the characteristics of the animal as it exists and is the result of both genetic and environmental effects. It is possible for animals to have the same genotype but different phenotypes owing to environmentally produced variation, i.e. how it is fed, housed and managed. Environment does not affect all characters to the same extent. For example, the normal homozygous black coat colour of the Aberdeen Angus is always dominant to the recessive red of the Hereford, thus all Angus × Hereford cattle are black with a white face. Environment has no effect as these characters are controlled by the presence of one pair of genes. The expression of many characters of economic importance – body growth, milk yield, milk quality, fertility – are controlled by many genes and environment also exerts an effect. Characters in which there is a close resemblance between parent and offspring, whatever the environment, are characters of high heritability (Table 18.38).

Greater progress can be expected if the breeding programme is concentrating on characters of high heritability.

Table 18.38 Heritability of breeding stock characters

Trait	*Heritability*
Birth weight	0.4
Weaning weight	0.2
Daily gain, weaning to slaughter	0.3–0.6
Food conversion	0.4
Mature weight	0.4
Wither height	0.5–0.8
Heart girth	0.4–0.6
Dressing percentage	0.6
Points of carcass quality	0.3
Cross-section of eye muscle	0.3
Bone percentage	0.5
Lactation yield	0.2–0.3
Fat content of milk	0.5–0.6

After Johansson & Rendel 1968 (reproduced with permission).

As the sire often serves many cows, a great deal of attention needs to be placed on his selection. The two main methods of evaluating bulls are:

(1) performance testing
(2) progeny testing.

Performance testing involves measuring an animal's individual performance – growth rate, feed conversion efficiency – and comparing them with animals from a comparable group which have been subjected to similar conditions of feeding and management. Performance testing can only be used to measure characteristics in the live animal. Assessing carcass composition used to necessitate slaughtering the animal. However, new techniques such as ultrasonics have to a large extent overcome this particular difficulty.

Progeny testing involves the examination of an animal's offspring. The characters are measured and compared against the progeny of other sires. Again progeny should be kept under similar environmental conditions. It is particularly useful for testing dairy bulls as milk production is only measurable in the female and the characters in question are often of low heritability. The method is used by the Milk Marketing Board for evaluating its dairy bulls for artificial insemination (AI) purposes. The bull's progeny can be compared with daughters sired by other bulls in the same herds. The record should reflect the differences attributable to the bull's genetic constitution and is referred to as a contemporary comparison. These contemporary comparisons can provide guides when selecting a bull for AI. The higher the contemporary comparison the greater the chances are that the bull will pass on genes resulting in improved daughters. The greater the number of daughters used to evaluate a bull in this way, the more reliable comparison figure. The number of daughters used is referred to as a 'weighting' which appears with the contemporary comparison figure.

References and further reading

Allen, D. & Kilkenny, B. (1980) *Planned Beef Production*. Granada, London.
Baker, H.K. (1975) *Livestock Production Science* **2**, 121.
Frood, M.J. & Croxton, D. (1978) *Animal Production* **27**(3), 285.
Hardy, R. & Meadowcroft, S. (1986) *Indoor Beef Production*. Farming Press, Ipswich.
HMSO (1984) *Dairy Herd Fertility*. Reference Book 259.
Johansson, I. & Rendel, J.L. (1968) *Genetics and Animal Breeding*. Oliver and Boyd, London.
MLC (1978) *Beef Improvement Services*. Data summaries on beef production and breeding.
MLC (1982) *Selecting Cattle for Slaughter*.
MLC (1986) *Beef Yearbook*.
MLC/MMB Joint Publication. *Rearing Replacements for Beef and Dairy Herds*.
MMB (1983) *The New Pricing Package*.
MMB (1986) *UK Dairy Facts and Figures*.
MMB (1986) *EEC Dairy Facts and Figures*.
Roy, J.H.B. (1980) In *The Calf*. Butterworths, London.
Scottish Agricultural Colleges (1979) *Feeding the Farm Animal – Dairy Cows*. Publication No. 42.
Thickett, W.S., Cutherbert, N.H., Brigstocke, T.D.A. &

Wilson, P.N. (1979) *BSAP Winter Meeting Paper* No. 18.
Warner, R.G. & Flatt, W.P. (1965) In *Physiology of Digestion in the Ruminant*, R.W. Dougherty, Ed. Butterworths, London.
Wood, P.D.P. (1972) *MMB Better Management* No. 7:1.

19

Sheep and Goats

R.A. Cooper

Sheep

International picture

There were 1111 million sheep in the world in 1993, and in most countries populations are increasing. Details of their distribution are given in Table 19.1. Production of sheepmeat by the world's major producers is shown in Table 19.2, with EU output detailed in Table 19.3. Mean carcass weight in most situations is 15–16 kg, but in areas where sheep are kept as dual-purpose milk–meat animals (e.g. in Italy and Greece) carcasses may only weigh 7–9 kg as lambs are slaughtered at six to eight weeks of age.

Sheep in the UK

The total ewe population of the UK is 20.65 million head. Data for breeding animals are given in Table 19.4, which also gives details of animals eligible for Hill Livestock Compensatory Allowance (HLCA) payments and thus indicates the importance of sheep in hill and upland areas. Table 19.5 gives a breakdown of the sheep industry by flock size. Sheep numbers increased in all geographic regions during 1992/93, with increases in flocks of more than 1000 head particularly noticeable.

The national sheep flock is made up of some 50 'pure' breeds and more than 300 crosses, but few of the pure breeds, and even fewer of the crosses, are of major significance except in localized circumstances. Details of individual breeds may be obtained from the National Sheep Association publication, *British Sheep*. Flock categories, important characteristics and major breeds are given in Table 19.6.

Sheep production systems in the UK are many and

Table 19.2 Sheepmeat production ($\times 10^3$ t), 1993

EU	1177	N. Zealand	500
USSR	500	N. America	202
Australia	643	Middle East	785
S. America	296	Far East	1387
Africa	880		

Table 19.3 Production and consumption of sheepmeat in EU 1992/3

	Production (10^3 t)	*Consumption (10^3 t)*
EU total	1177	1423
UK	355	386
France	172	321
Spain	247	235
Greece	132	153
Italy	86	106
Germany	44	82

Table 19.1 World distribution of sheep ($\times 10^6$) 1993

Africa 206.3		*N. America 18.1*		*S. America 99.9*		*Asia 341.6*		*Europe 133.5*		*Oceania*	
South Africa	30.0	USA	10.2	Argentina	24.5	China	109.7	E. Europe	80.6	Australia	38.1
Ethiopia	23.0	Mexico	6.0	Uruguay	25.7	India	44.6	UK	29.3	N. Zealand	51.0
Sudan	22.5	Guatemala	0.4	Brazil	19.7	Turkey	39.4	Spain	24.8		
Algeria	18.8	Cuba	0.4	Peru	11.9	Iran	45.2	France	10.4		
Nigeria	14.0			Bolivia	7.5	Pakistan	27.7	Italy	11.6		
Morocco	16.3					Afghanistan	14.2	Greece	9.7		
Somalia	6.5										

FAO Production Yearbook (1993), reproduced with permission.

Table 19.4 Numbers and distribution of breeding ewes in the UK ($\times 10^6$), 1992

		England	*Wales*	*Scotland*	*N. Ireland*	*Total*
Less-favoured areas	–SDA	3.55	4.34	4.39	0.71	14.53
	–DA	0.60	0.68		0.26	
Non-LFA		5.14	0.45	0.36	0.25	6.20
Total		9.29	5.47	4.75	1.22	20.73

Table 19.5 Distribution of breeding ewes by flock size (% of flocks, 1991)

Flock size	*England*	*Wales*	*Scotland*	*N. Ireland*	*UK*	
					% flocks	*% ewes*
1–49	3.5	1.7	1.8	8.5	2.9	0.9
50–99	6.2	3.5	4.4	16.0	5.7	1.9
100–199	14.7	8.4	9.6	26.9	12.7	5.2
200–499	34.4	28.6	22.2	33.3	30.0	16.9
500–999	25.0	31.8	26.4	11.0	26.2	23.5
>1000	16.2	26.0	35.5	4.3	22.5	51.6

Table 19.6 Breed types, important ewe characteristics and main breeds

Type	*Desirable characteristics*	*Main breeds*
Hill breeds	Regular lambing Milking/mothering ability Ability to rear 100% crop Hardiness Good wool weight and quality	Scotch Blackface Welsh Mountain Swaledale Cheviot Herdwick Gritstone
Upland breeds	Milking ability High fertility	Kerry Hill Clun North Country Cheviot
Longwools	High fertility and prolificacy Milking ability Growth rate	Blue-faced Leicester Border Leicester Teeswater
Down breeds	Growth rate Carcass quality	Suffolk Oxford Down Hampshire Down Texel Dorset Down

diverse, because of the varied conditions under which sheep are kept and the multiplicity of breeds and crosses available. They are linked, however, by an interdependence based on a substantial cross-breeding programme. A generalized outline of this programme, which is referred to as stratification, is shown in Fig. 19.1. Under the very harsh conditions experienced on many hill farms, ewes cannot thrive for more than four seasons, thus the hill farm retains many of its ewe lambs as replacement breeding stock while the sale of draft ewes provides an important part of hill farm income (14% of flock gross output at 1993 prices). Drafted into the better conditions on upland and marginal farms these ewes have several years of productive life left, and given better climate and nutrition are capable of lambing levels well beyond the 80–100% they have achieved previously.

On the upland farms to which these ewes go there is often little scope for finishing lambs satisfactorily. Put to a longwool ram these ewes produce a crossbred lamb with the benefits of heterosis evident in terms of fertility, milkiness and general vigour. There is a strong demand for such animals for lowland breeding flocks, where these attributes may be complemented by those of suitable Down rams for the production of prime lamb. Sixty per cent of ewes tupped under lowland conditions are derived from longwool × hill matings. Table 19.7 gives details of the most important of these crosses.

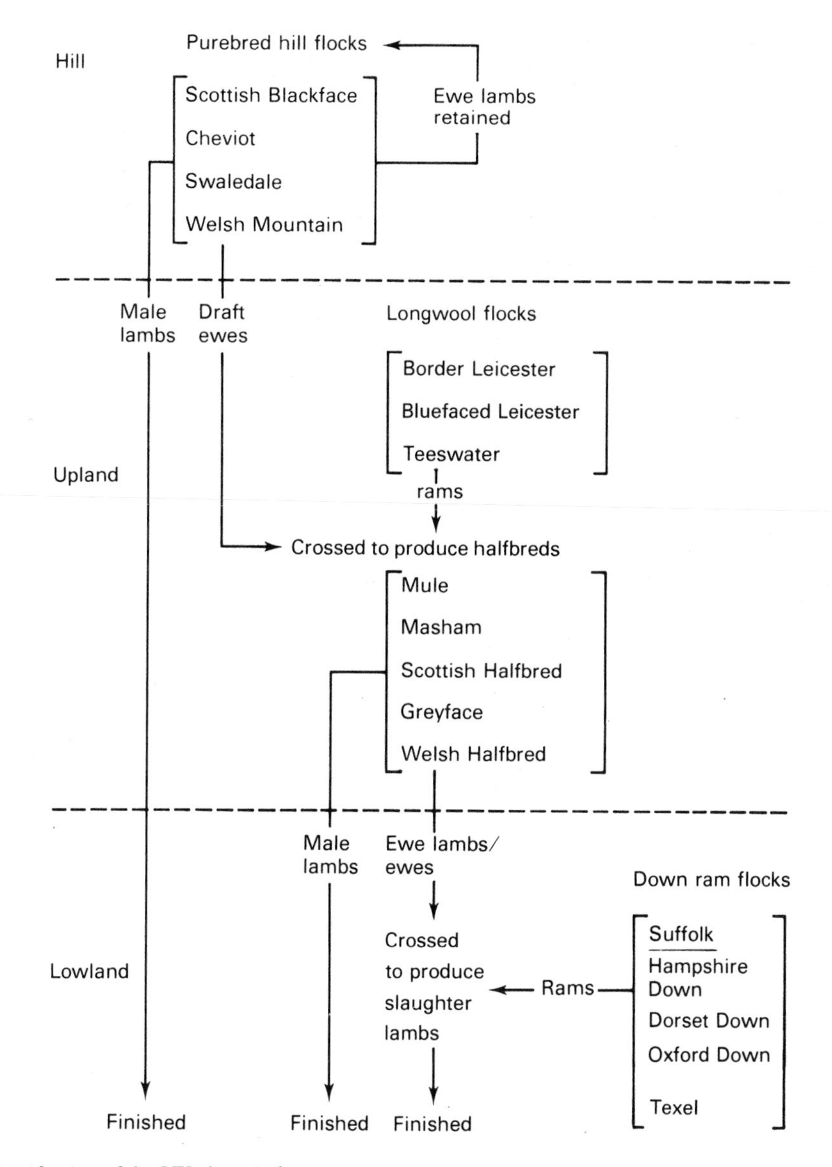

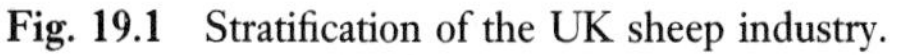

Fig. 19.1 Stratification of the UK sheep industry.

Table 19.7 Important crossbreeds and their derivation

Cross	*Sire breed*	*Dam breed*
Mule	Blue-faced Leicester	Swaledale, Blackface Speckledface, Clun
Greyface	Border Leicester	Scotch Blackface
Masham	Teeswater, Wensleydale	Dalesbred, Rough Fell Swaledale
Scottish Halfbred	Border Leicester	Cheviot
Welsh Halfbred	Border Leicester	Welsh Mountain

Factors affecting flock performance

Given the diverse nature of sheep production in the UK it is difficult to discuss flock performance in other than general terms. The standards which flocks recorded by the Meat and Livestock Commission (MLC) are achieving are shown in Figs 19.2 and 19.3. In each case target figures (top one-third) are given in parenthesis alongside achieved values. Under lowland conditions the top one-third producers achieved a gross margin 60% above average in 1993 (£880 vs £549 per hectare). The factor contributing most to this superiority was stocking rate (71%). For upland flocks the superiority of the top third was 50% (£821 vs £548 per hectare) with stocking rate contributing 80%. In hill flocks performance per ewe is more important with the number of lambs sold contributing 39% to top-third superiority and lamb value 27%.

It is apparent from these data that there is room for improvement in both hill and lowland flocks. Details on good grassland management can be found in Chapter 11. The improvement of lambing percentage, which should be defined as lambs reared/100 ewes to ram, is discussed below.

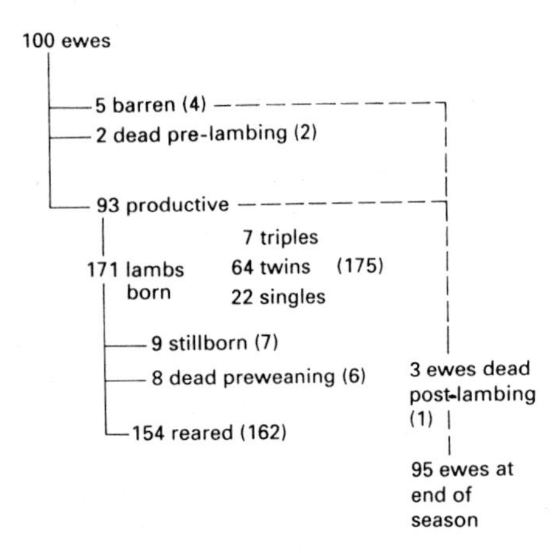

Fig. 19.2 Physical performance of lowland flocks (1993 data) (target values in parenthesis).

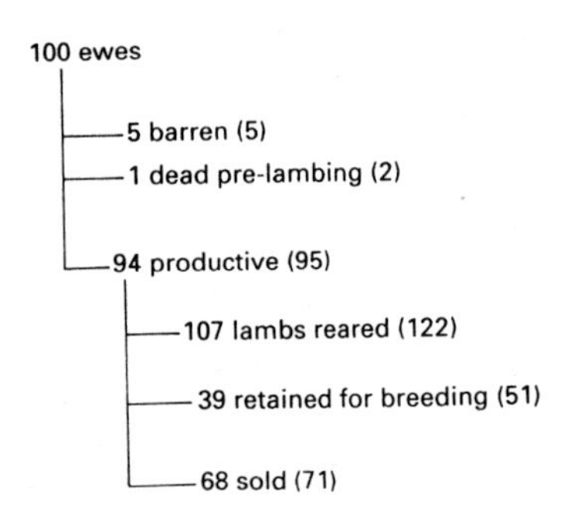

Fig. 19.3 Physical performance of hill flocks (1993 data) (target values in parenthesis).

Oestrus

The sheep is a species in which there is generally an annual rhythm of breeding activity, the onset of the breeding season corresponding with a period of decreasing daylength and cessation occurring as daylength increases in spring. There are, however, wide variations within as well as between breeds, and a much reduced cyclicity is evident at lower latitudes. Some idea of the range of dates involved is given in Table 19.8. These values can be taken as no more than guides because of the interactions of many other factors as well as daylength. The response of the ewe to photoperiod is known to be mediated by the production of a hormone, melatonin, by the pineal gland, during the hours of darkness. It has recently been demonstrated that the administration of exogenous melatonin, in the form of a subcutaneous implant, may be used to bring mating forward by up to six weeks. Implantation from mid-May (Suffolks) or early June (Mules) should be followed by the introduction of teaser rams 35 days later and of entire rams 17 days after that. Peak matings will occur 52–62 days post-implantation. This technique has also been shown to increase lambing percentage by up to 20%.

The sudden introduction of rams during the period immediately before the onset of oestrus may also bring forward breeding activity by a few days. More importantly it can stimulate a considerable degree of synchrony. When associated with the use of vasectomized, or 'teaser', rams the technique can be used to obtain a tighter lambing pattern when 'entire' rams replace the teasers after 14 d.

Ovulation and fertilization

One or more 'silent' ovulations often precede the onset of behavioural oestrus. Ovulations occur every 17 d (range 14–19 d) throughout the breeding season. Oestrus lasts for about 30 h (range 24–28 h) with ovulation occurring towards the end of this period. One of the prerequisites for a good lambing percentage is a high ovulation rate, and subsequently a high fertilization rate with minimal embryo mortality. Genotype will have influence here, although nutritional factors may confuse the situation. Within a flock ovulation rate will be influenced by the following factors.

Nutrition

It has been known for many years that putting ewes onto a rising plane of nutrition for two or three weeks prior to tupping has a beneficial effect. Traditionally ewes were weaned onto hard conditions, in part to assist drying off, in part to put them into lean condition before 'flushing'. This increasing body condition, or dynamic effect, gives some 6–8% more twins compared with the performance of ewes mated whilst on a maintenance diet, and 12–16% more twins when compared with ewes mated while in negative energy balance. More recently a 'static' effect has also been recognized, this being dependent on the level of body reserves in the ewe; ewes with greater body reserves having a higher ovulation rate. For ewes on a level plane of nutrition an extra 3.5 kg bodyweight at mating represents a potential 6% extra twins (comparisons within flock).

Recognizing the impracticability of weighing ewes on commercial farms, the former Hill Farming Research

Table 19.8 Breeding seasons for some British breeds

Breed	*Onset*		*Cessation*	
	Mean	*Range*	*Mean*	*Range*
Blackface	25 Oct	26 Sept–10 Nov	19 Feb	17 Jan–3 Apr
Welsh Mountain	25 Oct	11 Oct–11 Nov	17 Feb	1 Feb–25 Feb
Border Leicester	9 Oct	24 Sept–14 Oct	10 Feb	18 Dec–11 Mar
Romney	4 Oct	16 Sept–19 Oct	2 Mar	20 Jan–23 Mar
Suffolk	3 Oct	12 Sept–23 Oct	17 Mar	7 Feb–20 Apr
Dorset Horn	24 July	15 June–21 Aug	2 Mar	22 Jan–22 Apr

Adapted from Hafez, E.S.E. (1952) *J. Agric. Sci.* **42**, 189, with permission.

Organization devised a simple, rapid and accurate method of judging the body condition of ewes by palpating the spinal column over the loin area. Scoring, on a scale from 0–5, is possible to the nearest 0.5 (see Appendix 1). Body condition scoring may be used on sheep of any breed, and at any time of the year, as a guide to the adequacy of nutrition of the flock. It is particularly valuable in the pre-tupping and pre-lambing periods. Pre-tupping it can be used to divide the flock into groups according to their needs in terms of the duration and degree of flushing required. Lowland ewes should have a condition score of at least 3.5 at mating and hill ewes 2.0 to 2.5. In a normal flushing period (approximately three weeks), lowland ewes having an initial score of 3 or above can achieve this target easily. Those scoring 2.5 or less will need at least six weeks on improved grazing, or supplementary concentrates, if lambing percentage is not to suffer. The relationship between score at tupping and lambing percentage is shown in Table 19.9, but many breed differences exist and some of these are illustrated in Fig. 19.4. A scoring at least six weeks before tupping can identify those animals in need of preferential treatment and a second, three weeks later, can be used to monitor response. An improvement in condition from score 2.5 to score 3.5 is equivalent to a gain of some 6 kg liveweight. Body fat content (f) may be predicted using the equation $f = 8.69$ score $+ 2.69$.

Table 19.9 Relationship between condition at mating and lambing percentage

Breed	*Average score at mating*	*Live lambs/ 100 ewes to ram*
Scotch Halfbred	3.8	184
	2.6	120
Welsh Halfbred	3.5	158
	2.7	133
Masham	3.4	178
	2.2	127
Clun	3.6	156
	2.0	131

After MLC 1980, Sheep Improvement Services, Body condition score of ewes (reproduced with permission).

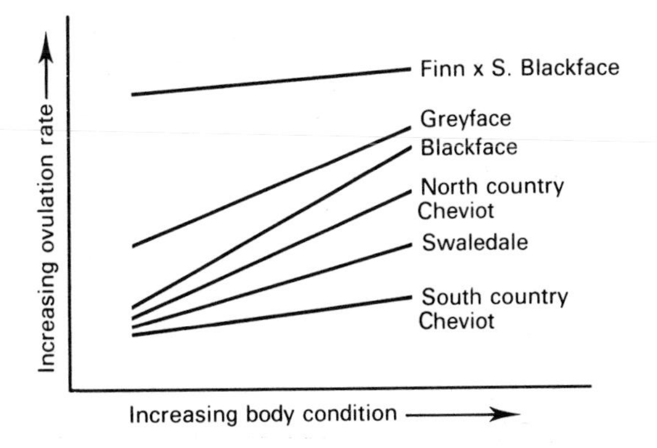

Fig. 19.4 Breed effects on the relationship between condition at mating and ovulation rate (From *Science and Quality Lamb Production* (1986). ARC with permission.)

Age

Ewe lambs may attain puberty in their first year. There is a nutrition/date of birth interaction, and with lightweight or late-born lambs oestrus may not be attained. Although most animals that attain puberty will rear one lamb at best, there is some evidence that animals which are bred as lambs may be more prolific in later life. Subsequently, mean ovulation rate will continue to rise to five or six years before falling off again. Hence the draft hill ewe has only just achieved her potential when she is sold.

Season

Many ewes begin the season with a 'silent' oestrus. Mean ovulation rate then rises for three or four cycles and will likewise drop at the end of the season. On average, therefore, lambing percentages will be lower in early or very late-lambing flocks and are very low in most out-of-season lambings. These problems are exacerbated by higher numbers of barren ewes at such times. The Finnish Landrace and its crosses do not appear to be affected in this manner and are thus widely used in 'out-of-season' production systems.

The relationship between tupping date (and hence lambing date) and lambing percentage in lowland flocks is shown in Table 19.10. Decisions on lambing date are likely to be influenced also by factors such as climate, altitude and politics. In hill situations the introduction of the ram into the flock will be dictated in part by tradition. Traditional dates in Scotland for example, are 10

Table 19.10 Effect of mating date on mean lambing percentage in lowland flocks

Lambing date	*Lambing %* *MLC data**	*WSCA† data*
January	148	137
February	155	145
March 1–15	160	—
March 16–31	167	159
April	168	—

* *MLC sheep facts* 1978 (reproduced with permission).
† West of Scotland College of Agriculture Research and Development Publication No. 7.

November for Cheviots and 25 November for Blackface. Under normal conditions all lowland sheep will ovulate and produce at least one ovum; the majority should produce two or more. Genotypically there is no reason why this should not be similar for hill breeds, the lower ovulation rates experienced in hill flocks being largely the result of environmental effects.

If all ova shed are to be fertilized, then rams must be both fertile and active. Low libido (sexual drive) may well be a problem in rams in early lambing or out-of-season situations. At the beginning of the breeding season all rams should be examined to ensure that they are healthy, sound in feet and legs and without any abnormality, especially of the testes. Microscopic examination of a semen sample, obtained by electro-ejaculation, is an additional check though results need careful interpretation, especially early in the season. Correct ewe to ram ratio will depend on circumstance. A strong ram should be able to cover 50–60 ewes adequately, but under extensive conditions the ratio may need to be lower and where an attempt has been made to synchronize oestrus it will need to be no more than 10:1. An inexperienced ram lamb may be used (carefully) in his first year, but only at a low ewe:ram ratio (20:1) and only with mature ewes. It is particularly important that ram lambs do not have to compete with experienced older rams at this time.

The practice of applying ochre to the breastbone of rams (keeling or raddling), or the use of suitable harness and crayon, is a helpful management tool. Rams treated in this way leave a mark on all ewes mated and this allows the monitoring of the progress of the breeding programme and, supplemented by more permanent markings, can form the basis of subdivision of the flock prior to lambing. In larger flocks this can lead to significant savings in concentrate costs. With such a system the colour of the raddle should be changed at least every 16 d, going progressively from lighter to darker colours.

Embryo mortality

A significant percentage of fertilized ova are lost in early pregnancy, especially in the first 30 d. Losses as high as 40% have been suggested but the mean is probably about 25%. Some loss of embryos is to be expected but steps can be taken to minimize it. In particular the avoidance of stress and the adequate nutrition of the ewe are critical; above-maintenance levels of feeding should be maintained for the first month of pregnancy. Embryo mortality also increases with increasing ovulation rate, so that ewes that have been flushed premating are likely to be more sensitive to postmating drops in feed intake. Embryo loss is also greater in early and out-of-season breeding and in ewe lambs. In many countries a further factor is likely to be heat stress, which can cause very high embryo losses if it occurs at the early cleavage stage.

During the middle part of pregnancy few losses appear to occur. The ewe is able to withstand periods of undernutrition – hill ewes often losing 5–10% of bodyweight during this period without significant effect. Losses in excess of this, or in the latter stages of pregnancy, may lead to fetal death, reduced birthweights and increased perinatal mortality.

Perinatal mortality

Perinatal mortality (including stillbirths and abortions) is a major source of loss in many sheep flocks. Surveys suggest that as many as four million lambs may be lost in the UK every year. Main causes of loss are: abortions and stillbirths 20–40%, starvation and exposure (hypothermia) 35–55%, infectious diseases 5–10%, misadventures 5–10%. Organisms of the *Pasteurella*, *Toxoplasma*, *Brucella* and *Rickettsia* groups may be implicated in so-called abortion storms in sheep. Vaccination can help avert the problem. Similarly, appropriate vaccination (e.g. against clostridia), given to the ewe prepartum, can help reduce postnatal lamb mortality.

The nutrition of the ewe in the six weeks prelambing is extremely important. During this period the fetus makes 70% of its growth. Undernutrition of the ewe at this time will lead to reduced birthweights and an increased susceptibility to hypothermia (see Fig. 19.5), or even to the death of some or all fetuses *in utero*. Additionally, correctly fed ewes are much less likely to suffer from pregnancy toxaemia (twin lamb disease), a metabolic disorder which may cause the death of both ewe and lambs. The accuracy with which pregnant ewes can be fed is increased if the number of fetuses carried is known. Pregnancy diagnosis using real-time ultrasonic scanning and carried out 50–100 d post-mating can detect pregnancy to an accuracy of 98% and fetus number to 95%. Such knowledge allows appropriate division of ewes, facilitating differential feeding and making lambing management easier. Savings of concentrates of up to 20 kg/barren ewe and 10 kg/single-

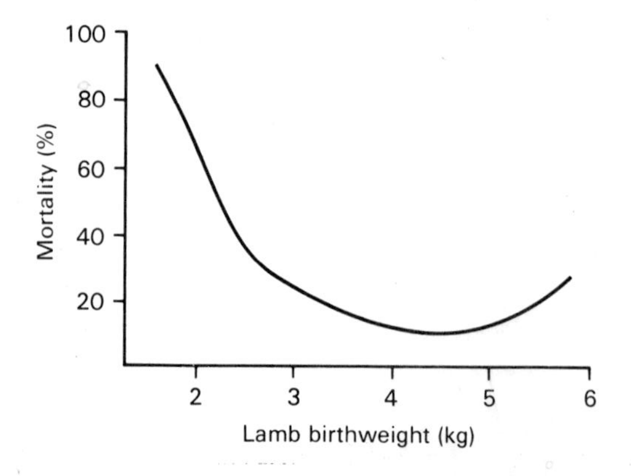

Fig. 19.5 Relationship between lamb birthweight and mortality. (From Maund B (1974) Drayton EHF Annual Review.)

bearing ewe are possible and will often cover the cost of scanning. Reduced ewe and lamb mortalities may further contribute to the cost-effectiveness of the technique.

Hypothermia

Once the lamb is born, careful shepherding can do much to ensure its survival. The lamb's ability to withstand the challenges of climate and infection is critical. A well planned care routine (see Table 19.11) can do much to reduce these problems. Hypothermia is one of the major causes of postnatal mortality. The newborn lamb has a coat with little insulating value, a large surface area:weight ratio, and is wet. It must produce heat as rapidly as it is losing it or it will die. Lambs have only small fat reserves, especially those that are small, and rely heavily on feed energy. The small lamb may quickly burn up body reserves in a cold, wet environment and thence be too weak to suckle properly. Such a situation will lead to death within a few hours of birth. The larger lamb may be able to utilize reserves for up to 24 h but must then replace them if it is to survive.

Body temperature is a good guide to the lamb's status. Workers at the Moredun Institute have developed a programme based on lamb body temperature (see Fig. 19.6) which, if rigorously applied, should dramatically reduce the number of lambs lost to hypothermia.

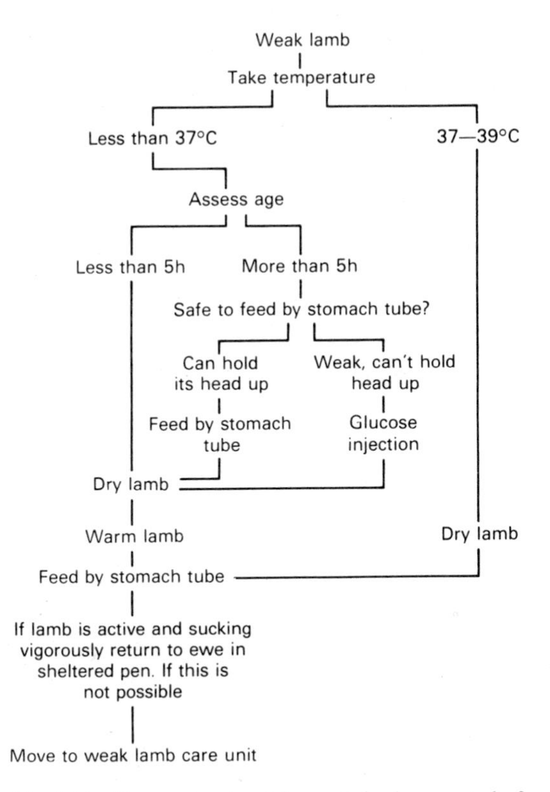

Fig. 19.6 Course of action taken with lamb suspected of being hypothermic. (From Eales, F.A. & Small, J. (1986) *Practical Lambing*. Longman, London, reproduced with permission.)

Table 19.11 Intensive care routine at lambing

(1) Check lamb breathing: clear mucus from airways
(2) Treat navel with chloromycetin or suitable alternative
(3) Individually pen ewe and lambs
(4) Draw milk from both teats of ewe
(5) Put lamb to teat and initiate sucking
(6) Use stomach tube to administer colostrum if necessary

Meat production

Growth and nutrition

Under UK conditions the main output of the sheep unit is in the form of lamb – either finished or store. The money obtained for each lamb is a function of weight, either live or carcass, quality and the date of sale. This section discusses the main factors affecting this output.

As can be seen from Fig. 19.7 the development of the components of the conceptus is variable, with most placental development having finished by six weeks prior to lambing while most fetal development is yet to take place (see Table 19.12). The importance of ewe nutrition during this period has been discussed in relation to the survival of the newborn lamb. It is also important in terms of the subsequent growth of that lamb, especially in the case of multiple births.

From a daily requirement of 12–18 megajoules (MJ) of metabolizable energy in late pregnancy, the energy demands of the ewe will increase to 20–30 MJ at peak lactation. Especially in the flock lambing before grass

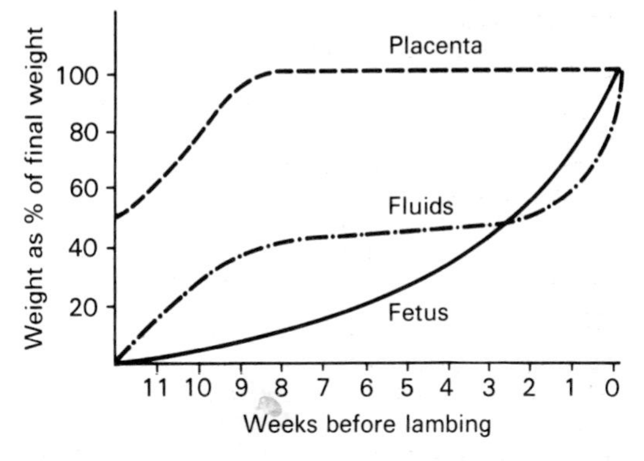

Fig. 19.7 Differential development of components of conceptus. (From Robinson, J.J. *et al. Journal of Agricultural Science*, 88, 539. Reproduced by courtesy of the editor and publisher.)

Table 19.12 Growth of fetus during twin pregnancy

Stage of pregnancy (days)	*Fetus weight (g)*
28	001
56	89
84	1 000
112	4 250
140	10 000

After Wallace, L.R. (1948) *Journal of Agricultural Science* 38, 93.

becomes available, this energy demand may not be met and the ewe will begin to 'live off her back'. Ideally the ewe will lamb down at condition score 3.5 if she is going to be able to milk satisfactorily. Ewes suckling more than one lamb respond by producing more milk; some 40% more if suckling twins, 50–55% more for triplets, and there may on occasion be a case for running ewes suckling triplets separately and feeding them accordingly. Notwithstanding this extra milk, lambs reared as multiples are unlikely to grow as fast as those reared as singles. Average figures for daily milk yield are given in Table 19.13. For the lowland ewe the average of 1.6 kg produced/d will support daily gains of 300 g in the single lamb. When producing 2.54 kg the ewe is supporting some 460–500 g of gain, but this is only 230–250 g/lamb. By the third month the relevant figures are 220 g and 130 g/d for singles and twins respectively. The effects of this on lamb liveweight are demonstrated in Fig. 19.8. For the single lamb, growth rate is almost linear, and with 'milky' ewes the lamb will need no supplementary feed. On the other hand triplet lambs start at lower weights, grow more slowly, become dependent on alternative feeds earlier and when grazing are thus much more prone to roundworm infestation. There are few advantages to be gained by leaving lambs with their dams beyond 12 weeks of age and indeed, in situations where grass is in short supply, there may well be advantage in weaning at ten weeks and allowing the lambs access to the best grazing.

A second factor affecting daily gain beyond the 10–12 week period is the shift, in terms of components of any gain, away from lean and towards fat. The characteristics of growth in the lamb are similar to those for other species, namely that initially bone growth predominates, followed by lean tissue and finally fat. Animals in which the waves of growth for bone, lean and fat are steep and close together are called early maturing. In sheep the tendency is for smaller breeds, such as the Southdown, to be early maturing and to begin to lay down fat at relatively low carcass weights, and for heavier breeds to be later maturing and capable of producing, given time, a heavier carcass at any given level of fatness.

The current demand is for leaner carcasses. The Meat and Livestock Commission (MLC) have developed a classification scheme for lamb carcasses which can be used to assess and describe any carcass in terms of its conformation (muscling) and fatness (see Fig. 19.9). On this basis, current demand is for a carcass falling into fat classes 2 and 3L. An indication of the relationship between carcass weight and classification is given in Table 19.14. It is possible to predict, with a reasonable degree of accuracy, the likely carcass weight of any breed or cross of lamb, at a fat score of 2–3L, using the formula $(WD/2 + WS/2)/2 \times$ estimated killing-out percentage (where WD = mature weight of dam breed; WS = mature weight of sire breed). Thus, for example, using a Suffolk ram (90 kg) on a Scotch Halfbred ewe (90 kg) will produce a 20-kg wether carcass at a killing-out percentage of 45. For ewe lambs this weight will be reduced by about 10%. Choice of sire breed is a matter

Table 19.13 Average milk yields during first 12 week of lactation

Type of ewe	*No. of lambs*	*Yield (litres)*		
		Month 1	*Month 2*	*Month 3*
Hill	1	1.35	1.30	0.95
	2	2.25	1.85	1.25
Lowland	1	1.60	1.55	1.25
	2	2.55	2.10	1.45

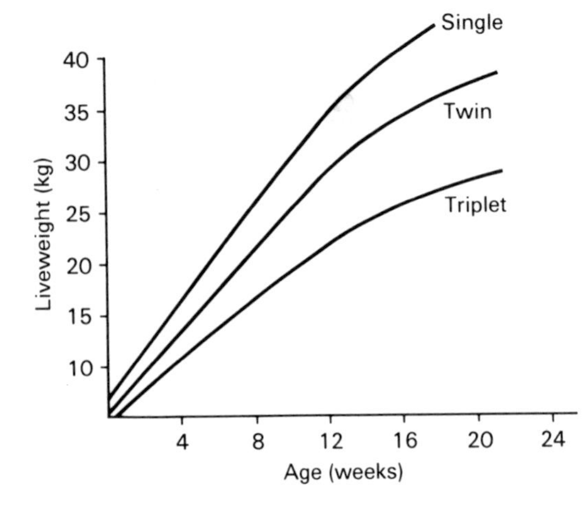

Fig. 19.8 Typical growth curves for lambs (From Spedding, C. (1970) *Sheep Production and Grazing Management*. Bailliere, London, reproduced by courtesy of the publishers.)

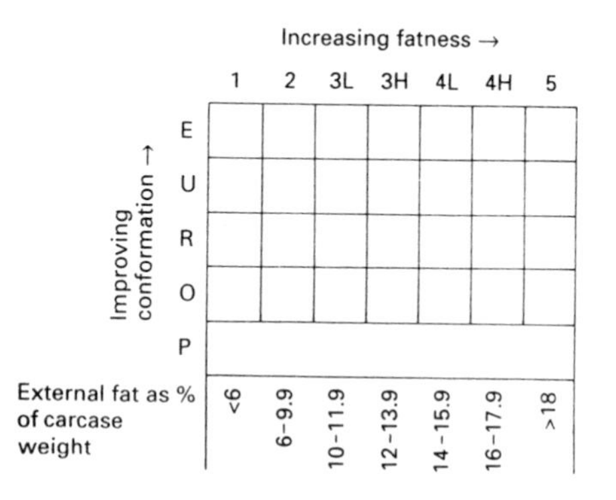

Fig. 19.9 Sheep carcass classification grid.

Table 19.14 Relationship between carcass weight and carcass classification

Carcass wt range (kg)	*% carcasses in fat class* 2	3	4	5
16–17	29	60	7	—
18–19	18	63	15	1
20–21	11	60	24	3
22–23	7	52	33	6
24–25	4	40	41	14

of individual judgement. It is essential that the breed of ram chosen suits the lamb finishing system to be used, and it must be remembered that a heavier lamb will take longer to produce and that this may be inconvenient. For example, lambs aimed at the early market need rapid growth and early maturity, and will tend to be sold at lower carcass weights, while animals to be finished on roots as late lambs or hoggets (animals sold after 1 January) need to be capable of producing a heavy carcass without laying down too much fat.

Systems of lamb production

If lambs are categorized according to date of sale, then in the UK early lambs, sold before the end of June, account for only 16% of total output; lambs finished off grass 35%, lambs finished on forage crops 30% and hoggets 19%. There are many systems of lamb finishing. Important elements of the main ones are as follows.

Early lamb

Aimed particularly at the Easter Market, prices peaking between mid March and late April as hogget numbers decline. Dorset Horns and their crosses are able to lamb regularly in the period September to December, as are selected strains of other breeds. The system is characterized by high lamb prices and high stocking rates for weaned ewes (25 ewes/ha), but these points tend to be offset by lower lambing percentages and higher concentrate costs.

Traditionally Dorset Horn lambs have been run on root crops, when only modest rates of gain are necessary. For later-born lambs the need is for concentrate feeding, with high quality feed (12.5 MJ, 170 g DCP) being fed from three weeks. On such diets an FCE (Feed Conversion Efficiency) of 3.5:1 is possible, with lambs gaining 350–400 g/day and going off at 34 kg liveweight in 12–14 weeks. This type of system fits well on to arable units where grass is limiting, and where 'off-peak' buildings and labour are available.

Lambs off grass

The traditional system of production in many areas. Such systems should aim for high lambing percentages (170+), high stocking rates (15–18 ewes/ha) and as many lambs sold 'off-the-ewe' as possible (75%). Lambs not sold by weaning may then be finished on silage aftermaths or forage crops, or sold as stores. Performance of lambs on such a system is improved by the adoption of forward-creep-grazing techniques. Two such systems, which have been well documented, are the 'Grasslamb' system developed by the Grassland Research Institute (now known as AGRI) (see GRI Farmers Booklet No. 2), and the Follow-N system of the National Agricultural Centre. Both these system utilize high levels of nitrogen (upwards of 250 kg/ha), the main difference being that the GRI system uses rotational grazing and forward creep grazes the lambs, while the NAC set stock and rely on anthelmintics to reduce worm burden. The stocking levels achieved may be compared with average values on more traditional systems of only 11–13 ewes/ha at nitrogen levels below 135 kg.

Lambs and hoggets off forage crops

Those lambs which have not been finished by 1 October in any year are generally termed store lambs. Many such animals change hands at this time of year but some will be finished on their home farms. Animals available will be wethers of the hill breeds, weighing 23–30 kg, wethers of the hill crosses (Mules, etc.) at 25–35 kg and the remains of fat lamb flocks at 30–40 kg. Feeds for these animals may be grass, silage, catch-crop roots, main-crop roots and arable by-products. It is important that lambs are matched to the feeding system proposed and that stocking rates are carefully monitored. For example, main-crop roots are expensive to grow and the aim must be for high yields and high stocking rates. On the other hand if stocking rates are too high, individual animal performance may suffer. Details of potential yields and stocking rates for the main crops are given in Table 19.15.

Grass finishing

Suited to Down crosses for finishing by December, halfbred wethers for February–March sale. Where dairy cow paddocks are being grazed, sheep must be off by December to avoid subsequent yield drop; potential silage ground may be grazed until March. Relatively low rates of gain (70–100 g/d) produce carcasses with acceptable fat levels, but Down cross lambs will become fat if held too long.

Silage finishing

Depends on availability of surplus silage of good quality (DM 25–35%, 'D' value 65 +, ME 10.5 MJ +, pH 4, NH_4N <10%). Needs lambs of 30 kg liveweight with good frame (e.g. Halfbred wethers). Growth rates are variable around 75 g/d. Where good silage is available (100–120 kg/lamb) it is only necessary to supplement with small quantities (30 g/d) of fishmeal to improve silage intake. For silage of 65 D up to 250 g/d of a 15% protein barley/protein mix will be necessary. This amount will increase to 450 g at 63 D and 750 g at 61 D. These quantities may need increasing further if silage intake is limited (e.g. for long-chop or poorly fermented material). At the same time, in some situations silage intake may be suppressed by concentrate feeding. A 10 cm feed face is required for *ad libitum* silage feeding and a 35–40 cm trough for concentrates. One of the problems of this system is that where lambs are held to take advantage of the higher prices in late February/March they may be occupying buildings needed for the breeding flock.

Finishing on roots

The performance of lambs on roots is very variable, between 50 and 200 g/d. Short-term systems, aiming at merely 'finishing' well-grown but lean lambs, will utilize

Table 19.15 Yields and stocking rates of forage crops

Crop	*Season of use*	*Yield (t/ha)*	*Stocking (lamb days/ha)*
Stubble turnips	Sept–Jan	40–50	2500 (i.e. 60 lambs for six weeks)
Rape	Oct–Feb	30–35	2500
Swedes	Dec–Mar	65–100	8000
Beet tops	Oct–Dec	?	1800
Grass	Oct–Mar	?	600

stubble turnips. Long-term systems, usually involving swedes or hardy turnips, aim to have animals putting on lean before finishing and depend on the rise in lamb prices in February/March for their profitability. One of the main problems with this system is that of balancing wastage and performance. High individual-animal performance is normally associated with high waste (up to 60% in a bad year). Where heavier stocking is practised, to reduce crop waste, lamb growth suffers. The feeding of up to 200 g barley/d can help maintain outputs and reduce variation between animals. Block grazing tends to produce lower wastage rates, without affecting lamb performance, and has lower labour requirement than strip grazing.

Finishing on cereals

The high cost of cereal-based diets means that this system is only applicable to February/March sales. Lambs must be capable of going to 40 kg liveweight without going to fat, but as with root-based finishing the production of over-fat lamb is a major hazard. Ideally lambs are introduced to cereals whilst still at grass, having been vaccinated against pulpy kidney (*C. welchii*), and preferably also against *Pasteurella*, at least two weeks earlier. Lambs may be housed when eating 500 g/d and gradually brought to *ad libitum* feeding, but shy feeders should not be brought in. The ideal cereal for lamb is whole barley, but addition of protein and minerals is then difficult. A compromise 13% protein ration might be as follows: whole barley 350 kg, rolled barley 300 kg, beet pulp shreds 230 kg, soyabean meal 100 kg, sheep minerals 20 kg. On such a diet lambs will gain 1 kg/week, at an FCR of 5–5.5:1, and will tend to get fat. Regular handling of all animals is necessary if fat class 4 carcasses are to be avoided.

The use of concentrate feeding may also be considered in systems involving 'out-of-season' lambing, when concentrates (12.5 MJ, 17% CP) are introduced at 7 d and fed *ad libitum* to slaughter, or when six-week weaning of one lamb from twins is practised (see later section headed 'Early weaning of hill lambs'). For such concentrate-fed lamb, milk substitute should be offered *ad libitum* for three to four weeks followed by one week of restricted access. A feed conversion of 1:1, on a dry matter basis, is possible for the liquid diet, with 4:1 being achieved on concentrates. Gains of 300 g/d are possible, with 10 kg milk substitute and 100 kg concentrates necessary per lamb.

Milk production

In recent years there has been an upsurge of interest in milking sheep in the UK. Elsewhere in the world the practice is widespread. Within Europe, Italy, Spain, Portugal, Greece and France have well established industries (Table 19.6). In Greece sheep's milk accounts for over 35% of all milk produced.

As may be seen from Table 19.16, yields of sheep's milk are extremely variable. In most of Europe, ewes are used as dual-purpose animals, producing both meat and milk, and are suckled for varying lengths of time. Under UK conditions it is likely that milking ewes will be weaned as soon as the lamb has taken colostrum. Many breeds may be milked satisfactorily but yields will vary. 'True' milking breeds, such as the Friesland and British Milksheep, will produce up to 450 litres in a 180–220 d lactation. More 'dual purpose' animals, such as the Clun or Mule, will yield upwards of 150 litres. The lactation curve of a sheep is similar to that of the cow. 'Letdown' occurs in two phases, with an initial flow of milk from the udder cistern followed some 20 s later by that from the alveoli. Udder washing may reduce this delay. Ewe's milk is very rich, having approximately 70 g fat, 50 g lactose and 50 g protein/kg. This creates a high energy demand (7.8 MJ/kg) as well as tending to create difficulties with milkstone deposits on equipment.

As with the dairy cow, the most difficult time in terms of feeding the ewe is in early lactation when appetite is limited. This is especially true of ewes yielding in excess of 3 litres. The nutrition of the ewe is discussed later. One of the major problems likely to be encountered when ewes are lambed in winter is that of low water intake. Ewes may be encouraged to drink by using a feed such as sugar beet pulp presented soaked and covered with water.

Table 19.16 Outline of sheep's milk production in Europe (1993)

Country	*No. of ewes* ($\times 10^3$)	*Av. flock size*	*Main breeds*	*Av. yield (kg)*	*Annual prod. ('000 t)*
			Sarda	100–250	
Italy	5000	10	Gentile di Puglia	30–35	653
			Sopravissana	50–55	
			Churra	120	
Spain	2500	60	Manchega	100	275
			Castellana	90	
			Chios		
Greece	2500	80	Vlahico	90	620
			Karagunico		
			Merino		
Portugal	1000	60–100	Churra	120	90
			Bordaliera		
			Manech	150	
France	1000	100–350	Corse	110	206
			Lacaune	160	

Wool

At one time wool was a major source of income for the sheep farmer. Today wool accounts for less than 5% of gross output. Yields of wool vary greatly between breeds, with longwools such as Teeswater and Devon longwool producing 5–6 kg/head while hill breeds such as the Swaledale and Herdwick will only yield 2 kg.

Wool is produced by follicles in the skin. Two distinct types of follicle are identifiable. Primary follicles, normally associated with an erector muscle and a sweat gland, produce coarser, medullated fibres of hair or kemp. In improved breeds many of these coarser fibres are shed soon after birth. Secondary follicles, associated only with a sebaceous gland, produce wool fibres. The ratio of primary:secondary follicles determines the fineness of the fleece and varies between breeds. For example the fine-woolled Merino may have 25 secondaries per primary, while most British breeds will have no more than eight.

Fineness is a major criterion in wool grading, the standard measure of fineness being the Bradford Count. Bradford Count is the number of hanks of yarn 510 m long that can be spun from 450 g wool. Values will vary from those of the Merino at 80+ to those of the Blackface at 27 to 30. Fibre diameters will vary from 23 μm at 60 to 38 μm at 45. There are also within-fleece differences, with breech wool being coarsest and shoulder wool finest. Breech wool also contains a higher percentage of kemp. A second quality factor in wool is its crimp; that is the number of corrugations per 25 mm. Crimp values vary from 8 to 28 and are closely correlated with fineness.

The growth of wool is photoperiodic and cyclic, 80% of growth taking place between July and November. Wool fibres are almost entirely keratin – a protein with a very high cystine content (12–14%). Notwithstanding this, the main relationship between nutrition and wool growth is in terms of energy input. Wool continues to grow even when the sheep is in negative energy balance. It does however grow more slowly and is finer under such conditions. While wool which is uniformly fine along its length is desirable, the development of a thinner area in an otherwise thicker fibre, a condition known as tenderness, is to be avoided since it reduces the value of the fleece and may even cause premature shedding. Tenderness develops in conditions of underfeeding, for example in ewes suckling triplets, or during bad attacks of parasitic gastroenteritis or liverfluke.

At shearing, care should be taken to preserve the quality of the fleece. Sheep must be dry when shorn; double cuts, which reduce effective staple length, should be avoided; organic matter such as straw and faeces should be kept out of the wool; and the fleece should be carefully packed and stored in the dry. Considerable penalties are incurred for badly presented or marked fleeces. Details of grades, prices and penalties are published annually by the British Wool Marketing Board, who have a monopoly on all wool sales.

No yield advantage accrues from shearing sheep more frequently than once yearly. Sheep are traditionally shorn in late spring or early summer, but recent years have seen the development of winter shearing of housed ewes. This technique allows increased stocking densities (up to 15% more) and tends to produce cleaner fleeces. It also reduces heat stress in the ewes, leading to increased dry matter intakes of the order of 10%. This extra intake is reflected in higher lamb birthweights and lower perinatal mortalities. In the prolific ewe flock at Drayton EHF, shearing increased birthweight by 600 g/lamb and reduced mortality from 16 to 7%.

Winter shearing does not fit all situations. To avoid problems the house must be well bedded and draught-free. A minimum of eight weeks must elapse before turnout, to allow some regrowth, and even then it is undesirable to turn shorn ewes out before early March in sheltered areas of the south or early April in the north. Ewes with condition score below 2.5 should not be shorn.

Flock replacements

Flock replacement costs are a major element in the profitability of a sheep enterprise. In 1993 the costs averaged £7.46/ewe in MLC recorded flock.

Keeping these costs down is a question of minimizing the number of ewes to be replaced each year and of keeping down the cost of each replacement.

On most hill farms some ewes will be lost and some will need to be culled. The majority of ewes leaving the farm will be drafted out at four to five years of age simply because they are no longer able to cope with conditions. Thus, on average some 25% of the flock will need to be replaced each year. Set against this is the fact that many of these ewes will be sold as draft breeding stock rather than culls, thus making a substantial contribution to the output of the flock. Traditionally, and of necessity, replacement ewes will be retained from within the lamb crop. This is essential if the ewe lamb is to become a productive member of the flock. Within a hill flock, groups of sheep become territorially organized or 'hefted'. This tendency is encouraged by carefully shepherding, for it reduces the need for fences and assists the survival of the sheep since they know of, and will seek out, areas of shelter and sources of food during bad weather. From a husbandry viewpoint one of the major problems created by the need to keep ewe lambs as replacements is what to do with them over their first winter. If they stay on the hill, then they add to feed requirements at a difficult time, and some indeed may not survive the winter. On the other hand, the traditional 'tacking' or 'agistment' of ewe lambs onto lowland farms has been made more difficult by a reduction in the number of farmers willing to cooperate and by escalating costs, both of agistment and of transport. One compromise which may become more common is the provision of housing for these animals on the home farms.

Under lowland conditions most ewes being sold off will be culls. Criteria taken into account when deciding on which animals to cull may include condition of udder, feet and mouth, previous history, condition and temperament. Many of these factors are influenced, to a greater or lesser extent, by management. It could be argued, for example, that culling because of bad udder or bad feet should only occur as an extreme measure, and that if many sheep are involved it may be that flock management is suspect. Culling on teeth condition will depend on two main management factors: stocking rate and winter feed policy. The need for a ewe to be 'sound' in mouth relates to her ability to feed. As stocking rates increase, so competition for grass increases and the ewe needs to be able to graze closer to the ground, a facility made more difficult if the ewe is 'broken-mouthed'. Similarly, if winter feeding involves the use of 'hard' roots, such as swedes or turnips, the broken-mouthed ewe will be unable to compete and will tend to lose condition. Additionally, the feeding of hard roots tends to increase teeth loss. By four and a half years of age up to 75% of ewes may still be full-mouthed if fed on hay and concentrates; in root-fed flocks the figure may be as few as 35%.

In addition to the above factors, opportunity plays a part in determining culling rate. In years when replacements are expensive, culling levels tend to be lower than in years when they are relatively cheap. The cost of each replacement can also be influenced by the age of the animal at purchase; ewe lambs being appreciably cheaper than shearlings. Having opted for the cheaper ewe lambs, the farmers must next decide whether to put them to the ram in their first year or not. Puberty in sheep is influenced by age, bodyweight and daylength, older heavier animals being more likely to attain puberty than younger or lighter ones. The advantages of breeding from ewe lambs are that it increases lifetime lamb production per ewe and decreases replacement cost, but against this must be set the likelihood of more mis-mothering problems and difficulties in getting their lambs away fat. In the end the decision may well depend on the source of the lambs. For homebred animals, whose management has been aimed at producing an animal suitable for tupping at seven to nine months, the system can work quite well. For the producer who is using longwool crosses, and who of necessity has to purchase all his replacements, it may be much less viable.

Feeding the ewe

Theory

As has been emphasized, the nutrition of the ewe is a major factor in determining flock performance. Feed requirements vary with liveweight, physiological state and environment. Details of these requirements, and suggested systems for meeting them, are discussed in the MLC publication *Feeding the Ewe*. The main points are discussed below.

Voluntary feed intake

Dry matter intake will vary with physiological state and with diet, as well as with bodyweight. Intakes will normally be within the range 60–70 g/kg $W^{0.75}$ (where $W^{0.75}$ metabolic bodyweight), but may be as high as 100 g/kg $W^{0.75}$ for concentrate diets or as low as 45 g/$kg^{0.75}$ on silage-based diets, especially if the silage is acid (pH below 4) or poorly fermented. Appetite is depressed by up to 20% in late pregnancy and very early lactation but recovers quickly to reach a peak some two to three weeks after peak milk yield is achieved. The increase in intake between late pregnancy and peak may be as high as 60%.

Energy requirements

Energy requirements vary according to bodyweight, stage of pregnancy and milk yield, and are reduced by 0.75–1.2 MJ/d in housed ewes. Maintenance requirements, based on $M_m = 1.8 + 0.1\,W$, are usually in the range 5–10 MJ/d. During pregnancy requirements increase significantly as fetal growth accelerates in the last six to eight weeks (see Fig. 19.7), being 20% higher six weeks before lambing and up to 150% higher immediately pre-lambing. During lactation energy requirement varies with milk yield, which is in turn influenced by the number of lambs suckling. Average milk yields are given in Table 19.13. Requirements for energy, based on the above considerations, are given in Tables 19.17, 19.18 and 19.19.

Protein requirements

Protein requirements also vary according to the physiological status of the ewe. Maintenance levels are generally between 65 and 100 g CP/d. For weight gains (increases in body condition) before tupping, these values increase by up to 100% (Table 19.17), but during mid-pregnancy maintenance levels are again adequate. During the last six weeks of pregnancy, requirements increase to between 110 and 235 g CP/d, depending on ewe weight and number of lambs carried (see Table 19.18). Throughout

Table 19.17 Mean daily allowances for ewes up to mid-pregnancy

*Liveweight (kg) and condition change**	*Pre-mating*		*Post-mating*					
			Month 1			*Months 2 & 3*		
	ME (MJ)	*CP (g)*	*ME (MJ)*	*CP (g)*	*MP (g)*	*ME (MJ)*	*CP (g)*	*MP (g)*
0.0	6	65	6	65	70	5.5	55	58
40 + 0.25	9	95						
+ 0.50	12	120						
0.0	8	85	9	90	91	7	70	71
60 + 0.25	12	125						
+ 0.50	17	165						
0.0	10	105	11	110	103	9	90	83
80 + 0.25	16	155						
+ 0.50	22	210						

* Level of body condition change in last 28 days pre-mating.

Table 19.18 Mean daily allowances for ewes in late pregnancy

Liveweight (kg)	*No. of fetuses*	*Weeks pre-lambing* 6			4			2		
		ME (MJ)	*CP (g)*	*MP (g)*	*ME (MJ)*	*CP (g)*	*MP (g)*	*ME (MJ)*	*CP (g)*	*MP (g)*
40	1	6.7	80	68	7.4	95	73	8.2	110	78
	2	7.1	95	72	8.2	120	78	9.5	150	87
60	1	8.8	110	81	9.8	125	86	10.8	140	92
	2	9.4	130	86	10.9	155	91	12.7	185	100
80	1	10.9	135	93	12.1	150	98	13.4	165	104
	2	11.8	155	97	13.7	190	103	15.8	235	112

Table 19.19 Mean daily allowances for lactating ewes (based on yield data from Table 19.13)

Liveweight (kg)	*No. of lambs*	*Month of lactation* 1			2			3		
		ME (MJ)	*CP (g)*	*MP (g)*	*ME (MJ)*	*CP (g)*	*MP (g)*	*ME (MJ)*	*CP (g)*	*MP (g)*
40	1	16.3	200	152	15.9	200	148	13.2	150	125
	2	23.3	300	215	20.3	240	188	15.6	190	146
60	1	20.3	210	183	19.9	205	180	16.7	170	151
	2	27.6	315	249	24.2	250	218	19.0	205	171
80	1	22.3	220	195	21.9	210	192	18.7	190	163
	2	29.6	335	261	26.2	265	230	21.0	220	183

this period the provision of 10 g CP/MJ ME in the diet should be adequate, but where ewes are expected to utilize backfat during late pregnancy there will be a need for supplementary protein, preferably in the form of a material of low degradability such as fishmeal or soyabean meal. For the lactating ewe protein requirements depend on stage of lactation and number of lambs suckling (see Table 19.19). Here, too, the use of an undegradable protein is advantageous, particularly in situations where ewes are utilizing back-fat reserves. In terms of metabolizable protein (MP), requirements for maintenance may be calculated as $2.57\,\mathrm{g}\,W^{0.73}$, that for wool growth may be taken as 20.4 g/day and that for lactation as 72 g/kg milk produced. (See Tables 19.17–19.) Calculation of requirements is complicated. Acceptable approximations (for maintenance and pregnancy) are given in Table 19.18.

Feeding in practice

In practice it may be neither practical nor economically desirable to meet the ewe's requirements in full at all times. Most flocks contain ewes with a range of bodyweights and carrying or suckling varying numbers of lambs. The aim should be to feed accurately at critical times in the production cycle, using the body condition of the ewes as a guide. The need for ewes to be in good body condition, and on a rising plane of nutrition, at tupping has already been discussed. The ewe should not be allowed to drop in condition for the first month of pregnancy, but during the second and third months a moderate degree of underfeeding, leading to weight loss of up to 5%, is acceptable. Forage-only diets, with an energy concentration of 8–9 MJ/kg DM, are adequate at this stage. During late lactation it is essential that ewes be brought back to good body condition (condition score $3-3\frac{1}{2}$). Pregnancy scanning can allow grouping of ewes according to number of fetuses and differential feeding. Otherwise lowland ewes should be assumed to be carrying twins and hill ewes singles. Concentrate feeding will normally begin, some six to eight weeks before lambing, at 100 g/head/d. This amount should gradually increase to a maximum of 600 g/d at lambing. With housed ewes and good quality silage, concentrates may only be fed for four weeks and to a maximum of 400 g. At all times the body condition of the ewe should dictate concentrate feeding level. Forage will usualy be offered *ad libitum*.

In many hill situations such feeding is economically more difficult to justify and practically often very difficult, not only because of the difficulty of reaching the ewes but also because it is undesirable to interrupt the normal grazing behaviour of the flock. In such conditions the judicious use of feed blocks may be considered. Such blocks should be sited at strategic points, allowing one block/30 ewes/week. These blocks should be offered from late January and where possible should be supplemented by small amounts of grain feeding in the last few weeks of pregnancy (up to 200 g/head/d). Additionally storm-feeding provision will often be necessary.

Adequate quantities of hay, ideally 1 bale/ewe, should be stored on the hill to be fed when grazings are snow-covered.

For the lactating ewe the amount of hand feeding will depend on date of lambing. Where grazing is immediately available, then no supplementary feeds may be needed. In early lambing flocks, root crops may be utilized or forage feeding continued. In these conditions the use of an undegradable protein source will maintain milk yields and lamb growth rates. Under normal circumstances the aim should be to restrict concentrate feeding to 45 kg/ewe per year, with about 0.5 t silage or 215 kg hay/ewe being needed during winter housing. Suggested rations for a 75 kg lowland ewe in late pregnancy are given in Table 19.20.

Grazing management

Good grazing management is one of the most important factors influencing sheep flock profitability. Variable costs will be affected by the extent to which the flock needs supplementary feeding, and by forage variable costs, while gross margin/ha is very dependent on stocking rate. Examination of MLC figures shows that 71% of the superiority of the 'top-third' flocks is accounted for by better stocking rates and only 6% by better lambing percentages. In 1993, average stocking rate was 12.3 ewes/ha while the top one-third of producers kept 16.9 ewes/ha. These higher stocking rates were achieved at higher levels of nitrogen/ha (91 versus 79 kg) but similar levels/ewe.

The use of ewes/ha as a measure of stocking rate can be misleading, since it takes no account of ewe size. In the MLC publication *Prime Lamb from Grass* an attempt has been made to overcome this problem by interrelating ewe weight, prolificacy and fertilizer usage, rather than ewe numbers. Representative values for average quality land are given in Table 19.21, but individual farm circumstances may markedly affect these values. Although there is a trend towards higher fertilizer N levels leading to higher stocking rates (+ 10 kg N = 0.3 ewes/ha), many farmers fail to fully exploit this potential. A second problem associated with the recording of grassland usage, especially on mixed farms, is that of allocating variable costs. The simplest way of standardizing figures is to use the livestock units system (see Chapter 11).

The times of the year when problems of high stocking rates are likely to be most apparent are in early spring and in the latter half of the grazing season. In early spring the provision of adequate grazing for the lactating ewe is important in terms of milk yield and lamb growth rates. It is easier to obtain this early grazing if the flock can be kept off the grass over winter. Where sheep are run at grass over winter, the effect on subsequent performance is variable and will depend on season and timing. Up to January there should be no effect on yield, but grazing in January and February will reduce early growth by 20%, or even more in a wet year. On the other hand, most swards have recovered from such an early grazing by silage or hay making. Wherever possible, therefore, sheep should be outwintered on fields destined for conservation.

In-wintering of sheep is becoming more common. More than two-thirds of lowland flocks are lambed indoors but only about 40% are housed throughout the winter. (The question of housing is examined thoroughly in Bryson, 1984.) In-wintering is expensive and unlikely to be justifiable in terms of increased ewe output alone. It can only be defended on grounds of easier shepherding and if grassland utilization is improved as a result.

In the latter half of the grazing season the main problem is competition between ewes and lambs, which can lead to reduced rates and increased parasitic gastro-enteritis in the lambs. This situation is most acute on all-grass farms where sheep are the only enterprise, since there is less scope for utilizing silage aftermaths or alternating grazing between species. In such conditions the forward creep grazing of lambs is worth considering and may give benefit of up to 500 g extra growth/lamb per week.

Table 19.20 Suggested rations for 75 kg ewe in late pregnancy (kg/d)

Ration	*Silage*	*Hay*	*Swedes*	*Kale*	*Concentrates**	*Barley*
1		0.6	4		0.6	
2		0.7		4		0.5
3	4				0.6	
4		1.0			0.6	

* 13 MJ/kg DM; 160 CP/kg DM.

Table 19.21 Relationship between ewe weight, nitrogen use and stocking rate on average land

Targets	*Nitrogen (kg/ha)* 0–75	75–150	150–225
Wt of ewe carried (kg)	750	900	1050
Wt of lamb produced (kg)	600	750	850
Stocking rates to achieve above targets (ewes/ha)			
Welsh halfbred	13	16	18
Mule	11	13.5	15.5
Scotch halfbred	10	12.5	14.5

After MLC *Prime Lamb from Grass* (reproduced with permission).

Increasing lamb numbers per ewe

As was shown in Figs 19.2 and 19.3, average flock performance is well below what is theoretically possible. Under most hill farm conditions a 100% lambing is considered satisfactory. Where higher lambing percentages are required, manipulation of ewe nutrition, as discussed earlier, can lead to increased numbers of lambs being born. There is then a need to adopt an alternative management strategy.

Early weaning of hill lambs

The main problems associated with twins from hill ewes are loss in body condition in the ewe, which may be severe enough to cause barrenness, and reduced growth rates in the lambs. Removal of one lamb, leaving a female where possible, can overcome these problems.

Weaning at 24 h, once the lamb has taken sufficient colostrum, allows the ewe to return immediately to the hill. The 'weaned' lamb must be fed on milk substitute and concentrates. A higher labour and feed cost, coupled with generally poor lamb performance, make this a difficult system to operate successfully. Where adequate inby grazing is available, it is more satisfactory to allow both lambs to suckle for five to six weeks before weaning one. The ewe may then be returned to the hill while the weaned lamb will either remain at grass, a technique demanding clean grazing and lamb-proof fencing, or be concentrate-fed indoors. This latter system is more likely to produce a finished lamb, with growth rates in excess of 200 g/d, and can improve the utilization of buildings used to house the flock through the winter.

Under lowland conditions there is often considerable scope for improving output using conventional management. A number of techniques are however available for those who wish to move further.

Use of more prolific ewes

Prolificacy in sheep is a trait with low heritability (0.1). The development of a flock of highly prolific ewes will generally involve the use of animals with some Finnish Landrace blood (e.g. the Finn–Dorset) or of the relatively new Cambridge breed. Where such animals are used as breeding females, then lambing percentages in excess of 250% may be expected. As mean lambing percentage in a flock increases, so the number of triplet and quadruplet births increases (see Table 19.22). A high level of stockmanship is necessary in such flocks. Chief features of the management in these situations are as follows:

(1) careful organization of mating to facilitate organized lambing;
(2) generous feeding in late pregnancy;
(3) closely supervised lambing, normally indoors in March early April;
(4) weaning by mid-July to allow ewes to regain body condition prior to tupping.

Additionally, the combination of Finnish Landrace blood and the smaller birthweights of triplet/quadruplet lambs means that careful selection of terminal sires is necessary if lambs are to be finished successfully.

Table 19.22 Relationship between lambing percentage and distribution of litter size

Mean no. born	*Litter size distribution (%)*				
	1	*2*	*3*	*4*	*5*
1.5	49	50	1		
2.0	12	76	12		
2.5	9	43	41	5	1
2.6	10	36	41	13	2

Increased frequency of lambing

Theoretically it is possible for ewes to lamb every six months and some may do this. However, ewes may not recover from one lambing in time to be mated again and the problems of operating such a system are many. In particular, low conception rates following the mating of lactating ewes necessitate very early weaning, and failure to conceive greatly increases barrener percentage or spreads lambing to an unacceptable degree. More commonly ewes will be lambed every eight months and often two flocks will be run, four months out of synchronization, so that ewes failing to conceive in one flock may be 'slipped' into the other and so given a second chance (see Fig. 19.10). In a frequent-lambing flock a variety of breeds may be used, but although exogenous hormones or day-light manipulation will generally be used, highly prolific ewes having a long breeding season are to be preferred. In this respect ewes of the Finn–Dorset type are probably ideal.

Although daylength may be manipulated to increase breeding frequency (see Fig. 19.11), the method is seldom commercially viable and exogenous hormones will be used. An intravaginal sponge, impregnated with progesterone, is left in place for 12 days. On its withdrawal an intramuscular injection of 500–750 units PMSG is given and ewes will show signs of oestrus 36–72 h later. A detailed calendar describing such a system is given in Appendix 2.

Critical to such systems are the management of the rams, the close supervision of lambing, the abrupt weaning of the lambs at one month of age and the careful nutrition of the ewe. Close supervision of lambing may be made easier if lambing is induced. The use of corticosteroid injection (e.g. 20 mg dexamethazone) on the evening of day 142 of pregnancy will result in ewes beginning to lamb some 36 h later. Lamb viability and growth rates, and ewe fertility, are not affected by such treatment.

Diagrammatic representation of the nutrition of a frequently lambing ewe is given in Fig. 19.12. Of particular importance is the protein input. In early lactation the inclusion of undegradable protein, such as fishmeal, facilitates backfat utilization and improves milk yield, while the reduction in total protein at the end of week 3 of lactation speeds the drying-off process and encourages backfat deposition.

The frequent lambing system is characterized by high variable costs. It has been suggested that an annual production in excess of 2.5 lambs/ewe is necessary for the system to be viable, but the high stocking rates possible with the system (25 ewes/ha) can help compensate for lower margins/ewe. A detailed discussion of this aspect of sheep production may be found in Littlejohn (1977).

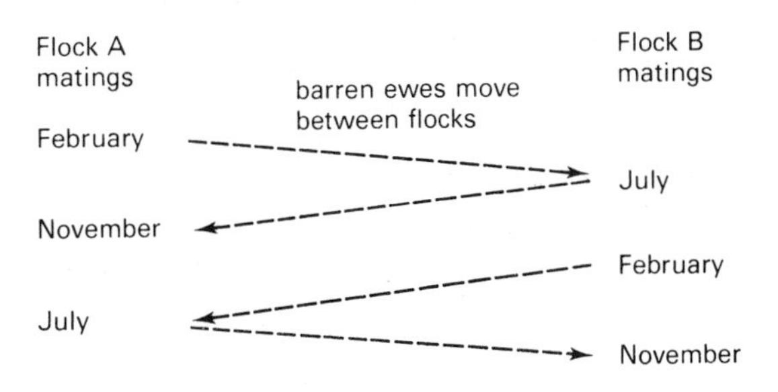

Fig. 19.10 Organization of split flock allowing 'slipping' of animals failing to conceive.

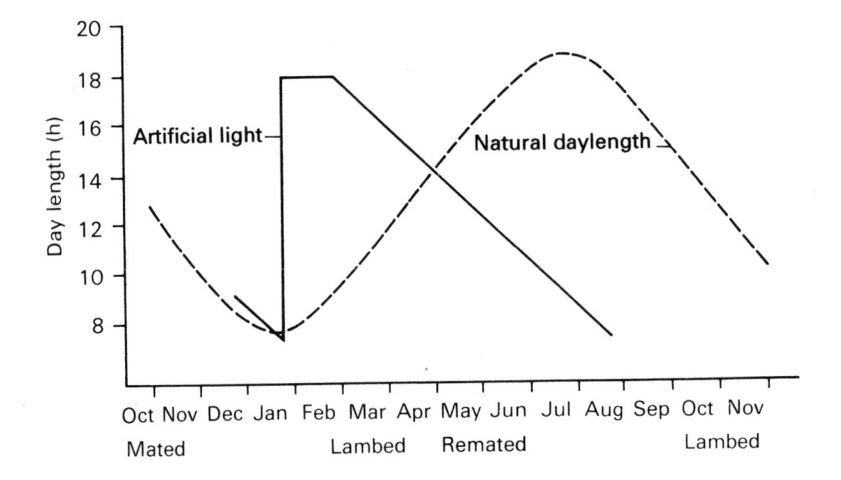

Fig. 19.11 Manipulation of daylength to control oestrus. (From Robinson, J.J. *et al.* (1975) *Annals Biol. Anim. Biochem. Biophys.* **15**, 345, reproduced by courtesy of the editor and publishers.)

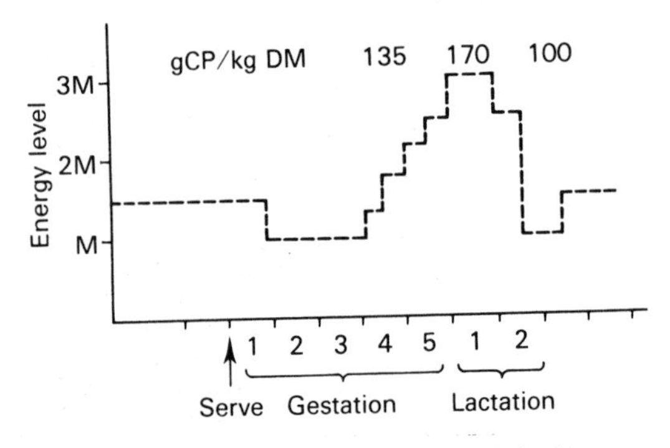

Fig. 19.12 Outline of nutrition of frequently lambing ewe.

Goats

There are more than 594 million goats in the world, largely in Asia and Africa (Table 19.23). Within Europe there are regional differences in the uses to which goats are put. In France and the UK the main output is milk and cheese. Here the most important breed is the Saanen (and its crosses), and average yields are 700 litres/year (France) and 800–1000 litres year (UK). In Spain and Italy approximately 30% of goats are milk-type, with the remainder dual-purpose or meat-producing, whereas in Greece almost all animals are dual-purpose and milk yields are lower. An outline of main breed types and milk yields is given in Table 19.24.

Goat production in the UK is still in its infancy, with approximately 100 000 breeding does of which some 20 000 are milked. The main output is milk, but interest in fibre production is growing. Estimated goat numbers, together with average milk production data, are given in Table 19.25.

Milk production

The lactation curve of the goat is similar to that of the cow. Potential lactation yield may be predicted by multiplying daily yield at peak by 200; beyond peak, yield may be expected to decline at 2–2.5% per week. In the non-pregnant goat, lactation can continue for up to two years. Whilst this will increase lactation yield, annual production is unlikely to equal that of two lactations in the same period. The seasonality of breeding activity in the goat makes it difficult to sustain milk production throughout the year. Many producers overcome this problem by manipulating day length so that does may be bred in April or May to kid in September and October (see Fig. 19.11).

Table 19.23 World distribution of goats ($\times 10^6$), 1993

Africa	172	*Asia*	358	*North America*	15.5	*Europe*	20.2	*S. America*	22.9
Nigeria	25	India	117.5	Mexico	11.1	Greece	5.9	Brazil	12.5
Ethiopia	19.5	China	97.5	USA	2.0	Spain	2.8	Argentina	3.3
Somalia	12.5	Pakistan	40.2	Haiti	0.9	France	1.0	Bolivia	1.5
Sudan	16.2	Iran	24			Italy	1.2	Peru	1.8
Kenya	7.3	Indonesia	11						
Tanzania	9.4	Bangladesh	26						

FAO Production Yearbook (1986) (reproduced with permission).

Table 19.24 Types and production of goats in Europe

Country	*Main products*	*Breed(s)*	*Yield (litres/year)*	*Ann. prod. (10^6 litres)*
France	Milk 50% Meat 50%	Alpine Saanen	770 770	417
Greece	Milk/meat	Not defined	120	466
Italy	Milk 30% Meat 70%	Sarda Maltese	150–200 450	155
Spain	Milk 30%	Murciana Malaguena Delas Mesetas Canaria	500–850 600 300 700	430

Table 19.25 Goat breed distribution and lactation data for UK

Breed	%	*Lactation yield (kg)*	*Fat (%)*
Crossbred	53	800+	3.5
Saanen	17	920	3.6
Anglo-Nubian	12	750	4.5
Toggenberg	12	880	3.3
Alpine	5	920	3.5

Table 19.26 shows that the analysis of goat's milk is similar to that of the cow. The fat in goat's milk has a higher content of small globules (28% versus 10% < 1.5 μm) and a higher percentage of short-chain fatty acids (15% versus 9%). It is some of these short-chain acids – caproic, caprilic and capric – which give goat milk its characteristic flavour. The protein in goat's milk is characterized by smaller casein micelles and by an increased β-casein content (67% versus 43%). The absence of carotene leaves the milk looking very white, although the level of vitamin A is higher than in cow's milk, as is that of nicotinic acid. On the other hand goat's milk has only 10% of the vitamin B_6 level found in cow's milk.

There is a specific demand for goat's milk and goat's milk products from those allergic to cow's milk. Estimates suggest that 7.5% of babies and 2.5% of adults may be allergic to cow's milk and that 60% of these are probably not allergic to goat's milk. There is no quota on goat's milk production and fewer regulations governing its production. The sale of goat's milk for human consumption is covered by Sections 1, 2 and 8 of the Foods Act (1984) which cover the questions of contamination and of health hazards.

Table 19.26 Typical composition of milk of different species

	Lactose (%)	*Fat (%)*	*Protein (%)*
Cow	4.5–5.0	3.5–4.0	3.0–3.5
Ewe	5.2–5.5	5.5–11.0	4.5–7.5
Goat	4.0–5.0	3.0–4.5	2.8–3.6

Meat production

In many parts of the developing world goats are an important source of meat. Total production exceeds 2.9 million tonnes. In the EU production is estimated at 92 000 t; Greece accounting for 42% of the total. Goat's meat may come from cull adults, from kids weaned and slaughtered at 6–12 weeks or from intensively finished animals usually surplus males.

On intensive systems, goats tend to have higher DMI than lambs of similar weight, but poorer efficiencies. Daily gains of up to 250 g may be expected at FCRs of 4–6:1. Killing-out percentages are similar to lamb but conformation is poorer and carcass composition is different, with goats having half as much subcutaneous fat but almost twice as much kidney fat. There is some evidence that Angora goats may be fatter than other types. Castration will tend to slow down growth rates and increase fat levels.

Fibre production

Most goats have a 'double-coat' fleece with long guard hairs covering a finer under-wool. It is this under-wool which is combed out to produce cashmere. In the Angora goat the under-wool is longer and coarser and is clipped twice yearly to produce mohair.

Cashmere fibres should be less than 6 cm long and 13–16 μm diameter. Yields vary between breed types and range from 50 to 250 g. Sixty per cent of world cashmere output comes from China. Mohair fleeces can weigh up to 2.5 kg, with fibres up to 15 cm long and a diameter of 20–40 μm. Major mohair producers are Turkey, South Africa and Texas.

Feeding goats

Detailed information on the nutrition of dairy goats is scarce. Dry matter intakes, at 80 g/kg $W^{0.75}$, are higher than for sheep, especially in lactating animals. Appropriate metabolizable energy inputs are of the order of 0.4 MJ/kg $W^{0.75}$ for maintenance and 0.7 MJ/kg $W^{0.75}$ during the last eight weeks of pregnancy. For a 60 kg goat these values equate to 9 MJ and 15 MJ/d respectively. During late pregnancy it is important to monitor body condition to avoid does becoming overfat and thus in danger of suffering from ketosis in early lactation. Mean energy requirement for milk production is 5.1 MJ/litre (6 MJ for Anglo-Nubians). During early lactation it is often impossible to meet energy requirements from the diet. Backfat utilization can provide 2.8 MJ/100 g backfat loss.

The protein requirement of a goat for maintenance is approximately 7 g DCP/MJ ME (equal to 60 g for a 60 kg doe). This amount should be increased by 60 g/d for the last eight weeks of pregnancy and by 55 g/litre of milk produced (see also Tables 19.27 and 19.28).

In lactating goats water intake is critical. In temperate areas intakes of 140 g/kg $W^{0.75}$ plus 1.4 litres/litre milk (i.e. 4–7 litres/d) may be expected. Goats do not like very cold water and intake may drop significantly in winter, adversely affecting yields. Warm water may need to be provided in such situations. Similarly, in concentrate-fed goats, reduced water intake can lead to the development of urinary calculi (bladder stones), a condition to which Angoras seem especially prone. Addition of 1% salt to the diet (as fed) reduces the danger of this condition developing.

Goats are browsers, but will graze when necessary. In grazing trials comparing sheep and goats, and using grass/clover swards and natural vegetation, goats have been shown to preferentially eat indigenous species such as bent grass (*Molinea caerulea*) and rush (*Juncus* spp), and to select against clover. This suggests that goats may have an important role in the improvement of hill grazing. On the other hand, problems with internal parasites (anthelmintics may not be used on lactating goats without withdrawing milk from sale) and difficulties with fencing have caused many goat's milk producers to house their animals year-round. The goat's tendency to produce tainted milk if fed on aromatic material is an additional hazard.

The most common feed for housed goats is hay, although good silage may also be used. Maize silage is particularly useful for raising milk yield and quality but needs careful balancing. Sugar beet pulp is a useful additional feed, especially early in lactation, when its palatability stimulates appetite and its digestible fibre helps maintain butterfat levels. Most producers favour concentrate feeds presented as a coarse mix, but there is no evidence that these are any better than pelleted diets; they are certainly more expensive.

Compared with silage, hay is generally lower in energy and much lower in protein. A switch to silage feeding, where possible, will thus lead to a saving in concentrate requirements. Yearly requirements for a dairy goat are given in Table 19.29. In fibre production the most important consideration is avoidance of nutritional stress, particularly in terms of energy input. Feeding at levels equal to those required by a dairy goat of similar size is unlikely to produce problems in this respect.

Table 19.27 Nutrient requirements of housed dairy goats

Yield (kg/day)	*Nutrient*	*Liveweight (kg)*		
		50	*60*	*70*
0	DM (kg)	1.5	1.8	2.1
	ME (MJ)	8.0	9.2	10.3
	DCP (g)	51	59	66
	MP (g)	43	50	56
1	DM (kg)	1.7	2.0	2.3
	ME (MJ)	13.1	14.3	15.4
	DCP (g)	106	114	121
	MP (g)	86	92	98
2	DM (kg)	1.9	2.2	2.5
	ME (MJ)	18.2	19.4	20.5
	DCP (g)	161	169	174
	MP (g)	129	135	141
3	DM (g)	2.1	2.4	2.7
	ME (MJ)	23.3	24.5	25.6
	DCP (g)	216	224	231
	MP (g)	171	178	184

Table 19.28 Nutrient requirements of growing goats

Liveweight (kg)	*Nutrient*	*Liveweight gain (g)*			
		50	*100*	*150*	*200*
10	DM (kg)	0.45			
	ME (MJ)	4.5	6.0	7.5	9.0
	DCP (g)	45	55	65	75
	MP (g)	40	55	71	86
30	DM (kg)	1.30			12.3
	ME (MJ)	8.3	9.8	11.3	90
	DCP (g)	60	70	80	95
	MP (g)	53	67	81	
50	DM (kg)	1.50			
	ME (MJ)	11.5	13.0	14.5	16.0
	DCP (g)	71	81	91	101
	MP (g)	66	79	92	105

Table 19.29 Annual feed requirements for a housed dairy goat (kg)

	Silage (10.5 ME)	*Hay (8.5 ME)*
Forage	3500	700
Beet pulp	110	110
Concentrate	200	450

Reproduction in the goat

Does are seasonally polyoestrous, but less so than ewes. Attainment of puberty depends on date of birth and level of nutrition and is generally at four to six months, with full sexual maturity reached at six to eight months in does and eight to ten months in bucks. Oestrous cycle length is very variable about at 21 d mean. Oestrous lasts 4–40 h, with ovulation 30–36 h after onset. Unmated does will have three to five cycles, with up to seven recorded for Angoras. Does in oestrus become vocal and restless, showing active male-seeking activity. Tail fanning tends to cause a wet area around the tail. Behaviour is more marked in the presence of an odorous male (the odour being produced by a secretion from a sebaceous gland at the base of the horns). Problems of oestrus detection in goats are less marked than in cattle because seasonality pre-ordains likely breeding periods and because breeding is unlikely to coincide with peak yield. Techniques for synchronizing oestrus, or for stimulating out-of-season breeding, are equally applicable to goats and sheep.

Gestation length varies between breeds, within the range 148–153 d, with Anglo-Nubians tending towards shorter and Saanens towards longer gestations. Litter sizes very between and within breeds, and are largely a reflection of environment ($h^2 = 0.1$). Mean litter size is approximately 1.75 under UK conditions.

Intersexuality

A well recognized problem in goats is that of reduced fertility resulting from intersexuality. The condition is related to the presence or absence of horns. Polledness is under the control of a single gene (designated P). Three genotypes and two phenotypes are identifiable:

PP = Homozygous polled
Pp = Heterozygous polled
pp = Homozygous horned

In females, PP genotypes are generally infertile while Pp genotypes have a 50:50 chance of being subfertile: these genotypes are indistinguishable. Fewer males are infertile if they are PP genotypes (up to 50%) but almost all will be subfertile. Careful inspection of the vulva of polled kids can often indicate whether or not they are likely to be infertile, the presence of a pea-like protrusion indicating a problem.

Appendix 1 Body condition scores

(0) Extremely emaciated and on the point of death. Not possible to detect any tissue between skin and bones.
(1) Spinous processes prominent and sharp. Fingers pass easily under ends of transverse processes. Possible to feel between each process. Loin muscles shallow with no fat cover.
(2) Spinous processes prominent but smooth; individual processes felt as corrugations; transverse processes smooth and rounder, possible to pass fingers under the ends with little pressure. Loin muscle of moderate depth but little fat cover.
(3) Spinous processes have little elevation, are smooth and rounded, and individual bones can only be felt with pressure. Firm pressure required to feel under transverse processes. Loin muscles full with moderate fat cover.
(4) Spinous processes just detected with pressure, as a hard line; ends of transverse processes cannot be felt. Loin muscles are full with thick covering of fat.
(5) Spinous processes cannot be detected, even with firm pressure; depression between layers of fat where processes would normally be felt. Loin muscles are full with very thick fat cover.

Appendix 2 Calendar for frequent-lambing flock

July
14 Insert sponges.
27 Withdraw sponges. Inject 750 units PMS.
29 Mating. Allow adequate rams (1:10) and rotate between paddocks every 8 h or use AI.

August
15 Mate repeats.
20 Remove rams.

December
23 Begin lambing.

January
9 'Repeats' lamb.
27 Wean first lambs. Insert sponges into all ewes.

February
10 Withdraw sponges. Inject 750 units PMS. Wean 'repeats'.
12 Mate.
28 Mate 'repeats'.

March
3 Remove rams.

July
9 First ewes lamb.
25 'Repeats' lamb.

September
20 Wean all lambs.

November
5 Introduce rams (normal mating, no hormonal treatment).

April
1 First ewes lamb.

June
14 Wean all lambs.

July
14 Insert sponges.
Etc.

Further reading

Alderman, G. (1993) *Energy and Protein Requirements of Ruminants*. CAB, Wallingford.
Bryson, T. (1984) *Sheep Housing Handbook*. Farming Press, Ipswich.
Eales, F.A. & Small, J. (1986) *Practical Lambing*. Longman, London.
Gall, C.G. (1981) *Goat Production*. Academic Press, London.
Littlejohn, L. (1977) *A Study of High Lamb Output Production Systems*. Scottish Agricultural Colleges.
MLC (1983) *Artificial Rearing of Lambs*. MLC, Milton Keynes.

20

Pig Production

M. A. Varley

Introduction

Pig farming has changed rapidly in recent years both in developed countries and in the less prosperous regions of the world. A principal reason for this is the high biological potential of the pig for converting feedstuffs into meat. Pigs are omnivorous and can utilize a wide range of food materials including plant proteins, bulky feeds, and human food waste. Modern hybrid pigs grow rapidly and efficiently to yield lean carcasses on energy-dense cereal-based feeds balanced with supplementary protein. Throughout Western Europe the nutrition of pigs is based on cereals such as barley, wheat or maize with soya and canola meal as protein sources. Some countries utilize imported cassava meal as an energy source and this is supplemented with soya protein or canola. In Third World countries the use of pigs for meat has taken criticism because the animals compete for available food energy which could be eaten directly by the human population. Much depends on the particular economic development of a country. If a given community has moved off the economic baseline, it will demand food in the form of animal protein. The use of pigs is then a viable proposition because of their innate capacity for efficient growth and production.

Reproductive characteristics of the pig are impressive when viewed alongside other species, and with today's systems of production it is possible to produce 25 piglets per sow per year. There is still large variation in what is actually achieved on farms, and even on the best of farms it is difficult to maintain constant production over a long time span. Research and extension work in recent years have given the industry sound working guidelines for reproductive management, and improved performance has been the result.

On a global scale there is now a great degree of uniformity in the methods used in production. Whether pigs are kept in Europe, North America, Australasia or in Africa, there has been a trend towards the use of prolific hybrid sows kept in individual stall houses, weaned at between three and four weeks, and the piglets are reared into flat-deck cages. It seems therefore that the pig industry is following the path that the poultry industry traversed around 20 years ago.

In Western Europe and in Britain particularly, another feature of pig farms has been the increasing specialization of pig units and the continued expansion of existing farms. The economies of scale for labour utilization dictate that viable units are much bigger than they were ten years ago and this process is not slowing down.

A few years ago an average commercial pig unit was 100 sows. Currently in the UK this figure is nearer 500 sows and herds of 1000 and 2000 sows are not uncommon. This rapid expansion has presented some problems, and one of these is the demise of the traditional family farm which was once the backbone of British agriculture. It is still possible to make a living from a smaller unit where the only labour used is family labour, but the smaller units are under increasing pressure from the larger operators with their smaller unit costs of production.

Another recent aspect of the European pig industry is the recent concern amongst the general public about the way in which animals are housed and managed. Legislation in some countries prohibits the use of certain practices such as early weaning or the use of sow tethers. Novel systems of production which minimize the stress imposed on animals seem likely to have a major impact on pig production in the next ten years or so. It may be that the structural changes seen in the past are gradually reversed as urban populations demand a more active interest in pig farming. A recent feature of pig production in the United Kingdom has been a very large swing towards outdoor production systems. It has been estimated that at the present time 25% of all sows are accommodated in outdoor systems. This move has in part been a response by the farming industry to public pressure for improved animal welfare.

It is the purpose of the ensuing text to give an overview of the pig industry, concentrating mainly on the UK. The intention is to cover important technological developments in the fields of genetics, reproduction, nutrition, housing and marketing to give the reader a basic working understanding of the pig industry.

The structure of the UK pig industry

The structure of the UK pig industry has altered significantly since the end of the Second World War as a result of social and economic pressures. In the early 1950s pigs were considered a secondary enterprise as part of a mixed farming pattern. Herds were small and only a small number of specialist producers existed. This scenario worked well in the context of integrated systems of livestock production. Arable farms growing their own cereals could process milled barley and wheat into pig meat, straw was freely available for bedding, and manure disposal was not a problem using cereal stubbles in the autumn months.

Today the pig industry is in relatively fewer hands and there is a continuing trend towards specialist production although as pig units have increased in size the large scale arable farmers of the eastern counties of England have found it easier to expand with minimal slurry disposal problems and readily available straw. There has been a gradual shift of production to the east of Britain partly because of the drier climate but mainly because of the logistics involved in the production, processing and transport of the raw materials involved such as corn and straw. In 1964 there were around 75 000 holdings in the UK which had sows on them. By the middle of the 1970s there were only about 22 000 holdings in the UK with pigs and at the present time about 80% of the nation's pigs are in the hands of 2000 producers. These structural changes allied to rapid increases in average herd size have meant that the cost of pig meat to the consumer has been kept as low as possible, and 33% of all meat eaten in Britain at the present time is pig meat.

Economic factors have been responsible for many of the changes, and within the European Union (EU) there is intense competition to secure and hold markets for pig meat. Denmark has traditionally had a large slice of the UK market for bacon and this seems likely to continue because of the vigorous marketing tactics used by the Danes. More recently, significant quantities of pig meat have been imported from The Netherlands and also from Ireland. In the future as part of a much larger economic community the marketing of pig meat will be increasingly complex and difficult for the home producer. The EU itself now contains Greece, Spain and Portugal, and these countries also have rapidly developing pig industries which in due course could provide major competition for the UK. Southern European countries can produce pig products significantly cheaper than in the North because of cheaper labour, feed and housing costs. When the final barriers to trade came down with 'harmonization' in 1993, it was also the case that there was a much more significant movement of fresh pig meat across Europe and greater difficulties for UK producers.

One of the facets of pig farming within the EU has been that the industry has not been supported financially by the same system of intervention prices and storage of products that has helped the milk, beef and sheep industries. Pig farmers have been price takers without the insurance of guide prices and financial subsidies. As a consequence of this, the industry in general has exhibited increasing technical skills and the uptake of new technology and research has been swift and effective.

A significant component of the financial burden in pig farming has been the increasing costs of feedstuffs. Whilst arable farmers have enjoyed strong support from the EU budget and cereal prices have been buoyant, the price of wheat and barley represents the major raw material cost to the pig farmer. Eighty per cent of the costs of producing pig meat is the cost of feedstuffs. This situation may change in the future and the EU administrators have recently changed their stance on the financial support of agricultural production to relieve the burden on the taxpayer. It may be that with the relaxation of intervention buying, the costs of production in pig farming could fall in real terms.

Another recent problem for the UK farmer in relation to feedstuffs stemmed from developments in our fishing industry. Surplus fish manufactured into fishmeal was until the 1975 'cod war' with Iceland a relatively cheap and high quality source of protein for most types of pig. At the present time it is difficult to justify the inclusion of fishmeal in the diets of growing pigs or sows because of its very high price. A benefit transpiring from this situation has been the stimulation of research and development work in non-ruminant nutrition. This has demonstrated clearly the value of alternative protein sources such as soya, canola and other leguminous protein meals.

The pig cycle

The so-called pig cycle has existed for many years now and its characteristics are illustrated in Fig. 20.1. Because the reproductive rate of the pig is so high it is possible for the national herd to expand and contract very quickly over a relatively short time span. At a time of high profitability and confidence in the industry, farmers expand and new producers enter the industry. The rapid rise in sow numbers inevitably leads to overproduction and the price of finished pigs falls. Profitability declines, sows are culled, some farmers go out of business and national production of pigs falls off. The national herd numbers return to baseline again and the cycle begins again.

The amplitude and frequency of the cycle vary considerably but the hope has always been that with large herds committed to steady state production and with heavy investment in buildings and equipment, the cycle will disappear. This so far has not been the case and larger producers seem prepared to expand continually. Current indications are that, after a number of very bad years, producers are being a little more cautious.

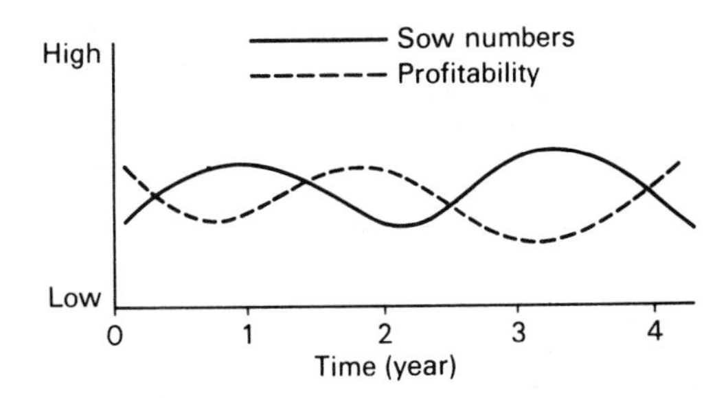

Fig. 20.1 A schematic representation of the 'pig cycle'.

Summary of the UK industry

In Table 20.1 are given data illustrating the current (1993) position of the UK pig industry. The industry is a

Table 20.1 UK pig industry

Parameter	*1983*	*1993*
Sow numbers	809 000	785 000
Bacon and ham production (t)	212 000	169 000
Pork production (t)	776 000	797 000
Consumption per capita: pig products, UK (kg)	22	24
Bacon production: self-sufficiency (%)	45	42

significant industry in its contribution to overall agricultural production, and in financial terms has a turnover around £1 billion per annum. In relation to the rest of the EU we have about 8% of the total number of pigs within the 12 nations of the EU. Pig production for the whole European Union is also a major meat industry and 43% of all meat produced is pig meat.

Trends in performance and profitability

In Table 20.2 data are presented from the Meat and Livestock Commission's (MLC) Feed Recording Service, showing the trends in physical performance in pig production over the years from 1970 to the present time. There has been a steady improvement in the national herd, as indicated from this large sample, and the process is not slowing down. What is also evident from MLC statistics is that a considerable gap exists between the 'average' producer in the recorded herds and the performance of the top 10% of herds. For example, in terms of breeding performance in 1993, the average herd achieved 2.3 litters per sow per year, 21.5 piglets reared per sow per year. The top 10% producers however achieved 2.4 litters per sow per year, 24.4 piglets per sow per year. Clearly there is scope for the forward-thinking producer through the application of good technical skills to generate a healthy financial return.

Table 20.2 Trends in pig performance

	1970	*1977*	*1987*	*1993*
Breeding herd				
Litters/sow/year	1.8	2.0	2.3	2.3
Pigs/litter alive	10.2	10.3	10.3	10.8
Mortality (%)	15.2	14.5	11.1	11.7
Pigs/sow/year	15.5	18.1	21.0	21.5
Feed/sow (t)	1.6	1.6	1.2	1.27
Feed price (£/t)	36	117	139	142
Finishing pigs				
Food conversion ratio	3.7	3.2	2.7	2.4
Daily gain (g/d)	—	—	588	591
Mortality (%)		3.9	2.7	2.3

Pig housing and animal welfare

The way in which pigs are housed has changed at a remarkable pace since the move to intensive systems. The rate of change has also not diminished and there has been a continuous development of new building systems to meet the changing needs of the industry. In the early days of intensification, criteria were established for the design of systems which could accommodate pigs efficiently, with a gain in productive efficiency for a reasonable life span. Pigs are also notorious amongst all the farm species for their capacity to damage buildings and equipment.

As indoor systems became more prevalent and the economic situation changed, it became more and more important to operate existing and new buildings as efficiently as possible. It was necessary to build them at minimum cost within the limits of the known environmental requirements of the pig. Consequently stocking densities in pig buildings have risen steadily. Most successful pig farmers, however, are well aware of the relationships which exist between stocking density and the incidence of disease. There is an optimum stocking density for any given building above which production falls off significantly as a direct result of increased bacterial contamination but also because of social stress caused by overcrowding. Unfortunately there are some farmers who increase stocking densities above the optimum and then solve problems with antibiotics. This is, of course, dangerous practice and potentially limits the effective clinical use of the same antibiotics against human diseases.

A more recent phenomenon in the development of new pig building systems is the involvement of public opinion in how animals are managed and housed. It would appear that the consuming public wish to have a say in the care of animals on farms, and the industry has begun to respond to these wishes. At the same time the general public needs education in order that they understand why animals are housed and managed as they are. There is at the moment a move to improve the channels of communication between the various interested parties. One aspect which is still a problem area is the economics of improved animal welfare. Animals can be produced in less intensive systems using high-welfare management. These systems do cost money, however, and inevitably unit costs of production and therefore the price of meat rise accordingly.

Breeding stock

The package-deal building is the standard housing unit. These are of prefabricated construction and are erected on-site by trained personnel. A very high standard of construction is available in terms of thermal insulation, ventilation, heating control and the internal fittings used. There is scope for small farms building their own facilities using timber and concrete, but for the larger units aiming at high levels of biological efficiency, the package-deal building is probably the best option.

Dry sows are housed principally in individual stalls within controlled environment buildings. These include slatted floors and many have completely automatic feeding systems. The labour input is minimal and the management of the sows can be of a high order. Problems arise because of the automation itself. If sows are placed in stalls in early pregnancy and are supplied with all of their needs by mechanical devices, then this reduces the necessity to observe animals as individuals on a routine and daily basis. Moreover, if sows and gilts are in stalls allowing only limited exercise for about 100 days at a time, then this may not be conducive to the maintenance of high health status. It is this last point which has brought so much criticism of the stall house, and many people believe they should be phased out of use. In particular, the use of neck tethers to reduce the constructional costs of the stall has been severely attacked by some pressure groups, and in many ways these tethers are difficult to defend. In the UK a decision has been made to phase out sow stalls in 1999 and this decision has been reached after much debate. This will put producers at a disadvantage compared with their EU counterparts and it remains to be seen whether this will become an adopted standard across Europe.

The well managed stall house does have many advantages, however. The system evolved in order to fulfil the nutritional, thermal and social needs of *all* sows. Sows in groups establish social hierarchies after much fighting and physical damage over a number of days after mixing. Sows at the bottom of the social order may be repeatedly attacked and prevented from obtaining food. It is also impossible as animals progress from one phase of the reproductive cycle to the next to avoid constant remixing of groups of animals. Sow stalls prevent such physical damage and allow all animals to receive the correct daily feed allowance. In addition, reproductive management is easier because it is possible to ensure the timely checking of animals at the different stages.

A proportion of sows are housed in semi-intensive building systems in groups of between 10 and 50. These are usually based on straw yards as in the eastern counties of England where they have been popular. The buildings themselves can be multipurpose, allowing producers to move in and out of pig production as the pig cycle progresses. There is a renewed interest in these systems and the use of electronic sow feeders may be partly responsible for this. These are devices where pigs live in a group of, say, 25 and have access to a special feeding stall. When a sow enters the stall, a gate closes preventing another sow from entering. Each sow wears an electronic transponder device which contains a sow identification code. This code is read electronically whilst the sow is feeding and the weight of feed eaten by the sow is monitored and recorded automatically by a dedicated computer. This allows group-housed sows to be fed individually and the system can be programmed to allow each sow a set daily allowance.

Many of the development problems of these feeders have now been resolved. Where these feeders are used, sows are still housed in a group for most of the day with all the problems this can entail, but the intense competition at feeding is eliminated and much of the severe aggression is precluded.

Sows are moved to farrowing crates about 3–7 days before delivery. Farrowing crates are designed to minimize the loss of baby piglets due to overlying by the sow in the early days after farrowing. This is the major cause of death in the neonatal period. Farrowing crates commonly include a 'creep' area where piglets have access to a supplementary dry food and a heat lamp to keep them warm. The crate is necessary to restrain the sow and prevent her overlying her piglets. Farrowing crates are not popular with some people and it has been suggested that restraining the sow during parturition slows down the process, but this has not yet been proven conclusively. Most stockmen and managers prefer the use of farrowing crates as they are used for a relatively short period of time when the sow is lactating.

There have been innovations in the design of farrowing crates recently. One of the more promising of these is the use of flexible hydraulic bars on either side of the crates. These bars slowly part when the sow lies down. When she stands up, the bars fold inwards, so preventing any sudden move downwards and hence protecting the piglets from being crushed. Another development is the use of blower fans around the legs of the sow. The fans are activated when the sow stands, helping to drive the piglets away from the place of maximum risk and into the creep area. A so-called high welfare crate has also recently become available called the 'Freedom Farrowing Crate'. This piece of equipment allows sows to move from a straw yard when they choose to do so at the end of pregnancy, and the crate allows them to leave and enter the crate at any time. When the piglets are born, they are prevented from leaving by roller bars and panels. Early results from this system have been variable although in welfare terms the equipment does offer potential.

When sow stall houses were first installed, it was usual to have boar pens in the dry sow house and sows and gilts were moved to these pens for heat detection and service. This worked well for some, but the arrangement is not ideal because of the logistical problems in moving animals in and out of stalls or tethers. Boar pens which are remote from weaned sows and unserved gilts do not provide the ideal pheromonal environment or even the correct physical environment for effective matings. Larger pig units now build specific houses for boars, weaned sows and gilts, and these buildings included purpose-built pens where mating can take place with the full supervision of the stockman. These service areas contribute to high conception and farrowing rates and management attention to this crucial phase of the cycle of production.

Sows under outdoor paddock conditions are given simple shelters usually made in a semicircular construction from galvanized corrugated iron and wood. Sows quickly identify their own huts and farrow down with the minimum of fuss. These systems are therefore extremely cheap in terms of capital requirement compared with indoor accommodation.

Weaner accommodation

When weaning was carried out at six or eight weeks of age, piglets could be transferred to large straw yards without special environmental controls and there were few problems. With the advent of early weaning it soon became clear that more careful consideration of the housing of post-weaned pigs was essential. Flat-deck cages were devised as a method of housing small pigs to provide an appropriate environment for minimizing the post-weaning growth check and promoting good health. The essence of flat-deck cages is the completely slatted floor area which ensures that piglets are in minimal oral contact with their own faeces. The walls and floors therefore carry less contamination and enteric diseases are minimal. The second feature of these houses is that they are well insulated and invariably have a heating system to keep the inside temperature at the optimum throughout the year. For newly weaned three-week-old piglets this is probably around 28°C. Flat-decks have also been improved over the years to provide plastic slotted floors which are kind to the piglets' feet, easy-clean surfaces so that each pen can be completely power-hosed in between batches, and better ventilation systems that give air change without draught. The outcome of these improvements is minimal mortality of post-weaned piglets and faster daily gain through to slaughter weight.

Flat-deck cages have also been criticized on animal welfare grounds. This is partly because they are associated with early weaning, but also because of the completely slatted floors which do not allow any bedding in the form of straw or wood shavings.

Another popular type of building for weaners is the verandah house. These give piglets access to an outside slatted dunging area and an indoor insulated kennel for groups of 20–40 piglets. These buildings can work well for weaning at four to five weeks but in adverse weather conditions (hot or cold) they can be difficult to manage.

Growing pigs

From 25 to 30 kg onwards piglets are transferred to the final accommodation where they remain until slaughter. These houses are variable in type, and design often depends on the availability of bedding. Many bacon houses are based on package-deal buildings with completely controlled environments and automatic feeding systems. Bedding is not used and most buildings have either a partly slatted floor as a dunging area or in some cases a fully slatted floor. Under these conditions the attainment of lean carcasses from pigs which have grown rapidly and efficiently is routine assuming a good degree of stockmanship.

The main alternative to the completely controlled environment house is a semi-intensive system based on straw yards and insulated kennels. These have become known as Suffolk houses because of the popularity of this type of house in that part of England. Good performance can be achieved with these houses but in the absence of a plentiful supply of cheap straw the system is not an economic proposition.

Associated with any large pig finishing enterprise is the disposal of large volumes of either slurry or farmyard manure. Straw-based systems have the advantage that muck can be stacked in heaps and spread when the land conditions are right. Slatted floor systems can present real problems due to the very large amounts of low dry matter slurry produced. Most buildings include underground tanks for short-term storage. Vacuum tankers are required to dispose of the slurry in dry conditions on stubbles and grassland and often on sacrifice areas. Because of the vagaries of British weather, regular spreading is not always possible and on many farms there are now large-capacity above-ground storage tanks. These are emptied at set times in the farming calendar or when weather conditions are suitable. The spreading of the slurry itself is still a major issue in some areas due to potential problems of run-off into waterways and problems of smell from farms adjacent to towns and villages. In some of the high-density pig areas of the UK, planning authorities are refusing to allow farmers to invest in new slurry-based systems. Currently, the handling and disposal of slurry constitute a major unresolved problem for the industry. It seems the UK may be going rapidly down the path of Dutch pig farmers who have been banned from expanding their industry in the south of The Netherlands because of persistent slurry problems. A further problem with slurry and its spreading on land in large quantities is the contamination of the soil with phosphates and possibly heavy metals. Nitrogenous material in large amounts also ends up in watercourses. As a result of very intensive farming practices, there have been problems in estuaries and in-shore areas where overgrowth of marine vegetation has occurred. This problem is currently being addressed throughout Europe, and legislation will be introduced to control the damage caused in this way. Research is also being carried out to look for optimum N and P inputs into the diets of pigs to minimize excretion rates without loss of performance.

Genetics and pig improvement

The size of the contribution of genetics to the improved performance of modern pigs cannot be overemphasized. Over the last 20 years the performance characteristics of pigs have altered enormously. A major factor was the foresight of pig breeders in using established principles of quantitative genetics and abandoning some of the traditional methodology of pedigree breeding. To an extent pig breeders followed a similar course to the poultry industry and adopted the techniques of mass selection for multiple objectives and used large populations to ensure statistical validity in the selection of parents to breed the next generation. A more traditional approach included the selection for traits of no economic merit and the progeny test as a selection tool. The drawback of progeny testing is the length of time before a selection decision can be made as this significantly reduces the rate of annual genetic improvement in a population.

As a result of the intense selection methods deployed over the last 20 years and also the contribution made by national organizations such as the MLC and the hybrid breeding companies, it is easy to see why British pigs are now in demand throughout the world. It is also interesting to note that Denmark, which prides itself on the production of quality pig meat, is now importing and using hybrid sows from a major UK hybrid company.

Breeds

The concept of a discrete breed is fast becoming irrelevant

and the use of commercial hybrids has superseded purebreds in commercial meat production. The hybrids themselves are created from the original parent purebreds and these are still maintained within nucleus populations. The bulk of UK slaughter pigs are produced with some degree of relationship to either one or both of the two most important breeds, the Large White and the Landrace. It is believed that the original ancestry of the Landrace breed belongs in Scandinavia. The Large White, related to the American Yorkshire breed, is more indigenous to Great Britain.

It was fortuitous that the Large White and the Landrace breeds came to the fore in Britain at a time when systematic breed improvement was being initiated. Both breeds are extremely prolific and can produce litters of 10 to 12 piglets consistently. Furthermore, both breeds produce an acceptable carcass even as a purebred. They have the length needed for bacon production and also a high lean content and low subcutaneous backfat thickness which is required by today's consumers. More specifically, they both have a white skin and this is needed by the meat trade in the UK industry. Both breeds are also noted for their ability to grow quickly and to show a high efficiency of food conversion. In short the Large White and the British Landrace are good all-rounders and are well fitted for meat production. It is easy to understand why the hybrid pig breeding companies initially selected these breeds as the foundation material for their selection programmes.

The other two breeds of any numerical significance in Britain are the Welsh and British Saddleback. These have been given resources over the years in the national improvement programmes. The Welsh breed is similar in some respects to the British Landrace and can be considered as an alternative in a crossing programme but this breed has been declining in numbers in recent years. The British Saddleback, however, is a black-coated breed with a white 'saddle' behind the front legs. This breed was established relatively recently by the merging of the herd books of a number of similar black-saddled breeds such as the Essex. The merits of the Saddleback are in its hardiness and its mothering ability which some strains of the white breeds seem to lack. British Saddlebacks are used in many parts of southern England for outdoor pig production either as a pure bred or more commonly now as a crossbred. The Saddleback is capable of producing strong healthy litters of weaners with the minimum of fuss. Some of the leading hybrid companies now market breeding females specifically designed for outdoor production and some of these have genes from the British Saddleback.

Minor breeds such as the Tamworth, the Large Black, the Gloucester Old Spots and the Middle White are still kept by a few enthusiasts and at rare breed society farms but the commercial performance of these breeds in terms of prolificacy, growth and carcass quality falls way below the breeds mentioned above. It is of incalculable value, however, that these breeds are still maintained both because of their intrinsic qualities as part of our heritage and also because of the need to keep a wide genetic base to allow for changing selection objectives in the future.

There has been periodic interest in imported breeds and some of these are now used in large numbers. The Hampshire breed from North America is used as a top crossing sire in pork production. This is because of the high lean content of these pigs and the very good conformation of the hams. Hampshire pigs also have a good eye-muscle area and a good killing-out percentage. The level of prolificacy in Hampshires is low, however, and as a breeding female they are not a viable proposition. There are also a number of Belgian Pietrain pigs in the UK. These have probably the best eye muscle and hams of any pig and a high lean content generally. They have been incorporated into sire-lines by the breeding companies to be sold as top-crossing sires for the production of slaughter pigs.

Throughout Europe the Duroc breed originally imported from North America has been popular for some years as a top crossing sire. These pigs are very hardy compared with our indigenous white pigs, and some strains of them can grow extremely quickly. The current interest in Durocs stems from the putative superior quality of the meat. It is said that Durocs have a high percentage of intramuscular fat and as a consequence Duroc meat may be juicier and may have more flavour than other breeds. Under British conditions to date the results of controlled experiments with Duroc meat have been rather equivocal. Throughout the whole of Europe and further afield there has been an increased use of Durocs as terminal sires both as purebreds and also as crossbreds. Durocs have also found a very useful niche for the outdoor pig industry. Duroc cross females can be highly productive on outdoor farms and they can also withstand the relatively harsh conditions in winter.

In 1987 after some years of negotiation, a sample of Chinese pigs was imported into Britain. The stimulus for this effort was the reports and first-hand accounts coming from China of the high prolificacy of these pigs. Some anecdotal accounts suggested that breeds such as the Meishan from central China could regularly produce 20 piglets in one litter. The objective was to incorporate the good genes of the Chinese pigs with British pigs to produce hyperprolific hybrids with an acceptable carcass. This last point is certainly a stumbling block, and research from France, where Chinese pigs have been investigated for some years, suggests that although the prolificacy is good it may not be good enough to cancel out the very poor carcasses that crossbred Chinese pigs produce. With the advent of sophisticated genetic engineering techniques it may be possible to identify the specific genes carried by the Meishan and others and then transfer these genes directly into the genetic constitution of our own indigenous pigs. Most British breeding companies now have populations of Meishans in the development stage but one hybrid company in 1992 introduced on to the market a product based on Meishan genes as a hyperprolific female. The company claimed to have reduced the subcutaneous fat levels using the principles of direct selection, and the final slaughter generation has only slightly increased backfat levels. It remains to be seen how successful this development will be. However, other companies at some future stage will be also releasing products.

Other pigs present in the UK in very small numbers include the Vietnamese pot-bellied breed, which is kept simply as an ornamental pig, and the Mexican Yucatan pigs formerly owned by the University of Leeds. These latter animals are farmed for the benefit of human infants with heart valve defects. It has been discovered that the heart valves of these pigs can be used as very effective replacements in infants with diseased valves. It may be that in future the pig may serve humans in many more ways than just for food.

Crossbreeding

Systematic crossbreeding has been used in pig production for some years and this seems likely to continue. Part of the problem in any breeding situation is that it is almost impossible to find the perfect blend of characteristics in a single breed. By merging together the good traits of two parent breeds in a crossing scheme it is possible to produce a slaughter generation containing all the good points. In pig breeding the major reason why crossbreeding is still so widely used is for the exploitation of hybrid vigour to improve reproductive performance. It was found in the 1960s that, by crossing together the Large White and the British Landrace, the first-cross female expressed greater prolificacy than either of the two parent breeds. This is a good example for heterosis or hybrid vigour. As the carcasses of both parent breeds are good, the use of the first-cross female as a breeding sow seemed the obvious choice. The use of LW × LR females has stood the test of time and now a very high percentage of female pigs are of this genotype.

Most hybrid females are based on this same first-cross animal. In the context of pig breeding, hybrid is thought of as an 'improved crossbred' where simultaneous genetic improvement is made in the parent lines as well as selection for crossing ability. Table 20.3 gives a schematic representation of the most popular system of crossbreeding used.

There are of course a number of variants on the scheme outlined in Table 20.3 but the majority of producers use bought-in first-cross females and purebred terminal sires from one of the hybrid companies. Some producers use their own gilt replacements and may buy in only a proportion of their annual needs. To give farmers flexibility in their replacement policies, many breeding companies sell grandparent females directly to commercial farmers who then carry out the crossing programme themselves to generate the bulk of their gilt replacements. Artificial insemination is also used in such schemes to replace the original grandparent gilts and as crossing sires on nucleus purebred females. As might be expected, there are advantages and disadvantages for the different breeding policies for individual farmers. Those who purchase all of their first-cross gilts every generation from a company or from a pedigree breeder will incur higher replacement costs and they also run a higher risk of introducing disease on to their farms. On the other hand, if they are buying in from a progressive company they will capitalize more quickly on rapid genetic improvement. If all replacement gilts are bought monthly from one source, then this simplifies the management process. In contrast, the management of grandparent gilts can be onerous and badly implemented.

Table 20.3 Crossbreeding to produce meat pigs

Generation	*Male*		*Female*
Grandparent	Purebred LW	×	Purebred LR
	Purebred LR	×	Purebred LW
		|	
Parent	Purebred LW	×	LW × LR
	Duroc sire-line, etc.	|	
Slaughter pigs	LW	×	(LW × LR)

The hybrid companies

There are about ten major hybrid companies in the UK in the business of selling boars and gilts to commercial farmers. The majority of them use proven scientific principles in their selection programmes allowing significant advances over time in the genetic quality of their stock. The origins of individual companies are quite diverse. Some were groups of collaborating pedigree breeders who effectively pooled their resources and expertise to maximize their breeding effort. Others have grown out of the ancillary industries such as the meat processing companies striving to develop products which fitted their own markets. There are also companies which developed by diversification from other breeding operations. All of the successful companies use essentially similar methods. They have large nucleus populations where direct selection is carried out under company control. These nucleus herds are owned predominantly by the companies themselves.

Improvements made at the nucleus level are then propagated by multiplying breeders who are usually leading pig farmers operating under contract with the hybrid company. Gilts are sold off these multiplication farms to commercial meat producers.

A major consideration in the selection of breeding replacements is health status. Many of the companies sell stock which are free from certain diseases such as virus pneumonia and atrophic rhinitis. They maintain a high health status by stringent controls at the nucleus level. Introduction to the nucleus herds may only be carried out using hysterectomy-derived litters which are specific pathogen free. Over long time spans, however, the health status of even the best companies can wax and wane, and this requires careful monitoring by commercial producers.

National involvement in genetic improvement

The UK has been fortunate to have had progressive organizations such as the Pig Industry Development Authority (PIDA) and later the MLC to coordinate a national programme of improvement. Whilst accepting that the role of the hybrid companies has taken over from pedigree breeders, it was the original work done by the PIDA and the MLC that focused the enthusiasm of the pedigree breeders into a collaborative effort.

The cornerstone of the MLC scheme was the rationalization of the existing structure of breeders into a clearly defined breeding pyramid. The best breeders were designated as nucleus or elite breeders and a second category of units was listed as multiplication units to supply commercial farms with breeding stock. Groups of test pigs were sent from all breeders within the scheme to central pig testing facilities. This breeding pyramid structure is illustrated in Fig. 20.2.

There were five testing stations throughout the country

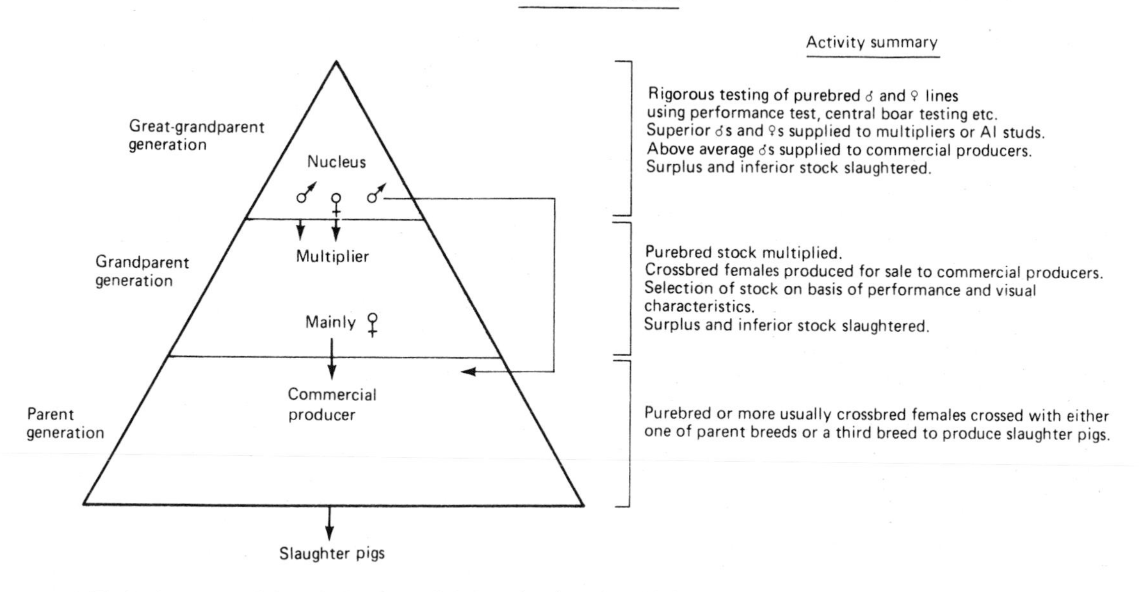

Fig. 20.2 The breeding pyramid.

where large numbers of potential young boars could be performance tested every year. A method of index selection was used on these stations to improve simultaneously a set of multiple selection objectives. These objectives were the economically important traits such as food conversion efficiency, growth rate, eye muscle area and muscle quality. In other words the scheme concentrated on traits of high heritability and therefore selected high quality boars which passed on their superiority to the next generation.

Each tested boar was given an index score which integrated his total economic value in terms of breeding value. For ease of understanding, the points score was designed so that an average boar in any given time period scored 100 points. Boars scoring 90 points or less were slaughtered after test. Boars scoring 120 points or more were called approved sires and were returned to the breeders who could then either use the boars themselves or sell them to other breeders. It was in the interest of nucleus breeders to use the highest pointed boars they could in order to stay ahead of the field. All the data were made public and multipliers could use only approved sires. By identifying good genes and disseminating them as widely as possible, rapid genetic progress was made. The hybrid companies also use similar testing methods.

The rewards for this effort and considerable expense have been enormous and it is estimated that over the two decades that the scheme operated the increase in profits to the commercial farmer were of the order of 50–60 pence annually on every pig sent for slaughter. Nationally therefore the scheme was an investment that has probably paid a tenfold return.

The last testing station in the MLC boar testing scheme closed in 1987. The remains of the scheme centres around a reduced super-nucleus list of pedigree breeders and the use of on-farm testing. These changes have been made as a result of the increased presence of the hybrid companies and also the escalating costs of operating central testing stations. The use of Best Linear Unbiased Prediction (BLUP) methodology has made the use of on-farm testing a much more accurate exercise. BLUP is now an integral component of the MLC scheme and the hybrid companies use it extensively to allow valid selection decisions amongst different nucleus farms.

Artificial insemination

One of the most efficient methods of spreading good genes widely and rapidly is the use of artificial insemination. Pig AI has not been quite the overnight success story that in many ways revolutionized cattle breeding. There are signs now that producers are at last aware of the benefits from using AI in their own breeding programmes. As a percentage of all matings carried out nationally in any one year, about 11% are artificial inseminations and this figure is rising steadily. In some neighbouring countries the figure is much higher. One of the reasons for not using AI is that there is less risk to human life and limb with boars on a farm than with bulls.

Pig farmers are familiar with the housing and handling of boars with the minimum of problems and costs. More important is the variable reproductive performance associated with the use of AI. Surveys in the past have shown that conception rates are often down when compared with natural service and in some cases the litter size is reduced. At the present time this should not be a problem and most users achieve very high levels of performance once their stockmen have been adequately trained. The third reason for not using AI is that farmers were concerned about the length of time required to carry out the AI procedure. This last point is quite contentious as the supervision of natural matings, done properly, involves care, attention and therefore time.

The factors in favour of using AI revolve around the superiority of the genetic material available at the AI studs. In practice there are boars at stud which have extremely high index scores and these are available to an average commercial producer at modest cost. The possibility therefore exists to upgrade an average herd very quickly using AI.

With regard to the problems of conception there is now a much better understanding of the relationships of the AI technique itself and the probability of failure to hold to service. There are sound and clear guidelines in the form of publications and consultancies and many producers achieve conception rates around 80–90%.

The principal AI scheme nationally was, until recently, the one operated by the MLC and this was based on a postal delivery service. The AI station was situated centrally and farmers with females in heat or expected in heat telephoned in to the station. Semen was despatched twice a day by post to arrive within 24 h of the farmer ringing in. This station is now owned by a commercial company but the scheme operates in just the same way. Most of the leading hybrid companies in fact own their own AI stations and operate similar programmes.

The AI process itself is carried out by the farmer or stockman using a disposable catheter and the insemination takes between 10 and 15 minutes. Many of the larger herds using AI place fixed orders on a weekly basis so that semen is available at the time animals are expected in heat.

Timing of the insemination is crucial to success and a rigorous heat detection programme must be used to determine the onset of oestrus. The first insemination should be carried out about 24 h after the onset of oestrus (not the first time that oestrus is actually observed) and a second insemination given between 8 and 16 h later.

There have been many attempts to use deep frozen semen in pig AI schemes but all have failed because of poor freezing and thawing characteristics of pig semen. In specialist breeding work there is still a place for the use of frozen semen and research is slowly improving the conception rates achieved with deep frozen semen. New diluents and cryo-preservatives are constantly being tried.

For the meat producer, AI offers the opportunity of replacing the existing herd with superior animals over a period. There are also an increasing number of meat producers who use AI widely in their main herd to gain from the significant financial advantage of using the progeny of AI boars as meat animals. Some of these producers use natural service for a first insemination followed by AI at the second. In addition to conserving 'boar power', this can give an overall improvement in fertility.

Sow and gilt reproduction

The breeding sow is the basic production unit for the production of as many weaned piglets as possible in a given time at least cost. The profitability of a breeding-finisher unit is intimately linked to maximum sow productivity. Where the statement falls down is that, in a changing price situation over long time spans, maximum profits are not always associated with maximum physical output. Furthermore, in the context of the public becoming increasingly interested in animal welfare, the hyper-producing sow may, in the future, be incompatible with the public's attitudes and understanding of animal welfare. Within normal bounds, sow management is directed towards farrowing as often as possible in a year and rearing the maximum number of piglets to weaning. It is intended in this section to outline female reproduction from puberty to weaning and illustrate the salient features.

Puberty and the gilt

Whatever the source of gilt replacements, it is imperative to introduce them to the main breeding herd as soon as possible. In other words, the sooner they are mated and produce their first litters, the lower the overall replacement costs. It was once thought that around a third of all sows a year needed to be replaced and every time a cull sow was removed from the herd a gilt was mated immediately as a replacement. This is probably not the case and in many herds replacement rates are around 50% per annum. Early weaning is a factor in this but not necessarily because the practice 'wears-out' sows rather quickly but because culling takes place after six parities and with early weaning this point is reached much more quickly. The annual culling rate therefore rises accordingly.

Puberty or first heat in the female pig occurs when the animal is about six months of age at 90 kg liveweight. There is enormous variation around this age and some gilts attain puberty as early as 130 days and some as late as 280 days. Much research has been carried out to establish the factors controlling puberty, and this is now quite well understood. Genotype, nutrition and season of the year are not the least of the factors that influence the time of first heat but the most potent influence is generally referred to as the 'boar effect'. If prepubertal gilts are suddenly exposed to contact with mature boars, this stimulates the first heat and normal cyclicity in a high proportion of gilts. There are a number of components of this effect such as sight, sound and tactile contact with a boar. The biggest component is that of boar odours and the pheromones (airborne hormones) contained within the boar's characteristic smell.

Gilts at 160–180 days of age are introduced to mature boars and given daily contact for at least 20 min/day in the same pen as the boar. It is highly desirable to house the gilts from the time of first introduction to males within the same air space or in the pen adjacent to the boar used for testing the gilts. This maximizes the exposure of the gilts to the pheromones. Stimulation with mature boars necessitates a new pen and perhaps relocation within the piggery. Gilts respond to this stimulus and it helps to initiate the first heat in many gilts. It is well known that gilts appearing on a farm from a breeding company exhibit oestrus within 7–10 days of delivery, and this is put to good advantage by farmers.

A final stimulus to early puberty is the mixing of groups of gilts into new social groups. This on its own will cause many heats within a few days of mixing. The use of this practice in a management programme may need careful control to avoid the physical trauma resulting from fighting in new social groups.

Despite the abundant data and research on gilts, failure to exhibit oestrus is a perennial and serious problem on some farms. There are now a number of pharmaceutical products which are used very effectively as cycle starters. These products are usually based on combinations of the gonadotrophin hormones (FSH and LH) or their analogues which cause follicles to grow and mature in the ovary.

Oestrous cycles

Following first heat, most gilts have regular oestrous cycles at approximately 21-day intervals. This varies from about 18 days to 23 days in individual animals. In some gilts, however, following first heat a number of so-called silent heats may occur. This is the situation where ovulation takes place and eggs are shed from the ovary under the control of the hormone LH, but the overt signs of oestrus including receptivity to the boar are absent. A similar problem is where gilts show only a weak oestrus response and mating is then difficult because of the behavioural problems encountered.

A major decision is at what stage, relative to puberty, mating is to be carried out. There is clear evidence that ovulation rate (the number of eggs shed) is very poor at pubertal heat. Mating gilts at this time therefore leads to an unacceptable litter size at the end of pregnancy. Most producers wait until the second or even third heat by which time the ovulation rate has increased to promote a reasonable level of prolificacy.

In larger herds where 20 gilts or more are mated in any one week, there may be a role for the use of synthetic materials to control oestrus in groups of gilts. The problem is that because of seasonal variations in the propensity to show first heat, it is difficult to ensure a regular supply of replacements. There is a product available which is based on an analogue of the naturally occurring hormone progesterone. This product (Regumate, Hoechst UK Ltd) can be given as a feed additive. On withdrawal of the product, animals show heat with a high degree of synchronization. Matings can be planned well in advance to maintain a steady throughput of gilts into the main breeding herd.

Oestrus and mating

When oestrus occurs in pigs, the outward signs are characteristic and clear. Gilts and sows in a group will tend to go off their food and in the pro-oestrus period will mount other females. They will also exhibit vulval reddening and swelling at this time under the influence of rising levels of the oestrogen hormones. At the onset of oestrus, animals will respond to the application of firm hand pressure on their backs and will stand rigidly often holding their ears erect. Once this response has been elicited, it can be extremely difficult to move the animals around. Stockmen often sit astride the backs of sows and gilts as a more vigorous method of heat detection and this is known as the 'riding test'.

In the absence of a boar for heat detection, it is possible to identify only a percentage of animals which are in heat. This figure may be up to 60%. With a boar in the same pen as pro-oestrus gilts or sows there is a maximum chance of detecting heats. Some farms use vasectomized young boars for heat detection. The sterile boars are used as teasers for the stimulation of first heat, without the risk of mating at the wrong time or with the wrong boar.

Oestrus normally lasts for 2–3 days and ovulation takes place in the middle of this oestrus period. The aim with natural service as with AI is to have fresh semen in the reproductive tract to await eggs being shed from ripe ovarian follicles. This is important because there are definite physiological processes whereby the sperm cells mature for a number of hours in the female's tract before they are able to fertilize viable eggs. If sperm are introduced into the female's tract either too early or too late, then the sperm and egg cells are out of phase in their physiological maturity. Conception could therefore fail or the litter size could be extremely small.

With natural service, it is best to carry out heat detection on a twice-daily basis to ascertain, as near as possible, the precise onset of oestrus. Service is carried out ideally 30 hours after the beginning of heat and in practice this is on the morning of the first time that oestrus is detected. Mating is then repeated the next day. For sows spotted in heat in the afternoon but not on the morning of the same day, it is better to wait until the next morning before the first mating is carried out. Figure 20.3 illustrates the timing of mating relative to the important reproductive events.

There are now some farmers who allow three matings through the course of the 2-day oestrus period. This does have some merit, and evidence has shown that conception rates and litter size are maximized by increasing the chances of placing fresh semen in the female's tract at just the right time. On the debit side, there is an increased requirement for the number of boars or semen samples needed. If sows or gilts are mated twice each and the optimum services-a-week for each boar is about four times, then this means that for every 20–25 females in the herd a farmer will need one boar. If triple serving is practised, this changes to around 15 females per boar to avoid the overuse of boars and to allow for the resting of boars periodically.

Feed allowances for gilts

Where gilts are reared on the farm where they are to remain as breeding females, there is more scope for the provision of a perfect nutritional and social environment. The evidence points to a medium plane of energy intake in the early rearing period to allow gilts to reach the threshold liveweight at an age where they are mature endocrinologically. If they are grown too quickly or too

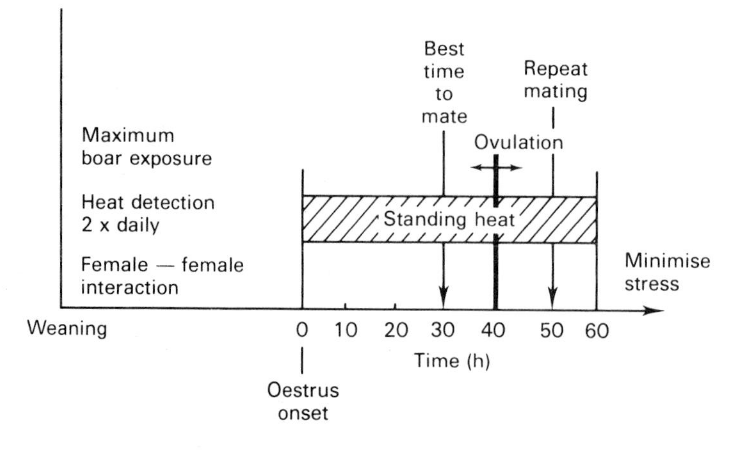

Fig. 20.3 Timing of mating.

slowly, then there is the possibility that puberty will be inhibited. In practice, up to the time gilts are ready for stimulation by boar contact, they are fed about 1.8 kg/d of a conventional breeder diet containing 14–15% crude protein (CP). The precise amount given may vary from farm to farm depending on the housing system, group size and so on.

From the time gilts are introduced to mature males, most farmers 'flush' their gilts. Flushing is the abrupt transition to a high plane of energy intake prior to ovulation, and gilts are commonly given around 2.5 kg/day or are even fed *ad libitum* on some farms until the time mating takes place. This practice ensures the maximum number of eggs are shed at ovulation. Immediately after service the feed scale should be adjusted again to a low plane at about 1.8 kg/d. This prevents the excessive loss of fertilized embryos which may happen if gilts are overfed in the first three weeks of pregnancy.

Pregnancy

Sows spend at leat two-thirds of their working lives in gestation and the bulk of feed inputs are given during this important perod. One of the problems with modern systems is that with automated housing and feeding systems, sows in pregnancy can easily be under-managed. If the whole unit is to operate successfully, sows should be monitored as individuals throughout pregnancy.

The period of gestation in the pig lasts about 115 days and the variation around this mean value is generally small although some sows will naturally give birth as early as 110 days after mating or as late as 119 days. The initial days of pregnancy are the most critical in terms of whether the pregnancy will last, and also it is in the initial weeks when the final litter size is determined. At first the fertilized eggs are free-living entities within the lumen of the uterus. By day 12 the initial attachment of embryonic membranes begins and this process of implantation continues until about day 20 of pregnancy. Because implantation is similar in some respects to the host-graft relationship seen in organ transplantation or in skin grafting, the process is a very delicate one. The developing embryos can be immunologically rejected by the uterus and development stops. Any stress on the sow or gilt at this time causes an imbalance in the hormone status of the sow and more embryos die. Similarly, any difficulty experienced by individual embryos in securing a supply of nutrients across the placental wall will also lead to embryonic death. Once the hurdle of implantation is passed, then the sow is more able to buffer embryos from an adverse external environment.

The final litter size is set initially by ovulation rate. Throughout gestation the number of potential piglets is reduced by losses at different stages. Under average conditions 25–30% of all fertilized eggs will be lost in the first three weeks of gestation. Another portion of loss occurs in the fetal stages when the developing fetuses compete for available space and nutrients within the uterus. This latter percentage of loss is less significant than early embryos losses. Figure 20.4 illustrates the general pattern of losses of potential piglets from ovulation onwards.

In the management of sows, the objective is to provide a minimum stress environment, particularly in the first weeks of pregnancy. The understanding of stress has a long way to go before firm recommendations can be given on housing and the ideal social environment. Some interesting work from Australia has shown the importance of social factors for reproductive performance. A novel finding was the impact that social factors in very early life can have on the ultimate lifetime performance of the animal. For example, ten-week-old piglets which were either reared in isolation or which were subjected to mishandling by humans in their early lives were especially prone to diminished reproductive performance at maturity. The notion that early stress can have profound influences on the physiology and therefore the physical performance of sows and gilts is an extremely important one. In the future we may see more attention given to the building of minimal stress farms where production systems are good for the animals' general well-being but also good for economical production. The Australian work carried out by Dr Paul Hemsworth and his colleagues has included a study of the way in which humans interact on farms with the animals in their care. This latter work

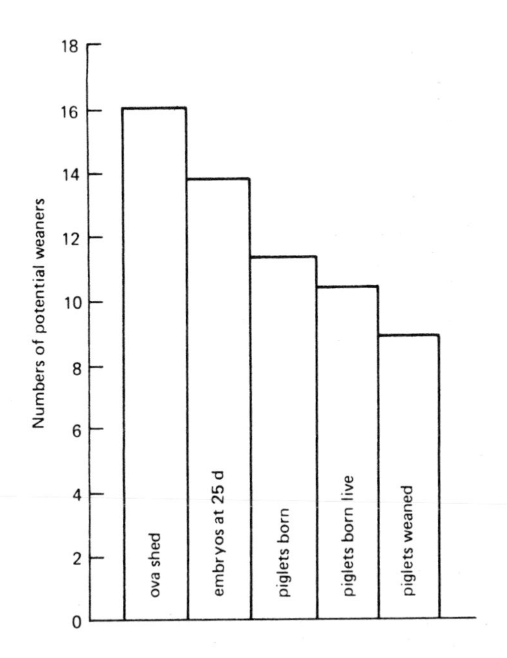

Fig. 20.4 Losses of potential weaners through pregnancy and lactation.

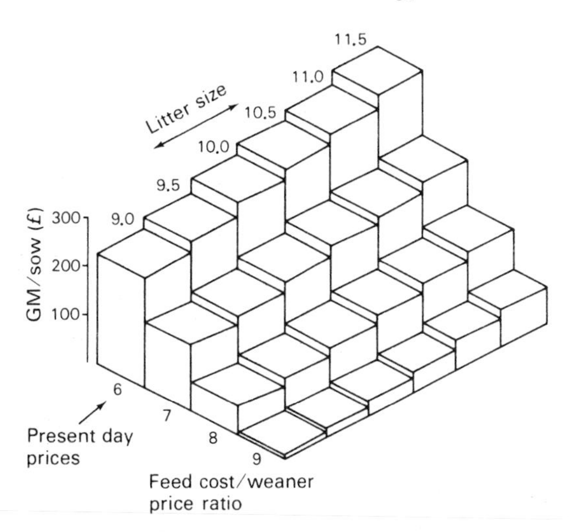

Fig. 20.5 Prolificacy and profitability.

has shown that the personality of the stockman looking after sows has a strong relationship with both litter size and the farrowing rate of individual farms. Where sows showed particular aversion to humans there was a detrimental effect on reproduction. The selection and training of competent stockmen may therefore turn out to be the biggest factor of all in determining the reproductive capacity of the whole herd. What is certain is that a high litter size is crucial for survival in the modern economic climate. At the University of Leeds we have recently confirmed the results of the Australian researchers and in addition we have shown that the physical environment in early life is also important. Piglets reared in a pleasant outdoor environment produced more piglets at maturity than their counterparts reared under indoor intensive conditions.

Figure 20.5 illustrates how, in a changing price structure, litter size affects profitability. It can be seen that those units operating below par in a bad year will quickly move into a loss situation.

The feeding of pregnant sows

Pregnant sows which are gaining body weight from one parity to the next need nutrients for maintenance and growth and also a supply of nutrients to sustain the developing conceptus. The factorial approach has been used in the past to calculate energy and protein requirements for sows, and recommendations are based on these calculations along with the results of controlled feeding trials. There are a number of difficulties with this approach. The first is that pregnant sows experience a hormonal environment which is stimulatory to growth in the same way that steroidal growth promoters work in a number of species of farm animals. The steroid hormone progesterone is at high concentrations in the peripheral circulation of pregnant females and this helps them to utilize their food more efficiently than they would otherwise. The second problem is that we have to take what may be an arbitrary decision as to how much weight we wish the sow to gain (or lose) through pregnancy.

Traditional feed scales allowed for high weight gain in pregnancy and much of the gain was catabolized in the ensuing long lactation. This proved an inefficient way to feed sows from one parity to the next. At the present time sows are offered a plane of energy intake in pregnancy only just above what would be a maintenance intake for a non-pregnant animal. At the end of gestation the sow has accrued modest weight gains and can then be fed directly in lactation for milk production. In this way sows gain weight at 15 kg from one parity to the next and do not waste energy in building up body reserves and then breaking them down again in lactation. In current commercial practice, sows are offered between 1.8 and 2.2 kg/day throughout pregnancy of diets containing around 12.5–13 MJ of digestible energy/kg. This scale of feeding is not altered throughout the whole of pregnancy as there is little merit overall in increasing the scale as parturition approaches. There has been some interest in the short-term increase of the energy concentration in the diets of sows in the last few weeks of gestation, and there have been claims that the average birthweight and the vigour of newborn piglets are improved. The results are often conflicting, however, and not convincing despite the fact that many feed companies offer high energy diets with an increased inclusion of fats.

Protein requirements for sows have also been well researched and it has been demonstrated that a sow will perform satisfactorily on a diet containing as little as 12% crude protein. This means that a diet composed of high protein barley plus synthetic lysine would be perfectly suitable for a pregnant sow. For sows expected to have a reasonably long productive life, these extremely low levels of protein may not be acceptable and a protein level of 14–15% is more commonly used for dry sow diets.

Pregnancy diagnosis

An array of ultrasonic devices is now available which can be used for the determination of pregnancy. The early detection of sows which are non-pregnant is a valuable management tool. Non-pregnant sows or gilts can be moved back to the service area or culled if required without the input of further feed and other costs. Diagnosis is possible as early as 28 days after service but repeated testing through pregnancy is necessary to pick up sows which resorbed their fetuses after initially conceiving. There is also a system of pregnancy detection based on a blood test for the hormone progesterone. In the future this method may be more accurate than ultrasonic machines and if it can be made to work on urine samples it is likely that this would be a preferred method.

Parturition

Parturition and the time that sows deliver their offspring are a crucial phase in the reproductive cycle. If things go wrong, piglets will either die before they are expelled from the uterus or they will die in the early hours or days after birth. The hormonal and physiological events taking place around parturition are now well known and this understanding has given us some valuable tools for the control of the process.

Labour is initiated 48 hours or so prior to any visible signs. The ovaries respond to a hormone of the prostaglandins series produced in the uterine wall and the net result is that the corpora lutea on the ovaries begin to regress and blood progesterone levels fall. Without the support of progesterone, uterine contractions begin and labour commences. A number of other hormones then control the frequency and the strength of the uterine contractions. Amongst these are oxytocin, relaxin and oestrogen. Perhaps partly because of the complexity of the hormonal events the whole process is prone to dysfunction. Any stress on the sow prior to parturition can cause delayed farrowing and prolonged delivery. This in turn leads to oxygen starvation in those piglets born last and they may be stillborn.

Sows should be transferred to their farrowing quarters at least 3 days before the expected time of farrowing. This allows them to acclimatize to the new environment before the litter is delivered. They should ideally be in a group farrowing within a few days of each other in order that they all farrow in a clean area away from the contaminants associated with older piglets. When a particular sow is about to deliver, she will exhibit 'nest building' activity or characteristic pawing with the front legs. Finally she will settle down and at the appearance of a bloody discharge she will begin to deliver piglets. There are no set rules as to the interval between piglets: in some cases it will be a few minutes or even as long as an hour or two. From beginning to end, the whole process may be from 1 to 12 hours or even longer. Above 6 hours the percentage of piglets which are stillborn rises exponentially and an obstetric examination may be required.

For a number of years now farmers have used synthetic versions of the naturally occurring prostaglandin hormones. These are extremely effective for inducing parturition and are now widely used. A single injection is given 26 hours before delivery is required and almost 100% of sows respond to this by going into labour at the prescribed time. It is therefore possible to ensure that night-time farrowings are avoided and also weekend farrowings when there may be no one around to supervise the deliveries.

The use of products such as prostaglandin analogues needs some caution as it is possible to induce farrowing at a time when the fetuses are not mature enough to survive the external environment. Piglets induced 2 days or more before the expected date can hence lack vigour. Postnatal care must therefore be of a high order.

Lactation and weaning

If all has gone well after the rigours of farrowing, lactation is initiated and suckling occurs more or less at hourly intervals. Daily milk production builds up to a peak at the end of the third week of lactation and then declines steadily. Figure 20.6 shows the pattern of daily milk output for the average sow.

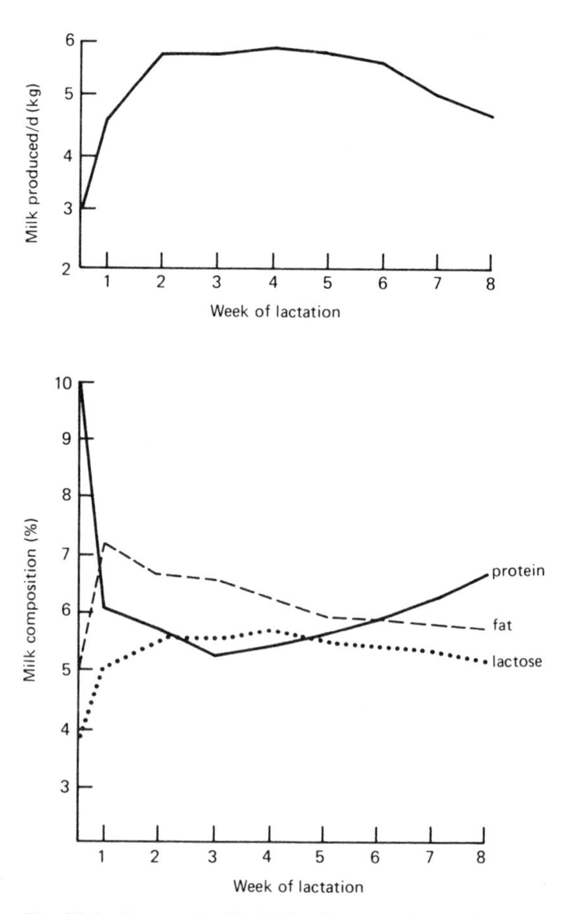

Fig. 20.6 Pattern of milk yield and composition in the sow. (After Salmon-Legagneur, E. (1961) *Ann. Biol. Anim. Biochem. Biophys.* **1**, 295–303. Reproduced by courtesy of the Editor and Publishers.)

Reproductive activity is at a low level generally throughout lactation and the ovaries are almost completely inhibited by the suckling stimulus. Feral pigs have been observed to show heats whilst they are still suckling but only after quite long periods of lactation. Experimentally it has been demonstrated that sows can be persuaded to escape the inhibition of the suckling stimulus and can exhibit concurrent pregnancy and lactation. Attempts have been made to exploit this commercially using groups of multisuckling sows and the introduction of boars to groups of sows and litters about 12 days after parturition. It has never, however, been possible to achieve successful conception before about four weeks into lactation. In view of the fact that a high percentage of producers wean at three weeks of age, and mating is carried out at 26–28 days *post partum*, the effort involved in inducing sows to ovulate in lactation does not seem worth-while.

The age at which piglets are weaned from the sow is a major decision and there is probably no single perfect time to wean which suits everyone. There are now well designed creep feeds for very young piglets and the environmental requirements are well worked out and have been translated into appropriate buildings. There is therefore no technical limitation to the age at which weaning takes place and even systems of weaning at birth have been explored experimentally. About 70% of UK producers now wean at between three and four weeks of age and the national weaning age has fallen steadily over a long time period. Economic pressure to maximize the annual sow productivity on farms is the biggest single reason why producers have moved in this direction. In theory, by reducing the time each sow spends in lactation, more reproductive cycles are fitted into a 12-month period and annual sow productivity rises. This is illustrated in Fig. 20.7 which shows the components of a sow's normal reproductive life. The faster each sow moves around the cycle, the higher is the output of piglets achieved. This relationship is not linear. As we reduce lactation length, the sow's physiology alters so that the litter size falls off and there is a prolonged interval from weaning to remating. In Table 20.4 a theoretical analysis is given to illustrate the expected outcome of weaning at different ages together with the real outcome as demonstrated in a number of large studies and surveys.

The data given in Table 20.4 serve to illustrate the point that, by continuing to reduce the lactation length below three weeks, there will be no further advantage in sow productivity because of the precipitous drop in litter size. On individual farms the precise outcome may vary but results in general will follow this same pattern. More recently another factor has come to bear in consideration of weaning age. It is still generally believed that annual sow profitability is directly proportional to annual sow productivity. Indeed this may be true in some situations where there is a high ratio between weaner price and feed costs. Through the 1980s we saw escalating feed costs relative to pig prices and this has shifted the balance towards weaning later. Early weaning necessitates the use of expensive creep diets and it is this that can severely deplete the profit of the system. On many farms careful consideration of all of the financial aspects of weaning age should be given as well as the likely output in terms of piglets per sow per year.

Feeding the lactating sow

The feeding of the lactating sow should always be assessed in relation to the feed scales adopted in pregnancy, but, in general, lactation is a time when relatively high energy intake should be allowed to promote milk production. Until recently most pig units used the same breeder ration in lactation as for pregnancy. The feed compounders now offer high energy diets with a high inclusion of fats to satisfy lactational requirements. In part this move has stemmed from the difficulties faced by gilts in their first lactation. These animals may have been mated relatively young and are still actively growing their first

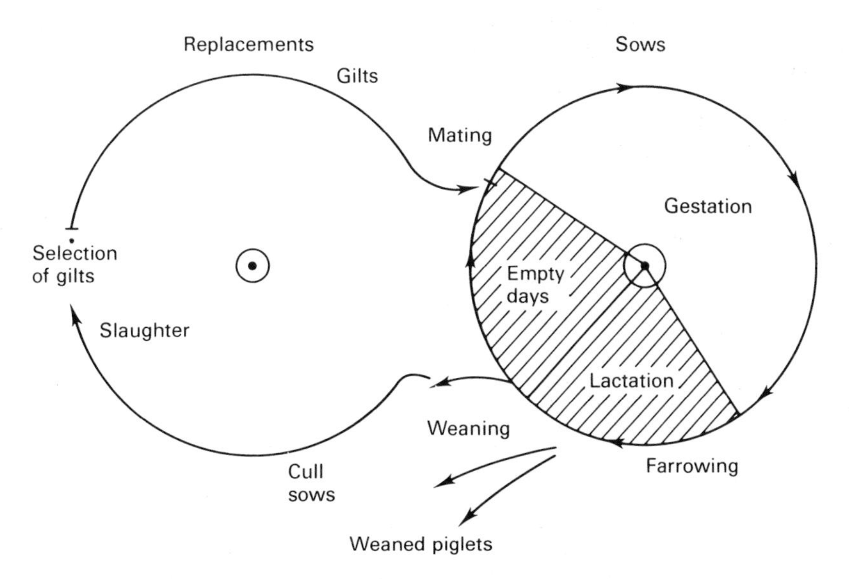

Fig. 20.7 The reproductive cycle.

Table 20.4 The effects of weaning age on sow productivity

	Lactation length (d) 14	21	28	35	42
Theoretical					
Pregnancy (d)	115	115	115	115	115
Lactation (d)	14	21	28	35	42
Weaning to mating (d)	5	5	5	5	5
Farrowing interval (d)	134	141	148	155	162
Litters/sow/year	2.72	2.58	2.46	2.35	2.25
Litter size	10.7	10.7	10.7	10.7	10.7
Piglets/sow/year	29.1	27.6	26.3	25.1	24.0
Actual					
Pregnancy (d)	115	115	115	115	115
Lactation (d)	14	21	28	35	42
Weaning to mating (d)	7.5	6.5	6.0	5.2	5.0
Farrowing interval (d)	137	143	149	155	162
Litter/sow/year	2.66	2.55	2.44	2.35	2.25
Litter size	9.4	10.6	10.7	10.7	10.7
Piglets/sow/year	25.0	27.0	26.1	25.1	24.0

parities. Particularly if they have been overfed in their first period of pregnancy, they will struggle to consume enough energy in their first lactation for milk production and growth. The net result is that these gilts reach the beginning of their second reproductive cycle in very poor condition, and reproductive performance in the second parity will be adversely affected. It is probable that the changing genotype of our pigs has contributed to this problem. By selecting pigs which grow fast and which are efficient converters of energy, we have also selected pigs with smaller appetites. As a consequence, a number of farmers give lactational diets to sows on an *ad libitum* basis.

For mature sows the daily allowance offered of a diet containing 13 MJ of digestible energy (DE) and 15% CP might be 1.8 kg plus 0.45 kg for each piglet suckling. A sow suckling a litter of ten piglets would therefore be offered 6.3 kg each day or around 82 MJ of DE. This allowance would not be offered at the beginning of lactation and from the gestational scale of, say, 1.8 kg/day an extra increment of 0.45 kg would be added each day until the set amount is reached.

The weaning to remating interval

After weaning and the removal of the suckling stimulus, the sow under normal circumstances exhibits oestrus and ovulates between 3 and 10 days later. The majority of animals show heat at either day 4 or day 5 after weaning but, as shown above, the early weaned sow is a little later at 6 or 7 days after weaning. When things are running smoothly and heats are expressed regularly and predictably, management is straightforward. All too often, however, these short periods in the sow's life can be the most difficult. Anoestrus or the complete absence of any signs of heat at all after weaning is one of the major reasons for culling sows from the herd and even the best herds will have 2–4% of these every cycle. At worst this figure can rise to 10–20% and leads to the complete disruption of the farrowing programme. The cause of anoestrus is multifactorial but sow body condition, season and health status are often implicated. A thorough heat detection policy is essential to ensure that there are no sows recorded as anoestrus which were simply missed by the stockperson.

There is no advantage in the withdrawal of either food or water in the first 24 hours after weaning. It was once thought that this practice might help to dry the sow off. It has been shown that this gives no benefit in terms of the time to return to oestrus or the percentage of sows showing oestrus within 10 days of weaning. The objective in feeding the sow during this empty period is to maximize the ovulation rate, and therefore a daily feed allowance of 4–6 kg is offered to this end. It is also a time when some sows need to begin to recover from the negative energy balance of lactation, and therefore feeding to appetite in this short period is good practice.

Weaner piglets

The management of small piglets has been made more straightforward with the development of flat-deck cages. These building systems help to reduce outbreaks of acute enteric disease which were commonly seen in earlier types of weaner accommodation. The understanding of the nutritional and immunological requirements of baby piglets is now well advanced and there are high quality diets available based on complex formulations to ensure minimum problems. Despite the technology available, the post-weaned piglet still represents real problems for some farmers.

The ability to withstand disease challenge is a principal attribute which piglets must acquire to survive. They are born devoid of any protective antibodies in their bloodstream and therefore are highly susceptible to even the simplest of bacterial types. Each piglet must therefore ingest as much of the sow's first milk or colostrum as possible. Colostrum contains a very high concentration of antibodies or immunoglobulins and immediately after birth these pass straight across the piglet's gut wall into the bloodstream. Within hours of birth the piglet's gut is

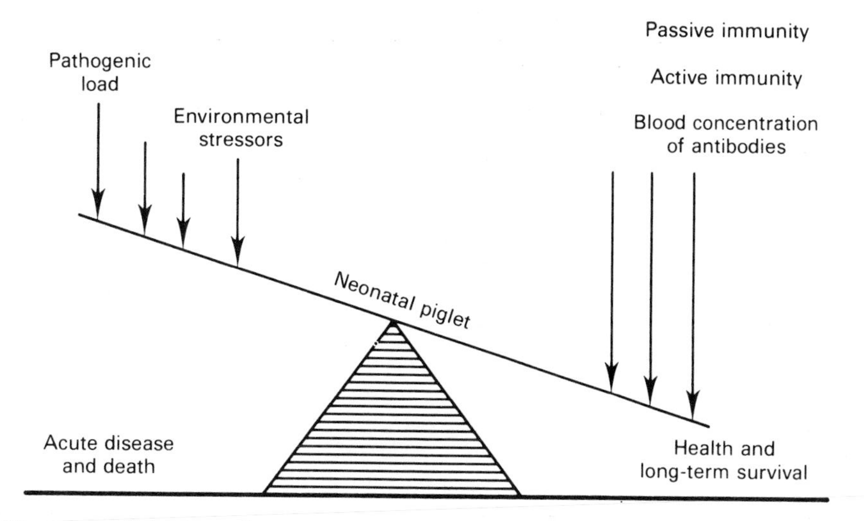

Fig. 20.8 The neonatal 'balancing act'.

said to close and it loses its capacity to transfer large immunoglobulin molecules straight into the bloodstream. It is also well known that piglets do not begin to manufacture their own antibodies in significant amounts until they are about two to three weeks of age. The amount of colostrum ingested in the first 12 hours of life determines the piglet's survival prospects for many weeks to come because after birth the blood concentration of antibodies falls at a regular rate. In Fig. 20.8 a schematic diagram is given of the way in which immunoglobulins are needed to balance out the prevailing pathogenic load. If the balance falls in favour of the bacteria and viruses, then the young animal will succumb to disease and may die.

From three weeks of age onwards the piglet's blood level of immunoglobulins (Igs) rises and the possibility of disease and death diminishes as complete immunocompetence is attained. There may, however, still be some problem piglets which only acquire enough colostrum to survive initially. These piglets may fall below a certain threshold of blood antibody at some later stage and they may die. Death could occur many weeks after birth but the time when the problem begins is immediately after birth.

It is important to supervise the postnatal period as much as possible to maximize the suckling performance of all piglets in every litter. Successful suckers within a litter will easily survive but poor suckers are prone to mortality for some time after birth. The relationship between suckling performance and the probability of death is illustrated in Fig. 20.9.

The use of prostaglandin analogues to induce farrowing during normal working hours may help to facilitate the supervision of suckling. Cross-fostering of piglets between a group of sows which have farrowed at the same time is also a useful technique.

The process of weaning itself is often made an abrupt event rather than a gradual transition from dependence on the dam to complete self-sufficiency. At weaning the piglet adjusts to and finds its new sources of food and water. It has also to adjust to the stress of relocation and of being mixed with new individuals from other litters. These combined stressors often lead to reduced immune competence which manifests itself as enteric or systemic diseases. At best post-weaning scours cause chronic loss in growth, and at worst mortality rates are significantly increased. One method to circumvent these problems is to spread the stressors as much as possible. The sow can be moved from the farrowing house leaving the litter of piglets *in situ* for as long as possible. It may then be possible to take away the dividing partitions between two or three litters and allow them to mix freely but still in familiar surroundings. Finally the new group is moved to weaner accommodation where they will find the same food and water delivery system with which they are already familiar.

Water delivery systems for weaned piglets need some care in their selection and operation. Nipple drinking systems have been used for some time and in general they are maintenance free and almost impossible for piglets to foul. Some types may not be easy for small piglets to operate, particularly in the immediate post-weaning period. It is becoming increasingly obvious that one of the first limiting factors to post-weaning growth may be water intake which in turn limits dry matter intake. There are many sweetening agents which can be added to water supplies to encourage intake but perhaps avoiding problems comes down to a high level of stockmanship.

Another factor known to be responsible for a high proportion of the outbreaks of enteric disease in the post-weaning phase is the type of protein source used in the feed. Even though a diet contains an ideal blend of ingredients to meet all the requirements for maintenance and growth, some types of protein may cause hypersensitivity reactions in the piglet's gut. Plant proteins are worse than animal proteins in this respect and soya proteins may be among the worst. It is easy to formulate a high energy diet with maize meal, wheat and corn oil and to balance this with skimmed milk powder as a protein source. These feed ingredients are highly nutritious and very palatable to young piglets, and will cause few problems. The trouble is that skimmed milk powder is now a

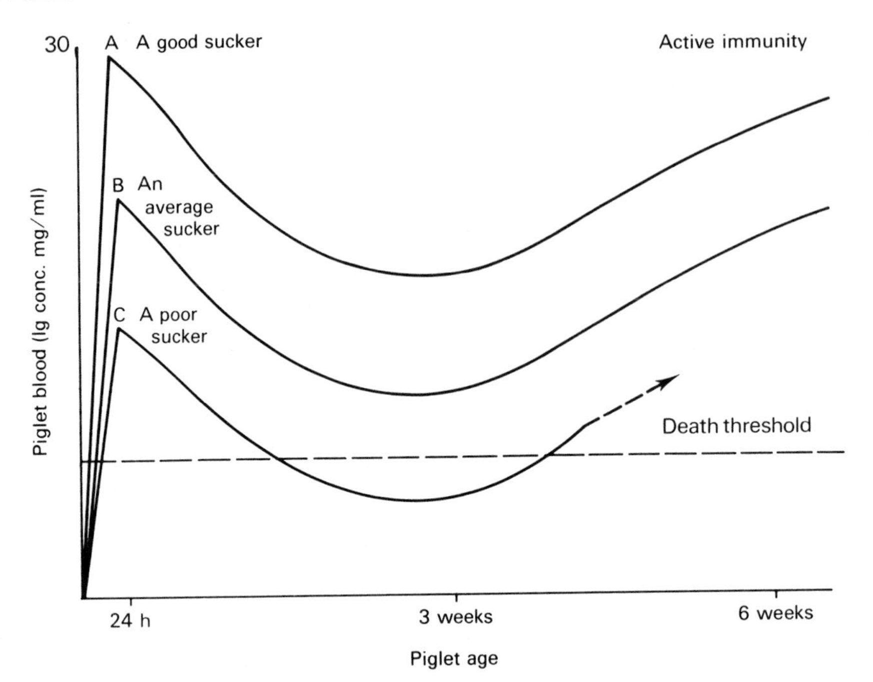

Fig. 20.9 Immunoglobulin concentrations after birth.

very expensive ingredient and will probably continue to be so. The inclusion of heat-treated soya bean meal in creep diets is highly attractive in keeping the price down. The use of these soya-based diets needs some caution although it is not impossible to use them. Many of the problems can be avoided by 'tolerizing' piglets to the protein source. This entails giving piglets the potentially harmful ingredients in small but regular amounts for as long as possible before weaning. When piglets are subsequently put on to a high level of the same feed ingredient after weaning, their immune system does not overreact with an allergic response because the animal is familiar with the source of protein. Another strategy is to offer no creep feed at all before weaning. The piglet is then presented with dry food for the first time after weaning. The sudden high level of intake of the protein source induces the tolerization process and the lining of the gut is not damaged. If it works, as it can on some farms, then this latter strategy may also cheapen feed costs considerably.

Amongst a variety of methods for minimizing post-weaning diarrhoea are the use of oral vaccines against the *E. coli* organisms which cause the disease. These can be given in the feed and although they may not work on every farm, they give protection against a spectrum of organisms. Probiotics or *Lactobacillus*-based agents are also used to minimize enteric diseases. *Lactobacillus* bacteria are normal inhabitants of the gut and a healthy flora of these organisms tends to inhibit the further colonization of the gut wall by potentially pathogenic bacteria such as *E. coli*. In addition, the *Lactobacillus* bacteria produce lactic acid which reduces the pH in the gut lumen creating a hostile environment for the proliferating pathogenic strains of bacteria.

A recent innovation in the management of very young piglets has been the development of synthetic colostrum products. Material derived from the plasma of growing pigs can be converted into a highly potent source of immunoglobulins. After suitable sterilization this can be given to neonatal piglets as a supplement to the colostrum of the sow. This is obviously of great use for orphan piglets or in situations where there are too many piglets and not enough sows for crossfostering.

Finishing pigs

From weaning to the time when pigs are slaughtered for meat, the management is relatively straightforward compared to the problems with sows and piglets although there is a large financial commitment at this time. In a well designed finishing house with a carefully controlled environment, the process is one of turning feed energy into saleable meat. The investment in terms of fixed equipment and working capital is such that great precision in technical management is needed to make a profit. Currently, compared with other animal enterprises, pork and bacon production can give a reasonable return on capital and a cash flow situation allowing regular receipts.

Traditionally there have been a number of different markets for finished pigs. The first of these is the pork market where the meat is sold as a fresh product. The slaughter liveweight for pork pigs varies from one part of the country to another but is usually in the range 50–90 kg. Jointing techniques for pork pigs also vary from one region to another depending on local customs. Consumer demand shows seasonal trends and the Christmas and Easter trades are the times when demand is greatest. The smallest demand is traditionally seen in the summer months but recently this is not so clear cut because consumers have responded to the competitive price of pork and historical fears about pig meat in hot conditions are not evident.

Bacon production tends to be the most profitable of all pig meat outlets, and large scale producers sell a large proportion of their output into this category. The slaughter-weight ranges between 80 and 95 kg. Individual bacon companies may specify a much narrower weight band to fulfil their own market requirements. In general the closer a producer can sell his pigs to the top of the weight range, the closer he is to maximizing his profitability. Pigs are therefore weighed carefully every week as they approach slaughter to ensure every individual attains the correct carcass weight.

In addition to pork and bacon pigs, there is a category of pigs known as cutters. The weight range is between 75 and 100 kg and historically these pigs were processed into manufactured products. There may be a better future for cutter pigs at around 80 kg liveweight. Production at this liveweight can be diverted to either quality bacon or to pork production depending on fluctuations in the markets of each. The hope is to balance one market against the other and to create a more stable price for each.

Heavy hogs or manufacturing pigs are slaughtered at 105 kg or more liveweight. These are destined to be processed into a large number of diverse products such as cooked meats, high quality hams, bacon products, meat pies, sausages and so on. The high slaughter weight means that, when the animals are killed, they are laying down body fat at a much greater rate than pigs at lighter weights. This can be somewhat inefficient production in biological terms because of the feed energy required to accrete fat. High quality bacon products sold at a premium price are produced by trimming off the excess backfat to yield rashers and joints with a high lean content. In the future there could be an increasing market for pigs of greater liveweight. This is because of the modern pig's capacity for lean tissue deposition and minimal fat cover. Heavy pigs fed on an *ad libitum* basis might offer better profitability. This strategy has yet to be proven and there is a reluctance on the part of the meat trade to move in this direction due to increased processing costs. Most male finishing pigs are not now castrated and this, in addition to the welfare advantages, offers significant improvements in lean tissue yield overall because of the reduced fat content of entire male carcasses.

Carcass quality and classification

The pig industry has been aware for some time of the need to generate products which consumers perceive as high quality and not just to produce what is easy to produce. As a consequence there has been a gradual reduction in the backfat scores of slaughter pigs. It is often said that in the UK we are still not capable of producing such consistent high quality products as our EC neighbours in Holland and Denmark. Bacon products which arrive on British supermarket shelves from these countries are uniform with a high lean content, and the packaging and marketing expertise are of a high order. Part of this apparent differential may be that UK producers sell almost all their produce on the home market. Danish and Dutch processors select the best pigs for the export trade and the meat sold on their own home markets is of a much lower quality. The image of British pig products on the home markets needs developing in the future is the industry is to remain competitive, even though farmers are capable of producing high quality pigs.

The criteria that determine quality in a finished pig revolve around the fat content of the carcass and in particular the amount of subcutaneous backfat. Most people prefer the taste of lean meat. Also, in the mid 1980s, a number of government reports on human nutrition pointed strongly to the need for the nation to reduce the number of calories eaten per person as saturated animal fats. The population has responded to this vigorously with much advertising effort given to low fat products. As a result there has been a continuing drive to produce leaner and leaner pigs. One point in favour of pig meat over most other meats is that the level of intramuscular fat is minimal and for a given weight of lean tissue the fat content is significantly lower.

The Meat and Livestock Commission operates a national scheme for the classification of carcasses. A very high percentage of pigs are classified within this scheme which is a valid EC-approved scheme.

Classification aims to describe carcasses objectively in a form which can be used by producers, meat wholesalers and retailers as a framework for payment. It is fortuitous that there is such a clear relationship between the depth of subcutaneous backfat on the pig and its saleable meat yield and lean content. This has meant that a relatively simple scheme of carcass measurements gives easily defined grading schedules based on what the meat trade and consumers want.

The three basic measurements of backfat used in the MLC scheme are taken at the last rib at set points from the mid-line. The names given to these positions are P1, P2 and P3 and these are shown in the illustration of a section through a pig carcass given in Fig. 20.10. All three measurements are not necessarily used in any one grading scheme. In pork production, for example, the P1 and P3 measurements are taken and added together to give a joint score. This score is then stamped on to the skin of the carcass. The measurements are made by an MLC officer after slaughter on the unsplit carcass. An intrascope device is used for this purpose. This is an optical tool pressed in at the appropriate points. The work involved in taking measurements can be automated, and computer interfaced systems are available which can identify individual pigs, take and record the necessary information and process the data for grading the pig and paying the farmer.

Grading schemes may be based on three or more discrete categories of carcass based on certain bands of backfat scores. The bands are designed to encourage the farmer to send the maximum number of pigs within the top grade for which the best price is paid. One of the elegant facets of the scheme is that, with ultrasonic machines, breeding companies and the meat trade can make the same measurements on live animals and use this information as a basis for selecting carcass characteristics.

Despite the fact that there is considerable variation in the backfat scores of classified pigs, in general pigs are now leaner than ever before. Some wholesale meat com-

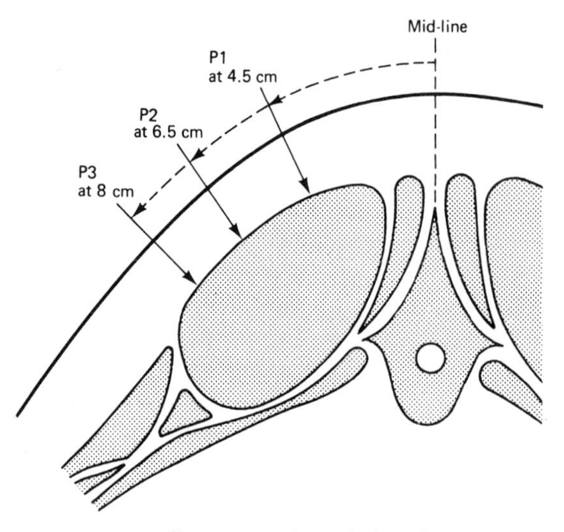

Fig. 20.10 Positions for the measurement of backfat thickness on the pig carcass. Note the P1, P2 and P3 probe measurements are taken 4.5, 6.5 and 8.0 cm respectively from the mid-line at the level of the head of the last rib. On carcasses these measurements are taken using an optical probe. On live animals measurements may be taken at similar points using an ultrasonic probe.

panies have built into their grading schedules a penalty for pigs which are too lean. This is partly because some extremely lean pigs have gone beyond the threshold of consumer acceptability, and partly because the texture of the meat is adversely affected and cutting becomes difficult. This seems a rather intractable dilemma: the producer is caught in the middle of consumers demanding increasingly lean meat and the meat trade seemingly reluctant to pay when very lean pigs are produced.

The industry has also been aware for some time of the need to carefully monitor meat quality. As pigs have become leaner they have also tended to show a high incidence of pale soft exudative muscle (PSE). Some breeds and crosses are known to have problems with PSE meat. Most hybrid companies therefore take steps to select against PSE in their breeding programmes, and use foundation stock with a low propensity to this problem. Breeds such as the Pietrain and some strains of Landrace pigs have a high incidence of PSE.

Pigs which are prone to PSE are often stress prone and possess the so-called halothane gene. They are extremely sensitive to halothane gas used in anaesthesia. Susceptibility to the gas is used to identify PSE genes and stress susceptibility. The genes can then be bred out of a population over a period of time. Many of the major breeding companies screen their pigs for halothane sensitivity.

Within the EC there have been strong pressures on member countries to bring together all classification schemes. There is now an agreed schedule which most nations have adopted. The scheme uses backfat measurements similar to the MLC scheme, but a combination of observations is used with prediction equations to assess the overall lean content of the pig. Individual carcasses are designated as one of six European standard classes (S, E, U, R, O, P) ranging from S containing 60% or more lean tissue to P containing 40% or less lean.

The nutrition of growing pigs

The energy and protein requirements of growing and finishing pigs are well established following research projects over many years. It is not intended to give an exhaustive account of this work here but merely to outline some of the salient features with regard to feeding and management. For comprehensive reports on the nutrition of pigs the reader is referred to the 1981 edition of the AFRC publication *The Nutrient Requirements of Pigs*.

Nutrient requirements are to a large extent determined by the stage of growth or age of the pig. Genotype, sex, thermal environment and social order are other important factors which have a direct bearing on the nutritional needs of the animal. As pigs grow older, because of their changing body composition and increasing propensity for fat deposition most feeding strategies are based on a changeover at around 50 kg or so from a high protein *ad libitum* schedule to a lower protein scale-fed system. Practically it would be more straightforward and convenient to use a single grower diet at about 16% crude protein and to offer this on an *ad libitum* basis through to 90 kg or slaughterable liveweight. This would lead to an increased number of overfat pigs which would fall into the lower categories of most grading schemes. Improved hybrids do have the capability to remain on *ad libitum* feeding to much greater weights without excess fat accretion. The majority of finishing pigs are still fed on a restricted basis in the final weeks up to slaughter to maximize the percentage of pigs achieving the top grade. One of the disadvantages in reducing energy intake in the last few weeks is that overall growth performance, usually expressed as 'average days to bacon', is made worse. As a consequence the throughput in a given finishing house is reduced. The production of finished pigs is therefore a balancing act, with feed inputs, slaughter weight and throughput being the principal variables.

Because of the complexities of the relationships involved in making on-farm decisions, many producers and nutritionists use computer simulation programmes which model their businesses to optimize inputs and output. These models are valuable tools in farm management, and as the many variables change it is possible to 'fine tune' the business to maximize profitability. Linear programming techniques are also widely available to farmers using their own microcomputers to enable them to formulate their own pig rations. As different samples of feed ingredients are available and prices change, it is possible to produce least-cost rations and minimize costs. For larger producers with their own mill and mixing plants, the ability to buy feed ingredients at the right price is crucial. Computers are necessary to evaluate quickly all the possible options before major decisions are taken.

Nutrient requirements for growing pigs are given in Tables 20.5–20.7. These requirements can be delivered to pigs in a variety of forms, at different nutrient densities within the final diet, and a large number of ingredients can be used depending on regional availability and season.

Table 20.5 Suggested energy and protein allowances for growing pigs*

	Approximate growth rate 600 g/d		*700 g/d*	
Liveweight (kg)	*Energy requirement (MJ/DE)*	*Protein requirement*[†] *(g/CP)*	*Energy requirement (MJ/DE)*	*Protein requirement*[†] *(g/CP)*
20	12.6[‡]	185[‡]	12.6[‡]	185[‡]
30	18.8	255	18.8[§]	255[§]
40	22.0	275	24.5	300
50	23.9	28.5	26.5	310
60	25.5	300	28.2	325
70	27.0	315	29.8	340
80	28.4	325	31.2	355
90	29.8	335	32.6	365

* Derived from ADAS (1978) Nutrient allowances for pigs. Advisory Paper No. 7, MAFF. Reproduced with permission.
† Assumes protein digestibility 80%, biological value 65. (Equivalent to an amino acid contribution in the protein of lysine 5.1%, methionine + cystine 2.8%, threonine 3.3% and tryptophan 1.0%.)
‡ Growth rate 400–500 g/d.
§ Growth rate 600 g/d.

Table 20.6 Suggested vitamin allowances for growing pigs

	Liveweights Up to 20 kg	*20–55 kg*	*55–120 kg*
Vitamin A (units/kg)	12 000–20 000	10 000–15 000	5000–10 000
Vitamin D (units/kg)	1500–2000	1500–2000	1000–2000
Vitamin E (mg/kg)	12–30	10–15	5–15
Vitamin K (mg/kg)	2–5	1–4	1–4
Thiamine (B_1) (mg/kg)	1–3	0–2	0–2
Riboflavin (B_2) (mg/kg)	3–6	3–6	3–10
Nicotinic acid (mg/kg)	20–25	10–20	10–15
Pantothenic acid (mg/kg)	10–15	10–15	5–15
Pyridoxine (B_6) (mg/kg)	2.5–6	1–4	0–4
Vitamin B_{12} (μg/kg)	10–30	10–30	10–20
Folic acid (mg/kg)*	0.8	0.5	–
Biotin (μg/kg)*	200	150	100
Choline (mg/kg)	150	150	100
Vitamin C (mg/kg)*	20	10	–

* Where only a single value is presented, there is a paucity of published evidence on which to suggest likely allowances.

Table 20.7 Suggested dietary allowances of minerals

	Liveweights Up to 20 kg	*20–55 kg*	*55–120 kg*
Calcium (%)	0.6–0.75	0.6–0.90	0.6–1.0
Phosphorus (%)	0.65	0.62	0.60
Salt (NaCl) (%)	0.35–0.60	0.35–0.50	0.35–0.50
Potassium (%)	0.22	0.22	0.22
Iron (mg/kg)	50	50	50
Magnesium (mg/kg)	–	–	–
Zinc (mg/kg)	160	160	160
Copper (mg/kg)	180	180	180
Manganese (mg/kg)	80	80	80
Iodine (mg/kg)	1.5	1.5	1.5
Selenium (mg/kg)	0.2	0.2	0.2

Note: With the exception of calcium, phosphorus and salt, values indicate the suggested level of micronutrient supplementation.

It is therefore impossible to present scales of intake for different liveweights.

The farmers who tend to make the most money from finishing pigs are those with a very flexible feeding system. They can meet the nutrient requirements of their pigs using a wide variety of ingredients, and as the price or availability of an ingredient changes, they are able to respond quickly. This is the basis of operation of the large scale producers around the major conurbations of the UK who have access to large quantities of cheap by-products and waste from the human food industry. They are able to operate with such low feed costs that the quality of their end product is of less consequence. For the majority of farms a blend of barley and soya bean meal with an added package of vitamins and minerals is still the most profitable combination for finishers.

There has been sporadic interest in alternative feed-stuffs for pigs such as fodder beet, potatoes and even grass silage. A minority of farms may be able to make some use of these materials where they are home grown, but for growing pigs appetite is a severe limiting factor. For sows some of these bulky feeds may offer an economic alternative to conventional rations.

General management of growing pigs

The general management and husbandry of growing-finishing pigs is to a large extent dictated by the type of housing available. In intensive systems the performance of pigs is highly temperature dependent. Normal growth can be sustained within a wide range of ambient temperatures given an adequate supply of nutrients. There may, however, be only a relatively limited range of temperatures where pigs will perform economically. This is because the pigs' biology strives to preserve its own internal environment. If external conditions (i.e. house temperature) change, then pigs use excess feed energy in temperature regulation. Growth rate and the efficiency of food conversion are adversely affected. The precise house temperature depends on factors such as the use made of straw and the group size. In modern intensive finishing house the temperature should be maintained at around 70°C. Some of the more sophisticated houses for growing pigs included separate rooms for each batch or group of pigs. Each room has its own control unit, and as the pigs reach higher liveweights the temperature settings can be reduced to provide each pig with as near an ideal environment as possible. These systems offer very high pig performance but at a high building cost.

Growing pigs are usually accommodated in groups of 10 to 20 in any one pen. Above this, it is difficult to monitor the progress of individuals and disturbances in the social order are often seen. As animals approach slaughter weight, they need to be weighed individually on a weekly basis and this is not easy with large groups. As particular groups reach the target liveweight, some individuals reach the target first and some may take another two or three weeks. In this situation there may be some pens with only one or two pigs housed because they cannot be mixed without considerable fighting and trauma. It is possible to use a tranquillizing agent to help mix a group of slow-growing individuals together, so to release some of the available pen spaces.

There are many methods of delivering feedstuffs to groups of finishers, but as labour costs have risen, the use of automatic systems has increased. Liquid feeding systems are popular with some farmers. The blend of ingredients in a meal form is mixed with water into a slurry which is then pumped around the farm from pen to pen. Some evidence has shown that these wet feeding systems are associated with improved feed conversion efficiency. However, one of the disadvantages is the need for more trough space within each pen and this tends to reduce the number of pigs housed. Dry feeding systems are usually based on pelleted feeds which reduce dust and waste. These are given on the floor of the lying area of the pen. This allows for increased stocking density and helps to persuade the pigs to dung in the allotted slatted area or dunging passage. There are now many automatic and semi-automatic systems for dispensing dry feeds to groups of pigs, and the reliability of these systems is much improved compared with earlier machinery.

The effects of sex on the performance of growing pigs is well documented, and it is known that, other things being equal, the lean tissue growth potential of boars is better than gilts and in turn gilts are slightly better than castrated males (often called hogs). In the past, when pigs grew much more slowly, entire males were slaughtered well after sexual maturity. Male pigs which are sexually mature produce steroidal products and a complex carbohydrate called skatole, which taint the meat giving it a strong flavour and a characteristic odour. The faster growing pigs of today are slaughtered at younger ages before a significant problem can be detected although a small proportion of entire males will exhibit boar taint at an early age. There seems little reason to continue the practice of castrating male piglets at about two weeks of age with all the stress this entails. Castration is associated with reduced lean tissue growth and a poor feed conversion ratio, and many consider it an unnecessary mutilation. There has been a great reluctance on the part of the meat trade to accept entire males as quality carcasses but in the last few years and in most regions of the country there are large numbers of entire males produced for meat. With this type of production it is then advantageous to separate males from females and to offer a different feeding regime to each. Boars can be offered an *ad libitum* scale of feeding through to slaughter with no loss in carcass quality. Females, on the other hand, still need a period of restricted feeding towards slaughter.

The use of steroidal growth promoters to modify the carcass composition of pigs was once practised throughout Europe and elsewhere. These are now banned in EC countries as a result of public concerns over meat residues. It seems likely that no other products will be used in the future for growing pigs. In the not too distant future it may also be possible, using transgenic animals, to design pigs with specific carcass characteritics and growth potential. These genetically engineered pigs could give the public meat products which are both nutritious and can promote good health. They may also be extremely efficient and profitable for producers. However, the technology involved is still a long way from commercial application. The few pigs that have been produced in this way, with extra genes for the production of endogenous growth hormone, also have associated problems of arthritic conditions and reproductive failure.

Conclusions

Those involved in pig production have seen many changes in the methods used in the last few years. The improvements in technical efficiency and management have been of a high order as producers have endeavoured to stay ahead of the economic field without the protective umbrella of a price support scheme. As a result the industry is lean and fit and the prospects are encouraging for those who are committed to production and vigorous marketing. It is still possible to step on to the farming ladder by producing pigs on a small scale, but this route is becoming more difficult to follow as the structure of the industry continues to change in favour of the large units. A major challenge for the future, apart from the production of low fat meat, is the accommodation of the views of the groups involved with animal welfare. In view of the industry's record for rapid change and forward thinking it is possible that within a short time the innovators within the industry will have created systems of pig farming that will provide for the welfare needs of the animals whilst still generating profits. It has been estimated that very soon up to 40% of pig farming in the UK will be back into outdoor production systems. This will depend on the availability of suitable soil types for outdoor herds, and this may limit the actual number of animals farmed in this way.

Further reading

Brent, G. (1986) *Housing the Pig.* Farming Press, Ipswich.

Cole, D.J.A. & Haresign, W. (1985) *Recent Developments in Pig Nutrition.* Butterworths London.

Meat and Livestock Commission. *Annual Yearbooks.*

Nutrient Requirements of Pigs (1981) Commonwealth Agricultural Bureaux, Slough.

Whittemore, C.T. & Elsley, F.W.H. (1979) *Practical Pig Nutrition.* Farming Press, Ipswich.

21

Poultry

J.I. Portsmouth

Poultry are kept in the UK primarily for meat and egg production. The two sectors are economically different, but naturally have physiological similarities, e.g. meat chickens are hatched from eggs produced by specially bred parent stock and commercial eggs are produced by specialist bred parents. The distinction is thus in the selection of meat or egg production characteristics. The utility fowl of the 1950s/1960s no longer exists. Chickens are specifically bred, managed and marketed to meet a very definite requirement of quality, quantity and price.

Meat production

For a summary of UK poultry meat production see Table 21.1 and Fig. 21.1. Total output increased by 43% in the ten-year period with turkey meat expanding by 47% and chicken by 50%. Waterfowl increased by 56%. A period of further expansion and also consolidation is anticipated over the next ten years as different sectors of the poultry and meat industry concentrate on developing even newer

Table 21.1 UK poultry meat industry (10^3 tonnes), 1981–1992

	1981	*1982*	*1983*	*1984*	*1985*	*1989*	*1991*	*1992*	*1993*
Poultry meat – output	746	812	804	845	874	993	1063	1069	1080
Imports	23.7	27.2	51.1	53.0	61.2	84.3	129.0	163.0	149.2
Exports	17.6	19.9	23.3	27.7	30.9	66.4	80.1	81.3	77.6
Consumption/head (kg)	13.5	14.4	14.5	15.5	16.0	18.9	19.2	20.0	21.5
Broilers – output	555.7	604.7	589.5	632.7	658.0	751.0	829.0	837.0	838.0
Slaughterings (10^6)	375.9	401.7	387.7	413.3	428.0	471.0	522.0	516.0	525.0
Placings (10^6)	423.9	439.3	434.8	465.1	487.8	593.5	644.7	640.3	669.2
Consumption/head (kg)	10.1	10.9	11.3	11.7	11.9	14.2	14.4	14.6	14.7
Producer price (p/lb)	23.1	24.7	25.7	26.6	26.0	25.9	25.4	25.8	27.9
Turkeys – output	116.7	130.3	136.6	138.5	140.4	177.0	174.0	172.0	182.0
Slaughterings (10^6)	22.2	24.7	26.0	26.3	26.7	34.0	33.0	33.0	35.0
Placings (10^6)	24.27	27.84	28.26	29.07	29.78	37.21	37.0	37.53	38.27
Consumption/head (kg)	2.2	2.3	2.5	2.7	3.0	2.8	2.9	3.2	3.4
Ducks – output	15.3	16.5	16.7	16.4	17.2	24.0	24.0	25.0	27.0
Slaughterings (10^6)	7.1	7.6	7.7	7.6	7.9	11.0	11.0	11.0	13.0
Placings (10^6)	7.98	8.48	8.77	9.13	10.67	14.89	14.89	NA	NA
Geese – output	2.0	2.0	2.0	2.0	2.0	3.0	3.0	3.0	3.0
Slaughterings (10^6)	0.4	0.4	0.4	0.4	0.4	0.6	0.6	0.6	0.6
Hens – output	57.0	59.1	59.9	57.0	57.8	38.0	33.0	32.0	30.0
Slaughterings (10^6)	36.5	38.3	38.6	36.2	36.9	25.0	21.0	21.0	19.0

Source: *Poultry World*, September 1992.

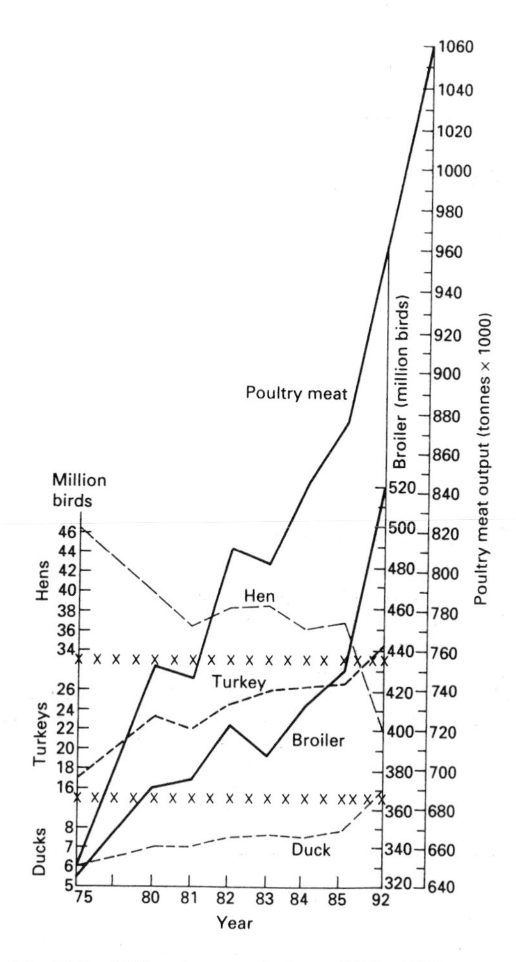

Fig. 21.1 UK poultry meat industry 1975–1992

Table 21.2 Poultry meat consumption (kg/head/year), 1981–1992.

Year	*Total*	*Turkey meat*	*Broiler meat*
1981	13.5	2.2	10.0
1982	14.4	2.3	10.9
1983	14.5	2.5	11.3
1984	15.5	2.7	11.7
1985	16.0	3.0	11.9
1986	17.3	2.6	13.2
1987	18.4	2.6	14.1
1988	19.4	2.7	15.0
1989	18.9	2.8	14.2
1990	19.3	2.9	14.5
1991	19.2	2.9	14.4
1992	20.0	3.2	14.6

Source: Poultry Meat Association reproduced with permission.

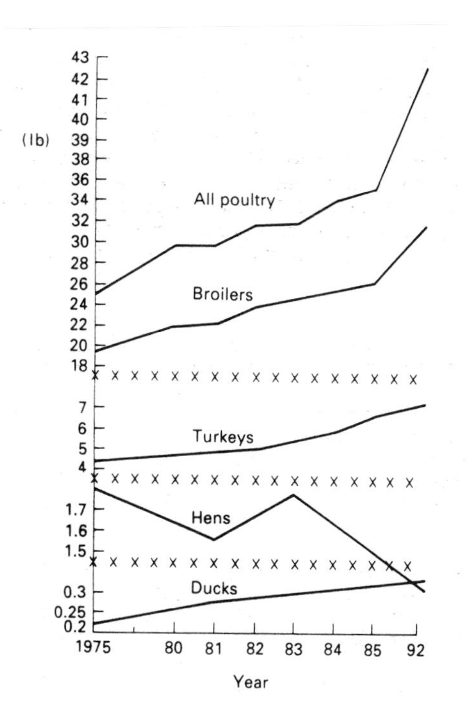

Fig. 21.2 UK poultry meat consumption per head.

markets for gourmet-type products as saturation point approaches for whole carcass sale. Such products include chicken rolls, sausages, portions, meat loaves and many gourmet dishes.

Figures in Table 21.2 and Fig. 21.2 show how poultry meat consumption has increased by 48% in the 11-year period, an annual increase of some 4.3%. Turkey meat consumption on the other hand has increased by 45% (4% per annum). A further increase in the consumption of poultry meat is almost certain to occur at the expense of red meats, the main reasons being price, flexibility and attractiveness of product, low animal fat content and high protein content. Duck meat consumption has shown a modest expansion since 1985 and this trend is likely to continue. Compared with some other countries in the EU such as France, where barbary (Muscovy) duck output is expanding, production in the UK is almost entirely 'Aylesbury' type. Geese consumption is increasing, as a percentage, more rapidly than that of duck, and this trend should continue as consumers look for traditional meat sources at Christmas time.

Broilers

Since the broiler industry began in the early 1950s it has developed rapidly and is now the most integrated form of livestock production. All the major companies, e.g. Premier, Marshalls, Sun Valley and Padley's, and many others, produce and market their own nationally advertised brands with total control of live bird growing, breeding, hatching, feeding and processing.

Only a handful of independent producers remain to supply the smaller packing stations with the high quality product demanded by the market.

Table 21.3 Recommended broiler ration specifications

Nutrient	*Feed type*			
	Starter	*Grower*	*Finisher 1*	*Finisher 2*
Protein (%)	23.0	21.5	21.0	19.0
Lysine (%)	1.38	1.27	1.17	1.08
Av. lysine (%)	1.21	1.12	1.03	0.96
Methionine (M) (%)	0.64	0.58	0.54	0.48
M + C (%)	1.00	0.95	0.90	0.84
Energy (ME kcal/kg)	3035	3166	3226	3250
Energy (ME MJ/kg)	12.70	13.25	13.50	13.60
Calcium (%)	0.9	0.9	0.9	0.9
Available phosphorus (%)	0.45	0.45	0.45	0.43
Salt (%)	0.34	0.36	0.36	0.36
Linoleic acid (%)	1.50	1.40	1.25	1.10

Source: JP Enterprises.

Of the total meat produced in the UK 75% is derived from broilers. A broiler is a young bird of either sex, marketed between 1.45 kg and 2.75 kg liveweight. It is slaughtered between 35 and 56 d (approximately) depending on the range of body weight needed and the most economical time of production. The broiler is a specifically bred hybrid meat bird derived from breeds and strains originally developed in the USA. The leading UK breeding companies are Cobb (Tyson) and Ross (Hillsdown Holdings).

Depending on killing age a broiler will convert food into meat with a ratio of 1.9:1. Optimum economic performance depends on correct nutrition and husbandry conditions. Work by ADAS at Gleadthorpe EHF shows that broilers respond optimally to high nutrient density (HND) when housed at 21°C and killed at 47–49 d. Tables 21.3–21.7 summarize the nutritional needs of broilers. It is anticipated that by the year 2000 the figures quoted in Table 21.5 will be achieved in 4 d less. By the end of the present decade a 2 kg liveweight broiler will be reached in 35 d.

Table 21.4 Broiler feed programmes (kg/1000 birds)

	Age at slaughter		
	35–42 d	*43–49 d*	*+53 d*
Starter	500	500	500
Grower	1200	1200	1000
Finisher 1	To finish	2050	2250
Finisher 2		750	1000

Source: JP Enterprises.

Table 21.5 Broiler performance guide

Age at slaughter (days)	35	42	45	52	59
Av. liveweight (kg)	1.7	2.1	2.4	3.0	3.5
FCR	1.6	1.8	1.9	2.0	2.1
Liveability %	97.0	96.5	96.0	94.5	93.0

Source: JP Enterprises.

Anti-coccidial drugs such as the ionophore narasin and chemical diclazuril are an essential part of an overall disease prevention and control programme. Without these anti-coccidial drugs modern broiler/turkey production under intensive husbandry methods would be impossible and the disease coccidiosis would cause considerable loss. Anti-coccidial drugs such as monensin and halofuginone are an important part of disease prevention in turkey production. The use of these medications is, however, relative to specific disease situations and together with anti-blackhead drugs the specific drug to use will depend upon the degree of disease and the type of management.

Growth enhancers such as the antibiotics are used in about 90% of UK broiler and turkey feeds. It is generally acknowledged that their presence probably enhances growth and improves feed conversion efficiency between 2 and 5%. Non-antibiotic digestive improvers, such

Table 21.6 Micronutrients recommended for inclusion in broiler diets

Nutrient	*Starter*	*Finisher*
Vitamin A (m i.u.)	12	12
Vitamin D_3 (m i.u.)	5	5
Vitamin E (k i.u.)	30	30
Vitamin K_3 (g)	3	3
Vitamin B_1 (g)	2	2
Vitamin B_2 (g)	8	8
Nicotinic acid (g)	60	30
Pantothenic acid (g)	12	12
Vitamin B_{12} (mg)	15	15
Vitamin B_6 (g)	3	2
Choline (g)	250	200
Folic acid (g)	1.5	1.0
Biotin (mg)	250	150
Manganese (g)	110	110
Zinc (g)	100	100
Copper (g)	20	20
Cobalt (g)	0.4	0.4
Iron (g)	30	30
Selenium (g)	0.25	0.25
Iodine (g)	1.0	0.5

Source: JP Enterprises.

Table 21.7 Additives commonly used in poultry production without veterinary prescription

Additive	*Stock type*	*Age (weeks)*	*Level of active ingredient in final feed (ppm)*
Growth and egg promoters			
Avilamycin	Broilers	0–16	5–10 avilamycin
Avoparcin	Broilers	0–16	7.5–15 avoparcin
Crinarom	Broilers	0–16	50 g crinarom 737
	Turkeys	0–26	50 g crinarom 737
Virginiamycin	Broilers	0–9	0–20 virginiamycin
Bambermycin	Broilers	0–16	2–5 bambermycin
	Turkeys	0–26	2–5 bambermycin
	Laying hens		2–5 bambermycin
Zinc bacitracin	Broilers	0–4	5–20 zinc bacitracin
	Broilers	4–16	5–20 zinc bacitracin
	Laying hens and breeders		15–100 zinc bacitracin
*Anti-coccidials**			
Monensin	Broilers, pullets	0–16	100–120 monensin
Narasin	Broilers	0–16	70 narasin
Lasalocid	Broilers	0–16	90 lasalocid
Maduramycin	Broilers	0–16	5 maduramycin
Salinomycin	Broilers	0–16	60 salinomycin
Halofuginone	Broilers	0–16	3 halofuginone
	Broilers	0–16	
Halofuginone	Turkeys	Unit 5 d before slaughter	3 halofuginone
Nicarbazin	Broilers	Until 5 d before slaughter	125 nicarbazin
Diclazuril	Broilers	Until 5 d before slaughter	1.0 diclazuril
Anti-blackhead medications			
Dimetridazole	Turkeys		125 dimetridazole

* Where an additive contains more than one active ingredient only the major constituent is shown.
Source: JP Enterprises.

as crinarom (made from extracts of essential oils), are gaining popularity in poultry raised on completely drug-free diets.

Commercial egg production

The UK laying flock has been declining since 1968. This is due to a fall in demand for eggs caused by such controversial issues as cholesterol and salmonella, and an increase in egg production per bird due to technological advances. The total number of poultry farms with laying birds has fallen from 125 258 in 1971 to 33 000 in 1992. The egg industry concentrates on fewer and larger units with more than 50% of the country's eggs produced from about 390 flocks averaging 20 000 or more laying birds each. The trend to fewer and larger units will continue unless intensified action by welfare interests moves the industry away from the battery cage and back to the less intensive management systems where husbandry practice favours the smaller unit.

In 1995 battery cage dimensions and the number of birds allowed in each cage change dramatically to allow each bird greater freedom of movement. This move will increase production costs and force many producers to quit the industry.

Free range production

Eggs produced by hens managed under free range and other alternative conditions account for approximately 15% of the total egg production. The proportion is highest in the southern and home countries of England where the claimed premium is also highest. It is apparent that the saturation point for free range eggs and eggs produced under other less intensive systems is around 20%. Beyond this the premium is eroded by a fall in demand and many free range units which began in the 1980s were out of business by 1990. This was caused partly by poor returns and partly by higher production costs incurred, especially in adverse weather conditions. Under extensive systems of management the need for a high degree of stockmanship is even more important than in the intensive systems.

Table 21.8 Selection of egg-producing strains available in UK

Commercial name	*Company*
Babcock 380	ISA Poultry Services
Hisex Brown	Euribrid
Shaver 579	Shaver
ISA Brown	ISA Poultry Services
Hy-line Brown	Hy-line Int. USA
Lohmann Brown	Lohmann Germany

Source: JP Enterprises.

In the late 1950s and early 1960s the laying flock were made up predominantly of white feathered birds producing white eggs. In 1975, 64% of the flock were brown and producing brown eggs. In 1979 almost 90% were brown egg layers and in 1992 this figure had reached virtually 100%.

The demand for brown eggs (including tinted eggs) has caused breeding companies to intensify efforts to increase food efficiency, which previously favoured white egg layers because of their smaller body size and therefore lower energy needs for maintenance. Table 21.8 shows the main 'breeds' and companies involved in breeding commercial laying stock in the UK.

Tables 21.9 and 21.10 show egg yields for hens housed in different husbandry systems and also per capita egg consumption since 1970 (see also Fig. 21.3). In 1960/1961 birds housed in cages on average produced 18.3 eggs more than the other flocks while in 1978/1979 this widened to 21.5 eggs per bird. The average of 249 eggs per bird is in fact insufficient to cover the production costs of the 1990s. Profitable flocks need to produce 285 eggs on low feed intakes. To obtain high and economic production, certain minimum amounts of nutrients must be provided. These are shown in Table 21.11.

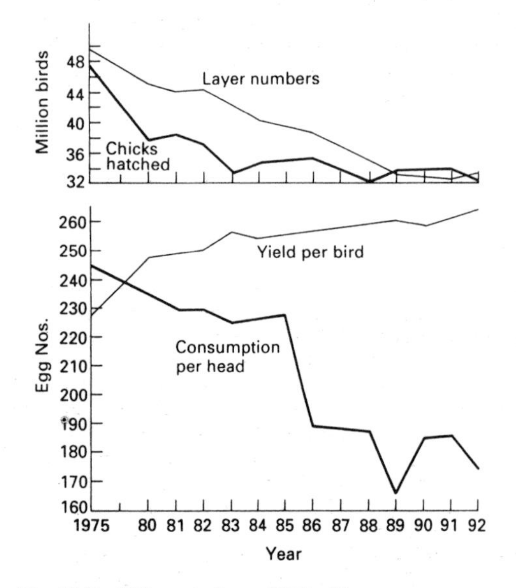

Fig. 21.3 UK egg industry 1975–92.

Table 21.9 Egg production/hen per annum by management system

Management system	*1972/23*	*1973/74*	*1974/75*	
Free range	181.2	172.9	175.7	
Deep litter	238.5	200.9	207.0	
Battery cage	214.0	232.3	236.4	
	1976/77	*1978/79*	*1984/85*	*1993*
Free range	192.1	196.8	258	260
Deep litter	224.5	227.8	264	270*
Battery cage	245.4	249.3	276	282

Source: ADAS.
* Refers to hens housed in perchery/barn systems.

Table 21.10 1970–92 annual egg consumption in the UK (no. eggs/person)

Year	*Total*	*Year*	*Total*
1970	275	1982	231
1971	273	1983	226
1972	273	1984	227
1973	262	1985	229
1974	256	1986	189
1975	246	1987	189
1976	248	1988	188
1977	248	1989	166
1978	250	1990	184
1979	251	1991	185
1980	236	1992	176
1981	227		

Source: *Poultry World*, with permission.

Table 21.11 Recommended intake of certain essential nutrients to produce 55 g egg output with house temperature of 21°C

Nutrient	*Minimum daily*	*% Nutrient in ration at specified average daily intake*		
		100 g	*110 g*	*120 g*
Protein	19.0 g	19	17.3	16
Methionine	420 mg	0.42	0.38	0.35
Methionine + cystine	640 mg	0.64	0.58	0.53
Lysine	900 mg	0.9	0.82	0.75
Calcium	4.2 g	4.2	3.8	3.5
Av. phosphorus	0.33 g	0.33	0.32	0.29
Linoleic acid (EFA)*	1.25 g	1.25	1.14	1.04
Xanthophylls†	14 mg/kg	14	12.75	11.70
Energy	325 kcal/day	325	325	325

* The level of linoleic acid influences egg weight. To maximize egg weight 3.0 g linoleic acid/bird/day is required.
† Recommended level of xanthophylls ensures a good yellow–orange colour (Roche fan colour 11).
Source: JP Enterprises.

Feed programmes for laying birds

Laying hens are invariably fed *ad libitum*. The hens' adjustment of feed intake to change in ration energy level is not precise but sufficiently accurate to make *ad libitum* feeding the most economical system. Restricted feeding, whereby a certain allowance is allocated once or twice daily, is hazardous. Variation in feed consumption within the flock may be as much as 40%, thus any physical restriction of feed penalizes the small appetite to the great detriment of egg output and profits. Free choice feeding, whereby cereals are fed whole with a pelleted protein/vitamin/mineral concentrate, allows the individual bird to select its own protein (amino acids) needs according to rate of lay, has shown some nutritional/physiological advantages. The system needs management perfection before it can become of practical value.

Feed restriction for replacement pullets

Excess body fat adversely affects subsequent rate of lay. Consequently, replacement pullets and replacement broiler breeding stock are fed controlled amounts of feed to reduce excessive energy intake. With commercial pullets the mid-term restriction during rearing proved to be popular in the early 1980s where it was required to bring them into lay at around 20 weeks of age. It was found that the mid-term restriction was more important than either early or late term restriction. In practice, this meant that the severest form of restriction took place during the period 9–15 weeks. More recently, in the rearing of commercial laying pullets the *ad libitum* feeding system has been favoured as it brings birds into lay at an earlier age and has been shown to produce superior egg production and egg size. Such pullets should not, however, be fat and the balance between additional body weight and earlier sexual maturity is a delicate one as it is important to optimize not only egg production but also egg weight.

Feed control with broiler breeders during rearing and laying is vital if excess energy consumption and overfatness are to be prevented. Broiler breeder companies have recommended feeding programmes and these should be followed closely for maximum economic performance. Table 21.12 details the essential micronutrients such as vitamins and trace elements which are used to supplement normal ingredients of poultry rations.

Table 21.12 Micronutrient recommendations for commercial layers, replacement and breeding stock

Nutrient	*Poultry breeder*	*Chick grower*	*Layer*
Vitamin A (m i.u.)	13	10	8
Vitamin D_3 (m i.u.)	5	3	3
Vitamin E (k i.u.)	80	50	15
Vitamin K_3 (g)	8	3	2
Vitamin B_1 (g)	3	1	0.5
Vitamin B_2 (g)	14	8	4
Nicotinic acid (g)	50	20	15
Pantothenic acid (g)	20	10	8
Vitamin B_{12} (mg)	25	18	10
Vitamin B_6 (g)	7	3	1
Choline (g)	400	200	–
Folic acid (g)	3	2	–
Biotin (mg)	300	100	–
Manganese (g)	120	110	110
Zinc (g)	100	100	100
Copper (g)	20	20	10
Iodine (g)	2	1	1
Cobalt (g)	0.5	0.5	0.4
Iron (g)	30	40	20
Selenium (g)	0.30	0.25	0.20
Molybdenum (g)	2.0	1.0	–

Source: JP Enterprises.

Broiler breeding industry

The two major breeding companies in the UK are Cobb and Ross, whilst in the EU Arbor Acre, Hubbard and Vadette also have good market shares. Both Cobb and Ross are so-called heavy meat strains yielding superior quantities of high value breast and thigh meat compared with the other strains. Broiler chick production of Cobb and Ross UK strains averages between 120 and 135 chicks per hen per 60-week period. The other strains may produce more chicks per hen but the growth rate and meat yield are generally lower. Because broilers are bred to grow fast, the replacement parent stock have to be reared under very strict management and nutritional conditions. Control of energy intake is of paramount importance in order to prevent excess body weight occurring.

Special feed programmes must be carefully followed with the emphasis on monitoring a pre-determined body weight for age guide. At 18 weeks increases in both nutrient quantity and quality are synchronized with an increase in the photoperiod. The effect of changing nutrition and day length serves to bring the flock into egg production with body weight still under careful control. Broiler breeding, like other sectors of the modern poultry industry, is highly sophisticated, demanding a great deal of skill and stock sense.

Environment of layers and broilers

Ventilation

Table 21.13 shows the ventilation rates needed for intensively housed poultry. The optimum temperature for adult layers is 21°C and for broilers, after initial brooding, 20–21°C depending on sex and ration nutrient density. Ventilation rate is also expressed according to the amount of food consumed. Thus, the optimum minimum for broilers is 2 mstd (cubic metres of air per second per tonne of feed per day). Maximum ventilation is ten times the minimum (that is 20 mstd). Note 1.0 mstd is approximately equal to 1.0 cfm per pound per day (cubic feet of air per minute per pound of feed per day). This standard therefore relates closely to the metric m^3/s/tonne/d, 25 cfm/lb/d being equivalent to 25 mstd, and 2 cfm matching 2 mstd.

Table 21.13 Ventilation rate requirements for intensively housed poultry

Stock	*Weight (kg)*	*Max. needed*		*Min. needed*	
		m^3/bird	*cfm/bird*	*m^3/bird*	*cfm/bird*
Layers	1.2	10	6.0	1.0	0.6
	2.0	12	7.0	1.2	0.75
	2.5	14	8.0	1.5	0.9
	3.0	14	8.0	1.7	1.0
	3.5	15	9.0	2.0	1.2
Broilers	0.05			0.1	0.06
	0.4			0.5	0.3
	0.9			0.8	0.45
	1.4			0.9	0.50
	1.8	10	6.0	1.3	0.75
	2.2	14	8.0	1.7	1.0
Turkeys	0.5	6	3.5	0.7	0.4
	2.0	12	7.0	1.2	0.7
	5.0	15	9.0	1.5	0.9
	7.0	20	12.0	2.0	1.2
	11.0	27	16.0	2.7	1.6

Source: MAFF, Gleadthorpe EHF. Reproduced with permission.

Lighting

Because nearly all intensive poultry units employ windowless controlled-environment housing, artificial lighting, both in extent and intensity, is extremely important. In laying birds changes in day length stimulate or depress the secretion of gonadotrophic hormones. Increasing day length stimulates and vice versa. Change in day length can be used to manipulate age of sexual maturity. Extending the day length in rearing advances sexual maturity. With a laying bird, day length is generally increased from about 8 h at 18 weeks to a maximum of 16 h achieved by weekly increases of 15 min each (see Fig. 21.4). Day length is then held at this level until the end of laying period. Egg output can be manipulated by so-called ahemeral lighting patterns where day length is more or less than the normal 24 h cycles. With a longer day length, i.e. 25–28 h, egg weight is increased, whilst with cycles less than 24 h the rate of production can be increased at the expense of egg weight. Ahemeral patterns are sometimes used in practice and may find application in selection of future breeding stock.

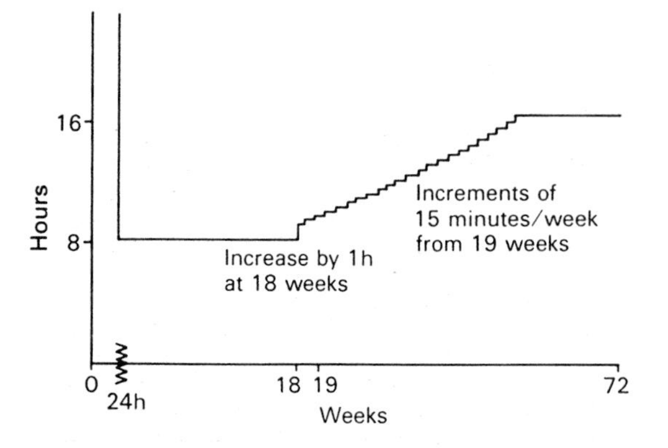

Fig. 21.4 Changing day length.

In broiler production the most popular lighting system is 24 h for the first 48–72 h followed by a 23 h day to slaughter. The 1 h dark period is used to accustom the bird to darkness in the event of a power failure.

Intermittent lighting programmes are also used. After the continuous 24 h for the first 72 h the programmes may be 2 h light followed by 2 h darkness or 1 h light and 3 h darkness. Intermittent programmes are thought to improve efficiency of digestion through the enforced rest of the dark period and also to reduce the incidence of leg weaknesses which is not uncommon in fast-growing broilers.

Light intensity (all poultry)

Under intensive housing systems a low light intensity is essential to avoid feather pecking and cannibalism. Additionally, birds are quieter, easier to manage and less active, thus using less energy for maintenance. Young chicks should be given 20–30 lux (lx) for the first 72–96 h to encourage early feeding. Thereafter, 6–7 lx is sufficient to allow normal feeding, optimum activity and ease of operations. For laying birds the optimum intensity is 5–7 lx. In the dark period one-tenth of the optimum figure is recommended, i.e. 0.5–0.7 lx.

Turkeys and waterfowl

Turkeys

Turkey production increased four-fold between 1965 and 1973. Disastrously low selling prices due to overproduction in 1973 saw drastic cutbacks in 1974 which were even more severe in 1975. Since then a better organized industry, with marketing at Easter, Whitsun and August Bank Holiday, has led to an almost speculative expansion. The market prospects to the end of the 1990s are good and, with a demand for high quality white meat, the future for the turkey industry during the 1990s is speculative. Predictions like this, however, can be severely affected by the output of our partners in the EU, particularly Italy and France who have relatively large turkey industries. Both of these countries are potentially large exporters of turkey meat which could seriously interfere with our own expansion plans. In 1971, 2300 flocks produced 14 000 000 turkeys whilst by 1985 this number had fallen dramatically for almost twice the output (28 000 000 birds) thus the trend of fewer and larger is seen also in the turkey industry. Table 21.14 shows how in 1985 some 92% of all turkeys were produced by units with in excess of 100 000 birds compared with some 4% produced by flocks of 5000–10 000 birds!

This trend has continued through to the present time. Four major companies – Bernard Matthews, Moorland Foods and Sun Valley – account for 85% of all turkeys produced in the UK. Eleven other companies producing some 3 million turkeys in total account for a further 8%.

Table 21.14 Number of turkeys in England and Wales and proportion of flock by size of flock

Size of flock	*1981* $\times 10^3$	%	*1984* $\times 10^3$	%	*1985* $\times 10^3$	%
1–25	3.6	0.1	2.7	0.1	2.8	0.1
26–99	6.5	0.1	3.6	0.1	3.9	0.1
100–499	34	0.4	22.3	0.3	19.0	0.2
500–999	38.9	0.5	24.0	0.4	32.7	0.4
1000–4999	238.3	3.0	253.8	3.7	258.2	3.4
5000–9999	340.9	4.3	272.4	4.0	299.6	3.9
100 000 +	7291.2	91.7	6254.7	95.0	6982.3	91.9
Totals	7953.4	100	6834.4	100	7598.5	100

Source: MAFF statistics. Reproduced with permission.

The remaining 7%, representing some 4 million birds, is produced by 1000 or so small seasonal turkey producers.

Turkeys are into a relatively new phase of marketing with expansion into the 'cut-up' portion, and further processed markets with such products as turkey rolls, sausages, patés, burgers, and a variety of easily prepared convenience foods competing very successfully with other sections of the meat market. This further process market requires a large bird with a high meat to bone ratio. Such birds are normally 20–26 weeks old and may weigh from 12 to 15 kg compared with the whole bird market which demands a smaller 4–6 kg oven-ready bird. Table 21.15 shows the carcass analysis of 9.3 kg male turkeys. Figure 21.5 shows the relationship between percentage meat of liveweight and turkey age. For a 22-week-old medium stag some 63% of total liveweight is meat. In the data of Table 21.15 the 9.3 kg bird is 14 weeks old and such birds have a 50% meat yield of liveweight. The larger the bird the greater its meat yield and its usefulness in the cut-up and meat stripping business. Consumption per capita of turkey meat is shown in Table 21.2. The average rate of increase is 2.9% per annum. In the last five years this has been due to greater consumption of turkey meat at times other than Christmas, for example Easter and Whitsun. Table 21.16 shows the growth rate, feed consumption and feed conversion tables for turkeys from day old to 22 weeks of age. These figures are just a guide to performance and can be wide due to strain differences, etc. Tables 21.17 and 21.18 deal with the nutritional requirements of turkeys at all ages.

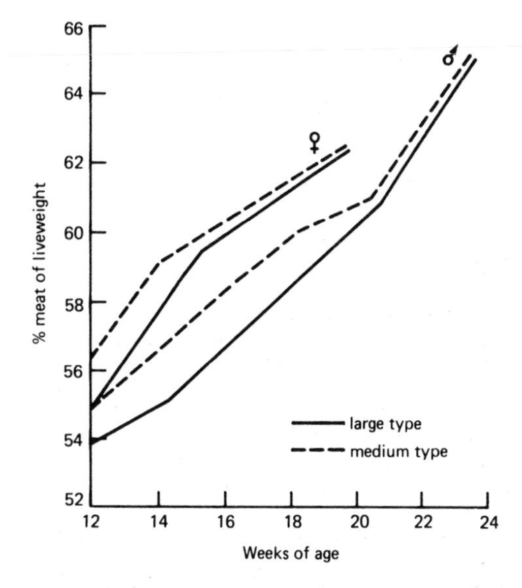

Fig. 21.5 Relationship between % meat/liveweight and turkey age (from *Poultry World* 26 June 1980, reproduced by courtesy of the editor and publisher).

Table 21.15 Turkey carcass analysis (male birds)

	kg	%
Weight	9.3	100
Breast meat	2.29	24.7
Leg meat	1.86	20.0
Total bone	1.72	18.5
Total meat	4.59	49.4
Total meat/skin	5.13	55.2

Source: British United Turkeys.

Waterfowl

Ducks

Meat production is small in comparison with turkeys and chickens. Table 21.19 shows the growth in the duck market from 1979 to 1993 in the UK and other EU countries. By far the largest output is from France at almost 53% of the total. The UK produces nearly 16% of the EU duckling production.

Based on 1983 figures consumption per capita is 0.87 kg. Duck meat is still regarded as a luxury. The majority of UK duck production is from the world's largest producer, Cherry Valley Farms Ltd, Lincolnshire. The two main breeds are Aylesbury and Pekin. The plumage of both is white. The Pekin hybrids commonly used in commercial production have better reproductive

Table 21.16 Management standards – turkey growth rate and food conversion (as hatched)

Age (weeks)	*Live weight (kg)*	*Total food consumed cumulative (kg)*	*Cumulative FCR*
1	0.14	0.13	0.94
2	0.32	0.38	1.21
3	0.60	0.81	1.36
4	0.97	1.42	1.47
5	1.44	2.20	1.53
6	2.01	3.24	1.61
7	2.66	4.47	1.68
8	3.37	5.93	1.76
9	4.14	7.53	1.82
10	4.93	9.32	1.89
11	5.72	11.27	1.97
12	6.51	13.28	2.04
13	7.3	15.55	2.13
14	8.05	17.79	2.21
15	8.78	21.24	2.30
16	9.50	22.70	2.39
17	10.17	25.32	2.49
18	10.81	27.99	2.59
19	11.41	30.80	2.70
20	11.97	33.75	2.82
21*	15.55	41.52	2.67
22*	16.40	45.28	2.72

* Figures in weeks 21 and 22 apply to stags only.
Source: British United Turkeys.

Table 21.17 Nutritional requirements of turkeys

Age (weeks)	*Pre-starter* 0–4	*Starter* 5–8	*Rearer* 8–12	*Early finisher* 12–16	*Late finisher* 16–22	*Heavy stag*
Protein (%)	30	27	24	21	15	15
Methionine (M) (%)	0.68	0.64	0.55	0.48	0.43	0.38
Lysine (%)	1.88	1.61	1.32	1.07	0.9	0.78
M + cystine (%)	1.22	1.13	1.00	0.85	0.77	0.68
ME (MJ/kg)	12.0	12.0	12.0	12.0	12.0	12.0
Calcium (%)	1.37	1.25	1.13	1.03	0.94	0.89
Phosphorus (%)	0.9	0.85	0.80	0.70	0.65	0.65
Av. phosphorus (%)	0.77	0.70	0.64	0.58	0.54	0.49
Salt (%)	0.34	0.34	0.34	0.34	0.34	0.34
Na (%)	0.16	0.16	0.16	0.16	0.16	0.16

Source: British United Turkeys.

qualities than Aylesburys, but hydrid crosses are superior to both of these pure breeds. The growth rate of duckling is far superior to chicken and turkeys as shown in Table 21.20. For a given age under eight weeks, duckling outgrows all other poultry. Its food conversion, however, is inferior. This is due to the higher feed consumption per unit of bodyweight gain and it is this which makes duckling a relatively expensive meat to produce. The yield of edible cooked meat is lower for duck compared with turkeys and broilers. Table 21.21 compares the relative cooking losses.

Geese

Goose production is the poor relation of the poultry industry but in recent years good attempts have been made to increase demand for the gourmet luxury meat. Few accurate records are available to show how many birds are produced annually. Many are raised on general farms and remain unrecorded. Sales mainly occur at the festive occasions, particularly Christmas. Being extremely good grazers and converters of grass into meat, the most profitable management of feeding systems utilizes this natural asset. Most popular breeds are the Embden, Toulouse, Roman and various hybrids involving crosses of these major breeds. Weights are dependent on feeding systems and the age of killing, with Embdens reaching 14–15 kg by five months, whilst the smaller Roman may only achieve 6 kg in this time. The Toulouse is slightly smaller than the Embden.

After feeding on a good proprietary chick pellet for three to four weeks after hatching, the growing goose is capable of surviving and growing on good quality

Table 21.18 Micronutrients recommended for inclusion in turkeys diets

Turkey supplements	*Turkey starter*	*Turkey grower/ finisher*	*Turkey breeder*
Vitamin A (m i.u.)	15	10	15
Vitamin D_3 (m i.u.)	5	3	5
Vitamin E (g)	50	40	60
Vitamin K_3 (g)	5	3	12
Vitamin B_1 (g)	5	1	2
Vitamin B_2 (g)	8	6	20
Nicotinic acid (g)	75	50	70
Pantothenic acid (g)	25	15	25
Vitamin B_{12} (mg)	20	20	30
Vitamin B_6 (g)	7	5	5
Choline (g)	400	150	200
Folic acid (g)	3	2	3
Biotin (mg)	300	300	300
Manganese (g)	120	100	120
Zinc (g)	100	70	100
Copper (g)	20	20	20
Iodine (g)	2	2	2
Cobalt (g)	0.4	0.4	0.5
Iron (g)	50	20	50
Selenium (g)	0.25	0.20	0.25
Molybdenum (g)	—	—	1.0

Source: British United Turkeys.

Table 21.20 Comparative growth performance

	Broiler strain	*Duckling*	*Turkey*
Age (d)	47	47	47
Live weight (kg)	2.40	3.73	2.66
FCR	1.90	2.20	1.68
Feed consumed per bird (kg)	4.56	8.20	4.47

Source: JP Enterprises.

Table 21.21 Comparison of cooking losses in ducks, turkeys and broilers

	Duck	*Turkey*	*Broiler*
(1) Live weight (kg)	2.7	3.6	1.8
(2) Bled/plucked weight of live weight (1)	82%	90%	87%
(3) Eviscerated weight of (1)	73%	70%	75%
(4) Cooking loss of (1)	37%	31%	22%
(5) Yield of edible cooked meat of (1)	25%	34%	27%

Source: After Hollows (1978) *Productivity of Ducks*. Society of Feed Technologists. Reproduced with permission.

Table 21.19(a) Duck hatchings ($\times 10^3$) in EEC countries

	1980	*1981*	*1982*	*1983*	*1984*	*1985*
West Germany	4 300	4 200	4 100	3 600	2 900	4 630
Netherlands	4 700	4 300	5 000	5 400	5 700	6 100
Belgium/Luxemburg	241	226	216	188	161	115
Italy	5 700	5 900	3 100	4 800	5 000	6 290
France	19 100	22 600	24 300	26 500	31 700	35 800
Denmark	3 200	3 500	3 100	3 700	3 500	3 940
Ireland	1 200	1 300	1 400	1 100	NA	NA
UK	8 290	7 980	8 480	8 770	9 130	10 670
Greece	—	—	—	—	—	25
EC total	46 731	50 006	49 696	54 058	58 091	67 570

NA, not available.
Source: After MAFF and AGRA Europe. Reproduced with permission.

Table 21.19(b) Duck hatchings ($\times 10^3$) in EU countries

	1988	*1989*	*1990*	*1991*	*1992*	*1993*
Germany (East and West)	NA	12 700	16 000	15 000	16 000	16 000
Netherlands	3 400	3 800	3 900	4 300	4 500	4 200
Belgium/ Luxemburg	110	110	150	200	200	200
Italy	5 800	5 800	6 000	6 000	6 300	6 400
France	47 000	52 000	55 000	59 000	64 000	68 000
Denmark	2 300	2 300	2 300	2 400	2 300	2 200
Ireland	1 200	1 300	900	1 000	1 200	1 500
UK	9 000	11 000	12 000	11 000	11 000	13 000
Greece	200	200	200	200	200	200
EU total	69 010	89 210	96 450	99 100	105 700	111 700

NA, not available.

grassland and small supplements of concentrate feed. (Recommended micronutrient supplements for both ducks and geese are given in Table 21.22.) This can be dispensed with after eight weeks and a satisfactory finish can be achieved on grass provided some supplementary grain is given during the last month of fattening. It is usual to restrict the range area at this time to conserve energy requirements. Penning the geese in straw yards is sometimes used in East Anglia, where straw forms a relatively cheap wall and bedding. The selling of day-old or part-grown goslings can be a possible side line. Breeding geese often mate for life and this strong 'pairing bond' is established in the autumn prior to the laying season in early spring. Normally one gander pairs with three to four geese. For best fertility the gander should be one year or older than the geese. Young geese show reproductive improvements with age through to ten years or more. Egg production varies, but may range from 30 to 75 eggs per goose per year. The eggs are best hatched artificially as geese make indifferent incubators and mothers.

Table 21.22 Recommended micronutrient levels for duck and geese rations

Supplement	*Starter/finisher*
Vitamin A (m i.u.)	12
Vitamin D_3 (m i.u.)	3
Vitamin E (g)	50
Vitamin K (g)	5
Vitamin B_1 (g)	2
Vitamin B_2 (g)	12
Vitamin B_6 (g)	4
Vitamin B_{12} (mg)	20
Biotin (mg)	110
Choline (g)	400
Folic (g)	2
Nicotinic acid (g)	80
Pantothenic acid (g)	15
Biotin (mg)	200
Cobalt (g)	0.5
Copper (g)	20
Iodine (g)	2
Iron (g)	15
Mangenese (g)	110
Selenium (g)	0.25
Zinc (g)	100

Breeding waterfowl should be given a vitamin supplement as specified for 'Poultry breeder' in Table 21.12.
Source: JP Enterprises.

Poultry welfare

The UK poultry industry is one of the most intensive in the world today. Intensive production developed in the mid 1950s to the mid 1980s as producers were forced to increase productivity and reduce costs. The 1980s saw a demand for free range and alternative-system eggs and meat. This developed well during the more prosperous 1980s but declined as the recession reduced spending power. Extensive production costs are almost double intensive costs. Anti-factory-farming lobbies and welfare groups have forced the poultry industry to provide more space per bird and a more acceptable environment, and to pay full attention to animal behaviour. From January 1995 new EC cage size regulations permitting more space and height in battery cages become law. The new regulations outlaw many older cage types and the necessary re-capitalization will force many UK egg producers to quit the industry. Although theoretically applicable to all EC member states their adoption by each one and policing by local authorities is questionable outside the UK. Imports from non-EC countries do not have to comply and the consumer wishing to influence animal welfare may well inadvertently condone the systems that anti-factory-farming lobbies are trying to outlaw.

In parts of Europe antibiotic performance enhancers (as listed in Table 21.7) are now banned. Such substances are licensed for use in the UK against Gram-positive as opposed to Gram-negative bacteria. Disease organisms which attack and spread in both intensively and extensively managed stock are controlled by products specifically licensed for this use. In countries not using UK permitted performance enhancers, more therapeutic drugs are used to keep the animals healthy. The development of non-antibiotic products to comply with the so-called 'green' issues will continue and may ultimately lead to their replacing performance enhancers.

Further reading

Sainsbury, D. (1983) *Animal Health*. Granada, St. Albans.
Scott, M.L., Nesheim, M.C. & Young, R.J. (1982) *Nutrition of the Chicken*. M.L. Scott and Associates, Ithaca, New York.
Poultry World (monthly magazine). Reed Business Publications Sutton, Surrey.
Poultry (monthly magazine). Misset, The Netherlands.

22

Animal Health

D.W.B. Sainsbury

Introduction

Good health is the birthright of every animal and it is the duty of the livestock keeper to do everything possible to ensure this. Considerable advances in the control of animal infections have led to the effective elimination of many of the more traditional causes of acute disease and it is no longer necessary to consider disease an inevitable part of a livestock enterprise. This situation has been achieved by a combination of good hygiene, the appropriate use of vaccines, good medicine therapy, the development of disease-free strains of livestock but above all by improved husbandry and management. These advances have made it possible to keep animals in much larger groups and more densely housed than hitherto but the results have by no means been a gradual disappearance of infectious and contagious diseases altogether.

On the contrary, a number of complex diseases have emerged, difficult to diagnose and induced by a multiplicity of pathogenic agents. Whilst these may cause an apparent or 'clinical' disease it is more likely that the effect will be less obvious and may only reduce the overall productivity of the livestock by, for example, slowing growth and reducing the food conversion efficiency. Animals may not die or even show any symptoms at all so that farmers may be unaware of what is happening unless they keep very careful records and use them with more than the usual degree of skill. It is also a very common phenomenon that intensive production on a livestock unit may start efficiently and effectively but deteriorate in time so gradually that it is not noticed until the consequences have become very serious and control becomes extremely difficult. The nature of these infections is of especial interest and concern because the environmental and housing conditions have a profound effect on their severity. In the field of contagious diseases the new problem of the 'viral strike' is emerging: many apparently new virus diseases, borne by wind or vectors, travel through areas causing some quite devastating effects for a time, especially in the larger livestock units. More virulent and mutating strains of pathogens appear to be produced the larger the 'biological mass' and the more vaccines are used to protect the animals from diseases.

A further major problem is that of the *metabolic diseases*. These are a group of diseases that are caused intrinsically by the animals being called to produce an end product faster than the body can process its intake of feed. Enormous efforts are made to provide the right nutrients in an easily assimilated form but as the metabolic disease is rather different from a deficiency disease, this does not necessarily work. In a way the metabolic diseases are the inevitable outcome of the success in conquering most of the acute virulent diseases together with the advances made in improved genetics, nutrition and housing. These improvements have led to much increased growth and productivity but the capability of the animals to keep pace with these has outstripped the normal functioning processes (or metabolism) of the body.

An example of this is provided by considering the growth of table chickens. When the industry started intensively it took about 13 weeks to produce a bird weighing 2 kg, at a food conversion efficiency of approximately 3:1 (that is, 3 kg of feed to produce 1 kg of liveweight). The hazards in growing the birds at that time were innumerable and were largely related to contagious and infectious diseases. Mortalities were high, often up to 30%, and the diseases played their part in slowing growth and damaging the food conversion efficiency. Now, some 30 years later, the position is quite different. It takes less than half the time for the birds to reach 2 kg, that is about $5\frac{1}{2}$ weeks. Mortalities normally average 3–4% and contagious and infectious diseases are controlled largely by good hygiene together with judicious use of vaccines and medicines. But now we have a number of the so-called metabolic or production diseases: for example, birds are affected with excess fatty deposition, causing degeneration of the liver and kidney and heart attacks. The skeletal growth may not keep pace with the rate of muscle development so that locomotor disorders occur, such as twisted, rubbery and broken bones and slipped tendons.

Types of diseases

An important factor influencing the incidence of disease in livestock today is the increasing immaturity of livestock. Improved performance has resulted in animals reaching market weight much earlier, whilst for genetic reasons breeding animals are also younger on average than previously. Thus, on a modern livestock farm, there is usually a high proportion of young animals that are in a state of susceptibility to infectious agents whilst they are still developing the ability to resist disease naturally or natural immunity to disease which will normally take place over a prolonged period. This difficult state of affairs is further exacerbated by the considerable size of many livestock units in which the young animals may have originated from various parents of very different backgrounds. In many cases the young or growing stock will have come in from widely separated areas. They may have no resistance to local infections and will therefore be totally susceptible to them and at the same time contribute a new burden of pathogenic micro-organisms to the unit they have entered. Altogether the modern livestock unit may present at any one time a confusing immunological state and the basic design and its management will influence the success or otherwise of disease control (see Table 22.1).

There are certain major groups of infections that account for most of these problems. The most widespread are probably respiratory diseases. Many of these are subclinical, have a pronounced debilitating effect on the animals and are caused by a large number of different infective agents even in any one disease incident. They may not respond satisfactorily to vaccines, antisera, antibiotics or any other drugs, so that the fundamental method of approach to their control is by managerial, environmental and hygienic measures.

Another significant group of diseases influenced similarly are the enteric infections. These, also, have many different primary causative agents, ranging from parasites to viruses and bacteria. The reasons for their increasingly harmful effects in recent years are not only those already listed but also the general trend with certain livestock to eliminate the use of bedding such as straw. The harmful effects of this may often be corrected by the use of good pen design, especially by the use of slatted or slotted floors, but it is nevertheless more difficult to separate animals from their urine and faeces when the flooring is without litter, as the latter has a diluting and absorbent effect on the excreta.

There are, in addition, several important bacterial infections that have tended to increase in large intensive units. Examples of these are *Salmonella* and clostridia

Table 22.1 The disease pattern on the farm

Poor husbandry	+ primary disease agent (often a virus)
	+ secondary disease agents (often bacteria or parasites)
	= subclinical or overt disease

Examples of bad husbandry
Too many animals on a site
Overcrowding within the buildings
Mixed ages within a house or site
Excessive movement of animals
Poor ventilation and environmental control
Bad drainage and muck disposal
Insufficient bedding
Lack of thermal insulation in construction
Unhygienic or insufficient food and watering equipment
Absence of routine disinfection procedures
Faulty nutrition

bacterial spores, together with *Escherichia coli*, *Pasteurella* and *Campylobacter*. Many forms of these organisms are normal inhabitants of the animals' intestines in small numbers but excessive 'challenges' causing disease may build up under unhygienic intensive conditions, encouraged by building with poorly constructed surfaces which cannot be cleaned.

Livestock unit size

In addition to the risks of the gradual build-up of disease-causing agents within livestock buildings there are serious dangers of livestock enterprises becoming too large. In all parts of the world where the development of large units has taken place there have emerged a number of viral diseases which tend to 'sweep' like a forest fire through areas with a high livestock population, sometimes leaving a trail of devastation. Often there is likely to be considerable loss over a concentrated period, after which the animal population may develop a natural immunity, at least for a time, or artificial immunities are promoted by the use of vaccines. It is now known that contagious virus and other particles can travel great distances from large infected sites. Certainly distances of 50 miles have been virtually proven, but they may well travel much further than this. If livestock enterprises continue to grow in size, then the dangers in this respect can only become greater. At the present time it is impossible to give soundly based objective advice on the optimal unit size and in any event the factors that would lead to making proposals are highly complex. However, there is clear evidence that animals thrive less efficiently in large numbers even in the absence of obvious clinical disease. Apart from this there is the difficulty of hygienic disposal of the dung when the unit is of excessive size.

A good example of the spread of virus disease was the sudden occurrence of 'blue-eared pig disease' (also known as the porcine reproductive and respiratory syndrome or PRRS), which swept through the most highly populated areas of the world with great rapidity. First recognized in 1987 in USA, it found its way to Europe in 1990 and to Britain in 1991.

It is normally possible to keep much greater numbers of *adult* animals together than young stock since there is nothing like the same number of contagious disease problems after the difficult growing stage and its immunological uncertainties are passed.

As to the maximum number of livestock that might be kept on a site, it will be appreciated that this will depend on a number of factors, apart from health considerations. Figures have been proposed which attempt to allow for all factors and those suggested are as follows:

Dairy cows 200
Beef cattle up to 1000
Breeding pigs 500
Fattening pigs 3000
Sheep 1500
Breeding poultry 3000
Commercial egg layers 60 000
Broiler chickens 150 000

Such figures can be no more than suggestions made in the light of practical experience. In due time more scientific evidence may be available to establish a better degree of accuracy. It is certainly likely that the figures will require constant adjustment in the light of new developments in husbandry, housing and disease control and especially with the anticipated trend towards the increasing provision of livestock free of specific disease.

The benefits of smaller livestock units are becoming appreciated more widely world-wide and there is evidence that 'small is beautiful' is becoming more fashionable.

Design essentials to minimize the disease challenge

In addition to the size of the livestock unit there are a number of other basic items which need to be considered in order to provide the bases of good health.

Depopulation

A fundamental concept in maintaining the health of animals is to ensure the periodic depopulation of a building or a site. The benefits of eliminating the animal hosts to potential disease-causing agents are well understood and the virtue of being able to clean, disinfect and fumigate a building when the animals have been cleared is also accepted. Periodic depopulation is especially important for young animals but is less so for groups of older animals which have probably achieved an immunity to many contagious diseases. Much also depends on whether the herd or flock is a 'closed' one with few introductions of fresh animals or an 'open' one with a constant renewal of the animal population. If the latter is the case, then constant depopulation is of greater importance as there is little or no opportunity for natural immunities to develop and the regular removal of the build-up of infection is of great assistance in ensuring the good health of the livestock.

The health status

The policy will also depend on the health status of the stock. At one extreme there are the so-called 'minimal disease' or 'specific pathogen-free' herds which have been developed to be free of most of the common disease-causing agents of that species. Here, depopulation is less critical than the protection of the animals from infections that come in from outside. Since this danger is very serious in most localities it is important to subdivide the animals in a unit into smaller groups, lessening the likelihood of a breakdown and/or enabling isolation and elimination of a group which may become infected. At the other extreme there are those units which have a constant intake of new animals from outside and of a totally unknown health status. In this case there is a high risk, and more usually a near certainty, that some will be either clinically infected with, or carriers of, disease-causing organisms. Design specifications for such units should be quite different from those of the closed herd or flock so that defined areas of the unit should have groups of animals put through them in batches after which the area can be cleared, cleaned and sterilized. Obviously, it is preferable if the whole unit can be so treated since it ensures an absolute break in the possible disease build-up cycle.

Between these two extremes is the more usual case in which a herd or flock is of reasonable health status, though certainly not free of all the common diseases, and in which new livestock are added only occasionally. In such cases the precautions in the housing against disease build-up and spread of infection can be rather more relaxed but there must still be a proper provision for the isolation of incoming and sick animals.

Group size

Animals thrive best in groups of minimal size. If groups are small, it is usually easier to match the animals within them for size, weight and age, and it is well established that growth under these circumstances is likely to be most even and economical. Behavioural abnormalities such as fighting and bullying, are also kept to a minimum – indeed they may be prevented altogether.

Fighting amongst animals is a highly contagious condition and under the most intensive husbandry systems an almost casual accident that may draw some blood can escalate into a blood-bath. Pens which keep the animals in small groups will tend to reduce the occurrence of such disasters and indeed with good management the removal of an animal that has accidentally injured itself or is off-colour and therefore prone to being bullied will stop the trouble before it ever has a chance to develop seriously.

There is yet another economic advantage in keeping animals in small groups. The farmer will achieve the best economic return if the stock are housed at the densest possible level for optimal productivity. If animals are kept in a house to allow a density such as this, it means that they should spread across the house evenly so they do in effect occupy and use this area. In practice, however, this is very difficult to achieve, especially when large numbers are housed together without any subdivision at all. Hence the birds, or any other livestock as may be at risk, crowd in certain parts of the building which can lead to grossly overstocked floor areas. If livestock crowd excessively in certain parts of a house, this area is likely to become more polluted with excreta and respiratory exhalations to an abnormal and harmful degree; the humidity becomes high, proper air movement is impeded and the animals soon may become ill. Sick animals feeling cold tend to huddle together more, so the vicious circle is perpetuated and there is seemingly no end to it unless some measures are taken to ensure a better distribution of the stock. When this problem arises under practical conditions it may be impossible to subdivide the animals at once. An immediate trend in the right direction, that is encouraging the animals to spread themselves more uniformly over the house, can often be achieved by introducing some artificial heat. There are excellent gas radiant heaters and oil-fired and electric blower heaters available.

When animals are penned in large numbers, the effects of a fright caused by an unusual disturbance can be extremely serious. It is almost impossible to guard against all the extraneous sounds and sights that may affect the stock. The best safeguard, therefore, is to have the animals housed in small groups so that the effect of a panic movement will be more limited and will never build up into highly dangerous proportions.

Floors

The profound effect of the floor surface on the health and well-being of livestock is well established. When bedding was almost invariably used in animal accommodation there were relatively few problems related directly to the flooring. Now that the farmer must frequently use housing systems without bedding, often with slatted or other forms of perforated flooring, to produce economically a comfortable and clean environment for the animals, new problems have emerged. Though we are still far from being able to advise the ideal floor, it is possible, by choosing a good combination of surface and bedding where used, to provide the animals with a comfortable, warm, hygienic and well-drained surface.

The best solid flooring is usually based on concrete because when properly made and laid it is hard wearing, hygienic and impervious to fluid. Where an animal lies directly on the floor, without bedding, there should be an area of insulated concrete, most often incorporated by using 100–200 mm of lightweight or aerated concrete under the top screed and above a damp-proof course.

The surface of the concrete must not be so smooth that the animals slip or injure themselves yet, on the other hand, if it is too rough it can cause abrasions and injuries. A happy medium is not easy to find but is usually achieved by using a wood float finish or by tamping the floor lightly with a brush. Also, very rarely are the 'falls' on the floor correct; again, there must be a nice balance between being steep enough to drain away liquids yet not so sloping that the animals slip. With some animals the floor is made more comfortable by placing rubber mats on top; these are used with some success with cattle, calves and pigs.

Some of the worst problems of injury, especially with cattle and pigs, have occurred with slatted or other forms of perforated floors. These have been particularly troublesome when the edges have been left too sharp or have worn leaving injurious protrusions. If bedding *can* be used for livestock it is usually better to do so. Good bedding is an insulator and warmer for animals and is probably the cheapest that there is. The air temperature of buildings can be lowered if bedding is used. When the bedding is straw it may form an acceptable, if not an essential, adjunct to the diet of the animal. The fibre may help digestion and create a sense of well-being. It keeps the animal occupied and assists in the prevention of vices.

Bedding also provides just about the safest flooring for the animal, so reducing the danger of injuries to the body, particularly the limbs. It is also much more difficult to keep the floor dry when no bedding is used and if the animal's surface is kept wet there is a further chilling effect created by contact with a wet floor. It is also important to emphasize the health risk from undiluted muck which does not pass through slats but accumulates at some points, and the great risk to the respiratory system of humans and animals from gases rising from slurry tanks or channels below the slats. Not surprisingly, therefore, there has been a trend towards the advocacy of bedding for most animals.

Isolation facilities

There is an urgent need for better consideration to be given by the farmer to the isolation of sick animals. Isolation removes the dangers of contagion to the normal animals. It also makes it more likely that the sick animal will recover without the unwelcome attentions of the other animals who will always tend to act as bullies. In nature the sick animal usually separates itself from the rest of the herd or flock but it is often impossible for the housed animal to do this. Also, under intensive management, more animals suffer from the aggression of their pen-mates and it is essential that such animals are removed to prevent them from suffering or being killed.

It is also easier with an isolated animal to give it such therapy as it requires and to give it any special environmental conditions, such as extra warmth. It is an interesting but important observation, known to most farmers, that if you take 'poor doers', perhaps animals suffering from subclinical disease, out of a group and pen them with more space and extra comfort, they may thrive without any further attention. Nevertheless, with the addition of suitable therapy, the cure may be accelerated and completed.

Health and the disposal of manure

The method used for manure disposal has potentially important effects on health, both human and animal. Whilst the smell from composted solid manure tends to be strong, it rarely travels far or creates a nuisance problem and there is little if any risk to the human or animal population from this form of manure under temperate climatic conditions.

Slurry, however, is quite a different matter. If placed straight on the land from the animal house or after holding in a tank anaerobically it has an extremely offensive smell. Whilst masking agents are possible they are too expensive at present to be considered economic. The worst smell comes from the pipeline and gun spreader, because the droplet size is small and light and particles may carry for considerable distances. Less smell arises from a tanker spreader because the slurry is much thicker and not spread by aerially dispersed small droplets. The most satisfactory way to prevent the slurry from causing offence is to treat it aerobically in some way before spreading. Many human and animal health problems may arise from the spreading of slurry. In surveys it has been found that potentially pathogenic bacteria were able to survive for up to three months in slurry kept under anaerobic conditions. Whilst the particular bacteria studied were *Salmonella* species and *E. coli*, there is little doubt that more resistant organisms, such as *Bacillus anthracis*, *Mycobacterium tuberculosis*, *Clostridium* species and *Leptospira* species could survive as long or probably very much longer. If the slurry should enter a river or stream, the pollution may have far-reaching and infinitely more serious effects. It is thus essential for all enterprises with slurry as the disposal system that either the slurry is placed on land where it cannot be a nuisance or health risk or it is so treated beforehand that the risks are removed.

The dangers from gases in farm buildings

Hazards associated with gases in and around livestock farms have been highlighted recently, particularly in association with gas effusions from slurry channels under perforated floors. Numerous fatalities have been recorded of livestock, and even in humans, due to gas intoxication. There is, in addition, mounting evidence that there may be concentrations of gases in many livestock buildings that may affect production adversely by reducing feed consumption and lowering growth rates and the animals' susceptibility to invasion by pathogenic micro-organisms.

The most serious incidence of gas intoxication arises from areas of manure storage in slurry pits or channels under the stock usually, but not always, associated with forms of perforated floors. The greatest risk arises when the manure is agitated for any reason, usually when it is removed. There is also an ever-present danger if a mechanical system of ventilation fails and this is the only method of moving air in the house. Several cases of poisoning have been reported when sluice gates are opened at the end of slurry channels and the movement of the liquid manure has forced gas up at one end into the building.

High concentrations of gases, chiefly ammonia, may also arise from built-up litter in animal housing. This is most likely in poultry housing since the deep litter system is the most commonly used arrangement with broiler chicken and poultry breeders. The danger has undoubtedly been exacerbated within the past few years, owing to the necessity of maintaining relatively high ambient temperatures in order to reduce food costs, while at the same time there has been good evidence that higher temperature than hitherto should be maintained for optimal productivity. Poultry farmers have often attempted to achieve such temperatures by restricting ventilation in the absence of good thermal insulation of the house surfaces, and the result can be generally harmful if not dangerous.

The most popular form of heating is by gas radiant heaters which are suspended from the ceiling. Well over half of all poultry housing is heated in this way and because the cost of gas as a fuel appears in some respects to be improving in relation to that of other fuels, the system is being used more for pig housing and especially for piglets in the early weaning system. There is a risk that with inexpert use such heaters may be improperly serviced and the house insufficiently ventilated to give complete combustion and that toxic quantities of carbon monoxide may be produced.

Disease and immunity

Organisms causing infections

There are different groups of organisms causing infectious disease in animals. The smallest of them are viruses, below 300 nm in size, but many are well below this. For example, the foot-and-mouth disease viruses are 10–27 nm (1 μm = 0.001 mm and 1 nm = 0.000 001 mm). Viruses cannot usually be seen under the ordinary microscope but can be photographed by the electron microscope. They are simple organisms that can only multiply within living cells and this property distinguishes them from bacteria. Viruses are classified into a number of different groups, e.g. reoviruses, adenoviruses and herpesviruses.

Bacteria are relatively large organisms compared with viruses and can multiply and grow outside living tissues. In size, for example, cocci (round bacteria) measure 0.8–1.2 μm, bacilli (rods) are 0.2–2 μm, and the spiral shaped spirella up to 50 μm. They are visible under the ordinary microscope. Many types, once outside the animal body, may sporulate to form a protective coat so that they can live many years in buildings, soil or elsewhere and can still be capable of infecting animal life. For example, the spores of anthrax and *Clostridium* can live for 20 years or more under favourable circumstances.

Mycoplasma are smaller organisms than bacteria, being about 0.25–0.5 μm (or 250–500 nm). They are rather like oversize viruses but, unlike viruses, they can be cultured on artificial media.

Rickettsiae are of somewhat similar size to *Mycoplasma*, being 0.25–0.4 μm, but they cannot be grown in ordinary culture media and will only multiply intracellularly, like viruses.

All these organisms are classified as from the vegetable kingdom and can be joined by the yeasts, moulds and other fungi which are larger than bacteria, several varieties of which cause diseases in animals and humans.

There are also many simple organisms from the animal kingdom, known as parasites, that cause disease in farm livestock. These range from the single-celled protozoa, such as the coccidial parasites, to parasites of ever increasing size and complexity, culminating in such relatively complicated organisms as the roundworms (helminths), lice, flies and ticks.

Organisms from these groups are the main causes of infectious disease in farm livestock. The only other agents causing disease, and which are not infections, are the metabolic disorders, poisons, deficiencies, excesses and injuries, examples of which are found throughout this chapter.

Immunity

In our understanding of the natural mechanisms that maintain good health it is essential to know the principles of the normal animal's immunological system and the way in which we make use of such products as vaccines and sera to boost it as necessary. This is best understood by studying the cycle of events in any animal's life (see Fig. 22.1).

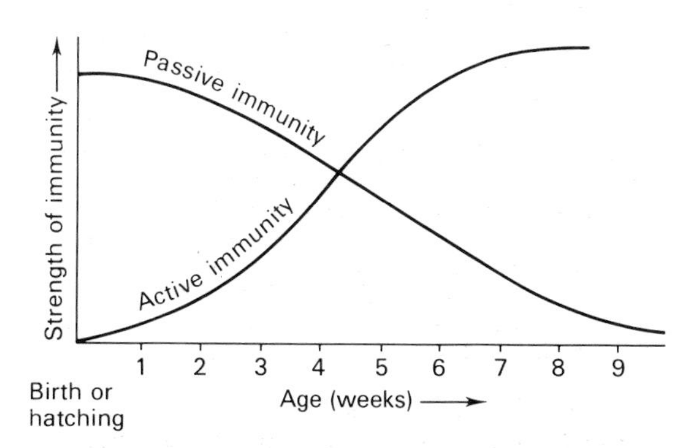

Fig. 22.1 Diagram showing how immunity develops in livestock. Curves represent antibody (i.e. immunity) produced by active or passive immunization. Animals derive passive immunity from their dams but it is short-lived. Active immunity comes either from vaccines or a disease challenge – which may be mild.

When an animal is born it is relatively free of disease organisms but it has what is known as *passive immunity*, which it receives from the mother whilst still *in utero*. If the mother has been vaccinated against a disease or has had an experience of the actual disease, then her immunological system will normally produce the antibodies that are capable of resisting the disease. These antibodies circulate in the blood and will also be transferred to the fetus. The antibodies themselves are effective only for about two to three weeks before they disappear.

The young animal can produce its own antibodies and thus develop what is known as *active immunity* only if it has experience of the organisms that cause the disease in question or has been vaccinated against it. It should be stressed that an animal will have some passive immunity only to those particular infections experienced by the mother. These may be quite limited and 'local'. This is a good argument for rearing an animal in the environment in which it is conceived and a good reason for not transporting young animals too early in life before they have some better ability to resist a 'foreign' disease challenge. Passive immunity in mammals is fortified immediately after birth by the young drawing the first milk (or colostrum) from the dam. Colostrum is especially rich in antibodies and also nutrients.

As the passive immunity the young animal receives from its mother fades, it may be replaced by an active immunity if the young animal is vaccinated or has some challenge with the organisms that can cause disease. Under practical conditions the aim is to make sure that any natural challenge of potentially pathogenic organisms is always mild or gradual, not massive and overwhelming, as in the latter case disease will arise.

Sometimes it is perfectly satisfactory to allow the animals to receive a 'natural' challenge from the environment but this is at the best inexact and at its worst ineffective. Thus, various methods of induced immunity are given, either *passive* or *active*. The passive immunity is induced by injecting into the animal hyperimmune antiserum prepared from the serum of animals which have experienced the actual infection and have recovered. This serum may also be used for the treatment of the disease but it has only a transitory effect lasting about three weeks, similar to the natural immunity received by the young animal from its mother.

For an active immunity vaccines are used. A vaccine is a way of activating the body's defence mechanism by challenging it with the pathogenic organisms so modified that they are active enough to produce an immunity to the disease but not the disease itself. The modification is done in very many ways, by adding chemicals, by growth of the organism in special media or livestock of a different species or by irradiation. Sometimes no modification is required as an organism can be used which is sufficiently closely related to induce immunity but not disease. Occasionally, in actual farming practice, if no vaccine is available for a disease which is almost certainly likely to challenge the animals later in life, the animals are deliberately exposed to the disease causing agent and at such a time when no actual disease will be caused. This is a venture which is hazardous and it is unwise to do it without proper veterinary supervision.

The production of vaccines is a very sophisticated process but can induce an enormously efficient protection for the animal. Vaccines must be used by skilled hands at the right time in the correct dose and repeated as necessary since the length and strength of the immunities developed vary according to the infection. It should be borne in mind that when a vaccine is given the immunity takes time to develop, somewhere in the order of three weeks, so it may be necessary to protect an animal during that time by antisera, medicines or simply by isolation. Some vaccines are of live organisms, others are dead. In general the live vaccine may give a stronger immunity more quickly but it is not necessarily any longer lasting and it may induce more reaction. Dead vaccines given by injection in oil or other special bases can be very effective for long-term immunities and are increasingly widely used. It should be emphasized that when a vaccine is administered to an animal it causes a degree of illness and during the period of immunological development the animal is experiencing stress and is *more* susceptible to infection than at other times. Thus it is necessary to practise even better management at the time of vaccination. Very often this factor is not appreciated and a failure that is unjustifiably ascribed to the vaccine is essentially due to mismanagement of the animals.

Medicines

Vaccines and sera are used to protect or treat animals for specific diseases. They should be used wherever they can be in the prevention of infections but in many cases no serum or vaccine exists and we must deal with illness using one or many of a wide choice of medicines. Also many of the disease conditions we are dealing with are caused by a number of different organisms and the identity of some or all may not be known definitely. Frequently the urgent need for treatment demands that a medicine is used before laboratory examinations have elucidated the cause, and we must rely on the experience of the farmer and the clinical skill of the veterinarian to judge the correct medicine to use. When the results of the tests are known, an appropriate change in treatment may be indicated.

Medicines can be given to have an almost immediate action. One administration may have an effect for only a few hours or in some cases a few days but usually, in order to keep the levels up within the blood stream of the animal, there must be a constant boosting of levels. It cannot be emphasized too strongly that the full course of treatment prescribed should be completed and levels maintained correctly or the benefits may be lost and medicine-resistant organisms produced; this possibility is one of the worst outcomes of the use of medicines. Care must also be taken in the agricultural use of medicines to observe certain conditions of use. For example, with many medicines a compulsory withdrawal period is required before animals can be sent for slaughter, varying from about 5–10 d and in some cases even longer, and milk from cows being treated for mastitis must also not be mixed with normal milk for specified periods.

The majority of medicines used for the prevention and treatment of disease are chemotherapeutic, the term implying the use of chemicals that may have been synthesized, such as the sulphonamides, or produced by the action of living organisms – these being the antibiotics.

The routes by which medicines are given vary from injection by the intravenous, intramuscular or subcutaneous routes to oral dosing and incorporation in feed or drinking water. With so many medicines available and the routes and methods of administration being so variable it is perhaps inevitable that careless and abusive use occurs. One cannot do more than emphasize that the prescribing of medicines requires a high degree of skill and their administration an abundance of diligence and competence.

Growth promoters

For many years are livestock industry has made use of antibiotics and other substances which, when incorporated in small amounts in the ration, may increase the growth rate of livestock and improve food conversion efficiency. In the early days of use of these materials, the same antibiotics were used for growth promotion as for the treatment of disease. Since medicines as growth promoters are used at a much lower level than for therapeutic use, the incorporation of antibiotics at a low level in the feed is unacceptable since resistant organisms will soon emerge and not only will the antibiotics no longer be of use as growth promoters but they will also have lost their value for therapeutic purposes. Because of this there is now a clear separation, legally enforced, between the use of antibiotics for these two purposes. Those antibiotics used as growth promoters (e.g. zinc bacitracin, flavomycin and virginiamycin) have no other use and are not used for any therapeutic purposes. The therapeutic antibiotics are used only for the prevention and treatment of disease and are available only on veterinary prescription to be used by or under the control of the veterinarian.

This problem of resistance is not the only one in connection with the use of the feeding of growth-promoting antibiotics. These substances have their greatest effect when the hygiene or husbandry system is less than ideal. Under really good management practices they may do little good at all. Thus they tend to 'mask' bad husbandry and this is a dangerous characteristic since such an effect will not last for ever; the bad practices, however, cannot only last indefinitely but will tend to have a cumulative effect on results.

There is still no absolute certainty as to the means by which growth-promoting antibiotics exert their effects but it seems most likely that these effects are due to the favourable action of the antibiotics on the balance of bacteria in the intestines, reducing the harmful ones and promoting the beneficial ones and thereby increasing the efficiency of food utilization and production.

The importance of disinfection

Livestock are generally housed much more intensively than hitherto. This involves not only the stocking of animals more densely but equally an increased size of units and the more efficient use of building space. It is rare for a livestock pen to be without stock; if there is a break between batches then it is usually a short-lived one. In the days before the use of antibiotic therapy, intensive methods tended to collapse under the burden of disease. Now they are again under threat because an increasing number of disease-producing organisms are resistant to drugs and also the cost of almost continuous antibiotic therapy is becoming an economically damaging burden to the stock keeper.

There is one viable answer to this and it is to institute a very efficient programme of cleaning and disinfection of the buildings. Let us be quite clear, however, disinfection needs to be done in a carefully organized way and must not be, as so often is the case, an application of any disinfectant chosen at random in the hope that it will kill the pathogens. First, it is imperative, at least with young livestock, to plan the enterprise in such a way that a building or even a site is totally depopulated of stock before the cleaning and disinfection takes place. This is necessary because the live animal, and it might be literally only one, is potentially a much greater reservoir of infection than the building itself. So the complete emptying of a house can be more important than the disinfection process itself. The disinfection process in sequence is as follows:

(1) Empty the pen and preferably the house of all livestock.
(2) Clean out all 'organic matter' – dung, bedding, old feed, any other material that could contain pathogenic micro-organisms. Remove totally from the area of the buildings.
(3) Remove all portable equipment for cleaning and disinfecting outside the building.
(4) Wash down with heavy duty detergent-disinfectant. Use a power washer wherever possible.
(5) Apply a disinfectant appropriate to the types of infection being dealt with. Usually required is a mixture active against viruses, bacteria, parasites and insects which may carry infection from one batch of livestock to another.
(6) Wherever possible fumigate the house with a disinfectant in aerosol or spray form after the equipment has been reassembled.
(7) Dry out and rest for a day or two before restocking.

Signs of health and disease

Early recognition of illness is particularly vital because the risk of spread of anything contagious is very great. Early recognition also increases the opportunity for successful treatment, enables the affected stock to be isolated and speeds any laboratory diagnostic work. It also helps in the good welfare of the animals since sick animals are often maltreated by their neighbours. All in all, it will also generally reduce losses to a minimum and thus make sound economic sense.

The following are the most important signs to look for.

Appetite

One of the earliest and surest signs of illness is a lack of interest in feed (anorexia) or a capricious appetite, though on occasions it may be the feed itself which is at fault and requires investigation.

Separation from the group

Such animals will usually separate from the others and hide in a corner or bury themselves in bedding, if available.

Excreta

This may be 'abnormal'. Sick animals often have scour (diarrhoea) or, more usually in the earliest stages of disease, they will be constipated, especially if there is a fever. In other circumstances the faeces may contain blood (dysentery) or may be of abnormal colour (which sometimes happens in cases of poisoning, for example).

Urine

This may show abnormalities such as the presence of blood or may be cloudy or yellow (jaundiced). Normal urine is pale and straw-coloured.

Posture
Sick animals may find it impossible to rise, or may hold themselves uncomfortably, and may be lame. The nature of the abnormality of posture may indicate the organ affected.

Appearance
Sick animals will often have a droopy head, with dull eyes and dry muzzle and possibly discharge from the nose and watery or catarrhal eyes. All mucous membranes may be discoloured.

Skin and coat
A healthy coat is clean and glossy. Abnormal coats may be 'hide-bound' which is when dehydration makes it difficult to move the skin over the underlying tissues. Sick animals may have bald, scurfy, staring, 'lousy', mangy or scabby skin, and the animals may scratch or rub. The skin, coat and feathers are very good indicators of disease.

Coughing
Most abnormalities of the respiratory system produce coughing in animals and the nature of the cough will often be diagnostic of certain conditions.

Pain
The presence of pain shows in a number of ways: grunts, groans, grinding of the teeth, squealing or crying, and arching of the back.

Mucous membranes
The normally moist membranous linings to the mouth, nose and other external orifices may show abnormalities such as discoloration, dehydration, discharge of a serous, mucous or purulent nature, or haemorrhages.

Temperature
An abnormally high temperature usually indicates an infection; an abnormally low one may be due to a metabolic defect, poisoning or simply the later or terminal stages of an infection. Normal temperatures and respirations are as follows:

	Temperatures °F	°C	*Respirations/ min*
Cattle	101.5	38.7	12–20
Sheep	103.0	39.4	12–30
Pig	102.5	39.2	10–18
Fowl	106.5	41.5	12–28

Respirations tend to be accelerated by infectious diseases and slowed by others though the nature of the breathing will also be vastly affected and may change from the deep to the shallow on the occurrence of illness.

Pulse rates
This is normally only studied by the veterinarian but the normal rates are as follows:

	Pulse rate/min
Cattle	45–50
Sheep	70–90
Pig	70–80
Fowl	130–140

Animal welfare and animal health

A recent phenomenon has been the widespread questioning of the humanity of certain methods of rearing livestock. Under critical attack from so-called 'welfare groups' have been systems of housing that greatly restrict the movement of animals, amongst these being cages for chickens, calves housed in crates for the production of veal and sows closely tethered in stalls during pregnancy. In the UK it is now illegal to rear calves for veal production in crates where they cannot turn round and have no bedding. Also, sow stalls and sow tethers will be illegal in the UK from 1997. Few other countries are copying such welfare legislation and there is a danger that unilateral welfare measures will merely encourage livestock production to expand in countries without welfare rules and regulations.

Animal welfare is very pertinent to the question of health since there is no dispute that one of the essentials for the provision of good welfare is the maintenance of health in the animals. It has also become apparent from recent activities in the research field that the eventual goal of establishing what constitutes good welfare in a scientific way is going to be very difficult indeed, if not impossible. Thus it will be necessary to continue to rely, at least in the foreseeable future, on the overall knowledge, expertise and even instinct of the persons concerned with the management and care of the animals, these being the farmers, stock keepers and husbandry and veterinary advisers. It is essential to keep in the forefront of consideration the overriding importance of good welfare and humanity to animals on all occasions. Anyone who has closely observed sick animals can understand what constitutes 'misery' in the animal kingdom and contrast this with the alert, buoyant and inquisitive appearance of an animal in good health.

A feature of this controversial field which has been beneficial is the way in which it has forced us to enquire whether it is necessary to house animals using methods which involve a high degree of restriction on their movement. There has been a vigorous search for alternatives which has been fruitful as it appears there is less need for the extremes of intensification than has been alleged and indeed it may be economically preferable to have systems which are more in harmony with the overall agricultural scene.

Large and intensive livestock enterprises may be unsuccessful for three principal reasons. First, the management of large numbers of *living beings* is not easily organized on the large scale. Stockmanship is partly science and partly art, and art must rely heavily on the individual. If only one small item goes wrong or is overlooked, it can lead to a very serious chain of events on the large scale.

The second reason is that disease is much more likely to have a serious deleterious effect in the large unit in subclinical or clinical forms. The failure of many units can be ascribed to their inability to control disease so that the economic viability of the unit has been totally destroyed.

The third reason is the heavy cost of buying in or bringing in fodder and moving out the muck in such enterprises. It is common for the highly intensive unit to be totally dependent on food from outside the farm or the locality and to be faced with great difficulty in disposing of all the waste products. In smaller units the crops can be used to feed the livestock and the 'muck' they generate is of great value to the land; energy is conserved and this is agriculture of a totally balanced sort.

It may be stated also that the welfare standards that are gradually emerging across the world are based on evidence which, so far as is possible, is not in conflict with economics. For example, if animals are housed at concentrations and intensities above those advised, it is likely that their productivity will fall. Particularly this is so when it comes to stocking density and its relationship to growth, production and the efficiency of food utilization. Good welfare standards can therefore, be used as a husbandry standard with some confidence.

The key to good welfare is a high standard of stockmanship. If this is to function, good facilities must be provided for the stockman. Many systems that have been developed in recent years completely ignore this. The essentials are as follows:

(1) plenty of room for the stockman to see the stock whenever he needs to;
(2) good lighting and no undue saving on passage space;
(3) facilities for isolation of sick stock;
(4) easy access by the stockman to his animals and, above all, a very high standard of handling facilities so that it is easy to medicate or treat the animals in any way necessary.

Disease legislation

In Great Britain, under the Diseases of Animals Act, there are a substantial number of Orders relating essentially to the State's interest in animal disease control. These regulations cover a wide spectrum of activity. For example, they deal with the importation of animals and animal by-products, the movement of animals, and the inspection of livestock and vehicles, and allow for the Ministry of Agriculture, Fisheries and Food to institute a large number of Orders relating essentially to specific diseases.

Under the Diseases of Animals Act certain specified diseases are designated notifiable. Those of current importance are:

Anthrax
Foot-and-mouth disease
Fowl pest
Bovine tuberculosis
Bovine spongiform encephalopathy ('mad cow disease')
Scrapie
Swine vesicular disease
Aujeszky's disease
Enzootic bovine leucosis
Rabies
Warble fly

There are also others, such as swine fever, which could well occur again as they are widespread elsewhere in Europe.

The most important feature about the notifiable diseases is that the owners or person in charge of an animal suspected as affected by a notifiable disease must report immediately his suspicion to a police officer or an inspector of the Ministry of Agriculture. Once the owner has notified the inspector or police, steps will be taken to diagnose the condition and measures may be taken to isolate the premises and treat, vaccinate or slaughter the stock. Different diseases have different procedures and these are dealt with later in the chapter under the respective diseases. Also, under the Zoonoses Order, the presence of conditions such as salmonellosis and brucellosis, which infect both animals and humans, must be notified.

The health of cattle

In this section health problems are dealt with on a life-cycle basis, dealing firstly with those health problems affecting the calf, continuing with growing stock and concluding with the mature beast and those diseases that have no special age incidence.

The health of the calf

Of the 5–10% mortality that occurs in calves, nearly two-thirds takes place in the first month of life and much of this is due to what is commonly called calf scours. The main cause is ultimately the ubiquitous bacterium *Escherichia coli* following initial environmental stress and usually a virus challenge.

Colostrum and protection against diseases in the calf

It is essential that a calf should have a good drink of colostrum very soon after birth. Colostrum contains the immunoglobulins that are the essential protection to infection. It also contains large quantities of vitamin A (six times as much as in ordinary milk). Immunoglobulins are the proteins which contain the antibodies which protect the calf from disease organisms, whilst vitamin A is the most important vitamin associated with the animal's resistance to infection.

Because calf scour is caused by a number of organisms, the use of specific vaccines and sera may have a limited chance of providing either good treatment or prevention, though sometimes this can be the case. More often the disease must be treated with antibiotics with a broad range of activity, such as the synthetic penicillins, together with replacement therapy for the enormous fluid depletion that takes place. This will include electrolytes, glucose, minerals and vitamins in easily assimilated liquid form.

Salmonellosis

The salmonella group of bacteria is a very large one with hundreds of different types, many potentially pathogenic to man and animals. Only relatively few are harmful, the majority being 'exotics' that come in via the feed and are quickly eliminated.

The two most important species in the condition in calves are *Salmonella typhimurium* and *S. dublin*. The symptoms with either of these infections or indeed with other virulent salmonellae can be largely similar. A percentage of calves may die so rapidly that no symptoms are seen; others scour profusely with dysentery, pneumonia, arthritis, jaundice and even nervous symptoms manifesting themselves in some cases. Urgent treatment will be essential to include antibiotics such as ampicillin, chemotherapeutics and fluid therapy, isolation and general nursing, especially warmth. Thorough disinfection of the quarters will be essential and preventive measures may include vaccination and possibly preventive medication. However, the use of the last measure is hazardous as it may generate drug resistance and encourage the establishment of the 'carrier' animal which can excrete salmonella organisms throughout its life.

Whilst salmonellosis is especially a disease of young stock, adult cattle do become infected and may not show symptoms, becoming carriers of great potential danger to man and other livestock. Abortions may be caused by salmonella, and salmonella can be a source of infection via milk. In the main the biggest problems from salmonellosis arise from farms where the calves are brought in from outside and infection can easily spread to the adults if they are on the same premises.

In Great Britain salmonellosis is a notifiable infection and official measures can be taken to control the disease as appropriate. In the event of an outbreak in adult cattle, measures that may be taken will include laboratory testing of the cattle to identify the carriers, treatment of animals with antibiotics, such as ampicillin, and appropriate care of the milk to remove the risk of human infection.

Calf diphtheria

The bacterium associated with foul-in-the-foot in cattle, known as *Fusibacterium necrophorum*, is also capable of causing a serious disease in calves known as calf diphtheria. The infection attacks very young calves causing inflammation within the mouth, which leads to soreness, ulcers and ultimately necrosis. The calf eventually becomes seriously ill. The condition responds well to treatment. Sulphonamides are effective as are a wide range of broad-spectrum antibiotics. It is a disease that is associated with bad hygiene.

Navel-ill or joint-ill

These two terms describe the disease affecting calves where bacterial infection causes inflammation in the region of the joints. It usually occurs because the calves are managed under dirty conditions and are challenged by a serious burden of bacteria before the navel has healed.

Treatment may be undertaken successfully in the early stages of the disease with sulphonamides or broad-spectrum antibiotics; local application of dressings to the navel will help to prevent infection.

The health of growing cattle

Internal parasites

There are three major groups of diseases in cattle caused by parasitic worms. There is 'husk' or 'hoose', which is due to worms which live in the lungs; parasitic gastroenteritis, which is caused by roundworms which live in the abomasum and/or intestines of cattle; and finally liverfluke disease, caused by flatworms or flukes, which live in the liver.

Parasitic worms do not multiply within the body of the animal in which they live as adults. The adults produce millions of eggs but none reaches maturity without first passing out of the animal in the faeces. They must then have a period of time outside the animal before they become infective, and this period varies from a few days to many weeks. Thus every worm in the animal has been picked up in the herbage while the animal was grazing. The significance of this knowledge is that it provides for two ways of attack, one within the body and one without. Removal of contact of animals with their own faeces is a vital factor. If the husbandry is good, with a clean environment and no overcrowding, the infestation may be so slight as to be unimportant. Mixed grazing is also useful since the parasitic worms are host specific and infect only one species. It is also important to know that it is chiefly the young and growing stock that are vulnerable to infection. Adults should have a naturally produced immunity due to earlier contact with the parasites. Nevertheless, the adults often have a worm burden and will produce some eggs. And even with adults they can become badly affected if they become stressed by poor conditions or nutrition. The worm eggs once voided can remain infective for up to or even more than a year.

Husk or hoose

This is an infection of the bronchial tubes and leads to the cough known as 'husk' or 'hoose'. It is caused by the lungworm *Dictyocaulus viviparus*. The white worms are threadlike, up to 75 mm long, and the females lay vast numbers of eggs. After hatching they produce minute larvae which are carried in the mucus up the windpipe and to the mouth. They are then swallowed and pass out onto the pasture with the dung. These larvae have to develop on the ground before they are able to infect cattle who take them in with the grazing. The development takes about 5 d in warm, moist weather and over a month when the weather is cold. Whilst developing in the dung the larvae go through a series of changes and moult their skin twice. The larvae have little resistance to drying and can die off rapidly under these circumstances.

When the infective larvae are swallowed by grazing cattle, they penetrate the intestinal wall and pass into the lymphatic system and by this route reach the blood stream and then pass to the lungs. Here they develop into the adult worms and begin to lay eggs after three weeks.

Cattle that have been exposed to lungworm infestation will in due course acquire a resistance. If the level of pasture infestation rises at a moderate rate, the gradual and increasing resistance of the calf will enable it to reject the effect of the worms and no disease will actually occur. But if the pasture infestation rises rapidly, as it may do if the climatic conditions are favourable, disease may occur.

High pasture infestation may occur in a number of circumstances. In early spring most pastures will be either clean or at least only very slightly infested. Contamination of the pasture by older animals, which have carried over some worms in their lungs from the previous season, may produce more significant levels of infestation on the herbage and may even give rise to dangerous infestation if the season and pasture conditions are suitable. Low levels of herbage infestation may be increased when susceptible calves become infected on the pasture and in consequence increased numbers of larvae are passed on to it in their dung.

Symptoms
The disease may be recognized, or at least suspected, if there is an increase in the respiratory rate, coughing, especially when the animals are moved, and some loss in body condition. If cases become severe, the coughing becomes louder, the respirations shallow with clear pain and distress, and the animal stands uncomfortably with head and neck outstretched and the tongue protruding.

Control
There are a number of useful medicines for lungworms, such as Oxfendazole and Wermectic. Wherever possible it is best to remove the animals from the pasture and house them in good buildings where they can be well fed and warmly maintained.

Prevention
It is wise to keep animals grazing their first year off pasture which has had animals on it recently. Strict systems of rotational grazing can always be of help since they will prevent heavy build-up of disease. The calves should be moved twice a week on to ground which has not had calves on it for about four months but even a shorter time would be of help in reducing infection. The vaccine is also a great help in preventing the disease but it is an *adjunct* to good management and not a substitute.

Liverfluke disease

Liverfluke disease tends to go quite undetected or unsuspected until animals start dying. It is admittedly more a disease of sheep than cattle but there is good evidence that liverfluke disease causes a very serious effect on growth and productivity. The liverfluke is a flat worm about the size and shape of a privet leaf, which lives and feeds on the liver of cattle. The flukes lay many thousands of eggs which pass on to the pasture in the dung. The eggs hatch and produce a very small swimming creature which has to enter a particular type of snail if it is to survive. The parasite grows in the liver of the snail, producing many young liverflukes. These look like the flukes which cause disease in the older stock except that they are much smaller. The young flukes leave the snail when the snail is in water and encyst on any herbage which they contact. They lose their tails and surround themselves with a protective covering. There they remain as cysts but are inactive for long periods, up to months on occasion, until eaten by the grazing animal. When eaten the protective coat disappears and the fluke emerges and makes its way into the liver. During the next two to three months the fluke grows to maturity and then lays eggs.

Symptoms
There are two forms which are clinically differentiated. The acute disease occurs when the grazing animal picks up many flukes in a short time. These damage the liver seriously and this can lead to sudden death in the apparently healthy animal. The chronic form is more common. It is due to the presence of flukes in the liver over a period of months, and affected animals waste progressively.

Control
Certain drugs now exist for treatment, namely oxyclozanide, nitroxynil and albendazole. Also recommended are rafoxanide, netobimin and triclabendazole. All will remove more than 90% of adult flukes from the bile ducts but have variable efficiencies against the immature stages migrating from the liver. Professional advice should generally be sought before use of these drugs in view of possible side effects, contraindications and withdrawal periods of milk for human consumption.

It is also effective to prevent the disease by controlling the secondary hosts, the snails. Infestation in flukey areas is worse in the months of October to December and hence fluke-free pastures should be grazed in this period wherever possible. Badly drained areas should be properly drained. The snails may also be dealt with by treating the pasture with either copper sulphate at 12 kg/acre or sodium pentaclorphenate at 4 kg/acre. As these are dangerous materials they should be applied with due caution.

Parasitic gastroenteritis

This is a disease, or infestation, caused by the presence in the abomasum or fourth stomach or the anterior part of the small intestine, of a group of small roundworms. The most important species are *Ostertagia ostertagi* and *Trichostrongylus axei* in the abomasum; *Cooperia oncophora*, *C. punctata*, *T. vitrinus* and *T. colubriformis* in the small intestine and *Oesophagostomum radiatum* in the large intestine. They cause considerable economic loss from both poor productivity and overt disease, chiefly in young animals.

The life-cycle of the worms is 'direct', there being no intermediate host. Eggs are laid by the adult worms passing out with the dung on to the pasture where they can hatch within about 24 h, producing minute larvae which feed on bacteria in the faeces. Within the next few days they moult twice and only then are they capable of infesting stock who ingest them in the herbage. In order to make the ingestion more certain they leave the dung and pass on to the grass, climbing upwards so that they will readily be consumed by the animals. All this can take not more than 4 d in warm, damp weather but may take weeks in cool conditions.

A limited number of worms do little or no harm to the beasts but a heavy infestation can be lethal. Thus infestation is assisted by overcrowding, moist, warm conditions and long grass in which moisture more readily persists.

Symptoms
These are nearly always indefinite and the usual signs are a progressive loss of condition possibly leading eventually to emaciation and death. There is usually a severe scour but the animals continue to eat voraciously. However, one should be aware of the fact that some animals will not scour and signs will be so indefinite as to go undetected. In every case it is wise to have positive identification by examination of the faeces to determine the degree of infection.

Control
The first course of action is to consider medication and there are a wide variety of materials that can be given, e.g. organophosphorus compounds, such as fenbendazole, albendazole and thiabendazole; also levamisole and systamex oxfendazole.

Preventive measures should consist of the following.

- Overcrowding of the young stock should be avoided.
- Grass should be kept short as larval worms tend to perish in such grass.
- Only adult stock should be put on badly infested land.
- Periodic treatment should be given to animals after laboratory examination of faeces has established the degree of infection.

Skin conditions and external parasites

Ringworm

This is an extremely common skin disease of cattle which tends to be a problem in winter when the animals are housed. It is caused by fungi belonging to the genus *Trichophyton*, two species of which occur in cattle in the UK and also attack horses, pigs, dogs and humans. *Trichophyton verrucosum* is the most common one and is responsible for most cases of ringworm in cattle, farmers, stock keepers and their families.

The fungi that cause ringworm are capable of surviving for very long periods in the farmyard, certainly for over a year and probably much longer. The spores that are picked up by the animals find their way into cracks on the skin and germinate to produce fine filaments which flourish in the surface of skin and hair. These fungal threads grow downwards inside the hair keeping pace with its growth from the hair bulb and weakening it so that it snaps off at the surface of the skin producing the characteristic bald areas. The skin reacts to the infection by inflammation followed by the formation of a grey to yellow white crust. The lesions of ringworm are most often seen on the head and neck, especially in calves, but in adults patches can occur on the flanks, back and rump, wherever rubbing takes place. Animals that are rundown are likely to suffer a severe infestation.

Control

Whilst ringworm will disappear spontaneously in the course of time, it is not advised to ignore treatment since the disease has a serious debilitating effect. Treatment may be effected by removing the crusts with a soft wire brush and applying one of several proprietary preparations containing iodine or quaternary ammonium compounds, also a 2% formalin solution, handling with care, wearing rubber gloves and overalls. A much better treatment, though expensive, is an antibiotic, griseofulvin. It is more vital to practise good hygiene measures by efficient cleansing and disinfection of the buildings and all objects that could harbour the fungal spores.

Mange

Certain forms of mites infest cattle. These are *Sarcoptes scabiei*, *Chorioptes bovis*, *Psoroptes communis* and *Demodex bovis*. All are extremely contagious conditions (sometimes called scabies). The result of the disease is very patchy coats and areas of bare skin with an extreme amount of irritation.

The parasitic mites spend their whole lifetime on the animals and only survive a few weeks off them. The symptoms are always at their worst during the final winter months. Diagnosis is made by microscopic examination of skin scrapings. The mites are visible to the naked eye.

Control

Control is rather difficult and requires professional veterinary advice. Any incidence should firstly be identified and once this is done measures are advised to eradicate it from the herd. Amitraz (Taktic, Smith, Kline and Beecham) is applied as a spray and Bromocyclen (Hoechst) as a spray or suspension. All animals should be treated and in order to eradicate the condition it will be necessary to re-treat two or three times at 10 d intervals.

It is also necessary to keep the cattle away from those parts of the building that could harbour the mites and/or disinfect these areas as effectively as possible to prevent re-infestation.

Blackquarter (blackleg)

The disease of blackquarter is similar to so-called blackleg of sheep. It is caused by a bacterium *Clostridium chauvoei*, the spores of which are common inhabitants of the soil in areas where sheep and cattle have long been grazed.

The disease occurs because the organism is able to multiply in the muscles of the leg. The symptoms are stiffness and lameness with swellings appearing in areas such as the loin, buttocks or shoulder. The swellings are hot and painful at first but later become cold and painless. The skin over the affected muscles will become hard and stiff and if pressure is applied there is a papery effect, a dry, crackling noise due to the stiffness of the skin and the movement of the gas underneath. In fatal cases the condition of the animal rapidly deteriorates and it will die quite quickly. Initially there is probably a bruise or injury to the muscle and it is in this area that the clostridia can multiply in the absence of oxygen. Then they produce the toxins which have the serious destructive effect on tissues.

Animals affected with the disease can be treated successfully with antibiotics, only if the disease is recognized very early. Penicillin preparations are satisfactory. Prevention can be effected by using a vaccine but the best measures of prevention rely on good pasture management, avoiding areas of known susceptibility and draining or cultivating those that must be used.

Coccidiosis in cattle

Coccidiosis is a form of dysentery that primarily affects young stock in the summer and autumn months. It is caused by a microscopic unicellular parasitic organism known as a coccidium. The coccidial parasite forms spores, known as oocysts, that can live outside the animal for many months. Adults are frequently carriers of the coccidial parasites and show no symptoms yet spread the disease to younger animals.

The oocysts are very resistant to destruction and can often remain infective for a year but if they are swallowed by a susceptible animal they hatch and the active forms of the coccidia are liberated. These penetrate the cells lining the intestine where they mature and multiply and break down areas of the mucosa causing diarrhoea together with bleeding.

Animals acquire a resistance to coccidiosis by receiving a light infestation early in life and developing an immunity. However, if the intake of oocysts is sudden and too great at one time before a resistance has been produced, then disease occurs. This is most liable to happen when animals are densely stocked and the weather conditions are warm and wet. Dirty walls of pens or boxes, as well as the floor, are sources of infection.

The disease is controlled in a number of ways. Affected animals should be treated with a suitable drug, such as sulphonamide, plus plenty of really good nursing, especially fluids. Overstocking must be avoided: good bedding and cleanliness are important and animals should be kept in good thriving conditions so that they can readily resist such infection that occurs.

Bovine virus diarrhoea (mucosal disease)

Bovine virus diarrhoea (BVD) or mucosal disease (MD) is an infectious disease of cattle caused by a virus. It was first recognized in the USA but in recent times has occurred in virtually all parts of the world. Originally the two diseases BVD and MD were distinguished but it is now clear that they are caused by the same virus so the best term is probably the BVD-MD complex.

Symptoms

The disease causes ulceration of the mouth and lips accompanied by foul smelling diarrhoea. There is often lameness and extreme scurfiness of the skin. By far the majority of cases occur in cattle between 4 and 18 months of age. Most cases can recover spontaneously, and only exceptionally, if there are many young animals affected, the death rate can be high.

Treatment and control

There is no specific treatment but supportive therapy in the form of intestinal astringents and electrolytes can be of value.

Infectious bovine rhinotracheitis (IBR)

This is an acute respiratory disease of cattle caused by a virus and is spread by contact or by the airborne route. It is a very contagious disease and can spread through a herd within 7–10 d.

Symptoms

The symptoms are of severe inflammation of the eyes and the upper respiratory tract. The disease can also cause a substantial fall in milk yield, abortion and infertility. There may be a fever, loss of appetite and drooling of the saliva. Sometimes there is a pronounced cough.

Control

This is not too serious a disease in Great Britain. IBR is treated with broad-spectrum antibiotics to help reduce pyrexia or the effect of secondary invaders. A vaccine is available to protect the herd.

Respiratory diseases in cattle

The number of pathogenic organisms that can infect the respiratory system of cattle is enormous: there are some 20 bacteria, six fungi and four parasites that *commonly* infect them, not to specify the underlying primary damage that may be wrought by viruses. The wide involvement of so many agents has made the production of satisfactory vaccines most difficult and the use of medicines complicated and expensive. Thus the prime responsibility for dealing with such conditions rests with getting the environment correct.

The general principles are given in Table 22.2.

Specific advice for cattle may be summarized as follows:

(1) Depopulate between batches and institute a rigid disinfection programme.
(2) Reduce numbers in one environment to a minimum.
(3) Keep different ages and sizes of animals separate.
(4) Ensure copious ventilation without draught as cattle are hardy animals and do not require cossetting.

Table 22.2 Essential action to counter respiratory disease in cattle

Expert attention required for correct therapy
Improve ventilation – more air flow is usually needed with less draught
Check stocking density – are the cattle overcrowded?
Is more bedding required?
Separate and if possible isolate those cattle that are badly affected
Provide warmth for the very sick
Attempt separation of different ages
Reduce movement of animals to a minimum
Consider insulation and other essentials of construction relevant to the building
Can serum, vaccine or drugs be used to protect the affected animals?

Wooden tongue and lumpy jaw

These are two distinct diseases which have rather similar symptoms and are caused by unrelated organisms. They both run a slow course and are characterized by a steady enlargement of the tongue and swellings in various parts of the animal but especially on the head and neck. Their presence may generally interfere with the normal functions of the animal. They can also form abscesses or ulcerate, discharging into or outside the body. Untreated they cause a steady wasting of the animal so that it becomes unproductive and will eventually die.

The causes of the two diseases are quite different. Actinomycosis or lympy jaw is caused by a ray fungus known as *Actinomycetes bovis*, whereas actinobacillosis or wooden tongue is caused by a bacterium *Actinobacillus ligneresi*. The main difference between the two diseases is that actinobacillosis tends to affect the soft tissues, such as the tongue, and actinomycosis usually causes diseases of the bones, especially those of the jaw.

Symptoms

The tongue is most commonly affected and becomes hard and rather rigid, hence the name 'wooden tongue'. There is a constant dribbling of saliva and food tends to be rejected. The disease will also affect the glands of the neck and swellings appear between the angles of the jaw. If this swelling is allowed to continue, it will affect breathing, and swallowing also becomes difficult. The lesions may burst and release a characteristic yellow and granular pus. In addition the bones of the jaw are often affected. Pus forms within the bone and these disease areas can eventually break through the skin to form an unpleasant discharging fistula. These are, however, the external signs of the diseases. They may also affect the internal organs and will then cause a progressive wasting.

Control and treatment

Other than in advanced cases, treatment can be very successful. If the disease does seem to have got a firm hold on the animal which is fast losing condition, it is probably best to slaughter it. *Actinobacillus* responds quite well to iodides and sulphonamides, and actinomycosis will be suitably treated with penicillin or broad-spectrum antibiotics.

Redwater

Redwater, also called bovine piroplasmosis or babesiasis, is a curious condition caused by the presence in the blood of minute parasites called piroplasms. These parasites live inside the red blood corpuscles where they can be seen in the early stages of the disease. It is a condition that is found in certain areas, for example in Wales, southwest England and East Anglia.

The organisms are especially associated with rough scrubland, heathland or woodlands where the ticks *Ixodes ricinus* are present as they transmit the organisms from animal to animal. The tick is infected by sucking the blood of an infected animal. Once the tick is engorged with blood it falls off and lies in the herbage. Later the tick may attach itself to other cattle and when it sucks from these it introduces the parasites into the blood. The two worst periods for redwater disease are from March to June and October to November.

Symptoms

The main effects of the disease are to cause loss of condition and a considerable loss of milk yield in the dairy cow, together with blood pigments in the urine. Animals affected also show a high fever and diarrhoea and later constipation. The animal also often becomes jaundiced, the mucous membranes becoming yellow. The heart beat becomes very loud and there are tremors of the muscles of the shoulder and legs. It should be noted that young cattle are less susceptible than adults which have not previously been infected. If contracted early in life, the disease is only mild and the animal is then immune against further infection. This indicates one of the great dangers; if adult cattle from a tick-free area are introduced into a redwater area, some of these may contract the infection and die quite rapidly without treatment.

Control

Treatment of infected animals will be twofold. First, the animal will require an early injection with one of a number of chemotherapeutic agents that are lethal to the parasite. Secondly, careful nursing is perhaps even more important. Feed well, house comfortably, do not drive or excite for fear of heart failure, and keep the animal warm.

Prevention

The disease could be eradicated if ticks were eliminated. The tick is extremely resistant to most measures of control. In addition it can feed off several species of animals and is not dependent on cattle alone. Sometimes sheep are put on pasture as 'tick collectors'. By placing them alone on the pasture at the worst tick seasons of the year and then adopting a heavy dipping programme, this will reduce numbers considerably. The only effective way of eliminating ticks is to clear the rough ground by cultivation and normal crop rotation.

Wherever this procedure is adopted and a part of the farm is freed from ticks, the importance of exposing young animals to tick infestation must be considered otherwise they will be highly susceptible as adults to a first infection. Great care must be taken not to put them on to tick infested ground for the first time in adulthood.

Diseases of mature cattle

Metabolic disorders

Bloat

In bloat, also known as blown, hoven or tympanitis, the rumen becomes distended with gas and the pressure from this on the diaphragm may lead to the animal dying from asphyxia or shock. Two types of bloat occur. The first is where the gases separate from the contents of the lower part of the rumen, and the second where the gases remain as small bubbles mixed with the contents in a foamy mass, giving rise to what is known as 'frothy bloat'.

Bloat most often occurs in the grazing of lush pastures which contain a high proportion of clover. It may be due to the fact that some kinds of herbage contain substances that may cause paralysis of the rumen muscles and of other organs concerned in the process of rumination. Or it may be due to the fact that more gas is produced than can be got rid of by eructation. Bloat may also be due to the result of foaming in the rumen caused by saponin and protein substances obtained from plants.

Symptoms Symptoms are obvious distension of the abdomen which is particularly pronounced on the left flank between the last rib and the hip bone. In the early stages the animal becomes uneasy, moves from one foot to another, switches its tail, and occasionally kicks at the abdomen. Breathing becomes very distressed and rapid. If the attack is a severe one, any movement intensifies the discomfort and so the animal stands quietly with legs wide apart. If relief is not given very quickly, death from suffocation or exhaustion can follow rapidly.

Control It is advisable to summon a veterinary surgeon as soon as possible. If time allows a stomach tube may be inserted into the rumen to allow the gas out. In acute cases the only remedy is to puncture the rumen on the left flank with the surgical apparatus, the trocar and cannula. In an emergency, where death seems imminent and specialized assistance is not available, the only course of action is to puncture the area with a short pointed knife with a blade which is at least 6 inches long, the blade being plunged into the middle of the swelling on the left side and twisted at right angles to the cut to assist the gas to escape.

Although this treatment will serve to give immediate relief, more treatment is required if the case is one of frothy bloat. Drenching with one of the substances known as silicones, or with an ounce of oil of turpentine in a pint of linseed oil, often helps in the less severe cases and some gentle exercising may help to relieve the pressure of the gas in the rumen. There are also a number of drug treatments available, such as Avlinox, which is an ethylene oxide derivative of ricinoleic acid.

Prevention Efficient pasture management can help to avoid the trouble. After a pasture is closely grazed and then rested, the clover recovers more quickly than the grasses so that when pastures are used intensively the clover becomes dominant in the sward. However, it is generally considered that if there is less than 50% of clover present, it is usually safe. A variety of seed mixtures have been tried to provide a suitable balance between clover and grass: one which has given promising results contains tall fescue. After it has been closely grazed this species of grass recovers almost as quickly as the clover and maintains a reasonably safe balance between grass and clover.

Another system which has been advocated is to allow the herd to graze a potentially dangerous herbage for a limited period and then turn it on to an old pasture for a time.

Controlled grazing can contribute to the prevention of bloat by ensuring that the grazing animal eats both leaf and stem, whether clover or grass, and that it cannot select clover or clover leaf which is thought to be more dangerous. If an electric fence is moved forward so that the cows have access to a succession of narrow strips of pasture during the course of the day, bloat may be reduced or eliminated but it may reduce the intake of the feed and milk yield.

The second method of controlling grazing is to mow strips of herbage, then allow it to wilt and let the herd graze it as it lies on the ground. Alternatively, the wilted herbage can be carted to the cattle and fed indoors. The feeding of roughage in the form of hay or oat straw shortly before the grazing of dangerous pasture has also been practised for many years. It is only completely reliable if the roughage is fed overnight.

Milk fever

Milk fever occurs principally in cows after calving, usually in dairy breeds of high milk yielding capacity. It is given various synonyms, such as hypocalcaemia, parturient hypocalcaemia, parturient paresis and parturient apoplexy. There is, in fact, no fever and indeed it is more usual for the cows to show subnormal temperatures.

Symptoms Usually within 12–72 h of calving there is a short period of uneasiness with paddling of the hind legs, swishing of the tail and convulsive movements, soon followed by depressed consciousness and paralysis. The animal usually falls on its side with its legs extended, rolling its eyes and breathing heavily. The characteristic picture is a twisting of the neck as it lies on one side, with spasm of the neck muscles, shallow laboured breathing, grunting, cessation of rumination and dryness of the muzzle. In straightforward milk fever the calcium and phosphate levels fall and the magnesium levels rise.

In practice, it is quite difficult to distinguish between the simple hypocalcaemia and the more complicated metabolic disorder where there are a number of deficiencies which have to be corrected.

Background to occurrence There does not appear to be any real breed incidence but it is usually well fed, high-yielding cows which are affected and the severity of the condition increases with age. The cause is usually ascribed to a temporary failure in the physiological mechanism which controls calcium levels in the blood. The stress of calving and the physical adjustment necessary for lactation will upset the balance of chemical regulators (hormones) produced by various glands. If the cow does not adjust itself quickly enough, it may suffer an attack of parturient hypocalcaemia.

Control The immediate administration of a solution of calcium borogluconate is effective in curing uncomplicated milk fever. It may be given intravenously for a quick response or subcutaneously for a slower effect. In addition, the cow should be propped up in the box by means of straw bales on each side to prevent it from becoming 'blown', and to prevent regurgitation of ruminant contents. Further injections of calcium may be required and in addition, in complicated cases, there will be the need for administration of phosphorus injections and even other minerals. Professional advice is certainly required for all but the most simple case of hypocalcaemia.

Injections of vitamin D increase the mobilization of calcium in the body and thereby may help to prevent levels falling too low.

Ketosis (acetonaemia)

Ketosis is a disease involving ketones which are chemical compounds which may be produced during the metabolism of fat. This condition may occur when an animal's food intake is inadequate for its needs and it is drawing on its own reserves in considerable amounts to make good the deficiency. The usual form of ketosis is ketonaemia which includes an excess of ketones in the blood. Since acetone is the simplest and most characteristic of the ketones, the disease is generally termed acetonaemia. It is also called 'post-parturient dyspepsia' as it often occurs soon after calving and is associated with indigestion. It often follows mild attacks of milk fever and almost appears to be a subnormally slow recovery, hence the traditional term for the disease of 'slow fever'.

Symptoms The symptoms of ketosis are usually seen within two weeks of calving and in well nourished stock of high milking capacity. But it may also occur much later than this, even several months after calving. It is most common during the tail-end of winter. Affected cows appear listless, and show a marked loss of condition, and there is a dull coat and a sickly sweet smell of acetone in the breath and also in the urine and milk. The milk production falls away quickly and the milk becomes tainted and undrinkable. Constipation with dark, mucus-covered dung is usual. Sometimes there is a licking mania, a rather wild looking appearance and champing of the jaws and salivation, and there may also be hyperexcitability.

Control First, tests are carried out to confirm the presence of ketones in the blood. It will be usual for the blood sugar to be low and injections of glucose are an early procedure. Corticosteroids may be administered to give a longer term beneficial effect. Other useful treatments are glycerine and preparations containing sodium propionate. Molasses and/or molassine meal are also helpful.

Prevention will depend principally on skilful management. The energy provided in the food must keep pace with the draining demands of production, especially in early lactation. In late pregnancy the cow should be kept in good condition but not over-fat. The roughage in the ration should also be carefully selected and kept up to a good level or at least reduced slowly if the cow's appetite and enthusiasm for a balanced diet are to be maintained.

Hypomagnesaemia
The amount of magnesium in the blood of normal animals keeps within a fairly constant range. However, should the magnesium level of the blood drop below this range, then the condition known as hypomagnesaemia, or grass tetany, can result. The condition occurs especially in areas of cooler climate where the pasture has been improved and production increased. In adult cattle it is called by various names, such as 'grass tetany', 'grass staggers', 'lactation tetany' and 'Hereford disease', whilst in calves it is known as 'milk tetany' or 'calf tetany'.

Underlying causes The occurrence of this disease is in the period of sudden change of lactating cows from winter housed conditions to those of the rapidly sown spring grass. But it is now known to occur in cattle under most feeding and management regimes, including winter feeding periods. It is not due to a simple deficiency of magnesium, for sometimes the condition occurs in cattle on pasture which has a very adequate magnesium content. Underfeeding appears to be a predisposing cause, especially in outwintered beef cattle, but overall the disease is due to physiological dysfunctioning that interferes with the absorption and utilization of the magnesium which is in the food.

Symptoms Symptoms vary according to the intensity of the attack. The first signs are often nervousness, restlessness, loss of appetite, twitching of the muscles, especially of the face and eyes, grinding of the teeth and, quite soon after, staggering. In less acute cases animals which are normally placid become nervous or even fierce. In severe cases paralysis and convulsions develop very soon after the onset of symptoms and if treatment is not given death can follow very quickly. In milking cows a sudden reduction in milk yield may occur just before an attack. Sometimes the attack comes on so rapidly that no symptoms are seen, merely a dead beast. There is also a chronic form of the disease in which cows show a gradual loss of condition although appetite and even milk yield do not show any drop. This chronic state can last several weeks and then develop into the more acute condition.

Control The veterinary surgeon will inject magnesium plus other solutions into the affected animal to cure the condition. To prevent the disease supplements of magnesium will be used. Two ounces (60 g) of magnesium oxide/head/d has been used for many years successfully in dosing adults to prevent the condition. The magnesium oxide is usually given as calcined magnesite in a granular form. This must be given daily to cattle under risk or blood magnesium levels can fall quickly.

The use on pastures of magnesium-rich fertilisers has been quite successful in preventing the disease. On some soils top dressing in January or February with 500 kg of calcined magnesite per acre can be a perfect preventative. Even 250 kg may be sufficient. Another method of preventing the disease is the magnesium 'bullet' in the reticulum with its slow release of magnesium, and attempts have also been made to give multi-vitamin injections to increase the absorption and utilization of magnesium.

Calf tetany
Many years ago it was established that calves could not be reared to maturity on milk alone because of the onset of deficiency conditions and in particular hypomagnesaemia or calf tetany during growth. But the condition does also occur under other conditions of feeding and it is an increasing problem in beef calves, especially if they are grazing at the time of the most active growth of spring grass.

The symptoms in calves are similar to those in older animals: hyperexcitability, irritability and twitching of the muscles. The walk often becomes spastic, the feet being carried well above the usual height. Thereafter in the progress of the disease convulsions may follow and eventually death.

Treatment consists of immediate administration of suitable magnesium injections. Prevention is effected by administering about 15 g ($\frac{1}{2}$ oz) of calcined magnesite daily, or 30 g (1 oz) of magnesium carbonate daily.

Mastitis

Mastitis means literally inflammation of the mammary gland or udder. It has always been an important disease of dairy cattle with potentially disastrous consequences but with the advent of antibiotic treatments there was a general optimism that the days of the disease were numbered. Regrettably this has not been the case, for organisms that cause mastitis have become increasingly resistant to the constant use or, perhaps more correctly, misuse and abuse, of antibiotic therapy. Also there has been a growth of large dairy cattle units with the greater opportunity for disease build-up and cross infection. Cow cleanliness has become less satisfactory and some of the most modern practices lead to dirtier cows and udders, for example, with loose housing and cubicles. There has also been the emergence of different forms of mastitis, called environmental because of their association with the newer forms of husbandry and the rather different balance of organisms involved.

Mastitis may affect as many as 50% of the cows in the National Herd in one form or another, many cases being subclinical and so not easily recognizable. Since a cow with mastitis may *readily* lose a quarter of its milk output, there is no doubt of the extreme economic importance of this condition. Whenever mastitis occurs it must be looked upon as a herd problem and should be dealt with urgently and thoroughly. Even a single cow may be a warning of a managemental error and should be carefully considered in this respect.

Subclinical mastitis means that the udder is affected mildy but the cow's health or even the milk yield are not obviously affected. It can be recognized by *cell counts*. If an inflammatory process is proceeding, then the milk will contain more white blood cells than normal. The best way for the farmer to deal with this is to arrange for a regular monitoring of the cells in the milk. The range of 'counts', with the significance given, is shown in Table 22.3.

The milking machine
A major predisposing cause of mastitis is the misapplication of milking machines, together with bad maintenance and poor hygiene routine during milking. One of the most serious of these is overmilking, i.e. the milking machine is left on much longer than it should be causing considerable damage to the delicate mucous membranes lining the teat canal allowing pathogenic micro-organisms to invade so that mastitis results. Teat liners must be chosen of the right size and of gentle fit. They must never be slack and should be carefully maintained to prevent them from causing damage. Pulsations of the machine must be maintained between 40 and 60/min; if the rate increases, squeezing may fail and mastitis is more likely to occur. The ratio between the release and squeeze phase should be between 2:1 and 1:1.

The pump needs to be maintained properly as it 'ages' or a surge of milk may result around the teats, such circumstances also being a predisposing cause of mastitis. The vacuum regulator prevents fluctuation in vacuum which can be very damaging to the udder and teats for, if it fails, a surge of milk results around the teat and clusters may fall off or, if it rises, it damages the teats causing extrusion of the lining of the teat canal.

Types of mastitis
The easily recognizable forms of clinical mastitis are the acute, sub-acute and chronic forms.

Acute mastitis The cow is obviously ill, feverish, fast breathing and depressed and there is no cudding. Examination of the udder will show that one or more quarters are tense, swollen, painful and possibly discoloured (blue). It may be possible to withdraw from the affected quarter(s) a small quantity of grossly abnormal fluid. Gangrene may set in causing the destruction of part of the udder. This 'acute' mastitis is often the same as 'summer mastitis' which, whilst it is most common in dry cows and heifers in the summer months, can occur in cows at other times. Such mastitis is caused primarily by a bacterium, *Corynebacterium pyogenes*, and must be treated vigorously with antibiotics and sulpha drugs. Because of its common association with 'dry' cows its incidence may be reduced by using long-acting antibiotic therapy as soon as the cow is dried off.

Sub-acute mastitis The symptoms are somewhat similar but milder and slower in their progress. The first sign of disease may be the appearance of small clots in the milk with some increased difficulty in extracting it. Pain also gradually increases and the affected quarter(s) swell. The milk eventually becomes yellowish and much decreased in amount.

Table 22.3 The significance of cell counts

Cell count ranges (cells/ml)	*Estimate of mastitis incidence in a herd*	*Estimate of milk (litres) production loss per cow*
Below 250 000	Negligible	–
250 000–499 000	Slight	200
500 000–749 000	Average	350
750 000–999 000	Bad	720
1 000 000 and over	Very bad	900

Chronic mastitis In this form of mastitis there is no pain and any general sickness in the cow is unlikely but there is a gradual hardening of the udder tissue and a decrease in the amount of milk.

Causes of mastitis
At one time the most frequent occurring bacterium was *Streptococcus agalactiae* but this has generally decreased in incidence with the advent of successful antibiotic treatment. Other streptococci, e.g. *S. dysgalactia* and *S. uberi*, are common and also *Staphylococcus* and *Corynebacterium*. There are, however, some serious new forms of mastitis that are of comparatively recent origin and are most difficult to treat.

Escherichia coli is usually associated with the so-called environmental mastitis. Also *Pseudomonas* is a common bacterial cause of mastitis and *Corynebacterium bovis*. Mastitis may also be caused by micro-organisms other than bacteria, such as mycoplasma (*M. agalactiae*) and moulds, the latter being termed mycotic mastitis.

Leptospira infections in cattle are also causing considerable concern, especially as such infections can cause serious illness in man. The most common organism is *Leptospira hardjo*, responsible for the so-called milk drop syndrome, and also a cause of abortion. The milk drop appears initially as a mastitis usually in all four quarters, falling almost to zero within 24 hours. The condition is accompanied by a high fever and the milk is thickened or clotted or even bloody. A very high percentage of cattle, as many as 50% of cows in a herd, may be affected.

The condition can be successfully treated by broad-spectrum antibiotics and can be prevented with a vaccine which needs annual administration.

Control of mastitis
The view is now taken that the control of mastitis must depend largely on the application of hygienic criteria of the highest standard. These methods are summarized in Table 22.4.

Disease problems associated with breeding in cattle

Lowered fertility results in lowered production and causes great economic loss. However, infertility is not a disease as such; it is only the symptom of one. It may be due to any condition that prevents a vigorous sperm from fertilizing a healthy ovum to produce a robust calf. The process involved is complicated and the different causes of infertility are numerous. Consider the whole physiological process. The bull must produce healthy sperm only, but the cow must not only produce a healthy egg but also provide the ideal condition for the fertilization of the ovum by the sperm and the establishment of the embryo in the uterus. The cow then has to provide the nutrients for the calf and then, when it is mature, she must expel the calf from the uterus.

Infertility has probably become more serious in recent years because it tends to be associated with cows giving an immensely good milk yield.

Table 22.4 Summary of measures aimed to prevent mastitis in the dairy cow

Basic design of cow accommodation must be right to ensure cow's bed is dry and clean and to minimize the chance of injury to udder
Replace fouled bedding frequently in stalls, yard or cubicles
Test milking machine regularly, check vacuum pressures, pulsation rates, air bleeds and liners daily
Monitor cell counts in milk. Keep detailed records for frequent reference
Teat-dip always to be used after milking. Use a sanitizing mixture with an emollient such as lanolin
Wash udders before milking with clean running water and dry with disposable towels. Milker should preferably wear smooth rubber gloves
Ensure early diagnosis by use of 'fore-milk' cup
Treat all cases of mastitis promptly
Cull chronic cases of mastitis
Correct laboratory diagnosis of causal organisms assists in taking proper remedial measures
Treat cows as they dry off with a long-acting antibiotic

Causes of infertility
Fundamentally the breeding potential of an animal depends on certain inherited genetic tendencies and the presence of any inherited abnormalities. Such genes may cause the animals either to be infertile or to produce faulty calves. However, some of the most important causes of infertility are bacterial or other infectious agents. Such an agent may be a specific organism producing a condition of which infertility is only one symptom. An example of this is contagious abortion (brucellosis), a disease which has now been virtually eradicated from the UK. Another example is trichomoniasis, the causal organism being *Trichomonas foetus*, which is spread at service by infected bulls which themselves show no obvious evidence of infection. A third example, vibriosis, is due to the organism *Vibrio foetus* and is carried and transmitted by apparently normal bulls. Both trichomonad and vibrio organisms multiply in the uterus and cause the death and abortion of the fetus. Infertility may also be caused by non-specific infections of the genital tract.

Control When a cow or heifer fails to conceive to a second service, it should be examined by a veterinary surgeon who, by carrying out tests, can establish if there is an infectious cause. If there is, it may well be treatable. Prevention depends on a number of factors. The herd should be kept as self-contained as possible and only maiden heifers and unused bulls should come into the herd. There should be no exchange of breeding animals. The diet must be adequate and balanced.

Bovine spongiform encephalopathy (BSE) or mad cow disease
This is a recently recognized progressive neurological disorder affecting adult cattle, mainly dairy cattle, in mainland Britain, Orkney, the Channel Islands, Northern Ireland, Saudi Arabia, France and Switzerland. Sporadic cases have also occurred in other countries.

Symptoms BSE is characterized clinically by behavioural changes and pathologically by vaccination of the neuropil and some nuclei of the brain stem. The pathological changes resemble those of certain other neurological disorders of animals and humans, collectively known as the transmissible spongiform encephalopathies and classified among the 'slow virus diseases'. It affects cattle of 3–6 years of age. The initial signs are subtle and include hyperaesthema, persistent grinding of the teeth, and repetitive, agitated, purposeless movements of the head and limbs. Affected animals become increasingly apprehensive, and may become aggressive when approached. A sudden noise may result in the animal collapsing into a tetanic spasm. The course of the disease varies between two weeks and several months. A presumptive diagnosis can be made on the clinical signs although in the early stages of the disease it may be confused with acetonaemia, hypomagnesaemia or listerial encephalitis. Confirmation of diagnosis can be made on a neuropathological examination.

Cause The agent causing BSE is not firmly established. It appears to be an infective particle, sometimes suggested to be a novel kind of infectious protein or 'prion', which is highly resistant to physical or chemical disinfection, such as boiling, exposure to formalin or ultraviolet radiation. The agent appears to have a close similarity to the agent causing scrapie, and evidence shows that the disease arose in cattle after the feeding of animal protein containing nervous tissue from scrapie-infected sheep which had been insufficiently sterilized. Animals so infected would have received massive doses of the infectious agent.

In the UK all suspicious cases in cattle are slaughtered after confirmation. The evidence is that there is very little vertical transmission from cow to calf and reported cases are now falling very significantly. No animal protein is fed to ruminants, and the brain and spinal cord are removed from beasts at the slaughterhouse and destroyed. Thus there is reasonable hope that the disease in cattle will be eliminated. Fears that BSE may be transmitted to man are not justified by the evidence.

Diseases affecting cattle of all ages

Foot-and-mouth disease

Foot-and-mouth disease is an acutely contagious disease which causes fever in cattle and all cloven hoofed animals, followed by the development of blisters or vesicles, which arise chiefly on the mouth and feet. The disease is an infection, world-wide in incidence, caused by at least seven major variants of the foot-and-mouth virus. Each type produces the same symptoms and they can only be identified by specialist examination. The disease is notoriously contagious, perhaps more so than any other disease, and it is known that it can spread some 50 miles down-wind from one outbreak to another. It has a fairly short incubation period of about 3–6 d so it can spread rapidly throughout a susceptible population.

The disease does not usually cause deaths, except in the young, but it leads to damaging after-effects. It will seriously lower productivity in all animals, especially the milking cow. Mastitis will possibly develop so milk production is permanently impaired and the infection of the hooves may cause lameness that leads to secondary infections.

Infection with the virus of foot-and-mouth disease can occur in many ways other than by wind-borne spread.

People, lorries, wild birds and other animals, wild life such as hedgehogs, markets and so on are just some of the ways in which the disease can spread. In addition, infective material can come in from abroad in imported substances such as hay, and meat and bones which have not been sterilized.

In this country foot-and-mouth disease is a notifiable disease and has long been dealt with by immediate slaughter of infected animals. This has been a very successful policy and has meant there have been few outbreaks. In fact, the last big outbreak occurred in 1967 and there have been just two small infections since then. The important thing is to be sure of recognizing a case as soon as possible so all the official measures can be taken to control the disease. Symptoms that would lead one to suspect foot-and-mouth disease are described below.

Infected cows suffer from fever and anorexia with a sudden drop in milk yield and there are blisters on the upper surface of the tongue and the balls of the heels. Animals prefer to lie down but when they are forced to walk they move painfully, occasionally shaking their feet. The lameness then becomes worse so the animal can barely move. Blisters also develop on the teats so that the cow cannot be milked.

Rabies

Rabies is a notifiable virus disease which affects nearly all animals and is especially dangerous to humans. It causes progressive paralysis and madness in most animals. The virus is present in the saliva so that the animals affected may bite one another, or humans, and in this way cause infection to spread. It is nearly always fatal. The greatest danger presented by this disease is that infection becomes endemic – virtually permanent – in wild animals, particularly foxes but also is badgers, deer and vampire bats, and these infect dogs, cats and farm animals. In this way is the particular danger to humans created.

The disease may be prevented in humans and animals by vaccination but it should be emphasized that no vaccinations are 100% effective and it is difficult to exaggerate the ghastly effects of this disease in the human.

One of the features of this virus is that after an animal becomes infected, perhaps via the saliva by being bitten by another rabid animal, several months may elapse before any signs of the disease are seen.

Symptoms

Symptoms in animals may show great variation, ranging from the classic mad, biting, salivating beast, to 'dumb' forms in which the animals are incoordinate, progressively paralysed and make no noise. Usually the 'dumb' form follows the mad stage.

Control

The British Isles have been free of rabies outside quarantine for 60 years, other than two cases in dogs which were quickly dealt with. Our rigid regulations prohibiting the importation of dogs and cats and certain other mammals without a six month quarantine period which includes vaccination is thoroughly justified. The fear remains that animals incubating the disease may be smuggled in, or could jump off a boat, and thereby rabies could become endemic in the wild animal population. This danger has increased with the spread of rabies in areas of Europe near the UK. In those countries where the disease is endemic, great strides are now possible towards eradicating rabies with the aid of the greatly improved vaccines now available.

Tuberculosis in cattle

Tuberculosis is a chronic infectious disease affecting virtually all species of animal and also humans and birds. It is due to a bacterium, *Mycobacterium tuberculosis*. There are three main strains of this organism. There is the human strain which affects primarily the human but can also affect cattle; the cattle (bovine) strain which is most prevalent in cattle but also affects humans, pigs and certain types of wild life; and finally we must be aware of the avian strains which primarily affect birds but also cattle, pigs and other animals at times. All types of cattle may be affected but it is in the dairy cow that the main risk occurs. The milk from an infected dairy cow may contain the organism that causes infection and this may infect calves or human subjects if the milk is not pasteurized or sterilized. Also, in advanced cases of tuberculosis, the uterus may become infected so that the calf is born with the disease.

Tuberculosis is essentially a slow, inflammatory action which produces almost anywhere in the organs of the body nodular swellings, known as tubercles, which are of fibrous tissue with a core of pus-like caseous (cheesy) material. Fortunately it is possible to test an animal for the presence of the disease long before it has reached the contagious stage.

Tuberculosis has been virtually eradicated from the UK though testing continues and some cases do occur. Animals may be infected by humans and wild life and there is, for example, much concern about the infection of cattle by badgers which, in some areas, are serious reservoirs of the disease. Those same measures that have generally been successful in eliminating tuberculosis in the UK are also being applied in a similar fashion elsewhere.

Anthrax

Anthrax is a serious bacterial disease which infects cattle, other animals and humans. It is rare to see symptoms in cattle as infection is so acute that it usually causes sudden death. The cause of the disease is a bacterium, *Bacillus anthracis*, which may persist in the soil of a farm, once present, almost indefinitely. However, few cases occur in cattle in the UK, around 50 a year, because of the means taken to prevent its spread and re-infection.

Anthrax should be suspected if an animal was found dead or there was a very high fever and swollen neck. At either point professional help should be sought. Diagnosis would be made by the veterinary surgeon taking a very small quantity of blood and examining this under the microscope. If the animal died from anthrax, or is infected with it, the blood will be teeming with the rod-shaped organisms that are distinctive features of this disease.

A major warning: do not allow the carcass to be cut or air to reach the blood since it is then that the organisms form spores which can survive for many years.

Some blood may be oozing from parts of the animal and this must be treated with disinfectant. Anthrax is a notifiable disease and measures that must be taken are mandatory. The carcass must be burned or buried deeply so that it is removed from any opportunity to cause re-infection or for the organisms to form spores. The area around which the animal was found dead should be disinfected and isolated for a short while in case of spillage of other organisms. It is most likely that the source of infection is from a feedstuff, possibly meat or bone meal, which has come from imported infected material.

Enzootic bovine leucosis (*EBL*)

This is a comparative newcomer to the list of notifiable diseases. It is a virus disease, slow and insidious in its effect, which may have been imported with Canadian Holstein cattle. It causes multiple tumours, known as lymphosarcomas, and whilst it is certainly not widely present in the UK, its presence has been confirmed. The symptoms in the live and adult animal are chronic ill health, anaemia, weakness and inappetance, which are not diagnostic. Only special laboratory examinations will give a definite diagnosis.

External parasites

Warble fly

There are two species of cattle warble fly that cause trouble, *Hypoderma lineatum* and *H. bovis*. Both are rather like bumble bees in appearance as their dark bodies are covered with white, yellow and black hairs. They have no mouths and only live for a few days. They have been found almost everywhere in the British Isles and most other parts of the world.

Adult female flies cause substantial losses to cattlemen, being responsible for 'gadding' during the spring and summer when the cattle can be so panicked by the approach of the flies that they may stampede in the fields. Such stampeding lowers milk yield and may cause serious injuries to the beasts and even broken limbs. There is also hide damage and this is by far the worst loss. The maggots of the fly make breathing holes and these can ruin parts of the hide, even after the maggot has left the hide and the hole has healed over with fibrous tissue: this remains a weak area in the leather so that it has a much reduced value. It also damages the meat itself as a yellowish jelly-like substance forms around each maggot and spoils the appearance of the meat.

The life-cycle of the warble fly is as follows. During the short life of the fly the female may lay some 50 or so eggs which are attached to hair on the legs and bellies of cattle, those of *H. bovis* being laid singly (like louse eggs) and those of *H. lineatum* in rows of up to about 20. In 4–5 d the eggs hatch and tiny maggots emerge, crawl down the hair and burrow through the skin of the beast by means of sharp mouth hooks.

The entry of the maggot causes some injury and it shows as a small pimple or scab. Under the skin the maggot moves along, feeding and growing. It takes seven to eight months to reach the skin of the back of the animal. When this point is reached the maggots make breathing holes and become isolated within a small abscess. It stays there for five to seven weeks, casting its skin twice and growing to a length of about 1 inch. The full grown maggot is dark brown, fleshy, barrel-shaped and covered with groups of tiny spines. When fully ripe they squeeze themselves out of their 'warbles' and fall to the ground. Here they slide into shallow crevices in the soil and pass underneath dead vegetation. The skin of the maggot darkens and it then becomes a puparium or chrysalis inside which the warble fly is formed. This usually takes about four weeks.

Control must be achieved by dealing with the maggot. These can be destroyed by treating the animal with one of the more recently produced systemic insecticides.

Warble fly is a notifiable disease. The disease had recently been eradicated from the UK but has been reintroduced in recent years due to the importation of infected cattle. Any suspicions must be reported to the divisional veterinary officer to the ministry who will supervise treatment of all cattle over 12 weeks old on the premises. The disease is nearly eradicated from the UK and only 34 cases occurred in 1986. Great care is needed in the use of the systemic insecticides as they are organophosphorus compounds and are given by pouring them on the backs of the animals or by injection. Sick animals must not be treated. Milking cows must be treated immediately after milking to allow at least 6 h before the next milking. The approved 'withdrawal' period – which is on the label of the product used – will specify that no animal may be slaughtered for meat under 14–21 d from treatment.

Lice

Lice infestation is a major problem only in cattle when they are housed. The thick winter coat, lack of sufficient air and sunlight and possibly poor nutrition can contribute to the occurrence of this problem. There are four species of lice infesting cattle, three being sucking lice and the fourth which is also the commonest, a biting louse. The sucking lice are *Haematopinus eurysternus*, *Linognathus vituli* and *Solenoptes capillatus*, and the biting louse is *Damalinia bovis*. The sucking lice cause the greater damage and irritation as their sharp mouthparts pierce the skin to draw blood on which they feed. The biting louse feeds on the scales of the skin and on the discharge from existing small wounds.

These four types of louse affect only cattle and cannot exist on other farm animals. The entire life-cycle takes place on the skin of the animals where the females lay eggs in small groups. They are attached to the base of the hair by a sticky secretion. There the eggs remain for up to three weeks after which the live young hatch out and very quickly start feeding and reproducing. Reproduction lasts for about five weeks; each female is able to lay about 24 eggs and if conditions are favourable the number of lice can increase very rapidly. Lice cannot survive for more than 3–4 d away from cattle so that transfer of infections is largely by direct contact.

It is not too difficult to recognize an infestation with lice. They tend to congregate on the animal's shoulders, the base of the neck, and the head and root of the tail, and examination of these areas will show them to be extremely scurfy and the lice should be obvious. As the infestation worsens more of the body can become affected. The affected animals rub and scratch themselves, greatly adding to the damage. The hair goes altogether in some areas. Rubbing thickens the skin which often breaks, so that bacterial infection gets in. All the effects of a bad infestation are ultimately to cause much loss of condition and production and should not be tolerated.

Insecticides must be applied to kill the lice. Materials

that are satisfactory include organophosphorus preparations. Liquid preparations can be applied either under pressure or with a scrubbing brush. As these preparations have a reasonable residual effect, it is feasible for one good application to kill the lice and the larvae that hatch out after the application.

After treatment, measures must be taken to prevent reinfestation by contact with untreated animals and also by putting in fresh bedding and probably cleaning equipment.

Never under-estimate the damage that can be done by lice both as mechanical irritators and as vectors of any livestock disease.

Foul-in-the-foot

Foul-in-the-foot is an infection with *Fusibacterium necrophorum* between the claws of the feet of cattle causing pus formation and necrosis. The organisms enter only through the broken skin, often caused by sharp stones. Cattle become lame and productivity can be severely affected. Treatment consists of trimming, cleaning and dressing the affected feet and giving an injection of a sulpha drug or antibiotic. As a routine preventive where the infection is widespread in a herd, the cattle may be routinely walked through a foot-bath containing 5–10% copper sulphate or 5% formalin.

A vaccine is now available for sheep with foot-rot and it is expected to be available soon for cattle with foul-in-the-foot.

The health of pigs

The sow and piglets around farrowing

Farrowing fever (metritis, mastitis, agalactia syndrome or MMA)

This is a complex condition that occurs in sows during the period of farrowing and which seriously affects the viability of the piglets and the health of the sows. The sow runs an elevated temperature (about 40.5–42°C) and the udder is often hard and tender. There will be little or no milk 'let-down' ('agalactia'), and there is often inflammation of the uterus ('metritis'), which may be associated with retained afterbirths.

It is usually possible to cure the disease by immediate administration of broad-spectrum antibiotic injections such as tetracycline or ampicillin, and posterior pituitary hormone (oxytocin) to 'let-down' the milk.

Transmissible gastroenteritis (TGE)

This is an alarming disease of pigs caused by a virus which causes a severe diarrhoea with many very young piglets rapidly dying of dehydration. Some piglets vomit; all tend to have a very inflamed stomach and intestines. The sows may also be ill and have little milk. Older piglets, above about three to four weeks, will also be affected to varying degrees and for a period will scour and make little progress but should recover. The only treatment is to administer fluids, provide good nursing and inject antibiotics to reduce the likelihood of secondary bacterial infections. (Also see later section for general advice against viral infections.)

Piglet anaemia

Every pig which is born in intensive housing requires some extra administration of iron to ensure that it does not become anaemic. The piglet is born with limited reserves of iron in its liver, iron being an essential constituent of haemoglobin, the oxygen-carrying element in the red blood cells. Some supplementary source of assimilable iron must be given. The preferred way of giving the iron is by injection of 200 mg of a soluble iron preparation at not more than 3 d of age. This is sufficient to carry the piglet through to the time it is eating solid food which should be liberally supplemented with iron.

Escherichia coli infection

There are various ways in which this ubiquitous cause of trouble on the livestock farm can do damage. In the young piglet during the first few days of life, *E. coli* may be a cause of an acute septicaemic condition. Within only a few days of birth a number of pigs in the litter can become very ill with no particular symptoms except extreme illness leading to death.

Neonatal diarrhoea

Those pigs that are infected with *E. coli* but less seriously than are the piglets killed with the septicaemic form of the disease show a most serious profuse watery diarrhoea.

Milk scours and post-weaning diarrhoea

Another form of diarrhoea due to *E. coli* but occurring later is that known as 'milk scours'. A pale, greyish scour affects piglets in the period between one to three weeks of age.

There are many strains of *E. coli* that cause these symptoms and almost certainly many of these strains will be living in perfect harmony within the intestines of the pig and will generally be present around the pigs' quarters. The *E. coli* appear to be able to multiply if there are certain stresses in the management.

Treatment and prevention

The treatment of *E. coli* infections is best achieved either by injections or medication incorporated in the feed or water. There are many antibiotics that can be used for this purpose, such as chlortetracycline, oxytetracycline, streptomycin, amoxycillin, lincospectin or the chemotherapeutic trimethoprim sulphonamide. Orally nitrofurans, framomycin or neomycin may be given. The recovery of pigs suffering with *E. coli* diarrhoea will also be helped by administering fluids fortified with minerals, electrolytes and vitamins.

Prevention may be achieved by using certain antibiotics in the feed. There are a number of alternatives to choose from including furazolidone, organic arsenicals, chlortetracycline or carbadox. An alternative approach is to attempt to create an active or passive immunity by the use of serum or vaccine. The former course makes use of *E. coli* hyperimmune sera. In the case of the vaccines, *E. coli* vaccine may be given to the sow during pregnancy.

Diseases of young and growing pigs

Bowel oedema

Also caused by *E. coli*, the condition invariably occurs about 10 d after weaning. The first indication of an outbreak is usually the appearance, in a pen of weaners, of one dead pig, often the biggest of the bunch. Others may show 'nervous trouble', staggering, incoordination, blindness, loss of balance, and falling about when roused. As the disease progresses in these pigs, they will lie on their side and paddle their legs and they then go into a coma and die within 1 d. In bowel oedema such pigs usually show swollen (oedematous) eyelids, nose and ears and may have a moist squeal resembling a gurgle. A post mortem, if carried out speedily, shows oedema of the stomach and in the folds of the colon and in the larynx.

The occurrence of this disease seems to be associated with the stress of weaning, with the addition of *ad libitum* dry feeding.

Treatment is rarely successful. To prevent the spread of the disease all dry foods should be withdrawn and replaced with a limited wet diet.

Streptococcal infections

In recent years streptococcal infections of young piglets have become very common. *Streptococcus suis* Type 1 infects the sow and she then transfers the infection to the piglets soon after birth. The organisms, after getting into the piglet, may cause a general infection.

The disease usually occurs at about 10 d of age. Affected pigs run a temperature, show painful arthritis, have muscular tremors, appear blind and cannot coordinate their movements. Some die suddenly with heart inflammation. Typically, some 20–30% of a litter are affected.

Affected pigs may be treated successfully with injections of suitable antibiotics such as penicillin or with sulphonamides such as trimethoprim.

The same organism, *Streptococcus suis* Type 1, can also cause serious disease in older pigs. Here, what seems to happen is that young pigs, already carriers of this disease, are mixed with older pigs around the early fattening stages, say 8–12 weeks. The carriers infect the pigs they are mixed with and after a few days of incubaton the affected pigs show symptoms similar to those displayed by the younger pigs.

Clostridial infections

Whilst the effects of clostridial infections on pigs are in no way comparable with those which affect sheep or even cattle, there are problems from time to time, particularly in outdoor reared pigs on 'pig-sick' land. For example, *Clostridium welchii (perfringens)* can produce a fatal haemorrhagic enteritis in pigs up to about a week old. The disease has a very dramatic effect, with profuse dysentery, becoming dark from the profusion of blood within it. Usually pigs so affected die very rapidly though a few take a more chronic turn. Most die within 24 h of the symptoms being noted. The litter affected should be treated with an antibiotic such as ampicillin, and those under risk of infection can be given antiserum. A more permanent answer may be needed and the sows can be given a vaccine for the infection, this being administered twice during pregnancy.

Piglets are also infected with *Clostridium tetani* in the same way as lambs by penetration of the organism from dirty conditions, possibly through the navel or by the injury caused during castration.

Greasy pig disease (exudative epidermitis; Marmite disease)

This is an acute skin inflammation of pigs two to eight weeks old with the production of excessive amounts of sebaceous secretion and exudation from the skin which does not cause irritation. It tends to be most common in hot weather.

For treatment inject a broad-spectrum antibiotic such as ampicillin or lincomycin. It is vital then to make sure that the cause is removed; in particular the best procedure is to provide plenty of soft bedding.

Swine influenza

The swine influenza virus causes serious losses amongst pigs in all parts of the world. In the normal way, if the animals are well housed and are not infected with other respiratory invaders, the course of the disease is acute but the animals quickly recover.

Atrophic rhinitis

Atrophic rhinitis is a respiratory disease caused principally by a bacterium, *Bordetella bronchiseptica* and other organisms, notably a virus – the inclusion body rhinitis virus. Atrophic rhinitis affects primarily the membranes lining the delicate bones of the nose and leads to inflammation, nose bleeding, sneezing and eventually general degeneration of the bones of the snout. This does profound damage to the growth of the pig and it also makes the animal more susceptible to enzootic pneumonia and any infections of the respiratory tract.

In order to prevent the condition a vaccine may be used or continuous preventive medication may be instituted with a suitable antibiotic such as Tylan or a chemotherapeutic drug such as a sulphonamide. It is better to improve the environment and especially ventilation and management than to rely on measures of vaccination and medication.

Diseases largely of fatteners

Salmonellosis

All forms of salmonellosis can occur in pigs. Symptoms vary tremendously depending on the type and virulence of the salmonellae. The worst outbreaks are usually after weaning, in large groups of 'stores'. This often shows itself in the septicaemic form, causing sudden death or acute fever with blue (cyanotic) discoloration of the ears and limbs. Another form of disease shows itself as an acute diarrhoea, also with a fever, but these are not all the signs. Pneumonia may occur with abnormal breathing, and also nervous abnormalities, such as incoordination and paralysis. Also the skin is often affected and parts may even 'slough-off' later in the disease.

Treatment will require the use of antibiotics with known activity against salmonella. Salmonellosis is a notifiable disease under the Zoonoses Order (see the earlier section headed 'Disease legislation').

The only form of salmonella that can be prevented by vaccination is *S. cholerae suis*.

Enzootic pneumonia

One of the most serious scourges and causes of economic loss on the pig farm is the disease of pneumonia caused by a species of mycoplasma *M. hyopneumoniae*. Though this is recognized as the principal cause of the disease, other organisms are common secondary invaders, including other types of mycoplasma and bacteria, such as *Pasteurella multocida* and *Bordetella bronchiseptica*. The lesions of enzootic pneumonia tend to be in the anterior lobe of the lungs which become consolidated (solid) and grey in colour.

Treatment should be instituted in pigs showing serious clinical signs. Broad-spectrum antibiotics, such as chlortetracycline and oxytetracycline, may be injected but may not have a permanent beneficial effect as no immunity follows an attack.

Whilst the disease may kill a proportion of pigs its most serious economic effect is to lower the growth rate and food conversion efficiency of nearly all the animals in the building. The longer-term measures to deal with the disease are to improve the housing and environment and/ or to consider re-stocking with enzootic-pneumonia-free pigs.

Swine dysentery ('bloody scours')

This is a disease largely of fattening pigs and the infection causes inflammation of the intestines which leads to dysentery (diarrhoea together with some blood effusions), and a general debilitation of the pigs. It is highly contagious, being spread via contact with infected faeces.

The main causal organism involved in swine dysentery is a bacterium, *Treponema hyodysenteriae*, though it is not the only one involved and others such as *Campylobacter*, *E. coli* and salmonellae of various species may always be present as secondaries.

Treatment

There are a number of drugs which may be given via the food or water such as the macralides, in the form of tylosin, erythromycin or spiramycin or lincomycin, which may be used as a feed additive or in combination with spectinomycin may be given in water. Further effective drugs are tiamulin and dimetridiazole.

Diseases of pigs largely without age incidence

Swine erysipelas ('the diamonds')

This disease is caused by a bacterium known as *Erysipelothrix insidiosa* or *E. rhusiopathiae*. Most pigs are under risk of this disease because the organism can live in and around piggeries and 'pig-sick' land for very many years in the sporulated form. In the peracute form there may be hardly any signs at all, and the pig may be found dead after a very brief period of severe illness. In the acute form the pig is also ill, has a high temperature and the skin is inflamed: it will not eat. Then there is a sub-acute form which is the more typical disease, showing characteristic discoloration of the skin with raised purplish areas said to be roughly diamond shaped throughout the back and on the flanks and belly. Finally, there are two 'chronic' forms of the disease. In one the joints are affected which causes the pig severe lameness, whilst in the other the organism causes erosion of the valves of the heart, leading to cauliflower-like growths, so that the pig will often show signs of heart dysfunction or even heart failure and death. A feature of swine erysipelas is that it tends to affect pigs during hot and muggy weather conditions.

The intestinal haemorrhage syndrome ('bloody gut')

This condition, often known as 'bloody gut', affects fatteners in the latter stages of growth, and after a short period of depression, with an appearance of paleness and an enlarged abdomen, they die rapidly. Post-mortem shows the small intestine full of blood stained fluid and also gas in the large intestine. There may also be a twisting of the gut. The condition is more particularly common in whey-fed pigs, but not exclusively so, yet its cause remains a mystery.

Mulberry heart disease

This is a curious disease of fatteners causing sudden death or the pigs become depressed and weak, collapse and die within about 24 h of the onset of the disease. Few recover. On post-mortem examination the signs are of an enlarged and mottled liver and the surface of the heart is streaked with haemorrhages running longitudinally and also occurring in the endocardium. The cause is not really known but there is a suggestion that a deficiency of vitamin E or selenium may be involved. There is little evidence that treatment does much good but the use of multi-vitamin injections and broad-spectrum antibiotics may be of some benefit.

Swine fever (hog cholera)

A very serious virus disease now eliminated from the UK. Affected pigs show many symptoms, especially a very high fever, great depression and discoloration of the skin and usually diarrhoea of a particularly foetid type. Pigs also have a depraved appetite and thirst. It is a highly contagious disease. If any cases occurred in the UK, the herd would be slaughtered.

Aujeszky's disease

This is a herpes virus which leads to nervous and respiratory symptoms with a fever. Mortality can be high in young pigs. It is a notifiable disease in the UK and affected herds are slaughtered.

Tuberculosis in pigs

Three forms of tuberculosis can infect the pig, the bovine, avian and human forms. Because of the much reduced incidence of tuberculosis in cattle and man, there is now very little in pigs but it does still exist as a clinical entity. Of the cases that occur, most are of the avian type doubtless infected from wild birds since tuberculosis is virtually non-existent in the domestic fowl.

The symptoms are rather non-specific and would rarely be suspected by the pigman. Loss of weight, coughing and discharge from the nose will hardly lead to suspicion so that it is unusual for the disease to be recognized before it is seen in the slaughterhouse by the occurrence of swollen and infected lymph nodes in the throat, chest and abdomen. If samples are taken the disease can then be diagnosed positively.

Treatment is rarely called for but careful note should be given to any lessons there may be in isolating stock from sources of infection and also in improving the hygiene of the premises. The organisms that cause tuberculosis in all species are very persistent and resist destruction by most means; it is therefore very important to take such measures that eliminate them. It is also pertinent to stress the risk to the human population and whilst it is not a highly contagious agent of disease, it could represent a very undesirable challenge to man if the infection became widespread in a pig herd.

A group of important virus infections affecting pigs of various ages

Pig enterovirus infections, including 'SMEDI'

A number of enteroviruses cause stillbirths, mummification, embryonic death and infertility – the initials SMEDI – and hence this name is given to this group of diseases.

Pig enteroviruses may also be associated with nervous symptoms such as incoordination, followed by stiffness, tremors and convulsions. The acute form of this disease, known as Teschen disease, is probably not present in the UK but it is a notifiable disease and would be dealt with by a slaughter policy to attempt to eradicate it. The mild form, known as Talfan disease, is certainly present in the UK but no special control policies are instituted and reliance is made on the pigs developing a natural immunity, as with many other virus diseases of rather mild and uncertain symptoms.

Rotavirus infection

This causes very severe scouring in piglets which can lead to quite a heavy mortality. The symptoms are anorexia, vomiting, and diarrhoea, yellow or dark grey in colour, and very profuse. The diarrhoea causes rapid dehydration, which can kill, otherwise the pigs can recover in about 7–10 d.

These clinical signs are very similar to other conditions of young pigs causing profuse diarrhoea, such as TGE and *E. coli* infection and only a laboratory diagnosis which identifies the virus will confirm the cause.

Vomiting and wasting disease

This is a disease of the newborn pig caused by a coronavirus similar to the TGE virus. Affected pigs firstly vomit, huddle together with general illness, are depressed, run a high temperature and show little interest in suckling. Only a proportion of piglets in a litter are affected and only a proportion of all litters may be affected at all.

Epidemic diarrhoea

This is a very contagious disease of pigs caused by a virus which produces a profuse diarrhoea, vomiting, wasting and inappetance. The disease is similar to TGE but affects largely the older pigs; younger pigs are affected very much less or not at all.

Parvovirus infection

Parvovirus infection of pigs is a cause of infertility, stillbirths, small litters and mummification. It is a relatively new condition that has been recognized in a number of large units. It tends to be most commonly seen in young gilts or newly introduced pigs which have no resistance to the 'local' infection. It may also infect boars which then act as spreaders of infection. It appears to be far more of a problem in those housing systems where the sows are kept as individuals (as in tethering or sow stalls) and where the lack of contact fails to produce a passing infection and then resistance. In this type of condition there is no treatment that is of any real use. Licensed vaccines have been produced in the UK. An alternative procedure is to 'infect' a young gilt before service and the most effective way is to 'feed' homogenized placenta from known infected cases. This is however, a crude method of 'vaccination' and has great dangers of spreading other infections indiscriminately.

Porcine reproductive and respiratory syndrome (PRRS or 'blue-eared' pig disease)

This condition has already been mentioned as a good example of a recent viral condition that has swept through heavily populated pig areas. Attempts to limit its spread by notification, isolation and slaughter have completely failed and it is now accepted as a condition that has to be lived with and which appears to be due to one or more viral infections.

Symptoms

The clinical signs are very variable but the following usually occurs when the disease first strikes: inappetance and listlessness; abortions; embryonic death leading to a substantial fall in total births; irregular oestrus or anoestrus; skin changes including blue coloration; poor milking; poor piglets; lameness and puffy eyes. A general increase in diseases occurs due to the immunosuppressive effect of the PRRS virus, and especially respiratory disease. One of the biggest problems that have occurred in the wake of PRRS has been infection in pig herds with the swine influenza viruses. Their incidence and effect appear to become much greater after a primary outbreak of PRRS. When a herd is affected with both PRRS and one or more of the strains of swine influenza virus, the effect can be quite devastating. This effect is especially serious in intensive units, and some farmers have reverted to less intensive methods in their determination to eliminate the effects of this range of infection.

General control measures for virus infections when no vaccines are available

The upsurge in the number of virus infections in pig herds in recent years is worrying and provides us with lessons of great importance. They appear to be very similar to those conditions that have affected poultry but the pig industry is much more vulnerable. It tends to be more careless about hygiene, has no general policy of depopulation of sites and moves pigs around during their lifetime much more generally than is the case with poultry.

To minimize the effects, the following preventive measures should be considered:

(1) Introduce a minimum of new stock from outside the unit.
(2) Depopulate the housing of young pigs as frequently as possible.
(3) Keep adult breeders closely but cleanly housed.
(4) Do not use sow stalls or tethers but kennels or yards which group the gilts, sows and boars.
(5) Limit the size of each self-contained unit to a minimum. This will ensure that the effects of any virulent virus take place as rapidly as possible and immunity may develop quickly and uniformly.

(6) Practice careful isolation of the site in terms of feed deliveries, visitors, collecting lorries for fat pigs and any other potential danger.

External parasites

Pigs can be quite badly infected with lice (*Haematopinus suis*), the mange mite (*Sarcoptes scabiei* var. *suis*) and the stable fly (*Stomoxys calcitrans*).

Lice

Lice tend to be most common on the folds of the skin of the neck, around the base of the ears, on the insides of the legs and on the flanks. The constant irritation causes the pig to rub and scratch and this reduces growth and food conversion efficiency. Also the lice may be vectors of other disease agents.

Mange

Mange is an infinitely more worrying problem than lice. The parasite which is most common in the pig (*Sarcoptes scabiei* var. *suis*) burrows in the skin and lives in so-called 'galleries'. The mange parasites are about 0.5 mm long and lay their eggs in the galleries and develop through larval and two nymphal stages to adults within a period of about 15 d. Whilst the parasites can only multiply on the pig, the mites can survive up to two to three weeks in piggeries.

Internal parasites

Ascariasis

Ascaris lumbricoides is the common large roundworm that lives in the small intestine. Some can achieve a length of up to $1\frac{1}{2}$ ft (450 mm). Infection can be so bad that the intestines are literally blocked with large numbers of these worms. The life-cycle is direct and after the eggs are laid and ingested by pigs kept under unhygienic conditions, the eggs hatch out. The larvae, in their development to the adult stage (which is always in the intestine) actually pass through the lungs and liver. Coughing and pneumonia result from the migration of the larvae in the lungs, and tracking of the larvae through the liver also causes damage and leaves small white areas known as 'milk spots'. Treatment of the pigs with anthelmintics will eliminate the worms and improved hygiene can destroy the eggs.

Other intestinal worms of pigs occur in various parts of the intestine. There are the two stomach worms *Ostertagia* and *Hyostrongylus rubidus* and the worm *Oesophagostomum*, which lives in the caecum and colon.

Finally there is a lungworm of importance known as *Metastrongylus*. The adult worm lives in the bronchioles of the lungs. In this case the life-cycle is indirect, infection being caused by ingestion of the earthworm which is the intermediate host of the larvae of the *Metastrongylus*. The lungworms undoubtedly cause damage to the lungs and pneumonia, but it is chiefly in the younger pigs that the symptoms are serious. Older pigs are not usually adversely affected but do remain as carriers of the infection.

Treatments can be carried out by in-feed medication or by injections using products such as fenbendazole, tetramisole, thiabendazole, piperazine and dichlorvos.

Pigs can be affected with anthrax; foot-and-mouth disease; a mild condition but rather similar to foot-and-mouth disease known as swine vesicular disease; and rabies, but all are only very rare occurrences if occurring at all in the UK. All, however, are notifiable diseases.

The health of sheep

Health problems at or near lambing

The first point to stress is the essential need for the highest standard of hygiene at all times. Lambing is the period of greatest risk from infection. Ewes must be carefully observed during lambing to reduce losses due to difficulties in the birth of the lamb (dystocia). If no progress is made within 3 h of the start of lambing, the competent shepherd should carry out an examination. He may find he can readily correct any malpresentation or may decide to call in a veterinary surgeon according to his judgement of the position.

E. coli infection (colibacillosis)

There are two types, the enteric and septicaemic forms. Symptoms in the enteric form usually manifest themselves at 1–4 d of age and the lamb becomes depressed, shows profuse diarrhoea or dysentery and dies, usually within 24–36 h of the onset of symptoms. Those affected with the septicaemic form are usually two to six weeks old. Affected lambs have a fever and become stiff and uncoordinated in their movements; later they lie down, paddle with their legs and become comatose.

To prevent this disease the ewes should be vaccinated during pregnancy. Treatment is by antiserum or broad-spectrum antibiotics.

There is a further condition affecting young lambs and also caused by *E. coli*. This is watery mouth or rattle belly. Lambs have a cold wet mouth with drooling of saliva – or even abomasal contents. Such lambs are depressed, usually scour and have an elevated temperature. Treatment is by nursing, warmth, antibiotics and vaccination for prevention of coliform infection in general.

Navel-ill

This occurs in lambs for the same reason as in calves (see 'The health of cattle'), and requires identical attention.

Mastitis

Mastitis in the ewe is a very damaging disease which often leads to the complete destruction of the parts of affected. Immediate treatment with a broad-spectrum antibiotic, given both locally into the teat and by injection, may save the udder and the ewe.

The clostridial diseases

The group of bacteria known as clostridia are spore-forming organisms found universally wherever sheep are kept. As soon as the bacteria pass out of the animal's body they form the tough spores (or capsules) which make them extremely resistant to destruction. Their mode of action is to manufacture toxins which make the animal ill or kill it. Clostridia will multiply only in areas where there is no oxygen, and they do this in wounds, especially deep ones, in the intestines of animals, or in organs within the body such as the liver. Clostridial diseases tend to affect animals in very good, thriving conditions. The main clostridial diseases are described below.

Lamb dysentery

Lamb dysentery occurs in lambs during the first few weeks of life and is caused by the organism *Clostridium perfringens* Type B.

The principal symptoms are of bloody diarrhoea (dysentery). There may be many deaths – up to 30% of the flock if it is unchecked.

Immediate treatment or prevention is by the use of antiserum. To protect from future attacks, the ewe should be given two vaccinations, the last one being about one month before lambing thus transferring a strong immunity to the lambs. In succeeding years ewes will only need one additional 'booster' injection.

Pulpy kidney disease

This is caused by *Clostridium perfringens* Type D. The first sign is often the sudden occurrence of a number of deaths in really good lambs at two to three months of age. The name 'pulpy kidney' is given because on post-mortem the kidneys show a high degree of destruction.

Antiserum or vaccination procedures will be similar to those that apply to lamb dysentery. A major predisposing cause is when sheep are placed on rather too good a pasture, and much can be done to prevent trouble by watching their condition and regulating their nutrition.

Blackleg

Caused by *Clostridium chauvoei* blackleg is similar to blackquarter in cattle which has been described earlier.

Braxy

Caused by *Clostridium septicum*. Usually the first sign is the sudden death of some young sheep in good condition on frosty autumn mornings. The predisposing cause is the eating of frosty food which damages the wall of the abomasum and allows the invasion of the clostridial organisms. Vaccination is totally effective as a preventive.

Black disease

The symptoms are sudden death, usually in adult sheep. The organism *Clostridium oedematiens* invades the liver after damage by the liverfluke. It usually occurs in the autumn and early winter when flukes migrate in their largest numbers from the intestine to the liver.

The same measures of control are used as in other clostridial diseases. It will also help to prevent black disease if the fluke infestation is controlled.

Tetanus (lockjaw)

Caused by the organism *Clostridium tetani* which finds entry through a wound which heals and the organism multiplies within this anaerobic atmosphere producing the toxin which causes nervous symptoms, spasms of muscular contraction, stiffening of the limbs and eventually death.

Control and prevention is on the same lines as other clostridial diseases. Proper treatment of wounds will greatly reduce the likelihood of tetanus.

The use of vaccines and sera with clostridial infections

All the clostridial diseases can be countered by the use of sera and vaccines – sera for immediate prevention or treatment or vaccines to build up an immunity over a period of weeks. Preparations combining several vaccines together (up to at least eight) covering all the common sheep diseases are now available.

Pneumonia

Acute pneumonia infections are especially serious in winter and autumn and are caused by infection with *Pasteurella* bacteria. The predisposing causes of infection can be intensification, harmful weather conditions, transportation under less than ideal conditions and infestation of the lungs with parasites.

Treatment is by appropriate antibiotics, and prevention by vaccination.

Orf (contagious pustular dermatitis)

This is a virus infection of sheep causing vesicles and scabs on the skin and especially over the mouth and legs. It can also cause an unpleasant disease to sheep handlers. After infection older animals develop an immunity.

Deficiency conditions

Swayback (enzootic ataxia)

Swayback affects the nervous system of young lambs causing incoordination and paralysis of the limbs of the body due to a degeneration of the nerve cells in the brain and spinal column; it is associated with a low level of copper in the blood and tissues of the lambs and ewes. The basic cause of the disease is a deficiency of copper, this element being essential in the formation of enzymes for the construction of nervous tissue. The critical level of copper necessary for the proper development of nervous tissue is 5 ppm and in swayback areas levels of copper are generally well below this. To prevent the disease it is necessary to supplement the copper intake of the pregnant ewe. Adult ewes require an average daily intake of 5–10 mg of copper as an adequate allowance. Mineral block mixtures containing 0.5% copper sulphate are satisfactory, or diet supplementation with 0.2 g of copper sulphate fed weekly to each sheep would be adequate.

Pine (pining)

Pine is a disease of sheep which leads to wasting. It is caused by a deficiency of cobalamin (vitamin B_{12}) due to a lack of sufficient cobalt in the diet.

A simple way of preventing pine is to add a supplement which has been enriched with cobalt to the normal concentrate feed. Animals may also be dosed individually, the cobalt 'bullet' being the best treatment since it is released gradually over a long period. The pasture may also be treated with 2 kg of cobalt sulphate per acre every five years.

Rickets

Rickets occurs in lambs and is a condition of poor bone construction when the calcification fails to take place efficiently. Proper formation of bones is due to a sufficient quantity and balance of the minerals calcium and phosphorus, together with usually vitamin D. In practice injections of vitamin D are successful in arresting the problem.

Internal parasites of sheep

The three main types of internal parasite that infest sheep are tapeworms, the flatworms represented by the liverfluke, and the roundworms, which infest both the intestines and the lungs.

Tapeworms

Tapeworms require an intermediate host, as well as sheep, for their existence. The adult tapeworm lives in the sheep's intestines, and lays eggs which pass out in the dung on to the pasture. Further development occurs in the intermediate host, which is a small pasture mite; lambs become infested when they consume the mites accidentally when grazing.

Sheep may also be infested with a number of bladder worms, infestation being picked up on grazing contaminated with the droppings of dogs and foxes which carry the adult tapeworms in their intestines. The bladder worms occur as thinly walled, fluid-filled bladders among the intestines, in the lung and liver and in the nervous system.

Liverfluke

Liverfluke can be a serious problem in wet years and has been fully described earlier in connection with cattle. The snail can exist only in areas where the soil is saturated with moisture for considerable periods though it is not found naturally in running water or ponds. In dry periods or in winter, it tends to burrow into the mud and can survive in the inactive state for a considerable time.

Roundworms

At least ten species of roundworms (nematodes) cause parasitic gastroenteritis. Heavy infestations are disastrous but even light ones can cause great loss of productivity. Principal worms are: *Haemonchus contortus*; *Ostertagia circumcincta*; *Trichostrongylus axei*; *Nematodirus spathiger*; *Trichuris ovis* and *Oesophagostomum columbianum*.

Control is tackled by treatment of the animal to eliminate the worm burden and by pasture and flock management to minimize infection.

Drugs used to control the infection are thiabendazole, levamisole hydrochloride and organophosphorus preparations. Suitable pasture management is based on rotational grazing allowing up to a week on pasture then closing it for three weeks.

Sheep may also be affected by lungworms. The reader is referred to the description of the similar condition in cattle.

External parasites

These include insects, mites and ticks. The easiest to kill are those which spend their whole time on the sheep – the mites, lice and keds. This group is effectively controlled by a single whole-body treatment with a preparation which is lethal to adults and young stages and persistent enough to kill the larvae as they hatch from resistant eggs, or by repeated use of a material which kills the adult only.

The blowfly, in comparison with the other external parasites, completes only one fairly short stage of its development on the sheep and is thus open to attack by an insecticide for only a short time. Larvae hatch rapidly and feed on the skin and tissues for several days, causing great damage. They fall on to the herbage and form the resting stage, known as the pupa, from which the adult fly emerges several weeks later.

The problem of fly-strike requires additional measures to be effective. All cases of scour should be treated and wounds should be dressed immediately.

In the north of England and in Scotland a serious problem has recently emerged, the *headfly*. This fly feeds on the fluid from the eyes, nose and mouth of sheep. As a result of the irritation the sheep rub their heads against fences and walls causing wounds which make them even more attractive to the unwelcome attention of the headfly. The result of this disturbance is a very restless animal and considerable economic loss.

The *sheep tick* attaches itself for only a relatively few days during its three-year cycle of development. The preparation that is used must therefore be highly active against all the developmental stages and should protect the sheep over a period of about two months.

Sheep scab, the commonly used name for a mange in sheep caused by the psoroptic mange parasite, was until recently a notifiable disease in the UK but has now been deregulated. The mange parasite can invade all parts of the body that are covered with wool and also the ears. The total effect is extremely damaging, which is the reason why the disease was notifiable. Although at one time the disease was eliminated from sheep in the UK, it has now returned to quite a worrying extent. Treatment and prevention are instituted by a dipping of the sheep with approved materials according to a planned programme. It is also possible to deal with the condition by double injections of Wermectin, an organophosphorus compound.

Infections of mature sheep

Scrapie

Scrapie is caused by an agent that acts like a virus but does not have the physical characteristics of one. The term 'prion' has been proposed. The disease affects the nervous system and the sheep becomes uncoordinated, develops a severe itchiness and rubs furiously against posts. No treatment is of any use and all sheep that are infected should be slaughtered.

The causal agent is believed to be the same as, or similar to, that which causes bovine spongiform encephalopathy (see section on this disease earlier in this chapter). It is believed that scrapie is transmitted vertically from ewe to lamb. It is also believed that some breed lines have a built-in susceptibility, so that where such lines are identified they should be culled from the breeding programme.

Foot-rot

This is one of the biggest causes of serious economic loss in sheep everywhere. It is primarily caused by a bacterium *Fusiformis nodosus* and spreads from sheep to sheep via infected soil. The harbourers of infection are in fact the sheep with bad feet in which the organism can survive indefinitely. The conditions that favour the spread of infection are wet weather and soil when the feet are softened and the organisms are released to invade other feet through any injured point.

Treatment and control have been described in the section on foul-in-the-foot in cattle.

Metabolic diseases of sheep

There are three most important metabolic diseases of sheep: pregnancy toxaemia (or twin lamb disease), milk fever (lambing sickness) and hypomagnesaemia (grass staggers).

Pregnancy toxaemia (twin lamb disease)

This is a disease of ewes which affects them in the last few weeks of pregnancy and nearly always occurs in ewes which are carrying more than one lamb. Affected ewes are incoordinate and then totally recumbent and comatose, and will invariably die unless treated.

To some extent the condition is due to excessive demands being made on the ewe's metabolism when associated with insufficient nutrition. Affected ewes may have the injections of intravenous glucose solutions together with dosing with glycerine and/or glucose solutions. However, most of the effort should go into prevention. The important and indeed essential requirement is that during the last six weeks of pregnancy the ewe is given a diet which is low in fibre, nutritious and easily digested.

Milk fever and hypomagnesaemia

Milk fever (lambing sickness) and hypomagnesaemia (grass staggers) both occur in sheep and are similar to the condition in cattle (see earlier section).

Diseases causing abortion

Enzootic abortion

This is caused by a chlamydial organism and tends to be common in certain well-defined areas; in the UK it is especially prevalent in NE England and SE Scotland. The infection is usually introduced into a flock by a 'carrier' which releases a great amount of infected material when it aborts. This infects the ewe lambs which will not abort the first season but will abort in the second. After abortion they are then immune but young sheep and brought-in animals will continue to become infected.

There is no specific treatment for infected animals but it is good policy to isolate the ewes which have aborted for the period when they may be discharging infected material. Thereafter a policy of vaccination may need to be instituted.

Salmonellosis

Salmonella abortion is an acute and contagious condition caused by the bacteria *Salmonella abortus ovis*, and can also be caused by the non-specific salmonellae, such as *S. typhimurium* and *S. dublin*.

Control can be instituted, including antibiotics or chemotherapy, vaccination and hygienic measures.

Vibriosis

Vibrionic abortion is caused by the bacterium *Vibrio foetus* var. *intestinalis*. Like salmonella infections, this causes late abortions. There is no specific treatment but vaccines are available to control the condition.

Brucella

The *Brucella* species of bacteria can also cause abortion in sheep, the effects being the same as in cattle. *Brucella abortus*, *B. mellitensis* and *B. ovis* are all capable of causing the problem. After positive diagnosis, control relies on hygiene and vaccination.

Toxoplasmosis

There is also a parasitic cause of abortion in sheep which is due to a toxoplasm, *Toxoplasma gondii*. The best procedure in an infected flock is to mix infected ewes with non-pregnant ones to stimulate natural immunity before pregnancy.

The notifiable disease of anthrax and foot-and-mouth disease also occur in sheep and the reader is referred to the section covering the diseases affecting cattle.

Maedi–Visna

This is a new 'slow' virus infection, also known as ovine progressive pneumonia and characterized by an insidious incurable respiratory disease. It is not a notifiable disease but there is an official M–V Accredited Flocks Scheme which enables farms to set up and maintain sheep known to be free after appropriate blood tests which are continued for member flocks.

Part 4

Farm Equipment

23

Services

R.P. Heath

There is a need for services in all areas of farm and rural pursuits. These may be power, heat, light, water and so on. Their suitability and effectiveness for the task in hand depend on the level and appropriateness of the service: this in turn relates to the adequacy and design of installation. The provision of services is usually vested in the 'utility' companies and similar bodies: indeed it may be a legal requirement to use their staff or their approved agents. In all cases of installation and use, it is essential to address safety considerations and legal requirements.

Energy

Compressed air

Compressed air can be a useful, flexible, easily installed source of power. The opportunity for safety is greater than with electricity provided that condensate does not cause deterioration of pressure vessels and installation. Maintenance must be of a high standard. Insurance company inspections of industrial premises set out to demonstrate the importance attached to the dangers. Leakage from the system must be stopped if the system is to be competitive on price with other power sources. Oilers should be fitted to hand-tool air supply lines. Clean air supply by use of filters and traps in the pipework is of benefit, particularly as water is detrimental to spray-paint operations. BS 1710 depicts the colour codes for pipework content. Mobile plant may offer a flexible power source for yard, barn and field work.

Electricity

Power generation is vested mainly with the generating companies: National Power plc, PowerGen plc and Nuclear Electric plc and, in Scotland, Scottish Power plc and Scottish Hydro-Electric plc. There is a movement by the distributors, i.e. the 12 regional electric companies such as SWEB, MEB, NORWEB, etc., towards generating and selling their own electricity. Likewise there is opportunity for high consumers of electricity to buy from any source as well as the franchise market. Non-Fossil Fuel Obligations have allowed smaller power generators to sell to the 'network'.

An excited direct current (DC) coil produces a magnetic field which is revolved inside a frame containing three sets of windings (coils). The magnetic field cutting the windings at speed produces a voltage in each of the separate windings. The speed of rotation is 3000 rev/min (50 rev/s). As the north and then the south poles of the solenoid affect the windings in turn at 50 cycles per second, an alternating voltage at 50 Hz frequency is produced. Voltage pushes current through a resistance (or impedance). Losses occur proportional to the transmission material, its length and area of section. The basic principle is to keep conductors as short and at as large a cross-sectional area as possible in order to minimize losses within the constraints of safety.

Generated electricity is transformed to 400 000 V for national distribution on the National Grid. At this high voltage the line losses are lower. The lines have to be high, on pylons, to give adequate insulation because of the surrounding air. At places through the grid the supply is transformed in step-down transformers to 295 kV, 132 kV, 33 kV, 11 kV and then 400 V as it gets to the consumer. The system is balanced with three power lines on three phase. At the last transformer, which may be a pole-type transformer by buildings on the holding, the neutral connecting all the number-two ends of the phase line windings is taken with the phase lines. Between any phase and neutral is 230 V supply.

Power for single-phase supply:

$$W = VApf$$

Power for three-phase supply:

$$W = \sqrt{3}\, VApf$$

Line amperage that a system demands:

$$A = \frac{kW \times 1000}{\text{Line voltage} \times 1.732 \times pf} = \frac{\text{Brake horse power} \times 746}{\text{Line voltage} \times 1.732 \times pf}$$

where W = watt; V = volt; A = ampere (amp); pf = power factor.

Power factor is the term used to describe the electrical inefficiency. It is the cosine of the angle by which the current lags behind the applied voltage, and occurs where windings and coils are used in alternating current (AC) electrical supplies, such as motors, welding sets and transformers. In such cases there is more current needed, with its inherent heating, for the same power output as displayed in the voltage expressions above. This increased current passes through the distribution system designed for normal flow and may cause failure of the conductor, switch-gear, etc. This adverse situation carries a cost penalty to the consumer because of these dangers. Electricity pricing is adjusted according to the pf level below the ideal of 1.0 but generally with a leeway to 0.8. Capacitors (or condensers) can be used to correct adverse pf. However, they do consume electricity, adding to running costs; so choice of level of correction is needed to balance capital costs and running costs. The capital cost may be as much as the piece of equipment to which it is connected.

Assessment of demand and routing prior to installation is essential. Mechanical, chemical or heat damage to wires, cables, fixtures and fittings must be avoided. Where not legally obliged to fit them, wiring diagrams and switch labelling associated with the installation are strongly recommended. New installations must be checked before a supply is connected. Since 1990, electricity companies contract in the marketplace and as such they will not check a service prior to connection as a matter of course as they did previously. Contractors installing electric wiring and equipment must ensure that it meets the current Institution of Electrical Engineers (IEE) regulation and recommendations.

Cables come in sizes characterized by the cross-sectional area of the conductor, which itself bears reference to the current-carrying capacity at the stated voltage and to the earth requirement (Table 23.1). Dimensions of cables have relevance to voltage drop from source to consumption point. This must not exceed 2.5% of nominal voltage. A 6.0 mm^2 conductor in a conduit has a voltage drop of 7.0 mV/m per ampere, thus 210 mV/m. Since 2.5% of a 230 V supply is 5.75 V, the nominal maximum length of cable run at maximum rating is 27.4 m for this case.

Fuses protect installations from electrical damage. Their action is too slow to protect life and a residual current device (RCD) should be employed. Such devices were known by various names, including earth leakage circuit breaker (ELCB) and residual current circuit breaker (RCCB). Their action is to sense an imbalance in the feed and return supply lines of a circuit. Any imbalance must mean a loss, possibly flowing to earth via a person. In the absence of protection via protective multiple earth (PME) systems, it is essential to fit an RCD. Full core balanced types offering better than 25 mA should be employed. Regular testing is essential to ensure that they trip under unbalanced load.

Electric motors must be fitted with a suitable isolator within reach of the operator and a starter if needed (typically over 3 kW and/or three phase (3 ph)). Three-phase motors may fail to revolve because of a breakdown in one phase; single-phase capacitor start motors because of a capacitor circuit failure. In such cases the other windings, circuits, etc., will be live. 'Electric' caution should be shown in attempting to resolve these situations. Incorrect direction of rotation of a three-phase induction motor is due to the phase wires being incorrectly connected. Reversing two wires will resolve the difficulty. Clearly, correct installation for the whole enterprise is to be recommended.

Table 23.1 Conductor data

Nominal area cross-section (mm^2)	*Rating (A)*	*Use*
Cable		
1.0	11	Lighting and small heaters
1.5	13	Ring circuits
2.5	18	Heaters to 3 kW
4.0	24	Radial circuits
6.0	31	Cooker circuits
10.0	42	Large heaters
Flexible		
0.05	3	
0.75	6	
1.00	10	
1.50	15	
2.50	20	

Colour codes

Dimension	*Old*	*New*	*Wire*	
Two-core flexible	Red	Brown	Live	
	Black	Blue	Neutral phase	
Three-core flexible, single-phase	Red	Brown	Live	
	Black	Blue	Neutral phase	
	Green	Green and yellow stripe	Earth	
Three-core flexible, three-phase	Red	Brown	Brown	Live
	Yellow	Brown	Yellow	Live
	Blue	Brown	Blue	Live

A good power output to frame size can be gained from the use of a three-phase supply. Where a three-phase supply is absent, electricity companies may fit two-phase. This supply has both phase circuits separate and effectively doubles the single-phase supply and makes 480 V not 415 V of three phase available.

A voltage of 110 V is used for hand tools and workshops, typically off the 230 V supply through a small mobile transformer. The reason for its use is that it is voltage that pushes current through a resistance. If the resistance is a human taking the current to earth, then the lower voltage will not have the same 'pushing' ability as 230 V; even though it is current that kills and for a given power the current of a system on 110 V is higher than that on a 230 V supply. The building industry uses this standard, which is only to be recommended for rural building and outside use.

Pricing of electricity is based upon overall and peak demands on a tariff structure designed to reduce peak demands in the system. Generators bid each day a rate and capacity for each half-hour period. Trying to get consumers to use electricity at a time when demand is low is going to be beneficial. Rural pricing is on a demand tariff on the overall capacity of the system, to try to cover the infrastructure costs, and then a price per kWh (unit) of power consumed. There are several pricing options for night, day/night, summer only (useful for summer holiday lets), etc. For larger farms, such as intensive livestock, where there is a more or less steady demand of >50 kVA, there are 'demand' charge systems. In situations where grain dryers can be switched out for certain seasons or where electricity is not used at certain times of the day, then seasonal time of day (STOD) tariffs are available.

Private supply

Alternators for private supply can cover for mains failure through a temporary fault. The units are stand-alone or tractor powered. When sizing a unit, the number and size of motors and other services are used to determine the capacity. What needs to be remembered is the current needed to start electric motors. Typically this can be some six times the full load current and may cause an alternator set not to have the capacity to be able to start them.

Fuel oil

Oil of 28-second viscosity is normally used for domestic-style burners though normal engine fuel oil (35-second) may be used if the appliance permits. Standard tank sizes are available and are typically self-supporting, being set upon piers or stands with suitable proof against corrosion and for contaminant separation. A provision to contain leaks to 110% tank capacity would lead to the recommendation that suitable under-cover provision is made which would also help corrosion and vandalism prevention. Gravity discharge is recommended and vehicular access on a load-bearing, fuel-resistant floor is needed. Advice from the National Rivers Authority (NRA) should be obtained from modification to existing facilities or the provision of new ones. (See Farm Buildings, Chapter 25.)

Liquefied petroleum gas (LPG)

As an oil-based product the price structure follows the fluctuations and fortunes of the oil industry. Convenient as a source of heat energy, being three times that of natural gas, it may be a plausible possibility for farm use. The gas is supplied in small bottles to 50 kg mass and large above-ground storage tanks. Storage rules are simple and based on the following: tap and outlet uppermost to prevent liquid and not gas into the system; freely ventilated as the gas sinks to the lowest level and may be in such a concentration that even away from the store it provides a hazard; the bottles must not be subject to excess heat nor cold and should not be insulated. Propane boils at −42°C and butane at −5°C. On pressure release when the control valve is opened, heat is removed from the surround. This may cause water vapour to freeze round the outlet. Clearly propane or several bottles convected in parallel may be the better answer in colder climes. No attempt should be made to use other than proprietary valves, taps, pipework, etc.

Pipes can be of smaller bore than for natural gas. Drawn copper tube to BS 1386 should be used for preference. Red lead paint should not be used as a pipe sealant at joints. Modern plastic will provide a more suitable installation medium and be safe from explosion in service. Solder joints should be replaced with brazed ones near the heat source. Blockage of the smaller jets may occur and flame failure is likely in draughts, which should be eliminated through good design of flues and through the position of their siting on building faces. British Gas plc, the supplier of LPG, or approved agents should install any fixtures and be called to give regular service. Refer to Building Regulations and amendments, the Gas Safety Regulations 1972 (SI 1972:1178), the Gas Act 1972, CP 331, CP 338, CP 339 and BS 5258.

Natural gas

Installation of 'mains' natural gas supply is the responsibility of British Gas or its approved contractors (CORGI registered under the Gas Safety (Installation and Use) Regulations 1984). Compliance with Regulations Parts I to VII is necessary. Specific responsibility and their property ends at the outlet from the meter terminating 2 m from the end of the underground service main. Service capacities and pressures are the subject of British Gas involvement. Typically a 6 m^3/hour meter is provided for domestic use. Gas safety regulations (1984 above) apply throughout and BS 6891 describes the materials, methods and design requirements for straight copper or iron pipe and fittings. Pipe corrosion can be a major problem. Ducting and protection are essential, but relatively straightforward with plastics sheaths. Sleeved protection through walls and inclusion in mass concrete floors is advised. Polyethylene (PE) pipe is commonly used for supply with exposed pipe in coated steel. All pipe runs should be well ventilated to prevent gas build-up from leakage. Flexible links to the appliance are usually fitted to aid service. Condensate may occur and pipework is laid to drain back to the meter or a drain point. Electricity and gas supplies must not be laid together in ducts, etc. Within 600 mm of the output side of the meter electrical cross-bonding must be clamped to the pipework and to the electricity earthing system with 10 mm^2 earth bond (or as amended by reference to IEE Regulations).

Gas is almost pollution free and has a calorific value of 37 MJ/m^3 and a supply pressure of 1−2 kPa. Burning is generally noisy due to the air supply volume for combustion.

Flues carry water vapour and carbon dioxide (on correct combustion) outside the building. Should failure occur, gas may also be ventilated out of the structure. Instantaneous heaters such as over the sink and shower heaters typically vent to the room. As with space and radiant heaters, water temperature raisers, etc., extreme care should be taken to ensure that appliances are regularly serviced by approved technicians.

Petrol

The Petroleum (Consolidation) Act 1928 controls the storage and licensing system. Enactment is through local authorities or equivalent, whose advice should be sought.

Chases and conduits

Chases and conduits provision allows specific access to the property and between-floors compartments. Attention must be paid to maintenance access to the service and necessary fire separation, noise control, moisture elimination and vermin control. Piped services have their own protection. Electrical service in particular needs protection, but in the domestic or public building situation some aesthetic role may be fulfilled. In commercial premises the need is more in the nature of damage protection than aesthetic. Maintenance of the service is an important consideration when at the design stage. Security of the installation on public access buildings poses particular problems, not least in relation to potential vandalism. It may be sensible to make a route from a service access passage through to the end use.

Heating

Human comfort relates to heat, draughts, CO_2 and oxygen concentrations in the air, etc. Relative humidity can influence ventilation air flow rates and exchanges of air, as can the amount of airborne pollution. Animals also show behaviour patterns in response to their environment. When temperatures fall below levels determined by human comfort, class of stock or plant, or to the detriment of infrastructure, then heating related to air flow rate and exchanges can be provided. Typically, heavy work areas require some 13°C, sedentary work 18°C and crop and animal units special conditions. For housed livestock some form of control is needed to ensure that welfare requirements are met. When designing a system the macro-considerations, such as resultant condensate, venting of hot water pipework or vibration in air ducts, should be taken into account.

A range of fuels is available (Table 23.2). Many are self stacking or have the advantage of inexpensive materials handling. The use of a particular one depends on the materials handling, security of supply, ease of management, cost and equipment costs. Efficiency also plays a part. Electricity is 100% efficient, oil burnt about 75%.

There are other considerations in the provision of a service. Infra-red heating has been used for animals in the form of a localized heat source with a built-in light attraction aspect. The heat is not always usefully used because of convection and evaporative cooling by water vapour, etc. Newer quartz linear heaters offer direct, efficient and directional heating. The penalty is some 400% capital cost.

Underfloor electrical heating is clearly directed whereas space heating using propane burners and the like is more general. Another aspect to consider is the presence of ammonia. This may corrode reflector fittings so rendering the function of the heater less efficient. Another less heat efficient but longer lived reflector unit may be chosen.

Low-grade heat from milk cooling has storage problems. Bacterial contamination must be controlled. The winter intake of water at an elevated temperature may have a benefit in feed productivity and performance. Biogas from the anaerobic digestion of animal and plant waste can form a useful source of energy. If burnt, the energy is not as great as if put through a spark engine with an alternator. The use to which the combined heat and power is put will affect overall efficiency. Where control-of-pollution systems have used anaerobic digestion, the gas is a useful source to offset capital and running costs. A regular throughput of material is essential with a residence time of 8, 17 and 17 days respectively producing 0.3, 0.2 and 0.28 m^3/kg waste from pig, cattle and poultry.

Table 23.2 Energy value of various fuels

Fuel	*Gross calorific value (MJ)*
Anthracite and good coal	35/kg
Biogas (60–70% methane)	22–26/m^3
Methane	37/m^3
Oil:	
Class C Kerosene	46.4/kg
Class G Heavy	42.5/kg
Peat (14% DM)	19/kg
Sawdust	18/kg
Wheat straw (10% DM)	17.7/kg

Lighting

Artificial light can give a safe, even level of light to buildings, plant and equipment where usually non-reflective agricultural building surfaces occur. Glare must be avoided and correct levels for detailed work or human access must be provided, though for many classes of livestock a 'night-light' standard is all that is required. The Bodmann ratio (the relationship between light level for the task, immediate surroundings and walls and ceilings) of 10:4:3 has been of great value in design to eliminate eye strain. Visual display equipment such as computers, display screen equipment (DSE), and the like have their own demands, requiring consideration of the ergonomics of posture, surrounding colours, etc. A cost–benefit analysis and reference to legislation will influence the level of provision. Some light fittings require additional equipment to 'strike' the luminaire. Correct choice of light source offers savings in running cost, colour rendering, safety in explosive atmospheres, corrosion resistance, water tightness, etc. Reference should be made to electricity companies and the Farm Electric Centre.

Illuminance intensity and luminaire life and efficacy are shown in Tables 23.3 and 23.4.

Table 23.3 Illuminance intensity

Area or task	*Intensity (lx)*
For selection by colour or detailed assembly	500–1000
For 'close' inspection	300
Farm workshop (general level)	100
Milking premises and passages	100
Others	50

Table 23.4 Luminaire life and efficacy

Nominal life (h)	*Lamp type*	*Efficacy (lm/W)*
7500	High pressure sodium (gold colour light)	70–100
7500	HID mercury vapour (+ halide colour renderer)	62–72
7500	HID mercury vapour (+ fluorescent coat)	35–50
7500	Fluorescent tubes ('white' colour)	54–67
7500	Fluorescent tubes (colour rendered)	33–40
2000	Tungsten-halogen	16–22
1000	Tungsten-filament	10–18

Assistance on fitting and suitability may be gained from the Farm Electric Centre, electricity companies and CP 324.101.

Shafts and ducts

Shafts and ducts are used where a large number or variety of services are needed. They may be in masonry, concrete or frame and board, etc. Typically, load-bearing capacity with suspended ceilings is confined to the ceiling itself; however, the void may provide useful service access to the floor above or to the room below. Hollow floors in agriculture are rare and thus ducting needs to be installed at the construction stage if expensive chasing is to be avoided. Unnecessary ducts in concrete are expensive but fitting at a later date has financial and aesthetic penalties. Inserting ducts between buildings at the time of casting concrete may be good insurance.

Gas, electricity or water service is contained in a plastics or vitreous duct entering the building at not less than 750 mm. Ducts are sealed at both ends with mastic or plastic to prevent water ingress and insulated for 600 mm at the ground-floor end. Where ducts penetrate the damp-proof membrane, there should be effective sealing prior to pouring the site concrete or the final screed. No building loads should be transferred from the foundation to the pipe. Ducts may enter the premises above damp-proof course level, and require the building to be sealed against the elements and mechanical and vermin damage.

During construction the main areas requiring attention are:

- entry of services at ground level;
- separation of services routes at ground level in particular;
- the correct sizing and positioning of holes to prevent structural damage;
- design for fitment of services later in the build sequence;
- repair and maintenance provision once up and running;
- the need to keep a permanent record of installation.

Drainage

Waste disposal, pollution of rivers legislation and local bye-laws must be observed. BS 5502:1978, particularly Section 3.1, is useful. Where a public sewer is involved, details of the size, discharge quantity and quality, etc. should be submitted to the authority. Approval must be gained prior to starting work. Diseased animal waste and high-risk wastes should be disposed of separately. Solids should be removed separately and soil/grease traps fitted to avoid blockage of the system.

Where workers occupy a building for 4 h or more, daily lavatory accommodation is required. Human waste must be disposed of separately from other wastes.

Natural water may be disposed of directly into a watercourse, or soakaway should the topsoil conditions permit. The soakaway should be away from a building boundary. A perforated wall may be used to contain porous fill, but a solid cover is needed. The capacity volume is taken as 13 mm over the drained area.

Water containing contaminants poses a particular problem. The best solution is to keep the water as clean as possible and dispose of the pollutant in as concentrated a form as possible. Livestock and vegetable wastes, organic matter, cleansing, disinfecting or chemical agents and fuel oils must be assessed in terms of biochemical oxygen demand, toxicity and physical handling properties.

No waste should be disposed of through areas of human habitation, food preparation or storage areas. Drains should be of stoneware, glazed, with no less than 150 mm of concrete surrounding them. This may prove insufficient where ground movement is a possibility: here flexible pipework may be needed. An adequate fall is necessary, but the old rules of thumb of 1 in 40 for 100 mm drain and such like are now unsatisfactory and have been replaced by falls commensurate with the content, pipe material and flow length. Excessive falls can be lessened by the use of backdrop inspection holes. Clean water velocity should not fall below 0.75 m/s. Foul water may have drain sizes based on the Colebrook and White formula. The maximum velocity of 1.5 m/s allows for cleaning to be maintained. In excess of this figure the liquid fraction is likely to leave the solid behind and blockage may occur. BS 5502:1978 contains tables from which soil pipe sizes may be determined.

Drains may be inspected for fall prior to backfill by the use of an angled mirror on a stick support offered to the end of the drain at the other end from a light source. This test will also be better than rodding prior to service in order to detect debris in the pipe run and ensure free egress of contents in service. Gas tightness using smoke generators in a closed pipe may be of value. Hydraulic tests using coloured fluid may be useful in detecting leaks under working loads. Caution is needed where support arising from the surrounding mass of the backfill is missing due to the open trench.

The pipes must be laid socket end uphill and joints made watertight. The joints should not interfere with the flow. Caulking the joint and masking with an even volume cement–sand mix can achieve this objective. The use of rubber and plastic gaskets, and steel, plastic or pitch-fibre pipe can overcome the brittle nature of the solid jointed glazed clay or iron pipes. Ventilation of drains should be provided and there should be free air passage along the pipe length. Discharges of foul water drains or stacks must in all cases pass through suitable 50-mm-seal water traps. Only air vents to positions not likely to cause a nuisance or health hazard are permissible.

Inspection chambers at no greater than 45-m centres should be provided with an extra rodding facility such as rodding eyes elsewhere. Inspection chambers are constructed in various materials but in any case it is important to check with the building control office of the local authority to determine if there are local material specifications. Covers should meet British Standards for the loads envisaged and may be grease or grease–sand mix sealed in the locating slot. Any screw fixing should be protected from corrosion. Drains adjacent to buildings, where the bottom of the trench is lower than the foundation of the wall, should be 'filled-in' with mass concrete. If the distance to the side of the trench is less than 1 m from the wall footing, the concrete must be to the level of the underside of the foundation. No continuous length of fill concrete should exceed 9 m.

A septic tank capacity of 0.1 m^3 per person daily is required (CP301:1971 and BS 5502:1978). Stores should withstand static and dynamic loadings and be of a material that will contain the deleterious substance likely to be encountered for the design life, which is typically at least 20 years. No store should be within 150 m of domestic dwellings. Animal access should be restricted and the provisions of the Health and Safety at Work etc. (1974) Act must be observed with respect to humans. Under Control of Substances Hazardous to Health (COSHH) requirements, assessment must be made of the risks involved to determine the precautions that must be taken should access to the facility be needed.

Rainwater goods

Design and installation must satisfy Building Regulations. Rainwater goods are designed to meet storm requirements. A storm is deemed to be 75 mm/h rain intensity, which occurs over a 5-minute period once every four years and over a 20-minute period once in 50 years.

The maximum roof drip should not exceed 50 mm to the gutter. Falls are needed to cope with this, bearing in mind the exit capacity of the downspout. Downspouts do not fill when working and thus their capacity will not be a simple volume relationship to the anticipated roof catchment volume. Gutters and downspouts are designed to the anticipated precipitation and the catchment area. The length of gutter should not exceed 6 m from a downspout. Protection of downspouts from animals or machinery accident may be aided by building them in or constructing from flexible materials. Catchment of the nominally clean water and removal separately from 'dirty' surfaces help to reduce the running costs of waste water disposal. It may form a useful source of water for fire fighting, for irrigation of animal housing to remove faecal material so reducing airborne pollution, or for the dilution of animal waste to aid materials handling and reduce costs. Clearly frost incidence will determine suitability.

Water supplies

Cold water supply

Water company supplies are provided to some farms but private supplies are more common. Reference must be made to the local water company if use of water from private sources is planned. Stock may drink from sources considered not 'potable' as far as humans are concerned, but increasing note is being taken of the importance of water quantity and quality in animal production. The demand for water by various classes of livestock, crop and agricultural tasks varies (Table 23.5). A useful rule of thumb for crop irrigation is 13 mm per week, but with variations about a typical 25% over-run in glasshouse cropping systems the vagueness of average consumption figures is drawn to attention. The volumes of water involved in agriculture make purification impractical and thus a suitable source of potable water must be found.

Soil chemical attack will be overcome by the use of correct bronze or plastics fittings. Communication piping from the main, called the service pipe, will typically be by polythene pipe to BS 1972 or 3284 (colour-coded sky blue) in roll lengths. (See Table 23.6.) The operating temperature of 70–115°C makes it suitable only for cold service. Grades vary for the system pressure to be contained (class B, C, D, E for, respectively, 600, 900, 1200 and 1500 kPa); though PVCu to BS 3505 may be acceptable. Mild steel (barrel) pipe to BS 1387 is cheap, strong and reasonably resistant to corrosion. Open systems should be galvanized; closed systems such as heating circuits should be of 'black' un-galvanized pipe. Where water is soft, corrosion may be high. As a guide to uses: class A, brown band, is used for waste; class B, yellow band, is used for distribution; and class C, green band, is used for rising mains. Wrought iron pipe (BS 788) is

Table 23.5 Daily water use for farm production and related storage

	Water required (litres)
Cow drinking, 22 litres milk per day	70
Cow drinking, dry	35
Cow allowance, dairy cleaning	20–50
Cow allowance, milk cooling	3 times yield
Beef animal drinking	25–45
Calves drinking, up to 6 months	15–25
Sows drinking, in milk	18–23
Sows drinking, in pig	5–9
Boars drinking	9
Pigs drinking, growing/fattening	2–9
Sheep drinking, growing/fattening	2.5–5
Sheep dipping, all sizes/dip	2.5
Poultry drinking, layers (100 birds)	20–30
Poultry drinking, fattening (100 birds)	13
Turkeys drinking, fattening (100 birds)	55–75
Sink, storage capacity allowance	90
WC, storage capacity allowance	90
Hot water system, storage capacity allowance	130
Farm office/canteen, storage capacity allowance	225
Farm workshop, storage capacity allowance	450

Table 23.6 Availability of polythene pipe to BS 1972 and BS 3284 (low and high density)

Pipe class and colour code:		B, Red		C, Blue		D, Green	
Working pressure head:		60 m		90 m		120 m	
Density type: (32 low, 50 high)		32	50	32	50	32	50
Nominal size (inch)	*Permitted range of outside diameter (mm)*						
$\frac{1}{4}$	17.0–17.3	×	×	×	√	√	√
$\frac{1}{2}$	21.2–21.5	×	×	√	√	√	√
$\frac{3}{4}$	26.2–26.9	×	√	√	√	√	√
1	33.4–33.7	√	√	√	√	√	√
$1\frac{1}{4}$	42.1–42.5	√	√	√	√	√	√
$1\frac{1}{2}$	48.1–48.5	√	√	√	√	√	√
2	60.1–60.5	√	√	√	√	√	√

Other sizes used in agriculture: 3, 4 and 6 inch.

little used. When metal was more normally used it was desirable to leave a goose neck to respond to ground movement. For long communication runs, plastics pipe has the advantage that it may be 'moled' into suitable ground. Use of polythene pipe offers more tolerance of adverse loads but indirect pipe run may still be used to advantage. In any case the pipe should not be less than 750 mm below ground surface level to avoid frost and mechanical damage (CP 99). Water authorities take varying views on the charge level for water loss from pipes, etc. through unplanned incidents. The communication pipe is the responsibility of the customer and regular checks on water meter readout and pipe run should be made. Ideally the pipe should be laid with a steady rise to the point of entry to the premises in order to vent fully. Where public supply is not available it will be necessary to use private supply. Pumping rather than sucking is more efficient and so energy provision is needed for the prime mover.

Stored water vessels are of galvanized steel (BS 417), polythene and polypropylene (BS 4123) or asbestos cement (BS 2777) with sizes coded C1 to C21 as in BS 417 Table 1. Plastics offer corrosion resistance but require support, rodent protection and caution to maintain an electric 'earth' bond.

In order to effect repairs a store of 24 hours' capacity is desired. Complying to 1985 Model Water Bye Laws the store should be vermin- and insect-proof. Bearings should be adequate to support the load and to avoid condensation damage to the structure. Low-temperature insulation is normally fitted to the vertical run of pipe in the building, the rising main, tank and supply pipe. Proprietary electric heating tapes or any heat source available such as chimney flues may be used to offset low-temperature conditions. It is suggested that the rising main be fixed some 750 mm away from an outside wall to avoid low temperature.

Distribution pipes from the main or header tank, the supply pipes, must be suitable for the pressure of the service. Half hard copper (BS 659), underground (coil) pipe (BS 1386) and galvanized iron pipe need support at 1.25 m horizontally and 1.75 m vertically for 15 mm or equivalent diameter pipe and so on according to BS 61. Modern plastics do offer cheaper pipe cost but, though light, strong and durable, need more regular support as they are not self supporting. Capable of handling hot water, plastics tend to have more expensive connectors and fittings and thus, as installed, tend to be as expensive as traditional metal materials. To install they do avoid the need for a heat source which may pose a fire hazard and for threading and bending equipment; are not prone to corrosion between dissimilar pipework materials; and do not require protection against lime-based building materials or need metal brackets. Control of the service is by stop valves, stopping the flow, and gate valves, controlling the flow as defined by BS 1010. Stored water must have its free surface at least 150 mm below the rising main valve and any feed water silencing devices should have anti-back-syphoning devices fitted. Overflows must clear the building and the outfall must be placed so that it can be observed. There need to be vermin guards on the pipework. Additional treatment to provide a 'clean' water supply may be needed to meet regulations, laws and bye-laws, such as Milk and Dairies Regulations.

Water will be drawn from not less than 50 mm from the base of the tank to avoid debris entering the pipe. Whilst desirable to size to allow adequate supply to all draw-off points at once, the cost–benefit must be addressed. Typically the flow rate is:

0.19 litre/s 12 mm draw-off
0.30 litre/s 18 mm draw-off
0.60 litre/s 25 mm draw-off

Clearly, however, these depend on available head, and pipe diameter should be adjusted to offer the desired flow rate.

Animal houses in particular require adequate water supply to all draw-offs to avoid restricting animals' intake which must be as and when they desire. There should be appropriate back-syphoning equipment fitted to the user end of the pipes. These devices may double in function as a valve to enable routine maintenance.

Corrosion prevention is imperative. Cement-based products should be isolated from copper. Electrolytic corrosion in pipework can be minimized by the use of correct fittings.

There are several types of water-level control valves. The traditional Portsmouth type have been replaced in recent years with units which have a better service life and allow for easier adjustment of water level. Many valves are made of plastics. When fitted to drinking facilities, it is imperative to protect the control from animal access, not least to prevent main contamination. Excess flow rate from incorrectly sizing volume flow rate and pipe cross-sectional area can cause velocity energy impact on the pipework (water hammer). This effect is especially a problem when associated with sharp changes in direction of water flow and poorly secured pipework. Damage may occur to the installation. Air cell 'capacitors' may be retro-fitted but good design is preferable. Larger pipe size may be desirable to improve flow where low height of store above point of use may restrict pressure head available to overcome pipe friction.

Hot water supply

Hot water is expensive to provide and the higher the

temperature the greater the rate of heat loss. Every effort should be made to offer the desired service with the minimum of pipe run and exposure to heat sinks and draughts. The stack effect for larger diameter non-pumped primary indirect heating situations may adversely affect the rate of heat rise. A 22-mm pipe has 3.5 times the capacity of a 15-mm pipe but clearly offers less frictional resistance. For domestic purposes the temperature rise to 40°C for 2-hour firing relates to about 10°C per hour average demand. In a similar manner the overall and peak demands for hot water must be calculated. Typically a one-hour demand store is needed with boiler capacity to replenish it in 1.5 to 2 hours. For example:

$$\text{Heat required} = \text{litres} \times 4.2\,\text{kJ/kg/°C} \times \text{temperature difference}$$
$$= 150 \times 4.2 \times 40$$
$$= 25\,200\,\text{kJ (or 25.2 MJ)}$$

Heating done in 3 hours (say on off-peak rate electricity) = kJ/time (h) × 3600 (seconds per hour)

$$= 25\,200/3 \times 3600$$
$$= 2.33\,\text{kW heater element (at 100\% efficiency)}$$
$$= 3.0\,\text{kW at about 30\% inefficiency}$$

Independent boilers, housed to provide half-hour fire protection, are used to supply heated water for farm, domestic and tourist/holiday-maker use. Oil is the main fuel, with 'bottled' gas used where a main gas supply is not available. The appropriate regulations, such as Approved Document J to the Building Regulations must be observed. 'Biogas' may be used with domestic boilers where an anaerobic animal waste digester is installed. Water quality and cost will determine whether a cast iron, mild steel or copper boiler is used. Condensing boilers recovering sensible and latent heat show impressive energy efficiency, which lowers running costs. Storage cylinders and pipe sizing must allow adequate draw-off at all points; likewise, correct length, section and insulation of pipework to match flow and temperature will add to energy efficiency and water volume consumption. Cylinder output pipe should travel for 0.5 m horizontally to prevent one-pipe circulation.

Any possibility of using low-grade heat as feedstock should be considered. Heat-pumped low-grade energy from aerobic slurry digestion, milk cooling, vegetable stores, etc. can be used. The capital and running costs must be investigated. Farm use of proprietary solar panels usually has pay-back periods far in excess of service life. Their economic use may be satisfactory to supplement electric direct heating for summer caravan and camping-site hot-water demands. Copper is the common material for pipework. ABS (acrylonitrile butadiene styrene) melts at 93°C and has been used with reusable quick-fit connectors as an alternative.

Instantaneous heaters may be divided into three classes:

- oversink: 20°C rise at 4.5 litres/min output;
- single-point boiling-water appliances: 1.5 litres/min domestic to 7.5 litres/min commercial;
- multi-point (bath, sink, shower, etc.) appliances: at 10 litres/min.

If the hot water storage is not vented to the atmosphere, the temperature must not exceed 100°C and safety device outlets must be visible and discharged so as not to injure people. Where heating is for industrial purposes or has less than 15 litres stored or is a space heater, the requirements are not necessary but sensible. Agrément approval of the package installed by an approved operator is required.

Fire fighting

When involved in a construction exercise, reference must be made to BS 5502:1978 and Building Regulations Parts G and L. Insulation, claddings or structural skins should not be of a material that produces toxic fumes, is highly flammable, produces dense smoke or constitutes a major hazard to the building concerned. Escape routes and building exits should be clearly marked, easily accessed, suitable for the occupancy and preferably illuminated. Where livestock are contained, the escape route needs careful thought to enable expeditious exiting of stock, preferably largely unaided by humans.

Chemical stores constitute a particular fire and safety hazard. The use of a lockable, separate, fireproof structure is recommended. Chemical content list should be kept away from the store, but readily accessible. (See Control of Pesticides Regulations 1986.) High fire-risk areas such as fuel stores should provide 2-hour fire resistance by walls, or have a 12-m separation from any other boundary. Boiler rooms, maintenance workshops, plant material drying facilities and potentially explosive fertilizer stores should provide 1-hour fire resistance by walls or 6-m separation. The separation of buildings physically to provide fire breaks can be extended to the inside of structures as well as by good design of the building layout. There should be fire breaks through compartmentalization, especially if the floor is greater than 500 m^2, ease of access for emergency services and ample water between 6 and 100 m away. Static tanks of 20 m^3 capacity for suction hose use should be provided if an adequate pipe supply is not available. Hose reels should comply with BS 5274 and hydrants to BS 3231. A list of portable fire-fighting apparatus is given in Table 23.7. Water or foam extinguishers should not be provided and used where there is a danger of electrocution from electrical services. Some extinguishers by their action reduce volume oxygen level and are a danger to fire fighters; others are toxic in confined spaces. Care is needed with provision. Regular maintenance of provision is a must. Alarm systems may be overlooked or seen as unnecessary or expensive, but their inclusion in buildings can only be to advantage. Bell systems should be replaced by solid-state electronic noise generators as these are more reliable. Smoke and flame detection, battery or mains or both sources of electrical supply are available and need regular testing.

Security

Small-capacity infra-red sensors are adequate to switch effective 500 W halogen lamps protecting the outside

Table 23.7 Portable fire-fighting appliances

Type	*Colour code*	*Best for, and use*
Water	Red	Class A involving solids. Mainly cools materials. Not electrical equipment nor burning oils. Aim at base of fire and keep moving across area of fire
Multi-purpose dry powder	Blue	Class A involving solids and class B involving liquids. Flame knock-down and melts to form skin. Does not penetrate spaces nor take heat out of fire materials. Beware hidden flame or restart of fire. Aim at base of fire and advance as flame is extinguished to cover the fire area
Standard dry powder	Blue	Class B fires involving liquids. Knocks down flame. Do not use where velocity of exiting powder stream will cause liquids to be blasted out over surrounds (i.e. chip pan). Does not penetrate spaces nor take heat out of fire materials. Beware hidden flame or restart of fire. Aim at base of fire and advance as flame is extinguished to cover the fire area
Halon 1211 (BCF)	Green	Class B fires involving liquids. Class A surface burning solid material fires. Chemical combustion inhibition effect. Do not use where velocity of exiting gas stream will cause liquids to be blasted out over surrounds (i.e. chip pan), nor where excess ventilation will dilute effect. Does not cool fire and is toxic in confined spaces or on hot metal
AFFF (Aqueous film-forming foam)	Cream	Class A and Class B involving liquids. Smother and cool. Do not use where velocity of exiting liquid stream will cause liquids to be blasted out over surrounds (i.e. chip pan). Use as water on Class A; use as foam on Class B fires
Foam	Cream	Class B fires involving liquids, but not all. Works by smothering the fire. Tend to be better in the hands of professionals. Allow the foam mat to build up in one place and spread over the burning liquid surface. Do not use where velocity of exiting liquid stream will cause liquids to be blasted out over surrounds (i.e. chip pan)
CO_2 (carbon dioxide)	Black	Class B fires involving liquids. Do not use where velocity of exiting liquid stream will cause liquids to be blasted out over surrounds (i.e. chip pan). Save and clean though caution in confined spaces. Noisy in operation
Fire blanket		For smothering small Class A and B fires

Class A fire: Solids (wood, cloth, paper, plastics, coal, etc.).
Class B fire: Liquid (grease, fats, oil, paint, petrol, etc.).

premises or providing courtesy lights on accesses. For security, relatively inexpensive infra-red sensors with microwave sensors will provide a better function as they rely on two input media. For farm buildings where fluctuations in heat are the norm, the problem of intruder detection is difficult to resolve. This can be especially the case in livestock buildings. The additional difficulty with remote premises is what signal to generate if there are intruders? Linking to a telephone line if one is nearby is an inexpensive option, as is a personal radio link if one is already in place. In any case lights and alarms should be triggered to offer some deterrence. For domestic premises the hallway and the living room are usually 'covered' first with similar provision for visitor and holiday accommodation. Remote signalling devices may also be used as part of systems failure sensors for water, feed or other welfare need.

Further reading

Foster, J.S. (1990) *Structure and Fabric*, Part 1. Mitchell, London.
Seeley, I.H. (1991) *Building Technology*, 3rd edn. Macmillan, London.

24

Farm Machinery

P.H. Bomford

The availability of suitable farm machinery gives the farmer the means to carry out farming operations, and the capacity to complete these operations within the time available. Successful management of machinery involves the selection of equipment of the correct function and capacity, and the supervision of its efficient and safe operation in order to achieve quality work at the lowest possible cost.

The agricultural tractor

Tractors account for about two-thirds of total farm machinery sales in the UK. On many farms, the tractor is the most costly machine (McKee, 1992).

The tractor provides power for almost all mobile operations on the farm, as well as for many stationary processes. Power is defined as the rate of doing *work*, and work is done when a *force* acts through a *distance*. One Newton metre (Nm) of work (or 1 Joule) is done when a force of 1 Newton acts through a distance of 1 metre. The same units define *energy*, which is the potential to do work. For the measurement of power, a time factor is included, and the unit of measurement is the joule per second, or watt (W). A rate of work of 1000 Newton metres per second is 1 kilowatt (kW).

Power for the tractor is produced by its engine, and this power is made available in two main forms: as pull at the drawbar, to operate trailed equipment, or as rotary power at the power take-off (PTO) shaft. A small proportion of the engine's power is available through the tractor's hydraulic system.

The 'size' of a tractor is generally described by quoting the power produced by its engine. Since the full power of the engine is not available to do work outside the tractor, power which is available at the PTO is a more useful indication of the work a tractor may be able to carry out.

The engine

The tractor's engine converts the chemical energy contained in diesel fuel into rotary power at the flywheel. Two-thirds or more of the energy value of the fuel is lost as waste heat via the exhaust and cooling systems. Flywheel power, often called brake power because it is measured by applying a braking load to the engine, is the product of the engine *speed* (N rev/min) and the *torque* (T Nm), or twisting effort that the engine can maintain at that speed. The relationship between these factors is:

$$\text{power (kW)} = \frac{2\pi\text{N (rev/min) T (Nm)}}{60\,000}$$

A typical tractor engine produces little torque below 500 rev/min. Maximum torque is reached at about 1400 rev/min, and torque then decreases with increasing speed to 90% or less of maximum at full speed, which is 2000–2800 rev/min. This torque reduction is called 'torque backup'. A large torque backup indicates a flexible engine with good 'lugging ability', and a reduced need for gear changing during work. Brake power, however, increases almost linearly with engine speed, and maximum power is produced only at maximum rated engine speed.

The conventional engine is the best mobile power source at present available in terms of efficiency, weight, availability of fuel and cost. However, it is by no means perfect. It is noisy, it vibrates and it produces exhaust gases which pollute the atmosphere. It is made up of many hundreds of individual components and has many points of wear. The life of a well maintained tractor engine, before it needs a major overhaul to renew worn parts, is between 4000 and 7000 h of work. The life of other mobile engines ranges from 200 h for small air-cooled engines, to 1500–2000 h for a car engine, to a maximum of 12 000 h for heavy duty industrial engines.

The faster an engine rotates, the greater the amount of power (and fuel) that is consumed in just keeping the engine turning at that speed, in relation to the power that is available at the flywheel. If it is not necessary to run an engine at full speed, because maximum power is not needed, operating the engine more slowly will save fuel. Slower operation also increases the engine's reliability and prolongs its working life.

Many diesel engines are fitted with turbochargers, to increase power output. A turbocharger is an exhaust-driven rotary compressor which forces more air into the engine. This allows more fuel to be burnt, releasing more energy and producing more power. A power increase of 30% or more may be achieved, but this will subject the engine to greater thermal and mechanical stresses. Mechanical components, and cooling and lubricating systems, must be upgraded accordingly. Since the increase in power is achieved with no increase in engine speed or size, frictional losses do not increase in proportion, and the increase in power also results in an increase in engine efficiency.

The transmission

In order to deliver a full range of engine power at a wide range of forward speeds, the engine is connected to the wheels by a transmission offering from 6 to over 30 gear ratios. With so many ratios available it is necessary for the driver to change gear often in order to match power and forward speed to changing conditions. Various devices are provided to make this task easier. Synchromesh transmissions synchronize the speeds of rotating parts, to allow quiet gear-changing on the move. Semi-automatic transmissions change gear hydraulically by releasing one clutch and engaging another. The commonest application of a semi-automatic transmission is in a 'high–low' change which inserts an extra ratio between each pair of existing gears; complete transmissions can operate this way, providing up to 15 forward and four reverse ratios. Semi-automatic transmissions deliver less of the engine's power to the wheels than conventional systems because power is lost due to friction and in operating the hydraulic control system of the transmission itself (Renius, 1992).

Some tractors are fitted with hydrostatic transmissions which provide an infinitely variable range of ratios with a stepless single-lever change even from forward to reverse. This system is ideal for the operation of trailed PTO-driven machines such as balers or forage harvesters. Forward speed can be continuously adjusted to match crop and ground conditions while maintaining a constant engine (and PTO) speed. For the same reason, hydrostatic transmissions are fitted to many combines and self-propelled forage harvesters. Where a major proportion of the tractor's power is to be used in traction, the lower efficiency of the hydrostatic transmission means that less power is available at the wheels from a given engine power (Tinker, 1993).

Drawbar power

A tractor develops drawbar (DB) pull as a result of the gross tractive effort developed between its drive wheels and the soil. After this tractive effort has overcome the rolling resistance of the tractor's own wheels moving the tractor along, any remaining force is available at the drawbar.

$$\text{DB pull} = \text{gross tractive effort} - \text{rolling resistance}$$

As will be seen, maximum drawbar pull is achieved both by maximizing the gross tractive effort, and by minimizing the 'parasitic' effect of rolling resistance.

Drawbar power is the product of drawbar pull and forward speed, as expressed by the equation:

$$\text{DB power (kW)} = \frac{\text{DB pull (kN)} \times \text{speed (km/h)}}{3.6}$$

Gross tractive effort

When the lugs of a drive wheel or track bite into the soil, the rearwards thrust of the lugs tends to push the soil back. As the soil trapped between the lugs is sheared from the underlying soil, a horizontal force is developed. The further the soil is displaced, the greater is this force.

As the tractor moves forward, exerting a pull on some following attachment, the soil is pushed backwards a little. The percentage the soil is moved back, in relation to the distance the tractor would move forward on a rigid surface with no drawbar load, is called the slip. A wheeled tractor develops its maximum drawbar pull at 20–25% slip, but maximum tractive efficiency occurs at about 10–12% slip (Gee-Clough *et al.*, 1982). Any slip at all means that some of the tractor's power is being lost in pushing soil backwards instead of pushing the tractor forwards, but since no pull can be generated without some slip, this must be accepted.

The shear strength of the soil under the tractor's wheel or track depends to a small degree on the rate of shearing; a tractor operating at a higher speed can generate a slightly greater pull because of this property. The major factors affecting soil strength are its coefficient of internal friction and its cohesion.

The more weight that is applied to the soil under the driving wheel, the greater will be its frictional strength and the greater will be the tractive effort generated at a particular level of slip. Coefficients of internal friction range from below 0.2 for a plastic clay, to over 0.8 for a coarse sandy soil. An 'average' figure is 0.6, which means that 60% of the vertical force applied to the soil by the driving wheels would be available as gross tractive effort. Excess water acts as a lubricant between the soil particles, and can reduce the coefficient of friction almost to zero.

Cohesion is the strength with which the soil clings together, even when no weight is applied to it. A typical value for a friable soil is $30\,kN/m^2$, with a range from zero for very coarse-textured soils to a maximum of $60\,kN/m^2$ in some clay soils. The greater the area of cohesive soil that is put in shear, the greater will be the tractive force generated. Compacting a loose soil, for example by running a wheel over it, will increase its cohesive strength so that a following wheel can generate a greater tractive effort than it could if running on uncompacted soil (Rackham & Blight, 1985). Cohesive strength is high in undisturbed soils.

Frictional tractive effort is increased by increasing the weight on the driving wheels or tracks. The loading may be by means of iron weights, or water ballast in the tyres, or the tractor can be made to carry part of the weight of the implement it is pulling. This last approach has the advantage that when the implement is detached from the tractor, so is the extra weight.

As the tractor pulls a load, the resistance of the load tends to tip the tractor backwards about a point on the ground beneath the rear axle. This has the effect of transferring weight from the front to the rear wheels, and thus increasing the available tractive effort. The height of the hitch point must be kept low enough to eliminate any risk of overturning the tractor. Front-end weights may be fitted to ensure that at least 20% of the tractor's weight remains on the front axle to give steering control. Four-wheel drive tractors, and crawlers, gain no benefit from weight transfer, since all the weight of the machine is carried on the driving members. On four-wheel drive versions of two-wheel drive tractors, weight transfer can remove much of the weight from the front axle under good tractive conditions so that the powered front axle contributes little to traction unless it is ballasted by front-end weights or front-mounted implements.

Cohesive tractive effort is increased by increasing the contact area between drive member and soil. The fitting of larger section rear tyres or dual wheels, or the use of four-wheel drive or crawler tracks all have this effect. Additional soil may be put in shear by the use of grousers, strakes or spade lugs, which can penetrate through a slimy surface layer into stronger underlying soil. However, traction in this case is increased at the expense of reduced tractive efficiency since power is lost in digging into the soil. The use of tyres at no more than the recommended inflation pressure for the load carried will ensure the maximum safe contact area between tyre and soil.

Ballasting

Correct ballasting is essential if a tractor's maximum tractive power output is to be realized on typical frictional-cohesive agricultural soils. A major research programme at the National Institute of Agricultural Engineering (NIAE) (now named the Silsoe Research Institute) has resulted in the production of very clear recommendations (Dwyer & Dawson, 1984). The total weight to be carried on the driving wheels (w) is found where

$$w \text{ (kg)} = \frac{650 - \text{PTO power of tractor (kW)}}{\text{working speed (km/h)}}$$

Much of this weight will be provided by the tractor itself and by any mounted implements that it carries, but additional iron weights and water-ballasting of tyres are normally required. In the case of four-wheel-drive tractors, front to rear weight distribution should be in proportion to the tyre-maker's recommended carrying capacity of front and rear tyres at equal inflation pressures. If ballasting is correct with the tractor stationary, the effect of weight transfer will not be large enough to reduce tractive performance.

When the correct ballasting has been established, the tractor must be equipped with drive wheels and tyres large enough to carry the necessary weight at a low inflation pressure (preferably no more than 1.0 bar) (Dwyer, 1983). In work the tractive load of the tractor should be adjusted so that wheel slip is 10–12%. This is not easy to judge, but it can be measured, and tractor-mounted slip indicators and slip control systems are available for this purpose.

Rolling resistance

In order to roll a wheel along a surface, a force must be applied to overcome its rolling resistance. Rolling resistance increases with the load carried by the wheel. In agricultural conditions the force ranges from 5% of the weight carried by the wheel when travelling over a dry field after a cut of silage, to 20% or more when harvesting root crops (Dwyer, 1985).

Rolling resistance is minimized by using the largest available tyre size inflated to the minimum pressure recommended for the load being carried, and by using more wheels to support the load in dual or tandem formation. Where radial tyres are available, a 5% reduction in rolling resistance can be attained by their use (Stadie *et al.*, 1989).

Tyres

Most tractors rely on rubber tyres to transmit the power of the engine to the soil and to generate tractive effort. So long as lug height is not less than 20 mm, tread pattern has little effect on overall tractive performance. There is negligible tractive advantage in increasing lug height above this value, as taller lugs will deform under load and lose their bite into the soil (Gee-Clough *et al.*, 1977a).

The familiar chevron tread pattern has the advantage that it is self-cleaning under quite sticky conditions so that the tread bars can continue to bite into the soil when other tread patterns would become completely clogged with mud. Because the tread is more rigidly supported, and the side walls are more flexible, radial tyres have less rolling resistance, give 5–15% better tractive performance, last longer and produce lower peak soil pressures than similarly loaded cross-ply tyres (Gee-Clough *et al.*, 1977b; Plackett, 1984).

The size of a tractor tyre is described by two dimensions (usually in inches). The first of these gives the maximum width of the tyre section, and the second indicates the diameter of the wheel rim on which the tyre is mounted. A 50 kW tractor can be fitted with 13.6 × 38, or 18.4 × 30 rear tyres, which are similar in overall diameter. However, the wider 18.4 × 30 tyre has a larger ground contact area and a greater carrying capacity and, when ballasted to take advantage of this, will give a 5–15% increase in tractive performance. It is not possible to use wide section tyres for row-crop work, but for most tillage and haulage operations there is no restriction on tyre width.

Compaction

Soil compaction by heavy machines is a problem not only of rutting the soil surface, but also of compression of the soil itself which reduces pore space, inhibits water movement, and increases the formation of clods. Compaction, under any particular combination of soil conditions, is largely a function of ground pressure and number of passes; the higher the pressure, the more dense the soil becomes, and the greater the depth to which compaction occurs. Ground pressure is reduced by spreading the weight of the machine over a greater ground contact area, using wider section tyres or dual wheels. Although the degree and depth of compaction are reduced, more soil is compacted by the wider contact surface. Minimum degree, depth and volume of compaction are achieved by a long, narrow contact patch, or by the use of wheels in tandem.

Compaction is also increased under a wheel which is operating at high slip, the maximum effect occurring at a slip range of 15–25% (Raghavan *et al.*, 1977). Since this is above the level at which maximum tractive efficiency occurs, it is advisable, for reasons of efficiency as well as reducing compaction, to operate a tractor at loads which only require a slip of 10–14%. Working rate is maintained by pulling these lighter loads at higher speeds.

The most effective method of controlling compaction is to use the lightest machines that will do the job, keep off the soil when it is wet, and reduce the number of passes over the field to the minimum.

To eliminate compaction of the cropped soil entirely, the tractor can be run in the same wheel-tracks for all operations, while the crop is grown in beds between the wheel-tracks. This system is known as 'controlled traffic' or 'zero compaction'.

Four-wheel drives

The majority of new tractors sold in Europe are fitted with four-wheel drive (Fig. 24.1). Driving all four wheels of a tractor offers a number of benefits:

(1) Greater soil contact area produces a greater cohesive pull, which is particularly advantageous under wet conditions when friction is low. When tractive conditions are good, the advantage is small.
(2) The pull is shared between four drive wheels, reducing each wheel's slip, compaction and sinkage. Ballast is also spread between all wheels, reducing total weight for a given tractor power. Two-wheel drive tractors above 50–60 kW cannot be ballasted sufficiently for maximum performance because of tyre limitations.
(3) The powered front wheels give improved steering control in wet conditions.
(4) The tractor's brakes are effective on all four wheels. Some tractors have front brakes, but normally only rear brakes are fitted, and front-wheel braking is only effective when front-wheel drive is engaged. Systems are available which automatically engage front-wheel drive (if not already engaged) when the breakes are applied.

The best tractive performance is produced by systems where all four wheels are the same size (Dwyer & Pearson, 1976). Smaller front wheels give a tighter steering lock but have smaller contact area and higher rolling resistance than full-sized front wheels. Equal-wheel tractors overcome the turning-circle problem with centre-pivot or four-wheel steering, which also allows front and rear wheels to run in the same tracks during turns.

If front wheels are to contribute fully to tractive performance, they must be correctly ballasted.

The power take-off (PTO)

Test reports show that between 80 and 94% of the power of a tractor's engine is available through the PTO shaft at standard speed, while only 50–70% is available through the wheels under average tractive conditions. The difference is mainly due to wheelslip and rolling resistance losses. There are two internationally standardized shaft sizes and speeds, a six-spline shaft turning at 540 rev/min and a 21-spline shaft turning at 1000 rev/min. Rotation is clockwise when seen from the rear of the tractor.

Since rotary power is the product of torque and rev/min, it can be seen that the faster PTO speed can transmit almost twice the power at any given torque, or through a shaft of a particular size since the power-

Fig. 24.1 The diesel-powered four-wheel-drive tractor is the most popular agricultural prime mover. Manoeuvrability problems in earlier four-wheel-drive designs have been overcome by means of revised front axle geometry. Courtesy John Deere Ltd.

carrying capacity of a shaft is limited by the torque it can withstand. Many tractors have dual PTO systems to accommodate all types of machine.

A fixed ratio between the engine and the PTO sometimes allows the engine to develop its maximum power at the standard PTO speed. Where this is not so, it is generally possible to over-speed the PTO so that engine power can be maximized.

The PTO is connected to the engine by its own clutch. If this can be operated quite separately from the transmission clutch, the system is called an 'independent' PTO. A two-stage clutch pedal controlling the transmission at half depression and disconnecting the PTO at full depression gives a 'live' PTO (see Fig. 24.5).

Hydraulic systems

According to model and specification, from 15 to more than 30% of a tractor's power is available through the hydraulic system. This is adequate for light work such as the operation of the three-point linkage, tipping trailers, most front-end loaders and a few light duty excavator attachments. Larger hydraulically operated machines such as high-lift loaders or hedge cutters must have their own hydraulic power units, driven by the tractor's PTO.

Hydraulic power (in kilowatts) is a function of fluid flow rate and pressure, and is represented by the expression

$$\frac{\text{flow rate (litre/min)} \times \text{pressure (bar)}}{600} = \text{power (kW)}$$

Tractor hydraulic systems operate at pressures of 140–200 bar, with flow rates of 20–100 litre/min.

The tractor's three-point linkage is of standard dimensions and pin sizes (Category I, II or III or combinations according to tractor size). It is able to carry mounted equipment for transport, and to control the working depth of many types of soil-engaging implements. Two alternative control systems are usually available, draught control and position control.

Draught control

This system adjusts the working depth of ploughs or other high-draught implements to maintain a constant draught or tractive load. Draught is sensed by springs or pins in the upper or lower linkage; changes in draught cause the hydraulic system to raise or drop the linkage. A 'response' adjustment controls the rate at which these hydraulic corrections are made; slow response gives the smoothest work, but fast response may be needed if the ground is uneven (Hesse & Withington, 1993).

Position control

This system will hold the linkage at a constant height relative to the tractor. This is generally used for transporting equipment in a raised position but may also be used to control the working position of some machines, sometimes in combination with draught control.

Front three-point linkages are available for many tractor models. The use of front-mounted implements can apply ballast to driven front axles, and can allow two light operations to be carried out simultaneously, at front and rear, saving time and labour.

Most tractors have available two or more pairs of external hydraulic couplings, controlled by separate double-acting control valves, and a coupling for the automatic operation of hydraulic trailer brakes. The valves may be used to control a front-end loader, rear fork-lift or other accessory, saving the cost of purchasing separate control valves for each attachment.

External hydraulic hoses are usually connected to the tractor by way of snap-on couplings, which will pull out and seal themselves if the machine becomes detached from the tractor. All couplings are a source of contamination to the tractor's hydraulic system, and care must be taken to avoid the entry of dirt when attaching and storing hydraulic accessories.

Fig. 24.2 This four-wheel drive tractor controls a reversible four-furrow plough by means of an electronic sensing system and an on-board computer. Courtesy Massey Ferguson (UK) Ltd.

Health and safety for the tractor driver

The tractor driver is subject to a number of potential risks due to his occupation. The major risks are suffering hearing loss from long exposure to noise, and injury from tractor overturning accidents.

Tractors are noisy, typically exposing an unprotected driver to a noise level of 95–105 dB(A). In the short term, such noise levels can cause fatigue and leave a ringing sensation in the ears when the noise has stopped. The long-term effect of exposure to high levels is to produce permanent hearing loss.

The ears can be protected by acoustic ear plugs or earphone-type protectors. Both are effective and cheap, but may not be comfortable, especially in hot weather. 'Q' cabs must reduce the tractor's noise to no more than 85 dB(A) inside the cab. Because of the logarithmic scale used, a reduction of 10 dB(A) represents a halving of the noise level to which the tractor driver is exposed.

Since much of the noise from a tractor is airborne, most of the effectiveness of the cab is lost if it is necessary to leave a door or window open, for access to implement controls or for ventilation in hot weather. Remote controls can eliminate the first problem, while the provision of refrigerated air-conditioning systems can (at a price) do away with the second.

Tractor overturns most commonly occur sideways, when operating on steep slopes or driving too close to gulleys, ditches or steep banks. Trailers, slurry tankers or other heavy, unbraked machines can push the tractor down hills. Backward overturns occur less frequently, usually from attempting to pull from a high hitch point (Hunter, 1992).

Since the introduction of BS approved safety cabs on all new tractors, the number of deaths from tractor overturns has diminished dramatically. There are almost no recorded instances of drivers being killed *inside* safety cabs. In an overturn accident, it is most important to hang on and stay inside the cab until the tractor has come to rest completely.

Cultivation machines

In order to convert the surface of a field carrying the remains of a previous crop into an environment tailored to the establishment and growing requirements of the next crop, a sequence of operations from a varying selection of machines must be employed. A detailed understanding of the actions and interactions of the full spectrum of cultivation machines, under a full range of soil situations, is essential if the aim is to achieve the desired result at least cost (Schaffer *et al.*, 1985). Primary cultivations are those involved in breaking up the soil initially, followed by secondary cultivations which refine the soil to produce a final smooth, level, firm seedbed.

Subsoilers

On heavy or poorly structured soils it is occasionally necessary to loosen the soil to a greater depth than that reached by normal cultivations, in order to improve drainage and root penetration. Subsoilers for this purpose can operate at depths from 300–600 mm, and at spacings as close as 1 m.

The subsoiler consists of one or more heavy vertical tines, with a replaceable point or foot. A knife-edged vertical tine is common, but a flat leading edge in front of a tapering tine requires a lower draught force, as does a tine which is angled forward. Both these latter alternatives have the disadvantage that subsoil will slide up the flat front of the blade and be left on the surface. All tines, but particularly those which are angled forward, lift the soil into a bulge or 'surcharge' ahead of the tine. It is important that no depth wheel or other component is positioned where it will prevent this action from taking place.

Below a particular depth, called the 'critical depth', at 200–400 mm according to soil conditions, the tine will just cut a slit through the soil, rather than bursting it upwards in a wide V. The width of soil loosened by the subsoiler can be considerably increased with only a small increase in draught force, by the addition to the foot of horizontal wings 300 mm wide. A pair of shallow leading tines, spaced at 0.5 m to either side of the main, winged, tine, loosen the upper soil layers and increase the critical depth for the deeper tine. The result is a large increase in soil disturbance with negligible increase in draft (Spoor & Godwin, 1978).

Ploughs

For centuries the plough has been the main implement for primary cultivation. As well as loosening and breaking-up the ground, it inverts the top soil to bury weeds and trash. The soil loosening function can be performed by many other machines, but if soil inversion and burying are required, the plough must be used. In many situations the value of inverting the top 200 mm of soil is being questioned, and some very successful cultivation systems are designed to keep the upper 100 mm, plus organic matter, on the surface by only cultivating to this depth. It was thought that the practice of straw-burning would further reduce the need for ploughing; as growers revert to straw incorporation, ADAS trials have re-emphasized the suitability of the mouldboard plough, without pre-cultivation, for this purpose (Fielder, 1986).

The plough is made up of a number of bodies, each of which turns one furrow, mounted on a rigid frame. Each body covers a width of 300–400 mm; this is normally fixed, but in a few cases furrow width can be adjusted to suit changing field conditions. The total width of all the furrow slices turned by the plough constitutes its effective working width, which can range from 300 mm to more than 4 m. Many different bodies are available to suit a full range of field conditions. A 'general purpose' body has a slow curvature which leaves the furrow slice intact, while a 'digger' body has a very abrupt curvature, to shatter the soil more effectively. 'High-speed' bodies reduce draught and leave neater work at high operating speeds.

Most ploughs can be fitted with pre-loaded release mechanisms which allow the whole body to fold back if it strikes an obstruction. This is particularly necessary where large ploughs are operated at high speeds. A small investment in protective devices can reduce the risk of long and costly delays and expensive repairs at a busy time.

The soil-engaging parts of the plough body are the coulter, a vertical knife or freely rotating disc, which makes a cut to divide the furrow slice from the unploughed ground; the share, which undercuts the furrow slice; and the mouldboard, which lifts and turns the furrow slice to leave it in its final inverted position. A small secondary body, or skim, may be used to shave off an upper corner of the furrow slice to ensure that no surface vegetation remains exposed. All soil-engaging parts are subject to wear, and are individually replaceable.

Since the plough body pushes soil to the side (to the right on conventional ploughs), it follows that the soil pushes the plough equally in the opposite direction. This side force is resisted by a flat plate, the landside, which bears against the vertical edge of the unploughed ground. A correctly adjusted plough exerts no side force on the tractor which pulls it.

While the construction of the plough sets the width of most of the furrow slices, the width of the front furrow slice is governed by the distance between the front body and the inside edge of the tractor rear wheel which is running in the previous furrow. The wheel must be set at the correct width from the tractor's centre line, as specified by the plough manufacturer. A further adjustment can be made to the plough itself, to steer it closer to or farther from the tractor wheel until the front furrow slice is at the correct width.

Ploughing depth, which should not exceed two-thirds of the width of the furrow slice if the slices are to turn satisfactorily, may be set by means of an adjustable depth-wheel, if one is provided. More commonly, the tractor's hydraulic draught control system is used. The use of a depth wheel gives more accurate control, and is not affected by changes in soil conditions, but the use of draught control without a depth wheel puts more weight on the tractor's rear wheels. This improves traction and also eliminates the extra rolling resistance of the trailed wheel. Long ploughs, of four furrows or more, often have a depth wheel at the rear to supplement the draught control system.

In addition to the side-force difficulty mentioned earlier, the one-sided action of the plough poses problems in working a field, since it cannot simply be drawn up and down like most other machines. The field must be marked out in 'lands', which are then worked separately. At the centre of the land a ridge is formed from the soil turned inwards by one or more passes of the plough in each direction, and then the plough works up and down each side of the ridge, 'gathering' the soil, until the land is ploughed. The finished field shows a ridge at the centre of each land, and a double furrow between adjoining lands. The furrows, in particular, can affect subsequent operations carried out in the field at least to the harvesting of the next crop.

The problems of unproductive time spent marking the field out into lands, time wasted travelling along the headland from one land to the next, and the uneven surface produced by the conventional plough have led to the development and widespread adoption of the reversible plough (see Fig. 24.2). In this machine a second set of bodies, which turn the soil to the left, is mounted on the beam in opposition to the right-hand set. The front end of the beam is attached to the tractor's three-point linkage by a headstock which can rotate the plough through almost 180 degrees, when it is raised, and thus transpose the two sets of bodies. By using alternate sets of bodies, the reversible plough can work across a field from one side to the other with no marking-out, no ridges or furrows, and a minimum of idle time on the headland. A higher rate of work can be expected as a result of the better use of time, although the heavier plough will be harder to pull and will cost about twice as much as a conventional plough with the same number of bodies. The extra weight gives good penetration into hard soil, but the extra complexity of the rotation mechanism and headstock can sometimes cause trouble.

Rates of work of (0.8–2.0 ha h)/100 kW rated tractor power can be achieved, depending on soil conditions and working depth. Plough and tractor should be matched so that the power of the engine is fully utilized at a speed of 5–8 km/h. Fully loading the tractor at slow speeds causes high losses due to wheelslip, while operating at very high speeds puts great strain on the plough and leads to high draught force, as draught increases with forward speed (Stafford, 1979).

The disc plough

The disc plough resembles a mouldboard plough in layout, but the bodies have been replaced by angled, inclined, free turning concave discs. Large stationary scrapers are fitted to prevent soil build-up on the discs and to increase the turning action on the soil.

The soil is loosened and mixed, rather than inverted; where erosion is a problem, partially buried crop residues can be very effective in binding and stabilizing the soil.

The machine is difficult to adjust, and the work produced does not look like conventional ploughing. Where obstacles such as roots or rocks abound, the discs avoid damage by rolling over the obstructions rather than catching under them.

Although not much used in this country, the disc plough and its derivatives are widely accepted in areas where their special properties can be used advantageously.

The rotary cultivator

The rotary cultivator may be used for primary or secondary cultivation work. The horizontal rotor, up to 3 m wide, carries right- and left-handed L blades and turns at 120–270 rev/min to produce either a coarse or a fine tilth. A rear hood may be raised to allow clods to be thrown out, or lowered to give a further shattering effect as the clods strike the inner surface of the hood. Working depth, to 200 mm, is controlled by a land wheel or a rear crumbler roller. An alternative spiked rotor may be used for seedbed preparation.

If the ground is hard or stony, some machines can be fitted with front loosening tines which reduce rotor power requirement and extend blade or spike life (Weise, 1993). The chopping and mixing action of the rotary cultivator is ideal for the incorporation of crop residues, fertilizers or chemicals into the soil. Repeated use of the rotary cultivator may break the soil down into fine dust, and there is some risk of polishing the underlying uncultivated soil if conditions are hard. The machine has a high power requirement of 15–30 kW/m of width, but where one pass of this machine can replace several passes with alternative machines this will be acceptable.

The spring-tine cultivator

This is a versatile machine for secondary cultivations. A grid frame 1.5–10 m wide carries a number of S-shaped spring tines distributed over the grid so that one tine passes through each 100–200 mm strip of soil. Replaceable points of various widths are available. The machine is normally tractor-mounted, but is carried in work by adjustable depth control wheels. Machines above 3 m in width are made up of a central section with two hinged wings which fold for transport.

The spring tines vibrate as the machine moves forward, which is very effective in shattering clods. Trash and hard clods are brought up to the surface. The machine leaves the soil furrowed at about 600 mm intervals, corresponding with the spacing of the last row of tines. Crumbler rollers or light harrows, available as extras to fit on the rear of the machine, will produce a smooth surface.

The spring-tine cultivator has a good mixing action at speeds above 7 km/h and may be used for the incorporation of soil chemicals. Two passes are recommended, at 45 degrees to each other. Heavy spring tines (or discs) are a faster-working alternative to the plough when dealing with chopped straw residues on the heaviest land. Work should be to a depth of 100 mm, and the land should be rolled immediately after cultivation to ensure good contact between soil and straw.

The disc harrow

The tandem disc harrow is made up of four 'gangs' of concave discs, each gang clamped to an axle which is free to turn in sealed bearings. The front pair of gangs is angled to turn soil outwards, while the rear gangs draw soil inwards. Seen from above, the four gangs form a wide X.

An alternative layout, known as the offset disc harrow, has only one front and one rear gang of discs, each of which extends across the full width of the machine. The gangs meet at one side and are separated at the other to achieve the same relationship of angles and disc orientation as one-half of the tandem machine.

The front gangs, which must penetrate firm soil, are often fitted with scalloped discs. The rear gangs use plain discs, which last longer and move more soil.

The action of the disc harrow is to cut downwards into the soil, and to turn and mix the material thus loosened. Clods and subsoil are not brought to the surface. Two fast passes of the machine can effectively incorporate straw or chemicals into the soil.

Penetration may be increased by increasing the angle of the discs to the direction of forward travel, but a more effective method of increasing penetration is to add weight to the frame of the machine.

Tractor-mounted disc harrows are available but, since much of the penetrating effect depends on weight, heavier trailed machines are better able to deal with tough or hard soil conditions. A total machine weight of 150 kg per disc is recommended for straw incorporation. Large units can perform primary cultivations, particularly in lighter soils.

These large machines are available in widths up to 4 m, with discs of 760 mm diameter, and weighing 1 t/m of width. A machine 4 m wide, pulled by a tractor of 120 kW, can cultivate at rates exceeding 3 ha/h.

PTO-driven secondary cultivation machines

In addition to the rotary cultivator, described above, other power-driven machines are available for secondary cultivation work. Many use vertical spiked tines rotating in pairs about a vertical axis, or reciprocating across the width of the machine. The speed of the tines is much greater than the forward speed of the tractor and their shattering action is thus more effective than that of rigid tines being pulled through the soil. The action of the tines also levels and compacts the seedbed, without raking up much buried trash or subsoil. Seed drills may be used in conjunction with some of these machines, to combine the operations of seedbed preparation and drilling.

Power requirement is 15–30 kW/m of width. Tine life can be short if soils are abrasive and this can increase the operating cost of the machine. Where one pass of the powered machine can replace several passes with conventional machines, this cost is likely to be justified.

Harrows

A harrow consists of a large number of small tines or spikes carried on rigid or flexible (chain) frames. As they are dragged along, the action of the tines is to sort, level and compact the seedbed, the degree of penetration depending on the weight of the frame and the size of the spikes. Harrows are also used after drilling to ensure seed is covered. Chain harrows may be used on grassland to break up matted swards or to spread dung after grazing.

The power requirement of harrows is very low, so it is uncommon for any tractor to be fully loaded by a set of harrows of normal width. Harrows may be used in combination with other machines – spring-tine cultivators, seed drills – to acheve two operations in one pass.

Rollers

Ridged, or Cambridge, rollers, made up of a number of ribbed cast iron wheels on a free turning axle, are often used in seedbed preparation to crush clods, compact the soil and leave a smooth finish. The compacting effect of a roller depends on its weight; it decreases with increasing roller diameter and decreases with increasing forward speed. Since Cambridge rollers weigh 300–400 kg/m of width and are used at fairly high speeds, their compacting effect is generally small and confined to the top few centimetres of soil only. Light, rigid-tined implements can be just as effective in increasing soil compaction, and do not leave a tight 'skin' at the surface.

Ridged rollers are available in widths to more than 7 m, the larger sizes being made up of a central section and two wings which sometimes fold hydraulically for transport.

Smooth rollers are commonly used for levelling grassland and pressing-in stones in spring to prepare for hay or silage harvesting later in the season. Rollers up to 3.6 m in width are available. By adding ballast to the hollow cylindrical rollers, weights from 0.5 to 1.3 t/m width can be applied.

Rollers weighing up to 5 t or more can sometimes take control on steep ground. Care must be exercised in matching the roller to a tractor of adequate weight and in operating safely on hillsides.

Some secondary cultivation machines may be fitted with a 'crumbler' roller. This is an open-cage roller, with a surface of spaced straight steel rods held in position by two or more integral wheels. The whole unit is free to turn in bearings at either end. As the roller moves forward, its weight is concentrated successively on each steel rod, which is quite effective in crushing clods at that point. There is also a good levelling effect.

Combination seedbed-preparation machines

Since many seedbed-finishing operations demand little power, a number of manufacturers have developed machines which combine several operations and can thus utilize a tractor's power more completely. Such items as ridged rollers, rigid or spring tines, crumblers or toothed rollers may be combined in such a way as to lift out and crush clods repeatedly, leaving a fine and level seedbed.

Reduced cultivations

Where a deep seedbed is not necessary (as it is for many root crops, for example), there can be savings in energy and increases in work rate if the soil is only cultivated to a depth of 100 mm. This system has been used successfully, even in heavy soils, for cereal production, without loss of yield (Ball, 1990). In many situations this practice has also led to long-term improvements in the structure of the upper soil layers. Problems with the build-up of annual grass-weed populations can occur unless this is controlled by herbicide applications or by occasional ploughing.

The soil may be worked to 100 mm depth by means of a heavy spring-tined cultivator or a rigid-tined cultivator, with tines spaced at 200 mm overall. Adjustable wheels control working depth. Three or four passes are needed before drilling in heavy soils.

Combination machines, generally using gangs of discs in conjunction with banks of heavy spring tines, can reduce the number of passes needed to produce a satisfactory seedbed. One example, 3 m wide, requires a tractor of 75–105 kW, and can cultivate up to 2.5 ha/h. Two or three passes are generally sufficient to produce a seedbed.

Fertilizer application machines

Ninety per cent of the nation's fertilizer is applied as dry granular, prilled, crystalline or powdered materials. In some areas liquid fertilizer solutions are available; they have the advantage of being handled easily by pumping, but the disadvantage is that they are generally less concentrated than solid fertilizers and thus more material must be handled in order to apply a given weight of nutrients to the land. Also, a storage tank must be installed on the farm to hold at least part of a year's supply of fertilizer.

Liquid fertilizer is applied by spraying machines very similar to those described below under the heading 'Crop sprayers'.

Dry fertilizer may be applied by a combine drill, simultaneously with planting the crop. In this situation it is not practicable to apply fertilizer at high rates, because the small carrying capacity of most combine drills would necessitate frequent refilling stops.

Broadcasters

Most dry fertilizer is applied by broadcasting machines. A hopper of 250–1000 kg capacity on mounted machines or up to 8 t on trailed machines discharges fertilizer on to a spreading mechanism comprising a single or double spinning disc, or an oscillating spout. If PTO speed is correctly set, the fertilizer can be spread to an effective width of 5–24 m, according to the particular machine and the material used. The material is distributed in a pattern which is heaviest directly behind the machine, gradually reducing to zero at a distance of 4–18 m to each side. As long as this pattern is symmetrical and the reduction of rate is constant with increasing distance from the path of the machine, a return bout at the correct spacing will produce an even application rate across the complete area. Light materials are not spread as widely as heavy ones, so that it is extremely important to follow the manufacturers' bout width recommendations for the type of fertilizer being applied.

Application rates of 20–2500 kg/ha are possible with broadcasting machines, at speeds up to 12 km/h. Application rate is controlled by the rate at which fertilizer is metered onto the spreading mechanism, and by forward speed. Some trailed machines have land-wheel driven metering devices which makes the application rate independent of forward speed.

The machine can be calibrated at a particular setting by operating it, stationary, indoors for a measured time, collecting and weighing the fertilizer delivered. The time needed to cover 1 ha at a known speed and spreading width (Table 24.1) can be used to convert the delivery rate/min to rate/ha. This test can also show up any difference in the amount of fertilizer thrown to right and to left, although the exact spread pattern cannot be determined.

In addition to the factors of correct PTO speed and forward speed, good condition of spreading mechanism, correct height above target and machine set level, the accuracy with which the operator maintains his required width of spread is critical to accuracy and evenness of application. Foam marker nozzles on a boom are very satisfactory, and the same attachment can also be used on a sprayer to spread the cost, which only amounts to a few pence/ha. Far more than this amount can be wasted by incorrect fertilizer application.

Fertilizer, particularly in combination with water from the atmosphere, can be very corrosive to mild steel, iron

Table 24.1 Time taken to cover 1 ha, according to working width and forward speed (min)

Effective width (m)	*Speed (km/h)*								
	5	*6*	*7*	*8*	*9*	*10*	*12*	*14*	*16*
3	40	33.3	28.6	25	22.2	20.0	16.7	14.3	12.5
4	30	25	21.4	18.8	16.7	15.0	12.5	10.7	9.38
5	24	20	17.1	15.0	13.3	12.0	10.0	8.57	7.50
6	20	16.7	14.3	12.6	11.1	10.0	8.33	7.14	6.25
7	17.1	14.3	12.2	10.7	9.52	8.57	7.14	6.12	5.36
8	15	12.5	10.7	9.38	8.33	7.50	6.25	5.36	4.69
10	12	10	8.57	7.5	6.67	6.00	5.00	4.29	3.75
12	10	8.3	7.14	6.25	5.56	5.00	4.17	3.57	3.13

Notes
(1) Non-productive time for such operations as filling hoppers, adjustment or turning is not included.
(2) Combination of width and speed not shown in the table may be evaluated by means of the formula $t = 600/(w \times s)$ where t is the time in min to cover 1 ha, w is the working width in m, and s is the speed in km/h.

or aluminium components. Manufacturers use corrosion-resistant materials such as plastics or stainless steel for some parts of the broadcaster, but it is important also that the machine be easy to dismantle and clean at the end of the season. Vulnerable components can be brushed or washed clean, and coated with oil to protect them in good condition for the following season.

Full-width spreaders

The broadcaster is an inexpensive machine which can be very accurate if calibrated often and operated correctly. However, the variation of application rate with speed, and the difficulty of maintaining the correct bout width, have led manufacturers to develop much more sophisticated machines which spread evenly over a known and constant width. Most of these machines take the form of a central hopper (mounted to 1 t, trailed to 5 t) with a metering mechanism, carrying a folding boom up to 30 m wide along which the fertilizer is conveyed by an air blast. The fertilizer emerges from a series of nozzles, spaced as close as 600 mm in some cases, to form an even overlapping pattern across the full width of the boom.

Where the metering is by a wheel-driven force-feed mechanism, the machine may be calibrated by rotating the mechanism by hand for a specified number of turns and collecting the fertilizer in the tray provided. These machines can be extremely accurate, at application rates from 1 kg/ha for certain granules or seeds, up to 2 t or more/ha. However, the machine must be in good condition throughout (MAFF, 1980), PTO speed must be correct, and spacing must be within 0.5 m of the correct width. Inaccurate driving can leave strips without any fertilizer, or apply double rates. Again, the advantages of using a marker device are very clear.

Planting machines

The objective of any planting operation is to produce the desired population of vigorous, healthy plants.

Grain drills

These machines can plant a range of seeds from grass or clover to beans. Seed is discharged in rows, in an even trickle, but not as individual seeds. Combine drills also deliver fertilizer, generally to lie close to the seed in the row.

In traditional machines, hoppers extend across the full width of the machine, which ranges from 2 to 6 m. Carrying capacity is 100–200 kg/m width on mounted machines, 220–300 kg/m on trailed models. Where a combined grain and fertilizer hopper is provided, it is often possible to reverse a central divider to give ratios of either 1:1 or 1:2 in weight of grain to weight of fertilizer, according to application rate. Removal of the divider allows the full hopper to be used for grain. Pneumatic grain drills have a large central hopper, generally carrying only grain (see Fig. 24.3).

Seed is dropped in rows 120–135 mm or 175–180 mm apart, the number of rows per machine being from 15 to 50. Narrower rows put seed further apart in the row (see Table 24.2), but make the machine more complicated and expensive. Closely spaced coulters are more prone to blockage, and less weight can be applied to each one if penetration is difficult.

The metering mechanism is driven from a ground wheel. There is normally one metering roller per row. As the roller turns in the bottom of the hopper, serrations grip the seeds and carry them out to a position where they are discharged down the seed tube. The seed rate is adjusted by changing the speed of the roller in relation to forward speed, or by changing the proportion of the roller exposed to the seed.

Some manufacturers can supply alternative sets of metering rollers with small or large serrations to deal with large or small-seeded crops.

Individual rows can be shut off by slides in the bottom of the hopper. Thus, a crop such as rape can be planted in wider rows than those used for corn, by shutting off two out of every three rows. Small internal hoppers can concentrate the seed over only those rows that are being planted. Automatic or semi-automatic 'tramlining' attachments can be fitted to close off certain pre-selected rows every two, three or four passes across the field, to leave unseeded strips for further passes of sprayers or fertilizer machines.

Fig. 24.3 Wheat being planted by a 32-row pneumatic grain drill. Disc markers ensure that bouts are accurately matched. Courtesy Massey Ferguson (UK) Ltd.

Most drills are supplied with calibrating trays to collect the seed, and handles to operate the mechanism, for the equivalent of a known area so that the correct setting for a particular batch of seed can be established before drilling begins. Static calibration does not however take into account the effect of forward speed or of wheelslip which can vary from 0.5 to 15% according to soil and tyre conditions. A final calibration where the drill is run at full planting speed over a measured area (see Table 24.3) can ensure that the amount of seed delivered is correct under field conditions.

Seed population can be further checked after drilling, and plant population after emergence, with reference to Table 24.2.

Seed is released by the metering mechanism up to 600 mm above the ground, and then falls or is blown to the ground through a seed tube. The tube spreads out the flow of seed from small groups which leave some types of metering device, into an even stream. Fertilizer or seed dressing can build up inside these tubes; the action of flexible tubes tends to dislodge these deposits as the coulters move up and down.

The seed is placed in the ground by the coulter. Ideally, each seed will be planted at the same depth, and surrounded by firm, moist, warm soil. Where the soil has been thoroughly prepared, either single-disc or Suffolk coulters are used.

The single curved-disc coulter uses an angled disc to cut a groove in the soil, and the seed drops into this groove. Some soil falls back onto the seed after the coulter has passed. The disc coulter can cut through loose surface trash, but stones can push it up out of the ground, resulting in uneven planting depth.

The Suffolk coulter is a fixed blade, shaped like the front of a boat. It presses a groove in the soil, into which the seed falls. It can work in stony ground, and can give very even planting depth in well cultivated soil. If loose surface trash is present, the Suffolk coulter will rake it up

Table 24.2 The number of seeds (or plants)/m^2 according to row width and spacing within the row

Within-row spacing (mm)	*Row width (mm)*										
	110	*120*	*125*	*130*	*135*	*140*	*165*	*170*	*175*	*180*	*190*
10	909	833	800	769	741	714	606	588	571	555	526
15	606	556	533	513	494	476	404	392	381	370	351
20	455	417	400	385	370	357	303	294	286	278	263
25	364	333	320	308	296	286	242	235	229	222	211
30	303	278	267	256	247	238	202	196	190	185	175
35	260	238	229	220	212	204	173	168	163	159	150
40	227	208	200	192	185	179	152	147	143	139	132
Metres of row/m^2	9.09	8.33	8.0	7.69	7.41	7.14	6.06	5.88	5.71	5.55	5.26

Table 24.3 Calibration distances for machines of different widths

Working width (m)	*Distance (m) to cover*	
	$\frac{1}{10}$ *ha*	$\frac{1}{25}$ *ha*
1	1000	400
1.5	667	267
2	500	200
2.5	400	160
3	333	133
3.5	286	114
4	250	100
5	200	80.0
6	167	66.7
7	143	57.1
8	125	50.0
9	111	44.4
10	100	40.0
11	90.9	36.4
12	83.3	33.3

and frequent blockages will reduce the rate of work.

Both types of coulter are carried on spring-loaded arms; increasing spring tension will increase planting depth. Individual coulters can move vertically to follow uneven ground. In both cases, there is advantage in using a following harrow to improve seed cover; a trailed harrow is ungainly to use, and many manufacturers can supply integral spring-tined harrows for use with their drills. Where additional compaction is required, a very few makes of drill can be fitted with a heavy narrow press wheel behind each coulter.

In order to match up adjacent passes of the machine, most drills are fitted with disc markers which leave a small furrow which is followed by the front wheel of the tractor at the next pass across the field.

Precision seeders

Where crops are grown as individual, separate, spaced plants, it is necessary to plant seeds singly rather than in a stream. This demands a metering mechanism which is capable of picking out individual seeds.

The most common metering device is the cell. Cells of the correct size to match the seed (which should itself be graded or pelleted to improve accuracy) are carried in the outer rim of a wheel, or as perforations in a flexible belt. The cells pass under the seed as it lies in a small hopper, and the seeds drop into the cells by gravity. After removal from the hopper, the seed falls from the cell to the ground by gravity, usually aided by an ejector. The seeds may be thrown rearwards as they leave the seeder, to counteract the machine's forward speed and reduce the tendency of the seeds to roll or bounce along the ground.

The most satisfactory results are achieved at low speeds (3–5 km/h), since this allows a sufficient time for the seeds to find their way into the cells. With increasing speed, more and more cells remain unfilled, and the number of gaps in the row of seeds increases.

An improved design of cell-wheel planter separates the two functions of selecting and feeding the seeds. A slow moving many-celled selector wheel turns in the bottom of the hopper. The seeds are delivered singly into a fast-turning feeder wheel, which ejects the seeds at ground level, with a rearwards velocity equal to the forward speed of the machine. Satisfactory precision planting is possible at speeds of 10–12 km/h.

More positive seed selection is achieved by the vacuum seeder. The vacuum is provided by a PTO driven fan. Single seeds are sucked into a ring of small perforated depressions on one side of a thin circular disc. As the disc rotates, the seeds are retained and lifted out of the hopper. An adjustable finger displaces any doubles. When the seed is directly above the coulter, the vacuum is cut off and the seed drops to the ground, again with a rearwards impetus in many cases. Some vacuum machines can plant seeds accurately at 8–10 km/h. One size of disc can meter a considerable range of seed sizes, so the time and cost normally involved in changing from one set of cells to another between crops is much reduced.

The metering mechanism is driven from a ground wheel on each seeder unit (the cheapest system to buy) or from a master landwheel which drives all units. Changing the drive ratio, by means of stepped pulleys or gears, changes the rotation of the mechanism in relation to the ground covered, and thus the seed spacing. On most machines, cell wheels or belts can be changed for ones with space for more, or less, seeds per turn, which will also change the spacing.

Once the machine has been prepared for work, by matching cells to seed size and by adjusting drive ratios, it is important to check performance in the field, at normal planting speed. A portion of every row should be uncovered so that spacing and evenness may be assessed, and corrected if necessary. Table 24.4 shows the relationship between spacing, row width and plant population.

Seed population must not be confused with plant population, which will be lower in proportion to the emergence percentage of the crop involved. The figure for sugar beet is 60–65%; extra seeds must be planted to allow for this loss.

Precision seeders are made up of a number of single-row units, each with its own seed hopper, attached by flexible links to a tractor-mounted toolbar. By moving the units along the toolbar, row widths down to 200 mm are possible on some machines. Closer spacings are possible by staggering units in two rows. The number of rows of the seeder should be a multiple of the number of rows to

Table 24.4 Plant (or seed) population (10^3/ha) according to row width and spacing

Spacing (mm)	*Row width (mm)*					
	400	*500*	*600*	*700*	*800*	*900*
100	250	200	167	143	125	111
125	200	160	133	114	100	88.8
150	167	133	111	95.2	83.3	74.1
175	143	114	95.2	81.6	71.4	63.5
200	125	100	83.3	71.4	62.5	55.6
250	100	80.0	66.7	57.1	50.0	44.4
300	83.3	66.7	55.5	47.6	41.7	37.0
400	62.5	50.0	41.7	35.7	31.3	27.8
km of row/ha	25.0	20.0	16.7	14.3	12.5	11.1

be harvested simultaneously, for example a crop to be harvested by a three-row harvester should be planted by a 6, 9, 12, 18 or 24 row planter. This avoids the risk of misalignment which might occur if the harvester had to overlap two passes of the seeder.

The seed is released very close to the ground, so that the spacing is not affected by the presence of a seed tube. This gives the machine very little ground clearance; soil must be smooth, trash free and well cultivated. A typical unit has a front depth wheel, followed by a boat-shaped coulter which presses a groove in the soil. The coulter may be lowered to increase planting depth. Behind the coulter is a covering device, followed by a second depth wheel. As coulters wear, they produce a wider furrow in which seeds can roll or bounce out of position; depth is less constant and seed–soil contact is reduced. Worn coulters must be replaced. Ceramic-tipped coulters last, on average, three times as long as the iron or steel components they replace.

Accessories are available on some machines to push aside loose dry clods, or to level and compact an uneven tilth. Angled discs or blades can draw soil over the seed, and heavy narrow press wheels can improve compaction round the seed. Use of the correct accessories can ensure fast, even germination under a range of soil conditions, but care must be taken not to disturb the seed spacing with any following treatment. Seeder units may also be equipped for the application of liquid or granular chemicals.

Seeder units are very compact and close to the ground, so that it is difficult for the operator to see whether or not they are working satisfactorily. Some machines can be fitted with simple electric monitors to show whether the mechanism of each unit is turning, or whether the seed hoppers are empty. A more sophisticated device causes a light beam to be broken by each seed as it is released. Any interruption in seed flow sets off an alarm, so that the fault can be identified and corrected immediately.

Potato planting machines

The potato planter must deal with a large amount of 'seed' material, typically 40 000 sets or 2.5 t/ha. The material must be handled carefully, especially if chitted seed is used.

Since the crop is grown in a ridge, a deep tilth must be prepared before planting to allow the ridge to be formed. In some cases, ridges are formed before planting but most modern potato planters form ridges from flat ground as part of the planting operation. It is important at all stages of production to avoid any activity which will press the soil into clods, as these are difficult to separate from potatoes at harvest time. If potatoes are to be grown in stony soil, it is desirable to pick up and remove stones from the topsoil, or to gather them into windrows which are buried away from the crop.

Hand-drop planters

These machines, which generally plant two rows at a time into ridged land, are based on the mouldboard type ridger. Seats are provided on the planter for one operator per row, and seed is carried in trays or a hopper. A bell, operated by cams on a ground wheel, sounds at regular intervals for forward travel, and the operators drop a potato into a tube at each ring. The potatoes fall into an open furrow and are then covered with soil as the machine moves slowly forward. A two-row machine can plant 1.25 ha/d.

Semi-automatic planters

The task of the operator is less onerous on this machine as he fills up a series of small trays or cups as they move. These containers subsequently release the tubers at the correct intervals as the machine moves along. One operator per row is still required.

Automatic planters

Operators are not required on automatic planters, as a series of cups or fingers picks up the tubers out of a bulk hopper (375–500 kg capacity for a two-row planter) and releases them close to the ground at the correct spacing. A coulter makes a small furrow to receive the seed, which is then covered by disc or mouldboard ridgers.

The metering mechanism is a single or double row of cups on an endless belt or chain, or a series of spring-loaded fingers on the face of a large disc. Some machines have a make-up device which can add a potato to the metering device if an empty cup is detected. Feeding rates of up to 500 tubers/min are achieved on some cup-feed machines, giving ground speeds to 6.5 km/h, or a rate of work of 1.4 ha/h for a four-row machine. A more typical figure for a four-row machine is 5 ha/d.

Not all automatic planters are gentle enough to deal with chitted seed, but most manufacturers produce special models for this purpose. Machine damage to chits typically causes a yield reduction of 1.8 t/ha (Maunder, 1983). Potatoes with short, green chits are most resistant to damage.

High-speed planting with gentle tuber handling can be achieved on machines where the 'seed' is handled by belts, and electronic control of spacing is incorporated in one design.

Because of the large weights of seed to be handled, as well as the fertilizer which is often applied by the planter, an efficient transport system is necessary if working rates are to be maximized.

Transplanters

Many vegetable and nursery crops are traditionally planted out as transplants. Bare-rooted transplants are planted by semi-automatic machines. The operator, seated on each single-row unit, selects plants and feeds them roots-up into a gripping device which may consist of a pair of flexible discs or a series of spring-loaded fingers. As this device rotates, it places the plant roots in a furrow which has been opened by a leading coulter, and releases the plants at the correct spacing. A pair of inclined wheels under the operator's seat gathers and firms the soil round the plant roots.

Spacing is adjusted by changing the gearing between the ground-wheel and the planting mechanism, or by changing the number of plant-places on the mechanism. Individual units can be spaced as closely as 600 mm; narrower rows can be planted by two or more banks of units. As with precision seeding, this operation demands a smooth, even soil surface with a good tilth.

Rates of work of up to 1500 plants/h per operator are

possible; a five-row machine operated by a team of six, and well serviced with supplies of plants, can plant 1 ha of cabbages in 7 h, or a 1 ha of lettuce in 13 h.

Block transplanters

To avoid the growth check which occurs when quite large plants are uprooted, handled in boxes for a while and then re-planted, many growers are considering transplanting very small plants growing in small blocks of peat compost (Fig. 24.4).

The transplanter presses square studs into the ground, and operators drop individual blocks into the depressions thus produced. Rain or irrigation washes the soil closely round the block, and growth continues without a check. Automatic block transplanting machines are reported to be able to work more than three times as fast as manual machines (Boa, 1984). Because the blocks are handled as well as the plants, materials handling becomes an even larger part of the operation. The use of smaller blocks combats the problem to some extent.

Crop sprayers

Most chemicals are applied as liquid solutions or suspensions, generally water-based. The liquid is applied as fine droplets to give an even cover (Table 24.5) and because small particles are more likely to be retained on the leaf than large drops, which can roll or bounce off (Spillman, 1984).

However, very fine drops fall slowly through the air, and are therefore subject to drift (Table 24.6).

Spray drift has been measured as far as 6 km from the spraying site (Sharp, 1984). The problem of drift is increased by the fact that water evaporates from droplets in dry weather, making them lighter and more drift prone.

Apart from the problem of drift, it can be seen that reducing droplet size offers two major advantages:

Table 24.5 Effect of droplet size on closeness of spray cover

Diameter of droplet (μm)	*Number of droplets/cm² at 45 litre/ha application rate*
400	13
250	55
150	254
50	1750

Table 24.6 Effect of droplet size on spray drift (after Spillman, 1984)

Diameter of droplet (μm)	*Horizontal drift (m) when released from a height of 0.6 m*		
	Wind 0.1 m/s	*Wind 1 m/s*	*Wind 5 m/s*
10	19.1	191	859
50	0.84	8.4	42
100	0.24	2.4	12.3
300	0.05	0.5	2.5
1000	0.01	0.1	0.6

(1) More close coverage of plant or soil. A more concentrated spray solution can be distributed evenly and a tankful of spray will treat a greater area.
(2) Better retention on leaves, so that a reduced application of chemical is necessry to give the required protection. There is less risk of the build-up of residues in the soil. Where very fine and uniform droplets can be produced, and with certain approved chemicals only, application rates can be reduced to as little as 0.1 litre/ha (see ULV, below). Typical field application rate is 200 litre/ha.

Fig. 24.4 Lettuces being transplanted in soil blocks. The tractor is fitted with narrow tyres and an optional slow-speed 'creeper' transmission. Courtesy John Deere Ltd.

Sprayer nozzles

The orifice of the sprayer nozzle meters the flow of spray solution according to fluid pressure; an increase in pressure gives an increase in flow rate through a particular nozzle. As the nozzle wears with use, its output will increase. If a nozzle is found to be delivering an excess of more than 5%, it should be replaced.

The nozzle breaks up the spray solution into droplets and discharges them in a pattern towards the target. With wear this pattern narrows, and concentrates more spray directly in line with the axis of discharge. Nozzles are classified according to their pattern, output and rated pressure. Patterns include fan (F), evenspray fan (FE), low pressure fan (FLP), hollow cone (HC) and deflector (D). The classification F/80/1.2/3 denotes a fan spray nozzle, the angle of spread of the fan being 80 degrees, and the nozzle producing an output of 1.2 litres/min at its rated pressure of 3 bar (BCPC, 1986).

The output of a nozzle is also classified as *very fine*, *fine*, *medium*, *coarse*, or *very coarse*. According to pressure, most nozzles can produce more than one grade of spray; manufacturers produce charts showing the pressures at which one or other grade is produced by a particular nozzle. Chemical manufacturers specify the spray 'quality' that should be used for each product, and this is shown on the product label. 'Fine' spray is used for foliar-acting weed control in cereals and for contact-acting fungicides and insecticides. Because of the risk of drift, fine sprays must not be used for products labelled 'toxic'. 'Medium' sprays are suitable for most purposes, while 'coarse' are used for soil-incorporated materials. 'Very fine' sprays are used for fogging, and 'very coarse' are used for liquid fertilizer application.

The nozzles are mounted along the boom at intervals of 0.5 m, and must be held at the recommended height above target if application is to be uniform.

The interchangeable nozzle tips may be made of brass, plastics, stainless steel or ceramics; durability of these materials increases in the same order. All nozzles should be replaced at least annually. The cost of a set of nozzles is no more than the cost of 1 litre of some spray chemicals, which is a very small price to pay for accuracy.

Many nozzle holders are equipped with check valves which shut off the flow of spray when pressure drops to 0.2–0.3 bar. This prevents dribble from the nozzles after the machine has been shut off at the end of a run. Corrosion-proof strainers may be fitted to protect the nozzle from blockage. It is important to match the screen size of the strainer to the aperture of the nozzle; if the strainer is too coarse, the nozzle is not protected; too fine, and the screen may clog up with small particles which could otherwise pass freely through the nozzle.

Conventional hydraulic nozzles produce a wide spread of droplet sizes, from those which are too heavy to stick to a leaf, to those which are light enough to be carried away by a breeze. Rotary atomizers have been shown to produce a very narrow spectrum of droplet sizes as the liquid spins off the edge of a rotating disc.

In the controlled droplet application (CDA) sprayer, droplets of 250–300 μm are produced. These resist drift and give good cover using less spray. Application rates of 20–40 litres/ha are recommended, which treats a very large area per tankful.

The ultra low volume (ULV) system uses a droplet size of 70 μm, at very low application rates. These droplets are carried by the wind to give a very thorough cover of the crop with excellent penetration into dense foliage. Herbicides and poisonous materials are not applied, because of the risk of drift, but satisfactory results can be achieved with approved fungicides and insecticides. Special oil-based formulations are used, to reduce the risk of evaporation while the tiny droplets are in the air. Only chemicals approved for this system of application may be used, and manufacturers' recommendations must be followed.

To propel the spray droplets positively toward the target, thus reducing drift and increasing the effective delivery of spray, a number of systems of electrostatic charging have been developed. A charge of up to 30 kV is applied to the droplets as they leave the nozzle. The voltage gradient from droplets towards the target at up to nine times the speed of similar sized non-charged particles (Lake *et al.*, 1980), much reducing the risk of drift (Sharp, 1984). The charged droplets are attracted on to the surface of the crop ensuring that a large proportion of the spray actually ends up on the crop. Some designs produce a blast of air with propels the spray droplets towards the target, reducing the risk of drift.

Other sprayer components

The tank, generally of a plastic material to resist corrosion, is of 200–1000 litres capacity on mounted machines, and up to 3 500 litres on trailed models. The contents can be agitated by a pressure jet or by the more positive mechanical paddle. Agitation is especially important where suspensions are applied. The tank has a filler opening and strainer on top. Since the top of the tank may be quite high, many machines also have a low level hopper from which the chemical can be washed into the main tank by the pump. The tank is fitted with a fine strainer at its outlet to protect other components from blockage. Direct injection sprayers have separate containers for water and for chemicals, which are metered separately and mixed continuously as the machine crosses the field. This development minimizes the need to handle the chemical concentrate and avoids the problem of disposing of dilute chemical should spraying have to stop unexpectedly (Landers, 1992).

The pump is of positive displacement, delivering 30–150 litres/min at 540 rev/min, at presures of 2–4 bar. Roller-vane pumps are cheapest but are subject to wear; diaphragm pumps are used on the majority of machines. Piston pumps are available, and can operate to pressures of 40 bar.

An adjustable pressure regulator maintains the correct spray pressure, and allows excess liquid to return to the tank once a pre-set pressure has been built up. Some pressure regulators contain a forward-speed measuring device which increases or decreases pressure in line with changes in forward speed. This reduces the possibility of errors in application rate because of speed variations.

The boom, which carries the spray nozzles, can be from 6 to 24 m in width; 12 m is a common size as it fits in with tramline systems. It is generally made in three sections, the outer two of which fold up for transport. Each section of the boom can be shut off independently.

The function of the boom is to carry the nozzles at a set height above the target, usually 0.5 m, with minimal vertical or longitudinal bounce. Boom height can be adjusted to maintain the desired clearance above a growing crop. Vertical bounce is countered by mounting the boom in a damped linkage to isolate it from the rocking of the machine. There may also be considerable longitudinal oscillations at the boom tip. The ground speed of the outer nozzles may vary from backwards, to forwards at 1.8 times tractor speed (Nation, 1982). This effect can also be reduced by appropriately designed flexible mountings (Nation, 1987).

Sprayer calibration

The value of spray chemicals handled by a sprayer in a year is usually far greater than the value of the sprayer itself. Time spent calibrating and checking the machine can yield large dividends in terms of more efficient use of chemicals. The sprayer should be calibrated at least as frequently as every 100 ha of use. The volume of water delivered/min by each nozzle at given gauge pressure can be measured using a special meter, or by collecting the liquid in a graduated 2 litre container for 1 min. This allows the output of the nozzle to be checked against the manufacturer's figure and for uniformity against performance of the other nozzles on the machine. A minimum of four nozzles should be checked, at least one from each section of the boom. The total output of the machine per minute can also be calculated. From Table 24.1, the time taken to spray 1 h at a given forward speed can be determined, and the output/ha of the sprayer can be calculated and compared with the desired application rate.

Correct forward speed is essential to precision spraying; tractor speedometers, if fitted, must be calibrated to take account of alternative tyre sizes and field conditions. Speed monitors, using independent ground wheels or radar sensors, are available, and some monitors will also display flow rates and application rate/ha. Control systems known as 'spatially variable technology' are capable of adjusting the sprayer's output of chemicals as conditions vary across the field. This will give greater precision in husbandry and better efficiency of pesticide use (Paice, 1993).

Servicing sprayers

Although the field sprayer is a wide machine, and high rates of work are possible, a typical work rate for a 12 m machine is only 2.5–3 ha/h. A survey (MAFF, 1976a) showed that over 70% of machines studied spent more than half their 'working' time on such tasks as filling the tank, or travelling to the water supply. By good organization some machines were able to spend 70% of their time spraying. Fast refilling was achieved by the provision of a water tanker on the headland, often fitted with a 450 litre/min centrifugal pump. This cut down travelling time almost to zero and severely reduced filling-up time in comparison to the time taken to fill the tank via a hose pipe from the mains. Where only a few days are available for the application of a particular chemical, this level of attention to detail can greatly increase the productive capacity of the spraying machine. The survey also pinpointed the importance of driving at the correct bout width, and demonstrated clearly the advantages of a bout-marker system.

Sprayer safety

The use of spraying equipment and spray chemicals is potentially hazardous, both to the operator and to members of the public, livestock, pets, beneficial insects and non-target crops.

The Food and Environment Protection Act 1985 (FEPA) and the Control of Substances Hazardous to Health Regulations control all aspects of the importation, advertisement, sale, supply, storage, handling and application of pesticides. An appropriate recognized certificate of competence is required by any person who sells, supplies or uses pesticides approved for agricultural, horticultural, forestry or amenity uses. Anyone applying these pesticides and not holding a certificate must be directly supervised by a certificate holder, unless they were born before 31 December 1964.

Only approved products may be advertised and sold, and users must comply with the conditions of approval relating to use.

The test for the certificate of competence in pesticides application (mandatory from 1 January 1989) is administered by the National Proficiency Tests Council (NPTC). Training is the responsibility of the Agricultural Training Board (ATB); courses are also available at some colleges. The following areas of expertise are included:

(1) Risk assessment.
(2) Selection of an appropriate approved pesticide, and the use of the correct application method and harvest interval as specified by the manufacturer.
(3) Correct storage of pesticides in a suitable store, with appropriate safety precautions and record keeping.
(4) Correct selection, deployment and use of gloves and other necessary protective clothing and equipment when handling or using pesticides.
(5) Safe handling, transport and mixing of pesticides.
(6) Correct application, machine calibration and operation, including measurement of forward speed, quantity and uniformity of output.
(7) Calculation of treatment rate.
(8) Correct procedures for warning neighbours and beekeepers prior to spraying, and for the protection of bees when spraying.
(9) Determination of safe wind conditions for spraying in order to minimize the risk of spray drift.
(10) Exclusion of people and animals from treated areas.
(11) Safe disposal of unused pesticides, surplus mixture, sprayer washings and empty containers.
(12) Decontamination and first aid procedures.

(NPTC, 1987; MAFF/HSE, 1990.)

Harvesting forage crops

Mowing and swath treatment machines

Whether they are to be conserved as hay or as silage, the majority of forage crops are first cut and then allowed to wilt or dry. The length of this drying period will vary from 24 h to several days, according to the conservation system.

Maximum drying rate of the cut swath depends on maximum use of the drying power of sun and wind. On one hot breezy day, sun and wind can remove more than 3 mm of water, 30 t/ha, from the leaves of a growing crop. If this moisture removal rate could continue after the crop was cut, hay would be made in a single day. As it is, the crop resists water loss, and conditions in the swath do not allow sun and wind to reach all parts of the crop.

The process of 'conditioning' the crop, normally carried out simultaneously with mowing, seeks to deal with both crop and swath factors by:

(1) Surface abrasion of the crop to allow water vapour to escape through the outer cuticle layer, once stomata have closed. The surface treatment, the severity of which should be adjustable to match crop requirements, is concentrated more on the thicker butts than on the more easily dried and fragile leaves.
(2) Production of a fluffy, durable swath for maximum wind penetration.
(3) Mixing of the crop to expose thicker butts to wind and sun.

This 'conditioning' must be achieved without the penalty of excessive fragmentation and leaf loss which can result from over-vigorous treatment. Losses increase with the length of time that the crop is drying in the field, so a treatment which gives a rapid drying rate for a 24 h wilted silage crop may well be too severe for a 5 d hay crop.

The ability to leave a very short stubble may seem to be advantageous but in fact a longer stubble, of 50 mm or more, has been shown to give faster drying rates than where the crop is shaved to the ground (Klinner, 1975). The risks of contaminating the crop with soil, and also of damaging the machine, are reduced considerably by leaving a longer stubble.

The cutterbar mower

This machine cuts by shear as the reciprocating knife sections move across the stationary ledger plates. Knives must be sharpened often. Work rate is 0.2–0.6 ha/h for a 1.5 m machine. A leaning or lodged crop reduces the work rate. The machine is very economical to use as it only requires 3–5 kW at the PTO and it is also the cheapest type of mower, being less than half the price of its nearest rival. The cut swath lies quite flat, with the butts covered; it must be tedded or conditioned immediately after cutting if a fast drying rate is to be achieved.

Impact mowers

All other mowers cut by impact. A cutting edge moving at 50–90 m/s smashes its way through the base of the crop. The crop is severed whether or not the machine is sharp, although blunt cutting edges may absorb 50% more power. Power requirement is much greater than for the cutterbar machine; the highest power is consumed when mature, stemmy crops are cut. There is danger both to the machine and to the user if it strikes a stone or other obstacle. Blades can be broken and fragments can be thrown out. Guards and shields must be kept in place.

Impact machines can offer very fast working speeds with little maintenance or downtime, and are little affected by the condition of the crop. Against these advantages must be set the machine's higher initial cost and greater power requirement.

Drum and disc mowers

Two or more rotors turn on vertical spindles. Small flat knives are attached to the periphery of the rotors, which are from 350 mm to 1 m in diameter. As the rotors turn, the knives cut a series of horizontal arcs as the machine moves forward. Tip speeds of 80–90 m/s ensure a clean cut.

The rotors of a disc machine are driven from below, and the cut material passes over them; the alternative is a drum-shaped rotor assembly driven from above. In the latter case the cut material passes back between adjacent pairs of drums. If the cut material does not pass quickly back off the machine there is a risk of double-cutting, where small fragments are severed from the butt end of the crop and are subsequently lost in the stubble.

Power requirement is 35 kW for a 1.6 m drum machine, with a 15% reduction for disc machines. High forward speeds are possible, and work rates of 0.5–1.5 ha/h/m can be achieved (Raymond *et al.*, 1986). Capital cost per unit width increases with width, with a large increase for trailed machines. For this reason, and as a means of avoiding the reduction in drying rate that is experienced in swaths produced by large machines, the use of twin smaller machines, one front and one rear mounted on the same tractor, may be considered (Tuck *et al.*, 1980).

The machines are complex; some have many fast-moving parts which may rotate at speeds in excess of 3000 rev/min. A high level of maintenance is vital if a long working life is to be achieved.

The cut swath is slightly fluffed-up, but the butts are covered. After an initial period of rapid drying, the overall rate is little better than that from a reciprocating mower. Tedding or conditioning is essential.

Mower-conditioners (Fig. 24.5)

As mentioned above, the swath produced by cutterbar or drum/disc machines is not 'set up' for rapid drying without further treatment to achieve the conditions described above. The combination of a mower with a 'conditioner' can perform both these operations in a single pass and leave the swath in an ideal state for fast drying. A mower-conditioner requires 35% more power than the equivalent mower and may cost 50% more to purchase. Work rate will be lower than that of the mower alone.

Two types of conditioner may be used. Roller machines have a pair of horizontal rollers which crush and bruise the crop before dropping it into a losse swath. These machines are best suited to stemmy crops such as lucerne, but deal less well with heavy, leafy, grass crops.

More recently, flail conditioners have been developed.

Fig. 24.5 A silage crop is cut by a 2.4-m PTO-driven mower-conditioner. The rear 'grouper' attachment places two swaths close together to suit the forage harvester. Courtesy John Deere Ltd.

The cut crop is lifted, mixed, turned and scuffed by a rotor carrying narrow fixed or hinged flails. This machine deals more thoroughly with thick, leafy crops, where the rigidly mounted resilient flails are able to penetrate throughout the material. It is less expensive but more power consuming than the roller machine. The severity of treatment can be easily adjusted by moving a baffle which can delay the passage of the crop so that the flails apply more impacts to it. The 'spoke' conditioner developed at NIAE (Klinner, 1975) has been widely adopted by mower-conditioner manufacturers.

The use of any steel-tined machine carries with it the risk that a broken tine may occasionally be left in the swath and cause serious damage to following machinery. Designs of conditioner which use platic bristles or toothed plastic sheet eliminate this problem, as well as offering greater improvements in crop drying rates (Klinner & Hale, 1984).

Swath treatment machines

Once cut, and possibly conditioned, further treatment to assist drying requires fluffing up (tedding) or spreading out the crop to maintain air circulation, and this must be followed by windrowing the crop to match the next stage of the harvest operation or to reduce the risk of spoilage if rain is expected. The crop should be tedded at least daily in good weather.

Spreading the swath produced by a mower or mower-conditioner allows the cut material to intercept most of the solar energy falling on the field, rather than half or less when gathered into rows. Drying breezes also have better access to all parts of the crop. The disappointing drying rates often experienced where crops are cut by machines over 1.8 m in width can be greatly improved if the crop is spread immediately after cutting (Jones, 1985). A Dutch system involving early morning cutting with a mower-conditioner at a stubble height of 75 mm, spreading immediately, and spreading again after 2 h has been shown to produce wilted material of 35% DM or more, by the end of the day of cutting (Bosma & Verkaik, 1986). Where spreading is used, crop losses are increased both by the scattering action and because a tractor must run over part of the crop in order to gather it back into windrows (Rees, 1982).

Most swath treatment machines use spring-steel tines to move the crop. If these are set too close to the ground the risk of breakage is increased. Farmers often modify their machines by adding small clips or ties so that broken tines are retained rather than being allowed to fall into the windrow. Machine designers have addressed this problem by the use of plastic or rubber crop-handling elements.

Combination machines

Some machines are able to perform all the functions of tedding, turning and windrowing. They cover a 3–5 m width, and generally consist of two PTO-driven tined rotors which rotate on vertical spindles, some with movable swath guides behind. The whole machine is tilted forwards to ted or scatter the crop, or levelled to produce a windrow.

Most machines can be made to rotate more slowly if gentle handling is required when the crop becomes dry and brittle or in difficult crops, such as lucerne, where leaf loss is a problem. It is not always easy to produce an even swath, but this is most important if a high harvesting rate is to be achieved by following machines.

'Specialized' machines

An alternative swath treatment system is to use separate machines for the different functions. Spreading machines of 4–7 m width use four or six PTO-driven tined rotors. Wide machines are articulated to follow ground contours.

A strip of spread material 3–5 m in width may be gathered into a windrow by a single-rotor machine on which groups of up to four tines sweep the ground as they pass across the front of the machine, and then fold back to release the crop and leave it in a windrow at the side. A guide, of canvas or of steel rods, assists in the formation of an even swath.

A gentle method of gathering spread material into a windrow, or of combining two or more rows into one, is by means of the finger-wheel side rake. A series of four to eight spring-tined wheels roll along the ground while held at an angle to the direction of travel. The tines sweep the ground and roll the crop from one wheel to the next across the width of the machine. The tines travel slowly in relation to the crop, and the rolling action retains small fragments of the crop that might otherwise be lost in the stubble. This action also tightens the swath and can reduce air circulation and the rate of drying if used at an early stage. The machine is inexpensive to buy and requires little power. Front mounting of the gathering rake can allow two rows to be put into one by the tractor which also powers the baler or the forage harvester.

Barn hay drying

When a hay crop has dried to 35% MC in the field,

further drying will only take place if the relative humidity of the air is low enough to extract moisture from the crop against the attraction of the plant material (Lamond & Graham, 1993). (See Table 24.7.)

This final stage can be speeded up, and exposure time in the field reduced by 1 or 2 d, if the hay is removed from the field at moisture contents of 50–35%, and dried with forced air in a barn or stack. The result will be a better quality product with reduced field losses and weather dependence, although it is produced at a higher cost/t than field-cured hay.

Conventional bales are generally dried in the store which they will occupy until used. Batch systems involve a second move of the crop which increases cost. The air is blown into the stack by a fan, via a false floor or large duct under the stack, or a vertical duct up the centre of the (square) stack. Mesh floor systems require an airflow of $(0.25\,m^3s)/m^2$ of floor area, while the vertical duct system demands $0.005\,m^3/s$ per bale, at pressures of 60–85 mm water gauge.

Where the air temperature can be raised by 5°C above ambient, drying can be continuous. Without heat, drying can only continue while the ambient air is dry enough to extract water. This requires constant monitoring of the relative humidity, and the drying process will be extended over a longer period.

Forage harvesters

The job of the forage harvester is to gather up crops which are to be conserved as silage or, occasionally, fed directly to livestock. The crop must be delivered into a trailer at an appropriate chop length to suit the conservation system and the feeding method. A short chop length requires more power and more expensive machines. Chopped material becomes more compact in load and store. Fine chopping improves fermentation, particularly of high dry-matter forage.

A $10\,m^3$ trailer can carry 1500 kg of fresh grass chopped to 200–250 mm, 2300 kg if chopped to 100 mm, and 2500 kg if chopped to 25 mm. For satisfactory compaction and fermentation is a clamp silo, a chop length of 200 mm is recommended for material up to 20% DM, 130 mm for material of 20–25% DM, 80 mm for 25–30% DM and 25 mm for material above 30% DM. For a tower silo a chop length of 25 mm or less is required.

Most forage harvesters discharge directly into trailers and cannot work at all unless a trailer is available. To keep the machine working at full capacity, careful organization of transport is of paramount importance.

Table 24.7 The equilibrium relationship between air relative humidity (RH) and hay moisture content (MC)

Air RH (%)	*Hay MC (%) below which moisture will not be removed*
95	35
90	30
80	21.5
77	20
70	16
60	12.5

Double-chop forage harvester

This machine combines a flail head, using edge-cutting flails, with a flywheel cutter and blower. This gives a more positive chopping action although chop length is not precisely controlled. The machine was originally designed for directcutting forage as part of a zero-grazing system, but is also suitable for picking up wilted material.

Typical chop length range is 50–150 mm and the machine requires 40–55 kW to achieve a net output of 8–12 t/h. This is a trailed offset machine which pulls the trailer directly in line with the tractor.

Precision-chop forage harvester

The precision-chop machine is normally fitted with a tined pickup for harvesting windrowed crops. Alternative headers for direct cutting and for harvesting maize are usually available.

'Precision' is attained by controlling the feed of the crop into the cutting mechanism so that the crop moves forward a set distance between each cut of the blades. This distance is known as the 'theoretical chop length', and can normally be adjusted between 5 and 50 mm. The average length of the chopped material will be greater than the set value, because of the way the crop is arranged in the swath. Chop length is adjusted by changing the rate of crop feed, controlled by two or one pair(s) of feed rollers, or by varying the number of knives on the rotary cutter. The shorter the chop length, the greater is the power consumption, or the lower the output of the machine.

Cutting is achieved by the shearing action of a sharp, fast-moving knife against a solid square-edged shear bar. Dull knives or excessive clearance between knife and shear bar waste power and reduce work rate. Machines are fitted with knife sharpeners and these should be used at least daily, followed by re-adjustment of clearances if necessary.

The cutting mechanism is particularly vulnerable to damage from stones or metal objects picked up in the windrow. Damage-resistant designs of cutter have been developed by several makers, and electronic metal detectors can stop the crop intake mechanism instantly if a piece of ferrous metal is detected. The particle of metal has to be picked out before harvesting continues. This device protects the machine from expensive breakdowns, and also prevents small metal fragments from being incorporated into the silage and later fed to livestock. Research workers have developed a pickup system which can reject the portion of a windrow containing a stone or a piece of metal, without the need to stop the harvester at all (Klinner & Burgess, 1982).

Most machines use cylinder cutters, which are similar to the mechanism of a traditional lawn mower. Up to 12 knives arranged round the periphery of a cylinder slice the crop against the shear bar. The direction of cut is normally downwards, and the cut material is dragged round under the cutter and then thrown up the chute. Upwards-cutting systems throw the cut crop directly upwards, eliminating drag, and releasing more power for cutting and throwing (Röhrs, 1986). A number of manufacturers use flywheel cutters. Power use is intermediate between the two types of cylinder systems, and the cutters are claimed to work more quietly and to resist damage better.

An experimental design of forage harvester which slices the crop with sharp-edged discs and then conveys it into the trailer with a belt conveyor has been shown to reduce cutting power requirement by 50% and conveying power by 70% and has achieved outputs of 60 t/h (Knight, 1986).

Trailed offset machines come in a range of sizes to match 50–150 kW tractors. Outputs range from 10 to 40 t/h, according to available power, crop density, chop length and field organization. Self-propelled harvesters (Fig. 24.6) generally contractors' machines, range in power from 110 to 350 kW. Outputs of 30 to over 100 t/h are possible, but this depends very much on the availability of large, even windrows and plenty of empty trailers. To reduce overhead costs, the machine's working season may be extended by the harvesting of a succession of crops throughout the summer and autumn.

Self-loading forage (SLF) wagons

An alternative one-person system of loading and transporting crops for clamp silage, the wagon, of 3.5–8 t capacity, loads itself by means of a reel pickup under the drawbar and a short vertical conveyor which packs the crop into the enclosed body. Up to 21 stationary knives may be positioned against the conveyor, to slice up the crop as it passes by. Chop lengths of 80–220 mm are achieved when knives are sharp. A chain and slat conveyor on the floor of the wagon can move the load rearwards as the machine fills and can discharge the contents through a rear gate in 3–10 min at the silo. When the trailer is loaded, the pickup is raised hydraulically and the combination is then driven to the silo.

The SLF wagon has a low requirement for power at the PTO, but a tractor of 35–50 kW is necessary to control the heavy trailer on hilly ground; three or four loads/h can be achieved over a transport distance of 500 m or less (MAFF, 1976b). The chop length is longer than that recommended for wilted silage.

Baled silage

Silage crops may be handled as unit loads if they are baled using large round balers or high-density large rectangular balers. These balers can also be used for packaging straw, but are not suitable for baling hay unless it is drier than 20% MC or treated with a preservative to prevent moulding. The heavy wet crop causes more wear to the baler than a similar amount of dry material.

Bales are handled by tractors with front and rear forks, and may be loaded onto flat trailers for haulage over greater distances. Individual bales may be enclosed in polythene bags of 80–125 μm thickness, or wrapped in four layers of thinner 'cling' material, or a stack may be built and covered with plastic, similar to the conventional clamp but without the cost of the walls. No further compaction is necessary, as the bales are compressed already. A high stack may be built from ground level without any need for a machine to be driven on to the material. Forage should be wilted to 30% DM or more to eliminate effluent problems and minimize settlement when stacked.

Individually wrapped bales offer the advantage of allowing extra forage to be ensiled without the need for additional silo space; batches of silage can be kept separ-

Fig. 24.6 The largest self-propelled forage harvester can achieve spot work rates of over 100 tonnes/h. Courtesy Claas UK Ltd.

ate, and unneeded silage can be sold off the farm as a cash crop. Care must be taken to protect the wrapping membrane from rodent, bird and wind damage. Stacks of baled silage must be carefully sited to avoid any risk of polluting a watercourse with the effluent (MAFF, 1991).

Silo filling

Handling at the silo must keep pace with harvesting equipment. A tractor with rear push-off buckrake can handle up to 30 t/h. Greater outputs are possible with a front-mounted buckrake, and more still from four-wheel-drive handling vehicles with fork capacities of 1–2 t. In addition to spreading the crop the machine also compresses it. Final densities of 700–850 kg/m^3, after fermentation, are common.

Silage may also be stored in tower silos, which reduce storage losses and permit every stage of silage handling to be mechanized. Tipping trailers deliver precision-chopped forage into a stationary dump box, which feeds it at up to 30 t/h into a forage blower driven by a tractor of 40–80 kW. The forage is blown up a vertical pipe 225 mm in diameter, over a curved section and against a spreading device in the centre of the roof of the silo. Spreading is essential if the silo is to be filled evenly and its potential storage capacity is to be fully utilized. A minimum filling rate of 2 m depth/d ensures satisfactory fermentation.

During the early stages of fermentation, gases such as carbon dioxide and oxides of nitrogen are evolved. If the silo is partly filled, these gases accumulate above the silage and can kill anyone climbing down on to the silage, perhaps to remove plastic sheeting after a weekend. Great care must be taken to clear any such hazards before entering the silo. *Never* enter a silo or other enclosed space (empty slurry tank, grain silo) without a rope attached, and the supervision of companions capable of giving assistance (without entering the silo themselves) should an emergency arise.

Baling and bale handling

A convenient system for handling hay (and straw) is to package it into bales. These may be moved by hand or grouped into larger unit loads for machine handling. Large bales must be handled by machine (Fig. 24.7). The process adds nothing to the final usefulness or value of the product; indeed this may be reduced. Handling as bales can offer a least-cost solution to the problem of moving bulky materials on the farm.

The conventional pickup baler

This machine produces bales of section 350 × 460 mm, and variable length. Larger sections such as 400 × 450 mm are used on some high-output machines. Bale density, which is adjustable, varies from 160 to 220 kg/m^3, and bale weights range from 15 to 40 kg.

The offset spring-tined pickup lifts the crop into the machine. Its height is controlled by skids or small wheels at either side.

The crossfeed moves wads of material from behind the pickup and feeds them laterally into the bale chamber each time the ram draws back.

The heavy ram slices off each wad from any trailing material, and forces it down the bale chamber towards the rear of the machine. As the compressed material moves along, driven by the ram working at 60–95 strikes/min, it carries with it two loops of string which will eventually form the securing bands of the bale.

When sufficient material has passed the bale length sensor, the tying cycle is initiated. This is timed to occur when the ram is forward. The needles bring up the ends of the strings to encircle the bale, place them in the knotters, and withdraw, trailling out a new string ready for the next bale. The knotters tie each bale string by forming a loop and pulling the two ends through. The strings are cut at the knotters to release the bale, which continues rearward past the adjustable restrictor that controls bale density, and is finally discharged via the bale chute.

Many balers have potential work rates of 15 t/h or more; typical field performance is 6–8 t/h. Large, even windrows allow the machine to operate steadily at high rates of throughput.

Handling conventional bales

The conventional bale was originally intended to be a suitable unit for manhandling, and can still be handled that way if labour is available. It remains a convenient unit for feeding to livestock.

Ideally, a bale handling system will deal with bales at all states – field stacking, transport from field to store, loading the store, and unloading from store to final use. It is very unusual for a conventional-bale handling system to fulfil all these requirements; many are only used for loading trailers in the field, which is a very small part of their potential.

Many mechanized bale-handling systems are available. Most make larger units by forming the bales into groups of 8, 10, 20 or multiples of these numbers. A collector sledge may be pulled behind the baler, and the bales released at intervals to form windrows. These windrows are formed into the desired groups by hand. Alternatively, the groups can be formed automatically on an 'accumulator', which releases each group when it is complete. These groups are dotted all over the field unless a further collector is drawn behind.

Groups of bales are picked up by specialized lifting attachments on tractor fore-end loaders or other handling machines, and may be built up into larger stacks or loaded directly on to trailers. Stacks of 40–144 bales may be picked up as units by specialized trailers and moved directly to store.

A survey has indicated the labour requirement for many systems of handling bales from windrow to store (MAFF, 1975, 1976b). The figures shown below are averages, and considerable reductions were achieved by the best or 'premium' operators.

- Bale/collect; hand-load trailers; hand stack: 83 man min/t
- Bale/accumulate 8 s; front-end loader to trailer: hand stack: 49 man min/t
- Bale/accumulate 10 s; field stack 100 s: specialized trailer to store: 26 man min/t
- Bale/self-loading wagon 88 s; place stack direct into store: 21 man min/t

Fig. 24.7 Moving round bales with a telescopic handler. Courtesy JCB Ltd.

It must be remembered that labour requirement is only one of a number of criteria on which a bale-handling system is chosen. Other factors may be overall rate of work, capital costs of the equipment, or the suitability of a system for an existing storage building. The final solution, as with other materials handling problems, must deal with bales at least cost when all forms of cost are taken into account.

Big balers

Handling crops in units weighing from 150 kg to 1 t allows fast rates of work, generally demanding only a single operator. All handling must be by machine, so store layouts have to allow this.

Large round baler

These machines produce a cylindrical bale, either 1.5 m wide × up to 1.8 m diameter or 1.2 m wide × up to 1.5 m diameter. Bale weights range from 150 to 800 kg, with densities from 90 to 140 kg/m^3. The bale is produced in a cylindrical chamber in the form of a roll. The machine has no crossfeed, so careful swath building and care in aligning windrow and pickup are essential for the production of good-shaped bales.

Fixed-chamber machines accumulate material until a predetermined tension is attained, while variable chamber systems, generally giving a higher density bale, allow any diameter of bale to be produced up to the full dimension of the bale chamber.

When the bale chamber is sufficiently full, the operator stops forward travel and initiates the wrapping of six or more turns of string round the bale as the baler continues to turn. The string is cut, but not knotted, and the bale is discharged through a rear gate. The bale rolls down a ramp and away from the baler to give room for the rear gate to close. If the windrows are close together it may be advisable to reverse the baler out of the row so that the baler can be dropped in a position where it will not interfere with the next pass of the machine.

The machine can be fitted with equipment for wrapping each bale with about two turns of a self-clinging plastic net. Because of reduced wrapping time, the output of the baler is increased by up to 50%. This advantage can often outweigh the extra cost of the attachment, and of the netting which is more expensive than conventional twine. Continuous baling, with no stops for wrapping, is achieved by the use of a 'feed chamber' where picked-up material can build up while wrapping and ejection are taking place.

Round bales can be carried by a grapple or prong attachment on the front or rear of a tractor, or may be loaded onto flat trailers. The smaller size allows two bales to be loaded across the width of a trailer or lorry, which is not possible with the larger size.

The bales may be self-fed, or can be unrolled on the ground. Mechanical unrollers, shredders and tub grinders may also be used to break up the bale.

Large rectangular high density baler

Packages weighing up to 0.5 t or more are produced by these very large machines, which require a tractor of at least 75 kW. Bale size varies: one popular machine produces a bale of 0.8 × 0.8 × up to 2.5 m, at a density of 300 kg/m^3 or more. This system is suitable for handling very dry hay, straw or silage crops (Fig. 24.8).

The bales are too heavy to be handled by smaller tractor front-end loaders, but are well within the capabilities of specialized handling vehicles fitted with grapple or squeeze-loader attachments.

Grain harvesting

The combine harvester

Grain crops (and also pulses, oil seeds and other seed crops) are now universally harvested by the combine harvester. This machine combines four major functions: it gathers the crop, threshes the seed out of the ear, separates the seed from the straw and finally cleans the seed of chaff, weed, seeds and dust. After temporary storage in a hopper on the machine the grain is delivered into a bulk trailer or lorry.

Most machines are self-propelled with cutting widths ranging from 2.1 to 6.7 m: most machines are offered with a choice of two or three alternative cutting widths. Engines of 32–220 kW are fitted. Although the 'size' of a combine is often described by quoting its cutting width, a more useful indication of its harvesting capacity is its power, or the width of the threshing cylinder. Under good working conditions, the factor which normally limits a combine's rate of work is its ability to shake out the grain from the remainder of the 'material other than grain' (MOG) in the separating mechanism. This ability varies from crop to crop according to maturity, moisture content, straw-length and threshing adjustments and so cannot be quoted as a constant figure.

A typical grain harvesting season contains 12–20 dry working days (up to 200 working hours), according to geographical location. While it is common to select a combine size which will provide approximately 1 m of cutting width for every 35 ha of grain (up to 60 ha in a few cases), an alternative approach is to specify a harvesting capacity which can deal with an expected crop within the working time available in the region. As explained earlier, no firm output figures can be given, but the following approximations are satisfactory for planning purposes; machines of 38–60 kW can harvest 4–7 t of grain/h, 60–80 kW, 6–10 t/h; 80–100 kW, 8–12 t/h; more than 100 kW, 11–16 t/h. The largest 'rotary' combines can attain rates in excess of 24 t/h in good conditions.

Fig. 24.8 This high-density baler produces bales 0.5 × 0.8 × up to 2.4 m long. A silage bale 1.5 m long will weigh up to 0.5 tonne. For straw, longer bales are made. Courtesy Claas UK Ltd.

Inadequate combine capacity leads to an extended harvest period with greater weather risk and rapidly rising levels of crop loss from shedding, both in the field and at the combine intake. A doubling of pre-harvest losses can occur after a delay of only two weeks in some crops (Klinner, 1980).

Most combines operate below full capacity. More than two-thirds of machines are used less than 200 h/year. The availability of excess harvesting capacity is a form of insurance against bad weather at harvest time, which also unfortunately leads to high overhead costs/ha harvested. A further form of insurance against a poor harvest season is the installation of grain drying equipment (see Chapter 14 and also Elrick, 1982).

Gathering

The crop is gathered into the machine at the 'table' or 'header'. It is drawn in by a rotating reel, which can be moved forward, lowered and speeded up to deal with a lodged crop. Unnecessary speed or reach can cause shedding, particularly in over-ripe crops. Crop lifters can be used to slide the crop gently up on to the table; in one trial the use of crop lifters in lodged barley was found to halve grain losses at this point. Header losses can also be reduced by cutting only against the lean of the crop.

The crop is cut by a reciprocating knife, with serrated blades to grip the hard stems. Control of the height of cut, which is adjusted hydraulically, is the most demanding task facing the combine operator. Many larger machines have automatic table-height controllers, and other machines have flotation devices which allow the table to follow ground contours.

The cut crop falls into the table and is drawn towards its centre by a double flight auger. Loss of grain from the table itself is minimized by the length and profile of the deck, which is shaped to retain loose grain should any be threshed out of the ears by the reel or the auger. The crop leaves the header via a conveyor in a band 0.8–1.6 m wide to match the width of subsequent mechanisms in the machine.

To increase the versatility of the machine, most tables can be detached quickly for transport or storage, and alternative gathering equipment is available for such crops as maize, windrowed crops, sunflowers or edible beans. A 'rotary stripper' which picks the ears off the standing straw has been shown to be more effective than a conventional header in gathering-up lodged crops. In addition, the amount of crop material entering the combine is halved, allowing faster working rates (Klinner *et al.*, 1987).

Threshing

The crop is threshed by a rasp-bar cylinder, or 'drum', working against an open grate concave which matches its curvature for almost one-third of its circumference. The impacts of the beaters on the crop at up to 30 m/s, and of the moving crop on the stationary concave, break the seeds loose. Severity of treatment is adjusted to match the type and condition of the crop by altering cylinder speed or varying the gap between concave and cylinder through which the crop is forced. Excessive threshing severity will break grains, while insufficient treatment will leave grains attached to the ears which will be discharged with the straw and lost.

In addition to the threshing action of the cylinder and concave, 70–90% of threshed grain is separated from the MOG by passing outwards through the spaces in the concave.

Separation

Separation of the remaining threshed grain from the straw is the function of the straw walkers, which offer a separating area of 1.8–7.9 m^2. To shake out the grain, the threshed mass is agitated as it moves rearwards, finally to be discharged on to the ground. The thicker the mat of straw on the straw walkers, the more difficult is the task of separation, and the situation is even worse if the straw is damp or broken up. Beyond a certain MOG throughput, different for every machine, satisfactory separation cannot be completed within the length of the straw walkers and grain is carried over the back of the machine in increasing quantities. Electronic loss monitors may be fitted at the rear of the straw walkers; with their aid (if regularly calibrated), the forward speed (and hence throughput) of the machine can be adjusted to maintain an acceptable level of loss at this point.

Most manufacturers have supplemented the action of the straw walkers on their conventional machines with rotary or reciprocating beaters, which shake out the mat of straw and allow trapped grains to escape. At the top end of the performance and price scale, 'rotary' models dispense with straw walkers in favour of one or several rotary separators. The crop is handled fast, in a thin layer, which gives more positive separation, at the cost of a greater power requirement and more broken-up straw behind the machine.

Cleaning

Impurities are removed from the crop by a combination of size and density separation. An adjustable air blast moves less dense particles to the rear of the machine, while two, or three, sieves of total area 2.0–6.5 m^2 prevent particles larger than those required from passing through. Adjustable sieves are convenient, but interchangeable fixed-size sieves are more precise in their dimensions, and also permit greater throughputs of smaller-sized seeds. Grain losses over the sieves can be detected by a grain-loss monitor in the same way as straw walker losses.

Grain losses

Losses of grain from the combine can occur at several points, for some of the following reasons:

(1) gathering losses – reel too aggressive; crop over-ripe; knife too high; crop lodged; inaccurate driving;
(2) threshing losses – cylinder too slow; concave clearance too large;
(3) separating losses – forward speed too high; straw walkers blocked;
(4) cleaning losses – too much wind; sieves insufficiently opened; sieves blocked.

Table 24.8 shows acceptable loss levels. In some cases, a greater straw walker loss may be acceptable, as this will allow harvesting to progress at a faster rate.

Grain losses are measured in the field by collecting and quantifying the lost grain from a known area. If the combine is stopped when harvesting an 'average' part of the field, losses may be assessed in the standing crop in

Table 24.8 Acceptable grain losses from a combine harvester in good harvesting conditions (Busse, 1977)

Gathering loss	0.2%	14 kg/ha in a 7 t/ha crop
Threshing loss	0.1%	7 kg/ha in a 7 t/ha crop
Straw walker loss	1.0%	70 kg/ha in a 7 t/ha crop
Sieve loss	0.2%	14 kg/ha in a 7 t/ha crop

front of the combine to give pre-harvest losses. After reversing the combine a short distance, lost grain from beneath the centre of the machine in its initial position is the sum of pre-harvest and gathering losses. Behind the machine lies the straw, with unthreshed grains still attached, and the sum of all types of loss. By subtraction, the extent of each individual type of loss may be determined, and the appropriate corrections can be made.

A convenient unit area is 1 m^2, which should extend across the full width of cut and can be marked out with a string and four pegs. All lost grain within this area is then gathered up. Table 24.9 shows the relationship between the number of grains found and the loss of grain/ha.

Table 24.9 A method of assesing grain losses

Crop	*Seeds/m^2 to equal 1 kg/ha*
Barley	1.7
Maize	0.2
Oats	1.6
Rape	20
Wheat	2

Effect of slopes on grain losses

Much grain is grown on sloping land. The combine works best on level ground, and inclination affects the operation of the straw walkers and sieves (Pascal & Hamilton, 1984). In particular, sieve losses increase very rapidly with increasing side slope (PAMI, 1977).

Since the sieves are the combine component most vulnerable to side slopes, side-levelling sieves are standard on one range of machines. Another manufacturer offers sieves that can oscillate laterally to counteract the effect of a side slope.

Hillside combines can automatically level the body of the machine on land sloping as steeply as 18 degrees up, 6 degrees down and 22 degrees to the side, a combination also being possible. Side-hill combines can accommodate side slopes only, to 11.5 degrees. Within these limits, the adverse effects of slopes are completely eliminated, and maximum throughput is possible.

Hillside combines cost about 70% more than the equivalent conventional model, while the side-hill type costs 20% more.

Monitors

Many combines are equipped with sealed sound-proof cabs, often air-conditioned. Even without this barrier it is very difficult for the operator to keep in touch with all the processes and components in the very complex machine. In addition to loss monitors, most of the larger machines have as standard, or can be fitted with, a range of 'function monitors' which detect blockages or malfunctions and alert the operator before a major breakdown can occur (Fig. 24.9).

Yield mapping

Some combines can be fitted with instrumentation that weighs the harvested grain as it enters the tank, and ground positioning systems. The information so gathered is processed into a detailed yield map of the field. Use of the yield map allows very precise control of subsequent husbandry inputs in order to optimize production with a minimum use of resources.

Fig. 24.9 A combine harvester working in wheat. Information is displayed on a TV screen in front of the driver. Courtesy Massey Ferguson (UK) Ltd.

Root harvesting

Potato harvesters

Lifters

Where labour is available, potatoes, particularly for the valuable early market, may be harvested by hand. The tubers are picked from the ground after the ridge has been broken up by a lifter or 'potato digger'. There are two types, a power-driven tined 'spinner', where the ridge is broken up and spread over previously picked ground to expose the tubers, or the elevator-digger on which the ridge is undercut by a flat share and then conveyed up and back by an elevator of round steel crossbars or 'webs'. Loose soil falls between the webs and the tubers are carried over the back to fall on top of the sifted soil.

Complete harvesters

In order to reduce the high labour demand at potato harvest, the elevator-digger has been developed into a complete harvester, which can deliver the potatoes directly into sacks, boxes or bulk trailers. More than 65% of the UK crop area is harvested by 'complete' machines (Statham, 1981). The web elevator has been extended, haulm-removing devices added, and most machines have a divided horizontal conveyor where a gang of four to eight pickers can remove stones and clods, or tubers. The tubers are then directed into the chosen handling system.

Many growers have adopted unmanned machines, typically harvesting two rows at a time, where any clods, stone or trash that cannot be separated by the machine are harvested and dealt with subsequently at the store by hand sorting, or by automatic sorting machines. Rates of work of 0.75–1.25 ha/row daily can be expected.

The mechanization of the potato harvest has brought with it the two problems of crop losses and crop damage. Most of the 1.25–2.5 t/ha that are left in or on the ground in a typical potato field are lost at the share of the machine. If the share is too shallow, tubers will be missed or sliced. If soil does not pass freely back over the share, perhaps under loose or wet conditions when working downhill, or when the share is trailing haulm, soil and tubers will be 'bulldozed' ahead of the share and some will be lost by spillage to either side. Improved harvester designs incorporate fine control of share depth, and may include automatic depth control.

Tuber damage, typically as high as 20% severe damage, occurs mainly on the web elevator which is also a frequent source of breakdowns. Forward speed of the machine and the amount of web agitation must be carefully matched to web speed to maintain a cushion of soil over the elevator. Designers are reducing the damage potential of machines by minimizing the length of web elevators and eliminating drops and changes of direction of the tubers.

A further site for tuber damage is the drop into the bulk trailer. This should be no more than 200 mm, which requires constant adjustment of discharge conveyor height as the trailer fills. Great care must also be taken when discharging the hopper of tanker-type machines.

Tubers can also be damaged by contact with stones during harvesting or transport. In stony ground the removal or windrowing of stones at planting time can significantly reduce tuber damage, as well as speeding the harvest operation (McRae, 1977).

Sugar beet harvesters

The entire sugar beet crop is harvested by machine. The harvesting process involves first cutting off the tops, which may be conserved for livestock feeding but are more usually discharged back on to the ground. The roots are then lifted from the soil, most commonly by a pair of inclined, sharp-edged, spoked wheels which cut into the soil to either side of the row and lift the root between them while soil can fall away through the spokes. A minority of machines used fixed shares. The beet is then raised by a web elevator which further cleans the crop, before discharging into an integral hopper or into a trailer running alongside.

Harvesters range from one-row trailed or self-propelled models, through three-row trailed machines, popular with larger growers, to self-propelled five- or six-row systems suitable for the largest producers and contractors. Rates of work of 0.125–0.2 ha/h per row can be expected.

Crop losses result from incorrect topping, from failing to gather whole roots into the machine, and from leaving the lower parts of roots in the ground. Typical losses are 2.5 t/ha, but over 30% of growers leave as much as 5 t/ha. One-third of these losses are visible as loose beet on the surface, the remainder being below ground. One 0.6 kg root lying on the surface in a 20 m length of row represents a loss of 2 t/ha (Davis, 1977).

The two most common faults in machine operation which lead to crop losses are failing to operate the lifters sufficiently deeply, and failing to position the lifters exactly on the row. The latter fault can be eliminated by the provision of automatic steering on the lifters, or by the use of quite inexpensive guide skids to centre the lifters on the row.

Faulty topping units can damage the beet, and may loosen the roots in the ground so that they are knocked out of the row and are not gathered in by lifting mechanism. Both the topping knife and the feeler wheels, which adjust the knife to the correct position below the beet crown and steady the root while it is topped, should be sharpened regularly. Lightweight topping units which locate the beet crown by feeler blades or spikes have been introduced. These can respond more rapidly to beet of different heights, and will work accurately at speeds up to 8 km/h. (Traditional toppers are not accurate at speeds beyond 4.5 km/h.) The skew-bar topper is a rotary, power-driven device which scrapes the leaves off the crown of the beet, leaving a domed shape. It is claimed to remove unwanted material thoroughly even at a fast forward speed, but takes off less of the actual root than a knife topper (Billington & Butler, 1984).

Harvested beet are often piled at the roadside for considerable periods before being transported to the factory. Losses are minimized if a concrete area is available for storing and loading the crop. The roots are removed from the pile by front-end loaders; industrial loaders can move up to 90 t/h, rough terrain fork lifts 60 t/h, tractor front-end loaders 20–30 t/h; slatted buckets allow some soil to escape, and this process can be continued by passing the beet through a cleaner-loader on its way to the lorry.

Irrigation

The use of irrigation equipment enables the farmer to prevent crop production from being limited by moisture stress at any time during the growing season. Benefits will include good early establishment, even development of the crop, and increased crop value through uniformity and enhanced yield and quality. Additional uses for irrigation equipment include softening the ground to aid seedbed preparation or root harvesting, the application or washing-in of fertilizers and other chemicals and the prevention of wind erosion (Willis & Carr, 1986). Irrigation systems may be installed specifically for the application of livestock wastes, or for the protection of sensitive crops from frost.

System capacity

Each irrigation treatment will typically involve the application of the equivalent of 25 mm of rainfall, and the capacity of the system should allow the whole area under irrigation at a particular time to be covered in 6–8 d to allow for a peak depletion rate of over 3 mm/d. With careful management, a modern irrigation system can be kept in operation for 20–22 h/d. The application of 1 mm of water to 1 ha (1 ha mm) represents 10 m^3 of water. Where an irrigation system applies 25 mm to 6 ha in a 22 h day, its water output is 68.2 m^3/h and it has the capacity to cover 48 ha in 8 working days.

With a traditional sprinkler system, a '1 acre setting' applying 25 mm to 0.4 ha in 3 h and covering 4 settings/d, water output will be 33.3 m^3/h and 12.8 ha can be irrigated in 8 working days.

Water supplies

Irrigation is impossible without a reliable supply of water of a quality appropriate to the crops being treated. The local Water Authority must always be consulted before abstracting water (MAFF, 1983a).

Direct abstraction

Because virtually no source works are required, the cheapest water supply is a river or stream from which water can be pumped directly into the irrigation system. Very few new direct abstraction licences are issued by Water Authorities because of the need to maintain flow in rivers during the dry part of the year. In areas where underlying geological structures are suitable, boreholes may offer a source from which water may be pumped directly on to the land.

Water storage

A significant component in the cost of most new irrigation systems is the construction of a reservoir in which winter water can be stored for summer use. Rivers, streams or field drainage systems may be used as sources of supply. Where the supply is irregular, measurements must be taken to ensure that an adequate volume of water will be available to service the planned irrigation system.

Total irrigation need for the season must be calculated, to arrive at the required storage capacity. Likely application quantities, ranging from 25 to 150 mm according to crop, are multiplied by the area of each crop to be irrigated, in order to give the total water requirement. An additional amount, normally equivalent to 500 mm depth in the reservoir, is allowed for seepage and evaporation from the reservoir.

The reservoir may be constructed by enlarging a hollow or closing-off a valley. In flat country, a rectangular bank is built up from the soil excavated from the area enclosed by the bank. Field drainage systems and any porous subsoil layers must be removed. If suitable clay is available, preferably from the excavation itself, this may be used to line the reservoir; if not, a lining of flexible sheet material must be used. The need for a lining of this type can double the cost of a reservoir.

Banks must be protected from the effects of weathering, and from wave action at the waterline. Points of water entry, and spillways to allow for overflow, must be reinforced to resist the erosive effects of concentrated water flow. The advice of a chartered civil engineer should be sought in the design of reservoirs and associated works, especially where reservoir capacity exceeds 20 000 m^3, or banks are higher than 5 m.

Ideally, the reservoir will be filled by gravity. A channel taken from far enough upstream can be led directly into the reservoir. Flow is controlled by a sluice at the upper end of the channel. Where this is not feasible, water is raised into the reservoir by a low-lift, high-volume pump positioned as near as possible to the supply water level.

Pumping

Centrifugal pumps

Because of their simplicity and reliability, centrifugal pumps are almost universally used in clean water irrigation systems. One or more vaned rotors spin the water to build up its kinetic energy, and this is converted to pressure energy due to the internal shape of the pump body. This pressure then causes the water to flow through the system. Increasing the rotational speed of the pump increases pressure and flow. Typical rotor speed is 1800–2800 rev/min; this allows direct drive from diesel engines or electric motors, but requires gearing-up if the pump is to be driven by a tractor PTO.

The maximum efficiency of a centrifugal pump is 65–80%, and the peak is at a very specific combination of flow and pressure. The pump for any particular irrigation system should be selected and then operated so that it will work under conditions at which its greatest efficiency will be realized.

In relation to its maximum delivery pressure of 9–15 bar, the inlet suction or lift capacity of a centrifugal pump is small. Most are fitted with a hand-operated priming pump so that they can be filled with water before start-up. The ideal position for a pump is on a floating raft, as close as possible to the water level, whether this is low or high in the reservoir. Where this is not convenient, the pump should be located on the bank as close as possible to the water.

Pump controls

In addition to manual start and stop controls, irrigation pumps are normally fitted with safety controls which will shut down the system if the motor becomes overloaded (electric drive) or if the engine temperature rises or oil pressure falls beyond a pre-set value (diesel drive). The system will also shut down if water pressure falls, indicating a loss of inlet prime.

To allow adjustments to be made to application equipment without a double visit to the pump, high- and low-level pressure switches can stop the motor when hydrants are closed prior to an adjustment, and start it up again once the applicators are re-set and the hydrant is re-opened. Timers can be used to set a given period of operation, allowing the final irrigation setting of the day to finish without supervision.

Pump selection

While operating efficiently, the pump must deliver a sufficient *flow* of water to meet the output requirements of the system, at *pressure* sufficient to overcome any resistance due to the height the water must be lifted from source to application point, and due to the length and narrowness of the pipes in the distribution system and the restrictions caused by valves, bends and other fittings (MAFF, 1982). This resistance is expressed as a 'head' or pressure; when this is added to the pressure required at the discharge appliance (typically 3–10 bar), the delivery pressure required from the pump may be computed.

Distribution of water

Where appropriate, water is distributed from the pump via a 'skeleton' of permanent underground 'mains', usually of 150 mm diameter, terminating in carefully positioned hydrants. From these points, portable aluminium mains, typically of 100 mm diameter in 9 m lengths, carry the water through the area to be irrigated, with further hydrants at spacings to suit the particular method of irrigation.

Aluminium irrigation pipes are good conductors of electricity, and there have been many serious accidents where pipes, being carried upright, have contacted overhead electricity cables. The greatest care should be taken when irrigating near overhead cables, both to avoid direct contact by pipes, and to avoid contact by solid water jets from large rain guns, which can also conduct electricity.

Application

Nozzles

When water is forced out through a nozzle, it forms a jet which atomizes, or breaks up into droplets in a range of sizes. At low pressure, there is little atomization, and a solid stream of water may result, while at high pressure there is very thorough atomization, resulting in a wide, feathery spray which does not carry far and is easily deflected by the wind. Between these extremes, there is an optimum range of pressures at which atomization and throw are satisfactory, and flow rate is sufficient. It is unusual for too-high pressures to be a problem, but many irrigation systems suffer from lack of pressure and this can result in poor distribution from nozzles, reduced water application and soil damage from the larger droplets which will be produced.

Larger droplets are produced by larger nozzles, even when operated at the recommended pressure, and the use of such nozzles carries with it the risk of soil damage from impact by droplets larger than 3–5 mm, especially when the soil is not fully protected by a growing crop. For this reason, 22 mm is the largest nozzle diameter recommended for use in beet or potato crops, while 28 or 30 mm nozzles can be used on grassland (Bailey, 1986a).

Rotating nozzles

Most irrigation systems incorporate one or more rotating, upward-facing, nozzles to discharge the water. Sprinklers, incorporating one, or two, nozzles, have a discharge rate of 0.5–5 m^3 h and are used at spacings of 12–24 m (Fig. 24.10). Rainguns, generally with a single large nozzle, have discharge rates of up to 150 m^3 h or more and may be spaced up to 140 m apart and irrigate over 1 ha from a single point (Fig. 24.11).

Rotation is by small increments, caused by reaction with the main water jet. Either a hammer is pushed back by the jet and then swings forward to strike the nozzle round, or an angled plate swings in and out of the jet, and the deflection of the water to one side causes the nozzle to move a small distance in the opposite direction. The resulting pattern of water distribution peaks directly below the nozzle, and decreases with increasing distance from that point. Satisfactory uniformity in still air is achieved by operating nozzles at their recommended spacing.

In windy weather, the circular pattern of a sprinkler or raingun is elongated into an egg-shape in the downwind direction, covering a smaller and narrower area and resulting in a slightly higher application rate within that area. Where possible, mobile irrigators should move across the prevailing wind direction; where this is not possible, spacing should be reduced to 70% of that recommended for still air conditions, in order to improve uniformity of application (Bailey, 1986a).

Many rotating nozzles are available with a 'sectoring' attachment, and can be set to cover only a segment of a circle after which the (single) nozzle flicks back to the start point and repeats the segment. This feature is useful when irrigating awkward areas, and is particularly suited to mobile irrigators. The water is sprayed away from the direction of movement of the irrigator so that the trolley carrying the raingun is able to move forward on to relatively dry ground.

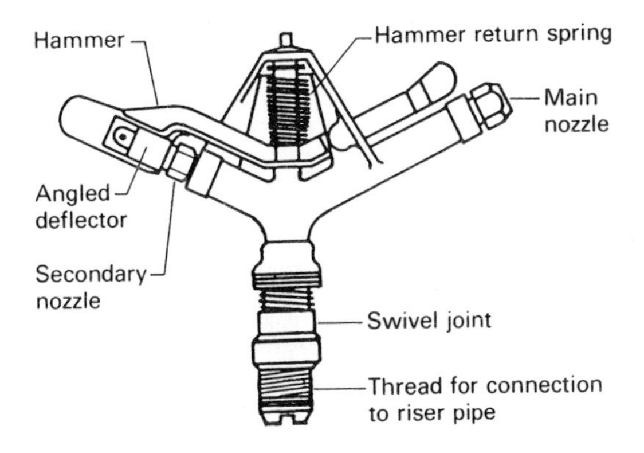

Fig. 24.10 A sprinkler (courtesy of Wright Rain Ltd).

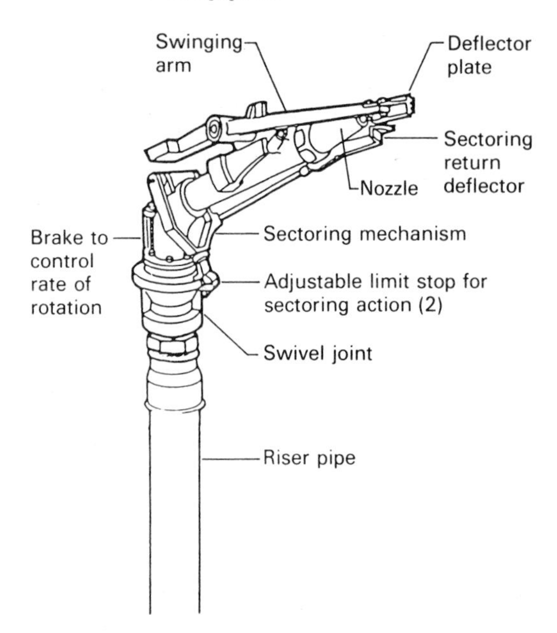

Fig. 24.11 A rain gun (courtesy of Wright Rain Ltd).

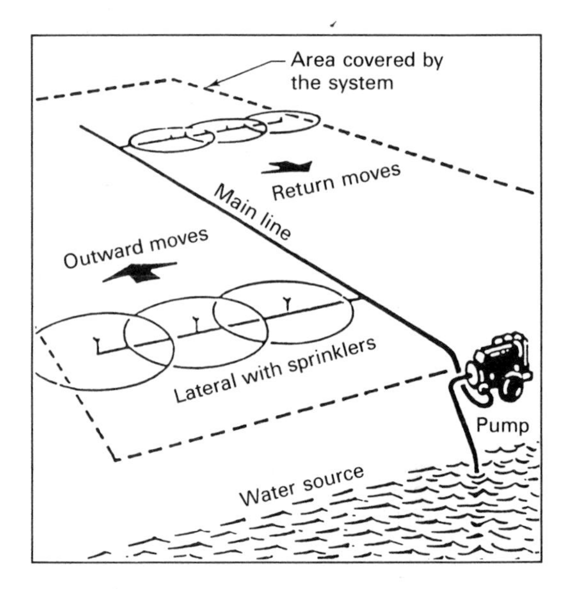

Fig. 24.12 Layout of a sprinkler irrigation system (courtesy of Wright Rain Ltd).

Fixed nozzles

Irrigation water can be discharged from many small nozzles spaced along a boom, which itself moves over the cropped area. The small droplets produced by these nozzles are gentle on soil and crop, even though high application rates are often involved. Some of the largest machines use downward-facing nozzles operating at pressures below 2 bar, carried 1–2 m above the crop. The result of this combination of features is to deliver a very uniform water application, little affected by wind, and using less energy for pumping than would be the case with a high-pressure system.

Sprinklers and laterals

A number of sprinklers (often 12), spaced along pipes of 75 mm diameter, is laid out to cover, typically, an area of 0.4 ha. The pipes, or laterals, are available in 6 and 9 m lengths to allow for different spacings, and sprinklers are normally held above crop height by vertical risers every 12–18 m (Fig. 24.12).

An application of 25 mm of water is delivered in 2–4 h, after which irrigation must stop and two or more people move the lateral to its next position. The work, in wet conditions after irrigation, is not pleasant, and involves the interruption of other activities at what is often a busy time of year.

'Self-moving' lateral systems have been developed in which the entire setting is moved on its own wheels from one position to the next between applications, generally powered by a small petrol engine. Another version comprises a wide rotating boom on a tractor-like self-propelled chassis. This irrigates a circular area, and is driven to its next position when the previous area has received sufficient water.

All these systems offer a reduction in labour over that necessary for moving conventional sprinklers, and allow the move to be completed more quickly, but attention is still necessary several times each day.

Solid-set sprinklers

The chore of moving sprinkler laterals three or four times a day can be eliminated where there are sufficient sprinklers and laterals to cover the entire area of a crop. The system is set out at the beginning of the life of the crop, and remains in place until no more irrigation is required.

Semisolid-set systems are also in use. Here, an extended area is laid out with pipes which offer many quick-coupling sockets into which standpipes and sprinklers can be inserted. Sockets not in use are sealed automatically. An 0.4 ha set of sprinklers can be moved from socket to socket across the area as irrigation proceeds, without the need to reposition the pipes. However, as with conventional sprinkler systems, attention is needed three or four times a day.

Mobile irrigators

The popularity of mobile irrigators is due mainly to the great savings in labour they offer over traditional sprinkler systems (see Table 24.10). The machine covers a large area per setting, and this allows a long period of operation before attention is required, to move it to a new position. Models are available to cover a great range of areas per setting, from less than 1 to more than 400 ha.

Because mobile irrigators are constantly on the move, they only actually apply water to a fraction of the set area at any one time. This fraction receives the full output of the machine, resulting in an extremely high spot application rate. Typical spot rates for hosereel boom machines are 60–80 mm/h (Bailey, 1986b), while the outer regions of centre-pivot systems can deliver more than 300 mm/h (PAMI, 1984). Such high application rates will cause ponding if the terrain is fairly flat, while runoff and possibly erosion will occur on sloping ground.

Table 24.10 A comparison of irrigation methods (after Curtis, 1980)

Type of system	*Approximate relative*		
	Cost	*Labour*	*Energy*
Hand moved sprinklers	100	100	100
Self-moving sprinklers	200	40	110
Solid set sprinklers	400	20	100
Rotating boom machine	160	60	180
Hosereel mobile	170	20	190
Hose trailing mobile	130	30	150
Centre pivot (low pressure nozzles)	250	15	75

Hosereel irrigators

Most new irrigation systems in the UK use hosereel machines (Stansfield, 1992). The main component is a large reel, carrying up to 400 m of up to 125 mm diameter hose. On the end of the hose is a wheeled trolley carrying a sectoring raingun capable of irrigating a strip of ground up to 140 m wide (Fig. 24.13). To prepare the machine for work, the reel is positioned at the headland of the field, and the trolley is pulled across the field by a tractor, unwinding the hose from the reel. The limit of operation is either the length of the hose or the width of the field.

When pressure is applied, the raingun irrigates a sector of ground facing away from the reel, and, after a delay, a water-powered motor begins to turn the reel, drawing in the trolley and raingun with a force of several tonnes. Application rate depends on nozzle output and rate of travel; it can take over 24 h to draw in the full length of the hose. Most machines have compensators to allow for the fact that the reel increases in diameter as the hose is wound on to it, but it is advisable to check the accuracy and consistency of speed settings from time to time.

When the trolley reaches the reel, a valve stops reel rotation; irrigation continues for a further pre-set period before that also is shut down. The machine must then be repositioned for another period of unsupervised work.

Some machines are fitted with a turntable so that the reel can be swung round to face in any direction without moving the chassis. This allows precise alignment with crop rows where these are not exactly at right angles to the headland, and it also facilitates irrigation to either side of a central laneway where a field is to wide to be irrigated fully from one side headland.

Where delicate crops are to be irrigated, or where poorly structured soils are not fully protected by crop cover, the raingun may be replaced by a boom up to 40 m in width carrying several sprinklers, or many smaller downward-facing nozzles, which will produce fine droplets and minimize impact damage.

An alternative and cheaper form of reel irrigator is one in which the reel is carried on the irrigator itself, but instead of winding in the hose a small diameter steel cable is used for propulsion and water is supplied by a flexible hose trailed on the ground behind the machine. The 'hose-trailing' machine is less power-consuming than the hosereel type, but requires more labour when moved.

Because of the long runs possible, reel irrigators can be made to operate for almost 24 h/d. Larger machines can work two 11 h runs, while smaller machines can be set for a single 22 h run. Where awkward fields do not allow this, short runs can be worked in the daytime, and long runs saved for night operation (Bailey, 1986a).

Centre pivot and linear-move irrigators

These machines allow a very uniform application of water to be applied automatically to a large area of land. Coefficient of uniformity for centre-pivot machines is 90–95%, compared with 50–60% for hosereel machines, and this allows more precise control of water

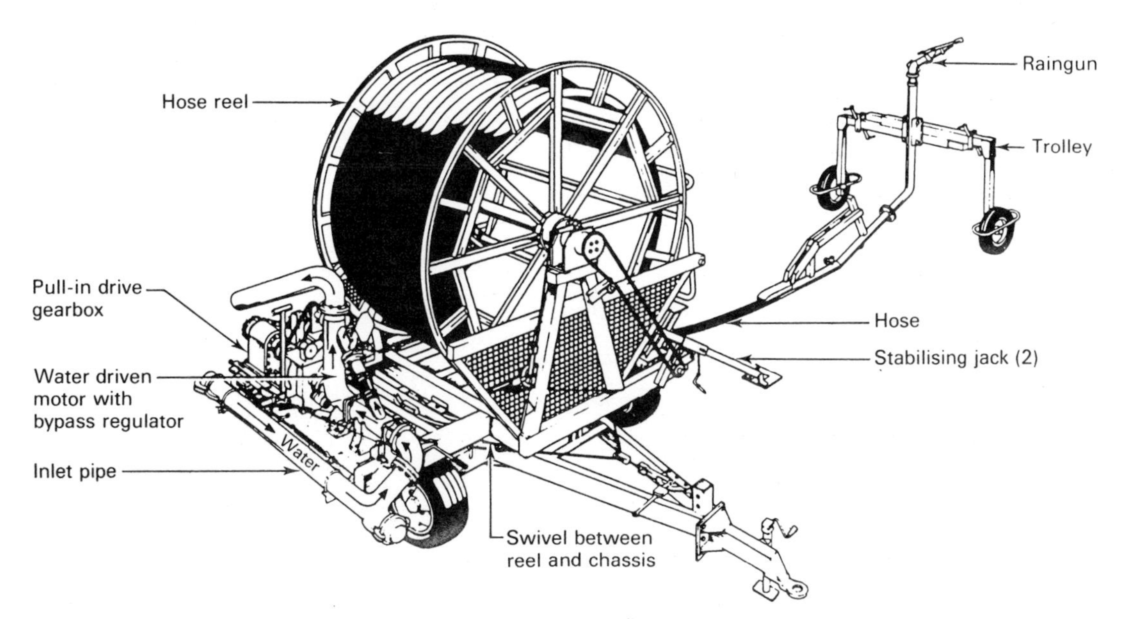

Fig. 24.13 A hosereel mobile irrigator (courtesy of Wright Rain Ltd).

use (Willis & Carr, 1986). Power requirement is also less, although capital cost is greater than for hosereel machines.

Both types of irrigator consist of a long boom often several hundred metres in length made up of a number of segments, each approximately 30 m in length. Wheeled towers at the outer end of each segment carry the boom above crop height. Sprinklers or downward-facing low pressure nozzles are spaced along the length of the boom.

Centre-pivot machines are anchored to a central mast, through which water and electrical power (for the propulsion system) are fed in. Driven by the wheeled towers, the boom then rotates around the mast, covering a circular area of land which can range from less than 20 to more than 400 ha. Rotational speed of 1–100 h/rev (Kay, 1983) is set by the travel speed of the outermost drive unit; sensors at each tower ensure that the boom is kept straight. Most makes can be fitted with a hinged outer extension, which will unfold to reach into the corners of a square area, guided by a low-voltage cable buried in the soil below plough depth.

Flat ground is ideal, both for the drive-system and because of the heavy application rate, but the drive system can negotiate undulating ground if necessary. Circular tracks are left by the wheels; hedges and banks must be opened up to allow the towers to pass. The area lost to wheel tracks is less than 1% of the whole (PAMI, 1984).

The size and spacing of output nozzles along the boom varies so that all parts of the cropped area receive the same application of water.

The linear-move irrigator follows a straight path, at right angles to the boom itself, and covers a rectangular area. Water enters the boom at one end, either from a trailing or reeled pipe or pumped directly from a supply ditch running down the field. Combination of both principles gives a machine which irrigates a rectangular area, plus a semicircle at either end.

Irrigation scheduling

During the growing season, the soil is depleted of water at a rate depending on the amount of water taken out by the crop (transpiration) and the amount returned by rainfall. When transpiration exceeds rainfall, a *deficit* develops. This is expressed either as the amount of rainfall or irrigation that would return the soil to field capacity (mm), or the tension with which the remaining water is held in the soil.

As water is depleted, crop growth begins to be affected. The tension or deficit (SMD) at which this takes place is known as the critical SMD, and will depend on the crop itself (see Table 24.11), its rooting depth and the available water storage capacity (AWC) of the soil in which it is growing (see Table 24.12). The scheduling of the timing and size of irrigation treatments must be based on an understanding of these variables. A deep-rooting crop on a soil with a high AWC and a high infiltration rate (see Table 24.13) can be given heavy, infrequent water applications which are only necessary when a large SMD has developed, while a shallow-rooted crop on a soil of low AWC must be irrigated frequently with smaller applications of water, to maintain a small SMD. Some crops are particularly sensitive to soil water conditions at certain stages of their growth, and this must also be taken into account (see MAFF, 1982).

Soil moisture tension may be measured directly by means of tensiometers buried in the ground. This system has the great attraction that no other measurement and

Table 24.11 Soil moisture deficit criteria for the major irrigated crops (MAFF, 1984)

Crops to be irrigated	*Recommended water per application (mm)*			*Critical SMD (mm)*			*Start SMD* (mm)*		
	Soil class			*Soil class*			*Soil class*		
	A	*B*	*C*	*A*	*B*	*C*	*A*	*B*	*C*
Early potatoes and canning: emergence onwards	25	25	25	25	30	30	20	25	25
Second early potatoes from tuber size 6 mm	25	25	25	25	30	30	20	25	25
Main crop potatoes from tuber size 10–20 mm	25	30	30	35	50	50	20	30	30
Sugar beet	25	25	25	25	30	50	20	25	25
May } 75% crop cover	25	25	25	35	35	50	20	30	30
June } 75% crop cover	25	40	50	45	50	100	35	40	90
July	25	50	50	55	75	125	50	60	100
August	25	—	—	50†	—	—	35	—	—
Cereals	25	25	25	25	35	35	20	30	30
Grass-grazing (May onwards)	35	50	50	35	50	50	30	40	40
Grass-conservation (May onwards)									

Start SMD is on assumption that it will take 2 to 4 days to irrigate an area and one has to anticipate the deficit building up.
† Up to anthesis.

Table 24.12 Classification of the available water capacities on common soil textures in approximately increasing order (MAFF, 1982)

A. Low	
AWC not more than 12.5% by volume (less than 60 mm/500 mm soil depth)	Coarse sand Loamy coarse sand Coarse sandy loam
B. Medium	
AWC more than 12.5% but not more than 20% by volume (60–100 mm/500 mm soil depth)	Sand Loamy sand Fine sand Loamy fine sand Clay Sandy clay Silty clay Clay loam Sandy loam Sandy clay loam Silty clay loam Fine sandy loam Loam
C. High	
AWC more than 20% by volume (greater than 200 mm/500 mm soil depth)	Very fine sand Loamy very fine sand Very fine sandy loam Silt loam Silty loam Peaty soils

no calculation is necessary; some irrigation systems are controlled automatically on the basis of tensiometer measurements. However, several tensiometers must be distributed in each field to give accurate results, and much care is necessary in their use (MAFF, 1983b).

More commonly, a running balance sheet of SMD is kept, either on paper or is a computer. Potential transpiration data are available, based on regional weather measurements, and these figures are combined with local rainfall, crop cover, and irrigation records to arrive at the weekly SMD for each field. Soil information for each field is needed, to indicate the critical SMD and the point at which irrigation should commence. Many consultancy services are available to assist the farmer in scheduling his irrigation applications.

Frost protection by irrigation

Some fruit crops are particularly vulnerable to damage from late frosts. In some cases, the entire crop can be lost. During a frost, developing buds can be prevented from falling below 0°C if they are kept wet by a continuous light application of irrigation water. The latent heat given up by the water as it freezes maintains the temperature at 0°C, and a layer of ice forms over the bush. An application rate of 2.5 mm/h will protect down to −3.3°C; 3.2 mm/h to −5.6°C (Ingram, 1984).

Quite a large outlay is involved in providing a solid-set sprinkler system and frost controls to protect the whole area of a vulnerable crop, but this could be recovered in a single season if the crop would otherwise have been lost.

Since there is no conflict with conventional irrigation, much of the equipment used for frost protection can also be used for this purpose.

Organic irrigation

Irrigation equipment may be used to transport and spread livestock wastes on the land. This can make manure disposal almost into a 'push button' operation, and also avoids the need to move heavy tankers on the land with attendant risks of soil damage and safety hazards on slopes.

Because of the fibrous nature of manure solids and the probability of hay, silage or straw being present, special pumps, often fitted with chopping attachments, are specified. Also to avoid blockage, rainguns with large nozzles are used, either static or as part of a hosereel irrigator. Only machines recommended for organic irrigation should be used, as liquid manure can affect the water motor and valves of conventional irrigators.

The suitability of liquid manure for handling in this way depends very much on its solids content: generally, slurries containing up to 8–10% DM are suitable for organic irrigation (Schofield, 1984). This corresponds to the dilution of faeces and urine combined with an equal quantity of water, or to the liquid fraction produced by a slurry separator.

When planning and using organic irrigation, it is essential to follow the guidelines set out in the Code of Good Agricultural Practice for the Protection of Water (MAFF, 1991).

Table 24.13 Infiltration capacities associated with some common soil textures (MAFF, 1982)

Category	*Equilibrium infiltration capacity range (mm/h)*	*Textures*
Very high	Greater than 100	Coarse sands, sands, loamy coarse sands, loamy sands
High	20–100	Sandy loams, fine and very fine sandy loams, loamy fine sands and loamy very fine sands
Moderate	5–20	Loams, silt loams, silty loams, clay loams
Low	Less than 5	Clays, silty clays, sandy clays

Feed preparation machinery

Grinding

Cereals and other ingredients of livestock feeds are ground into fine particles for a number of reasons:

(1) to improve digestion by breaking up the protective outer skin and increasing the surface area of the material;
(2) to increase intake;
(3) to aid mixing of ingredients into homogeneous final diets.

Desirable features of a grinding machine include:

(1) an end product of uniform particle size, with a minimum of dust;
(2) fineness adjustable to suit different classes of livestock;
(3) minimum power requirement in relation to output;
(4) ability to operate reliably without supervision.

Hammer mills

Grain is metered into the mill where it is sheared by sharp-edged beaters rotating at up to 100 m/s. Moving particles from this action are sheared again by the sharp edges of the perforated screen which surrounds the rotor. The process continues until the particles are small enough to pass through the perforations of the screen. A range of screens from 1 to 9.5 mm caters for all requirements. Most mills are equipped with an integral pneumatic conveying system which aids the flow of grain into the mill and also delivers the meal into a hopper or mixer. Less power is needed to convey the meal by auger, and some manufacturers use this alternative system. Throughput of a particular mill decreases with fineness and with increasing moisture content of the grain; 14–16% MC is considered ideal. Edges on rotor and screen become blunt with wear, and must be turned or renewed to restore performance. Typical output would be (45–85 kg/h)/kW of power available; machines range in size from 2.2 to 37 kW, those up to 7.5 kW being the most popular on farms.

The high working speed of the rotor makes the mill vulnerable to damage if metal objects enter it, so most machines are protected by magnets, which will retain ferrous metals. A considerable amount of fine dust is produced by this method of grinding, necessitating the use of cyclones and dust socks.

Hammer mills can operate unattended for several hours; many small machines are run at night, drawing corn from an overhead hopper which has been filled to the required level for making up a batch of feed. When the hopper is empty, a switch in its base is released, and this shuts down the mill.

Plate mills

The plate mill grinds by abrasion and shear. Grain up to 18% MC is broken up between the serrated surfaces of a pair of cast iron discs, one of which rotates as it is pressed against the other. Reduced clearance between the discs results in a finer product and a lower throughput. A more uniform meal is produced, with less dust than the hammer mill. Discharge is by gravity only, so a conveying system must be added unless the mill is positioned directly above a receptable for its output. The mill produces (60–110 kg of meal h)/kW of power available.

Roller mills

A roller mill crushes grain between a pair of flat surfaced or serrated cylindrical rollers. The rollers are spring-loaded against one another, and spring tension can be adjusted. Dry grain is cracked to produce a coarse meal, while grain of higher moisture content is flattened and crushed into flakes. Moisture content of 18% or more is recommended and the machine lends itself well to a system of high-moisture grain storage. Dry grain can be moistened to improve its rolling characteristics.

Flaked grain is almost dust-free and hence more palatable to livestock than dusty materials. Break-up of the fibrous portion of the grain can be minimized, making a suitable product for feeding to ruminants.

An output of (90–120 kg/h)/kW can be expected, and machines are available in sizes to 7.5 kW with electric drive, as well as larger PTO-driven models. Discharge is by gravity, so that further handling or receiving equipment must be installed. Flaked material is fragile; handling and mixing equipment must be selected and operated with this in mind if the product is to be preserved in this desirable form.

Mixing

When all ingredients of a ration are available in ground form they may be mixed into a homogeneous final product. Mixes are generally prepared in batches of 0.5–1 t. Ingredients are weighed into the mixer in most cases, but some ingredients may be added by volume into a calibrated container. Since different ingredients are of different densities, the container must be calibrated with care if the final diet is to be as required. The volume occupied by 1 t of various ingredients is shown in Table 24.14 (see also Table 24.15).

Larger operators may find it more convenient to produce diets continuously rather than in batches; equipment for this purpose is produced by several manufacturers.

Table 24.14 The volume of 1 t of various livestock feedstuffs

	m^3
Wheat	1.29
Oats	1.96
Barley	1.43
Beans	1.21
Wheat meal	2.13
Oat meal	2.38
Barley meal	1.96
Bean meal	1.85
Grass meal	2.85–3.90
Dry beet pulp	5.67–3.90
Fish meal	1.83–2.08
Soyabean meal	1.48–1.83
Pelleted ration	1.60–1.69
Crumbed ration	1.83

Table 24.15 Typical annual concentrate consumption of various types of livestock

	Tonnes
Dairy cow	1.25–2.0
Sow or gilt	1.0–1.8
Baconers (from 8 weeks)	0.7–1.0
Porkers (from 8 weeks)	0.5–0.7
1000 laying hens	35–45
1000 broilers	15–20

Vertical mixers

The vertical mixer is an upright cylinder, narrowing to a funnel-shaped base. Capacities of 0.5, 1 and 2 t are available. Ingredients may be added through the top, often from an upper floor or granary, or via a small hopper at the base. A central vertical auger circulates and mixes the ingredients. Adherence to the recommended mixing period of 12–20 min is essential; too short a period results in incomplete mixing, while an excessive time may result in separation of some ingredients. Automatic timers are available.

The mix is discharged from an outlet or sacking spout just above the base of the machine.

Horizontal mixers

The horizontal mixer is a rectangular box with a sloping bottom. A loading hopper is positioned at the low end, while discharge spouts are located under the high end of the machine. The flat top carries the drive motor and can accommodate a grinding machine discharging its output directly into the mixer.

The ingredients are mixed by the action of a chain and slat conveyor which moves up the sloping bottom of the mixer and returns across the top. Mixing time is short (2–3 min) and the gentle action is very suitable for rations containing flaked ingredients.

Blenders

A blender prepares the diet as a continuous process. All ingredients are metered out simultaneously in a stream which is mixed as it is carried away by an auger conveyor.

The metering devices are generally short auger conveyors, the speeds of which are adjusted to give the desired flow rate of each ingredient. Since the metering is in fact volumetric, each auger must be carefully and regularly calibrated to ensure that the proportions, by weight, of each ingredient in the mix remain correct.

Mill mixers

Complete mill mixer units are sold by several manufacturers. They normally consist of a small mill, served by a corn hopper and delivering into a mixer. Controls are pre-wired to the motor, so that the unit can be installed on the farm at minimum cost.

An alternative to fixed equipment, which still permits the farmer to include his own corn in his livestock rations, is the use of a contractor-operated mobile mill mixer. This is a large, high-output machine mounted on a lorry which can be brought on to the farm at intervals to prepare batches of feedstuffs to the farmer's requirements.

Cubers

Meal may be made more palatable to livestock if it is pressed into 'cubes', 'crumbs', 'pellets', 'nuts', 'pencils' or 'cobs'. This is achieved by forcing the meal into a funnel-like 'die', of a size appropriate to the desired product. The pressure, plus the heat which is generated by friction, forms a durable cube which may be further strengthened by the use of additives such as molasses, water or beet pulp. Freshly made cubes must be cooled before being stored in bulk, and cubed material must be handled gently.

Cubing is a slow, power-consuming process, and die life is relatively short, making the cuber a rather expensive addition to a farm-scale provender plant.

Milking equipment

Milking parlours

Most dairy animals are milked in parlours, a system in which the milking equipment and operator(s) are static and the cows move from a collecting yard to be milked, and then depart, to be replaced by further batches of cows. This system gives economies in labour over the more traditional cowshed milking, where the cows are tethered in a large building and operators move the milking equipment from cow to cow as milking proceeds.

The most popular form of parlour is the 'herringbone', where a row of standings is constructed on each side of a sunken operator's pit. The cows face slightly outwards, so that their rear ends are in towards the operator. A feeder is provided for each cow to allow individual rationing. Milking units and milk handling pipelines are centrally installed. In many cases, only one line of cows can be milked at a time, while the other batch is being installed, fed and washed. The provision of a second set of milking units allows milking to begin on all cows as soon as udder washing is complete, regardless of the progress of the previous batch, and can increase the throughput of cows. Herringbone parlours are described by quoting the number of milking units and then the number of standings, for example an 8/16, where there are two rows of eight standings and one set of eight milking units, or a 20/20.

Milking performance in a conventional herringbone parlour is 50 cows/man-h, with up to 75 possible. Further refinements in feeding and milking equipment, and in the shape and layout of the parlour, can reduce the man-time/cow still more. Outputs of 120 cows/man-h are being achieved, while research workers quote figures of 150 or more as being attainable.

Milking machines

Milk is extracted from the udder by applying a vacuum of 50 kPa to the teats. A higher vacuum will extract the milk faster but the risk of teat damage is also increased. Although a constant vacuum is applied to the interior of the flexible teat-cup liner, this vacuum is only applied to the teat for two-thirds or half of the time. The alternation of vacuum with atmospheric pressure in the space between the liner and the teat-cup, controlled by the pulsator, allows the liner to collapse round the teat and shield it from the vacuum for 0.33 s, at 1 s intervals. This allows blood to circulate more freely, and maintains udder health.

The pulsation ratio, the ratio of the time the teats are exposed to vacuum to the time the liner is collapsed, is typically 2:1 as described above, although some equipment operates at 1:1. Pulsation rate is normally 60 cycles/min. Milking rate can be increased to some extent by widening the pulsation ratio or by increasing the pulsation rate.

The four teat-cups are connected to the claw, which provides pulsation and vacuum connections to each teat-cup. This whole assembly is the cluster. Some claws have a built-in shut-off valve to prevent air entry when the cluster is removed from the cow or if it falls off during milking. Alternatively, a pinch valve on the long milk tube can perform the first function. A small hole in the claw allows air bubbles to enter the milk, making it easier for the vacuum to raise it up the long milk tube.

Most parlours are provided with graduated glass recorder jars of 23 litres capacity, either at operator's head level or below the cows, where low-level pipelines are used. The milk from each cow accumulates in the corresponding jar, so that yield can be recorded or a sample taken.

An individual batch of milk can be rejected, if necessary, without contaminating the entire system. The recorder jar has in some cases been replaced by a milk meter. When installed in the long milk tube, the milk meter indicates milk yield and collects a representative sample of the milk for butterfat analysis. When yields are not being recorded, the meters are not installed and milk passes directly into the pipeline system. A milk flow indicator may be used to let the operator know when a cow has finished milking, or a transparent section of tube or a transparent milk filter may be included in the long milk tube.

A further refinement is to couple the automatic detection of the end of milk flow with the automatic removal of the cluster from the cow (ACR). The cluster is raised by a cord powered by a vacuum-operated piston, while the milking vacuum is shut off to release the udder and prevent air entry. If an individual cow has an erratic milk output, ACR can terminate milking before all her milk has been removed.

Milk enters the pipeline either from a transfer valve at the base of the recorder jar, or directly from the milk meter or the long milk tube. An air bleed into the claw or the transfer valve helps the milk to flow towards the source of vacuum until it reaches the releaser vessel (capacity 25 litres) which is normally mounted on the milk room wall. An electric milk pump is actuated by a high level of milk in the vessel to withdraw milk against the vacuum and discharge it through a filter into the bulk tank.

Pipeline milking systems and all their components are designed to be cleaned and sterilized in place. Circulation cleaning is a routine by which the system is first rinsed through with hot water then a hot solution of detergent and sodium hypochlorite is circulated through the system for 5–10 min followed by a final water or hypochlorite rinse. The various liquids are drawn from a trough or tank into the wash line, and enter each cluster via a set of four jetters which fit over the ends of the teat-cups. From here, the liquid follows the route of the milk described above, until it reaches the releaser pump which either returns it to the trough for recirculation or discharges it from the system.

A saving in chemicals, but an increased consumption of electricity for water heating, is achieved by the process of acidified boiling water (ABW) cleaning. Sterlization is mainly by heat, as the entire system is raised to 77°C for at least 2 min; 14–18 litres/unit of near-boiling water (96°C) acidified by a small quantity of sulphamic or nitric acid, is drawn once through the system at a rate controlled by an inlet orifice, and discharged by the releaser pump. In addition to a mild sterilizing action, the acid helps to keep the glass tubes clear of hard-water salts.

The milking plant is used 365 d of the year; regular maintenance is necessary to ensure that rubber parts are in good condition, pulsators and vacuum regulator are operating correctly and vacuum pump is lubricated. There should be an annual machine test by the machine manufacturer's agent. Such care will help to ensure that the installation gives reliable, efficient milking, good milk hygiene and minimum machine-induced mastitis.

Robotic milking

Systems are under development for the robotic milking of cows. The milking machine is available round the clock, and the animals may present themselves for milking at any time (Street, 1993). The unit automatically carries out all the necessary operations: identifying the cow, putting on the teat cups, milking, recording the yield, removal and cleaning of teat cups, and udder disinfection. One unit is required for every 20–30 cows.

Many benefits are envisaged:

- Welfare. Animals choose when and how often to be milked.
- Yield. More frequent milking generally results in increased milk yield per cow.
- Better working conditions. The robot can perform tasks that are unpleasant and monotonous for workers (Mottram, 1992, 1993).

Milk cooling and storage

Milk is normally cooled, and stored prior to collection, in a stainless steel bulk tank of 600–4000 litres capacity. When holding 40% of its nominal capacity (from the evening milking) which has been cooled to 4.4°C overnight, the remaining 60% of capacity must be cooled from 35°C to the same level within 30 min of the end of filling, at an ambient temperature of 32.2°C.

The milk is cooled by contact with the chilled inside surface of the tank; an agitator paddle keeps the milk in circulation to assist in this process. The agitator must also be able to homogenize a sample of milk of 4.5% butterfat so that after 2 min operation the milk may be sampled to give an accuracy of ±0.05 units of butterfat percentage. The tank is insulated so that the chilled milk will not rise in temperature by more than 1.7°C over an 8 h period when the ambient temperature is 32.2°C.

Most bulk tanks are cooled by an ice bank. A medium sized refrigeration unit, the evaporator coils of which are submerged in water in the space between the inner and outer walls of the tank, operates for many hours daily to build up an ice bank. A control will stop the refrigerator when the ice bank is sufficiently large to cope with the expected milk yield and ambient temperature. At milking, the circulating warm milk inside the tank gives up its heat which is absorbed as latent heat by the melting ice. This system allows a relatively low-powered refrigeration unit to give a rapid rate of milk cooling.

Where very large quantities of milk are handled, it may be more economical to cool the milk in a separate plate-type heat exchanger before it is stored in an insulated tank.

The conventional bulk tank will cool about 45 litres of milk for each kWh of energy consumed. This figure may be almost doubled if the milk can be pre-cooled in a heat exchanger before it reaches the bulk tank. If a free source of cold water, such as a spring, is available, or if large quantities of purchased water are to be used elsewhere, this water can be used to carry away as much as half the heat from the milk. A heat pump may also be used to recover some of the energy from the milk and use it to heat washing water.

Bulk tanks are always cold, so that heat sterilization is impossible; chemical sterilization must be used. Most tanks are equipped with a programmed cleaning unit which, by means of a spray head, rinses, washes, sterilizes and finally rinses the tank. This cycle is set in motion by the tanker driver after he has collected the milk.

Parlour feeding

Feeding of the cows in the parlour occupies them during milking and also allows individual rationing of each cow. In a small parlour, the operator can recognize each cow, remember her ration, and operate a volumetric feeder a certain number of times to deliver the correct ration into a trough in front of her.

In order to take up less of the operator's time, parlour feeding systems have been developed in a number of ways:

(1) vacuum operation of the feeders from a central control panel, to reduce the distance walked;
(2) a programmable unit which releases the correct ration at the correct position when the cows' identification numbers are manually dialled or punched in;
(3) machine recognition of individual cows at each standing by means of a coded transponder worn around the neck, followed by the release of the correct quantity of feed. The memory unit is updated regularly, to adjust the ration according to yield and stage of lactation.

As milking routines speed up, the cow has less and less time to eat her ration. While a typical 6/12 parlour gives her time to eat 5.5 kg, a 12/12 reduces her waiting time, and she only has time to consume 3.5 kg of feed. In order to satisfy the requirements of high-yielding cows, a number of manufacturers offer out-of-parlour feeders, which respond only to cows wearing transponders with a particular coding. A batch of feed is delivered to that individual – and then no further batch can be delivered to her – for a pre-set period, in order to control her concentrate intake. Some systems also record the amount of food delivered to each cow in this way.

Information technology in the milking parlour

In addition to the recording of individual milk yields and the dispensing of feed, information technology can gather and display valuable herd management data such as milk conductivity (indicator of mastitis), milk temperature (indicator of ill health or 'heat'), activity of the animal (indicator of 'heat') and animal weight (indicator of condition) (Peiper *et al.*, 1993). Such measurements can be made automatically at every milking, and are then immediately available on-screen to the herd manager. Any necessary medication, nutritional change or other action can be implemented at once.

Manure handling and spreading

Farmyard manure

The traditional method of dealing with manure is to provide straw or other absorbent material as bedding, and to handle the combined product of faeces, urine and bedding as farmyard manure. This increases the volume of material to be handled by 10–25% in comparison with the figure mentioned in Table 24.16).

Where there is sufficient clearance, farmyard manure may be stored in place until it can be spread. Alternatively, a double-handling system is used, involving removal to a temporary storage site followed by spreading onto the land at a convenient time of year.

Using a tractor mounted front-end loader, a rough terrain fork lift with manure bucket, or an industrial loader similarly equipped, farmyard manure can be loaded into trailers or spreaders at a rate of 20–60 t/h.

Spreading on the land is by rear discharge or side discharge spreaders, which have carrying capacities of 2–10 t.

Rear discharge machines resemble trailers. They are equipped with a slatted floor conveyor to move the load rearwards into a beating and shredding mechanism. This distributes a swath of manure behind and to either side of the machine. Some of these spreaders can be adapted to handle other materials.

Side discharge spreaders are cylindrical in shape, with an opening along the top portion for loading and discharge. A shaft through the centre of the cylinder carries

Table 24.16 Daily production of excreta by livestock

Beef Bullock	27 litres
Dairy cows	57 litres
1000 turkey growers	124 litres
Pigs 45–75 kg	4 litres
Pigs 16–20 weeks	7.5 litres
Sow and litter	12 litres
1000 broiler chickens	36 litres
1000 laying hens	115 litres

Source: MAFF, 1991.

numerous chain flails, and a rigid flail at either end. As the shaft rotates, manure is discharged first from the ends of the load, and then progressively towards the centre of the load. Spreading is to one side of the machine only; a wide range of materials from strawy farmyard manure to semi-liquids can be carried and spread.

Spreading of manure in any form must follow the guidelines set out in the Code of Good Agricultural Practice for the Protection of Water (MAFF, 1991).

Liquid manure

In livestock housing systems where little or no bedding is provided, the manure consists only of faeces and urine. This is removed from contact with the animals by allowing it to fall through a slatted floor into a space below, or, as in the case of cubicle housing for dairy cows, dunging areas are scraped daily by a tractor rear-mounted scraper.

It is rarely possible to spread manure daily, so a storage tank must be included in the system to hold, typically, at least three months' production. A longer storage period may permit most of the manure to be applied to stubbles in autumn, or some other seasonal priority.

Size of store

When calculating the size of store necessary for a particular length of storage, it is advisable to add a factor of 40% to the figure given in Table 24.16, to allow for split drinking water and any washing water which will find its way into the store. If rainwater cannot be excuded, an additional volume V litres must be added, where $V \times$ (mm of rainfall over the storage period) $\times$ (m^2 of area draining into the store + m^2 surface area of the store).

Size of tanks

While liquid manure stores can be excavated and lined, or constructed of concrete or masonry to any size, many farmers install prefabricated stores which are bolted together from curved coated steel panels, mounted on a ground level concrete base. Stores are available in diameters of 4.5–25 m, heights of 1.2–6 m, and capacities up to 3000 m^3. When planning a manure handling system, provision must always be made for increasing its capacity, even if no immediate increase in livestock numbers is envisaged. An increase in store height to give greater capacity is only feasible if the original store was designed to accept this increase.

The storage tank may be below the level of the yard, in which case it can be filled directly by the scraper. Alternatively, partially or completely above-ground tanks may be used. A small below-ground sump, covered by a heavy grid, receives the scraped manure from the yard. This is transferred into the main tank at intervals by a pump. The size, design and siting of any manure store must follow the guidelines set out in the Code of Good Agricultural Practice for the Protection of Water (MAFF, 1991).

Successful operation of a liquid manure system demands care and good management at all stages, to include the following points.

- Exclusion of any fibrous material: hay, silage, straw. This will otherwise block up pumps, and contribute to crust formation.
- Pumping of the manure can be facilitated by the addition of water to the dung and urine, although this also increases the volume of material to be handled. Dung and urine together form a slurry at about 12–15% solids content which is barely pumpable; adding water to produce a less viscous consistency at 6–7.5% solids makes the resultant slurry very easy to pump (Schofield, 1984).
- If the manure is to be handled by pumps and tanks, the contents of the tank must be thoroughly agitated at least fortnightly to prevent the formation of surface crusts in the case of cattle manure, or sludge deposits in the case of pig manure. The pump which is used for handling the manure can also be used for this purpose. Movable jets can be provided at various levels in the tank, and the output of the pump is concentrated through any jet to break up the surface crust, dislodge sludge deposits, or just mix the contents of the tank. Where regular mixing has been neglected, or mixing power is inadequate, and crusts or deposits have built up, propeller agitators can be used to homogenize the contents of the tank.

Liquid manure is unloaded from the store directly by a centrifugal or positive-displacement pump, or by vacuum into a tanker of 3500–13 500 litres capacity. After transport to the field, a built-in spreading mechanism distributes the manure as the machine moves forward. Considerable odour may be produced, particularly if the liquid manure is thrown high into the air. As an alternative, some tankers can be fitted with heavy tine 'injectors' which can place the manure below the soil surface (Godwin *et al.*, 1990). This minimizes odour and also reduces the loss of nutrients by run-off.

Care must be exercised when operating tankers across slopes, particularly when partly full. The load shifts to the downhill side, moving the centre of gravity nearer to the downhill wheel. Overturns will occur on side slopes where the full tanker, or a trailer, could be used quite safely (Hunter, 1992).

If the manure store can be entered by a ramp down from ground level, it can be emptied by a self-loading tanker which is backed down the ramp. Alternatively, a four-wheel-drive loader with a large-capacity bucket may be used to unload the manure at rates of 60–80 m^3/h into tankers or tipping trailers.

Slurry separation

Many farmers have installed animal waste handling systems including shurry separators. The slurry is separated into solid (30–60% DM) and liquid (1.5–8% DM) fractions, generally by pressure against a perforated screen, but sometimes by gravity or vacuum. The solid fraction may be stored for further processing or subsequent spreading. The liquid fraction is retained in a tank (see above), and applied to the land by means of an organic irrigation system when conditions permit (MAFF, 1991).

Slurry separators

By passing liquid manure over a series of sieves, rollers or vacuum devices it is possible to remove about half the solids, leaving a liquid that can be stored without problems and pumped easily. The 'solids' can be stacked for later spreading with FYM equipment (Pain *et al.*, 1978).

References

Bailey, J. (1986a) Choosing and improving an irrigation system. *Br. S. Beet. Rev.*, **54**, 3–4.

Bailey, J. (1986b) Report on an irrigation demonstration. *Br. S. Beet. Rev.*, **54**, 37–38.

Ball, B.C. (1990) Reduced tillage for energy savings with cereals: practical and research experience. *Agr. Engr*, **45**, 2–6.

Billington, W.P. & Butler, J. (1984) The design and development of a skew-bar topper for sugar beet. *J. agric. Engng Res.*, **29**, 329–335.

Boa, W. (1984) The design and performance of an automatic transplanter for field vegetables. *J. agric. Engng Res.*, **30**, 123–130.

Bosma, A.H. & Verkaik, A.P. (1986) Developments in forage conditioning. *Ag. Eng. 86* Conference Papers, 181–2, Wageningen.

British Crop Protection council (1986) *Nozzle Selection Handbook.*

Busse, W. (1977) The design and use of combine harvesters for minimum crop loss. *Agr. Engr.*, **32**, 7–9.

Curtis, G.C. (1980) Application equipment – present and future. *Soil & Water*, **8**, 17–18.

Davis, N.B. (1977) The minimisation of crop losses associated with sugar beet harvesting. *Agr. Engr*, **32**, 10–13.

Dwyer, M.J. (1983) Soil dynamics and the problems of traction and compaction. *Agr. Engr*, **38**, 62–68.

Dwyer, M.J. (1985) Power requirements for field machines. *Agr. Engr*, **40**, 50–59.

Dwyer. M.J. & Dawson, J. (1984) Tyres for driving wheels. Proc. BSRAE Assoc. Members Day, 4 Oct. 1984.

Dwyer, M.J. & Pearson, G. (1976) A field comparison of the tractive performance of two- and four-wheel drive tractors. *J. agric. Engng Res.*, **21**, 77–85.

Elrick, J.D. (1982). How to choose and use combines. The Scottish Agricultural Colleges, Publication No. 88.

Fielder, A. (1986) Straw incorporation – a practical guide. Progress August, 2–4.

Gee-clough, D., McAllister, M. & Evernden, D.W. (1977a) Tractive performance of tractor drive tyres I. The effect of lug height. *J. agric. Engng Res.*, **22**, 373–384.

Gee-Clough, D., McAllister, M. & Evernden, D.W. (1977b) Tractive performance of tractor drive tyres II. A comparison of radial and cross-ply carcass construction. *J. agric. Engng Res.*, **22**, 385–395.

Gee-Clough, D., Pearson, G. & McAllister, M. (1982) Ballasting wheeled tractors to achieve maximum power output in frictional-cohesive soils. *J. agric. Engng Res.*, **27**, 1–19.

Godwin, R.J., Warner, N.L. & Hann, M.J. (1990) Comparison of umbilical-hose and conventional tanker-mounted injection systems. *Agr. Engr*, **45**, 45–60.

Hesse, H. & Withington, G. (1993) Digital electronic hitch control for tractors. *Agr. Engr*, **48**, 1–17.

Hunter, A.G.M. (1992). A review of research into machine stability on slopes. *Agr. Engr*, **47**, 49–53.

Ingram, J.A. (1984) Frost protection of fruit crops. *Irrigation News*, No. 6, 17–22.

Jones, L. (1985) The effect of ground area cover on the drying rate of mechanically conditioned grass crops. Grassland Research Institute Final Ann. Rep. 1984–85, 102–3.

Klinner, W.E. (1975) Design and performance characteristics of an experimental crop conditioning system for difficult climates. *J. agric. Engng Res.*, **20**, 149–165.

Klinner, W.E. (1980) Reducing field losses in grain harvesting operations. *Agr. Engr*, **35**, 23–27.

Klinner, W.E. & Burgess, L. (1982) Separating heavy objects from forage harvester windrows. Proc. BSRAE Assoc. Open Day, April.

Klinner, W.E. & Hale, O.D. (1984) Design evaluation of crop conditioners with plastic elements *J. agric. Engng Res.*, **30**, 255–263.

Klinner, W.E. Neale, M.A. & Arnold, R.E. (1987) A new stripping header for combine harvesters. *Agr. Engr*, **42**, 9–14.

Knight, A.C. (1986) Energy requirement to comminute forages by slicing. *J. agric. Engng Res.*, **33**, 263–271.

Kay, M. (1983) *Sprinkler Irrigation. Equipment and Practice.* Batsford, London.

Lake, J.R., Frost, A.R. & Wilson, J.M. (1980) The flight times of spray drops under the influence of gravitational, aerodynamic and electrostatic forces. Proc. BCPC Symposium 'Spraying Systems for the 80s', 119–125.

Landers, A. (1992) Direct injection systems on crop sprayers. *Agr. Engr*, **47**, 9–12.

Lamond, W.J. & Graham, R. (1993) The relationship between the equilibrium moisture content of grass mixtures and the temperature of the air. *J. Agric. Engng Res.*, **56**, 327–335.

Maunder, W.F. (1983) Planting and mechanical handling of seed (potatoes). *Agr. Engr*, **38**, 38–41.

McRae, D. (1977) The design and operation of potato harvesters for minimum damage and loss. *Agr. Engr*, **32**, 17–19.

MAFF (1975) Bale handling methods 1972, 1973 and 1974. Farm Mechanization Studies No. 28.

MAFF (1976a) The utilisation and performance of field crop sprayers 1976. Farm Mechanization Studies No. 29.

MAFF (1976b) Self-loading forage trailer performance Pwllpeiran 1975. Mechanization Dept, ADAS Wales.

MAFF (1980) The handling and application of granular fertilizers – 1980. Farm Mechanization Studies No. 35.

MAFF (1982) *Irrigation*. Reference Book, 138. HMSO, London.

MAFF (1983a) *Water for Irrigation*. Reference Book 202. HMSO, London.

MAFF (1983b) Methods of Short Term Irrigation Planning. Booklet 2118. MAFF Publications, Alnwick.

MAFF (1984) Daily Calculation of Irrigation Need. Booklet 2396. MAFF Publications, Alnwick.

MAFF (1991) Code of Good Agricultural Practice for the Protection of Water. MAFF Publications, London.

MAFF/HSE (1990) Pesticides: Code of Practice for the Safe Use of Pesticides on Farms and Holdings. HMSO, London.

McKee, A. (1992) The role of the conventional tractor. *Agr. Engr*, **47**, 41–47.

Mottram, T.T. (1992) The design and management of automatic milking systems. *Agr. Engr*, **47**, 87–90, 115–118.

Mottram, T.T. (1993) The design and management of

automatic milking systems. *Agr. Engr*, **48**, 6–12.
Nation, H.J. (1982) The dynamic behaviour of field sprayer booms. *J. agric. Engng Res.*, **27**, 61–70.
Nation, H.J. (1987) The design and performance of a gimbal-type mounting for sprayer booms. *J. agric. Engng Res.*, **36**, 233–260.
NPTC (1987) *Pesticides Application* National Proficiency Tests Council, Stoneleigh.
Paice, M. (1993) Patch spraying. *Agr. Engr*, **48**, 24–25.
Pain, B.F., Hepherd, R.Q. and Pittman, R.J. (1978) Factors affecting the performances of four slurry separating machines. *J. agric. Engng Res.*, **23**, 231–242.
PAMI (1977) Evaluation Report No. E 0576C – John Deere Sidehill 6600 Self Propelled Combine. Prairie Agricultural Machinery Institute, Humboldt.
PAMI (1984) Evaluation Report No. 388 – Lockwood Model 2265 Centre Pivot Irrigation System with Flexspan Corner System Attachment. Prairie Agricultural Machinery Institute, Humboldt.
Pascal, J.A. & Hamilton, A.J. (1984) Grain losses associated with combine harvesters operating on sloping land – performance changes over the past fifteen years. *Agr. Engr*, **39**, 124–130.
Peiper, U.M., Edan, Y., Devir, S., Barak, M. & Maltz, E. (1993) Automatic weighing of dairy cows. *J. agric. Engng Res.*, **56**, 12–24.
Plackett, C.W. (1984) The ground pressure of some agricultural tyres at low load and with zero sinkage. *J. agric. Engng Res.*, **29**, 159–166.
Rackham, D.H. & Blight, D.P. (1985) Four-wheel drive tractors – a review. *J. agric. Engng Res.*, **31**, 185–201.
Raghavan, G.S.V., McKyes, E. & Chassé, M. (1977) Effect of wheelslip on soil compaction. *J. agric. Engng Res.*, **22**, 79–83.
Raymond, F., Redman, P. & Waltham, R. (1986) *Forage Conservation and Feeding*, 4th edn. Farming Press, Ipswich.
Rees, D.V.H. (1982) A discussion of sources of dry matter loss during the process of haymaking. *J. agric. Engng Res.*, **27**, 469–479.
Renius, K.T. (1992) Developments in tractor transmissions. *Agr. Engr*, **47**, 44–48.
Röhrs, W. (1986) Investigation on energy consumption of cylinder type forage harvesters. *Ag. Eng. 86* Conference Papers, 84–5. Wageningen.
Schaffer, R.L., Johnson, C.E., Elkins, C.B. & Hendrick, J.G. (1985) Prescription tillage: the concept and examples. *J. agric. Engng Res.*, **32** 123–129.
Schofield, C.P. (1984) A review of the handling characteristics of agricultural slurries. *J. agric. Engng Res.*, **30**, 101–109.
Sharp, R.B. (1984) Comparison of drift from charged and uncharged hydraulic nozzles. Proc. 1984 British Crop Protection Conference, 1027–1031.
Spillman, J.J. (1984) Spray impaction, retention and adhesion: an introduction to basic characteristics. *Pesticide Science*, **15**, 97–106.
Spoor, G. & Godwin, R.J. (1978) An experimental investigation into the deep loosening of soil by rigid tines. *J. agric. Engng Res.*, **23**, 243–258.
Stadie, A.L., Dawson, J.R. & Neuyen, N.V. (1989) A comparison of the rolling resistance of single, dual and tandem wheels. *Agr. Engr*, **44**, 34–37.
Stafford, J.V. (1979). The performance of a rigid tine in relation to soil properties and speed. *J. agric. Engng Res.*, **24**, 41–56.
Stansfield, C.B. (1992) Irrigation surveys by ADAS. *Agr. Engr*, **47**, 6–8.
Statham, O.J.H. (1981) Maincrop potato production in Great Britain. *Agr. Engr*, **36**, 25–28.
Street, M.J. (1993) Robotic milking. *Agr. Engr*, **48**, 24–25.
Tinker, D.B. (1993) Integration of tractor engine, transmission and implement depth controls. Part 1, Transmissions. *J. agric. Engng Res.*, **54**, 1–27.
Tuck, C.R., Klinner, W.E. & Hale, O.D. (1980) Economic and practical aspects of high-capacity rotary mower and mower-conditioner systems. *Agr. Engr*, **35**, 11–14.
Weise, G. (1993) Active and passive elements of a combined tillage machine: interaction, draft requirement and energy consumption. *J. agric. Engng Res.*, **56**, 287–299.
Willis, S.T. & Carr, M.K.V. (1986) Centre pivot irrigators. *Irrigation News*, Spring 16–36.

Further reading

Culpin, C. (1992) *Farm Machinery*, 12th edn. Blackwell Science, Oxford.
Witney, B.D. (1988) *Choosing and Using Farm Machines*. Longmans, London.

25

Farm Buildings

J.L. Carpenter & R. Coates

Design criteria (BS 5502, Part 21, refers)

Introduction

The great variety of purpose and intensity of use, together with alternative construction methods, results in considerable differences of building investment on individual farms. Even when there are similarities of production, it may range from basic weather protection for stock up to elaborate environmental control systems, producing variations from 10 to 90% of a farm's capital. Bearing this potentially heavy, possibly crucial, expenditure in mind it is imperative that construction projects are well conceived, thoroughly thought out and carefully managed to a satisfactory completion for use. Thus, a strong case can be made in the present economic climate (particularly with regard to protection of the environment, the present agricultural industry 'shrinkage' and the consequent cost-cutting) for more care and 'state of the art' advice to be used in the planning and purchase of farm building systems than has been exercised hitherto.

Agricultural building design is currently under considerable pressure from many external factors. Among these are a rapidly awakening environmental consciousness, a need for economic stringency, brought about by general market and legislative forces and the demand for countryside-based (but non-agricultural) industrial and leisure activities. As a result of these forces, farm construction now has a British Standard (BS 5502) to guide (in practice control) its use of materials and methodology and a series of Welfare Codes for livestock to provide recommendations (and in some cases legislate) for 'good' buildings. Additionally through planning legislation (e.g. Town and Country Planning Acts), pollution control (various 'Water' Acts) and public health regulations (e.g. Health and Safety at Work Act) it must now conform to strict environmental parameters. One of the results of this recent increase in external influence has been a potential increase in cost. This implies that it is more important than ever that the farmer/investor should ensure optimum value for money from any farm building installation. This requires, in turn, that a greater degree of professionalism be used in all phases of the design and construction process. Although, clearly, a hired specialist designer, whether an architect, surveyor or other expert, should be in a position to best advise, they are relatively rarely employed to do so, partly due to their 'added cost' but also because there are comparatively few of them. However, to be of maximum use they must be consulted as early in the design process as is possible – before mistakes are made, rather than after. The next sections outline the principles and procedures to be used and the general order in which they should be adopted in this process. Where appropriate in the text, 'agencies' are mentioned which may provide help with specific design problems.

NB. There are many valuable sources of research and advisory information on topics included in this section. The most relevant are summarized later in 'Further reading' and are mainly of a 'general text' nature. These in turn will provide pointers to particular references as necessary.

Planning and design

Despite the great variation implied in the Introduction above, derived from type and scale of farm operations, there are some considerations which are common to the planning processes of all farm buildings. These may be summarized as follows:

- Suitability for purpose and use (including conformation to requirements of BS 5502, see later).
- Appropriate cost, relative to permanence and maintenance.
- Flexibility of design to allow for
  - expansion,
  - produce and method of production changes,
  - alternative (or alternate) use.
- Minimal labour and energy use.
- Minimal risk
  - to health of crop, stock or manpower,
  - of fire,
  - of polluting the surroundings.
- Utilization of common/standard/local materials (with account taken of DIY element).
- Harmony with surrounding environment.

- Appropriate use of site topography for aspect/drainage/shelter and gravity movement of materials.
- Finance available.
- Impact of legislation and codes of practice.
- Potential modification or extension of existing buildings allowing for age and technological differences.

Specific design

Whilst the general points above must be borne in mind at all times in any building development, specific use structures will have particular requirements concerned with their productive functions, e.g. precise environmental control. The route by which they may be included in the overall design is indicated below in the scheme of work represented by steps A–H and the chart in Fig. 25.1.

Where possible, design should conform to a British Standard laying down firm guidelines for the construction of farm buildings which was first published in 1980. This Standard, BS 5502, is intended to embrace all agricultural structures and impose upon them precise structural and environmental requirements which will result in satisfactory operation for a predicted and assured lifetime whilst providing safe working conditions during that lifetime. All buildings conforming to the BS should carry a plate showing their classification, manufacturer and date. Buildings not conforming to BS should not be acceptable. Classification (see Table 25.1) relates to occupancy, predicted safe lifespan, type of activity intended for, and

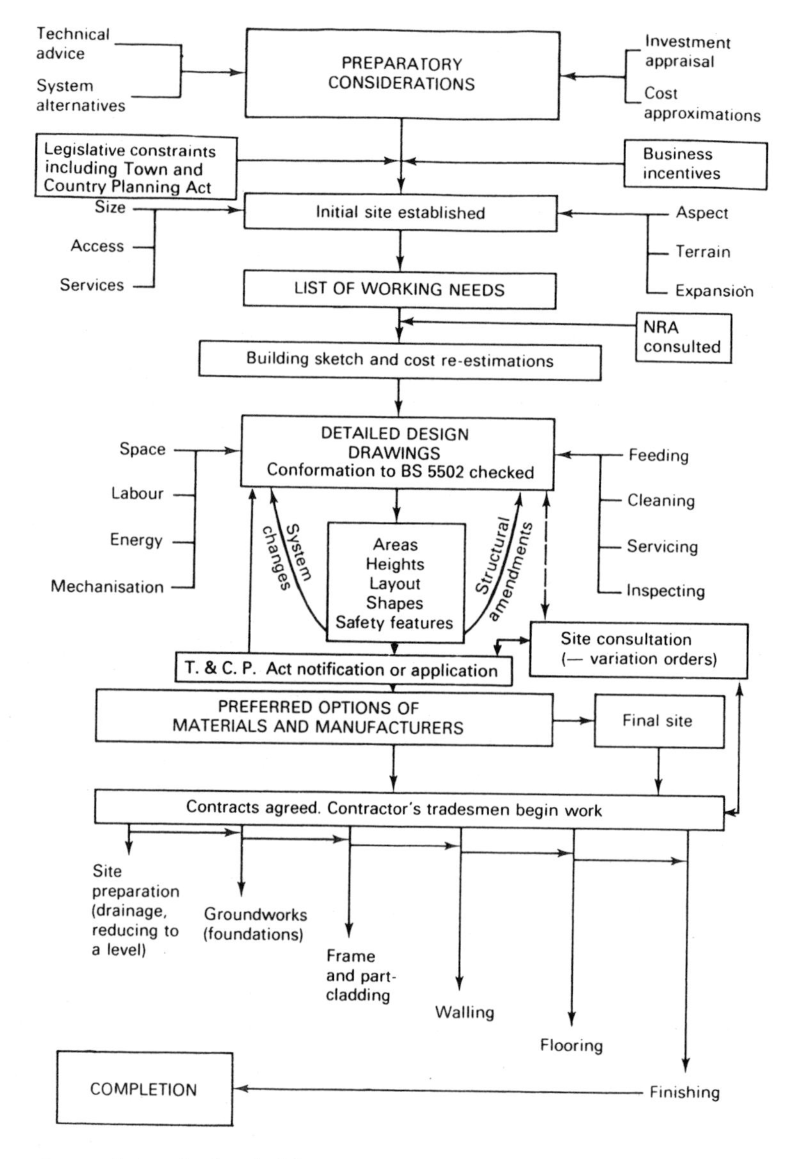

Fig. 25.1 Steps to the completion of a farm building.

Table 25.1 Summary of provisions for BS 5502 classification

Class (and potential example)	*Expected structure life (minimum) – reflecting strength of components (years)*	*Human occupancy h/d at population density (persons/m²)*	*Situation (m minimum from classified road or non-owned dwelling place)*
1. (Vegetable processing shed)	50	Unrestricted	Unrestricted
2. (Milking parlour)	20	6 at 2	10
3. (Pig fattening)	10	2 at 1	20
4. (Plastic envelope sheep house)	Temporary	1 at 1	30

the construction material's safe stressing limitations together with its general suitability for the purpose and use.

BS 5502 incorporates animal welfare guidance and lays down environmental standards desirable for healthy livestock in humane production systems. It does this by recommending space allocations, target temperatures, ventilation rates, confinement methods, etc. Another important function of the standard is to ensure that the fabric and equipment, as well as the layout, of farm buildings is adequate for their purpose and use. It thus lays down rules for types, qualities, installation methods and required long-term abilities of material to be used in the construction of all elements of a farm structure.

All parts of this section of the *Agricultural Notebook* should be read with the guidelines laid down in BS 5502 in mind. Where appropriate, reference is made to specific parts of the Standard as and when they are relevant to the text content.

A. Initial preparation

(Confer at all appropriate stages with appropriate advisory organization.)

(1) Establish the business motives for the building investment and the level of acceptable cost relative to the return on the capital expended.
(2) Consider alternative systems to achieve similar results and consult other existing users (if any) who have experience of them.
(3) Estimate the overall size of the completed structure(s) and ancillaries required and hence assess the *approximate overall* cost (see 'Cost ranges', page 601 and following).
(4) Obtain information on

- grant aid potential
- tax concessions, etc.,
- need to conform to legislative requirements such as planning/building/safety/pollution/ health regulations including BS 5502 (see above).

(5) Examine possible sites for the building(s), considering their interactive effects on the farm business and the rural environment both during and after construction. Take particular account of:

- size (area and height – note planning permission restrictions);
- access (for vehicles, animals, operatives, building and other materials handling);
- circulation space (and expansion/change potential);
- drainage (within, from and around building – note anti-pollution requirements);
- aspect (climate and its environmental effects – note also visual effects in areas of particular sensitivity);
- availability of services (provision in remote areas may be very costly);
- utilization of slopes (gravitational movement of materials, use of storage space and drainage);
- potential for expansion or change (space of surroundings, layout of services).

(6) Confirm the building concept by sketching and listing the requirements of the building. Include provision for all necessary activities within and surrounding it.
(7) Re-estimate the cost in the light of decisions taken about siting and building system/methods, which may modify the original concepts or facilities required (see 'Cost ranges', page 601 and following).

B. Design criteria establishment

(1) Consider the space, labour, mechanization and energy inputs required to operate the building system within and around the building. Note and list requirements for activities such as feeding, cleaning, handling, emptying, filling, heating, cooling, weighing, lighting.
(2) Draw up a list of relevant criteria to be met such as living or storage space requirement, insulation standards, slurry capacity, mechanical handling clearances, ventilation rates, cleanability of materials, circulation areas, etc.

C. Concept detailing

(See later for details of specific building types.)

(1) Calculate

- *Areas* required (of floor, pens, circulation, access, slats, handling equipment, feeders, fixed equipment, etc.) Remember that Net essential area = Gross building area – Structure occupancy.
- *Heights* (of ceilings, doorways, walls, mechanical equipment, services, etc.). Remember to allow for essential clearances, e.g. trailer tipping, pig reach.

(2) Arrange the *interior layout* for stock control, materials movement, equipment installation, and environmental effects.
(3) Derive the *external shape* from (1) and (2) taking account of weather resistance (roof shape and wall presentation) and impact on the surrounding rural environment, including any adjacent buildings (odours, contrasts, textures, etc.).
(4) Check and include *safety* elements which are intended to limit hazards (to operatives, stock and goods) from fire, structure failure, poor atmospheric composition, and themselves.
(5) Decide on *preferred* methods, types and manufacturers of materials and structures. Note that individual preference for quality standard, dimensional system, localization of supply source and workability of various materials will give a very great range of options.
(6) Consider relevant requirements imposed by the need to conform to BS 5502 (see specific designation for details).

D. Design enhancement

There is an increasing awareness of the need for thought to be given to the quality of design in the countryside in order to give buildings character and local distinctiveness. However, in view of the infinite variety of sites and circumstances, the following ground rules can only be a general guide to stimulate quality.

(1) Siting

- Reduce impact by letting into slope – should avoid made-up ground especially for load-bearing structures.
- Avoid skyline.
- Position in relation to existing buildings following the pattern of the farmstead (also bearing in mind access, and circulation requirements).
- Utilize existing landscape features to provide backcloth: retain as much as possible when preparing site.
- Allocate space for additional landscaping – remembering a natural shelter belt is the best wind filter.
- Try to avoid 'all under one roof'. (Will probably improve access and ventilation in doing so.)
- Step floor and roof levels down steep site.
- Consider 'near' and 'far' viewpoints, especially from the farmhouse, approach and nearest highway.
- Orientate to suit use in relation to sun/prevailing wind.
- Consider breaking up building to create second structure at right angles (instead of lean-to) to provide shelter and create courtyard with existing buildings.
- Utilize tower silos, elevator penthouses, etc. to create focal point – but ensure non-reflective and harmonize colour.
- Where traditional ranges are unspoilt or have potential for diversification, consider forming second farmstead.

(2) Roof

- Colour (new buildings) must be dark and relate to existing surfaces – slate or tile, etc. (Note: 'Plastisol' coated steel sheet is not recommended for roofs in dark colours and is therefore not suitable.)
- Reroofing should be carried out in materials as close as possible to the original (also necessary for environmental grant).
- Try to relate the facing roof of new buildings to those adjacent by offsetting ridge from centre so that a slope can be created to match in height and pitch, and then clad in the same material.
- Surfaces should be non-reflective. (Manufacturers please note! Currently not available in sheet surfaces.)

E. Dimensional formulation

Dimensional formulation of plans should now be possible. This entails accurate drafting, preferably to a standard scale (1:10, 20, 50, 100, 200 as appropriate), all the external and internal aspects, in detail, which will enable the builder to construct without further instruction. At this point it may be advisable to employ specialist help with knowledge of drawing interpretation. However, it should always be remembered that experts in the building trade are not experts in agriculture and so thoroughgoing instructions should be given wherever possible.

(1) Drawings (see Fig. 25.2) should be prepared to scales of 1:50 or 1:100 showing:

- The layout in plan view indicating position of doors, barriers, feeders, circulation space, drainage, water and electricity supplies in relation to frame members, walls and other structural features at or above ground level. (Some of these might be more clearly shown on separate additional drawings to avoid confusion of lines and shading.)
- The building in elevation to demonstrate shape, volume, cladding materials and external features likely to be obtrusive.
- Cross-sections of the structure demonstrating material details and thicknesses, positions of service fixings and the nature of space usage in the building.

(2) A scale map of 1:2500 should be used to show the location and general arrangement of the building relative to its surroundings. (Permission may be granted for a copy to be made for personal use from an Ordnance Survey Map of this scale.)
(3) Further drawings to scales of 1:10, 1:20 and 1:50 may be prepared to show constructional features such as foundations, framing and fixings, drainage schemes, damp and vapour proofing, insulation methods and mechanical services. These are generally the province of professional experts, surveyors and architects or may be provided by commercial building firms as part of a sales service.
(4) Planning and grant-aiding authorities may need to see some or all of these drawings to ensure that the

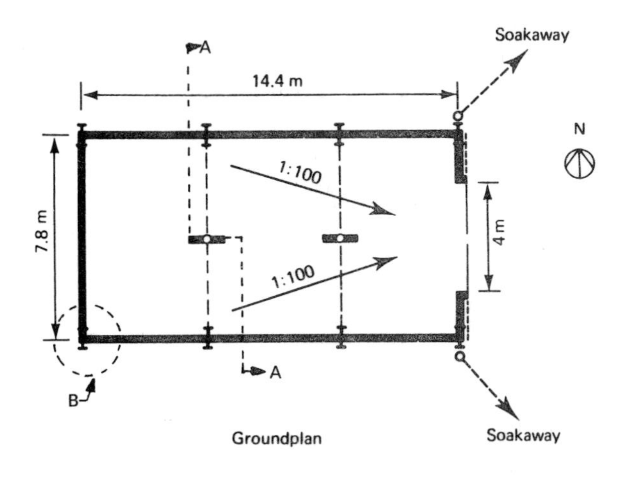

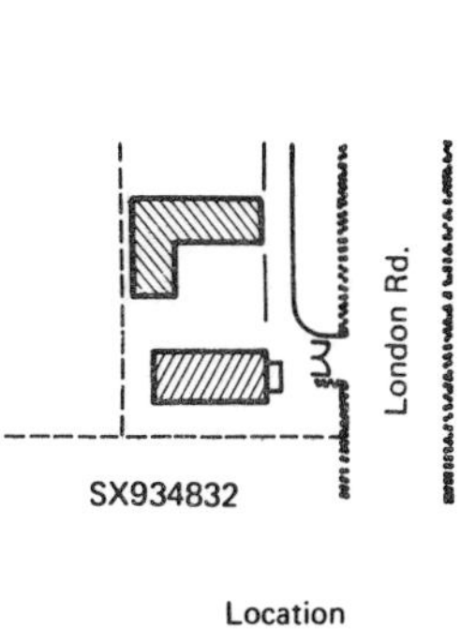

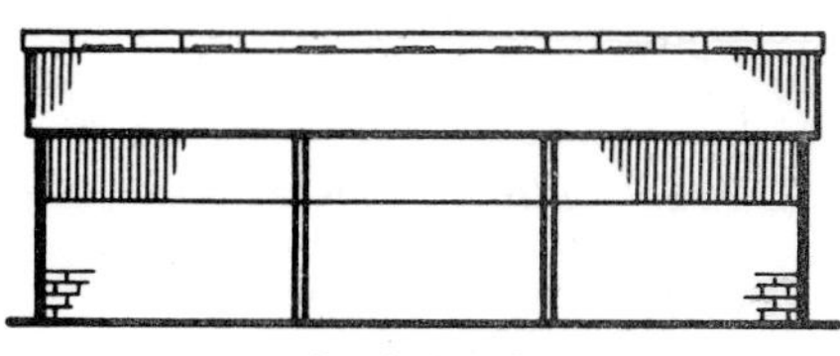

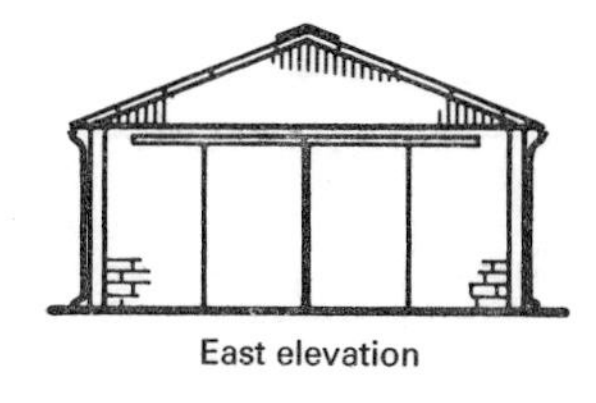

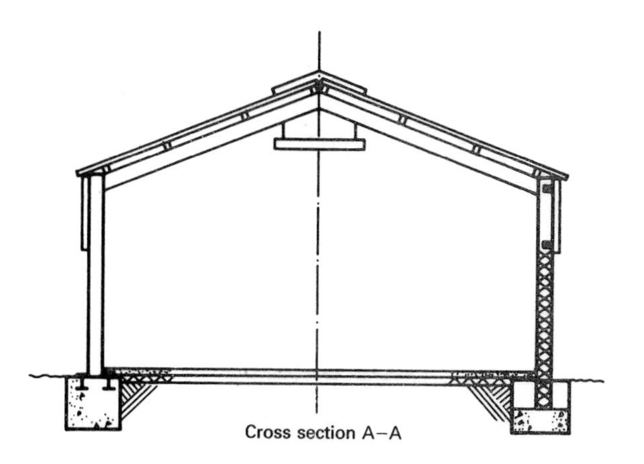

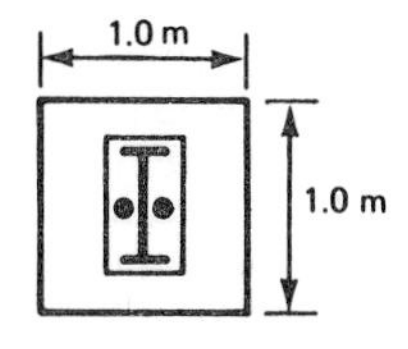

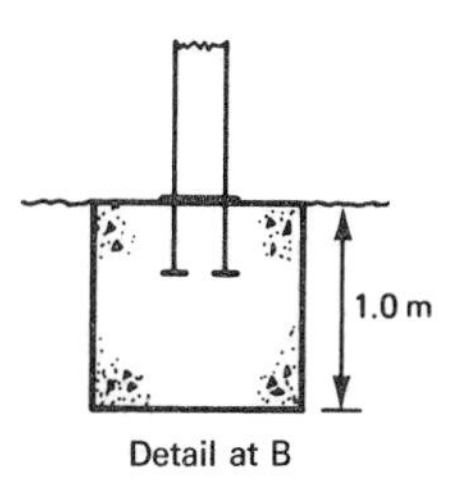

Not to scale

Fig. 25.2 Building drawings.

law is being complied wih in every respect and that the building conforms to the requirements of BS 5502. (In practice it is advisable to consult the appropriate bodies at an early stage of design thinking. It is not good, or safe, procedure to request retrospective approval.)

F. Final site preparation

This should now be carried out by way of a *survey* using the dimensions of the building established previously. Account should be taken of the area required, the existing slopes and drainage pattern together with the soil characteristics governing the foundation type and sizes. From the results of this survey estimates may be made of soil quantities to be removed and the problems if any, which might arise in the installation of services on the site. At this point the precise position of the building outline can be marked by pegs and string or chalk lines (described in the next section) as a first step in the construction process.

G. Design amendment

Although the above stages in the design have been taken in logical order, one by one, it is likely that there will be a continual need throughout to readjust ideas and objectives in attempts to match conflicting (even contradictory) requirements. Compromise between farm business aims and building techniques is often unavoidable, as for instance in the situation where the drainage of slurry may create foundation design problems. Even when construction has started, cases arise where the three-dimensional nature of the building work demonstrates that the two-dimensional planning process is inadequate and necessitates changes in the instructions to the builder. As this will increase the cost of the work significantly, a contingency sum should be set aside to cover these eventualities, but they should be avoided wherever possible.

H. Contracting for the construction site work

This can proceed as soon as the drawings and building instructions (with, perhaps, a 'bill of quantities') are drawn up. Negotiations can be entered into with potential contractors or building/material suppliers, and formal agreement can be reached as to the precise conditions under which the contract will be carried out including, hopefully, a fixed and final cost. Builders and their estimators may prefer to operate on a 'cost plus' (cost of materials and labour plus a reasonable addition for expenses and profit) or a 'price, ex-works, at date of dispatch' basis but it is more usual to negotiate a contractual sum for the provision of completed buildings, including all site works required. This latter element often provides some nasty surprises for the unwary if anything but a straightforward total contract sum is agreed beforehand. (It is also worth noting that there may be extra administrative complications through VAT when building costs for materials, etc. are obtained piecemeal.)

Agricultural construction is very commonly carried out in separate stages by different types of commercial organization. For instance, frame-members and part-cladding for the structure might be provided first, by a large-scale manufacturer. This would be erected by a subcontracted site crew, whilst the rest of the work is subsequently carried out by a local building firm who may, in their turn, subcontract specialist jobs, such as electrical installations. This system of construction is 'flexible' an allows for elements of DIY or the use of 'labour-only' subcontracting. It gives potentially lower cost but may lead to lack of contract responsibility and site control. An alternative arrangement (commonly used outside agriculture) in which a 'main-contractor' is responsible for most of the site works and usually for all supervision, generally leads to less confusion and controversy over things such as trades' attendance and timeliness. This militates also for higher quality in workmanship. The work expected of all parties on a site should be made clear before contracts are signed and work commences. Above all the temptation to make adjustments or improvements to the design during construction should be resisted, unless they correct obvious failings which will impair the building system's efficiency later. Amendments to the work in progress are nearly always done hastily, without sufficient regard to their total impact and are relatively costly, as the resulting 'variation orders' are not constrained by the original pricing for the contract tender.

Stages in constructing a building (BS 5502, Parts 21 and 22, refer)

Preparation – setting out a simple rectangular building

(1) The turf and surface soil are stripped from the site, along with trees, shrubs and bracken, complete with roots, if possible. The subsoil thus exposed is 'levelled' over an area exceeding that of the building. If the site slopes, then the surface must be 'reduced' to a satisfactory gradient as required by use of spirit-, water-, or surveyor's levels (usually to the horizontal).

(2) Divert existing drains which cross the site or re-lay in a straight line, surrounded by 150 mm of concrete – see later section on concrete – deeper than the proposed works and on suitable falls, or construct a 'barrier' drain trench up-slope from the site (this latter is sometimes known as a 'french' drain). It is a legal requirement that any such drain is positioned at least 10 m away from certain types of structure, e.g. dung steads. Thus it is important to check with the appropriate authority before installation.

(3) On the prepared subsoil surface stretch a line between pegs to represent the position of the longest side of the building but much greater in length such that the pegs are well clear of the proposed corners as at (a) in Fig. 25.3.

(4) Mark one of the corner positions with another peg (b, Fig. 25.3) and at right angles from the first line stretch another to mark a second building side (c, Fig. 25.3). An accurate right angle may be made by using a builder's sqaure, a surveyor's site square, or the method of Pythagoras (3:4:5 triangulation giving 6 m along one side, 8 m along a second and 10 m across the measured hypotenuse).

(5) In a similar manner mark the third and fourth sides (d). The dimensions and position of the building are now shown in plan by the lines and corner intersections. It is usual for these to be the outside

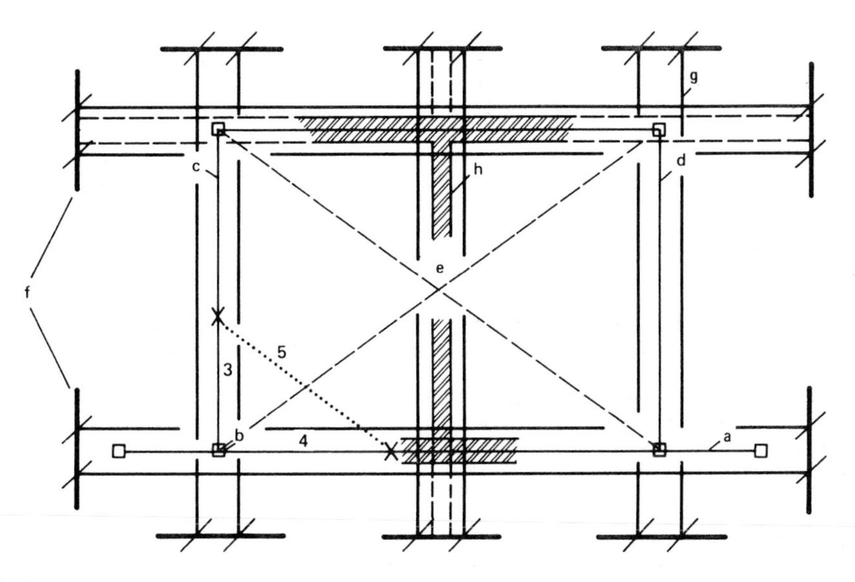

Fig. 25.3 Setting out.

edges of the building above ground but they may represent centrelines or inside surfaces. It is important to plan according to fixed reference surfaces and levels throughout the construction and to avoid changing them unless it is absolutely necessary.

(6) Check the diagonals (e) of the building outline: they should be the same (for a rectangular building).

(7) Profile boards (Fig. 25.3 and Fig. 25.4) are now sited beyond each corner, well clear of the building arrea. These are pieces of timber which have permanent markers (nails or sawcuts) indicating the position of one or more aspects of construction, for example the edges of the foundation trench, the position of the wall to be built on the foundation and internal features such as doorways and walls (h, Fig. 25.3).

(8) Lines (g) stretched from marker to marker on the profile boards provide an accurate facsimile of the ground plan full size on the surface to be worked on. Before construction begins, they are replaced by trickled silver sand, chalk dust, paint lines or other easy to follow methods of ground marking. The boards, however, are retained until such time as the reference marks on them have no value.

(9) The heights and levels over the site should be set out and checked against a vertical reference stake driven firmly into the ground, clear of all site activity. Marks on the stake should indicate ground level when the stake was driven in, a general reference height, say 1 m, above ground level and other points necessary to the construction such as damp course level. (Clearly visible reference lines painted on a nearby permanent structure might be preferable in some cases.)

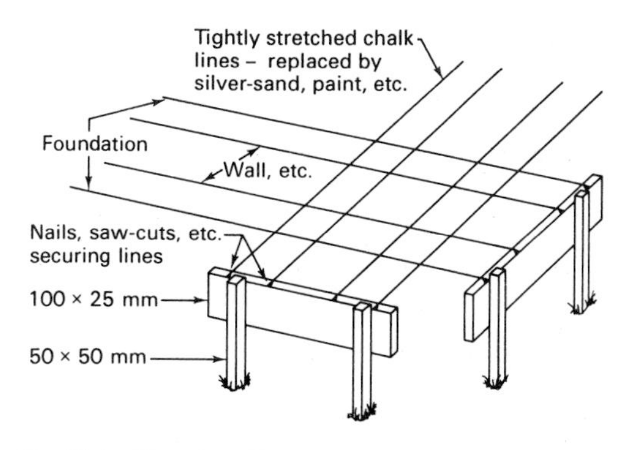

Fig. 25.4 Use of profile boards in marking out.

Construction – groundworks (BS 5502, Part 22, refers)

(See sections on Foundations and Drainage.)

(1) Trenches and/or pad support areas are dug out to profiled width and correct depth with vertical sides and firm bottom. Any soft areas should be stabilized with stone ballast or concrete and rammed down until firm. Water and water-softened soil should be removed prior to the foundation concrete being poured. In very soft soil conditions it may be necessary to support the trench sides with struts and board temporarily, not least to ensure safety of workers.

(2) Foundations should always be horizontal; on sloping sites a 'stepped' foundation is required. The depth of each step should, for convenience, be a multiple of whole building units, e.g. block(s) or brick(s). Thus:

$$\text{Distance between steps} = \frac{\text{Depth of step} \times 100}{\text{Slope \%}}$$

Each step must project over the slab below substantially more than the step 'depth'.

(3) To ensure horizontal levelling and the correct thickness of materials used, drive pegs into the trench bottom so that the tops represent the finished foundation surface. Use a spirit level and a straight board over short distances but a surveyor's level, either telescopic 'automatic' or watertube type (see 'Further reading' for a surveying reference text).
(4) The concrete mixture used for most farm works will normally be that specified in BS 5328 (e.g. as type RC 35) as described later. NB: Liquid-retaining structures are also covered by BS 8007, which normally specifies type RC 45 for that purpose. In foundation trenches with substantial thicknesses of concrete to be laid down, lesser specification concrete may be used. Again refer to BS 5328 for details. (Reference may be made by contractors and others to the use of mixes described as 1:2:4, 1:2½:4, 1:3:6. These outmoded descriptions relating to volumetric proportions do not imply satisfactory quality standards and should be avoided. Concrete mixes may be described by the also outmoded designations of C20P, C7.5P, etc. These are more reliable descriptive guides to concrete quality, but still may mislead the unwary and are not as defined as those under BS 5328. They will not be used in specifications henceforth.)
(5) The top surface of the foundation concrete must be level, smooth and horizontal to facilitate the laying of brick, block or framework.
(6) Drainage trenches are dug out and pipes laid where appropriate, at the same time as the foundations are excavated. Any work which affects both is carried out first. For instance any pipes to be positioned under or through strip foundations should be placed and fixed by, for instance, encasing in 150 mm of concrete (to BS 5238) before the rest of the mass is poured around the pipework.
(7) Services are laid onto the site and led to the main distribution points for connection on a temporary basis during construction.

Construction – frame and related cladding (*see* Figs 25.5 and 25.6)

(1) Steel, concrete and timber frames have different shapes and material characteristics which require different assembly and fixing techniques (see Fig. 25.6). The manufacturer of the frame will give specific guidance on the fixing and foundation requirements of particular structures.
(2) Each foundation pad supports and locates a frame member. The location system, of which there are many, using sockets and bolt assemblies, should be positioned exactly so as to receive the stanchion of the frame without distortion. Some frame manufacturers may wish to cast pad foundations themselves to avoid potential site difficulties. Small frames of 10 m span or less may not require a fixing system but rely on their weight to hold them in position.
(3) The frame members will arrive on site, sectionalized, for assembly prior to lifting into position by mobile crane. The span sections are loosely bolted in place and stabilized spatially by use of longitudinal purlin and wind-bracing members until the whole structure is complete at which time the joints are tightened up and the structure becomes rigid.
(4) Once the frame is secured, roof cladding is fixed, starting from one (usually leeward) gable end. Care is needed to align the corrugated sheets squarely with the verge and eaves of the building, otherwise there may be a danger of too little or too much overhang as the laying of sheets progresses down the length of the building.
(5) After both slopes are covered, the ridge capping is fixed in place. The overlap on both sheets and capping should be such as to minimize penetration by the prevailing wind.
(6) The gable ends and side sheeting are positioned, cut as necessary and secured, followed by the guttering, at a suitable slope, and the 'down' pipes to channel the surface water to ground level.

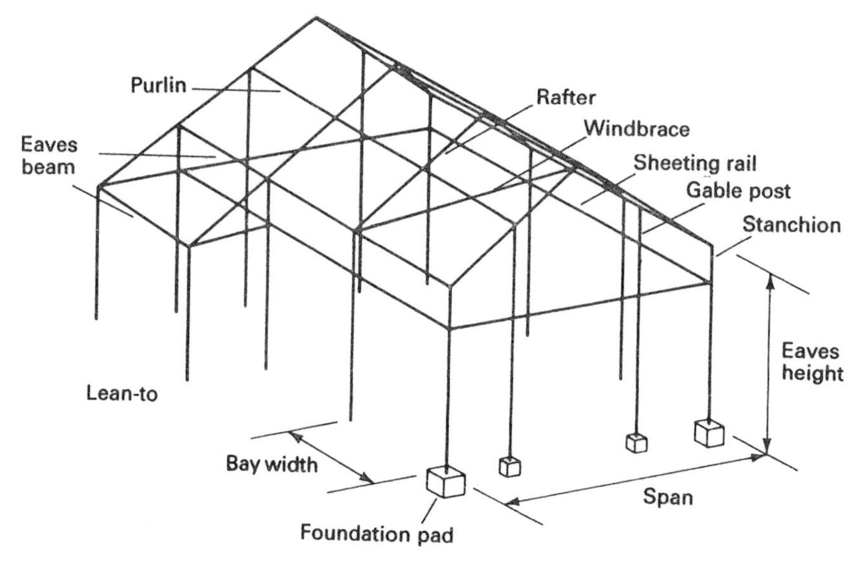

Fig. 25.5 Structure elements.

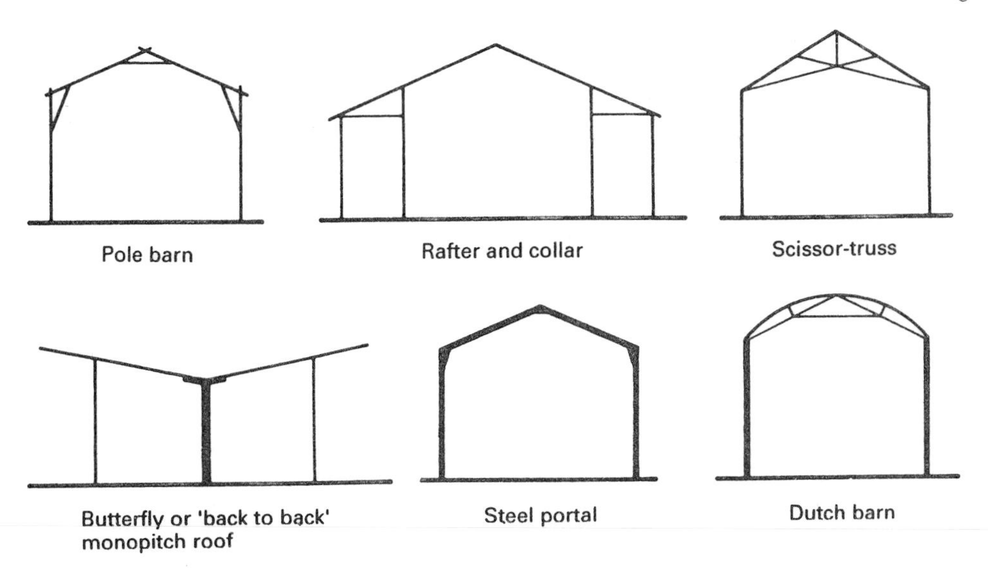

Fig. 25.6 Frame systems.

(7) Special features like translucent sheeting, eaves fillers, ventilation ducting, insulation or anticondensation materials may be incorporated as the roof is clad.

Construction – masonry cladding and floor laying

(1) With the frame in place, blocks or bricks may be laid on 'strip' foundations between stanchion members up to 'damp-course' (DPC) level which should be at least 150 mm above the outside ground surface. NB. No DPC may be used where there is any possibility of side thrust being applied to a wall. They are not therefore commonly used in agricultural situations, even though they are mandatory in other spheres.
(2) Doorways and other gaps are left in the wall, as appropriate, with timber 'liners' or with fixing blocks left in the masonry for subsequent connections.
(3) The walls will be built according to the design specification and may incorporate support piers and/or other reinforcements, insulation, waterproofing or a cleanable surface.
(4) When the walls have reached damp-course level (see (1) above), it is often convenient to lay the concrete floor. (For lightweight agricultural buildings the method of construction sometimes used is the 'ring-beam' foundation system, in which the floor is cast at the same time as the frame and wall supports with reinforcement incorporated at the edges.) The ground surface should already be level and free of topsoil so pegs are placed to indicate the finished floor surface. Hardcore is laid on to the soil to the required thickness (75–125 mm) and vibration-rolled into a hard, dense mass. The top of this is then 'blinded' by sand to fill all surface cracks and provide a smooth plane on which to place the damp-proof membrane. Polythene sheet, 125 μm thick, or damp-proof 'building' paper, is laid, with a minimum overlap of 150 mm at the sheet edges, to cover the floor and is bonded to any adjacent vertical wall and its DPC by bitumen-based paint or mastic. (The sheeting also serves as a 'slip' plane for the concrete mass.)
(5) The floor concrete (see Mass concrete later) should be poured in convenient shuttered bays, not exceeding 10 m × 3 m in area, for manageability and timeliness. It should be to BS 5328 specification and of a suitable thickness, usually a minimum of 100 mm. It must be tamped and surfaced by floating or other method, preferably after the use of vibration equipment for consolidation. It should be left to 'cure' for at least 7 d and preferably 14 d, before the enclosing shuttering is removed or it is required to bear a load. (If the curing period is very warm or dry, then a gentle re-wetting of the surface or spraying with resin sealant is recommended to preserve the mix moisture level.)
(6) An additional surface may be called for in some buildings, for instance in a farm dairy where a granolithic 'screed' may be necessary to provide anti-slip characteristics. This is a thin (up to 50 mm) top layer of fine granite aggregate concrete mix, applied to the base mix before curing has finished, and preferably within 72 h of laying. This is a skilled job, usually best left to specialist tradesmen.

Construction – internal finishing and services

The internal fittings such as water supplies, electrical installations, joinery work and masonry finishing (plastering, etc.) are normally the province of professional tradesmen, whose skill can make or mar a building. In this work there is a flexible element of design which is not possible in other areas of construction. 'On-site' decisions are quite usual and in any case craftsmen would be expected to think for themselves in respect of details like pipe and wire runs and fixing methods on all but the largest farm construction. However, there are some pitfalls on which fixing advice needs to be given, for example on door security in piggeries, moisture proofing of electrical equipment and drinking bowl positions.

The complexity of the work to be done and the need for a careful order of procedure to be followed means that the timeliness of the attendance of the specialists is often important to the smooth transition in the stages of building construction to avoid time wasting and mistakes. The employment of an experienced professional to provide overall supervision may be crucial to the successful completion of a complicated building.

Building materials and techniques (BS 5502, Part 21, refers)

Mass concrete

There are various types of concrete in general use but that for farm construction is usually a mixture composed of ordinary Portland cement (as opposed to rapid hardening Portland cement and other types), fine aggregate, coarse aggregate and water. Aggregate may consist of gravel or crushed stone, of a light or heavy nature, and may be round or sharp (angular). The fine aggregate (or sand) fills in the voids between the coarse particles (stone, gravel), the cement coating adhering to both. This should produce an almost solid matrix when set if air is expelled by good mixing and consolidation by vibration and tamping. All aggregates must be clean and free from vegetable matter, salt and clay. They must be dry (for accurate volume or weight measurements to be made prior to mixing). 'All-in' or 'as-dug' aggregates should not be used. Always use clean water, preferably from a drinking source. Although it is possible to mix concrete by hand in small batches, it is not recommended practice. The most reliable way of apportioning mix ingredients is by 'weigh batching', that is, weighing out enough materials to mix with 1 bag (50 kg) of cement, or a multiple number of bags, and to load the quantities into a suitably sized mixing machine in one whole batch.

The time of mixing should be long enough to ensure that all the aggregate particles are coated evenly with cement to give a consistent greyish coloration, for which, normally, a minimum of 2 min is required after the last ingredient has been placed in the drum. Most mixers work best when the coarse aggregate is placed in the drum first, followed by the water, then half the sand, then the cement and finally the second half of the sand. Some 'sticky' aggregates may require that this order be varied to achieve the best possible mixture.

The quantities of cement, aggregate and water will vary according to the job to be done.

Table 25.2 shows the normal proportions by volume, for the work loads expected in three types of circumstance, on farm sites. (A recent reclassification of concrete mixes under BS 5328:1991 has altered the designated coding for all uses: see Table 25.3.)

The size of aggregate must be taken into account when selecting an appropriate mix. Thin section or reinforced concrete requires stone which is relatively small in size, graded from 6 to 20 mm, whereas large mass concrete can satisfactorily enclose aggregate of up to 40 mm size grade. Fine aggregate is 5 mm or less in size usually, but must not contain too much dust.

Water quantities are critical – almost as important to the ultimate strength of the concrete as the cement itself. Too much weakens the concrete by leaving cracks when it dries out, too little will give ineffective bonding due to lack of chemical reaction. Workability is also governed by

Table 25.3 Designated mixes for premixed concrete according to BS 5328: 1991. Sample of types and applications (see British Concrete Association Fact Sheet No. 16 for more information)

Typical application	*Designated mix*
Concrete blinding under floor	GEN 1
Oversite concrete to be finished with screed, etc.	GEN 2
Unreinforced foundations – strip, stanchion bases, etc.	GEN 3
Reinforced foundations, crop store and livestock floors	RC 35
Slurry floors (and walls), workshop and sugar beet floors	RC 40*
Silage floors (and walls), stable floors, brewers' grain stores	RC 45*
Parlour and dairy floors (50 mm slump plus carborundum dusting)	RC 45*
External yards and roads – air entrained mix	PAV 1

*Equivalent to C35A current.

Table 25.2 Concrete mixes for various purposes proportional by volume

Purpose and use	*Large bulk volumes, e.g. deep foundations*		*Average use, floors, roadways*		*Heavy use and thin sections*	
Cement	2 vol	1 bag	3 vol	1 bag	3 vol	1 bag
Fine aggregate (damp)	5 vol	190 kg	5 vol	115 kg	4 vol	90 kg
Coarse aggregate	8 vol	270 kg	9 vol	195 kg	7 vol	170 kg
Water	1–1.25 vol	20 litres	1.25–1.5 vol	23 litres	1 vol	17 litres
Approx. yield	7–8 vol	200 litres	9–10 vol	170 litres	7–8 vol	140 litres
Previous designations	1:3:6/C7.5P		1:2:4/C20P		1:1.5:3/C25P	
BS 5328 Code	GEN 1/2		ST4		ST5	
Table 25.3 near practical equivalent			GEN 3/PAV1		RC 35	

the water content of a mix and as a consequence very often concrete is mixed and laid much too wet in order to increase the facility of placement. The moisture content can be checked by use of the 'slump' test. Special testing cones can be purchased or hired but a steel bucket may be used as follows instead to provide a rough guide. The bucket is filled with concrete and is then inverted on to a flat surface (steel/wood plate). It is then drawn gently away from the contained concrete leaving a cone-shaped heap, the height of which has 'slumped'. A slump of more than 25% of the height of the bucket shows the mix is too wet. Using the correct test equipment 50 mm ± 25 mm should be attained at the correct water content. If such a mix is made it will require vibrating equipment to place and consolidate it satisfactorily. This equipment may be hired.

All concrete should be laid as soon as possible after mixing and always before 1 h has elapsed. It should be carefully positioned from a low height avoiding any movement which might cause desegregation of the constituents, for instance rough barrowing. There is a brief period of 'setting' (about $\frac{1}{2}$ to $2\frac{1}{2}$ h) followed by 'hardening' during which it gradually achieves greater strength. During this latter 'curing' period, which is indefinite but for practical purposes is said to be 28 d, the concrete should be undisturbed, kept from drying out and in a frost-free environment. Under no circumstances should concrete be mixed when the aggregates are frosted, but in a very cold conditions (less than 4°C) a solution of builder's antifreeze (*not* motor-car type) may be used to enable concreting to be carried out.

NB Pulverized fuel ash (pfa) cement should be considered for use where alkali hoof erosion is potentially possible on new concrete. Also ensure that no limestone aggregate is used for silage bunker casting.

Pre-cast concrete

Many concrete items used in construction are 'pre-cast', that is, they are made away from the point of use, for example, concrete blocks, drain covers, frame members.

Concrete blocks

These are supplied in various common forms, standard sizes (see Fig. 25.7 for a typical range) and aggregate materials to suit the requirements of the structure. The majority of those used in farm construction work are of 'dense' aggregate, although 'cellular' blocks are used from time to time for insulation purposes. These latter have a bath sponge appearance, due to the foaming method of manufacture. Although of low heat conductivity they are relatively weak and are porous to water passage. (Note, however, that all concrete blocks are poor moisture barriers compared with bricks.)

When calculating quantities to be purchased it should be noted that the actual sizes of the blocks may be less than the 'nominal' size by 10 mm, the mortar gap allowance.

Concrete block walling

Most walls constructed in farm buildings are of 140 mm solid or 215 mm wide hollow or solid units laid in 'stretcher' bond pattern (see Fig. 25.8). (Lighter 'boundary' walls of 100 mm thick block may be used for the enclosure of small stock.) For some purposes however,

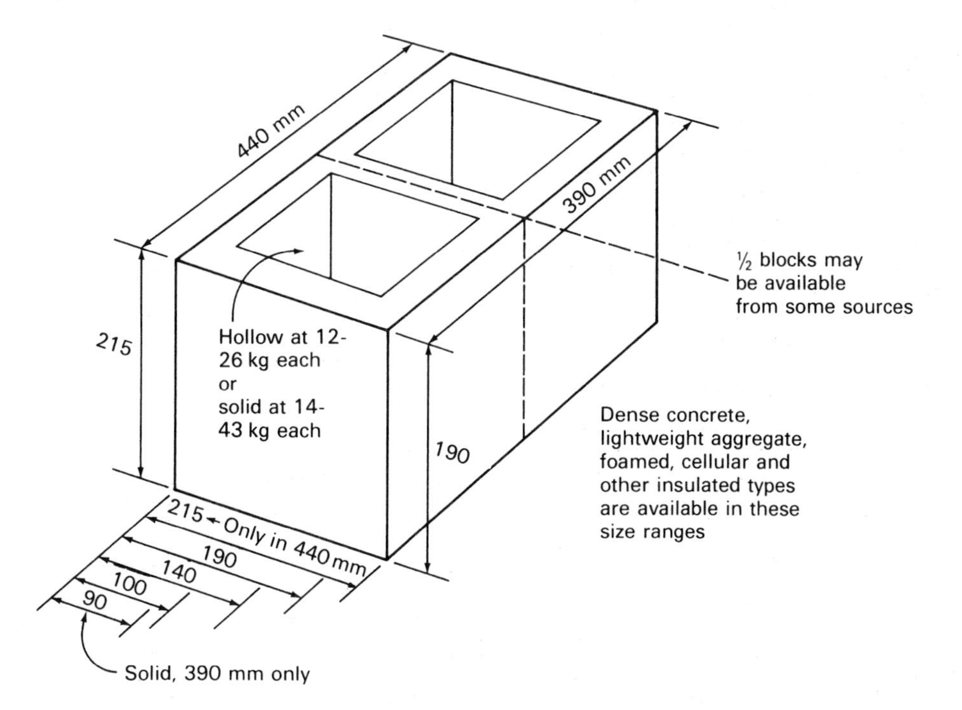

Fig. 25.7 Concrete block dimensions.

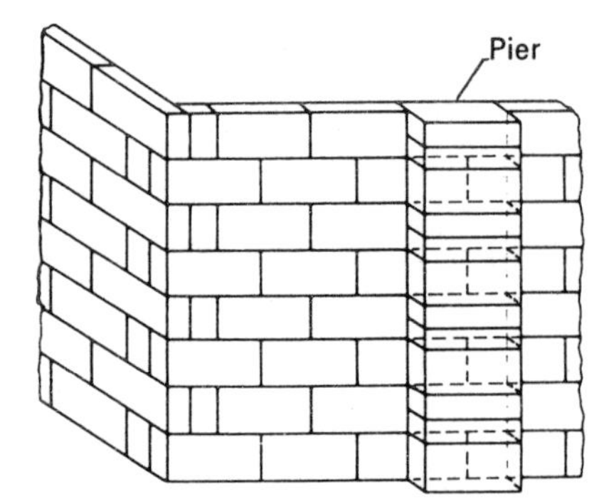

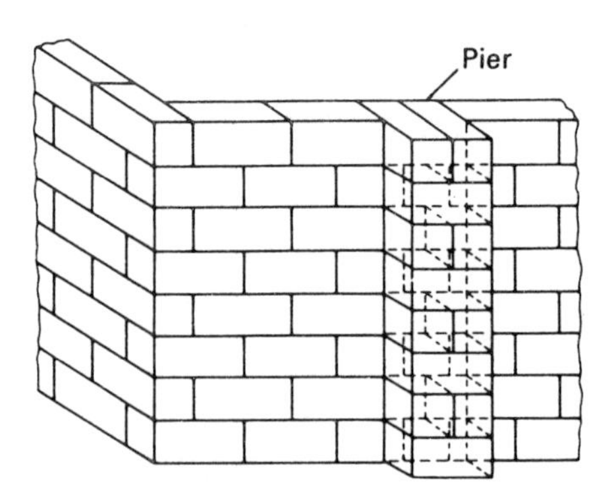

Fig. 25.8 Wall bonding.

all types of wall may require strengthening. This is carried out, traditionally, by either the incorporation of piers or the insertion of steel rods or netting into the bond pattern of the wall. Piers (see Fig. 25.8) are part of the wall which are thickened by the inclusion of blocks such that a vertical 'ridge' is formed, normally one block wide up the wall. The various forms of steel reinforcement provide strength in tension which the concrete block-work requires but lacks under bending loads. The steel must be bonded into the wall such that it all behaves as one composite material and yet is correctly positioned to resist the tension forces. In practice, piering and reinforcement are used together to achieve maximum strength, one method supplementing the other.

See Table 25.4 regarding wall stability and 'rule of thumb' relationships of wall width, strength and use.

NB Many agricultural buildings carry wall loads which are lateral and quite severe. 'Rules of thumb' are not recommended for these and expert advice should be sought for their design.

'Cavity' walls are not common in agriculture, although they are good insulators, as they are complicated to build, are weak under lateral pressures and can be expensive. They consist of two 'skins' (or leaves) of 100 mm concrete blocks 'tied' together by means of steel or plastic connectors (ties) across a 50–60 mm gap left between the skins which may be partially filled by insulation board adjacent to the inner leaf. The ties are placed in the wall, as it is built, about 1 m apart horizontally and 0.5 m vertically. It is important to ensure that the ties are kept scrupulously clean and free from mortar droppings. The internal skin may be of cellular block to increase the thermal insulation level yet further.

Table 25.4 Block thickness for unreinforced walls (but which may include piers)

	Block size (mm)	*100*	*140*	*215*
Walls lacking any support at top	Max length	3.6 m	4.5 m	9 m
	Normal max. related height	1.8 m	3 m	3.6 m
With continuous top support, e.g. roof framing, wall-plates, eaves beams	Max length	4.5 m	6 m	7.5 m
	Normal max. related height	4.5 m	6 m	7.5 m

NB Walls designed to withstand wind and stock pressures; not to retain stored grain or dung. For larger stock use the large block sizes.

Where the length of a wall exceeds its height by $1\frac{1}{2}$–2 times or where there is no natural bond break within 6 m, an expansion break must be included to accommodate thermal and moisture movement to which concrete blocks are particularly prone. This may be provided by a slot cut in the outer block surface at least 50 mm deep or, possibly, completely through the wall. In the latter case the stability is retained by the use of steel dowels set in the mortar on one side of the break and greased before laying in the other side. The slot should be filled with mastic at the outer surface to provide a weather-seal

The mortar used in concrete block laying should be in the volume proportions of 1:2:9 or 1:1:6 mixed from cement, lime and builder's sharp sand. The lime may be replaced with proprietary ingredients known as plasticizers, which make the mortar 'fatty', that is, slip easily into place between the blocks. These mixes are weak, relative to block strength, so that if there is a tendency to crack it will appear less unsightly and be of less significance to wall stability when it is confined to the mortar gaps.

Concrete block walls may, depending on their purpose, require waterproofing. This is primarily done by use of render coatings of mortar-like material, usually in two coats; the first, at 10 mm thick of 1:1:6 mix as described before and the second at 6 mm thick, of $1{:}\frac{1}{4}{:}3$ mix. Alternatively, the block pores can be filled in and the mortar gaps flush pointed to give a flat surface which can be painted with chlorinated rubber, epoxy-resin or acrylic-resin bonding paints.

A new approach to block laying combines the simplicity of 'dry-laying' (with no mortar), and render-coating for water proofing, by the use, as a wall surface fixative, of a mix of cement, sand, specially prepared glass fibre and bonding chemical laid on to produce a hard, durable coat which stabilizes and creates the strength of the wall from the outside surfaces.

Reinforced concrete

Contains steel reinforcement at a position determined by the need to carry tensile loads, for which concrete is relatively weak, having only one-tenth of its capacity to resist compression. The steel is usually below the centre line, towards the bottom surface. Frame members of reinforced concrete are not very often installed nowadays owing to their relatively high cost, but many structures built in the 1960s and 1970s were of this material. Simple structures, e.g. door lintels, may be cast *in situ*, but the commonest use of reinforcing is where strong flooring is required for resistance to heavy imposed loads such as the weight of stored grain. In situations where very high strength is needed, together with extreme economy of concrete use (to minimize cost and space wastage), a technique known as pre-stressing may be used. This method creates an apparently high tensile resisting ability in the mass of concrete by pre-compressing it through tensile load imposed on the steel in advance of setting.

Steel for all concreting purposes may be in the form of wires, bars or net depending on the situation and load type, but whereas rods or wires are usually used for walls or frames, net reinforcement is more commonly incorporated into flooring and is less often utilized in walls. The steel must be covered by a substantial layer of concrete wherever it is used so that it will resist corrosion and fire whilst performing well as the tension element in the combined mass.

NB For the case of liquid-retaining structures see also Mason, P.A., CIRIA 126 (Further reading).

Timber

Timber used in construction should be well seasoned, pressure treated with biocides and fire retardants, and free from warping, knots, sapwood, bark and shakes (splits). Softwood (see Fig. 25.9 for available size ranges) is used for nearly all work other than decorative or highly specialized functions due to the high cost and difficult working of most hardwoods.

Lengths of timber are generally purchased by the 'metric-foot' – actually 300 mm but for large quantities it may be sold by the cubic metre with additional sawing charges. Pieces in excess of 5.4 m long may exact a premium price. Sawing and planing 'allowances' must be taken into account when sizing timber sections as these may reduce the dimensions by up to 2 mm.

Probably the biggest drawback to the use of timber in construction is biotic decay. In the UK it is affected by three main agents, 'woodworm' (actually the larvae of a beetle, *Anobium punctatum*) 'dry rot' (a fungus, *Merulius lacrymans*) and 'wet rot', which has several possible causal fungi. All these problem pests can be prevented from doing damage by treating the timber initially with a cocktail of potent chemical biocides, but the life-span of timber can, in any case, be prolonged by the avoidance of poor climatic or environmental conditions.

Repair of timber which is already infected is difficult and sometimes, as in the case of dry rot, almost impossible without wholesale replacement.

Recently the use of 'stress-graded' timber has become a requirement in some types of structure. This is timber which has been inspected visually for flaws such as knots, size imperfections, shakes and other defects and is stamped with a grade category which indicates the limitations to its use in structural work. GS (general structural) has been passed as acceptable for general use, for example as floor joisting; SS (special structural) implies a more particular usage (see BS 4978 on stress grading of timber).

For information about screw and nail sizes for timber work see BS 1210 and 1202. Recently there have been a number of new types of fixing method devised, most of which combine the driving ease of the nail with the firm hold of the screw. A review of these innovations is recommended before extensive timberworks are carried out.

There are two further general design concerns expressed about structural timber. The first is its apparent susceptibility to fire damage which, owing to its characteristic of charring rather than burning, is not as threatening as

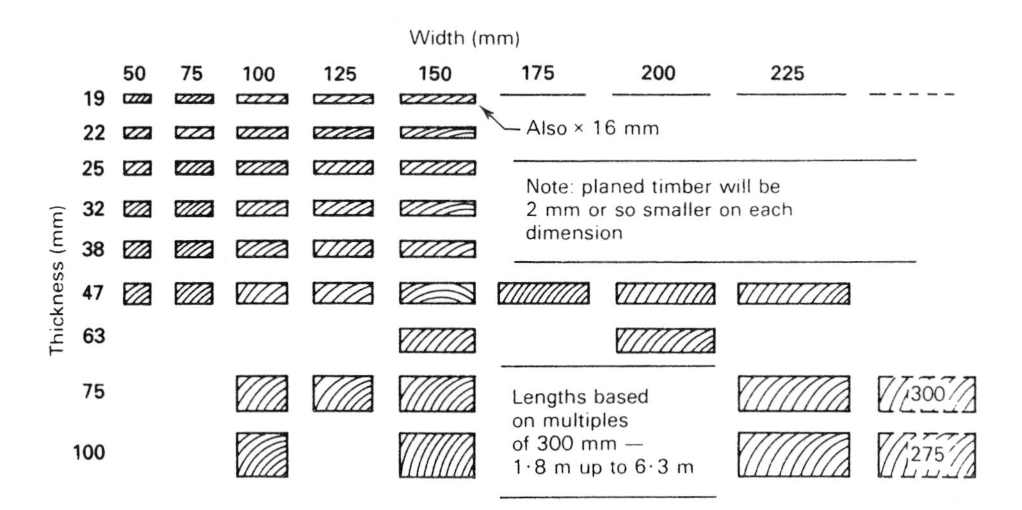

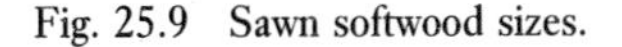
Fig. 25.9 Sawn softwood sizes.

might at first be thought. If proper safety factors are used in design considerable parts of the cross-section of timber beams must be burnt away before failure occurs.

The second concern relates to its flexibility. Timber lacks 'stiffness' compared with structural concrete and steel so its use is restricted in simple beam or lintel framing to short spans. In larger spans the timber must be used in a lattice girder or similar pattern, to achieve the required strength and rigidity.

Timber boards, particle boards and composites

The cost of timber has risen rapidly in recent years and will continue to do so in the foreseeable future. As a consequence composite materials have been developed which utilize the lesser quality portion of the timber previously thrown to waste, and at the same time make better use of the high quality material. The original form these took was that of 'plywood' or 'chipboard' but now they may be made partly or even mostly from sources other than wood, for example, recent building boards which are combinations of plasterboard and glass- or mineral-fibres, also plastic-based insulation combined with wood particle board.

The structural framework used for mounting these boards (usually supplied in 2.44 m × 1.22 m sheets) is generally based on the traditional 'studwork' used for timber planking (see Fig. 25.10a) but they are sometimes supported on rails in a similar manner to cement-fibre corrugated sheeting (see Fig. 25.10b). External grade (WBP, weather and boil proof) plywood sheets make excellent cladding for walls and roofs (the latter with a felt covering) though care must be taken when fixing if warps are to be avoided. All edges should be secured with galvanized nails or screws, with intermediate nailing for hardboard (or other types of 'particle' board) at 600 mm centres. Generally, the main studs should be at 1.2 m centres, equal to the sheet widths with noggins at 600 mm vertical centres. Oil-tempered hardboard, an inferior but cheap alternative, should be nailed every 200 mm or screwed every 300 mm to a frame specially provided with studs and noggins to accommodate it. NB: Chip (particle) board is not normally suitable for 'weather' cladding but there are some special resin-impregnated varieties available which can withstand deleterious water, solar and biotic effects. Traditional timber boards are used as an outside wall cladding for poultry houses, barns, etc., either horizontally or vertically (with tongued and grooved joints). The boards may be horizontal and parallel-faced with lapped joints (e.g. 'shiplap' boarding) or, preferably, horizontal and taper-faced with lapped joints (weather boarding). All boards used should be 25 mm thick though 19 mm is not unusual on cheaper work. In the latter case they should be 'clout', galvanized nailed to vertical studs placed at 450 mm centres. For proofing against wind and snow the boards should be lined with building paper trapped to the studwork.

Vertical boards may be used to provide ventilation to cattle yards (space-boarding/Yorkshire boarding), being nailed to horizontal rails, with about 13–18 mm gaps between. These boards, 125 or 150 × 19 or 25 mm, should be above stock height and should be associated with ridge ventilation to assist air circulation. Yorkshire boarding is also used to roof cattle yards, being an alternative to slotted steel or steel or aluminium or gapped cement-fibre sheets. Boards used in this manner should be 150 × 25 mm with two longitudinal drainage grooves, one near each topside edge. They should be nailed over a spacer which lifts the boards 6 mm above the purlins. Gaps of 6–12 mm are left between them.

Various materials are available for internal lining boards, including flat cement-fibre sheets (2.4 × 1.2 m) in several qualities and ranging from 3 mm up to 12.5 mm thick. These are suitable for lining timber frames or panelling gates and are often used for controlled-environment housing. Similarly, oil-tempered hardboard in thicknesses of 3–8 mm and sheet sizes up to 1.2 × 3.6 m may be a useful, cheap cladding for this purpose as it resists water absorption. Another common material is manufactured from compressed plastic waste to form a strong, dense but flexible and waterproof board of 12 mm thickness or more.

Additionally, there are many insulation boards formed from single materials such as foamed PVC or combinations of timber, plastic and metals bonded with plastic glues.

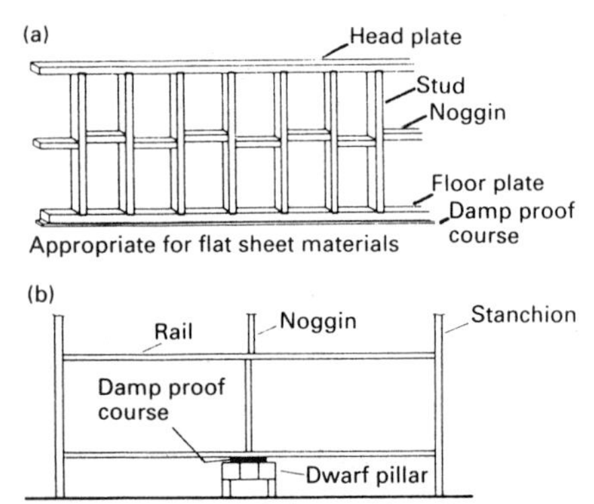

Fig. 25.10 Wall framing.

Non-timber lightweight cladding

Frames may be clad in simple fashion with uninsulated sheets of cement-fibre, steel, plastic and bitumen-coated hessian, or combinations of these (aluminium rarely) and fixed with nails or special bolts to wall rails or roof purlins. Corrugated cement-fibre (replacing asbestos-cement), together with plastic (PVC) coated steel are the usual claddings preferred.

The former, however, has low impact strength, becomes brittle with age and requires 'crawling' boards for maintenance procedures. The most common type of this material, so-called Big Six, with standard lengths in increments of 300 mm from 1.5 m to 3 m, will, when laid with a single corrugation lap, have an effective width of 1 m. The minimum end lap should be 150 mm and for roofs of under 15 degree pitch should be sealed with a bitumen cord. Maximum purlin and rail spacing for this sheet type should be 1.35 and 1.8 m, respectively. These sheets are available with slots in the crown of the corrugation which can assist ventilation of roofs for cattle. Other types of sheet are available witn smaller 'Standard' dimensions as shown in Fig. 25.11. It should be noted

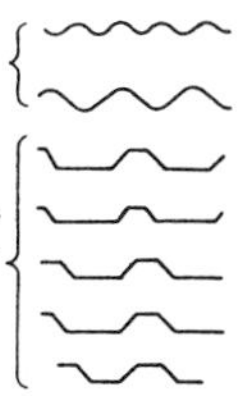

	Depth (mm)	Thickness (mm)	Pitch (mm)	Cover width (mm)
Cement-fibre	25	5,5.5	73	648
	54	6,7.5	146	1016
Examples of plastic-coated steel	30	-	200	1000
	19	-	152	762
	26	-	150	900
	24	-	143	1000
	23	-	125	1000

Fig. 25.11 Common corrugated sheet profiles.

that different manufacturers do not observe the same profile for similar sheet sizes, thereby making a match difficult for assembly or repair.

Translucent plastics/glass fibre sheets for rooflights or wall panels follow the same profile and sheet size as cement-fibre. Clear plastic ('Perspex') sheets are also available and increase light transmittance, but being much more expensive are rarely used.

Sheets are fixed with hook bolts over steel and concrete rails or purlins or with drive screws into timber, the latter being cheaper. Bolts and screws should be galvanized and protected externally with plastic washers and caps. Holes for fixings should always be drilled in the top of the sheet corrugation (except in the case of steel sheet – see later). It is usual to support each sheet in at least three places. Where four cement-fibre sheets coincide (and overlap), it is necessary to cut two of the sheets on a mitred angle to lay them as flat as possible. There are many specially formed accessory panels available to cover awkward areas and difficult shapes at, for example, roof edges and on curved roofing.

Corrugated steel, available in galvanized, plastic-coated, painted or, very rarely, plain versions, may be supplied in lengths from 2–3 m, gauges of 22, 24 and 26, with a net covering width of 610 mm. It is a lighter and more impact-resistant alternative but however well protected is still subject to rusting processes after ageing. The great majority of these sheets are used coated with flexible co-polymer paint (trade name Colorfarm AP) to match the requirements of owners or local planners. Plastic-coated sheeting should not be used in dark colours on the roof. Exposed sites should be accommodated by use of double corrugation laps. In addition sheet lengths of up to 7.5 m are obtainable to special order, obviating the need for frequent end laps. Fixing methods are similar to those of cement-fibre sheets except that corner mitring is unnecessary and holes may be punched through the sheet. They must also be at the bottom of the corrugation troughs.

Aluminium, although not common owing to cost, is useful for its lightweight characteristics and solar reflectance.

There has been a recent revival of interest in slotted and slatted roofing for ventilation in extensive livestock housing. New, cheaper methods have been devised taking the form of corrugated sheeting laid in normal fashion from ridge to eaves but with no side overlap and leaving a small gap (15–20 mm wide) between the sheet 'columns'. The sheets are laid upside-down with the edges of the corrugations upwards at the slots. Roofing costs are reduced and the cost of a purpose-built ventilation system let into the roof may also be saved.

Frames and simple roof support structures

The frequently used term 'portal' should specifically be applied to support frames providing clear, unimpeded headroom to the ridge inside the building, but it is often corrupted to mean any factory made, site assembled, large span system (Fig. 25.6). Common components (Fig. 25.5) include the upright supports (stanchions), the arms (rafters) running from the eaves to the ridge, purlins, sheeting rails, eaves beams and ridge boards. Many types of connector are used in assembling the structural members, including bolts, clamps, nails, staples, welds and even glue, each relating to an appropriate material and structure system.

The common and 'preferred' size ranges for major frame dimensions to which most manufacturers adhere are shown in Fig. 25.12. Eaves height is comparatively simple to adjust during manufacture or assembly and so provides many options which are suited to individual requirements. As a general rule wider buildings have shallower slopes to their roofs to reduce overall height but the framework elements must become stiffer to compensate. Thus it is often more economical to create very wide structures from main central spans with lean-tos at one or both eaves or, alternatively, multi-span arrangements with roof valleys between slopes. A lean-to may extend as much as 11 m from the main structure.

Normally purlins will be of the same material as the frame except that timber may be supplied with steel frames of smaller bay widths and where easy sheet fixing (by nailing) is required. Purlin spacing and size depend more on cladding sheet length than on frame rigidity considerations, as each sheet must be supported at the top, centre and bottom, resulting in a normal arrangement at approximately 1.35 m intervals down a roof.

Occasionally, in a narrow building, it may be more appropriate and economic to use a roof truss supported by the walls instead of a frame system. This may be of steel, bolted or welded type, or of timber in a number of patterns, from the simple traditional nailed or bolted variety to the sophisticated design of glued or stapled latticework. The simplest type of frame of all is the 'beam and post' construction of which there are many versions from the rolled steel joist (RSJ) to the box beam in 'glue-lam' plywood. The choice of these is usually left to structural engineers and should not be attempted by the non-professional.

Foundations

Foundations are intended to sustain building structures against three main loading problems in agriculture (note: reference to 'soil' implies subsoil, i.e. below the agricultural soil horizon):

(1) Sinkage – when the building weight (its own structure load and that imposed upon it by contents) exceeds the support capacity of the soil beneath, downward or sideward movement may occur. This can be differential in nature, varying from place to

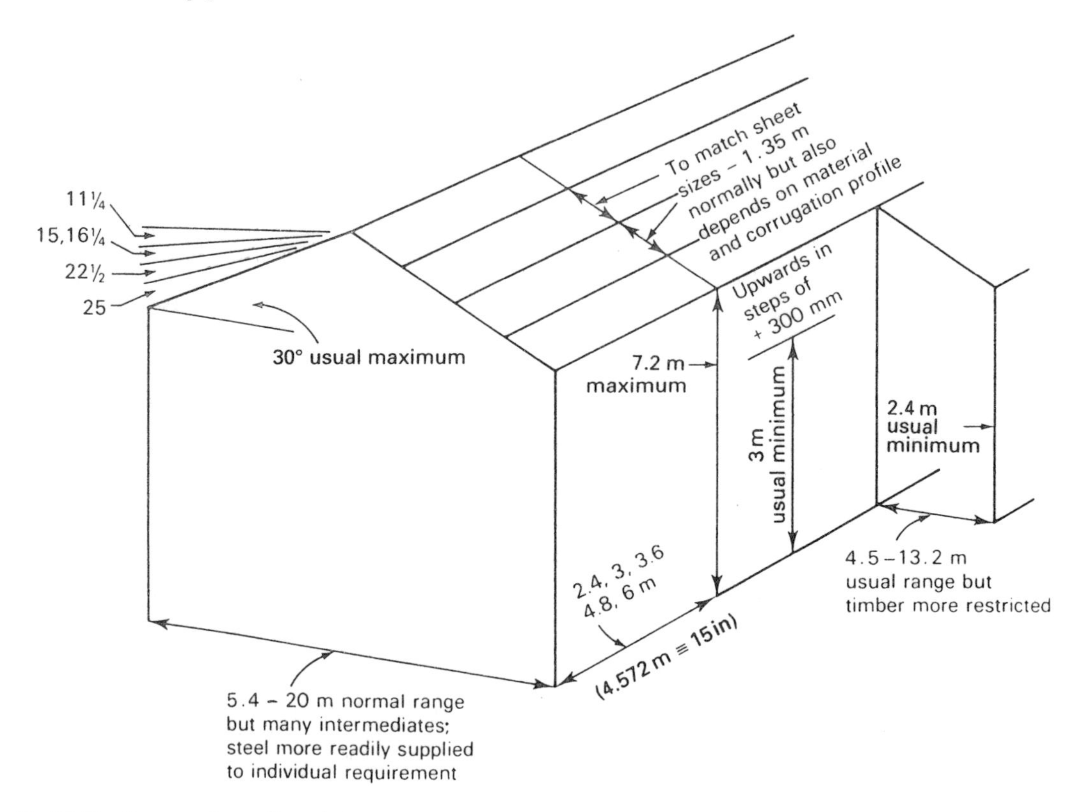

Fig. 25.12 Common frame dimensions.

place, creating bending and twisting loads on the structure.

(2) Water effects:

   (a) Volume changes due to soil moisture content variation (sand soils best, clay and peat worst) cause structure stress and movement by soil 'heave'.
   (b) Frost action giving volume and structure changes in soil (minimum depth for foundations related to climatic considerations), by expansion/contraction.
   (c) Leaching and erosion undermining the base for support, caused by poor drainage (from soil or inadequate piping configurations).
   (d) Tree root influences, creating changing moisture levels in the soil horizon, especially after felling.

(3) Cartwheeling – where lateral (side) loads applied by building contents, e.g. stored grain, turn the structure base through 90 degrees.
(4) Foundations must resist other forces such as vibration and load cycling but fortunately these are comparatively rare in agriculture and can be ignored for practical design purposes. (They must also resist chemical attack from soil acids and sulphates – but these are not directly structural matters, although important.)

Thus foundations must be:

(1) correct in area (load imposed ÷ soil bearing capacity × safety factor) which sometimes, for simplicity, results in a standard recommended size;
(2) at a correct depth (about 600 m to 1.2 m according to soil and climate, see 2a and b above);
(3) of a correct width to prevent cartwheeling and provide for easy construction practices, e.g. trench and hole sizes are dictated partly by the skill and ease of digging, also masonry units require a margin of measurement accommodation;
(4) of a thickness to be strong enough to resist failure in compression, i.e. to avoid a hole being punched through by the load applied (it may be necessary to reinforce with steel to achieve this);
(5) of sufficient material quality to resist failure by chemical reaction and premature loading during curing.
(6) below any 'fill' material on 'made-up' ground area.

The actual size and depth are therefore inherently related to particular sites and building circumstances, and there can be no safe 'rules of thumb'.

Foundation design advice should always be obtained from an expert source familiar with the site and the proposed building before commitment to a type of construction is made.

Strip foundations (Fig. 25.13)

As agricultural buildings are usually lightweight with few load complications and offer minimal risk to the general public, some restricted soil situations can be provided for in the limited case where simple blockwalls are to be erected. The recommendations in Table 25.5 apply only where the walls supported are for boundary purposes, carry no imposed load from above and there is no risk of human injury. (Again, to be safe, have the dimensions checked by an expert.)

Pad foundations

These provide support for frame uprights and locate them laterally. Manufacturers of frames will provide detailed drawings of the socket, bolt or dowel fixings with their positions, in pads of recommended size. It is not possible to give guidelines for these as they vary tremendously with shape, dimension and type of frame and the properties of the soil on which they rest.

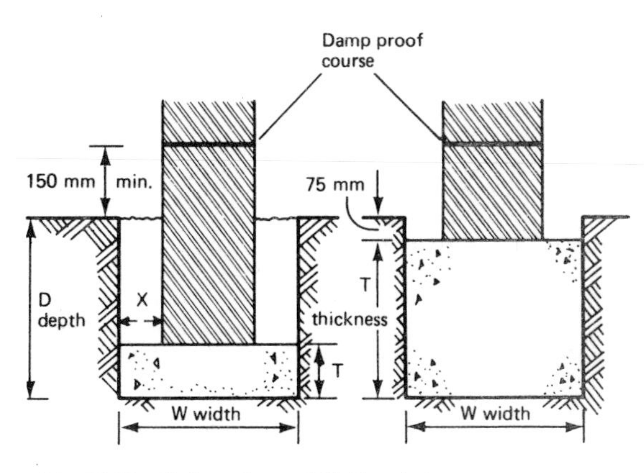

Fig. 25.13 Strip and trenchfill foundations.

Table 25.5 Foundation dimensions

Strip foundations	*Clay*	*Soft subsoils*	*Sand, gravel, chalk*
Depth (normal climate)	1 m	750 mm	600 m
Depth (frost prone)	1.5 m	1.1 m	1 m
Minimum width (100 mm wall)	300 m	500 m	300 m
Width with working space (100 mm wall)	400 mm	500 mm	400 mm
Minimum width (270 mm wall)	600 mm	700 mm	450 mm
Width with working space (270 mm wall)	600 mm	700 mm	600 mm
Minimum thickness	200 mm	300 mm	150 mm (200 mm in soft conditions)
(For strip foundations extension X must never exceed thickness T – Fig. 25.13)			

For 'trench-fill' systems read as strip foundations for depth; always use minimum width. Thickness is depth minus 75 mm. To be safe in uncertain soil conditions use minimum width required in soft soil.
NB This table should only be used where the soil on which the foundation rests is firm, undisturbed, not subject to frost or water action and can support the loads implied by the table conditions.

Drainage

Two categories of drainage must be considered (and treated) separately in practice (see Chapter 23 on Services, and 'Water pollution control' later in this section for further information). They are:

- Surface water, that is, water from rainfall run-off, which should be 'clean' and may be discharged directly into a watercourse for disposal. In effect this implies water from roofs only.
- Foul water (or soiled water) including sewage and slurry flows from all sources and other 'dirty' water such as cow yard washings. This category must in no circumstances be discharged into watercourses without purification which will ensure that it conforms with the various anti-pollution requirements before release.

Roadways and external works

Within a farm there are usually three types of road: the access road connecting the farmstead with the highway, the roads which link buildings together and the lanes connecting with the fields. Each serves a different purpose and requires a different investment approach to be economic.

The access road is the main artery of the farm and needs a good layout and surface, justifying the use of concrete which is normally too expensive for other road purposes. The width should be not less than 3 m to prevent concentrated wear and should allow for the passage of wide vehicles. Passing bays should be provided at not more than 150 m intervals. The 'weight' of traffic on a farm road is light, being related to the frequency of use rather than wheel loading, as the majority of vehicles travelling along it have comparatively low ground contact pressures. However, fast moving traffic will wear surfaces rapidly so it is advisable to limit permissible road speeds for vehicles such as bulk tankers, which have high weight-carrying capacities. Particular problems for all farm roads result from silage effluent attack and the combination of slurry droppings and the passage of cloven hoofs, but it is difficult to see how the latter can be minimized without handicapping the farm transport network. Concrete, being the most resistant common surfacing material, may be used to combat these effects but only at a cost which may be inappropriate.

Roadways around the buildings are an important factor in the planning process and will vary in width according to the amount of traffic expected to use them. They should be laid well clear of the buildings, except where they serve as direct links. Large vehicles entering a cul-de-sac must be provided with space to turn. All bends and turning circles should be not less than 16 m diameter and should be strengthened to accept the greater loads induced.

Internal roads usually serve to link fields and buildings together and take mostly tractors, trailers and stock, none of which justify the building of an elaborate road. Thus, to keep down cost, local material should be used wherever possible: chalk, quarry waste, ashes, hardcore from old buildings, flints, burnt shale, limestone chippings, are all materials which make satisfactory internal roads. The finished surface may then be treated with a coat of tar or bitumen plus chippings or a more costly layer of precoated stone (Macadam). These will keep the water out of the subsoil, but may not be very hard wearing. (NB These latter, coated, materials are not generally recommended for surfaces over which livestock will walk frequently.)

Keeping the subsoil dry is one of the key factors in successful roadmaking.

It is not usual therefore to excavate into the ground to make a road, though fibrous/humus soils should be removed; it should, instead, be built up from the ground surface. Where a road is made down into the subsoil because of poor support conditions the lower 'in-soil' part is made of introduced stone or hardcore and the surface should still be above the surrounding ground. Fabric specially designed to sustain roads over soft under-surfaces (e.g. the product trade-named 'Terram') may be incorporated below roads which are likely to carry heavy imposed loads frequently, or where the going is naturally 'boggy'.

On sloping sites, a land drain should be provided on the rising gradient side of the road. All work should be carried out in dry weather, preferably during the summer months. Any junction with a highway must be carried out to the satisfaction of the County Council, possibly in pre-coated stone laid on a suitable stone base. It is often feasible to have this work done by the highway authority when a roadmaking gang are working in the area. An existing farm road can be given a wearing surface of pre-coated stone laid between 50 and 75 mm thick and then well rolled. The work should be completed whilst the stone material is moist. Alternatively, uncoated aggregate can be laid first and grouted with hot tar afterwards. All sharp bends should be constructed in concrete as the greatest amount of wear takes place on corners.

Wheel track roads in concrete cost almost as much as a full width road, owing to the labour required in forming the track. They require a lot of maintenance, can lead to tyre-wall damage and are therefore not usually recommended.

Concrete for roadways should normally be from 100 to 150 mm thick on at least 150 mm of compacted hardcore and of an appropriate quality to BS 5328, usually PAV 1, with air entrainment for maximum wear life, with a mesh reinforcement when bulk tankers use the road.

Reinforcement is only necessary in soil of low bearing capacity such as peat, fenland silt and the like. Expansion joints are usually provided every 5 m length of road when laying continuously.

To reduce expense, passing bays can be constructed of hardcore: they should be 2.5 m wide by 12 m long.

Very light use internal farm roads may be made from 100 to 150 mm of hardcore laid on top of the subsoil surface after spraying with herbicide and rolling well. Larger stone should be used at the edges with smaller stone filling the middle. A drainage furrow may be ploughed alongside the edge which also acts as a support against the surface spreading. Any wet spots should be drained before the stone is laid down and consolidated.

Environmental control

The term 'environmental control' in agriculture is generally understood to mean the manipulation of temperatures (and occasionally humidities) by ventilation and heating. Its main use is in the optimization of livestock housing climates, but in limited form 'conditioning' in crop stores

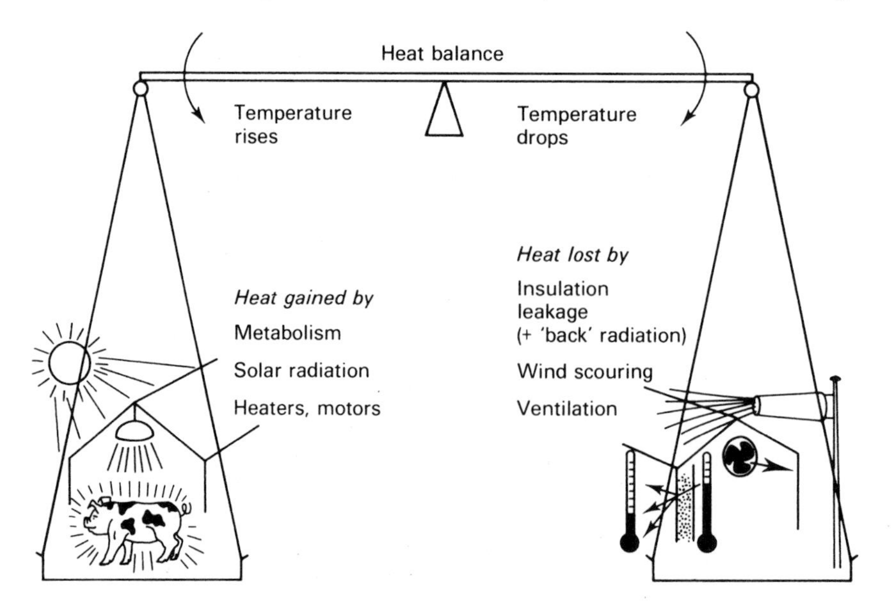

Fig. 25.14 Major elements in environmental control.

may also be described as such. Horticulturally, the term has a much wider implication and may imply, among other things, adjustment of gas levels (CO_2, O_2, N_2), supplementation of light and 'natural' control of pests. These activities call for a precision of monitoring and control which is not usually justifiable economically for less sensitive agricultural produce such as potatoes and pig-meat. However, environmental control is still appropriate to these latter, though on a lesser scale, because it is only by the achievement of a reasonable energy flow relationship between input and output that satisfactory profits can be made.

The major elements in the balancing act between input and output of energy, which is said to be 'environmental control' in livestock buildings, are represented in Fig. 25.14. As can be seen, some of these factors are the result of management decisions, whilst others are established by the nature of the building's design. However, by far the biggest influences brought to bear are those of the weather. Thus, whereas we loosely use the word 'control' to denote our ability to change conditions within the environment, often our attempts to do so are rendered difficult or even ineffective by the rapidly altering climate outside. Nevertheless, by manipulating the two 'mechanical' pieces of the equation, ventilation and heating (or cooling), some semblance of temperature (and humidity) control may be achieved. The process is made much easier, and less costly, if there is a 'buffer' between the outside elements and the internal environment. This takes the form of an insulative envelope, which can be simple as in the case of 'Yorkshire board', or complicated like the composite cladding around a pig fattening house.

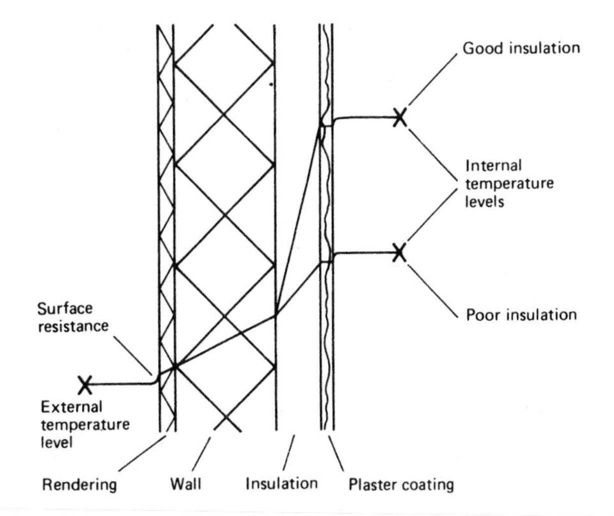

Fig. 25.15 Temperature gradients through insulation.

Any of these systems has the objective of providing initial weather protection. Therefore, it is obvious that the starting point for all environmental control systems should be consideration of the standard of insulation to be used. In addition, it is now being realized that animal welfare is often prejudiced by inadequate insulation, particularly in flooring. This reinforces the need to review this factor first.

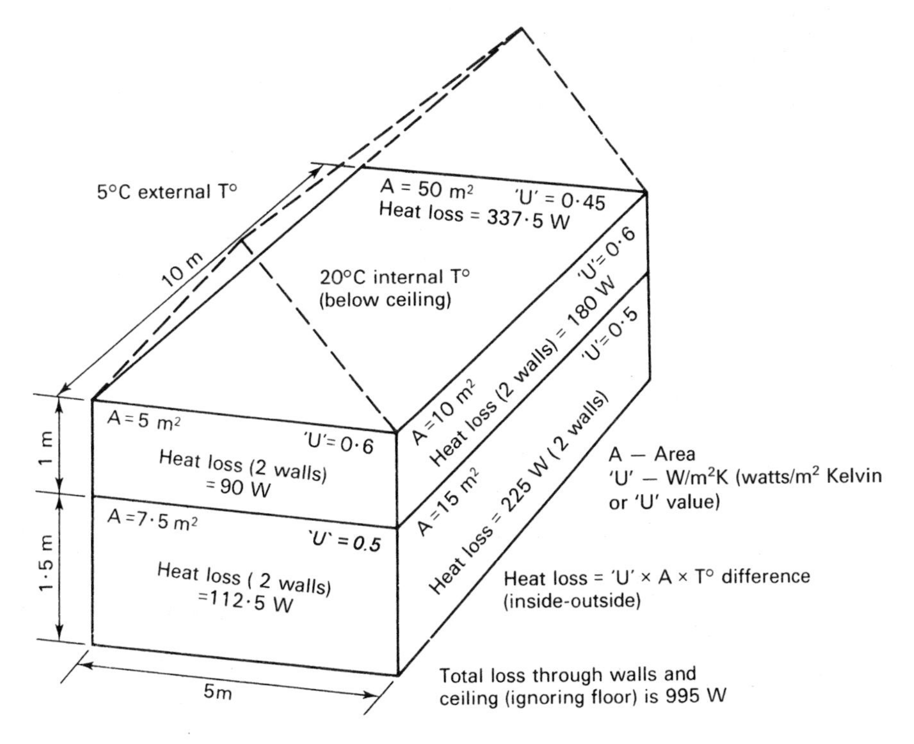

Fig. 25.16 Example of heat losses from a warm environment.

Insulating methods

Virtually any structure which encloses an environment is acting as an 'insulator' but the term is generally reserved for materials which have the specific property of resisting, or opposing, heat transfer. This implies that they are capable of preserving a high (relative) temperature difference between external and internal environments, irrespective of which of these is the higher. The practical effect of a high temperature gradient may be seen from the example shown in Fig. 25.15. Some insulators are clearly better than others (e.g. timber as opposed to dense concrete) but the great majority of specific purpose materials cannot be judged by instinct or on 'common knowledge' grounds. They need, therefore, to be compared by numerical values, of which there are several standard alternatives. The most frequently used of these is the 'U' value, a measure of the rate of heat flow expected through a structure under given temperature conditions; better insulators are associated with lower values. Table 25.6 gives some examples and Fig. 25.16 shows how they may be applied to calculate heat losses. Perhaps more logically, the 'R' value – for resistivity – is quoted for some materials. This is the inverse of 'U' and as such gives larger numbers to represent greater insulating effect. Both 'U' and 'R' are affected by their working 'conditions'. For instance, on exposed sites the 'U' will be greater and the 'R' smaller than normal, due to heat being scoured away from the insulating surfaces, creating more rapid transmission effects. In still air, however, the values will be appreciably better. Thus, they reflect the insulation properties for typical situations of material, climate and aspect rather than indicating absolute values which hold good at all times. Care should therefore be exercised in their interpretation for unusual situations.

Two further factors should be accounted for in the practical use of insulators. First, the majority of the better ones contain large quantities of air, usually in tiny bubble formations; these cavities should not be allowed to fill with water by condensation or wetting otherwise their insulating properties will be lost. Secondly, radiant heat (positive solar gain and negative 'back' radiation loss to sky) can be an embarrassment which is difficult to counteract. Insulators which inherently have anti-radiation properties, being light in colour or having silvered surfaces, will help to limit heat transfer from this cause.

A good insulator, therefore, usually takes the form of a lightweight, hydrocarbon-plastic mass, filled with minute, discrete air bubbles and covered with a shiny, preferably silvered, surface which is also a water-vapour proofing coat.

Table 25.6 'U' and 'R' (= 1/U) values of common insulators

Wall	'U' (W/°C m^2, also W/m^2 K)	'R' (°C m^2/W, also m^2 K/W)
1. 215 mm dense, hollow, concrete blockwork	2.05	(0.49)
2. As (1) + outside rendering + 12 mm expanded polystyrene slab	1.14	(0.88)
3. 275 mm 'cavity wall' of 100 mm dense concrete blocks + 25 mm expanded polystyrene slab	0.68	(1.47)
4. 215 mm solid wall of foamed blockwork + rendering both sides	0.45	(2.22)
5. 6 mm cement-fibre sandwich on 50 × 50 mm timber frame with expanded polystyrene core	0.45	(2.22)
Roof		
6. Corrugated 'Big Six' cement-fibre sheet	6.53	(0.15)
7. Corrugated sheet steel + fibre 'insulation' board	4.82	(0.21)
8. Corrugated 'Big Six' cement-fibre sheet + 60 mm mineral wool (+ vapour barrier) + 4.5 mm cement-fibre lining board	0.60	(1.67)
9. Corrugated 'Big Six' cement-fibre sheet + 55 mm extruded, foil-faced, polystyrene sheet	0.59	(1.69)
10. Corrugated 'Big Six' cement-fibre sheet + 40 mm foil-faced polyurethane board	0.50	(2.0)

NB Mandatory minimum requirement for 'U' of industrial walls and roof is 0.45; 0.45 and 0.25 respectively for dwellings.

Floor	'R_{f45}'
11. Dense concrete floor	0.042
12. Concrete slatted panel	0.086
13. 18 mm screed on 150 mm lightweight aggregate	0.17
14. Wooden slats, 58 × 70 mm with 10 mm gaps	0.23
15. 60 mm dry straw of dense concrete	0.66

Note that R_{f45} is not directly equivalent to R(= 1/U) above as it takes into account sideways heat movement from 'point' body contacts, etc.

Ventilation methods

Whereas insulation is normally an inseparable part of a building, ventilation systems are often 'add-ons' or at least intrusions (e.g. ridge vents) into the structural envelope. They are therefore much more variable in pattern, but can broadly be divided into two types, although there is overlap between them. The simplest of these from the point of view of equipment and installation is 'natural' ventilation, the power source of which is derived from wind force or warm air buoyancy (or more generally a combination of the two). 'Forced' ventilation, the alternative, is more complicated, needs mechanical/electrical power to operate and may involve the use of fans, ducts, diffusers, filters and controllers. Its advantage in contrast to 'natural' ventilation systems, which may outweigh these complications, is that it provides positive air movement and exchange irrespective of climatic or housing conditions. In livestock housing each of these systems has its appropriate place directly related to economic levels of investment, whereas in crop storage only the more positive powered methods are likely to be successful.

Natural ventilation

Natural ventilation (Fig. 25.17), created from 'stack' effects (buoyancy) and wind flow, is generally used for housing animals that have low 'critical' temperatures and where stocking rates are sufficient to produce satisfactory flows/exchange rates. Two general problems arise from the use of this system. The first occurs where there are no wind flows of any consequence as for instance in anticyclonic weather conditions. The second is where the building is so wide from eaves to ridge that air flow patterns through the building are sluggish due to inertial effects. Both of these situations result in stagnant air masses which, once established, are difficult to move or exchange, and thus create health and structure hazards from condensation and general dampness. The problem is made worse by a reluctance to provide an adequate width of opening at the ridge, mainly due to the apparent risk of weather penetration. A protected/unprotected ridge (see Fig. 25.18) which is open overcomes this problem.

It is acknowledged that natural ventilation is, at best, unpredictable and at worst impossible to control, the situation not being improved by common variables such as building shape and size. Many structures, having the traditional slotted or open ridge, are under-ventilated much of the time. In recognition of this there has been a resurgence of interest in slatted (or slotted) roofing for large livestock buildings. A simple way of creating this effect is to leave gaps of 10–20 mm between each column of sheets when covering the roof (having first turned them upside down so that the edge corrugations conduct rain satisfactorily to the gutters!). This provides good ventilation rates and patterns, balancing adverse wind forces on the roof, but nevertheless still lacks the essential element of controllability.

Automatically controlled natural ventilation (ACNV), a fairly recent development based on improved electrical control equipment (electronics), is now commonly used for small volume, high stocking density situations such as weaner housing, where expensive external heat sources are often required. This system uses a combination of wind flow and stack effect to achieve good air flow rates (and control) by mechanically actuated ventilation flaps arranged in optimal structure positions. It may also be used in conjunction with fan power assistance, as a backup, when the wind flow fails. This circumvents some of the drawbacks inherent in a purely 'natural' system.

Forced ventilation

Forced ventilation which relies solely on fan power for air exchange and distribution can be divided into 'suck' (exhaust) and 'blow' (pressurizing) methods. Each has advantages for specific types of building but generally those that exhaust are simpler and cheaper but less effective, particularly in achieving good air distribution. In contrast, pressurizing methods will exchange air efficiently, enabling better scouring and control. They also offer the possibility of air recirculation, filtration and silencing but almost invariably cost more. The potential for creating draughts will also be less with pressurized air flow, so long as it is conducted in properly designed ducting at appropriate (low) velocities. Figure 25.17 demonstrates some of the air-flow pattern variants which are possible with both these systems.

Ventilation rates (see next section)

The need to achieve a reasonable temperature stability in a building at a set level can result in very different (and variable) air flow needs, possibly requiring alteration from the minimum to remove noxious gases in winter up to a high rate sufficient to reduce heat build-up in still-air summertime conditions. This ratio of demand may be as much as 30 or 40:1, making successful ventilation very difficult to achieve, even without interference from adverse wind flows or the inadequacies of control equipment.

The actual air-flow rates required can be calculated on two bases, either whole building exchange (house volumes/h) or the needs of the particular species concerned (m^3/kg liveweight, m^3/t, litres/bird). Both these methods have had their merits but recently the latter approach has been more popular.

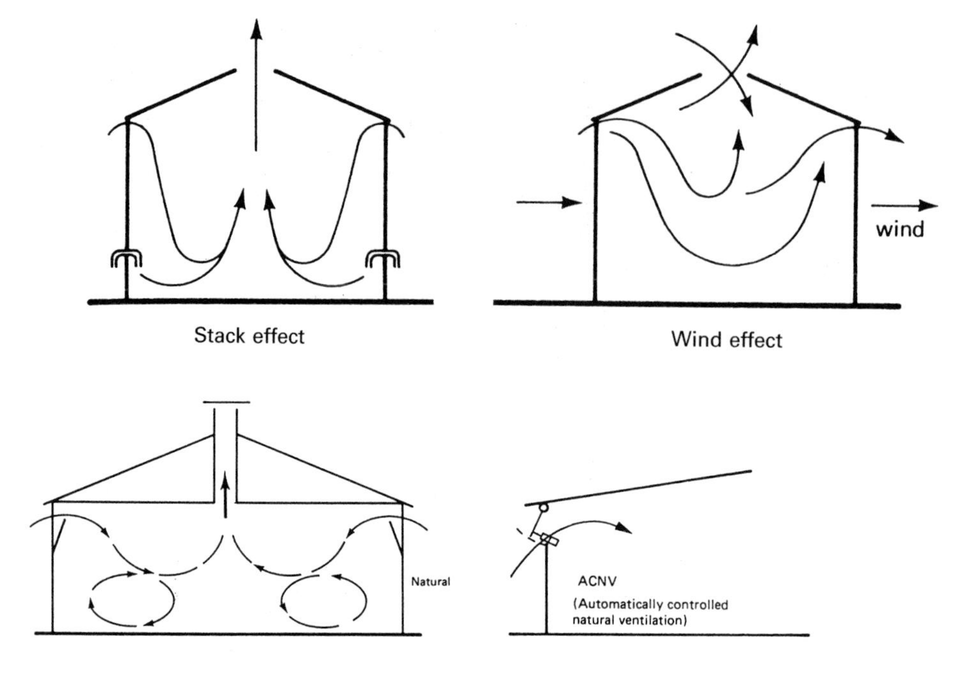

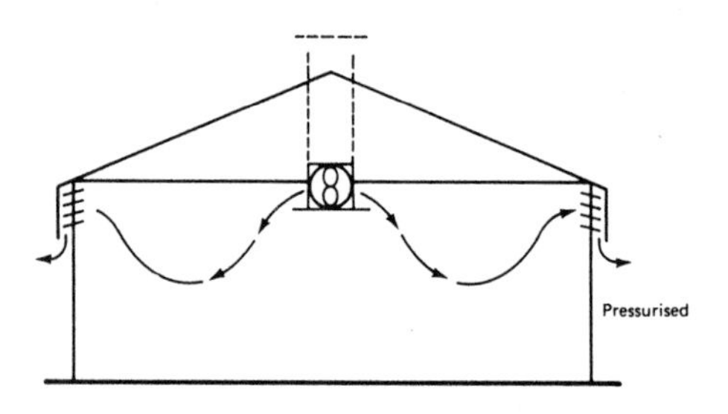

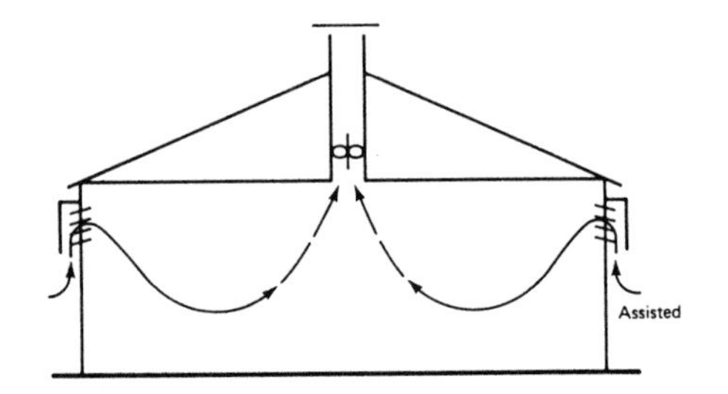

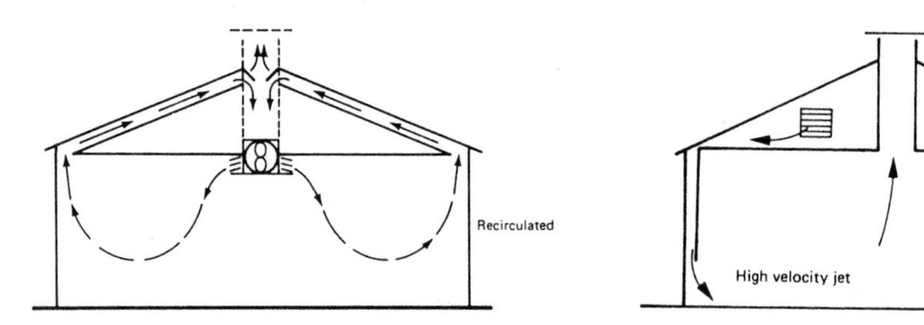

Forced ventilation

Fig. 25.17 Ventilation systems.

(See also BS 5502, Part 52, for alarms and emergency ventilation.)

Humidity control

This has not hitherto figured very largely in environmental control design, due mainly to the difficulty of sensing any change in conditions with reasonable precision. Added to this, the reaction of stockmen to humidity tends to be anthropomorphic and so controllers are often over-ridden when the atmosphere is felt by them to be 'uncomfortable'. However, as there is a growing awareness of the effects of 'damp' conditions in disease transmission and general animal welfare, more emphasis is bound to be placed on this in future. One particular problem of high humidities lies in the implied risk to structures from condensation effects, both on cold structure surfaces and within insulation under 'dew-point' conditions. These situations if permitted over long periods will depreciate building materials and reduce their effectiveness, ultimately affecting stock and crop health.

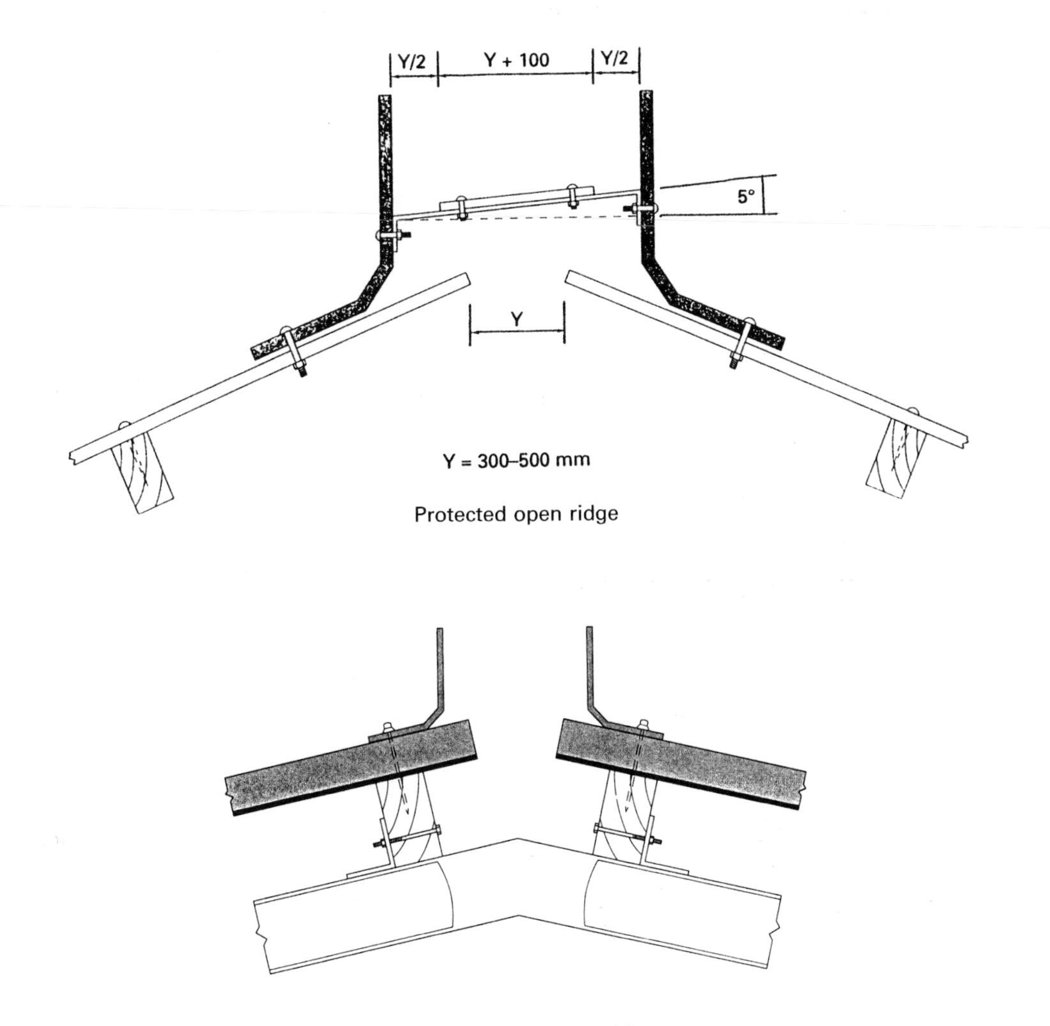

Fig. 25.18 Ridge ventilation (taken from Eternit Detail Handbook).

Water Pollution Control (BS 5502, Part 50, refers)

The problem of watercourse pollution from farms is one of the main issues on the environmental agenda. It has been tackled by means of a comprehensive Code of Practice, backed up by specific legislation that affects every holding.

(1) Code of Practice

No installation of control should be attempted without knowledge of the requirements laid down in the MAFF: Code of Good Agricultural Practice for the Protection of Water, and the detailed CIRIA Report 126, Farm Waste Storage: Guidelines for Construction. (See Further reading.)
The requirements summarized here are mandatory for England and Wales. Scotland and Northern Ireland have equivalent regulations and codes which require the same high standards, but there are important differences in interpretation so the correct Codes for the location should be applied.

(2) Legal requirements

(a) Adherence to the above Code and Guidelines is not legally binding. The legal requirements are set out in Statutory Instrument No. 324 (1991), The Control of Pollution (Silage, Slurry and Agricultural Fuel Oil) Regulations 1991 (under the Water Act 1989). Contravention is a criminal offence.
(b) Air pollution must also be considered: now covered by MAFF: *Code of Good Agricultural Practice for the Protection of Air.*
(c) See also relevant sections in the Agricultural Notebook on health and safety (Chapter 5) (Health and Safety at Work Act 1974) and animal health (chapter 22) (Code of Recommendations for the Welfare of Livestock and the supporting regulations (1990).

(3) Planning permission and building regulations

All structures or engineering works will require either full planning permission or a planning notice, although the extent of applicable works may vary from one authority to another. There should be no problem, however, with complying with the exemption requirements as prescribed for Class III of the Building Regulations 1985. See under 'Design criteria'.

(4) Consultation

Prior consultation with the NRA is always advisable. Free general advice is currently available from ADAS. Neither of the above consultations will be sufficient to draw up a scheme, which should be done by ADAS, consultants, equipment suppliers, or a combination.

(5) Notice of Construction

It is a legal requirement that notice must be served on the NRA of the relevant works at least 14 days before use. Having served the notice you can proceed with the use *after the 14 days have elapsed.* There is no need to wait for an inspection. If it has to be brought into use, even in unfinished state, notice should still be served in time with an explanation.

(6) Construction standards (see 'Specific purpose buildings')

It is essential for farmers to stipulate that the installation must be constructed to:

(a) BS 5502 Class I (50-year life) *or* Class II (20-year life) (regulations require 20 year min. design life).
(b) The relevant part of BS 5502 for that structure, e.g. Part 50:1989 for slurry storage.
(c) The relevant BS and manufacturer's instructions for each material, e.g. BS 8007 for concrete.

(7) Applicable works (summary only)

(a) All forms of silage storage except:

  (i) baled silage in sealed bags at least 10 m from any unsealed water, e.g. land drain, ditch, pond;
  (ii) temporarily exempt (until regulations reviewed on or after 1 September 1996), if built before 1 March 1991 or contract for construction placed before 1 March 1991 and completed by 1 September 1991 – providing prior notice of intention to continue to use was served before 1 September 1991, *and* the above 10 m rule applies. If an exempted structure is substantially altered or extended, then it loses the exemption.

(b) All forms of slurry (which includes water contaminated with excreta) storage except:

  (i) slurry temporarily stored in mobile tanker, maximum capacity of 18 000 litres;
  (ii) exempt structures as above;
  (iii) solids that cannot be pumped, i.e. farmyard and poultry manure, and solids from a separator.

(c) All forms of fuel oil storage on farms except:

(i) fuel temporarily stored in a mobile tanker;
(ii) fuel stored in an underground tank;
(iii) exempt structures as above;
(iv) total quantity on farm does not exceed 1500 litres;
(v) fuel oil used exclusively for heating a dwelling.

(8) Notice requiring works

Where NRA satisfied there is a significant *risk* of pollution from an existing installation, it may serve notice on the person having custody or control of relevant substance requiring him to carry out such works as it considers appropriate to reduce that risk to a minimum within a reasonable period (minimum 28 days). There is also provision for negotiation and appeal.

(9) General design calculations

(a) Calculate total livestock excreta in each age group held in buildings/yards and period held. Assess maximum daily flow.
(b) Add total silage effluent and maximum flow rate.
(c) Add total wash-down water and any other effluent, plus maximum daily flow.
(d) Add rainfall – under the Farm and Conservation Grant Scheme a report from the Meteorological Office is required (1993, cost £50) giving the average maximum (on a 5-year return period) for: 1–3 hours, 24 hours and 4 months. The Meteorological Office figure should be multiplied by the following areas:

   (i) any external uncovered area where livestock is held or areas polluted from such holding locations or polluted by silage effluent (it is reasonable to exclude roads and tracks where stock is passing through but not held);
   (ii) any roof area discharging on to (i);
   (iii) any surface water discharging on to (i) from fields or roads;
   (iv) the total surface area of the uncovered storage facilities (in the case of a slurry lagoon this could double the required storage capacity);
   (v) addition for future expansion and/or contingencies.

(e) Calculate total low-risk land available for effluent available for distribution (after taking out land within 10 m of any watercourse – increase the 10 m according to risk (e.g. sloping towards watercourse and to 50 m from a spring, well or borehole).
(f) Consider options for spreading and relate to spreading rates (maximum rate 50 m^3/ha per application) and maximum available nitrogen, and other limiting factors (see Code). Relate to volume above.
(g) It is likely that the main problem will be excess surface water especially in high rainfall areas. It is therefore essential that a great deal of thought be put into the design of the system to eliminate pollution of 'clean' areas and separation/diversion of surface water running on to the collection areas or into the drains.
(h) The minimum total capacity of the system is laid down in the regulations as set out under 'Specific purpose buildings' with one important exception, the *pumping* chamber in a low volume irrigation system: it merely states that the pump should be able to cope with a maximum 24-hour volume, whereas it is also essential that at least 24 hours' storage is available to allow for breakdown, freeze-up etc. (with back-up such as a tanker spreader) or 48/72 hours without. In many situations, especially higher altitudes, it will also be necessary to provide for emergency additional storage for much longer periods of freezing and/or snow covered conditions: use adjusted figures omitting most of the rain water.

(10) Special areas

If the farm is sited in a National Park, ESA, NSA or SSSI, you must approach the appropriate authority first before designing any scheme as the regulations are different especially for the last three. Some restrictions may also be imposed by the NRA on aquifer zones.

(11) Site ground conditions

As with all farm buildings it is essential to know the soil bearing capacity. Ignoring it in the design not only contravenes the British Standards but is courting financial disaster and prosecution. Likewise in the construction of any earth-sided bunker (not allowed for silage in England and Wales) or lagoon, a soil analysis is required unless a lining is to be used. (Excellent guidance in CIRIA Report 126.)

(12) Environmental statement

May be required to support a planning application on high sensitive sites and large scale pig/poultry enterprises especially with low areas of land.

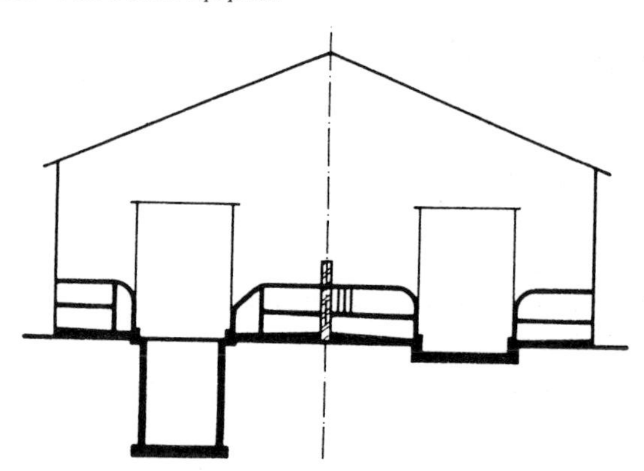

Fig. 25.19 Section: slurry cellar/solid floor cubicle building.

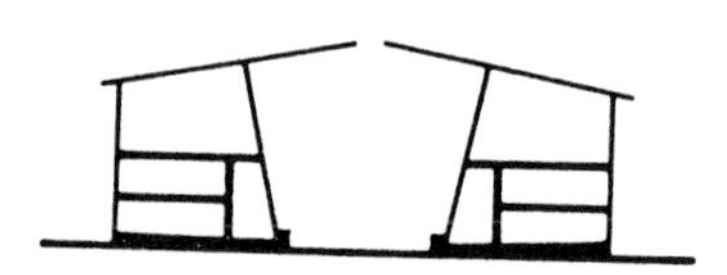

Fig. 25.20 Typical kennel-type cubicle building.

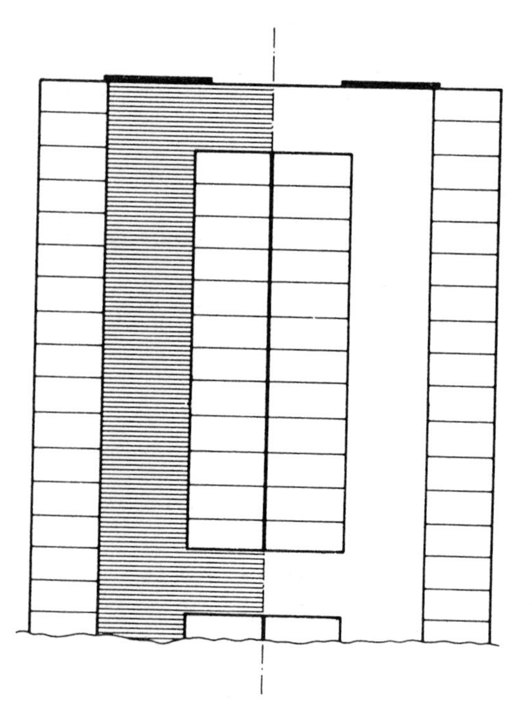

Fig. 25.21 Plan of the above (Fig. 22.19).

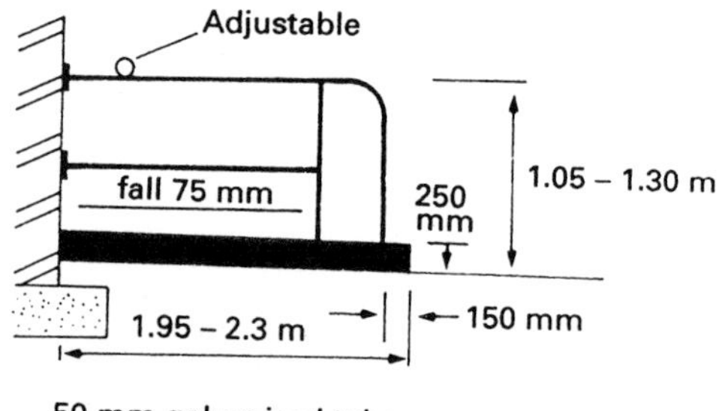

50 mm galvanized tube or timber posts 100 × 50 mm and 150 × 47 mm rails, (bottom rail better if of twisted rope).

Fig. 25.22 Section of typical cubicle.

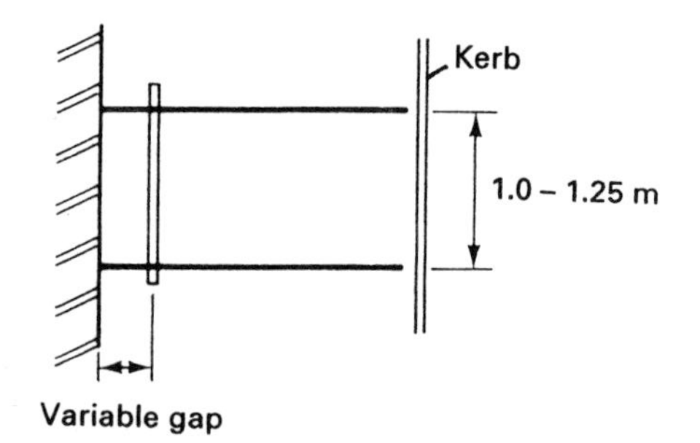

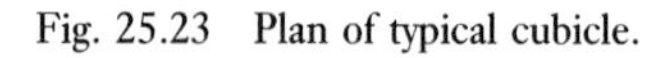

Fig. 25.23 Plan of typical cubicle.

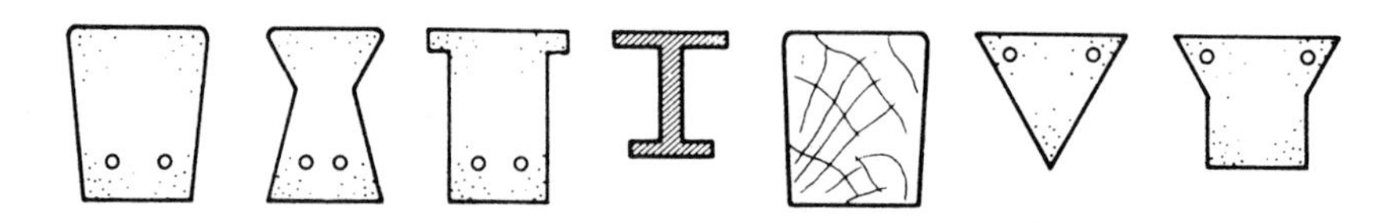

Fig. 25.24 Slats: some alternative sections (BS 5502, Part 51, refers).

Specific Purpose Buildings

Cattle accommodation (BS 5502, Part 40, refers.)

Dairy cows

Space for loose housing Cost range: £120–£140/m^2	Between 5.0 m^2/small animal, and 8.0 m^2/large animal (of which 75% strawed) + 1.0–3.0 m^2 when housing not completely covered
Space for cattle on slats Cost range: £170–£180/m^2	3.0–4.0 m^2. Slats 125 mm wide with 40 mm gap (see Fig. 25.24).
Cubicle/kennel dimensions (see Figs 25.19–23)	1.1–1.3 m wide × 2.0–2.4 m deep (for small and large cows) + passageway (2.4–2.7 m)
Feeding and circulation space	2.0–5.0 m^2/cow
Minimum height to eaves	2.2 m (for cubicle systems) to 3 m for loose housing plus maximum bedding depth
Temperature and insulation	Not important down to 5°C, no control of environment usually. 2°C is 'lower critical' temperature of small (50 kg) animal. Insulation limited to bedding or, occasionally, flooring
Ventilation	Natural; by open ridge or slotted/spaced roof and spaced boards in upper wall area. Usually underestimated, with too little ridge vent provided.
Bedding, water and feed consumption	Straw 1–4 tonnes/cow/winter for loose housing, 1–200 kg/cow/winter chopped straw for cubicle building. Wood shavings 500–1000 kg/cow per winter Silage 20–40 kg/d per cow Hay 5–10 kg/d per cow Concentrates 5–7 kg/d per cow (for 15–20 litres of milk) Water 35–70 litres/d per cow (from trough with allowance of 20 cm^2/cow or one drinking bowl/10 cows)
Effluent output, drainage and storage	Urine 25–60 litres/d per cow } depends on feed and housing system Dung 10–35 litres/d per cow } 57 litres/d excreta adopted by MAFF for pollution potential calculations Drainage slopes on floors from 1:50 to 1:100 Slats may be part or whole of floor, may have drainage channel or tank under, to contain 4–6 months' accumulation, may be raised 2+ m to provide total storage for permanently housed stock. Slats may be of timber, concrete or steel with general dimensions of 125 mm wide with 38 mm gap, 150 mm deep and 25 mm taper (see Fig. 25.24)
Feed space	600–760 mm trough length/cow 125–200 mm self-feed silage face/cow with unrestricted access 250–300 mm self-feed silage face/cow with restricted access 600–700 mm silage and hay manger/cow
Collecting area	1.3–1.5 m^2/cow.
Cost range: £30–£50/m^2 Other considerations	Races (700–800 mm wide, internal), and handling equipment should be adjacent to and integrated with the housing Non-slip grooved floors and steps required. Where steps provided, rise should be 250 mm/step and 'going' or tread 760 mm/step. Need an isolation box separated from rest of buildings

	Provide calving boxes (min. 15 m^2 each), number to suit calving pattern and herd size (min. 1 box/30 cows)
General	Consider cow management and handling, effluent problems, feed store siting, cow comfort, access for mechanical equipment

Calves

Space for individual pen housing up to 4 weeks old Cost range: £100–£120/m^2	1.1 m^2–1.5 m^2 in single pens of approximately 0.8 m × 1.5 m (minimum depth) for small calves
Group housing space, up to 3 months	1.5 m^2 increasing to 2.5 m^2
Space in open-fronted housing, 3–6 months	2.5–3.0 m^2 with three or four per pen (see Fig. 25.25)
Height recommendations	2.5 m to ceiling, minimum of 6 m^3/calf. Barrier height 1.0 m minimum, up to 1.4 m for larger calves to provide isolation between pens
Temperature	20°C for young calves, 15°C at 1 month old and 12°C at 3 months. Note that lower critical temperature varies greatly with bedding type and dryness. Thus, if kept in 'non-controlled' accommodation may be very close to LCT, particularly if draught-prone position
Insulation of small calf environments	Floor damp-proofed and having 'U' value of 0.5 W/m^2°C. Roof, double-skinned or with ceiling, of 'U' value 0.6 W/m^2°C maximum. Walls, cavity or other, with 'U' value of 0.9 W/m^2°C maximum
Ventilation	Adequate, draught-free ventilation essential for good health. Natural ventilation in housing for stock over 3 months old, with hopper windows or space boarding and ridge ventilation
Feed and water consumption	Feed from bucket or machine at pen front, or trough for older calves (350 mm/animal), all near passageway. Milk substitute and small quantity of concentrate (0.7 kg/d) then up to 2.5 kg/d to 3 months of age. 15–25 litres/day per calf from water bowls or bucket near passage
Effluent output	10–30 litres/d depending on food, size and age
Flooring	Hard, cleanable, durable, resistant to disinfectant. Surfaced with wood float in exposed areas if concrete. Drain to 'centre dome' passageway at 1:20, and via step of 100 mm at edge of pen to drain channel at passage-edge. Slats at 3 months and above only. 100 mm wide with 30 mm gap
Walling	As for floor with hard cement render and waterproofing additive. Form curved junction with floor
Other considerations	Solid or 'see-through' pen divisions? Opinions vary as to value of each because of hygiene and social aspects Care with surfaces as calves chew and lick everything

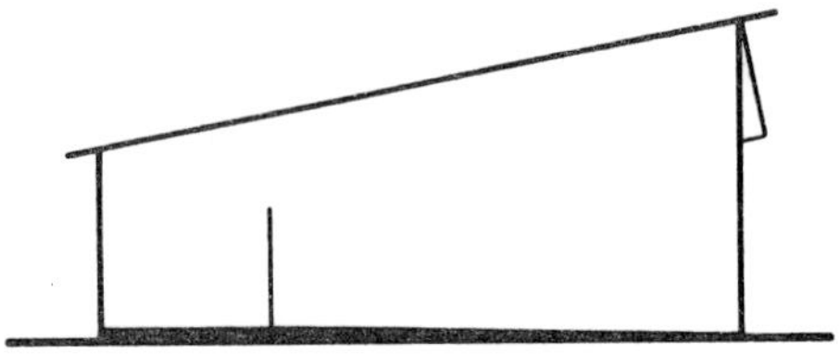

Fig. 25.25

Drainage must never interconnect pen to pen

Relative humidity should be kept down to 70% or less. (May be more easily achieved with 'open' than 'controlled environment' accommodation.)

Beef animals

Space on straw (see Figs 25.26 and 25.27)
Cost range: £90–£120/m²

2.0–3.5 m² per animal at 12 months, 4.5–6.0 m² per 2-year-old bedded animals (4.0 m² with self-feed silage systems)

10–20 in group to one court or pen

Space for animals on 'sloped' ('Orkney'), (lightly littered floors)

2.5–3.5 m² per animal 12–24 months (sloped area at 1:16)

Space for beasts on slats
Cost range: £150–240/m²
(the latter when with deep cellars)

1.4–1.9 m² per animal at 12 months, 100 mm wide with 30 mm gap
2.0–2.5 m² per 2-year-old. 125 mm wide with 40 mm gap
20–25 animals per court or pen (see Fig. 25.24)

Height to the eaves

3.0 m minimum (allow extra for over-winter dung build-up of 1.2 m in strawed yards, or slurry depth under slatted floor)

Roofless units

3.0–5.0 m² per animal, strawed; 1.0–2.0 m², slatted; 1.7–3.0 m², with sloping concrete and litter. (May have cubicles.)

Temperature and insulation

Not critical within range 2–20°C
Insulation not required

Ventilation

Natural, with space-boarded upper walls and ventilated ridge. Avoid draughts. (Slatted or slotted roofing appropriate.)

Bedding and fodder consumption

Straw 1–3 tonnes/beast per winter
Silage 30–40 kg/beast daily
Rolled barley 7–10 kg/beast daily

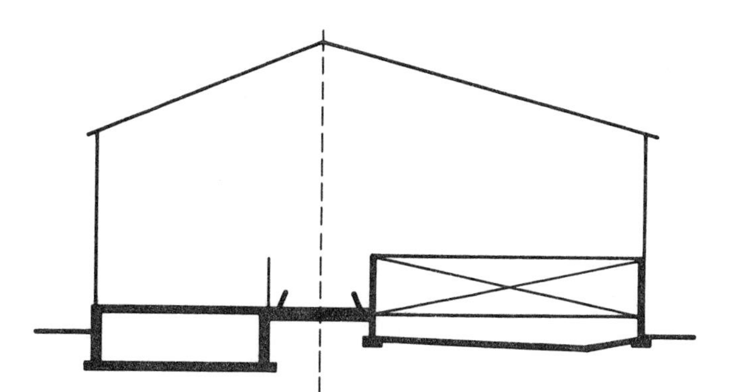

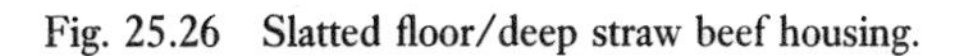

Fig. 25.26 Slatted floor/deep straw beef housing.

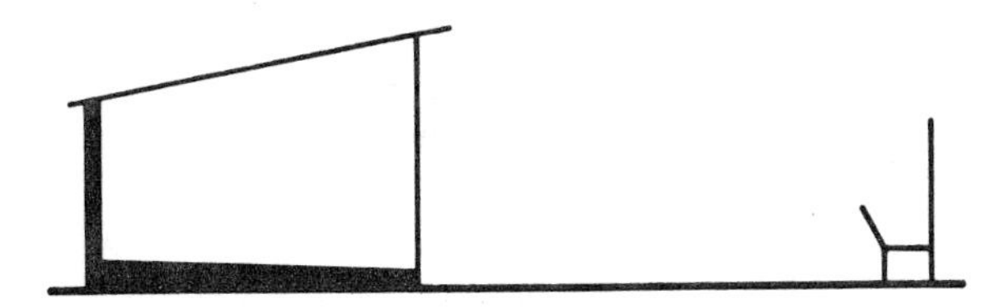

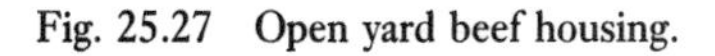

Fig. 25.27 Open yard beef housing.

Trough space	500 mm length/9 to 12-month-old beast. 600 mm length/18-month-old beast. 700 mm length/2-year-old beast. 800 mm/mature bull beef animal. 150 mm/animal for *ad libitum* fed young stock
Water consumption	25–45 litres/d per head Allow 20 cm^2 trough per head or one drinking bowl per ten animals
Other considerations	Kennels and cubicles may be appropriate (see Dairy cows) Safety of personnel (especially with bull beef animals): requirement for sturdy barriers and doors. Vertical rail partitions and gates at 150 mm centres for bull beef. Feed barrier at 400 mm centres. (Normal beef centre barrier at 300 mm.) Heights to be a minimum of 1.3 m

Milk production buildings and bull pens (BS 5502, Part 49, refers)

Note: Many aspects controlled by the Milk and Dairies Regulations.
Cost range: Extremely variable but not likely to be less than £140/m^2

Abreast parlour (six standings) (see Fig. 25.28)	Overall depth 4.8–5.8 m Overall width 6.9 m Side exit passage 0.3 m for personnel Minimum headroom over standings 2.0 m (plus 0.45 m step to standings gives total height minimum 2.45 m)
Tandem parlour (six standings) (see Fig. 25.29)	Overall width 5.1 m Overall length 9.3 m Passage widths 1.0 m (one end, across, and two exit passages) Minimum headroom over standings 2.0 m Depth of operator pit 0.8 m Width of operator pit 1.8 m
Herringbone parlour (eight standings) excluding 'quick-exit' parlours (see Fig. 25.30)	Overall width 4.8 m Overall length 6.9 m (Passageways where required 1.0 m minimum) Depth of operator pit 0.8 m Width of operator pit 1.8–2.4 m Minimum headroom over standing 2.0 m Width of cow standing 1.5–1.7 m May be open at entry end
Chute parlour (Fig. 25.29)	As tandem, less end and exit side passages and with narrower pit. Six standing measures 7.4 m × 3.1 m overall
Rotaries, tandem (Fig. 25.31)	Space occupied overall from 5.5 m × 5.5 m for six milking points up to 15.5 m × 15.5 m for 18 points
abreast	Space occupied overall from 8.1 m × 8.1 m for 12 milking points up to 14.0 m × 14.0 m for 30 points
herringbone	Space occupied overall from 7.0 m × 7.0 m for 12 milking points up to 14.0 m × 14.0 m for 28 points
Polygon parlours (see Fig. 25.32)	Four ranks (or three in 'trigon') of standings with up to eight cows per rank, in flattened diamond or triangle shape arrangement, 16–40 cows at one time in parlour, 16–18 m wide × 18–33 mm long, with 7–9 m at pit centre
Collecting yards and circulation space	Circular or rectangular, open or covered (in which case ventilation should not be through the parlour) allowance of 1.3–1.5 m^2 per cow
Drainage	Smooth floated channels or vitrified pipe. Overall falls 1:100 or 150 over lengths of buildings. 1:25 for standings and falls to outlets and gullies. Channels to have 50 mm high minimum rise at edges

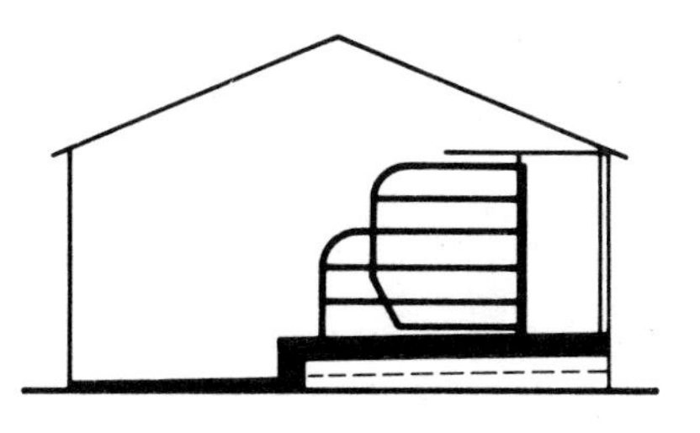

Fig. 25.28

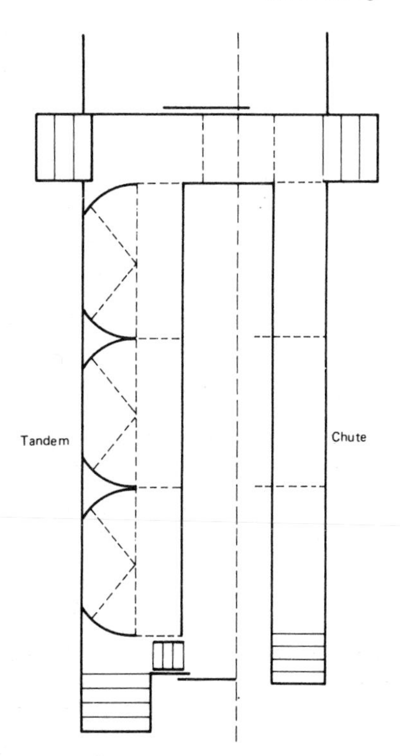

Fig. 25.29

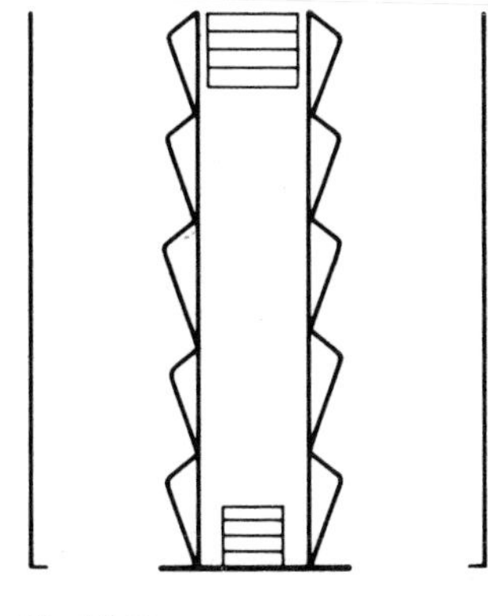

Fig. 25.30

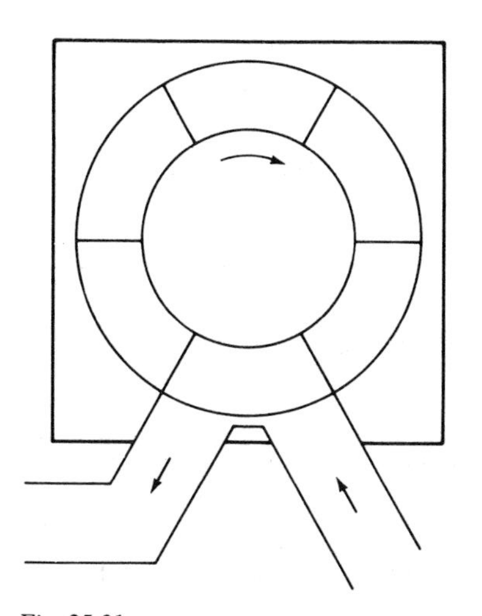

Fig. 25.31

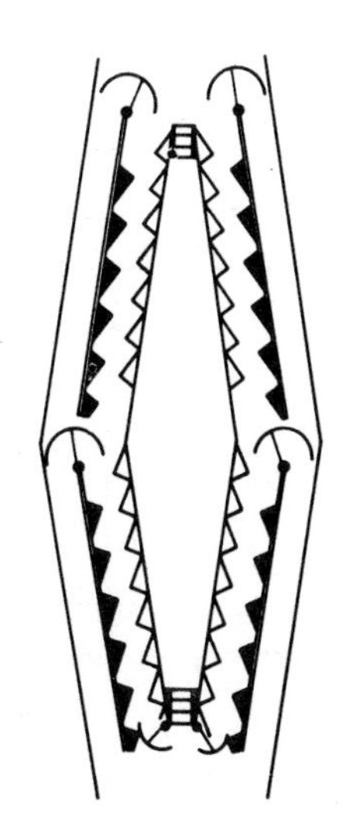

Fig. 25.32

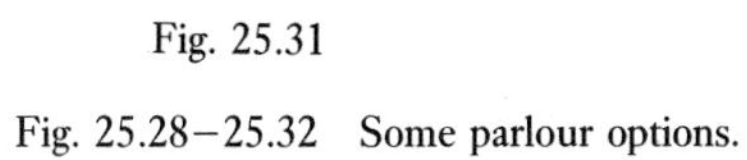

Fig. 25.28–25.32 Some parlour options.

Water supplies	0.5–1.0 litres/d/cow udder wash allowance 10–20 litres/milking unit per d circulation cleaning allowance; 0.5–1.5 m^3/d for pressure hose facility (18 litres/d/cow used for overall calculations)
Services	Pipes and fittings required for vacuum, milk transfer, cleaning systems; storage facilities for hot water, warm water, milk All accessible electricity supplies and fittings should be low voltage (110, 24 or 12 V) or leakage protected. All metal fittings to be 'earth' bonded. Consider heating installations against freezing of pipes and for operator comfort
Light levels and sources	Natural preferred from above for general illumination. Local high intensity artificial light at udder level. Visual contact to be maintained between parlour, dairy and collecting yard
Flooring	Hard, durable, resistant to detergents and disinfectants with non-slip surface (granite chippings and carborundum dust) which is cleanable by pressure hose
Walls	Rendered with waterproof, cleanable, smooth surface to a height of 1.37 m minimum (1.7 m preferable) painted with epoxy resin or chlorinated rubber or other acceptable finish. No insulation provided. Curved filler in render at junction between floor and walls
Other considerations	Steps for cows up and down should be non-slip with 150–250 mm 'rise' and 760 mm 'going'. Heelstones on pit edges should be a minimum of 100 mm above standing level. Steps for operatives should not exceed 220 mm rise and should not be less than 250 mm going Sliding doors or rubber/plastic flap doors required with automatic opening and closing as appropriate Provide ventilation for high summer working, e.g. variable skylight opening
Dairies	See Fig. 25.33. Many features controlled by Milk and Dairies Regulation for essential hygiene
Bull pens	See Fig. 25.34. All housing elements must be designed to resist aggressive behaviour. Provide escape routes from areas used by operatives

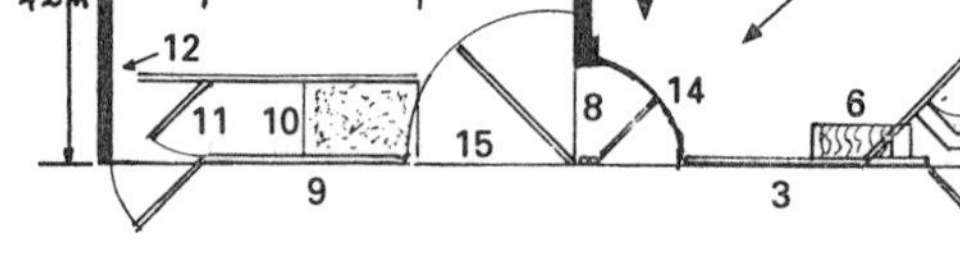

Fig. 25.34 Bull box and exercise yard.
1, Box – covered (insulated or heated floor?)
2, yard – covered or open (both with non-slip floors).
3, dairy collecting or dispersal yards (usual site in relation to).
4, large manger for silage with yoke.
5, childproof access door – width to suit feed system.
6, water trough – protected service box and rails over.
7, reinforced solid walls – rendered if concrete block.
8, can add further gate or slide in rails.
9, vertical rails at 150 mm centres to ground level.
10, service pen – 1 m internal width with sand bed.
11, cow yoked gate with feed area and exit gate.
12, squeeze gap, 300 mm wide – must not be accessible to children.
13, refuge with vertical ladder rails – 300 mm gap either side.
14, all gates double latched – capable of adult operation only.
15, steps to suit, especially if outside pen area is scrapped.

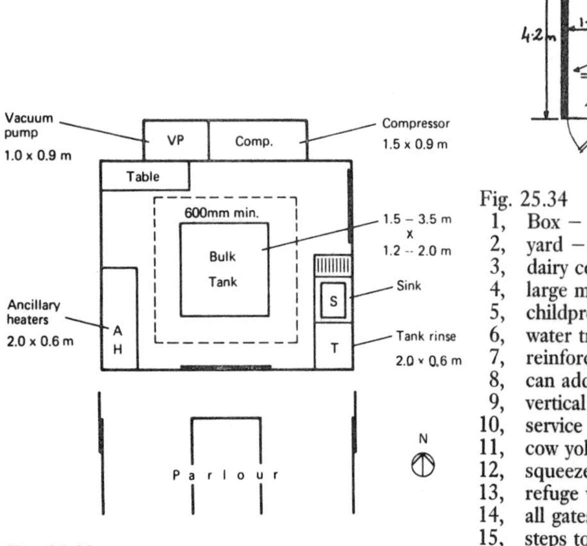

Fig. 25.33

Pig housing (BS 5502, Part 42, refers)

General

The last few years have seen very rapid changes in both the structure and the technology of pig-producing systems in Europe. Among these changes has been a major move towards freedom of movement and away from restrictive enclosure or tethering. Additionally, there has been a trend towards liquid and/or automatic feed dispensing systems which have had a radical effect on building design. Straw has been deemed to be the most appropriate bedding method for pigs long term, and ways have been sought to incorporate the material in housing methods which do not bedevil the problems of manure handling and labour use. A further factor has been the advancement of group (family) housing theories among the leading exponents of 'natural' pig production.

All these causes have allied with the recent financial problems of the industry to ensure that pig housing design was, and is, subject to considerable revisionary influences. Some of the text below describing housing types therefore refers to systems as they are, rather than 'state of the art' technique as they should be if the world was perfect (for instance in the case of the 'flatdeck' still portrayed here). The above remarks should be borne in mind and appropriate advice sought so that intended new systems do conform to the latest strictures on the health and welfare of pigs as the descriptions below cannot encompass all the potential variations – or pitfalls for the unwary!

Pig fattening accommodation

Space within controlled environment housing
Cost range: £200–£240/m^2

Absolute minimum of 0.5 m^2/'grower' pig place

	Total area (m^2)	*Dunging area (m^2)*	*Alternate slatted (m^2)*
Porkers	0.6–0.7	0.25	0.12
Baconers	0.75–0.9	0.3	0.15
Heavy hogs	0.9–1.0	0.35	0.2

(May have slightly less than above with some straw-based systems where there is a definite strawed sleeping area.)
10–20 pigs/pen normal; some systems 30–40

Pen dimensions and arrangement (Figs 25.35–38)

Optimum depth of lying area 1.8 m. Trough width 300–400 mm. Trough length/pig 250, 300, 350 mm porkers, baconers and heavy hogs respectively. *Ad libitum* feeding systems only require 225 mm for all classes of fattening pig). Some systems have no trough allowance – floor feeding, etc. (Trough length usually governs length and shape of pen and thus the house.)
Square (or nearly square) pens are now favoured, particularly for larger groups

Height of building

As low as possible. Passageways for operatives give minima; 2.15 m normal, 2.7 m minimum with scraping, 3.0 m with catwalk over pen walls for feeding, etc. Eaves height 1.7 m over slatted areas or 1.1 m with strawed/open yards and 'pop-hole' access. (If liquid feed systems are to be used, the height must provide for convenient siting and fixing of pipes and controls to give sufficient drop out of the pigs' reach.)

Passage widths

Dung passages 1.05–1.35 m slatted outer, 1.15–1.65, slatted central, 1.15–1.5 m concrete solid, outer, 2.1 m minimum for tractor. Feed passages, 1.0–1.2 m

Temperature and insulation

Minimum of 21°C mean throughout environment. NB Nurtinger kennels described later under 'Weaner housing', measuring up to approx. 2.4 × 1 × 0.9 m. Straw bedding affects lower critical temperature, to reduce thermal risks. 2°C less, generally, for pigs over 4 months old. Roof and walls should have 'U' value of 0.5 and 1.0 W/m^2°C maximum respectively. Floor in lying area should be insulated and vapour proofed to a particularly high standard

Ventilation

2.5 litres/s/50 kg liveweight in winter
12.5 litres/s/50 kg liveweight in summer

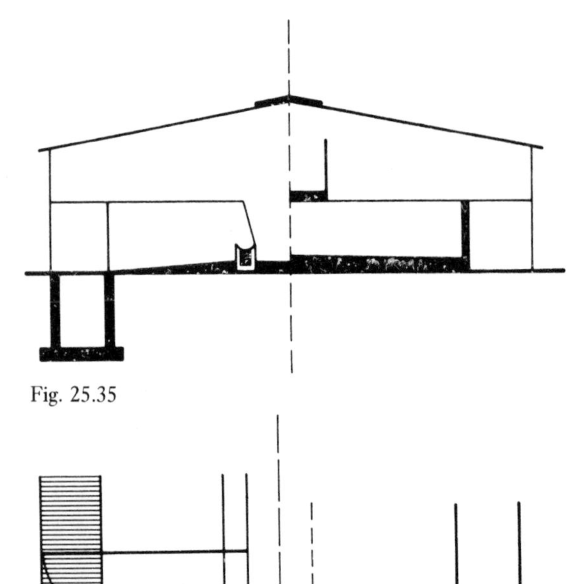

Fig. 25.35

Fig. 25.36

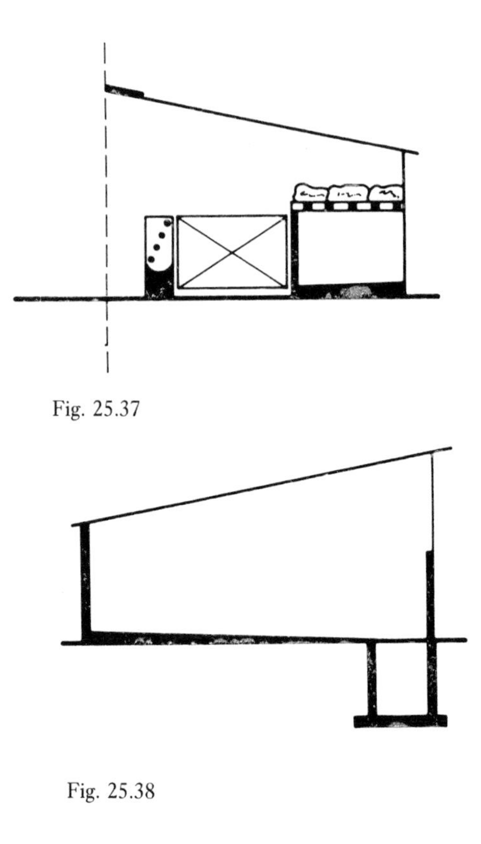

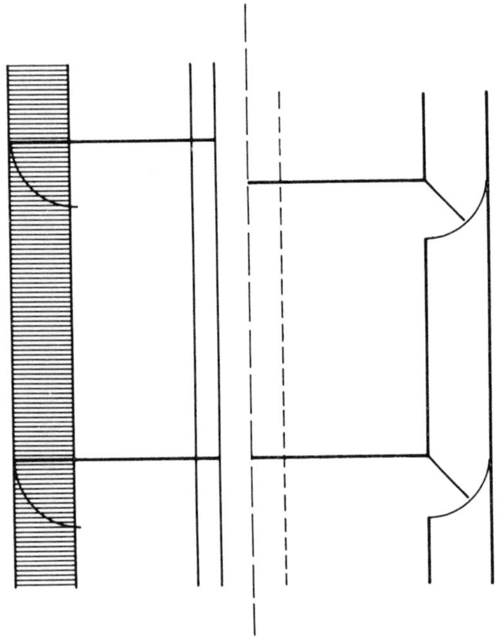

Fig. 25.37

Fig. 25.38

Feed and water consumption	2–4 kg/pig per d of meal 4.5–9 litres/pig per d through trough with food, + bowls or drinkers in dunging area or over trough at one per six pigs. (Minimum of two drinkers per pen.) Pipe-line feeders supply through delivery pipe at high level down to troughs in conventional position
Effluent and drainage	7 litres/d from baconer on dry feed. 14 litres/d from baconer on *ad libitum* wet feed. Floor falls 1:15 in lying area; 1:50 to 1:100 for dung passages or channels Slatted area – part or whole. Slats–concrete, 75–100 mm top width + 15–20 mm gaps. Steel T bars, 35–50 mm top width + 10–12 mm gaps. Step of 50 mm to slatted area sometimes. 'Straw-flow' systems create a natural movement of straw from the back of the pen to the front and to waste clearance through a much more pronounced floor slope
Other considerations	Chew-proof doors and partitions with secure latching and fixings

Farrowing accommodation

Cost range: Very variable but likely to be £220–£280/m^2

Space for crate (see Fig. 25.39)	2.4–2.95 m × 0.6 m wide (+2 × 400 mm side clearance) × 1.4 m high. (Plinth may be provided 200–300 mm high.) Crates vary considerably both in their construction and ultimate effects on the pigs, so it is important that a thorough review of current options is carried out before any contemplated investment is made. In general they are tending to provide greater and even unrestricted movement, positioning the sow according to her natural tendencies rather than forcing her into an unnatural posture
Space for circulation space round crate	Passage at rear minimum of 1.4 m. Feed passage at head end minimum of 0.8 m. Eaves height 2.4 m from passage floor
Space for creeps surrounding crate	Up to 0.1 m per piglet allowance. Maximum is likely to be 0.4–0.5 m wide × 1.4 m long (length of crate). Covered with lid and heat source above piglets below lid. Bottom rail or pophole access for piglets. Rail height 250–300 mm or three to four popholes along creep base
Traditional farrowing places/pens with weaning space and creep (see Fig. 25.40)	2.4 m deep × 3.0 m long (pens either side of feed passage). Eaves height 2.4 m. Creep rails on rear and side wall nearest creep, set 250 mm out from walls and 250 mm above floor NB The development of 'family' or other 'natural' group sow (and other pig class) housing in batches of 5–10 occupying one very large space Creep at pen front, 0.9 m × 0.8 m area with access from pen via popholes and creep rail. Lid over, with heat source below
Open front pen (see Fig. 25.41)	1.5 m wide × 4.5–5.5 m deep × 1–1.5 m at rear eaves (mono-pitch with front eaves of 2.1–2.4 m), against rear wall. Removable farrowing crate rails linked to creep, arranged along the pen at centre rear
Temperature and insulation	General areas 16–20°C. Creep at 30°C initially, reducing to 24°C over 3–4 weeks from birth. 'U' values of 0.5 and 1.0 W/m^2°C for roof and walls respectively. Crate and creep areas to be well-insulated concrete, may have in- or under-floor heating. NB Nurtinger kennels described later under 'Weaner housing'
Ventilation	Natural in open systems and in small, pen-type housing (up to five sows). 5 litres/s per sow in winter and 15 litres/s per sow in summer in controlled environment houses

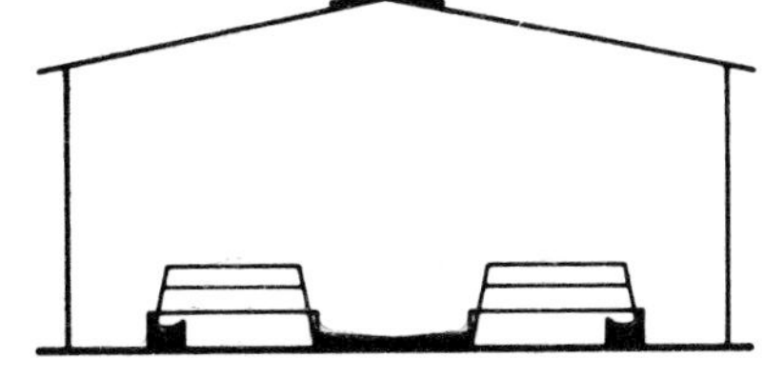

Fig. 25.39

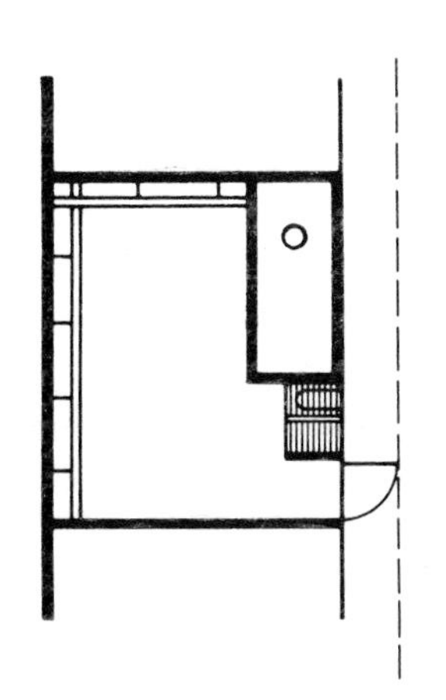

Fig. 25.40

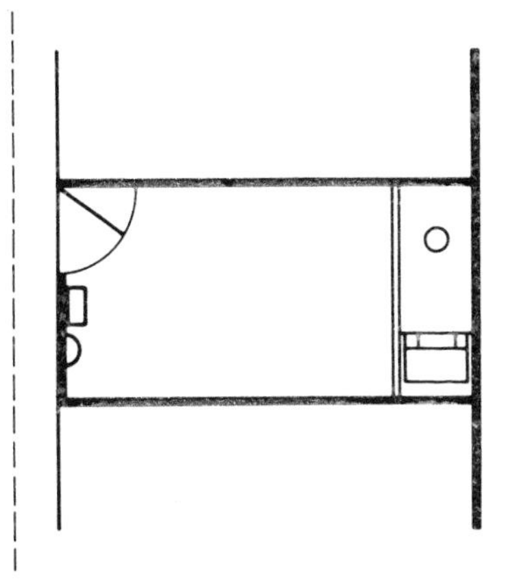

Fig. 25.41

Feed and water consumption

5–8 kg/d per lactating sow. 0.6 m long trough with drinker over. 250 mm above floor level. 18–25 litres/per lactating sow. Supplementary water for piglets provided by 'tube' drinkers or trough away from creep feeder, over slats if present. Use of liquid feed is reducing the need for service passageways at the sides of houses, etc.

Effluent and drainage

20–25 litres/d per sow.
Drainage falls in lying and creep areas to be 1:20, 1:40 in general areas. Steps to drain or dung areas not to exceed 50 mm rise. Slats or mesh area in rear half of crate. Slats consist of slabs of concrete with 10 mm slots and 75–100 mm slat tops. Steel tube or cast iron, 15–50 mm with 10–12 mm gaps. Meshes of steel, plastic, plastic-coated steel of various conformations and aptitudes, care needed for foot comfort of piglets. Crate raised 400–500 mm above surrounding levels at rear to accommodate slurry build-up and drainage flows.

Materials of walls and floor

Easily cleaned and resistant to strong disinfectants. Floor flat but not smooth trowelled – no tamping grooves to impede drainage

Other considerations

Robust fittings, including water bowls required. Strong post and frame fixings. May have crates in an 'open' arrangement with or without low partitioning between each, situated on a plinth or on a well sloped floor. All crates should be removable for easy cleaning of crate sections and floor

Dry sows and boar accommodation

Space allowances
in open yards
Cost range: Variable – around £130/m^2

2.5–3.7 m^2/sow of which 1.2–1.5 m^2 is covered and possibly strawed. 65% lying/feeding and 35% dunging area. Group size normally up to 20–40/yard. Automatic feeders require 4–5 m^2 +

sows in pens
Cost range: £200/m^2

2.1 m × 1.5 m/individual sow. Trickle feeder space – 400 + mm width

in boar pens (see Fig. 25.42)

4.5 m × 3.0 m pen of which 4.5 m^2 is lying area (2.4 m × 1.8 m minimum)

Temperature and insulation

21°C maximum, 14°C minimum. Roof and walls of 0.5 and 1.0 W/°C m^2 'U' value respectively

Ventilation

Natural usually adequate but forced ventilation in some totally enclosed houses at 50 litres/s per pig maximum

Feed and water consumption

2–3 kg/d per pregnant sow in individual feeders
3–5 kg/d per boar from a trough of 0.6 m length
9–18 litres/d per sow or boar

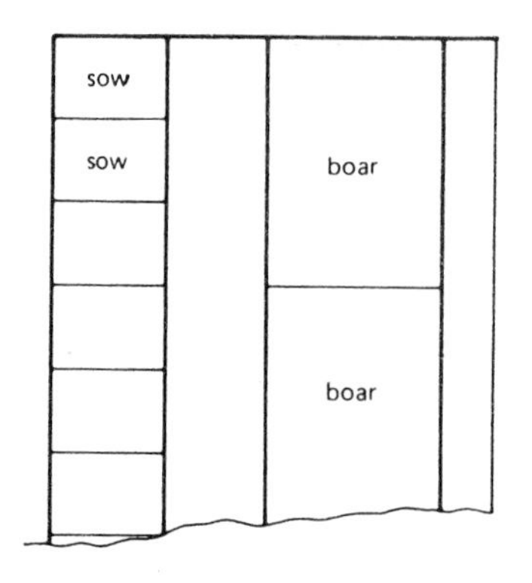

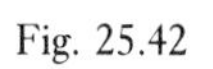
Fig. 25.42

Effluent and drainage

20–25 litres/d per animal. Floor falls laid to 1:40 generally; 1:20 to discharge into open gullies. House end falls 1:50–100. (Slats may be used, see Farrowing accommodation.)

Other considerations

Boars kept within visual contact of sows in extensive houses. Sow restraint may reduce bullying, can produce sores

Weaner housing

Space requirements
Cost for verandah type: £145/m^2

0.5–0.6 m^2 per pig in a strawed yards of which up to 0.3 m^2 is lying area if open fronted or 'verandah' type (see Fig. 25.43 for slatted version). Eaves height 0.9 m to 2 m depending on whether kennel or framed type. Groups of 10–20/pen are usual

0.15–0.25 m^2 per pig in 'flat deck' accommodation. Eaves or ceiling height 2.0 m. Groups of 10, 15, 20/pen are common (see Fig. 25.44). NB Current production of this type of house is now very limited and is not likely to survive much longer, but there are many still operating successfully.

Temperature and insulation

23–27°C inside kennels and in controlled environment housing. Roofing or kennel lid 'U' value should be 0.3–0.5 W/m^2°C. Wall 'U' value should be 0.5 W/m^2°C. Well insulated floors with in-floor heating using electric cable or hot water piping are strongly recommended. Extra heat usually needed in winter – gas heaters common for air flow systems

Nurtinger kennels may be used to provide 'an environment within an environment', to give a near-natural pig microclimate. This may be an asset for all pig-producing systems but it is particularly useful for weaners. The kennels take the form of small-section but relatively long 'hutches' with one long side curtained in flexible plastic. Their dimensions vary widely according to the class of pig accommodated, but the range is: length 1.5–2.4 m, width 0.35–1 m, height 0.45–0.9 m.

Ventilation

2.5 litres/s per 50 kg liveweight in winter.
15.0 litres/s per 50 kg liveweight in summer
Care to ventilate avoiding passage of noxious gases from dung through house in total control systems

Feed and water consumption

Ad libitum feeding from 50 kg capacity hoppers with 50 mm/pig access. 1–2 kg/d per pig. Some rationing systems use troughs at 75–100 mm/pig; should be spill-proof. 1–2 litres/d per pig from nozzle-drinkers (2/pen minimum) over well-drained area. Very young pigs might need temporary ramp, etc. to reach drinkers

Effluent and drainage

1–2 litres/d per pig. Either open dunging area outside kennel or weldmesh/expanded metal/plastic mesh part or whole floor, e.g. Weldmesh is 76 mm × 13 mm × 8 swg (10 swg for smaller) over slurry pit 1–2 m deep or raised 0.6 m in total control housing. Many alternatives available.

Materials

Prefabricated buildings with easy erect, easy clean and control facilities

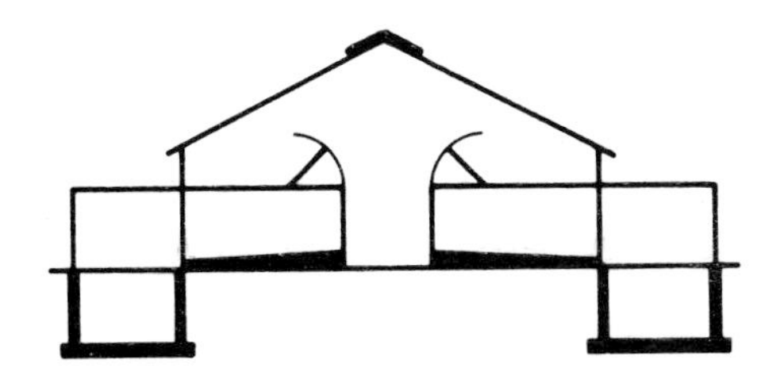

Fig. 25.43

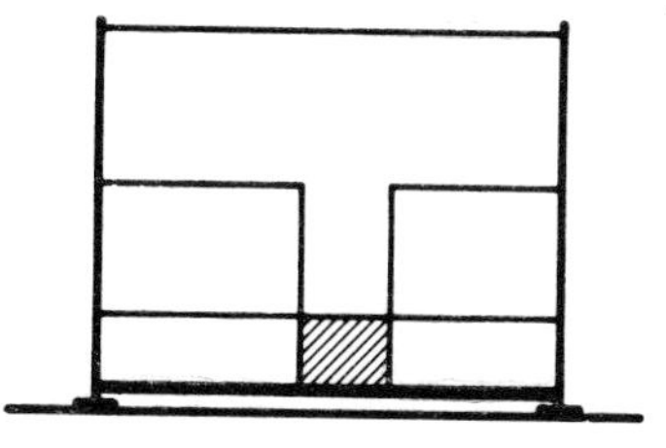

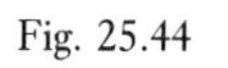

Fig. 25.44

Sheep accommodation (BS 5502, Part 41, refers)

Sheep housing (Figs 25.45 and 25.46)

Space requirements	(See Table 25.7)
Cost range: £50–£70/m^2	
Temperature and insulation	Ambient of surroundings, down to 2–5°C. No insulation necessary
Ventilation	Adequate air exchange essential to maintain health of stock. Natural from space boarding walls and ridge ventilation (open 300–600 mm, relative to building span): spaced or slotted roofs possible. Must avoid low level draughts. Building may be open at one side
Bedding, feed and water consumption	0.8–1.5 kg/d of straw/ewe in yards. 0.25–1.0 kg/d per ewe of concentrate. 1–1.5 kg/d per ewe of hay. Up to 5 litres/d per animal from a water trough 0.3 × 0.6 m in each pen or one bowl to 50 ewes. (Running water to overflow desirable, particularly with hill breeds.)
Floor and drainage	4.5 litres/d per adult animal on average. Slats are wood, 38–60 mm top width × 25–30 mm thickness, with 12–20 mm gaps according to size of breed, etc. Leave 0.65 m clearance underneath for one season's holding capacity, 0.2–1.0 m common. Floor falls, when concrete, to be 1:100 to drainage collection point Solid floor of 150 mm stone (50 mm to dust grade) on geotextile fabric, vibration roll consolidated
Special areas (lambing, shearing)	Clean area, isolated from general holding area Adequate services. Pens of 0.75–1.2 m × 1.2–1.8 m 0.5–0.6 m^2/animal holding area. 1.5 m^2/shearer
Other considerations	Polythene or polyester 'tunnels' provide low-cost short-term accomodation. Provide races and satisfactory entry and exits to dipping and handling areas. Ramps to and from house to be 1:10 maximum. Provide long-term fodder storage within building envelope for provision against poor weather periods

Sheep handling areas

Space allocations Cost range £30–£40/m^2	Gathering and dispersal pens to contain groups of up to 100 sheep at 0.5–0.6 m^2/animal Catching pens to contain 10–15 sheep at 0.3–0.4 m^2/animal; may be circular with centre pivot gate to crowd into race, dip, etc., or rectangular Central races 3.0–6.0 mm long × 0.45 m wide × 0.8–1.1 m high. Footbath and race 3 m long minimum × 0.25 m wide at bath (0.15 deep with 0.1 m fluid depth) and 0.45 m wide at top of boards, about 1.0 m high

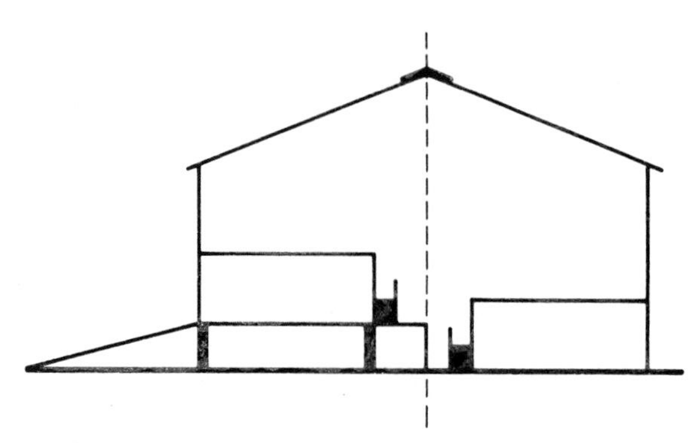

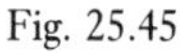

Fig. 25.45

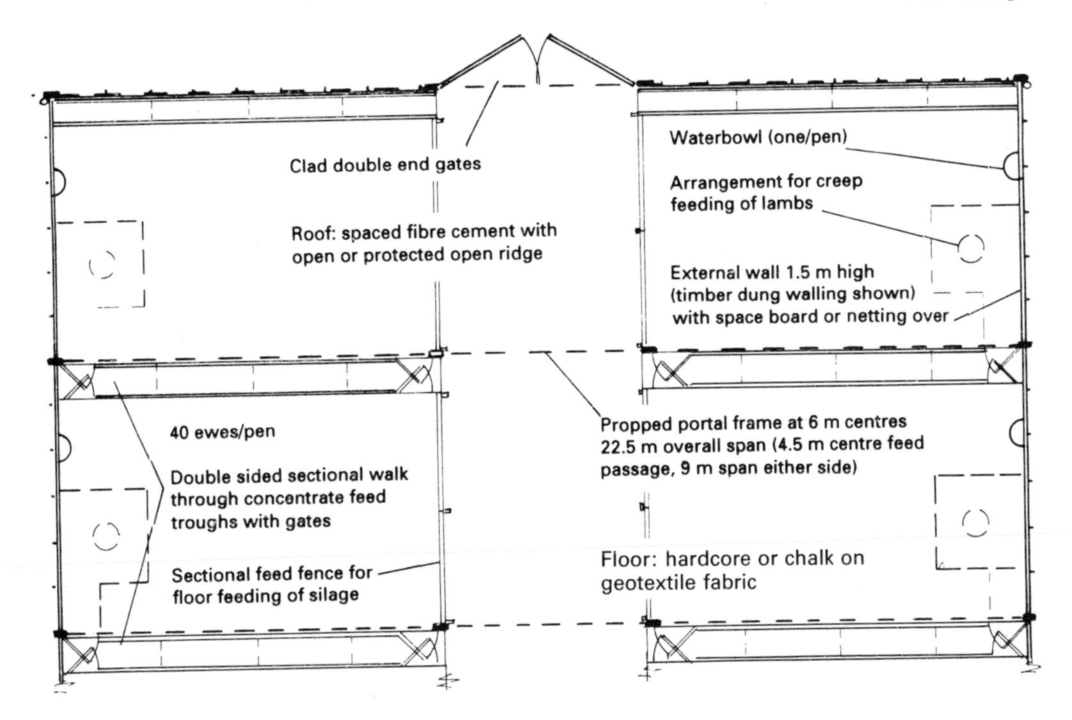

Fig. 25.46 End two bays of typical sheep housing.

Table 25.7 Sheep housing: special requirements (from Astley Cooper *et al.*, 1991, with permission)

Type of housing	*Area on straw* (m^2)	*Area on slats* (m^2)	*Trough space (access from outside pen)*	
			Concentrates (mm)	*Ad-lib hay or silage (mm)*
Large ewe 60 kg up to 90 kg in lamb	1.2–1.4	0.9–1.1	450–500	200–225
Large ewe 60 kg up to 90 kg + lamb	1.4–1.8	1.2–1.7	450–500	200–225
Small ewe 45 kg up to 60 kg in lamb	1.0–1.3	0.7–0.9	400–450	175–200
Small ewe 45 kg up to 60 kg + lamb	1.3–1.7	1.0–1.4	400–450	175–200
Hoggs 32 kg up to 45 kg	0.7–0.9	0.5–0.7	350–400	150–175
Hoggs 23 kg up to 32 kg	0.15	0.4–0.5	300–350	125–150
Lamb creep (2 weeks)	0.4	–	–	–
Lamb creep (6 weeks)	2.2	–	–	–
Ewe plus lamb in pen	1.5	–	–	–
Lambing pen		–	–	–

* On self-feeding 100–125 mm is adequate for all sheep
NB For clipped ewes the figures above may be reduced by at least 10%.

	Dip tank (see Fig. 25.47) capacity from 0.75 m^3 for 300 ewe flock, to 2.5 m^3 for flocks of 1000+ (see below for general advice)
	Ancillary holding pens for up to 50 sheep, short term, at 0.4 m^2/animal
Other considerations	Allow 2.5 litres/sheep (minimum 350 litres, then 1000 litres/500 ewes up to 4000 litres) of dip fluid. Circular dip best for large flocks. Floors should be roughened concrete with slope to drain channels of 1:50–100, or may be hard, dry soil in well drained areas, or of introduced stone. Slope from bath should be 30 degrees
	Drain baths, tanks to an approved holding or disposal point. (Care to avoid pollution.)
	Pen posts for rails should be away from sheep, either on non-sheep side or with shoulder protector boards

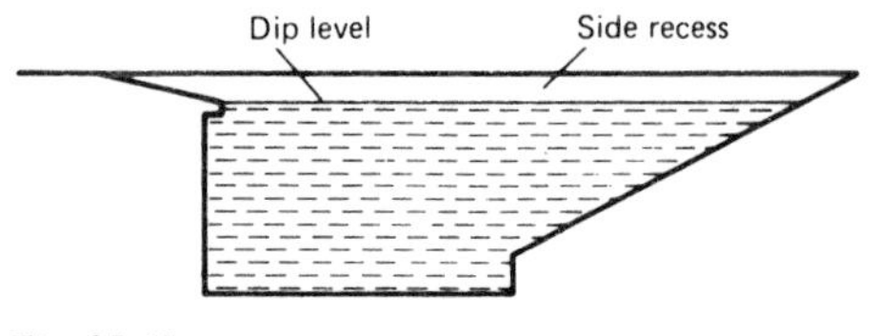

Fig. 25.47

Crop storage (BS 5502, Parts 70 and 74, refer)

Grain storage (Table 25.8)

Table 25.8 Grain storage

		Barley	*Beans*	*Herbage seeds*	*Linseed*	*Maize*	*Oats*	*Peas*	*Rape*	*Rye*	*Wheat*
Space requirements (Base on conventional storage MC and 20–30° repose angle)	(m^3/t)	1.45	1.2	3.0–5.0	1.45	1.35	2.0	1.3	1.6	1.45	1.3
	(kg/m^3)	705	833	200–500	705	740	513	785	625	705	785
Capacity of level, 2 m fill, of loose grain, per 4.8 m bay and 6 m span	t	42	51		42	47	29	44	40	42	44
Capacity of fill when heaped above 2 m, per 4.8 m bay and 6 m span	t	57				64	39				
Tonnes of product 2 m high, level fill/m^2 of floor area	t	1.34	1.67	0.4–0.6	1.34	1.48	1.0	1.54	1.33	1.34	1.54
Capacity, of 4.5 m diameter silo 9 m high	t	102					76	220		98	110

Note: Further moisture adds weight/m^3 at 1–1.5% extra per 1% addition of moisture over standard conventional.

Cost range: £125–£170/m^2; towers on concrete £60–£80/m^3; 'on-the-floor £40–£70/t; bin and drying £70–£120/t

Store types (see also Chapter 14 on grain storage)

Circular bins 2–6 m in diameter, up to 6 m high (height not to exceed twice the diameter) – see Fig. 25.48 and 25.49. Materials may be hessian and wire mesh, oil-tempered hardboard, galvanized corrugated steel, galvanized slotted steel, plyboard, welded steel plate, glass- or enamel-surfaced steel, or concrete stave with steel hoops. Floor may be plain concrete, aerated mesh or sloped steel (45 degrees). Emptied by 'air-sweep', slope, or auger conveyor

Square or rectangular bins 2.4–4.5 m with heights up to 9 m (6 m commonly for drying purposes). Materials may be corrugated steel sheet (square profile), galvanized steel pressed panels or reinforced concrete block with substantial piering and frame support. Flooring will be plain, ventilated or sloped as with circular type

Loose bulk ('on-the-floor') storage uses standard building types and frames; with or without simple retaining walls for containment at the edges or at dividing partitions (see Fig. 25.50). Variants include systems for air passage through grain from main ducts and laterals, under-floor ducting, floor perforations, and radial distributors buried from top (see Chapter 14 on grain storage). Duct/outlet type and positioning are dictated by grain type, moisture levels, suck or blow system, high or low ventilation rates. The building size and shape are also critical to the system used. Refer to specialist texts for details. It is important to appreciate that the building width in particular is critically dependent on the duct system chosen.

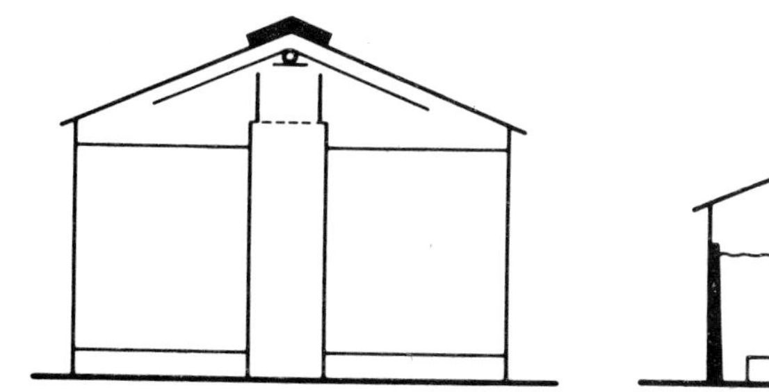

Fig. 25.48

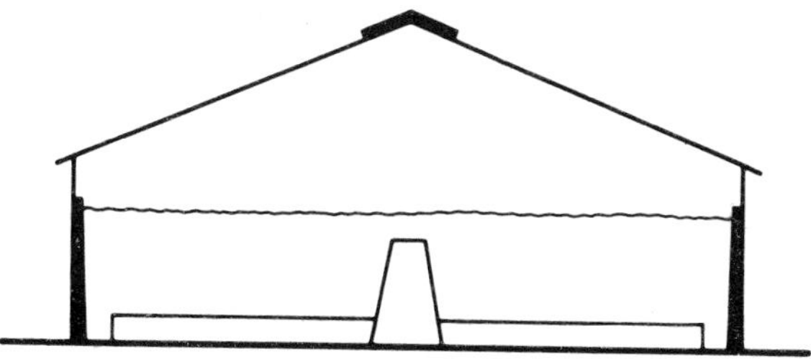

Fig. 25.50

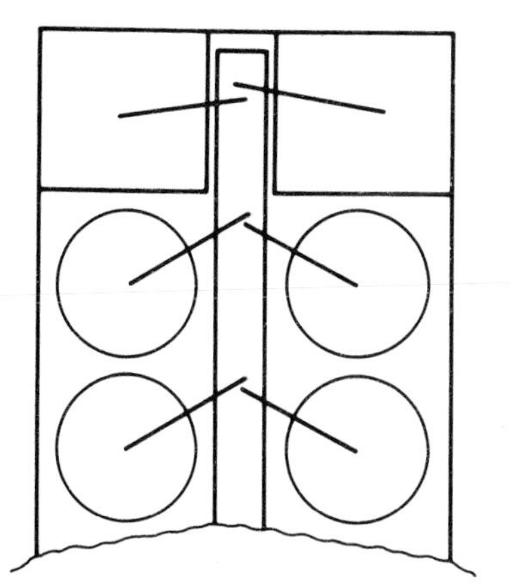

Fig. 25.49

Wet grain storage, in sealed silos (or with non-sealed roofing), using circular, galvanized, glass fused or enamelled steel. (Allow for density of product at 14% MC plus 5–10%.) Danger from CO_2 build-up

Wet grain heaped 'on-the-floor' in clear span framed buildings, preserved by propionic acid application. (Density as above.)

Refrigerated grain stored, at 18–22% MC in bins at a density of 3–5% more than at 14% MC

Floor

Hard, durable, smooth, dust-free and dry. 150 mm of concrete, reinforced on soft, difficult soils and at edges when supporting bins; steel floated and incorporating a damp-proof membrane. May have to accommodate air ducts or conveyors

Walls

Thrust resistant and air tight. Must be designed with adequate foundations. Building frame may support thrust *if designed to do so*

Other considerations

High power consumption (e.g. 30–60 kW) likely in some stores

Reception pit needed for bin systems, 2.5 m × 1.5 m × 1.5 m deep, with a sloping delivery floor at 60 degrees to the horizontal

Overhead hoppers to fill lorries rapidly, with high delivery rate conveying systems to match. Large headroom and strong framing required

Must be vermin-proof (mandatory control legislation against *Salmonella*) thus totally enclosed and no natural light – to discourage birds (See BS 5502, Part 30, for general guide to infestation proofing.)

Doors should provide clearance for tipped trailers (up to 6 m). Provision should be made for heavy vehicles in the flooring in the vicinity of grain stores. Special consideration should be given to fire hazards in grain drying areas. Deep-pit elevator equipment and temporary storage prior to drying may be required. (Service access needed below ground level.)

Hay, straw, grass. (See next section for silage)

		m^3/t
Space required for bulky low density materials	Hay, loose	9.0
	Hay, medium density baled	6.0
	Hay, barn dried	7.5
	Barley straw, medium density baled	11.5
('Average' figures provided: there will be considerable variation from sample to sample, dependent on moisture content, species, etc.)	Wheat straw, medium density baled	13.0
	Grass, wilted	2.2
	Grass, silage, consolidated	1.4
	Grass, tower silage, high dry matter	1.25
	Grass, high temperature dried, 75 mm long	8.5
	Grass, high temperature dried, ground	1.5

Cost range: £50–£60/m^2

Volumes and tonnages — A 4.8 m bay of a building will contain:

35 t of baled hay at a height of 4.8 m

6.9 t of consolidated silage at a height of 2.4 m

20 t of medium density baled barley straw at a height of 5.4 m

Store types — Standard frame with open or closed sides, may have special walling and floor system (see Fig. 25.51)

Flooring — Earth or rammed hardcore on well drained soils, for simple storage of baled hay and straw. Raised floor weldmesh, etc. to provide plenum chamber for drying of hay in batch or storage driers. (See also Chapter 24 on Farm Machinery – Barn hay drying.)

Walls — Hay barn walls may be removable in the form of wooden baulks or steel sheet or they may, in the case of loose material be of wire netting. Some types of walling must be air tight in storage drying. Top of wall near eaves should be slatted for up to 1 m to allow air movement during drying

Roofing — Hay barn roofing should be ventilated and of non-condensation-drip forming type

Other considerations — Fire risk should be minimized. Ease of access to farm roads and livestock buildings. Power requirement may be high in storage driers or tower silo systems (20–100 kW). Drying fans tend to be noisy in operation

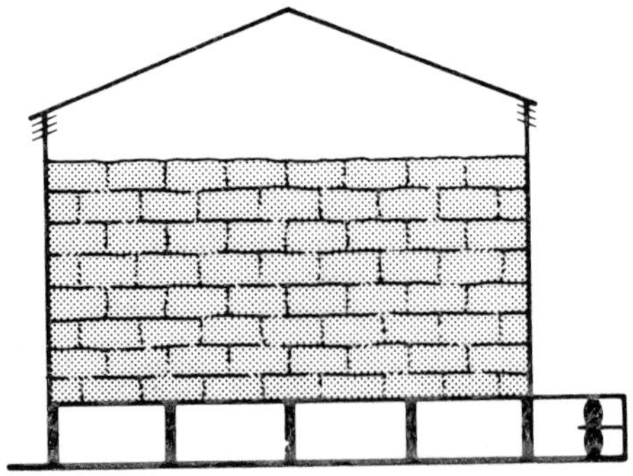

Fig. 25.51

Silage bunkers (See Fig. 25.52)

Cost for 1000 tonnes £30–40/tonne

BS Walls	BS 5502 Part 22:1987 (legal requirement)
BS Concrete floor	BS 8007:1987 (to comply with legal requirement to be impermeable)
Space	1.4 m^3/tonne grass silage @ 20% dry matter (minimum) 1.5 m^3 tonne grass silage @ 25% dry matter (optimum) 1.6 m^3 tonne maize silage @ 30% dry matter (maximum)
Depth	Optimum 2.4 m or 3 m. Maximum height for self feed 1.8 m unless top removed daily. Must *never* increase height by surcharging – consolidated height must not project above top of wall for at least 600 mm from edge and only slightly domed in centre
Width	Optimum face, for grass, to prevent spoiling, 7.2–9.0 m or up to 12 m if can use fast enough. For maize up to 18 m. Conflict if self-feeding as must allow min 150/200 mm of face width/cow (or large beef) for 24-hour access
Length	No limit except practical considerations. Long bunkers much more flexible if open both ends – if site permits. Avoid demountable front walls if possible
Falls	From back to front: 1 in 100 (CIRIA), or 1 in 75 (Code) From centre to sides: 1 in 75 (CIRIA), or 1 in 50 (Code) This means that the walls must either be stepped or sloping – if sloping they will no longer be in the vertical plane. Confirmation that this is acceptable within the design of prefabricated panels must be obtained
Roof	Not justified on economic grounds, but does eliminate rainwater from effluent system and provides for straw storage over, cover for self feed, and scope for other uses

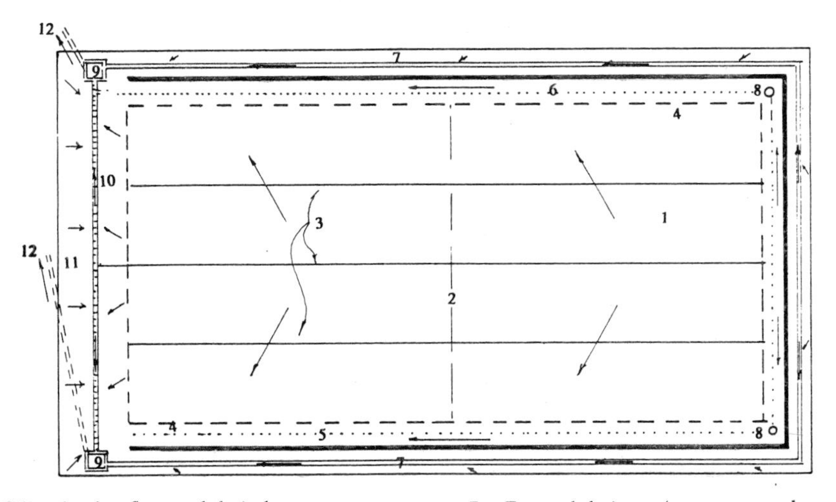

Fig. 25.52 Silage bunker floor and drain layout.
1, Concrete floor to BS 8007 (very severe rating).
2, Contraction joint (see Fig. 25.53).
3, Construction joints (see Fig. 25.54).
4, Expansion joints (see Fig. 25.55).
5, Perimeter concrete (to design for wall).
6, Internal drainage (e.g. Porcupipe).
7, External drainage (e.g. concrete channel).
8, Rodding chamber.
9, Catch pit.
10, Front drainage (e.g. polymer concrete with grating).
11, Front apron with reverse fall to drain.
12, To effluent tank.

Walls

Options (all of which must have an engineer's certificate confirming that the wall type and foundation requirements conform to BS 5502):

- No walls -- (still needs impermeable floor with upstanding curb and outside channel unless only used for bagged silage)
- Stub or building stanchions with horizontal rail(s) and:
  - Concrete or timber (usually external ply) vertical panels
  - Sleepers (vertical)
- Stub or building stanchions and:
  - Concrete horizontal prestressed panels
  - Cast in situ concrete walling
  - Sleepers (horizontal) would need stanchions at max 2.6 m c/c.
- Cantilevered vertical prestressed concrete panel
- Cantilevered cast in situ concrete (with engineer's supervision)
- Precast concrete 'L' shape retaining wall sections (height up 3.6 m)

NB. Reinforced hollow concrete block walls no longer recommended.

Floor

Concrete floors must be laid with expert supervision that has a working knowledge of how to meet the very high standards of the 'very severe' category of BS 8007, which could include the following:
Mix: RC 45 (previously C35A) 0.55 water/cement ratio and 325 kg/m^3 minimum cement content, 75 mm slump, use PFA (pulverized fuel ash) aggregate if available; never use limestone
Thickness: 150–250 mm depending on ground bearing and loading factors, with min. 50 mm cover over reinforcement
Reinforcement: Weldmesh fabric to suit jointing and bay size
Membrane: 1200 gauge polythene (used mainly as slip surface for contraction, but also additional pollution protection)
Joints: See Fig. 25.52, or lay single slab laid with pumped concrete
Finish: Retamp approx 20 min after laying across normal traffic direction
Temperature: Do not lay in extremes of temperature – hot or cold

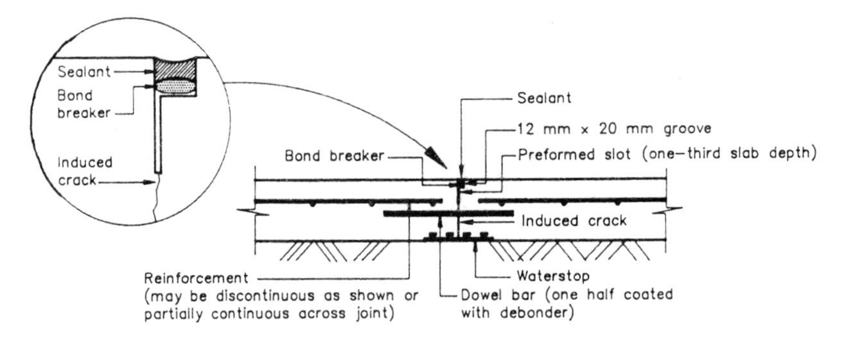

Fig. 25.53 Contraction joint detail (adapted from BCA Farm Concrete Fact Sheet and CIRIA Report 126).

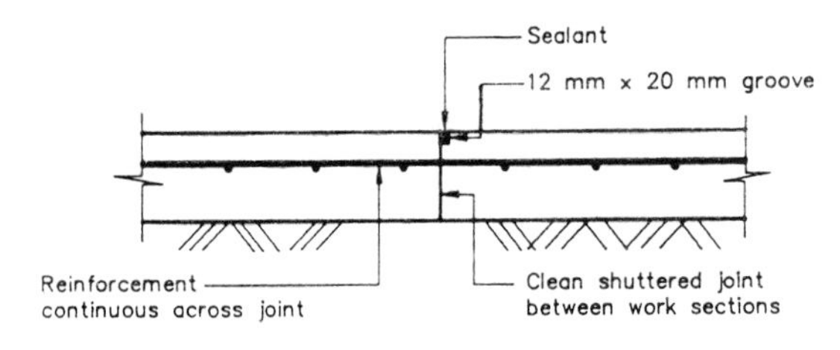

Fig. 25.54 Construction joint detail (adapted from BCA Farm Concrete Fact Sheet and CIRIA Report 126).

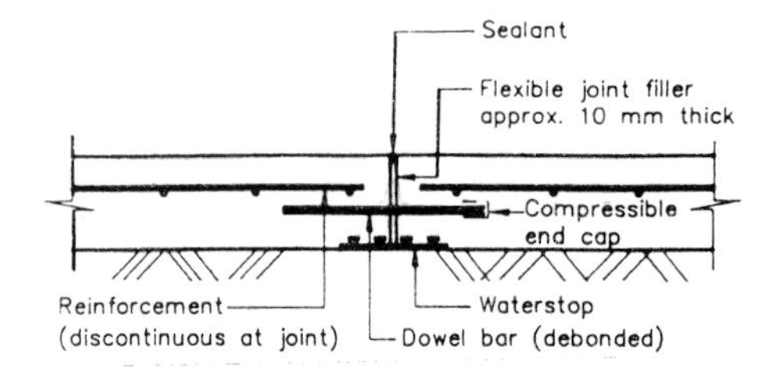

Fig. 25.55 Expansion joint detail (adapted from BCA Farm Concrete Fact Sheet and CIRIA Report 126).

Surface coating: Use purpose-designed acid-resisting coating – also required for inside face of walls (two coats at base)
Joint sealant: Use only purpose-designed one- or two-part polysulphide or polyurethane – not bitumen. See CIRIA Report section 7.4.4 about the precautions necessary

Hot rolled asphalt (HRA)

Although offered in the Code as an alternative with a bitmac base, it is not recommended at present until the problem of the expansion joint with the perimeter concrete has been overcome. It does, however, have a very useful role to play in upgrading existing concrete floors (and for surfacing new floors providing there is sufficient time lapse), using 50 mm hot rolled asphalt laid by mechanical paver and 8-tonne roller using 40% hardstone type aggregate – *not* limestone – to BS 594 column 21 Table 5 modified. Preparation is the secret to success. The tack coat must be applied to a very clean and textured surface

Channel external

There is a legal requirement to install an impermeable channel around the outside with adequate fall (min. 1/100) in an extension of the base of the silo to discharge into the effluent system; usually formed by insetting half-round salt-glazed or polymer pipe or dishing the concrete

Drain internal

The design of the walls will be dependent on the provision of an internal drain within 500 mm of the inside face to relieve the hydraulic pressure. This can be omitted if the wall design allows for the effluent to pass through or underneath. There are three options:

- temporary flexible 75 mm perforated PVC land drain – often tied to 75 mm timber batten to prevent crushing;
- permanent 100 mm drain – slot or Porcupipe with rodding access points;
- permanent channel – with engineering brick or timber infill or heavy duty narrow slot grating.

NB With the second and third options thicken concrete accordingly and maintain 50 mm cover over mesh.

Front drain

Must be set approx. 1 m beyond front edge of retaining wall and have further 1 m of apron reverse falling towards bunker so that even if drain blocked the effluent will pond. Use same drainage system as internal with letterbox slot and sump at end to collect ponding effluent.

Effluent tank legal requirement to provide to following capacity:

up to 1500 m^3 = 20 litres/m^3
over 1500 m^3 = 30m^3 + 6.7 litres/m^3 over 1500 m^3
It must also have a *maintenance-free* life expectancy of 20 years. The alternative is to discharge direct into the dirty water system providing that system conforms and precautions are taken (and notices erected) about the danger of the production of highly toxic hydrogen sulphide (bad-egg smell). Both the Code and CIRIA recommend GRP bottles or cylinders with the special silage coating may be used but access to get out heavy sand deposits which cannot be pumped is difficult. (They *must not be entered* without full breathing apparatus and safety harness, etc.)

Precast concrete tanks are offered by several manufacturers which overcome the access problem with slat covers, but must be designed and installed with instructions certified by an engineer, and tested on completion. (See Box 51, CIRIA Report.) Direct engineer's supervision is required for a cast in situ tank, and blockwork is not suitable

Notice

Legal requirement to display notice showing the following:

- Moisture content after rolling not to exceed ... %.
- Internal drainage provided.
- Gross vehicle not to exceed ... (usually 8 t).
- Maximum depth of silage when rolled not to exceed ....
- Minimum distance of centre of wheel of rolling.
- Vehicle to edge of bunker 0.6 m.
- (Usually added) Complies with BS 5502 Part 22: (and date of construction).

Concentrated and root crop feed material

Space for 'concentrates' ('Average' values provided – there will be considerable variation from sample to sample, dependent on moisture content, consolidation) Cost range: £70–£120/m^2	Meal, loose, 2.0 m^3/t or 500 kg/m^3 Pellets, loose, 1.7 m^3/t or 600 kg/m^3 Nuts, loose, 1.4 m^3/t or 700 kg/m^3 (50 kg bags stack 2 high, occupy 2.0 m^3/t or 500 kg/m^3) Beet pulp 1.5 m^3/t or 650 kg/m^3 Fodder beet 1.7 m^3/t or 600 kg/m^3 Turnips 1.8 m^3/t or 550 kg/m^3 Brewers' grains (dry) 2.0 m^3/t or 500 kg/m^3
Space for equipment used in milling and mixing	Horizontal mixer, 4–10 m^2 × 2–3 m high Vertical mixer, 3–5 m^2 × 3–6 m high Crushing mill, 2–3 m^2 × 1–1.5 m high Hammermill 1–6 m^2 × 1–1.5 m high (5 m with cyclone above) Allow 1–1.5 m headroom for overhead conveying
Other considerations	Self-emptying bins should have floors at 60 degrees to the horizontal. Noise and dust are potential health hazards. Risk of explosion in a dusty atmosphere. Very high power requirement by most barn machinery (5–40 kW). Mechanical handling requires smooth running, hard flooring. Use of gravity desirable in conveying – overhead hoppers for lorry and trailer filling with 10–20 t capacity for big units. High clearance needed (up to 6 m)

Vegetable storage (BS 5502, Part 71, refers)

		m^3/t	*Mean temperature to store long term*	*Maximum height*
Space requirements and environment				
Cost range: Insulated, £170–£200/m^2 Non-insulated, £125–£170/m^2	Potatoes, bulk (see Figs 25.56 and 25.57)	1.5–1.6	Ware 4–5°C	4.0 m (6 m if air flows can reduce temperature gradient)
	Potatoes, in pallet boxes (1.2 m × 1.4 m × 0.8–1.0 m) (see Figs 25.58 and 25.59)	2.4	Ware – as above. Crisping and processing, 6–8°C	4–6 boxes (4.0–5 m)
	Onions, dry (see Figs 25.56 and 25.57)	1.5–2.0	2–3°C	4–5 m
	Onions, green topped	3.0	Ambient temperature –3–6°C until dry	2.5 m
	Onions, refrigerated	2.0	0.5–1.5°C	3.0 m
	Onions dry, in boxes (1.1 m × 1.3 m × 1.05 m high)	3.0	Ambient temperature +3 to 6°C	Boxes stacked to eaves

Store types	Standard, clear span frames
Floor	As under 'Grain storage' plus smooth-running heavy-duty floor for pallet handling equipment and which may incorporate air-flow and conveying ducts
Walls	Thrust resistant (see under 'Grain storage') – angle of repose for potatoes is 30–40 degrees). In pallet box stores cheaper walling may be used. 'U' value of 1.0 maximum W/m^{2}°C for walls (and roof) is necessary to protect potatoes from frost. Care must be taken to avoid 'cold bridging' across the walls via reinforcing or framing systems. Walls should be smooth and cleanable

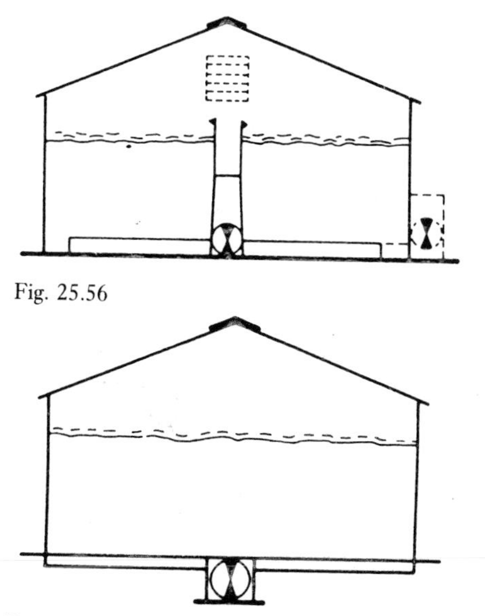

Fig. 25.56

Fig. 25.57

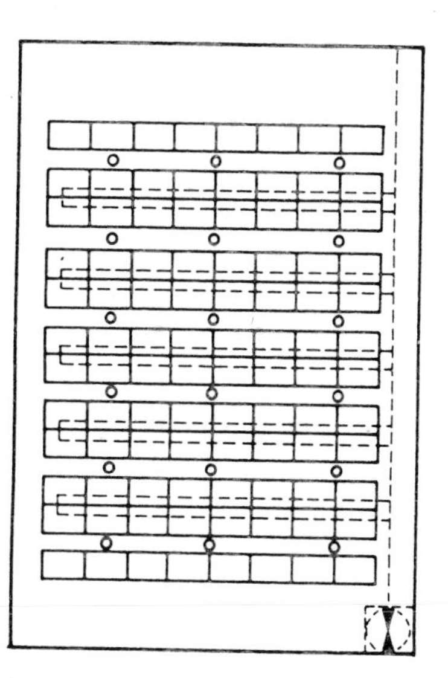

Fig. 25.58

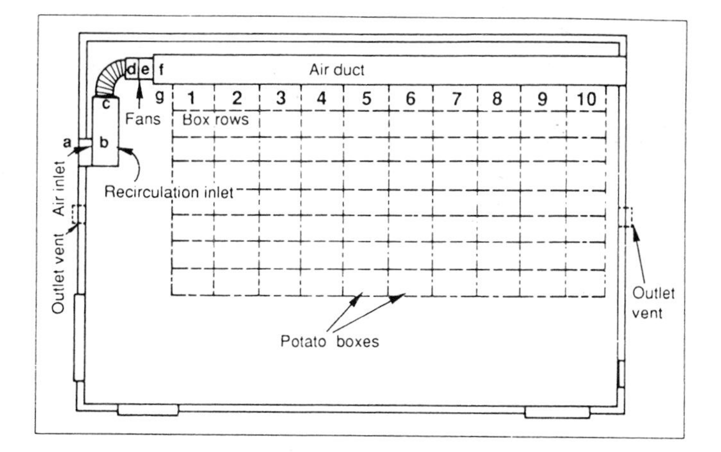

Fig. 25.59 Plan of potato store showing arrangement of boxes.

Ventilation	Main purpose to control temperature rather than moisture. Insulation of structure greatly assists but does not remove need for ventilation except in short-term box storage or shallow (max 1.8 m) on-the-floor storage. Even then a low volume ventilation system is desirable to dissipate the heat and moisture given off by the crop. The systems illustrated above are usually controlled by automatic heat and humidity sensors placed in the crop and measuring the ambient air temperature and humidity. This gives the option to recirculate the air if the need arises. Note also the laterals are wider spaced than for grain storage and can exceed 10.5 m in length if tapered. Essential to provide pressure release flaps in a controlled environment with input fans
Other considerations	Avoid materials in structure with a high thermal conductivity. Use straw as insulation layer on top of stack, and bales as temporary wall insulators. Should be a concrete apron of 2.5 m minimum width in front of loading doors

Porch desirable to enable store to be filled to end wall, potatoes rest on removable boards supported on porch side walls which carry stack weight as it is built up

If only used for storage exclude natural light but if building used for sorting and grading, roof lights preferable – keeping crop covered with straw. Strong artificial light, 300 lx, is required in all instances of working machinery.

Chitting houses for potatoes (see Fig. 25.58)
(BS 5502, Part 66 refers)

Potatoes stacked in trays, with access to illumination from movable lighting columns suspended from ceilings. Trays are 750 × 450 × 150 mm on pallets 1.5 × 0.9 m in area, stacked to depth of 3.5 m maximum. Alleyways 0.5 m minimum between rows. 1 m side alleys. Generous end space in building. Should be frost-proof and force ventilated, with heating equipment available when necessary

Servicing and general buildings

Fertilizer storage

Space required
Cost range: £70–£120/m^2

Loose, bulk, 0.9–1.1 m^3/t average for prilled fertilizer

50 kg bags, stacked 1.0–1.2 m^3/t (at 180 mm thickness per bag, with maximum of 10 high)

'Big' bags containing 0.5 t, stack two or three high; are nearly equivalent to loose

Other considerations

Stored in standard frame building with weather protection and anti-condensation provisions to prevent dripping on to bags. Thrust-resistant walls for bulk fertilizer with corrosion-resistant surfacing. Ends and top of heap covered with plastic sheeting. Walls and floor clearance of 0.5 m and 0.15 m respectively for bags (on pallets or boards). Provide temporary divisions to isolate different fertilizer types

Slurry and farmyard manure storage

BS 5502: Part 50:1989 (legal requirement) and BS 8007:1987 for impermeable concrete floors where applicable

Space required

Dirty water: 1 m^3/tonne or 1000 litres/tonne
Slurry: 1–1.4 m^3/tonne (3–12% dry matter)
Farmyard manure: 1.5–2.0 m^3/tonne ('solid' manure)

Cost range: £7/m^3 (earth lagoon) – £80/m^3 (weeping wall store)

Minimum capacity

Dirty water: *Either* total containment – 4 months (legal) *or* with distribution system (usually low volume irrigation): no specific obligation except to be adequate. Recommend three-chamber with 'H' pipes (see Fig. 25.60)

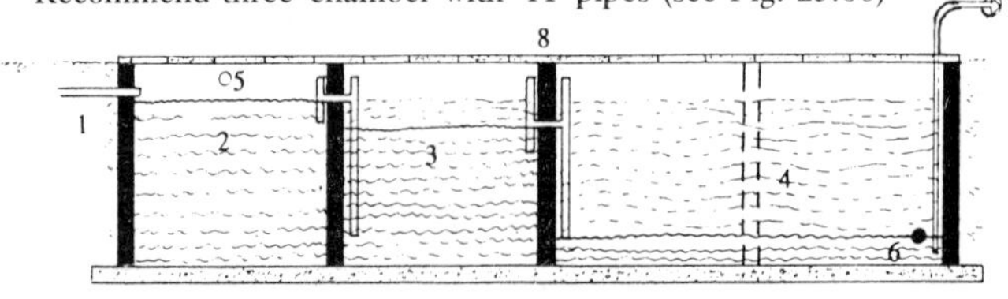

Fig. 25.60 Dirty water settlement tank.
1, Inlet (to be kept as high as possible).
2, primary settlement tank (usual actual minimum 7500 litres*).
3, secondary settlement tank (usual actual minimum 5000 litres*).
4, pumping tank (with/without intermediate wall, including top section of secondary settlement tank and excluding level below float switch (300 mm) – usual minimum capacity 24 hours *without pump functioning and* providing emergency overflow arrangements – subject to NRA agreement (if no overflow, recommend min. 48 hours capacity)
5, overflow as above.
6, pump intake 150 mm above base and float switch 150 mm above intake.
7, to low volume irrigation system. (Pump kept well above tank in case of flooding and provided with froststat control heaters with warning light – usually on adjacent buildings).
8, contruction to regulations – need easy access for desludging – slats shown.

* Will not be sufficient for units with large open yards.

Slurry: *Either* unseparated total containment: 4 months plus 300 mm min 'freeboard' or 750 'freeboard' for earth bank lagoons (legal) *or* separated by weeping wall or separator: as above for remaining solids – i.e. less all liquids and 10% (weeping wall), 20% (separator), of slurry. In latter case the separated materials become dirty water and farmyard manure in classification. NB The minimum capacity for a Slurry reception pit: 48 hours (legal)

Farmyard manure: No minimum requirements. Permanent stores should have impermeable floor. Can be temporary stored in field providing does not cause pollution from run-off

Options for storage

- Under slat storage cellars or channels leading to other storage
  Floors: min. 150 mm RC 40 reinforced with membrane laid horizontal with 100/150 mm minimum water containment so dung solids will float
  Walls: Impermeable and double load bearing (for outside earth thrust) in reinforced concrete (BS 8007) or masonry (BS 5628) waterproofed inside and outside. Cellars with ramp or reinforced panel access for emptying. Channels with weir every 18 m; 150 mm high, usually approx. 1 m wide.
- Above-ground circular store: proprietary steel or concrete on concrete base and ring beam to manufacturer's requirements up to 33 m dia. and 7.2 m high with access platform and steps, mixing system. Used in conjunction with reception pit (minimum capacity 48 hours with two locked sluice valves – both legal requirement) and tanker filling pump
- Dirty water below-ground tanks for low-volume irrigation: three chamber as above usually prefabricated concrete panels or rings, but can be constructed as underground cellars above in purpose-designed cast in situ or masonry. GRP not suitable. Must also have adequate locked access for pumping out, preferably with removable cover on settlement chamber for mechanical removal of silt. Accessible 'H' pipes between chambers. Can use earth lagoon (see below) but need extra filtration for pump.
- Weeping wall store – used with a dirty water system
  Construction as option 1 with level floor with 150 mm water-holding upstand, or with floor horizontal around perimeter, but with slight fall *towards* fill area. Outside the walls the floor must be extended by 300 mm incorporating a channel as per silage (legal requirement)
  Walls: 2.4/3 m high with 25/30 mm gaps at 200/450 centres on some or all faces using concrete panels or wooden sleepers in steel uprights
  Filled by channel or ramp. Emptied by either ramp or removable panels
- Impermeable wall store – usually used as above or with slurry pump.
  Construction of floor as above except that the external channel is not required, and walls as per silage bunker. Usually too expensive for 4-month storage. Can be used with internal weeping wall strainer used in conjunction with drain or pump to remove dirty water
- Earth bank lagoon (sometimes called compound when used for slurry). Suitable for either slurry or dirty water, short term, emergency overflow or 4-month minimum. The 'freeboard' minimum goes up to 750 mm and the soil must be suitable (legal requirement). The criteria for the construction are set out in CIRIA Report 126. (Note that the code does not give sufficient guidance.) In very simple terms, providing the soil samples taken have passed a detailed analysis, it can be used without lining as shown in Fig. 25.61. It must also be constructed in accordance with the standards laid down including compaction in layers to the table provided, e.g. 150 mm using eight passes of a vibratory roller between 1300 and 1800 kg. If there is a shortage of suitable material, it is possible to use an impermeable core, or, if the material is not suitable, then a liner can be used in one of the approved materials – usually with a guarantee, but a leakage monitoring system is required so it is a last resort. All lagoons must be protected with child-proof fencing, locked gates and a hazard warning sign. Earth bank lagoons can be used with low volume irrigation systems with separating banks to subdivide the three tanks with T-pipe overflows between each (usually two to each). Large lagoons must be designed to be emptied economically – preferably with a ramp access. If to be emptied from the top by swing shovel, the width must be restricted to suit the machine with a 4 m top access strip. A strainer box to ease the extraction of liquid is also essential with provision for operator access.

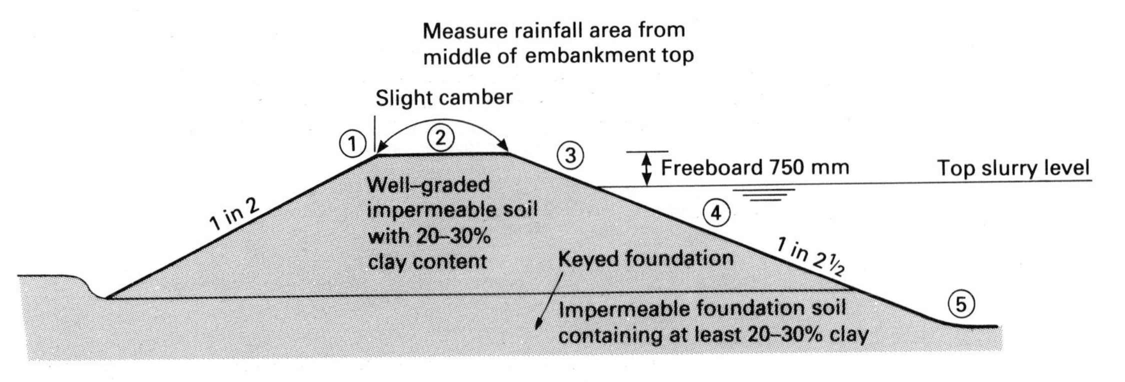

Fig. 25.61 Lagoon embankment on impermeable soil (one of four options from CIRIA Report 126).
1, Suggested position of child-proof fence.
2, top width varies with depth but allow minimum 3 m for vehicle access, 4 m if emptied from top.
3, seed exposed faces to stabilize.
4, place 'ladders' around bank – often made out of tyres.
5, floor of lagoon should have shallow fall towards fill position.

Machinery storage

(BS 5502 Part 80 refers)

General — Height and width are often more of a problem with storage than area or volume occupied, due to doorway restrictions, eaves heights and turning circles. Site building near good hard road, central to farm operations, adjacent to workshop and cleaning facilities. Provide sufficient light, natural and artificial, for inspection and manoeuvring

Cost range: £70–120/m^2

Clearance requirements for tractors and implements — Medium size tractor – 3.0 m high × 1.85 m width (may be extended to 2.5 m wheel track), area 6.75 m^2, with a turning circle of 7.5 m diameter (exhaust should have 0.5 m minimum height clearance from structure). Medium size tractor plus loader, turning circle 10 m diameter

Combine harvester, 3.0–6.0 m wide × 7.5–10 m long × 3.0–4.0 m high

Trailers 2.0–2.43 m wide × 4.0–5.5 m long × up to 4.5 m high (tipped)

Lorries 2.43 m wide × 4.5–9.1 m long (rigid chassis) or 10.0–12.5 m long (articulated) × up to 4 m high (unusually 5 m high by 16 m long)

Plough, reversible, three-furrow, 1.5 m wide × 2.5 m long. Disc harrow, 2.5–5 m wide × 1.9 m long. Seed and fertilizer drills, 2.5–4 m wide (trailed) or 2.5–6.5 m wide (mounted) × 2.0–3.0 m long

Balers, medium density, 2.5–3.0 m wide × 4.5–5.5 m long

Beet harvesters 2.5–4 m wide × 5.5–7 m long × 3.5 m high

Other considerations — Hard, well-drained, floor (but not necessarily concrete). High eaves at doorway. Consider upstand in roof at this point for better access). Translucent sheeting in roof to give high level of natural light. Concrete at point of entry to building with generous hard standing outside, adjacent. Limited number of socket outlets for electricity supply (low voltage output desirable)

Fuel storage

Diesel store only (petrol storage restricted and licensed)

Typical tank size (volumes approximate): *Steel*
1 m × 1 m × 1 m (1000 litres)
2 m × 1.25 m × 1.25 m (3125 litres)
Plastic
1.15 m dia. × 1.3 m (1000 litres)
1.60 m dia. × 1.32 m (2000 litres)

Steel may be supported on walls; plastic on a 'solid' base (paving slabs on concrete lintels). All high enough to discharge to highest tank filler on vehicles when almost empty. Bottom slope from delivery pipe to sludge drain cock. Dipstick and ladder access or sight-gauge fitted. Only approved materials to be used. If 1500 litres or more is stored, then tank should be mounted in enclosing 'bund' walling to hold 110% of tank fill. Flexible hose should have auto-close valve. All valves kept locked until use. Tanks are not installed within 10 m of watercourse. Minimum life of 20 years

Workshop accommodation (BS 5502, Part 80, refers)

Area required and facilities	One or two bays of standard frame building of 9–15 m span. Pit, 2.0 m × 0.8–1.0 m wide × 1.7 m deep with steps at each end, waterproofed, with drain sump: placed centrally in one of the bays with access from main door to workshop. Clear wall length of 5 m for benching with electric power (110 V) and spot lighting. Concrete area of 50 m^2 minimum outside main workshop door with good drainage falls to sump or trapped catchpit. Floor inside of 150 mm concrete, reinforced to take heavy equipment. Overhead gantry for crane desirable

Farm office accommodation

Space required (may be controlled by Offices, Shops and Railway Premises Act, 1973)	10 m^2 minimum, 15–20 m^2 preferred, with washroom area and WC of 3 m^2 adjacent, all served by porch lobby of 3 m^2 minimum. Ceiling 2.5 m high. Natural light required through one wall of office, preferably with view of main farm access road. Electric power, heating, water and telephone services essential. Need insulation and background heating if computer terminal to be installed. Large area of clear wall space desirable. Drainage and other building features are regulated by Building Regulations and inspected during construction by Local Authority Building Control Officer.

Further reading

General

Astley Cooper Sir P. (Ed.), *et al.* (1991) *Farm and Rural Buildings Pocketbook*. Farm and Rural Buildings Centre, NAC, Stoneleigh, Kenilworth.

Barnes, M. & Mander, C. (1992) *Farm Building Construction: the Farmer's Guide*. Farming Press, Ipswich.

BSI (1989–94) BS 5502 *Buildings and Structures for Agriculture*. Parts 10–19, General data series. Parts 20–39, General design series. Parts 40–59, Livestock building series. Parts 60–79, Crop building series. Parts 80–99, Ancillary buildings. BSI.

MacCormack, J. (Ed.) (1995) *Farm Buildings Cost Guide, 1995*. Centre for Rural Building, Aberdeen.

Noton, N.H. (1982) *Farm Buildings*. College of Estate Management Reading, distributed by Spon, London.

Specific – livestock

Baxter, S. (1984) *Intensive Pig Production, Environment Management and Design*. Granada, London.

Brent, G. (1986) *Housing the Pig*. Farming Press, Ipswich.

British Veterinary Association (1987) *Farm Animal Housing*. BVA, London.

Bryson, T. (1984) *The Sheep Housing Handbook*. Farming Press, Ipswich.

Clarke, P.O. (1980) *Buildings for Milk Production*. British Cement Association, Slough.

Clarke, P.O. (1985) *Buildings for Beef Production*. British Cement Association, Slough.

Farm Electric Centre (1990) *Controlled Environments for Livestock*, EA 1011. NAC, Stoneleigh, Kenilworth.

Farm and Rural Buildings Centre (1992) *Planning Sheep Handling Units*. NAC, Stoneleigh, Kenilworth.

Farm and Rural Buildings Centre (1992) *Planning Sheep Housing*. NAC, Stoneleigh, Kenilworth.

MAFF (1986) *Guidelines for Housed Livestock*. HMSO, London.

Meadowcroft, S. and Hardy, R. (1990) *Indoor Beef Production*. Farming Press, Ipswich.

Mitchell, D. (1976) *Calf Housing Handbook*. Farming Press, Ipswich.

Sainsbury, D. & P. (1988) *Livestock Health and Housing*. Baillière Tindall, London.

Wathes, C.M. & Charles, D.R. (1994) *Livestock Housing*. CAB International, Slough.

Crop storage and effluent control

Barnes, M.M. (1986) *Concrete Floors in Crop Stores*, Farm Note 15. British Cement Association, Slough.

Farm Electric Centre (1985) Hay Drying, EC4744. NAC, Stoneleigh, Kenilworth.

Farm Electric Centre (1985) Potato Storage, EC4457. NAC, Stoneleigh, Kenilworth.

Farm Electric Centre (1990) *Bulk Grain Drying and Conditioning*, EA 1032. NAC, Stoneleigh, Kenilworth.

MAFF and WOAD (1991) Code of Good Agricultural Practice for the Protection of Water. Code of Good Agricultural Practice for the Protection of Air. MAFF Publications, London.

Mason, P.A. (1992) Farm Waste Storage: Guidelines for Construction. CIRIA Report No. 126. Construction Industry Research and Information Association, London.

Building and surveying

British Cement Association, Farm Concrete Fact Sheets and Booklets, BCA Publications, Crowthorne.

Irvine, W. (1987) *Surveying for Construction*. McGraw-Hill, London.

Mitchell, G.A. & A.M. (1988) *Building Construction* (in five volumes). Batsford, London.

Glossary of units

Metric units and 'imperial' conversion factors

The established metric unitary system used in the UK is the Systeme Internationale d'Unites (SI) which replaces imperial measures. The system consists of basic units from which all measurements are specified by use of multiplying factors. The most important of these factors in common usage are:

Multiplying factor		*Prefix*	*Symbol*
0.000 000 000 001	(10^{-12})	pico	p
0.000 000 001	(10^{-9})	nano	n
0.000 001	(10^{-6})	micro	μ
0.001	(10^{-3})	milli	m
0.01	(10^{-2})	centi	c
1 000	(10^{3})	kilo	k
1 000 000	(10^{6})	mega	M
1 000 000 000	(10^{9})	giga	G

Examples of the use of these factors are:
millilitre (ml), i.e. 0.001 × 1 litre and kilogram (kg), i.e.
1000 × 1 gram. (The latter is often shortened to 'kilo' in common usage.)

In general, measurements should be specified in terms of units such that the 'whole number' part of the measurement is between 1 and 1000, e.g. 100 kg not 100 000 g, and 20 mm not 0.020 m.

The seven primary units used in SI are:

	Unit	*Symbol*	*Conversion factor SI to imperial*	
Length	metre	m	to ft	3.281
Weight	kilogram	kg	to lb	2.204
	tonne	t	to ton	0.984
Time	second	s		1.000
Temperature	degree kelvin or Celsius	K or °C	to °F	1.8 (+32)
Luminous intensity	candela per square metre	cd/m^2	to cd/ft^2	0.0929
Electric current	ampere	A		1.000
Radiation	becquerel	Bq (≡ 1 spontaneous nuclear transition per second)		

The above are the basis of many derived practical units which include:

	Unit	*Symbol*	*Conversion factor SI to imperial*	
Volume	litre	l (≡ 1000 cm^3)	to gallons	0.218
			to pints	1.744
Force	newton	N ≡ kg m/s^2	to lbf	0.2248
Work energy	joule	J ≡ Nm	to ft lbf	0.7376
			to Btu	0.9479×10^{-3}
			to Therms	9.479×10^{-9}
			to Wh	2.7×10^{-3}
Power	watt	W ≡ Nm/s	to HP	1.341×10^{-3}
Pressure	pascal	Pa or N/m^2	to in H_2O	4.015×10^{-3}
			to in Hg	2.953×10^{-4}
			to lbf/in^2	1.45×10^{-4}
			to $tonf/ft^2$	9.324×10^{-6}
			to $tonf/in^2$	6.475×10^{8}

Pressure is also expressed in the 'non-preferred' unit of the millibar (1 mbar is equivalent to 100 N/m^2)

(1 atmosphere ≡ 1 bar ≡ 100 kN/m^2 ≡ 100 kPa ≡ 10.197 mH$_2$O ≡ 33.025 ftH$_2$O ≡ 760 mmHg ≡ 14.7 lbf/in^2)

Other units commonly used in agriculture are:

	Unit	*Symbol and conversion factor*	
Calorific value	kilojoules per cubic metre	kJ/m^3 to Btu/ft^3	2.684×10^{-2}
U-value	watts per square metre per degree Celsius	W/m^2 °C to Btu/ft^2 h°F	0.176
Specific heat capacity	kilojoules per kilogram per degree Celsius	kJ/kg °C to Btu/lb°F	0.239
Moisture capacity	grams per cubic metre	g/m^3 to grains/100 ft^3	43.6
Illumination level	lux	lx to 1 m/ft^2	0.0929
Land area	hectare	ha to acres	2.47
Fuel consumption	kilometres per litre	km/litre to miles/gal	2.825
Rate of usage	grams per square metre	g/m^2 to oz/yd^2	0.0295
	kilograms per hectare	kg/ha to lb/acre	0.892
	litres per hectare	litre/ha to gal/acre	0.089
Crop yield	tonnes per hectare	t/ha to ton/acre	0.398
		to cwt/acre	7.97
Density	tonnes per cubic metre	t/m^3 to ton/yd^3	0.753
	kilograms per cubic metre	kg/m^3 to lb/ft^3	0.0624
Speed	kilometres per hour	km/h to miles/h	0.621
	metres per second	m/s to ft/min	196.9
Flow rate	cubic metres per second (cumecs)	m^3/s to ft^3/min	2.118×10^3
	litres per second	litre/s to gal/min	13.2

With considerable amounts of imperial standard equipment still in use it will be necessary to convert from metric to imperial measure for many years to come. Accurate conversion will always be the best solution, though quick approximation may be of value to the farmer.

The following near equivalences are useful to remember and do not incur great error.

Length
25 mm = 1 in
300 mm = 1 ft
1 m = 1.1 yd
8 km = 5 mile

Area
1000 mm^2 = 1.15 in^2
1 m^2 = 10 ft^2
4 ha = 10 acre

Volume
100 cm^3 = 6 in^3
3 m^3 = 100 ft^3
3 m^3 = 4 yd^3
4 m^3 = 100 bushel
9 litres = 1 gal
1 m^3 = 220 gal

Rate of use and yield
4.5 kg/ha = 4 lb/acre
125 kg/ha = 1 cwt/acre
5 t/ha = 2 ton/acre
11 litre/ha = 1 gal/acre
25 mm over 1 ha = 250 m^3 =
1 acre in = 22 600 gal

Rate of flow
1 m^3/s = 2000 ft^3/min
1 litre/s = 13 gal/min

Common metric equivalences
1 litre = 1000 cm^3
1 ha = 10 000 m^2
1 hamm = 10 m^3
1 m^3 = 1000 litres
1 km^2 = 100 ha
1 MN/m^2 = 1 N/mm^2 (= 1 MPa)

Mass
30 g = 1 oz
1 kg = 2 lb
50 kg = 1 cwt
1 t or 1000 kg = 1 ton

Force and pressure
4.5 N = 1 lbf
10 kN = 1 tonf
7 kN/m^2 = 1 lbf/in^2

Heat energy
1 kJ = 1 Btu
3 W or J/s = 10 Btu/h

100 MJ = 1 therm

Index

abattoirs, 392
abortion, enzootic, 524
access to countryside, 57–9
accounts, 26–30
acetic acid, metabolism, 350
acetonaemia (ketosis), 358, 511–12
actinobacillosis, cattle, 509
actinomycetes, soil, 108–9
actinomycosis, cattle, 509
active immunity, 502
activity plans, 50
Acts of Parliament
  Agricultural Holdings Act (1986), 55
  Agriculture (Miscellaneous Provisions) Act (1976), 56
  Criminal Justice and Public Order Act (1994), 57
  Data Protection Act, 51
  Diseases of Animals Act, 505
  Employment Protection (Consolidation) Act (1978), 61–2
  Environmental Protection Act (1990), 53
  Food and Environment Protection Act (1985), 69
  Highways Act (1980) S164, 58
  insecticide, 297
  Occupiers Liability Act (1984), 57
  Petroleum Act (1928), 529
  Public Order Act (1986), 56
  Rent (Agriculture) Act (1976), 54
  Rights of Way Act (1990), 53, 58–9
  Town and Country Planning Act (1990), 59
  Trade Union Reform and Employment Rights Act (1993), 61–2
  Wild Life and Countryside Act (1981), 58–9
  see also Health and Safety at Work
Acyrthosiphum pisum (pea aphid), 295
agalactia syndrome, farrowing fever, 517
Agricultural Holdings Act (1986), 55
agricultural land
  access, 58–9
  classification, 78–9, 112
    crop suitability maps, 112
  development and use, legal constraints, 58–61
  occupation with/without security of tenure, 55–7
  occupier's liability, 57–8
  ownership, possession, occupation, legal aspects, 55–8
  reduced cultivations, 544
  see also agricultural law; alternative enterprises
agricultural law, 53–63
  agricultural law, 53–63
  development and use of land, 58–61
  earliest, 64
  employer's liabilities, 61–2
  English legal system, 53–4
  farm tenencies, 55–7
  occupier's liabilities, 57–8
  ownership, possession, occupation of agricultural land, 55–8
  see also tenure
Agriculture (Miscellaneous Provisions) Act (1976), 56
Agriotes (wireworms), 279, 284–95
Agrostis, grass, 195
Agrotis (cutworm), 285, 288
Allium ameloprasum see leeks
Allium cepa (dry bulb onions), 182
alsike, 202–3
Alternaria see black/sooty moulds; damping-off; leaf blight
Alternaria dauci (damping-off and leaf blight, carrots), 318–19
alternative enterprises, 71–85
  ancillary resources, 77
  defined, 71–2
  establishment, 79
  examples, 80–2
  management, 78–80
  marketing, 78
  monitoring and control, 79–80
  policy climate, 72–4
    economic, 72–3
    environmental, 73
    food and health, 73–4
    political, 73
  resource base, 78–9
  types
    alternative crops and stock, 76
    ancillary resources, 77
    horse enterprises, 82
    public goods, 77–8
    tourism and recreation, 74–6, 80–2
    value-added enterprises, 77
  see also enterprise planning; farm business management
amino acids
  characteristics, 339
  metabolism, 351–2
ammonium nitrate, 138
ammonium phosphate, 140
ammonium sulphate, 139
anaemia, piglets, 517
animal breeding *see* animal reproduction
animal diseases *see* diseases of animals
animal growth and development, 379–84
  body function regulation, 335–8
  chemical composition and food, 337–43
  endocrine system, 335–6
  environmental influences, 123–7
  genetic effects, 380–1
  growth promoters, 503
  hormonal control and manipulation, 382
  influencing factors, 380–2
  metabolism, 348–58
  nature, 379
  nervous system, 335–7
  nutrient requirements, 418–19
  nutritional effects, 381
  sigmoid curve, 379
  temperature, 123–4, 383–4

animal growth and development (*cont.*)
 thermal balance, 383
 see also animal nutrition
animal health, 497–524
 vs disease, signs, 503
 disposal of manure, 500
 floors, 499–500
 group size, 499
 health status, 499
 isolation facilities, 500
 livestock unit size, 498–9
 manure disposal, 500, 503
 unit depopulation, 498
 see also animal nutrition; diseases of animals
animal nutrition
 amino acids, 339, 351–2
 blood chemistry, 359
 carbohydrates, 340–1, 344–6, 348
 digestion biochemistry, 344–8
  carbohydrates, 344–6
  proteins, 346–8
 digestive tract
  anatomy, 343–4
  monogastric, 344
  pre-ruminant, 344
  ruminant, 344
 fatty acids, VFA, 350
 growth effects, 381
 lipids, 341–3, 350–1
 minerals (micronutrients), 343, 352–4, 419–20
  cobalt, 354, 522
  copper, 354–5
  essential elements, 352
  iodine, 355
  iron, 107, 133, 354
  licks and blocks, 414
  manganese, 355
  molybdenum, 355
  selenium, 355–6
  sodium, potassium, and chloride, 354
  sulphur, 354
  zinc, 355
 nutrient requirements, 414–20
 UME (utilized metabolizable energy), 211–12
 vitamins A–K
  B vitamins, 193
  metabolism, 356–8
 voluntary food intake, 360–1
 water, 343
 see also carbohydrates; fats; feeds and feeding; proteins
animal physiology
 environmental physiology, 382–3
 see also animal nutrition
animal reproduction, 361–73
 advances in reproductive technology, 373
 embryo transfer, 373
 fertilization, 370
 mating, 369
 nutrient requirements, 417–18
 pregnancy, 370–1
 reproductive development, 122, 370–1
animal reproductive system
 female, 361–6
 male, 366–9
animal welfare, 385–93, 504–5
 acceptable level, 392
 assessment, 389–90
 behavioural indicators, 391–2
 definition, 385–6
 environmental physiology, 384
 farm buildings, BS-5502, 576–7
 physiological and biochemical indicators, 390–1
 pig housing, 466
 poultry, 496
 production indicators, 391
anthrax, cattle, 515–16
antibiotic performance enhancers, poultry, 489, 496
antisera, 502
aphids
 cereal pests, 281–2
 see also pea and bean pests
Aphis fabae (black bean aphid), 286, 296
Apodemus sylvaticus (woodmouse), 286
appetite, sign of health and disease, 503
appraisal benefits, 50
Arable Area Payments Scheme, 152, 169
arable cropping, 150–93
 field constraints, 154
 see also crop physiology; *specific crops*
Arion hortensis (garden slug), 284
artificial insemination, 368–9
 genetics, pig improvement, 471–2
ascariasis, pigs, 521
Ascochyta see damping-off, seedling blight and root rots
Ascochyta fabae (bean leaf, stem and pod spot), 316
Ascochyta pisi (pea leaf, stem and pod spot), 316
ascorbic acid, 357–8
Atomaria linearis (pygmy beetle), 287
ATP, 348
atrophic rhinitis, 518
Aujeszky's disease, 519

babesiasis, 510
Bacillus anthracis, 500, 515
bacteria
 soil, 108
 types, 501
badgers, 515
balance sheet projections, assessment, 26–7
balers
 bale handling
  conventional bales, 556–7
  conventional pickup baler, 556
 baled silage, 555–6
 large rectangular high density, 558
 large round, 557–8
barley
 diseases, 305–7
  of ear, 307
  of leaf and stem, 305–7
  of root and stem base, 307
 eyespot, 307
 in feeds and feeding, 410
 halo spot, 307
 net blotch, 306
 quality standards, 157
 see also cereals; barley yellow dwarf virus
barley yellow dwarf virus, 161–2, 299
 barley, 306
 oat, 308
 vector (bird cherry aphid), 282
 wheat, 303
beans *see* pea(s) and bean(s)
beans, pests and diseases *see* pea and bean pests and diseases
beef production, 432–7
 Beef Special Premium, 12
 CAP price mechanisms, 12
 cereal beef, 434–6
  18–month grass beef, 435–6
 grass silage beef, 436
 selection for slaughter, 433–4
 suckled calf production, 436–7
 systems, 434–7
 see also calves; cattle
beet
 sheep stocking rates, 453
 weed beet, 175, 262–8
 see also sugar beet
beet cyst nematode, 175, 289–90
beet flea beetle, 287
beet rust, 315
Beta vulgaris see beet; sugar beet
bird cherry aphid, vector of BYDV, 282
black bean aphid
 pea and bean pests, 296
 sugar beet pests, 289
black bean aphid (*Aphis fabae*), 286, 296
black disease, 522
black medick *see* trefoil
black rot, carrots, 318
black scurf and stem canker, potato, 312
black/sooty moulds, 307
 wheat, 305
blackleg
 cattle, 508
 potato, 310
 sheep, 522
 sugar beet, 315
blackquarter, cattle, 508
blenders, 569
bloat, mature cattle, 510–11
blood chemistry, nutritional status, 359

'bloody gut' (intestinal haemorrhage syndrome), 519
'bloody scours' (swine dysentery), 519
blowfly, fly-strike, 523
blown, 510
'blue-eared' pig disease (porcine reproductive and respiratory syndrome), 520
body condition scores, 463
bone meal, as animal protein supplements, 413
borage, 193
Bordetella bronchisepta, 518, 519
boron, 107, 132–3
  deficiency symptoms, 132–3
  sources, 107
  toxicity symptoms, 133
borrowing, capital costs, 33–4
Botrytis allii (neck and storage rots, onions and leeks), 318
Botrytis cinerea (storage rots, carrots), 318
Botrytis fabae (leaf, stem and pod spots, peas and beans), 316
Botrytis (grey mould/pod rot, peas and beans), 317
bovine piroplasmosis, 510
bovine rhinotracheitis (IBR), 509
bovine somatotrophin, 377
bovine spongiform encephalopathy (BSE), 413, 514
bovine virus diarrhoea (mucosal disease), 509
bowel oedema, growing pigs, 517–18
brain and nervous system, 335–7
Brambell Committee, 385, 386
brashing, new farm woodland, 245
Brassica campestris (turnips), 192
Brassica juncea (condiment mustard), 193
Brassica napus
  forage rape, 191–2
  swede, 181–2, 191
Brassica oleracea
  calabrese, 185–6
  cauliflowers, 183–5
  kale, 191
  see also cabbage
brassicas
  diseases and pests, 291–4, 312–15
    brassica pod midge, oilseed rape, 293
    cabbage aphid, 292
    canker, 313
    cauliflower mosaic virus, 314
    clubroot, 314
    flea beetles
      kale, 293–4
      oilseed rape, 291
    powdery mildew, 312–13
braxy, 522
Brevicoryne brassicae (cabbage aphid), 292
brewing by-products, 411
bridle-paths, 58–9
British Standard (5502)
  farm buildings, 576–7
  water pollution, 598–9
broad bean rust, 317
broadcasters, 544–5
broadleaves (non-woody)
  annuals, 253
  herbicidal weed control, 262
  perennials, 253
broilers *see* poultry
bromes, forage, 202
brown foot rot and ear blight
  barley, 307
  wheat, 304
brown rust
  barley, 306
  wheat, 303
brucellosis
  cattle, 514
  sheep, 524
BSE (bovine spongiform encephalopathy), 514
BST (bovine somatotrophin), 377
budgets
  capital, 25–6, 29–30
  cash flow, 23–5, 27–9
  enterprise, 19–22
  outputs and costs estimation, 22–3
  profit, 19–23, 29–30
  profit assessment, 23
bulls, Wild Life and Countryside Act (1981) S59, 58
bunt, wheat, 161, 305
business management *see* farm business management
butyric acid, metabolism, 350

cabbage
  nutrient requirements, 186
  spring, summer, autumn and winter, 186–7
  see also brassicas
cabbage aphid, oilseed rape pests, 292
cabbage ringspot, brassicas, 314
cabbage seed weevil, oilseed rape pests, 293
cabbage stem flea beetle
  kale pests, 294
  oilseed rape pests, 291
cabbage stem weevil, oilseed rape pests, 292
cables, 528
calabrese, 185–6
calciferols, 357
calcium, 132
  fertilizers, 209
  and magnesium, 106–7, 132
    soil reactions, availability and losses, 106–7
    sources, 106
  metabolism, 353
calcium ammonium nitrate, 139
calcium borogluconate, 511
calcium cyanamide, 139
calves
  accommodation, 602–3
  colostrum, protection against diseases, 505
  dentition, 424
  diseases
    calf scours, 505
    diphtheria, 506
    navel-ill, 506
    salmonellosis, 505–6
    tetany, 512
    see also cattle, diseases
  rearing, 424–5
    early weaning, 425–6
    single, double, multiple suckling, 426
    suckled calf production, 436–7
    veal production, 12, 426–7
canker, brassicas, 313
Cannabis sativa (hemp), 192
CAP *see* Common Agricultural Policy
capital
  budgets, 19–30
  costs of borrowing, 34
  recording and analysis, 30
  sources, 32–3
carbohydrates
  animals, 341
  biochemistry in animal digestion, 344–6
  characteristics, 340–1
  classification, 340
  glucose metabolism, 348–50
  plant tissues, 340
carbon:nitrogen ratio, 110–11
carbon cycle, 109–10
carcass quality and classification
  changes in, 380
  pig, 481–3
  stress before slaughter, 390
carotenoids, 356
carrot diseases, 317–18
  carrot motley dwarf virus, 318
  storage rots, 318
carrots, 181
cash flow
  budget assessment, 25
  monitoring, 27–9
  see also budgets
cassave, 409
cattle, 421–44
  accommodation, 601–6
    beef animals, 603–4
    calves, 602–3
    dairy cows, 601–2
    milk production buildings and bull pens, 604–6
  breeding, 361–73
    infertility, 513–14
    management of stock replacement, 427–9
    target weights for two-year calving, 427
  breeds and breeding
    disease problems, 513–14

cattle (*cont.*)
heritability of characters, 443
progeny testing, 443
stock replacement management, 427–9
susceptibility to disease, 513–14
target liveweights, 427
CAP price mechanisms, 12
definitions and terminology, 422–3
diseases
calf scours, 505
coccidiosis, 508
dysentery, 508
enzootic bovine leucosis (EBL), 516
external parasites and skin conditions, 508–10
internal parasites, 506–7
lumpy jaw, 509–10
mange, 508
mastitis, 512–13, 514
milk drop syndrome, 513
parasitic gastroenteritis, 507–8
rabies, 515
redwater, 510
respiratory diseases, 509
ringworm, 508
salmonellosis, 505–6
tuberculosis, 515
wooden tongue, 509–10
feeds and feeding
beef cattle, 429–32
breeding stock, 429–30
dairy cattle, 429–32, 437–43
diet formulation, 431–2
flat-rate feeding, 439
growing/finishing, 430–1
production systems, 432–7
health, 505–17
heifer calving
autumn-born, two years old, 428
spring-born, two years, 428–9
infertility, 513–14
normal body temperature, 504
normal respiration rate, 504
pests, warble fly, 516
population (1991), 423
slaughter, selection, 433–4
Suckler Cow Premium, 12, 60
trends, 421–2
UK output 1984, 1992, 421
see also beef production; calves; dairy cattle
cattle lice, 516–17
cauliflower mosaic virus, 314
cauliflowers, 183–5
centrifugal pumps
irrigation, 562
controls, 562–3
selection, 563
cereal pests
bird cherry aphid, 282
cereal cyst nematode, 281
control, 165
dead heart, 161
establishment pests, 161
field slug, 161, 280–2
frit fly, 161, 278–9
grain aphid, 281–2
leatherjackets, 280
rose grain aphid, 282
wheat bulb fly, 161, 277–8
wireworms, 279
yellow cereal fly, 161, 278
cereals, 157–66
desiccation, 165
disease control, 164
establishment diseases, 161–2
grain and by-products
feeds and feeding, 409–12
markets, 157
minor cereals, 166
straw, 408
growth cycle, 157–8
herbicidal weed control, 258–62
marketing, 157
CAP price mechanisms, 9–10
nutrient requirements, 162–6
plant growth regulators, 164–5
preparation and processing, 410
seeds
dressings, 161
population and seedrate, 158–9
seedbed, 158
sowing date, 160–1
sowing depth, 161
silage from, 220
soil types and rooting, 158
weed control, 162, 258–62
Zadok's decimal code for growth, 159
Zeleny index, 9
see also named crops
Ceutorhynchus assimilis (cabbage seed weevil), 293
Ceutorhynchus picitarsis (rape winter stem weevil), 291
Ceutorhynchus quadridens (cabbage stem weevil), 292
Chaetocnema concinna (beet flea beetle), 287
chases and conduits, 530
chemical weed control *see* herbicidal weed control
chlamydial infections, cattle, 524
chlorine
animal nutrition, 354
plant nutrition, 134
chlormequat, 167
Chorioptes, 508
Christmas trees, 230
citrus pulp, high energy feeds, 412
Cladosporium herbarum (black/sooty moulds), 307
clear felling, 228
clostridial diseases
cattle, 508
growing pigs, 518
sheep, 521–2
vaccines and sera, 522
clover
and bloat, 511
crimson clover, 203
nitrogen additions, 206–7
red clover, 202–3
white clover, 200, 202
clubroot, brassicas, 314–15
coaching, 50
cobalt
animal metabolism, 354
deficiency, 522
soils, 134
coccidiosis, cattle, 508–9
cocksfoot grass, 201
colibacillosis *see Escherichia coli* infection
Colletotrichum (damping-off, seedling blight and root rots), 316
Colletotrichum lindemuthianum (leaf, stem and pod spots, peas and beans), 317
colostrum
for calves, 505
protection against diseases, 424–5, 505
Columba palumbus (pigeon), 291–2
combinable crops, 157
combination seedbed-preparation machines, 544
combine harvester, 558–60
Common Agricultural Policy
background, institutions and legislative process, 1–4
environment, 15–16
European Union, 1–15
formation and development, 5–6
monetary arrangements, 13–14
structural and environmental policy, 14–15
origins, 6
price mechanisms
beef and veal, 12
cereals, 9–10
eggs and poultrymeat, 13
fruit and vegetables, 13
milk and milk products, 11–12
oilseeds, 10–11
original, 6
other products, 13
pigmeat, 13
protein crops, 11
sheepmeat, 13
sugar, 11
problems, surpluses, 7–8
problems and developments, 16
reformation, 8–9
common scab, potato, 311
composts, 138
compressed air, 527
conifers, pruning, 245–6
conservation
and drainage, 100
forest management, 249
Wild Life and Countryside Acts

(1981 and 1985), 60
contagious pustular dermatitis (orf), 522
Contarinia pisi (pea midge), 296
continuous grazing, 214–15
Control of Pesticides Regulations, 60, 256
conversion factor, metric units and 'imperial', 626–7
Cooperia, 507
copper
animal metabolism, 354–5
deficiency symptoms, 133, 522
metabolism, swayback, 355, 522
plant nutrition, 133
in soils, 107
toxicity symptoms, 133
toxicity to sheep, 137
coppice, 229
short-rotation for firewood and biomass, 229–30
with standards, 229
stored, 229
corpus luteum, 363–6
Corynebacterium mastitis, 513
COSHH Regulations, 60
coughing, in disease, 504
covered smut, barley, 307
craft products, 76–7
Criminal Justice and Public Order Act (1994), 57
crimson clover (trifolium
crop diseases *see* diseases of crops
crop health, 252–320
diseases, 298–320
pests, 277–98
crop nutrition, 130–49
see also crop physiology; soil
crop physiology
competition, 127–9
germination, 121–2
growth, 120
light effects, 124–7
pH tolerance, 147
photosynthesis and respiration, 118–20
plant growth regulators, 129
processes, 118–20
seed development, 122–3
senescence, 123
temperature effects, 123–4
vegetative development, 122
yield, 120
Zadok's decimal code for growth, 159
see also soil; *specific nutrients*
crop rotation, 153–4
crop sprayers, 549–51
components, 550–1
safety, 551
servicing, 551
sprayer calibration, 551
sprayer nozzles, 550
crop storage, 614–19
cropping systems, 152–3
crops
alternative range, 76
energy, 192–3
fibre, fuel and non-food, 192–3
medicine, flavouring and minor, 193
crown rust
oat, 308
rye grasses, 319
cubers, 569
cultivation, strength of soil, 93
cultivation machines, 541–3
cultivators, 542–3
cutworms
potato pests, 285
sugar beet pests, 288
Cydia nigricana (pea moth), 295
Cylindrosporium concentricum (light leaf spot, brassicas), 313
cyst nematodes
beet cyst nematode, 175, 289–90
cereal cyst nematode, 281
pea cyst nematode, 294–5
potato cyst nematode, 283–4

dairy cattle *see* cattle
dairying
milk production buildings, 604–6
trends, 437
Damalinia, 516
damping-off
forage maize and sweet corn, 309
and leaf blight, carrots, 317
peas and beans, 316
seedling blight and root rots, linseed, 317
Dasyneura brassicae see brassica pod midge
Data Protection Act, 51
Daucus carota see carrots
deaths, fatal injuries (1986–92), 65
decisions, European Union, 5
definitions and terminology, cattle, 422–3
Delia coarctata (wheat bulb fly), 277–8
Demodes, 508
Deroceras reticulatum (field slug), 161, 280–2
development cycle, 121–3
development of land, legal constraints, 59–60
'diamonds' (swine erysipelas), 419
diarrhoea
calf scours, 505
mucosal disease of cattle, 509
post-weaning, pigs, 517
Dictyocaulus viviparus, lungworms, 506–7
diet *see* feeds and feeding
digestibility measurements, 394–5
digestion, 343–8
see also animal nutrition
digestive tract, anatomy in animals, 343–4
diphtheria, 506
Directives, European Union, 5
disc harrow, 543
disc plough, 542
disciplinary procedures, 45
dismissal, 46
fundamental steps, 46
investigation, 45
retrospective action, 46
warning
conduct, 46
performance, 46
diseases of animals
challenge, minimizing by design, 499–500
control
depopulation, 499
floors, 499–500
group size, 499
health status, 499
isolation facilities, 500
and immunity, 501–5
legislation, 505
types, 497–8
diseases of crops, 298–320
air-borne, 298
seed-borne and inflorescence diseases, 301
soil-borne, 300–1
vector-borne, 301–2
disinfection, importance, 503
distillery by-products, 411
diversification *see* alternative enterprises
Diving Operations at Work Regulations (1981), 68
docking disorder, 290
dogs, attacking livestock, 56
downy mildew
brassicas, 313
onions and leeks, 318
peas and beans, 316
sugar beet, 315
drainage, 94–100, 531–2
and conservation, 100
economics, 100
irregular/random systems, 99–100
maintenance of drains, 100
materials, 96–7
needs and benefits, 94–5
secondary drainage, 97–9
systems
design, 99–100
site investigation, 99
moling, 97–8
open ditch, 95
pipe drainage, 95–9
subsoiling, 98–9
see also soil
drugs (medicines), 502–3
dry bulb onions, 182
dry rot, potato, 311
ducks, 493–4, 495
durum wheat, 166

earthworms, 109
EBL (enzootic bovine leucosis), 516
ECU
  defined, 13
  switchover coefficient, 14
education, learning aids, 49
egg production
  CAP price mechanisms, 13
  commercial, 489–91
  free range, 489–90
electricity
  conductor data, 528
  farm services, 527–9
  private supply, 529
ELISA test, milk progesterone, 373
embryo transfer, 373
embryonic and fetal mortality, 371
  sheep, 450
employers, Health and Safety at Work (etc.) Act (1974), 66–7
employer's liabilities, 61–2
Employment Protection (Consolidation) Act (1978), 61–2
endocrine system, 335–6
  see also hormones
energostasis, limits and mechanisms, 360
energy *see also* animal nutrition; feeds and feeding
energy fermentable metabolizable energy, 347, 396
English legal system, 53–4
enterovirus infections, pig, 520
enterprise budgets, 19–22
enterprise planning, 154–6
  see also alternative enterprises
environment
  animal welfare, 384
  Common Agricultural Policy, 15–16
  control, horticulture, 592–3
  environmental statement, water pollution control, 599
  grassland farming, 223–4
    effect on botanical composition of swards and soil, 224
    nitrate losses, 224
Environmental Protection Act (1990), 53
Environmentally Sensitive Areas (ESAs), 77–8, 153
enzootic abortion, sheep, 524
enzootic ataxia, sheep, 522
enzootic bovine leucosis (EBL), 516
enzootic pneumonia, in pigs, 518–19
epidemic diarrhoea, pigs, 520
Erwinia (blackleg), 310
Erysipelothrix, 519
Erysiphe cruciferarum (powdery mildew, brassicas), 312–13
Erysiphe graminis (powdery mildew, grasses and herbage legumes), 320
Escherichia coli infection
  cattle mastitis, 513
  pigs, 517
  sheep, 521
ethephon, 167
ethnic products, 76
European Agricultural Guidance and Guarantee Fund, 6, 14–15
European Commission, 4
European Council, 4
European Monetary System, Exchange Rate Mechanism, 14
European Parliament, 4–5
European Union
  decision making, 4–5
  origins and development, 5–6
  see also Common Agricultural Policy
evening primrose, 193
Ewe Premium Rights, 60
excreta
  in disease, 503
  disposal, 500
exudative epidermitis (greasy pig disease), 518
eyespot
  barley, 307
  wheat, 303–4

factory farming, 385, 496
Farm Animal Welfare Council, 385
farm buildings, 575–600
  British Standard (5502), 576–7, 580–91
  construction
    frame and related cladding, 582–3
    groundworks, 581–2
    internal finishing and services, 583–4
    masonry cladding and floor laying, 583
    stages, 580–1
  design criteria, 575
  environmental control, 592–7
  gases, 500–1
  insulating methods, 594
  lighting, 530–1
  materials and techniques
    concrete block walling, 585–6
    drainage, 591
    foundations, 589–91
    mass concrete, 584–5
    non-timber lightweight cladding, 588–9
    particle boards and composites, 588
    pre-cast concrete, 585
    reinforced concrete, 586–97
    timber, 587–8
  planning and design, 575–6
  planning permission and regulations on water pollution control, 598
  servicing and general, 622–4
  site preparation, 580
  specific design, 576–7
  specific purpose, 601–25
  storage
    concentrates and root crop feed material, 620–2
    fuel storage, 624–5
  ventilation methods, 596–7
farm business management, 17–37
  accounts, 27
  balance sheet projections, assessment, 26–7
  budgets, 19–30
  cash flow, budget assessment and recording, 25–9
  definition, 17
  financial measures, 18
  management by objective, 50
  marketing, 36–7
  planning change, 30–2
  pricing, 37–8
  primary objectives, 17–18
    planning, 18–19
  profit and capital recording, 29–30
  promotion, 37
  records, 30
  risks, 34–6
  shared ownership, 34
  see also alternative enterprises; budgets; capital
farm machinery, 536–74
  balers and bale handling, 555–8
  broadcasters, 544–5
  centrifugal pumps for irrigation, 562–3
  crop sprayers, 549–51
  cubers, 569
  cultivators, 542–3
  disc harrow, 543
  feed preparation machinery, 568–9
  grinding machinery, 568
  hammer mills, 568
  harrows, 543
  harvesting machinery
    grain harvesting, 558–60
    green forage crops, 552–8
    root crops, 561
  mixers, 569
  mower-conditioners, 552–3
  mowing and swath treatment machines, 552
  planting machines, 545–9
  plate mills, 568
  plough, disc, 542
  ploughs, 541–2
  potato harvesters, 561
  potato planting machines, 548
  PTO-driven secondary cultivation machines, 543
  roller mills, 568
  rollers, 543–4
  rotary cultivator, 542–3
  seeders, 547–8

spreaders, full-width spreaders, 545
spring-tine cultivator, 543
storage of farm machinery, 624
subsoilers, 541
sugar beet harvesters, 561
swath treatment machines, 553–4
transplanters, 548–9
work days, 93–4
see also irrigators; milking; pumps; tractor
farm office accommodation, 625
farm roadways, and external works, 591–2
farm staff
appraisals, 49–50
discipline, 45
discrimination, 44
dismissal, 46
fatal injuries (1986–92), 65
feedback, 50
gang-work day charts, 40
grievances, 47
house rules, 45
ill health, 46
incentives, 51
industrial relations, 47
industrial tribunals, 46
interviewing, 42–7
job analysis and design, 39–40
management, 39–51
control, guidance and negotiation with staff, 44–7
maintaining and increasing performance, 49–50
poor standards, 51
poor standards, 51
recruitment and pay, 41–2
retirement, 46
skills audit, 47–8
teams, 51
training and development, 47–8
Work Equipment Regulations (1992), 68
farm tourism, alternative enterprise model, 80–2
farm woodland *see* forest and woodland
Farm Woodland Premium Scheme (MAFF), 250
farmyard manures *see* manures
farrowing fever (mastitis), 517
fats
FGAF, 411
metabolism, animals, 350–1
and oils
high energy feeds, 411–12
oilseed residues, 412
fatty acids, 342–3
VFA metabolism, 350
VFA in rumen liquor, 346
feeds and feeding, 394–420
chemical composition and nutritive value, common foodstuffs, 404–7
complete diets, cattle, 439
concentrate and compound feeds, 399–401
annual consumption by type of stock, 569
storage, 620
diet formulation
daily gains target, 431–2
raw materials, 401–2
digestibility, neutral detergent cellulase + gammanase, 394–5
energy
energostasis, 360
UME (utilized metabolizable energy), 211–12
energy measurement, 395
gross value, 395–6
loss, 375
metabolizable and fermentable metabolizable, 347, 396
net energy value (NE), 396–7
energy requirements, 414–16
evaluation, 394–401
feed preparation machinery, 568–9
feed supplements, 414
micronutrients, 399–401
chemical analysis and prediction, 399
near infrared reflectance analysis, 401
nutrient requirements, 414–20
protein evaluation of feeds, 397–9
rations, performance prediction, 431
raw materials, 401–2
voluntary food intake, 360–1, 456
see also animal nutrition; cereals; grain; haymaking; *other specific feeds*; silage, grass and grasslands, grazing
FEOGA *see* European Agricultural Guidance and Guarantee Fund
FEPA (Food and Environment Protection Act) (1985), 69
fertility, male, 369
fertilization, 370
fertilizers
application machines, 544
calculations, 147
compound, 143
environmental considerations, 144
grassland, 206–9
inorganic, 138–43
forms available, 141–3
liquid manure, 572
N additions, 104–5
optimum rates, 144–5
organic, 134–8
placement, 143–4
recommendations, 146–7
storage, 622
see also manures; *specific mineral nutrients*; *specific types*
Festuca
meadow fescue, 201
red fescue, 195
sheep's fescue, 194
tall fescue, 511
fetal development, 370
fetal mortality, 371
fibre crops, 192–3
financial management
accounts, 27
budgets, 19–30
see also farm business management
fire
fire regulations, 51–2
fire-fighting, 534
forest protection, 246
portable appliances, 535
first aid, 52
Health and Safety (First-Aid) Regulations (1981), 68
fish meal
animal protein supplements, 413
price, 465
flavouring crops, 193
flax *see* linseed
floors, and animal health, 499–500
flora and fauna, Habitats Directive (1994), 60
flukes, 507, 523
fly larvae, grain damage, 161
fly-strike, blowfly, 523
fodder beet diseases *see* sugar beet, diseases
food
meat consumption per head (UK), 422
public concerns, 74
see also feeds and feeding
Food and Environment Protection Act (1985), 69
foot-and-mouth disease, 514–15
foot-rot
plant crops *see* damping-off
sheep, 523–4
footpaths, 58–9
forage crops, 174–81, 187–93
beet and mangels, 190–1
harvesting machinery, 552–8, 561
SLF wagons (self-loading forage wagons), 555
see also green forage crops; root crops
forage harvesters, 554–5
double-chop, 554
precision-chop, 554–5
forest and woodland
coppice, 229–30
diseases, 247
grants
Farm Woodland Premium Scheme (MAFF), 250
Woodland Grant Scheme (Forestry Authority), 16, 240
harvesting, 234–5

forest and woodland (*cont.*)
  inspection, new farm woodland, 245
  integration with agriculture, 249–50
  landscape design, 247–8
  management, 227–51
    grants, 250
    investment, 249
    non-timber uses, 248–9
    objectives, 227
  measurement, 230–2
    volume of a stand, estimating, 231–2
    volume of standing tree, estimating, 231
  natural regeneration, 228–9
  neglected
    enrichment, 230
    improvement, 230
    replacement, 230
  new, 235–46
    aftercare, 244–6
    beating up, 243
    brashing, 245
    cleaning, 245
    climate and topography, 235
    drainage, 241
    fencing, 241
    fertilizer application, 243–4
    herbicides, 244–5
    inspection racks, 245
    inter-row vegetation management, 245
  plant handling, 242
    plant origin, 241
    planting stock, 241–2
    pruning, 245–6
    pure crops and mixtures, 240
    site preparation, 240, 241
    site and species selection, 235–40
    soil and ground vegetation, 235
    spacing, 242
    time of planting, 242
    tree shelters, 243
  products and markets, 227–8
  protection, 246–7
  species characteristics, 236–9
  thinning, 232–4
    cycle, 233
    intensity, 232–3
    no-thin systems and delayed thinning, 234
    type, 233
    yield, 233
  weed control, 270–6
    frill girdling, 275
    handwork, 270–4
    herbicidal treatments, 274
    large tree removal, 275–6
    mechanical, 274
    methods, 270–6
    new plantings, 244
  weed types, 270
  see also trees
foul-in-the-foot, 517
free range egg production, 489–90
frit fly, 278–9
frost protection, by irrigation, 567
fruit crops, frost protection, 567
fuel
  electricity, 527–9
  energy values, various fuels, 530
  fuel crops, 192–3
  fuel oil
    farm services, 529
    Regulations, 61
  liquefied petroleum gas (LPG), 529
  natural gas, 529
  petrol, 529–30
  storage, 624, 624–5
  farm buildings, 624–5
fungi, soil, 109
fungicides, cereals, 166, 300
Fusarium
  brown foot rot and ear blight, 304, 307
  damping off, 317
  stalk rot, 309, 316
Fusarium oxysporum (pea wilt), 169, 316
Fusarium solani (f.sp. *caeruleum*, dry rot), 311
Fusiformis nodosus, foot rot in sheep, 523–4
Fusibacterium necrophorum
  calf diphtheria, 506
  foul-in-the-foot, 517

Gaeumannomyces graminis (take-all whiteheads, barley), 307
gang-work day charts, 40
gangrene, potato, 311
gas, liquefied petroleum gas (LPG), 529
gases, farm buildings, danger, 500–1
gastroenteritis
  cattle, 507–8
  pigs, 517
gates, 58
geese, 495–6
General Agreement on Tariffs and Trade (GATT), 7–8, 73
germination, 121–2
Gibberella (stalk rot), 309, 316
gleys, 111
Globodera pallida and *G. rostochiensis* (potato cyst nematodes), 283–4
glossary of units, 626–7
glucose metabolism, 348–50
goat, breeding, 462–3
goats, 460–3
  body condition scores, 463
  breeding, 462–3
    intersexuality, 463
  feeds and feeding, 461–2
  fibre production, 461
  meat production, 461
  milk production, 460–1
  and sheep, 445–63
grain aphid, 281–2
grain handling, 329–30
  auger conveyors, 329
  belt conveyors, 329
  bucket elevator, 329–30
  chain and flight conveyors, 329
  grain cleaners, 328
  grain thrower, 330
  harvesting machinery, 558–60
  loading lorries, 330
  pneumatic conveyors, 330
  weighers, 330
grain harvesting, 558–60
grain preservation and storage, 321–31, 614–15
  aeration, 328
  agencies causing deterioration, 321–2
  chemical preservatives, 331
  cooling, 328
  drying, 323–7
    air supply and distribution, 326
    axial-flow fans, 327
    centrifugal fans, 327
    continuous flow driers, 324–5
    ducts and drying floors, 327
    high-temperature driers, 324–5
    low-temperature driers, 325–6
    measurement of air relative humidity, 326
  handling grain, 329–30
  hot spot elimination, 328
  management, 328
  moisture content and measurement, 322–3
  monitoring temperature, 328
  safe conditions, 321–2
  sealed containers, 330–1
  storage, 614–15
  time factor, 322
grain sowing, drills, 545–7
grain storage *see* grain preservation and storage
grants, farm woodland management, 16, 240, 250
grass staggers (hypomagnesaemia tetany), 359
grass storage, 616
grasses and grasslands, 194–226
  agricultural, characteristics, 200–2
  characteristics of grasses, 200–3
  conservation methods, 216–23
    acidification, 216
    conserved grass feeds, 403–8
    dehydration, 216–17
    finishing of stores, 437
    freezing, 217
    grass digestibility, 212
    see also feeds and feeding; haymaking; silage
  diseases, 319–20
  distribution and use, 195–6

efficiency of use, 212
establishment, 198–9
fertilizers, 206–9
Festuca spp., 194, 195, 201, 511
glossary of terms, 224–6
grazing, 402–3
animal output experiments, 213
block grazing systems, 215
buffer grazing and feeding, 216
continuous, 214–15
full, 214–15
grazed herbage, 402–3
management for sheep, 458
monitoring availability, 213–14
mountain, 194
nutrient requirements, 206–9
output from grazing animals, 213–14
paddock, 215–16
rough, 194, 195
set-stocking, 214
for sheep, management, 458
strip, 216
systems, 214–16
three-field system, 215
zero, 216
see also green forage
herbage seed production, 203
improvement, 196–200
nature of growth, 196–8
permanent, 195
production
effect of site class, 210–11
output expression, 211–12
patterns, 210–11
stocking rate method, 211
utilized metabolizable energy, 211–12
rotational, 195
seed mixtures, 205
seed production, 203–5
seed rates, 204
sheep stocking rates, 453
weed control, 268–9
see also green forage
greasy pig disease, 518
green forage crops
drying, 223
feeds and feeding, 409
harvesters, precision-chop, 554–5
harvesting machinery, 552–8
herbage legumes, 202–3
lucerne, 189–90
maize, 187–9
seed production, 203–5
sheep stocking rates, 453
see also grass and grasslands; legumes
green manures, 154
green pound, 14
grey mould/pod rot, peas and beans, 317
grinding machinery, 568
growth *see* animal growth and development; crop physiology

habitats, public goods, 77–8
Habitats Directive (1994), 60
Haematopinus (lice), 516
Haemonchus, 523
halo blight, dwarf and navy beans, 317
halo spot, barley, 307
hammer mills, 568
harrows, 543
harvesting machinery
grain harvesting, 558–60
green forage crops, 552–8
hay, straw, grass storage, 616
haymaking, 221–3
barn hay drying, 222
facts and figures, 223
hay additives, 222–3
see also grass and grasslands, conservation methods
headfly, 523
'health products', 76
health and safety, 52, 64–9
tractor driver, 541
Health and Safety at Work (etc.) Act (1974), 64–6
agriculture activities, regulations, 67–8
enforcement, 67
septic tanks, 532
statutory requirements, 66–7
Health and Safety at Work Regulations (1992)
Display Screen Equipment Regulation (1992), 68
management, 68
Health and Safety Commission, policy, 66
Health and Safety Executive
operational staff, 66
technological, science and medical support, 66
Health and Safety (First-Aid) Regulations (1981), 68
health of stock *see* animal health
heart disease, mulberry, in pigs, 519
heating, 530
see also fuel
hedges, 73
Helicobasidium brebissonii (violet root rot, carrots), 318
Helminthosporium solani (silver scurf, potato), 312
hemp, 192
herbage seed production, 203
herbicides, 253–8
application
foliar, 254
soil or residual applications, 254–5
spray application, 256
choice, 255–7
annual weeds, 259–60
broad spectrum, 255
established grassland, 268–9
forest and woodland, 270–3
major broadleaved crops, 265–7
non-crop situations, 275–6
perennial weeds, 258
root crops, 262
selective control, 261, 263–4
volunteer crops as weeds, 267
cost of treatment, 256
farming system and cropping programme, 255
integration with those used on other crops, 255
programmes, 255
purchase and use, 256–7
selective treatments, 254
stage of growth of crop and weed, 255
trees
farm woodland, 244–5
forest and woodlands, 274
see also weed control
herpes virus, pigs, 519
Heterodera avenae (cereal cyst nematode), 281
Heterodera goettingiana (pea cyst nematode), 294–5
Heterodera schachtii (beet cyst nematode, 289–90
highways, 58–9
Highways Act (1980) S164, 58
barbed wire, 58
hog cholera (swine fever), 519
hoose (*Dictyocaulus viviparus*), 506–7
hormones
animals, reproductive, 364–6
control and manipulation of growth, 382
endocrine system, 335–6
list, 338
mammary development, 375
plants, 129, 164–5
pregnancy, assay, 371
stress indications, 390
horse enterprises, alternative enterprise model, 82
house rules, 45
housing of stock *see accommodation* under *named animals*
hoven, 510
HRG *see* ryegrass
HSW *see* Health and safety
HSW fatal injuries (1986–92), 65
HSW Act *see* Health and Safety at Work (etc.) Act (1974)
husk, 506–7
hybrid companies, genetics, pig improvement, 470
Hyostrongylus, 521
hypocalcaemia, 511
Hypoderma, warble fly, 516
hypomagnesaemia, 132, 136, 359
calves, 512
mature cattle, 512

IBR (infectious boving rhinotracheitis
immunity, animal health, 501–2

immunization, ovarian steroids, 373
implantation, 370
incentives, 51
industrial relations, negotiation and bargaining, 47
industrial tribunals, 46
infections, causative organisms, 501–4
infectious bovine rhinotracheitis (IBR), 509
influenza, pigs, 518
injuries, fatal injuries (1986–92), 65
(near) infrared reflectance analysis, 401
insecticides, schemes, Regulations, and Acts of Parliament, 297
insects, forest protection, 246–7
insulating methods, farm buildings, 594
insurance, employer's liabilities, 61–2
integrated crop management systems, 153
Intervention Board for Agricultural Products (IBAP), 6
interviewing
  and discrimination, 42–7
  documentation, 43
  information acquisition, 43
  information supply, 43
  preparation, 42–3
  questions, 43–4
  selection, 42
  structure, 43
  telephone screening, 42
  welcome, 43
intestinal haemorrhage syndrome ('bloody gut'), 519
iodine, animal metabolism, 355
IRG (Italian ryegrass) *see* ryegrass
iron, 107, 133
  animal metabolism, 354
  deficiency symptoms, 133
irrigation, 562–7
  frost protection, 567
  organic liquids, 567
  scheduling, 566–7
  solid-set sprinklers, 564
  system capacity, 562
  water storage, 562
irrigators
  centre pivot and linear-move, 565–6
  hosereel, 565
  mobile, 564–5
isolation facilities, 500
Ixodes (ticks), 510

job analysis and design, farm staff, 39–40
joint-ill, 506

kale, 191
  brassica flea beetles, 293–4
  cabbage stem flea beetle, 294
ketosis, 358
  mature cattle, 511–12

lactation, 373–8
  lactation curves, cattle, 438
  nutrient requirements, 415–17
  and weaning, pigs, 476–7
  see also milk
lactational anoestrus, 372–3
land, *see* agricultural land
law *see* agricultural law;, *see also* Common Agricultural Policy
leaf blotch
  barley, 306
  rye grasses, 320
leaf growth, 122
leaf spot
  and seedling blight, oak, 308
  stem and pod spots, peas and beans, 316
  sugar beet, 315
  wheat, 303
leaf stripe, barley, 161, 306
learning aids, induction, 49
leatherjackets
  potato pests, 280
  sugar beet pests, 288
leeks, 182–3
  diseases, 318–19
    leek rust, 319
legal system, English, 53–4
legumes, 170–4, 187
  CAP price mechanisms, 11
  diseases, powdery mildews, 320
  herbage, diseases, 319–20
  lucerne, 189–90
  nutrient requirements, 187
  seeds, feeds and feeding, 412–13
  silage from, 220
  see also peas and beans
Leptosphaeria maculans (canker, brassicas), 313
Leptosphaeria nodorum syn. *Septoria nodorum*, 303
Leptospira, survival, 500
Leptospira infection mastitis, 513
lice
  cattle, 516–17
  pigs, 521
light, photoperiod, 124
light capture by crop canopies, 124–5
  canopy structure, 125–6
  depression problems, 126–7
  factors affecting, 125
  nitrogen fertilizers, 125
  seedrate increasing, 125
light leaf spot, brassicas, 313
light sensitivity, yellow ears, 409
lighting, 530–1
  security, 534
liming, 101–2, 103, 147–9
  cation exchange, 101–2
  cereals, 162–3
  nutrient availability, 102–3, 147–9
  overliming, 149
  pH and, 102–3
Linognathus, 516
linseed, 169–70, 192
  diseases, 169, 317
Linum usitatissimum see linseed
lipids
  characteristics, 341–3
  classification, 342
  digestion, 348
  fatty acids, 342–3
liquefied petroleum gas (LPG), farm services, 529
liquid manure
  slurry separators, 572
  tanks size, 572
Liscombe star system, silage, 218
liverfluke
  cattle, 507
  sheep, 523
livestock
  alternative ranges, 76
  unit size, 498–9
livestock manures *see* manures
lockjaw (tetanus), 522
Longidorus see nematodes
loose smut
  barley, 307
  wheat, 304–5
LPG (liquefied petroleum gas), 529
lucerne, 189–90
lumpy jaw, cattle, 509–10
lungworms (*Dictyocaulus viviparus*), 506–7
lupins, 413
lysine, legumes, 412–13

Maastricht Treaty, 6, 15
McConnell, P, biographical sketch xiii-xiv
machinery *see* farm machinery
macronutrients, 103–4
macroorganisms, soil, 109–10
MacSharry Package, 8, 16
mad cow disease (bovine spongiform encephalopathy, BSE), 514
Maedi–Visna virus, 524
magnesium, 106–7, 132
  administration, 512
  availability and losses, 106–7
  deficiency symptoms, 132, 136, 209, 359, 512
  fertilizers, 209
  grass staggers (hypomagnesaemia tetany), 209, 359, 512
  index, 114, 115
  metabolism, 350
  sources, 106
maize
  by-products, 411
  in feeds, 410
  forage maize, 187–9
  grain maize, 166
  nutrient requirements, 188
  silage, beef production, 435

silage from, 220
temperature requirements, 188
see also cereals
maize diseases, 308
smut, forage maize and sweet corn, 309
mammary gland
anatomy, 373–4
development, 374–5
management *see* farm business management
manganese, 133
animal metabolism, 355
deficiency symptoms, 133
soil, 107
toxicity symptoms, 133
mange
cattle, 508
pigs, 521
sheep, 523
mangold diseases *see* sugar beet, diseases
manioc, 409
Manual Handling Operations Regulations (1992), 68
manures, 134–8, 571
disposal, health, 500
effects, 136
farmyard, 571
green, 154
handling and spreading, 571–2
liquid, 572
management aspects, 137–8
nutrient value estimates, 135–6
risks from application, 136–7
storage, 622–4
marketing
crops, 152
cereals, 157
farm business management, 36–7
Marmite disease (greasy pig disease), 518
mastitis
cattle, 512–13, 514
causes, 513
cell counts, 513
control schedule, 514
pigs (farrowing fever), 517
sheep, 521
mating, 369
see also animal reproduction
meadow fescue, 201
meat and bone meal, animal protein supplements, 413
meat production
goats, 461
poultry, 486–7
sheep, 451–4
see also beef production
Medicago sativa (lucerne), 189–90
medicines, 502–3
animal health, 502–3
crops, 193
melatonin, 364
out-of-season breeding, 373
Meligethes (pollen beetles), 292
mentoring, 50
metabolic disorders, 358–9, 497
rumen-related, 359
metabolism in animals, 348–58
see also animal nutrition
Metapolophium dirhodum (rose grain aphid), 282
Metastrongylus, 521
methionine, legumes, 412–13
metric units, *glossary of units*, 626–7
'imperial' conversion factors, 626–7
metritis (farrowing fever), 517
mice, seedling pests, 286
micro–organisms, soil, 108–9
micronutrients
animal requirements, 419–20
soil, 107, 131–4
mildews *see* downy mildew; powdery mildew
milk
biosynthesis, 375–8
fat, 376–7
lactose, 377
nutritional influences, 377
protein, 375–6
site, 375
composition, 375
cooling and storage, 570–1
economics, average returns (by products), 423
milk ejection mechanisms, 377–8
milk quality, 439–40
production, 442
age at first calving, 427
CAP price mechanisms, 11–12
goats, 460–1
sheep, 454
progesterone, rapid test, 373
sheep, European production, 454
trends, 421, 422
see also lactation
milk by-products, animal protein supplements, 413
milk drop syndrome mastitis, 513
milk fever (parturient paresis), 358–9
hypomagnesaemia, sheep, 524
mature cattle, 511
Milk Quotas, 60
milk scours
pigs, 517
milking machine, and mastitis, 513
milking parlour, 569
equipment, 569–71
feeding, 571
information technology, 571
milking machines, 569–70
robotic milking, 570
milking process, 442–3
millipedes, sugar beet pests, 285
minerals, plant requirements *see* crop physiology; soil; *specific minerals*
minerals in diet *see* animal nutrition
mites, 508, 523
mixers
horizontal, 569
mill, 569
vertical, 569
mixing machinery, 568–9
MMA (farrowing fever), 517
molasses, high energy feeds, 412
molybdenum, 133–4
animal metabolism, 355
deficiency symptoms, 134
moors and heaths, 194–5
mortality
embryonic and fetal, 371
fatal injuries (1986–92), 65
mountain grazing, 194
mower-conditioners, 552–3
mowing and swath treatment machines, 552
mucous membranes, in disease, 504
mulberry heart disease, 519
mustard, *Brassica juncea* (condiment mustard), 193
Mycobacterium hypopneumoniae, pigs, 519
Mycobacterium tuberculosis, cattle, 500, 515
mycoplasma, 501
Mycosphaerella brassicola (cabbage ringspot), 314
Mycosphaerella graminicola syn. *Septoria tritici* (leaf spot), 303
Myzus persicae, (peach potato aphid), 283, 288
Myzus persicae, (potato leaf roll virus), 310

natural gas, farm services, 529
nature conservation
production methods, 60
public goods, 77–8
Wild Life and Countryside Acts (1981 and 1985), 58, 60
navel-ill
calves, 506
sheep, 521
neck and storage rots, onions and leeks, 319
negotiation and bargaining, types, 47
nematodes, 284, 290
pea cyst nematode, 294–5
potato cyst nematode, 283–4
see also roundworms
Nematodirus, 523
neonatal diarrhoea, pigs, 517
nervous system, 335–7
net blotch, barley, 161, 306
nitrogen
cereals, 163–4
deficiency symptoms, 130
fertilizers, 125, 138–40, 207–8
animal-recirculated, 134–5, 206
clover, 206–7
nitrogen index, 207–8
sources, 206–8

nitrogen (*cont.*)
  nitrogen index, 114
  non-protein nitrogen sources (NPN), animal protein supplements, 413–14
  soil, 103–4
  soil analysis, 114
  utilization by plant, 130
  utilization by ruminant, 347
  see also proteins
nitrogen cycle, 104
notching, 276
nutrition
  amino acid metabolism, animals, 351–2
  energostasis, limits and mechanisms, 360
  energy requirements
    growth, 418–19
    lactation, 415–16
    maintenance, 414–15
    reproduction, 417–18
  high energy feeds
    citrus pulp, 412
    fats and oils, 411–12
    molasses, 412

oak, seedling blight, and leaf spot, 308
oat, in feeds and feeding, 410
oat, diseases, 307–8
  barley yellow dwarf virus, 308
  ear, 308
  leaf and stem, 307–8
  oat mosaic virus, 308
  powdery mildew, 307
  root, 308
  smuts, 308
  see also cereals
occupier's liability, agricultural land, 57–8
Occupiers Liability Act (1984), 57
Oenothera (evening primrose), 193
Oesophagostomum radiatum, 507, 521, 523
oestrone sulphate test for pregnancy, 371
oestrous cycles, 362–4
  pigs, 473
  synchronization, 364–6
oestrus
  detection, 440–2
  and mating, pigs, 473
  sheep, 448
oilseed rape, 166–9
  forage rape, 191–2
  nutrient requirements, 168
  residues, feeds and feeding, 412
  soil and site, 168
  spring types, 169
  variety selection, 168
  winter types, 168–9
oilseed rape pests, 290–3
  brassica flea beetles, 291
  brassica pod midge, 293
  cabbage aphid, 292
  cabbage seed weevil, 293
  cabbage stem flea beetle, 291
  cabbage stem weevil, 292
  field slug, 290
  pigeon, 291–2
  pollen beetles, 292
  rape winter stem weevil, 291
  slug, 280–2, 290
  winter stem weevil, 291
oilseeds, CAP price mechanisms, 10–11
onions and leeks
  diseases, 318
  neck and storage rots, 319
Onobrychis vicifolia (sainfoin), 190
Onychiurus (springtails), 286
open ditch drainage, 95
Opomyza florum (yellow cereal fly), 278
orf (contagious pustular dermatitis), 522
organic irrigation, 567
organic matter
  carbon cycle, 109
  decomposition, 109
  fertilizers and manures, 134–8
  impact of agricultural practices, 110–11
  role, 109
organic/sustainable systems, 116
Oscinella frit (frit fly), 278–9
Ostertagia circumcincta, 523
Ostertagia ostertagi, 507, 521
ovarian steroids, immunization, 373
ovulation and fertilization, sheep, 448
ownership, possession, occupation of agricultural land, legal aspects, 55–8

P, K and Mg index, 114
pain
  checklist, 388
  definition, 386–7
  in disease, 504
parasites
  external, 508, 521, 523
  internal, 506–8, 521, 523
Paratrichodorus, 284
  see also nematodes
parsnips, 182
parturient paresis, 358–9, 511
parturition
  gestation length, 372
  induction, 372
  initiation, 372
  post-parturient dyspepsia, 511
  stages of labour, 372
parvovirus infection, pigs, 520
passive immunity, 502
Pasteurella, 519
Pastinaca sativa (parsnips), 182
paths, 58–9
pea(s) and bean(s), 170–4, 187
  CAP price mechanisms, 11
  field beans, 170–1
  field peas, 172–4
pea and bean pests and diseases, 294, 316–17
  black bean aphid, 296
  broad bean rust, 317
  halo blight, 317
  leaf, stem and pod spots, 316
  pea aphid, 295
  pea and bean weevil, 294
  pea cyst nematode, 294–5
  pea midge, 296
  pea moth, 295
  pea wilt, 316
  powdery mildews, 320
  seed-borne diseases, 316
peach potato aphid, 283
  sugar beet pests, 288–9
peats, 112
Pellicularia filamentosa syn. *Rhizoctonia cerealis* (sharp eyespot), 304, 307
perennial ryegrass, 200–1
Peronospora destructor (downy mildew, onions and leeks), 318
Peronospora farinosa (downy mildew, sugar beet), 315
Peronospora parasitica (downy mildew, brassicas), 313
Peronospora viciae (downy mildew, peas and beans), 316
Personal Protective Equipment at Work Regulations (1992), 68
pesticides
  Control of Pesticides Regulations, 60, 256
  interactions with soil, 107–8
  and pollution, 60–1
pests of crops, 277–98
petrol, farm services, 529–30
pheasants, forest management, 248–9
Phoma betae (leaf spot, sugar beet), 315
Phoma exigua var. *foveata* (gangrene), 311
Phoma lingam (canker, brassicas), 313
phosphate, phosphorus, 105–6, 130–1
  cereals, 162
  deficiency symptoms, 131
  index, 114, 115
  inorganic fertilizers, 140–1, 208–9
  losses, 106
  organic fertilizers, 135
  soil reactions and availability, 106
  sources, 105–6
phosphorus, metabolism, 350
phosphorus cycle, 105
photosynthesis and respiration, 118–20
Phyllotreta (flea beetles)
  kale, 293–4
  oilseed rape, 291

physiology, and nutrition, 335–84
Phytophthora
P. infestans (potato blight), 180, 309–10
(pink rot and watery wound rot), 312
pigeons, 291–2
pigs, 464–85
accommodation, 466–8, 607–11
animal welfare, 466–7
dry sows and boars, 610–11
farrowing, 467, 609–10
paddock shelters, 467
pig fattening, 607–8
sow stalls, phasing out, 467
weaner housing, 467–8, 611
breeding, 361–73
artificial insemination, 471–2
breeding stock, 466–7
crossbreeding, 470
hybrid companies, 470
parturition, 576
pregnancy, 474–5
diagnosis, 576
feeding, 475
puberty
female, 362, 472–3
male, 366–7
sow and gilt reproduction, 472–8
lactation and feeding, 477–8
timing of mating, 473, 474
weaning age and sow productivity, 478
weaning to remating interval, 478
breeds, 469–70
carcass quality and classification, 481–3
diseases
ascariasis, 521
'bloody gut' (intestinal haemorrhage syndrome), 519
'bloody scours' (swine dysentery), 519
'blue-eared' pig disease (porcine reproductive and respiratory syndrome), 520
bowel oedema in growing pigs, 517–18
enzootic pneumonia, 518–19
Escherichia coli infection, 517
exudative epidermitis (greasy pig disease), 518
farrowing fever, 517
hog cholera (swine fever), 519
lice, 521
mange, 521
milk fever, 517
neonatal diarrhoea, 517
parvovirus infection, 520
piglet anaemia, 517
porcine reproductive and respiratory syndrome (PRRS), 520
rotavirus infection, 520
salmonellosis, 518
SMEDI, 520
transmissible gastroenteritis, 517
tuberculosis, 519–20
vomiting and wasting disease, 520
feeding
gilts, 473
growing pigs, 482–4
pregnancy, 475
finishing, 480–1
genetics, pig improvement, 468–72
artificial insemination, 471–2
breeds, 468–70
crossbreeding, 470
hybrid companies, 470
national involvement, 470–1
growing stock, 468
gilts, feed allowances, 473–4
management, 484
nutrition, 482–4
growth cycle, 465
health, 517–21
immunoglobulins, after birth, 479
industry, 464–85
CAP price mechanism, 13
carcass quality and classification, 481–3
European industry, 465
performance and profitability trends, 466
size of unit, 464
UK structure, 464–6
normal body temperature, 504
normal respiration rate, 504
performance, trends, 466
weaner accommodation, 467–8
weaner piglets, 478–80
pine (pining), sheep, 522
pink rot and watery wound rot, potato, 312
pipe drainage, 95–7
Pisum sativum see peas
planning, 59–60
plant breeding, 129
plant growth *see* crop physiology; plant hormones
plant hormones (plant growth regulators), 129, 164–5
timing, cereals, 167
plant populations, competition, 127–9
planting machines, 545–9
Plasmodiophora brassicae (clubroot), 314–15
plate mills, 568
Pleospora bjoerlingii (blackleg, sugar beet), 315
ploughs, 541–2
pneumonia
in pigs, 518–19
sheep, 522
podzols, 111
pollen beetles, oilseed rape pests, 292
pollution
grassland farming, watercourses and drinking water, 223–4
and pesticides, 60–1
silage effluent, 219
see also water pollution
Polyscythalum pustulans (skin spot), 312
posture, in disease, 504
potassium, 106, 131–2, 354
cereals, 162
deficiency symptoms, 132
fertilizers, 208–9
index, 114
inorganic fertilizers, 141
losses, 106
soil reactions and availability, 106
sources, 106
potassium nitrate, 139
potato, 176–81
harvest and storage, 180
markets, 176
nutrient requirements, 180
planting machines, 548
hand-drop planters, 548
semi-automatic planters, 548
seed certification, 177
seed treatment, 177–8
seedrates, 178
potato, diseases, 309–12
black leg, 310
dry rot, 311
gangrene, 311
leaf and stem, 309–11
mild and severe mosaic, 310
potato blight, 180, 309–10
potato cyst nematodes, 283–4
potato leaf roll virus, 310
powdery scab, 311
silver scurf, 312
skin spot, 312
spraing, 310
tuber and storage, 311–12
virus X, 310
virus Y, 310
wart disease, 312
potato harvesters, 561
potato, pests, 282–5
cutworms, 285
free-living nematodes, 284
garden slug, 284
peach potato aphid, 283
potato cyst nematodes, 283–4
wireworms, 284–95
poultry, 486–96
accommodation
layers and boilers, 491–2
lighting, 492
ventilation, 491–2
poultry, light intensity, 492
antibiotic performance enhancers, 489, 496

poultry (*cont.*)
  broiler breeding industry, 487–9, 491
  CAP price mechanisms, 13
  egg production, 489–91
    free range, 489
    laying birds, 491
    strains available, 490
  feed programmes
    laying birds, 491
    micronutrients, 491
    nutrient requirements (RDAs), 490
  normal body temperature, 504
  normal respiration rate, 504
  replacement pullets, feed restriction, 491
  UK meat industry, 486–7
  welfare, 496
powdery mildew
  barley, 305
  brassicas, 312–13
  grasses and herbage legumes, 320
  oat, 307
  sugar beet, 315
  wheat, 302
powdery scab, potato, 311
pregnancy, 370–2
  diagnosis, 371–2
  maintenance, 370–1
pregnancy toxaemia (twin lamb disease), 524
PRG *see* ryegrass
prion disease, 514, 523
profit and capital recording, 29–30
profit *see also* budgets
progesterone, 364, 365–6
  milk, rapid test, 373
progesterone test for pregnancy, 371
propionic acid, metabolism, 350
prostaglandins, 364–6
protective equipment, Regulations, 68
protein crops, CAP price mechanisms, 11
proteins, 339–40
  biochemistry in animal digestion, 346–8
  chemical composition of animals, 339–40
  evaluation of feeds, 397
    crude protein (CP), 397
  measurement of quality in feed, 397–9
  non-protein nitrogen sources, 413–14
  protein hydrolysis, 413
  protein supplements, 413
  requirements
    growth, 418–19
    lactation, 415–16
    maintenance, 414–15
    reproduction, 417–18
  supplements
    fish meal, 413
    meat and bone meal, 413–14
  see also nitrogen
PRRS (porcine reproductive and respiratory syndrome), 520
pruning, new farm woodland, 245
Pseudocercosporella herpotrichoides (eyespot), 303, 307
Pseudomonas phaseolicola (halo blight, dwarf and navy beans), 317
Pseudoseptoria stomaticola syn. *Selenophoma donacis* (halo spot), 307
psoropteric mange, 523
Psoroptes, 508
Psylliodes chrysocephala (cabbage stem flea beetle), 291, 294
PTO-driven secondary cultivation machines, 543
public goods, 77–8
Public Order Act (1986), 56
  Criminal Justice and Public Order Act (1994), 57
public rights of way
  access to countryside, 58
  barbed wire, Highways Act (1980) S164, 58
  bulls, Wild Life and Countryside Act (1981) S59, 58
  gates and stiles, 58
  misleading notices, 58
  obstructions, 58
  overhanging vegetation, 58
Puccinia allii (leek rust), 319
Puccinia coronata, (crown rust), oat, 308
Puccinia coronata, (crown rust), ryegrass), 319
Puccinia recondita (brown rust, cereals), 303, 306
Puccinia striiformis (yellow rust), 302, 305
'puffy' seedbeds, 93
pulpy kidney disease, 522
pulse rates, 504
pumps, centrifugal pumps, 562–3
pygmy beetle, sugar beet pests, 187
Pyrenophora avenae (leaf spot and seedling blight), 303, 308, 315, 316
Pyrenophora brassicae (light leaf spot), 313
Pyrenophora graminea (leaf stripe), 161, 306
Pyrenophora teres (net blotch), 161, 306
Pythium (damping-off, seedling blight and root rots), 317
Pythium (pink rot and watery wound rot), 312

quotas, 60

rabies, cattle, 515
rainwater goods, 532
Ramularia beticola (leaf spots, sugar beet), 315
rape *see* forage rape; oilseed rape
rape, sheep stocking rates, 453
recreation, 75–6
recruitment
  advertising, 41
  disabilities, 44
  pay and reward systems, 41–2
  production/performance rates, 41–2
  service and loyalty rewards, 42
  time rate, 41
rectal palpation, 372
red clover, 202–3
red leaf (barley yellow dwarf virus), 306
redwater, cattle, 510
Regulations
  Control of Pesticides Regulations, 60, 256
  COSHH Regulations, 60
  Diving Operations at Work Regulations (1981), 68
  European Union, 5
  fire regulations, 51–2, 534
  insecticides, 297
  Manual Handling Operations Regulations (1992), 68
  Personal Protective Equipment at Work Regulations (1992), 68
  Reporting of Injuries, Diseases and Dangerous Occurrence Regulations (1985) (RIDDOR), 68–9
  Safety Signs Regulations (1980), 68
  silage, slurry and fuel oil, 61
  water pollution control, planning permission and building regulations, 598
  Work Equipment Regulations (1992), provision and use, 68
  Workplace (Health safety and Welfare) Regulations (1992), 68
  see also Acts of Parliament; Health and Safety at Work
Rent (Agriculture) Act (1976), 54
Reporting of Injuries, Diseases and Dangerous Occurrence Regulations (1985) (RIDDOR), 68–9
reproductive system *see* animal reproduction
respiration rate, normal, 504
retirement, farm staff, 46
rhizomania, 176, 315
  vector, 315
Rhopalosiphum padi (bird cherry aphid), 282
Rhynchosporium orthosporum (leaf blotch, ryegrass), 320
Rhynchosporium secalis, (leaf blotch, barley), 164, 306

Rhynchosporium secalis, (leaf blotch, ryegrass), 320
rice bran, 411
rickets, sheep, 522
rickettsiae, 501
RIDDOR (Reporting of Injuries, Diseases and Dangerous Occurrence Regulations (1985))
Rights of Way Act (1990), 53, 58–9
  chief obligations, 58–9
ringworm, cattle, 508
roller mills, 568
rollers, 543–4
root crops, 174–81, 190–1
  in feeds, 408–9
  forage, 190–1
  harvesting, 561
  harvesting machinery, 561, 552–8
  herbicides, 262
  storage, 620
rose grain aphid, 282
rotary cultivator, 542–3
rotavirus infection, pigs, 520
rough grazing, 194, 195
roundworms, 507
  cattle, 506
  pigs, 521
  sheep, 523
  see also nematodes
row crops, herbicides, 262
rumen degradable crude protein, 346
rumen-related metabolic disorders of livestock, 359
ruminant digestive tract, 344
rushes and sedges (perennial), 253
rusts
  beet rust, 315
  broad bean rust, 317
  brown rust
    barley, 306
    wheat, 303
ryegrass
  diseases, 308
    leaf blotch, 320
  hybrid, 201
  Italian ryegrass, 201
  perennial, 200
  Westerwolds, 201

SACs (special areas of conservation), 60
Safety Signs Regulations (1980), 68
sainfoin, 190
salmonellosis
  cattle, 505–6
  pigs, 518
  sheep, 524
Sarcoptes scabiei, 508, 521
scabies, 311, 508, 521
Sclerotinia sclerotiorum (storage rots), 169, 314
sclerotinia stemrot, brassicas, 314
Sclerotium cepivorum (white rot, onions and leeks), 319
scours, 'bloody scours', 519
  calf, 505
  pigs, 517
scrapie, sheep, 514, 523
Scutigerella immaculata (symphylids), 286
seaweed, 137, 138
security, 534–5
seeding
  direct drilling, 199
  scatter and tread, 200
  seeders, precision, 547–8
  slit-seeding, 199–200
  without a cover crop, 199
seeds
  crop establishment, 155–6
  development, 122–3
  herbage seed production, 203
  mixtures, 203–5
    check-list for studying, 204–5
    purpose, 204
    types, 204
  seed-borne diseases, 316
  seedling protection, 156
  undersowing, 199
selenium, animal metabolism, 355
self-loading forage (SLF) wagons, 555
Selye, general adaptive syndrome, 388–9
senescence, 123
septic tank, 532
septoria diseases, wheat, 303
services, 527–35
set-aside, Common Agricultural Policy, 10, 15–16
set-aside cover crops, 154
set-stocking, 214
sewage sludge, 137, 138
sex determination, 361–2, 373
sexual development
  female, 362
  male, 366
shafts and ducts, 531
shared ownership, 34
sharp eyespot
  barley, 307
  wheat, 304
sheep
  accommodation, 612–13
    handling areas, 612–13
  body condition scores, 463
  breed types, 446–7
    crossbreeds, 447
    ovulation rate, 449
  breeding, 361–73
    age, 449
    embryo mortality, 450
    nutrient requirements, 448–9
    oestrus, 448
    ovulation and fertilization, 448
    perinatal mortality, 450–1
    season, 449–50
  breeding season, 449
  calendar, frequent-lambing flock, 463
  carcass classification grid, 452
  copper toxicity, 137
  culling, 455–6
  deficiency conditions, 522
    pine (pining), 522
    rickets, 522
    swayback (enzootic ataxia), 522
  diseases
    black disease, 522
    blackleg, 522
    braxy, 522
    brucella, 524
    causing abortion, 450, 524
    clostridial infections, 521–2
    contagious pustular dermatitis, 522
    enzootic ataxia, 522
    Escherichia coli infection, 521
    external parasites, 523
    foot-rot, 523–4
    lamb dysentery, 522
    liverfluke, 523
    Maedi–Visna virus, 524
    mange, 523
    metabolic diseases, 524
    navel-ill, 521
    orf, 522
    pneumonia, 522
    pregnancy toxaemia (twin lamb disease), 450, 524
    pulpy kidney, 522
    rattle belly, 521
    salmonellosis, 524
    scrapie, 523
    sheep scab, 523
    tapeworms, 523
    tetanus (lockjaw), 522
    vibriosis, 524
    watery mouth, 521
  energy requirements of ewe, 456
    milk yields, 452
  flock performance
    affecting factors, 448–51
    feeding the ewe, 456–7
    frequent-lambing flock calendar, 463
    growth curves, 1–3 lambs, 452
    organization of barren ewes, 459
    replacements, 455–6
    twin pregnancy, 451
  growth and nutrition, 451
  health, 521–4
  international production, 445
  lamb hypothermia, 451
  lamb production systems
    early lambs, 453
    finishing on cereals, 454
    finishing on roots, 453–4
    forage crops, 453
    grass finishing, 453
    hill, early weaning, 458–9
    prolificacy of ewe, 459
    silage finishing, 453

lambing
health problems, 521
increased frequency, 459–60
numbers per ewe, 458–60
M–V Accredited Flocks Scheme, 524
meat production
CAP price mechanism, 13
tonnes, 445
mortality
embryo, sheep, 450
hypothermia, sheep, 451
perinatal, sheep, 450–1
normal body temperature, 504
normal respiration rate, 504
numbers
UK, 446
world, 445
nutrition, 448–9
feeding ewes, 456–8
protein requirements, 456–7
voluntary feed intake of ewe, 360–1, 456
performance, lowland vs highland, 448
season, 449–50
shearing, 455
UK production, 445–7
sheep milk, Eurpoean production, 454
sheep wool, 455
shelterbelts, 248
Shoard, M., *The Theft of the Countryside*, 73
silage
arable mixtures, 220
baling and bale handling, 555–6
bunkers, 617–19
crops for, 220–1
effluent, 219
effluent tank, 619
facts and figures, 221
feeding big-bale, 220
legume, 220
Liscombe star system, 218
maize, 220
making, 217–18
big-bale, 219
in clamps, 219
in towers, 219–20
mechanical removal, 220
Regulations, 61
self-fed silage, 349
self-feeding, 220
silage bunkers, 617–19
storage, 218
whole-crop cereals, 220
silicon, 134
silo filling, 556
silver scurf, potato, 312
silviculture *see* forest and woodland; trees
Sitobion avenae (grain aphid), 281–2
Sitona lineatus (pea and bean weevil), 294
skills audit, 47–8
questions, 48
training, 48
skin and coat, disease, 504
skin spot, potato, 312
SLF wagons (self-loading forage wagons), 555
slow fever, 511
slow virus diseases, 514
sludge, analysis, 137
slugs, field, 161, 280–2
slugs, garden, potato pests, 284
slurry
irrigation, 567
Regulations, 61
separation of liquid manure, 572
storage, 622–4
SMEDI, pigs, 520
smuts
barley, 307
onions and leeks, 319
wheat, 304–5
snails, liverfluke transmission, 507
sodium, 131–2, 354
deficiency symptoms, 132
sodium nitrate, 139
soil
aeration, 93
analysis, 113–14
nitrogen, 114
bacteria, 108
biological properties, 108–9
chemical properties, 100–8
crop requirements, 100–1
components, 87
constraints, 151
density, 91
bulk, 91
particles, 92
drainage status, 89
elements, 130–4
formation, 111–12
fungi, 109
infiltration capacities, 567
ion exchange processes, 101
machinery work days, 93–4
macronutrients, 103–4, 130–4
removal by some crops, 101
see also specific nutrients
macroorganisms, 109–10
management, 87–117
texture, 89–90
micronutrients, 107, 131–4
removal, 101
microorganisms, 108–9
actinomycetes, 108–9
bacteria, 108
fungi, 109
mineral components, 88
nitrogen, 103–4
nutrient index system, 114
nutrient retention, 89
cation and anion exchange, 101–2
nutrients (N, P, K etc.) *see* soil, macronutrients
organic matter and the carbon cycle, 109
pesticide interactions, 107–8
adsorption, 107–8
degradation, 108
pH, nutrient availability and liming, 102–3, 147–9
physical properties, 87–8
podzols, 112
pore space, 92
profile description, 113
sampling, analysis and nutrient indices, 113–14
soil moisture deficit criteria, 566
strength and cultivation, 93
structure, 90–1
formation, 90
importance, 90–1
modification, 91
surface compaction, 93
surfaces, electrical charges, 101–2
sustainable production, 114–16
texture, 88–90
laboratory/field determination, 88
trace elements *see* soil, micronutrients
types
brown earths, 112
classification and analysis, 111–12
gleys, 112
peats, 112
podzols, 112
water capacities, 567
water retention, 92–3
workability and trafficability, 89
see also drainage; liming
Soil Association, permitted fertilizers, 135
Solanum tuberosum see potatoes
solar panels, 534
Solenoptes, 516
solid-set sprinklers, irrigation, 564
sorghum, feeds, 410
sowing
protection, 156
transplants, 156
ungerminated seed, 115–56
SPAs (special protection areas), 60
sperm production, 367–9
data, 368
Spongospora subterranea (powdery scab), 311
spraing, potato, 310–11
spreaders, full-width spreaders, 545
spring cabbage, 186
spring-tine cultivator, 543
springtails, sugar beet pests, 286
sprinklers and laterals, irrigation, 564
squirrels, forest protection, 247
staggers *see* magnesium deficiency (tetany)
stalk rot, forage maize and sweet corn,

308
Staphylococcus mastitis, 513
Stemphylium radicinum (black rot, carrots), 318
steroids, 343
ovarian, 373
stiles, 58
Stomoxys, stable fly, 521
storage rots, carrots, 318
straw
as feed, 408
in manures, 137–8
storage, 616
streptococcal infections, growing pigs, 518
Streptococcus, mastitis, 513
Streptomyces scabies (common scab of potatoes), 311
stress, definition, 387–8
stubble cleaning, 258
subsoilers, 541
Suckler Cow Premium, 12, 60
suffering, definition, 387, 504
sugar, CAP price mechanisms, 11
sugar beet, 174–6
harvesters, 561
pests and diseases, 315
rhizomania, 176, 315
rust, 315
weed beet, 175, 262–8
sugar beet pests and diseases, 285–90
beet cyst nematode, 175, 289–90
beet flea beetle, 287
black bean aphid, 289
cutworms, 288
free-living nematodes, 290
leatherjackets and diseases, 288
millipedes, 285
peach potato aphid, 288–9
pygmy beetle, 187
springtails, 286
symphylids, 286
wireworms, 287
woodmouse, 286
sugar beet pulp, 409
sulphur, 132
animal metabolism, 354
deficiency symptoms, 132
fertilizers, 209
soil reactions, availability and losses, 107
sources, 107
superovulation, 373
surpluses, 7–8
swath treatment machines, 553–4
barn hay drying, 553–4
combination machines, 553
'specialised' machines, 553
swayback (enzootic ataxia), sheep, 355, 522
swedes, 181–2, 191
sweet corn, diseases *see* maize
swine dysentery ('bloody scours'), 519
swine erysipelas ('the diamonds'), 419
swine fever (hog cholera), 519
swine influenza, 518
symphylids, sugar beet pests, 286
Synchytrium endobioticum (wart disease), 312

take-all
barley, 307
forage maize and sweet corn, 309
oat, 308
wheat, 304
tall fescue, 202
tapeworms, sheep, 523
tapioca, 409
temperature
in animal production, 383–4
normal body, 504
plant growth and development, 123–4
temperature-induced stress, 124
tenancies
occupation with/without security of tenure, 55–7
occupier's liability, 57–8
terminology, cattle, 422–4
tetanus (lockjaw), sheep, 522
Thanatephorus cucumeris (black scurf and stem canker), 312
thyroxine, 355
ticks, 510, 523
Tilletia caries (bunt), 161, 305
timber
species characteristics, 236–9
see also forest and woodland; trees
timber measurement, 230–2
diameter, end uses, 228
volume, yield class system, 232
timothy grass, 201
Tipula (leatherjackets), 280, 288
tissue growth, animal, 379–80
tocopherols, 357
tourism, 74–5
farm, alternative enterprise model, 80–2
Town and Country Planning Act (1990), 59
toxic elements, soil, 134
toxoplasmosis, sheep, 524
tractor, 536–41
ballasting, 538
compaction, 538–9
draught control, 540
drawbar power, 537–8
engine, 536–7
four-wheel drives, 539
gross tractive effort, 537
hydraulic systems, 540
position control, 540–1
power take-off (PTO), 539–40
rolling resistance, 538
transmission, 537
tyres, 538
tractor driver, health and safety, 541
Trade Union Reform and Employment Rights Act (1993), 61–2
training and development, farm staff, 47–51
transmissible gastroenteritis (TGE), pigs, 517
transplanters, 548–9
block, 549
transplants, 156
travellers, 56
trees
basal bark spray, 275
broadleaves, pruning, 246
diseases, 247
epicormic branches, pruning, 246
species characteristics, 236–9
stump treatment, 275
tree felling, control, 235
tree shelter tubes, 243
unwanted
basal bark spray, 275
frill girdling, 275
injection, 276
notching, 276
see also forest and woodland; timber
trefoil, 203
trespass, 57
tricarboxylic acid pathway, 349
Trichodorus *see* nematodes
trichomoniasis, 514
Trichophyton, 508
Trichostrongylus, 507, 523
Trichuris ovis, 523
Trifolium (crimson clover), 203
triticale, 166
diseases, 308
tuberculosis
cattle, 515
pigs, 519–20
TURERA (Trade Union Reform and Employment Rights Act) (1993), 61–2
turkeys, 492–3
turnip moth, cutworms, 285, 288
turnips
quick-growing 'continental/stubble', 192
sheep stocking rates, 453
twin lamb disease, 524
tympanitis, 510

ultrasonic techniques, 372
UME (utilized metabolizable energy), 211–12
urea, 138–9, 413–14
urine, examination in disease, 503
Urocystis cepulae (smut, onions and leeks), 319
Uromyces betae (beet rust, sugar beet), 315
Uromyces fabae (broad bean rust), 171, 317
Ustilago avenae (oat smut), 308
Ustilago hordei (covered smut, barley), 307
Ustilago hordei (oat smut), 308

Ustilago maydis (maize smut), 309
Ustilago nuda (loose smut), 307, 340–5

vaccines, 502
value-added enterprises, 77
marketing, 77
processing, 77
veal production, 426–7
see also calves
vectors (of cereal viruses) *see* aphids; *named cereals*
vegetable crops, 181–7
CAP price mechanisms, 13
pests, 296–7
storage, 620
ventilation, danger from gases, 500–1
ventilation methods
farm buildings, 595–7
forced, farm buildings, 595
humidity control, farm buildings, 597
natural, farm buildings, 595
rates, farm buildings, 595–7
VFA metabolism, animals, 350
vibriosis
cattle, 514
sheep, 524
Vicia faba see beans
violet root rot, carrots, 318
viral strike, 497
virus infections, pigs, control measures, 520–1
virus yellows, sugar beet, 315
viruses, 501
vitamins
deficiencies in sheep, 522
metabolism and nutrition of animals, 357–8
vitamin B, animal nutrition, 193
vomiting and wasting disease, pigs, 520

warble fly, cattle, 516
wart disease, potato, 312
water
availability to plants, 89
centrifugal pumps, 562–3
chemical composition of animals, 343
cold supply, 532–3
distribution, 563–7
hot supply, 533–4
irrigation, 562–7
irrigators, 564–6
and nutrient uptake, 118
Regulations, 61
soil capacities, 567
soil moisture deficit criteria, 566
storage, 562
see also pollution
water pollution control, 598–600
applicable works, 598–9
code of practice, 598
constructions standards, 598
consultation, 598
environmental statement, 599
farm waste management plan, 599
general design calculations, 599
legal requirements, 598
notice of construction, 598
notice requiring works, 599
planning permission and building regulations, 598
site ground conditions, 599
special areas, 599
water supplies, 532–4
irrigation, 562
application nozzles, 563
direct abstraction, 562
distribution, 563
fixed nozzles, 564
rotating nozzles, 563
sprinklers and laterals, 564
waterfowl, 492–3
weed beet, 175, 262–8
control, 268
weed control, 252–76
annual cultivation, 257
arable crops and seedling leys, 257–8
cereals, 258–68
foliar application, 254
forest and woodlands, 270–6
general control measures, 253
grassland, established, 268–9
groundkeepers, 262
new farm woodland, 244
non-crop situations, 275, 276
perennial cultivation, 257–8
rotation, 257
shed crop seed, 262–8
soil application, 254
unselective, non-crop situations, 276
weed types
annuals, 252, 253, 257
perennials, 252–3, 257–8
woody, 242, 253
see also herbicides
welfare *see* animal welfare
Westerwolds ryegrass, 201
WGS *see* Woodland Grant Scheme (Forestry Authority)
wheat
by-products, 411
in feeds, 410
quality standards, 157
see also cereals
wheat, diseases and pests, 302–5
barley yellow dwarf virus, 303
brown foot rot and ear blight, 304
brown rust, 303
bunt, 161, 305
ear diseases, 304–5
eyespot, 303–4
leaf and stem, 302
powdery mildew, 302
root and stem base, 303–4
septoria diseases, 303
sharp eyespot, 304
take-all, 304
wheat bulb fly, 277–8
wheat bunt, 305
yellow rust, 302
white clover, 202
increasing in existing swards, 200
white rot, onions and leeks, 319
Wild Life and Countryside Acts (1981 and 1985), 58, 60
willow carrot aphid, vector for carrot motley dwarf virus, 318
wind, forest protection, 246
wireworms, 279
cereal pests, 161
potato pests, 284–95
sugar beet pests, 287
wooden tongue, cattle, 509–10
Woodland Grant Scheme (Forestry Authority), 16, 240
woodlands *see* forest and woodland; trees
woodmouse, seedling pests, 286
wool, 455
Work Equipment Regulations (1992), provision and use, 68
Workplace (Health safety and Welfare) Regulations (1992), 68
workshop accommodation, 625
worms (earthworms), 109

yellow cereal fly, 278
yellow ears, light sensitivity, 409
yellow rust
barley, 305
wheat, 302
yield, 120

Zadok's decimal code for growth, 159
Zea mays see maize
zinc
animal metabolism, 355
plant nutrition, 134
zinc sulphate turbidity test, 424